DICTIONARY OF

PHYSICS

EDITED BY

H. J. GRAY

C.M.G., M.Sc., LL.B., M.P.A(Harvard),
M.Inst.P., F.R.S.A.

ALAN ISAACS

Ph.D., B.Sc., D.I.C., A.C.G.I.

LONGMAN

This new edition has been prepared and typeset by Market House Books Ltd

The work of revision and the preparation of new entries has been carried out by:

VALERIE ILLINGWORTH, M.PHIL., B.SC.
JOHN DAINTITH, PH.D., B.SC.
J W WARREN, PH.D.

Additional contributions have been made by:

RICHARD BATLEY, PH.D., M.INST.P.
COLIN BRETT, PH.D., B.A.
H M CLARKE, M.A., M.SC.
PROFESSOR A J H GODDARD, PH.D., M.I.MECH.E.,
    F.INST. NUC.E., F.INST.P., M.S.R.P.
RICHARD RENNIE, PH.D., M.SC., B.SC.

NOTES

An asterisk indicates a cross reference.
An entry having an initial capital letter is either a proper name or a trade name.
Syn. is an abbreviation for "synonymous with".
*All other abbreviations will be found in the Tables of SI Units (page 625) and the
Table of Symbols for Physical Quantities (page 633).*

ACKNOWLEDGEMENTS

The Table of Fundamental Constants has been selected from CODATA
Bulletin No.63 (Nov. 1986)

Longman Group UK Limited,
*Longman House, Burnt Mill, Harlow, Essex CM20 2JE, England
and Associated Companies throughout the world.*

FIRST PUBLISHED 1958
SECOND EDITION 1975
THIRD EDITION 1991

BRITISH LIBRARY CATALOGUING IN PUBLICATION DATA
DICTONARY OF PHYSICS. — 3RD ED.
  1. PHYSICS
  I. ISAACS, ALAN II. GRAY, H.J.
  530

ISBN 0–582–03797–2

Produced by Longman Singapore Publishers (Pte) Ltd.
Printed in Singapore

PREFACE

TO THE NEW EDITION

The Longman DICTIONARY OF PHYSICS was originally edited by
H J Gray and first published in 1958. In 1975 a second, fully revised
edition was published, edited by H J Gray and Alan Isaacs. Now
the editors believe that the time has come to produce a new
edition of the dictionary.

In preparing this third edition of the dictionary the editors
have had the same two objectives as for the second (1975) edition.
First to bring the book up to date and secondly, to maintain its
usefulness to the potential readers for whom it was originally
compiled.

In the 15 years since the last revision of the book a large
number of advances have been made in the theory and techniques
of physics — in particular, in such fields as particle physics, nuclear
physics, solid-state physics, electronics, astrophysics, cosmology,
and computer science. These changes have been reflected in the
new edition, both in changes to existing entries and in the
inclusion of a large number of new entries. At the same time, the
editors have kept the entries of historical interest in the belief that
the dictionary should cover the whole of physics — not only
modern physics.

The potential readership of the book remains the same. We
hope that it will prove useful to practising physicists and to
students and teachers of the subject. It should also be of use to
physical chemists, engineers, and other scientists working in
related fields.

Finally, the editors would like to record their appreciation
for the contributions and help given by a number of people. For
the first edition these were Professor Allan Ferguson, S G Starling,
Professor J A Crowther, Professor W Wilson, Professor Kathleen
Lonsdale, Dr E G Richardson, Dr C Dodd, W Swaine, C J G
Austin, W Ashworth, and J R Barker. For the second edition they
were John Young, Carol Young, Stephen Dresner, Paul Collins,
and John Illingworth. Contributors to the present edition are listed
opposite.

1990   THE EDITORS

CONTENTS AND TABLES

A

Å Symbol for angstrom.

ab- A prefix that, when attached to the name of a practical electrical unit, denotes the corresponding unit in the CGS-electromagnetic system. This system of units is no longer employed.

abampere *See* ab-.

Abbe, Ernst (1840–1905) German physicist who played a great part in the development of the important optical firm of Carl Zeiss, Jena. He was responsible for many improvements in optical instruments, introducing the substage condenser for microscopes (*see* Abbe condenser), the oil-immersion objective, and greatly improved achromatic systems. He also designed the Abbe refractometer (*see* refractive index measurement) and many other devices. (The name Abbe is frequently rendered incorrectly, as Abbé.)

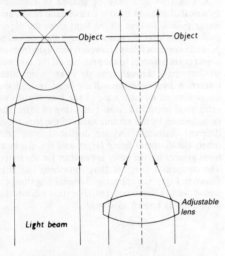

a Abbe condenser *b* Variable-focus condenser

Abbe condenser A simple two-lens *condenser that has good light-gathering ability, the *numerical aperture being 1.25. (Fig. *a*.) It is therefore extensively used in general microscopy. Aberrations are not well corrected. A modified Abbe condenser called a *variable-focus condenser* is used to obtain a greater illuminated field area. (Fig. *b*.) The lower lens can be adjusted to bring light to a focus between the lenses. *See also* achromatic condenser.

Abbe criterion *See* resolving power.

Abbe number *Syn.* constringence; V-number. Symbol: *V*. The reciprocal of the dispersive power. *See* dispersion.

Abelian group *See* group theory.

aberration Of lens or mirror. In a general sense this refers to a defect of the image revealed as blurring or distortion, and is classified as *chromatic aberration (due to *dispersion) and *spherical aberration (due in the main to curvature of surface). In a particular sense it refers to a measure of the amount rays fail to focus accurately, distortion of the wave front, etc. (*See* aberrations of optical systems; lateral aberration; longitudinal aberrations.) Aberrationless points and systems in which all rays from an object point pass through a point image over a wide aperture (*Cartesian systems*) should be distinguished from *aplanatic systems* (*see* aplanatic), although frequently the terms have been used synonymously. *See also* Seidel aberrations.

aberrationless systems *See* aplanatic.

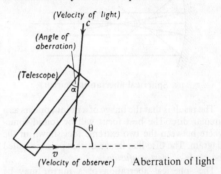

Aberration of light

aberration of light The seasonal small displacement of stars, attributable to the effect of the orbital motion of the earth round the sun on the direction of arrival of the light. If the observer has velocity *v*, i.e. the earth's velocity in its orbit, and *c* is the speed of light a telescope must be inclined forward by an amount α, the *angle of aberration*, to receive the starlight, where

$$\sin \alpha = (v/c) \sin \theta$$

This has a maximum value for $\theta = 90°$ when $\sin \alpha = (v/c)$ and $\alpha = 20.5''$. *Bradley, who first observed the phenomenon (1727), used it to estimate the speed of light. Modern explanations are given in terms of *relativity.

aberrations of optical systems 1. *General.* The simple theory of *centred optical systems holds good only for paraxial rays, i.e. those passing near the axis. When the angles involved become so large that it is no longer accurate to replace the sine of the angle by the angle itself (as is done in the simple theory), the point, line, and plane correspondence between object and image no longer holds good, and certain defects in the image occur.

By expanding the sine terms to two or more terms in the series

$$\sin \theta = \theta - \theta^3/3! + \theta^5/5! - \dots ,$$

the deviations of the path of a ray from that predicted by the simple theory can be expressed in terms of five sums called the *Seidel sums* or the *Seidel terms*. The presence of one or more of these terms can be

aberrations of optical systems

linked with certain recognizable defects in the image, i.e. those called *spherical aberration, *coma, *astigmatism, *curvature of field, and *distortion. In addition, if light of more than one colour is involved, false colour effects may be introduced in the image, a defect known as *chromatic aberration.

Of these six defects only spherical and chromatic aberrations are found in the images of axial points. The other four aberrations occur only when extra-axial points are involved.

2. *Spherical aberration.* The rays meeting a spherical concave mirror near the periphery are brought to a focus nearer the mirror than are those meeting the mirror near the pole (Fig. *a*).

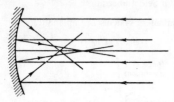

a   Spherical aberration (mirror)

The result is that the image of a point appears as a circular disc. The best focus will be found somewhere between the two extreme foci shown in the diagram. The disc-like image at this point is called the *circle of least confusion*.

The spherical aberration of a mirror may be eliminated by grinding the surface of the mirror to give it the shape of a paraboloid (for use with distant objects) or of an ellipsoid (where the object is at a finite distance).

A second method is to place at the centre of curvature of the spherical mirror a transparent plate known as a *Schmidt corrector*. This plate has the shape of a convex lens near the centre and that of a concave lens near the periphery. The central rays thus converge to a focus nearer the mirror than normally, while the outer rays diverge to a focus further away from the mirror than normally, so that it is possible to bring all rays together into one focus. The use of this plate has the advantage over a paraboloidal mirror in that a larger field of view is sharply focused. It is also more useful when a mirror of short focal length is required.

In the case of lenses, spherical aberration is encountered as indicated in Fig. *b*. The rays refracted at the periphery of the lens cross the axis nearer to the lens than do those refracted at the centre. Such a lens is said to be *undercorrected* for spherical aberration. An *overcorrected* lens will show the opposite effect.

The aberration is reduced by choosing a lens of the right shape for the particular work required. In general, the aberration will be small if the angles of incidence of the rays on the lens surfaces are small. The solution is therefore to share the necessary deviation of the ray equally between the two faces of the lens. Thus a telescope *objective, which is used

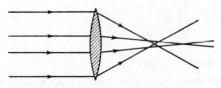

b   Spherical aberration (lens)

to form an image of distant objects is roughly planoconvex with the convex face pointing towards the object. The deviation of the rays is then about the same at the two faces of the lens. In a microscope objective, a number of lens components are used so that the necessary deviation of the rays is shared between a number of refracting surfaces.

For a particular zone of the lens, the *longitudinal spherical aberration* is the axial separation of the images formed by rays passing through that zone and those through the centre of the lens. The *transverse spherical aberration* is the height of the intercept of a ray through the zone on a screen placed at right angles to the axis at the paraxial image.

3. *Coma.* A lens may be shaped to eliminate spherical aberration for one pair of axial object and image points, so that all the light from the object point passes after refraction through the one image point. It does not follow, however, that light from an object point adjacent to the first, but off the axis, will similarly pass through a single image point after refraction. In general this will not be so. Instead, the image formed by each zone of the lens of such an extra-axial point will consist of a ring of light. The rings formed by the various zones of the lens are of different diameters and are displaced from each other, the diameter being larger and the displacement greater for the outer zones than for the inner. The composite image is thus something like that shown in Fig. *c*, the appearance resembling that of a comet with a luminous tail. For this reason the aberration is known as coma.

c   Coma

It is found that, provided spherical aberration has first been eliminated, the condition for absence of coma for points near the axis is that the ratio of the sines of the angles made with the axis by the incident and emergent rays must be the same for all zones of the lens system. This is known as the *optical sine condition*. The object and image points for which this condition is satisfied are known as the *aplanatic points* of the system. The term "aplanatic" signifies freedom from both coma and spherical aberration.

4. *Astigmatism.* With a simple lens it is found that a point object off the axis gives rise to two images each in the form of a line. The tangential fan of rays OAB (Fig. *d*) is brought to a focus at T′, while the so-called sagittal fan OCD is focused at S′. If the object were a wheel centred on the axis the spokes would be seen at S′ and the rim at T′. This defect is called astigmatism.

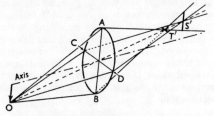

d   Astigmatism (lens)

A similar effect is found in the images of extra-axial points formed by spherical mirrors.

The eye may produce astigmatic images of axial points because of nonspherical curvature of refracting surfaces. This is corrected by using spectacle lenses with cylindrical surfaces.

5. *Curvature of field.* In general, a plane object at right angles to the axis of an optical system will not give rise to a plane image. Instead, in the absence of astigmatism, the image would lie on a paraboloidal surface known as the *Petzval surface. This aberration is called curvature of field (or of image). The effects of astigmatism will be superimposed on those of curvature of field with the result that the tangential and sagittal focal planes T and S are displaced from the Petzval surface P. Two possible cases are indicated in Figs *e* and *f*.

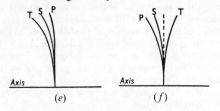

(*e*)   (*f*)

Curvature of field

In the case shown in Fig. *f*, the effects of astigmatism have been used to offset the curvature of the Petzval surface, and a flat screen placed in the position indicated by the dotted line would show a reasonably focused image. Variations of this sort are achieved in practice using optical systems with more than one lens, altering the spacing of the various component lenses, and adding suitably positioned stops.

In *anastigmat photographic lenses, this procedure has been carefully applied so that the image obtained is substantially free from the defects of astigmatism and curvature of field.

6. *Distortion.* If the magnification produced by an optical system varies over the field of view it covers, then a plane object placed normally to the axis will give rise to an image of different geometrical shape. Thus, if the magnification increases with the distance of the object point from the axis, the diagonals of a square object will be magnified more than the

sides, and the resulting image will have the defect called *pinchushion distortion.* In the converse case, the defect is called *barrel distortion.* (*See* Fig. *g*.)

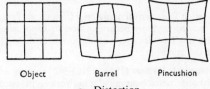

Object   Barrel   Pincushion

g   Distortion

Distortion is controlled by introducing suitably placed stops into the system.

7. *Chromatic aberration.* The *refractive index of all transparent substances varies with the colour of the light. A simple lens will therefore form a series of images, one for each colour of light present (Fig. *h*.). The result is that coloured haloes are seen round the focus.

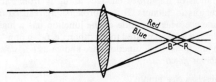

h   Chromatic aberration

Moreover, because the focal length alters, the magnification will also vary with the colour. The effect of this *chromatic difference of magnification* will be that the image of, say, a star off the axis of the lens will consist of a short radial spectrum.

Chromatic aberration such as that indicated is reduced in practice by employing a composite lens. Two or more lenses, made of different types of glass, are cemented together to form an *achromatic doublet* or *triplet*. A typical doublet telescope objective is shown in Fig. *i*. The general idea is that a "powerful" convex lens (one of short focal length combined with a weak concave lens) made of strongly dispersing flint glass. In this way it is possible to bring two colours, say the blue and the red, to the same focus. For thin lenses, the condition is that the focal lengths of the two component lenses are numerically proportional to the dispersive powers of their respective glasses. There will still be some residual colour effects, known as a *secondary spectrum*, but these will be slight. If desired, a triplet lens can be used which will bring three colours together at the same focus and reduce the aberration still further.

i   Achromatic lens

In the case of *eyepieces, the chromatic difference of magnification can be avoided by using separated components of the same glass. For thin lenses the condition that the focal length of the combination should be the same for all colours is that the separation should be equal to the mean focal length of the two components.

ablation The removal of material from the surface of a moving body by decomposition or vaporization resulting from friction with the atoms or molecules of the atmosphere.

Abney, Sir William (1843–1920) Brit. chemist and physicist, noted especially for his work on the chemistry of the photographic processes (including colour photography and infrared photography) and on printing processes. He invented a form of flicker photometer. *See* photometry.

abnormal glow discharge *See* gas-discharge tube.

A-bomb *See* nuclear weapons.

abrupt junction A *p-n junction in which the concentration of impurities changes abruptly from *acceptors to *donors. It is approximately realized in practice when one side of the junction has a high doping level compared to the other, i.e. p^+-n or n^+-p junctions. Such junctions are known as *one-sided abrupt junctions*.

abscissa *See* graph.

absolute electrometer *See* attracted disc electrometer.

absolute expansion *See* coefficient of expansion.

absolute humidity *See* humidity.

absolute magnitude *See* magnitude.

absolute temperature Former name for *thermodynamic temperature.

absolute unit If a quantity y is uniquely defined in terms of quantities $x_1, x_2, \ldots$ by
$$y = f(x_1, x_2, \ldots),$$
the unit U_y of y can be obtained from the units U_{x_1}, U_{x_2} of x_1, x_2 from the equation
$$U_y \propto f(U_{x_1}, U_{x_2}, \ldots).$$
In any given system an absolute unit is one for which the constant of proportionality is unity. All *SI units are absolute.

absolute value *Syn.* modulus. The magnitude of a quantity without regard to its sign.

absolute zero The unattainable lower limit to temperature. It is the zero of *thermodynamic temperature: $0 \text{ K} = -273.15\,°\text{C} = -459.67\,°\text{F}$.

According to thermodynamics, if an ideal heat engine working in a *Carnot cycle has its lower-temperature isothermal process at absolute zero, no heat will be discharged and the efficiency of the engine will be one.

The ideal-gas absolute scale can be shown to be equivalent to the thermodynamic scale. Temperature on this scale is defined to be proportional to the product of pressure and volume in the limit as pressure tends to zero and volume to infinity. Abso-

lute zero is that temperature at which this product is zero.

In quantum theory, absolute zero is interpreted as the temperature at which all particles are in the lowest-energy quantum states available. Generally the available states do not have zero energy, so there is still molecular energy at absolute zero (the *zero-point energy*). The kinetic energies of the molecules of an ideal gas would be zero at 0 K, but not those of a real substance. In statistical theory, negative thermodynamic temperatures are considered. *See also* vacuum state.

absorbance *See* internal transmission density.

absorbed dose *See* dose.

α ($\%$) for metal film boundaries

$\lambda/(nm)$	silver	aluminium	gold
450	10	13	67
600	7	11	16
700	5	13	8
800	3	15	5
4000	2	6	3

absorptance *Syn.* absorption factor. Symbol: α. A measure of the ability of a body or substance to absorb radiation as expressed by the ratio of the absorbed *radiant flux or *luminous flux to the incident radiant or luminous flux. For radiant heat the absorptance of a body, measured against a vacuum, depends on the thermodynamics temperature T of the body receiving the radiation and on the wavelength. The absorptance at a fixed frequency of radiation is called the *spectral absorptance*. Typical values of α (expressed as a percentage) are shown in the table. The absorptance of a body was formally called its *absorptivity; it is equal to its *emissivity. *See* Kirchhoff's law.

absorption 1. *Syn.* sorption. The process in which a gas or liquid is taken up by another substance, usually a solid. Thus the absorbed material permeates into the bulk of the solid, as opposed to *adsorption which involves accumulation of material at a surface. Charcoal, for example, contains many small pores and any foreign substance finding its way into these may be said to be adsorbed on their surface. Since the substance permeates into the bulk of the charcoal the process may also be regarded as absorption.

2. Reduction in the flux of *electromagnetic radiation or other ionizing radiation on passage through matter. *See* absorption of electromagnetic radiation.

3. Reduction in the intensity of sound on passage through matter. *See* absorption of sound.

absorption bands (and **lines**) Dark bands or lines present in a *spectrum as a result of absorption by some intervening medium. The absorption lines of the solar spectrum were first observed by Wollaston but rediscovered by Fraunhofer who labelled the more prominent ones *A*, *B*, *C*, etc.

absorption coefficient 1. Of electromagnetic radiation. *See* linear absorption coefficient.

$\alpha(\%)$ for air-material boundaries

$f/(Hz)$	carpet pile	concrete	3-ply wood
250	14	1	28
500	37	2	26
1000	43	2	9
2000	27	2	12
4000	25	3	11

2. Of sound. Symbol: α. The ratio of the absorbed sound energy at a boundary to the incident sound energy. Its value depends on the material and on the frequency (*f*) of the sound. Some typical values of α (expressed as a percentage) are shown in the table. A perfect absorber, for which $\alpha = 1$, is often called an *open window* and the *open window unit* has been used as a unit of absorption coefficient, being the area of an open window having the same degree of absorption as the whole area of the boundary under consideration.

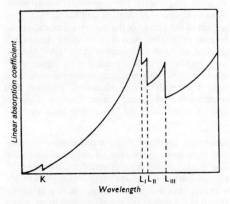

Absorption discontinuity

absorption edge (discontinuity or **limit) 1.** An abrupt discontinuity in the graph relating the amount of absorption of X-rays in a given substance with the wavelength of the radiation. At certain critical absorption wavelengths the absorption shows a sudden decrease in value. This occurs when the quantum energy of the radiation becomes smaller than the work required to eject an electron from one or other of the quantum states in the absorbing atom, and the radiation thus ceases to be absorbed by that state. Thus, radiation of wavelength greater than the K absorption edge cannot eject electrons from the K states of the absorbing substance.

2. A similar discontinuity in the absorption spectra of nonmetallic solids using infrared, visible, or ultraviolet radiation. It occurs when the quantum

absorption of electromagnetic radiation

energy is equal to the energy gap between the valence and the conduction bands.

absorption factor *See* absorptance.

absorption hygrometer *See* chemical hygrometer.

absorption of electromagnetic radiation When *electromagnetic radiation is incident on the boundary between two media, three processes can occur: reflection, transmission, and absorption. Reflection is the return of radiation into the original medium without any change in frequency of the components of which the radiation is composed. Transmission is the passage of radiation through the material without any frequency change. Absorption is the transformation of the energy of the radiation into a different form of energy. The nature of this process depends on the frequency of the radiation and on the substance involved. Radiation of all frequencies will increase the internal energy of an absorbing medium by various mechanisms. If the quantum energy is more than a few electronvolts, chemical changes may be caused. (*See also* photoelectric effect; photoionization; fluorescence; Compton effect; pair production).

The relative extents to which these three processes occur are given by the *reflectance, *transmittance, and *absorptance of the specimen or medium. The absorptance is the ratio of absorbed *luminous flux or *radiant flux to the incident luminous or radiant flux. (In the absorption of sound an analogous quantity is the *absorption coefficient). Similar definitions are given for transmittance and reflectance.

When a beam of radiation passes through a specimen the amount transmitted depends not only upon the amount absorbed but also on the amount reflected at the interfaces. It is also possible that diffuse transmission may occur, that is transmission in which the laws of refraction are not obeyed on the macroscopic scale and part of the radiation is scattered and lost from the beam without being converted into another form of energy (absorbed). The *internal transmittance is the ratio of the flux reaching the exit surface to that leaving the entry surface, and it is assumed that diffuse transmission does not occur. Similarly the *internal absorptance is the ratio of the flux absorbed between the inlet and exit faces to the flux leaving the exit face. Both these quantities are thus properties of the specimen under conditions in which reflection and diffuse transmission can be neglected. They depend on the thickness of the specimen and, for a given specimen, on the path length of the radiation, which depends on the angle of incidence. The *transmissivity is the internal transmittance of a layer of material for which the path length is unity. It is thus a property of the material itself and does not depend on the dimensions of the specimen. *Absorptivity is similarly defined. It is often convenient to use the *transmission density, which is the common logarithm of the reciprocal of the transmittance. This quantity is sometimes called *optical density*. The *internal transmission density is the common logarithm of the internal transmittance. It has also been called *absorbance*.

In its passage through a material the flux of radiation diminishes because of absorption and because of diffusion. If a collimated beam of monochromatic radiation is considered and diffusion and absorption occur, then the flux (Φ) falls off with distance according to the equation:

$$\Phi = \Phi_0 e^{-\mu x},$$

Φ_0 being the initial flux and Φ that after the radiation has traversed a distance x; μ is the *linear attenuation coefficient. A similar equation holds when absorption is the only process occurring:

$$\Phi = \Phi_0 e^{-ax},$$

a being the *linear absorption coefficient, sometimes called the *absorption coefficient*. This law is known as the *Lambert* or *Bouger law of absorption*.

absorption of sound When sound falls upon the boundary between two media, part is reflected and part enters the second medium and may be regarded as being absorbed. The ratio of the absorbed energy to the incident energy is called the *absorption coefficient. (A corresponding quantity in the *absorption of electromagnetic radiation is called the *absorptance.)

The loss in sound energy as it passes through a medium is given by the equation:

$$E = E_0 \exp(-\mu_\alpha x),$$

where E_0 is the incident energy, E the energy after a distance x, and μ_α a constant called the *absorption coefficient* or, to avoid confusion with the absorption coefficient mentioned earlier, the *linear absorption coefficient*.

The absorption of energy in sound waves has several causes:

(1) The viscous forces opposing the relative motion of the particles as the sound passes. This involves the transformation of acoustic energy into internal energy.

(2) The heat conducted from the compressed particles, which were at higher temperature, to the rarefied ones at lower temperature.

(3) The heat radiated from compressions to rarefactions, similar to conduction but with less magnitude, also causes some dissipation of energy at low frequencies.

(4) Mutual diffusion of the gas molecules (especially if the gas is admixed with some impurities) also produces an additional absorption by reason of the lighter molecules being driven in the sound wave from a position of compression to a position of rarefaction.

(5) The exchange of energy from translational to internal *degrees of freedom of the molecules.

The principal effects are viscosity and heat conduction and they are summed up in the *Stokes–Kirchhoff equation*:

$$\mu_\alpha = \frac{4\pi^2}{v\rho}\left(\frac{4}{3}\eta + \frac{(c_p - c_v)}{c_p c_v}\lambda\right)f^2,$$

where v is the speed of sound, ρ the density of gas, η its viscosity, c_p and c_v its principal specific heat capacities, λ its thermal conductivity, and f the frequency of the sound.

Absorption of sound by gases is basically due to viscosity, the conduction and radiation effects becoming more important for waves of larger amplitude. Water vapour content (humidity) also affects the absorption, μ_α taking a maximum value at about 15% relative humidity, and decreasing rapidly for higher and lower values; at its maximum μ_α tends to vary more directly with frequency f, rather than with f^2 (as the Stokes–Kirchhoff equation implies).

Absorption of sound in architectural materials (carpets, draperies, etc.) is essentially due to viscosity and conduction, since the attenuation is greatly increased when the particles of air are brought into contact with the walls of the narrow pores in the material. According to classical acoustics the degree of absorption in a narrow tube depends on its radius r:

$$\mu_\alpha = \frac{2}{vr}\left(\frac{\eta c_p}{\rho c_v}\right)^{1/2} f^{1/2}.$$

In fact, μ_α is generally rather larger than predicted by this formula, particularly for small values of r and very low frequencies. The experimental determination of values (*see* absorption coefficient) show that, as well as the nature of the material, attention must be paid to the treatment of its surface (whether it is painted, etc.) and the thickness of the layer. Absorption can also vary with the angle of incidence of the sound waves, being smallest for normal incidence; thus it depends on the mounting, position, and orientation of the sound source. Its value is important for determining the *reverberation characteristics of a room.

absorption spectrum When electromagnetic radiation from a source producing a continuous emission spectrum is passed through a medium, the spectrum reveals dark (or low-intensity) regions where absorption has taken place (continuous, line, and band types). In general, the medium absorbs at wavelengths at which it would emit radiation. Solids and liquids show broad continuous absorption spectra; gases give more discontinuous types (line and band). *See* spectrum.

absorptivity 1. A measure of the ability of a substance to absorb radiation, as expressed by the *internal absorptance of a layer of substance when the path of the radiation is of unit length and the boundaries of the material have no influence. (*See* absorption of electromagnetic radiation.)

2. Symbol: a. Formerly the fraction of radiant energy, incident from a vacuum on a body at temperature T, that is absorbed by the body. This term has now been replaced by *absorptance.

abundance *Syn.* relative abundance. Symbol: C. The number of atoms of a given isotope in a mixture of the isotopes of an element; usually expressed as a percentage of the total number of atoms of the element:

(1) *Natural abundance* Symbol: C_0. The abundance in a naturally occurring isotopic mixture of an element.

(2) *Cosmic abundance* The abundance of a nuclide or element in the universe expressed as a fraction of the total.

abvolt *See* ab-.

a.c. Abbreviation for *alternating current.

accelerating electrode Any electrode in an electronic valve or tube that serves to accelerate electrons in the electron beam. *See also* electron gun.

acceleration 1. Linear acceleration. Symbol: a; unit: m s^{-2}.

(1) In popular use and in scientific studies in one dimension it is the rate of increase of speed. If speed is decreasing, there is said to be a deceleration or a negative acceleration.

(2) In general scientific use, the rate of change of velocity. This is a vector quantity, having both magnitude and direction. The terms deceleration and negative acceleration are inapplicable in this sense.

2. Angular acceleration. Symbol: α; unit: rad s^{-2}. The rate of change of angular velocity. This is a pseudovector quantity having magnitude and the direction of orientation of an axis. In simple cases in which the axis of rotation is fixed, angular acceleration may be regarded as a scalar having a positive sign for increasing angular speed and a negative sign for decreasing angular speed.

acceleration of free fall *See* free fall.

accelerator A device for increasing the kinetic energy of charged particles or ions, such as protons or electrons, by accelerating them in an electric field. A magnetic field is used to maintain the particles in the desired direction. The particles can travel in a straight, spiral, or circular path. In a *linear accelerator the particles travel once through a long succession of radio-frequency cavities. Typical rates of energy gain are 7 MeV per metre for electrons and 1.5 MeV per metre for protons. The two-mile-long electron linear accelerator at Stanford (US) has a maximum energy of almost 50 GeV.

To achieve high energies in linear accelerators, the accelerating path must be extremely long. To overcome this disadvantage, cyclic accelerators were developed in which the particles completed hundreds or thousands of orbits of a circular path. The *cyclotron was the first such device, in which the accelerated particles follow a spiral path. The maximum energy is limited to tens of MeV due to the relativistic increase in mass.

The invention of the *synchrotron, which uses a different magnetic-field system from the cyclotron, results in far greater energies being possible. Other high-energy machines include the *betatron and the

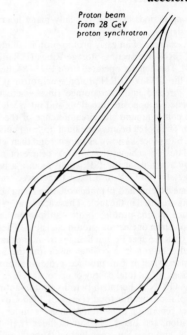

Proton beam from 28 GeV proton synchrotron

a   Intersecting storage ring at CERN

*synchrocyclotron. At present, the highest energies are obtained in the *proton synchrotron. Before acceleration, the kinetic energy of a particle is very small compared to its rest energy. Modern high-energy accelerators can produce kinetic energies many times greater than the rest energy. A typical rate of energy gain in a proton synchrotron is 40–50 MeV per metre. The rest energy of the proton is about 1 GeV. The *Super Proton Synchroton* (*SPS*) at *CERN (Geneva) accelerates protons to 450 GeV. (From 1990 CERN became known as the European Laboratory for Particle Physics.) In the USA, the Fermi National Acceleration Laboratory proton synchrotron (the TEVATRON) gives protons of 900 GeV. Superconducting magnets are used in the TEVATRON to produce large magnetic fields (about 4 T) thereby allowing larger particle energies.

In *fixed target* experiments the accelerated particles are extracted from the accelerator and directed into stationary targets where interactions can occur and be studied. In this case, conservation of momentum requires that the system of particles resulting from an interaction must have a large amount of kinetic energy. The total energy available for the production of new particles is therefore reduced, some of the energy being 'wasted' as kinetic energy. This problem is circumvented in the *colliding beam accelerator* where two beams of particles (usually of equal energy) moving in opposite directions are made to collide head-on. In a collider, the accelerator operates as a *storage ring* in which the accelerated particles are kept circulating around the machine

7

for long periods of time, typically many hours or even days.

An example of an early proton-proton collider is the CERN Intersecting Storage Rings (ISR) illustrated. This collider operated from 1972–84. Beams from the 28 GeV CERN proton synchrotron (CPS) were directed into two storage rings where they circulated in opposite directions and intersected at eight points around the circumference of the machine. The stored beams were built up from pulse to pulse by a process known as *stacking* and then left to circulate. To achieve the 56 GeV centre of mass energy available at the ISR would require a beam energy of 1700 GeV in a fixed target accelerator.

Some existing and planned colliding beam accelerators are listed in the table. These accelerators can produce proton-proton, proton-antiproton, electron-positron or electron-proton collisions. The 450 GeV CERN Super Proton Synchrotron for example can also be run in a collider mode (as the Spp̄S Collider) when it can produce proton-antiproton collisions with a total energy of up to 900 GeV. The TEVATRON at Fermilab permits proton-antiproton collisions with a total energy of 1800 GeV. A further generation of hadron colliders is under consideration, the Large Hadron Collider (LHC) at CERN and the Superconducting Super Collider (SSC) in Texas, which, if built, will reach total energies of about 16 TeV and 40 TeV respectively.

The Large Electron Positron (LEP) Collider at CERN accelerates electrons and positrons to 50 GeV and by 1994 the beam energy will probably be increased to about 90 GeV. Building a circular electron accelerator with energy greater than this is extremely difficult because of the radiation produced in bending the beams. Higher energy d^+e^- colliders will therefore require the use of linear accelerators. A step in this direction has been made at SLAC (at Stanford in California) where the existing electron linac has been modified to produce the Stanford Linear Collider (SLC). Electron and positron pulses are accelerated in the linac and then directed around two semi-circular arcs at the end of which they collide head-on with a total energy of about 100 GeV, similar to the energy of LEP. Unlike LEP however where the stored beams circulate for many hours, repeatedly colliding with each other, only a single collision is possible at the SLC before the two beams are deliberately dumped. To achieve an interaction rate comparable to that at LEP, the beams at the SLC must be squeezed down to very small size, a few microns across, and sophisticated focusing is then required to ensure that the two fine beams do indeed meet each other head-on.

In low-energy machines the particles are produced from heated filaments, ion sources, etc. The beams from these machines, including linear accelerators, cyclotrons, and *Van de Graaff accelerators, are injected into and further accelerated by high-energy accelerators. The final beam emerging from an accelerator can be magnetically deflected onto any of several targets, such as a *bubble cham-

ber. The resulting collisions give information concerning the basic nature of matter. New *elementary particles, requiring enormous energies for their formation, have been detected and their structure and properties determined.

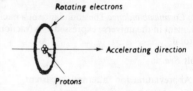

b Collective-effect accelerator

There are technological limits for both the electric field and the magnetic field, although the development of superconducting magnets and superconducting R.F. cavities has extended these limits. However, a more basic limit is imposed by the electric and magnetic fields generated by an intense beam of charged particles. These collective *self-fields* cause considerable instabilities even when they are less than 1% of the externally applied field. They will be very noticeable in the intense beams contained in particle-storage rings. At present, research is in progress to put the collective self-fields to practical use in a *collective-effect accelerator*. A beam of 10–20 MeV electrons can be deflected into a ring-shaped cluster by a suitable magnetic field. The magnetic attraction of the parallel electron currents cancel about 99.9% of the electrostatic repulsion. The introduction of a small number of protons results in a cancellation of this repulsion and a self-stabilized cluster is obtained with protons trapped in a deep potential well by the strong electric forces of the cluster (Fig. *b*). Acceleration of the cluster by an electric field of x V m^{-1} at right angles to the cluster plane produces relativistic effects which increase the electron mass by a factor of about 23. For a given velocity change, the energy change is directly proportional to the mass of the particles. The protons carried along inside the electron ring gain energy at a rate of xm_p/m_e eV m^{-1}, where m_p/m_e is the proton mass relative to the apparent electron mass. With an applied field of 10 MV m^{-1} the peak enhancement is expected to be about $10 \times 1836/23$ MeV m$^{-1} \cong 800$ MeV m^{-1}. This is much greater than the energy increase achieved in the present generation of cyclic accelerators.

accelerometer Any device used to measure acceleration. An *integrating accelerometer* is capable of performing one integration to obtain the velocity and a subsequent integration to obtain the distance travelled.

acceptor 1. The *impedance of a circuit comprising *inductance and *capacitance in *series has a minimum value at one particular frequency (the frequency to which the circuit is tuned; *see* tuned circuit). Such a circuit is an acceptor for that frequency. In practice, the effective resistance of such a circuit cannot be made zero and hence the impedance at the

Name	Start Date	Type	Beam Energy (GeV)	Circumference (km)
TRISTAN	1987	e^+e^-	30	3.02
SLC	1988	e^+e^-	50	1.46
LEP	1989	e^+e^-	50	26.66
Spp̄S	1981	pp̄	315	6.91
TEVATRON	1987	pp̄	900	6.28
LHC	1996(?)	pp	8000	26.66
SSC	1998(?)	pp	20000	83.63
HERA	1991	ep	26(e)+820(p)	6.34

Leading accelerators

frequency to which the circuit is tuned cannot be zero either. *See also* rejector.

2. *See* semiconductor.

access time The mean time interval between demanding a particular piece of information from a *storage device and obtaining it. In a given *computer system, the *memory will have the shortest access time, while the access time for information stored on *magnetic tape is much greater; *direct-access devices have an intermediate value.

accommodation The ability of the eye to alter its focal length and to produce clear images of objects at different distances. (*See* far point; near point.) Through the action of the ciliary muscle and the elasticity of the crystalline lens, the power of the latter and of the whole eye is varied. (*See* amplitude of accommodation.) With age, the lens becomes more rigid and the accommodation decreases. (*See* presbyopia.) The *Helmholtz theory assumes that the ciliary muscle acts so that the lens under its own elasticity becomes more convex; the *Tscherning theory* assumes that the ciliary muscle pulls on the lens making it more convex centrally.

accommodation coefficient Symbol: α. The ratio of the change in temperature of gas molecules occurring on collision with a surface to the difference between the temperature of the surface and that of the incident molecules. The coefficient, α, is therefore a measure of the extent to which the gas molecules leaving the surface of a solid in an atmosphere of the gas accommodate themselves to the surface temperature of the solid T_s. It is defined as:
$$\alpha = (T'_g - T_g)/(T_s - T_g),$$
where T_s is the surface temperature, T_g the temperature of the gas, and T'_g the temperature corresponding to the mean kinetic energy of gas molecules leaving the surface.

accretion disc A rotating disc-shaped mass of matter around a black hole or neutron star.

accumulator 1. (electrical). An electric cell that can be recharged after use. The common 'lead' accumulator consists in principle of two plates coated with lead sulphate immersed in aqueous sulphuric acid. If connected to a suitable d.c. supply, current is sent through the cell and the anode is converted to lead

(IV) oxide (lead peroxide) and the cathode is reduced to metallic lead. If the two plates are then connected through an external circuit, the chemical action is reversed and current flows round the external circuit from the brown peroxide plate to the grey lead plate. The action may be summarized in the equation:
$$PbO_2 + Pb + 2H_2SO_4 \leftrightharpoons 2PbSO_4 + 2H_2O,$$
with the right-to-left reaction occurring on discharge and the left-to-right reaction occurring on charge. Accumulators employing electrodes of nickel and iron (or cadmium) immersed in a solution of potassium hydroxide are also used. (*See* Edison accumulator.)

2. *See* hydraulic accumulator.

achromat *See* optical glass; achromatic lens.

achromatic colours Colours having no hue or saturation but only lightness. White, greys, and black are examples.

achromatic condenser A *condenser corrected for chromatic and spherical *aberrations, usually by having four elements, two of which are *achromatic lenses. It has a *numerical aperture of 1.4. It is used in *microscopes when high magnification is required. *See also* Abbe condenser.

achromatic fringes The spacing of most systems of fringes formed by *interference is dependent on the wavelength of the light. If the source is white, overlapping systems of coloured fringes are produced and usually only a few fringes are distinguishable. The first fringe of a *Lloyd's mirror system and the central fringe of a Fresnel's *biprism system are examples of achromatic (black and white) fringes.

A complete system of achromatic fringes is obtainable with a Lloyd's mirror if the "source" is a narrow projected spectrum, about 10^{-4} m long, with the violet end nearest to the mirror and suitably displaced from it.

achromatic interval *See* achromatic threshold of vision.

achromatic lens A combination of two or more lenses using, if necessary, different kinds of glass, designed to remove the major part of *chromatic aberration. (*See* achromatism.) First successfully

9

attempted by *Dolland (1757). The elementary theory of simple achromatic doublets assumes two lenses of powers P_1 and P_2 placed in contact (with total power $P = P_1 + P_2$) made of glasses of *dispersive powers ω_1 and ω_2 so that the condition for achromatism ($\omega_1 P_1 + \omega_2 P_2 = 0$) is satisfied. To produce an achromatic converging lens, e.g. for telescope or photographic objectives, the dispersive power of the higher-power convergent lens must be less than that of the divergent lens of the combination. Before the development of the wide selection of optical glasses at Jena, low dispersive powers (crowns) were generally associated with low refractive indexes. Photographic lenses made with such types of glass are called *old achromats*. The term *new achromat* refers to the type in which glasses of inversely related refractive index and dispersion are used. (*See* optical glass; anastigmat.) The elementary theory of eyepieces (Huygens, Ramsden) shows that two thin lenses made of the same type of glass and separated by a distance equal to half the sum of their focal lengths provide correction in which the power referred to principal points (*see* principal planes) is the same for two main colours. This provides approximately lateral achromatism (*see* lateral aberration). The Kellner eyepiece provides a better colour correction than the Ramsden design. More complete corrections for objectives and eyepieces require the use of more types of glass. *See* apochromatic lens.

achromatic prism A combination of two or more prisms that produces the same deviation of two or more colours so that objects viewed through them will not appear coloured (*see* chromatic aberration). As with thin lenses, "narrow angle" prisms are placed in contact in opposition, so that the *dispersive powers of the two glasses are inversely proportional to their angles of *deviation.

achromatic threshold of vision The smallest light stimulus detectable by the *adapted eye* (*see* adaptation of the eye). At low illumination, all colours lose their hue. There is a much bigger *achromatic interval* (difference between colours and brightness limits) for short waves.

achromatism The removal of *chromatic aberration, or *chromatic differences of magnification, or both, arising from dispersion of light. Owing to nonlinearity of dispersion the correction is attempted for two colours in the first approximation, and for three colours in higher corrections. (*See* apochromatic lens.) Visual or optical achromatism corrects for C and F lines (*see* Fraunhofer lines). Actinic achromatism corrects for D and G lines for photographic lenses and for F and G lines for astonomical lenses (*FG achromatism*). The necessary use of thick and of separated lenses of different kinds of glass makes the conditions of achromatism somewhat involved, especially when considered in relation to other factors: object position, spherical aberration, and degree of correction. In some instruments and in telescopes and microscopes, more perfect achromatism is effected by the use of compensating eyepieces.

aclinic line *Syn.* magnetic equator. A curve drawn in such a manner that all places on the curve have zero magnetic dip. *See* isoclinal.

acoustic absorption coefficient *See* sound absorption coefficient.

acoustic bridge The practical measurement of acoustic *impedance is rather more difficult than the measurement of electrical impedance in which resistance, inductance, and capacitance are usually separable quantities. The direct method is to measure the pressure gradient using a calibrated manometer capsule and the displacement current (or more precisely the strength of sound) by a calibrated hot wire; then the impedance is the ratio between them.

There are, however, some bridge methods similar to those used in electricity. The most suitable one is a form of acoustic Wheatstone bridge. It consists of two similar tubes forming the ratio arms in which sound waves are propagated by a loudspeaker. One of these tubes forms a known reactance standard and the other is the unknown impedance. The sound pressures at points equidistant along the ratio arms are balanced by adjusting the standard reactance. Balance is indicated by a differential microphone, which corresponds to the galvanometer or headphone in electrical circuits.

acoustic capacitance The imaginary component of acoustic *impedance due to the stiffness or elasticity (k) of the medium; it is equal to S^2/k where S is the area in vibration. In a *Helmholtz resonator the elasticity is seated in the air of the cavity. The capacitance (C) is given by $-\delta v/\delta p$ where δv is the change of volume of air in the cavity caused by an increase of pressure δp. For an adiabatic change:

$$\frac{\delta v}{\delta p} = -\frac{v}{\gamma p} = -\frac{v}{c^2 \rho},$$

where c is the speed of sound, γ the ratio of specific heat capacities, and ρ is the mean density of the medium. The capacitance is thus a function of the volume of the vessel of the resonator.

acoustic delay line *See* delay line.

acoustic filters As in the case of electrical filters, lines of acoustic *impedances can be made by proper adjustment to transmit high frequencies only (high-pass filters) or low frequency only (low-pass filters) or any given band of frequency (band-pass filters).

The theoretical treatment of impedances shows that if any simple harmonic motion is impressed on equal impedances Z_1 connected in a conduit and separated by branches containing other equal impedances Z_2, it will not pass through unless the ratio Z_1/Z_2 for the frequency of this S.H.M. lies between certain values; i.e. all other frequencies which do not satisfy this condition will be rapidly attenuated and only those covering this range will get through. The conditions corresponding to the three types of filter are:

(1) For low-pass filters: Let Z_1 consist of inertance only (L) and Z_2 of capacitance only (C); then:

$$Z_1 = \omega L = Z_2 = -1/\omega C.$$

The filter passes low frequencies between 0 and $1/\pi\sqrt{LC}$. This is satisfied in practice by a tube having cavities of considerable size set close together along the sides, as branches.

(2) For high-pass filters: Let the series impedances be capacitance only (C) and the shunt impedances be inertances only (L); then:

$$Z_1 = -1/\omega C \text{ and } Z_2 = \omega L.$$

The filter passes high frequencies between $1/4\pi\sqrt{LC}$ and ∞. In practice, this is satisfied by a wide tube with holes which may have short necks surrounding them.

(3) For medium or band-pass filters. These have capacitances C_1 in the main circuit but both capacitances C_2 and inertances L_2 in the branch line. The limits of frequencies are given by:

$$\omega = 1/\sqrt{(C_2 + 4C_1)L_2}$$

and

$$\omega = 1/\sqrt{L_2 C_2}.$$

The main tube of such a filter has side holes with necks, which are broken at the centre and communicate with closed vessels which surround them. *See also* cavity absorbent.

acoustic grating A series of objects, such as rods of equal size, placed in a row a fixed distance apart constitute an acoustic grating having similar properties to an optical *diffraction grating. When a sound wave is incident upon an acoustic grating, secondary waves are set up that reinforce each other or cancel out according to whether or not they are in phase. The result for a sinusoidal sound wave is a series of maxima and minima spaced round the grating. When the incident sound is normal to the grating, the condition for a maximum diffracted sound at an angle θ to the normal is $\sin\theta = m\lambda/e$, where λ is the wavelength of the sound, e is the width of a rod plus the space between it and the next, and m is an integer. e must be greater than λ for a diffraction pattern to be formed and this condition necessitates very large gratings for low-frequency sounds. Concave acoustic mirrors are generally used to make the incident sound parallel and also to focus the diffracted beam on to a sensitive detector such as a radiometer.

acoustic impedance *See* impedance.

acoustic inertance The imaginary component of acoustic *impedance, due solely to inertia. It corresponds to inductance in electric circuits. In the case of a mass (m) of gas in a conduit of cross section S, the inertance (L) is equal to m/S^2.

In a *Helmholtz resonator it will be comprised of the mass of the air or gas in the neck. Therefore in this case $L = \rho l/S$ where ρ is the density of gas and l is the length of the neck including the end-correction. In the case of a pipe, the inertance cannot easily be separated from the other components of impedance.

acoustic interferometer An instrument in which *standing waves are set up to measure the characteristics of a medium or its boundaries, by studying the *interference patterns produced.

acoustic levitation *See* levitation.

acoustic mass *See* reactance (acoustic).

acoustic power *See* sound-energy flux.

acoustic pressure *See* sound pressure.

acoustic reactance *See* reactance.

acoustics 1. The science concerned with the production, properties, and propagation of sound waves.
2. The characteristics of a room, auditorium, etc., that determine the fidelity with which sound can be heard within it. *See* auditorium acoustics.

acoustic shadow An acoustic shadow may be produced by an obstacle placed in the path of a sound wave in a similar manner to the formation of an optical shadow. There are great practical differences in the two cases, however, due to the fact that the wavelength of sound is much larger than that of light. As a result, sound waves tend to bend round obstacles, shadows are smaller than the geometrical shadows, and their edges are not clearly defined. This is a diffraction effect in which secondary waves originating near the edge of an obstacle penetrate into the geometrical shadow and may add together or cancel out at any point depending on their relative phases. In the case of a circular disc normal to a plane sound wave it can be shown that in any plane parallel to the disc there is a central maximum surrounded by a series of concentric rings of maximum and minimum intensity. This effect has been used in direction finding since if a large disc is rotated until it is normal to the sound the central maximum will be on its axis. An obstacle must be very large compared with the wavelength of the sound to produce a clear image. This has been achieved in the case of high-frequency sounds, which have correspondingly short wavelengths. When the dimensions of an obstacle are small compared with the wavelength of the sound, no shadow of any kind is produced. The waves appear to pass round the obstacle and join together immediately behind it.

acoustics of buildings *See* auditorium acoustics.

acoustic spectrum A representation of the distribution of sound intensity emitted by a source with respect to frequency.

acoustic stiffness *See* reactance.

acoustoelectronics The study and use of devices in which electrical signals are converted into acoustic waves by *transducers and the acoustic signals are propagated through a solid medium. Since sound travels more slowly than electric signals, acoustic *delay lines are much lighter and more compact than purely electronic devices of comparable performance.

actinic A term referring to the ability of radiation, such as light or ultraviolet radiation, to produce a chemical change in exposed materials. In photography, ultraviolet and blue light are more highly actinic than other parts of the spectrum, an ordinary emulsion being most sensitive to this radiation. Ultraviolet radiation is actinic as regards its sun-tanning effect.

actinic achromatism *See* achromatism.

actinium series *See* radioactive series.

actinometer 1. An instrument used to measure intensity of radiation. Acintometers usually depend on determining the extent to which a screen fluoresces or the extent to which a substance is decomposed by the radiation.

2. An instrument designed to measure the intensity of solar radiation. Acinometers are now usually called *pyrheliometers.

actinon A former name for radon-219. It decays by α-particle emission with a half life of 3.92 s, and is found as a member of the actinium *radioactive series.

action 1. The product of a component of momentum, p_i, and the change in the corresponding positional coordinate, q_i; or more precisely the integral $\int p_i dq_i$. (*See* Hamiltonian function; quantum of action.)
2. Twice the time integral of the kinetic energy of a system, measured from an arbitrary zero time. (*See* least-action principle.)

activation The process of heating an absorbent material, such as charcoal, and pumping off the gases so released. Charcoal activated in this way is capable of absorbing large quantities of gases.

activation analysis A sensitive analytical technique in which the sample is first activated by bombardment with slow neutrons, high-energy particles, or gamma rays and the subsequent decay of radioactive nuclei is then used to characterize the atoms present. For example, stable sodium nuclei can be activated by neutron capture:
$$^{23}Na + n \rightarrow {}^{24}Na + \gamma.$$
The ^{24}Na nuclei decay to give γ-rays, electrons, and neutrinos:
$$^{24}Na \rightarrow {}^{24}Mg + \gamma + e^- + \bar{\nu}.$$
The electrons have a characteristic energy spread and the γ-rays have energies of 2.75 and 1.37 MeV. Sodium can thus be detected by the presence of lines at these energies in the *gamma-ray spectrum of the irradiated material.

The usual method of irradiation is by neutron bombardment in a nuclear reactor. The technique is highly sensitive and can be used for a large number of elements. If the intensity of γ-ray emission is compared with that from a similarly treated standard, a quantitative analysis can be made.

activation cross section *See* cross section.

activation energy Energy that must be supplied to initiate a reaction. Thus, although the fission of a nucleus of ^{235}U yields nearly 180 MeV of energy, fission does not occur unless the nucleus first absorbs about 5 MeV of energy.

activator An impurity atom that will produce or increase the efficiency of *luminescence of a solid. Copper atoms are added to zinc sulphide as activators and thallium atoms can be used with potassium chloride.

active *See* radioactive.

active aerial *See* directive aerial.

active component An electronic component such as a *transistor or *thermionic valve, that can be used to introduce *gain into a circuit. *Compare* passive component.

active current The component of an alternating current that is in *phase with the voltage, the current and voltage being regarded as vector quantities. Alternative terms: active component, energy component, power component, in-phase component, of the current. *Compare* reactive current.

active voltage The component of an alternating voltage that is in *phase with the current, the voltage and current being regarded as vector quantities. Alternative terms: active component, energy component, power component, in-phase component, of the voltage. *Compare* reactive voltage.

active volt-amperes The product of the current and the *active voltage or the product of the voltage and the *active current. It is equal to the power in watts. Alternative terms: active component, energy component, power component, in-phase component, of the volt-amperes. *Compare* reactive volt-amperes.

activity 1. Symbol: A. The number of atoms of a radioactive substance that disintegrate per unit time ($-dN/dt$). It is measured in *becquerel, or, formerly, in *curie (3.7×10^{10} Bq).
2. A quantity used mainly in chemical thermodynamics to express the effective concentration of a substance in a solution or mixture. Absolute activity (λ) is defined by the equation:
$$\lambda = \exp(\mu/RT),$$
where μ is the *chemical potential of the substance. The relative activity (a) is given by λ/λ^*, where λ^* is the absolute activity of the pure substance at the same temperature and pressure as the mixture. The law of *mass action is valid if concentrations are replaced by activities.
3. *See* optical activity.

adaptation of the eye The physiological process that enables the eye to adjust its sensitivity for different levels of illumination. The most sensitive state is that of dark adaptation (scotopic vision) and requires over an hour's rest in total darkness (after exposure to daylight) to reach it. The process of light adaptation is more rapid. Adaptation is measured by the *adaptometer*, which provides a measure of the lowest brightness of an extended area that can just be detected.

adaptometer *See* adaptation of the eye.

ADC Abbreviation for analogue/digital converter.

Adcock direction-finder *Syn.* Adcock antenna. A radio direction-finder employing a number of spaced vertical aerials. It is designed so that any horizontally polarized components of the received waves have the minimum effect upon the observed bearings. In this manner, the errors due to such horizontally polarized components are effectively eliminated.

A/D converter *See* analogue/digital converter.

additive process A process by which almost any colour can be produced or reproduced by mixing together lights of three colours, called *additive primary colours*, usually red, green, and blue, the proportions of which determine the colour obtained; white light is obtained from approximately equal proportions of red, green, and blue light; yellow from a mixture of red and green light. *Colour television uses an additive process for final colour production. *See also* chromaticity. *Compare* subtractive process.

address A unique identification number assigned to a unit of computer *memory. The smallest addressable entity can be either a *word or a *byte. Addresses are often written in *hexadecimal notation or *octal notation.

adhesion The interaction between the surfaces of two closely adjacent bodies that causes them to cling together. *Compare* cohesion.

adhesional work *See* spreading coefficient.

adiabatic demagnetization A process used for the production of temperatures near *absolute zero. A paramagnetic salt is placed between the poles of an electromagnet and the field switched on, the resulting heat being removed by a liquid-helium bath. The substance is then isolated thermally and on switching off the field the substance is demagnetized adiabatically and cools. Gadolinium sulphate, cerium ethyl sulphate, and potassium chrome alum have all been used.

adiabatic lapse rate The rate of change of temperature with height of a large mass of air rising in the atmosphere, without any condensation of water vapour. When air rises, the pressure upon it decreases so it expands, doing work at the expense of its internal energy. Air is a very poor conductor of heat, so any large quantity can be regarded as perfectly insulated and the expansion is an *adiabatic process. Atmospheric processes are sufficient-

ly slow to be nearly *reversible. For a change of height dx the change of temperature dT is given by:
$$dT = -g \, dx/c_p,$$
where g is the acceleration of free fall and c_p is the specific heat capacity at constant pressure. This gives the lapse rate of approximately 0.01 °C per metre.

For fine weather in temperate regions the temperature gradient is typically about two-thirds of the adiabatic lapse rate. Thus if air rises for any reason its temperature soon falls below that of its surroundings, so it becomes denser and a resultant downward force acts on it. Conversely, if air falls it becomes warmer than its surroundings and the resultant force is upward. Hence the atmosphere is generally stable, despite being heated from below by solar radiation absorbed on the ground. With intense heating the temperature gradient becomes greater than 0.01 °C m^{-1} and the atmosphere is unstable, causing violent storms, tornadoes, etc.

adiabatic process A process in which no heat enters or leaves a system. For example, if gas is compressed or expanded in a cylinder by a piston it undergoes an adiabatic change if the cylinder does not allow transfer of heat between the gas and the surroundings. An adiabatic expansion results in cooling of a gas whereas an adiabatic compression has the opposite effect. If an *ideal gas undergoes a reversible adiabatic change in volume, the pressure (p) is related to volume (V) by the equation: $pV^\gamma = K$, where K is a constant and γ is the ratio of the *heat capacities C_p/C_V of the gas. Any reversible adiabatic change (*see* reversible change) is *isentropic*, i.e. during the change the entropy of the system remains constant. *Compare* isothermal process.

adiathermic or adiathermanous Not transparent to heat, as opposed to *diathermic*.

admittance Symbol: Y. The reciprocal of *impedance; it is related to *conductance (G) and *susceptance (B) by:
$$Y^2 = G^2 + B^2$$

adsorbate Any substance adsorbed on the surface of an *adsorbent.

adsorbent A substance on which *adsorption takes place.

adsorption The formation of a layer of foreign substance on an impermeable surface. *Compare* absorption.

advanced gas-cooled reactor (AGR) *See* gas-cooled reactor.

advection A process of transfer of atmospheric properties by horizontal motion in the atmosphere, such as the movement of cold air from polar regions. Advection is concerned with large-scale motions in the atmosphere; vertical, locally induced motions are *convection processes. In oceanography, advection is the flow of sea water as a current.

aeoleotropic *See* anisotropic.

aeolian tone A cylindrical obstacle (e.g. a wire) placed in a stream of fluid sets up a series of vortices on each side of the obstacle, revolving in opposite directions (*see* vortex street). The stream is set into transverse vibration as the eddies are formed and break away. If the wire dimension and flow are suitably adjusted, the alternating pressure waves so formed become audible. It is found approximately that the frequency of the sound is $0.0185 \, v/d$ where v is the speed of air flow and d the obstacle thickness. Such sounds in air are called aeolian tones. The aeolian harp (Aeolus – God of Winds), dating in its present form from the end of the sixteenth century, used this principle for the production of vague musical sound. A number of strings of varying diameter are mounted on a wooden resonance board, and all tuned to unison, the unison frequency being so chosen that a large number of harmonics lie in the audible range. The wind sets the strings in vibration, the harmonics more than the fundamental being generally heard. Since the strings are of different thickness, a given wind speed stimulates different harmonics in the different strings and an impression of chords is heard, which varies considerably with wind-speed changes. The different wire diameters employed also ensure that at least one of the strings will be excited to resonance. Aeolian tones are also heard when the wind passes telegraph wires and the smaller twigs and branches of trees.

aerial *Syn.* antenna. That part of a radio system from which energy is radiated into (*transmitting aerial*), or received from (*receiving aerial*) space. An aerial with its *feeders and all its supports is known as an *aerial system*. The most important types of aerial are the *dipole aerial and the *directive aerial.

aerial array *Syn.* beam aerial. An arrangement of radiating or receiving elements so spaced and connected that directional effects are produced. With suitable design, very great directivity can be obtained and also, as a consequence, large *aerial gain. An array of elements along a horizontal line, which has marked directivity in the horizontal plane in a direction at right angles to the line of the array, is referred to as a *broadside array*. One that has directivity in the horizontal plane along the line of the array is called an *end-fire array* (or *staggered aerial*). Arrays are commonly designed for directivity in both horizontal and vertical planes. The horizontal directivity is influenced by the number of aerial elements that are arranged horizontally, whereas the vertical directivity depends upon the number of elements that are stacked in tiers (or stacks), one vertically above the other.

aerial feed-impedance The effective impedance that an aerial, considered as a load, presents between the two terminals at the point where it is fed.

aerial gain The ratio of the signal power produced at the receiver input by the aerial to that which would be produced by a standard comparison aerial under similar conditions (i.e. similar receiving conditions and the same transmitted power). If the type of standard comparison aerial is not specified, a *half-wave dipole is implied.

aerial radiation resistance Fictitious resistance that takes into account the energy consumed by an aerial system as a result of radiation. For any element of the aerial, it is that resistance that, when carrying a current I_0, would dissipate the same energy as that radiated from the element, I_0 being the root-mean-square value of the current at a point in the element at which the current is a maximum (i.e. at a current antinode).

aerial resistance The resistance that takes into account the energy consumed by an aerial system as a result of radiation and losses (e.g. I^2R losses in aerial wires, dielectric losses, earth losses, etc.). It is equal to the power supplied to the aerial divided by the square of the current at the aerial supply point.

aerial system *See* aerial.

aeroballistics The study of the interaction between the earth's atmosphere and high-speed projectiles or spacecraft.

aerodynamics The study of the motion of gases (particularly air) and the motion and control of solid bodies in air.

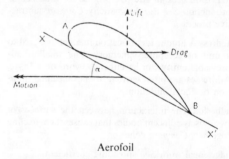

Aerofoil

aerofoil A body for which, when in relative motion with a fluid, the resistance to motion (drag) is many times less than the force perpendicular to motion (lift). The flight of aircraft depends on the use of aerofoils for wing and tail structure. The essential features of the aerofoil are the rounded leading edge A and the sharp trailing edge B. The projection of the section on to the common tangent XX′ is called the *chord*; the angle α is the *angle of incidence* or *attack*.

Due to viscosity, the layers of fluid passing over the upper and lower surfaces of the aerofoil arrive at the trailing edge with different velocities to form a *surface of discontinuity. This leads to the production of an eddy or vortex at the trailing edge accompanied by a counter-circulation around the aerofoil. This circulation is essential for the production of the lift force. *Compare* Magnus effect.

aerometer *See* hydrometer.

aeronautics The scientific study of flight, including the design of aircraft and the theory and practice of aerial navigation.

aerophysics The physics of the earth's atmosphere, especially as it interacts with bodies travelling at high speeds or high altitudes.

aerosol A dispersed system of solids or liquids in a gas: it includes smoke, dust, fog, clouds, mist, haze, and fumes.

aerothermodynamic duct *See* ramjet.

aether *See* ether.

a.f.c. Abbreviation for *automatic frequency control.

afterglow *See* persistence.

afterheat The heat generated in a nuclear reactor after it has been shut down. It is caused by radioactive substances that form in the fuel elements.

agate Any one of a group of amorphous and pseudocrystalline forms of silica. Apart from its wide use for ornamental purposes (cameos are cut from banded agates), agate is used in the manufacture of knife edges of delicate balances, small mortars and pestles, burnishers, etc.

a.g.c. Abbreviation for *automatic gain control.

age equation or theory *See* Fermi age theory.

age hardening The progressive hardening of an alloy with time, under given conditions of temperature, due to the separation of a differently oriented crystalline phase.

age of the earth *See* dating.

age of the universe A time of between 10×10^9 to 20×10^9 years as determined by the *Hubble constant. This value is clearly uncertain and depends on current theories of *cosmology. *See also* expanding universe.

agglomerate cell *See* Leclanché cell.

agonic line A curve drawn in such a manner that all places on the curve have zero magnetic *declination. *See* isogonal.

AGR Abbreviation for advanced *gas-cooled reactor.

Aharonov–Casher effect An effect in which an electrically neutral particle with a magnetic moment is influenced by an electric field. It can be demonstrated by diffraction of neutrons by a line of electric charge. There is a magnetic field associated with a moving neutron in the vicinity of an electric field, and this affects the neutron's phase. A similar effect on electrically charged particles moving close to a magnetic field is called the *Aharonov–Bohm effect*.

air Normal dry air has the following composition by volume:

Nitrogen	78.08%
Oxygen	20.94%
Argon	0.9325%
Carbon dioxide	0.03%
Neon	0.0018%
Helium	0.0005%
Krypton	0.0001%
Xenon	0.000 009%
Radon	6×10^{-18}%

For dry air:

Specific heat capacity at constant volume	718 J kg^{-1} K^{-1}
Specific heat capacity at constant pressure	1006 J kg^{-1} K^{-1}
Ratio of specific heat capacities	1.403
Boiling point (at atmospheric pressure)	$-193\,°C$ to $-185\,°C$ depending on age

Liquid air is produced by strong cooling under high pressure (*see* liquefaction of gases). It has a pale blue colour due to the presence of liquid oxygen.

For air at 0 °C and 1 atmosphere pressure the refractive index is given by the *Cauchy dispersion formula:
$$n - 1 = A + B/\lambda^2,$$
where n is the refractive index for a wavelength λ, A $= 2.875\,66 \times 10^{-4}$, and B $= 1.3412 \times 10^{-18}$ m^2. At 15 °C and 1 atmosphere, A $= 2.726\,43 \times 43 \times 10^{-4}$ and B $= 1.2288 \times 10^{-18}$ m^2, giving for the sodium D line $n = 1.000\,2765$.

air-break (Of a switch, *circuit breaker, etc.) Having contacts that separate in air. *Compare* oil-break.

air-cooled (Of an electrical machine, transformer, or engine) Having air as the sole medium for cooling the *windings, *core, or other working parts.

air equivalent A measure of the efficiency of an absorber of nuclear radiation, expressed as the thickness of a layer of air at standard temperature and pressure that causes the same amount of absorption or the same energy loss.

air gap A break in the ferromagnetic portion of a *magnetic circuit. Its length, measured in the direction of the *magnetic flux, is usually short compared with that of the remainder of the magnetic circuit. An air gap may be necessitated by mechanical considerations as, for example, in the case of the air gap between the *armature and *pole faces of a d.c. machine. Here the armature must be able to rotate without fouling the pole faces. *See also* section gap.

air-insulated (Of an electrical apparatus or machine) Having air as the whole, or the major portion, of the insulating medium between the live parts, their supports, and their surrounding casing.

air mass A part of the lower atmosphere in which the horizontal temperature gradient at all levels is very small. At the margins of an air mass the temperature gradients become steep; such a transition zone is called a *front*.

air pumps Air-exhausting pumps are used for withdrawing air or other gases from a closed chamber. Many of them are based upon the ordinary piston principle; others, e.g. the *Sprengel air pump and the *filter pump, use a jet of mercury or water to trap and remove air from the vessel to be exhausted. In the *Gaede molecular air pump*, a grooved cylinder rotates in a casing with very little clearance; a fixed comb projects into the grooves in the cylinder, the inlet for gas being on one side of the comb and the outlet on the other. On rotation of the cylinder, the gas is dragged from the inlet to the outlet; low pressures of the order of 0.001 mm of mercury can be achieved with a speed of revolution of the cylinder of 8000 to 12 000 revolutions per minute. *See also* pumps, vacuum.

air terminations *Syn.* air-termination networks. The parts of a lightning protective system from which it is intended that lightning discharges shall emanate into the atmosphere or at which the discharges shall be collected from the atmosphere. *Compare* earth terminations.

air wall *See* ionization chamber.

Airy, Sir George B. (1801–1892) Brit. astronomer. In addition to substantial contributions to astronomy, including the origination of photographic records of sunspots, discovery of irregularities in the motions of Venus and the earth, and the design of transit instruments, he estimated the mass of the earth by observations of *g* (the acceleration of free fall) at the top and bottom of a mineshaft and made many optical discoveries. *Airy's spirals* arise in the polarization of light by quartz crystals. *Airy's disc* is the diffraction pattern that constitutes the "image" of a distant point-object given by a telescope. The larger the aperture of the telescope, the smaller is the disc. *See* diffraction of light.

albedo 1. The fraction of incident light diffusely reflected from a surface. Some typical values are: fresh clean snow 0.8 to 0.9; fields and woods 0.02 to 0.15; mean for the whole earth 0.4; moon 0.073; Mercury 0.07; Venus 0.72, Mars 0.17; Jupiter 0.70. Saturn 0.63; Uranus 0.63; Neptune 0.73.
   2. The probability that a neutron entering into a region through a surface will return through that surface.

alcoholometry The determination of the strength of spirits. Sikes's *hydrometer is usually used and its readings are expressed in terms of over- or underproof strengths. *See* proof.

Alcomax *See* Alnico.

Alfvén, Hannes Olof Gösta (*b.* 1908) Swedish physicist who became professor at the University of California (San Diego) in 1967. Specializing in plasma physics he predicted (1942) the existence of waves in plasmas (now called Alfvén waves) and shared a Nobel prize (1970) with *Néel for his work on magnetohydrodynamics. His work on stellar plasmas led to a theory (1950) on the origin of the solar system.

Alfvén wave A magnetohydrodynamic wave propagated in a plasma. Alfvén waves are transverse waves; the plasma particles vibrate perpendicular to the direction of propagation, which is along the lines of magnetic flux density *B*. They travel at the *Alfvén speed*, equal to $B/\sqrt{(\rho\mu)}$, where ρ is the density and μ the magnetic permeability.

ALGOL (ALGOrithmic Language) A language designed on a rigorous syntactical basis for the unambiguous expression of *algorithms. It is used as a scientifically oriented high-level *programming language.

algorithm A method or procedure of computation, involving a series of steps or instructions. If no algorithm exists, a *heuristic solution has to be sought.

aligned-grid valve A *thermionic valve in which the wires or other conductors forming the screen grid and (usually) the control grid are arranged so that only a very small fraction of the total space current is intercepted by the screen grid. For example, in the *beam pentode tetrode* the wires of the screen grid are placed in the shadow of those of the control grid, as far as the electron stream is concerned.

alive *Syn.* live. Applied to an electrical conductor or circuit that is not at earth potential. *Compare* dead.

allobar A mixture of the isotopes of an element in proportions that differ from the natural isotopic composition.

allochromy The emission of electromagnetic radiation from atoms, molecules, etc. induced by incident radiation of a different wavelength, as in *fluorescence or the *Raman effect.

allotropy The existence of a solid, liquid, or gaseous substance in two or more forms (*allotropes*) that differ in physical rather than chemical properties. Solid allotropes often result from *dimorphism.

allowed band *See* energy bands.

allowed transition A transition between two states of a system that is consistent with the *selection rules. If an excited state can go to a less excited state by an allowed transition, the mean life of the state is very short unless the energy difference is very small. *Compare* forbidden transition.

alloy A material, other than a pure element, that exhibits characteristic metallic properties. At least one major constituent must be a metal. A definite chemical compound of a metal with a nonmetallic element, e.g. an oxide or a sulphide of a metal would not be counted as an alloy even if it showed some metallic properties.
   Most engineering alloys are still made by the procedure of melting a mixture of the constituents and then casting, but alloys are also produced by compacting metal powders, simultaneous electro-

deposition, diffusion of constituents in the solid state, and by the condensation of vapours.

An alloy examined under the microscope after suitable polishing and etching may appear to be homogenous, consisting of grains of identical composition and nature, or may consist of two or more *phases, having, for example, grains of one composition embedded in a matrix of grains of a different composition. The individual grains are crystalline but do not usually betray their crystalline nature by external form because of the way in which they are packed together.

Homogenous phases may consist of a *solid solution* of one element in another. This implies that the parent metal (the solvent) has absorbed the solute element into its crystalline lattice, either by the substitution of solute atoms (on a random pattern or in a regular manner) or by the accommodation of solute atoms in the interstices of the lattice. The latter process occurs only when the solute has relatively small atoms. Solid solutions may also occur with two or more solutes simultaneously present and, as with 'ordinary' (liquid) solutions, there may be a limited range of solubility of a given solute or the solvent and solute may be miscible in all proportions.

Intermetallic compounds may have characteristic crystal lattices that differ from those of the ingredient elements. Unlike ordinary chemical compounds, the proportions of the various atoms in the compound may often be varied over appreciable ranges and even the basic ratios need not be simple. This arises from the special nature of the metallic bonding, which differs from the types of valency bond occurring in conventional chemical compounds.

Most alloys solidify from a melt over a range of temperature, in contrast to the well-defined freezing point of typical elements or ordinary pure chemical compounds, and the first crystals to form do not have the same composition as the melt from which they separate. Consequently, the residual liquid progressively changes its composition during solidification. Even with *binary alloys*, with only two constituents, there are various possibilities. Where the range of solid solubility is limited, it is commonly found that the progressive change in composition of the residual liquid, as the temperature drops and more solid separates, is arrested by the complete solidification of the residual liquid as a fine intermixture of distinct phases. The material has now the *eutectic composition* and an alloy of this particular composition will have a definite freezing-point, which will be a minimum for the series of alloys. The eutectic alloy behaves in some respects as though it were a definite compound and it was at one time so regarded.

Eutectic alloys are of considerable importance. For example, they are of value in low-melting-point solders (*see* soldering), and in many foundry alloys and in fusible safety plugs.

During the solidification of a melt, the first crystals to form frequently grow as *dendrites* or tree-like structures and in the final alloy, these dendritic growths are often still discernible, enmeshed in a mass of grains of different composition. In the newly formed solid, especially if solidification has taken place rapidly, the whole mass is unlikely to be in equilibrium. There will, in general, be a reshuffling of the atoms if conditions permit. This may, for example, involve a recrystallization of part of or all the material and may involve migration of some of the atoms through the structure. Such rearrangements are often facilitated by maintaining the alloy for a prolonged period at a fairly high temperature as in *annealing. The equilibrium arrangement at the higher temperature may differ appreciably from that at room temperature, so that rapid cooling (e.g. by *quenching in water or oil) may produce a very different result from slow cooling. The *temper* of steel is controlled in this way.

Some of the physical properties of an alloy are very susceptible to changes in internal structure and, if the results of measurements are to be generally applicable, the physical condition of an alloy, or the treatment to which the alloy has been subjected, should be stated as well as its chemical composition. It should also be noted that where the individual grains or crystals are markedly *anisotropic, the specimen as a whole will also be anisotropic if there is any *preferential orientation* of the crystallographic axes of the grains. Such orientation frequently arises in industrial processes such as rolling (to make a sheet) and wire-drawing.

Steels are alloys of iron with the nonmetallic element carbon, often with other elements deliberately added; there is a wide range of alloy-steels used for special purposes, e.g. rustless chromium steels and high-permeability silicon steels. *Amalgams* are alloys involving the metal mercury. For the composition and special properties of some important alloys, *see* under separate headings, e.g. brass; bronze; constantan; duralumin; Manganin; Nichrome; phosphor bronze; type metal.

alloyed junction A *semiconductor junction, used in *transistors, that is formed from a wafer of semiconductor, which forms the *base region. Metal contacts are bonded onto the wafer and the structure is heated. The metal contacts alloy with the semiconductor to form the *emitter and *collector regions. This method is now rarely used except for the manufacture of germanium transistors, although some micro-alloy transistors are still in use.

Alnico A series of proprietary permanent-magnet alloys typically consisting of 18% Ni, 10% Al, 12% Co, 6% Cu, balance Fe. Typical values for *remanence are 0.6–1.0 tesla, coercive force 7.5×10^4–3.5×10^4 amperes per metre, and relative density 6900–7400 kg m^{-3}. On heating to 1200 °C and cooling in a unidirectional magnetic field, the materials become magnetically anisotropic with directional $(BH)_{max}$ up to 4.2×10^4 J m^{-3}. These treated alloys are marketed as a series called *Hycomax* and *Alcomax*.

alpha counter An electronic system, such as a *spark counter, including a special counter chamber, for detecting *alpha particles.

alpha decay A radioactive disintegration process whereby a parent nucleus decays spontaneously into an *alpha particle and a daughter nucleus. The *mean life of the parent nucleus varies from 10^{-7} seconds to 10^{17} years. On classical theory the maximum height of the *potential barrier between the α-particle (atomic number Z_a) and the daughter nucleus (Z) is given by:

$$V_R = Z_a Z e^2 / R,$$

where R is the radius of the nucleus. If a particle is inside a potential well with an energy $E_1 < V_R$, then the α-particle will be unable to escape. *Wave mechanics is necessary to explain the movement of α-particles through the barrier in terms of the *tunnel effect. The mean life, τ, of the parent depends on E_0, the energy available for reaction in a particular nucleus, and can be expressed approximately as:

$$\frac{1}{\tau} = 10^{21} \exp\left[\frac{-4(V_R - E_0)^{3/2}}{E_0}\right].$$

For $Z = 80$, V_R is 25 MeV. If $E_0 = 4$ MeV, $\tau = 3 \times 10^{12}$ years; if $E_0 = 8$ MeV, $\tau = 10^{-6}$ s.

alpha particle (α-particle) The nucleus of a helium (^{4}He) atom, carrying a positive charge of $2e$. Its *proton number and *neutron number are both 2, a *magic number, so that it is a very stable particle. It has a *relative atomic mass of 4.001 506.

alpha rays (α-rays) A stream of *alpha particles ejected from many radioactive substances with speeds (of the order of 1.6×10^6 metres per second) characteristic of the emitting substance. Alpha rays have a penetrating power of a few centimetres in air but can be stopped by a thin piece of paper, the maximum range varying as the cube of the velocity. They produce intense *ionization along their track and can be detected by an *alpha counter, *Geiger counter, or *bubble chamber, by their effect on a photographic plate, by the scintillations they produce on a fluorescent screen, etc.

alphatron An ionization-type vacuum gauge in which a small quantity of radium provides a constant source of alpha rays as the ionizing source. The number of ions formed in the gauge depends directly on the number of gas molecules with which this constant radiation can collide. The positive ions formed are collected by a grid and cause a current flow that is amplified and measured.

Alpher, Ralph Asher (b. 1921) Amer. physicist who, with Hans Bethe and George Gamow, produced the Alpher–Bethe–Gamow theory (1948; sometimes known as the α-β-γ theory) of the origin of the chemical elements. This theory assumes that the universe consisted initially of neutrons, which decayed into protons, and into heavier elements by subsequent neutron capture. He also predicted

(1948) the *microwave background radiation that originated with the big bang.

Alter, David (1807–1881) Amer. physicist who was the first to suggest that each element had a characteristic *spectrum. This idea laid the foundation for the development of spectrum analysis, mostly by *Bunsen and *Kirchhoff.

alternating current (a.c.) An electric current that periodically reverses its direction in the circuit, with a *frequency, f, independent of the constants of the circuit. In its simplest form, the instantaneous current I varies with the time t in accordance with the relation:

$$I = I_0 \sin(2\pi f t),$$

where I_0 is the peak value of the current. *See* cycle; period.

alternating-current commutator motor An a.c. motor that has an *armature provided with a *commutator and in which the latter is included in the a.c., circuit.

alternating-current generator A *generator for producing alternating e.m.f.s and currents. Examples are: *induction generator and *synchronous alternating-current generator.

alternating-current motor A motor that operates with *alternating current.

alternator *See* synchronous alternating-current generator.

altimeter An aneroid *barometer measuring the decrease in atmospheric pressure with height above ground and calibrated to read the height directly.

altitude 1. The vertical distance above sea level.
2. One of a pair of coordinates, the other being *azimuth, giving the position of a star. (*See* astronomy; celestial sphere.)

Aludure An alloy of aluminium and magnesium with good electrical conductivity and a higher tensile strength than that of pure aluminium. It is used in overhead transmission-line cables.

alumina Aluminium oxide (Al_2O_3), occurring widely as *corundum, which when coloured by traces of other metallic oxides forms a variety of natural gems such as ruby and sapphire. Corundum is widely used as an abrasive, emery being a form of it. It is also used as an insulator in thermionic valves because of its excellent properties at high temperatures. In solid state electronics it is used as a dielectric in some thin-film *capacitors and as the *gate dielectric in some M.I.S. transistors. *See* field-effect transistor.

Alvarez, Luis Walter (1911–1988) Amer. physicist. A professor at the University of California, he discovered orbital electron capture by a nucleus and with Felix Bloch made the first measurement of the neutron's magnetic moment. He constructed (1947) the first linear proton accelerator, with which he detected many new particles. For this work he received a Nobel prize (1968).

amalgam A solution of a metal in mercury. *See also* alloy.

amalgamation (Of the zinc pole of a primary cell) To reduce wastage of the metal by *local action the surface of the zinc is rubbed with mercury to form an *amalgam.

amber A fossil resin found in tertiary strata. The physical characteristics of amber are those of a resin. It is perfectly *amorphous, and occurs in the form of rods, drops, plates, etc. When rubbed on cloth, amber becomes strongly charged with negative electricity. It is a very good electrical insulator.

ambient 1. Surrounding, encompassing; e.g. ambient temperature is the temperature of an immediate locality.
   2. Freely moving, circulating; e.g. ambient air.

ametropia Optical defects of the eye including *myopia, *hypermetropia, and *astigmatism. *Compare* emmetropia.

Amici, Giovani Battista (1786–1863) Italian biologist responsible for several improvements to optical instruments.

Amici prism 1. *Roof prism. See* prism.
   2. *Direct-vision prism*, a compound prism consisting of oppositely directed prismatic components that produce a net dispersion without deviation of light in the middle of the spectrum. *See* direct-vision prism; spectrometer.

ammeter An instrument for measuring electric current. Common types are (*a*) the moving coil, (*b*) the moving iron, and (*c*) the thermoammeter. In (*a*), the current passes through a coil, pivoted with its plane parallel to the lines of a radial magnetic field produced by a permanent horseshoe magnet. The rotation is controlled by hair springs, and the deflection is directly proportional to the current. This is the most accurate type of ammeter, but only measures direct current. It can be adapted for alternating-current measurement by embodying a rectifier in the circuit. In (*b*), the current passing through a fixed coil magnetizes two specially shaped pieces of soft iron, the mutual repulsion of which causes the rotation of a pointer over a scale. Since the effect depends on the square of the current, the instrument can be used both for a.c. and d.c. measurements. It is less accurate than (*a*) and the scale is nonuniform. The control may be either by gravity or by a spring. In (*c*), the current passes through a thin resistance wire (usually in a vacuum) that, in consequence, becomes heated. The rise in temperature is measured by a thermojunction soldered to the wire and connected to a sensitive moving-coil instrument. It is particularly useful for measuring high-frequency currents.
   For many purposes, electronic instruments using digital display have replaced the earlier pointer instruments.

ammonia clock *See* clocks.

amorphous Devoid of crystalline form. True amorphous solids lack regular arrangement of their atoms. They are often supercooled liquids like glass and they may crystallize slowly in suitable conditions. Polycrystalline materials (e.g. metals) should not be confused with amorphous substances.

amount of substance Symbol: *n*. A dimensionally independent physical quantity that is not the same as *mass. It is proportional to the number of specified particles of a substance, the specified particle being an atom, molecule, ion, radical, electron, photon, etc., or any specified group of any of these particles. The constant of proportionality is the same for all substances and is known as the *Avogadro constant. Amount of substance is measured in *moles.

ampere Symbol: A. The *SI unit of electric current, defined as the constant current that, if maintained in two straight parallel conductors of infinite length, of negligible circular cross section, and placed one metre apart in a vacuum, would produce between these conductors a force equal to 2×10^{-7} newton per metre of length. This unit, at one time called the absolute ampere, replaced the international ampere (A_{int}) in 1948. The latter was defined as the constant current that, when flowing through a solution of silver nitrate in water, deposits silver at a rate of 0.001 118 grams per second. $1 A_{int} = 0.999 850$ A.

Ampère, André-Marie (1775–1836) French physicist. A pioneer in electrodynamics, he studied the forces between currents and investigated the magnetic field strength due to an electric current (*see* Ampère–Laplace theorem). He suggested that magnetism is due to minute circulating currents and attributed terrestrial magnetism to currents flowing round the earth. The unit of current is named in his honour. *See also* Ampère's rule; Ampère's theorem.

ampere balance An instrument for determining the size of the ampere by balancing the force between two current-carrying conductors against the force of a mass in a gravitational field. *See* current balance.

ampere-hour The quantity of electricity conveyed across any cross section of a conductor when an unvarying current of one ampere flows in the conductor for 1 hour. The term is employed in stating the capacity of accumulators. One ampere-hour = 3600 coulombs.

ampere-hour efficiency Of an accumulator. The ratio of the quantity of electricity available during discharge to the quantity of electricity used to charge the cell.

Ampère–Laplace theorem A theorem investigated experimentally by Ampère and in the context of electrical theory by Laplace for the magnetic flux density due to the current in a conductor. It is probably most useful in the form:

$$d\boldsymbol{B} = \frac{\mu_0}{4\pi} \frac{I \sin \theta}{r^2} dl$$

Ampère's rule

where dB is the elemental magnetic flux density at a point distance r from an elemental length dl of conductor carrying a current I; θ is the angle between the direction of I and the radius vector r; μ_0 is the *magnetic constant, i.e. $4\pi \times 10^{-7}$ henry per metre. Because the formula was confirmed (for a straight conductor) in an experiment devised by *Biot and Savart, it has sometimes been called *Biot and Savart's law.

Ampère's rule One of the numerous mnemonics for recalling the relation between the direction of an electric current and that of its associated magnetic field. To an observer looking along a wire in the direction of the current, the magnetic lines of force are clockwise.

Ampère's theorem If the *Ampère–Laplace theorem is applied to an infinitely long straight conductor carrying a current I, the flux density B at a point a distance r from the conductor is found to be given by $B = \mu_0 I / 2\pi r$, where μ_0 is the *magnetic constant. From this Ampère showed that along any closed path around N conductors of elemental length dl, each of which carries a current I,

$$\oint B \mathrm{d}l = \mu_0 NI.$$

This is Ampère's (circuital) theorem. He expressed it originally in terms of the work done in taking a unit magnetic pole (*see* unit pole) completely around a circuit, but since this latter quantity is now considered to have doubtful validity, the equation has been updated to the form quoted above.

ampere-turn Symbol: A or At. The *magnetomotive force produced when a current of one ampere flows through one turn of a coil.

ampholyte ions *See* zwitterions.

amplidyne A special form of *metadyne generator designed so that the power required by the controlling field winding is reduced to an extremely small value. It has an important use as an automatic regulator (voltage or speed) since it can control the excitation of a large generator or motor while requiring only very small control power (commonly less than 1 watt).

amplification factor 1. The limiting value of the ratio of a small change in anode voltage of an electronic tube to the change in grid voltage that would restore the anode current to the value it had before the anode voltage change.

$$\mu = -\left[\frac{\delta V_A}{\delta V_G}\right]_{IA \text{ const.}}$$

2. *See* gas amplification.

amplifier A device for reproducing an electrical input at an increased intensity. If an increased e.m.f. is produced operating into a high impedance, the device is a *voltage amplifier*, and if the output provides an appreciable current flow into a relatively low impedance, the device is a *power amplifier*. The

most commonly used amplifiers operate by *transistors or *thermionic valves. Types of amplifier include: *class A, *class AB, *class B, *class D, *emitter follower, *feedback, and *operational. The symbol for an amplifier is:

See also power amplification.

amplitude The *peak value of an alternating quantity in either the positive or negative direction. This term is applied particularly to the case of a sinusoidal vibration.

amplitude distortion *See* distortion.

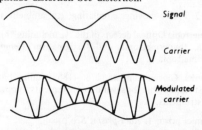

Amplitude modulation

amplitude modulation (a.m.) A type of *modulation in which the amplitude of the *carrier wave is varied above and below its unmodulated value by an amount that is proportional to the amplitude of the modulating signal and at a frequency equal to that of the modulating signal. An amplitude-modulated wave in which the modulating signal is sinusoidal may be represented by:

$$e = [A + B \sin pt] \sin \omega t,$$

where A = amplitude of the unmodulated carrier wave, B = peak amplitude variation of the composite, $\omega = 2\pi \times$ frequency of carrier wave, $p = 2\pi \times$ frequency of modulating signal. The modulation factor is defined by the following expression: modulation factor, $m = B/A$, so that the expression for the modulated wave may be written:

$$e = [1 + m \sin pt] A \sin \omega t.$$

Compare frequency modulation.

amplitude of accommodation A numerical figure expressed in *dioptres that describes the maximum change of power of the eye as the focus changes, by accommodation, from the far point to the *near point. If A is the amplitude in dioptres, and R and P are the reciprocals of the distances in metres of the far point and near point respectively from the eye, $A = (R - P)$. (R is positive when behind the eye as in *hypermetropia, P is negative when the near point lies in front of the eye.)

a.m.u. Abbreviation for atomic mass unit.

anallatic telescope A telescope used in surveying, fitted with crosswires, from which heights of buildings, etc. can be readily determined: the *telecentric principle is employed.

analogous pole The end of a pyroelectric crystal that becomes positively charged with rising temperature. *Compare* antilogous pole.

analogue circuit *See* linear circuit.

analogue computer *See* computer.

analogue/digital converter *Syn.* A/D converter; ADC. A device for converting a continuously varying signal, normally voltage or frequency, into a series of numbers on a medium suitable for use by a digital *computer.

analogue delay line *See* delay line.

analogue gate *See* gate.

analyser A device (crystal, *Nicol prism, etc.) by which the direction of polarization of a beam of light can be detected; usually the light has been passed through a *polarizer before arriving at the analyser.

analysis line The particular spectral line that is used to determine the concentration of an element in spectrographic analysis (*see* spectrometer). It has sometimes been called the *impurity line*, the *element line*, and the *minor-element line*. *Compare* internal standard line.

analysis of polarized light The examination of light using a *Nicol prism and a *quarter-wave plate to recognize the existence of unpolarized or polarized light (linearly, circularly, or elliptically polarized and its possible mixtures).

anaphoresis Migration to the anode of fine particles suspended in a liquid when an electric field is applied. *See* cataphoresis; electrophoresis.

anastigmat A photographic objective in which correction of spherical aberration, coma, radial astigmatism, curvature of image, and chromatic aberration, are attempted for large apertures and wide fields. The first anastigmats (P. Rudolph, 1889) employed combinations of new and old achromats (*see* achromatic lens) so arranged that their astigmatism and field curvature opposed one another. These were followed by numerous other designs, symmetrical and otherwise, involving the use of three- or four-lens cemented or uncemented objectives.

anchor ring *Syn.* toroid. The solid traced by the rotation of a circle of radius a about an axis in the plane of the circle at a distance b from its centre, where $b > a$. Its volume is $2\pi^2 a^2 b$.

Anderson, Carl David (*b.* 1905) Amer. physicist. His experimental researches on gamma rays and cosmic radiation revealed the positron for the first time (1932) and the muon (1938) (with Neddermeyer).

Anderson, Philip Warren (*b.* 1923) Amer. physicist. A professor at Princeton, he has worked on many solid-state problems, including ferroelectrics, superconductors, superfluids, and the migration of impurities within a crystal (the so-called *Anderson localization*). In 1977 he shared a Nobel prize with John *Van Vleck and Nevill *Mott for their work on the electronic structure of magnetic and disordered systems.

Anderson's bridge A *bridge for comparing an *inductance with a *capacitance. It is shown in the

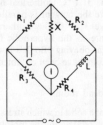
Anderson's bridge

diagram, where I is an indicating instrument such as a microphone or oscilloscope, C the capacitance, and L the inductance. R_4 and X are adjusted until the bridge is balanced, when:
$$R_2 R_3 = R_1 R_4$$
and
$$L = C[R_2 R_3 + (R_3 + R_4)X].$$

AND gate *See* logic circuit.

Andrade, Edward N. da C. (1887–1971) Brit. physicist. His researches cover a wide range, notably on viscosity and in metal physics. He wrote many well-known books, including *The Structure of the Atom*, orginally published in 1923. *Andrade's formula* for the coefficient of viscosity of a liquid is often quoted in the form $\eta v^{1/3} = A e^{b/vT}$, where A and b are constants for the liquid of viscosity η and v is its specific volume at thermodynamic temperature T.

Andrews, Thomas (1813–1885) Irish scientist. His investigation of the behaviour of carbon dioxide over a wide range of pressures and temperatures established the existence of a critical point and demonstrated the impossibility of liquefying substances at temperatures above their critical temperatures, no matter how great a pressure may be used. *See* Andrews's curves; critical temperature.

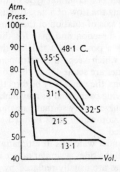
Andrews's curves

Andrews's curves A series of *isothermals for carbon dioxide showing the variation of volume with pressure at a series of temperatures. At temperatures below 31.1 °C (the *critical temperature), the curves have a region in which they are parallel to the volume axis, corresponding to the formation of liquid with subsequent large decrease of volume, and a region in which they are almost parallel to the pressure axis, corresponding to the liquid state where a large pressure increase causes a very small decrease in volume. The 31.1 °C isothermal has a point of inflection where the tangent is parallel to the volume axis. Above 31.1 °C the isothermals more nearly approximate to the hyperbolic form applicable to an ideal gas obeying *Boyle's law: pv = constant. Similar results have been obtained for many other substances.

anechoic chamber *See* dead room.

anelasticity A property of a solid in which stress and strain are not uniquely related in the preplastic range.

anemograph A recording *anemometer.

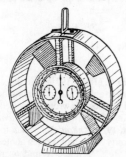

Vane anemometer

anemometer A device for measuring the velocity of a fluid, specifically, of the air. Velocimeters are of three types: (1) those that depend for their measurement on the difference of pressure between two points in the flow, such as the *venturi tube and the *Pitot tube; (2) those that use the cooling experienced by a heated body exposed to the fluid (*see* hot-wire anemometer); (3) those that use the momentum of the fluid to drive a small windmill, set of cups, or waterscrew facing the direction of flow.

The vane anemometer is a common type in the third group for recording the flow of a gas. It must be so mounted that the axis of rotation of the vanes points along the wind direction and may for this purpose be pivoted on a post with a tail fin, the aerodynamic force on which keeps the instrument facing upwind. The vanes are set at 45° to the axis. To record the revolutions, the vanes may be directly geared to a pointer moving over a dial or may make electric contacts to a buzzer every so many revolutions. In either case a stop-watch must be used to calculate the wind speed.

For measurements under water, as in estimating the speed of a river or or of the tides, a similar device is employed, but the vanes are usually replaced by a screw, which must be furnished with fins to keep it horizontal and pointing upstream. The instrument must also be loaded or anchored to keep it under water. Contacts connected through a cable to a buzzer on deck or on shore enable the revolutions to be counted.

The cup anemometer is often used in meteorology. Three cup-shaped bodies are attached by radial supports to a vertical axle. The axle rotates at a speed related to the wind speed. The rate of rotation may be recorded mechanically or electronically. This instrument is not very accurate for low wind speeds or for rapidly fluctuating winds.

aneroid Not containing a liquid, e.g. aneroid *barometer; aneroid manometer. *See also* Bourdon tube.

angle of contact *See* contact angle.

angle of friction The angle whose tangent is the coefficient of *friction. In measuring the coefficient of friction a body is placed on a plane and the latter tilted until the body will just slide down when gently tapped. The angle of the plane to the horizontal is then the angle of friction. *See* angle of repose.

angle of repose **1.** The maximum angle of inclination assumed by the surface of a heap of loose material, such as sand, when in equilibrium under gravity.
2. A less common name for *angle of friction.

angstrom (or **ångstrom**) A length unit of 10^{-10} metre (0.1 nm), used in spectroscopy and to measure intermolecular distances and variously abbreviated as A.U., Å.U., or Å. The visible spectrum extends roughly from 4000 A.U. to 7000 A.U. The use of this unit is discouraged.

Ångström, Anders Jonas (1814–1874) Swedish physicist. His work in spectroscopy is recognized by the name chosen for the spectroscopic unit of length (*see* angstrom). He recognized the existence of hydrogen in the solar atmosphere and he made the first investigations of the spectrum of the aurora borealis.

Ångström pyrheliometer *See* pyrheliometer.

angular acceleration *See* acceleration.

angular dispersion *See* disperson.

angular displacement The angle through which a point, line, or body is rotated in a specified direction and about a specified axis.

angular distance The apparent distance between two celestial bodies, measured in terms of the angle subtended by the bodies at the point of observation.

angular frequency Symbol: ω. The *frequency of a periodic quantity expressed as the product of the frequency in hertz and the factor 2π.

angular impulse The time integral of the torque applied to a system, usually when applied for a short time. It is equal to the change in *angular momentum that it would cause on a free mass acting about a principal axis.

angular magnification *See* convergence ratio.

angular momentum *Syn.* moment of momentum about an axis. *Symbol: L.* The product of *moment of inertia and *angular velocity ($I\omega$). Angular momentum is a *pseudovector quantity. It is conserved in an isolated system. For atomic and subatomic systems interacting only by central forces, components of the angular momentum are quantized in integral or half-integral multiples of the rationalized Planck constant. *See* moment.

angular velocity Symbol: ω. The rate at which a body rotates about an axis expressed in radians per second. It is a *pseudovector quantity equal to the linear velocity divided by the radius.

angular wavenumber *See* wavenumber.

angular wave vector *See* wavenumber.

anharmonic motion The motion of a body subjected to a restoring force that is not directly proportional to the displacement from a fixed point in the line of motion. *Compare* simple harmonic motion.

anhysteretic The magnetic state of a specimen when, in a constant magnetic field H, it has been subjected to an alternating field progressively reduced from a value greater than H to zero.

anion An ion that carries a negative charge, and in electrolysis moves towards the *anode. *Compare* cation.

anisotropic *Syn.* aeleotropic. Not *isotropic.

anisotropy The variation of physical properties with direction.

annealing The process of heating a substance to a specific temperature lower than its melting point, maintaining that temperature for some time, and then cooling slowly. Slow crystallization thus takes place in the solid state under controlled temperature conditions. Annealing generally softens metals and stabilizes glass articles by allowing stresses produced during fabrication to disappear. *See* alloy.

annihilation An interaction between a particle and its *antiparticle in which the two bodies disappear and photons or other elementary particles or antiparticles are created. Energy and mass are conserved in the process.

At low energies, an *electron and a *positron annihilate to produce electromagnetic radiation (*annihilation radiation*). Usually the particles have little kinetic energy or momentum in the laboratory system before interaction; hence the total energy of the radiation is very nearly $2m_0c^2$, where m_0 is the rest mass of an electron. In nearly all cases two *photons are generated, each of 0.511 MeV, in almost exactly opposite directions to conserve momentum. Occasionally three photons are emitted, all in the same plane. Electron–positron annihilation at high energies has been extensively studied in particle *accelerators. Generally the annihilation results in the production of a *quark plus an antiquark (for example, $e^+e^- \rightarrow u\bar{u}$) or a charged *lepton plus an antilepton ($e^+e^- \rightarrow \mu^+\mu^-$). The quarks and antiquarks do not appear as free particles but convert into several *hadrons, which can be detected experimentally. As the energy available in the elec-

tron–positron interaction increases, quarks and leptons of progressively larger rest mass can be produced. In addition, striking resonances are present, which appear as large increases in the rate at which annihilations occur at particular energies. The *J/psi particle and similar resonances containing a *charm quark plus an anticharm antiquark are produced at an energy of about 3 GeV, for example, giving rise to abundant production of charmed hadrons. Bottom (b) quark production occurs at energies greater than about 10 GeV. A resonance at an energy of about 90 GeV, due to the production of the Z^0 gauge boson (*see* gauge theory) involved in *weak interactions is currently under intensive study at the LEP and SLC e^+e^- colliders (*see* accelerator).

A *nucleon and an *antinucleon annihilating at low energy produce about half a dozen *pions, which may be neutral or charged. (An equal number of positive and negative pions is produced to conserve electric charge.) *See also* pair production.

annular eclipse A solar *eclipse that occurs when the moon is approaching its furthest distance from the earth (near apogee), so that it cannot completely obscure the sun's disc but leaves a bright ring exposed.

annular effect The phenomenon in fluid motion analogous to the *skin effect in alternating electric currents. With steady direct flow at low velocity in a tube, the velocity falls steadily from the centre towards the walls (*see* Poiseuille flow), but when the motion is alternating, as it is, for instance, when sound waves are being propagated in the tube, the mean alternating velocity rises from the centre towards the walls and finally falls within a thin laminar *boundary layer to zero at the wall itself. This is known as a *periodic boundary layer*. Its thickness increases as the square root of the frequency of the alternation. Similar conditions hold if the alternating flow overlays a direct flow in the tube.

anode The positive electrode of an electrolytic cell, discharge tube, valve, or solid-state rectifier. It is the electrode by which electrons leave a system. *Compare* cathode.

anode drop (or **fall**) The difference of potential, of the order of 20 volts, between the anode in a *gas-discharge tube and a point in the gas close to the anode.

anode feed resistance A resistance that is connected in series with the anode supply to a *thermionic valve and forms part of a *decoupling circuit.

anode load The total impedance in the anode circuit of a *thermionic valve external to the valve itself.

anode modulation A form of *amplitude modulation in which the output amplitude of a suitably designed radio-frequency amplifer (or oscillator) is made proportional to the anode supply voltage. If a voltage proportional to the modulating signal is superimposed on the anode supply voltage of the amplifier, the output of the amplifier (or oscillator) is amplitude-modulated. Such an amplifier (or oscillator) usually operates under *Class C conditions. *Compare* grid modulation.

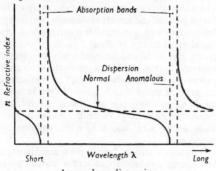

Anomalous dispersion

anode rays *See* gas-discharge tube.

anode saturation A condition arising in a valve or tube when electrons are no longer attracted by the anode. This is due to a build up of electrons around the anode preventing further discharge. *See also* space-charge region.

anode stopper *See* parasitic oscillations.

anodizing The process of thickening the natural oxide film on aluminium, and some of its alloys, by electrolytic action, using the aluminium as the anode in a suitable electrolytic cell. The process confers improved resistance to corrosion and increased surface hardness. The oxide film will absorb dyes enabling tinting of the surface to be achieved.

anomalous absorption of sound Taking into account the two main factors for absorption, namely: (1) viscous drag; and (2) heat conduction, it was found that only a few perfect gases under very restricted conditions of purity, temperature, pressure, and frequency behave in accordance with the classical theory of absorption as given by the Stokes–Kirchhoff equation (*see* absorption of sound). In many cases, however, dispersion of velocity as well as abnormal values of absorption occur. These anomalies are most marked in certain frequency ranges in each gas. There are various hypotheses to account for such anomalies:

(1) *Relaxation theory.* A delay in the change of translational energy into vibrational energy of molecules. (*See* absorption of sound.)

(2) *Selective absorption.* The molecular system encountered by the acoustical radiation is considered as a set of resonators tuned to a common frequency. If the size of an obstacle is approaching the wavelength of radiation, there will be a scattering of radiation, that is to say, a diminution in forthright intensity. The gases that show the sound dispersion may be in certain conditions of pressure, temperature, humidity, and frequency so as to produce such selective absorption.

(3) *Viscous absorption at high frequencies.* At high frequencies the classical theories of viscosity and heat conduction may not be applicable.

See dispersion of sound.

anomalous dispersion Rapid changes of refractive index with wavelength when the wavelength lies in the neighbourhood of *absorption bands of the material. On the longer wavelength side of the absorption band, the refractive index is high, and on the shorter wavelength side, low. Normal *dispersion is such that shorter wavelengths are associated with higher refractive index.

anomalous viscosity The coefficient of viscosity of a normal homogeneous fluid at a given temperature and pressure is a constant for that fluid and independent of the rate of shear or velocity gradient; fluids obeying this rule are said to be *Newtonian*. (*See* viscosity.)

In colloids, and in fact all fluids that consist of two or more phases present at the same time, this coefficient is not a constant but is a function of the rate at which the fluid is sheared as well as of the relative concentration of the phases; these fluids are said to be *non-Newtonian*.

Usually the viscosity diminishes as the velocity gradient increases, particularly in *gels in which a structure of a fibrous nature interleaves the continuous phase but tends to break down as the liquid is sheared. (*See* thixotropy.)

The existence of this property of anomalous viscosity confuses the measurements made in conventional viscometers, for the velocity gradient or rate of shear varies in them from place to place. It is usual in such cases to specify the value of the viscosity derived from the application of the usual instrumental equations (which assumes it to be independent of rate of shear) as an 'apparent viscosity'.

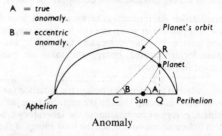

Anomaly

anomaly An *angular distance describing the position of an orbiting body such as a planet. There are three types of anomaly:

(1) *True anomaly.* For a planet orbiting around the sun, this is the angle between the *perihelion, the sun, and the planet, measured in the direction of motion of the planet.

(2) *Mean anomaly.* A planet's velocity is not constant but is greater at perihelion than at aphelion. The mean anomaly is the angle between the perihelion, the sun, and a fictitious planet having the same period as the real planet but moving with a constant velocity.

(3) *Eccentric anomaly*. If a circle is drawn, centred at the midpoint, C, of the major axis of a planet's elliptic orbit, and a perpendicular line, RQ, is drawn from the circle through the planet's position to the major axis, then the angle RCQ is the eccentric anomaly.

antenna *See* aerial.

antibonding orbital *See* molecular orbital.

anticathode A metal block in an X-ray tube upon which the cathode rays are focused, and from which the X-radiation is thus emitted. It either forms the anode of the tube, or is connected to it.

anticlastic *See* principal radii.

anticoagulant A substance that stabilizes a *colloid and prevents its separation into distinct phases.

anticoincidence circuit A circuit with two input terminals, designed to produce an output pulse if one input terminal receives a pulse, but not if both terminals receive pulses within a specified time interval. An electronic counter incorporating such a circuit will record specific events occurring singly and is termed an *anticoincidence counter*. *See also* coincidence circuit.

anticyclone A rotary atmospheric disturbance with a spirally outward flow of air from a centre of high pressure. In the northern hemisphere the rotation is clockwise. *Compare* cyclone.

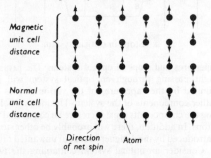

Crystal lattice of an antiferromagnetic material

antiferromagnetism The property of certain materials that have a low positive magnetic *susceptibility (as in *paramagnetism) and exhibit a temperature dependence similar to that encountered in *ferromagnetism. The susceptibility increases with increasing temperature up to a certain point, called the *Néel temperature*, and then falls with increasing temperature according to the *Curie–Weiss law. The material thus becomes paramagnetic above the Néel temperature, which is analogous to the *Curie point in the transition from ferromagnetism to paramagnetism.

Antiferromagnetism is a property of certain inorganic compounds such as MnO, FeO, FeF_2, and MnS. It results from interactions between neighbouring atoms leading to an antiparallel arrangement of adjacent magnetic *dipole moments. The diagram shows a crystal lattice, with the net *spins of atoms represented by arrows. The unit cell of the normal lattice, i.e. that determined by the positions of atoms, is half the size of the unit cell in the magnetic lattice. This can be demonstrated by *neutron diffraction. *See also* ferromagnetism.

anti-interference aerial system An *aerial system intended for use with a receiver and designed so that energy from any incident radiation is collected only by the aerial itself and not in any degree by the *feeder connecting the aerial to the receiver. An aerial system of this type can be used to reduce the effects of *interference caused by electric motors, motor-car ignition systems, etc. (i.e. "man-made" interference) by locating the aerial itself in a position remote from the cause of such interference.

antilogous pole That end of a pyroelectric crystal (*see* pyroelectricity) that becomes negatively charged with rising temperature. *Compare* analogous pole.

antimatter Matter composed entirely of *antiparticles. An antihydrogen atom would consist of a nucleus containing an antiproton and an orbiting positron. The existence of antimatter in the universe has not been detected. *See* early universe.

antineutrino The *antiparticle of the *neutrino.

antineutron The *antiparticle of the *neutron. It has the same mass as a neutron, but its *magnetic moment is opposite in sign, relative to its spin.

antinodal points *Syn*. negative nodal points. The pair of conjugate points of a thick lens or centred optical system for which the *convergence ratio is -1. *Compare* nodal points. *See* centred optical systems.

antinode A position in a *standing wave at which one of the types of disturbance in the wave has a maximum value. Generally there is another type of disturbance, the value of which will be a minimum at this point, and this position is said to be a *node of the other disturbance. Thus in standing electromagnetic waves there will be electric antinodes (i.e. places where the electric field has maximum values), which are nodes for the magnetic field and vice versa. In acoustic standing waves pressure antinodes are generally particle-displacement nodes and vice-versa.

antinucleon An antineutron or antiproton.

antiparallel Parallel but pointing in opposite directions.

antiparticle To each elementary particle, except for the photon and π^0, there exists another particle having charge (Q), *baryon number (B), *strangeness (S), *charm (C), and *isospin quantum number (I_3) of equal magnitude but opposite sign. These particles are called antiparticles. Examples of a particle and its antiparticle include the electron and the positron, the proton and the antiproton, the positive pion and the negative pion, and the up quark and the up antiquark. The photon and π^0 are their own antiparticles. The antiparticle corresponding to a particle a is usually denoted ā. Some examples of particles and their antiparticles are given in the table below.

For fermions the concept of an antiparticle has a special significance, which does not apply in the case of bosons. According to the relativistic wave mechanics of Dirac, there are states of negative energy that are normally all occupied by fermions of a particular type. These particles are unobservable in ordinary circumstances as there are no empty states available into which they can be displaced by any interaction unless the energy available is more than twice the rest energy of the particle. If sufficient energy is provided, a particle may be displaced into a state of positive energy thereby becoming detectable. The empty state of negative energy is then observable and is regarded as the antiparticle. *See* annihilation; pair production.

	Q	B	S	I_3
P	+1	+1	0	$\frac{1}{2}$
P̄	−1	−1	0	$-\frac{1}{2}$
π^+	+1	0	0	+1
$\bar{\pi}^+ \equiv \pi^-$	−1	0	0	−1
K°	0	0	1	$-\frac{1}{2}$
K̄°	0	0	−1	$+\frac{1}{2}$

antiprincipal points *Syn*. negative principal points. The pair of conjugate points of a thick lens or *centred optical system for which the *lateral magnification is −1. Corresponding principal and antiprincipal points are equally spaced from the appropriate focus. *See* principal planes and points.

antiproton The *antiparticle of the proton. Protons and antiprotons are created in pairs when protons with kinetic energy above about 4 GeV hit matter. They undergo *annihilation on interaction with protons or neutrons giving about half a dozen pions.

antiresonance The condition in which a vibrating system responds with minimum amplitude to an alternating driving force, by virtue of the inertia and elastic constants of the system.

aperiodic 1. Nonperiodic (*see* period). Applied to a system (e.g. an electric circuit, instrument, etc.) that is adequately *damped.
2. Without frequency discrimination. Applied to a circuit designed for use at a frequency (or over a range of frequencies) sufficiently far removed from any of its natural or *resonant frequencies for its characteristics to be substantially independent of frequency within the required limits.

aperture The part of a lens through which light is allowed to pass, or the part of a mirror or other reflecting surface from which light or other radiation can be reflected. The aperture is also the diameter of

such an area. *See also* apertures and stops in optical systems; f-number.

aperture angle The semi-angle subtended by the *entrance pupil of an instrument at the object. *See* numerical aperture.

aperture distortion Loss of definition in a *television image due to the size of the scanning spot or aperture. Thus, it is impossible to reproduce any details in the original object that are finer than the area represented by the aperture. For example, if the aperture travels across a line which, in the original marks an abrupt change between black and white, the light intensity of the area represented by the aperture will change gradually instead of suddenly from black to white.

aperture ratio When a light beam passing through a lens comes from a near object, the beam is not parallel and the light-passing power of the lens depends on the aperture ratio. This is given by $2n \sin \alpha$, where n is the refractive index of the medium in the image space and α is the angle, measured in the same medium between the optic axis and a ray from the axial object point passing through the edge of the mechanical aperture. *Compare* f-number.

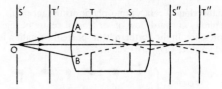

a Stops in optical system

apertures and stops in optical systems The rays of light passing through an optical system will be limited by the *apertures of the various lenses or other components of the system. The mounts holding the components must therefore be considered as stops. In addition, there will probably be other stops introduced by the designer to cut out unwanted rays.

Consider an optical system containing the two stops (or lens apertures) S and T (Fig. *a*). Let S′ and T′ be the images of these stops formed in the "object space" – that is by rays of light passing through the system from right to left. If light is to pass through the aperture in T, it must pass through the corresponding image T′ before entering the system. Thus the light from the axial point O that will succeed in passing through the system will be limited to the cone of rays OAB. A possible subsequent course for these rays has been indicated by the dotted lines.

The amount of light from O which passes through the system is in this case limited by the aperture in the stop T. T is known as the *aperture stop* of the system. Its image in the object space, T′ is called the *entrance pupil*. T″, the image of the aperture stop in the image space (rays passing from left to right), is called the *exit pupil*.

Now consider light coming from the extra-axial point P (Fig. *b*). The ray PQ which passes through the centre of the entrance pupil is called the *principal ray*. The field of view in the plane of S′ will be limited by the diameter of the aperture in S′, and no light from points further from the axis than the point R will pass through the system. The actual stop S that limits the field of view in this way is known as the *field stop*. Its image in the object space, S′, is the *entrance port* or *entrance window*. Its image in the image space, S″, is the *exit port* or *exit window*. The *field of view* is measured by the angle subtended by the entrance window at the centre of the entrance pupil.

If the plane of the object does not coincide with a stop image, there will be no sharp cut-off at the edge of the field of view. Instead, the field seen will have a central well-illuminated portion – sometimes called the *true field* – and an outer portion in which the illumination gradually falls off (the *partial field*).

In a telescope or in binoculars of magnifying power *m*, the rays leaving the instrument make angles with the axis *m* times as great as those for the incident rays. Thus if the angular field of view is θ, the extreme rays leaving the instrument make an angle $m\theta$ with each other. This product of field of view and magnifying power is called the *apparent field of view*. The apparent field of view that can be provided by the manufacturer is limited in practice by difficulties in keeping *aberrations within limits for wide-angle rays.

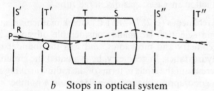

b   Stops in optical system

aperture stop *See* apertures and stops in optical systems.

aperture synthesis A procedure for obtaining full-size aperture results using a series of *radio telescopes of small aperture. The large aperture is divided into units equal in size to the actual small-aperture aerials, which are moved into every combination of position and orientation. Although it clearly takes a long time to complete the observations, the information gained enables a map to be constructed that extends over a large area of the sky. A map of this size, if constructed from the pencil-beam observations of a single conventional instrument, would take longer to complete.

apex *See* refracting angle and edge.

aphelion *See* perihelion.

apical angle *See* prism.

Apjohn's formula A formula for calculating the pressure of water vapour in the air using readings of a *wet and dry bulb hygrometer:

$$P_w - P = 0.000\,75\,H(t - t_w) \times$$
$$[1 - 0.008\,(t - t_w)],$$

where P_w is the saturated vapour pressure at the temperature t_w of the wet bulb, P is the actual pressure at temperature t, and H is the barometric pressure.

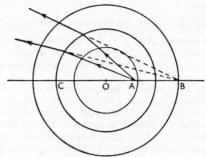

Aplanatic points

aplanatic Denoting a point or surface for which the image is free from spherical *aberration and coma for the rays. In the case of the spherical surface OC, there exist two points A and B along the one radius, so that OA = OC/*n* and OB = *n*. OC, so that all rays from A appear to pass through B after refraction. A and B are said to be *aplanatic points* of the surface. It is possible to build up aplanatic lenses and systems by arranging spherical surfaces in succession so that the above condition holds for each succeeding image (microscope objectives, aplanatic menisci). The term has also been applied in the sense of freedom from spherical aberration (aberrationless or Cartesian systems) as well as to systems in which true aplanatism is only approximately realized.

aplanatism *See* sine condition.

apochromatic lens A lens with a very high degree of correction of *chromatic aberration. The residual secondary spectrum of a lens achromatized for two colours is further reduced by using lens combinations with three or more different kinds of glass with appropriate partial dispersions. It has been applied to microscope objectives (Abbe, 1886), and special photographic lenses.

apocynthion The time or the point at which a spacecraft launched from the earth into lunar orbit is furthest from the surface of the moon. *Compare* pericynthion; apolune.

apogee *See* perihelion.

apolune The time or the point at which a spacecraft launched from the moon into lunar orbit is furthest from the surface of the moon. *Compare* perilune; apocynthion.

apostilb A unit of *luminance defined as the luminance of a uniformly diffusing surface that emits 1 *lumen per square meter. It is equivalent to a luminance of $1/\pi$ *candela per square metre or 10^{-4} *lambert.

apparent expansion *See* coefficient of expansion.

apparent magnitude *See* magnitude.

apparent resistance An obsolete name for *impedance.

apparent viscosity *See* anomalous viscosity.

appearance potential 1. The potential difference through which an electron must be accelerated from rest to produce a given ion from its parent atom or molecule.

2. This potential difference multiplied by the electron charge, giving the least energy required to produce the ion. A mass spectrometer (*see* mass spectrum) is focused on the ion of interest and a beam of electrons is passed through the sample gas, the accelerating potential difference being increased until the ion appears. A calibrating gas giving a known appearance potential is required to eliminate certain possible causes of error. A simple ionizing process gives the *ionization potential of the substance; for example:

$$Ar + e \rightarrow Ar^+ + 2e.$$

Higher appearance potentials may be found for multiply charged ions:

$$Ar + e \rightarrow Ar^{++} + 3e.$$

A molecular gas may give various ions, for example methane, CH_4:

$$CH_4 + e \rightarrow CH_4^+ + 2e,$$
$$CH_4 + e \rightarrow CH_3^+ + H + 2e,$$
$$CH_4 + e \rightarrow CH_2^+ + H_2 + 2e,$$
$$CH_4 + e \rightarrow CH_2^+ + 2H + 2e.$$

The existence of the last two processes shown is indicated by a considerable increase in the slope of the yield curve for the CH_2^+ ion at a value above the appearance potential corresponding nearly to the dissociation energy of the H_2 molecule.

Appleton, Sir Edward Victor (1892–1965) Brit. physicist. His researches on the ionized layers of the atmosphere (*see* ionosphere), studied by means of their effect on propagation of radio waves, are commemorated by the naming after him of one of these layers. He played a very important part in the development of radar.

Appleton layer *See* ionosphere.

apple tube A form of *cathode-ray tube for colour television. The screen consists of vertical strips of red, green, and blue *phosphors, which are excited independently in response to the transmitted signal. A black-and-white picture is produced by exciting the three colours in coordination to produce white light.

apsis (*plural*: apsides) *Syn*. apse. Either of the two extreme points on the major axis of the orbit of a planet or comet. The point located nearest the sun is the *perihelion, the one farthest from the sun is the aphelion. The *line of apsides* is the line joining the two apsides, i.e. the major axis of the orbit.

Arago, Dominique (1786–1853) French physicst. Among many important researches in optics, sound,

and magnetism, mention may be made of his cooperation with *Fresnel to establish the *wave theory of light. He demonstrated the existence of a bright spot in the centre of the shadow of a circular disc (shown by *Poisson to be a logical prediction of Fresnel's theory) and he and Fresnel jointly showed that the state of polarization of interfering beams is important, leading *Young and Fresnel to conclude that light is transverse. He suggested (1838) the rotating-mirror method for measuring the speed of light, later adopted by Foucault.

Arago's disc A horizontal disc of copper that can be rotated about a vertical axis, and is contained in an air-tight box. A horizontal bar magnet suspended above the disc, but outside the box, is deflected when the disc is rotated, and will eventually rotate with a smaller velocity but in the same direction as the disc, if the rotation of the latter is sufficiently rapid. The effect is due to the eddy currents induced in the disc by its motion in the magnetic field of the bar magnet, and is an illustration of *Lenz's law.

arc A luminous electrical gas discharge characterized by high current density and low potential gradient. The intense ionization necessary to maintain the large current is provided mainly by thermionic emission from the cathode, which is raised to incandescence by the discharge. Thermal ionization of vapour may contribute to a mercury arc. *See* conduction in gases; gas-discharge tube.

Archimedes (*c.* 287–212 B.C.) Greek mathematician and scientist. He made considerable advances in geometry and mechanics, and was the founder of hydrostatics. In geometry, he evaluated the circumference of the circle in terms of the radius, arriving at a good approximation to π. He made a number of inventions including the *Archimedian screw, applied to drain the holds of ships. He established the basic theory of the centre of gravity and the principle of the lever.

Archimedes' principle A body floating in a fluid displaces a weight of fluid equal to its own weight. *See* buoyancy.

Archimedian screw An apparatus for raising water, consisting of a tube shaped like a corkscrew and inclined at an angle to the surface of the water. When the tube is rotated, the water flows up the tube and out at its upper end.

arcing contacts Auxiliary contacts in any type of *circuit-breaker switch, designed to close before and open after the main contacts thereby protecting the latter from damage by an *arc.

arcing horn *Syn*. protective horn. A conductor shaped like a horn and fitted to an insulator to prevent damage to the latter by a power arc. Also, an element of a *horn-gap.

arcing ring A metal ring fitted to an insulator to prevent damage to the latter by a power arc.

arcing shield *See* grading shield.

arc lamp A type of lamp that utilizes the brilliant light accompanying an electric arc. The major portion of the light comes from the incandescent crater formed at the positive electrode.

arcover *See* flashover.

arc-suppression coil *Syn.* Petersen coil. An *earthing reactor used in a.c. power transmission systems and so designed that under fault conditions the *reactive current to earth balances the capacitive current to earth flowing from the lines so that the current at the fault to earth is limited almost to zero. In such a system, an arc forming between one line and earth is rapidly extinguished.

areal velocity The area swept out in unit time by the *radius vector to a point describing a plane curve. *See* Kepler's laws.

Argand, Jean Robert (1768–1822) Swiss mathematician who introduced the geometrical representation of a complex number (1806; *see* Argand diagram) and applied it to show that every algebraic equation has a root.

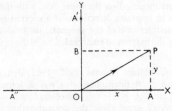

Argand diagram

Argand diagram A representation of complex numbers with reference to perpendicular axes: the horizontal axis of real quantities and a perpendicular axis of imaginary quantities. Representing a real quantity x by a length OA along OX, multiplication by $\sqrt{(-1)}$ is interpreted as a rotation through a right angle to OA'. Repetition gives $(\sqrt{-1})^2 x = -x$ shown by OA'', which is OA reversed, as it should be. The complex quantity $x + iy$, where $i = \sqrt{(-1)}$, is represented by the line OP (or the point P). The length of OP is the *modulus* of the complex quantity $z = x + iy$ and is thus $r = \sqrt{(x^2 + y^2)}$, while the angle between OX and OP is the *amplitude* or *argument* θ of z and it can be shown that z may alternatively be written as:
$$re^{i\theta} = r\cos\theta + ir\sin\theta = x + iy.$$
Addition and subtraction follow the same geometric procedure as for vectors but multiplication is different. The product of complex quantities z_1, z_2 (represented respectively by OP_1, OP_2) is a line of length equal to the product $OP_1 \times OP_2$ and its argument is the sum of the argument of z_1 and z_2:
$$z \equiv z_1 z_2 = (r_1 e^{i\theta_1})(r_2 e^{i\theta_2})$$
$$= (r_1 r_2)e^{i(\theta_1 + \theta_2)}.$$

Aristarchus of Samos (3rd century B.C.) Greek philosopher. He was the first to suggest that the earth revolved round the sun rather than *vice versa*. He was also the first to attempt to use geometry to calculate the distances from the earth to the sun and the moon.

Aristotle (384–322 B.C.) Greek philosopher. A great collector and systematizer of knowledge, his work dominated medieval thought up to the Renaissance. Much of his biological work was sound but his physics and astronomy involved many erroneous ideas. The four "elements", earth, air, fire, and water were produced by action of opposing "principles": hot and dry produced fire, cold and dry produced earth, etc. A fifth element, the "aether", constituted the incorruptible heavens. Bodies were supposed to seek their natural levels, light bodies rising and heavy bodies sinking. The earth was assumed to be the centre of the universe; continual effort was needed to maintain motion; and "Nature abhors a vacuum".

arithmetic mean *Syn.* average. *See* mean.

arithmetic series A mathematical series in which each term is equal to the sum of the preceding term and a constant:
$$a + (a + d) + (a + 2d) + \ldots$$
$$+ (a + (n-1)d).$$
The sum of n terms is:
$$\tfrac{1}{2}n(2a + (n - 1)d).$$

armature 1. *Syn.* rotor. The rotating part in an electric *motor or *generator.
2. Any moving part in a piece of electrical equipment in which a voltage is induced by a magnetic field or which closes a magnetic circuit, e.g. the moving contact in an electromagnetic relay.
3. *See* keeper.

armature reaction Distortion of the magnetic field of a dynamo caused by rotation of the armature. It is a function of the *reluctance of the magnetic circuit, the mechanical design of the field structure, and the *phase angle between the voltage and current in the armature. The main effects are: an alternation in the effective excitation of the machine; distortion of the air-gap flux distribution; and, in an a.c. synchronous machine, an influence upon the *power factor at which the machine operates.

armature relay *See* relay.

Arrhenius, Svante (1859–1927) Swedish chemist and physicist, who worked largely on *electrolytic dissociation in solution and on atmospheric electricity.

arrow of time An effect that gives a direction to time, despite the fact that physical laws do not distinguish forward and backward directions. The direction of the psychological arrow is given by the observer's impression that time is something that passes, and the fact that one can remember the past but not the future. The direction of the thermodynamic arrow is given by the second law of *thermodynamics, i.e. the passage of time is always associated with an increase in entropy in a closed system. The direction of the cosmological arrow of time is given by the direction in which the universe is expanding. Some physicists believe that these apparently separate arrows may be manifestations of the same phenomenon.

Arsonval, J. A. d' *See* d'Arsonval.

artificial aerial *Syn.* dummy antenna. A device that simulates an actual aerial in all its electrical characteristics except that substantially all the energy supplied to it is dissipated in a form other than electromagnetic radiation (usually heat in a resistor). Certain adjustments made to a transmitter or receiver while an artificial aerial is connected to it are adequate when the actual aerial is connected in place of the artificial aerial.

artificial horizon A device employed when finding the altitude of a heavenly body if no natural horizon is visible. A telescope directed to a star and then turned to view the light reflected in the surface of a pool of mercury must be rotated through twice the angular altitude.

artificial line An electrical *network comprising resistance, inductance, and capacitance elements, designed to behave in exactly the same way at any particular frequency as an actual *transmission line as far as the terminals are concerned.

artificial radioactivity *See* radioactivity.

	Thermal conductivity k $J\,m^{-1}\,s^{-1}\,K^{-1}$
Copper (18° C)	385
Mercury (50° C)	7·90
Crown glass	1·04
Asbestos	0·125

asbestos A class of fibrous silicate of magnesium, calcium, sodium, and iron. It is a good heat insulator and is used for this purpose in sheets or bricks, its insulating properties being illustrated by the table. Its melting point lies in the range 1280–1310 °C. Exposure to large amounts of asbestos can cause the lung disease *asbestosis*.

asdic An acronym for *a*llied *s*ubmarine *d*etection *i*nvestigation *c*ommittee. *See* sonar.

aspect ratio 1. The ratio of the width of a *television picture to its height. An aspect ratio of 4:3 has been adopted in most countries including Britain and America.
2. In *MOS integrated circuits of transistors and resistors, the ratio of the width to length of the conductivity channel.

asperity *See* friction.

aspherical lens or mirror A lens or mirror whose surface is part of a parabola, ellipse, hyperbola, etc., rather than part of a sphere, thus reducing optical *aberration, especially spherical aberration, to a minimum. A spherical surface can only form aberra-

tionless and *aplanatic images over a very restricted range. The use of aspherical surfaces also leads to more simplified optical systems as the number of optical elements required is reduced.

A *bispherical lens* has a spherical surface but with different curvatures in the centre and at the edge. It has the same advantages as an aspherical element.

Assmann psychrometer *See* psychrometer.

astable multivibrator *See* multivibrator.

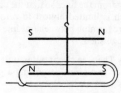

Astatic galvanometer

astatic system A system of magnets so arranged that there is no resultant directive force or couple on the system when placed in a uniform magnetic field. The simplest form consists of a pair of equal and parallel magnets mounted on the same axis, with their polarities in opposite directions. If a current-bearing coil encircles one of the magnets, its field affects mainly the magnet so encircled. This is the principle of the *astatic galvanometer*.

Astbury, William Thomas (1898–1961) Brit. physicist and crystallographer who was a founder of molecular biology. He was one of the first to use *X-ray crystallography for the determination of the structure of polymers.

asteroids (minor planets) A multitude of small planets that orbit the sun primarily between the planets Mars and Jupiter. The largest, Ceres, has a diameter of 1003 kilometres.

astigmatic difference A term sometimes used to describe the linear distance between focal lines, but more commonly the difference in dioptres between the principal convergences. If OC and OD are measured in metres (*see* diagram under astigmatism), then the astigmatic difference is $1/OC - 1/OD$.

astigmatic interval *Syn.* interval of Sturm. The portion between the focal lines of an astigmatic pencil. *See* astigmatism.

astigmatic lenses Planocylindrical, spherocylindrical, and spherotoric lenses (*see* toric (toroidal) lenses) used to correct *astigmatism of the eye (first used by *Airy). Stokes proposed the use of two cylindrical lenses with axes crossed in different directions to produce a lens of variable power.

astigmatic (focal) surfaces Due to *astigmatism of oblique incidence, the image-side focus of a plane consists of two curved surfaces coinciding at the axis (tangential and sagittal surfaces).

astigmatism An aberration in which, instead of rays converging to a single focus, they are caused to focus

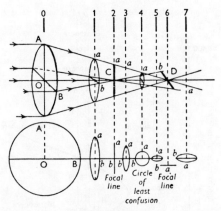

Astigmatism

to two lines at right angles, separated by an interval known as the *conoid of Sturm* (*see* diagram). This effect is produced by astigmatic systems by reflection or refraction, even when light is incident normally, due to meridian differences of curvature (toroidal, cylindrical, etc.). A slightly different form of astigmatism is produced by oblique centric or eccentric reflection and refraction at most forms of surface. (*See* radial astigmatism.) The term is generally restricted to narrow pencils. *Astigmatism of the eye is caused either by curvature differences of the cornea or to a lesser extent because the crystalline lens is somewhat tilted. *See* aberrations of optical systems.

astigmatism of the eye The main cause of this form of *astigmatism is the toroidal shape of the cornea, which in youth commonly has a greater curvature in the vertical *meridian and tends with age to become greater in the horizontal. Correction involves the use of sphero-cylindrical or *toric lenses to bring the foci of the two meridians together at the retina, when accommodation is at rest. The optical effect of astigmatism is to blur vision and to make lines, in directions associated with the astigmatism, appear clearer than those at right angles.

Aston, Francis Williams (1877–1945) Brit. physicist. His main work was on the mass spectrometer (*see* mass spectrum), developed from Sir J. J. Thomson's method of positive-ray analysis, and on the determination of the exact masses of isotopes using this technique.

Aston dark space The nonluminous layer (about 1 mm thick) that, when an electric discharge is passed through certain pure gases, separates the surface of the cathode from the negative glow. *See* gas-discharge tube.

astrolabe 1. An arrangment of rings representing the equator, prime meridian, ecliptic, etc., used by astonomers in the middle ages.
2. An instrument used by mariners of the sixteenth century for ascertaining the *altitude of the

sun, consisting of a graduated disc with a movable arm (the *alidade*) pivoted on the centre.

astrometry *See* astronomy.

astronautics The scientific study of the scientific, technical, and medical aspects of space flight. The first artificial earth *satellite was launched in 1957 by the USSR, who also put the first manned spacecraft into an orbit round the earth in 1961. Lunar space probes from both the USSR and USA, in following years, sent back photographs and valuable scientific information enabling the first moon landing to be achieved by the USA in 1969. Unmanned spacecraft are investigating the physical conditions of planets, satellites, and other bodies within the solar system.

astronomical telescope 1. *Syn.* Kepler telescope. *See* refracting telescope.
2. A *telescope designed for astronomical use. It can be mounted in any of several ways. In an *equatorial mounting* the telescope rotates about a polar axis, which points towards the celestial poles and is parallel to the earth's axis. The telescope turns at the same rate as the *diurnal motion of celestial bodies by means of a clock drive. It has two graduated circles reading right ascension and declination (*see* celestial sphere). The *coudé system* consists of a reflector or refractor having an equatorial mounting. The image is formed, after an additional reflection, at a point on the polar axis (the *coudé focus*). As this point remains fixed with respect to the earth, light can be analysed with a permanently installed *spectrometer, etc.
The altazimuth mounting is one in which the telescope can move in *azimuth about a vertical axis and in *altitude about a horizontal axis.
In the *meridian circle*, the telescope is mounted on an east-west axis so that it turns in the meridian plane. It is used to give the altitude of a star at the moment it crosses the meridian and for deducing right ascension and declination.

astronomical unit (AU) A unit of length used in astronomy. It is defined as being equal to 149 597 870 km, which is very nearly equal to the mean distance between the centre of the earth and the centre of the sun (the original definition of the term).

astronomy The study of the universe and its contents. The main branches of the subject are:
(1) *Astrometry*, positional measurements of the stars and planets on the *celestial sphere;
(2) *Celestial mechanics*, relative motions of systems of bodies associated by *gravitational fields;
(3) *Astrophysics*, the internal structure of planets and stars and their consequent external features and positions on the *Hertzsprung–Russell diagram.
See also radio astronomy; cosmology.

astrophysics *See* astronomy.

asymmetrical breaking-current *See* breaking-current.

asymmetrical deflection *See* symmetrical deflection.

asymptote A straight line that is tangential to a curve at an infinitely great distance from the origin.

asymptotic freedom A property of the strong interaction by which, at high particle energies (i.e. shorter distances), the force is weakened and quarks and gluons behave almost as free particles. As the distance between particles tends to zero, the force tends to vanish. Asymptotic freedom is a consequence of certain *gauge theories with unbroken symmetries (i.e. non-Abelian gauge theories; *see* group). *See also* quantum chromodynamics.

asynchronous logic *See* synchronous logic.

asynchronous motor An a.c. motor whose actual speed bears no fixed relation to the supply frequency and varies with the load. An *induction motor is a typical example.

atmolysis A method of separating the constituents of a gas mixture by making use of their different rates of diffusion through a porous material.

atmometer An instrument for measuring the rate of evaporation of water into the atmosphere. In the simplest form, evaporation is followed by measuring the water level in a cylindrical container at frequent intervals.

atmosphere 1. The *air. *See also* atmospheric layers.
2. Any gaseous medium.
3. *See* standard atmosphere.

atmospheric absorption *See* atmospheric windows.

atmospheric electricity The general electrical properties of the atmosphere, both under normal conditions and at the time of electric discharge (i.e. a lightning flash).

The following data characterize the properties of the atmosphere at, or just above, sea level. They are mean fine-weather values:

direction of field	downward
potential gradient	130 volts per metre
total conductivity	3×10^{-4} siemens per metre
small ion mobility	1.4×10^{-4} $(m\ s^{-1})/(V\ m^{-1})$
air-earth current density	2×10^{-14} ampere per metre^2

An average lightning flash has a potential of about 4×10^9 volts. It provides a charge of 15 coulombs, and possesses about 2×10^{10} joules of energy. The average upward current is rather less than one ampere, but the peak current in a flash may reach several tens of kiloamps.

atmospheric layers The gaseous layers into which the earth's atmosphere can be divided according to the change in physical properties, especially temperature. The altitude figures given in the diagram are approximate, as they vary over the earth's surface, and also show seasonal and diurnal changes. There are five main layers – the *troposphere, *strato-

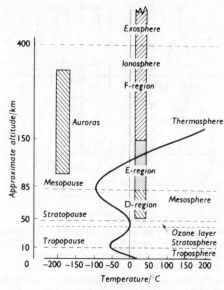

Atmospheric layers

sphere, *mesosphere, *thermosphere, and *exosphere. The *ionosphere extends from the stratosphere into the exosphere. *See also* upper atmosphere.

atmospherics Electromagnetic radiation produced by natural causes such as lightning. The term is also used to describe the disturbing effects that such radiation produces in a radio receiver.

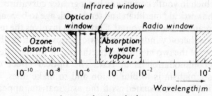

Atmospheric windows

atmospheric windows The gaps in atmospheric absorption that allow *electromagnetic radiations of certain wavelengths to penetrate the earth's atmosphere from space.

(1) *Optical window.* This allows through the whole visible spectrum (approx. 760–400 nm) and ultraviolet wavelengths down to about 320 nm. Radiation of wavelengths less than 320 nm is absorbed by atoms and molecules in the atmosphere.

(2) *Infrared window.* This allows through wavelengths of 8–11 μm, corresponding to the region in which there is no absorption of infrared radiation by water vapour in the atmosphere. There are additional narrow-band infrared windows at other micrometre wavelengths (1.25, 1.6, 2.2, 3.6, 5.0, and 21 μm). *See* infrared astronomy.

(3) *Radio window*. This allows through short-wave radio waves of wavelengths of approximately 1 mm to 30 m. *See* radio astronomy.

atom In the fifth century BC Leucippus and *Democritus asserted that matter is composed of atoms – that is very small indivisible particles. This idea was rejected by most philosophers, but was occasionally revived by scientists in the seventeenth and eighteenth centuries despite strong criticism by *Descartes. At the beginning of the nineteenth century *Dalton developed chemistry using the concept of an atom as the smallest part of an element that can take part in a chemical reaction. The value of this concept in chemistry, and the related developments in *kinetic theory in physics, led to widespread, but not universal, acceptance of atomicity. Towards the end of the century *Kelvin proposed a number of arguments, all giving approximately the same values for the sizes of atoms, although it was not possible to determine accurate values until the publication of *Planck's formula in 1900.

Following the discovery of the electron by *Thomson in 1897 it was recognized that the atoms of science had structure and thus differed from the indivisible atoms of the philosophers. Since electrons are negatively charged it is apparent that a neutral atom must have a positive component. To explain the stability of atoms, Thomson supposed that the electrons were disposed within a continuous distribution of positive charge. This model had limited success and failed completely to explain the large-angle scattering of alpha particles by thin metal foils observed by Marsden and Geiger. In 1912 *Rutherford argued that to account for these observations it is necessary to assume that nearly all the mass of the atom is concentrated in a positively charged *nucleus (radius of the order 10^{-15} m), while the electrons occupy the surrounding space to a radius of 10^{-11} to 10^{-10} m (in the ground state). According to *classical physics, such a system must emit electromagnetic radiation continuously and consequently no permanent atom would be possible. The development of *quantum theory eventually solved this problem. According to Rutherford the nucleus has a positive charge of Ze surrounded by Z electrons. The *Bohr theory of the atom and its application by Moseley showed (1913) that Z was the atomic number – that is the number of the element in the *periodic table. At about the same time evidence was found for the existence of *isotopes. Since the chemical properties depend upon Z it is concluded that all atoms of an element have the same value of Z but may differ in other respects.

The Bohr theory introduced the concept that an electron in an atom is normally in a state of lowest energy (*ground state*) in which it remains indefinitely unless disturbed. By absorption of electromagnetic radiation or collision with another particle the atom may be excited – that is an electron is moved into a state of higher energy. Such excited states usually have short lifetimes (typically nanoseconds) and the electron returns to the ground state, commonly by emitting one or more quanta of electromagnetic radiation. The original theory was only partially successful in predicting the energies and other properties of the electronic states. Attempts were made to improve the theory by postulating elliptic orbits (*Sommerfeld 1915) and electron *spin (Pauli 1925) but a satisfactory theory only became possible upon the development of *wave mechanics after 1925.

According to modern theories, an electron does not follow a determinate orbit as envisaged by Bohr but is in a state described by the solution of a wave equation. This determines the *probability* that the electron may be located in a given element of volume. Each state is characterized by a set of four *quantum numbers, and, according to the *Pauli exclusion principle, not more than one electron can be in a given state. (*See* atomic orbital.)

An exact calculation of the energies and other properties of the quantum states is only possible for the simplest atoms but there are various approximate methods that give useful results. (*See* perturbation theory.) The properties of the innermost electron states of complex atoms are found experimentally by the study of X-ray spectra. The outer electrons are investigated using spectra in the infrared, visible, and ultraviolet. Certain details have been studied using microwaves (*see* Lamb shift). Other information may be obtained from magnetism, chemical properties, *appearance potentials, or *photoelectron spectroscopy.

atomic bomb *See* nuclear weapons.

atomic clock *See* clocks.

atomic force microscope A type of microscope, similar in operation to the *scanning tunnelling microscope but using mechanical forces rather than electrical effects. In the atomic force microscope the probe is a tiny chip of diamond held on a spring-loaded cantilever. The probe, which is in contact with the sample surface, is slowly moved and the tracking force between its tip and the sample is measured by deflections of the cantilever. The probe is raised and lowered in order to maintain the force constant, and a contour of the surface is thus produced. A raster scan of the whole sample allows a computer-generated contour map of the surface to be obtained. The atomic force microscope, unlike the scanning tunnelling microscope, can produce images of nonconducting materials, such as biological specimens. The sample does, however, have to be fairly rigid.

atomic fountain A technique for the spectroscopic study of the hyperfine structure of atoms. It involves forcing atoms upwards by means of a laser beam and allowing them to fall back under gravity. Because the atoms move slowly near the top of their path, precise measurements of energy levels can be made.

atomic heat The former name for the *molar heat capacity of an element. *See* Dulong and Petit's law.

atomic mass unit

atomic mass unit (unified) *Syn*. dalton. Abbreviation: a.m.u. Symbol: u. A unit of mass equal to $1/12$ of the mass of an atom of carbon-12. It is equal to 1.6605×10^{-27} kg or approximately 931 MeV/c^2. It should not be confused with the *atomic unit of mass.

atomic number *Syn*. proton number. Symbol: Z. The number of protons in the nucleus of an atom or the number of electrons revolving around the nucleus. The atomic number determines the chemical properties of an element and the element's position in the *periodic table. All the isotopes of an element have the same atomic number although different isotopes have different *mass numbers.

a The probability distribution of the electron in the hydrogen atom

atomic orbital An allowed *wave function of an electron in an atom obtained by a solution of *Schrödinger's wave equation. In a hydrogen atom, for example, the electron moves in the electrostatic field of the nucleus and its potential energy is $-e^2/r$, where e is the electron charge and r its distance from the nucleus. A precise orbit cannot be considered as in Bohr's theory of the *atom but the behaviour of the electron is described by its wave function, Ψ, which is a mathematical function of its position with respect to the nucleus. The significance of the wave function is that $|\Psi|^2 d\tau$ is the probability of locating the electron in the element of volume $d\tau$. Schrödinger's equation for the hydrogen atom is:

$$\frac{\partial^2\Psi}{\partial x^2} + \frac{\partial^2\Psi}{\partial y^2} + \frac{\partial^2\Psi}{\partial z^2} + \frac{8\pi^2 m}{h^2}\left(E + \frac{e^2}{4\pi\varepsilon_0 r}\right)\Psi = 0,$$

where m is the mass of the electron and E its total energy. When the equation is solved and boundary conditions applied to make the solutions meaningful in physical terms (*see* wave function), it is found that the electron can only have certain allowed wave functions (eigenfunctions). Each of these corresponds to a probability distribution in space given by the manner in which $|\Psi|^2$ varies with position. They also have an associated value of the energy E. These allowed wave functions, or orbitals, are characterized by three quantum numbers:

n, the *principal quantum number*, can have values of 1, 2, 3, etc. The orbital with $n = 1$ has the lowest energy. The states of the electron with $n = 1, 2, 3$, etc., are called shells and designated the K, L, M shells, etc.

l, the *azimuthal quantum number*, which for a given value of n can have values of 0, 1, 2, ... $(n-1)$. Thus when $n = 1$, l can only have the value 0. Electrons in the L shell of an atom with $n = 2$ can occupy two "subshells" of different energy corresponding to $l = 0$ and $l = 1$. Similarly the M shell ($n = 3$) has three subshells with $l = 0$, $l = 1$, and $l = 2$. Orbitals with $l = 0, 1, 2$ and 3 are called s, p, d, and f orbitals respectively. The significance of the l quantum number is that it gives the angular momentum of the electron. The orbital angular momentum of an electron is given by:

$$\sqrt{[l(l+1)(h/2\pi)]}.$$

The component of the angular momentum in any defined direction is given by $mh/2\pi$, where m is the *magnetic quantum number*, which for a given value of l can have values $-l, -(l-1), \ldots, 0, \ldots (l-1), l$. Thus for a p orbital for which $l = 1$, there are in fact three different orbitals with $m = -1, 0$, and 1. These orbitals, with the same values of n and l but different m values, have the same energy in the absence of external fields. If the atom is in a magnetic field, the states have slightly different energies (*see* Larmor precession; Zeeman effect). According to wave theory the electron may be at any distance from the nucleus but in fact there is only a reasonable chance of it being within a distance of $\sim 5 \times 10^{-11}$ metre. Indeed the maximum probability occurs when $r = a_0$ where a_0 is the radius of the first Bohr orbit (*see* Bohr theory of the atom). It is customary to represent an orbital by a surface enclosing a volume within which there is an arbitrarily decided probability (say 95%) of finding the electron. These are drawn in Fig. *b* for some simple orbitals. Note that although s orbitals are spherical ($l = 0$), orbitals with $l > 0$ have an angular dependence.

Finally, the electron in an atom has a fourth quantum number, M_s, characterizing its *spin direction.

This can be $+\frac{1}{2}$ or $-\frac{1}{2}$ and according to the *Pauli exclusion principle, each orbital can hold only two electrons. The four quantum numbers lead to an explanation of the periodic table of the elements.

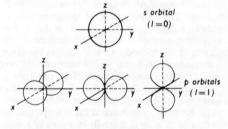

b Atomic orbital

atomic pile The original name for a *nuclear reactor.

atomic stopping power *See* stopping power.

atomic theory of magnetism *See* paramagnetism; ferromagnetism.

atomic unit of energy 1. *Syn.* hartree. The magnitude of the potential energy of an electron in the first orbit in Bohr's theory of the hydrogen *atom. It is given by $e^2/4\pi\varepsilon_0 a_0$, where e is the electron's charge and a_0 the atomic unit of length. It is equal to 27.190 electronvolts or 4.356×10^{-18} joule.

2. *Syn.* rydberg. The atomic unit of energy is sometimes defined as one half of this value. This is the ionization potential of the hydrogen atom.

atomic unit of length *See* Bohr radius.

atomic unit of mass A unit of mass equal to the rest mass of the electron. This unit is sometimes used in atomic physics. 1 atomic unit of mass is equal to $9.1084 \pm 0.003 \times 10^{-31}$ kg. It is not to be confused with the *atomic mass unit.

atomic volume The volume in the solid state of one mole of an element. Thus, atomic volume = relative atomic mass $\div$ density of the solid.

atomic weight *See* relative atomic mass.

attenuation The reduction of a radiation quantity, such as *radiant intensity, particle *flux density, or energy flux density, upon the passage of the radiation through matter. It may result from any type of interaction with the matter, such as absorption, scattering, etc. In an electric circuit it is the reduction in current, voltage, or power along a path of energy flow. *See* linear attenuation coefficient; attenuation constant.

attenuation band *See* filter.

attenuation coefficient *See* linear attenuation coefficient.

attenuation constant Symbol: α. For a plane progressive wave at a given frequency, the attenuation constant is the rate of exponential decrease in amplitude of voltage, current, or field-component in the direction of propagation of the wave. For example,
$$I_2 = I_1 e^{-\alpha d},$$
where I_2 and I_1 are the currents at two points (I_1 being nearer the source of the wave) a distance d apart. α is usually expressed in *nepers or *decibels. *See* propagation coefficient.

attenuation distortion *See* distortion.

attenuation equalizer An electrical *network designed to provide compensation for attenuation *distortion throughout a specified band of frequencies.

attenuator An electrical *network or *transducer specifically designed to attenuate a wave without distortion. The amount of attenuation may be fixed or variable. A fixed attenuator is also called a *pad. Attenuators are usually calibrated in *decibels.

atto- Symbol: a. The prefix 10^{-18}, e.g. 1 am = 10^{-18} metre.

attracted disc electrometer An early type of electrometer in which a circular metal disc A is held parallel to a larger metal disc B, by an extensible

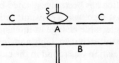

Attracted disc electrometer

spring S. The force per unit area on A, when a potential difference V is maintained between A and B, is equal to $\varepsilon_0 V^2/2d^2$ where d is the distance between the discs. A metal annulus, CC, known as the guard ring, ensures the uniformity of the field over the area of the attracted disc A. Since the potential difference is measured in terms of fundamental mechanical quantities the instrument is a form of *absolute electrometer*. Variations of this instrument have been used for measurements at very high potential difference.

attractor A set of points in *phase space to which the representative point of a dissipative system (i.e. one with internal friction) tends as the system evolves in time. The attractor can, for example, be a single point or it can be a closed curve (a *limit cycle*), which describes a system with periodic behaviour, or it can be a *fractal (or *strange attractor*), in which case the system exhibits *chaos.

Atwood, George (1746–1807) Brit. mathematician, chiefly remembered for his invention of *Atwood's machine.

Atwood's machine (1784) An educational device for measuring approximately the acceleration of a body falling under gravity and for illustrating the laws of motion. It consists basically of two equal masses M connected by a string passing over a pulley. A small mass, m, is placed on one of the masses M, which then moves from rest and is accelerated downwards; after a certain distance has been moved the mass m is automatically removed and the system continues to move with a constant velocity, which is measured.

audibility Ease of detection of a sound by ear. The sensitivity of the ear to a note depends on both its intensity and frequency. The intensity of a pure tone that is just audible is known as the *threshold of audibility*. Above this, the ear can detect greater intensities up to the threshold of feeling, beyond which the sensation changes to one of pain rather than hearing. These threshold intensities vary with frequency, the minimum threshold of audibility corresponding to an rms pressure amplitude of 0.0008 pascal at a frequency of 3500 hertz. Nothing would be gained if the ear were more sensitive than this since thermal noise in the air would then become audible. At a frequency of 1000 hertz the maximum intensity that the ear can detect is about 10^{14} times the minimum. The average range of frequencies to which the ear is sensitive is from about 20 to 20 000 hertz at an rms sound pressure of 1 pascal. The frequency range becomes smaller at greater or lower sound intensities and also diminishes as a person becomes older. The differential sensitivity of the ear to both intensity and pitch has been investigated by Knudsen who found that there was a constant ratio between the smallest detectable change in intensity and the original intensity except at high and low frequencies. A similar relation was found in the case of pitch sensitivity.

audiofrequency Any *frequency to which a normal ear can respond; it extends from about 20 to 20 000 hertz. In communication systems, satisfactory intelligibility of speech (i.e. of commercial quality) can be obtained if frequencies lying between about 300 and 3400 hertz are reproduced, and any frequency within this range is described as a *voice frequency*.

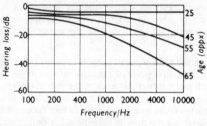

Audiogram

audiogram A graph showing hearing loss or threshold intensity shift in decibels plotted against frequency. This type of graph is useful for showing the full extent of deafness and its variation throughout the audible range. In such audiograms it is convenient to sketch in the normal thresholds of audibility and of feeling. An audiogram of hearing loss caused by noise shows the actual effects on the listener and is often more useful than a spectrum of the noise itself. The data for plotting such curves are obtained with an *audiometer. The hearing loss can be correlated with age. (*See* graph.)

audiometer An instrument for producing a sound of known frequency and intensity in a telephone earpiece. It is used to measure the hearing loss due to deafness and the masking produced by noise. Many forms of audiometer exist, the most common type consisting of an oscillator and amplifier feeding into a telephone earpiece. The frequency is variable throughout the audible range and an attenuator enables the output intensity to be cut down by specified amounts by rotation of a dial. The attenuator control can be calibrated in decibels so that hearing loss is read directly when the output is reduced to the threshold intensity of the person under test. This calibration is done by passing the sound from the earpiece through an artificial ear canal to the diaphragm of a calibrated microphone. For hearing-loss tests, the earpiece is held against the ear but for measuring the masking due to noise, it is kept a fixed distance away from it and the threshold of the sound in the presence of the noise is observed. Instead of the earpiece, a vibrating rod can be substituted. This is pressed against the skull for cochlear tests of the ear by *bone conduction. In

some audiometers the sound generator is a buzzer having component frequencies throughout the audible range. Sometimes a special recording is used as the source of a sound. Usual test tones are 2^n hertz, where $n = 7$ to 13. *See also* noise level.

auditorium acoustics The following are the main conditions that should be satisfied for a room to be acoustically good: (1) there should be no noticeable echoes; (2) loudness should be adequate and uniform throughout the room; (3) the reverberation time should be near the optimum value for the room; (4) resonance should be avoided; (5) the room should be sufficiently sound-proof. Echoes may be reduced by breaking up large plane surfaces, by avoiding sweeping curves, which tend to focus the sound, and by the use of absorbent material. Suitably placed reflectors may increase the loudness in certain parts of the room. The walls of a room can act as good reflectors but the path difference between direct and reflected sound must not be so great as to produce distinct echoes. Thus, a fan-shaped room with a sloping ceiling behind the source of sound and an absorbent back wall opposite to it tends to increase loudness without producing unwanted echoes. In large halls electronic amplifiers may be used successfully, provided echoes are avoided. Both echo and reflection effects can be studied in a model by spark photography or ripple-tank methods before the actual building is constructed. A common acoustical defect is due to a room having a long reverberation period. This causes poor articulation since the sound from one syllable does not die away appreciably before the next commences. Optimum reverberation times vary between about 1 and 2.5 seconds, being near the lower value for speech and having a high value for orchestral music, which is improved by the "body" that reverberation gives to it. The optimum time also depends on the size of the room, being longer the larger the room. In considering reverberation times, allowance must be made for absorption due to the audience. This considerably reduces the reverberation period observed in an empty room. By estimating the number of absorption units due to each part of a room, its reverberation time can be calculated before it is built. If necessary, the time may be reduced by avoiding or partitioning off recesses or by increasing the amount of absorbing material in the room. Resonance in the audible range of any part of a room or of the air contained in it should be heavily damped to avoid distortion due to amplification of a narrow band of frequencies. Some sound proofing of a room may be necessary and this entails massive walls, tightly fitting doors and windows, insulated heating and air pipes, and possibly a "floating" floor. *See* reverberation; sound insulation.

auditory acuity *See* audibility.

auditory perspective The ability of a person to detect the location of a sound, viz. the *azimuth, elevation, and distance of the sound from the observer. The latter quantity is most difficult to estimate aurally. At one time there was believed to be a velocity dispersion of the frequency components of a given sound and consequent change of the quality of a complex note with distance. Experiments show that there is no dispersion but rather an absorption with distance that is greater for the higher frequencies than for the lower. There is, therefore, a mellowing of the sound with distance, which, even when coupled with a general reduction of sound intensity, gives only a poor estimation of the range of the sound from the hearer. The use of the two ears of the observer permits a fairly accurate value of the azimuth and elevation of a sound source. (*See* binaural location.) Slight movements of the head, probably imperceptible to the observer, eliminate the ambiguities in angles that would be observed with the head stationary. Observations inside a room are complicated by multiple echoes. Information received by the other senses, particularly the eyes, is added to that received by the ears, and materially helps in the location of the sound.

Auger, Pierre Victor (*b.* 1899) French physicist who has made several contributions in the field of nuclear and atomic physics.

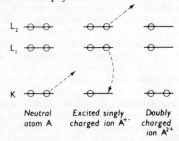

Neutral atom A   Excited singly charged ion $A^{+\cdot}$ Doubly charged ion A^{2+}

The dotted arrows indicate movement of electrons.

Auger effect initiated by the removal of a K electron

Auger effect *Syn.* Auger ionization. The spontaneous ejection of an electron by an excited positive ion to form a doubly charged ion: i.e.

$$A - e \rightarrow A^{+*} \rightarrow A^{2+} + e^{-},$$

where A^{+*} represents an excited state of a singly charged ion and A^{2+} a doubly charged ion that may or may not be in its ground state. The first step may result from *internal conversion. Alternatively, it may be induced by an external stimulus such as bombardment by electrons or photons. If this ion was formed by removal of a K electron (*see* diagram) , the ion would usually become de-excited by an electron in an outer shell (say L_1) moving into the vacancy in the K-shell, the energy released by this process being emitted as *characteristic X-radiation. However there is a probability (especially for elements of low atomic number) that the energy released may be taken up in the simultaneous emission of a further electron from another shell (say L_2). This process would be written K – $L_1 L_2$ and the

electron ejected would have an energy determined by the relative energy levels of the shells involved.

*Autoionization is very similar to the Auger effect and the terms are sometimes used synonymously. The diagram is a schematic representation of the Auger effect.

Auger electron An electron ejected by the *Auger effect.

Auger shower *Syn.* extensive air shower. A *shower of elementary particles produced by a primary *cosmic ray entering the atmosphere.

aurora An intermittent electrical discharge occurring in the rarefied *upper atmosphere. Charged particles in the *solar wind become trapped in the earth's magnetic field and move in helical paths along the lines of force, oscillating between the two magnetic poles. On entering the upper atmosphere the charged particles excite the air molecules. The resulting *emission of light takes many beautiful forms from slight luminosity to large streamers moving rapidly across the sky. Since the lines of force converge at the magnetic poles, which are close to the N and S poles, the intensity of the aurora is greatest in polar regions although it is sometimes seen in temperate zones. During high solar activity the number of particles in the solar wind, and their velocity, increases and consequently the aurora intensity increases. *Aurora borealis* denotes the northern aurora, *aurora australis* the southern one, and *aurora polaris* is a general term for either phenomenon. *See also* Van Allen belts.

autocollimation An autocollimating eyepiece includes a crosswire that can be illuminated from the side by using a small reflecting prism. (*See* Gaussian eyepiece.) When placed in the tube of a telescope, the light is reflected through the objective on to a plane mirror, which returns the light back through the objective, and a person looking through the eyepiece focuses the reflected image in the same plane as the original crosswires. The telescope is thus focused for parallel light and can then be used to focus a collimator. The theory is illustrated by the autocollimating method of measuring the focal length of a lens, i.e. by reflecting the light passing through a lens both ways to focus in the original source position. The *littrow prism uses the autocollimation principle together with its dispersing properties. The term is also applied to a small-aperture convex mirror with its pole at the focus of a concave mirror and their two centres coincident. Incident parallel light on the concave mirror will, after double reflection, emerge parallel to the incident light but opposite in direction of travel.

autodyne oscillator *See* beat reception.

autoemission *Syn.* autoelectronic emission. *See* cold cathode.

autoionization A form of ionization involving two steps, first the excitation of an atom (or molecule) into a state with an energy above its *ionization potential and second, de-excitation from this state to give a positive ion with a lower energy and an ejected electron. The process is similar to the *Auger effect with the difference that the initial vacancy in an electron shell is caused by transfer of an electron from one orbital to another empty orbital. In the Auger effect the vacancy is formed by complete removal of the electron to give an ion. Autoionization results in a singly charged positive ion:

$$A \rightarrow A^* \rightarrow A^+ + e,$$

where A^* is an excited atom. The electron ejected has a characteristic energy equal to the difference between the energy of the excited atom and that of the ion.

automatic frequency control (a.f.c.) A device that automatically maintains the frequency of any source of alternating voltage within specified limits. The device is "error-operated" and usually consists of the following parts: (1) a *frequency discriminator, which produces a d.c. voltage output approximately proportional to the frequency difference, with sign controlled by the direction of drift; (2) a *reactor, which forms part of the *oscillator tuned circuit and is controlled by the output of the frequency discriminator in such a way that it reduces the oscillator frequency.

automatic gain control (a.g.c.) *Syn.* automatic volume control. A method of automatically holding the output volume constant in a radio receiver, despite variations in the input signal. A control voltage is derived from the input signal and applied to a variable gain element in the receiver. Variations in the magnitude of the input signal cause changes in gain in the receiver so that the output is maintained constant.

automatic tuning control A type of *automatic frequency control incorporated in a radio receiver; it adjusts the tuning to a correct setting for a given received signal when the manually operated adjustment has only been set approximately to the correct position for that signal.

automatic volume contractor *Syn.* compressor. *See* volume compressors (and expanders).

automatic volume control *See* automatic gain control.

automatic volume expander *Syn.* expander. *See* volume compressors (and expanders).

autoradiograph A photograph of the distribution of a *radioisotope within a thin specimen of metal, biological tissue, etc. The specimen is *labelled with a *radioisotope and placed in contact with a photographic plate for a suitable exposure time, and an image is produced by the action of the radiation emitted. On developing the film, the autoradiograph can be seen.

autotransductor A transductor in which the same windings are used for the main and control currents.

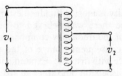

Connections for single phase autotransformer

autotransformer A *transformer with a single winding, tapped at intervals, instead of two or more independent windings. The voltage drop across each tapped section is related to the total applied voltage in the same proportion as the number of turns of the selection is related to the total number of turns of the winding.

autotransformer starter A *motor starter for an a.c. motor in which a reduced voltage is applied to the motor at starting by means of tappings on an *autotransformer.

autumnal equinox *See* equinox.

auxiliary circle The circle that has the major axis of an ellipse or elliptical orbit for its diameter.

avalanche A cumulative ionization process such as that occurring in a *Geiger counter when a single particle or photon ionizes several gas molecules. Each electron liberated is accelerated towards the anode and gains sufficient kinetic energy to ionize other molecules, giving more free electrons and ions. The positive ions move towards the cathode where they may release further electrons by *secondary emission. Excited atoms or molecules and recombining ions may emit radiation with sufficient quantum energy to cause *photoemission from the cathode, thereby extending the discharge rapidly. Thus a small initial amount of ionization may cause a large surge of current, the magnitude of which is independent of the original number of liberated electrons.

avalanche breakdown A type of *breakdown in a *semiconductor diode caused by the cumulative multiplication of free charge carriers under the action of a strong electric field. Some free carriers gain enough energy to liberate new hole-electron pairs by collision.

average *Syn.* arithmetic mean; mean. **1.** The sum of a number of observations divided by their number.

2. In the case of a continuous function $f(x)$, the mean value over the range x_1 to x_2 is

$$\frac{\int_{x_1}^{x_2} f(x)\, dx}{x_2 - x_1}.$$

Physically this corresponds to the mean ordinate of the graph of $f(x)$ (as ordinate) against x. Similar definitions exist for functions of more than one variable.

3. *See* weighted mean.

average life or lifetime *See* mean life.

Avicenna (Abdallah ibn Sina) (980–1037) Arab scientist who anticipated Newton's first law of motion and proposed a corpuscular theory of light.

avionics The application of electronics to *aeronautics and *astronautics.

Avogadro, Amedeo (1776–1856) Italian physicist. His most important discovery, now called *Avogadro's hypothesis, enunciated in 1811 was ignored until brought to general notice by *Cannizzaro in 1854.

Avogadro constant Symbol: L or N_A. The number of molecules contained in one mole of any substance. *Amount of substance is proportional to the number of specified entities of that substance, the Avogadro constant being the proportionality factor. It is the same for all substances and its value is $6.022\ 136\ 7 \times 10^{23}$ mol^{-1}. *See also* Loschmidt constant.

Avogadro's hypothesis Equal volumes of all gases measured at the same temperature and pressure contain the same number of molecules, i.e. the volume occupied at a given temperature and pressure by a mole of a gas is the same for all gases (22.4 $\times 10^{-3}$ m^3 at STP). *See* ideal gas.

axes of strain *See* homogeneous strain.

axes of stress *See* stress components.

axial ametropia A defect of eyesight due to length of eyeball. When the eyeball is too long, it is known as axial *myopia; when too short, axial *hypermetropia.

axial chromatic aberration *See* longitudinal aberrations.

axial hypermetropia *See* axial ametropia.

axial modulus *Syn.* modulus of simple longitudinal extension. *See* modulus of elasticity.

axial myopia *See* axial ametropia.

axial ratio The relative lengths of the three edges of the *unit cell of a crystal lattice, taking that of the b axis as unity.

axial vector *See* pseudovector.

axiom A self-evident proposition, not requiring demonstration.

axion A hypothetical subatomic particle postulated to account for the fact that theories of *quantum chromodynamics predict the violation of *CP invariance in the strong interaction, but this has not been observed. Axions, if they exist, are light particles (less than 10^{-12} times the mass of the proton). Some physicists believe that large numbers of them may exist in the haloes of galaxies and that they may contribute to the *missing mass in the universe.

axis of lens For *spherical lenses, a straight line perpendicular to each face and joining the centres of the two surfaces. For *cylindrical and sphero-cylindrical lenses, a line (on the surface) parallel to the geometrical axis of the cylinder surface. In *astigmatic lenses, generally a line parallel to one of the principal meridians passing through the optical centre.

axis of permanent magnet A line through the centre of the magnet such that the torque exerted on the *magnetic field vanishes when this line coincides in direction with the lines of the field.

Ayrton, William Edward (1847–1908) Brit. physicist who made several contributions to the theories of dielectrics and terrestrial magnetism. He also invented several measuring instruments, including a spiral-spring ammeter and wattmeter.

Ayrton–Jones current balance *See* current balance.

Ayrton–Mather shunt *See* universal shunt.

azimuth 1. Position as measured by an angle round some fixed point or pole. (*See* astronomy; celestial sphere; coordinate.)
  2. In relation to elliptically polarized light, the direction of the vibration plane of the light if the phase difference corresponding to the ellipticity is reduced to zero without altering the amplitude of the components.

azimuthal quantum number Symbol: l. A *quantum number introduced by Sommerfeld in his theory of elliptical electron orbits to specify the angular momentum of the motion. *See* atom; atomic orbital.

B

Babbage, Charles (1792–1871) Brit. mathematician, pioneer in the field of calculating machines.

Babinet, Jacques (1794–1872) French physicist. The originator of the suggestion that a particular spectral line should be used as a standard of length (1829). His researches on the analysis of polarized light produced the *Babinet compensator. *See also* Babinet's principle.

Babinet compensator An optical device that introduces a *phase difference of variable magnitude between the ordinary and extraordinary rays (*see* double refraction). It comprises two narrow-angle quartz wedges with parallel refracting edges and hypotenuse faces adjacent; the optic axes of the prisms are mutually perpendicular, and aligned parallel to the refracting edges. A light ray passing vertically downwards through some point on a refracting edge will traverse a distance d_1 in the upper wedge and d_2 in the lower wedge. The total relative phase difference is then proportional $(d_1 - d_2)$. This value can be varied by sliding one wedge over the other, maintaining parallel edges, and hence the desired relative phase difference can be obtained. *See also* half-wave plate; quarter-wave plate.

Babinet's principle Two complementary diffracting screens (i.e. screens in which the transparent portions of the one occupy the same positions as the opaque portions of the other) produce the same illumination at a point which would show zero illumination if both screens were absent.

Babo's law The lowering of the *vapour pressure of a solvent by addition of a nonvolatile solute is proportional to the concentration of the solution.

back electromotive force An e.m.f. that opposes the normal flow of current in an electric circuit. Thus, in a water voltameter a back e.m.f. is generated by the evolution of oxygen and hydrogen on the plates (which act as a voltaic cell). In an electric motor the rotation of the armature in the field magnets induces a back e.m.f. in the armature coils.

back focal length The distance from the last surface of an optical system to the second principal focus. In practice the power of a *spectacle lens corresponds with the *back vertex power*, the reciprocal of the back focal length expressed in metres.

background counts In a radiation *counter, counts registered when the source of radiation under consideration is not present. Such counts may arise from sources other than the one being measured, naturally occurring *background radiation, contaminations in the counter itself, or spurious signals in the electronic circuits of the counter.

background noise Syn. random noise. See noise.

background radiation 1. The low-intensity radiation resulting from the bombardment of the earth by cosmic rays and from the presence of naturally occurring *radionuclides (such as ^{40}K, ^{14}C) in rocks, soil, air, building materials, etc. When measurements of radiation are being carried out a correction must be made for the background radiation. See also background counts.
2. See cosmic background radiation.

back heating See magnetron.

backing pump See pumps, vacuum.

backing store Large-capacity computer storage devices, usually *disks, in which programs and data are kept when not required for processing by a *computer. A program and its associated data is copied from backing store into *main store only when the program is about to be executed.

backlash Looseness in a mechanical system, e.g. a screw head can be turned through a small angle before a thrust is developed on the nut; the driving shaft in a gear train may be turned through a small angle before a torque is developed at the final driven shaft. Some backlash is essential if the system is to be free to move but its effect may be reduced by springs which keep the driven member in contact with the driving member.

back layer photocell See photovoltaic cell.

back-reflection photography A crystal-diffraction method in which the photographic film is placed between the source of incident X-radiation and the specimen; used chiefly for studies of texture and for precision measurements of spacing of surface atoms.

back scatter The scattering process by which radiation emerges from the same surface of a material as that through which it enters. The term also applies to the radiation undergoing such a process. Back scatter must be taken into account when a source of *ionizing radiation is calibrated.

back vertex power See back focal length.

backward lead See brush shift.

backward-wave oscillator See travelling-wave tube.

Bacon, Francis (Baron Verulam) (1561–1626) Brit. philosopher and statesman, a vigorous exponent of inductive science.

Bacon, Roger (Friar Bacon) (c. 1214–1292) Brit. medieval scholar who is regarded as the first modern scientist. Although he believed that experience consists of a combination of an inner knowledge of divine origin and practical knowledge gained by observation, his greatest contribution to the scientific method was his insistence on the value of experiment. He regarded speculation as useless unless backed by measurements or observations. He showed that air is required to support combustion and he used lenses to correct vision.

baffle A partition used with a sound radiator to increase the path difference between sound originating from the front and back of the radiator. It is most commonly used to improve the frequency response of a loudspeaker. A cone speaker on its own acts as a double source, there being 180° phase difference between sound originating at the front and the back. When the speaker is surrounded by a baffle whose dimensions are small compared with the wavelength of the sound, it continues to act as a double source and its power output is proportional to the square of the frequency. When the baffle is large compared with the wavelength, however, the two sides of the cone act separately and the power output is independent of the frequency. If the wavelength of the sound is equal to the acoustic path length between front and back of the speaker the two sounds will be exactly out of phase and there will be a sudden drop in output at this frequency. Thus, the output from a speaker mounted at the centre of a circular baffle rises with frequency until the baffle dimensions are just under half a wavelength. Above this the response levels out except for a sharp dip when the wavelength is equal to the baffle diameter. This dip can be eliminated by mounting the speaker asymmetrically in a rectangular or circular baffle so that the path difference between front and back is not constant. To give a level response down to a frequency of 100 hertz a baffle about 2 metres square is needed. In most commercial sound reproducers the speaker is housed in a cabinet that acts as the baffle.

A baffle is also placed round a *ribbon microphone to increase the path difference between the two sides. The ribbon is actuated by the pressure difference caused by the phase change arising from this path. A baffle plate has also been used with a *hydrophone. When placed near one of the diaphragms it causes a bidirectional hydrophone to give a unidirectional response. Its action is not quite clear since it is not large enough to cast a real shadow of the sound.

Baird, John Logie (1888–1946) Brit. electrical engineer who gave the first demonstration of the transmission of a picture by radio waves in 1926. The Baird Television Development Co. provided the first BBC television programme in 1929. Baird's mechanically scanned 240-line system was replaced in 1935 by the electrically scanned 405-line system developed by EMI. At present a 625-line system is used. *See also* Zworykin, Vladimir Kosma.

balance An instrument whose primary function is to compare two masses. The balance with the widest usage is the *beam balance*, in which objects to be weighed and known weights are placed on opposite sides of a pivoted bar. If the arms of the balance are of equal length, and P and Q are two slightly different masses, placed one in each pan of the balance with arms of length a, and the balance comes to rest with the arms making a small angle θ with the horizontal then

$$\tan \theta = \frac{(P - Q)a}{(P + Q + 2w)h + Wk},$$

where w is the mass of a scale pan, and W is the mass of the balance beam, h is the height of the central knife-edge above the line joining the outer knife-edges, and k is the distance of the centre of gravity of the beam below the central knife-edge. Tan $\theta/(P - Q)$ is a measure of the sensitivity of the balance for a load P.

If the balance arms are of unequal lengths a and b, a method of weighing first used by Gauss may be employed, in which a substance of mass W is weighed successively in the two pans of the balance and the indicated masses are W_1 and W_2; then $W = \sqrt{W_1 W_2}$. This manner of weighing enables one to calculate the arm ratio of the balance, for the above equation is derived from the equations $Wa = W_1 b$ and $W_2 a = Wb$, which gives also $W_1/W_2 = a^2/b^2$, and finally the arm ratio a/b.

A second method of weighing was devised by Borda. A body whose mass need not be known (called the tare) is placed in one of the scale pans together with the object whose mass is to be measured, and masses are placed in the other scale pan until equilibrium is attained. The object of unknown mass is removed from the scale pan and masses are put in its place until equilibrium is restored. These masses are, obviously, the mass of the body.

Other forms of balance are: (1) the *decimal balance*, which has arms in the ratio 10:1 and thus avoids the necessity for using heavy weights; (2) the *spring balance*, which consists essentially of a helical spring with its axis vertical. It may be used either by extending the spring or by compressing it as in the familiar use of household scales; (3) the *torsion balance*, which consists essentially of a vertical straight torsion wire, the upper point of which is fixed while the lower carries a horizontal beam. This balance was used by Cavendish to measure the attraction between a large massive sphere and a smaller sphere attached to the end of the horizontal beam (*see* gravitation); (4) the *hydrostatic balance*. The equal-armed balance may be used as a hydrostatic balance by placing a wooden bridge across, but not touching, one of the balance pans and hanging a piece of the substance whose *relative density is required from a hook at the top of the pan support and weighing this first in air and second in a beaker of water standing on the bridge; (5) *Jolly's balance*, which consists of a helical spring whose axis is vertical and the upper point is fixed. The lowest point of the spring carries a horizontal pointer which moves over a vertical graduated scale. A specimen of the solid whose density is required is placed on a pan connected to the lowest point of the spring and the ·extension of the spring is measured (*a*) when the substance is in air, (*b*) when it is immersed in water; (6) the *chain-drum balance*, which obviates the necessity for the use of small fractional weights. It consists of a fine chain, one end of which is attached to the beam of a balance and the other end passes over a vertical drum which is attached by means of a clamp to the upright and hangs in a loop between the drum and the plane of the balance beam. Then, instead of using small fractional weights, the chain is wound from the drum in such a way that the length of the loop is altered until equilibrium is attained. The additional weight is obtained from the scale on the drum; (7) the Roman *steelyard*, which consists of a simple lever of which one arm is short and the other arm long. The object to be weighed is hung from a hook at the end of the shorter arm of the beam and equilibrium is obtained by moving a weight along a graduated scale which is marked on the beam; (8) the Danish *steelyard*, which is similar to the Roman steelyard, but is not so convenient since equilibrium is obtained by sliding the fulcrum along the bar which is graduated to read the required wieight directly; (9) the *modern balance* in which the deformation of a supporting system is measured using *strain gauges, usually with a digital display. *See also* Eötvös torsion balance; microbalance; and for Odén's balance *see* sedimentation.

balanced amplifier A push-pull amplifier. *See* push-pull operation.

balanced polyphase load A load on a symmetrical *polyphase system is balanced when the currents in the several circuits, or phases, are equal and have the same *power factor. *See* symmetrical components.

balance of a rotating body A rotating body is said to be balanced if its rotation does not cause radial forces to act on the bearings holding its centre. *See* centrifugal moment.

balancer *Syn.* static balancer. **1.** Direct current balancer. An auxiliary d.c. *motor generator or *dynamotor used with a multiple wire d.c. system (e.g. *three-wire system) for the purpose of dividing the total voltage substantially equally between the wires. **2.** Alternating-current balancer. A *reactor or *autotransformer designed for use with a multiple wire d.c. or a.c. system for the purpose of dividing the total voltage substantially equally between the wires. When used with d.c. systems it is connected via brushes and slip-rings to points in the armature winding, so that it is in fact operating with an alternating voltage. An a.c. balancer is sometimes used with a *three-phase a.c. system for obtaining a neutral point connection when four-wire working is required.

balancing speed *See* free-running speed.

ballast resistor A resistor constructed from a material having a high *temperature coefficient of resistance in such a way that, over a range of voltage, the current is substantially constant. It is connected in series with a circuit to stabilize the current in the latter by absorbing small changes in the applied voltage. The most common types are the *barretter and the *thermistor.

ball-ended magnet A magnet made of this steel rod with ball ends. The poles are situated at the centre of the spheres and if the magnet is long experiments can be carried out as if with point poles.

ballistic Of an instrument. Designed to measure an impact or brief flow of charge. *See* pendulum; ballistic galvanometer.

ballistic galvanometer A galvanometer adapted to measure the charge, Q, flowing through the instrument during the passage of a transient current, where

$$Q = \int_0^\infty I.\mathrm{d}t.$$

The period of the moving part of the galvanometer must be long compared with the duration of the current; the electromagnetic impulse due to the passage of the transient can then be deduced from the ballistic "throw", θ. For a suspended-magnet type galvanometer, $Q \propto \sin \frac{1}{2}\theta$; for a moving-coil instrument, $Q \propto \theta$. Damping of the movement should be small; correction can then be made for it by substituting $\theta(1 + \lambda/2)$ for θ in the above relations, where λ is the *logarithmic decrement. The instrument is calibrated by means of a standard capacitor or inductor. *See also* fluxmeter.

ballistic missile Any missile which is propelled or guided only during the first part of its flight, especially a ground-to-ground missile.

ballistics The study of projectiles.

ball lightning A rare form of lightning appearing as a slow-moving incandescent globe, which often explodes and may set fire to objects it touches. Its cause is unknown.

balloon When a balloon floats at rest in a still atmosphere, the upthrust, which is equal to the weight W_1 of the air displaced, just balances the weight W_2 of the gas within the balloon envelope, i.e. $W_1 = W_2 + W_3$, or $W_2 = W_1 - W_3$. The weight lifted (W_2) can therefore be increased by increasing the difference between W_1 and W_3. For a given size of envelope and at normal temperatures and pressure, W_1 is fixed so that ($W_1 - W_3$) can only be increased by using as light a gas as possible for filling the envelope. Hydrogen is the lightest gas and is cheap and easily obtainable; hence it has been widely used. Unfortunately it burns fiercely in air and mixtures of air and hydrogen are dangerously explosive. Helium is the second lightest gas giving values of W_2 about 9% less than for hydrogen. It is completely safe but is relatively expensive. Following a period of about forty years in which airships were rarely used, there has been a revival of interest in helium-filled airships for commercial purposes. The first balloon was the hot-air balloon of the brothers Montgolfier (1783) in which the air in the envelope was heated and became lighter on expansion, so causing the balloon to rise and stay up so long as sufficient heat is applied. Balloons are used for atmospheric and cosmic-ray research and for military use. Hot-air balloons are used for sport. *See* pilot balloon; sounding balloons.

Balmer, Johann J. (1825–1898) Swiss physicist. He took the initial step in the unravelling of atomic spectra by his discovery of a simple formula to represent the wavelengths of a set of spectral lines. *See* hydrogen spectrum.

Balmer series *See* hydrogen spectrum.

balun Acronym for balanced unbalanced. An electrical device used to couple a balanced impedance, such as an aerial, to an unbalanced *transmission line, such as a coaxial cable.

banana tube A special type of *cathode-ray tube used as a *colour picture tube. The name apparently follows the fashion set by the American *apple tube. The line scan is produced inside the tube, and the field scan produced by cylindrical lenses mounted arund the tube. The *phosphor is in the form of stripes of the three primary colours.

band 1. In communications, a range of frequencies within specified limits used for a definite purpose, e.g. *medium wave band* in radio is a range of frequencies, certain of which are assigned to various transmitting stations, and the radio may be tuned to any frequency in the band.

2. A closely spaced group of energy levels in atoms. (*See* energy band.)

3. *See* band spectrum.

band-edge energy *See* semiconductor.

band-pass filter *See* filter.

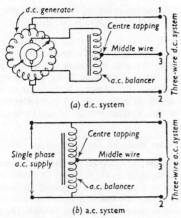

(a) d.c. system

(b) a.c. system

Connections for a.c. balancer in three-wire systems

band pressure level The *sound pressure level of a sound within a specified band of sound frequency.

band spectrum *Syn.* molecular spectrum. Under low resolution, the spectra of compounds under normal experimental conditions appear as fluted bands separated by dark spaces. With higher resolution, the bands appear as fine lines, sharp and closer together at the head of the band, and becoming more separated amd more diffuse farther away from the head. *See* spectrum.

band-stop filter *See* filter.

band structure of solids *See* energy bands.

bandwidth 1. In a communication system. The band of frequencies, in the transmitted signal, taken up by the modulating signal on each side of the frequency of the carrier signal. **2.** In an antenna array. The deviation in frequency that the system can handle without a mismatch. **3.** In an amplifier. The band of frequencies over which the *power amplification falls within specified limits of the maximum value (usually one half).

bar Symbol: bar. A CGS unit of pressure equal to 10^5 pascal. It may be used with SI units, and SI prefixes may be attached to it. The millibar (Symbol: mbar or mb) is a commonly used unit of pressure in meteorology.

Bardeen, John (*b.* 1908) Amer. physicist who, with Walter Brattain and William Shockley, first developed the point-contact *transistor (1947). He is also noted for his work on the BCS theory of *superconductivity (1957), in collaboration with Leon Cooper and John Schrieffer. Bardeen is the only scientist to have been awarded two Nobel prizes for physics (1956; 1972).

Barkhousen, Heinrich (1881–1956) German physicist who made substantial contributions to the theory of magnetism. *See* Barkhausen effect.

Barkhausen effect The magnetization of a ferromagnetic substance does not increase or decrease steadily with steady increase or decrease of the magnetizing field but proceeds in a series of minute jumps. The effect can be demonstrated by winding a magnetizing coil and secondary coil on a ferromagnetic specimen and connecting the secondary to an oscillograph which can show the rapid fluctuations as the primary current increases steadily. The effect gives support to the domain theory of *ferromagnetism.

Barkhausen–Kurz oscillations Sustained oscillations, in which the frequency is controlled by the applied voltages, occurring in a plasma of electrons and ions in a discharge tube. *See* plasma oscillations.

Barkla, Charles Glover (1877–1944) Brit. physicist who showed that when elements are exposed to X-rays there is a secondary emission of X-rays, the penetrating power of which increases with the atomic number of the irradiated element.

Barlow lens A planoconcave lens placed between the *objective and *eyepiece in a *telescope to increase magnification.

Barlow's wheel A primitive form of electric motor. It consists of a copper disc, free to rotate in a vertical plane, between the poles of a horseshoe magnet, and dipping at its lowest point into a pool of mercury. If a potential difference is maintained between the axle and the mercury, the wheel rotates.

barn Symbol: b. A unit of *nuclear cross section. 1 barn = 10^{-28} m^2.

Barnard, Joseph Edwin (1870–1949) Brit. amateur physicist who developed ultraviolet microscopy using quartz lenses and a photographic technique.

Barnes *See* Callendar and Barnes's continuous flow calorimeter.

Barnett effect A long iron cylinder rotating at high speed about its longitudinal axis develops a slight magnetization proportional to the angular speed of rotation. This magnetization is due to the effect of the rotation on the electronic orbits in the atoms of the iron and on the electrons themselves, which have their own intrinsic *spin. An iron cylinder freely suspended rotates slightly when suddenly magnetized – the reverse of the Barnett effect. *See* Einstein and de Haas effect.

barograph A recording *barometer. The common type consists of an aneroid barometer operating a pen that traces a line on a sheet of graph paper mounted on a slowly revolving drum.

barometer An instrument for measuring atmospheric pressure.
   1. *Mercury barometers* consist of a glass tube about 80 cm long, closed at one end. The tube is filled with mercury and then placed, open end downward, in a reservoir of mercury as in Fig. *a.* As the atmospheric pressure changes, the level of the mercury changes and, to a lesser degree, so does the level

barometric formula

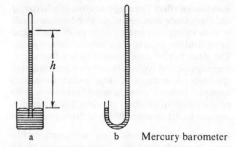

a   b   Mercury barometer

in the reservoir. The pressure is measured by the difference h between the levels (*see* pressure). The space at the top of the tube is known as a *Torricellian vacuum* after the inventor of the instrument.

In *Fortin's barometer*, the scale for measuring the height is fixed and the level of the mercury in the reservoir is adjusted (by moving the flexible bottom of the reservoir) to be at the zero of the scale. The difference in height can then be read directly. There are also instruments in which the scale is moved so that its zero coincides with the lower mercury level. In the *Kew-pattern barometer*, the reservoir and scale are both fixed, but the scale is calibrated so as to read the barometric height h directly. The *syphon barometer* is shown in Fig. *b*; it can be used as a recording instrument (*barograph) since a float on the mercury in the open limb can be made to move a pen.

2. The *aneroid barometer* consists basically of an evacuated flat cylindrical closed metal box with corrugated flexible faces that are kept from collapsing together by a spring. As the external pressure varies, the distance between the flat faces alters and a pointer is operated. Such an instrument is calibrated by comparison with a mercury barometer. It is also used as an *altimeter* to measure heights by means of the decrease in atmospheric pressure with height about sea level.

barometric formula A formula indicating the variation of pressure with height for the gas in the earth's atmosphere. It is based on the assumption that the temperature does not vary with height and has the form:

$$p = p_0 e^{-mgx/kT},$$

where p is the pressure at a height x, m the mass of a molecule, g the acceleration of *free fall, k is the Boltzmann constant, and p_0 the pressure when $x = 0$. The formula is a special case of *Boltzmann's formula, in which mgx is the potential energy of a molecule.

barostat A constant pressure device or pressure regulator, especially one that compensates for changes of atmospheric pressure, as in the fuel-metering system of an aircraft engine.

barrel distortion Distortion of the image of a square arising from the lateral magnification decreasing with size of object (*see* aberrations of optical sys-

tems). The real image formed by a simple converging lens with front stop shows barrel distortion.

barretter A device used for stabilizing voltage, consisting of a sensitive metallic resistor whose resistance increases with temperature. The resistor is usually enclosed in a glass bulb, e.g. an iron wire in an atmosphere of hydrogen. When used in series with a circuit the voltage drop is kept constant over a range of variations in current. *See also* ballast resistor.

barrier-layer photocell *See* photovoltaic cell.

bars in transverse vibration A vibrating bar can be regarded as an extension in theory of a *transverse vibration in a string whose *stiffness has been increased so that the restoring forces are due to the *bending moment rather than the tension. The theory of the bar assumes it to be uniform in thickness and density and not subject to tensional forces. The equation of motion of a transversely vibrating bar is:

$$\frac{\partial^2 y}{\partial t^2} + K^2 V^2 \frac{\partial^4 y}{\partial x^4} = 0,$$

where $V = \sqrt{E/\rho}$, the velocity of a longitudinal wave in the bar (E is Young's modulus and ρ the density), and K is the radius of gyration of the section considered about the neutral surface. The complete solution of this is:

$$y = (A \cosh mx + B \sinh mx + C \cos mx + D \sin mx) \cos \omega t,$$

where $m^4 = \rho \omega^2 / EK^2$. The unknowns in this equation are determined by the end conditions. Three different end conditions are possible: (1) free end. No curvature or shear; $\partial^2 y/\partial x^2 = 0$, $\partial^3 y/\partial x^3 = 0$; (2) clamped end. No displacement or slope; $y = 0$, $\partial y/\partial x = 0$; (3) supported end. No displacement or curvature; $y = 0$, $\partial^2 y/\partial x^2 = 0$. The frequency of vibrations in a bar of length l is given by:

$$f = \frac{q\pi}{8} \cdot \frac{K}{l^2} \sqrt{\frac{E}{\rho}}.$$

For a bar free at both ends or clamped at both ends, q is 3.001 2 for the fundamental and $3^2, 7^2, 11^2$, etc., for the partials. For a bar clamped at one end only q has values 1.194, 2.988, $3^2, 5^2, 7^2$, etc. For a bar supported at both ends, $q/4 = 1^2, 2^2, 3^2$, etc. It will be seen that in general the partials are not harmonics and also since the frequency is inversely proportional to the square of the length of the bar the velocity of transverse vibrations depends on the frequency. As in the case of a string, a bar may be excited by bowing, striking, or by electromagnetic methods. A thin bar clamped at one end and free at the other is used as a *reed in many musical instruments with a column of air acting as a resonator to reinforce the fundamental of the reed. In the xylophone, wood or metal bars are supported near their ends and excited by striking with a hammer. The *tuning fork is an extension of a transversely vibrating bar and may be considered either as a bent free-free bar or as two clamped-free bars.

bar suspension *Syn.* yoke suspension. A method of mounting an electric traction motor. One side of the motor frame is axle hung, i.e. is supported by special bearings on the axle. The other side is bolted to a bar that lies transversely across the truck, the bar being supported from the truck by springs. The method ensures that the axis of the armature is always parallel to and at a fixed distance from the axis of the axle, and is therefore suitable for use with geared drive. *Compare* nose suspension.

Bartholin, Erasmus (1625–1698) Danish mathematician who discovered the *double refraction of light in a crystal of Iceland spar, giving the names ordinary and extraordinary rays to the two beams.

Bartlett force *See* exchange force.

barycentre *See* centre of mass.

barye A former CGS unit of pressure. 1 barye = 0.1 pascal.

baryon A collective name given to *hadrons (*elementary particles that can undergo strong interactions) with half integer spin. Baryons are thus *fermions. All baryons have a mass equal to or greater than the mass of the proton. An additive *quantum number, called the baryon number (B), may be defined such that baryons have baryon number $B = +1$, antibaryons $B = -1$, and all other particles $B = 0$. The total baryon number, which equals the number of baryons minus the number of antibaryons is conserved in all particle interactions. Baryons are composed of three *quarks while antibaryons contain three antiquarks. For example, the proton contains two up quarks (u) and a down quark (d) while the neutron contains an up quark and two down quarks. *See* Appendix, Table 8.

barytes Barium sulphate or "heavy spar". Relative density 4.5. Used in special cement for protection from *ionizing radiations such as X-rays.

basal plane The plane normal to the unique axis in trigonal, tetragonal, or hexagonal crystals. It is also the plane normal to the axis of a crystal prism or forming the base of a pyramid.

base The region in a *junction transistor separating the *emitter and *collector. *See also* transistor; semiconductor.

baseball bars Current-carrying bars used in experimental fusion devices to increase *plasma stability. *See* fusion reactor.

base curve of a lens *See* toric lenses.

base electrode The electrode attached to the *base in a *junction transistor.

base exchange *See* ion exchange.

base magnification *See* total relief.

base units *See* SI units.

Basov, Nikolai Gennediyevich (*b.* 1922) Soviet physicist. His research into amplification of electromagnetic radiation by excited atoms and molecules culminated in the development of the *maser (1955), with the cooperation of Aleksandr Prokhorov. For this work they shared a Nobel prize (1964) with Charles Townes, who had developed the maser independently in America. Basov later developed semiconductor *lasers (1958).

batch processing A system by which the programs of individual users are collected together and submitted as a single unit to a computer, either at one central site or at a small number of *remote job entry stations. *Compare* time sharing. *See also* interactive.

Bateman equations A set of equations that describe the decay of a chain of radioactive nuclides. The first member of the chain decays into the second with a *decay constant λ_1, and at a time t there are N_1 atoms left where:
$$dN_1/dt = \lambda_1 N_1.$$
The rate of appearance of the second nuclide is governed by two processes, its rate of production from the first nuclide and its rate of disappearance by decay into the third nuclide. Consequently:
$$dN_2/dt = \lambda_1 N_1 - \lambda_2 N_2,$$
where N_2 is the number of atoms of this second nuclide and λ_2 is its decay constant. In general the rate of appearance of the nth nuclide is given by:
$$dN_n/dt = \lambda_{n-1} N_{n-1} - \lambda_n N_n.$$
If only the parent nuclide is initially present and there are N_1^0 atoms, then the number of atoms of the nth nuclide after a time t is given by the equation:
$$N_n(t) = \sum_1^n \frac{\lambda_1 \lambda_2 \cdots \lambda_{n-1} N_1^0 e^{-\lambda_n t}}{(\lambda_1 - \lambda_n)(\lambda_2 - \lambda_n) \cdots (\lambda_{n-1} - \lambda_n)}$$

battery Two or more secondary cells, primary cells, or capacitors, electrically connected and used as a single unit.

battery, floating An accumulator battery connected to a discharging circuit which it supplies, and also to a charging circuit, the mean charging current being adjusted so that the state of charge of the battery remains constant. It is used to provide a constant e.m.f. in the discharge circuit, in spite of fluctuations in the main supply.

baud Symbol: Bd. A unit of signal speed in a computer system or a communication system such as a telephone link. It is equal to the number of times per second that the signalling element changes state. When the signal is a sequence of *bits, one baud is equal to one bit per second (1 bps).

BCS theory *See* superconductivity.

BDV Abbreviation for *breakdown voltage.

beam aerial *See* aerial array.

beam balance

beam balance *See* balance.

beam coupling The production in a circuit of an alternating current between two electrodes, when an intensity-modulated electron beam is passed.

beam coupling coefficient The ratio of the alternating current produced to the *beam current applied in *beam coupling.

beam current The current consisting of the beam of electrons arriving at the screen of a *cathode-ray tube.

beam hole A hole made through the shield and usually through the reflector of a nuclear reactor to permit the escape of a beam of neutrons for experimentation purposes.

beam of radiation A narrow stream of approximately unidirectional electromagnetic radiation (e.g. a light ray) or of particles (e.g. an electron beam).

beam pentode and tetrode *See* aligned-grid valve.

beat-frequency oscillator An apparatus for generating electrical oscillations, the frequency of which can usually be varied over a range of audiofrequencies or *video frequencies. It incorporates two radio-frequency oscillators, one of which has a fixed frequency while the other has a frequency that can be varied at will. The output is obtained by the beating (*see* beat reception) of the two radio-frequency oscillations. A particular advantage of this type of oscillator is that the output voltage remains substantially constant at all output frequencies within the range covered.

beat oscillator *See* beat reception.

beat reception *Syn.* heterodyne reception. A method of radio reception in which beating is employed. The *beats (usually at an audiofrequency) are produced by combining the received radio-frequency oscillations with radio-frequency oscillations generated in the receiver in a separate oscillator (called a *beat oscillator*). The combined oscillations are then detected, amplified, and rendered audible. A beat oscillator that also functions as a detector and amplifier is called an *autodyne oscillator*. *Compare* superheterodyne receiver.

beats Fluctuations in sound intensity observed when two tones very nearly equal in frequency are sounded simultaneously. The phenomenon may be compared with that of *interference. At certain equal time intervals the wavetrains are in phase and reinforce each other; at intermediate periods they are in opposite phase and tend to neutralize each other. Combining the two tones of frequencies m and n, a tone of frequency $(m + n)/2$ is observed whose amplitude varies from $2A$ to 0 (A being the amplitude of each of the primary tones) at a frequency of $(m - n)$. Mathematically, the addition of two tones of frequencies m and n where m approximately equals n, may be expressed:

$$y - y_1 + y_2$$

$$= A \sin 2\pi m(t - x/v) + A \sin 2\pi n(t - x/v)$$
$$= 2A \cos \pi(t - x/v)(m - n)\sin 2\pi N(t - x/v),$$

where $2N = m + n$; $N = m = n$; and y is the displacement at any point x in time t. This equation is that of wave motion of frequency N and of varying amplitude. The beat frequency is $(m - n)$. If $(m - n)$ exceeds 20 per second the beats merge into a tone called *difference tone.

*Consonance and dissonance receive adequate explanation using the idea of beats between the upper partials of two notes of an interval. This theory is based on the observation that at any given audiofrequency the number of beats reaches a maximum of harshness to the ear, e.g. at middle C (256 hertz) mamimum harshness is observed at about 30 beats per second. A greater or less number of beats per second reduces the unpleasantness.

In tuning a musical instrument the beats between upper partials are observed and a note is altered in pitch until the required number of beats flat or sharp or zero beats are obtained. Two tones of notes may be tuned to the same frequency to an accuracy better than 1 beat in 30 seconds.

The electrical counterpart of beats is often said to occur in the *heterodyne* detection of radio frequency signals where one signal is beat with another to produce a third frequency – the difference or summation frequency. It would appear more appropriate in some respects to couple the idea of heterodyne with that of combination tones. (*See* beat reception.)

On organs there are generally two ranks of very softly voiced pipes one of which is tuned to beat slowly with the other. Occasionally three ranks are provided, one tuned to the pitch of the organ, one rank tuned slightly high and the other slightly low – to give the same number of beats. Generally the tuning is arranged to increase the number of beats as the frequency of the pipes increases. Such sets of pipes are known as Vox Celeste, Vox Angelica or Unda Maris depending upon the type of pipes used in the stop.

Beattie–Bridgman equation of state An empirical equation of state given by:

$$pV^2 = RT\left(1 - \frac{c}{VT^3}\right)\left[V + B\left(1 - \frac{b}{V}\right)\right] - A\left(1 - \frac{a}{V}\right),$$

where A, B, a, b, and c are experimentally determined constants for each fluid. *See also* equations of state.

Beaufort scale A numerical scale used in meteorology in which successive values of wind velocities are assigned numbers ranging from 0 (calm) to 12 (hurricane) thus indicating wind forces. Numbers 13–17 are often added to indicate specific hurricane speeds. For a scale 0–12, force 12 applies to winds over 75 m.p.h. See table.

Becher, Johann Joachim (1635–1681) German scientist who originated the *phlogiston theory.

Beaufort number	Wind description	Wind m.p.h.
0	calm	0–1
1	light air	1–3
2	light breeze	4–7
3	gentle breeze	8–12
4	moderate breeze	13–18
5	fresh breeze	19–24
6	strong breeze	25–31
7	moderate gale	32–38
8	fresh gale	39–46
9	strong gale	47–54
10	whole gale	55–63
11	storm	64–72
12	hurricane	73–82
13	hurricane	83–92
14	hurricane	93–103
15	hurricane	104–114
16	hurricane	115–125
17	hurricane	126–136

Becker and Kornetzki effect The lowering of *internal friction of ferromagnetic materials when subjected to a magnetic field of saturating value.

Beckmann, Ernst Otto (1853–1923) German physical chemist who devised a method for determining the elevation of boiling point and the *Beckmann thermometer.

Beckmann thermometer A mercury-in-glass thermometer used for the accurate determination of small temperature changes. The lower bulb is much larger than that of an ordinary thermometer and the scale behind the capillary tube, which is about 30 cm long, is divided into hundredths of a degree and covers only about 5–6 °C. The temperature change to be measured can take place about any mean temperature in the range, say, 0 °C to 100 °C, by varying the amount of mercury present in the lower bulb. This is made possible by running in more mercury from the small reservoir bulb at the top of the capillary, or conversely by running some mercury from the lower bulb into the reservoir where it plays no further part in the production of the thermometer reading. The variable amount of mercury present in the bulb means that the scale graduations will only exactly represent true degrees Celsius at the setting for which the scale has been calibrated.

becquerel Symbol: Bq. The derived SI unit of *activity equal to one disintegration per second.

Becquerel, Antoine Henri (1852–1908) French physicist. In 1896 he discovered *radioactivity by the effect of rays emitted by uranium salts on a photographic plate. He also made contributions to the study of phosphorescence, inventing a *phosphoroscope to estimate the duration of the glow after illumination, and to magnetism. His father, A.

E. Becquerel (1820–1891), also a physicist, had worked in similar fields and had also cooperated with his father, **A. C. Becquerel** (1788–1878), in electrochemical and other investigations.

Becquerel effect The e.m.f. produced by illuminating the surface of one electrode in an electrolytic cell.

Becquerel membrane A semipermeable membrane produced *in situ* by a chemical reaction, e.g. by the contact of solutions of silver nitrate and sodium sulphide.

Bednorz, J(ohannes) Georg (b. 1950) German physicist. Working at IBM's Zürich research laboratory on *superconductivity, he achieved superconductivity in a barium-lanthanum-copper oxide at 35 K, 12 K higher than previously achieved in any substance. For this work he shared a Nobel prize (1987) with Karl Müller.

Beilby layer An amorphous layer about 5 nm thick produced by polishing a surface; the ordinary crystalline material is present below this layer. It has been found that this surface layer is produced by sliding friction dissipating enough energy to melt the surface. A substance can thus be polished by one of higher melting point. *Running in of mechanical parts produces a deep Beilby layer.

bel Symbol: B. A logarithmic unit used particularly in telecommunications for comparing two amounts of power. Two amounts of power, P_2 and P_1, are said to differ by N bels when:

$$N = \log_{10}(P_2/P_1).$$

If P_1 is the power input of an electrical network and P_2 the corresponding power output, the above expression gives the gain of the network in bels. Note that if $P_2 < P_1$, N is negative, i.e. the gain is

Beckmann thermometer

negative, or is in fact a loss. A more commonly used unit is the *decibel (symbol: dB), which is one tenth of a bel. Thus:

$$N = 10 \log_{10}(P_2/P_1) \text{ dB}.$$

If the two powers are dissipated in impedances which have either (i) equal resistance R then, since

$$P_2 = |I_2|^2 R \quad \text{and} \quad P_1 = |I_1|^2 R,$$

$$N = \log_{10}\left|\frac{I_2}{I_1}\right|^2 = 2 \log_{10}\left|\frac{I_2}{I_1}\right| \text{ bels,}$$

or $\quad N = 10 \log_{10}\left|\frac{I_2}{I_1}\right|^2 = 20 \log_{10}\left|\frac{I_2}{I_1}\right| \text{ dB};$

or (ii) equal conductance G then, since $P_2 = |E_2|^2 G$ and $P_1 = |E_1|^2 G,$

$$N = \log_{10}\left|\frac{E_2}{E_1}\right|^2 = 2 \log_{10}\left|\frac{E_2}{E_1}\right| \text{ bels,}$$

or $\quad N = 10 \log_{10}\left|\frac{E_2}{E_1}\right|^2 = 20 \log_{10}\left|\frac{E_2}{E_1}\right| \text{ dB}.$

Note that if the two impedances are identical the conditions in (i) and (ii) above are both satisfied. *See* neper.

Bell, Alexander Graham (1847–1922) Scottish-born Amer. scientist and inventor. Invented the telephone (1876). The *bel is named in his honour.

bell sounds The bell exists in many different shapes and forms. The sounds produced are dependent to a great extent on the design and material used; however, in general the frequency is inversely proportional to the diameter multiplied by the cube of the mass. There is no adequate mathematical treatment. The bell may be regarded either as a development of the hollow tube closed at one end, or as a bent plate. The bell is generally of nonuniform thickness. Bell metal generally consists of about 3 parts of copper to 1 of tin. The bell may be excited either by a clapper hung loosely inside or by an external hammer, both hitting the sound bow – the thickened portion of the bell near the open end. (*See* Fig. *a*.) The three ways of sounding a bell with a clapper produce sounds varying little except in intensity: (*a*) chiming – the rope moves the bell just sufficiently for the clapper to strike its side; (*b*) ringing – the bell is swung upside down and allowed to return, the clapper hitting the bell; (*c*) the clapper of the bell is agitated by a rope – a method which may crack the bell. Experiments on bells have shown that the following *partials are heard:

(*a*) lowest partial: 4 nodal meridians, no nodal circle.

(*b*) second partial: 4 nodal meridians, 1 nodal circle.

(*c*) third partial: 6 nodal meridians, no nodal circle.

(*d*) fourth parttial: 6 nodal meridians, 1 nodal circle.

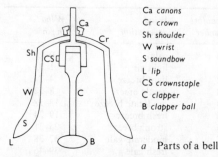

Ca canons
Cr crown
Sh shoulder
W wrist
S soundbow
L lip
CS crownstaple
C clapper
B clapper ball

a Parts of a bell

(*e*) fifth and highest partial: 8 nodal meridians, no nodal circle. The bell founders know these various partials as nominal, fifth, tierce, fundamental, and hum note respectively. By suitable design of the bell, the founder aims at hum note, fundamental, and nominal as successive octaves. Another partial which gives its name to the bell, but which has doubtful existence, is the *strike note. The various tones of a bell have different intensities and are attenuated differently with time. It is thought that bells having partials with a near uniform attenuation with time give the best sound quality. The characteristic waxing and waning of the intensity of sound from a bell, which may sometimes be attributed to air turbulence, is mainly caused by a slow rotation of the nodal lines as the bell vibrates, the nodes and antinodes being presented in turn to the observer. (*See* Fig. *b*.)

N_1, N_2 nodes
A antinodes
NANA nodal circle: gives rise to longitudinal vibrations
$N_1 \, N_1 \, N_2 \, N_2$ nodal meridians: give rise to transverse vibrations
[N for longitudinal vibration must be A for transverse vibration]

Clapper causes A at point where it strikes soundbow

b Nodes and antinodes of a bell

belted cable Multicore electric cable in which the conductors are each lapped with impregnated paper and then wormed together, the interstices being filled with packing material (worming) to form the cable into a circular section. A belt of impregnated paper surrounds the cores and the whole is enclosed in a lead sheath to render the cable waterproof. Three-core belted cables are suitable for use on *three-phase a.c. systems up to about 22 kV.

bending moment The algebraic sum of the moments, about any cross section of a beam, of all the forces acting on the beam on one side of this section. It is immaterial which side of the section is considered.

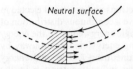

Neutral surface

Bending moment

The diagram shows the forces due to bending acting on the boundary of the shaded section, the resultant couple being anticlockwise and (conventionally) positive. The forces shown arise from the stretching of the filaments of the beam below, and the compressing of those above the *neutral surface.

Benedicks effect An e.m.f. produced in a closed circuit, composed of one metal only, under asymmetrical temperature distribution. The effect is not present if the metal is spectroscopically pure and free from internal strain.

Bennet, Abraham (1750–1799) Brit. physicist who invented the gold-leaf *electroscope and a simple induction machine.

Bernal, John Desmond (1901–1971) Brit. physicist who made important contributions to the study of the structure of proteins and viruses by *X-ray crystallography.

Bernouilli (or **Bernoulli**), **Daniel** (1700–1782) Swiss mathematician. He has been called the founder of mathematical physics, his work covering a wide range of topics. He is, however, best known for his work on fluids, and in particular for the formulation of the principle known as *Bernouilli's theorem or equation. He was a pioneer of the *kinetic theory of gases; he also made considerable contributions to the theory of vibrating strings. Many other members of the family achieved distinction in mathematics and physics.

Bernouilli's constant See Bernouilli's theorem.

Bernouilli's theorem In the steady frictionless motion of a fluid acted on by external forces which possess a gravitational *potential (V) then

$$\int \frac{\mathrm{d}p}{\rho} + \frac{1}{2} v^2 + V = C,$$

where p and ρ are the pressure and density of the fluid; v is the velocity of the fluid along a stream line; and C is a constant, depending on the particular stream line chosen, called *Bernouilli's constant*. The equation can be shown to agree with the principle of conservation of energy and may be expressed in a generalized form:

$$\int \frac{\mathrm{d}p}{\rho} - \frac{\partial \phi}{\partial t} \pm \frac{1}{2} v^2 \pm V = A,$$

where ϕ is the velocity potential and A is a function of the time (t). For steady motion this reduces to the original equation.

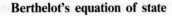

Berthelot, Pierre E. M. (1827–1907) French chemist. He was the founder of thermochemistry.

Berthelot's apparatus for latent heat Berthelot used a condensation method of mixtures in order to determine the specific latent heat of vaporization of a liquid. A gas ring r burning under a metal plate m, boils the liquid in the flask, the vapour passing down the tube T into spirals immersed in a water calorimeter where it condenses and collects in the reservoir R. The calorimeter and its water jacket are protected from the radiation of the burner by a slab of wood covered by a wire gauze sheet. The quantity of heat (H) gained by the calorimeter, during the distillation, through conduction down the tube T, is allowed for by observing the rate of rise of temperature of the water in the calorimeter both before the distillation begins and after it has finished. Then:

$$W(\theta_2 - \theta_1) - H = M\left[l + c\left(\theta - \frac{\theta_1 + \theta_2}{2} \right) \right]$$

where W is the water equivalent of the calorimeter and contents, M is the mass of liquid distilled, θ_1 and θ_2 are the initial and final temperatures of the water in the calorimeter, corrected for radiation losses, and θ is the boiling point of the liquid whose specific latent heat is l and whose specific heat capacity is c. The chief inaccuracy of the method lies in the fact that the vapour becomes superheated when passing through the tube T near to the gas ring. Modern modifications of the apparatus include a small electric heating coil immersed in the liquid, this being generally preferable to the use of gas heating.

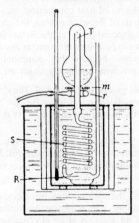

Berthelot's apparatus for determining specific latent heat of vaporization

Berthelot's equation of state The equation:

$$\left(p + \frac{a}{Tv^2} \right)(v - b) = RT.$$

Berthollet, Claude Louis, Count

This gives better agreement with experiment than the *Van der Waals equation at moderate pressures, but fails at the critical point. An empirical modification:

$$pv = RT\left[1 + \frac{9}{128}\frac{T_c}{p_c}\frac{p}{T}\left(1 - \frac{6T_c^2}{T^2}\right)\right]$$

is useful when correcting temperatures on a gas scale to those on the thermodynamic scale and is used when experimental results on the compressibility of the gas are not available. See equations of state.

Berthollet, Claude Louis, Count (1748–1822) French chemist. He was the founder of physical chemistry.

Berzelius, Jöns Jakob, Baron (1779–1848) Swedish chemist, discoverer of several elements (selenium, silicon, thorium, zirconium). He invented (about 1815) the modern system of chemical symbols.

Bessel, Friedrich W. (1784–1846) German astronomer and mathematician. He improved theory and practice of pendulum observations and made substantial advances in the treatment of atmospheric refraction. He introduced *Bessel functions into analysis.

Bessel functions Certain functions that can be expressed as power series in x, which are solutions of linear differential equations of the form:

$$x^2\frac{d^2y}{dx^2} + x\frac{dy}{dx} + (x^2 - n^2)y = 0.$$

They have many applications in physics, for example in problems of heat conduction, etc. There are various kinds of Bessel functions, the most important being denoted by $J_n(x)$, the Bessel function of order n. Bessel functions are sometimes called *cylindrical functions* because they commonly arise in problems involving symmetry about an axis.

beta current gain factor Symbol: β. The short-circuit current-amplification factor in a bipolar *transistor with *common-emitter connection. It is given by:
$$\beta = (\partial I_C / \partial I_B),$$
the collector voltage, V_{CE}, being constant; I_C is the collector current and I_B the base current. β is always greater than unity and in practice takes value up to 500.

beta decay The spontaneous transformation of a nucleus into one of its neighbouring *isobars accompanied by the ejection of an electron or positron. The nucleus produced always has the same *mass number as the initial nucleus but differs in *atomic number by one. If an electron is emitted the number of nuclear protons increases by one, if a positron is emitted it decreases by one. Two examples of beta decay are:
$$^{14}_{C} \rightarrow\ ^{14}_{N} + e^- + \bar{\nu},$$
$$^{11}_{6}C \rightarrow\ ^{11}_{5}B + e^+ + \nu.$$

It is found that electrons and positrons ejected in beta decay always have a continuous distribution of energies and not a single energy equal to the energy released. The "missing energy" is carried away by the neutrino ν or antineutrino $\bar{\nu}$. The total energy carried by the electron and antineutrino or by the positron and neutrino is constant for a particular decay and thus the neutrino enables energy to be conserved. The neutrinos also enable *angular momentum and linear momentum to be conserved. Beta decay is a *weak interaction. *Parity is therefore not conserved. In addition to the continuous distribution of electrons or positrons, a line spectrum is sometimes superimposed. These electrons are ejected from shells in the atom by internal conversion. See also radioactivity.

beta particle (β-particle) An *electron or *positron emitted by the nucleus of a *radionuclide during *beta decay.

beta rays (β-rays) A type of *ionizing radiation consisting of a stream of *beta particles. The most energetic emitters give particles with energies up to several MeV, which penetrate matter up to a few kg m^{-2}. Low-energy emitters may have spectrum limits of 10^5 or even 10^4 eV, which give very low penetration.

beta-ray spectroscopy See electron spectroscopy.

beta-ray thickness gauge An instrument for determining the thickness of sheet metal, paper, etc., by measuring the intensity of beta radiation from a radioactive source, after it has passed through the material. As the thickness varies the absorption of the particles varies accordingly.

betatron A cyclic *accelerator for producing high-energy electrons by means of magnetic induction. If an electron is describing a circular orbit of radius r in the magnetic field between the poles of an electromagnet, an increase in the magnetic flux through the orbit produces an acceleration of the electron, the momentum p of the electron being related to the flux ϕ by the equation $dp/d\phi = e/2\pi r$. It can be shown that if the field at the circumference of the orbit is equal to half the average field inside the orbit, the radius r is unaltered, i.e. the particle continues in the same path. This is achieved by shaping the pole pieces.

In the betatron the magnet is excited by alternating current, and the electrons are injected into the field when the current is beginning to rise from zero. They are deflected out of the field just before the current reaches its peak value, having completed several hundred thousand revolutions. They travel in an evacuated annulus known as a *doughnut*. The angular velocity ω of a particle moving in a fixed orbit in which the magnetic flux density is B is given by $\omega = eB/m$, where m is the mass. To maintain a constant angular frequency and orbit, B is increased by the same factor as m increases (due to the relativistic velocity of the particle). The functioning of the machine is therefore not affected by the relativistic mass increase. Energies up to 310 MeV have been produced; the electron beam is used to produce high quantum energy X-rays and also in particle research. *See* Kerst, D. W. *See also* synchrotron.

betatron synchrotron *See* synchrotron.

Bethe, Hans Albrecht (*b.* 1906) German-born Amer. physicist who first proposed that *fusion reactions were responsible for stellar energy. *See also* Alpher, R. A.; Gamow, G.

beva- (prefix). *See* giga-.

bevatron The name of the *proton synchrotron at Berkeley, US, that accelerates protons to an energy of 6 GeV.

Bhabha, Homi Jehangir (1909–1966) Indian physicist who carried out extensive research into the nature and origin of *cosmic rays.

B/H loop *See* hysteresis loop.

bias *Syn.* bias voltage. A voltage applied to an electronic device to determine the portion of the *characteristic of the device at which it operates.

biased automatic gain-control *Syn.* delayed automatic gain-control. *Automatic gain-control which is designed to operate only when the magnitude of the received signal exceeds a predetermined value.

biaxial crystal A crystal in which there are two directions along which the polarized components of a ray of light will be transmitted with the same speed. Such crystals belong to the orthorhombic, monoclinic, or triclinic systems.

biconcave *See* concave.

biconvex *See* convex.

bifilar suspension A type of suspension used in electrical instruments in which the moving part is suspended on two parallel threads, wires, or strips.

bifilar winding A method of winding a wire to form a resistor or coil with negligible *inductance: the wire is doubled back on itself and wound double from the looped end.

Bifilar winding

big-bang theory The theory in cosmology that all matter and radiation in the universe originated in a cataclysmic explosion that occurred $10–20 \times 10^9$ years ago. Since this initial state of extreme density and temperature, the universe has expanded and cooled.

Within a tiny fraction of a second from the big bang, a variety of elementary particles and antiparticles had been created and were undergoing interactions. Photons of radiation were produced by their *annihilation. Deuterium and helium nuclei were synthesized some 100 seconds after the big bang

when the temperature was about 10^9 K. After about 10^4 years, when the temperature had fallen to about 10^4 K, the matter in the universe was composed mainly of an ionized gas of free electrons, protons, and helium nuclei.

Neutral hydrogen formed, by combination of free electrons and protons, when the temperature had dropped to about 3000 K. This process took place about 3×10^5 years after the big bang. Prior to this, the scattering of photons on electrons coupled matter and radiation so that they shared a common temperature. Since the combination of free electrons and protons, the radiation has cooled from 3000 K to 3 K, i.e. to the observed temperature of the microwave background (*see* cosmic background radiation). The matter, no longer coupled to the radiation, has interacted to form stars and galaxies.

The big-bang theory has been successful in explaining the expansion of the universe (detected from the observed *redshift of distant galaxies), the measured *cosmic abundance of helium, and the microwave background. *See also* early universe; inflationary universe.

Billet split lens A lens cut in two, so that the optical centres of the semi-lenses are slightly displaced laterally; in consequence, two real images of a slit are formed and in the overlapping region in front of these (coherent) images *interference takes place.

billion One thousand million (10^9). In American and French usage the term has always been used for 10^9. In Britain, a billion was formerly one million million (10^{12}). In the 1960s, the British Treasury started using the American sense (a thousand million) in its economic statistics, and this usage has now largely supplanted the original meaning.

bimetallic strip Two metals having different coefficients of expansion riveted together: an increase in temperature of the strip causes the strip to bend, the metal having the greater coefficient of expansion being on the outside of the curve. One end is rigidly fixed and movement of the other end can serve to open or close an electric circuit of a temperature control device, or to move the pointer of a pointer-type thermometer.

bimorph cell A device for converting electrical signals into mechanical motion using the *piezoelectric effect. It consists of two piezoelectric crystals (such as Rochelle salt) cut and joined together so that an applied voltage causes one to expand and the other to contract. The composite crystal thus bends as a result of the voltage across it. The converse effect of generation of an electric voltage by bending the cell is also used. Bimorph cells are used in record-player pickups and loudspeakers.

binary alloy *See* alloy.

binary notation A method of expressing numbers by two digits, 0 and 1, rather than the ten digits used in decimal notation. The binary equivalents of some decimal numbers are as follows.

Decimal	Binary
0	0
1	1
2	10
3	11
4	100
5	101
6	110
7	111
8	1000
9	1001
10	1010
100	1100100

Since only two symbols are used they can be represented by either of two alternatives, for example the presence or absence of a pulse in an electronic circuit; for this reason binary notation is used in *computers. See also bit.

binary star A system of two stars that revolve around a common centre of gravity. In *visual binaries* the pairs of stars can be seen to be separate with an optical telescope. However, even when the distance between the stars is too small to enable them to be separated visually, their presence can sometimes be detected by other methods. *Spectroscopic binaries* can be resolved by the *Doppler shift of their spectral lines as the stars, revolving about their common centre of gravity, approach and recede from the earth. In these cases the spectral lines appear to double. If one of the stars is considerably brighter than the other the spectral lines appear to oscillate about a mean position. In *photometric binaries*, one star eclipses the other: this occurs when the major axis of the orbit is close to the line of sight. In these cases variations of brightness indicate the progress of the eclipse and enable the orbit and masses of the components to be calculated.

binaural location An important function of the ears – the sense of aural direction – is only possible when both ears are used together. A normal person can estimate the direction of a sound to within a few degrees (except for a possible error of 180° when the sound comes from directly in front or behind). The obvious explanation of this is that there is a difference of intensity at the two ears, the sound being judged to be on the side of maximum intensity. Rayleigh has shown, however, that owing to diffraction, the difference in intensity at the front and rear of the average head is only 10% for a note of frequency 256 hertz and less than 1% for a note of half this frequency. It therefore appears that location by intensity difference is only an important factor above 1200 hertz. It has been suggested that for frequencies below this limit the sense of direction is based on the difference of phase between the sound arriving at the two ears. The phase difference effect can only be of importance at low frequencies since if the phase difference amounted to half a wavelength there would be no difference between an advance or retardation in phase of this amount and thus no direction could be associated with it. It seems, then, that the intensity effect predominates above about 1200 hertz and the phase effect is the important factor below this. The mechanism of detection of a phase difference at the two ears is not yet understood. It has been suggested that the effect may be due to the time difference of arrival of the two sounds rather than the phase difference. The method of binaural location has been extended to determine the direction of distant sounds. The sound intensity at the ears is increased by using two large horns, and the direction sensitivity is improved by increasing the distance apart of the collectors. Rotation of the horns varies the path difference of the sound reaching the two ears. When the sound appears to cross over from one ear to the other, the source is in line with the axis of the horns.

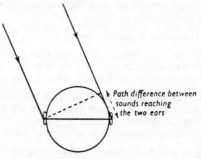

Path difference between sounds reaching the two ears

Binaural location

binding energy 1. Symbol: E_B. The mass of a nucleus is slightly less than the mass of its constituent protons and neutrons. By Einstein's law of the *conservation of mass and energy ($E = mc^2$), this mass difference is equivalent to the energy released when the nucleons bind together. This energy is the binding energy. The graph of binding energy per nucleon, E_B/A, against *mass number, A, shows that as A increases E_B/A increases rapidly up to a mass number of 50–60 (iron, nickel, etc.) and then decreases slowly. There are therefore two ways in which energy can be released from a nucleus, both of which entail a rearrangement of nuclei occurring in the lower half of the curve to form nuclei in the upper, higher-binding-energy part of the curve. Fission is the splitting of heavy atoms, such as uranium, into lighter atoms, accompanied by an enormous release of energy. Fusion of light nuclei, such as

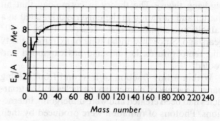

The binding energy for nucleon in MeV, as a function of the mass number A

deuterium and tritium, releases an even greater quantity of energy.

2. The work that must be done to detach a single particle from a structure. *See* ionization potential.

Bingham fluid An idealized fluid that does not begin to flow until the shear stress exceeds a certain fixed value (the *yield value*). The fluid then behaves as a *Newtonian fluid and its flow is proportional to the excess rate of shear over and above the yield value. This makes the coefficient of *viscosity, as usually defined, diminish as the velocity gradient increases and the fluid thus exhibits *anomalous viscosity.

No real fluid fills these conditions precisely, but the concept is a convenient approximation to the behaviour of a number of colloids and gels. *See* plasticity.

Binnig, Gerd (*b.* 1947) German physicist. Working at the IBM Research Laboratory in Zürich with Heinrich Rohrer, he invented and built the first *scanning tunnelling microscope. For this work they shared half a Nobel prize (1986), the other half being awarded to Ernst Ruska.

binocular Permitting the simultaneous use of both eyes (e.g. binocular *microscope, binocular vision). Sometimes used as a contraction of binocular telescope (e.g. *prismatic binoculars). There are various grades of binocular vision of which stereoscopic vision is the highest grade.

binomial distribution In a trial that can only have one of two results (say success or failure), the binomial distribution of probabilities, P_r, for obtaining r successes in n independent trials is given by:

$$P_r = \frac{n!}{r!(n-r)!} \, p^r q^{n-r},$$

where p is the probability of success in any one of the trials, and q is that of failure ($p = 1 - q$).

If c coins are thrown in one trial, the probability of a specific success (say heads only) is obtained from the binomial expansion of $(x + y)^c$, where x represents heads only and y tails only. The probability p of 1 head and 2 tails when 3 coins are tossed ($c = 3$) is the quotient of the coefficient of xy^2 and the sum of all the coefficients. Thus $p = 3/8$ and $q = 5/8$.

The probability P_r of obtaining 1 head and 2 tails once in tossing 3 coins 4 times is:

$$\frac{n!}{r!(n-r)!} \cdot p^r q^{n-r} = \frac{4!}{1!3!} \cdot \left(\frac{3}{8}\right)\left(\frac{5}{8}\right)^3 = 0.37.$$

binomial expansion If $|x| < 1$, the expression $(1 + x)^n$, where n may be any positive or negative number and not necessarily an integer, is equal to

$$1 + nx + \frac{n(n-1)x^2}{2} + \frac{n(n-1)(n-2)x^3}{3 \cdot 2}$$
$$+ \frac{n(n-1)(n-2)(n-3)x^4}{4 \cdot 3 \cdot 2} + \cdots$$

Thus, for example, if x and y are both *much* less than 1 only the first two terms are appreciable,

$$\frac{(1+x)^{1/2}}{(1-y)^2} = (1+x)^{1/2}(1-y)^{-2}$$

$$= (1 + \tfrac{1}{2}x)(1 + 2y)$$

$$= \left(1 + \frac{x}{2} + 2y\right).$$

(The condition that $|x| < 1$ is not necessary in those special cases where the series is not an infinite one, i.e. when n is a positive integer.)

binormal *See* normal.

biological half-life The time required for half the concentration of a particular substance to be removed from the body, or from a specific region in the body, by biological means when the rate of removal is approximately exponential.

biological shield A massive structure surrounding the core of a *nuclear reactor, provided to absorb most of the neutrons and gamma radiation in order to protect the operating personnel. Such shields are commonly of concrete and iron.

biomass energy *See* renewable energy sources.

biophysics Physics applied to biology. Biophysics is concerned with the physics of biological systems, with the use of physical methods in the study of biological problems, and with the biological effects of physical agents.

Biot, Jean Baptiste (1774–1862) French physicist who first recognized the rotation of the plane of *plane-polarized light; this led to the development of polarimetry as an analytical method.

Biot and Savart's law The magnetic field due to current flowing in a long straight conductor is directly proportional to the current and inversely proportional to the distance of the point of observation from the conductor. The law is derivable from the *Ampère–Laplace theorem but was obtained experimentally by the authors.

Biot–Fourier equation An equation for heat conduction through a solid:

$$\partial T/\partial t = -(\lambda/c\rho)\nabla^2 T,$$

where $\partial T/\partial t$ is the rate of change of temperature, ∇^2 is the *Laplace operator, λ the *thermal conductivity, c the *specific heat capacity and ρ the density. For heat flow in one dimension $\nabla^2 T$ becomes $\partial^2 T/\partial x^2$.

Biot's law The degree of rotation of the plane of *polarization of light propagated through an optically active medium, is inversely proportional (approximately) to the square of the wavelength of the light and proportional to the path length in the medium and to the concentration if the medium is a liquid.

bipolar electrode A metal plate in an electrolytic cell through which the current, or part of it, passes, but which is not connected either to the anode or the cathode of the cell. Since the current passes through the plate one face serves as a subsidiary cathode, the other as an anode.

bipolar integrated circuit A type of monolithic *integrated circuit based on *bipolar transistors. A section of a typical circuit is shown in the diagram. A *semiconductor wafer forms the substrate (p-type substrate is shown). Buried n^+ (highly doped) regions are selectively diffused into the substrate before the n-type *epitaxial layer is grown. These regions reduce the *collector series resistance of the completed transistors. The epitaxial layer, typically about 7 μm thick is grown on the substrate and a layer of oxide grown on this surface. The insulating oxide layer is then etched, using *photolithography, in the desired positions and isolating diffusions of the same type as the substrate are made to isolate individual components from each other. The process of oxide growth and photo-etching is carried out between each step to ensure the correct positioning of the diffused regions. The individual components are formed by diffusing in turn the appropriate type of impurity into the epitaxial layer. A final passivating oxide layer is grown and windows etched in this layer to enable contacts to be made to the semiconductor. Finally a metal layer (usually aluminium) is deposited by evaporation and etched to form the desired interconnection pattern.

Isolation of the individual components in a circuit is ensured by taking the substrate to the most negative potential which makes the collector–substrate reverse biased and prevents current flowing across the junction. The resistor box (n-type region surrounding the resistor) is held at the most positive potential, which ensures that the resistor–resistor box junction is reverse biased also.

bipolar machine An electrical machine with a field magnet that has only two poles. *Compare* multipolar machine.

bipolar transistor A *transistor in which both electrons and holes play an essential part, e.g. a *junction transistor. A bipolar junction transistor is commonly referred to simply as a transistor. *Compare* field-effect transistor.

biprism A prism with a very obtuse angle acting virtually as two narrow angle prisms placed base to base thereby splitting a beam into two parts with a small angle between the parts. A doubled image (small separation) of a single object can be formed. It can be used to produce *interference and has various other uses where image duplication is of value. The angle of duplication is the sum of the deviations produced by the constituent narrow prisms.

biquartz Two pieces of quartz, cut perpendicular to the axis, rotating the plane of *polarization equally in opposite directions and placed alongside one

another. It is useful to determine accurately the plane of polarization.

birefringence *See* double refraction.

bispherical lens *See* aspherical lens or mirror.

bistable *Syn.* bistable multivibrator. A type of circuit having two stable states. *See* flip-flop.

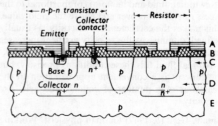

A: Metal interconnections   D: n-type epitaxial layer
B: Insulating layer   E: p-type subsrate
C: Isolating diffusions

Typical bipolar integrated circuit

bit A contraction of *bi*nary digi*t*. Either of the digits 0 or 1, used in computing for the representation within a *computer of numbers (*see* binary notation), of letters, punctuation marks, and other characters (encoded in binary form), and also of machine instructions. The bit is thus the smallest unit of information in a computer. It is also the smallest unit of storage, since it can be represented as a physical state of a two-state system. Examples include the two possible directions of magnetization of a spot on a magnetic disk or tape, and the high or low voltage that can be fed to a logic or memory circuit.

Bitter patterns Patterns demonstrating the presence of *domains in ferromagnetic crystals (F. Bitter, 1931). They can be observed by coating the polished surface of the material with a colloidal suspension of ferromagnetic particles. The particles tend to gather at the domain boundaries where there is a strong magnetic field. The technique can also be used for detecting cracks and imperfections in ferromagnetic materials.

Black, Joseph (1728–1799) Brit. chemist and physicist, responsible for clarifying ideas on *specific heat (capacity) and for introducing the concept of (specific) *latent heat. He made fairly accurate measurements with a simple ice calorimeter consisting of an ice block with a hollow, covered by a slab of ice as a lid.

black body *Syn.* full radiator. A body or receptacle that absorbs all the radiation incident upon it; i.e. a body that has both an *absorptance and *emissivity of 1 and that has no reflecting power. The radiation from a heated black body is called *black-body radiation. Stellar radiation can be described by assuming that stars are black bodies. While a black body is in fact only a theoretical ideal, it is in practice most nearly realized by the use of a small slit or hole in the wall of a *uniform temperature enclosure.

black-body radiation The thermal radiation from a *black body at a given temperature. Experimental investigation of the spectral distribution of energy in black body radiation was made by a number of workers, notably *Lummer and Pringsheim. Their results show a distribution of the form shown in the diagram. The most striking feature of a set of distribution curves is that each has a definite maximum and that this shifts towards the region of shorter wavelengths as the temperature rises. The intensity of radiation of any given wavelength increases steadily as the temperature rises.

Many theoretical and empirical attempts were made to find a general formula to represent the black-body spectrum. Thermodynamic reasoning does not give a complete answer, though it does predict two characteristic features of the radiation, firstly that for curves at different temperatures the value of $\lambda_{max}T$ is constant. This statement is known as the *Wien displacement law*. The second deduction that may be made from thermodynamics is that the heights of corresponding ordinates vary directly as T^5. These two rules permit the construction of complete curves for all temperatures once any one is accurately known, but no further deductions can be made without making assumptions that are independent of any purely thermodynamic foundation. Wien deduced that:

$$M_{e,\lambda} = c_1\lambda^{-5}\exp(-c_2/\lambda T),$$

where $M_{e,\lambda}$ is the *radiant exitance per unit wavelength range for wavelength λ and c_1 and c_2 are constants. This is known as the *Wien radiation law*. It is successful for short wavelengths, but it gives values somewhat too low for long wavelengths.

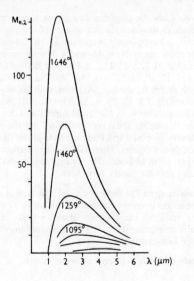

Rayleigh and Jeans applied the principle of equipartition of energy to a system of electromagnetic vibrations of different frequencies and gave:

$$E = CT\lambda^{-4},$$

which is in agreement with experiment only for long wavelengths. (*See* Rayleigh–Jeans formula). Planck gave, by reasoning which formed the starting point of the quantum theory,

$$M_{e,\lambda} = C\lambda^{-5}/[\exp(hc/\lambda kT) - 1],$$

in which C, c, k and h are constants, which agrees with experiment for all wavelengths. (*See* Planck's formula.)

The total amount of energy of all wavelengths emitted by a black body is given by the *Stefan–Boltzmann law, namely $M_e = \sigma T^4$ where σ is the Stefan–Boltzmann constant.·

black-body temperature The temperature at which a *black body would emit the same radiation as is emitted by a given body. It is the temperature of a body as measured by a (radiation) *pyrometer. It is usually appreciably less than the true temperature of the body. For a temperature T_0 observed by a *total radiation pyrometer, the true thermodynamic temperature T is given by $T_0^4 = \varepsilon T^4$, where ε is the emissivity of the source.

Blackett, Patrick M. S., Lord (1897–1974) Brit. physicist. His chief field of study was cosmic radiation and he developed the *cloud chamber technique to permit automatic photographic recording; for this he was awarded a Nobel prize (1948). He was the first to photograph a nuclear transmutation (1925) and he discovered the positron in 1932, independently of Anderson. He propounded a theory of the magnetism of the earth and the sun and other stars, postulating an inherent magnetization of massive rotating bodies.

black hole An astronomical body with so high a gravitational field that the relativistic curving of space around it causes gravitational self-closure, i.e. a region is formed from which neither particles nor photons can escape, although they can be captured permanently from the outside.

The most promising candidates for black holes are massive stars that explode as *supernovae, leaving a core in excess of three solar masses. A core of this mass must undergo complete gravitational collapse because it is above the stable limit for *white dwarfs and *neutron stars. (*See also* Schwarzschild radius.) It has been suggested that black holes are the unseen components of certain binary systems. They could be detected through the gravitational fields.

Supermassive black holes, of 10^6 to 10^9 solar masses, probably lie at the centre of some galaxies. They may give rise to the *quasar phenomenon and the phenomena of other highly active galaxies. *See also* Hawking radiation.

black-out point *See* cut-off.

Blagden's law

Blagden's law The lowering of the freezing point of a solvent is proportional to the concentration of the solute.

blanket A layer of *fertile material surrounding the *core of a *nuclear reactor either for the purpose of breeding new fuel or to reflect some of the neutrons back into the core.

blazed grating A reflective diffraction grating so ruled that the reflected light is concentrated into a few orders, or even into a single order, of the spectrum by inclining the grooves at an angle to the grating surface so that radiation is reflected in the required direction. *See also* echelette grating.

blink microscope or comparator An instrument used to detect small differences in the *luminosity or position of stars between two photographs of the same part of the sky. The photographs are viewed alternately in rapid succession using a mechanical device.

Bloch, Felix (1905–1983) Amer. physicist, born in Switzerland, who made substantial contributions to *wave mechanics and the theory of magnetism and *energy bands in solids.

Bloch functions The solutions of the *Schrödinger wave equation for an electron moving in a potential that varies periodically with distance (F. Bloch, 1928). They have the form:
$$\psi = u_k(r) \exp(i\mathbf{k}.\mathbf{r}),$$
where u is a function depending on k, the wave vector, and k varies periodically with distance r. k has the same period as the potential and the lattice. Bloch functions are used in the mathematical formulation of the band theory of solids (*see* energy bands).

Bloch wall The transition layer between adjacent ferromagnetic *domains magnetized in different directions. It allows the spin directions to change slowly from one orientation to another, rather than abruptly.

blocked impedance If the motion of an electromechanical or acoustical *transducer is blocked the *impedance measured is called the blocked impedance. *Compare* motional impedance.

blocking capacitor A *capacitor included in an electric circuit for the purpose of preventing the flow of direct current and low-frequency alternating current while permitting the flow of higher-frequency alternating current. Its *capacitance is usually chosen so that its *reactance is relatively small at the lowest frequency for which the circuit is intended to be used.

blocking layer photocell *See* photovoltaic cell.

blocking oscillator A type of oscillator in which, after completion of (usually) one cycle of oscillation, blocking (i.e. cessation of oscillation) takes place for a predetermined period of time. The whole process is then repeated. It has applications as a *pulse generator or as a *time-base generator and is fundamentally a special type of *squegging oscillator.

Bloembergen, Nicolaas (*b.* 1920) Dutch-born Amer. physicist. A professor at Harvard, he developed a *maser that worked on a continuous cycle. He later worked on the use of lasers to break chemical bonds, sharing a Nobel prize (1981) for this work with Arthur Schawlow and Kai Siegbahn.

Blondel–Rey law The apparent brilliance, B, of a light source, flashing at a frequency less than 5 hertz, is given by:
$$B = B_0 t / (t + \alpha),$$
where t is the duration of the flash, B_0 is the actual brilliance, and α is a constant. *Compare* Talbot's law.

blooming of lenses The process of depositing a transparent film (about one quarter wavelength) of lower refractive index on the surface of a lens of higher index, whereby, through destructive interference, surface reflection is reduced. Calcium or magnesium fluoride is deposited by evaporation in a vacuum. Bloomed (or *coated*) surfaces have a pale purplish hue when examined by reflected light as the interference only occurs in the middle part of the spectrum.

blur circle If an image screen is not at the correct focus, a point object produces a patch of light, in general circular in shape, of radius determined by the aperture and distance from the focus. The term is also used for the patches produced by *aberration or *astigmatism, whether circular or not.

Board of Trade unit The former unit for the supply of electricity in Britain; one kilowatt-hour. It is equivalent to 3.6×10^6 joules.

Bode, Johanne E. (1747–1826) German astronomer remembered for his empirical rule, now known as *Bode's law.

Bode's law An empirical law giving a sequence of numbers to which the major axes of the planetary orbits are supposedly proportional. The formula for the sequence is $3n + 4$ where n denotes successive numbers $0, 1, 2, 2^2, 2^3, \ldots$ Thus the Bode sequence is $4, 7, 10, 16, 28, 52, 100, 196, 388, 772$ which agrees well, except for the last two terms, with the observed values $3.9, 7.2, 10, 15, -, 52, 95, 192, 301, 396$. The gap occurs between Mars and Jupiter and this position is roughly speaking "occupied" by the minor planets. Neptune and Pluto are both closer than the Bode's-law figures. There is no accepted theoretical explanation of Bode's law and its approximate validity may be purely fortuitous.

body-centred The form of crystal structure in which the atoms occupy the centre of the lattice as well as the vertices. The diagram shows a body-centred cube arrangement typified by crystals of iron. *Compare* face-centred.

body centrode *See* instantaneous centre.

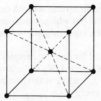
Body-centred cube

body cone *See* instantaneous axis.

body force *See* force.

Bohr, Aage (*b.* 1922) Danish physicist and son of Niels Bohr. Professor at the Institute of Theoretical Physics (later called the Niels Bohr Institute), his work on the quadrupole moment of the nucleus in conjunction with Ben Mottelson led to the collective (or unified) model of the nucleus. For this work they shared a Nobel prize (1975) with L. J. Rainwater.

Bohr, Niels (1885–1962) Danish physicist. He applied quantum conditions to the nuclear atom proposed by *Rutherford, obtaining (1913) the first model capable of explaining satisfactorily the origin of spectral lines and especially the significance of spectral terms in spectrum analysis (*see* Bohr theory of the atom). The model later, in conjunction with the *Pauli exclusion principle gave a satisfactory interpretation of the main features of the *periodic table. He introduced the concept of *complementarity and later in his life contributed to the theory of *nuclear fission.

Bohr–Breit–Wigner theory The theory that a *nuclear reaction occurs in two stages. In the first the colliding particle is captured by the nucleus to produce a new, highly-excited, compound nucleus. In the second this compound nucleus emits one or more particles or gamma-ray photons to form a different nucleus.

Bohr magneton *See* magneton.

Bohr radius *Syn.* atomic unit of length. Symbol: a_0. The radius of the electron orbit in the ground state of a hydrogen atom in the *Bohr theory of the atom. In the nonrelativistic approximation *wave mechanics interpets this as the most probable distance of the electron from the nucleus. It is given by:
$$a_0 = \varepsilon_0 h^2/\pi m e^2.$$
Taking m as the actual mass of the electron (not the *reduced mass in the atom) this gives:
$$a_0 = 5.291\,7725 \times 10^{-11} \text{ m}.$$

Bohr theory of the atom (1913) The first significant application of the quantum theory to atomic structure. Although the theory has been replaced (*see* quantum mechanics) it introduced several concepts that have remained as essential features of later theories.

The theory was applied in particular to the simplest atom, that of hydrogen, consisting of a nucleus and one electron. It was assumed that there could be a *ground state* in which an isolated atom would remain permanently, and short-lived states of higher

energy to which the atom could be excited by collisons or absorption of radiation. It was supposed that radiation was emitted or absorbed in quanta of energy equal to integral multiples of $h\nu$, where h is the Planck constant and ν is the frequency of the electromagnetic waves. (Later it was realized that a single quantum has the unique value $h\nu$.) The frequency of radiation emitted on capturing a free electron into the nth state (where $n=1$ for the ground state) was supposed to be $nh/2$ times the rotational frequency of the electron in a circular orbit. This idea led to, and was replaced by, the concept that the angular momentum of orbits is quantized in units of $h/2\pi$. The energy of the nth state was found to be given by:
$$E_n = -me^4/8h^2\varepsilon_0^2 n^2,$$
where m is the *reduced mass of the electron. This formula gave excellent agreement with the then known series of lines in the visible and infrared regions of the spectrum of atomic hydrogen and predicted a series in the ultraviolet that was soon to be found by Lyman. (*See* hydrogen spectrum.) Atoms in excited states were shown to be much larger than in the ground state, which explained the absence of certain lines in the spectra of laboratory sources that were found in stellar spectra. It was proved that certain lines previously attributed to hydrogen were derived from singly ionized helium atoms.

The extension of the theory to more complicated atoms had some successes but raised innumerable difficulties, which were only resolved by the development of *wave mechanics. *See also* atom; atomic orbital; Bohr radius; magneton.

boiling point The temperature of a liquid at which visible evaporation occurs throughout the bulk of the liquid, and at which the vapour pressure of the liquid equals the external atmospheric pressure. It is the temperature at which liquid and vapour can exist together in equilibrium at a given pressure. The variation of boiling point with pressure may be obtained from the *Clausius–Clapeyron equation. The term is commonly restricted to that temperature at which liquid and vapour are in equilibrium at standard atmospheric pressure ($1.013\,25 \times 10^5$ Pa).

boiling-water reactor (BWR) A type of thermal *nuclear reactor in which water is used both as *coolant and *moderator, being allowed to boil by direct contact with the fuel elements. *See* pressurized-water reactor.

bolide A large bright *meteor: some bolides explode with a loud detonation on entering the earth's atmosphere.

bolograph The graph obtained by allowing a pencil of light reflected from the galvanometer of a *bolometer to fall on a moving strip of photographic material.

bolometer An instrument for the measurement of the total energy flux of electromagnetic radiation, especially that of microwave and infrared radiation.

Boltzmann, Ludwig

There are many types of bolometer, including semiconductor and specially cooled devices, but each is essentially a small resistive element capable of absorbing radiation. The resulting rise in temperature, which leads to a change in the resistance, is a measure of the power absorbed.

Boltzmann, Ludwig (1844–1906) Austrian physicist who developed statistical mechanics, applying it to the *kinetic theory of gases and to *thermodynamics. He enunciated the principle of *equipartition of energy and discovered the law of *distribution of velocities among the molecules of a gas, known as the Maxwell–Boltzmann relation since Maxwell derived it by another approach. Boltzmann connected entropy with probability in about 1877 (*see* Boltzmann entropy theory). He gave a thermodynamic derivation of the result, previously enunciated by his former teacher Stefan as a generalization from experiment, that the energy of a *black-body radiation is proportional to f_4, T being the thermodynamic temperature of the emitter, the result being sometimes known as the *Stefan–Boltzmann law. Boltzmann also played an important part in supporting and developing Maxwell's electromagnetic theory.

Boltzmann constant Symbol: k. When the entropy of a system (a gas, for example) is divided by the natural logarithm of its statistical probability, the result is the constant named after Boltzmann. It is equal to R/L, where R is the *molar gas constant and L is the *Avogadro constant. It has the value $1.380\,658 \times 10^{-23}$ J K^{-1}.

Boltzmann distribution *See* Boltzmann's formula.

Boltzmann engine An ideal thermodynamic engine operating in cycles and employing enclosed *black body radiation as a working substance, conceived for the theoretical derivation of the *Stefan–Boltzmann law of black body radiation.

Boltzmann entropy theory The entropy S of a system of particles is a linear function of the logarithm of the statistical probability W of the distribution; i.e. $S = k \log W$, where k is the *Boltzmann constant. *See* entropy.

Boltzmann's formula *Syn.* Maxwell–Boltzmann law. The formula giving the distribution of energy among the particles in a body in thermal equilibrium, according to classical physics. It has the form:

$$N_E\,dE = Ag_E e^{-E/kT}dE,$$

where N_E is the number of particles per unit energy range at energy E, k is the Boltzmann constant, T is the thermodynamic temperature, g_E is a *statistical weight factor, and A is a constant. A and g_E are characteristics of the system and type of energy.

It can be shown that at low concentrations and high temperatures the distribution laws given by *quantum statistics tend to the form of Boltzmann's equation. In this case $g_E\,dE$ represents the number of quantum states with energy in the range E to $E + dE$ and $Ae^{-E/kT}$ is the average number of particles per state. For this reason classical physics gave

certain results, which were in agreement with experiment in certain conditions but which were found to be incorrect in other cases, especially at low temperatures.

The formula can be used for discrete energies. The infinitesimal range dE is omitted, N_E is the number of particles with energy E, and g_E is the number of quantum states at this energy.

bomb calorimeter A device used for measuring the heat evolved by the combustion of a fuel. A strong metal bomb A with a gas-tight lid B, contains a platinum crucible C, held by a framework D. A known mass of solid or liquid fuel is placed in C and ignited electrically, in an atmosphere of compressed oxygen, by passing a current through a fine iron wire F. The oxygen is admitted at G through a valve H. The whole bomb is immersed in a water calorimeter K of known thermal capacity and fitted with stirrers L. A water jacket surrounds the calorimeter so that radiation losses may be allowed for. After ignition, the stirrers are operated until the thermometer registers the maximum temperature of the water.

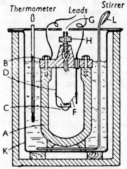

Bomb calorimeter

bond angle The angle between two lines joining the nucleus of an atom in a molecule to the nuclei of two other atoms to which it is bound, taken as the angle between the two chemical bonds. In water molecules, for example, the angle between the two oxygen–hydrogen bonds is 104.45°.

bond energy The energy required to break a chemical bond between two atoms in a molecule. The bond energy depends on the type of atoms and on the nature of the molecule. In molecules that contain only one type of bond, the bond energy can be uniquely defined. For example in water it is simply one half of the energy required to convert the molecule into atoms. In a molecule containing more than one type of bond the energies can be estimated from thermochemical and spectroscopic data and by comparison with data from similar molecules.

Bondi, Sir Hermann (*b.* 1919) Brit. physicist and cosmologist who, with Sir Fred Hoyle and Thomas Gold, proposed the *steady state theory of the universe.

bonding orbital *See* molecular orbital.

bonding pad Metal pads usually arranged around the

edge of a semiconductor *chip to which wires may be bonded to make electrical connection to the component(s) or the circuit(s) on the chip.

bond length The length of a chemical bond in a molecule as determined by the mean distance between the atomic nuclei. This distance depends on the types of the two atoms and on the nature of the molecule. For example, the carbon–carbon bond length in ethane is 0.1543 nanometre whereas in benzene it is 0.1395 nm.

bone conduction Vibrations in the audible range may be transmitted from one side of the head to the other through the bones themselves; this is called bone conduction. For example, if the stem of a tuning fork is pressed on the skull, the vibrations are conveyed direct to the cochlea through the bones. An important application of this medium of sound conduction is found in some types of *hearing aid. If deafness is due to a defect in the middle ear only, the output from a hearing aid is produced in a diaphragm which presses against the mastoid bone just behind the ear. The defective part of the ear is thus by-passed and the sound transmitted direct to the cochlea. The method of pressing a tuning fork on the skull has been used to test whether defective hearing is due to the middle ear or the cochlea. Another application of bone conduction is found in the phenomenon of masking of a sound applied to one ear by another sound applied to the other. This binaural masking has the same characteristics as monaural masking except that the masking tone must be raised by about 50 deibels in the former case. This is believed to be the intensity loss caused in its transmission by bone conduction from one ear to the other. For a similar reason "objectibe" beats may be heard between two notes, each being applied to a different ear.

booster 1. A generator or transformer inserted in an electric circuit to enable the voltage acting in the circuit to be increased (positive booster) or decreased (negative booster), or to change the phase of the voltage.
2. In broadcasting, a repeater station that receives the signal transmitted from a main station, amplifies it, and then retransmits it, sometimes with a change in frequency.

bootstrap A means or technique enabling a system to bring itself into some desired state. For example, it may be an electronic circuit in which positive *feedback of the output signals is used to control conditions in the input circuit.

bootstrap theory (From the phrase, "pulling oneself up by the bootstraps".) A theory that leads to, or is concerned with, the self-consistency of a more enveloping theory.

Bordini effect An effect observed in a metal of *close-packed structure containing *dislocations. If it is subjected to an oscillating stress of variable frequency there is a peak in the *internal friction at a particular frequency. This is due to the dislocations in the metal oscillating between two equilibrium positions at this frequency.

Born, Max (1882–1970) German-born British physicist who has made substantial contributions to the mathematical development of *quantum mechancis.

Born equation An equation derived by Max Born giving the heat of solvation (ΔH) of an ion, i.e.
$$\Delta H = (Lz^2 e^2/8\pi\varepsilon_0 r)(1 - 1/\varepsilon_r),$$
where, L is the *Avogadro constant, z is the valency and r the radius of the ion, ε_r is the relative *permittivity of the solvent and ε_0 is the electric constant.

boron chamber An *ionization chamber lined with boron or boron compounds or filled with boron trifluoride gas.

boron counter A proportional counter that uses a nuclear reaction with boron–10 for the detection of slow neutrons.

Bose, Satyendra Nath (1894–1974) Indian physicist who collaborated with Einstein in working out Bose–Einstein statistics (*see* quantum statistics). *Bosons are named after him.

Bose condensation *Syn.* Bose–Einstein condensation. A phenomenon that occurs at low temperatures in systems consisting of large numbers of *bosons whose total number is conserved in collisions. Used in the explanation of *superfluidity, this phenomenon enables a significant fraction of the particles to occupy a single quantum state. No analogous phenomenon occurs for two or more *fermions, which are prohibited by the *Pauli exclusion principle from occupying the same quantum state.

Bose–Einstein statistics and distribution law *See* quantum statistics.

boson Any particle having integral *spin. Bosons obey Bose–Einstein statistics (*see* quantum statistics). All particles are either bosons or *fermions: *photons, *pions, and *kaons are bosons. Bosons are not subject to any law of conservation of particle number; they can be created or destroyed freely.

Bothe, Walther (1891–1957) German physicist who in 1930 made the first experimental discovery of the particle later identified by *Chadwick as the *neutron.

Bouguer's law of absorption *See* linear absorption coefficient.

boundary conditions Conditions that must be satisfied at a surface separating two media differing in electrical or magnetic properties. For *dielectrics*, the tangential components of the electric intensity must be the same in the two media, and the difference between the normal polarizations in the two media must equal the surface density of electrification. For two *magnetizable substances*, the tangential component of the magnetic field must be the same in both media, and the magnetic induction at right angles to

boundary layer

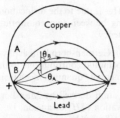

Boundary conditions

the surface must be the same in both media. At the surface separating two *conductors* of *resistivity ρ_A and ρ_B, the directions of the lines of flow of a current passing from one medium to the other will be deviated in accordance with the relation $\rho_A \tan \theta_A = \rho_B \tan \theta_B$, where θ_A and θ_B are the angles made by the line of flow with the normal to the surface in the media A and B respectively (see diagram). The current is thus deflected away from the normal on passing into a medium of greater conductivity.

boundary layer If a fluid of low viscosity (e.g. air or water) has a relative motion with respect to solid boundaries, then at a large distance from the boundaries the frictional factors are negligible with regard to the inertia factors while near the boundaries the frictional factors are appreciable.

The theory, first propounded by Prandtl, is that the fluid may be divided into two parts: first, a thin layer of fluid close to the solid boundaries in which the viscosity of the fluid is of major importance – this layer is called the *boundary layer*; secondly, the portion of the fluid that remains outside this boundary layer within which the fluid may be considered as nonviscid. It can be shown that the thickness of the boundary layer is directly proportional to $\sqrt{\nu}$ where ν is the *kinematic viscosity and that the normal pressure on the solid boundary is unaltered by the presence of the boundary layer.

The approximate pressure distributions outside the boundary layer can be ascertained from the assumption of irrotational motion of the perfect fluid, only if the boundary layer does not separate from the solid body. Otherwise a wake is formed in the rear of the body and the pressure distributions are affected, resort then being made to experiment.

Bourdon tube (and **gauge**) A curved tube of oval cross section with the longer diameter of the oval perpendicular to the plane in which the tube is curved. If the volume of the tube is made to increase (by an excess pressure inside), the oval cross section becomes more nearly circular and the tube tends to straighten out.

The Bourdon tube can be used as a recording thermometer. The tube is closed at both ends and completely filled with a liquid. The liquid expands more than the tube material with increase in temperature and causes the tube to straighten out. One end of the tube is fixed and the other, connected to a tracing point, draws a graph of temperature against time on a slowly moving surface. The Bourdon tube is also used as a *pressure gauge. The tube is closed

60

at one end and the pressure applied at the other. One end of the tube is fixed and the other operates a pointer that indicates the pressure on a calibrated dial. A series of these Bourdon gauges can be used from vacuum to several tens of megapascals.

Boyle, Robert (1627–1691) Irish scientist and philosopher. Often known as the father of chemistry, he played a great part in dissociating chemistry from the mystical mixture of science and fancy involved in alchemy, his outstanding publication being *The Sceptical Chymist* (1661). He established that air has weight, was a pioneer in Britain in the use of a sealed thermometer, and made advances in heat, in crystal optics, and in electricity. *Boyle's law, established by experiment and published in 1663, states that at constant temperature, the volume of a fixed mass of gas is inversely proportional to its pressure. In France it is often called *Mariotte's law* although Mariotte did not publish his statement of the law until 1676. *See* Boyle temperature.

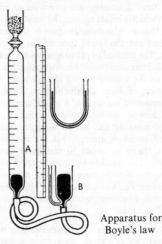

Apparatus for Boyle's law

Boyle's law If a given mass of gas is compressed at constant temperature, the product of the pressure and volume remains constant. The law may be investigated experimentally by containing the gas in the graduated vessel A and the pressure is varied by adjusting the height of the reservoir B. The law is found to be only approximately true for real gases, being exactly fulfilled only at very low pressure. An *ideal gas by definition obeys Boyle's law exactly under all conditions.

Boyle temperature The *equation of state of one mole of a real gas can be written in the form:

$$pV = RT + Bp + Cp^2 + Dp^3 + \dots,$$

where RT, B, C, etc., are the virial coefficients (*see* virial expansion). R is the *molar gas constant and T the *thermodynamic temperature. For all gases the quantity B is negative at low temperatures and positive at high temperatures. Higher-order terms, which are generally positive, only become significant at very high pressures. At the Boyle temperature T_B the quantity B is zero so the gas obeys *Boyle's law almost exactly over a wide range of pressure.

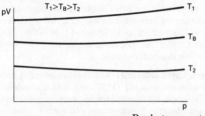

Boyle temperature

Boys, Sir Charles Vernon (1855–1944) Brit. experimental physicist who introduced quartz fibres into instrument suspension (*see also* Boys' radiomicrometer). He also used a rotating-drum camera to photograph lightning, and spark photography to photograph bullets in flight. *See also* Boys' method.

Boys' method A method of measuring the refractive index n of the glass of a lens. It depends on the formula:
$$1/f = (n - 1)\{(1/r) + (1/s)\},$$
where f is the focal length. To obtain the radii of curvature (r and s) of the faces (here both positive if convex outwards), reflection from a face is used to give an image coincident with the object, the reflection being reinforced if necessary by floating the lens on mercury, as shown. The measured distance x and

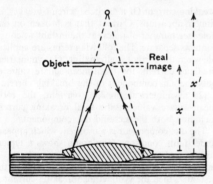

Boys' method for measuring n

the focal length f (previously determined, e.g. with a plane mirror behind the lens) give x' in the formula
$$1/f = (1/x) + (1/x')$$
and the radius of curvature of the lower face $r = -x'$. The lens is turned over to find the other radius s, and n can then be calculated as shown.

Boys' radiomicrometer A thermocouple and moving coil galvanometer combined to make a sensitive detector of radiant energy.

A single antimony–bismuth junction is connected directly to a moving coil. Usually a small blackened platinum disc is fixed to the junction as a collector for radiation.

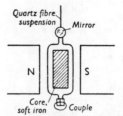

Boys' radiomicrometer

brachistochrone A curve along which a smooth wire joining two points A and B must lie in order that a bead may take the shortest possible time to slide along the wire from A to B under gravity.

Brackett series *See* hydrogen spectrum.

Bradley, James (1693–1762) Brit. astronomer. He discovered the phenomenon of stellar aberration (*see* aberration of light), a periodic change in the apparent position of stars due to the changing velocity of the earth in its orbital motion, and applied this to obtain a good estimate of the speed of light.

Bragg, Sir William Henry (1862–1942) Brit. physicist. In cooperation with his son, Sir (William) Lawrence Bragg (1890–1971), he developed the X-ray spectrometer, experiments with which led to great advances in the knowledge of atomic and crystal structures.

Bragg angle Symbol: θ. *See* Bragg's law.

Bragg curve A curve on a graph produced by plotting the specific ionization caused in air by alpha particles (as ordinate) against the distance from the source (as abscissa). It has a characteristic shape in which the specific ionization increases with distance (as the energy falls) followed by a sudden drop at the point at which the energy of the alpha particles becomes so low that they capture electrons to become neutral helium atoms.

Bragg–Pierce law The mass absorption coefficient (*see* linear absorption coefficient) of a beam of monochromatic X-rays of wavelength λ, in material of atomic number Z, is given by $\mu = CZ^4\lambda^{5/2}$, where C is a constant which changes abruptly at each *absorption edge. The law is, at best, only a very rough approximation to the facts, and may not even be of the right form. $\lambda^{2.88}$ gives better agreement with the experimental data.

Bragg's law If a parallel beam of *X-rays, wavelength λ, strikes a set of crystal planes it is reflected from the different planes, *interference occurring between X-rays reflected from adjacent planes. Bragg's law states that constructive interference takes place when the difference in pathlength, BAC, is equal to an integral number of wavelengths:
$$2d \sin \theta = n\lambda,$$
where n is an integer, d is the interplanar distance, and θ is the angle between the incident X-ray and the crystal plane. This angle is called the *Bragg angle* and a bright spot will be obtained on an interference

Brahe, Tycho

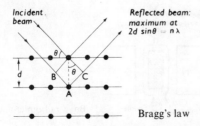

Incident beam

Reflected beam: maximum at $2d \sin\theta = n\lambda$

Bragg's law

pattern at this angle. A dark spot will be obtained if
$$2d \sin\theta = m\lambda,$$
where m is half-integral. The structure of a crystal can be determined from a set of interference patterns found at various angles from the different crystal faces.

Brahe, Tycho (1546–1601) Danish astronomer, whose accurate observations provided *Kepler with the material on which he formulated his laws of planetary motion.

brake dynamometer *See* torquemeter; dynamometer.

brake horsepower The *horsepower of an engine or motor as measured by a brake *dynamometer.

braking an electric motor *See* electric braking; magnetic braking.

braking radiation *See* bremsstrahlung.

Bramah, Joseph (1749–1814) Brit. engineer. Inventor of hydraulic machinery, especially the *Bramah press. See* hydraulic press.

branch Of an electrical network. *See* branching; network.

branching The occurrence of competing decay processes (*branches*) in the *disintegration of a particular radionuclide. *See also* branching fraction.

branching fraction The fraction given by the number of disintegrating nuclei that follow a particular branch of a decay process to the total number of disintegrating nuclei of a radionuclide. It is usually expressed as a percentage.

branching ratio The ratio of two specified *branching fractions.

brass The generic name for an *alloy in which copper and zinc are the dominant constituents. Most grades of brass are somewhat denser (about 5%) than steel; strength and melting point tend to be lower. A typical *yellow brass* (relative density 8.4, M.P. 940 °C) may contain about 67% of copper, 33% zinc; *spring brass* has some lead and iron in addition. *Red brass* (relative density $\sim$ 8.8, M.P. 1050 °C) is usually about 90 Cu: 10 Zn but some grades are nearer 85 Cu: 15 Zn. The coefficient of thermal expansion is usually about 18×10^{-6} per °C (linear) but may be appreciably changed by compar-

atively small additions of other elements such as tin. *Compare* bronze.

Brattain, Walter Houser (1902–1987) Amer. physicist. Working at the Bell Telephone Company on the properties of semiconductors, with John Bardeen and William Shockley he invented (1947) the point-contact transistor. The three shared the 1956 Nobel prize for this achievement.

Braun, Carl Ferdinand (1850–1918) German physicist who constructed (1897) the first cathode-ray oscilloscope. He is noted for his work on measuring the wavelength of radio waves and for his construction of improved radio transmitters.

Bravais lattice *Syn.* space lattice. An indefinitely repetitive arrangement of points in space that fulfils the condition that the environment of each point is identically similar to that of every other point. There are fourteen such arrangements. *See* crystal systems.

brazing *See* soldering.

breakdown 1. A sudden disruptive electrical discharge through an insulator, or between the electrodes of an *electron tube.
2. A sudden transition from high dynamic resistance in a semiconductor device to substantially lower dynamic resistance.
In both cases, the voltage at which breakdown occurs is called the *breakdown voltage*.

breakdown voltage (BDV) *See* breakdown.

breaking current Of a *switch, *circuit-breaker, or similar apparatus. Current that is broken on one pole (*see* number of poles) at the instant when the contacts separate. The following terms are applicable when the current is alternating: (1) symmetrical breaking current, the *root-mean-square value of only the a.c. component of the breaking current;
(2) asymmetrical breaking current, the root-mean-square value of the total breaking current including both the a.c. and d.c. components.
The d.c. component is a transient which appears under fault conditions. The graph shows a typical current under fault conditions in an a.c. system. If the contacts of the circuit-breaker, etc. open at the instant A then the symmetrical breaking current is $BC\sqrt{2}$ and the asymmetrical breaking current is $\sqrt{[(BC)^2/2 + (AB)^2]}$. *Compare* making current.

breaking stress The tension required to separate two neighbouring crystal planes normal to a rod of unit cross section: observed values for ionic crystals are found to be only a small fraction of calculated values.

Bredig arc A method for producing colloidal metal sols by striking an arc between two electrodes of the metal, held below the surface of water kept cool by an ice bath.

breeder reactor A *nuclear reactor in which more *fissile material is produced from *fertile material than is consumed. In strict usage the term is restrict-

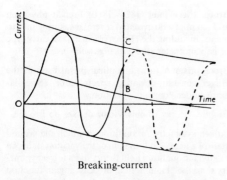

Breaking-current

ed to reactors in which the nuclide produced is the same as that which is consumed. If they are different it is a *converter reactor. *See also* fast breeder reactor.

breeding A process of nuclear transformation in which the *breeding ratio is greater than one.

breeding gain *See* breeding ratio.

breeding ratio The ratio of the number of fissile atoms produced in a *breeder reactor to the number of fissile atoms of the same kind that are destroyed. The breeding ratio minus one is sometimes called the *breeding gain*.

Breit–Wigner formula An equation giving the absorption *cross section, σ, of a particular nuclear reaction when the intermediate excited nucleus can decay in any of several ways. The cross section is a function of the energy, E, of the bombarding particle and when E is close to the energy E_c of the *compound nucleus then:

$$\sigma = \frac{\sigma_0 \, \Gamma^2 (E_c/E)^{1/2}}{\Gamma^2 + 4(E - E_c)^2}$$

where σ_0 is the resonance *cross section and Γ is the width of the excited quantum state.

bremsstrahlung Electromagnetic radiation produced by the rapid acceleration of an electron during a close approach to an atomic nucleus. The radiative loss due to the "braking effect" increases rapidly with the energy of the electron, and for energies exceeding 150 MeV is responsible for most of the absorption of the electron's energy. The energy lost at each encounter is radiated as a single photon. Bremsstrahlung radiation forms an important constituent of secondary *cosmic rays and is the continuous radiation that occurs in the production of *X-rays by electron bombardment. *See also* synchrotron radiation.

Brewster, Sir David (1781–1868) Brit. physicist. His main research work was in optics, in which field he was a pioneer in spectral work and described (1832) *absorption spectra due to coloured glass and to some gases; he studied the polarization of reflected light and the properties of biaxial crystals and dis-

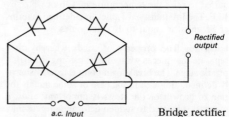

covered double refraction accompanying strain. He also discovered fringes formed by multiple reflections between glass plates (*Brewster fringes*, 1817) and invented the kaleidoscope (1816). He was one of the main organisers of the British Association for the Advancement of Science (founded 1831).

Brewster angle *See* polarizing angle.

Brewster's law *See* plane-polarized light; polarizing angle.

Brewster windows Reflecting surfaces, used in certain gas *lasers, to reduce reflection losses that would arise from using external mirrors. The surfaces are set at the Brewster angle to the incident light. *See* polarizing angle.

bridge A circuit made up of electrical elements (e.g. resistors, inductors, capacitors, rectifiers, etc.) arranged in the form of a quadrilateral, or its electrical equivalent. Two opposite corners of the quadrilateral are made the input and the other pair the output of the circuit. (*See* bridge rectifier.) Bridges are most commonly used in a variety of measuring instruments in which the output is connected to a current detector and the circuit adjusted until the bridge is balanced and no current is detected. In this way an unknown resistance, capacitance, or inductance can be compared with known standards. *See* Wheatstone bridge; Anderson's bridge; Campbell's bridge; de Sauty bridge.

bridge rectifier A *full-wave rectifier consisting of a bridge with a rectifier in each arm, as shown in the diagram.

Rectified output

a.c. Input   Bridge rectifier

bridge transition A method of *series-parallel control of a pair of traction motors in which the motors and their control rheostats are connected in a *Wheatstone bridge circuit during transition from series to parallel connection. Its advantage over *shunt transition is that the whole tractive effort is maintained during transition.

Bridgman, Percy Williams (1882–1961) Amer. physicist. A leading investigator in the field of high-pressure physics and an authority on thermodynamics.

Bridgman effect The absorption or liberation of heat arising from a nonuniform current distribution that occurs when an electric current passes through an *anisotropic crystal. *See also* Peltier effect; Thomson effects.

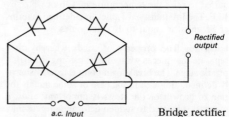

Brillouin function

Brillouin function Symbol: B_J. The magnetization of a paramagnetic substance can be expressed by the equation: $M = NgJ\mu_B B_J(x)$, where N is the number of atoms per unit volume, g the *Landé factor, J is the magnetic moment *quantum number, and μ_B the *Bohr magneton. x is given by the expression $gJ\mu H/kT$, where H is the magnetic field strength and T the thermodynamic temperature. The Brillouin function is expressed by the equation:

$$B_J = \frac{2J+1}{2J} \coth\left[\frac{(2J+1)x}{2J}\right] - \frac{1}{2J} \coth \frac{x}{2J}.$$

Brillouin zone If the *Schrödinger equation for electron energies is solved with a periodic function $u(k)$ to give the energies of an electron in a solid, the solutions fall into permitted bands (see energy bands). If the solutions are plotted in the *reciprocal lattice of the crystal being considered, the zones enclosing the solutions for $k = 1, 2, \ldots n$, are called Brillouin zones.

Brinell, Johann August (1849–1925) Swedish engineer who devised the Brinell *hardness scale.

Brinell hardness See hardness.

Britannia metal A hard silver-white alloy containing approximately 80–90% tin, 5–15% antimony, 0–3% copper, and sometimes some lead or zinc. It is used as an antifriction material and for domestic purposes.

British thermal unit (Btu) The amount of heat required to raise the temperature of 1 lb of water by 1 °F. The International Tables Btu, defined in terms of SI units, is equal to 1055.06 joules.

brittleness The property of solids whereby they separate into pieces without first passing through a plastic stage. The breakdown point is usually sharply defined, although in the case of some metals the rate of growth of the stress is important; thus, a material subjected to sudden deforming forces may be brittle but it may yield plastically when the stress is built up slowly. The value of the stress at breaking point as determined by experiment is usually less than the value calculated from a knowledge of the properties of the crystals from which the substance is formed. See elasticity; hardness.

brittle rupture The sudden separation of a single crystal into two parts; this may depend on the rate of application of the force, there being no sharp dividing line between brittle rupture and *plastic deformation.

broadside on A magnet is said to be broadside on to a point, if the point lies on a straight line drawn through the centre of the magnet at right angles to its magnetic axis; i.e. on the magnetic equator of the magnet. Also known as the Gauss B position. See Gauss positions.

Broca, Pierre Paul (1824–1880) French physician and scientist who invented several measuring instruments, including the *Broca prism and an early type of galvanometer (the *Broca galvanometer*).

Broca prism A form of optical prism used in the Hilger constant-deviation spectroscope. (See diagram under constant deviation.)

Broglie, (Prince) Louis Victor de See de Broglie.

broken symmetry A condition in which the ground state of a many-body system or the vacuum state of a relativistic quantum field theory has a lower symmetry than the *Hamiltonian or *Lagrangian function defining the system. Examples in condensed-matter physics include antiferromagnetism and superconductivity. The Weinberg–Salam model (see electroweak theory) is a physically important example of a relativistic quantum field theory with broken symmetry.

Bronson resistance An *ionization chamber enclosing a constant source of ionization, usually a layer of uranium oxide. If the p.d. between the electrodes is small compared with that required for saturation, the ionization current is approximately proportional to the applied p.d., and the system acts as a high resistance. The resistance can be increased by decreasing the activity of the source employed.

Brönsted, Johannes Nicolaus (1879–1947) Danish physical chemist who contributed to the theory of electrolytes and introduced a new definition of acids and bases.

bronze The generic name for *alloys in which copper and tin are the dominant elements; it has been extended by usage to include many other copper-rich alloys other than the *brasses, which are predominantly copper–zinc alloys, but even some of these are often described as "bronze". The percentage of tin may vary over a wide range (1 to 30%) and zinc and lead may also be present. Typical relative density 8.7, coefficient of thermal expansion 18×10^{-6} per °C (linear); M.P. ranges from 745 °C (67 Cu, 33 Sn) to cover 1000 °C (e.g. 1000 °C for 90 Cu, 10 Sn). Many of the bronzes are resistant to corrosion, they are often very suitable for casting and they make good bearing materials. See also gunmetal; phosphor bronze.

Brown, Robert (1773–1858) Brit. botanist, discoverer of the motion of the suspended particles of a colloidal solution now known as *Brownian movement.

brown dwarf A type of astronomical object with a mass between that of a large planet and a small star. Brown dwarfs are faint objects with a mass too small to sustain fusion but great enough to generate energy by gravitational pressure. Their mass can be several times the mass of Jupiter (up to a limit of about 80 times the mass of Jupiter). It has been suggested that brown dwarfs (sometimes called 'jupiters') are significant contributors to the *dark matter in the universe.

Brownian movement Robert Brown in 1827 observed that small particles about 1 μm in diameter, such as pollen grains, when held in suspension in a liquid exhibited unceasing and irregular motion in all directions. Initially the phenomenon was ascribed to a vital force but subsequent work showed it to be a visible demonstration of molecular bombardment by the molecules of the liquid. (*See* kinetic theory.) The smaller the suspended particles, the less likely are the molecular impacts on opposite sides to balance out simultaneously and the more noticeable the motion. It can also be observed in particles of smoke.

*Perrin carried out experiments between 1900 and 1912 on suspensions of gamboge or mastic and was able to determine the value of the *Avogadro constant with considerable accuracy.

Brownian motion sets a theoretical limit to the sensitivity of a chemical balance at 10^{-9} g and similarly sets a limit to galvanometer measurements of currents at 10^{-11} A.

brush A conductor that serves to provide electrical contact with a conducting surface moving relatively to the brush, usually between the stationary and moving parts of an electrical machine.

brush contact resistance The *contact resistance between a *brush and the moving surface. In electrical machines the contact resistance between a carbon brush and the copper *commutator bars is not constant but descreases in a very marked manner as the current density at the contact surface is increased. The result of this is that the voltage drop at the contact surface is very roughly constant at one volt (common value) over a considerable range of normal working current densities.

brush discharge A luminous discharge from a conductor that occurs when the electric field near the surface exceeds a certain miniumum value but is not sufficiently high to cause a true spark. It appears as a large number of intermittent luminous branching threads, penetrating some distance into the gas surrounding the conductor; the distance is greater for an anode than for a cathode at the same potential. A nonuniform field is essential for the effect.

brush rocker *Syn.* brush-rocket ring. Part of an electrical machine to which are fixed the brush holders (*see* brush) and which is provided with means whereby the whole assembly can be moved circumferentially. The adjustment enables the most satisfactory brush position to be obtained as, for example, the position for sparkless commutation in a d.c. machine.

brush shift *Syn.* brush lead. The amount by which the brushes of an electrical machine fitted with a *commutator are displaced circumferentially from the *neutral position. It is expressed in commutator bars or *electrical degrees. If the brushes are moved from the neutral position in the direction of rotation of the commutator they are said to have a forward shift (or forward lead) and if displaced in the opposite direction, backward shift (or backward lead). In a d.c. machine not fitted with *compoles, the brushes are usually given forward shift in the case of a generator (backward shift or a motor) to improve commutation but when compoles are fitted, the brushes are fixed in the neutral position.

bubble chamber An instrument in which the tracks of an ionizing particle (*see* ionizing radiation) are made visible as a row of tiny bubbles in the liquid inside a large chamber. The liquid, usually hydrogen, helium, or deuterium, is maintained under pressure so that it can be heated without boiling to a temperature slightly above its normal boiling point. Immediately before the passage of a particle, the pressure is reduced, and the liquid would normally boil after about 50 milliseconds. However, the release of energy resulting from ionization of atoms along the track of the moving particle causes rapid localized boiling along this path. After about 1 ms, the bubbles are big enough to photograph and a record is obtained of the particle's track and that of any decay or reaction products. The pressure in the chamber is then increased again to prevent the bulk of the liquid from boiling.

Buchholz protective device A protective relay fitted to an oil-filled transformer tank and operated either by an accumulation of gas produced by faults or by oil surges due to explosive faults within the tank. The device is arranged to operate an alarm or to trip (*see* tripping device) the transformer out of circuit.

bucket brigade *See* charge-transfer device.

buckling A constant that is a measure of the curvature of the *neutron flux distribution inside a nuclear reactor. In this context the curvature is that of a plot of neutron flux at a particular point against radial distance from the centre of the reactor. The *geometric buckling*, which depends only on the dimensions of the reactor core, is the curvature required to keep the reactor in a steady state. The *material buckling* depends only on the materials of construction, and is equal to the geometric buckling when the system is in a steady state.

Bucky diaphragm A grid consisting of narrow strips of lead placed between the specimen and film in radiography. It allows primary rays to pass, but cuts off secondary rays and thereby increases the sharpness of the image (see diagram).

buffer An isolating circuit used to minimize reaction between two circuits. Usually it has a high input *impedance and low output impedance. An *emitter follower is an example.

bug An error or fault in computer programs or equipment. *See* debugging.

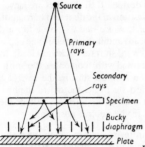

Bucky diaphragm

build-up time The time taken for a current in an electronic circuit or device to rise to its maximum value.

bulk lifetime The average time interval between the generation and recombination of *minority carriers in the bulk material of a *semiconductor.

bulk modulus *See* modulus of elasticity.

bulk strain *See* strain.

bumping In the absence of nuclei, bubbles do not form until the temperature of a liquid is above the boiling point so that when formed the vapour pressure inside the bubbles greatly exceeds the applied pressure. The consequent rapid expansion of the bubbles causes violent motion (bumping) of the containing vessel. Small pieces of porous pot placed in the liquid, by providing nuclei, prevent this occurring.

bunching *See* velocity modulation.

Bunsen, Robert W. E. (1811–1899) German chemist. He was jointly responsible with *Kirchhoff for the introduction of spectrum analysis (1859), and the discovery by this means of the elements caesium (1860) and rubidium (1861). Bunsen was also responsible for the *filter pump, for the *grease-spot photometer and for the *Bunsen burner, *Bunsen cell, and a type of calorimeter (the *Bunsen ice calorimeter*).

Bunsen burner A gas burner in which a regulated amount of air mixes with the gas stream at the bottom of the tube of the burner, the flame being at the top. The air is drawn in by the suction effect of the fine gas jet, a consequence of *Bernouilli's theorem. The flame does not proceed back down the tube in normal working because the velocity of the unburnt gas mixture exceeds the velocity of propagation of the flame-front but if the mixture contains too much air, and especially if the light is applied before sufficient gas has entered the tube, the burner may "burn back" and the gas burn usually with incomplete combustion, at the jet.

Bunsen cell A primary cell, much used before the introduction of accumulators. The negative pole is an amalgamated zinc rod immersed in dilute sulphuric acid in a porous pot. The positive pole is a plate of hard gas carbon immersed in strong nitric acid, which serves as the depolarizer. The e.m.f. is about 1.9 volt.

buoyancy The tendency of a fluid to exert a lifting effect on a body wholly or partly immersed in it. *Archimedes principle states that such a body experiences an upward force equal to the weight of the fluid that would fill the space occupied by the immersed part of the body. This force acts through the centre of gravity of that fluid which would replace the immersed part of the body and this point is the *centre of buoyancy* of the body. The plane in which the liquid surface intersects the stationary floating body is the *plane of flotation*. For the body to be in equilibrium: (*a*) the upthrust must be equal to the weight of the body; and (*b*) the centre of gravity of the body and the centre of buoyancy must be in the same vertical line.

The equilibrium is obviously stable with respect to sideways translation of the floating body and also for vertical displacements. For small rotations without change in the volume displaced, the equilibrium is only stable if the centre of gravity is below both *metacentres.

burden The load connected across the secondary terminals of an *instrument transformer under specified conditions, usually expressed in *volt-amperes. Thus, if a *current transformer is designed for a burden of 15 volt-amperes at a full rated secondary current of 5 amperes, then, under these conditions, the secondary terminal voltage will be 3 volts and the impedance of the connected load 0.6 ohm. The term is sometimes, but less correctly, used for the total *impedance connected across the secondary terminals.

burial *Syn.* graveyard. A place where highly radioactive products from *nuclear reactors may be safely deposited and stored, usually in noncorrosive containers.

burn-up 1. The significant reduction in the quantity of one or more *nuclides arising from neutron absorption in a *nuclear reactor. The term can be applied to fuel or other materials.

2. *Syn.* fuel irradiation level. The total energy released per unit mass of nuclear fuel.

bus Contraction of busbar. A set of conducting wires connecting several components of a computer and allowing these components to send signals to each other. It is thus a pathway for signals.

busbar 1. *Syn.* bus. Generally, any conductor of low *impedance or high current-carrying capability relative to other connections in a system. It is usually used to connect many like points in a system, as with an *earth bus*. Busbars frequently feed power to various points.

2. *See* bus.

button microphone *See* carbon microphone.

BWR Abbreviation for *boiling-water reactor.

bypass capacitor A shunt capacitor connected in a circuit in order to provide a path of comparatively low *impedance for alternating current. The frequency of the alternating current passed depends on the magnitude of the capacitance. Such a capacitor is commonly used to prevent a.c. signals from reaching a particular point in a circuit or to separate out a desired a.c. component.

byte A fixed number of *bits (now almost always 8 bits) that can be handled and stored as a single unit in a computer.

C

cadmium cell *See* Weston standard cell.

cadmium ratio Symbol: R_{Cd}. The ratio of the neutron-induced radioactivity in a sample to the radioactivity induced under identical conditions when the sample is covered with cadmium, which has a high capture cross section for thermal neutrons and has a very low cross section for energies above 0.3 eV. For large values of the ratio, it is therefore a measure of the ratio of *thermal neutrons to *fast neutrons.

cadmium sulphide cell A compact photoconductive cell (*see* photoconductivity) consisting of a layer of cadmium sulphide sandwiched between two electrodes. The high electrical resistance of the CdS drops when light falls on the cell. A current flowing through the cell will vary according to the amount of incident light. A battery is required to provide the current in the cell. It is used in *exposure meters and can be built into cameras. It has a much higher sensitivity than the *selenium cell.

caesium clock *See* clocks.

Cailletet, Louis Paul (1832–1913) French scientist who was the first to produce liquefied air, hydrogen, and nitrogen.

calcite *Syn.* Iceland spar. A crystalline form of calcium carbonate easily breaking along cleavage planes into rhombohedrons, each face being a parallelogram with angles 78° 5′ and 101° 55′. The crystal exhibits *double refraction and is used in the construction of *Nicol prisms. The ordinary and extraordinary rays show the light polarized at right angles. Optically it is classed as a negative *uniaxial crystal.

calculator In general, a device for carrying out logical and arithmetical operations of any kind. There are many kinds of calculators, ranging from a simple hand-operated device, such as an abacus, to large *computers capable of extremely fast and complex data handling. The term is usually applied to small electronic machines capable of performing simple arithmetical tasks such as addition, subtraction, multiplication, division, extracting square roots, etc. These electronic machines consist of integrated logic circuits that perform the calculations, a keyboard

for feeding data in, and an illuminated display consisting of for example, *digitrons, or *light-emitting diode arrays.

calculus of variations A mathematical method for solving those physical problems that can be stated in the form that a certain definite integral shall have a *stationary value for small changes of the functions in the integrand and of the limits of integration. Such problems include: (1) Determination of conditions of equilibrium from the *least-energy principle; (2) Determination of the path of a ray of light from *Fermat's principle of stationary time; (3) Solution of dynamical problems by means of *Hamilton's principle.

calendar year *See* time.

calibration Determination of the absolute values of the arbitrary indications of an instrument.

Callendar, Hugh L. (1863–1930) Brit. physicist. His work was mainly in the field of heat, especially investigations on the electrical method for measuring the mechanical equivalent of heat (with H. T. Barnes) and on the specific heat capacity of water, on continuous-flow calorimetry (used in the above), on the expansion of mercury (with Moss), on thermometry, developing the platinum resistance thermometer into an instrument of precision (and, with Griffiths, developing a suitable bridge for use with it) and on the equation of state and thermodynamic properties of steam.

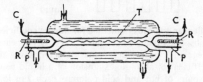

Continuous flow calorimeter

Callendar and Barnes's continuous flow calorimeter (1900) A calorimeter for the determination of the *mechanical equivalent of heat and for determining the variation with temperature of the specific heat capacities of liquids, especially water. A steady stream of water is heated in a fine flow tube T by an electric current passing through a central platinum resistance wire, so that differential platinum resistance thermometers R placed in thick copper tubes registered a steady difference of temperature θ between the inflowing and outflowing water (*see* diagram). If mass m of water of specific heat capacity c flow through in time t then, $IVt = mc\theta + H$. The potential difference (V) across the heater is obtained in terms of the Clerk standard cell by taking leads PP to a potentiometer while the current (I) is similarly measured by noting the potential difference across a standard resistance placed in series with the batteries and the current terminals CC of the heater. H represents the heat lost during the experiment and is made small and definite by surrounding the apparatus with a vacuum jacket inside a water bath at

Callendar and Moss's method

constant temperature. By varying the flow in a second experiment so that mass m of water is heated through the same rise of temperature (θ) as before, by a different current I' and p.d. V' it is possible to eliminate H; thus:

$$(I'V' - IV)t = (m' - m)c\theta.$$

In the original apparatus, by the use of platinum resistance thermometers, the rise in temperature θ could be measured to closer than one-thousandth part of a degree. The Clark cells mainly employed in the original experiment were of the hermetically sealed type and maintained their relative differences constant to one or two parts in 100 000 for two years. The time of flow of the total mass of water collected was generally 15 to 20 minutes and was measured on an electric chronograph reading to 1/100 of a second. The mass of water was collected and weighed in such a way as to eliminate the possibility of loss by evaporation.

Values of c can be determined with this type of apparatus to an accuracy of three significant figures. The experiment was originally used to determine the mechanical equivalent of heat; using a value of c for water of 1 calorie per gramme, J was found to be 4.182×10^7 ergs per calorie.

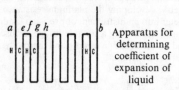

Apparatus for determining coefficient of expansion of liquid

Callendar and Moss's method for finding coefficient of absolute expansion of liquid (1911) They repeated *Regnault's heat experiments for determining the coefficient of absolute expansion of a liquid using six pairs of hot and cold columns each 2 m long instead of a single pair. The hot and cold columns, marked H and C respectively, were arranged in series but with the portions ef, gh, etc. doubled back so that all the hot columns were together in an oil bath while all the cold columns were in an ice bath. The coefficient was measured within the range 0 ° to 330 °C to an accuracy of 1 in 10 000 since the difference in height of the levels a and b is six times that due to a single pair.

Callendar's compensated air thermometer (1891) A constant pressure *gas thermometer in which the error due to the gas in the connecting tube and the manometer being at a different temperature from that in the bulb of the thermometer is eliminated. The pressure of the air in the thermometer bulb B is kept equal to that of the air in the bulb D by altering the amount of mercury in the reservoir S until the levels of the sulphuric acid in the manometer G are equal. The bulb D carries a tube equal in size and in close proximity to the tube connecting B to S. If D and S are immersed in melting ice then the thermodynamic temperature of the bulb B is given by:

$$T = T_0[b/(d - s)],$$

where b, d, and s, are the volumes of air in B, D, and S respectively.

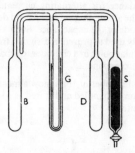

Compensated air thermometer

Callendar's equation of state The equation:

$$V - b = \frac{RT}{p} - c_0 \left(\frac{T_0}{T}\right)^n$$

which is used in dealing with the case of steam, and may be applied to gases and vapours at moderate pressures, although it fails at the critical point. *See* equations of state.

Callendar's radio balance *See* radio balance.

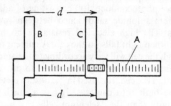

Sliding callipers

callipers *Syn.* calipers. **1.** A pair of dividers with bowed legs for measuring the diameter of convex bodies; also a similar instrument for measuring the bore of tubes, etc. **2.** *Sliding callipers* consist of a scale A graduated in cm carrying a fixed arm B. An arm C, parallel to B, slides on A. The two distances marked d are equal and indicated by the scale; the instrument will thus measure either the distance between objects (top arms) or the thickness of objects (lower arms). The scale may be fitted with a *vernier for accurate reading (*vernier callipers*). **3.** *Micrometer callipers. See* micrometer screw.

calomel electrode A *half-cell consisting of a mercury electrode in contact with a solution of potassium chloride saturated with calomel (mercury(1) chloride, Hg_2Cl_2). It is used as a reference electrode in physical chemistry.

caloric theory The theory of the nature of heat widely held up to about 1850, according to which heat was an imponderable, self-repellant fluid, caloric. It was unable to account for the production of an unlimited supply of heat by friction as occurred in the experiments of *Rumford and was abandoned when *Joule showed that heat was a method of transfer of energy and determined the value of the *mechanical equivalent of heat.

calorie A unit of heat and internal energy no longer employed in scientific calculations. Formerly defined as the quantity of heat required to raise the temperature of one gram of water from 14.5 °C to 15.5 °C at standard pressure, the calorie (symbol: cal_{IT}) is now formally defined as 4.1868 joules.

calorific value The amount of heat liberated by the complete combustion of unit mass of a fuel, the water formed being assumed to condense to the liquid state. The determination is carried out in a *bomb calorimeter and the value is usually expressed in $J\,kg^{-1}$ or similar units. See calorimetry.

calorimeter Any vessel or apparatus in which quantitative thermal measurements may be made. The simplest form, used for the *method of mixtures, consists of a copper can containing water and resting on insulating feet inside a water jacket at a definite temperature, which enables the radiation correction to be calculated. Through an insulating lid, used to prevent evaporation, passes a thermometer, to record the temperature changes of the water in the copper can, together with a stirrer. Other modifications used for special purposes are the *bomb calorimeter, the *Bunsen ice calorimeter, the *Joly steam calorimeter, the *Nernst and Lindemann vacuum calorimeter, the copper block calorimeter, the *continuous flow calorimeters of Callendar and Barnes and of Scheel and Heuse, the *Dewar liquid oxygen calorimeter, the *microcalorimeter, and *Simon and Lange's adiabatic vacuum calorimeter. See calorimetry.

calorimetry The study of the measurement of quantities of heat. (1) *Units*. The traditional thermal units were based on the heating of a standard mass of water. The *calorie and the *British thermal unit have now been replaced for scientific purposes by the SI unit of energy, the *joule.
(2) *Specific heat capacity determination*. An object gives out as much heat in cooling through a given range of temperature as it absorbs in being heated through the same range. The heat evolved while a sample cools may be made apparent by the rise of temperature it causes in a calorimeter (*see* method of mixtures) or by the amount of material it can melt (as in an *ice calorimeter) or boil away (as in *Dewar's liquid oxygen calorimeter). Alternatively, the sample may be heated electrically, as in the *Nernst and Lindemann calorimeter or by condensation on it, as in *Joly's steam calorimeter. For a discussion of these possibilities, *see* specific heat capacity.
(3) *Specific latent heat determinations*. Specific latent heats of sublimation (direct solid-vapour transition) may be obtained by the evaporation type of method as used for *specific latent heats of vaporization: the condensation methods are not usually applicable owing to blockage occurring on

deposition of solid. *See also* specific latent heat of fusion.
(4) *Heats of solution and of reaction*. The heat evolved in a chemical reaction or in the solution of a solid in a liquid solvent may be measured by initiating the reaction or solution by mechanical means (tipping a tube of reagent or smashing a thin glass vessel) inside a suitable calorimeter. Sometimes a *Dewar vessel is used as a calorimeter especially if the reaction is slow. Since the heat capacity of such a vessel is not easy to estimate, it is preferable in such cases to reproduce the rise of temperature in a second experiment by using electrical heating. If the heat of solution or of reaction is negative, i.e. there is a cooling during the experiment, it is desirable to make the experiment a null one (see 9) and adjust electrical heating so that the cooling is compensated at all stages of the experiment.
(5) *Fuel calorimetry*. To find the *calorific value of a fuel, a weighed amount of fuel is ignited in conditions that ensure complete combustion, usually in compressed oxygen in a *bomb calorimeter. Gaseous fuels are fed to a burner mounted inside a vessel in which an elaborate system of tubing leads the exhaust fumes away in counterflow to a current of water. This is an example of a *heat exchanger. There is close thermal contact between the two streams, through the metal walls of the tubes, so the exhaust gases give their excess internal energy to the water-flow and the steady rise in temperature, coupled with a knowledge of the rates of flow of water and gas, serves to give the evolution of heat per unit volume of gas. Tests are regularly made on supplies of coal-gas and other fuel gases in this type of instrument.
(6) *Heats of combustion*. The study of energy changes in chemical reactions is intimately related to the measurement of the heats of combustion of pure chemical substances, for the latter gives the heat liberated when the substance combines with oxygen. (*See* Hess's law.) Thus, to find the *heat of formation* of 1 mole of carbon dioxide CO_2, from its elements, carbon (C) (assumed to be in the form of graphite) and oxygen (as molecular oxygen, O_2):
heat of combustion of carbon to carbon dioxide:
$$C + O_2 = CO_2 + 393\ kJ\,mol^{-1};$$
heat of combustion of carbon monoxide to dioxide:
$$CO + \tfrac{1}{2}O_2 = CO_2 + 285\ kJ\,mol^{-1};$$
whence by subtraction,
heat of formation of carbon monoxide:
$$C + \tfrac{1}{2}O_2 = CO + 108\ kJ\,mol^{-1};$$
i.e. the formation of 1 mole, or 28 grams, of CO is accompanied by the evolution of 108 kJ. In thermodynamics the convention is to express this as a negative quantity, i.e.
$$\Delta H_{CO} = -108\ kJ\,mol^{-1}.$$
(*See* enthalpy.)
The experimental measurements follow the same lines as in 5.
(7) *Continuous-flow calorimetry*. It is sometimes possible to arrange an experiment so that there is a

69

steady evolution of heat. This may then be collected in a steady stream of fluid, usually water, as in the fuel calorimetry of gases (5). This method eliminates the necessity of knowing the heat capacities of the solid parts of the apparatus, because these all have steady temperatures and heat losses are made definite and reproducible. It is then usually possible to eliminate heat losses, if necessary, by repeat runs with the same rise of temperature in the water stream, attained by altering the rate of generation of heat and adjusting the water flow. *See* Callendar and Barnes; Scheel and Heuse.

(8) *Heat losses in calorimetry.* The method of mixtures and many other calorimetric measurements are subject to uncertainty because of difficulty in assessing the amount of heat lost to the surroundings during the experiment. Various methods are available of applying a *cooling correction* (*see also* radiation correction) to the observed value of the change in temperature of the calorimeter, mostly based on the assumption that the rate of heat loss is, for a given system (roughly) proportional to the excess in temperature above the surroundings. This is not valid for large temperature excesses.

(9) *Null methods and adiabatic calorimetry.* A cooling effect, i.e. an *endothermic process, can be compensated by an adjustment of current through an immersed heating coil and the total supply of heat computed from a record of the electrical supply. An exothermic process can likewise be compensated by adjustable cooling, e.g. by circulating chilled brine through tubing in the calorimeter. Such null methods eliminate (or greatly reduce) the need for a cooling correction and they obviate knowledge of heat capacities of the calorimeter and accessories.

An *adiabatic calorimeter* has an independently heated jacket surrounding the calorimeter and this is maintained, manually or automatically, at the same temperature as the outside of the calorimeter, there being usually a thermocouple connected differentially between the two. Heat losses are thus eliminated.

(10) *Miscellaneous methods.* In the *microcalorimeter very small rates of production of heat are measured, e.g. the liberation of heat by a small sample of radioactive material. This is usually accomplished by differential methods in which the temperature of the calorimeter is constantly compared with that of a dummy calorimeter under identical conditions but without the sample. The calorimeters may be provided with resistance-thermometer windings, used in adjacent arms of a Wheatstone bridge, or thermocouples may be used.

Useful information can often be obtained by calorimetric measurements on magnetic phenomena and also when a substance changes structure to an allotropic form. When this happens at a well-defined temperature, the corresponding amount of heat evolved (or absorbed) over a narrow range of temperatures leads to an anomalous specific heat capacity.

calutron An electromagnetic separator of isotopes, based on the principle of the *mass spectrometer. It has been used to separate the fissile nuclide uranium-235 from uranium-238.

camera 1. A light-tight chamber containing a *photographic objective, a shutter, and a film or plate at the back. Usually the shutter has a variable diaphragm so that the *f-number can be altered. The exposure time is also a variable quantity. Film is sensitive to light, ultraviolet, and X-radiation, and can be made sensitive to infrared radiation. (*See also* panchromatic film; colour photography.) The lens-system material must be transparent to the radiation incident on the film. Glass or plastic lenses are used in cameras for normal use. Additional lens systems are the *zoom lens and *telephoto lens. Various types of camera are used in scientific work for recording and measuring purposes. Fast-moving events, such as a spray of liquid, are photographed using a camera in which the lens is left permanently open and the light is provided by an electrically controlled flash (*see* photography) of brief but known duration.

For the recording of some processes a *drum camera* is used. In this case the film is rotated on a drum and the event or process is recorded on the film during one revolution of the drum. This process is used in photographing the trace on a *cathode-ray oscilloscope, with the *time-base switched off. The *Polaroid (Land) camera is a means of obtaining an immediate scientific record of an oscilloscope trace, microscope image, etc.

2. *See* television camera.

3. *See* Schmidt telescope.

camera lucida A microscope accessory for attachment at the eyepiece end, whereby a virtual image is formed in the plane of drawing paper, permitting the simultaneous view of a hand-manipulated pencil to draw the outline.

camera obscura A dark chamber provided with a large aperture lens for projecting an image on to a flat surface. Frequently it is a room provided with a horizontal table on which several spectators can view an outside landscape, the objective and deflecting mirror being situated in the roof.

camera tube The *transducer device in a *television camera that converts the optical image of the scene to be transmitted into electrical video signals. Most camera tubes are *electron tubes: the two basic types are the *image orthicon* and the *vidicon*, from which many other tubes have been developed. There is also a solid-state camera tube in which the transducer is an array of *CCDs (charged-coupled devices), and is thus much smaller and lighter than devices containing electron tubes.

In the image orthicon (Fig. *a*) light from a scene is focused on the *photocathode, which consists of a light-sensitive material deposited on a thin sheet of glass. Electrons are emitted from the photocathode in proportion to the intensity of the light and fo-

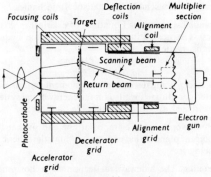

a Image-orthicon tube

cused onto a target consisting of a thin glass disc with a fine mesh on the photocathode side of the disc. The impact of electrons from the photocathode causes secondary emission of electrons from the target, greater than, but proportional to, the original electron density from the photocathode. These secondary electrons are collected by the mesh screen and drained off to a power supply. The target is left with a positive static-charge pattern corresponding to the original light image. The reverse side of the disc is scanned with a low-velocity electron beam produced from an *electron gun. The beam is deflected by magnetic coils. The electrons are turned back at the target glass into the multiplier section. Areas of positive charge on the target are neutralized by electrons from the beam, so the beam returning to the multiplier section varies in density in proportion to the charge on the plate: the return beam is thus modulated with video information.

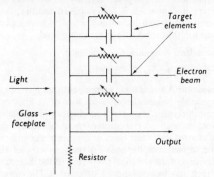

b Photosensitive target area of the vidicon

The vidicon type of camera tube (Fig. *b*) is widely used in closed-circuit television and as an outside broadcast camera as it is smaller, simpler, and cheaper than the image-orthicon type. The photosensitive target area of the vidicon consists of a transparent conducting film placed on the inner surface of the thin glass faceplate, and a thin photoconductive layer deposited on the film. The photoconductive layer may be considered as an array of discrete elements consisting of a light-dependent

resistor with a parallel capacitor (Fig. *b*). A positive voltage is applied to the conducting layer and this has the effect of charging the capacitive elements. The amount of charge in each element will depend on the value of the parallel resistor – the lower the resistance the more charge will be stored. Normally the resistors have a high value, but when light strikes the target area, the resistance drops. If the target is scanned with a low-velocity electron beam, the capacitors will be discharged and a current will flow in the conducting layer (Fig. *c*). The magnitude of the current developed is a function of the charge on the target area and hence of the illumination from the optical lens system.

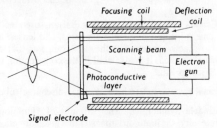

c Vidicon tube

In the *plumbicon*, a modern development of the vidicon tube, the photoconductive material is replaced by a layer of semiconductor material. The mode of operation is similar to the vidicon tube, but the target elements may be considered as semiconductor current sources controlled by light energy. When not illuminated, the target elements are essentially reverse-biased diodes with a very low current. The principal advantages of these tubes compared to the vidicon are the low dark current and good sensitivity and light-transfer characteristics.

The performance of camera tubes depends very greatly on the scanning system. Beam alignment is achieved by using small coils to ensure that the electron beam emerging from the electron gun is central. Deflection of the beam is provided by deflection coils controlling the horizontal and vertical directions; these coils are supplied with a sawtooth voltage causing a linear scan with very rapid return to the start of the scanning position. Focusing coils are also provided to ensure a small cross section when the beam reaches the target, and the decelerating electrode ensures that the beam is essentially stationary and perpendicular to the target surface. *See also* colour television.

Campbell's bridge A *bridge for measuring a mutual inductance by comparison with a standard capacitance *C*. The usual arrangement is illustrated in the diagram. I is an indicating instrument, such as a microphone or *oscilloscope, *L* the self inductance of the coil between A and B, and *M* the mutual inductance of the pair of coils. The resistances are varied until the bridge is balanced, when:

$$\frac{L}{M} = \frac{R + R_1}{R} \quad \text{and} \quad \frac{M}{C} = RR_2.$$

Campbell's bridge

Canada balsam A transparent balsam used as a cement for optical elements, prisms, and lenses, as its refractive index is in the same range as that of glass, i.e. 1.55. It is also used in the construction of *Nicol prisms.

canal rays (kanalstrahlen) *See* gas-discharge tube.

candela Symbol: cd. The *SI unit of *luminous intensity; the luminous intensity in a given direction of a source that emits monochromatic radiation of frequency 540×10^{12} Hz and that has a radiant intensity in that direction of $(1/683)$ watt per steradian.

This definition replaces the former definition: the luminous intensity, in the perpendicular direction, of a surface of $1/600\,000$ m^2 of a *black body at the temperature of freezing platinum at standard atmospheric pressure.

candle *See* international candle; photometry.

candle power Former name for *luminous intensity.

Cannizzaro, Stanislao (1826–1910) Italian chemist who devised a new method for determining atomic weights (relative atomic masses) making use of *Avogadro's hypothesis, which had been completely neglected until revived by Cannizzaro.

canonical (math.) Furnishing, or according to a general rule or formula.

canonical distribution A term introduced into statistical mechanics by J. Willard *Gibbs. It is expressed by:
$$f = A \exp(-\text{energy}/\Theta)$$
$$dp_1 \ldots dp_n dq_1 \ldots dq_n,$$
in which f means the fraction of the systems in an assemblage (molecules in a gas for example) whose momenta lie between $p_1 \ldots p_n$ and $p_1 + dp_1 \ldots p_n + dp_n$ and whose associated coordinates lie between $q_1 \ldots q_n$ and $q_1 + dq_1 \ldots q_n + dq_n$. A is a constant and Θ is the modulus of the distribution; Θ can be identified with kT, the product of the *Boltzmann constant and the thermodynamic temperature. Maxwell's law of distribution of velocities among the molecules in a gas is a limiting case of a canonical distribution.

canonical ensemble *See* statistical mechanics.

canonical equations Equations of classical mechanics as expressed in the form of *Hamilton's equations, namely
$$dp_i/dt = -\partial H/\partial q_i;$$
$$dq_i/dt = \partial H/\partial p_i;$$
p_i and q_i being respectively the momenta and associated coordinates, while H is the energy of the system expressed as a function of p_i, q_i, and time t.

cantilever A beam clamped at one end so that the tangent plane at that end is horizontal, the other end being free.

canting The sideways deflection of a horizontal beam built in at one end and subject to a critical vertical load at the other.

capacitance Symbol: C. The property of an isolated *conductor, or set of conductors and *insulators, to store electric charge. If a charge Q is placed on an isolated conductor, the voltage is increased by an amount V. The capacitance of the conductor is defined as Q/V; for a given conductor it is a constant and depends on the size and shape of the conductor. Two conductors, or a conductor and a semiconductor, together form a *capacitor, and the capacitance C is defined as the ratio of charge on either conductor to the potential difference between them. The unit of capacitance is the *farad. *See also* mutual capacitance.

capacitive coupling *See* coupling.

capacitive electrometer *Syn.* condensing electrometer. A type of electrometer (due to Volta) in which the p.d. is first applied to two large plates of a capacitor, with the plates initially almost touching. The p.d. connection is then removed and the plates separated, which increases the p.d. between them (up to 100 times) to a measurable level.

capacitive reactance *See* reactance.

capacitive tuning *See* tuned circuit.

capacitor An electronic component that has an appreciable *capacitance. It consists of at least one pair of conductors, or of a conductor plus semiconductor, each pair separated by a *dielectric (an insulator). For most types of capacitor, the value of the capacitance depends on the geometry of the device and the electrical properties of the dielectric, which may be solid, liquid, or gaseous. The capacitance may be a fixed or a variable value. Symbols:

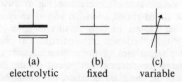

| (a) | (b) | (c) |
| electrolytic | fixed | variable |

capacitor microphone A type of microphone consisting essentially of a diaphragm forming one plate of a

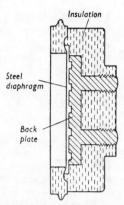

Capacitor microphone

capacitor, the other plate being fixed. Movement of the diaphragm caused by sound-pressure variations alters the capacitance of the capacitor. The diaphragm is of metal and is about 0.025 mm thick. This is tightly stretched by a large steel ring so that its resonant frequency is above the audible range. The other plate of the capacitor is formed by an insulated steel disc that is radially grooved to provide acoustic damping. The spacing between the plates is of the order of 0.025 mm and should be made as small as possible so that a movement of the diaphragm will produce a large variation in capacitance. A high potential is applied across the capacitor so that all dust must be removed from the narrow air gap to avoid breakdown of the insulation. The air gap is usually sealed from the outside air. The microphone is connected in series with a high resistance across a steady potential difference of about 300 volts. Changes in capacity thus produce corresponding changes of potential difference across the capacitor. The microphone has no background noise and is a high-quality instrument with a good frequency response. Its output, however, is low and it has directional effects. An unusual type of capacitor microphone that eliminates diaphragm resonance consists of two thick metal plates with a free passage of air between them. Changes of air density due to the sound cause changes of dielectric constant and thus variations in the capacitance of the two plates.

capacitor motor A type of single-phase a.c. *induction motor somewhat similar in construction to a *split-phase motor. In common with the latter it has two stator windings separated by 90 *electrical degrees. Both windings are connected to the a.c. supply at starting and when running at normal speed, but one of them has a capacitor connected in series with it so that the necessary *phase difference exists between the currents in the two windings. The motor is usually provided with a squirrel-cage rotor. Compared with split-phase motors, capacitor motors have better starting and running characteristics and in particular have a better *power factor and are quieter (less vibration). In some types, the winding

capillary electrometer

having the capacitor in series with it is in circuit only at starting and during acceleration, and is disconnected for normal running. These are known as *capacitor-start* motors.

capacity 1. The amount of information that can be held in a computer storage device.

2. Obsolete name for capacitance.

capillarity An obsolete term to describe the effects of surface tension. The term is derived from the most prominent of these effects – the rise or fall of liquids in vertical capillary tubes (Latin – *capilla*, a hair).

capillary Having a minute or hair-like bore.

capillary constant *Syn.* specific cohesion. The quantity $2T/g(D - d)$, denoted by the symbol a^2, where T is the interfacial surface tension for two immiscible fluids of densities D and d, and g is the acceleration of free fall. (If one phase is air, T is the surface tension of the liquid and d is the density of the air.) It has the dimensions of area and is usually given in mm^2 and has been widely used. It is equal to the rise of the interface in a narrow capillary tube times the radius of the tube, if the contact angle with the tube wall is zero. Its square root, a, simplifies many formulae in surface-tension problems. A few writers have used the symbol a^2 for a capillary constant of half the size of that defined here.

capillary curve The curve in which a vertical plane perpendicular to a glass plate dipping vertically into a liquid intersects the liquid surface.

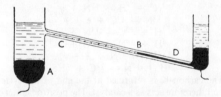

Capillary electrometer

capillary electrometer An electrolytic cell, one electrode of which is a pool of mercury A, while the other is the meniscus B of a thread of mercury in a capillary tube, CD. If a potential difference is applied to the electrodes, a small charging current flows through the cell, producing *polarization at the meniscus B. (The polarization at A is negligible on account of its much greater area.) The electric field due to this polarization alters the *surface tension of the mercury and the meniscus, therefore, moves to a new position of equilibrium. The apparatus is very sensitive, but cannot be used to measure potential differences greater than 0.9 volt. Leakage of charge across the instrument is also troublesome. The sensitivity can be increased by decreasing the angle of inclination of the tube to the horizontal. The movement of the mercury is not strictly proportional to the applied p.d., and the electrometer is most conveniently used as a null instrument.

capture Any process in which an atom, ion, or molecule absorbs a particle from outside. *Radiative capture* occurs when a neutron is captured by a nucleus, with emission of a gamma ray, for example:

$$^{59}_{27}\text{Co} + {}^{1}_{0}\text{n} \rightarrow {}^{60}_{27}\text{Co} + \gamma,$$

which is usually written $^{59}_{27}\text{Co}(\text{n},\gamma)^{60}_{27}\text{Co}$. In *electron capture* the nucleus acquires an electron from an inner orbit, changing a proton into a neutron, for example:

$$^{55}_{26}\text{Fe} + {}^{0}_{-1}\text{e} \rightarrow {}^{55}_{25}\text{Mn} + \nu.$$

Since the electron usually comes from the innermost K-shell, this process was formerly called *K-capture*. Electrons from outer orbits fall into the vacant shell with emission of characteristic X-rays. The capture of an electron into an outer orbit of an atom or molecule, to form a negative ion is called *electron attachment*.

carat 1. The fineness of gold expressed in parts (by mass) of gold per 24 parts of the alloy. Thus 22 carat gold is 22/24 pure gold by mass.

2. A measure of mass (200 mg) for diamonds and other gems.

Carathéodory's principle A theorem in thermodynamics that can be used to derive the second law, without making reference to thermodynamic cycles. It states that it is impossible to reach every state in the neighbourhood of any arbitrary initial state by means of adiabatic processes only.

carbon cycle A cycle of *nuclear reactions resulting in the formation of one helium nucleus from four hydrogen nuclei, with the emission of gamma rays (γ), positrons ($\bar{e}$), and neutrinos (ν):

$$^{12}\text{C} + {}^{1}\text{H} \rightarrow {}^{13}\text{N} + \gamma$$
$$^{13}\text{N} \rightarrow {}^{13}\text{C} + \bar{e} + \nu$$
$$^{1}\text{H} + {}^{13}\text{C} \rightarrow {}^{14}\text{N} + \gamma$$
$$^{1}\text{H} + {}^{14}\text{N} \rightarrow {}^{15}\text{O} + \gamma$$
$$^{15}\text{O} \rightarrow {}^{15}\text{N} + \bar{e} + \nu$$
$$^{1}\text{H} + {}^{15}\text{N} \rightarrow {}^{12}\text{C} + {}^{4}\text{He}$$

The carbon-12 is reformed at the end of the cycle and therefore acts as a catalyst. The positrons interact with electrons giving annihilation radiation. This cycle is believed to be the major source of energy in hot massive stars. *See also* proton–proton chain; thermonuclear reaction.

carbon fibre A material consisting of silky fine threads of pure carbon produced by heating and stretching organic textile fibres. The carbon has an orientated crystal structure and a higher strength and stiffness than other materials of comparable weight. The tensile strength may be as high as 220 meganewtons per square metre. Carbon fibres are used to reinforce metals, ceramics, and plastics. They are used in applications where lightness and strength are required at high temperatures, such as in components of jet engines.

carbon-14 dating *See* dating.

carbon microphone A device making use of the variation of the resistance of an assembly of carbon granules with stress. Sound waves cause oscillation

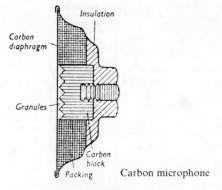

Carbon microphone

in a diaphragm, which is in contact with the granules, causing oscillation of the resistance, which produce an electric signal in the circuit.

carbon resistor *See* resistor.

Cardew voltmeter *See* hot-wire ammeter.

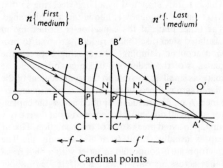

Cardinal points

cardinal planes (and **points**) Gauss, in 1841, developed a theory that reduced the calculation of the image-forming properties of a succession of centred spherical refracting surfaces for the paraxial region to the consideration of a single simple system of six points and planes, viz. *focal, *principal, and *nodal points and planes. The *principal points* P and P', are two conjugate points for the whole system where the lateral magnification is unity (*unit points*). An object PB is equal in size to the image P'B'. The *nodal points* N and N' are two conjugate points for which a ray, e.g. AN, emerges through N' in a parallel direction N'A'. Angle ANO = Angle O'N'A'. The *focal points*, F and F', possess the properties that a ray AB parallel to the axis emerges in the final medium to pass through F', and a ray in the reverse direction parallel to the axis, passes through F. Once the positions of F, F'; P, P'; N,N'; are calculated, any object OA will form an image determined thus: ray AB parallel to the axis, emerges from B' at an equal distance from the axis to pass through F'. Ray AF produced to C, emerges through C', at an equal distance from the axis, to pass along C'A' parallel to the axis. A' is thus determined. Other relations that hold are (a) $\text{FP}/n = -\text{F'P'}/n'$, where n, n' are the refractive indices of the media, and (*b*) FP = N'F'; FN = P'F'. *See also* centred optical systems.

cardioid A heart-shaped curve with polar equation $r = 2a(1 + \cos \theta)$.

cardioid condenser An aplanatic microscope condenser for dark-ground illumination, using reflection first at a spherical surface and then at a cardioid surface; the latter is sometimes replaced by a spherical surface since only a small portion is used at any time (Siedentopf-Zeiss).

Carey–Foster bridge A modification of the *Wheatstone bridge designed to measure the difference in resistance between two nearly equal resistances in terms of the resistance per unit length of the bridge wire. Two resistances are placed in the ratio arms of a Wheatstone bridge and the balance point found on the resistance wire. The resistances are then switched and a new balance point found; the distance between the points on the resistance wire is proportional to the difference between the resistances. If the resistance wire has been precalibrated, the difference in length is equal to the difference between the resistances.

Carnot, Sadi Nicholas Léonard (1796–1832) French soldier-scientist. Although handicapped by his belief in the *caloric theory of heat, Carnot's acute mind enabled him to obtain fundamental results in the theory of heat engines (*Réflexions sur la puissance motrice du feu*, 1824), introducing the idea of reversible operations and ideal cycles. This work, modified by recognition of the kinetic nature of internal energy and the conservation of energy by William Thomson (Lord *Kelvin), *Clausius, and others, formed the basis of the second law of thermodynamics, first enunciated in 1850.

Carnot–Clausius equation For a reversible closed cycle (*see* reversible change), an equation giving the total change in the *entropy of the system as $dq/T = 0$ where dq is the quantity of heat taken in by the system during an infinitesimal reversible change of state and T is the thermodynamic temperature of the system during this change.

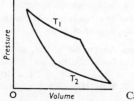

Carnot cycle

Carnot cycle An ideal thermodynamic cycle for a heat engine in which every step is a *reversible process. The working substance expands isothermally and reversibly at T_1 doing work and drawing heat from a furnace. It then continues to expand adiabatically so that the temperature falls to T_2. Work is done upon it in an isothermal compres-

sion at T_2, heat being given out to a "heat sink". The substance is then compressed adiabatically until the temperature is T_1. The pressure, volume, and temperature now have their initial values. Because every process is reversible the change of *entropy of the whole system comprising the engine and its surroundings is unchanged. The cycle gives the highest efficiency theoretically conceivable for a heat engine. (No real engine can attain this.) Kelvin defined his *thermodynamic temperature scale so that the efficiency of an engine working in a Carnot cycle is equal to $(T_1 - T_2)/T_1$:

Carnot's theorem No heat engine can be more efficient than a reversible engine working between the same temperatures. Hence all reversible engines working between the same temperatures are equally efficient, the efficiency being independent of the nature of the working substance, depending only on these temperatures. *See* thermodynamics.

carrier 1. An electron or *hole that can move through a metal or *semiconductor. Carriers enable charge to be transported through a solid and are responsible for conductivity. (*See also* majority carrier; minority carrier.)
  2. A substance used to provide a bulk quantity of material containing traces of *radioisotopes for use in physical and chemical operations. It is used in *radioactive tracer studies and in the preparation of chemical compounds containing radioisotopes. Usually it is a stable isotope or one of its compounds that is identical or similar to the radioactive material used. In some applications the carrier has to be similar to the tracer but finally separable from it. Carriers often depend on selective adsorption of the radioactive substance or on the formation of mixed crystals or complexes.
  3. A carrier wave. *See* modulation.

carrier concentration The number of *carriers – electrons or holes – per unit volume of *semiconductor.

carrier mobility *See* semiconductor.

carrier storage *See* storage time.

carrier system Of telegraphy and telephony. *Carrier transmission of telegraphic or telephonic signals along lines or cables. A special application in telephony is the multichannel carrier system. This latter system provides many (up to 200 or more) communication channels, which may be used simultaneously, and it requires only two cables between the sending and receiving points.

carrier transmission Electrical transmission by means of a modulated carrier wave. *See* modulation.

carrier wave *See* modulation.

Cartesian coordinates *See* coordinate.

Cartesian diver A device illustrating simultaneously the transmission of pressure by liquids, Archimedes' principle, and the compressibility of air. A float containing air is made to rise or sink in water by the removal or application of pressure to the water surface.

Cartesian geometry Geometry using Cartesian *coordinates.

Cartesian ovals Surfaces, first investigated by *Descartes (1637), of such a shape that rays emanating from an object point are refracted to an aberrationless image point (not aplanatic in the Abbe sense). Each object point requires its own surface.

Cartesian systems *See* aberration; aplanatic.

cascade 1. A number of *capacitors are said to be connected in cascade (or in series) if the outer plate of the first capacitor is connected to the inner plate of the next, and so on. All the plates are insulated except the last. The plates of the compound capacitor so formed are the inner plate of the first condenser and the outer plate of the last. The *capacitance of the compound capacitor is given by

$$1/C = 1/C_1 + 1/C_2 + \ldots.$$

The charges on all the capacitors are equal.

2. A chain of electronic circuits or elements connected in series, so that the output of one is the input of the next.

3. *See* isotope separation.

cascade control 1. A method of connecting two a.c. *induction motors to provide an economical change of speed. The two motors are mechanically coupled and the supply is connected to the stator winding of the main motor. The main rotor *slip rings are connected to the stator winding of the auxiliary motor. The latter may have a squirrel-cage rotor or a wound rotor with slip-rings and external starting rheostat. If the torques of the two machines are in the same direction, the *cascade synchronous speed* in revolutions per second is obtained by dividing the supply frequency in hertz by the sum of the number of pairs of poles of the two machines, i.e.

$$n = f/(p_1 + p_2).$$

If $p_1 = p_2$, the cascade synchronous speed is half the synchronous speed of either machine used alone.

2. A system of automatic control in *computers, in which each control unit is controlled by the preceding unit, and in turn, controls the unit following it.

cascade liquefaction First used by *Pictet in 1878. A gas with a high critical temperature is liquefied by increase of pressure, and evaporation of this liquid cools a second gas below its critical temperature so that it, too, may be liquefied by increase of pressure, and so on. In the diagram, machine 1 uses chloromethane, CH_3Cl, whose critical temperature is 143 °C and which can therefore be liquefied by applying a pressure of a few atmospheres. The liquid chloromethane evaporates under reduced pressure at –90 °C in the jacket 2, in which ethene, C_2H_4, is cooled and liquefied by increase of pressure. The liquid ethene evaporates under reduced pressure at about –160 °C in the jacket 3, in which oxygen is

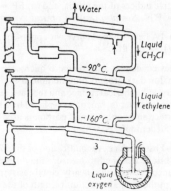

Cascade process of refrigeration

cooled and liquefied under pressure, being collected in the Dewar vessel D. Neither hydrogen nor helium can be liquefied by this method since their critical temperatures (33 K and 5.2 K) cannot be reached in this way.

cascade process of isotope separation *See* isotope separation.

cascade shower *See* cosmic rays.

cascade synchronous speed *See* cascade control.

Casimir effect An attractive force between two closely spaced metal plates in a vacuum, caused by radiation pressure generated by the zero-point energy of the background electromagnetic field.

Cassegrain, N. (17th century) French physician and astronomer who developed the Cassegrain telescope.

Cassegrain telescope *See* reflecting telescope.

casting The process of making a shaped part by pouring molten metal into a suitably shaped cavity (usually made in moulding sand) and allowing it to solidify.

Diecasting is a similar process in which the cavity is in a metal of higher melting point and, unlike the sand cavity, does not have to be remade for each casting. Zinc alloy diecastings are common.

Castings can also be made from other materials, e.g. certain plastics, plaster of paris.

catadioptric system An optical system, such as a telescope, that uses both reflecting and refracting components to form the final image. Examples include the *Schmidt and *Maksutov telescopes.

cataphoresis *Syn.* electrophoresis. The movement of colloidal particles in an electric field. The suspended particles in a *sol carry electric charges that are due to (a) selective adsorption of ions from the liquid, (b) superficial ionization of the colloid particles, or (c) entrainment of ionizable impurity in the substance during preparation. *See* anaphoresis; electrophoresis; isoelectric point.

catastrophe theory A theory of dynamic systems based on analogy with topographical form. If a system depends on n variables, a state of the system can be represented by a point in n-dimensional space, possible states being represented by a region in this space. Catastrophe theory considers the topological classifications of such regions, and, in particular, the conditions under which a discontinuous 'catastrophic' change can occur. Originally developed for biology, the theory has applications in physics (e.g. in optics and mechanics) as well as uses in social sciences.

Catching diodes for voltage limitation

catching diode *Syn.* clamping diode. A *diode used to limit the voltage at some point in a circuit. A diode will start to conduct at the *diode forward voltage, V_d, typically 0.7 V, and will therefore prevent the voltage applied in the forward direction from rising above this value. Two diodes may be used together, as in the illustration, to keep the voltage within specific limits, i.e. within $\pm V_d$.

If V rises to the diode voltage, B will conduct and prevent any further rise in voltage. If V falls to the diode voltage (say –0.7 V), A will conduct and prevent any further fall in voltage.

catenary The curve assumed by a uniform, perfectly flexible chain hanging in equilibrium between two supports. Its equation is $y = k\cosh x/k$, where k is the y intercept.

Soap film forming catenoid surface

catenoid The surface formed by rotating a catenary about its directrix. A soap film formed between the plane, circular ends of two open coaxial tubes has this shape.

cathetometer A device for measuring vertical heights consisting of a vertical scale along which a horizontally mounted telescope or microscope may be moved.

cathode The negative electrode of an electrolytic cell or electron tube. It is the electrode by which electrons enter a system. *Compare* anode.

cathode dark space *See* gas-discharge tube.

cathode fall or drop *See* gas-discharge tube.

cathode follower *See* emitter follower.

cathode glow *See* gas-discharge tube.

cathode-ray oscilloscope (CRO) Usually shortened to oscilloscope. An instrument that enables a variety of electrical signals to be examined visually. Any variable that can be converted into an electrical signal can be studied, making the oscilloscope an extremely valuable tool. The signal of interest is fed, after amplification, to one set of deflection plates of a *cathode-ray tube, usually the vertical deflection plates. The beam is moved horizontally across the screen by the voltage from a sweep generator (usually called a *time-base generator) incorporated in the oscilloscope. The resultant trace seen on the screen is a composite of the two voltages, and suitable choice of sweep speed in the horizontal direction allows easy visualization of the input signal. The simplest type of time base is a constantly variable sweep generator producing a *sawtooth waveform, so that the trace moves slowly and uniformly across the screen, then returns almost instantaneously to the starting point. A more sophisticated type of sweep-trigger circuit may be employed when the sweep is initiated by an external trigger pulse (often the presented signal), so that each sweep is started in synchronism with the trigger pulse.

Extra facilities usually found on a modern oscilloscope include a delayed trigger, access to the X-deflection plates allowing an external time base or other modulating signal to be used, and often facilities for beam-intensity modulation.

cathode rays *See* gas-discharge tube.

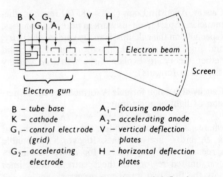

B – tube base
K – cathode
G₁ – control electrode (grid)
G₂ – accelerating electrode
A₁ – focusing anode
A₂ – accelerating anode
V – vertical deflection plates
H – horizontal deflection plates

a Electrostatic focusing and deflection

cathode-ray tube (CRT) A funnel-shaped *electron tube that permits the visual observation of electrical signals. A CRT always includes an *electron gun for producing a beam of electrons, a grid to control the intensity of the electron beam and thus the brightness of the display, and a luminescent screen where the electron beam is focused to produce a spot source of light. *Focusing of the beam of electrons and the deflection of the beam according to the electrical signal of interest, may be done either electrostatically or electromagnetically (*see* Figs. *a* and *b*), or by a combination of both methods. In general, electrostatic deflection is employed when high-frequency waves are to be displayed, as in most *cathode-ray oscilloscopes, and electromagnetic deflection is employed when high-velocity electron beams are required to give a bright display, as in *television or *radar receivers.

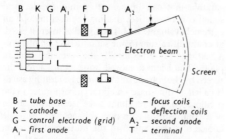

B – tube base
K – cathode
G – control electrode (grid)
A_1 – first anode

F – focus coils
D – deflection coils
A_2 – second anode
T – terminal

b   Electromagnetic focusing and deflection

cathode-ray voltmeter An electrostatic voltmeter consisting of a cathode-ray tube of known sensitivity of deflection.

cathodoluminescence The emission of light by substances under bombardment by *cathode rays.

cation An ion that having lost one or more electrons has a net positive charge and thus moves towards the cathode of an electrolytic cell. *Compare* anion.

catoptric power Of a mirror. *See* power.

catoptric system An optical system in which the principal optical components are reflecting surfaces. *See also* catadioptric system.

Cauchy, Augustin Louis (1789–1857) French mathematician. Apart from his work in pure mathematics, especially on the calculus, Cauchy studied elasticity, and suggested a model, based on the elastic solid *ether theory, to explain dispersion (*see* Cauchy dispersion formula).

Cauchy dispersion formula A formula for the dispersion of light of the form:

$$n = A + (B/\lambda^2) + (C/\lambda^4),$$

where n is the refractive index, λ the wavelength, and A, B, and C are constants. It gives a reasonable agreement with experiment for many substances over limited regions of the spectrum. Sometimes only the first two terms are necessary.

Cauchy's relations Relationships between the *elastic constants depending on the condition that the forces exerted by atoms in a crystal on one another depend only on their natures and on their distance apart. This condition appears to be fulfilled only for the alkali halides at room temperatures (and then incompletely).

causality The principle that every effect is a consequence of an antecedent cause or causes. For causality to be true it is not necessary for an effect to be predictable as the antecedent causes may be too numerous, too complicated, or too interrelated for analysis by man or machine. Roulette is a game of chance in which the outcome of a single spin of the wheel is not predictable, but the law of causality is

not broken because the result could be calculated given sufficiently accurate information regarding the momentum of the ball and the wheel and the friction forces between them.

According to *quantum mechanics events on the atomic scale are subject to laws that are not entirely determinate but which predict the probabilities of certain events. Thus the eventual decay of a radioactive nucleus may be regarded as certain, but the actual time at which decay occurs is not determined. If some phenomenon ensures that an electron is in a particular place at a certain time the subsequent motion is completely indeterminate.

When the exact form of the relevant law of quantum mechanics is unknown, or the solution is too difficult, a useful approximate calculation of the degree of indeterminacy is given by the *uncertainty principle.

Attempts have been made, so far unsuccessfully, to find a principle of determinacy underlying the apparent indeterminacy of quantum mechanics. Another unsolved problem is that of establishing the degree to which macroscopic phenomena may deviate from determinacy as a result of microscopic disturbances causing large-scale effects (*see* chaos theory).

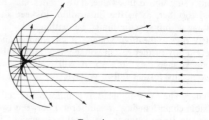

Caustic curve

caustic curve (and **surface**) Rays in a meridian plane from an object after reflection or refraction at spherical surfaces in general do not focus at one point. Consecutive rays as one moves away from the axis, intersect at points lying on a curved line (the caustic), possessing an apex or *cusp* lying at a paraxial focus. (*See* spherical aberration.) Reflected and refracted rays are tangential to the caustic.

Cavendish, Henry (1731–1810) Brit. physicist and chemist. He investigated and identified hydrogen in 1776, and in 1781 outlined the composition of the atmosphere and he showed that water is a compound. He was the first to measure Newton's gravitational constant in the laboratory, using a torsion balance of his own design, thus calculating the mean density of the earth. He provided an experimental proof (1772) of the inverse square law in electrostatics, and showed that no charge remains on an inner conductor after it has been once connected to a surrounding conductor. The Cavendish Laboratory at Cambridge University is named after him.

cavitation The formation of cavities in liquids. Cavitation will take place in liquid flow when at some point the velocity becomes so large that the pressure approaches zero. In the flow of a liquid the lowest equilibrium pressure reached is the vapour pressure; any further increase in velocity is associated with a change in condition of the flow; *Bernouilli's theorem breaks down and cavities are formed.

If the fluid is streaming, the vapour-filled cavities are carried downstream to regions of high pressure where the cavities will collapse. The quick collapse of the cavity gives rise to a large impact pressure that will cause pitting or deformation of any solid surface close to the cavity at the time of collapse. This effect of cavitation is very noticeable with high-speed hydraulic machinery such as propellers and pumps.

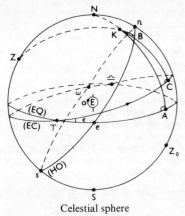

Celestial sphere

cavity absorbent A device that may take the form either of a narrow tube or cavity through which sound is passed, or a hollow resonator placed in the sound field. In the case of a resonant cavity, large vibrations are set up in it at its natural frequency and sound energy is absorbed from the field in the neighbourhood of the resonator. Brittain and others have investigated the absorption of sound in air ducts lined with resonators that were covered with thin membranes. When the resonators were empty, sharp absorption peaks were found in the low-frequency region but these broadened out considerably when the cavities were filled with rock wool. Absorption by small nonresonant cavities is due to greatly increased viscous forces and heat-conduction losses caused by the large surface area of the material in contact with the air. Most absorbent materials are porous so that they actually consist of large numbers of small cavities. Some commercial materials making use of both these methods of absorption consist of small cavities loosely filled with absorbent material, each cavity being partially covered by a membrane with a hole in the centre.

cavity resonator *Syn.* resonant cavity. When suitably excited by external means, the space contained within a closed or substantially closed conducting surface will maintain an oscillating electromagnetic field, and the complete device, the cavity resonator, displays marked electrical *resonance effects. It has several resonant frequencies that are determined by its dimensions. Cavity resonators are used in place of tuned resonant circuits for high-frequency applications.

CCD Abbreviation for change-coupled device. *See* charge-transfer device.

CD Abbreviation for *compact disc.

celestial mechanics *See* astronomy.

celestial sphere A sphere of infinite radius, with its centre at the centre of the earth E, that rotates once in 24 hours of sidereal *time. It is used for positional astronomy.

N	north celestial pole: point of projection of earth's north pole
S	south celestial pole
EQ	celestial equator: circle of projection of earth's equator
EC	ecliptic: circle of projection of apparent path of the sun around the earth; the sun moves anticlockwise as viewed from N
♈	vernal equinox: point of intersection of equator and ecliptic, where sun crosses from south to north of equator
♎	autumnal equinox: point of intersection of equator and ecliptic, where sun crosses from north to south of equator
ε	obliquity of ecliptic: 23.4°
O	observer on earth
Z	zenith: point of projection of O
Z_0	nadir
HO	horizon: great circle having Z, Z_0 as poles
n	north point (of horizon): point of intersection of ZN extended and horizon
s	south point
e	east point
w	west point
K	celestial object
BK	altitude (a) of K
ZK	zenith distance ($z = 90° - a$) of K
nB	azimuth (k) of K: measured in degrees east of the north point
eB	amplitude ($j = 90° - k$) of K
AK	declination (δ) of K: regarded as positive if K is north of the equator
NK	north polar distance of K ($90° - \delta$)
♈ A	right ascension (α) of K: measured in hours and minutes (24 hours = 360°) anticlockwise from ♈
ZNn	meridian for observer at O: when K lies on meridian it is said to transit and to have an *hour angle* H of zero; H increases after transit, and is equal to the difference between local sidereal time and the right ascension of the body

CK   celestial latitude (β) of K: regarded as positive if K is north of the ecliptic

Υ C   celestial longitude (λ) of K: measured in degrees anticlockwise from Υ

nN   altitude of north celestial pole, equal to terrestrial latitude of observer

For almost all astronomical observations, right ascension and declination coordinates are employed. Azimuth, altitude observations and longitude, latitude observations can be converted to the standard coordinate system:

$$\sin \delta = \sin a \sin \phi + \cos a \cos \phi \cos k$$
$$\cos H = (\sin a - \sin \delta \sin \phi) \cos \delta \cos \phi$$
$$\sin \delta = \sin \beta \cos \varepsilon + \cos \beta \sin \varepsilon \sin \lambda$$
$$\cos \alpha = \cos \beta \cos \lambda \sec \delta.$$

cell 1. A pair of plates in an electrolyte from which electricity is derived by chemical action; a unit of a battery. A *primary cell* (or *voltaic cell*) is one in which the current is produced directly from chemical action by the solution of one of the plates. Current can be drawn at once from a primary cell as soon as it is made. A *secondary cell* has to be "charged" by passing a current through it in the reverse direction to its discharge, the chemical actions in the cell being reversible (*see* accumulator). The potential difference between the poles of a cell in a closed circuit depends on its internal resistance, and on the external resistance through which it is maintaining a current. The p.d. (U) between the terminals of a cell is given by:

$$U = ER/(r + R),$$

where E is the e.m.f., r is the internal resistance, and R the external resistance. (*See* polarization.)

2. A unit of a lattice in a *nuclear reactor.

3. A shielded compartment for storing or processing radioactive materials.

4. *See* unit cell.

cell constant The area of the electrodes in an electrolytic *cell divided by the distance between them.

cellular telephone A radiotelephone network linked to the main telephone system, for mobile subscribers (e.g. in cars). In this network the country is divided into adjacent cells, each containing a receive/transmit station. As the user moves from one cell to the other, the signals are automatically switched from cell to cell. In the UK the network, known as Cellnet, is operated jointly by British Telecom and Securicor.

Celsius, Anders (1701–1744) Swedish astronomer. He originated the first scale of temperature, which was "centigrade" in the sense that there were 100 degrees between the ice point and the steam point although in his original scale Celsius had 100° for the ice point and 0° for the steam point. He is honoured by the adoption of his name officially for describing the normal *centigrade scale.

Celsius scale The official name of the centigrade temperature scale with the ice point as 0° and the boiling point as 100°. The degree Celsius (symbol: °C) is equal in magnitude to the *kelvin. A Celsius temperature t, in °C, is converted to a *thermodynamic temperature T, in kelvin, by the relationship:

$$t = T - T_0 = T - 273.15,$$

where T_0 is a thermodynamic temperature fixed as 0.01 K below the triple point of water (273.16 K).

cent 1. The interval equal to $1/1200$ of the interval of two frequencies having the ratio 2:1, viz. the octave. The ratio of the two tones of an interval may be expressed as the number of cents in the interval. For example, in the case of two tones A and B, the number of cents in the interval, $I = k \log_{10} A/B : k$ is defined from the number of cents per octave that when equal to 1200, gives

$$k = 1200/\log_{10} 2$$

Algebraic addition of the number of cents in two different intervals, e.g. D to C and E to D, gives the number of cents in the interval E to C.

2. *See* dollar.

centi- Symbol: c. A prefix denoting 10^{-2}, e.g. 1 centimetre (cm) $= 10^{-2}$ metre.

centigrade scale The temperature scale due to *Celsius on which the ice and steam points are defined as 0 °C and 100 °C respectively. The °C is defined as one-hundredth of this temperature interval. The name is no longer used for scientific purposes, although it is still sometimes used in meteorology. *See* Celsius scale.

centimetre-gram-second system *See* CGS system of units.

centimetric waves Radio waves having wavelengths between 1 cm and 10 cm.

centrad A small angle unit, one-hundredth of a radian, used to specify angles of deviation of narrow-angle prisms. *See* prism dioptre.

central force A force on a moving body that is always directed towards a fixed point or towards a point moving according to known laws.

central orbit The orbit traversed by a body moving under the action of a *central force.

central processing unit (CPU) *Syn.* central processor. The principal operating part of a *computer, in which program instructions are interpreted and executed. In larger and more complex computers, processing tasks and functions are now distributed among various units, each acting independently. These units are simply referred to as *processors*.

centration *Syn.* centring. The process of placing (or the condition of placing) the optical centres of lenses, etc. at the geometrical centres of mounts. Alternatively, the process of arranging the centres of curvature of refracting and reflecting surfaces along the same axis, e.g. objectives and eyepieces of instruments. Also the adjustment of an object or eyepiece graticule to lie in the centre of the field, or of light sources in projection systems to secure uniform illumination. *See* optical centre.

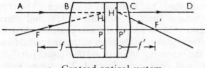

a Centred optical system

centred optical system A system consisting of a number of spherical refracting or reflecting surfaces having their centres on a common axis. The theory of such systems has been developed by *Gauss, *Maxwell, and others. The Gaussian treatment elucidated the properties of such systems in terms of certain constants – the so-called *Gaussian points* or *constants*. Maxwell considered a perfect system in which there was complete point, line, and plane correspondence between object and image. Such correspondence is found in practice provided the rays forming the image are restricted to those passing near the axis of the system.

A ray AB (Fig. *a*) entering the (converging) system parallel to the axis will in general cross the axis after emergence at some point F′ (the *second focal point*). Similarly the ray CD, at the same height above the axis as AB, will have passed into the system after crossing the axis at some point such as F (the *first focal point*). The two incident rays shown fix an object point H for which the corresponding image point must be H′. The plane HP drawn through H perpendicular to the axis is the *first principal plane*. The plane H′P′ similarly drawn through H′, is the *second principal plane*. The principal planes thus have the properties of being *conjugate (object and image planes), and of yielding unit magnification (since HP = H′P′). P and P′ are the first and second *principal points*.

The *first focal length* of the system, *f*, is defined as the distance from the first principal point to the first focal point. The *second focal length*, *f′*, is the distance from the second principal point to the second focal point. It can be shown that if there is the same medium on both sides of the system, the two focal lengths are equal. If the media on the object and image side have different refractive indices n and $n′$, then the relation between the numerical values of the focal lengths is that $n/f = n′/f′$.

Provided object distances are measured from the first principal plane and image distances from the second, the results of simple thin-lens theory can be applied to any centred optical system. In a thin lens the principal planes coincide in the plane of the lens.

In Fig. *b*, the rays entering the system directed towards the axial point N leave as though from the axial point N′ and make the same angle with the axis. The points N and N′ that have this property are called the first and second *nodal points of the system. They are conjugate points of unit angular magnification.

It can be shown that the distance from the first focal point to the first nodal point is equal to the second focal length, while the distance from the second focal point to the second nodal point is equal

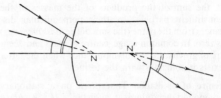

b Nodal points of centred optical system

to the first focal length. This is illustrated in Fig. *c*. With the same medium on both sides of the system the two focal lengths become equal. The principal points must then coincide with the nodal points.

The three pairs of points – focal, principal, and nodal – are called the *cardinal points of the system. If the positions of the cardinal points of a system are known, the position, nature, and size of the image of any object can be calculated without reference to details of the system. The most complicated system then becomes amenable to simple calculations. *See* cardinal planes and points.

c Focal lengths of centred optical system

centre of area *See* centroid.

centre of buoyancy *See* buoyancy.

centre of curvature (of a **lens** or **spherical mirror**) The centre of the sphere of which a lens face or spherical mirror is part. The *radius of curvature* is the radius of this sphere.

centre of figure *Syn.* centre of volume. *See* centroid.

centre of gravity 1. Of a body in a uniform gravitational field (e.g. a body small compared with the earth in the earth's gravitational field). The force on the body (its weight) is the resultant of the forces on the individual particles. As these forces are all parallel, their resultant passes through a particular point fixed with respect to, but not necessarily on, the body; however it may be turned relative to the field. This point is the centre of gravity and it coincides with the *centre of mass of the body.
2. Of a body in a nonuniform gravitational field. The forces on the particles of the body (no longer a system of parallel forces) are reducible, in general, to a single force and a couple. This single force does not, in general, pass through a single point fixed with respect to the body, as the body is turned in the field. If the matter in the body is distributed with spherical symmetry, the couple reduces to zero and the force always passes through the centre of mass; only such a body has a centre of gravity in a nonuniform field, and is said to be *centrobaric* or *barycentric*.

centre of inertia *See* centre of mass.

centre of mass *Syn.* centroid; barycentre; centre of inertia. A point such that if any plane passes through

81

it, the sum of the products of the masses of the constituent particles by their perpendicular distances from the plane (the sum of the *mass moments*) is zero. In common usage, centre of mass and *centre of gravity are synonymous since when the latter exists it coincides with the former.

centre of oscillation That point on a stationary compound pendulum at a distance $(k^2 + h^2)/h$ vertically below the *centre of suspension*. (The centre of suspension is where the axis of rotation intersects a perpendicular plane passing through the *centre of mass; k is the *radius of gyration about the centre of mass and h is the distance of the point of suspension from the centre of mass.)

If the pendulum axis is moved to the centre of oscillation, the point of suspension becomes the new centre of oscillation, and the period is unchanged; the two points are said to be *conjugate*.

centre of percussion If a compound pendulum is given an impulse at the *centre of oscillation P, there is no impulsive reaction at the centre of suspension S; in this case P is called the centre of percussion. If the pendulum is struck at P in the absence of any constraint at S, it begins to turn about S, i.e. S is the *instantaneous centre*. (A hammer is so shaped that P is in the head and S in the handle.) P and S are interchangeable and, in general, there are any number of pairs of points related to one another like S and P.

centre of pressure The point on a plane surface, immersed in a fluid, at which the resultant force on the surface may be taken to act. If the surface is horizontal in the liquid, the centre of pressure coincides with the *centre of gravity; otherwise it is below the centre of gravity but gets nearer to it as the liquid depth increases. (It is defined with respect to a plane area because the system of forces on a curved area is not always reducible to a single force.) *See* buoyancy.

centre of suspension *See* centre of oscillation.

centrifugal force *See* force.

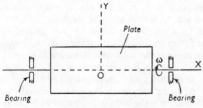

Centrifugal moment

centrifugal moment The total moment of the fictitious centrifugal forces of all the particles of a rotating body about any line. For example, suppose a flat plate coincides with the XY plane of Cartesian coordinates OXYZ. Let the plate rotate with uniform angular velocity ω about the axis OX and be

constrained by bearings as shown. The total centrifugal force is parallel to OY and equals $\omega^2 \Sigma my$. This is zero if the axis passes through the centre of gravity of the plate so that the radial forces on the bearings if any exist are then equal, parallel to OY, and oppositely directed. These constitute a couple that is equal to the total moment of the centrifugal forces

$$= \omega^2 \Sigma mxy = \omega^2 H$$

about the axis OZ where H is the *product of inertia with respect to OX and OY. The centrifugal moment about OZ for unit angular velocity of rotation about either OX or OY is equal to the product of inertia H. The centrifugal moment is zero if a body is rotating about one of its *principal axes* (*see* product of inertia) and these axes are therefore also called *free axes*.

The rotating system is said to be *balanced* if no radial forces act on the bearings.

centrifugation potentials An electric field generated in a colloidal solution when centrifuged. The *disperse phase is generally denser than the continuous phase and moves away from the rotational axis though retaining its charge. The smaller counterions remain behind. This gives rise to a p.d. between points in the solution at different distances from the axis. It can be used as a means of determining the *zeta-potential in colloidal solutions.

centrifuge A rapidly rotating bar or flywheel on a vertical axle from the rim of which a series of tubes are suspended so that their lower closed ends are free to tilt upwards and outwards. At high speed the centrifugal force outwards is far greater than gravity, and suspensions put in the tubes settle out much more quickly than in the ordinary way. It is also used for measuring sizes, shapes, and weights of particles. *See also* ultracentrifuge.

centring *See* centration.

centripetal force *See* force.

centrobaric *See* centre of gravity.

centrode *See* instantaneous centre.

centroid 1. *See* centre of mass. **2.** *Syn.* centre of area. The centroid of a surface is the centre of mass of a thin uniform sheet of matter with the same shape as the surface. **3.** *Syn.* centre of figure. The centroid of a volume is the centre of mass of a uniform solid with the same shape as the volume.

centrosymmetry Symmetry with respect to a point. Crystals that are centrosymmetrical have their faces arranged in parallel pairs that are alike or enantiomorphous in surface characteristics (*see* enantiomorphy). Centrosymmetry is equivalent to twofold rotation about an axis plus reflection across a plane perpendicular to the axis.

ceranuograph A device working on the principles of a radio receiver for recording the occurrence of a lightning discharge that is too far away for it to be seen or its thunder heard.

Cerenkov (Cherenkov), Pavel Alekseevich (*b.* 1904) Russian physicist. He was awarded the Nobel prize, jointly, in 1958 for the explanation of *Cerenkov radiation, which he first observed in 1934.

Cerenkov detector A sensitive device in which the *Cerenkov radiation, produced by a fast particle usually moving through water, is detected by one or more *photomultipliers, so allowing the path of the particle to be reconstructed.

Cerenkov radiation The bluish light emitted by a beam of high-energy charged particles passing through a transparent medium at a speed, v, that is greater than the speed of light c' in that medium. The light is emitted at all angles, θ, to the direction of motion of the particles, thus forming a conical wave front of angle 2θ, where $\cos \theta = c'/v$; $c' = c/n$, where n is the refractive index of the medium and c is the speed of light in a vacuum. The angle θ can be used to measure the speed of the particles, and hence the kinetic energy if the nature of the particles is known. Cerenkov radiation is caused by the rapid changes in the polarization of the medium as a charged particle moves through it.

cermet A contraction of *cer*amic and *met*al. A hard material formed by sintering a powdered mixture of a ceramic and a metal. It is used when resistance to high temperature, corrosion, and abrasion is required.

CERN The European Laboratory for Particle Physics, previously known as the European Organization for Nuclear Research. The research centre for high-energy physics situated at Geneva, Switzerland. In 1990 the member countries were: Austria, Belgium, Denmark, France, W. Germany, Greece, Holland, Italy, Norway, Portugal, Spain, Sweden, Switzerland, and the United Kingdom. *See* accelerator.

CerVit A proprietary glass-ceramic material that alters very little in size or shape when subjected to normal temperature changes. It has thus been used in telescope optics.

CGS system of units A system of units based on the centimetre as unit of length, the gram as unit of mass, and the second as unit of time. Although strictly applicable to mechanical measurements only, the system was extended to cover thermal measurements by the addition of the inconsistently defined *calorie. In extending the system further to enable electrical measurements to be carried out, it was recognized that a further fundamental quantity needed definition. This idea gave rise to two alternatives:
(a) the CGS-electromagnetic units, based on the *permeability of free space having unit size;
(b) the CGS-electrostatic units, based on the *permittivity of free space having unit size.
Because, as Maxwell proved, the product of the permeability and permittivity of free space is c^{-2}, where c is the speed of light, systems (a) and (b) are mutually exclusive.

The *Gaussian* (or symmetric) *system of units* uses units from system (a) to measure magnetic quantities and those from system (b) to measure electric quantities. In consequence, some equations of electromagnetic relationships contain c explicitly. All versions of the CGS system have now been superseded by *SI units for general scientific purposes; however, Gaussian units are still used in particle physics and in relativity. *See also* Heaviside–Lorentz units. *See* Conversion Factors (Appendix, Table 1).

Chadwick, Sir James (1891–1974) Brit. nuclear physicist, the first to realize the significance of observations of penetrating radiations derived in bombarding some light elements with alpha particles and thus to establish the identity of the neutron. Nobel prize for physics 1935.

Chadwick–Goldharber effect The disintegration of atomic nuclei by the action of *gamma-radiation.

chain-drum balance *See* balance.

chain reaction 1. Any self-sustaining chemical reaction in which the products of the reaction are necessary to keep the reaction going.
2. A series of nuclear transformations initiated by a single nuclear fission. For example, the fission of a ^{235}U nucleus is accompanied by the emission of one, two, or three neutrons, each of which is capable of causing further fission of ^{235}U nuclei.
When each transformation causes an average of one further transformation, the reaction is said to be *critical*. If the average number of further transformations is less than one, the reaction is *subcritical*; if it exceeds one, it is *supercritical*. *See also* critical mass.

chain-structure A crystalline arrangement in which the forces between atoms in one general direction are greater than those in any other direction, so that the atoms tend to form chains.

Chamberlain, Owen (*b.* 1920) Amer. physicist who with Emilio G. Segrè, first produced *antiprotons by bombarding copper with protons (of energy 6 GeV) in the *bevatron (1954). They shared a Nobel prize (1959) for this work.

Chandrasekhar, Subrahmanyan (*b.* 1910) Indian-born Amer. astrophysicist, who worked in England and America. His *Introduction to the Study of Stellar Structure* (1939) explained many problems in stellar evolution and set the so-called *Chandrasekhar limit*. *See* white dwarf.

Chandrasekhar limit *See* white dwarf.

channel 1. A route along which information can be sent in a computer or communications system. It may be, for example, a telephone link between two computers or a band of frequencies assigned for transmission and reception of radio or TV broadcasts.
2. In a *field-effect transistor, the region between *source and *drain, whose conductivity is modulated by the voltage applied to the *gate.

3. In a p-n-p junction *transistor, extension of the n-type *base region across the surface of the *collector to the edge of the die. This is caused by a layer of n-type material forming on the surface of the high-resistivity p-type collector due to the inevitable presence of positive changes at the interface with the passivating oxide and results in excessive leakage current flowing in the collector. It can be overcome by using a *channel-stopper.

channel spin The vector sum of the *spins of either the initial particles in a *nuclear reaction (*entrance channel spin*) or the resulting particles (*exit channel spin*).

channel stopper 1. In a p-n-p junction *transistor, a ring of low resistivity (i.e. highly doped) p-type material entirely surrounding the n-type *base region, to limit the extent of *channel formation.
2. In *MOS integrated circuits spurious *field-effect transistors are liable to be formed if the voltage on interconnections passing over adjacent drain regions exceeds the field *threshold voltage. The channel stopper is a region of highly doped material of the same type as the lightly doped substrate; it increases the field threshold voltage and inhibits the formation of unwanted conducting-channels.

chaos theory The theory of the unpredictable behaviour that can arise in systems obeying deterministic laws as a result of their sensitivity to variations in the initial conditions or to an excessive number of variables. Although deterministic laws enable the condition of a system to be predicted at any time in the future, to do so often depends on an ability to specify with great precision a set of parameters at an exactly specified moment. An example of chaos theory occurs in long-term weather forecasting. The meteorological laws may be well understood, but obtaining exact parameters to use with them may not be possible. In the *butterfly effect*, for example, it is postulated that the flap of a butterfly's wings can so upset the sensitive meteorological dynamics that an unforecast tornado may be set off by it. In fact, simulation of weather systems has played an important role in the development of chaos theory. Other fields in which this form of unpredictability occurs include turbulent fluid flow, oscillations in electric circuits, reaction kinetics in chemistry, and many situations in biology, astronomy, and economics.

Chapman, Sydney (1888–1970) Brit. mathematician. Professor at Imperial College, he made advances in the kinetic theory of gases, predicting (1917) thermal diffusion in gases. His other work was concerned with geomagnetism, in particular variations in the earth's magnetic field and the theory of magnetic storms.

Chappuis and Harker's constant volume gas thermometer (1900) A gas thermometer incorporating a barometer so that the absolute value of the pressure exerted by the gas (hydrogen) can be obtained directly in terms of the height of a column of

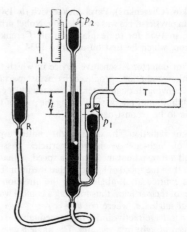
Constant volume gas thermometer

mercury. A litre bulb T (*see* diagram) of platinum-iridium, whose expansion is allowed for, is joined to the manometer by a capillary tube so that the dead space is small. The height of the mercury reservoir R is adjusted until the mercury just touches the fixed pointer p_1 and the barometer tube carrying its vernier is adjusted until the mercury just touches the pointer p_2 fixed in the vacuum. The vertical distance between p_1 and p_2 gives the total pressure, (H + h) cm of mercury, of the gas and is read off directly from the scale and vernier. For temperatures above 500 °C nitrogen is used in place of hydrogen since the latter diffuses through platinum at these temperatures.

characteristic A relation between two magnitudes that characterizes the behaviour of any device or apparatus. The relations are usually plotted in the form of a graph and are most frequently used for *transistors.

The most usual characteristics are: (1) the static characteristic showing, for example, *collector current against *base voltage, with all other voltages kept constant; (2) the dynamic characteristics showing the current from one electrode against the voltage of another under dynamic conditions, such as (a) a specified load impedance or (b) a sinusoidal voltage superimposed on the initial constant supply voltage.

characteristic curves 1. Families of graphs plotted for electronic devices, relating the currents obtained to the voltages applied and other operating conditions. The curves have value in designing uses for the devices. (*See* characteristic.)
2. A graph relating the optical density of a photographic film to the logarithm of the *exposure for a particular set of conditions.

characteristic equations *See* equations of state.

characteristic function One of a set of functions satisfying a particular equation with specified boundary conditions. For example, the functions $A_n \sin n\pi x/l$ (where n is any integer) all satisfy the differential equation of transverse wave motion on a uniform flexible string of length l with both ends fixed.

The corresponding values of some parameter associated with these solutions (such as the frequencies of vibration of the string in the example given) are known as "characteristic values". The general motion of the string can be regarded as a superposition of these characteristic functions, which are also known in acoustics as *normal modes of vibration*.

In *wave mechanics, characteristic functions are *well-behaved* (i.e. physically possible) solutions of the *Schrödinger wave equation for an atomic particle, and the corresponding values of the energy of the particle are known as *characteristic values*. If there is more than one solution of the differential equation corresponding to a particular characteristic value, the system is said to be *degenerate*.

Characteristic functions and values occur in *matrix mechanics also. In quantum mechanics particularly, characteristic values and functions are often called *eigenvalues* and *eigenfunctions* (from the German).

characteristic impedance 1. Of a uniform *transmission line. The limiting value that the impedance between the sending-end terminals approaches, under steady-state conditions, as the length of the line is indefinitely increased. In complex form it is given by:
$$Z_0 = \sqrt{(R + i\omega L)/(G + i\omega C)},$$
where Z_0 = characteristic impedance when R = total series resistance of the line, L = total series inductance of the line, G = parallel conductance of the line, C = parallel capacitance of the line, $\omega = 2\pi \times$ frequency. (Note, R, L, G, and C must all be for the same length of line, e.g. for unit length.) As $\omega \to 0$, $Z_0 \to \sqrt{R/G}$, and as $\omega \to \infty$, $Z_0 \to \sqrt{L/C}$, and at both limits, Z_0 is in the nature of a pure resistance. The second of these is a very useful approximation at high (e.g. radio) frequencies. *Compare* surge impedence.

2. Of a *quadripole. A special case of *iterative impedance.

characteristic temperature See Debye theory of specific heat capacities.

characteristic value See characteristic function.

characteristic X-radiation A series of discrete-wavelength X-rays characteristic of the element emitting them. See X-rays; X-ray spectrum.

charge Symbol: Q. A property of some *elementary particles that causes them to exert forces on one another. The natural unit of negative charge is that possessed by the electron and the proton has an equal amount of positive charge. The terms positive and negative are used because a body that carries equal quantities of each type of charge is electrically neutral. A convention adopted in early work on electrostatics led to the definition of the electron

charge as negative. Like charges repel and unlike charges attract each other. The charge of a body or region arises as a result of an excess or deficit of electrons with respect to protons. Charge is the integral of electric current with respect to time and is measured in coulombs. The electron has a charge of $1.602\ 177\ 33 \times 10^{-19}$ coulomb.

charge conjugation parity *Syn.* C-parity. Symbol: C. A quantum number associated with those elementary particles (such as π^0 and η) that have zero charge, baryon number, and strangeness. It is conserved in *strong and *electromagnetic interactions. In simple terms, it shows whether the *wave function describing the particle is unchanged ($C = +1$) or changes sign ($C = -1$) when the particle is replaced by its *antiparticle. *See also* CP invariance; CPT theorem.

charge-coupled device See charge-transfer device.

charge density 1. *Volume charge density*. Symbol: ρ. The electric charge per unit volume of a medium or body. It is measured in coulombs per metre cubed.

2. *Surface charge density*. Symbol: σ. The electric charge per unit area of a surface. It is measured in coulombs per metre squared.

charge exchange The transfer of electric charge between two colliding particles, such as the transfer of an electron from a neutral atom or molecule to an ion.

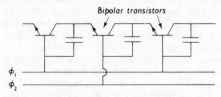

a Bipolar bucket-brigade device

charge-transfer device A *semiconductor device in which discrete packets of charge are transferred from one location to the next. Such devices can be used for the short-term storage of charge in a particular location, provided that the storage time is short compared with the *recombination time in the material.

Applications of charge-transfer devices include short-term memory systems, *shift registers, or imaging systems. Information is usually only available for serial access.

Several different types of charge-transfer device exist, the main classifications being *bucket-brigade* devices, *charge-coupled* devices, and *surface-charge transistors*.

1. *Bucket-brigade devices*. These devices consist of a number of capacitors linked by a series of switches that in practice consist of *bipolar or *field-effect (MOS) transistors (*see* Figs *a* and *b*). These circuits consist of discrete components but are invariably manufactured as *integrated circuits. *Clock pulses

Charles, Jaques A. C.

are applied to close the switches, a two-phase system (ϕ_1 and ϕ_2) being used. As each switch is closed, charge is transferred from one capacitor to the next.

Bucket-brigade devices are frequently used as *delay lines in both *digital and *analogue systems, since the amount of charge stored may vary continuously from zero to the limit set by the magnitude of the capacitance and the operating voltage. The capacitors are provided in practice by using the collector-base capacitance or drain-gate capacitance of the transistors used in the circuit.

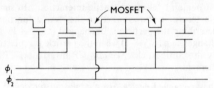

MOSFET

ϕ_1
ϕ_2

b　MOS (metal-oxide silicon) bucket-brigade device

2. *Charge-coupled devices* (CCD). Charge-coupled devices consist of arrays of MOS capacitors (i.e. MOS transistors without source or drain diffusions). (*See* Fig. *c*.) If a suitable potential is applied to the gate electrodes of such structures, a depletion region will be formed in the semiconductor beneath the gates. An inverted channel cannot be formed since no source of carriers of the opposite type exists (because there are no source or drain regions). If minority carriers are injected into the system by some means (e.g. a forward-biased p-n junction or light-induced electron-hole pairs) they will enter the depletion region and be held at the surface of the semiconductor until they recombine. If the adjacent capacitor has a more depleted region than the first, and the physical separation is sufficiently small, the charge stored in the first capacitor will be transferred to the second and so on. Packets of charge may be transferred along a line of capacitors by suitably adjusting the gate potentials and hence the depletion layer. Some loss of charge occurs at each transfer, so that regenerative devices must be included at regular intervals to restore the original charge packet (e.g. every 8 or 16 stages).

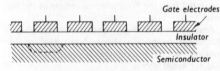

Gate electrodes

Insulator

Semiconductor

c　Charge-coupled device

Charge-coupled devices may be used as shift registers or in imaging systems. In the latter application the array may be exposed to a light image and the capacitors become charged according to the light-intensity distribution. This charge pattern may then be systematically shifted out of the array to provide a suitable video signal.

3. *Surface-charge transistor* (SCT). The surface-charge transistor is a modification of the charge-coupled device. The essential difference between the SCT and CCD is that charge-transfer does not rely on the physical proximity of the electrodes. An extra electrode (the transfer gate) is provided between each storage location, overlapping the storage electrodes (source/receiver electrodes) that, when a suitable potential is applied to it, provides a transfer path between the storage locations. A single SCT is shown in Fig. *d*. The surface-charge transistor requires a two-level electrode system in manufacture, but allows more efficient transfer of charge than the CCD. CCDs and SCTs exist only as integrated circuits, and are not formed from discrete components.

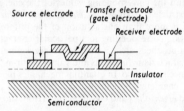

Source electrode　Transfer electrode (gate electrode)

Receiver electrode

Insulator

Semiconductor

d　Surface-charge transistor

Charles, Jaques A. C. (1746–1823) French physicist. He was responsible for the first balloon ascents with hydrogen-filled balloons (1783), the preceding ascents having been carried out with hot air. He did not publish the results of his experiments on the expansion of gases and hence the honour of the discovery of *Charles's law is more properly given to *Gay-Lussac, who published his work in 1802.

Charles's law (Also known as Gay-Lussac's law.) All gases and unsaturated vapours have the same mean thermal expansivity at constant pressure over the range 0 °C to 100 °C. The law in fact applies approximately over a much wider range of temperatures but is not exact except at very low pressures.

The law is often stated incorrectly in the form of a linear relationship between volume and temperature, and this is sometimes extended to a direct proportionality between volume and thermodynamic temperature. Such statements cause difficulty because of failure to specify the temperature scale used and to distinguish between real and *ideal gases.

charm A *quantum number used in the theory of *quarks and *hadrons. The charm quark (c) has charm +1 and its antiquark c̄ has charm –1. All other quark flavours have charm 0. The charm of a particle is the sum of the number of charmed quarks minus the number of anticharmed quarks.

chart-recording instrument *See* graphic instrument.

Chattock gauge A form of *micromanometer. A and B are two water vessels attached to the two almost equal sources of pressure whose difference is required. The central vessel C is completely filled with an oil lighter than and not miscible with water. An excess of pressure in B will raise a bubble of water into the oil on the end of the tube D. The instrument is attached to a rigid framework that can be tilted slightly from the horizontal by means of a micrometer screw M. Initially the vessels A and B are open to the atmosphere and the meniscus at D is adjusted so that its image is at a suitable position on the eyepiece graticule of a low-power observing micro-

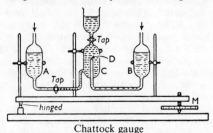

Chattock gauge

scope. The small pressure difference to be measured is applied to A and B and a displacement of the meniscus occurs that is reduced to zero by tilting the framework by the micrometer screw. Displacement of the micrometer screw is directly proportional to the applied pressure difference.

chemical constant The constant of integration i, occurring in the expression for the *vapour pressure of a substance. It is determined by evaluating all other terms in this equation. Stern and Tetrode applying statistical mechanics to the equilibrium between a solid and its monatomic vapour obtain:

$$i = \log \frac{(2\pi m)^{3/2} k^{5/2}}{h^3},$$

where m is the mass of an atom, k is the Boltzmann constant and h is the Planck constant. This gives $i = 10.17 + 1.5 \log_e M$, where M is the molecular weight, and the value thus given agrees well with the constant in the vapour-pressure equation.

chemical dosimetry *See* dosimetry.

chemical energy Energy released when chemical bonds are formed. Usually, in a chemical reaction, some bonds are broken while others are formed. Energy is given out if, in total, the new bonds are stronger than the old.

chemical hygrometer *Syn.* absorption hygrometer. A *hygrometer that measures the actual mass of water vapour present in a given volume of air together with the mass to saturate the same volume at the same temperature, the ratio of these masses giving the relative *humidity of the atmosphere. Water flowing from the aspirator D causes a known volume of air to be drawn through E, any moisture contained in the air being removed in the drying tubes A, B, and C, which contain phosphorus pentoxide. Tubes A and B are then disconnected at H and their increase in weight is due to the amount of water vapour absorbed. The tubes are replaced after refilling D,

and the tube RL containing broken pumice and water is joined to the system at E. Any air passing through RL becomes saturated with water vapour so that the increase in weight of the drying tubes A and B, when the same volume of air is drawn through the system, gives the mass of water to saturate that volume at the same temperature as before.

chemical potential Symbol: μ. A property of substances used in chemical thermodynamics defined by

$$\mu_A = \left(\frac{\partial G}{\partial n_A}\right)_{T, P, \ldots},$$

where G is the *Gibbs function and n_A is the amount of substance (in moles) of the component A. The temperature, pressure, and amounts of other substances are constant and μ_A measures how G depends on n_A. Two phases are in equilibrium when their chemical potentials are equal. *See also* activity.

chemical shift A change in the position of a spectrum peak resulting from a small change in energy of a given state caused by a chemical effect. Thus, in the *Mossbauer effect there is a difference between the energy of a given state for nuclei in the pure element and the energy when the element is combined with other elements in a compound. This appears as a shift in the spectrum. For the Mossbauer effect it is sometimes called the *isomeric shift*. A similar effect occurs in *nuclear magnetic resonance where the energy difference between two nuclear states of different spin depends, to a small extent, on the chemical compound used. The term is also applied to changes in energy of states of orbital electrons in inner shells as determined by X-ray *photoelectron spectroscopy.

chemiluminescence *See* luminescence.

chemisorption Atoms of gas, held to a solid by covalent interactions, are said to be chemisorbed, and need high temperatures and pumping to remove them. For example, oxygen adsorbed by charcoal can only be removed by pumping at high temperature and comes off as oxides of carbon. *Compare* physical adsorption.

chemosphere *See* thermosphere.

Cherenkov, Pavel Alekseevich *See* Cerenkov.

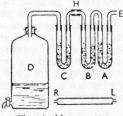

Chemical hygrometer

Cherwell, Viscount *See* Lindemann, Frederick Alexander.

cheval-vapeur *Syn.* force de cheval; pferdekraft; metric horsepower. A metric unit of horsepower equal to 75 metre-kilograms force per second. It is equivalent to 735.7 watts.

chief ray The central or representative ray of a pencil of rays from an object point on or off the axis, to the centre of the entrance pupil. It therefore passes through the centre of the actual aperture stop and the centre of the exit pupil (aberration for the stop position may introduce complications). The intersection of a chief ray with an image screen determines the position of the image or the centre of the blur circle, if the screen is not at the focus position.

Child's law Under space-charge-limited conditions, where the voltage drop V across an electronic tube is less than that which would give the maximum electron emission possible, the anode current follows the law:

$$J = 2.331 \times 10^{-6} \ V^{3/2}/s^2,$$

for flat surfaces, where s is the cathode-to-anode distance in m, and J is the current density in A m^{-2}.

chip A small piece of a single crystal of *semiconductor material containing either a single electronic component or device or an *integrated circuit. Chips are commonly sliced from a large single crystal or semiconductor – a *wafer* – that is used as the substrate on which the integrated circuits, etc., are fabricated. The chips are then packaged in various ways, e.g. as a *dual in-line package (DIP).

chirality In general, the property of existing in left- and right-handed forms, so that one form is a mirror image of the other. In a molecule, an atom attached to four different groups is a *chiral centre*; molecules that have such atoms can show *optical activity. The idea of chirality is extended to elementary particles. For any spin ½ (Dirac) particle u, two *chiral states*, u_L and u_R (known as 'left-handed' and 'right-handed' states), can be defined. For a massless particle, such as the *neutrino, the left-handed and right-handed chiral states correspond to the particle having helicity –1 and helicity +1 respectively, i.e. to the particle spin being oriented opposite to or along the particle's direction of motion. For a particle with non-zero mass, such as the *electron, this is true only in the limit of very high energy. The chiral states are useful in theories of the *weak interaction and in *electroweak theories. Weak interactions involving the charged W$^\pm$ bosons can be expressed purely in terms of left-handed particles and right-handed antiparticles; photon and Z^0 interactions involve both chiral states. Only left-handed neutrinos and right-handed antineutrinos are found to exist in nature, i.e. the neutrino always has helicity –1 and the antineutrino always has helicity +1.

Chladni, Ernst F. F. (1756–1827) German physicist. One of the founders of the science of acoustics. He discovered the longitudinal vibrations in a rod, and investigated the vibrations in flat plates (*see* Chladni's plates). He obtained the speed of sound in various gases by using the change of pitch of an organ-pipe filled with the gas; and he invented the euphonium, a musical instrument of the saxhorn family.

Chladni's plate An experimental method of demonstrating displacement nodal lines in transversely vibrating plates. Chladni's work preceded the theoretical work on the subject but his simple method gives such good results that it is still widely used. Chladni fixed his plate to a central pillar thus making a displacement node at the centre of the plate. He held the plate at one point along its edge and bowed it at another. By sprinkling fine sand on to it the nodal lines are rendered visible since the sand collects on these lines. A great variety of patterns can be obtained by bowing and clamping at different points. Chladni used both square and circular plates of brass and glass. Lissajous showed, with a Y-shaped interference tube, that the phase of vibration changes on crossing a nodal line. Many different methods of excitation have been used to produce Chladni's figures. In the case of steel plates an electromagnet fed from an oscillator can be employed. At high frequencies small plates have been made to vibrate by touching them with a magnetostrictive rod. Another method of excitation is to touch the plate with the edge of a piece of solid carbon dioxide. The study of vibrating plates is important in view of their applications in telephone diaphragms and piezoelectric oscillators.

choke 1. An *inductor that presents a relatively high impedance to alternating current. It is often used in audiofrequency and radio-frequency circuits, to impede the audiofrequency or radio-frequency signals, or to smooth the output of a rectifying circuit.
2. A groove cut into the metal surface of a *waveguide, approximately one-quarter of a wavelength deep, to prevent the escape of microwave energy.

chroma The attribute of a visual sensation by which the amount of pure colour can be judged, irrespective of the amount of white or grey present.

chromatic aberration *Syn.* chromatism. Since the refractive index of a refracting medium depends on the wavelength (*see* dispersion), the focal length of a lens varies according to the colour of the incident light. The image of a point source of white light is therefore blurred and appears coloured; tinged with a surround of blue or violet at the focus for red, and with red at the blue focus. At an intermediate position a white circle AB occurs – the *circle of least confusion. For standards of comparison, the colours corresponding with the C (red) and F (blue-green) lines of hydrogen are chosen. The distance between the foci for these colours is the *longitudinal chromatic aberration*. The reciprocals of the principal focal lengths are the powers; the difference of these powers is commonly referred to as the *chromatic aberration*. For a thin lens, the last-mentioned

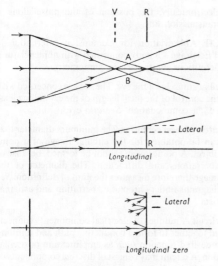

Chromatic aberration

chromatic difference of magnification Chromatic effects produced by the sizes of images being different for different colours (sometimes referred to as lateral chromatism). *See* chromatic aberration.

chromaticity An objective description of the *colour quality* of a visual stimulus, such as a coloured light or a surface, irrespective of its *luminance (see colour system).* Chromaticity and luminance completely specify a colour stimulus. The colour quality is defined in terms of its *chromaticity coordinates.* These three coordinates, x, y, z are equal to the ratio of each of the *tristimulus values* of a light to their sum. The tristimulus values, X, Y, Z, are the amounts of the three reference or matching stimuli required to match exactly the light under consideration in a given trichromatic system. Hence

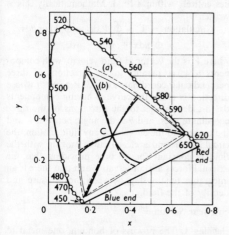

Chromaticity diagram and two colour triangles. The position of the spectral colours

chromatic aberration is ωP, where ω is the dispersive power of the glass and P is the power of the lens for yellow (sodium D) light.

The sizes of the images for different colours will be different; the difference in size is called the *lateral chromatic aberration* for the object considered. For thick systems, the principal points for the different colours will occupy different positions even if the longitudinal aberration has been reduced to zero, so that lateral aberration may still be present. Lateral aberration is sometimes referred to as *chromatic difference of magnification* or lateral chromatism. When chromatic aberrations have been corrected for two colours, owing to nonlinearity of dispersion there is a residual chromatic aberration referred to as the *secondary spectrum. See* aberrations of optical systems.

$$x = \frac{X}{X + Y + Z} \text{ (redness)}$$

$$y = \frac{Y}{X + Y + Z} \text{ (greenness)}$$

$$z = \frac{Z}{X + Y + Z} \text{ (blueness)}$$

Thus all colour can be reduced to a common function since $x + y + z = 1$. When x, y, and z all approximately equal $\frac{1}{3}$, the colour is almost white.

A *chromaticity diagram* is obtained when x is plotted against y, the graph being horseshoe-shaped and the locus of all monochromatic colours (*see* diagram). The straight line joining the ends is the locus of pure purple, i.e. combinations of the extreme red and blue monochromatic colours. All colours lie within these loci. White lies at the point, C, (the white point), having coordinates $x = y = \frac{1}{3}$. Any colour lies on the line joining a spectral colour (on the horseshoe) to C. The wavelength of the spectral colour used is the *dominant wavelength* for the colour under consideration. The position of this colour on the line depends on the proportions of the spectral colour and white required to obtain the colour. The *excitation purity* of the colour is the ratio of the distances of the colour and the spectral colour from the white spot. The dominant wavelength is roughly equivalent to *hue, and excitation purity to *saturation. Dominant wavelength, excitation purity, and *luminance are frequently used as the set of coordinates in an objective *colour system. The colour resulting from the addition of two colours lies on the straight line joining them.

One or more *colour triangles* can be drawn on the chromaticity diagram. They represent the entire range of chromaticities that can be obtained from a combination, by a *subtractive process, of three dyes of the *primary colours cyan (blue-green),

89

magenta (red-blue), and yellow. Triangle (a) shows the colours obtained by a combination of the three dyes above; triangle (b) is obtained by mixing the three dyes and a white dye.

chromaticity diagram *See* chromaticity.

chromatic resolving power Of a prism. The ratio:
$$\lambda/\Delta\lambda \ (= b\Delta n/\Delta\lambda),$$
where λ is wavelength, $\Delta\lambda$ the limiting difference of wavelength detectable by the use of a prism of width of base b with refractive index n at wavelength λ. *See* resolving power.

chromatic scale *See* musical scale.

chromatic sensitivity Of the eye. The smallest change of wavelength that produces a just detectable difference of hue. Across the visible spectrum there are three or four maxima and minima of sensitivity, the sensitivity decreasing rapidly at the extremes.

chromatism *See* chromatic aberration.

chromatron A type of *colour picture tube, used in *colour television, with four screens. The colour phosphors are arranged as vertical stripes on the screen.

chrominance signal *See* colour television.

chromosphere A layer of a few thousand kilometres thickness in the sun between the *photosphere and the *corona. It is visible during a total eclipse. At the lower levels the temperature is about 4000 K and the pressure about 10^{-6} atmosphere. At the top the temperature is about 40 000 or 50 000 K.

Because of the low density, ions and electrons only recombine very slowly, thus the absorption of radiation from below causes the amount of ionization to rise far above the equilibrium values at these temperatures. Elements in the upper chromosphere are generally very highly ionized, as is shown by examination of the *flash spectrum.

chronograph A mechanism for recording time signals on either a paper tape or a cylindrical drum moved by a clock mechanism. It is used for the precision measurement of time intervals. *See also* clocks.

chronometer An exact type of portable timepiece, used for example in the determination of longitude on board ship. Its rate of motion is controlled by a balance wheel and hair spring, but there are several technical differences between a chronometer and a watch, making the former the more accurate. *See also* clocks.

chronon A hypothetical particle of time; the time taken for a photon to traverse the diameter of an electron. It is approximately equal to 10^{-24} seconds.

chronoscope An electronic instrument for measuring very short time intervals.

chronotron A device for measuring the time interval between events. Each event initiates a pulse, and the time between events is measured by determining,

electronically, the position of the pulse along a transmission line.

CIE Abbreviation for Commision Internationale d'Éclairage – a body concerned with nomenclature, definitions, units, etc., in photometry.

ciliary muscle Of the eye. The muscle concerned with the control of the focal length of the crystalline lens, i.e. *accommodation. *See also* eye.

circle of least confusion The minimum diameter that can be obtained for the circular image of a point object. A point image cannot be formed, because of aberrations and diffraction. The diameter of the image therefore measures the limit of definition. *See* diagrams under chromatic aberration and astigmatism.

circuit A number of electrical components connected together to form a conducting path and fulfilling a desired function such as amplification or oscillation. A circuit may consist of discrete components or may be an *integrated circuit.

If the components form a continuous closed path through which a current can circulate, the circuit is said to be *closed*; when the circuit is broken, as by a switch, it is said to be *open*. *See also* magnetic circuit.

circuital *See* curl.

circuit-breaker A device for making and breaking an electric circuit under normal or under fault conditions. A circuit-breaker usually breaks the circuit automatically, the making operation being by hand. *See* contactor; switch; tripping device.

circular mil A unit of area, equal to the area of a circle of which the diameter is one-thousandth of an inch.

circular polarized radiation *See* polarization.

circulation The flow round a closed curve (C) that lies entirely within a fluid. Mathematically this is written
$$\text{Circulation} = \Gamma = \oint_C V.\mathrm{d}s = \oint_C V\cos\theta\mathrm{d}s$$
$$= \oint_C u\mathrm{d}x + v\mathrm{d}y + w\mathrm{d}z,$$
where V is the vector of fluid velocity with components u, v, w, parallel to the directions of three perpendicular coordinate axes; $\mathrm{d}s$ is a vector whose direction is along the curve C having components $\mathrm{d}x$, $\mathrm{d}y$, $\mathrm{d}z$, parallel to the directions of the three coordinate axes; and θ is the angle between V and the tangent to the curve. For any fluid motion, the circulation round a closed curve moving with the fluid is constant for all time providing that the external forces are conservative and that there is an integrable relationship between the pressure and density of the fluid.

civil year *See* time.

cladding 1. The process of bonding one metal to another to prevent corrosion of one of the metals. It is used in *nuclear reactors to prevent corrosion of a *fuel element by the *coolant and the escape of fission products.

2. *See* fibre-optics systems.

clamping diode *See* catching diode.

Clapeyron, B. P. E. (1799–1864) French physicist who developed and publicized *Carnot's work in thermodynamics. His form of the first latent-heat equation was improved by *Clausius. *See* Clausius–Clapeyron equation.

Clark cell A voltaic cell, formerly adopted as a standard of e.m.f. It consists of a mercury electrode surrounded by a paste of mercury sulphate, the negative electrode being a rod of pure zinc in a saturated solution of zinc sulphate. Its e.m.f. was defined to be 1.4345 volts at 15 °C. It has now been superseded by the *Weston standard cell.

class A amplifier A *linear amplifier operated under such conditions that the output current flows over the whole of the input cycle. The output wave shape is essentially a replica of the input wave shape. Class A amplifiers have low distortion and low efficiency.

class AB amplifier A *linear amplifier operated so that, in general, the output current flows for more than half but less than the whole of the input cycle. Class AB amplifiers tend to operate as *class A for low-input signal levels and *class B at high-input signal levels.

class B amplifier A *linear amplifier operated to produce a half-wave rectified output, i.e. the output current is cut off at zero input signal. In order to duplicate the input waveform successfully, two transistors are required each conducting for one half of the input cycle. Class B amplifiers have high efficiency but suffer from crossover distortion.

class C amplifier An *amplifier in which output current flows for less than half of the input cycle. The output waveform is not a replica of the input waveform for all amplitudes and class C amplifiers are therefore nonlinear. Class C amplifiers are more efficient than other types, but introduce more distortion.

class D amplifier An *amplifier operating by means of pulse-width modulation. (*See* pulse modulation.) The input signal is used to modulate a square wave with respect to its *mark-space ratio. The modulated square wave then operates *push-pull switches (*see* push-pull operation) so that one switch operates when the input is high, and the other when the input is low. The resultant current in the output load is proportional to the mark-space ratio and hence the input signal. Class D amplifiers are theoretically highly efficient, but to avoid distortion the switches must be operated faster than is generally practicable.

classical physics The long-established part of physics, excluding *relativity and *quantum theories.

clathrate structure A crystal structure in which one kind of molecule forms a cagework that imprisons another kind of molecule, so that the latter cannot escape except by fusion, solution, or drastic mechanical treatment of the substance as a whole.

Claude, Georges (1870–1960) French chemist and physicist. He devised commercially successful methods for the liquefaction of air and other gases and a rectification process for separating the constituents of the air.

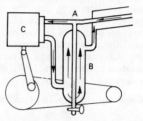

Claude process of liquefying air

Claude process for liquefying air A process that depends on the cooling produced when a gas undergoes adiabatic expansion and performs external work. The gas under pressure divides at A (*see* diagram) into two parts, one part going to the expansion chamber C where it performs external work (used in the compressor) and cools during the adiabatic expansion. This cooled gas cools the other part from A in the interchanger B and as this gas is under pressure it eventually liquefies. The early difficulty of lubrication of moving parts at these low temperatures was overcome first by the use of ether and later by the use of the liquid air itself as lubricant.

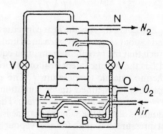

Claude's process for separating air into constituents

Claude's process of rectification A process used to separate air into its constituents. The diagram represents the process schematically. In steady working conditions, bath A contains nearly pure oxygen. Air entering at 20–30 atmospheres pressure circulates in tubes immersed in this bath (much simplified in the diagram) and an oxygen-rich fraction condenses in B while a nitrogen-rich liquid collects in C, under the combined action of pressure and cooling. The oxygen liquefies more readily because of its higher boiling-point. Valves VV control the supply of liquids descending the column. Oxygen progressively replaces nitrogen in the liquid stream and almost pure nitrogen gas is drawn off at the top (N) and pure oxygen from above the liquid surface (at O).

Clausius, Rudolf J. (1822–1888) German physicist. He was responsible for the first adequate treatment of the *kinetic theory of matter (1857), giving an expression for the mean free path of a molecule in 1858. He introduced the *virial in 1870. He played an important part in laying the foundations of thermodynamics, improving on the work of *Clapeyron on the *Carnot cycle (1850), giving a formulation of the second law of *thermodynamics; he introduced the concept of entropy (1854) and later, its name. Clausius also suggested the hypothesis of dissociation in electrolytes (1857), afterwards more fully developed by *Arrhenius.

Clausius–Clapeyron equation The equation:

$$\frac{dp}{dT} = \frac{L}{T(v_2 - v_1)},$$

where v_1 and v_2 are the specific volumes of the substance in two different phases, L is the specific latent heat for the change from one of these phases to the other at temperature T and pressure p. The equation gives the slopes of the boundaries between the regions representing solid, liquid, and vapour phases on a p–T diagram. Thus, it gives the variations with pressure of the freezing, boiling, and sublimation points. *See also* triple point.

Clausius's equation The equation:

$$c_2 - c_1 = T \frac{d}{dT}\left(\frac{L}{T}\right),$$

where c_1 and c_2 are the specific heat capacities of the liquid and vapour respectively, measured under conditions such that the two phases remain in equilibrium, and L is the specific latent heat of vaporization at the thermodynamic temperature T. The specific heat capacities are defined under conditions such that the two phases remain in equilibrium. The value for the liquid is almost the same as that measured at constant pressure but that for the vapour differs greatly and may even be negative for some substances, including steam.

Clausius's equation of state For a real gas,

$$\left(p + \frac{a}{T(V + c)^2}\right)(V - b) = RT.$$

This is a purely empirical equation. *See* equations of state.

Clausius's virial law The mean kinetic energy of a system is equal to its *virial equation. It is a general

theorem that can be used to obtain, for example, the *equation of state for a solid.

cleavage The easy separation of a crystal into two parts owing to a relative weakness of bonds along a particular direction. Normal to this direction the two parts of the crystal show clean good surfaces parallel to the *cleavage plane*.

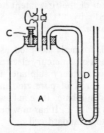

Clément and Desormes' method for determining γ

Clément and Desormes' method for determining γ The ratio (γ) of the principal specific heat capacities of a gas is determined by allowing an adiabatic expansion of the gas from a known pressure (p_1) to atmospheric pressure B. The gas is then allowed to attain room temperature at constant volume, the final pressure p_2 being noted. Air is pumped into a large vessel A (*see* diagram) through a valve C and left to stand until the levels in the manometer D are steady at a difference h_1. The stopcock is opened for a while and then closed. The pressure falls to atmospheric but slowly rises as the gas warms up to room temperature giving a difference of levels h_2. Then

$$\gamma = (\log p_1 - \log B)/(\log p_1 - \log p_2),$$

where $p_1 = B + h_1$ and $p_2 = B + h_2$. If both h_1 and h_2 are small compared with B, the formula simplifies to: $\gamma = h_1/h_1 - h_2$. The chief error in the method is that oscillations in pressure occur when the tap is opened and the pressure may not be atmospheric when the tap is closed. This difficulty is removed in the methods of *Partington and of *Lummer and Pringsheim.

clepsydra A water clock used by the Greeks and Romans, in which time was measured by the rate of flow through small holes at the bottom of a globe.

Clinical thermometer

clinical thermometer A mercury in glass thermometer used for the accurate determination of the temperature of the human body and graduated from 35–46 °C (95–115 °F). The mercury in the thin-walled bulb B expands past the constriction S into the capillary tube. On removing the thermometer from the patient, the mercury beyond S cannot recede into the bulb because of the constriction and so the meniscus remains, except for a slight contraction of the small volume of the mercury thread, in the same position as before removal. The reading can thus be taken at leisure, the fine mercury thread being magnified by shaping the glass round the capillary. The thermometer is re-set by shaking until the mercury thread is forced past the constriction into the bulb.

clinographic projection A method of crystal drawing in which the eye, or point of projection, is at infinity and is above and somewhat to the right of the crystal.

clinometer An instrument for measuring angles of inclination by reference to a plumb-line, spirit level, etc.

clock frequency *Syn.* clock rate. The master frequency delivered by an electronic device, called a *clock*, at fixed intervals to synchronize operations in a *computer. The clock, generally a stable oscillator, generates an extremely regular series of fixed-width pulses – *clock pulses*. The reciprocal of the pulse repetition rate is the clock frequency, normally given in megahertz. Because of its constant rate, the clock signal can be used to synchronize the operations of related pieces of computer equipment so that events take place in sequence at fixed times. For example, the clock signal is used to initiate actions within a *logic circuit and to synchronize the activities of a number of such circuits.

clock pulses *See* clock frequency.

clock rate *See* clock frequency.

clocks The earliest clocks were based on processes that take place at a constant rate, such as the apparent movement of the sun, the rate at which a candle burns, or the fall of sand in an hourglass. More advanced devices use periodic processes of constant frequency.

(1) *Pendulum clock.* The pendulum clock employs Galileo's discovery that the period of a pendulum is a function only of its length and not its mass or initial displacement. Each swing should take place under the same conditions as its predecessor; good clocks have *compensated pendulums and are kept in airtight cases under constant temperature conditions. The highest precision pendulum clocks are accurate to about 0.01 seconds per day.

(2) *Crystal clock.* For precise scientific measurements a higher degree of accuracy is obtained from the crystal clock in which a quartz crystal is made to oscillate at about 100 000 hertz by *electrostriction. Such clocks are accurate to about 0.001 seconds per day.

(3) *Atomic clock.* Even greater accuracy can be obtained from atomic clocks in which the periodic process is a molecular or atomic event associated with a particular spectral line. In the ammonia molecule clock the umbrella inversion of the ammonia molecule provides the basic oscillation. In the *ground state the molecule has the shape of a pyramid with the nitrogen atom at the apex and the three hydrogen atoms at the corners of the triangular base. If energy is supplied, the molecule can exist in a vibrationally excited state in which the nitrogen

atom passes through the plane of the hydrogen atoms to an equivalent position on the opposite side. This oscillation has a frequency of 23 870 hertz and ammonia therefore strongly absorbs electromagnetic radiation of this frequency, which is in the microwave region. Early forms of ammonia clock have a quartz oscillator supplying energy to ammonia gas at this frequency. When the oscillator supply varies from this value, the energy is no longer absorbed and is used in a feedback circuit to correct the crystal oscillator.

An ammonia *maser can also be used as a frequency standard. A beam of ammonia molecules from an oven passes through an electric field. In the ground state the molecules have a dipole moment and are deflected away by the field whereas the excited molecules have no dipole moment (on average the nitrogen atom is in the same plane as the hydrogen atoms) and are therefore undeflected by the field. These molecules then enter a *microwave cavity to which the *resonant frequency of the oscillation is supplied. The microwave signal is amplified by *stimulated emission and the oscillations of the amplifier can be used as a frequency.

(4) *Caesium clock.* A similar device is the caesium clock in which the frequency is defined by the energy difference between two different states of the caesium nucleus in a magnetic field (*see* nuclear magnetic resonance). A beam of caesium atoms is split by a magnetic field as in the *Stern–Gerlach experiment. Atoms in the lower energy state are directed into a cavity and fed with radio-frequency radiation at a frequency of 9 192 631 770 hertz, which corresponds to the energy difference between the two states. Some caesium atoms are raised to the higher energy state by absorption of this radiation and the mixture of caesium atoms is analysed by a further magnetic field. A signal from the atom detector is fed back to the r.f. oscillator supply to prevent it from drifting from the resonant frequency. In this way the supply is locked to the spectral line frequency and the accuracy is better than one part in 10^{13}. The caesium clock is used in the international (SI unit) definition of the *second.

See also synchronous clock; clock frequency.

closed circuit *See* circuit.

close-packed structure A crystalline arrangement in which similar atoms, supposed spherical, are packed as economically of space as is possible. The two common arrangements are the *face-centred cubic and hexagonal close-packed structures (*see* crystal systems), but combinations of these also occur. The essential condition is that each atom shall be symmetrically surrounded by twelve others.

cloud chamber *Syn.* Wilson cloud chamber. An apparatus for making visible the tracks of ionizing particles. It consists of a gas-filled chamber C containing a saturated vapour, which can be made supersaturated by the sudden cooling produced in an adiabatic expansion. The excess moisture is deposited in drops on the trail of ions left behind by

cloud-ion chamber

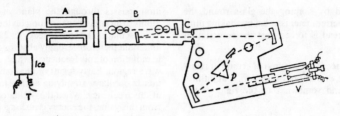

Coblentz apparatus for determining C_2 (Displacement law)

the passage of a particle. Adiabatic expansion can be produced by having a well-fitting movable piston P as the base of the chamber; alternatively the piston may be replaced by a rubber diaphragm. *See* diffusion cloud chamber.

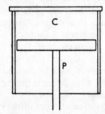

Cloud chamber

cloud-ion chamber An instrument combining the functions of an ionization chamber (utilizing free-electron collection) and the Wilson *cloud chamber, in the same gas volume. Isoamyl alcohol and argon mixture can be used to satisfy the necessary conditions for the gas.

Clusius column *See* isotope separation.

cluster An aggregation of stars that move together through space. *Galactic* or *open clusters* are loosely packed groups of young stars found principally in the plane of the spiral arms of the Milky Way. *Globular clusters* are densely packed groups of old stars found in the spherical halo around the nucleus of the Milky Way. About 120 have been identified, each containing about a million stars, although the existence of about 500 has been predicted.

CMOS *See* complementary transistors.

coated lenses *See* blooming of lenses.

coaxial cable *Syn.* coax. A cable that consists of two or more coaxial cylindrical conductors, insulated from each other. The outermost conductor is often earthed. Coaxial cables do not produce external fields and are not affected by them. They are thus frequently used for transmission of high-frequency signals, as in television, radio, etc.

cobalt bomb 1. An apparatus used in the radiotherapy of cancer. It has a source of radioactive cobalt-60 (a γ-ray emitter) surrounded by a shield with a collimating system to restrict the radiation to a particular part of the patient's body.

2. A potential nuclear weapon, with a shell of cobalt around it that would become radioactive if the weapon were exploded.

Coblentz, William Weber (1873–1962) Amer. physicist and astronomer who made substantial advances in the field of radiation measurements (*see* Coblentz radiation experiments), especially in connection with radiation from the stars and planets.

Coblentz radiation experiments 1. He used a high-temperature black-body source to confirm the *Stefan–Boltzmann law and determine the *Stefan–Boltzmann constant. The radiation was received by a modified Angstrom *pyrheliometer containing a thin blackened manganin strip placed just in front of a thermopile that produced a deflection of a galvanometer. The strip was then heated electrically to produce the same deflection so that the rate at which energy was received by the strip could be evaluated.

2. He verified Wien's formula experimentally and determined the value of the constant C_2 in the radiation formula by determining the position of the maximum of the energy-distribution curve. Radiation from a *black-body furnace A (*see* diagram), whose temperature was measured by a thermocouple T, is focused on the spectrometer slit C by the use of silvered mirrors in an airtight box B. The prism p of fluorite or quartz forms the spectrum and the intensities for different wavelengths are measured by the vacuum bolometer V. The constant C_2 is then calculated from the wavelength λ_m corresponding to the maximum energy for a source at thermodynamic temperature T from the equation $C_2 = 4.965 \ 1 \lambda_m T$. *See* displacement law.

Cockcroft, Sir John Douglas (1897–1967) Brit. nuclear physicist. With E. T. S. *Walton, he effected the first purely artificial disintegration of an atomic nucleus (1932), for which they shared the 1951 Nobel prize. (*See* Cockcroft and Walton experiment.) Cockcroft played a leading part in radar research and in work on nuclear bombs and power.

Cockcroft and Walton experiment The first purely artificial nuclear disintegration. High-velocity protons, accelerated by a potential of some 300 kV, bombarded lithium, thus producing two alpha-particles:
$$^1_1\text{H} + ^7_3\text{Li} \rightarrow ^4_2\text{He} + ^4_2\text{He}$$
A replica of the original apparatus can be seen in the Science Museum (London).

Cockcroft–Walton generator or accelerator A high-voltage direct-current *accelerator especially for the acceleration of protons. The d.c. voltage is produced from cascaded rectifier circuits and capacitances to which a low a.c. voltage is applied.

Coddington lens A powerful magnifying glass; in effect, a complete sphere with a central stop.

coefficient of absorption *See* absorption coefficient.

coefficient of contraction The ratio of the area of the *vena contracta of a jet of fluid to the area of the orifice through which it is discharging; values lie between 0.5 and 1.

coefficient of coupling The ratio between the actual mutual inductance between two coils and the maximum possible. It depends on the relative position of the coils and the distance between them. If air-core coils are wound closely over each other, their coefficient of coupling is about 0.6 or 0.7. It approaches unity only when a closed iron core is used.

coefficient of expansion 1. For a solid, the coefficient of expansion is given by the expression $\Delta X / X \times 1/t$, where t is the rise in temperature producing an increase ΔX in the magnitude of the quantity X. The unit is simply $(°C)^{-1}$. If X is the length of the solid, the coefficient of linear expansion is obtained; if X is the area of the surface of the specimen, the coefficient of superficial expansion is obtained; and if X is the volume of the specimen, the coefficient of cubic expansion is obtained. In general,
$$X_t = X_0(1 + \alpha t),$$
where X_0 is the original value of the quantity and α the appropriate coefficient of expansion. Since the coefficients are small, the coefficient of superficial expansion may be taken as twice the coefficient of linear expansion and the coefficient of cubic expansion may be taken as three times the coefficient of linear expansion. The coefficient of linear expansion for material in the form of a long bar may be found by the *comparator method. For small crystals, *Fizeau's method may be used.
   2. For a liquid, there are two coefficients of cubic expansion. The coefficient of apparent expansion is the coefficient calculated from $\Delta V / V \times 1/t$ without account being taken of the expansion of the containing vessel. The coefficient of real or absolute expansion is the coefficient obtained when allowance is made for the expansion of the containing vessel, and is equal to the sum of the coefficient of apparent expansion and the coefficient of cubic expansion of the material of the containing vessel. The coefficient of apparent expansion of a liquid may be found experimentally by using (*a*) a *dilatometer, (*b*) a *pyknometer, (*c*) a *weight thermometer, (*d*) by *Mathiessen's sinker method. The coefficient of absolute expansion of a liquid may be obtained directly by the methods of (*a*) *Dulong and Petit, (*b*) Regnault (*see* Regnault's heat experiments), (*c*) *Callendar and Moss.
   3. For a gas, the expansion is considerable, and the expansion per degree centigrade must be ex-pressed as a fraction of the volume at 0 °C and not of the original volume. The coefficient of increase of volume of a gas at constant pressure is thus the ratio of the change in volume per degree change in temperature to the volume at 0 °C, the pressure remaining constant. (*See* constant pressure gas thermometer; Callendar's compensated air thermometer.) The coefficient of increase of pressure of a gas at constant volume is the ratio of the change in pressure per degree change in temperature to the pressure at 0 °C, the volume remaining constant. For an ideal gas both these coefficients (α) are equal to 0.003 6608 per °C (*see* Heuse and Odo's experiments) so that:
$$V = V_0(1 + \alpha t).$$
This means that the volume V would become zero at a temperature of $-1/\alpha$ or -273.15 °C. This temperature is also the absolute zero of *thermodynamic temperature.

coefficient of friction *See* friction.

coefficient of restitution Newton's experimental law states that if two spheres collide directly, the relative velocity after impact is in a constant ratio to the relative velocity before impact, and in the opposite direction. If the bodies collide obliquely, the same result holds for the relative velocity components along the line of centres at the instant of collision. This constant ratio, which depends on the materials of the spheres, and is unity for perfectly elastic bodies and zero for completely inelastic ones, is called the coefficient of restitution. The coefficient becomes slightly dependent on the relative velocity if this is very high.

coefficient of turbulence *See* eddy viscosity.

coefficient of viscosity *See* viscosity.

coercive force The reversed magnetic field required to reduce the magnetic induction (*magnetic flux density) in a substance from its remanent value to zero. It is represented by OC on the *hysteresis loop. *See* ferromagnetism.

coercivity The value of the coercive force for a substance that has been initially magnetized to saturation.

coherent radiation Radiation in which the waves are in *phase both temporally and spatially. If the phase and amplitude at any particular time or position are known, then the phase and amplitude at a subsequent time or position can be determined. Most practical radiation sources are not coherent over an appreciable length of time since *wave trains of limited length are emitted at random intervals. The *laser is a source of coherent radiation. Radiation that is not coherent is said to be *incoherent*. Since the *interference of light normally requires two coherent sources, a single source is generally employed with a device to split the light into two beams. Two laser beams can produce much better interference patterns.

coherent sources Light sources vibrating in identical phases necessary for the production of interference fringes of stable character. Generally, such sources are produced by a device that doubles a single source, e.g. a *biprism.

coherent units A system of units, such as *SI units, in which the quotient or product of any two units gives the unit of the resultant physical quantity. For example, in SI units the unit of length is the metre and the unit of time is the second, the coherent unit of velocity is therefore the metre per second. The *base units* of a coherent system (such as the metre and second in SI units) are an arbitrarily defined set of physical quantities: all the other units in the system are derived from the base units by defining relationships and are called *derived units*.

cohesion The property of a substance that enables it to cling together in opposition to forces tending to separate it into parts; the tendency of the different parts of a body to maintain their relative positions unchanged. *Compare* adhesion.

cohesional work *See* spreading coefficient.

coil A conductor or conductors wound in a series of turns. Coils are used as *inductors and in the windings of *transformers, *motors, and *generators.

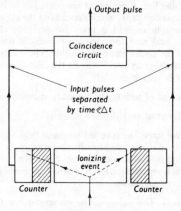

Coincidence counter

coincidence circuit A circuit with two input terminals that is designed to produce an output pulse only when both input terminals receive a pulse within a specified time interval, Δt. Such a circuit may be incorporated into a *coincidence counter*, the outputs of two radiation counters being fed into it. This device can be used to determine the direction of radiation or of *cosmic-ray showers. A *delayed coincidence circuit* gives an output pulse only when one input terminal receives a pulse within a given time interval beginning a specified time after the other terminal has received a pulse. *See also* anticoincidence circuit.

coincidences, method of The method of measuring lengths, timing periodic phenomena, etc., by observing coincidences between the lengths to be measured or the phenomena to be timed and those of some standard. For example, the use of the *vernier scale for measuring lengths: *interference fringes for comparing wave lengths: the use of the *stroboscope or of a pendulum whose period of vibration is known for timing a periodic phenomenon.

cold cathode A cathode of an electron tube that emits electrons by a process other than thermionic emission (i.e. emits electrons at low temperature). Electrons may be ejected from the cathode by *secondary emission caused by positive ions produced from residual gas in the tube, or in the case of very highly evacuated tubes *field emission may occur.

cold emission *See* field emission.

cold fusion Nuclear fusion occurring at normal temperatures, rather than at the high temperatures necessary to overcome electrostatic repulsive forces between nuclei. There have been two main approaches to producing fusion at low temperatures. One is an electrolytic method; it has been suggested that, under certain conditions, electrolysis of deuterium oxide using a palladium cathode can produce low-temperature nuclear fusion. Deuterium ions liberated at the cathode are absorbed in the crystal lattice of the electrode, where they are forced together, thus overcoming the repulsive electrostatic force. However, claims that high-energy outputs using this method have been obtained have not been reproduced; the necessary output of neutrons for a genuine fusion reaction has not been detected.

The other approach to cold fusion has been to shield one of the deuterium atoms by binding it with a negative muon. In this technique a muon replaces an electron in a deuterium atom. Because the muon is 207 times heavier than the electron, the resulting muonic atom of deuterium is much smaller and is able to approach another deuterium atom more closely, allowing nuclear fusion to occur. The muon is then released to form another muonic atom, and so on; i.e. the muon acts as a catalyst for the fusion reaction (known as **muon-catalysed fusion**). One problem with this approach is the short lifetime of the muon, which restricts the number of fusion reactions it can catalyse.

cold trap A tube, cooled with liquid air, or dry ice (frozen carbon dioxide) in acetone, that will condense vapour passing into it. It may be used in a pumping line between a mercury diffusion pump (*see* pumps) and the system being evacuated, and also to keep vapours from entering a *McLeod gauge. Liquid-air cooling is necessary to trap water but dry ice in acetone suffices for mercury.

collective excitation A quantized mode in a manybody system that arises when cooperative motion of the system as a whole occurs. This type of *excitation arises as a result of the interactions between particles. *Plasmons and *phonons in solids are examples of collective excitations. Collective excitations obey Bose–Einstein statistics (*see* quantum statistics).

collector The region in a bipolar junction *transistor into which *carriers flow from the *base. The electrode attached to this region is called the *collector electrode*.

collector-current multiplication factor In a junction *transistor, minority *carriers entering the *collector from the *base region can carry sufficient energy to create electron-hole pairs in the collector thus causing an increase in minority carrier current in the collector. The collector-current multiplication factor is the ratio of enhanced current flow in the collector as a result of the flow of minority carriers from the base into the collector, to the current carried by the minority carriers at collector voltage. Under normal operating conditions this factor is unity, but under high-field conditions it rapidly increases to infinity as the *avalanche breakdown voltage is reached.

collector ring *See* slip ring.

colligative property A property of a system that depends on the number or concentration of molecules present rather than their nature. The *osmotic pressure of a solution and the *pressure of a gas at low pressures are colligative properties.

collimation 1. The alignment of an optical system or the parts of an optical system, often in order to produce a parallel beam of light.
2. The limiting of a beam of radiation or of particles so that it has the required dimensions. This process has to be carried out with great accuracy when radiation is used for diagnostic and radiotherapeutic purposes. *See also* collimator.

collimator 1. An optical system that produces a beam of parallel light. It is used in *spectrometers, *telescopes, etc.
2. *Syn.* finder. A small fixed telescope attached to a larger one in order to set the line of sight of the large instrument.
3. An apparatus, usually in the form of a cylindrical tube of a heavy material, that is used to limit the size of a beam of charged particles or X- or gamma-radiation to the required dimensions.

collision In *kinetic theory, the mutual action of molecules, atoms, etc., when they encounter one another. A collision is thought of as being one of three kinds, i.e. elastic, inelastic, or superelastic.
(1) *Elastic collision.* One in which the total kinetic energy of translation is unchanged after the collision, none being translated into other forms. In nuclear physics, an elastic collision is one in which the incoming particle is scattered without exciting or breaking up the struck nucleus.
(2) *Inelastic collision* (or *inelastic collision of the first kind*). One in which the total kinetic energy of translation is decreased by the collision while some other form of energy is increased. For example, a neutron may undergo an inelastic collision with a

nucleus, which is thereby raised to an excited state that decays by gamma emission. The most extreme case of an inelastic collision is one in whch the colliding particles do not separate after impact. Some writers (especially engineers) reserve the term "inelastic" for this special case.
(3) *Superelastic collision* (or *inelastic collision of the second kind*). One in which the total kinetic energy of translation is increased by the collision while some other form of energy is decreased. For example, molecules of a gas in contact with a solid at higher temperature on the average recoil with higher kinetic energy, at the expense of the vibrational energies of the molecules of the solid.

In all kinds of collision total energy, mass, momentum, and angular momentum are conserved. Interactions of electromagnetic radiation with atomic particles can be regarded as collisions of particles with the photons. *See* Compton effect; Raman effect.

collision density The total number of a specified type of collision occurring per unit time per unit volume of material. It is often used for systems in which a gas or solid is bombarded with a beam of particles or photons. In nuclear physics it usually refers to the total number of collisions of a given *flux of neutrons per unit time passing through unit volume of a given material.

colloid A substance consisting of particles of ultramicroscopic size, intermediate between those of a true solute and those of a suspension. They exhibit Brownian movements, and owing to their electrical charge are subject to *cataphoresis. Though frequently used, the term is a loose one since most substances can be brought to the colloid state by a suitable technique. *See* colloidal state; crystalloid.

colloidal solution or suspension *See* colloidal state.

colloidal state Matter in the colloidal state consists of a suspension of fine particles whose diameters may vary between 1 and 100 nm. Such suspensions usually do not settle out and will not pass a fine filter (*see* dialysis). The dispersion medium may be a gas or a liquid, e.g. smoke is dispersed in air, and starch can be dispersed in water. In the latter case, the particles are intermediate in size between those of coarse suspensions (like muddy water or milk of lime, which slowly settle) and those of solutions (in which the particles are of molecular sizes).

Colloidal solutions may be produced by (*a*) grinding; (*b*) striking an electric arc between rods of the metal; (*c*) chemical methods (e.g. adding a drop of acid to a solution of sodium thiosulphate in water). *See* colloid; sol.

colorimetry The science that aims at specifying and reproducing colours as a result of measurement. Colorimeters of three types: (*a*) colour album or filter samples for comparison – essentially empirical; (*b*) monochromatic colorimeters that match colours with a mixture of monochromatic and white

colour

lights; (c) trichromatic colorimeters in which a match is effected by a mixture of three colours.

colour 1. The sensations of whiteness and colour depend very much on conditions including intensity, field of view, and contrast. Thus normal daylight has an energy distribution corresponding roughly to that of a black body at 6000 K and is seen as white. Similarly a lamp with a filament temperature of 2000 K gives white light to a room. If the lamp is observed in full daylight the filament appears yellow, while if a small beam of sunlight enters the room lit by the lamp the beam appears blue-green. Thus when the eye perceives light with a spectral distribution significantly different from that to which it is habituated, there is the sensation of colour.

The colour sensation arises most strongly from radiation covering only a small part of the visible spectrum. Besides possessing *luminosity*, colours have *hue* and *saturation*. Saturation is the degree to which a colour departs from white and approaches a pure spectral colour. Hue is determined by wavelength – a pure continuous spectrum shows a continuous variation of saturated hues. When a hue is diluted with white light (desaturated, impure), the colour is classed as a *tint*. The *shade* of a colour refers to its *luminosity*.

Observers with normal *colour vision group the hues of the spectrum in the six colours: red, orange, yellow, green, blue, violet. Such subjects observe typically 100 steps of hue difference across the spectrum. A few exceptional individuals observe about 130 steps and distinguish a distinct colour indigo between blue and violet. Persons with defective colour vision distinguish considerably less than 100 hue steps.

By mixing colours, other colours emerge and a sharp distinction must be drawn between combining coloured lights, the mixing of pigments, and the transmission of light by colour filters. Mixing coloured lights is an *additive process. The other two are *subtractive processes. When lights of different colours are arranged to illuminate a white screen viewed by eye, or if they are arranged to produce overlapping illumination on the retina of the eye, the ensuing sensation has another colour whose hue and saturation depend on the relative proportions of the mixing colours (Newton) and its luminosity is the sum of the separate luminosities (Abney). Pairs of pure spectral hues produce colours of different hues with differing degrees of saturation: certain pairs, called *complementary colours, combine to form white light.

By using lights of three *primary colours in various proportions it is possible in general to produce any other colour, i.e. a colour match can be made with another colour, and from a measure of the relative quantities of the constituent colours, the *chromaticity coordinates can be found. Slightly fewer colours can be obtained by mixing dyes and inks as the three colours used do not correspond exactly to the three light primaries. Colour triangles

(*see* chromaticity) indicate the possible colours. *See also* surface colour; colour system.

2. A quantum number of *quarks. *See* quantum chromodynamics.

colour blindness *See* colour vision.

colour centre A point *defect that causes absorption of light over a range of wavelengths in an otherwise transparent crystal.

colour equation An algebraic equation that expresses the results of an additive mixture (*see* additive process) of three *primary colours in terms of another colour or white. If quantities of colour are expressed in convenient units (i.e. of such magnitude that equal quantities of the three primaries produce white), then it is possible to write, say:
$$m_1 C = r_1 R + g_1 G + b_1 B,$$
where R, G, and B represent units of (say) red, green, and blue proportioned in amount by the quantities r_1, g_1, and b_1; the result is an amount of colour C represented by m_1. Again, white light may be produced according to:
$$nW = rR + gG + bB.$$
On equal sensation units, the equation is commonly written with $n = 1$ and $(r + g + b) = 1$. With luminosity units,
$$m_1 = r_1 + g_1 + b_1.$$
If C is highly saturated, it may be necessary to have a negative coefficient for R, G, or B. In such cases, a test colour must be mixed with a primary before the other two colours can be made to effect a colour match, e.g.
$$m_1 C + b_1 B = r_1 R + g_1 G.$$
See also colour system; chromaticity.

colour mixture *See* additive process; subtractive process.

colour photography A process based on the fact that any colour can be formed from a suitable combination of three *primary colours. Modern colour photography is basically a *subtractive process. *Kodachrome* is a colour transparency film that consists of three layers of emulsion on a plastic base, usually cellulose acetate. The top emulsion layer is an ordinary photographic emulsion, sensitive to blue-violet light. The second layer is made sensitive to green as well as to blue light. The third layer is made sensitive to red as well as blue light. Blue light is prevented from reaching the second and third emulsion layers by a yellow filter sandwiched between the first and second layers, which absorbs blue light (the *complementary colour of yellow). Following exposure of the film a blue object will be recorded as a latent image (*see* photography) in the top layer, a green object will be recorded in the second layer, a yellow object in the second and third layers, a white object in all three layers, etc.

The latent images in all three layers are then developed to give silver images in a normal black-and-white developer. The film is next exposed to red light so that a latent image is formed in the residual silver halide of the red-sensitive layer. The film is developed in a cyan-forming colour developer and an image is formed of the complementary colour to red (cyan) in those areas not originally exposed to the red light of the scene being photographed. The film is then exposed to blue light, from the top surface. Due to the yellow filter this only forms latent image in the residual silver halide in the top layer. Development in a yellow-forming developer gives a yellow dye image in those areas not exposed to the blue light of the original scene. Development of the film in a magenta developer converts the remaining silver halide in the middle green-sensitive layer to a magenta (complementary to green) image. The silver images and the yellow filter layer are bleached and fixed leaving only the positive dye images in the film. The colours of the final image are therefore made up of the correct proportions of the three colours yellow, magenta, and cyan. A red image will be formed from a combination of yellow and magenta in the top and middle emulsion layers by a subtractive process. (Yellow dye absorbs blue light, magenta dye absorbs green light, and only red light passes through the transparency). A black image is formed by a combination of yellow, magenta, and cyan dyes, and so on.

In colour prints a colour negative is produced in complementary colours to the original scene. A red bus will appear cyan, blue sky will be yellow. The negative is used to reproduce the scene in its original colours as a coloured print on a paper base.

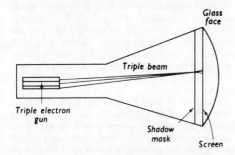

a Colourtron, with triangular arrangement of electron guns

colour picture tube A type of *cathode-ray tube designed to produce the coloured image in *colour television. The coloured image is produced by varying the intensity of excitation of three different *phosphors, which produce the three primary colours red, green, and blue, and thus reproduce the original colours of the image by an *additive process.

The conventional colour picture tube consists of a configuration of three *electron guns tilted slightly so that the electron beams intersect just in front of the screen. Each electron beam has an individual electron lens system of focusing and is directed towards a different colour phosphor. There are several different types of colour picture tube, the main differences being in the configuration of electron

guns and arrangement of the phosphors on the screen.

One main type of tube (e.g. the *colourtron*) has a triangular arrangement of electron guns, and the

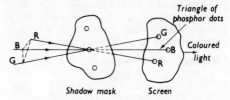

b Light production in colourtron

phosphors arranged as triangular sets of coloured dots. A metal shadow mask is placed directly behind the screen, in the plane of intersection of the electron beams, to ensure that each beam hits the correct phosphor, the beams being blanked out by the mask while moving from one position to the next (Figs. *a*, *b*).

The other main type has the three electron guns arranged in line horizontally, an aperture grille of vertical wires, and the phosphors arranged as vertical stripes on the screen (Fig. *c*). The latter type has advantages in focusing the beams but has a smaller field of view than the former.

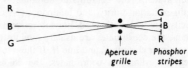

c Horizontal arrangement of electron guns

The *Trinitron* is a type of colour picture tube that has certain advantages over conventional tubes. It has a single electron gun with three cathodes aligned horizontally, an aperture grille, and vertically striped phosphors. The cathodes are tilted towards the centre so that the electron beams intersect twice, once within the electron-lens focusing system and once at the aperture grille (Fig. *d*). This allows a single electron-lens system to be used for all three beams, needing fewer components and thus making the system much lighter and cheaper. It also gives a greater effective diameter of the lens system than is found in conventional tubes, hence sharper focusing of the beams is possible. Horizontal alignment of the cathodes means that misconvergence of the beams only occurs in the horizontal direction; in the triangular configuration misconvergence occurs both horizontally and vertically, but without the associat-

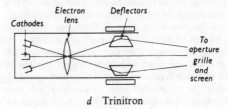

d Trinitron

colour quality

ed reduction in lens aperture found with a three-gun tube. The single electron gun allows the diameter of the tube to be smaller for a given screen size. The Trinitron has considerable potential for future development, particularly the development of a wide-angled colour picture tube, making colour television receivers relatively smaller, and also for uses where multiple electron-beam tubes are required.

colour quality *See* chromaticity.

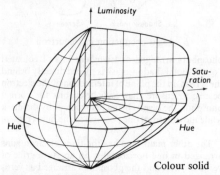

Colour solid

colour system The representation of a colour in terms of a specific set of coordinates. For objective colour systems, the coordinates dominant wavelength, excitation purity, and luminance (*see* chromaticity) are frequently used. For subjective colour systems, the coordinates are usually luminosity L, saturation or chroma S, and hue H. If these form a system of cylindrical coordinates, the colour is found inside a roughly elliptical solid.

colour television Colour television operates by an *additive process. Red, green, and blue may be combined to produce a wide variety of colours, and are the three colours on which colour television is based. Colour camera tubes produce three separate video signals (*see* television camera); these are then combined into a composite signal that is broadcast and received by colour receivers. These have a special type of *cathode and a screen that has three *phosphors producing red, green, or blue when excited (*see* colour picture tubes). Video information for each primary colour is extracted from the broadcast signal by a colour coder that feeds the appropriate electron gun, and in turn excites the appropriate phosphor. The image is produced by a combination of the three primary colours.

The composite signal transmitted in colour television needs to be compatible with the large number of black-and-white (monochrome) receivers in use. It is therefore in two parts, called the *luminance signal* and the *chrominance signal*. The luminance signal contains brightness information obtained by combining the outputs of the three colour channels. This produces the black-and-white image. The colour information is contained in the chrominance signal, which is transmitted as a subcarrier signal at a frequency chosen to cause the least interference on a

monochrome set. In colour receivers the chrominance circuits are disabled by the colour killer when a black-and-white signal is being received. This ensures that only luminance information reaches the tube, and prevents colour fringing on the image.

colour temperature Of a nonblack body. The temperature of a *black body that has approximately the same energy distribution as occurs in the spectrum of the body. *See also* selective radiation.

colour triangle *See* chromaticity.

colourtron *See* colour picture tube.

colour vision The *retina of the human eye contains two sorts of light-sensitive cells – rods and cones. The rods are distributed over the greater part of the retina but are absent from the central fovea. They are sensitive to a range of wavelengths over a great range of intensities, but do not distinguish colours. Hence, in very dim light objects appear only in black, white, and grey. The cones are concentrated in the fovea and over a region extending for a few millimetres around this. They are insensitive to low intensities but discriminate fine detail and distinguish colours, provided there is sufficient illumination. Colour vision can be explained by the *trichromatic theory* (originated by Young in 1807, developed by Helmholtz in 1852). It assumes that there are three separate systems of cones sensitive to either red, green, or blue light. Incident light will therefore stimulate one or more of these systems to an extent depending on its colour. Red light will stimulate the red cones, yellow light the red and green cones. The cones are linked to the optic nerve and electric impulses resulting from the stimulation are sent to the brain. Some insects are known to have four separate systems, giving exceptional colour vision.

About 5% of human males and 0.5% of females have *daltonism*, that is defective colour vision. (The popular term *colour blindness* is misleading as total incapacity to distinguish colour is very rare.) In such individuals the number of cones of one or more kinds is much less than normal and may be zero in some cases. Deficiency of red, green, or blue cones is described as *protanopia*, *deuteranopia*, or *tritanopia* respectively. Daltonism is hereditary and colour vision is controlled by the X-chromosome. Human females have two X-chromosomes, one derived from each parent. Males have only one, derived from the mother. Generally the genes giving defective colour vision are recessive, so a female with one normal and one abnormal chromosome will have normal colour vision. *See* additive process; duplicity theory; Purkinje effect.

columnar ionization The production, in a gas, of ionization confined to one or more paths of very small area of cross section; e.g. the ionization along the track of an alpha-particle.

column of air The vibrations of air columns are the sources of sound in the organ as well as in the

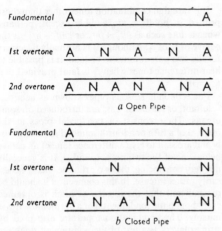

a Open Pipe

b Closed Pipe

Vibrating air column

3*c*/4*l* and so on. In general, the frequencies bear the ratios 1:3:5: . . . etc., representing the odd harmonic series of wavelengths. These deductions can be obtained by applying the end conditions to the mathematical expression of the acoustic impedance of a pipe.

The theory of the column of air, as applied to cylindrical pipes, is not valid for conical tubes where spherical waves are diverging from the tip of the cone. In this case, it is immaterial whether the vertex is open or closed. For a cone having the wide end open at a distance *l* from the vertex, the full harmonic series occurs and the fundamental has the frequency *c*/2*l*, which is the same as that for an open cylindrical pipe. For a closed cone the fundamental frequency is given by 1.43*c*/2*l* and the partial tones are harmonic, the antinodes being equidistant while the nodes are unequally spaced.

In the above treatment it is assumed that the nodes and antinodes are formed exactly at the end of the tube. This is true at a closed end if the unyielding nature of the material ensures that the displacement of the particles is nil. For the open end it is not true since the waves are propagated in spherical type outside the free air. There is then an *end correction.

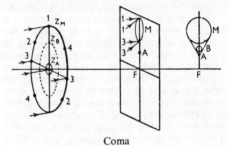

Coma

different types of wind instruments. The phenomenon has been studied theoretically since the ancient Egyptian and Greek civilizations.

The simplest case of a vibrating mass of air is that in a hollow cylindrical pipe, the ends of which may be open or closed. To put forward a simple theory, the following assumptions must be made: (1) The motion in the tube is uniform, i.e. the viscosity of the medium inside the tube is neglected so that only plane waves need be considered. To satisfy these conditions the diameter of the tube should be sufficiently great yet small with respect to the length of the pipe and with the wavelength of the sound. The walls of the pipe are assumed to be rigid. (2) Vortices or rotatory motions are not set up inside the tube. (3) Oscillations are so rapid that the changes may be considered adiabatic.

Under these conditions, when a cylindrical air column is set into resonant vibration by some means, *standing waves are set up due to *progressive and retrogressive waves. In the open pipe (that having both ends open, Fig. *a*), there must be a displacement antinode A at each end and therefore, in the fundamental mode of vibration, a *node N in the middle. The wavelength of the sound is therefore 2*l*, where *l* is the length of the pipe. The frequency of the fundamental is then *c*/2*l*, where *c* is the speed of sound in the medium inside the pipe. In the next possible mode, there must be two nodes in the pipe and an antinode at the centre. The wavelength is now *l* and the frequency *c*/*l*. In the next possible mode, three nodes occur in the pipe and the frequency is 3*c*/2*l*. The frequencies of the different modes are in the ratio 1:2:3: . . . etc. . . . representing a full *harmonic series of wavelengths.

For a closed pipe (that having one end closed and the other open, Fig. *b*), there must always be by the same reasoning, a displacement node at the closed end and an antinode at the open end. In the fundamental mode, the wavelength is 4*l* and the frequency *c*/4*l*. In the next mode of vibration, the frequency is

coma An aberration of a mirror or lens in which the image of a point lying off the axis presents a comet-shaped appearance. While the rays from the central zone Z_A (*see* diagram) focus to a point A, the zones Z_B and Z_M form annular rings or *comatic circles*, B and M of progressively varying diameter and centration. The overlap of these circles produces the comatic patch AM, which is called the *tangential coma* and the radius of the comatic circle M the *sagittal coma*: the latter is practically one-third of the former. For freedom from coma, the lateral magnification for all zones should be constant, which demands the fulfilment of the *sine condition. For a single lens, the lens with least spherical aberration has the least coma. *See* aberrations of optical systems.

comatic circles *See* coma.

combination tones *Syn.* resultant tones. If two tones of frequencies f_1, f_2, are sounded together very loudly, at least two other rather faint tones, of frequencies $(f_1 + f_2)$ and $|f_1 - f_2|$ (the absolute value), may be heard.

The first is an example of a summation tone and the second an example of a difference tone; collectively they are known as combination tones. Although originally the formation of these tones was confused with the formation of *beats, two alternative theories have been propounded to explain them.

(1) The Helmholtz theory of objective combination tones. Since intense primaries are required it is assumed that the restoring forces of the medium are not proportional to displacement, but that a term involving the square of the displacement should be introduced. Neglecting damping forces the equation of motion of a particle in the transmitting medium, is of the form:

$$m\ddot{x} + l\dot{x} + kx^2 = A_1\cos\omega_1 t + A_2\cos(\omega_2 t - d),$$

where A is the amplitude factor, x the displacement, m the mass, ω the angular frequency, t the time, d the phase factor. The solution is expressed in terms of

$$\cos 2\omega_1 t + \cos 2\omega_2 t, \cos(\omega_1 - \omega_2)t,$$

and

$$\cos(\omega_1 + \omega_2)t,$$

viz., the combination tones consist of the octave of the primaries and also the sum and difference tones. The amplitude factors for appreciable combination tones require A_1 and A_2 to be large. The theory explains the principal experimental observations.

(2) Subjective combination tones. The ear is a nonlinear device, its tissues moving through a distance that is not proportional to the change in pressure that takes place when a sound is generated. Thus the vibrations set up by a simple harmonic sound are not in themselves simple harmonic. Although described as subjective, it must be recognized that the combination tones are real, and may be picked out with the aid of sensitive *Helmholtz resonators tuned to the appropriate frequencies; they are not, as some investigators have implied, subjective in the sense that they are formed by the brain rather than the ear. The most obvious means of producing combination tones is to sound a harmonically pure tone of gradually increasing loudness. At a certain level the *octave will be recognized, and in due course the *interval of a twelfth. These are the summation tones $2f(f + f)$ and $3f(f + 2f)$. In principle, if tones of frequencies f_1 and f_2 are employed, the following families of combination tones occur:

simple summation

f_1	$2f_1$	$3f_1 \ldots$
f_2	$2f_2$	$3f_2 \ldots$

combined summation

$f_1 + f_2$	$f_1 + 2f_2$	$f_1 + 3f_2 \ldots$
$f_2 + f_1$	$f_2 + 2f_1$	$f_2 + 3f_1$

simple difference

$\lvert f_1 - f_2 \rvert$	$\lvert f_1 - 2f_2 \rvert$	$\lvert f_1 - 3f_2 \rvert \ldots$
$\lvert f_2 - f_1 \rvert$	$\lvert f_2 - 2f_1 \rvert$	$\lvert f_2 - 3f_1 \rvert \ldots$

complex summation

$$2f_1 + f_2 \ldots \quad 3f_1 + f_2 \ldots$$
$$\text{etc.}$$

complex difference

$$\lvert 2f_1 - f_2 \rvert \ldots \quad \lvert 3f_1 - f_2 \rvert \ldots$$

It is rare that combination tones other than those printed in boldface are audible except, perhaps, when a note such as $mf_1 + nf_2$ or $\lvert mf_1 - nf_2 \rvert$ is the octave of f_1 or f_2. Although it is normally necessary for the two parent tones to be loud, it is possible to hear difference tones when f_2 is faint provided that $f_2 \simeq nf_1$.

For economic reasons pipes known as acoustic, resultant, or harmonic bass are introduced on some organs. These consist of two pedal pipes at the interval of a fifth or a fourth sounded simultaneously and are said to give a difference tone of an octave below one of the two pipes sounded. It is generally not very convincing, the two primaries being necessarily rather intense. In this connection it should be remembered that the illusion of pitch is obtained even though the fundamental of a harmonic series is missing. The combination tones are also used by string players as a test of the accuracy of doubling stopping.

comma A small interval between the larger and the lesser whole tones on the scale of just temperament. The frequency ratio may be stated to be 81/80. It is sometimes defined as the small interval by which a major tone exceeds a minor tone in the same scale. See musical scale.

common-base connection A method of operating a *transistor in which the *base (usually earthed) is common to the input and output circuits; the *emitter is the input terminal and the *collector is the output terminal. This type of connection is frequently used as a voltage amplifier stage.

common branch Of an electrical network. See network.

common-collector connection A method of operating a *transistor in which the *collector (usually earthed) is common to both input and output circuits; the *base is the input terminal and the *emitter is the output terminal. See also emitter follower.

common-emitter connection A method of operating a *transistor in which the *emitter (usually earthed) is common to the input and output circuits; the *base is the input terminal and the *collector is the output terminal. This type of connection is used for power amplification with a nonsaturated transistor and for switching with a transistor in saturation.

common-impedance coupling See coupling.

communications channel A *channel used for the transfer of data. See also communications system.

communications line or link Any physical medium, such as a telephone line, cable, radio beam, or optical fibre, used to carry information between different locations.

communications network See communications system; network.

communications satellite See satellite.

communications system Any system whereby information can be conveyed efficiently and reliably from source to destination. With more than one source and/or destination, the system is called a *network. Information in a communications system is sent by means of a *communications channel. It is encoded in a digital form before transmission and decoded at its destination(s). This reduces to a minimum the errors arising in the information as a result of *noise in the channel.

commutating zone The zone containing a group of armature conductors being short-circuited by a brush on the *commutator of an electrical machine. During rotation of the commutator, a brush short-circuits at any instant two or more commutator bars.

commutator 1. A device for reversing the direction of the current in an electric circuit or in some part of a circuit.
   2. A device employed in electrical machines to connect in turn each of the sections of an armature winding with an external electric circuit. It may be used as a simple current-reverser, or to convert alternating current into direct current (or vice versa). In a d.c. machine it is typically a cylindrical assembly of copper commutator bars, each of which is provided with a commutator lug for connection to the armature winding. The commutator bars are insulated from one another by means of mica or a mica compound (e.g. micanite) and they are clamped in position by metal *V-rings. The latter are insulated from the commutator bars by V-rings of mica or micanite. The complete assembly rotates with the armature and its winding, and connection is made with the external circuit by means of carbon *brushes, which are kept in contact with the outer surface of the commutator.

compact disc (CD) A 120 mm metal disc on one side of which information is digitally recorded. The recording is protected by a layer of clear plastic. The most common application is the high-quality recording and reproduction of sound. The audio information to be recorded is encoded, using an *analogue/digital converter, into digital information. This is impressed on the disc as a series of minute bumps that lie in a track spiralling outwards from the centre of the disc. The information is retrieved by means of a small low-power laser. Destructive *interference occurs between the laser light reflected from a bump and that reflected from the unraised surface, with a consequent attenuation of the reflected light. The modulated reflected light is detected by a *phototransistor and the resultant electrical signal converted into the equivalent audio signal.
   A sophisticated error-control system is used to keep the laser beam both focused and centred on the track, and to ensure that the disc is rotated at the correct speed: the rotational speed is not constant but is a function of the track radius.
   Compact discs are also used in computing as high-capacity storage devices, mainly in the form known

as *CD-ROM* (*see* ROM) and *CD-I* (the interactive form).

compandor *See* volume compressors (and expanders).

comparator 1. A device for measuring the difference in length between two line standards or for measuring horizontal distances by comparison with a standard scale.
   2. A circuit, such as a differential amplifier, that compares two signals and produces an output that is a function of the result of the comparison.

Determination of coefficient of linear expansion

comparator method The standard precision method for determining the coefficient of linear expansion of a specimen in the form of a long rod or tube. The specimens have a scratch near each end and are mounted horizontally in a constant temperature bath so as to be able to expand freely. Two rigidly mounted travelling microscopes M are focused one on each scratch and the temperature of the bath is then raised and maintained at a higher temperature. The distance each microscope has to be moved to obtain the image of the scratch on the cross-wires of the eyepiece is noted, the expansion being given by the sum of these two movements. The expanding length between the scratches is noted by means of a cathetometer.

compass A magnet freely pivoted horizontally so that it can set itself along the lines of force of the earth's magnetic field. It usually carries a scale divided into degrees and marked with the cardinal points, or these may be printed on a circular card to which the magnet, or system of magnets, is fixed. In a liquid compass, the compass card is enclosed in a case completely filled with alcohol. The card bears a float that helps to take the weight off the point of suspension and also provides a damping effect. *See also* gyrocompass.

compensated pendulum A pendulum so constructed that the distance between the support and the centre of gravity of the bob is independent of temperature so that the time period does not vary with temperature. In *Graham's pendulum* some mercury is contained in a hollow bob B (Fig. *a*), attached to a long metal rod A. Expansion of this rod lowers the centre of gravity of the system but the expansion of the mercury in the bob raises the centre of gravity and if the correct amount of mercury is present, the resultant motion of the centre of gravity due to change of temperature is nil. In *Harrison's gridiron pendulum*, the bob B (Fig. *b*), is supported by five iron rods C and four brass rods D attached to cross bars as

compensating eyepiece

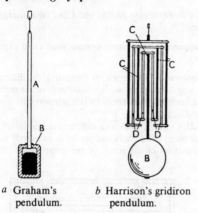

a Graham's pendulum.
b Harrison's gridiron pendulum.

shown so that expansion of the iron rods tends to lower the bob, while expansion of all the brass rods tends to raise the bob. The downward motion of the bob equals the expansion of three iron rods, while the upward motion of the bob equals the expansion of only two brass rods, but since the coefficient of linear expansion of brass is about 1.5 times that of iron, the net movement of the bob is zero.

compensating eyepiece An eyepiece that corrects the chromatic differences of magnification of an objective (microscopes, telescopes).

compensating leads An extra pair of leads, similar to the working leads of a resistance thermometer, which run alongside them almost to the point of working. They are connected together at the coil end and at the other are connected in series with the balancing resistance of the bridge circuit. Any resistance changes caused by temperature variations in the working leads are also produced in the compensating leads and balance them exactly.

With thermocouples, compensating leads join the working head of the instrument to an indicator or to an ice box when they join copper leads. They must be such that the thermoelectric e.m.f. they develop is the same as would be produced by the genuine thermocouple wires extended over the same range. Base-metal wires may thus serve with platinum or platinum-alloy thermocouples.

compensating winding Of an electrical machine. Winding housed in axial slots in the *pole shoes and designed to prevent the distortion of the magnetic flux distribution caused by *armature reaction by neutralizing the magnetizing action of the armature currents as far as possible, over the whole of the pole pitch.

compensators *See* quadrantal deviation. *See* also Babinet compensator; Soleil compensator.

compiler A *computer program taking a high-level *programing language as its input and converting it to a set of machine instructions.

complementarity The principle that a system, such as an electron, can be described either in terms of particles or in terms of wave motion (*see* de Broglie equation). According to Bohr these views are complementary. An experiment that demonstrates the particle-like nature of electrons will not also show their wave-like nature, and vice versa.

complementary acceleration *See* Coriolis theorem.

Colour	$\lambda/10^{-10}$ m	Complementary	$\lambda/10^{-10}$ m
Red	6562	Green-Blue	4921
Orange	6077	Blue	4897
Golden Yellow	5853	Blue	4854
Yellow	5671	Indigo Blue	4645
Green Yellow	5636	Violet	4330

complementary colours Two pure spectral *colours that when mixed produce white light. The negative of a colour photograph is in the complementary colours of the original scene. The table gives the complementary colours of lights (not pigments). There is no complementary spectral colour to green.

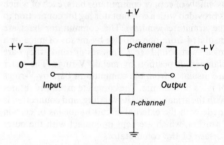

CMOS circuit

complementary transistors A pair of transistors of opposite type, i.e. n-p-n and p-n-p bipolar junction *transistors. *Class B push-pull amplifiers often employ complementary bipolar junction transistors.

Complementary MOS *field-effect transistors (abbreviation: *CMOS*) are used for *logic circuits with low heat dissipation and power consumption. The basic *inverter is shown in the illustration. When the input is low, the p-channel device conducts and the output is high. If the input is high, the n-channel device conducts and the output is low. The output of such a device will, in general, be driving like stages with essentially capacitive input impedance and the d.c. current flowing will therefore be zero.

complexion Any given set of values of the positional and momental coordinates of the molecules or other particles making up a system, treated as a particle distribution among the phase-space elements.

compliance Symbol: *C*. The reciprocal of *stiffness.

compole An auxiliary magnetic pole mounted midway between the main poles of an electrical machine having a commutator, to produce an auxiliary magnetic flux in the region occupied by the conductors of the coils undergoing commutation. As a result an e.m.f. induced in these coils, with correct design, neutralizes the e.m.f. of self-induction and gives rise to sparkless or "black" commutation. A compole is excited by a winding that carries the load current of the machine.

component The number of components in a system in equilibrium is the number of independently variable constituents by means of which, either directly or in a chemical equation, one may define the composition of each *phase, e.g. the water, water-vapour, ice system has only one component: H_2O. *See* phase rule.

component of a vector *See* resolution.

compound microscope Any *microscope with objective and eyepiece, to distinguish it from the simple microscope or single lens magnifier.

compound nucleus A highly excited nucleus, of short lifetime, formed immediately after a nuclear collision. This then ejects a particle or particles, completing the reaction. *Compare* stripping.

compound-wound machine A d.c. machine in which the *field magnets are provided with both series and shunt excitation windings. The series windings carry the load current of the machine and if the series field assists the shunt field, the machine is said to be *cumulatively compound-wound*; if the series field opposes the shunt field, the machine is said to be *differentially compound-wound*. A cumulatively compound-wound d.c. generator in which the series field is so proportioned that the full-load terminal voltage is equal to that at no load is termed *flat-compounded* or *level-compounded*. If the terminal voltage increases with load it is termed *overcompounded*. Generators are often over-compounded to give a rise in terminal voltage of between 5% and 10% from no load to full load in order to compensate for voltage drop in the *feeders. Cumulatively compound-wound motors are used in situations where a *series-characteristic motor is desirable but where it is possible for the mechanical load to be thrown off completely and, therefore, where it is essential to limit the no-load speed to a safe value. A differentially compound-wound motor can be designed to have an almost constant-speed characteristic. *See also* shunt-wound machine and series-wound machine.

compressibility Symbol: κ. The reciprocal of the bulk modulus (*see* modulus of elasticity); it is the negative of the volume strain divided by the pressure change:

$$\kappa = -(1/V)(\partial V/\partial p),$$

under specified conditions, usually constant temperature (isothermal) or constant entropy (adiabatic). For solids, values are typically of the order 10^{11} Pa^{-1}, for liquids 10^{-9} Pa^{-1}, and for gases the value at constant temperature is the reciprocal of the pressure (10^{-5} Pa^{-1} at about atmospheric pressure).

compressible flow As long as the velocities and their fluctuations in a fluid are small compared to the *speed of sound, changes of density so induced in the medium may be neglected. Otherwise the equation of continuity (*see* continuity principle) must include such changes of density. The *Bernouilli theorem has also to be modified in the light of the "gas equation", which is assumed to connect the initial and final states of the motion. If this equation is taken to be the one characteristic of adiabatic change, it can be seen how the speed of sound (c) is involved, for then, $c^2 = \partial p/\partial \rho$, p being the pressure and ρ the density of the gas.

To obtain the equations of a potential flow (*see* laminar flow) in a compressible medium, the three component velocities may be written:

$$U + u = U + \frac{\partial \phi}{\partial x}; \quad v = \frac{\partial \phi}{\partial y}; \quad w = \frac{\partial \phi}{\partial z};$$

where u, v, w but not U, are small compared to c ($U < c$). It can then be shown that, with the substitution of a new coordinate:

$$x' = \left(1 - \frac{U^2}{c^2}\right)^{1/2} x \quad \text{for} \quad x,$$

the potential satisfies the same conditions that characterize incompressible flow.

compression 1. (sound) Occurring in *longitudinal waves traversing an elastic medium and resulting in a greater density than for normal pressure. The diagram illustrates at (a) the particle displacement at a given instant for a sinusoidal *progressive wave traversing a homogeneous medium. If the displacements are longitudinal, i.e. in the same direction as the wave propagation, changes of density due to the relative juxtaposition of neighbouring strata occur as shown at (b), the number of lines in unit distance representing in highly exaggerated degree the local density. Points such as c and r represent compressions and rarefactions respectively. **2.** *See* ellipticity.

compression cable A cable in which the *dielectric is maintained at a pressure in excess of that of the atmosphere by means of a gas under pressure external to the cable sheath. For *three-phase working, the normal type is a three-core cable having a single sheath of triangular shape. The cable is inserted in a welded gas-tight steel pipe that is filled with nitrogen at a pressure of between 12 and 14 atmospheres.

compressor *See* volume compressors (and expanders).

Compton, Arthur Holly (1892–1962) Amer. physicist. A leading authority on nuclear physics, he played a notable part in the study of cosmic and nuclear energies. His name is associated with many discoveries in the field of X-rays, including the scattering process that bears his name, i.e. the

Compton effect

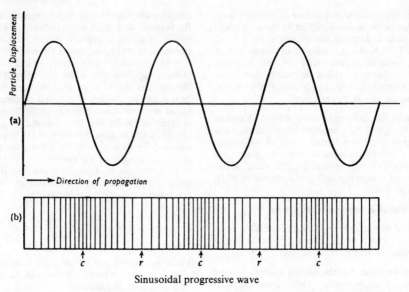

Sinusoidal progressive wave

*Compton effect. *See also* Compton electrometer; Compton equation; Compton recoil.

Compton effect An increase in wavelength that occurs when radiation is scattered by loosely bound (i.e. effectively free) electrons. The change $\Delta\lambda$ in wavelength is given in picometres by the relation, $\Delta\lambda = 4.85 \sin^2 \frac{1}{2}\phi$, where ϕ is the angle between the directions of the incident and scattered radiation (*see* Compton equation). The outer electrons in all elements and the inner ones in those of low atomic number have *binding energies negligible compared with the quantum energies of all except very soft X- and gamma rays. Thus most electrons in matter are effectively free and at rest and so cause *Compton scattering*. In the range of quantum energies 10^5 to 10^7 electronvolt this effect is commonly the most important process of attenuation of radiation. The scattering electron (*Compton electron*) is ejected from the atom with large kinetic energy and the ionization that it causes plays an important part in the operation of detectors of radiation.

In the *inverse Compton effect* there is a gain in energy by low-energy photons as a result of scattering by free electrons of much higher energy. As a consequence, the electrons lose energy.

Compton electrometer A type of *quadrant electrometer devised by A. H. *Compton and his brother K. T. Compton (1919). It has the needle slightly tilted and one pair of quadrants slightly above the other; this allows the sensitivity to be controlled and up to 60 000 mm per volt to be attained.

Compton electron *See* Compton effect.

Compton equation The theoretical equation for $\Delta\lambda$, the Compton change in wavelength (*see* Compton effect):

$$\Delta\lambda = (2h/mc)\sin^2 \frac{1}{2}\phi,$$

where h is the Planck constant, m the rest mass of the electron, and c the speed of light.

Compton recoil When an X-ray photon is scattered by an electron at an angle ϕ, the electron recoils at an angle θ using some of the photon's energy and hence reducing its frequency. The incident photon has a momentum of $h\nu/c$, and the scattered photon $h\nu'/c$, while the recoiling electron has a momentum of mv, or $m_0 v/\sqrt{(1-\beta^2)}$, where $\beta = v/c$. (*See* Compton effect; Compton equation.) The scattered ray is of greater wavelength than its parent primary ray and a corpuscular radiation accompanies the phenomenon made up of the recoil electrons.

Compton wavelength 1. For any particle of rest mass m_0 it is $h/m_0 c$, where h is the Planck constant and c the speed of light. The constant in the entry for Compton effect is twice this quantity for the electron.

2. The quantity defined above, divided by 2π. For the electron this has value $3.861\,5906 \times 10^{-13}$ m.

computer Any device by which data, received in the appropriate form, can be manipulated so as to produce a solution to some problem. In general the word now refers to a *digital computer*, the other basic form, the *analogue computer, being far less versatile.

The digital computer accepts and performs operations on discrete data, i.e. data represented in the form of combinations of characters. Before being fed into the computer, the characters consist of letters, numbers, punctuation marks, etc. Once inside the computer, the characters are represented as combinations of *bits. Arithmetic and logic operations can then be performed on these groups of bits according to a set of instructions. The instructions form what is known as a *program, which is stored along with the data in the computer *memory. The program instructions are interpreted and executed by one or more *processors. Data and programs are generally input to a computer by means of a *keyboard. The information is usually output on a *VDU, *printer, or *plotter. *See also* logic circuit; microcomputer; mainframe.

computer graphics *See* graphics.

concave Curving inwards. *Concave mirrors* are converging in action. As applied to lenses, it describes those lenses that are thinner at the centre and diverging in action – *biconcave, planoconcave*, and *convexoconcave*, although the last mentioned is generally designated in technical optics as a concave or diverging meniscus. (*See* diagrams under *lens.*) The *concave grating* (Rowland, 1883) was a diffraction grating ruled on a spherical metal reflecting surface, eliminating *chromatic aberration and ultraviolet absorption by the media (glass, etc.) otherwise required to produce dispersion, to focus the light, etc. *See* convex.

concentration Symbol: c_A (indicating concentration of solute A). The amount of solute present in a given volume of solution, usually expressed in mol m^{-3} or mol dm^{-3}. Concentration can also be measured by the amount of solute present in a given mass of solvent; it is then usually expressed in mol kg^{-1} and is called the *molality* (symbol: m_A). The term *molarity* has sometimes been used for a concentration expressed in mol m^{-3}, but this usage is deprecated in view of possible confusion with molality.

A solution containing x_A mol dm^{-3} of A is sometimes called an *x*-molar solution of A. This usage is also deprecated as it conflicts with the agreed usage of the word *molar to mean divided by amount of substance.

concentration cell A cell in which two electrodes of the same metal are immersed in solutions of different concentrations of some one salt of the same metal. The solutions may be separated by a porous partition. Metal dissolves in the weaker and is deposited from the stronger solution. The e.m.f. depends on the substances and the concentrations, but is usually a few hundredths of a volt.

concentric lens A convexoconcave lens whose surfaces have the same centre of curvature; its central thickness $t = r_1 - r_2$.

condensation 1. The process in which a vapor or gas is transformed to a liquid.

2. The ratio of the instantaneous excess of density to the normal density at a point in a medium transmitting longitudinal sound waves: $s = \delta\rho/\rho$. In plane longitudinal waves it is equal to minus the gradient of particle displacement along the direction in which the wave is travelling, i.e. in terms of the symbols used under *progressive wave, $-\partial\zeta/\partial\chi$.

condensation pump *See* pumps, vacuum; air pumps.

condenser 1. (heat) A device for the continuous removal of heat, e.g. a stream of cold water that removes the latent heat evolved when a vapour condenses to a liquid as in distillation. In a heat engine it is the system or reservoir to which the working substance rejects that part of the heat not converted into work.

2. (light) A mirror or lens combination used in optical instruments (e.g. projectors, compound microscopes) to concentrate light from a source into a defined beam so that the light source can be focused on to an object (which may be an opaque object or a transparent slide). It is usually a planoconvex lens or a pair of planoconvex lenses with plane sides facing out. The design differs widely according to the purpose. A *Fresnel lens is often used as a condenser in a projector. It is arranged to focus an image of the source without serious *aberration within the aperture of the projector. The transparency is placed immediately in front of the condenser. In episcopic projection (opaque reflecting object), a wide-aperture mirror condenser is used. For cinematographic projection, since the film picture is small, an image of the source is projected by a condenser (combination of ellipsoidal mirror and diverging lens) on the plane of the film. The term *microscope condenser* refers to the more elaborate substage lens or mirror combinations, for use with higher-power objectives, so that rays converge without aberration and fill the aperture of the objective uniformly. Its *numerical aperture should not be less than that of the objective. The term is also applied in microscopy to special designs for dark-ground illumination and for simple condensing lens devices. *See* Abbe condenser.

3. (electric) Obsolete term for *capacitor.

condensing electrometer *See* capacitive electrometer.

conditioned observations A class of observations in which all systems of values are not equally possible owing to the existence of conditions that must be exactly satisfied; e.g. when the three angles of a triangle are measured, their sum must be 180°. Such a relationship is known as an "equation of condition".

condition of aplanatism *See* sine condition.

Condon, Edward Uhler (1902–1974) Amer. physicist. He is remembered for his early work on quantum theory, especially in predicting the nuclear motions in molecules that accompany electronic transitions (the *Franck–Condon principle*), for developing a tunnelling model for the emission of alpha particles, and for an analysis of proton–proton scattering experiments, which led to the concept of isotopic spin in the theory of nuclear structure.

conductance Symbol: G. The reciprocal of resistance (for a direct current) or the real part of the *admittance Y, where $Y = G - iB$ and B is the *susceptance (for an alternating current). It is measured in siemens.

conductimetric titration The end-point in an acid–alkali titration can be determined by measurement of the *conductance, which shows a sharp discontinuity at the end point. The method is useful with solutions that are very dilute, or strongly coloured, with which normal indicators are unsuitable.

conduction band *See* energy bands.

conduction current *See* current.

conduction electrons Electrons in the conduction band of a solid. *See* energy bands.

conduction in gases The passage of an electric current through a gas. Because of *background radiation, gases normally contain small numbers of free electrons and positive ions, with some negative ions in a few cases. In the absence of any applied field, the electrons and positive ions recombine at the same rate as they are generated, giving a steady low concentration. If a small potential difference is applied between two electrodes in the gas, a current flows, which increases with p.d. until it reaches a steady value (the *saturation current) at which nearly all the ions and electrons are collected before they can recombine. The magnitude of this current is proportional to the product of the mass of gas between the electrodes and the intensity of the ionizing radiation, values of the order of 10^{-15} A being typical for normal background. (*See* ionizing chamber.)

At higher potential differences, electrons acquire sufficient kinetic energy to create further positive ion–electron pairs on collision, so the current rises above saturation. Positive-ion bombardment of the cathode causes *secondary emission of electrons, further increasing the current. Radiation from excited molecules on recombining ions may contribute by photoelectric emission from the cathode. At a sufficiently high p.d. known as the *breakdown potential* there is an *electric discharge*. According to the conditions there will be an *arc discharge, *glow discharge, *corona, or *spark.

Immediately after the onset of arc or glow discharges the p.d. across the tube falls to a definite value (the *maintenance potential*) and the current becomes constant. The curve relating current to applied p.d. for a given separation of electrodes in a gas at a particular pressure is known as the *discharge characteristic*.

At pressures about one pascal, the mean free path of the electrons becomes comparable with the dimensions of a *gas discharge tube, and the streams of electrons are called *cathode rays. For potential differences of 10^4V and above these rays generate *X-rays on hitting the anode on the walls of the tube.

conduction of heat The transfer of heat through a body not involving radiation or the flow of the material. In gases transfer is caused by collisions between the higher-energy molecules in regions at higher temperatures with the lower-energy molecules in adjacent cooler regions. In dielectric solids the vibrations of molecules are transmitted through the body as waves as a result of the elastic bonding between the particles. These waves are of the same nature as sound but have very much higher frequencies (of order 10^{12} Hz), and their energies are quantized as *phonons. On balance more energy is transferred by phonons from the hotter to the colder parts than is transferred the opposite way. Theories of heat conduction treat the phonons as if they were gas molecules moving within the space occupied by the solid, being scattered by irregularities in the material. Conduction in nonmetallic liquids is intermediate in character between that in solids and that in gases.

In solid and liquid metals conduction is normally almost entirely by means of the valence electrons. These are treated theoretically as gas molecules. *See also* thermal conductivity; Wiedemann–Franz–Lorenz law.

conductivity 1. Symbol: γ or σ. The reciprocal of *resistivity. It is also defined as the current density divided by the electric field strength: this definition is often more useful when considering solutions; it is then known as the *electrolytic conductivity*, symbol: κ. Conductivity is measured in siemens per metre. The term specific conductance was formerly used but is now deprecated owing to the meaning given to *specific.
2. *See* thermal conductivity.

conductor (electrical) A substance, or body, that offers a relatively small resistance to the passage of an electric current.

conduit A tube that carries and protects insulated cables and wires. Types in common use are *plain steel conduit and screwed steel conduit.

cone *See* colour vision.

confinement 1. *See* containment.
2. *See* quark confinement.

conic sections *Syn.* conics. The family of curves formed as sections when a right circular double cone is cut by a plane. The possible sections are: the *ellipse* (with the circle as a special case), the *parabola*, and the *hyperbola* (with a pair of lines as a limiting case). Alternatively, a conic is the locus of a point moving so that its distance from a fixed point (the *focus*) bears to its distance from a fixed line (the *directrix*) a constant ratio (the *eccentricity*, symbol e). If $e < 1$, the conic is an ellipse ($e = 0$ for a circle); if $e = 1$, it is a parabola; if $e > 1$, it is a hyperbola.

The parabola has only one finite focus and corresponding directrix; the other conics have symmetrical pairs of foci and corresponding directrices and a centre of symmetry and hence are *central conics*. In coordinate geometry, an equation of second degree always represents a conic. If the terms of second degree have real factors, the conic is a hyperbola (including the degenerate case of a pair of lines) and has real asymptotes. The simplest forms of the equations are:

Ellipse, $\dfrac{x^2}{a^2} + \dfrac{y^2}{b^2} = 1$, where $b^2 = a^2(1 - e^2)$;

Hyperbola, $\dfrac{x^2}{a^2} - \dfrac{y^2}{b^2} = 1$, where $b^2 = a^2(e^2 - 1)$;

Parabola, $y^2 = 4ax$.

In polar coordinates, the conics may all be represented by:
$$l/r = 1 \pm e\cos\theta,$$
with the pole at one focus and with the appropriate value of e.

conjugate 1. (general) Joined in a reciprocal relation, as two points, lines, quantities, things that are interchangeable with respect to the properties of each. (*See* centre of oscillation, for example.)
2. (light) Of foci, planes, and points. Relating to interchangeable properties of object and image. Thus, if I is the image of O, then if I is made the object, its image would be at O. *Conjugate focus relations* are algebraic equations in which the relation between object and image positions is expressed in terms of the optical constants of the system, e.g.
$$1/v - 1/u = 1/f,$$
where u and v are the object and image distances from the mirror (or lens) and f the focal length. The principal points, nodal points, symmetric points of a system are in turn conjugate pairs; the principal focal points are conjugate with infinity. The points at which object and image are coincident (centre of curvature of a mirror; pole of a mirror or thin lens) are *self-conjugate*. *See* Newton's formula.

conjugate complex quantities Complex quantities that differ only in the signs of their imaginary parts, e.g. $z = x + iy$ and $z' = x - iy$, where $i = \sqrt{-1}$. Such quantities have the same modulus. *See* Argand diagram.

conjugate impedances Impedances that have equal resistance components, and also equal reactance components, the latter having opposite signs. For example, two impedances,
$$Z_1 = R + iX \text{ and } Z_2 = R - iX$$
are conjugate impedances.

conjugate particle *See* antiparticle.

conjunction *See* opposition.

conoid of Sturm *See* astigmatism.

conpernic A magnetic alloy of nickel (50%) and iron (50%), used in magnetic shielding.

consequent poles Magnetic poles in excess of the usual pair in any magnetized body.

conservation of areas A general principle of the kinematics of central orbits, as illustrated by Kepler in his second law of planetary motion. *See* Kepler's laws.

conservation of charge The principle that the total net charge of any system is constant.

conservation of energy *See* conservation of mass and energy.

conservation of mass and energy The *principle of conservation of energy* states that the total energy in any system is constant. The *principle of conservation of mass* states that the total mass in any system is constant. In classical physics these two laws were independent. Although conservation of mass was verified in many experiments the evidence for this was limited. In contrast the great success of theories assuming the conservation of energy established this principle with the highest degree of certainty, and Einstein assumed it as an axiom in his theory of *relativity. According to this theory the transfer of energy E by any process entails the transfer of mass $m = E/c^2$, hence the conservation of energy ensures the conservation of mass.

In Einstein's theory inertial and gravitational mass are assumed to be identical and energy is the total energy of a system. Some confusion often arises because of idiosyncratic terminologies in which the words *mass* and *energy* are given different meanings. For example, some particle physicists use "mass" to mean the rest-energy of a particle and "energy" to mean "energy other than rest-energy". This leads to alternative statements of the principles, in which terminology is not generally consistent. *See* Einstein's law.

conservation of matter *Syn.* conservation of mass. Matter can be neither created nor destroyed. One of the basic principles of 19th-century chemistry. *See* conservation of mass and energy.

conservation of momentum 1. In any system of mutually interacting or impinging particles, the linear momentum in any fixed direction remains unaltered unless there is an external force acting in that direction.
2. Similarly, the angular momentum is constant in the case of a system rotating about a fixed axis provided that no external torque is applied.

conservative field A field of force in which the work done in taking a test particle from one point to another is independent of the path taken between them (e.g. a scalar potential field such as an electrostatic or gravitational field). *See* potential energy.

consistency 1. A quality of a material that depends on its viscosity, plasticity, etc. **2.** A qualitative measure of the agreement between different determinations of the same physical quantity.

consolute *Syn.* critical solution temperature. If two partially miscible liquids, e.g. aniline and hexane, are in contact, a state of equilibrium will be reached with each layer of liquid having dissolved some of the other. If the temperature is increased the relative proportions of the liquids in the layers will alter. At a particular temperature, called the *consolute*, the two layers become identical in composition. In Fig. *a*, the continuous line is a graph of the consolute temperature for varying proportions of the two

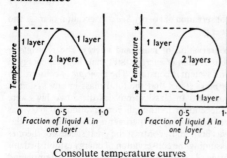

Consolute temperature curves

components. For conditions represented by points above the line, the two liquids are completely miscible and are said to be consolute; below the line, the system forms two distinct layers. Some systems, e.g. nicotine and water, have two consolute temperatures (known as the upper and lower) and the graphs of these against composition meet to form a closed loop as in Fig. b. For systems represented by points inside the loop, there are two distinct layers and for points outside there is only one.

consonance A combination of two or more notes generally accepted as giving a satisfying effect in itself, i.e. irrespective of context. Dissonance describes any combination of notes not forming consonance. Although these terms are mainly subjective, it appears that the degree of consonance is greatest when the two integers that define the interval ratio are small. The following list of intervals (with the ratios given in parentheses) is arranged in approximately decreasing order of consonance: unison (1:1), octave (2:1), perfect fifth (3:2), major sixth (5:3), major third (5:4), perfect fourth (4:3), minor sixth (8:5), minor third (6:5).

Consonance is explained in terms of the nature of the combined sound wave, and this depends on three factors.

(a) Prominence of beat frequencies. Dissonance combinations result from the roughness that is attributed to clearly observed beats. Maximum irritation appears to occur at beat frequencies of between about 25 and 40 Hz.

(b) Harmonic content. This is measured by the number and intensity of the upper partials, since these too can introduce beats.

(c) Blending due to the separation of the component notes. Very close notes (i.e. small intervals such as the semitone or second) and widely separated notes (i.e. large intervals such as the seventh) do not appear to be consonant. See also interval.

constancy of angle, law of Angles between corresponding faces of all crystals of the same chemical substance are always the same and are characteristic of the substance.

constantan An alloy of about 50% copper and 50% nickel with a comparatively high resistivity and low temperature coefficient of resistance. It is used extensively in winding electrical resistors, and with

copper, iron, silver, etc. in forming thermocouples with a comparatively large e.m.f. Because of the large thermoelectric e.m.f. with copper connecting-wire, constantan is unsuitable for high-precision resistors used in d.c. circuits and the more expensive alloy manganin is used. Constantan is suitable for a.c. work and for low-precision d.c. measurements. See also manganin.

constant current source A circuit designed to produce a specified current that is independent of voltage, i.e. a circuit with an infinitely high output *impedance. In practice such high output impedances are only approached over a limited range of output voltages. Symbol: ─∞─

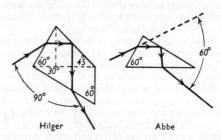

Constant deviation prisms

constant deviation 1. (Reflectors) A ray lying in a principal section of two inclined plane mirrors after successive reflection is deviated through a constant angle equal to twice the angle between the mirrors. When the angle between the mirrors is 45°, the deviation is 90°. Although refraction occurs in the case of the *pentagonal prism, the same deviation of 90° is produced for different angles of incidence on the first face.

2. (Spectroscope, spectrograph, prism-dispersion.) For some types of *spectroscope and *monochromator, the collimator and telescope are fixed in relative position (e.g. 90° or 120°) and by rotation of a dispersing prism (involving also a reflection), light of different wavelength is caused to travel down the telescope axis, showing thereby a constant deviation, which is also the effective minimum deviation of the prism for the light in question.

constant pressure gas thermometer A thermometer in which the volume occupied by a given mass of gas at a constant pressure is used for the measurement of the temperature of the bath in which the bulb containing the gas is immersed. A temperature of t_p on this scale is defined as:

$$t_p = \frac{V_t - V_0}{V_{100} - V_0} \times 100°C \qquad (1)$$

where V_t, V_{100}, and V_0 are the volumes occupied by the gas at the temperature t_p, the steam point, and the ice point respectively.

The scale given by employing an actual gas in the thermometer will differ from the *ideal gas scale, due to the fact that the gas does not obey Boyle's law exactly except at infinitely low pressure. In order to correct the temperatures obtained on this scale to those with an ideal gas use is made of the fact that the behaviour of a gas at constant temperature may be represented by the *virial expansion:

$$pV = A + Bp + Cp^2 + Dp^3 + \ldots \quad (2)$$

Consider a constant pressure thermometer in which the pressure employed is p_0. Then we have from equation (2):

$$p_0 V_0 = A_0 + B_0 p_0,$$
$$p_0 V_{100} = A_{100} + B_{100} p_0,$$
$$p_0 V_t = A_t + B_t p_0,$$

as at the pressures employed in gas thermometry C and D are not significant.

Thus, substituting in (1) we have:

$$t_p = 100 \frac{(A_t - A_0)}{(A_{100} - A_0)} \left\{ \frac{1 + \dfrac{p_0(B_t - B_0)}{(A_t - A_0)}}{1 + \dfrac{p_0(B_{100} - B_0)}{(A_{100} - A_0)}} \right\}.$$

Then using the binomial theorem and neglecting quantities of the order $(Bp_0)^2/(A)^2$ we have:

$$t_p = 100 \frac{(A_t - A_0)}{(A_{100} - A_0)}$$

$$\times \left\{ 1 + \frac{p_0(B_t - B_0)}{(A_t - A_0)} - \frac{p_0(B_{100} - B_0)}{(A_{100} - A_0)} \right\}.$$

Now $100 \left[(A_t - A_0)/(A_{100} - A_0) \right]$ is the temperature t' that would have been obtained had the gas been ideal or the initial pressure infinitely small, so that the correction to be applied to the reading t_p of the gas thermometer in order to reduce it to t' on the ideal gas scale is given by:

$$(t_p - t') = t' \left\{ p_0 \frac{(B_t - B_0)}{(A_t - A_0)} - p_0 \frac{(B_{100} - B_0)}{(A_{100} - A_0)} \right\}. \quad (3)$$

Now $$(A_t - A_0) = \frac{A_{100} - A_0}{100} t'$$

and $$A_t = A_0 \left\{ 1 + \frac{(A_{100} - A_0)}{100 A_0} t' \right\}.$$

Putting $$\frac{(A_{100} - A_0)}{100 A_0} = a,$$

then $$A_t - A_0 = A_0 a t'$$

i.e. in (3)

$$(t' - t_p) = p_0 \left\{ \frac{(B_{100} - B_0)}{100 A_0 a} t' - \frac{(B_t - B_0)}{A_0 a} \right\}. \quad (4)$$

This gives the correction that must be applied to the readings of a constant-pressure thermometer working at pressure p_0 in order to obtain the temperatures that would be given by a thermometer employing the same gas at infinitely low pressure, or an ideal gas. As t' occurs in the correction term, and yet

is not known until after the correction is applied, it is necessary to use the method of successive approximations, the value t_p first being used in the right-hand side of equation (4) and an approximate value of the correction thus obtained used to give a value of t', which is again substituted in the right-hand side of (4), and so on. In actual practice the first approximation is sufficient as the actual correction is very small.

For details of a constant pressure thermometer often used in practice, see Callendar's compensated air thermometer. See also gas scales of temperature.

constant volume gas thermometer A thermometer in which the pressure exerted by a constant volume of gas is used for the measurement of the temperature of the bath in which the bulb containing the gas is immersed. A temperature of t on this scale is defined as:

$$t = \frac{p_t - p_0}{p_{100} - p_0} \times 100°C \quad (1)$$

where p_0, p_{100}, and p_t are the pressures exerted by the gas when at the ice point, the steam point, and at the temperature t respectively. Using hydrogen or nitrogen gas in a platinum-iridium or platinum-rhodium bulb, temperatures from -260 °C to 1600 °C may be standardized. (See Chappuis and Harker's thermometer.) There are two chief errors causing the temperature on the gas scale to differ from the *thermodynamic temperature: (1) the gas is not ideal so that the product pV is equal to $(A + Bp)$ and only becomes independent of the pressure if the pressure is very small. From compression experiments at constant volume carried out at the ice and steam points and at t we have:

$$\left. \begin{array}{l} p_0 V_0 = A_0 + B_0 p_0 \\ p_{100} V_0 = A_{100} + B_{100} p_{100} \\ p_t V_0 = A_t + B_t p_t \end{array} \right\}$$

so from (1):

$$t = \left(\frac{A_t - A_0}{A_{100} - A_0} 100 \right)$$

$$\times \left\{ 1 + \frac{B_t p_t - B_0 p_0}{A_t - A_0} - \frac{B_{100} p_{100} - B_0 p_0}{A_{100} - A_0} \right\}.$$

Now

$$\left(\frac{A_t - A_0}{A_{100} - A_0} \times 100 \right)$$

where t' is the temperature that would have been obtained had the gas been ideal or the initial pressure infinitely small so that the correction to be applied to the reading t of the gas thermometer in order to reduce it to t' on the work scale is:

$$(t - t') = t' \left(\frac{B_t p_t - B_0 p_0}{A_t - A_0} - \frac{B_{100} p_{100} - B_0 p_0}{A_{100} - A_0} \right).$$

111

constringence

The correction is determined at a series of fixed points and the correction at intermediate points obtained by interpolation on the correction/temperature graph thus obtained. (2) The volume of the gas is not constant and not all at the same temperature. Let V be the volume of the bulb and H (in mmHg) be the pressure of the gas at 0 °C. If v is the volume of the *dead space that is at a temperature θ throughout the experiment and α is the pressure coefficient of the enclosed gas, then the total volume of enclosed gas measured at STP is:

$$V_0 = \left(V + \frac{v}{1 + \alpha\theta} \right) \frac{H_0}{760}.$$

When the bulb is heated to t the pressure becomes $(H_0 + h)$ so that:

$$V_0 = \left(\frac{V(1 + \gamma t) + bh}{1 + \alpha t} + \frac{v}{1 + \alpha\theta} \right) \frac{H_0 + h}{760}$$

where y is the coefficient of cubical expansion of the material of the bulb and b is the increase in volume of the bulb for a pressure increase of 10 mmHg. Equating these two expressions for the total volume V_0 we have:

$$\alpha t = \frac{H_0 + h}{H_0} \cdot \frac{V(1 + \gamma t) + bh}{V} + \frac{hv}{H_0 V} \frac{1 + \alpha t}{1 + \alpha\theta} - 1$$

in which t, the temperature to be determined, ocurs on the right-hand side also. The uncorrected value of $t = h/H_0\alpha$ is first used on the right-hand side to give a more correct value for t and by the method of successive approximations the corrected value may be obtained. *See* gas scales of temperature.

constringence *See* Abbe number.

contact angle The angle, lying between 0° and 180°, between the normal to the liquid surface, drawn into the substance of the liquid, and the normal to the solid, directed towards the substance of the solid. It is zero if the liquid wets the solid perfectly. (*See* wetting.) For mercury and glass it is about 140°. The value of the contact angle is determined by the relative amounts of *adhesion (liquid to solid) and *cohesion (liquid to liquid). If the contact angle is zero the adhesion is as strong as the cohesion; if the angle is 180° there is no adhesion whatsoever – a case not met in practice. The contact angle is often different according to whether the liquid has advanced over the surface or receded back along it (e.g. if a raindrop moves down a window pane, the lower edge has a larger contact angle than the upper). This is called *hysteresis* of the angle of contact.

A contact angle between two immiscible liquids can be similarly defined.

contactor A type of switch for making and breaking an electric circuit, designed for frequent use as, for example, in the control equipment of traction motors. Its operation is electromagnetic, electropneumatic, or mechanical.

contact potential The difference of potential that arises when two conductors of different material are placed in contact. Thus a metal will be at a different potential from a conducting liquid in which it is immersed; an electric field will exist in the space between plates of different metals, when the plates are electrically connected. The contact potential is usually of the order of a few tenths of a volt. It results from the difference between the *work functions of the two metals.

contact resistance The resistance at the surface of contact of two conductors. An important example is the *brush contact resistance between the carbon brushes and copper commutator segments in a d.c. machine.

containment 1. *Syn.* confinement. The process of preventing the *plasma from coming into contact with the walls of the reaction vessel in a controlled *thermonuclear reaction. The time for which ions are trapped in the plasma is called the *containment time.

2. The prevention of the release of unacceptable quantities of radioactive material beyond a controlled zone in a nuclear reactor.

3. The containment system of a nuclear reactor.

continuity principle For continuous motion, the increase of mass of fluid in any time interval δt within a closed surface drawn in the fluid is equal to the difference between the mass flow in and the mass flow out through the surface. This statement is expressed mathematically in the *equation of continuity*:

$$\frac{\partial \rho}{\partial t} + \frac{\partial(\rho u)}{\partial x} + \frac{\partial(\rho v)}{\partial y} + \frac{\partial(\rho w)}{\partial z} = 0$$

where ρ is the density of the fluid at time t and (u, v, w) are the Cartesian components of the vector velocity V at the space point (x, y, z). The continuity equation may be expressed in other coordinate forms.

continuous flow calorimeter A type of calorimeter in which heat is supplied at a constant rate to fluid flowing at a constant rate. A steady state is eventually reached when all temperatures remain constant with time so that small temperature differences may be accurately determined without any error due to lag of the thermometers. The heat capacity of the calorimeter does not enter into the thermal equation and since all temperatures are steady the external loss of heat by radiation or other means is more regular and certain. *See* Callendar and Barnes; Laby and Hercus. *See also* calorimetry.

continuous loading *See* loading.

continuous phase The medium in which a substance is dispersed in a *collidal solution or suspension. *See* disperse phase.

continuous spectrum *See* spectrum.

continuum A continuous series of components or elements that together form a reference system. The three dimensions of space and the dimension of time together form a *four-dimensional continuum.

contraction coefficient *See* coefficient of contraction.

contrast *Object contrast* is the ratio of the difference between background brightness and object brightness to the background brightness. Another definition is the ratio of the difference to the sum of background and object brightness. *Contrast sensitivity* of the eye is expressed as the *Fechner fraction (or its reciprocal), i.e. the least detectable difference of brightness as a fraction of the mean brightness of two adjacent fields.

control electrode An electrode to which the input signal voltage is applied to produce changes in the currents of one or more of the other electrodes. In a bipolar *transistor with *common-emitter connection the control electrode is the base electrode, in a *field-effect transistor it is the gate electrode, in a *cathode-ray tube it is the *modulator electrode, in a *thermionic valve it is the control grid.

control grid Of a *valve. A grid used as a *control electrode.

controlled-carrier modulation *See* floating-carrier modulation.

controlled thermonuclear reaction *See* nuclear fusion.

controller *See* motor controller.

control rod One of a number of rods that can be moved up or down along its axis into the *core of a *nuclear reactor to control the rate of the *chain reaction. The rods usually contain a neutron absorber, such as cadmium or boron.

convection (of heat) The process of transfer of heat in a fluid by the movement of the fluid itself. There are two distinct types of convection:
(1) *Natural* (or *free*) *convection*, when the motion of the fluid is due solely to the presence of the hot body in it giving rise to temperature and hence density gradients, the fluid thus moving under the influence of gravity.
(2) *Forced convection*, in which a relative motion between the hot body and the fluid is maintained by some external agency (e.g. a draught), the relative velocity being such as to make the contribution of the gravity currents negligible.

A theoretical treatment of convection is best achieved by means of dimensional analysis using mass, length, time, and temperature as primary dimensions to obtain expressions for h – the rate of heat transfer per unit area of surface of the hot body.

For dynamically similar bodies it is shown that for natural convection:

$$\left(\frac{hl}{\lambda\theta}\right) = f_1\left(\frac{l^3 g\alpha\rho^2\theta}{\eta^2}\right)f_2\left(\frac{C\eta}{\lambda\rho}\right).$$

where θ is the temperature excess, l is a particular linear dimension of the body, λ is the thermal conductivity of the fluid, α is the density temperature coefficient, C is the heat capacity of unit volume of the fluid, η/ρ is the kinematic viscosity of the fluid, and g is the acceleration of free fall.

The expression contains three dimensionless groups, namely: $(hl/\lambda\theta)$, the *Nusselt number; $(l^3 g\alpha\rho^2\theta/\eta^2)$, the *Grashof or free convection number; $(C\eta/\lambda\rho)$, the *Prandtl number. The form of the functions f_1 and f_2 must be assumed to be dependent on the shapes of the bodies, etc. involved.

In the case of forced convection the expression takes the form:

$$\left(\frac{hl}{\lambda\theta}\right) = F_1\left(\frac{lv\rho}{\eta}\right)F_2\left(\frac{C\eta}{\lambda\rho}\right),$$

introducing in addition to the Nusselt and Prandtl numbers, the *Reynolds number $(lv\rho/\eta)$, where v is the relative velocity of the fluid to the body.

In the case of a solid surface losing heat by natural convection in a fluid in contact with it, there is always a layer of fluid at rest relative to the solid, heat being transferred through this layer by thermal conduction into the bulk of the moving fluid. In free convection this film is stationary while for forced convection it is continuously being removed and renewed.

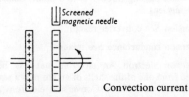

Convection current

convection current 1. (heat) A stream of fluid, warmer or colder than the surrounding fluid and in motion because of the buoyancy forces arising from the consequent differences in density.
2. (electricity) A moving electrified body constitutes an electrical convection current, and this type of current can flow without potential difference or energy change, and produces no heat. It can nevertheless produce a magnetic effect, as is shown by rotation of the plates of a charged capacitor near an electrostatically screened magnetic needle that is deflected when rotation occurs (*see* diagram).

convectron An instrument to give electrical indication of deviation from the vertical. The convection cooling of a straight, fine wire is much greater when the wire is horizontal than when vertical. In the tube shown in the diagram, an electric current heats the

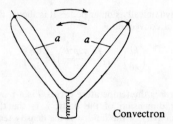

Convectron

two wires (*a*, *a*) that are in a bridge circuit arranged to be balanced in the position shown. If the tube is displaced in the direction of either arrow, the cooling effects differ in the two limbs with consequent increased temperature of one wire, and decreased temperature of the other, which alters the resistance of the wires and throws the bridge out of balance.

conventional current The concept of a current that flows from positive to negative, i.e. in the opposite direction to the electron flow. This convention, which originated before the electronic nature of a current was understood, is still used by some authors.

convergence A reciprocal distance used to describe the state of convergence of rays, regarded as positive vergence. Reduced convergence is the reciprocal of a reduced distance. (*See* dioptre.) The convergence of the visual axes of the two eyes is necessary for binocular single vision; this use has a physiological context.

convergence ratio *Syn.* angular magnification. The ratio of the tangents of the slopes of conjugate rays to the axis.

converging lens A *lens that can bring a parallel light beam passing through it to a point. *Compare* diverging lens.

conversion *See* converter reactor.

conversion conductance *See* mixer.

conversion electron An electron that has been ejected from one of the shells of an atom as a result of *internal conversion. Conversion electrons are often responsible for discontinuities in the electron spectrum (*see* electron spectroscopy) of nuclei that are undergoing *beta decay.

conversion factor 1. A factor by which a quantity expressed in one set of units must be multiplied in order to express that quantity in different units. **2.** *Syn.* conversion ratio. In nuclear physics, the ratio of the number of *fissile atoms produced from the fertile material in a *converter reactor to the number of fissile atoms of fuel destroyed in the process.

converter In general, a machine for converting electric current of one kind into current of another kind. An example is the *rotary* (or *synchronous*) converter that is used for changing polyphase alternating currents (*see* polyphase system) into direct currents or, less frequently, the reverse (inverted running). This type of converter is a synchronous machine in which

there is a single armature winding connected to a *commutator. Tappings from the winding are also brought out to *slip rings. As normally used, the a.c. supply is connected to brushes on the slip rings and the d.c. output is taken from the brushes on the commutator. Another example is the *motor converter*, which consists essentially of an *induction motor coupled to a rotary converter both mechanically and electrically. The rotor of the induction motor is coupled mechanically and the a.c. input to the rotary converter is obtained from the rotor winding of the induction motor.

converter reactor A *nuclear reactor in which *fertile material is transformed by a nuclear reaction into *fissile material. This process is known as *conversion*. A converter reactor can also be used to produce electric power. *See also* conversion factor; fast breeder reactor.

convex Curving outwards. A *convex mirror* is diverging in action. A *convex lens* is thicker at the centre. Thin lenses are classed as *biconvex, planoconvex*, and *concavoconvex* (the last mentioned is generally described as *convex meniscus*). (*See* diagram under lens.) While thin convex lenses are converging in action, thicker lenses may be telescopic, diverging, or convergent according to thickness. *See* concave.

coolant A fluid used to reduce the temperature of a system by conducting away heat produced by the system.

In a *nuclear reactor the coolant transfers heat from the *core to the steam-raising plant or to an intermediate heat exchanger. In *gas-cooled reactors the coolant is usually carbon dioxide. In *boiling-water and *pressurized-water reactors, water acts as both coolant and *moderator: in *heavy-water reactors, *heavy water fulfils this dual role. In *fast breeder reactors the need to transfer a large quantity of heat through a small surface area necessitates a liquid-metal coolant (e.g. sodium).

Coolidge tube An X-ray tube invented by W. D. Coolidge (1910) in which the electron stream is derived from a heated cathode by thermionic emission (as distinct from early tubes using ionic bombardment of a cold cathode) and focused on the target by a metal cylinder surrounding the cathode.

cooling correction *See* radiation correction; calorimetry.

cooling curve A temperature/time curve used for the determination of melting points or for applying a radiation or cooling correction. In the diagram, the portions AB for the liquid and CD for the solid are roughly exponential curves for which *Newton's law of cooling holds. The horizontal portion BC indicates that the latent heat liberated by solidification of the liquid is being given to the surroundings without any fall of temperature occurring until at C all the liquid has solidified. The temperature indicated by *t* is thus the melting point.

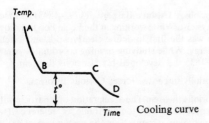

Cooling curve

cooling method The determination of the specific heat capacity of a liquid by comparing the time taken for the liquid and an equal volume of water to cool in identical vessels through the same range of temperature. Liquids that have the same cooling surface and the same temperature-excess over their surroundings lose heat at the same rate. Hence, the surface area and radiating power being the same for the two vessels, one containing the liquid and the other water, then:

$$(Mc_2 + m_1c_1)/t_1 = (Mc_2 + mc)/t,$$

where M and c_2 are the mass and specific heat capacity of each vessel; m_1 the mass of liquid of specific heat capacity c_1; m the mass of water and c its specific heat capacity; and t_1 and t are the times taken for the liquid and the water to fall through the same range of temperature, say, θ_1 to θ_2.

Cooper, Leon N. (b. 1930) Amer. physicist, who collaborated with John Bardeen and John Schrieffer in the formulation of the BCS theory of *superconductivity (1957).

Cooper pair See superconductivity.

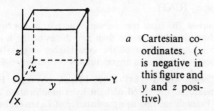

a Cartesian coordinates. (x is negative in this figure and y and z positive)

coordinate 1. One of the quantities used to define the position of a point relative to a *frame of reference. There are three main coordinate systems:

(1) *Cartesian coordinates.* Three mutually perpendicular lines OX, OY, OZ (Fig. a) are drawn through a point O known as the *origin*. These lines and O are fixed in the frame of reference, and in abstract work may themselves be the frame of reference. The position of a point P relative to these axes is given by the perpendicular distances of P from the three coordinate planes ZOY, XOZ, YOX; these distances x, y, z, are the coordinates of P. The axes are "right-handed" or "left-handed" according to whether turning OX towards OY (through the smaller angle) would drive a right-handed screw along the z-axis in a positive or negative direction respectively. Right-handed axes are most often used.

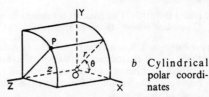

b Cylindrical polar coordinates

(2) *Cylindrical polar coordinates.* The position of a point P is specified by three coordinates: radial distance r, azimuthal angle θ, and axial distance z (Fig. b), these being related to the Cartesian system by

$$x = r\cos\theta, \ y = r\sin\theta, \ z \equiv z$$

These coordinates are especially useful if the system has some degree of symmetry about OZ, the *polar axis*.

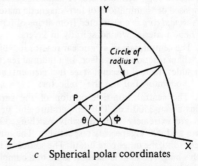

c Spherical polar coordinates

(3) *Spherical polar coordinates.* The coordinates of P (Fig. c) are the radius r, the angle of colatitude (or azimuthal angle) θ, and the angle of longitude ϕ. These are related to the Cartesian system by

$$x = r\sin\theta\cos\phi, \ y = r\sin\theta\sin\phi, \ z = r\cos\theta$$

This system is useful if the system has some symmetry about the point O.

In two-dimensional problems only two coordinates need to be specified; these are (x,y) in Cartesian, or (r,θ) in plane polar coordinates. r is often called the *radius vector* or *radius* in spherical and plane polar coordinates.

Curvilinear coordinates are any sets of parameters that define the position of a point as the intersection of curves or of curved surfaces. Thus, (2) and (3) are special cases. Latitude and longitude are curvilinear coordinates for the surface of the earth.

2. Generalized coordinates. *See* degrees of freedom.

coordinate bond See covalent bond.

coordination lattice A crystal lattice in which each ion bears the same relation to the neighbouring ions in all directions, so that the identity of the molecules becomes ambiguous.

Copernican system See Copernicus.

Copernicus, Nicolaus (1473–1543) Polish astronomer (original name Koppernigk). He affirmed, contrary to the accepted theory of his day, that rotation of the heavenly bodies took place round the sun and not round the earth. This concept is referred to as the *Copernican system*. *Compare* Ptolemy.

copper loss *Syn.* I^2R loss. The power loss in watts due to the flow of electric current in the windings of an electrical machine or transformer. It is equal to the product of the square of the current and the resistance of the winding.

Corbino effect If a current is passed from the centre to the circumference of a metal disc, the surface of which is normal to a magnetic field, a current will flow round the circumference.

core 1. The ferromagnetic portion of the magnetic circuit of an electromagnetic device. A simple *ferrite core* is a solid piece of ferromagnetic material in the shape of a ring, cylinder, etc. A *laminated core* is composed of *laminations of ferromagnetic material. A *wound core* is constructed from strips of ferromagnetic material wound spirally in layers.

2. The central part of a *nuclear reactor in which the *chain reaction takes place. In a thermal reactor it includes the fuel assembly (*see* fuel element) and the *moderator, but not the *reflector.

3. The central iron-rich portion of the earth, radius about 3500 km. The temperature and pressure are extremely high, possibly reaching 5000 kelvin and 400 gigapascals respectively. The inner core is solid, the outer core liquid. The source of the earth's magnetic field is thought to lie in a complex dynamo action in the outer core. *See* geomagnetism.

4. A small ferrite ring formerly used in a computer *memory to store one *bit of information. The direction of magnetization of the core was sensed and read as 0 or 1.

cored electrode A metal electrode used in electric welding that has a core of flux.

core loss *Syn.* iron loss. The total power loss in the iron core of a magnetic circuit when subjected to cyclic changes of magnetization such as, for example, that which occurs in the core of a transformer. The loss is due to magnetic *hysteresis and *eddy currents. It is usually expressed in watts at a given frequency and value of the maximum flux density.

core plate *See* lamination.

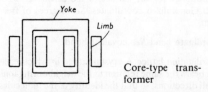

Core-type transformer

core-type transformer A *transformer in which the windings enclose the greater part of the laminated *core. The windings are formed around the *yoke*, which is built up from a stack of laminations (*see* diagram). Additional laminations form the *limbs* around each winding and complete the core. *Compare* shell-type transformer.

Coriolis, Gustave-Gaspard (1792–1843) French physicist. While working at the Ecole Polytechnique he was the first to define clearly work and kinetic energy. While studying rotating coordinate systems in 1835, he developed his *Coriolis theorem.

Coriolis force *See* force; Coriolis theorem.

Coriolis theorem The acceleration of a particle with respect to a Newtonian frame of reference (*see* Newtonian system) N, is the vector sum of (*a*) its acceleration with respect to some other frame of reference S, which is in motion relative to N, (*b*) the acceleration of N relative to S, and (*c*) the *Coliolis acceleration* (sometimes called the *complementary acceleration*), which equals twice the vector product of the angular velocity of S with respect to N and the linear velocity of the particle with respect to S.

The quantity (mass × Coriolis acceleration) has the dimensions of force and the Coriolis force has this magnitude but is oppositely directed from the Coriolis acceleration. It is an inertial force. *See* force, for a simple example.

corkscrew rule A rule for determining the direction of lines of magnetic force around a wire carrying a current. If a corkscrew is imagined to be turned in the manner necessary to drive it along in the direction of the current, then the lines of force are in the same sense as the rotation of the head of the corkscrew.

Cormack, Allan Macleod (*b.* 1924) South-African-born physicist, who became professor at Tufts University in the USA. Independently of the British scientist Godfrey Hounsfield, he developed the mathematical basis for computed-assisted *tomography (CAT).

cornea The front transparent outer coat of the *eye that contributes more than two-thirds of the total converging power of the eye. Higher degrees of ocular astigmatism are commonly corneal in origin.

Cornu, Alfred (1841–1902) French physicist. He measured the speed of light by a modification of *Fizeau's method using a distance of 23 km (1874).

Cornu double prism *See* quartz.

Cornu–Hartmann formula *See* Hartmann formula.

Cornu's spiral *See* diffraction of light.

corona 1. An electric discharge appearing round a conductor when the potential gradient at the surface is raised above a critical value, so that a partial breakdown of the surrounding gas takes place.

2. The outermost region of the sun's atmosphere: it is visible as a faint halo during a solar *eclipse. It has an approximate temperature of one million degrees and contains highly ionized atoms (iron can lose up to 13 electrons). *See* chromosphere.

corpuscular theory The theory, which has been proposed in various forms from time to time, that light consists of particles. A luminous body was supposed to emit small elastic particles with the speed of light.

They travelled in straight lines in isotropic media, were repelled on reflection, suffered change of direction by attraction on refraction. Since the corpuscular theory required a faster rate of travel in optically denser media, which is at variance with the Foucault experiment (1850) (*see* Foucault's method), the wave theory supplanted it. Although Newton's name is commonly associated with the corpuscular theory, he only threw out the idea as a suggestion. The *wave theory* offers a readier explanation of interference, diffraction, and polarization but fails to explain the interaction of light with matter, the emission and absorption of light, photoelectricity, dispersion, etc.; these can only be explained by a quasi-corpuscular theory involving packets of energy – light quanta or photons. It thus appears that two models are required to explain the phenomenon of light, according to Bohr's principle of *complementarity. *See* quantum theory.

correcting plate *Syn.* corrector. A thin lens or lens system used to correct *spherical aberration in spherical mirrors and *coma in parabolic mirrors. It is used especially in reflecting telescopes, such as the *Schmidt telescope.

correlation A reciprocal relationship between two variables *x* and *y*. Sometimes due to inability to control the conditions of an experiment, measurements may only indicate in a vague way that *x* and *y* are related. Statistical methods may then be applied to the results to determine whether or not this apparent connection is significant. A quantity called the *correlation coefficient r* is evaluated; $r = 0$ corresponds to no connection whatever, and $r = 1$ to perfect correlation.

correspondence principle The principle, due to *Bohr, that since the classical laws of physics are capable of describing the properties of macroscopic systems, the principles of *quantum mechanics, which are applicable to microscopic systems, must give the same results when applied to large systems. For example, the electrons in Bohr's theory of the *atom can only occupy certain orbits. It is found that for larger orbits the behaviour of the atom becomes more like the behaviour expected from classical mechanics.

corresponding states States of different gases all possessing the same values for the reduced pressure, reduced temperature, and reduced volume. The law of corresponding states maintains that if two of these reduced quantities are the same for each gas, then the third must be the same for each. *See* reduced equation of state.

corundum (Al_2O_3). A crystalline variety of alumina, of which three varieties are known, namely, ruby, sapphire, and emery. Corundum is one of the most refractory materials known, its melting point being over 1950 °C. It is nonmagnetic and a very poor conductor of electricity. It is widely used as an abrasive, emery being an impure corundum, mixed with magnetite and haematite.

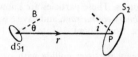

Intensity of illumination (cosine law)

cosine law 1. (light) (i) *See* Lambert's law; photometry. (ii) *Of the intensity of illumination* on an inclined surface. The intensity varies as the cosine of the angle of incidence; the intensity is independent of the nature of the surface and has a purely geometrical connotation. If *B* is the normal brightness of an element dS_1, delivering light in a direction θ to strike the surface S_2 at *P*, at an angle of incidence *i*, the intensity of illumination at *P* is equal to

$$\cos i(B.dS.\cos \theta)/r$$

2. (due to *Knudsen) A law determining the probability of a gaseous molecule leaving a surface in a given direction. The number of molecules leaving unit area in the solid angle $d\omega$ making an angle θ with the normal to the surface, per second is proportional to $\cos \theta d\omega$.

cosmic abundance The relative proportion of each element found in the universe, measured in terms of mass or numbers of atoms. In terms of mass, there is approximately 73% hydrogen, 25% helium, 0.8% oxygen, 0.3% carbon, 0.1% neon and nitrogen, 0.07% silicon, 0.05% magnesium.

cosmic background radiation Diffuse radiation from space, detected at many wavebands throughout the electromagnetic spectrum. At radio, infrared, X-ray, and possibly γ-ray wavelengths, it is thought to be the cumulative contribution of many unresolved and individually weak sources. The most important form, however, is the *microwave background radiation*. This background peaks at a wavelength of about 1 mm and is *black-body radiation characteristic of a temperature of 2.9 K. It is very nearly equal in intensity from all directions of the sky, i.e. it is very nearly isotropic. It is now assumed to be the remnant of the radiation content of the very early hot phase of the universe. *See* big-bang theory.

cosmic dust Fine particles of matter, smaller than 10^{-2} mm, that are distributed throughout space and occur in an extensive belt in our galaxy.

cosmic rays Highly energetic particles that move rapidly through space and continuously bombard the earth's atmosphere from all directions. They were originally discovered because they ionize the air to a small extent and cause the charge to leak from an electroscope. Experiments in balloons showed that cosmic radiation increases with height and thus that it must originate outside the earth's atmosphere. It can be detected by *counters, photographic emulsions, *bubble chambers, *cloud chambers, etc. Cosmic rays consist mainly of nuclei of the most abundant elements in the universe, primarily protons. Also present are a small number of electrons, positrons, antiprotons, neutrinos, and gam-

cosmic strings

ma-ray photons. These particles are known collectively as *primary cosmic rays*, and have a wide range of energies, from about 10^8 to over 10^{20} electronvolts.

The primary particles collide with oxygen and nitrogen nuclei in the atmosphere and these events, together with subsequent decays and interactions of the resulting particles, lead to large numbers of *elementary particles and photons constituting *secondary cosmic rays*. A large number of particles can be formed from one primary particle and this is known as an *air shower*. The initial products are principally charged and neutral *pions. Muons and neutrinos are formed from the subsequent decay of charged pions:

$$\pi^+ \rightarrow \mu^+ + \nu, \, \pi^- \rightarrow \mu^- + \bar{\nu}.$$

Muons do not interact strongly with matter and a large proportion are detected at the earth's surface and even in deep mines. A neutral pion (π^0) decays into two very energetic gamma-ray photons with energies of about 70 MeV. These each can produce one electron and one positron in passing near the nucleus of an atom (*see* pair production). Each particle then loses energy and emits *bremsstrahlung radiation – more photons – which produce yet more electrons and positrons, and so on. Thus the original neutral pion leads to a large number of electrons and positrons, called a *cascade* or *cascade shower*.

Occasionally a single particle with an energy greater than 10^{15} eV may enter the atmosphere and lead to a large number of secondary cosmic rays, extending over a large area. This is called an *extensive air shower* or *Auger shower*.

The flux of secondary particles at sea level is very low, being greatest at lower energies – several thousand particles per square metre per second. The flux varies with latitude because charged particles are affected by the earth's magnetic field, and is a minimum at the equator. Particles must have a minimum energy before they can overcome the earth's field and enter the atmosphere. There is also an *east-west effect* in that more particles approach from the west than the east. The maximum effect is at the equator where there is a 14% excess. The effect shows that cosmic rays have a positive charge.

The origin of cosmic rays is still uncertain. It is thought that almost all cosmic rays with energies less than 10^{18} eV are generated by sources within our Galaxy: *supernova explosions may produce medium- and low-energy particles, solar flares produce only very low-energy particles. It is also thought that these particles are confined within the Galaxy for millions of years by the weak magnetic field of the Galaxy, their directions of travel becoming almost uniformly scattered.

cosmic strings Entities that are solutions of certain types of *grand unified theories. Cosmic strings are associated with non-trivial *topology and may have important consequences in *cosmology, including the formation of galaxies. *See also* string theory.

cosmogeny The study of the origin and development of the universe or of a particular system in the universe, such as the solar system or a satellite system of a planet.

cosmology The branch of astronomy concerned with the evolution, general structure, and nature of the universe as a whole. *See also* big-bang theory.

cosmotron The name of the *proton synchrotron at Brookhaven, US, that accelerates protons to an energy of 3 GeV.

Cotton–Mouton effect Some isotropic transparent solids and liquids show slight double refraction towards light when in a strong magnetic field. *See* Kerr effects.

coudé system *See* astronomical telescope.

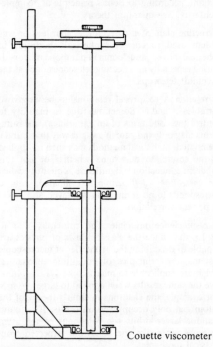

Couette viscometer

Couette viscometer Suppose two vertical coaxial cylinders of unit height stand with the space between them filled with a liquid of viscosity η. The outer one is rotated at constant angular velocity Ω and the inner is suspended by a torsion wire. Consider an annulus of liquid δr from the common axis. If the angular velocity of the inner periphery of this annulus is ω and that of the outer ($\omega + (\delta\omega/\delta r)\delta r$), then $(r(\delta\omega/\delta r).\delta r)$ is the difference of *linear* velocity and $r(\delta\omega/\delta r)$ is the gradient of this velocity across the interspace.

The moment about the axis of the viscous force is therefore $2\pi r^2 \mu(\delta\omega/\delta r)$ per unit length of the cylinders. It can be deduced from this, taking a to be the radius of the inner and b that of the outer cylinder, that the moment is:

$$M = 4\pi\eta\Omega \frac{a^2b^2}{b^2 - a^2}.$$

In Couette's form of the apparatus, the cylinder of liquid has a depth l, of the order of 20 cm and the moment on the inner cylinder is therefore Ml (see diagram). The outer cylinder is rotated by a motor through gearing and the moment on the inner measured through the twist of its suspension wire.

A similar formula applies to Searle's form of the apparatus in which the inner cylinder is rotated while the outer is twisted against its suspension. The Couette instrument has been modified for use with gases.

coulomb Symbol: C. The *SI unit of electric *charge, defined as the charge transported in one second by an electric current of one ampere.

Coulomb, Charles Augustin (1736–1806) French physicist. From researches on the torsional elasticity of wires and hairs, he was led to the invention of the torsion balance (1777), which he used to establish that the law of inverse squares held for electric charges (1785) (see Coulomb's law). He also investigated the effects and magnitude of electrostatic fields (see Coulomb's theorem). He discovered also the laws of friction.

The unit of charge (*coulomb) is named in his honour.

Coulomb field The *electric field around a point charge.

Coulomb force A force of attraction or repulsion resulting from the interaction of the *electric fields surrounding two charged particles. The magnitude of the force is inversely proportional to the square of the distance between the particles.

coulombmeter An instrument in which the electrolytic action of a current is used for measurement of the quantity of electricity passing through a circuit. Also called *voltameter*, though the term is now rarely used because of its similarity to voltmeter. A silver coulombmeter was formerly used in defining the international units of quantity of electricity.

Coulomb scattering The *scattering of charged particles, such as alpha particles, by nuclei as a result of the electrostatic forces between them. If the incident beam contains one alpha particle per unit area, then the number of particles, w, per unit solid angle that suffer a deflection ϕ is given by:

$$w(\phi) = \left(\frac{Z_1 Z_2 e^2 m}{4\pi\varepsilon_0 p^2}\right)^2 \frac{1}{\sin^4(\phi/2)}$$

where $Z_1 e$ and $Z_2 e$ are the charges of the scattered and scattering particles and m and p are the mass and momentum of the scattered particle. Measurements of the angles of scattering and their variation with particle energy give information on the nature

of nuclei and on interactions between particles and nuclei.

Coulomb's law From researches on the torsional elasticity of wires Coulomb designed a torsion balance, which he used to establish the inverse square law associated with his name. This states that the mutual force F exerted by one electrostatic point charge Q_1 on another Q_2 is proportional to the product of the charges divided by the square of their separation d:
$$F = Q_1 Q_2 / 4\pi\varepsilon d^2,$$
where ε is the absolute *permittivity of the medium in which they are situated.

Coulomb's theorem The intensity E of an electric field near a surface possessing a surface density of charge σ is given by
$$E = \sigma/\varepsilon,$$
where ε is the absolute *permittivity of the medium in which they are situated.

counter 1. Any device for detecting and counting individual particles and photons. The term is used for the detector and for the instrument itself. Most detectors work by multiplication of the number of ions or electrons formed by a single particle or photon; each ionizing event leads to a pulse of current or voltage and these are electronically counted. (See Geiger counter, crystal counter, proportional counter, semiconductor counter, and scintillation counter.)
2. Any electronic circuit that records and counts pulses of current or voltage. See also counter/frequency meter.

counter/frequency meter An instrument containing a frequency standard, usually a piezoelectric oscillator, that can be used as a counter or frequency meter by counting the number of events, or cycles, in a specified time. It may also be used to measure the time between events by counting the number of standard pulses occurring during a given number of events or cycles.

couple A system composed of, or equivalent to, two equal and antiparallel forces. The *moment is equal to the product of either force by the perpendicular distance between them and is the same about any axis perpendicular to the plane of the forces. It is an axial *vector.

coupled systems If two or more mechanical vibrating systems are connected so that they react on one another, the complete system is known as a mechanically coupled system. In such systems there is a transfer of energy from one system to another involving a change in the natural frequency of the individual systems. There are conditions, particularly at *resonance, when the energy drain from one system to another is sufficiently great for one system to be unable to maintain maximum amplitude. The usual example quoted consists of a heavy beam of mass M free to move in the direction of its length, carrying two simple pendulums, lengths l_1 and l_2,

coupled systems

119

masses m_1 and m_2, free to vibrate in a vertical plane through the axis of the beam. If one pendulum is set in vibration, the energy of motion will be transferred almost completely to the other pendulum, and then the process is reversed. This is particularly so as the natural periods of the two pendulums are altered towards the same value. The coefficient of coupling of such a system is defined as

$$\sqrt{\frac{m_1 m_2}{(M + m_1)(M + m_2)}}.$$

The coupling is diminished as the mass of the beam is increased. This should be compared with the electrical case of two inductances L_1 and L_2, coupled through the mutual inductance M_i. The coefficient is then $\sqrt{(M_i{}^2/L_1 L_2)}$. (See coupling.)

The frequency of an accurately electrically maintained tuning fork may be seriously affected by the manner of mounting of the fork, for the mounting and fork constitute a coupled system.

Most musical instruments may be considered to be coupled systems, some elements being sharply tuned, e.g. the strings of a piano, and some having a very broad resonance curve, e.g. the soundboard of the piano. In all applications, the design – which has been mainly achieved empirically – should be such that the energy transfer from the source is not too great to damp the source before a certain interval of time has passed.

The flue and the reed organ pipes may be considered to be examples of two coupled systems observed in wind instruments. The former mainly consists of a double coupled system, the latter of a triple system. The organ pipe is so designed that at the normal blowing pressure, the *edge tone has the same frequency as the fundamental of the pipe, viz. coupled at resonance. The air column is the less strongly damped component and therefore governs the frequency produced. The edge-tone frequency may be pulled a considerable way from its natural frequency but the air column alters very slightly with a change of edge tone. To a much less extent the material of the pipe and the surrounding air may be considered as elements of the coupled system. The reed organ pipe consists mainly of a triple element coupled system, (a) the boot in which the reed is mounted (roughly equivalent to the mouth in orchestral wind instruments), (b) the reed, and (c) the air column. The reed is the main frequency-deciding element. Much of the sound produced by the vibrating reed is inharmonic – although the reed vibrations themselves are almost entirely simple harmonic. The air column, however, does not present any adequate impedance to the inharmonic components. For optimum sound production by the reed pipe, the boot length should be an odd multiple of a quarter of a wavelength. Reeds vibrate most freely when placed near a node of the column of air in the supply tube and the air column proper.

The double *Helmholtz resonator consisting of two resonators coupled by a narrow neck and tuned to nearly the same frequency, is used to obtain a flattened response at a particular band of frequencies. By suitable proportioning of the resonators a large transfer of air through the neck may be obtained. This property is used in the design of sensitive detectors.

The human vocal apparatus may be considered as a coupled system of the pipe organ variety, the main elements consisting of the vocal chords and adjacent air cavities. There is some doubt as to the exact mechanism in operation. See vowels.

couple on a magnet If a magnet of moment M is inclined at an angle θ to a field of strength H, the couple acting on it is $HM \sin \theta$.

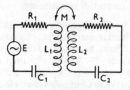

a   Mutual-inductance coupling

coupling 1. Of two oscillating electric circuits. The means by which the circuits interact so that energy is transferred from one to the other. In Fig. *a*, the circuits are coupled by mutual inductance between their individual inductances. This is *mutual-inductance coupling*, which is sometimes made by means of an impedance that is common to both circuits, examples of which are shown in Figs *b* and *c*. This is *common-impedance coupling*. The *coupling coefficient*, K, may be defined by:

$$K = X_m / \sqrt{X_1 X_2},$$

where X_m is the reactance common to both circuits, X_1 and X_2 are respectively the total reactances of the two circuits, both of the same kind as X_m. Thus, in Fig. *a*:

$$K = \omega M / \sqrt{(\omega L_1 \times \omega L_2)} = M / \sqrt{L_1 L_2}$$

In Fig. *b*:

$$K = \omega L_m / \sqrt{[\omega(L_1 + L_m) \times \omega(L_2 + L_m)]}$$
$$= L_m / \sqrt{[(L_1 + L_m)(L_2 + L_m)]}$$

In Fig. *c*:

$$K = \sqrt{\{C_1 C_2 /[(C_1 + C_m)(C_2 + C_m)]\}}$$

Sometimes the coupling is mixed, e.g. the types shown in Figs *a* and *c* are applied simultaneously.

2. An interaction between different properties of a system or an interaction between two or more sys-

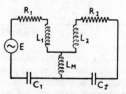

b   Inductive coupling

tems. There are two extreme types of coupling for atomic or nuclear particles:

In *Russell–Saunders coupling* (or *L–S coupling*), the resultant, *L*, of the *orbital angular momentum of all particles interacts with the resultant, *S*, of the *spin of all particles. In *j-j coupling*, the total angular momenta (orbital + spin) of individual particles interact with each other.

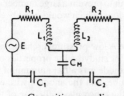

c Capacitive coupling

coupling coefficient 1. *See* coupling.

2. The numerical assessment between 0 and 1 characterizing the degree of coupling between two mechanically vibrating systems or electrical circuits. Maximum coupling is 1, no coupling 0. *See* coupled systems.

covalent bond A type of chemical bond formed by sharing of electrons. A hydrogen atom, for example, has one electron; two atoms can form a molecule (H_2) in which each atom contributes one electron to the bond. Each atom thus has two electrons orbiting around it and has the electron configuration of an inert gas (helium). Covalent bonds differ from *electrovalent bonds in that the latter result from electrostatic attraction between ions. When a covalent bond is formed in which one atom provides both electrons, the bond is called a *dative* or *coordinate bond*. *See also* molecular orbital.

C-parity *See* charge-conjugation parity.

CP invariance The *symmetry generated by the combined operation of *charge conjugation (*C*) and *parity (*P*). *CP violation* occurs in weak interactions of kaon decays in which CP symmetry is violated to one part per thousand. *See also* CPT theorem; *T* violation.

CPT theorem The theorem that the simultaneous operation of *charge conjugation *C*, *parity *P*, and time reversal *T*, is a fundamental symmetry of relativistic *quantum field theory. If *C*, *P*, or *T* are violated singly or in pairs the principles of relativistic quantum field theory are not affected; however, violation of *CPT invariance* would make a fundamental difference to relativistic quantum field theory although no experimental evidence exists of its violation.

CPU Abbreviation for *central processing unit.

cradle guard *See* guard wires.

creep The slow permanent deformation of a crystal or other specimen under sustained stresses.

crest factor *See* peak factor.

crest value *See* peak value.

crith *Syn.* krith. The mass of 1 litre of hydrogen at 0 °C and 760 mmHg. It is sometimes used to express the density of a gas relative to hydrogen. It is equal to 0.896 g.

critical angle The angle of incidence of light, proceeding from a denser medium towards a less dense one, at which grazing refraction occurs (angle of refraction = 90°). Light incident at a greater angle suffers total internal reflection. The critical angle, *C*, is given by $\sin C = n'/n$ in which $n > n'$ where *n* is the refractive index of one medium and *n'* of the other.

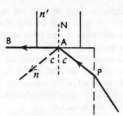

Critical angle refractometer

critical angle refractometers *Refractometers in which grazing incidence (*see* critical angle) is arranged between a medium whose refractive index is required and another of known index; the position of the boundary of light transmitted enables the refractive index to be calculated or to be observed directly on a scale, e.g. the Pulfrich and Abbe refractometers (*see* refractive-index measurement). The incident ray PA proceeds along AB. In refractometers, the light passes in direction BAP and the direction of the emergent light is observed.

critical constants *See* critical pressure; critical temperature; critical volume.

critical damping *See* damped.

critical isothermal The isothermal curve relating the pressure and volume of a gas at its critical temperature.

critical mass The minimum mass of a *fissile material that will sustain a *chain reaction. *See also* nuclear weapons.

critical point *See* critical state.

critical potential *See* excitation energy.

critical pressure The saturated *vapour pressure of a liquid at its critical temperature.

critical reaction *See* chain reaction.

critical solution temperature *See* consolute.

critical state The state of a substance when it is at its *critical temperature, pressure, and volume. Under these conditions the density of the liquid is the same as that of the vapour. The point corresponding to this state on an isotherm, is the *critical point*.

critical temperature The temperature above which a gas cannot be liquefied by increase of pressure. *See* Andrews's curves; equations of state.

critical velocity The velocity of fluid flow at which the motion changes from *laminar to turbulent flow. This velocity is sometimes called the *higher critical velocity*, the term *lower critical velocity* being given to the velocity at which the eddies in originally turbulent flow die away and the motion becomes laminar. It was first investigated by *Reynolds.

critical volume The volume of a certain mass of substance measured at the critical pressure and temperature.

CR law If a capacitor and resistor are in series, the rate of rise of the potential of the plates of the capacitor depends only on the product of the resistance and the capacitance. This relationship is important in the transmission of electrical signals in submarine telegraphy as the speed of operation depends on the product of the resistance and capacitance of the cable and cannot be increased except by lowering one of these factors.

CRO Abbreviation for *cathode-ray oscilloscope.

Cronin, James Watson (*b.* 1931) Amer. physicist who became a professor at Princeton and later at Chicago. He shared a Nobel prize (1980) with Val Fitch for their work on *CP invariance.

Crookes, Sir William (1832–1919) Brit. chemist and physicist. A pioneer in the field of vacuum-tube discharge, his name being given to the nonluminous region adjacent to the cathode, Crookes dark space (*see* gas-discharge tube). He studied extensively the properties of cathode rays and favoured their particle nature, confirmed by others. He was the inventor of the *Crookes radiometer intended for the measurement of the radiation pressure of light. He discovered the element thallium and separated thorium from uranium. *See also* Crookes glass.

Crookes dark space *See* gas-discharge tube.

Crookes glass A type of spectacle glass originally designed by *Crookes to protect the eyes of industrial workers from intense radiation. It is now used to guard against injury from bright sunlight. It contains cerium and other rare earths and has a low transmission of ultraviolet radiation.

Crookes radiometer

Crookes radiometer An instrument used to detect the presence of heat radiation. Four vertical vanes V, each blackened on one side, are free to rotate about a vertical axis in an evacuated glass vessel. When radiation falls on the blackened face of a vane it is absorbed and the temperature of the black face rises above that of the clear face. Molecules striking the blackened face carry away on an average, more momentum than those rebounding from the clear face, and thus the vane tends to rotate in such a direction that the black face continually recedes from the source of radiation. The gas pressure must be low enough to prevent a large number of intermolecular collisions, which would quickly equalize the velocities.

Crookes tube A low-pressure discharge tube as used by Crookes for studying the properties of *cathode rays. The cathode was a flat aluminium disc at one end of the tube, and the anode a wire electrode at one side of the tube outside the line of the cathode stream. *See* conduction in gases.

cross coupling *See* decoupling.

crossed cylinder A thin lens with cylindrical surfaces whose axes are crossed obliquely or at right angles. More particularly, a weak lens that has the effect of equal concave and convex cylinders crossed at right angles. It is always possible to calculate a spherocylindrical form of lens, which is equivalent to a pair of crossed cylinders. *Stokes's lens is a variable power combination.

crossed lens A form of spherical lens that shows minimum spherical *aberration in parallel light. For glass of refractive index 1.5, this occurs in the biconvex form in which the second surface has about six times the radius of curvature of the first.

crossed Nicols Two Nicol prisms arranged so that their vibration planes are perpendicular (*crossed*). The combination is opaque as the first Nicol polarizes the light in one plane and the second (analyser) prevents the polarized vibrations perpendicular to its own standard direction from passing on. *See* Polaroid.

cross modulation *Syn.* intermodulation. An effect occurring when a complex signal is fed into a nonlinear amplifier. Additional unwanted components are found in the output signal due to parts of the signal being modulated by other parts within the amplifier.

crossover network A type of filter circuit designed to pass frequencies above a specified value through one path, and frequencies below that value through another. The value of the frequency at which the output crosses from one path to the other is the *crossover frequency*, and the circuit is so designed that at that frequency the output of the two channels is equal. Such a network is widely used in high-fidelity systems, to feed the bass and treble components to the appropriate speakers.

cross section Symbol: σ. A measure of the probability of a particular collision process, stated as the effective area particles present to incident particles for that process. For example, if a beam of neutrons is passed through matter containing N atoms per unit volume one possible reaction is neutron *capture in which a nucleus retains the neutron on collision.

It is supposed that capture occurs when the neutron is some minimum distance, d, from the centre of the nucleus. For capture, the nuclei appear to present an effective cross-sectional area (called the *capture cross section*) σd^2 to the incident neutrons. If the neutron beam travels a distance dx through the substance the probability of a neutron being captured is $N\sigma dx$. If the beam intensity (in neutrons per unit area, say) is I and it falls by dI in travelling dx, then $-dI = IN\sigma dx$. Thus $I = I_0 e^{-N\sigma x}$, where I_0 is the initial intensity and I that after a distance x. $N\sigma$ is often represented by μ, and called the *linear absorption coefficient* for the process.

σ is not the physical cross-sectional area of the nucleus but the effective cross section for neutron capture. It is sometimes called the *activation cross section*. Its value depends on the energy of the incident neutrons as well as the nuclei considered. In particular, when the kinetic energy in centre of mass coordinates equals the energy difference between the *ground state of the nucleus and some higher state, the cross section is high and is called the *resonance cross section*.

The use of cross sections is not confined to neutron capture but is also applied to other nuclear reactions, as well as to interactions between atoms, electrons, ions, etc. Cross section has units of m^2 and is often measured in *barns. In nuclear-reactor theory, the quantity σ is sometimes called the *microscopic cross section* to distinguish it from the *macroscopic cross section*, which is the product of σ and the number of nuclei per unit volume. The macroscopic cross section is equal to the reciprocal of the *mean free path for the reaction in the material.

crosstalk Interference in the form of an unwanted signal in part of a circuit due to the presence of a signal in an adjacent circuit. It is a very common type of interference occurring in telephone, radio, and many data systems.

Crova's hygrometer A portable dew-point *hygrometer modified to be easily portable and usable in the open air.

crown glass See optical glass.

CRT Abbreviation for *cathode-ray tube.

cryogenics The study of the production and effects of very low temperatures. A *cryogen* is a refrigerant used for obtaining very low temperatures.

cryohydric point See eutectic.

cryometer A thermometer designed for the measurement of very low temperatures.

cryostat A vessel that can be maintained at a specified low temperature; a low-temperature thermostat.

cryotron A type of switch that operates at very low temperatures and depends on *superconductivity. One form consists of a wire surrounded by a coil in a liquid helium bath. Both the wire and the coil are superconducting and a low voltage can produce a current in the wire. If a current is also passed through the coil, its magnetic field alters the superconducting properties of the wire and switches off the current, thus the presence or absence of a current in the coil determines the ability of the wire to conduct. Cryotrons can be made very small and have low current requirements.

crystal 1. A three-dimensionally periodic arrangement of atoms in solids. Partial crystallinity can exist in two or one dimensions. See crystal structure; crystal systems; crystal texture.

2. In electronics, an element specially cut from a *piezoelectric crystal, such as quartz or barium titanate.

crystal analysis See X-ray crystallography.

crystal base The entire content of the *unit cell, whether considered as a symmetrical arrangement of atoms, or as a symmetrical distribution of electron density.

crystal class Crystals that are brought to self-coincidence by the operations of a *point group.

crystal clock See clocks.

crystal-controlled oscillator An electromechanical oscillator employing a mechanically vibrating quartz crystal. The link between the mechanical vibrations and the electric circuits is provided by the *piezoelectric effect of the quartz crystal. A particular feature of this type of oscillator is that it can be designed to have a very high degree of frequency stability. See clocks.

crystal counter A device for detecting and counting subatomic particles that depends on their ability to increase the conductivity of a crystal. If a potential difference is applied across a crystal that is struck by a particle or photon, the electron-ion pairs produced by the impact cause a transient increase in its conductivity. The pulses of current resulting from successive impacts are electronically counted. The operation of a crystal counter is analogous to that of a *Geiger counter in which the radiation induces conductivity in a gas.

crystal cut Crystal sections (usually thin plates or bars) may be cut in particular crystallographic directions, e.g. for use as *piezoelectric oscillators; these directions are specified as *cuts* and designated by letters such as X-cuts, AT cut, and so on. See Curie cut.

crystal detector A *detector that depends for its action upon the rectifying (see rectifier) properties of certain semiconducting crystals when placed in contact with one another, or of a crystal in contact with a metal. It was used extensively in the earliest types of radio receiver, and more recently as a detector and mixer of microwaves.

crystal diffraction The constructive and destructive interference of waves scattered by the periodic arrangement of electrons, nuclei, or field of force in a crystal, to give a pattern of discrete spectra.

crystal dynamics The study of the movements of atoms or of the variations of electron density in crystals.

crystal filter A filter that uses one or more piezoelectric crystals to provide its resonant or antiresonant circuits.

crystal grating A crystal in which the planes of atoms act like a diffraction grating for X-rays, electrons, thermal neutrons, etc.

crystalline lens The elastic lens of the eye that lies between the aqueous and vitreous humours by which the focal power of the eye can be altered. *See* accommodation; eye (human).

crystallites Very small crystalline particles that together make up a single (mosaic) crystal. (*See* crystal texture.) In petrology, they are the incipient crystals that occur in many glassy igneous rocks.

crystallography The science of the forms, properties, and structure of crystals, that is, of solids in which physical properties may vary regularly with direction, being the same along all parallel directions. *See also* X-ray crystallography.

crystalloid A substance that forms a solution as distinct from a sol. They are usually substances that crystallize, whereas those forming colloidal solutions are amorphous. The terms *colloid* and *crystalloid* were suggested by Graham who studied *dialysis between 1861 and 1864, but the distinction is not now maintained as, for example, common salt (a crystalloid) can, if suitably treated, be obtained as a colloid. *See* colloid.

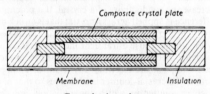

Composite crystal plate

Membrane   Insulation

Crystal microphone

crystal microphone *Syn.* piezoelectric microphone. A device making use of the *piezoelectric effect to convert the mechanical strain produced by sound pressure into electrical signals. P. and J. Curie discovered this effect when they found that crystals of quartz or Rochelle salt cut in a certain direction showed electrical charges on their faces when pressure was applied. Thus, the varying pressures caused by a sound wave striking the crystal face will produce alternating e.m.f.s across the crystal. One type of crystal microphone consists of two composite plates of Rochelle salt with an air gap between them

(*see* diagram). Pressure variations produced by the sound bend the plates and corresponding e.m.f.s appear across them. The connections are made so that the effect of the crystals is additive, the output being then used to drive an amplifier. The crystal plates and assembly are kept small so that the resonant frequency is above the audible range; this also makes the microphone practically nondirectional. In general, it has a good frequency response and no background noise, but it is relatively insensitive and needs subsequent amplification. In another type of microphone a separate diaphragm is used that is mechanically coupled to the centre of a double sheet of Rochelle salt. This type is more sensitive but its frequency response is not so good and it has directional properties.

crystal parameter *See* parameter.

crystal structure The specification both of the geometric framework (*see* unit cell; space group) to which the crystal may be referred, and of the arrangement of atoms or electron-density distribution relative to that framework.

crystal systems A classification of crystals based on their *unit cells. The 14 *Bravais lattices, and 32 *point groups, can be referred to seven crystal systems of three axes: (1) triclinic, in which the axes need be neither equal nor mutually perpendicular; (2) monoclinic, in which the axes need not be equal, but one is perpendicular to the other two; (3) orthorhombic, in which the axes need not be equal, but they are all mutually perpendicular; (4) tetragonal, in which two axes must be equal, and all are mutually perpendicular; (5) rhombohedral, in which all axes are equal, and equally inclined to each other at an angle of less than 120°; (6) hexagonal, in which two axes are equal and inclined to each other at 120°, being both perpendicular to the third unique axis; (7) cubic, in which all axes are equal and are mutually perpendicular. Since a crystal built on a rhombohedral lattice can in fact be referred also to hexagonal axes, it is sometimes the practice, particularly in the USA, to amalgamate (5) and (6) and to distinguish 6 crystal systems only.

crystal texture A crystal may be, in theory, an ideally periodic arrangement of atoms throughout the whole of its volume. In practice, however, such perfect crystals almost never occur. Even apparently single crystals consist of a conglomerate or mosaic of smaller crystallites that may be parallel (but with discontinuities at their mutual boundaries) or slightly disorientated. A massive crystalline specimen may consist of crystallites of larger or smaller grain size, partially or completely disorientated in one, two, or three directions. All these questions of crystal perfection, crystallite size, and orientation are covered by the word texture, and profoundly affect many crystalline properties.

cube-surface coil A system of five equally-spaced square coils designed to give a uniform magnetic field over a large volume. The uniform region is easily accessible from outside the coils.

cubic (isomeric regular) system *See* crystal systems.

cumulatively compound wound *See* compound-wound machine.

cup electrometer If a metal cup is attached to the plate of an electroscope or electrometer, any charged body touching the inside of the cup gives up the *whole* of its charge, instead of merely sharing it with the instrument. This is because electrostatic charges reside on the outside of conductors.

curie Symbol: Ci. A former unit of *activity of a radioactive nuclide corresponding to 3.7×10^{10} disintegrations per second (approximately equal to the activity of 1 g of radium). It has been replaced by the *becquerel, where $1 \text{ Ci} = 3.7 \times 10^{10} \text{ Bq}$.

Curie, Madame Marie (*née* Marya Sklodowska) (1867–1934) Polish-born physicist who married Pierre *Curie and worked in France. She discovered the radioactivity of thorium, following Becquerel's discovery of the radioactivity or uranium, and showed natural radioactivity to be unaffected by physical change or chemical combination. She discovered (in collaboration with her husband) the new elements polonium (named after Poland) and radium (1898) and isolated radium from the mineral pitchblende. The unit of radioactivity (the *curie) is named after her. She was the mother of Irène *Joliot-Curie.

Curie, Pierre (1859–1906) French physicist. He made important investigations on paramagnetism and on piezoelectricity (with his brother Jacques) as well as on radioactivity, most of which work was in association with his wife Marie *Curie.

Curie balance An instrument for the measurement of the *susceptibility of feebly magnetic materials. The specimen is placed in a glass tube at the end of an arm with a balance weight, the system being suspended by a silk fibre. It is deflected by a strong permanent magnet, and the deflection is compared with that produced with distilled water or some other substance of known permeability in the tube instead of the specimen.

Curie constant The product of the magnetic *susceptibility per unit mass and the thermodynamic temperature; this quantity is approximately constant for many paramagnetic substances. *See* Curie's law.

Curie cut A piezoelectric quartz crystal cut normal to one of the three electric axes connecting the opposite corners of the hexagon.

Curie point *See* Curie temperature.

Curie scale of temperature A temperature scale, used for the measurement of temperatures near the absolute zero, based on the assumption that a paramagnetic substance continues to obey Curie's law $\chi T =$ const. at the temperatures measured (below 1 K). Thus if χ_1 is the susceptibility at a known tempera-

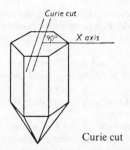

Curie cut

ture T_1 in kelvins and χ_2 the susceptibility at an unknown temperature, then this temperature T^* on the Curie scale is given as:
$$T^* = (\chi_1 T_1 / \chi_2).$$
The Curie law cannot be expected to hold at temperatures approaching absolute zero, but above 0.07 K, it was reliable to within 0.01 K.

Curie's law The principle that the susceptibility (χ) of a paramagnetic substance is universely proportional to the *thermodynamic temperature (T): $\chi = C/T$. The constant C is called the *Curie constant* and is characteristic of the material. This law was based on experimental measurements. Langevin later developed a classical theory of *paramagnetism in which he assumed that the magnetic behaviour was the result of each molecule having an independent magnetic *dipole moment. The tendency of the applied field was to align these and this was opposed by the random motion due to the temperature. He was thus able to explain the law.

A modification of Curie's law is known as the *Curie–Weiss law.

Curie temperature *Syn.* Curie point. Symbol: T_C or Θ_C. *See* Curie–Weiss law.

Curie–Weiss law A modification of *Curie's law, followed by many paramagnetic substances (*see* paramagnetism). It has the form
$$\chi = C/(T - \Theta)$$
The law shows that the susceptibility is proportional to the excess of temperature over a fixed temperature Θ: Θ is known as the *Weiss constant* and is a temperature characteristic of the material.

For ferromagnetic solids (*see* ferromagnetism) there is a change from ferromagnetic to paramagnetic behaviour above a particular temperature and the paramagnetic material then obeys the Curie–Weiss law above this temperature; this is the *Curie temperature*, Θ_C, for the material. Below this temperature the law is not obeyed. Some paramagnetic substances, such as gadolinium, obey the Curie–Weiss law above the temperature Θ_C and do not obey it below, but are not ferromagnetic below this temperature. The value Θ in the Curie–Weiss law can be thought of as a correction to Curie's law reflecting the extent to which the magnetic dipoles interact with each other. In materials exhibiting *antiferromagnetism the temperature Θ corresponds to the *Néel temperature*.

curl *Syn.* rotation. A vector quantity associated with a vector field, *F*. It is the vector product (*see* vector) $\nabla \times F$, where ∇ is the differential operator *del. Thus:

$$\text{curl } F = \nabla \times F =$$
$$i \times \partial F/\partial x + j \times \partial F/\partial y +$$
$$k \times \partial F/\partial z,$$

where *i*, *j*, *k* are *unit vectors along the *x*-, *y*-, and *z*-axes respectively.

The magnitude of curl *F* at a point is the maximum value of the line integral of the vector, per unit area, taken round the bounding edge of an infinitesimally small area at the point when the small area is oriented to produce the maximum line integral. The direction of the vector is that of the normal to the area and its sense is such that when looking along the vector one sees the line integral taken clockwise round the area boundary. (E.g. in the motion of liquids, if the curl of the velocity is not zero the particles are in rotation.)

A field that has a curl is generally described as *rotational*, *vortical*, or *circuital*; if the curl is zero everywhere, the field is irrotational or nonvortical. *Compare* gradient; divergence.

current Symbol: *I*. A flow of electric *charge in a substance – solid, liquid, or gas. The charge carriers may be electrons, *holes, or ions. The magnitude of a current is given by the amount of charge flowing in unit time; it is measured in *amperes. The direction is by convention from a point of higher potential to one of lower potential.

A *conduction current* is a current flowing in a conductor, the electricity being conveyed by the motion of electrons or ions through the material of the conductor. A conduction current of 1 ampere is equivalent to the flow of about 10^{18} electrons per second. A *displacement current is due to a change in the electric flux density in a dielectric; e.g. the current through a capacitor when connected in series with an alternating p.d. A current flowing always in the same direction in a circuit is called *unidirectional*; it may or may not be pulsating. A *direct current* flows always in the same direction, without sensible pulsations. *Compare* alternating current.

current balance An instrument for accurately determining a given current or, more fundamentally, the size of the *ampere by measuring the force between current-carrying conductors. The design in use in the UK is the *Ayrton–Jones balance*; it is based upon Neumann's law (*see* electromagnetic induction) for the energy *E* of two circuits of mutual inductance *M* carrying currents I_1, I_2:

$$E = M I_1 I_2.$$

If ds_1, ds_2 are vector elements of the circuits when at a separation *r*,

$$M = 10^{-7} \int_{s_1} \int_{s_2} ds_1 . ds_2 / r.$$

The force *F* between the circuits is

$$F = dE/dx$$
$$= I_1 I_2 dM/dx,$$

and d*M*/d*x* can be calculated for an inductance of coaxial helices. Small cylinders hang from the arms of a very sensitive beam balance, with coils R consisting of a single layer but many turns of bare copper wire. Each coil hangs in the centre of a larger cylinder wound with two identical single layer coils P, Q, one above and one below the middle. The same current passes through all three coils but in different directions through P and Q. R thus experiences an upward force (see diagram). The directions of flow of the current through the coils on the other side of the balance are such that R experiences a downward force. The forces are counterpoised by weights, and thus the sizes of the force per unit current on each coil can be ascertained. By various current reversals the effect of cross forces between the outer coils on one side and the suspended coils on the other can be eliminated, as well as the effects of external magnetic fields. Provided that all corrections are carried out, as well as dissipating the heat generated, an accuracy of better than 4 parts in 10^6 can be achieved. A modification of the general design (known as a *Rayleigh current balance*), which eliminates the exhausting procedure of accurately determining the diameter and precise position of the wire in the turns, is sometimes employed. *See also* Kelvin balance.

current circuit *Syn.* series circuit; main circuit. The circuit in an electrical instrument carrying either the current flowing in the circuit upon which measurements are being made, or a current proportional to that being measured. The term enables a distinction to be made between the two circuits (current circuit and *voltage circuit) of certain types of instrument (e.g. electrodynamic watt-meter).

current density Symbol: *j* or *J*. The ratio of the current to the cross-sectional area of the current-carrying medium. The medium may be a conductor, or a beam of charged particles. The ratio may be specified either as a *mean current density* or as density at a point.

current-limiting reactor A *reactor connected in series with an a.c. circuit to limit the current to a predetermined value especially under fault (e.g. short-circuit) conditions.

current transformer *Syn.* series transformer. An instrument transformer utilizing the current-transformation property of a transformer. The primary winding is connected in series with the main circuit and the secondary winding is closed through an instrument (e.g. ammeter) or other device. The ratio of primary and secondary currents is approximately the inverse of the primary to secondary turns-ratio of the transformer. Current transformers are extensively used to extend the range of a.c. instruments, to isolate instruments from high-voltage circuits and to operate protective relays in a.c. power systems.

curvature The reciprocal of the radius of curvature of a circle or sphere. It is applied to mirror and lens surfaces and wave fronts. Capital letters are com-

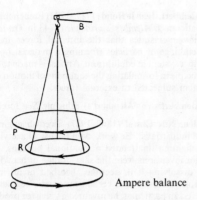

Ampere balance

monly used, e.g. $R = 1/r$. If r is in metres, R is in *dioptres. The sign given depends on convention. *See* principal radii.

curvature of image When *astigmatism for oblique pencils traversing a lens has been removed, the image of a plane object lies on a curved surface (the *Petzval surface). Curvature of the image is thus one of the aberrations of sphericity. For an infinitely distant object and a combination of thin lenses in air, the curvature of the image is given by $\Sigma(F/n)$, where F and n are the powers and corresponding refractive indices of the lenses. *See* aberrations of optical systems.

curve of buoyancy *See* metacentre.

curvilinear coordinates *See* coordinate.

cusp *See* caustic curve or surface.

cut-off *Syn.* black-out point. The point at which the current flowing through an electronic device is cut off by the *control electrode. In a transistor, for example, the cut-off current is the minimum base current at which the device conducts. In a cathode-ray tube the cut-off bias is the bias voltage that just reduces the electron-beam current to zero. In all cases, the values are dependent on the conditions at the other electrodes, which must be specified.

cut-off frequency Of a passive electrical or acoustical network as a nondissipative system – both characterized by possessing no internal source of energy. The frequency, reached by varying the frequency of the applied voltage or driving force, at which the attenuation quickly changes from a small value to a much higher value. For example, a typical loudspeaker with a cut-off frequency of 6000 hertz will fail to respond to frequencies above this value.

An active electrical or acoustical network as a dissipative system has the same cut-off frequency as a passive network with the same inductance (or inertia) and capacitance (or elastic) components.

The term is also applied to the limiting frequencies of acoustic and electric filters. *See* filters.

cyanogen absorption Bands occurring at 418 nanometres in the absorption spectra of stars due to

atmospheric cyanogen (C_2N_2). The absorption is more intense for *giant than for *dwarf stars and serves as a measure of the absolute *magnitude of stars.

cybernetics The theory of control systems first introduced by the American mathematician Norbert Wiener in 1948. It is concerned with the common characteristics of diverse systems including computers, automated factory processes, and the nervous systems of living organisms (such as man). It enables a comparison to be made of problems of control, communication, and feedback of information in biological and engineering systems. The theory has been applied to the design of automatic control mechanisms.

cybotaxis The special arrangement of molecules in a *liquid crystal.

cycle 1. An orderly set of changes regularly repeated.
2. One complete set of changes in the value of a periodic function, e.g. an alternating current passes through its cycle of values once in every period; one vibration, one oscillation, etc.
3. In computing, a set of operations that may be treated as a unit and repeated as a unit.

cycle of operations A series of operations starting with given conditions and bringing the substance back again to the same conditions. *See* Carnot; Rankine; Otto; Diesel.

cyclic currents *See* network.

cyclone A large area of atmospheric disturbance with an inward spiral rotation of air about a point of low pressure. In the northern hemisphere its rotation is anticlockwise. *Compare* anticyclone.

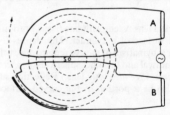

The magnetic field is perpendicular to

the plane of the paper

Cyclotron

cyclotron An *accelerator in which a beam of energetic particles is produced. Charged particles describe a spiral path of many turns at right angles to a constant magnetic field, and are given an acceleration, always in the same sense, from an alternating electric field at the beginning and end of each half turn of the spiral. The cyclotron depends on the fact that the time t taken by a particle of mass m and charge e to describe a semicircle in a plane at right angles to a uniform magnetic flux density B is $\pi m/Be$ and is thus independent of the velocity. Thus if an ion such as a proton, deuteron, alpha particle, electron, etc., is projected at S in the space between a pair of semicircular metal boxes A and B placed between the poles of a powerful electromagnet, it will, if the plates are connected to an alternating source of potential frequency $1/2t$, be accelerated each time it crosses the gap between the two conductors (known, from their shape, as *dees*). The radius r of the semicircle described by the particle increases as the velocity v increases: $v = Ber/m$. Thus large pole pieces and very powerful magnets are required. As the beam approaches the circumference of the dees, an auxiliary electric field deflects the particles from the circular path and they leave through a thin window. The energy that can be obtained from such a device is limited by the relativistic increase in mass of the particle as the velocity increases. Also it is necessary that B decreases slightly as r increases to prevent the beam hitting the dees. As the mass increases, the time taken to complete one semicircle increases and the particles get out of step with the alternating electric field. *Compare* betatron; synchrocyclotron; synchrotron.

cyclotron frequency The frequency at which an electron moves round a circular orbit in a plane perpendicular to a magnetic field of flux density B, given by $Be/2\pi m^*$, where m^* is the *effective mass of the electron in the substance. The term is sometimes used specifically for the value when the electron is a free particle and the mass has the normal value.

cyclotron resonance Resonance absorption of energy from an attenuating electric field by electrons moving in a magnetic field at the *cyclotron frequency. By measuring the frequency at which absorption of energy is a maximum, the *effective mass of the electrons in a substance can be determined.

cylindrical lens A lens with one face a portion of the curved surface of a cylinder. Thin lenses for correcting astigmatism of the eye require one surface to be cylindrical and the other spherical – a *spherocylindrical lens*. (*See also* toric lens.) Reflecting surfaces may also be a portion of the curved surface of a cylinder.

cylindrical polar coordinates *See* coordinate.

cylindrical winding A type of winding used in *transformers. The coil is helically wound and may be single-layer or multilayer. Its axial length is usually several times its diameter. *Compare* disc winding.

D

Dainton, Sir Frederick Sydney (*b.* 1914) Brit. physical chemist. A professor at Oxford and Leeds, he later became an administrator. His early work was on the chemical changes produced by high-energy radiation, particularly the formation of hydrated electrons in liquids.

d'Alembert, Jean le Rond (1717–1783) French mathematician. *D'Alembert's principle* (1743 in *Traité de dynamique*) states that the internal forces in an assemblage of particles (forming a material body) form a system in equilibrium. Appeal is made to the principle in formulating the equations of motion of a system subjected to external forces.

d'Alembert's (or **Alembert's**) **principle** *See* force.

Dalén, Nils Gustaf (1869–1937) Swedish engineer and industrialist. He worked on the use of acetylene for lighting unattended navigational beacons. His main inventions were the safe storage of acetylene by dissolving it in acetone absorbed in an inert porous medium, and certain automatic regulating valves. In particular, he invented a valve for producing light flashes of a given pattern to identify the source and a 'sun valve', which cut off the acetylene during the daytime. His development of these regulators won him the 1912 Nobel prize for physics. It has been suggested that the selection committee had, as their first choice, Thomas Edison and Nikolai Tesla, but that Tesla refused to share the award with Edison.

DALR Abbreviation for dry adiabatic *lapse rate.

dalton *See* atomic mass unit.

Dalton, John (1766–1844) Brit. chemist and physicist, the founder of the modern atomic theory. He approached the theory from the chemical side, in terms of combining weights and without any concept of the kinetic theory of matter and he did not fully appreciate the distinction between atoms and molecules but he succeeded in establishing the basic ideas. He established the law of partial pressures and the law of multiple proportions. Being himself colour-blind, he was able to make valuable comparisons of his vision with others and published (1794) the first known scientific paper on the subject. *Daltonism* or *dichromatism* is the condition in which green and red are confused.

Daltonism *Syn.* dichromatism. *See* colour vision.

Dalton's law of partial pressures The total pressure of a mixture of gases is equal to the sum of the *partial pressures that would be exerted by the gases if they were present separately in the container. *See* ideal gas.

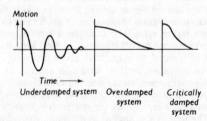

Underdamped system   Overdamped system   Critically damped system

damped Of a free oscillation. Progressively dying away due to an expenditure of energy by friction, viscosity, or other means. The word *damping* is used for both the cause of the energy loss and the progressive decrease in the amplitude of oscillation. If the damping is such that the system just fails to oscillate, the system is *critically damped* and for greater or lesser degrees of damping than this, it is *overdamped* or *underdamped* respectively. In electrical indicating instruments, three systems of damping are in common use: (1) air friction, (2) fluid friction (oil), and (3) eddy-current, and for such instruments the damping is usually designed to be slightly less than critical. (1) and (2) rely on viscous forces, and in all three systems, resistance is nearly proportional to velocity of motion. *See* damper; deadbeat.

damped vibrations Vibrations that decrease in amplitude with time. This decay is due to the resistance of the medium to the vibration. For small amplitudes of vibration the resistive force is approximately proportional to the velocity.

The force equation for a damped simple harmonic motion is:
$$m\ddot{x} = -kx - \mu\dot{x},$$
where x is the displacement, and m, k, and μ are inertia, elastic, and resistive terms respectively. The solution of this equation is:
$$x = a\,e^{-\alpha t}\sin(\omega t - \delta),$$
where the *decay* or *damping factor* $\alpha = \mu/2m$, and the frequency $n = \omega/2\pi$ where
$$\omega = \sqrt{\left(\frac{k}{m} - \frac{\mu^2}{4m^2}\right)}.$$

damper 1. A device incorporated in an electrical or other indicating instrument to provide the necessary damping (*see* damped). Common types are: (*a*) Air friction. Two forms are shown in Fig. *a*.

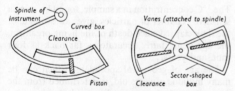

a Typical air dampers

Between the piston (or pairs of vanes) and the surrounding box is a very small uniform clearance through which the air passes when the piston, etc. moves. Air dampers are commonly employed in *moving-iron instruments. (*b*) Fluid friction. The fluid is usually oil. Two forms are shown in Fig. *b*. Fluid friction dampers are sometimes used in electrostatic instruments. (*c*) Eddy current. The instrument spindle (Fig. *c*) carries a metal disc or drum (usually made of aluminium or copper) situated in the *air gap of a permanent magnet. Movement of the disc or drum causes *eddy currents to be induced in it and these react with the magnetic flux to produce a force (and hence torque) that opposes the motion. Dampers of this type are commonly fitted to *hot-wire ammeters and *induction instruments.

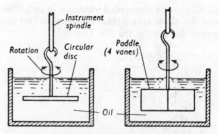

b Typical fluid friction dampers

This form of damping is also used in *permanent magnet moving-coil instruments but in this case, a separate damper is not used. The coil is wound upon an aluminium or copper former, which moves with the coil, in the air gap of the permanent magnet. Eddy currents induced in the metal former then give rise to the damping torque. **2.** A *damper winding.

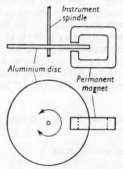

c Typical eddy current damper

damper winding *Syn.* damping winding. A special winding normally fitted to a.c. synchronous machines to damp out any cyclic speed irregularities and reduce the possibility of *hunting. It consists of a number of bare copper bars that are housed in slots or holes in the *pole shoes, the ends of the bars being brazed or welded into copper end-rings at both ends. The winding resembles that of a *squirrel-cage rotor in an *induction motor.

damping factor 1. *Syn.* decrement. The ratio of the amplitude of any one of a series of damped oscillations to that of the following one.

2. *Syn.* decay factor. *See* damped vibrations.

Daniell, John Frederick (1790–1845) Brit. physicist and chemist, who invented the hygrometer and the variety of two-fluid primary cell known by his name.

Daniell cell A primary *cell, now used only for demonstration purposes. The positive pole is of copper immersed in a saturated solution of copper (II) sulphate and the negative an amalgamated zinc rod in a solution of dilute sulphuric acid. The two solutions are separated by a porous partition. The cell has a fairly constant e.m.f. of about 1.08 volts and an internal resistance of a few ohms. This form of cell must be dismantled when not in use, otherwise the electroytes diffuse into one another and copper deposits on the zinc electrode.

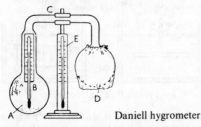

Daniell hygrometer

Daniell hygrometer A dew-point *hygrometer consisting of two bulbs A, D, joined by a tube C from which all air has been removed. Bulb A contains liquid ether whose vapour occupies the rest of the space. Ether is poured over the muslin round D and evaporates taking its heat from D, which cools. Some ether vapour inside D condenses and its place is taken by vapour evaporating from the liquid in A. The bulb A consequently cools slowly to the *dew point when dew appears on its surface. The dew point is measured by the thermometer B dipping into the liquid ether in A, the air temperature being measured by an external thermometer E. *See* humidity.

dark-field illumination *Syn.* dark-ground illumination. *See* microscope.

dark matter *See* missing mass.

dark space The comparatively nonluminous portion of an electrical discharge through a gas. *See* glow discharge; gas-discharge tube.

d'Arsonval, Jacques A. (1851–1940) French physicist, inventor of the moving-coil galvanometer provided with a mirror for giving the deflections. He also worked on high-frequency alternating current, especially on therapeutic applications.

d'Arsonval galvanometer An instrument in which the current to be measured is passed through a narrow rectangular coil, suspended so as to be free to turn about a vertical axis in the magnetic field between the poles of a permanent horseshoe magnet. A cylinder of soft iron is usually supported inside the coil. The instrument combines reasonably high sensitivity (currents of the order nanoamp may be detected) with a comparatively low resistance and high degree of damping. The name is often applied to any form of moving-coil galvanometer.

dart leader stroke *See* leader stroke.

dash-pot A mechanical device that prevents any sudden or oscillatory motion of a moving part of any piece of apparatus. It depends for its action upon the viscous resistance of air or or of a liquid (e.g. oil), and in a simple form consists of a piston loosely fitted in a cylinder filled with oil. A dash-pot is sometimes

used to provide a time lag in the operation of, for example, an electric circuit-breaker.

dasymeter An instrument for measuring the density of gases. It consists of a thin glass globe that is weighed in the gas under test and then in a gas of known density.

DAT Abbreviation for *digital audio tape.

database A computer data file containing information that can be organized, changed, and accessed according to certain rules.

dating Any of several methods for determining the age of archaeological and fossil remains, rocks, etc., by measuring some property of the organic or inorganic material that changes with time. This property may be dependent on some aspect of nuclear decay, such as the decay of radiocarbon or the uranium series, *thermoluminescence, or *electron spin resonance. These aspects are studied by *radiometric dating* techniques. The property may alternatively be dependent on a chemical change with a time-dependent rate constant, such as amino acid racemization.

(1) *Radiocarbon dating* (or *carbon-14 dating*) is a method for determining the age of objects up to 35 000 years old containing matter that was once living, such as wood. Atmospheric carbon consists mainly of the stable isotope ^{12}C and a small but constant proportion of ^{14}C, a radionuclide of half-life 5730 years resulting from the bombardment of atmospheric nitrogen by neutrons produced by the action of *cosmic rays. All living organisms absorb carbon from atmospheric CO_2, but after death, absorption ceases and the once-constant ratio $^{14}C/^{12}C$ decreases due to the decay of ^{14}C:

$$^{14}C \rightarrow {}^{14}N + e + \bar{\nu},$$

The ^{14}C concentration in a sample, found by using a sensitive *counter of particles, gives an estimate of the time elapsed since death of the living organism, the age being fairly accurate as far back as four thousand years.

The age of geological specimens, which can be many millions of years old, is determined from the proportion of a natural radionuclide (with a very long half-life) and its daughter nuclide contained in a sample of rock or mineral.

(2) *Potassium-argon dating.* Potassium, in combination with other elements, occurs widely in nature especially in rocks and soil. Natural potassium contains 0.001 18% of the radioisotope ^{40}K, which decays, with a half-life of 1.28×10^9 years, partly to the stable isotope of argon, ^{40}Ar. Determination of the ratio of $^{40}K/^{40}Ar$ gives an estimation of ages up to about 10^7 years.

(3) *Rubidium-strontium dating.* Natural rubidium, a much rarer element than potassium, contains 27.85% of the radioisotope ^{87}Rb, which decays with a half-life of 4.8×10^{10} years into the stable isotope of strontium, ^{87}Sr. Determination of the ratio $^{87}Rb/^{87}Sr$ gives an estimate of age of up to four thousand million years.

(4) *Uranium-lead* and *thorium-lead dating*. The long-lived nuclides thorium-232, uranium-238, and uranium-235 decay through radioactive series to lead-208, lead-206, and lead-207 respectively. Determination of the abundances of uranium and/or thorium and the isotopic composition of lead in rocks permits dating up to about 4×10^9 years.

The age of the earth has been estimated by radioactive dating, to be about 4×10^9 years.

dative bond *See* covalent bond.

daughter product Any nuclide that originates from a given *nuclide, the *parent, by radioactive *decay.

Davisson, Clinton Joseph (1881–1958) Amer. physicist, who worked at the Bell Telephone Laboratory. He shared a Nobel prize (1937) with George Thomson for their independent discovery of electron diffraction by crystals, which provided experimental evidence for de Broglie's wavelength equation for electrons. *See* Davisson–Germer experiment.

Davisson–Germer experiment (1927) The first experiment to demonstrate *electron diffraction. A narrow pencil of electrons from a hot filament cathode was projected *in vacuo* onto a nickel crystal; the distribution of the electrons scattered back from the crystal was investigated by a *Faraday cylinder. The experiment showed the existence of a definite diffracted beam at one particular angle, which depended on the velocity of the electrons. Assuming this to be the Bragg angle (*see* Bragg's law) the wavelength of the electrons was calculated and found to be in agreement with the *de Broglie equation. *See also* Thomson, Sir George Paget.

Davy, Sir Humphry (1778–1829) Brit. chemist and physicist. A prolific and notable scientist, he was responsible for the discovery or identification of many elements (sodium, potassium, aluminium, calcium, chlorine, iodine, barium, strontium, and magnesium). In this work he made great use of electrolysis. He was a pioneer in the use of nitrous oxide as an an anaesthetic and first suggested the electrical nature of chemical affinity. He invented the safety lamp still used in mines.

Michael *Faraday started his scientific career under Davy at the Royal Institution and in due course became his successor.

Dawes' rule An empirical rule for calculating the *resolving power of a *telescope in seconds of arc, given by $1.8/d$ where d is the diameter of the objective in centimetres.

day A unit of time equal to 24 hours, i.e. 86 400 seconds. *See* time *for* sidereal day.

DBS Abbreviation for *direct broadcast by satellite.

d.c. Abbreviation for direct current.

dead Applied to a conductor or circuit that is at earth potential. *Compare* alive.

deadbeat Indicating an instrument that is *damped so that any oscillating motion of its moving parts dies away very rapidly.

dead room *Syn.* anechoic chamber. A room that absorbs practically all the incident sound. For a room to be completely dead it must be made soundproof by insulating its floor, walls, and ceiling from the rest of the building and by using heavy soundproof doors. All the surfaces, including the floor, are then covered with several centimetres of highly absorbent material such as rock wool. The possibility of the formation of standing waves is further reduced by using an asymmetrical room or by placing absorbent deflectors in suitable places in the room. Most modern dead rooms employ large numbers of inward-pointing pyramids covered with an absorbing material to minimize possible reflections of sound. The subjective effect of the complete absorption found in a dead room is very oppressive and speech becomes tiring. Such rooms are used in acoustic laboratories for the production of free progressive waves and for experiments on reflection of sound. The transmission of sound through materials can be found by sealing up a space between two dead rooms with a partition of the material under test. Sound from a source in one room is directed on to the material and the intensity is measured in the other room both with and without the material in position.

dead space Space containing gas at a temperature different from that of the main volume of gas, e.g. the volume of gas in the capillary tube of a *constant volume gas thermometer.

dead time In any electrical device, the time interval immediately following a stimulus during which it is insensitive to another stimulus. Any device used as a *counter must have a dead time correction applied to the observed counting rate to allow for probable events occurring during the dead time.

de Broglie (Prince), Louis Victor (1892–1987) French physicist. He created *wave mechanics and was a leading authority on electron optics and particle and wave aspects of matter.

de Broglie equation A particle of mass m moving with a velocity v will under suitable experimental conditions exhibit the characteristics of a wave of wavelength λ given by the equation $\lambda = h/mv$, where h is the Planck constant. The equation, advanced on theoretical grounds, has been confirmed experimentally for electrons, protons, neutrons, α-particles, and some atoms and molecules. The equation is the basis of *wave mechanics. *See* de Broglie waves.

de Broglie waves *Syn.* matter waves; phase waves. A set of waves that represent the behaviour, under appropriate conditions, of a particle (e.g. its diffraction by a crystal lattice). The wavelength is given by the *de Broglie equation. They are sometimes regarded as waves of probability, since the square of their amplitude at a given point represents the

debugging

probability of finding the particle in unit volume at that point. These waves were predicted by de Broglie in 1924 and observed in 1927 in the *Davisson–Germer experiment.

debugging The process of discovering, diagnosing, and curing errors in computer *software or *hardware.

debye A former unit of electric dipole moment equal to 10^{-18} e.s.u. Equal and opposite charges, each equal to the electronic charge (4.80×10^{-10} e.s.u.), displaced 10^{-8} cm, produce a dipole moment of 4.80 debyes. 1 debye = $3.335\,64 \times 10^{-30}$ coulomb metre.

Debye, Peter Joseph Willem (1884–1966) Dutch scientist (later in U.S.A.). Much of his work has been in physical chemistry, particularly in dealing with electrolytes and with dipole moments. His theory of specific heat capacities (*see* Debye theory of specific heat capacities) was an advance on Einstein's and represents closely the behaviour of a wide range of substances. For his pioneer work in X-ray powder crystallography, *see* Debye–Scherrer ring. *See also* Debye–Sears effect.

Debye effect Debye predicted that alternating potentials should accompany the passage of ultrasonic waves through an electrolyte. The effect was confirmed experimentally and shown to depend on velocity variations within the waves, to be a function of the electrolyte, and to be independent of molar concentration in the range 0.005 to 0.0005 for univalent electrolytes.

Debye length The maximum distance at which the *Coulomb field of charged particles in a *plasma can interact.

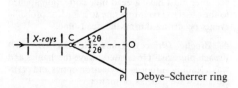

Debye–Scherrer ring

Debye–Scherrer ring The circular diffraction ring, concentric with the undeflected beam, formed when a narrow pencil of monochromatic X-radiation is passed through a mass of finely powdered crystal. Since the orientation of the powdered fragments is entirely random, the incident radiation must fall on some of them at an angle, which is the Bragg angle (*see* Bragg's law) for some one given set of the crystal planes, and hence a diffracted beam will be formed. Since everything is symmetrical about the axis of the incident pencil, the diffracted rays will lie on the surface of a cone (*see* diagram). Each set of crystal planes will form its own diffracted cone, and on intercepting the emergent radiation by a photographic plate, PP, the resulting negative will show a series of dark rings surrounding the central spot O,

due to the undeflected beam. The vertical angle of the cone is 4θ, where θ is the Bragg angle for the wavelength of the radiation employed, and the given set of planes. This method was devised, independently by Debye and Scherrer, and by Hull. It is used in *X-ray crystallography.

Debye–Sears effect An effect used for studying and measuring the speed of sound waves in a transparent liquid. If a *piezoelectric crystal is placed in the liquid and vibrated at a fixed frequency it sets up acoustic waves – alternate regions of compression and rarefaction in the liquid with nodes every half wavelength. The liquid is held in a parallel-sided glass cell through which a beam of light of known wavelength is passed. The regions of compression and rarefaction in the liquid act as a *diffraction grating with a grating interval equal to the wavelength of the acoustic waves. This wavelength can be measured from the position of the diffracted light beam and the speed of sound in the liquid can thus be obtained from the product of its frequency and wavelength.

Debye theory of specific heat capacities (1912) Debye applied the quantum theory to the independent vibrations of a solid considered as a continuous elastic body with an atomic structure such that the frequencies stop abruptly at a maximum frequency (ν_m) given by:

$$3L = 4\pi V\left(\frac{1}{C_L^3} + \frac{2}{C_T^3}\right)\int_0^{\nu_m} \nu^2 \, d\nu$$

where C_L and C_T are the velocities of longitudinal and transverse waves respectively in the body and L is the *Avogadro constant. The molar heat capacity at constant volume is then given by the *Debye function*:

$$C_v = 9R\left(\frac{4}{x^3}\int_0^x \frac{\xi^3}{e^\xi - 1} \, d\xi - \frac{x}{e^x - 1}\right)$$

where $x = h\nu_m/kT$; $\zeta = h\nu/kT$, k is the Boltzmann constant and h is the Planck constant. The *Debye characteristic temperature* Θ_D is defined as $h\nu_m/k$, so that C_V is a function of (Θ_D/T), which is in fair agreement with experimental results. *See also* quantum theory.

Debye T³ law At low temperatures the specific heat capacity of a dielectric solid is proportional to the cube of the thermodynamic temperature. At such temperatures the Debye function becomes
$$C_V = \tfrac{12}{5}\pi^4 RT^3/\Theta_D{}^3,$$
so that, since R and Θ_D are constants for the substance, $C_V \propto T^3$.

Dielectric solids generally agree well with this law up to a few kelvin. Metals have the relationship $C_V = AT^3 + BT$ where A and B are constants. In this case the term BT is attributed to the free electrons while the lattice vibrations give the Debye term AT^3. Departures from these laws are observed, for example, in the region of the *Néel temperature.

deca- Symbol: da. A prefix meaning 10. For example one decameter (1 dam) = 10 metres.

decalescence A phenomenon that may be exhibited by a heated metal. It consists of a sudden decrease in the rise of temperature due to an endothermic structural change within the metal. *Compare* recalescence.

decametric waves Radiowaves having wavelengths between 10 m and 100 m.

decay 1. The transformation of a radioactive *nuclide, the parent, into its *daughter product by disintegration, resulting in the gradual decrease in the *activity of the parent. (*See also* alpha decay; beta decay; radioactivity.)
2. The gradual decline of brightness of an excited *phosphor.
3. *See* damped vibrations.

decay constant *Syn.* disintegration constant. Symbol: λ. The probability per unit time of the radioactive decay of an unstable nucleus. It is given by the formula:
$$\lambda = -dN/dt.1/N,$$
where $-dN/dt$ is the *activity, A, of the nuclide and N is the number of undecayed nuclei present at time t. The exponential decrease with time of the activity of a radionuclide in which there are N_0 nuclei at time $t = 0$ is found from the formula $N = N_0 e^{-\lambda t}$. The time required for half the original number of nuclei to decay ($N = \frac{1}{2}N_0$) is the *half-life, $T_{1/2}$, given by $T_{1/2} = 0.693\ 15/\lambda$. The reciprocal of the decay constant is the *mean life.

decentred lens A lens in which the optical centre does not lie at the geometrical centre of the rim of the lens. Decentration has the effect of combining a centred lens with a weak prism ($P = F \times C$), where P is the prismatic effect in prism dioptres, F the principal power in dioptres, and C the decentration in prism dioptres.

deci- Symbol: d. A prefix meaning 0.1. For example one decimeter (1 dm) = 0.1 metre.

decibel Symbol: dB. A logarithmic unit for comparing two amounts of power. Thus, the logarithm to the base 10 of the ratio of two sound intensities is defined as the relative magnitude of the intensities in *bels. More often the decibel, equal to one-tenth of a bel, is used in practice. Thus if I_1 and I_2 are the intensities of two notes, the difference in intensity level is given by $\log_{10}(I_1/I_2)$ bels or $10\log_{10}(I_1/I_2)$ decibels. This logarithmic scale is convenient since between the thresholds of audibility and feeling the sound intensity increases in the ratio 1 to 10^{12} (*see* audibility). One decibel represents an increase in intensity of 26%, which is about the smallest change that the ear can detect. In using the decibel scale the intensity with which a note is compared is usually the threshold intensity of a note of the same frequency (*see diagram under* noise); in this case the relative magnitude of the note is called its sensation level. The *perceived noise decibel* (PN dB) is defined as the

sound pressure in decibels (above a datum level of 2 $\times$ 10^{-5} pascal root-mean square), as judged by an otologically normal binaural listener, of a band of random noise of width one-third to one octave centred on a frequency of 1000 hertz. Although the loudness of a note to an observer is related to the logarithm of its intensity, the decibel is not a measure of loudness, since the sensitivity of the ear to changes of intensity varies with frequency. The equivalent loudness of a note is measured in *phons. *See also* neper.

decimal balance *See* balance.

decimal system 1. The number system in general use in which numbers are expressed by combinations of the ten digits 0–9, the units of counting (ones, tens, hundreds, etc.) being related by powers of 10. **2.** A system of measurement, such as the metric system, in which the multiple and submultiple units are related by powers of 10.

decimetric waves Radiowaves having wavelengths between 10 cm and 1 m.

declination 1. (magnetic) The angle between the magnetic meridian and the geographical meridian at a particular point. Its value depends on the position of the point on the earth's surface, and at a given point changes slowly with time.
2. (astronomical) *See* celestial sphere.

declinometer An instrument for determining the magnetic *declination. It consists essentially of an arrangement by which the angle between the magnetic axis of a compass needle and the direction of some heavenly body can be read on a horizontal circular scale. An approximate value can be obtained with a prismatic compass: for accurate measurements the *Kew magnetometer is used.

decomposition voltage The maximum potential difference that can be applied to the electrodes of an electrolytic cell without giving rise to a permanent current through the cell. For a cell with platinum electrodes containing dilute sulphuric acid as the electrolyte, its value is 1.67 volts.

decoupling The removal of any a.c. components from a circuit or circuit element. These unwanted components are sometimes caused by *coupling between circuits, particularly those with a common power supply (in which case it is known as *cross coupling*). Decoupling is usually achieved by using a series inductance or shunt capacitor.

decrement *See* damping factor.

decrement gauge *See* molecular gauge.

de-emphasis *See* pre-emphasis.

dees *See* cyclotron.

defect All crystalline solids consist of regular periodic arrangements of atoms or molecules. Departures from regularity are known as *defects* and they can be classified in two types. (1) *Point defects* are defects

defect conduction

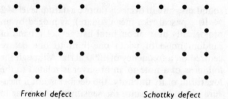

Frenkel defect   Schottky defect

a   Point defects in crystal lattice

involving one single atom or molecule. A *Frenkel defect* is a vacant lattice site with an associated interstitial atom – i.e. an atom that is not in a normal lattice position. Such atoms are called *interstitials* and the vacant lattice site is called a *vacancy* (Fig. *a*). The Frenkel defect is both the vacancy and the interstitial and it can be formed by movement of an atom from its normal lattice position. A *Schottky defect* is simply a vacant lattice point. It requires energy to create a point defect by moving an atom from its position but, since the defects introduce disorder into a crystal, this is offset by an increase in configurational *entropy (i.e. entropy caused by disorder). Consequently, all crystals above absolute zero have at least a certain number of such defects. The equilibrium number of Schottky defects in a crystal is:

$$n = N\,e^{-E/kT},$$

where N is the total number of lattice sites and E the energy required to produce the defect. The number of Frenkel defects is given by:

$$n = \sqrt{(NN'\,e^{-E/2kT})},$$

N' being the number of interstitial sites. The equilibrium number of point defects rises exponentially with temperature; typically, for a metal at about 700 °C, 1 in 10^5 sites are vacant. Point defects are responsible for *diffusion in solids. They can be produced in high concentrations by heating the solid to a high temperature and cooling it, by straining it, or by treatment with *ionizing radiation. In some chemical compounds point defects are responsible for the existence of *nonstoichometric crystals*. These are crystals in which there is more of one component than the other, so that the compound does not have a true chemical formula. A typical example is zinc oxide, which often contains slightly more zinc than oxygen ions. This excess acts as an electron donor and causes zinc oxide to be an n-type *semiconductor. Another example is nickel oxide, which frequently has slightly more oxygen ions than nickel ions. The absence of nickel ions causes nickel oxide to be a p-type semiconductor.

(2) *Line defects* are extended departures from regularity in crystals and they are often called *dislocations*. There are two basic types, *edge dislocations* and *screw dislocations*. An edge dislocation is shown in Fig. *b*. It corresponds to an extra plane of atoms introduced in one part of the crystal. The *dislocation line* extends into the crystal perpendicular to the plane of the paper at P. Screw dislocations are more difficult to visualize and illustrate. If a cylinder is taken, as in Fig. *c*, and cut along a plane ABCD and

134

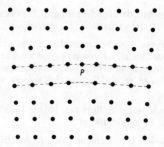

b   Edge dislocation in crystal lattice

displaced as shown, then a screw dislocation results. The atoms thus have a helical arrangement around the axis of the cylinder, which is the dislocation line. Many dislocations in crystals have features of both edge and screw dislocations. The dislocations in solids are responsible for plastic deformation above the *elastic limit. They are also formed by deformation. A good natural inorganic crystal may have 10^{12} dislocations crossing 1 m^2; the best organic crystals have fewer (perhaps 10^{10} per m^2). Annealed metals have 10^{13} to 10^{14} dislocations per m^2, while heavily worked metals may have as many as 10^{16} per m^2.

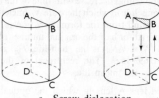

c   Screw dislocation

defect conduction In a *semiconductor, conduction due to the presence of *holes in the *valence band.

deflection defocusing An increase in the size of the luminous spot in a *cathode-ray tube that occurs when the latter is deflected from its undeflected position, and becomes greater as the deflection is increased. See symmetrical (and asymmetrical) deflection.

deflection sensitivity Of a *cathode-ray tube in which the electron beam is deflected electrostatically. A linear displacement of the spot on the screen when unit potential difference is applied to a pair of *deflector plates. The displacement is usually expressed in millimetres and the p.d. applied to the deflector plates in volts so that the deflection sensitivity is expressed in millimetres per volt. For a given tube, the deflection sensitivity is inversely proportional to the voltage applied to the final anode and it is necessary either for this voltage to be specified, or (more usually), for the deflection sensitivity to be expressed in the following form:

Deflection sensitivity = a/V mm/volt, where a is a constant for a particular tube, and V is the voltage of the final anode. A typical value for a is 750.

For electromagnetic deflection, the deflection sensitivity is expressed in metres per tesla of the deflecting magnetic field.

deflector coils Of a *cathode-ray tube employing electromagnetic deflection of the electron beam. The coils through which a current is passed to produce a magnetic field for the purpose of deflecting the beam of electrons. Two pairs of coils mounted close to the outside of the tube are usually employed and the pairs are described as X or Y according to whether they produce horizontal or vertical deflections of the beam respectively. The linear displacement of the luminous spot on the screen produced by one pair of coils is approximately proportional to the current in the coils. *Compare* deflector plates.

deflector plates Of a *cathode-ray tube employing electrostatic deflection of the electron beam. Electrodes to which a voltage is applied to produce an electrostatic field for the purpose of deflecting the beam of electrons. The electrodes are usually plates of metal and there are two pairs of them, the pairs being described as X or Y according to whether they produce horizontal or vertical deflections of the beam respectively. The deflector plates are mounted inside the tube. The linear displacement of the luminous spot on the screen produced by one pair of plates is approximately proportional to the voltage applied between the plates. *Compare* deflector coils.

de Forest, Lee (1873–1962) Amer. scientist who invented the three-electrode thermionic valve (triode) in 1907.

deformation potential The electric potential caused by mechanical deformation of the crystal lattice of *semiconductors and conductors. *See* piezoelectric effect.

degas *See* outgassing.

degaussing 1. Neutralization of the magnetization of a ship by surrounding it with a system of current-carrying cables that set up an exactly equal and opposite field.
2. In colour television, the use of a system of coils to neutralize the earth's magnetic field thus preventing the formation of colour fringes on the image.

degeneracy 1. A condition that arises when an atom or molecule has two or more quantum states with the same energy. The states are said to be *degenerate*. *See* statistical weight.
2. The condition of matter at high density, particularly at low temperatures, when the energy distribution of one or more types of particle departs greatly from the classical form, so that the exact equations of *quantum statistics must be used. This applies to the valence electrons in solids, to the electrons (and sometimes other particles) in highly condensed stellar interiors, and to superfluid liquid helium. *See* degenerate gas.

degenerate *See* characteristic function; statistical weight.

degenerate gas A gas in which the concentration of particles is sufficiently high for the Maxwell–Boltzman distribution (*see* distribution of velocities) not to hold; the behaviour of the gas is then controlled by *quantum statistics.

The pressure in a degenerate gas consisting of *fermions is called the *degeneracy pressure*; this exceeds the thermal pressure because according to the *Pauli exclusion principle particles very close together must possess different momenta and according to the *uncertainty principle the difference in momentum is inversely proportional to the distance between them. Thus, in a high-density gas the relative momentum of the particles is high and unlike thermal pressure does not tend to zero as the temperature tends to absolute zero. *White dwarfs and *neutron stars are thought to be supported against collapsing gravitationally by the degeneracy pressure of their electrons and neutrons, respectively.

degenerate level An *energy level of a quantum mechanical system that corresponds to more than one quantum state.

degenerate semiconductor A *semiconductor with the *Fermi level located inside either the *valence band or conduction band (*see* energy bands). The material is essentially metallic in behaviour over a wide temperature range.

degradation 1. The decrease in the availability of energy for doing work, as a result of the increase of *entropy within a closed system. (*See* thermodynamics.)
2. The loss of energy of a beam of particles or an isolated particle passing through matter as a result of the interaction of the particles with the matter.

degree 1. A unit of temperature difference. The Celsius and Fahrenheit degrees were formerly defined as 1/100th and 1/180th respectively of the temperature difference between the ice and steam points, so that $1 \,°C = \frac{9}{5} \,°F$. The unit of *thermodynamic temperature, no longer called a degree, is the *kelvin.
2. (math.) The rank of an equation or expression as determined by the highest power of the unknown or variable quantity. The degree of a curve or surface is that of the equation expressing it.
3. *See* electrical degrees.

degree Celsius Symbol: °C. A unit used in expressing temperatures on the *Celsius scale. It is now an *SI unit, defined in terms of thermodynamic temperature, one degree Celsius being equal to one *kelvin. (*See also* degree.) It was formerly called the *degree centigrade*.

degrees of freedom 1. The number of degrees of freedom of a mechanical system is equal to the number of independent variables needed to describe its configuration; e.g. a system consisting of two particles connected by a rigid bar has 5 degrees of freedom since 5 coordinates (3 of the mass centre or

degrees of variance

of either particle, together with 2 angles) are needed to specify its state. The smallest number of coordinates needed to specify the state of the system is called its *generalized coordinates* and since these specify the state of the complete system, they also specify the state of any individual particle of the system. The generalized coordinates may be chosen in more than one way – as in the example quoted. The number of degrees of freedom depends only on the possibilities of motion of the various parts of the system and not on the actual motions. For a monatomic gas the number is 3. For a diatomic gas with rigid molecules it is 6, made up of 3 degrees of freedom of the centre of gravity to move in space, 2 degrees of freedom of the line joining the two atoms to change its direction in space and 1 for rotation about this axis. In applying the principle of *equipartition of energy the number of degrees of freedom is taken as the number of independent squared terms in the expression for the energy of a system, subject to certain reservations. (*See* equation of energy.)

2. *Syn.* degrees of variance. In the *phase rule, the variable factors, such as temperature, pressure, and concentration, needed to define the condition of a system in equilibrium.

degrees of variance *See* degrees of freedom.

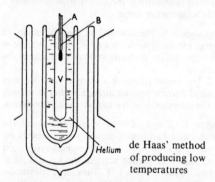

Helium   de Haas' method of producing low temperatures

de Haas' method of producing low temperatures de Haas used the adiabatic demagnetization of cerous fluoride (1935) to produce temperatures in the neighbourhood of absolute zero. A tube of the substance was held inside an almost completely exhausted space V by a carrier B and a rod A fixed to a balance. The vessel V was immersed in liquid helium boiling in a Dewar vessel at 1.26 K and the salt placed in the position where the vertical component of the nonhomogeneous magnetic field was a maximum. A strong magnetic field was applied and after four hours the salt attained the temperature of the bath (1.26 K). On removing the field a temperature of 0.27 K was attained. Much lower temperatures have been obtained by development of this method.

de Haas–van Alphen effect The diamagnetic susceptibility of bismuth, which is independent of field strength at ordinary temperatures, becomes a com-

plicated function of the field at the temperature of liquid hydrogen and lower. The susceptibility also becomes anisotropic.

Dehmelt, Hans (*b.* 1922) German-born Amer. physicist. His early work in Germany on *nuclear quadrupole resonance was followed by work using radio-frequency spectroscopic techniques to establish the magnetic moments of free electrons (1958).

dekatron A type of cold-cathode *scaler with usually ten sets of electrodes that function in turn. When an impulse is received, a glow discharge is transferred from one set of electrodes to the next. The tubes may be used for switching or for visual display of counts in the decimal system.

del *Syn.* nabla. Symbol: ∇. The differential operator
$$i(\partial/\partial x) + j(\partial/\partial y) + k(\partial/\partial z),$$
where *i*, *j*, and *k* are *unit vectors along the *x*-, *y*-, and *z*-axes respectively. The *Laplace operator is ∇^2.

delay cable A length of coaxial cable used in connection with high-voltage impulse (or surge) testing to delay the arrival of the impulse. According to requirements the cable may be connected between the impulse generator and the apparatus under test or between the latter and a remote point of measurement.

delayed automatic gain-control *See* biased automatic gain-control.

delayed neutrons Neutrons arising from nuclear *fission that are not directly emitted in the fission process but are ejected from fission product nuclei, which are left in excited states following *beta decay. The delayed neutron abundances then fall off after fission according to the half-lives of the beta-emitting nuclides, which range from 55 s to a fraction of a second. From 0.25% to 0.75% of the fission neutrons are delayed, the fraction depending on the fissile nuclide and the energy of the incident neutron. The presence of delayed neutrons makes it far easier to control reactors than would otherwise be possible. *See also* prompt neutron.

delay line A transmission line or any other device that introduces a known delay in the transmission of a signal. An *acoustic delay line* is a device in which the signal is converted to acoustic waves, usually by means of the *piezoelectric effect, and these waves are delayed by circulating them in a liquid or solid medium. They are then reconverted to electrical signals. Fully electronic *analogue delay lines* are based on *CCDs while *digital delay lines* use CCDs or *shift registers.

delay time In general, the time taken for a pulse to traverse any device or circuit. In a switching *transistor, it is the time taken between the application of a pulse at the input terminal to the appearance of a pulse at the output terminal. The time is usually measured between points corresponding to 10% of the peak amplitude.

Delbrück scattering The scattering of *photons by an atomic *nucleus. It is thought to occur for *gamma rays but the effect is small and has not been conclusively demonstrated.

delta connection A particular example of the *mesh connection employed in *three-phase a.c. circuits in which three conductors, windings, or phases are connected in series to form a closed circuit, which may be represented by a triangle (Δ), the main terminals of the circuit being the junctions between the three separate circuits. *Compare* star connection; double-delta connection.

delta function (δ-function) *See* Dirac function.

delta radiation (δ-radiation) Secondary electrons emitted by the impact of *ionizing radiation on matter. Their energies are of the order of only 10^3 eV; they can cause further ionizations.

delta voltage *See* voltage between lines.

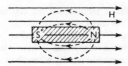

Demagnetizing field

demagnetizing field The magnetic field due to the free poles developed on a specimen of ferromagnetic material during the process of magnetization. The demagnetizing field (*see* diagram), in the medium between the free poles, opposes the applied magnetic field, H, the effective strength of which is thus reduced. Thus, if H is the applied field and I the existing intensity of magnetization of the specimen, the effective magnetizing field $H' = H - NI$, where N is a constant that depends on the geometry of the specimen, and can be calculated in some simple cases. *See also* keeper.

demand indicator *See* maximum-demand indicator.

Democritus (c. 460–357 B.C.) Greek philosopher who advocated a form of atomic theory and an idea of cosmic evolution.

demodulation The reverse of *modulation, i.e. it is the extraction or separation of the modulating signal from a modulated carrier wave. Circuits or devices used for this purpose are called demodulators or detectors.

demodulator *See* detector.

Dempster, Arthur Jeffrey (1886–1950) Canadian-born Amer. physicist, who became professor at Chicago. He developed the Aston mass spectrograph (*see* mass spectrum) into a more sensitive instrument and showed (1935) that uranium has several isotopes, of which U–235 was later found to be capable of sustaining a chain reaction.

demultiplexer *See* multiplex operation.

denaturant An isotope added to a *fissile material to make it unsuitable for use in nuclear weapons.

dendrites Tree-like crystalline growths that arise frequently when material crystallizes from the molten state. *See* alloy.

densitometer An instrument for measuring the optical transmission or reflection of a material. It is often used for converting the images on a photographic plate into quantitative form. For example, it may be used for converting a photographic record of an *emission spectrum into a graphical record of the intensities of the lines.

density 1. Symbol: ρ. The mass per unit volume of a substance. In SI units it is measured in kg m^{-3}; in other systems it is measured in g cm^{-3}, g ml^{-1}, and 1b in^{-3}. The relative density (symbol: d) is the density of a substance divided by the density of water; being a ratio, it has no units. This quantity was formerly called specific gravity but this usage is now deprecated because of the special meaning given to the word *specific (i.e. divided by mass).

At the *maximum density of water, $\rho = 1000$ kg m^{-3}; therefore the relative density of any substance is one-thousandth of its density. (*See* pyknometer, relative-density bottle, hydrometer, for some methods of measuring relative density. *Archimedes' principle may also be applied for determining relative density.)

2. Vapour density. The density of a gas or vapour divided by the density of hydrogen, both being at *STP. This may be measured by variations of the basic methods:

(i) Gay-Lussac's method. A known mass of substance is introduced into a vacuum over a mercury column and vaporized. The fall in the mercury column is noted.

(ii) *Dumas's bulb.

(iii) *Victor Meyer's apparatus.

(iv) *Dasymeter.

3. Density, in general, expresses the closeness of any linear, superficial, or space distribution, e.g. electron density = number of electrons per unit volume; *see also* charge density.

4. *See* reflection density; transmission density.

density of states curve *See* energy bands.

depletion layer A *space-charge region in a *semiconductor in which there is a net charge due to insufficient mobile charge carriers. Depletion layers are formed, for example, at the interface between a p-type and n-type semiconductor in the absence of an applied field. They are also formed at the interface of a metal and a semiconductor.

depletion mode 1. An operating mode of *field-effect transistors such that increasing the magnitude of the *gate bias decreases the current. **2.** A type of *field-effect transistor in which conduction occurs at zero *gate bias. All junction field-effect transistors are depletion mode devices. An n-channel insulated gate field-effect transistor (IGFET) tends to be a depletion mode device; a p-channel IGFET is difficult to fabricate.

depolarizer *Syn.* depolarizing agent. A substance used for removing the effects of *polarization in a primary cell, by reacting either chemically or electrolytically with the hydrogen ions liberated at the positive pole. For example, in the Leclanché cell, the hydrogen is oxidized by manganese dioxide; in the Daniell cell, the hydrogen ions are replaced by copper ions from a solution of copper sulphate.

deposition potential The minimum potential difference that must be applied to an electrolytic cell to produce electrolytic deposition.

depression of freezing point The lowering of the freezing point of a solvent by the addition of solute: it is used in the determination of relative molecular masses (molecular weight). It obeys *Blagden's law.

depression of zero Of a mercury-in-glass thermometer. The temporary depression of the zero occurring while the thermometer is being heated.

depth magnification The ratio of a small segment of the axis on the image side of an optical system to its conjugate segment on the object side. If f and f' are the principal focal lengths and x the distance of an object from the first focal point, depth magnification = $-ff'/x^2$.

depth of field If the lens in a camera or other optical instrument is focused on a particular object, the image of the object will be in focus but the images of objects on either side will be slightly out of focus. The zone in which the blurring of the image cannot be noticed is the depth of field of the lens. It depends not only on what standard of sharpness is acceptable, but also on the aperture and focal length of the lens and the distance from object to lens. A small aperture and short focal length produce a large depth of field. *Compare* depth of focus.

depth of focus Because of the finite resolving power of the eye (geometrically sharp point images cannot occur), the existence of indistinctness in any image whether due to incorrect focusing, aberration, or diffraction cannot always be recognized. Consider two conjugate points, object and image; the former may be moved over a range beyond or within the conjugate position, while the fixed-image screen does not show detectable blurring (*object space depth of focus*). Again the object may be fixed and the image screen displaced over a range (*image space depth*). Special interest attaches to its application to photographic lenses, etc. Assuming that the eye's resolving power is 1 minute (0.000 294 radians), the retinal blur of a point is about 0.005 mm, so that a photograph of circles separated by a distance of 0.07 mm would appear sharp when placed 250 mm from the eye. Frequently, the depth is expressed as a linear range or as a difference of the reciprocals of the two extreme distances mentioned, i.e. as an equivalent power effect at the lens. It is directly proportional to *f-number and object distance and

inversely proportional to the focal length of the lens. *Compare* depth of field.

derivative *Syn.* differential coefficient. If a quantity x is dependent in a continuous manner on another quantity t, then the rate of increase of x with t is known as the derivative of x with respect to t and denoted by dx/dt or $\dot{x}$ (e.g. velocity is the derivative of displacement, and acceleration the derivative of velocity, with respect to time).

derived units *See* SI units.

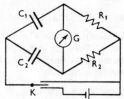

de Sauty bridge

de Sauty bridge A method for the comparison of capacitances. The capacitors to be compared occupy adjacent arms of the bridge, the other two arms being noninductive resistances. If there is no kick in the ballistic galvanometer, G, when the capacitors are either charged or discharged by means of the key K, then:

$$R_1 C_1 = R_2 C_2.$$

See also bridge.

Descartes, René (1596–1650) French mathematician and philosopher who worked mostly in the Netherlands. He applied algebra to geometry, using *Cartesian* coordinates. As a result of his development of geometrical optics, *Snell's law of refraction is sometimes named after him. He explained correctly the positions of the central rays of the primary and secondary rainbows but did not recognize dispersion or understand the action of a prism. His *Principia Mathematica* (1644) attempted to explain the universe in terms of mechanics. In this he adumbrated some of the concepts later developed by *Newton, but his theory was stultified by his belief that volume, not mass, was the primary property of matter and that a vacuum was impossible. He attempted to explain *Kepler's laws by assuming that planets were moved by *cortices* in interplanetary matter. Descartes was a convinced supporter of *Copernicus but was greatly disturbed by the condemnation of *Galileo. This led him to express some of his ideas ambiguously.

Descartes gave a very extensive account of phenomena, involving both accurate and faulty observations. His analyses are sometimes logical and sometimes very illogical. His work was greatly esteemed for many years despite the ultimately sterile nature of his methods. *See also* Cartesian diver; Cartesian ovals.

desorption The removal of adsorbed gas from a solid surface during which process heat is taken from the surface. The method is used for the liquefaction of helium. *See* Simon's method of liquefying helium.

Despretz's dilatometer experiment Despretz used a *dilatometer to determine the temperature of the

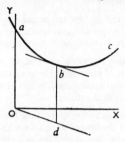

Despretz's determination of temperature at which water has maximum density

*maximum density of water. The apparent volume Y of water in a dilatometer is determined at a series of temperatures X and the curve *abc* plotted. (*See* graph.) Knowing the coefficient of expansion of the glass, the change in the volume of the vessel may be calculated at various temperatures and the line *Od* drawn. The true volume of the water at any temperature is given by the vertical distance between the curve *abc* and the line *Od*. This distance is at least at the temperature corresponding to *b* where the tangent to *abc* is parallel to *Od*, so this temperature corresponds to the minimum volume and therefore maximum density of the water.

Destriau effect *See* electroluminescence.

destructive interference *See* interference.

detailed balancing The principle that every process of transformation or exchange of energy that occurs in a system in thermodynamic equilibrium is invariably accompanied by an analogous reverse process, the two processes occurring with the same frequency.

detector 1. *Syn.* demodulator. In communications, a circuit or apparatus used to separate the original information from the modulated *carrier wave.
2. Any device or apparatus used to detect or locate the presence of a physical property, e.g. radiation detector, particle detector.

detergent A substance, such as soap, that can be used as a cleansing agent. Many synthetic detergents exist; all owe their properties to the fact that they are surface-active and modify the surface properties of the dirt that they remove.

Nonpolar substances, such as grease, are insoluble in pure water. Detergent molecules have a *lyophilic portion that has an affinity for nonpolar substances and a *hydrophilic portion that has an affinity for water. The detergent is adsorbed on the surface of the "dirt" by its lyophilic part and the hydrophilic groups then allow the dirt to be carried into solution.

determinant A collection of algebraic symbols written in a rectangular array. The expression

$$\begin{vmatrix} a_1 & b_1 \\ a_2 & b_2 \end{vmatrix}$$

is a determinant of the second order and is equal to $(a_1 b_2 - a_2 b_1)$. The expression

$$\begin{vmatrix} a_1 & b_1 & c_1 \\ a_2 & b_2 & c_2 \\ a_3 & b_3 & c_3 \end{vmatrix}$$

is a determinant of the third order, of value

$$a_1 \begin{vmatrix} b_2 & c_2 \\ b_3 & c_3 \end{vmatrix} - b_1 \begin{vmatrix} a_2 & c_2 \\ a_3 & c_3 \end{vmatrix} + c_1 \begin{vmatrix} a_2 & b_2 \\ a_3 & b_3 \end{vmatrix}$$

or

$$a_1(b_2 c_3 - b_3 c_2) - b_1(a_2 c_3 - a_3 c_2) + c_1(a_2 b_3 - a_3 b_2).$$

The expression

$$\begin{vmatrix} a_1 & b_1 & c_1 & d_1 \\ a_2 & b_2 & c_2 & d_2 \\ a_3 & b_3 & c_3 & d_3 \\ a_4 & b_4 & c_4 & d_4 \end{vmatrix}$$

is a determinant of the fourth order, and so on. Determinants arise in the solution of simultaneous linear equations; they are also of great value in higher mathematics.

deuteranopia *See* colour vision.

deuterium The isotope of hydrogen with mass number 2.

deuteron A nucleus of an atom of deuterium; i.e. an atom of deuterium that, having lost its single extra-nuclear electron, has a unit positive charge. It is regarded as a close combination of a proton and a neutron.

developable surfaces *See* principal radii.

deviation 1. *Syn.* variation. The difference between an observation and its true value. The latter has often to be replaced by its nearest known value, i.e. the mean or *average of all the observations, in which case the difference is often called the *residual*. The *mean deviation* is the average of the deviations when all are given a positive sign. The *standard deviation* is the square root of the average of the squares of the deviations of all the observations. (*See* frequency distribution.) **2.** In *frequency modulation, the amount by which the carrier frequency is changed by modulation.

deviation (angle of) The angle between the incident ray and the reflected (or refracted) or emergent ray. A ray after a single reflection is deviated $(\pi - 2i)$, where i is the angle of incidence; after successive reflection at two plane mirrors the deviation is $(2\pi - 2A)$, where A is the angle between the mirrors. The angle of deviation by a prism depends on the angle of incidence and the angle of the prism. *Minimum deviation* by a prism occurs when the refraction is symmetrical. The minimum deviation D appears in the equation:

$$n = \sin \tfrac{1}{2}(A + D)/\sin \tfrac{1}{2}A,$$

where A is the principal refracting angle of the prism and n the refractive index. For narrow-angle prisms:

$$D = (n - 1)A,$$

(approximately), and over a moderate range of angle of incidence there is an approximately constant deviation.

deviation ratio *See* frequency modulation.

Dewar, Sir James (1842–1923) Brit. chemist and physicist. In his low-temperature work, he invented the *Dewar vessel, he was the first to liquefy hydrogen (1898) and to solidify it (1899) and he studied many properties of substances at low temperatures including magnetic properties and phosphorescence. For low-temperature measurements of specific heat capacities he designed the *Dewar liquid oxygen calorimeter.

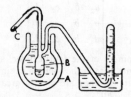

Liquid oxygen calorimeter

Dewar liquid oxygen calorimeter A calorimeter used to determine the mean specific heat capacity of solids between room temperature and the boiling point of oxygen (–183 °C). The bulb B contains liquid oxygen and is immersed in a *Dewar vessel, A, also containing liquid oxygen. The specimen is held in a tube at C until the amount of gas issuing from the delivery tube under water is small and constant. The specimen is then tipped into B by bending the rubber connection to C and the oxygen evolved is collected over water in a graduated jar, its volume being corrected to the value (v) for dry oxygen at STP when the density is p. Hence

$$mc(\theta + 183) = vpL,$$

where L is the specific latent heat of vaporization of liquid oxygen at its normal boiling pont; m and c are the mass and mean specific heat capacity respectively of the specimen and θ its original temperature. Allowance is made for the amount of gas that would have been evolved in the time occupied by the experiment if the specimen had not been added. Lead, whose mean specific heat capacity is known, is used to calibrate the calorimeter so that the value of L need not be known.

Dewar vessel *Syn.* vacuum flask; *UK tradename*: Thermos flask. A glass vessel consisting of a double-walled flask with the interspace completely evacuated to prevent gain or loss of heat by the contents of the flask through gaseous conduction and convec-

tion. Transfer of heat by radiation is reduced by silvering the inside walls so that a body placed in such a vessel is thermally isolated from the atmosphere outside and its temperature will remain practically unchanged for long periods.

dew point The highest temperature a surface may have in order that dew may condense on the surface from a humid atmosphere. *See* humidity.

dextrorotatory Capable of rotating the plane of polarization of polarized light in a clockwise direction, as viewed against the direction of motion of the light. *See* optical activity.

dialysis The separation of a colloid from a solute by diffusion of the latter through a suitable filter such as parchment, animal bladder, or a clay diaphragm (known as *semipermeable membranes). The apparatus used is called a *dialyzer*. In *haemodialysis*, waste materials or poisons in blood are removed by means of dialysis in a dialyzer, often called an artificial kidney.

dialyzer *See* dialysis.

diamagnetism A property of substances that have a negative magnetic *susceptibility. It is caused by the motion of electrons in atoms around the nuclei. An orbiting electron produces a magnetic field in the same way as an electric current flowing in a coil of wire. If an external magnetic field is applied, the electron orbits are distorted so as to produce a magnetic field that opposes the applied field, in accordance with *Lenz's law. Thus the relative *permeability of a diamagnetic substance is less than that of a vacuum and lies between 0 and 1.

Lines representing the flux in a uniform magnetic field become more separated when passing through the material; similarly, if a diamagnetic substance is placed in a nonuniform field, there is a force acting from the stronger to the weaker part of the field. If a bar of diamagnetic material is placed in a uniform magnetic field, it tends to orientate itself so that the longer axis is at right angles to the flux.

Diamagnetism is a very weak effect: the relative permeability is only slightly less than 1, for example, in bismuth it is 0.999 99. All substances possess diamagnetism but in some cases it is totally masked by stronger *paramagnetism or *ferromagnetism. Examples of purely diamagnetic substances are copper, bismuth, and hydrogen. The diamagnetic properties of materials are not affected by their temperature. A substance in the superconducting state can be regarded as strongly diamagnetic as the magnetic flux is excluded from the material except for a very thin surface layer (*see* Meissner effect).

diametral voltage *See* voltage between lines.

diamond A crystalline form of carbon and one of the hardest and most infusible substances known. Imperfect and discoloured diamonds are widely used as abrasives and cutting tools in drills, etc. In the diamond crystal lattice, each carbon atom is at the centre of a tetrahedron with its vertices at its nearest four neighbours. Diamond is an excellent electrical insulator in marked contrast to the other crystalline form of carbon, namely *graphite. It is a very good conductor of heat.

diaphragm An opaque screen containing a circular aperture centred and normal to the axis of an optical system; it controls the amount of light passing through the system.

diathermic or diathermanous Transparent to heat, as opposed to *adiathermanous*.

diathermy A method for the treatment of disease by the use of undamped high-frequency current oscillations, usually of the order of 10^6 hertz. The absorption of energy from the currents, owing to the electrical resistance of the tissues causes a considerable, but controllable, rise in temperature.

diatonic scale *See* musical scale.

dichroism The property possessed by some crystals of selectively absorbing light with electrical vibrations in one plane, while allowing the vibrations at right angles to pass through, e.g. tourmaline, Polaroid, etc.

dichromate cell A primary cell in which poles of carbon and amalgamated zinc are immersed in a solution of potassium dichromate ($K_2Cr_2O_7$) in dilute sulphuric acid. A two-fluid form, in which the zinc is in dilute sulphuric acid and the carbon in an aqueous solution of dichromate, the two being separated by a porous partition, has also been used. The e.m.f. is 2.03 volts.

dichromatic vision *See* colour vision.

Dicke, Robert Henry (*b.* 1916) Amer. physicist. A professor at Princeton, he predicted (1964) independently of others that there would be observable residual microwave radiation as a result of the big bang (*see* big-bang theory). The subsequent observation of 7 cm radiation helped to establish the big-bang theory. Dicke's work on gravitation established the equivalence of inertial and gravitational mass to one part in 10^{11} and with Carl Brans he proposed that the gravitational constant is slowly decreasing (the Brans–Dicke theory).

Dickinson, Harper, and Osborne's determination of the specific latent heat of fusion of ice (1913) They found the specific latent heat of fusion of ice at low temperatures by an electrical method. The electrical work E required to raise the temperature of unit mass of ice from T_1 to T_2 was determined in a *Nernst and Lindemann calorimeter containing ice below 0 °C. If c_1 and c_2 are the specific heat capacities of ice and water respectively and T_0 is the melting point:
$$E = \int_{T_1}^{T_0} c_1 \, dT + L + \int_{T_0}^{T_2} c_2 \, dT,$$
from which L may be obtained.

die *See* integrated circuits.

diecasting *See* casting.

dielectric A substance that can sustain an electric field and act as an insulator.

dielectric constant *Syn.* relative permittivity. Symbol: ε_r. The ratio of the absolute *permittivity of a substance or medium to the *electric constant (the permittivity of free space).

dielectric heating The heating effect that occurs when a high-frequency alternating electric field is applied across a nonconducting material, i.e. a dielectric. It is a result of dielectric hysteresis. Normally the power is provided by some form of oscillator. Dielectric heating is used extensively for the preheating of plastic materials.

dielectric hysteresis A phenomenon, akin to magnetic hysteresis, as a result of which the *electric displacement in a dielectric depends not only on the applied electric field strength, but also on the previous electrical history of the specimen. It entails a dissipation of energy from the field (the dielectric hysteresis loss) when the specimen is subjected to an alternating electric flux.

dielectric loss The total dissipation of energy that occurs in a dielectric when it is subject to an alternating electric field.

dielectric polarization *Syn.* electric polarization. Symbol: P. Stress set up in a dielectric owing to the existence of an electric field, as a result of which each element of the dielectric functions as an electric dipole. It measures the increased flux present in the dielectric due to the presence of the latter, and is defined as the product of the *electric constant ε_0 and the *electrization, or as the function $(D - \varepsilon_0 E)$, where E is the applied field strength and D is the electric displacement. *See also* displacement current.

dielectric strength *Syn.* disruptive strength. The maximum electric field that an insulator can withstand without breakdown, under given conditions. It is usually measured in $V \, mm^{-1}$.

Diesel, Rudolf (1858–1913) German inventor of the Diesel engine. *See* Diesel cycle.

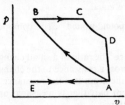

Diesel cycle

Diesel cycle A *heat engine cycle in which air is the working substance and the fuel a heavy oil. From A to B the air is compressed adiabatically to a very high temperature. From B to C the burning fuel causes expansion at constant pressure, while CD is the remainder of the working stroke, being an adiabatic expansion. At D a valve opens and the pressure

Dieterici equation of state

falls to atmospheric. AE and EA represent the exhaust and charging strokes. A separate fuel pump is necessary to inject the oil into the cylinder at high pressure.

Dieterici equation of state An *equation of state for real gases:

$$p\,e^{a/RTV}(V - b) = RT,$$

which leads to *van der Waals' equation as a first approximation. It predicts a value for (RT/pV) at the critical point of 3.69 and gives $V_c = 2V_0$.

difference tone If two tones are sounded simultaneously, other tones may be heard. One consists of a tone whose frequency is the difference of the frequencies of the primaries and is known as the difference tone. See combination tones.

Differential air thermometer

differential air thermometer A simple instrument for the detection of radiant heat. Two equal closed bulbs A and B, one clear, the other blackened, contain air at atmospheric pressure. Radiation falling on the apparatus is more readily absorbed by the blackened bulb and so the pressure inside B rises so that the liquid stands at a higher level in the left-hand connecting tube than in the right-hand tube.

differential amplifier A type of *amplifier with two inputs, whose output is a function of the difference between the inputs.

differential analyser A type of *analogue computer used to solve differential equations. It can also be a mechanical or electrical device.

differential coefficient See derivative.

differential galvanometer A *galvanometer of the moving magnet type, having two separate coils through which currents can be passed in opposite directions. If the two coils are identical, the galvanometer will show no deflection when the currents in the two coils are equal.

differentially compound-wound See compound-wound machine.

differential resistance The ratio of a small change in the voltage drop across a resistance to the change in current producing the drop. It is the resistance of a device or component part that is measured under small-signal conditions.

differential steam calorimeter A form of steam calorimeter in which the small amount of steam condensing on a body of small heat capacity is capable of precise measurement. In Joly's apparatus for the determination of the specific heat capacity of a gas at constant volume (c_v) two identical hollow copper spheres SS are suspended in the same steam cham-

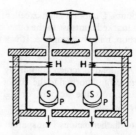

Joly's differential steam calorimeter

ber from the arms of a chemical balance. The spheres are both evacuated and counterpoised at room temperature (θ_2). Steam at a known temperature is admitted to the chamber for some time, some condensing on each sphere and being caught on the pans PP attached to the bottom of the spheres. The spheres are then counterpoised by the addition of weights to allow for any dissimilarity of the spheres. On cooling, one sphere is filled with gas under pressure, the mass of the gas m being found by recounterpoising. Steam is then passed through as before and the extra mass of steam M condensing on the sphere containing the gas is found by weighing. Then:

$$ML = mc_v(\theta_1 - \theta_2),$$

where L is the specific latent heat of vaporization of steam at θ_1. Condensation round the suspension wires where these leave the steam chamber is prevented by small electric heating coils HH surrounding these regions and heated to redness. A cover placed over each sphere prevents water condensed on the top of the chamber from falling on to the collecting pans.

differential thermal analysis A technique for investigating changes of state, heats of reaction, etc., in a sample by comparing it with a standard of identical *heat capacity. The temperature of both specimens is increased slowly and the difference between their temperatures is followed. The standard is chosen so that it does not change over the temperature range studied – its temperature can be varied linearly with time. A graph is plotted of this temperature (T) against the temperature difference. This difference (ΔT) will be zero until T reaches a value where the change occurs, when ΔT will change because of release or absorption of energy in the sample. Thus the graph gives information on the temperatures at which changes occur and on the associated energy changes. The results obtained may depend on the rate at which the temperature is varied.

differentiator A circuit designed so that the output is the differential with respect to time of the input; e.g. $v_o \propto dv_i/dt$, where v_o and v_i are output and input voltages respectively.

diffraction analysis The study of crystal structure by means of the diffraction of a beam of electrons, neutrons, or X-rays. See electron diffraction; X-ray analysis; neutron diffraction.

diffraction grating A device for producing spectra by diffraction and for the measurement of wavelength. Commonly it consists of a large number of equidistant parallel lines (of the order 7500 per cm) ruled with a diamond point on glass, speculum metal, or an evaporated layer of aluminium (ruled gratings) or of a plastic cast taken from a ruled surface (replica grating). Diffracted light after "reflection" or "transmission" produces maxima of illumination (spectral lines) according to the equation:

$$d(\sin i + \sin \theta) = m\lambda,$$

where d is the grating interval, i.e. the distance between corresponding points of adjacent lines ($= 1/N$, in which N is the number of lines per unit distance), i the angle of incidence, θ the direction of the diffracted maximum with the normal corresponding with the "order" m of the spectrum ($m = 0$, for the central image). *Concave gratings* are ruled on the front surface of concave mirrors and can be used to focus spectra without the use of lenses. This is an advantage for higher energy radiation, such as ultraviolet, which is strongly absorbed by matter. Plane gratings are used at grazing incidence for X-rays. *See* blazed grating; crystal grating; echelette grating; echelon grating; Merton grating; Rowland grating. Transmission gratings can be used in the visible, near ultraviolet and near infrared. Reflection gratings are needed for the far ultraviolet. Reflective gratings used at grazing incidence permit absolute measurements of X-ray wavelengths.

diffraction of light 1. *General.* If the shadow of an object cast on a screen by a small source of light is examined, it is found that the boundary of the shadow is not sharp. The light is not propagated strictly in straight lines, and peculiar patterns are produced near the edges of the shadow, which depend on the shape and size of the object. This breaking up of the light, which occurs as it passes the object, is known as *diffraction* and the patterns observed are called *diffraction patterns*. The phenomenon, which arises as a consequence of the wave nature of light, was first commented on by Grimaldi in a book published in 1665.

It is usual to distinguish between two classes of diffraction phenomena. In the first class – *Fresnel diffraction* – only the simple arrangement of source, diffracting object, and screen are involved. In *Fraunhofer diffraction*, a parallel beam of light passes the diffracting object and the effects are observed in the focal plane of a lens placed behind it.

2. *Fresnel zones.* Certain of the effects observed in Fresnel diffraction can be explained if one considers the wavefront falling on the obstacle divided into a number of concentric annular zones, the distances of the peripheries of the zones from the observation point on the screen increasing by one half wavelength from zone to zone. The zones are called *Fresnel zones*. Each point on the wavefront can be considered as the source of a secondary wave (*see* Huygens's principle; wave motion) and each of these secondary waves will make its own contribution to the light reaching the observation point. It

can be shown that all the Fresnel zones have roughly the same area, so that each zone can be considered to contain the same number of secondary sources, and hence each zone will contribute the same amount of light to the observation point. However, because each zone is a further half wavelength away than the next, the contributions from adjacent zones will be out of phase. Thus the total contribution can be represented as the sum of a series of terms alternately positive and negative, each term representing the contribution from one Fresnel zone.

Although all have the same area, the central zones are pointing squarely at the observation point, whereas the outer zones are directed more obliquely. The result is that the magnitude of the terms representing the contributions of the zones falls off steadily from term to term. The sum of the series can now be shown to be equal to half the first term, so that the amplitude of the light reaching the observation point from the whole unrestricted wavefront is half the amplitude that would result if all but the first Fresnel zone were blocked out. The intensity of the light is proportional to the square of the amplitude and will thus be one quarter of that due to the first zone alone.

If the obstacle has a circular aperture, the amount of light reaching the central point of the diffraction pattern will depend on the number of Fresnel zones that fill the aperture. If all but the first zone is blocked out by the obstacle, the intensity at the central point is four times greater than that observed if no obstacle were in place. If the first two zones are effective, the resulting intensity at the central point will be very small, since the contributions from the two zones are nearly equal but are out of phase. In this case, the diffraction pattern consists of a bright circle of light with a central dark spot. In general, if an odd number of zones is effective, the centre of the pattern is bright, while an even number results in a dark central point. The general pattern consists of concentric light and dark rings.

If a circular object is used, the central Fresnel zones will be blocked out. The sum of the series of terms that represent the effective zones will still be equal to half the first term, so that this resulting amplitude at the centre of the pattern is half the amplitude due to the first effective zone. It follows that there will always be some light reaching the central point of the diffraction pattern so there is always a relatively bright spot at the centre of the shadow of a circular obstacle.

If the obstacle consists of alternately opaque and transparent annular zones, it is possible to arrange that every other Fresnel zone is effective for a particular observation point. The result will then be a high intensity of illumination at that point, since the light from the effective zones will all arrive in phase. Such an obstacle (called a *zone plate*) will produce at the observation point a bright "image" of the point source and will in this sense act as a lens. It is capable of producing, by diffraction, an image of any small bright object.

diffraction of light

In practical cases the radii of the Fresnel zones are very small. Thus diffraction effects are seen only with small obstacles or apertures and at the edges of the shadows of larger obstacles. In other cases the effects predicted are exactly the same as those that would follow from the *rectilinear propagation of light.

3. *Strip division of the wavefront.* In considering the diffraction patterns produced by obstacles consisting of a straight edge, a slit, or a wire, it is convenient to divide the wavefront into strips parallel to the edges of the obstacle rather than into circular zones. Again we take the edges of the strips at distances from the observation point increasing by one half wavelength from strip to strip. The diffraction patterns can then be predicted by considering the total effects at the observation point of all the half-period strips not blocked out by the obstacle.

If the obstacle is a straight edge, then at the edge of the geometrical shadow the obstacle will block out all the strips over one half of the wavefront. Passing into the shadow, more and more strips of the other half of the wavefront will be blocked out and there is a gradual diminution of light received until almost complete darkness prevails. Outside the shadow, the full effect of one half of the wavefront is seen and in addition more and more strips of the other half are uncovered as one moves outwards. Since adjacent strips are out of phase with each other, the total effect will be a minimum if an even number of additional strips is uncovered and a maximum if the number is odd. Thus outside the geometrical shadow there are alternate bright and dark bands of decreasing contrast.

In the case of a slit there will be a bright central line to the diffraction pattern if an odd number of strips of each half of the wavefront is uncovered, and a dark line if the number is even. In general the pattern consists of an unsharp shadow of the slit crossed by dark lines.

The case of the wire is rather similar to that of the circular obstacle in **2**. There is always a relatively bright line in the centre of the shadow.

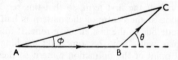

a Resultant of two light beams of differing phase and amplitude

4. *Graphical methods in diffraction problems.* Instead of using the zone treatment outlined in **2** and **3**, problems in diffraction may often be elucidated by graphical methods.

Provided the light is of uniform frequency (i.e. of the same colour) the resultant amplitude of two beams can be determined by drawing two sides of a triangle equal on some suitable scale to the amplitudes of the two separate beams, with the angle between the sides equal to the phase difference

between the beams. The resultant is then represented in amplitude and phase by the magnitude and direction of the third side of the triangle.

Thus, if two beams have amplitudes proportional to AB and BC in Fig. *a* and they differ in phase by θ, then the resultant amplitude is proportional to AC and there is a phase difference of ϕ between the resultant and the beam represented by AB.

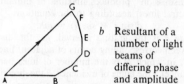

b Resultant of a number of light beams of differing phase and amplitude

If there are more than two beams (AB, BC, CD, . . . , FG in Fig. *b*) a vibration polygon can be drawn to find the resultant amplitude (AG).

Each point on the wavefront is emitting secondary waves, but since different points are at differing distances from the observation points, there is a phase difference between secondary wavelets arriving at that point. Thus if the wavefront is split up into smaller annular zones (similar to, but smaller than, the Fresnel zones of **2**) the effect of the zones can be represented by lines such as AB, BC, etc., and the final resultant obtained in Fig. *b*. If the size of zones chosen is progressively smaller and smaller, then in the limit the vibration polygon turns into a smooth curve as in Fig *c*.

The vibration diagram for the complete wavefront is illustrated in Fig. *c*. In this diagram the first Fresnel zone is represented by the curve from A to B, since the phase of B (represented by the direction of the tangent) is 180° different from that at A. Similarly the part of the curve from B to C represents the contributions of the second Fresnel zone, C to D the third, and so on. Since the contributions of the zones become progressively smaller as we pass out from the centre (because of the obliquity), the curve never circles back exactly to the starting-point; instead, it spirals in to the point J.

If the obstacle blocks out all but the first Fresnel zone, the resulting amplitude is represented by a line AB. The amplitude due to two Fresnel zones is AC, due to three is AD, and due to the whole wave is AJ. Thus the amplitude produced by the first zone is twice that due to the whole wave as was mentioned in **2** and indeed, all the facts deduced from the zone treatment can be deduced from the study of the vibration diagram of Fig. *c*.

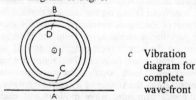

c Vibration diagram for complete wave-front

In the case of strip division of the wavefront, there is the difference that the areas of the strips (and

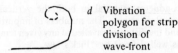

d   Vibration
    polygon for strip
    division of
    wave-front

therefore the amplitudes produced) decrease rapidly away from the centre. Thus, if finite strips are considered the vibration polygon is similar to Fig. *d*, which in the limit gives the curve of Fig. *e*.

A similar curve is obtained for each half of the wave but it is convenient to draw these curves in opposite quadrants so that the vibratior. diagram for the whole wave is represented by Fig. *f*. This graphical treatment was first conceived by *Cornu, and the cnrve of Fig. *f* is called *Cornu's spiral*.

The amplitude due to the whole wave is represented by J'J. If a straight edge is interposed to cut out

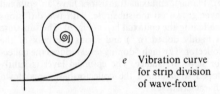

e   Vibration curve
    for strip division
    of wave-front

half the wave, the amplitude falls to OJ. A slit that is wide enough to let through the part of the wavefront represented by the curve along AOB will yield a resultant amplitude of amount AB, and so on. The problem of diffraction by straight-edges, slits, and wires can be considered in detail using Cornu's spiral.

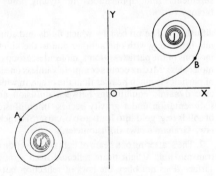

f   Cornu's spiral

5. *Fraunhofer diffraction*. In this class of diffraction, a parallel beam of light falls on the diffracting object and the effects are observed in the focal plane of a lens placed behind it. Thus in Fig. *g*, AB represents a slit whose length is perpendicular to the plane of the paper, and on which falls a parallel beam of light. According to Huygens's principle, each point in the slit must be considered as a source of secondary wavelets that spread out in all directions. Now the wavelets travelling straight forward along AC, BD, and so on, will arrive at the lens in phase and will produce a strong illumination at O. On the other hand, the secondary wavelets spread-

ing out in a direction such as AE, BF, and so on will arrive at the lens with a phase difference between successive wavelets, and the effect at P will depend on whether this phase difference causes destructive *interference or not.

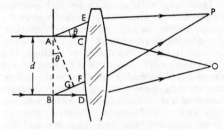

g   Fraunhofer diffraction

For example, if the distance BG in the diagram is equal to one wavelength of the light used, then there is a path difference of one whole wavelength between light reaching the lens from opposite ends of the slit. There is accordingly a path difference of half a wavelength between light from any point in the upper half of the slit and the corresponding point in the lower half, and light from these two points will interfere destructively. The result will be that the light from the whole of the upper half of the slit interferes destructively with that from the whole of the lower half, and no light reaches P. A similar effect occurs if BG is two wavelengths, three wavelengths, and so on. Between these positions of zero illumination, there will be bright regions, and the resulting pattern seen in the plane OP is of alternate dark and light bands running parallel to the length of the slit (i.e. perpendicular to the plane of the paper). It will be noticed that (unlike the corresponding case of Fresnel diffraction) there is always a bright line at the centre of the diffraction pattern. The separation of the diffraction bands increases as the width of the slit is reduced; with a wide slit the bands are so close together that they are not readily noticeable. The separation also depends on the wavelength of the light, being greater for longer wavelengths. If white light is used, a few coloured bands are therefore produced.

In the case of a circular aperture, the diffraction pattern consists of a central bright patch (called the *Airy disc*) surrounded by alternate dark and light circular bands. Again the bands are close together if a large aperture is used but their separation increases as the diameter of the aperture used is reduced.

In the case of the slit shown in Fig. *g*, the first dark line at P is in a direction θ such that BG is one wavelength, λ. If *d* is the width of the slit, then $\theta = \lambda/d$ (since θ is small). In the case of a circular aperture it can be shown that the direction of the first dark circle is given by a similar expression, $\theta = 1.22\lambda/d$.

diffraction of sound Sound waves (or indeed waves of any type) only cast sharp shadows when intercepted

diffraction pattern

by a solid or when made to pass through a slit, if the solid or slit is of large dimension compared with the wavelength. In such cases, the rays are lines normal to the wavefront and the laws of geometrical optics apply.

These laws, while adequate for such practical purposes are not strictly correct in theory, since to a certain degree the rays are bent or "diffracted" giving an interference pattern of regions of varying intensity just inside the shadow or shadows. This diffraction is more pronounced as the dimension of the solid or slit is made smaller, and in the extreme case where the dimension is less than the wavelength the sound is reradiated from the obstacle with uniform intensity in all directions; the slit radiates only on the side away from the incident rays and thus produces hemispherical waves, but the solid radiates in all directions and therefore produces spherical waves.

The analysis of the general case is facilitated by Huygens' construction, i.e. by regarding the wavefront as consisting of an infinite number of point sources radiating spherical waves. These secondary waves interfere with each other to produce the interference patterns obtained.

The wavelength of audible sound in air varies from approximately 20 mm to 10 m depending on the frequency, and therefore, since the obstacles usually met are within this range, sound shadows are not well defined in practice. A further contribution to the elimination of shadows is made by the presence of reflecting surfaces.

diffraction pattern *See* diffraction of light.

diffraction spectrum A spectrum produced by diffraction, particularly one produced by a *diffraction grating.

diffractometer An instrument used in *diffraction analysis to measure the intensities of diffracted beams of X-rays or neutrons at different angles. An *ionization chamber or *counter is usually used and the beam to be diffracted is usually monochromatic.

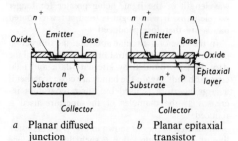

a Planar diffused junction   *b* Planar epitaxial transistor

diffused junction A junction between two regions of different conductivity within a *semiconductor, formed by diffusion of the appropriate impurity atoms into the material. The material is heated to a predetermined temperature in an atmosphere containing the desired impurities in gaseous form. Atoms that condense on the surface diffuse into the

semiconductor material in both the vertical and horizontal directions. The numbers of impurity atoms and the distance travelled at any given temperature is well defined by *Fick's law.

Originally, nonselective diffusions were performed, the entire surface of the semiconductor wafer being exposed to the impurities and unwanted areas being etched away (as in *mesa transistors). Surface-leakage problems were common to these transistors. Modern diffusion techniques use the *planar process; selective diffusions are made through holes in an oxide mask, which are formed using *photolithography. The structure is made passive by the oxide layer and surface-leakage currents are much reduced.

Planar diffused transistors have two successive diffusions forming the base and emitter regions (Fig. *a*). Planar *epitaxial transistors have an epitaxial layer grown on the substrate, with two diffusions made into the epitaxial layer (Fig. *b*). The substrate is highly doped (n^+) and forms the bulk of the collector. The high *doping level reduces the collector series resistance. The epitaxial layer is lightly doped to maintain the collector-base breakdown characteristics.

diffusiometer A device for measuring diffusion in liquids. If the diffusion is measured over microscopic distances, the time taken and the stirring effects of vibration and temperature changes are much reduced. The apparatus is then called a *microdiffusiometer*. Apart from theoretical interest, these measurements find applications in dyeing and in biology.

diffusion 1. The process by which fluids and solids mix intimately with one another due to the kinetic motions of the particles (atoms, molecules, groups of molecules). Mixing occurs completely unless one set of particles is much heavier than the other, in which case dynamic equilibrium between diffusion and sedimentation under gravity occurs. Interdiffusion of solids (e.g. gold into lead) also occurs. (*See* Fick's law; Graham's law; diffusiometer.)

2. The scattering of a beam of light on reflection or transmission. A light beam reflected from a rough surface does not obey the laws of reflection but is scattered in many directions (diffuse reflection). Similarly, light transmitted by certain materials does not obey the laws of refraction and is scattered in the medium (diffuse transmission).

3. The degree to which the directions of propagation of sound waves vary over the volume of a reverberant sound field.

diffusion cloud chamber A type of *cloud chamber in which supersaturation is achieved by diffusion of a vapour from a hot to a cold surface through an inert gas. As the vapour supply is continually replenished by diffusion, the chamber can be made almost continuously sensitive to ion tracks. There are no moving parts.

diffusion coefficient Symbol: D. The mass of substance transported in unit time across unit area of a fluid, divided by the concentration gradient across this area. It is measured in $m^2\,s^{-1}$. (*See* Fick's law.) This term is not synonymous with *diffusivity.

diffusion current *See* limiting current.

diffusion length In a *semicondcutor, the average distance travelled by *minority carriers between generation and recombination.

diffusion pump *See* pumps, vacuum; air pumps.

diffusion theory A theory concerning the *diffusion of particles, especially neutrons, based on the assumption that, in a homogeneous medium, the *current density is proportional to the negative gradient of the particle flux density.

diffusion transfer *See* Polaroid (Land) camera.

diffusivity Symbol: α. A measure of the rate at which heat diffuses through a substance. It is equal to the thermal conductivity divided by the specific heat capacity at constant pressure and the density, i.e. $\alpha = \lambda/\rho c_p$. It is measured in $m^2\,s^{-1}$. This term is not synonymous with *diffusion coefficient.

digital audio tape (DAT) Magnetic tape used for digital recording of sound and also for storing computer information. In the case of digital recording, the audiofrequency signal is normally sampled 48 000 times per second and the characteristics of the sampled signal are converted into 16-bit words – as occurs in *compact disc systems. The recording method for DAT is derived from video recording, the tape being wrapped helically around a rotating drum. This enables one or more tape heads in the drum to record the digital signal on slanted tracks on the slow-moving tape at a very high density. The sound reproduced on replay of DAT is of very high quality.

digital circuit Any circuit designed to respond to discrete values of input voltage and produce discrete output voltage levels. Usually only two values of voltage are recognized, as in binary *logic circuits. *Compare* linear circuit.

digital computer *See* computer.

digital delay line *See* delay line.

digital inverter *See* inverter.

digital recording A means of recording sound in which the audiofrequency signals are sampled up to 30 000 times per second and the characteristics of the sampled signal are converted to a digital form – a series of discrete numbers – that is then stored on a system such as a *compact disc or *digital audio tape. The digits are then reconstituted in the player, which enables them to be played back without interference or distortion.

digital voltmeter (DVM) A voltmeter that displays the measured values as numbers composed of digits. The voltage to be measured is usually supplied as an analogue signal, and the voltmeter samples the signal repetitively and displays the voltage sampled. A record of the output may be obtained with such voltmeters, and they function as a form of analogue to digital converter.

digitron *Syn.* Nixie tube. A type of cold cathode scaling tube in which the cathodes are shaped into the form of characters, usually the digits 0 to 9. A switching connection to one side of the power supply selects the cathode required. These tubes have been used for display purposes in *counters, calculators, etc.

dilatancy A name given by Osborne Reynolds to certain unexpected phenomena depending on the normal piling of granules. For example, a number of small shot contained in an indiarubber bar are in normal piling when the volume of their interstices is at a minimum value. Thus, if the indiarubber bag be filled with shot and water and the shot is in normal piling, then any deformation by squeezing of the indiarubber bag (which is supposed to be fixed to a cylindrical glass tube) will cause the level of the water in the tube to sink when one would have ordinarily expected it to rise. This affords an obvious explanation of the fact that pressure of the foot on the damp sand of a beach causes the sand to dry slightly in the neighbourhood of the foot. *See* thixotropy.

dilation *Syn.* dilatation. A change of volume.

Dilatometer

dilatometer An apparatus for studying thermal expansion. A *volume dilatometer* is used for the variation of the volume of a liquid over a range of temperature. The liquid is contained in a bulb B of known volume, joined to a graduated capillary tube, T, the top being closed by a ground stopper to prevent loss of liquid by evaporation. The volume of a known mass of liquid at any temperature can be read off from the position of the meniscus in the capillary tube, but allowance must be made for the expansion of the material of the bulb.

dilution law *See* Ostwald's dilution law.

dimensional analysis A technique whose main uses are: (*a*) to test the probable correctness of an equation between physical quantities; (b) to provide a safe method of changing the units in a physical quantity; (c) to assist in recapitulating important formulae; (d) to solve partially a physical problem whose direct solution cannot be achieved by normal methods; (e) to predict the behaviour of a full-scale system from the behaviour of a model; (f) to suggest relations between fundamental constants.

The basis of the technique is that the various terms in a physical equation must have identical *dimensional formulae if the equation is to be true for all consistent systems of units. E.g. in the well-known equation $s = ut + \frac{1}{2}at^2$ (which applies to uniformly accelerated motion in a straight line), all the terms have the dimension of length. The method does not check pure numbers (e.g. the $\frac{1}{2}$ in the last term).

The equation is said to have dimensional homogeneity and is true if the units for s, u, a, and t are respectively m, m s^{-1}, m s^{-2}, and s, or any other set of units having these dimensions. Thermal quantities require temperature Θ as an additional dimension, e.g. thermal conductivity has dimensions $MLT^{-3}\Theta^{-1}$. See also dynamic similarity.

dimensional constant A factor in a physical equation whose magnitude depends on the size of the fundamental units but not on the particular system to which the equation is applied. E.g. the gravitational constant G in the Newtonian formula $F = Gm_1m_2/r^2$ has the dimension of Fr^2/m_1m_2 (namely $M^{-1}L^3T^{-2}$).

dimensional formula A symbolic representation of the definition of a physical quantity obtained from its units of measurement.
(M = mass, L = length, T = time).
E.g. area = L^2, velocity = LT^{-1}, force = MLT^{-2}, energy = ML^2T^{-2}. See dimensional analysis.

dimensions (electrical and magnetic) Mechanical quantities can be expressed in terms of power of mass M, length L, and time T only. Electrical and magnetic quantities are found to require four independent dimensional fundamentals. Present-day users of *dimensional analysis prefer to include current I or charge Q as the fourth dimension. Earlier workers tended to use absolute permeability μ for magnetic analyses or absolute permittivity ε for electrical analyses, despite the fact that the powers of the resulting dimensional expressions were more often than not fractional. (The dimensions of μ and ε are related through Maxwell's equation connecting the free-space values: $c^2 = \mu_0^{-1}\varepsilon_0^{-1}$, where c is the speed of light.) Using the definition of power P in terms of I and voltage V, $P = IV$, the dimensions of voltage are shown to be $ML^2T^{-3}I^{-1}$ or $ML^2T^{-2}Q^{-1}$. The dimensional forms in terms of μ and ε are, via Coulomb's inverse square law, shown to be $M^{1/2}LT^{-2}\mu^{1/2}$ and $M^{1/2}L^{1/2}T^{-1}\varepsilon^{-1/2}$.

dimorphism The existence of a substance in either of two possible crystalline forms. Carbon, for example, can crystallize as graphite or diamond. See also allotropy.

dineric Having two liquid layers (phases).

Dines hygrometer A hygrometer, similar to *Regnault's hygrometer. A can with its axis vertical communicates at its lower end with a horizontal pipe that, in its turn communicates with a horizontal rectangular chamber containing a delicate thermometer. Its upper surface is closed by a thin horizontal plate of silver or of blackened glass. Water, which is cooled by dropping fragments of ice into it, is contained in the can and flows into the horizontal chamber whose upper surface is therefore progressively cooled until dew forms on the silver plate.

dineutron An unstable system consisting of a pair of *neutrons; it is assumed to have a transient existence in certain *nuclear reactions.

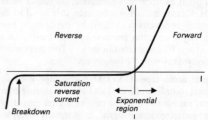

V/I curve for semiconductor diode

diode Any electronic device with only two electrodes. There are several different types of diode and their applications depend on their voltage characteristics. Diodes are usually used as rectifiers. Although *thermionic valves were originally used, diodes are now semiconductor devices. Semiconductor diodes consist of a single p-n junction. Current flows when forward voltage is applied to the diode (see diode forward voltage), and increases exponentially with voltage, becoming substantially constant after a few tenths of a volt (see illustration). If voltage is applied in the reverse direction, only a very small leakage current flows until the *breakdown voltage is reached. See also Gunn diode; IMPATT diode; light-emitting diode; photodiode; tunnel diode; varactor; Zener diode.

diode forward voltage Syn. diode drop; diode voltage. The voltage across the terminals of a semiconductor *diode when current flows in the forward direction. Because of the exponential nature of the semiconductor-diode current characteristic, the diode voltage is approximately constant over the range of currents commonly used in practical circuits. A typical value is about 0.7 V at 10 mA.

diode laser See semiconductor laser.

diode transistor logic (DTL) An early family of integrated *logic circuits, with the inputs through *diodes and the output taken from the collector of an inverting transistor. The basic DTL gate circuit is a *NAND gate and is shown in the illustration. If

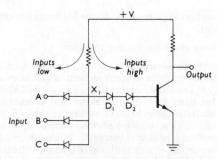

DTL gate circuit (NAND gate)

any one of the inputs A, B, or C is low, current flows through the input diode, and the point X_1 is low. This has the effect of shunting current away from the base of the inverting output transistor, and the output of the gate is high. If all the inputs are high, current flows through the diodes D_1 and D_2, towards the base of the transistor, and the output goes low. Two DTL NAND gates may be combined into an *OR gate, by connecting their outputs together.

The output transistor is designed to operate under saturated conditions, and the speed of DTL logic circuits is therefore less than *emitter-coupled logic circuits, because of the longer delay time. DTL circuits have been largely replaced by *transistor-transistor logic circuits.

diode voltage *See* diode forward voltage.

dioptre A unit used to express the power of a spectacle lens, equal to the reciprocal of the focal length in metres. It is also applicable to *vergence and curvature, convergence being regarded as positive.

dioptric system An optical system in which the principal optical components are refracting elements, such as lenses. *Compare* catoptric system.

dip *Syn.* inclination. The angle made with the horizontal by the direction of the earth's local magnetic field. It varies from 0° at the magnetic equator to 90° at the magnetic poles. A magnet freely suspended at its centre of gravity would set with its magnetic axis in the magnetic meridian and inclined to the horizontal at the angle of dip.

DIP Abbreviation for *dual in-line package.

dip circle An instrument for determining the angle of *dip. It consists essentially of a thin magnet M supported so as to be free to rotate about a horizontal axis A through its centre of gravity, its inclination to the horizontal being read on a vertical circle C, divided into degrees. For accuracy numerous adjustments are required, including the reversal of the needle on its bearings, and the reversal of the magnetism. These are provided for in the Kew standard dip circle (*see* illustration).

diplopia Double vision arising because corresponding points of the two eyes do not receive their

appropriate images. It can be produced by pressure against one eyeball; a prism before one eye, especially when deviating in the vertical direction; binocular instruments out of parallelism, etc.

dipolar ions *See* ampholyte ions.

dipole A system of two equal and opposite charges placed at a very short distance apart. The product of either of the charges and the distance between them is known as the *dipole moment* (symbol: *p*). A small magnet constitutes a magnetic dipole. Also known as a doublet.

dipole aerial An *aerial very commonly used for frequencies below 30 MHz. It consists of a centre-fed horizontally mounted conductor, the length of which is usually one-half wavelength of the transmitted or received radio wave (a *half-wave dipole*). The ends may be folded back and joined together at the centre (a *folded dipole*).

Dip circle
(*By courtesy of Messrs. W. & J. George & Becker, Ltd.*)

dipole moment *See* dipole; debye.

diproton An unstable system consisting of a pair of *protons; it is assumed to have a transient existence in certain *nuclear reactions.

Dirac, Paul Adrien Maurice (1902–1984) Brit. physicist. A leader in the field of quantum mechanics. Dirac predicted on theoretical grounds the existence of the positive electron (positron) in 1930, well before its discovery by Anderson and by Blackett in 1932. The form of statistics now known as

Dirac constant

Fermi–Dirac statistics was introduced independently by Fermi and Dirac. Dirac's quantum theory predicts the existence of electron *spin.

Dirac constant *See* Planck constant.

Dirac equation An equation for the wave functions of fermions used in relativistic *quantum mechanics. It can be regarded as a version of the *Schrödinger wave equation that takes relativity into account. There are a number of ways of expressing the equation; one form is:

$$i\alpha.\nabla\psi + (mc/\hbar)\beta\psi = (i/c)\partial\psi/\partial t,$$

where m is the mass of a free particle, c the speed of light, t the time, and $\hbar$ the rationalized Planck constant. The wave function is ψ, i is $\sqrt{-1}$, and α and β are square matrices satisfying certain symmetry rules. The Dirac equation can be used to show that fermions must have spin 1/2 and it also predicts the existence of antiparticles.

Dirac function *Syn.* δ-function. A function of x defined as being zero for all values of x other than $x = x_0$ and having the definite integral from $x = -\infty$ to $+\infty$ equal to unity. It is much used in quantum mechanics and can be used, for example, to represent an impulse in dynamics. The graph of the function is that of a single, infinitely high and infinitesimally wide peak of total area unity placed at $x = x_0$.

direct access *Syn.* random access. A method of data organization in which any part of a computer *file may be reached without starting at the beginning and working through. A direct access device, such as a *disk, must be used for this access method. *Compare* sequential access.

direct broadcast by satellite (DBS) A method of broadcasting in which a communications *satellite in geostationary orbit is used as the main transmitter. The signal to be broadcast is transmitted from its point of origin on the earth to the satellite, where it is amplified and retransmitted to cover a wide area of the earth's surface. It is detected directly by individual receivers using a suitable *dish aerial tuned to the DBS signals.

direct-coupled amplifier *Syn.* d.c. amplifier. An *amplifier in which the output of one stage is coupled directly to the input of the next stage, or through a chain of *resistors. It is capable of amplifying *direct current and is sometimes called a direct-current amplifier.

direct current Symbol: d.c. An electric current that flows in one direction only and is substantially constant in magnitude.

direct-current generator A generator for producing *direct current.

direct-current restorer *Syn.* d.c. restorer. A device to restore or impose a given d.c. or low-frequency component to a signal after passing through a circuit that has low impedance to fast variations in current, but high impedance for d.c. or low-frequency current.

direct effect *See* ionizing radiation.

direct-gap semiconductor A semiconductor, such as gallium arsenide, in which an electron with an energy E_g can make a direct transition across the forbidden band between the valence and conduction bands (*see* energy bands), the energy gap being E_g; it does so by absorbing a photon of energy E_g or by emitting a photon of energy E_g. In an *indirect-gap semiconductor*, such as silicon, an electron with energy E_g cannot be excited directly across the forbidden band but requires a change in momentum.

direction angles and cosines The direction of a line passing through the origin of a set of Cartesian axes, can be specified by the angles made by the line with positive parts of three axes. These angles are called the *direction angles* and the values lying between 0° and 180° are used.

The cosines of the direction angles are called the *direction cosines* and the sum of their squares equals one. Quantities proportional to the direction cosines are called *direction ratios*.

direction ratios *See* direction angles and cosines.

directive aerial An aerial that, as a radiator or receiver of radio waves, is more effective in some directions than in others. The directivity is often obtained by employing a *passive* (or parasitic) aerial in conjunction with an *active* aerial. The latter is an aerial connected directly to the transmitter or receiver. The former is an aerial that influences the directivity but, in transmission, is excited by the e.m.f. induced in it by the nearby active aerial, or, in reception, reacts with the active aerial by virtue of the mutual impedance between them. A passive aerial is called a *reflector* or a *director* according to whether it is placed respectively behind or in front of the active aerial.

directivity factor The ratio of the intensity of the sound radiated from a source (at a specified frequency) at any remote point on a reference axis, to the average (for all directions) of the intensity of the sound at the same distance from the effective centre of the source. The *directivity index* is ten times the logarithm to the base ten of the directivity factor.

direct lightning surge *See* lightning surge.

director (aerial) *See* directive aerial.

directrix *See* conic sections.

direct stroke *See* lightning stroke.

direct vision Central vision in which the image falls on the fovea centralis of the eye.

direct-vision prism A prism combination that produces dispersion without deviation of the central part of the spectrum (yellow D line). It is used in the construction of the *direct-vision spectroscope*. *See* Amici prism; prism; spectrometer.

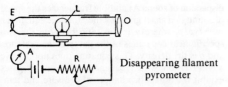

Disappearing filament
pyrometer

disappearing filament pyrometer An optical pyrometer in which an image of the hot source is focused by a telescope objective lens O onto the filament of an electric lamp L, which is viewed through a red filter by the eyepiece E. The observer varies the current through the lamp filament by means of the rheostat R until the filament becomes indistinguishable against the background of the image of the source. If the current is too large, the filament appears brighter than its background, while if the current is too small, the filament appears dark. The red filter enables the matching to be made for a small band of wavelengths and the instrument is calibrated by noting the reading of the ammeter A for various sources whose temperatures are measured by thermocouples. The range can be extended beyond the cold point by the use of a *rotating sector placed before the objective. If T' is the apparent temperature obtained from the calibration curve using a rotating circular sector whose angle of opening is α, the true temperature T of the black-body source is given by

$$\log_e\left\{\frac{2\pi}{\alpha}\right\} = \frac{1\cdot432}{\lambda}\left(\frac{1}{T'} - \frac{1}{T}\right),$$

where λ is the mean wavelength transmitted by the filter and 1.432 is the value of the constant C_2 in the *Wien radiation law.

discharge 1. To remove or reduce an electric charge from a body.

2. The passage of an electric current or charge through a *gas-discharge tube or dielectric, usually accompanied by luminous effects. (*See* conduction in gases; brush discharge; glow discharge; spark; arc; corona.)

3. The maintenance of a current by a cell, particularly an *accumulator. This involves chemical changes such that the cell eventually ceases to operate. It is then said to be *discharged*.

discharge coefficient The ratio of the actual discharge of fluid from an orifice to the discharge calculated from the velocity given by *Torricelli's law. The discharge coefficient is given by $Q/A\sqrt{2gh}$, where Q is the actual discharge through an orifice of an area A under a static head h.

discharge tube *See* gas-discharge tube.

discharge-tube rectifier A rectifier that consists of a *gas-discharge tube in which the electrodes are arranged in a manner that ensures that the discharge current in one direction is greater than that in the opposite direction when the same voltage applied between the electrodes is reversed.

discomposition effect *See* Wigner effect.

discriminator 1. An electronic circuit for changing a frequency-modulated or phase-modulated signal into an amplitude-modulated signal.

2. A circuit that delivers output pulses only for input pulses of greater than a certain chosen amplitude.

discriminator frequency *See* frequency discriminator.

disc winding A type of winding used in transformers. It consists of a number of flat coils, each wound in the form of a disc. The winding may be (i) single-disc, (ii) double-disc, or (iii) continuous-disc. In (ii) each coil is wound with a continuous length of conductor in the form of two discs side by side, the coil ends being at the outer periphery. In (iii) the whole winding consists of a continuous length of conductor. Disc windings are usually employed for the high-voltage windings of power transformers. *Compare* cylindrical winding.

dish A type of aerial used in radio astronomy, satellite communications, etc., and consisting of a sheet-metal or mesh reflector, spherical or paraboloid in shape. The reflected radio waves or microwaves are brought to a focus above the dish, where they are collected by a secondary aerial, called a feed.

disintegration Any process in which a nucleus emits one or more particles, such as *beta particles, *alpha particles, and *gamma rays, either spontaneously or following a collision.

disintegration constant *See* decay constant.

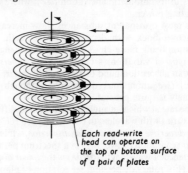

Each read-write
head can operate on
the top or bottom surface
of a pair of plates

Computer disk

disk *Syn.* magnetic disk. A *storage device, used in computing systems, that consists of a circular plate coated on one or more usually both sides with a magnetic film. Data is stored on a series of concentric tracks in the film; there are several thousand tracks. The disk substrate may either be rigid (in which case it is known as a *hard disk*) or it may be flexible (a *floppy disk*). The amount of data that can be stored on a disk depends mainly on its type and size, the number of tracks per disk, and the recording density along the tracks.

dislocation

Data is stored on and retrieved from disks by means of a device called a *disk drive*. A single disk may be used, or a stack of disks mounted on a common spindle. The disk or disk pack is rotated in the disk drive at constant high speed. A *read-write head* can be instructed to move radially over each coated surface to select a particular track. The disk's rotation brings a particular location on the track to the read-write head. Items of data can thus be found directly and within a very short time, usually some tens of milliseconds.

dislocation *See* defect.

disperse phase The suspended substance in a *colloidal solution or suspension. For example, in a suspension of small carbon particles in air the carbon is the disperse phase and the air is the *continuous phase.

dispersion The decomposition of a beam of white light into coloured beams that spread out to produce spectra, or *chromatic aberration. More precisely, it is concerned with descriptions of the variation of refractive index (n) with wavelength (λ). When n is written as a function of λ a *dispersion equation* is formed. (*See* Sellmeier equation; Cauchy dispersion formula.) The differential coefficient ($dn/d\lambda$) describes dispersion at a particular region of colour. Of more common reference is *mean dispersion*, which is the difference of refractive index for light of the F and C lines of hydrogen, i.e. ($n_F - n_C$). *Dispersive power* (ω) is the ratio:
$$(n_F - n_C)/(n_D - 1).$$
For intermediate differences, *partial* dispersions are quoted. The dispersion of a narrow-angle prism refers to its chromatic aberration (ωP), where P is deviating power.

As most transparent substances show increasing refractive index with decreasing wavelength, the variation being more rapid at shorter wavelengths, this type of variation is called *normal dispersion*. Near an absorption band this normality apparently ceases (*see* anomalous dispersion) although the anomaly is quite general.

The rate at which an angle of refraction or diffraction varies with wavelength ($d\theta/d\lambda$) is also referred to as *dispersive power* or *angular dispersion*, while the linear separation of two lines in a spectrum per unit difference of wavelength ($dl/d\lambda$) is the *linear dispersion*. (For rotary dispersion, *see* rotation of plane of polarization.)

Dispersion, in general, is a manifestation of the dependence of wave velocity on the frequency of the wave motion, and is a property of the medium in which the wave is travelling. It is not only light that is dispersed. Radio waves for example are slowed down when they travel through an ionized medium: the lower the frequency the greater the delay. *See also* dispersion of sound.

dispersion forces *See* van der Waals forces.

dispersion medium Fluid, liquid, or gas, in which colloidal particles are dispersed.

152

dispersion of sound At audible frequencies the speed of sound in a gas is given by the *Laplace equation $c = \sqrt{\gamma p/\rho}$, where γ is the ratio between the two specific heat capacities of the gas, p is the pressure, and ρ is the density. This equation does not involve any variation with frequency. Air and all gases exhibit the adiabatic changes of pressure and temperature when audible sound passes through as implied by this formula.

At higher frequencies, however, Pierce found that the speed of sound in certain cases, notably carbon dioxide, varies with frequency in a peculiar manner. In carbon dioxide, the speed of sound at 0 °C was found to be 258.52 metres per second at a frequency of 42 kilohertz increasing to 260.15 at 200 kHz. At higher frequencies carbon dioxide was found to be completely opaque to sound waves.

These variations of velocity with frequency (or dispersion) are always accompanied by abnormal values of absorption. This fact cannot be explained on the classical theory of speed of sound in gases. The *relaxation theory* as developed by Herzfeld and Rice is the most acceptable one to account for such dispersion phenomena. (*See* anomalous absorption of sound.)

It seems that at higher frequencies there is a lag in the interchange of translational and vibrational energy of the gas molecules. In other words the oscillatory degrees of freedom no longer have time in these rapid acoustic vibrations to adapt themselves completely to the adiabatic changes of temperature.

Kneser derived the following formula for the velocity in terms of *pulsatance (ω) and molecular constants:

$$c = \sqrt{\frac{p}{\rho}\left(1 + R\,\frac{\omega^2\beta^2 C_{va} + C_v}{\omega^2\beta^2 C_{va}^2 + C_v^2}\right)},$$

where C_v is the molar heat capacity at constant volume; C_{va} is the molar heat capacity of translational degrees of freedom; R is the gas constant; β is the mean life of an energy quantum (or relaxation time) and is expressed by the relation
$$C_{vi}^t = C_{vi}^{st}\,(1 - e^{st/\beta})$$
where C_{vi}^{st} is the total heat of vibration and C_{vi}^t is the effective heat of vibration at this instant. When ω approaches zero, i.e. at low frequencies, the speed of sound becomes

$$c_0 = \sqrt{\frac{p}{\rho}\left(1 + \frac{R}{C_v}\right)},$$

but at high frequencies (ω approaching ∞),

$$c_\infty = \sqrt{\frac{p}{\rho}\left(1 + \frac{R}{C_{va}}\right)}.$$

If the square of the speed of sound is plotted as a function of log n (where n = frequency) the *dispersion curve* is obtained. It has a point of inflection at a frequency $(1/2\pi\beta)(C_{va}/C_v)$.

By means of this relation, the relaxation time β can be determined. Thus the measurement of the dispersion of sound in a gas throws light on important intramolecular processes. For example, a region of dispersion exists for each natural frequency of the molecule, the regions being liable to overlap. The relaxation time depends upon pressure, temperature, admixed impurities, or chemical dissociation and therefore sound dispersion occurs by variation of any of these factors.

dispersive medium A medium in which the *phase speed of a particular kind of wave depends upon the frequency. All substances are to some degree dispersive for electromagnetic radiations. Sound is not dispersive in any materials except at high ultrasonic frequencies (*see* dispersion of sound). Waves on the surface of liquids have speeds independent of frequency at about a certain wavelength (roughly 20 mm for water at normal temperatures); for lower frequencies (long wavelengths) the speed tends towards proportionality to the square root of the wavelength; for shorter waves the speed increases in proportion to the inverse three-halves power of wavelength. *See* anamalous dispersion; dispersion; group speed.

dispersive power *See* dispersion.

displacement 1. A vector representing in magnitude and direction the difference in position of two points. It is the basic vector used in physics. Velocity is defined as rate of change of displacement, acceleration is rate of change of velocity. Thus mechanical quantities such as *momentum and force, and electrical quantities such as *electric field depend upon the concept of displacement.
  2. The quantity of fluid displaced by a submerged or partially submerged body.
  3. *See* electric displacement.
  4. *See* Wien displacement law.

displacement current The rate of change of electric flux through a dielectric when the applied electric field is varying. When a capacitor is charged, the conduction current flowing into it is considered to be continued through the dielectric as a displacement current so that the current is, in effect, flowing in a closed circuit. Displacement current does not involve motion of the current carriers (as in a conductor) but rather the formation of electric dipoles (a phenomenon known as *dielectric polarization*), thus setting up the electric stress. The recognition by Maxwell that a displacement current in a dielectric gives rise to magnetic effects equivalent to those produced by an ordinary conduction current is the basis of his *electromagnetic theory of light.

displacement law *See* Wien displacement law.

displacement rule *See* Soddy and Fajan's rule.

disruptive discharge The passage of an electric current through an insulating material when the latter breaks down under the influence of a dielectric stress equal to or greater than the *dielectric strength of the particular insulating material. *See* spark.

disruptive strength *See* dielectric strength.

dissociation 1. The breakdown of molecules into smaller molecules or atoms. Some compounds dissociate at room temperature. For example, dinitrogen tetroxide (N_2O_4) exists in equilibrium with nitrogen dioxide (NO_2) at normal temperature: $N_2O_4 \rightleftharpoons 2NO_2$. All molecules can be broken down into atoms at sufficiently high temperatures. Most dissociations of simple molecules are reversible; the equilibrium constant for dissociation is often called the dissociation constant. (*See* mass action, law of.)
  2. The breakdown of molecules into ions in solution. *See* electrolytic dissociation.

dissociation constant *See* Ostwald's dilution law.

dissonance Any combination of two notes not forming a *consonance.

distance of (most) distinct vision Conventional distance of 25 cm or 10 in for use in comparing magnifying powers of microscopes.

distorted waveform Usually a *waveform that is not *sinusoidal. *See* distortion.

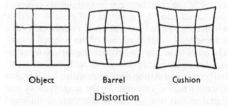

Object   Barrel   Cushion

Distortion

distortion The extent to which a system, or part, fails to reproduce accurately at its output the characteristics of the input.
  (1) (electrical) The modification of a *waveform of voltage, current, etc., by a transmission system or network. It involves the introduction of features that do not appear in the original or the suppression or modification of features that are present in the original. The main classifications are: (i) Attenuation distortion (or frequency distortion). Distortion produced when the gain or loss of a transmission system varies with frequency. (ii) Phase distortion. Distortion produced when the phase change introduced by the transmission system is not a linear function of frequency. (iii) Harmonic distortion. Distortion due to the production of harmonics that are not present in the original. (iv) Amplitude distortion. Distortion, due to variation with the amplitude of the input, in the ratio of the rms value of the output of the system to the rms value of the input, on the basis that both input and output waveforms are sinusoidal. If other harmonics are present in the output, the rms value of the output is for that of the fundamental component only. (v) Nonlinear distortion. Distortion produced by a system in which the transmission properties vary with the instantaneous magnitude of the input. Nonlinear distortion gives rise to the types listed under (iii), (iv), and (vi). (vi) Intermodulation distortion (or combination-tone

distortion). Distortion resulting from the introduction of spurious combination-frequency components in the output of a nonlinear transmission system when the input consists of two or more sinusoidal voltages applied simultaneously. Distortion is particularly important in telecommunications.

(2) (sound) In its application to sound transmission or reproduction, the percentage of first, second, third, and higher harmonics produced in the output for a given single sine-wave input is frequently quoted. The main types of distortion are (i) nonuniform transmission of amplitude at different frequencies for a constant amplitude input (frequency distortion); (ii) nonlinear relation between input and output at a given frequency at different amplitudes of the input (amplitude distortion); (iii) a phase shift between different components on transmission. This is largely immaterial when a system is handling steady notes, but is of great importance when considering transients; (iv) transient distortion, giving a duration of certain components of a note in excess of the duration of the input; (v) nonlinear distortion resulting from a lack of constancy of the ratio between instantaneous values of the output and of the input; (vi) scale distortion, which may be due to operation of a loudspeaker at a volume level other than that of the original sound. The amplitude relation between the various frequencies for equal perception varies with volume level; (vii) cross-modulation distortion with a loudspeaker system where a variation in the amplitude of one signal affects the output of another of different frequency but of constant output; (viii) the production of spurious combination tones when an input of two or more frequencies is applied to a nonlinear part of a system.

(3) (light) If the image formed by an optical instrument is not geometrically similar to its object, the *aberration in the image is called *distortion*. Instead of the lateral magnification (y'/y) being constant, its value depends on the size of the object (y). When the magnification decreases with object size, a square object is imaged with *barrel-shaped distortion* (*see* barrel distortion); the reverse case is *cushion-shaped distortion*. In general, a front stop yields barrel-shaped distortion; a rear stop yields pincushion type; symmetrical doublets with central stop for which position the lens is spherically corrected, are free from distortion. *See* aberrations of optical systems; trapezium distortion.

distributing main *See* distributor.

distributing point *See* feeding point.

distribution function The mathematical expression of a *frequency distribution.

distribution law *Syn.* partition law. (Berthelot, 1872). If to a system of two immiscible or slightly miscible liquids, a third substance is added that is soluble in both liquids, this will distribute itself so that its concentrations in the two liquids are in a constant

ratio at a constant temperature. This ratio is known as the *distribution* or *partition coefficient*.

distribution of velocities Maxwell's law of the distribution of velocities, based on classical statistics,

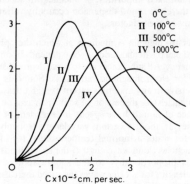

Distribution of molecular velocities for hydrogen

states that for a gas in equilibrium the number of molecules whose total velocity lies in the range $c \to (c + dc)$ is given by the expression

$$dN_c = 4\pi N \left\{ \frac{hm}{\pi} \right\}^{3/2} (e^{-hmc^2})c^2 \, dc,$$

where N is the total number of molecules, m is the mass of a molecule, h is a constant and equal to $1/(2kT)$, where k is the *Boltzmann constant, and T the thermodynamic temperature.

The distribution yields values as follows:
C, root mean square velocity $= \sqrt{(3/2mh)}$,
$\bar{C}$, the mean velocity $= \sqrt{(3/\pi hm)}$,
C_p, the most probable velocity $= \sqrt{(1/hm)}$.
The distribution is represented graphically in the illustration, plotted at four temperatures for hydrogen.

The units for the ordinate axis are chosen to make the area of the curve equal to N, the total number of molecules. The maxima of the four curves occur at the points $C = 1.50, 1.76, 2.53,$ and 3.24×10^3 m s^{-1} respectively.

The higher the temperature the more scattered is the distribution, but at all temperatures there is theoretically a small number of molecules whose velocity approaches infinity. The law, which has been verified experimentally by *Eldridge, is also known as the Maxwell–Boltzmann relation. *See also* canonical distribution.

distributor *Syn.* distributing main. The electric line supplying electrical power to the *service lines of the consumers. The generating station is connected to the distributor by one or more *feeders.

diurnal motion The apparent motion of celestial bodies across the sky from east to west, caused by the rotation of the earth.

divergence 1. (div) The *flux per unit volume leaving an infinitesimal element of volume at a point in a vector field; e.g. in an electrostatic field, the divergence of the field is zero unless the volume element contains an electrostatic charge and the vector field is therefore *solenoidal. The divergence of a vector field F is the scalar product (*see* vector) $\nabla . F$, where ∇ is the differential operator *del:

$$\text{div } F = \nabla . F = i . \frac{\partial F}{\partial x} + j . \frac{\partial F}{\partial y} + k . \frac{\partial F}{\partial z}$$

$$= \frac{\partial F_x}{\partial x} + \frac{\partial F_y}{\partial y} + \frac{\partial F_z}{\partial z} \quad \text{(scalar)}$$

where $F = iF_x + jF_y + kF_z$, and i, j, and k are *unit vectors.

2. (light) The spreading out of light rays from a real object point or a virtual image point. In technical and ophthalmic optics, divergence is reckoned as negative and is measured (in *dioptres) as the reciprocal of the distance of the point of divergence from the optical element (expressed in metres). Concave lenses and convex mirrors are diverging in action.

diverging lens A lens that causes a beam of parallel light to diverge. *Compare* converging lens.

diversity factor The ratio of the sum of the *maximum demands of all the separate consumers connected to a power station to their maximum simultaneous demand on the station. A common value in generation is about 25.

diversity system A system of communication employing two or more paths or channels so that their combination produces a single received signal. In radio-communication, *fading can be minimized by the use of frequency, space, or polarization diversity.

diverter A resistor connected in parallel with a winding of an electrical machine in order to divert a fraction of the current from that winding. A diverter is commonly connected in parallel with the field winding of a d.c. series motor to provide control of speed.

D-layer or region *See* ionosphere.

D-lines of sodium Two yellow lines very close together in the *emission spectrum of sodium. D_1 has a wavelength of 589.6 nm and D_2 is 589.0 nm. Because these lines are bright and easily produced, sodium light is used as a reference line in spectrometry. When the refractive index of a substance is quoted, it is often understood that the sodium D_1 line has been used although the helium line at 587.6 nm has been recommended as a standard in refractometry.

Dolby *Tradename*. A system for reducing *noise in magnetic and photographic sound recording and reproduction.

Dolezalek electrometer An improved form of *quadrant electrometer, giving greater sensitivity by a reduction in the size of the quadrants and the weight of the needle. The instrument is used hetero-statically, the needle, suspended by a fine phosphor bronze strip, being charged to a constant potential of about 100 volts. Used with a lamp and scale, deflections of the order of 1000 mm at a distance of 1 m can be obtained for a difference of potential of 1 volt between the quadrants. The instrument was widely employed in early investigations on the ionization of gases and radioactivity.

dollar A unit of *activity equivalent to the amount of activity required to make a *nuclear reactor critical (*see* chain reaction) using only *prompt neutrons. A *cent* is equal to one-hundredth of a dollar.

Dollond, John (1706–1761) Brit. optician who invented the achromatic lens.

domain A bounded region of ferromagnetic body throughout which the magnetization is uniform except for slight fluctuations in direction or intensity. The direction of magnetization depends on the magneto-crystalline and strain anisotropy, and on the effective field. The number, shape, and size of the domains in a body change with its magnetic state. For the domain theory of ferromagnetism, *see* ferromagnetism. *See also* Bitter patterns.

dominant wavelength *See* chromaticity.

Donders reduced eye In order to simplify calculations of image size and position in the case of the human eye, the latter is regarded as a convex refracting surface of radius 5 mm separating air in front from water (refractive index 4/3) behind. The anterior focal distance is 15 mm, the posterior focal distance 20 mm. The surface is placed at a position about 1.5 mm behind the pole of the actual cornea. *See* eye.

donor *See* semiconductor.

doping The addition of impurities (*dopants*) to a *semiconductor to achieve a desired n-conductivity or p-conductivity.

doping compensation The addition of a particular type of impurity to a *semiconductor to compensate for the effect of an impurity already present.

doping level In a semiconductor. The number of impurity atoms added to the material to achieve the desired polarity and resistivity. Low doping levels yield a high-resistivity material: high doping levels yield a low-resistivity material. *See also* semiconductor.

Doppler, Christian Johann (1803–1853) Austrian mathematician and physicist. He attempted to explain the colorations of stars in terms of an apparent change in frequency due to motion similar to the change in *pitch in sound from moving objects. In fact this *Doppler effect is too small with stars for visual detection but the rotation of the sun was demonstrated, using spectroscopic examination, in 1871 and since then the effect has been much used in astrophysics.

Doppler broadening

Doppler broadening An effect observed in *line spectra when radiation forming a particular spectral line has a spread of frequencies because of the Doppler effect, caused usually by the thermal motion of molecules, atoms, or nuclei. The apparent frequency of radiation from a single molecule of gas depends upon the component of the velocity of the molecule in the line of sight with respect to the observer. The number of molecules with components of velocity V per unit range of velocity is proportional to $\exp(-mV^2/2kT)$ according to the *Maxwell distribution. The fractional change of frequency given by the Doppler effect is V/c when c is the speed of light, hence the spectrum line is spread according to the factor: $\exp(-mV^2/2kTc^2)$, giving the line a *Doppler width*. Thus to obtain a sharp spectral line, especially with light molecules, it is necessary to keep the temperature of the gas as low as possible, or to use a *molecular beam moving transversely to the line of sight. The effect can be used to determine approximately the temperature of a *plasma. *Absorption spectra are similarly affected.

There is a related phenomenon in the gamma-ray spectra of solids using the Mössbauer effect. The low-excitation energies of the states studied in this way have lifetimes large compared with the period of vibration of the atoms, hence the mean component of velocity is zero averaged over the time of emission or absorption. The mean-square value is, however, nonzero, hence there is a frequency change dependent on the second-order Doppler effect.

Doppler effect

Doppler effect The change in apparent frequency of a source (of light, sound, or other wave motion) due to relative motion of source and observer. Originally advanced for the optical case (*see* Doppler) its most familiar application is to sound.

It is illustrated simply by assuming the velocities are in the line joining the source and observer; suppose C is the speed of sound, u_s the velocity of the source, u_0 the velocity of observer, n the true frequency of the source, and W the velocity of the medium. If S is the initial position of source and S′ its position one second later, SA $= C + W$ and SS′ $= u_s$. The waves emitted by the source in one second occupy the distance S′A $= C + W - u_s$, which contains n waves. Similarly let O be position of observer; in one second the waves received occupy the distance O′B $= C + W - u_0$, which contains n waves. Then, the apparent frequency is:

$$n' = n \times \frac{C + W - u_0}{C + W - u_s}.$$

When the medium is still:

$$n' = n \times \frac{C - u_0}{C - u_s}.$$

Starting with the speed of the source increasing from zero, n' will increase until $u_s = C$ where $n' = \infty$; if $u_s > C$, then n' decreases as u_s increases but in a negative sign, i.e. the waves emitted by the source are received in the reverse order. If u_s is negative, i.e. the source is going away from the observer, n' diminishes as u_s tends to ∞.

The same applies to the motion of the observer. As u_0 tends to C, n' diminishes to zero. If $u_0 > C$, n' increases but with a negative sign. If u_0 is negative, i.e. the observer is approaching the sound, n' increases indefinitely as u_0 increases.

The principle is applicable to all types of wave motion. Thus, for sound waves it is noticed that the pitch of a whistling locomotive drops suddenly as the locomotive passes an observer standing on the platform of a station; a similar effect occurs when the observer is moving and the source of sound is fixed. For electromagnetic radiation *relativity gives the formula:

$$\lambda = \lambda_0 (1 + V_r/c)(1 - V^2/c^2)^{-1/2},$$

where λ_0 is the wavelength measured by an observer at rest with respect to the source and λ is that measured by an observer with relative velocity V. V_r is the component of the velocity of the source away from the observer in the line of observation; c is the speed of light. The term V_r/c gives the *first-order Doppler effect* and that in V^2/c^2 gives the *second-order Doppler effect*. *See* wavelength; Doppler broadening; Doppler shift.

Doppler radar *See* radar.

Doppler shift The magnitude of the change in frequency or wavelength of waves that results from the *Doppler effect. *See* redshift.

Dorn effect *See* migration potential.

dose A quantity of radiation or absorbed energy.
  1. *Absorbed dose* (symbol: D) is the energy absorbed per unit mass in an irradiated medium. The SI unit is the *gray.
  2. *Exposure dose* (symbol: X) is a measure of X- or gamma-radiation to which a body is exposed. It is equal to the total charge collected on ions of one sign produced in unit mass of dry air by all *secondary electrons liberated in a volume element by incident photons stopped in that element. The unit is the coulomb per kilogram, which replaces the roentgen.
  3. *Dose equivalent* (symbol: H) is used for protection purposes. The unit is the *sievert. It is defined by:

$$1 \text{ sievert} = 1 \text{ gray} \times \text{QF},$$

where QF is the *quality factor* for a particular type of radiation and is a means of relating absorbed doses of different radiations to give the same biological effect. For example, 1 gray of neutrons has a much greater biological effect than 1 gray of gamma-rays.

Typical quality factors are: X-rays, gamma-rays, and high-energy beta-rays, QF = 1; low-energy beta-rays, QF ≃ 1.8; neutrons, QF = 10.

4. *Maximum permissible dose* is the recommended maximum dose that a person, exposed to *ionizing radiation, should receive during a specified period. *See also* dosimetry; dosemeter.

dose equivalent *See* dose.

dosemeter *Syn.* dosimeter (now deprecated). Any instrument or material used for measuring radiation *dose. Commonly used dosemeters include the *ionization chamber, film dosemeter, and lithium fluoride dosemeter. *See* dosimetry.

dosimeter *See* dosemeter.

dosimetry The measurement of radiation *dose, the choice of method being determined by the quantity and quality of radiation delivered, the rate of delivery (*dose rate*), and the convenience. The most common method is to measure the *ionization caused by the radiation, as in an *ionization chamber.

Film dosimetry is a means of measuring dose using photographic film. The degree of blackening on the film after exposure to radiation and development under controlled conditions gives a measure of the dose received. The usual method of using film dosemeters is to compare the blackening with that caused by a known dose, developed under the same conditions.

A small piece of film is used in *film badges to measure the dose received by personnel exposed to radiation.

High-energy radiation induces changes in the mechanical, electrical, and optical properties of polymers, such as perspex and PVC. In *perspex dosimetry* a piece of perspex is irradiated producing an increase in *optical density, which is proportional to dose over a certain dose range. The absorption maximum of clear perspex is in the ultraviolet and the optical density is read from a spectrophotometer (*see* spectrometer). Red perspex is a more useful dosemeter since radiation induces absorption peaks in the visible region, the dose being proportional to the degree of blackening. The irradiated perspex can be scanned by a *microdensitometer (which determines optical density in the visible and near ultraviolet region), giving a recording of the dose distribution over the area occupied by the perspex in the radiation field. A stack of perspex pieces can indicate dose distribution in a particular volume of a radiation field. The molecular weight of perspex is about equivalent to that of biological tissue so that perspex is a convenient biological dosemeter.

Doses of radiation can be measured by *chemical dosimetry*. A specific number of molecules of a particular type are formed or changed by a particular dose. The *G-factor* (*syn.* G-value) is that number formed or changed by 100 eV of absorbed energy. The G-factor of a molecule is dependent on the type of radiation, dose rate, and solute concentration. The optical density of a solution is proportional to

solute concentration. Measurement of the optical density of a particular solution and knowledge of its G-factor under the prevailing radiation conditions give a reproducible and accurate determination of dose (optical density or dose × G-factor). The *Fricke dosemeter* uses the radiation-induced oxidation of aerated ferrous ion to ferric ion. Another reaction used is the oxidation of the ferrocyanide ion $(Fe(CN)_6)^{4-}$ to the ferricyanide ion $(Fe(CN)_6)^{3-}$.

Lithium fluoride dosimetry involves the measurement of the *thermoluminescence from the irradiated phosphor, lithium fluoride. Following irradiation the lithium fluoride is heated and the thermoluminescent output (light) is determined by using a *photomultiplier. This output is proportional to integrated dose.

dot and cross notation *See* vector.

double-base diode *See* unijunction transistor.

double bridge *See* Kelvin double bridge.

double-current system A telegraph system in which the transmission of signals is carried out by reversing the direction of the electric current. *Compare* single-current system.

Double-delta connection

double-delta connection A method of connection used in six-phase a.c. circuits in which the six conductors, windings, or phases are connected to form two three-phase *delta connections displaced 180° in relation to each other.

double refraction *Syn.* birefringence. When near objects are viewed through Iceland spar, they appear doubled (Bartholinus, 1669). The light is split into two parts: an *ordinary ray* (*o-ray*), which obeys the ordinary laws of refraction and an *extraordinary ray* (*e-ray*), which follows a different law. The light in the ordinary ray is polarized at right angles to the light in the extraordinary ray. Because of the crystalline nature of the medium, two groups of Huygens wavelets (*see* Huygens principle) progress; the ordinary wavefront is developed by spherical wavelets and the extraordinary wavefront is developed by wavelets that are ellipsoids of revolution. Along an optic axis these travel with the same velocity. The measurement of the double refraction of any crystal is given by the difference of its greatest and least refractive indices. Some crystals are uniaxial (calcite, quartz, ice, tourmaline); others are biaxial (mica, selenite, aragonite). Uniaxial crystals are divided into two classes: positive crystals (e.g. quartz), in which the ellipsoid lies inside the sphere, and negative crystals (Iceland spar), in which the sphere lies inside the ellipsoid. *See* Voigt effect.

double resonance

double resonance A spectroscopic technique in which a sample is irradiated with electromagnetic radiation at two different frequencies, so that the excited atoms or molecules formed by absorption of one frequency absorb further radiation at the second frequency. *ENDOR (electron-nuclear double resonance) is a widely used example. *See also* Kastler, Alfred.

double-sideband transmission Transmission of the two *sidebands that are produced by the *amplitude modulation of a carrier wave, the latter being suppressed at the transmitter so that it is not transmitted. At the receiver, it is necessary to reintroduce the carrier artificially by combining the sidebands with a locally generated oscillation. The practical difficulties encountered in ensuring that the local oscillation is of correct frequency and phase are so great that double-sideband suppressed-carrier systems are little used. *Compare* single-sideband transmission.

doublet 1. A pair of closely spaced lines in a *spectrum. (*See* spin.) **2.** *See* dipole.

doughnut A toroidal vacuum chamber as used in circular particle-accelerators, such as a *betatron or *synchrotron, or in some types of *fusion reactor.

Dove prism *See* prism.

drag coefficient When a body and fluid are in relative motion the body experiences a *drag force* (D) parallel to the direction of relative motion but in the opposite direction. The magnitude of this drag force will depend on ρ, the fluid density, V the relative velocity, l some characteristic length of the body, and on ν the coefficient of kinematic viscosity. By dimensional analysis it is shown that $D = k_0 \rho l^2 V^2$, where k_0 is a dimensionless number being a function of *Reynolds number lV/ν. The ratio of the drag force to $\rho l^2 V^2$ is called the drag coefficient (k_0). This is not a unique definition, the term $\frac{1}{2}\rho V^2$ being used sometimes instead of ρV^2. For an *aerofoil the drag coefficient is $D/\rho S V^2$, where S is the area of the aerofoil projected onto the chord. *Compare* lift coefficient.

drain The electrode in a *field-effect transistor through which *carriers leave the interelectrode region.

drift mobility Symbol: μ. In a *semiconductor, the average velocity of excess *minority carriers per unit electric field. In general, the mobilities of *holes and *electrons are different.

drift transistor *Syn.* graded-base transistor. A transistor in which the impurity concentration in the *base varies across the base region. A high *doping level at the emitter-base junction reduces across the base to a low doping level at the collector-base junction. Drift transistors have a good high-frequency response.

drift tube *See* linear accelerator.

driver A circuit that provides the input of another circuit or controls the operation of that circuit. The term is commonly applied to the amplifier stage preceding the output stage of a transmitter or receiver. In general, a term applied to circuits with low output impedance that are capable of driving one or many other circuits.

driving point impedance 1. The ratio of the rms value of the sinusoidal voltage applied to two terminals of an electrical network to the rms value of the current that flows between the terminals as a result of the applied voltage.

2. When a telephone receiver is driven by alternating current, a counter-alternating e.m.f. is generated by the oscillation of the diaphragm. This counter e.m.f. is proportional to the velocity of the diaphragm and hence to the driving current at any given frequency. It manifests itself as a component of the impedance of the machine, which component is the difference between the impedance when the moving membrane is free to oscillate and the impedance when the motion is prevented. This component is called the *motional impedance* or the *driving impedance*. It can be calculated by the vector difference between the input impedance measured under a specified load and the blocked impedance (i.e. when the impedance of the load is inifinite).

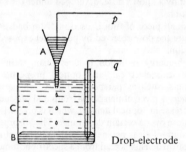

Drop-electrode

drop-electrode A device introduced by Lord *Kelvin for determining the difference of potential between the earth and any point in the atmosphere, and subsequently used for measuring the single surface potential difference between a mercury surface and an electrolyte with which it is in contact. If a conducting liquid is falling in a series of small drops into a dielectric medium that is not at the same potential as the main body of liquid, each drop will carry away an induced charge until the potential of the liquid becomes equal to that of the dielectric. Similarly, if a stream of drops of mercury fall from the nozzle of a funnel A, which is below the surface of an electrolyte, the mercury drops will carry away charges until the potential of the mercury in the funnel becomes equal to that of the electrolyte. The difference of potential between the platinum electrodes p and q then measures the single surface potential difference between the pool of mercury at the bottom of the vessel and the electrolyte C.

dropweight method A method of measuring the surface tension of a liquid by weighing the drops forming slowly and falling from the end of a capillary tube. The method can be used for interfacial surface tensions if drops of the heavier liquid form in and fall into the lighter liquid. The method is a comparative one as a calibration curve has to be drawn from the dropweights of liquids of known surface tension. *See* stalagmometer.

Drude, Paul (1863–1906) German physicist. He extended the uses of Maxwell's electromagnetic theory of light in optics, dealing especially with the optical properties of metals. He developed the first theory of the conduction of heat and electricity in metals by free electrons, assuming classical statistics (1900). This theory suggested an explanation of the *Wiedemann–Franz–Lorenz law and gave a value for the constant in this theory in agreement with experiment. In some other respects the theory was found to be in very serious disagreement with experiment, causing a major problem in theoretical physics. This was solved by *Sommefeld (1927) using *Fermi–Dirac statistics.

drum armature An armature with a *drum winding in an electrical machine.

drum winding A type of *winding used in electrical machines. It consists of coils, usually former-wound, which are housed in slots either on the outer periphery of a cylindrical core or on the inner periphery of a core having a cylindrical bore. Modern machines usually have this type of winding. *Compare* ring winding.

dry adiabatic lapse rate (DALR) *See* lapse rate.

dry cell A primary cell in which the active constituents are absorbed in some porous material so that the cell is unspillable. The usual form consists of a zinc container (forming the negative electrode) lined with a paste of ammonium chloride and plaster of Paris, and having in the centre a carbon rod surrounded by a mixture of ammonium chloride, powdered carbon, zinc sulphate, and manganese dioxide made into a stiff paste with glycerine. Its action is the same as that of the *Leclanché cell; e.m.f. about $1\frac{1}{2}$ volts.

dry ice Solid carbon dioxide, used as a refrigerant.

DTL Abbreviation for *diode transistor logic.

dual in-line package (DIP) A standard form of package for *integrated circuits in which the circuit is encapsulated in a rectangular plastic or ceramic package with a row of metal legs down both of the longer sides. The legs are terminating pins. The legs can be soldered into holes on a printed circuit board or inserted into a chip socket.

duality The principle that certain entities exhibit some of the characteristics of both waves and particles.

(1) *Electromagnetic radiation*. The propagation through space is fully described by the wave equation of electromagnetism. Einstein showed that interaction with matter appears to take place as if the radiation were propagated as quanta with energy $h\nu$, where h is the *Planck constant and ν is the frequency of the wave. The quantum of electromagnetic radiation or *photon* is regarded as a quasiparticle, with mass $h\nu/c^2$, momentum $h\nu/c$ and angular momentum most usually equal to the Debye constant.

(2) *Atomic particles*. The particle nature of such bodies as electrons, protons, atoms, etc. is fully established. According to some formulations of *quantum mechanics the motion of such particles is described in terms of equations analogous to those used for waves. Beams of particles exhibit *diffraction described in terms of the *de Broglie equation. *See also* de Broglie waves; complementarity; wave mechanics.

Duane–Hunt relation The shortest wavelength (λ_{min}) generated in an X-ray tube is inversely proportional to the potential difference (V) applied to the tube. If e, h, and c are, respectively, the electronic charge, the Planck constant, and the speed of light, then:
$$Ve = hc/\lambda_{min} = h\nu_{max},$$
where ν_{max} is the maximum frequency emitted. This is a special case of *Einstein's law.

ductility A combination of properties of a material that enables it to be drawn out into wires.

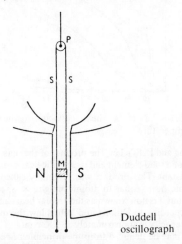

Duddell oscillograph

Duddell oscillograph Essentially, a high-frequency moving-coil galvanometer, formerly used for investigating the waveform of alternating currents or e.m.f.s. A single strip, ss (*see* diagram) of phosphor bronze has its two ends fixed to two terminals in the base of the instrument, and is looped over a suspended pulley, p, by means of which a considerable tension is applied to it. The strip passes between the poles, N, S, of a powerful electromagnet, and its deflections are observed by light reflected from a light mirror M, bridging the two parts of the strip. The reflected light describes a horizontal line, the instantaneous deflection being proportional to the

159

Dufay, Charles François de Cisternay

instantaneous value of the current. To investigate the waveform the light can be focused on a photographic plate moving vertically, or reflected from a rotating mirror driven by a synchronous motor from the a.c. supply under investigation. The instrument is now only of historical interest, having been replaced by the *cathode-ray oscilloscope.

Dufay, Charles François de Cisternay (1698–1739) French chemist who first recognized the existence of two kinds of electricity, that causing attraction and that causing repulsion.

Duhring rule The rule that if two substances A and B have boiling points T_A and T_B at a pressure P and T'_A and T'_B at a pressure P', then
$$(T_A - T'_A)/(T_B - T'_B) = C,$$
where C is a constant.

Dulong, Pierre (1785–1838) French chemist. He was responsible in 1819, with Petit, for the law concerning the molar heat capacities of solids. Dulong and Petit originated the balancing-columns method for the measurement of the absolute coefficient of expansion of mercury.

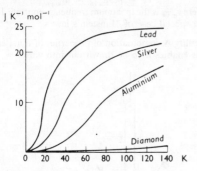

Curves of molar heat capacity against temperature

Dulong and Petit's law The product of the mass per mole of a solid element and its specific heat capacity is constant. This product was originally called the *atomic heat (equal to atomic weight × specific heat) but it is now known as the *molar heat capacity. According to Dulong and Petit's law the molar heat capacity is approximately 25 J K^{-1} mol^{-1}. This value may be deduced from the principle of *equipartition of energy: the movement of the lattice units, involving both kinetic and potential energies, requires an energy of RT per mole per degree of freedom (where R is the *molar gas constant and T the thermodynamic temperature). Thus the molar heat capacity for three degrees of freedom will be $3R$ (25 J K^{-1} mol^{-1}).

This value holds only for simple substances that crystallize in the regular or other simple systems and at high temperatures. At lower temperatures (*see* illustration) the value falls below $3R$, tending to zero as T tends to zero. This was explained in principle by *Einstein's theory of specific heat capacities (1903).

Einstein's theory does not give the correct form of the law at very low temperatures. The *Debye theory (1912) and later improved analyses all predict heat capacities proportional to T^3 near absolute zero, in agreement with experiment for nonmetals. The values for metals are affected by the free electrons, as explained by *Sommerfeld (1927).

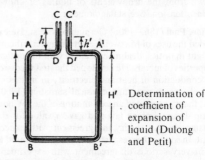

Determination of coefficient of expansion of liquid (Dulong and Petit)

Dulong and Petit's method for determination of coefficient of absolute expansion of a liquid A method that depends on balancing columns of the liquid in the arms of a U-tube, which are at different temperatures. Regnault modified the apparatus by bringing the upper ends of the tube closer together as shown in the diagram, so that the difference in height of the two columns can be easily read. The colums AB, CD, and C'D' are maintained at 0 °C by a bath of melting ice, whilst A'B' is kept at any temperature t °C by an oil bath. If H, H', h, h' denote the lengths of the liquid in the various columns as shown and d, d_0 the densities of the liquid at t °C and 0 °C respectively, then, since BB' is horizontal, the hydrostatic pressure at these points is the same so that $h'd_0 + H'd = (h + H)d_0$. But $d_0 = d(1 + \alpha t)$, where α is the coefficient of absolute expansion of the liquid so that
$$\alpha = \frac{H' - H + (h' - h)}{H - (h' - h)} \times \frac{1}{t},$$
the difference of levels $(h' - h)$ being obtained using a cathetometer. *See also* Regnault's heat experiments.

Dumas, Jean Baptiste André (1800–1884) French chemist who devised a method of measuring vapour densities and hence relative molecular masses. *See* Dumas's bulb.

Dumas's bulb A form of bulb, used on the principle of the *relative-density bottle for measuring the density of vapours and hence their relative molecular masses.

Duperrey's lines Lines traced out on the earth's surface by a traveller, starting at any point and always moving in the direction in which the compass points; first used by Duperrey in 1836.

duplexer A two-channel *multiplexer commonly used in *radar, in which a transmit-receive switch functions during the finite time between transmitting a pulse and receiving the return echo, so that the transmitter and receiver are connected in turn to the same aerial system.

duplex operation The operation of a communications channel between two points in both directions simultaneously. When the operation is limited to either direction but not both at once, it is called *half-duplex operation*.

duplicity theory A theory of vision (*see* colour vision), now generally accepted, that allocates to the rods of the retina the function to appreciate light and movement (without colour), being particularly active at low illuminations. The cones of the retina are active in form and colour vision, where the eye is adapted to higher illuminations.

duralumin An alloy of aluminium containing 4% copper, 5% manganese, 5% magnesium, which develops its hardness as a result of the precipitation of copper on ageing. A lightweight material, it is extensively used in aircraft construction.

Dushman, Saul (1883–1954) Russian-born Amer. physicist, noted for work on high vacua and on thermionic emission.

Dushman's equation *See* Richardson–Dushman equation.

dust core A core for magnetic devices consisting of a powdered magnetic material, such as *ferrite, sintered or cemented into a compact block. It is used for minimizing *eddy-current loss in high-frequency equipment.

DVM Abbreviation for *digital voltmeter.

dwarf star Former name for main-sequence star. *See* Hertzsprung–Russell diagram. *See also* white dwarf.

dynamic Changing, capable of being changed, or taking place over a period of time, usually while a device or system (electrical, electronic, or computing) is in operation.

dynamic braking *See* rheostat braking.

dynamic capacitor electrometer *Syn.* vibrating-reed electrometer. An instrument for measuring very small steady potential differences and direct currents (typically 10^{-15} A) by generating and amplifying an a.c. signal. The d.c. signal is applied to a small capacitor, one plate of which is mounted on a vibrator. The varying distance between the plates causes a periodic change in the capacitance hence generating an a.c. signal proportional to the applied steady p.d. The instrument has low drift and high sensitivity of up to 10 000 divisons per volt.

dynamic characteristic *See* characteristic.

dynamic equilibrium 1. *See* force.
2. A balanced state of constant change, e.g. if water is sealed in an exhausted vessel and kept at constant temperature, although molecules are constantly being exchanged between the ice, water, and water vapour phases, it will be in equilibrium in so

far as the pressure and volume of the phases are concerned.

dynamic friction *See* friction.

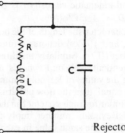

Rejector circuit

dynamic impedance The impedance of a parallel *tuned circuit (rejector circuit) at the resonant frequency. For the circuit shown, dynamic impedance $Z_d = L/CR$ ohms, where L = inductance in henries, C = capacitance in farads, R = resistance in ohms. *Note*: Z_d is purely resistive (i.e. it has no reactance component).

dynamic range The range over which a useful output is obtained from a device. For an electronic device, it is often expressed as the difference in *decibels between the noise level of the system and the overload level.

dynamics The branch of *mechanics concerned with forces that change the motions of bodies. *Statics is sometimes considered as a separate branch of dynamics.

dynamic similarity *Syn.* similarity principle. The dimensions of all mechanical quantities (velocity, acceleration, force, etc.) can be expressed uniquely in terms of the fundamental dimensions of mass (M), length (L), and time (T), certain combinations of the dynamical quantities producing nondimensional numbers. Two systems in motion possess *dynamic similarity* when, for equal values of some dimensionless grouping of the dynamic quantities, they pass through geometrically similar configurations. E.g. for all geometrically similar pendulums whose mass is similarly distributed, the time required to swing through an angle α is $t = k\sqrt{l/g}$, where l and g are the pendulum length and acceleration of free fall respectively, and k is a nondimensional constant depending on the value chosen for α. Thus, the quantity $t\sqrt{g/l}$ is dimensionless (and equal to k) so that pendulums (1) and (2) will be dynamically similar if $t_1\sqrt{g_1/l_1}$ equals $t_2\sqrt{g_2/l_2}$, and will, after time intervals t_1 and t_2, have swung through the same angle α.

In the motion of fluids two systems are dynamically similar when the body boundaries and the corresponding flow patterns are geometrically similar, the nondimensional groupings consisting of a combination of one or more of the dimensionless *Reynolds, *Froude, and *Mach numbers.

In *hydrodynamics and *aerodynamics, the principle of similarity is used extensively in calculating the effect of similar flow upon a scale model. For example, if gravity and elastic forces are neglected, by the method of dimensions it is shown that the drag force on a body in a fluid with relative velocity V, density ρ, and kinematic viscosity v is given by:

$$D = A\rho l^2 V^2 \phi(lV/v),$$

where l is some characteristic length of the body, A is a pure number, and $\phi(lV/v)$ denotes a function of Reynolds' number (lV/v). Similarly the direction of flow lines is dependent solely on Reynolds' number. Thus, providing the values of Reynolds' number for each system are equal, then the flow patterns will be geometrically similar for each system. Furthermore, equal values of Reynolds' number imply that the function $\phi(lV/v)$ has the same value in each system and so the magnitude of D for the full-scale prototype can be calculated from the observed value of D for the scale model providing the relationships between the quantities p, l, V are known. The design and practical requirements of aircraft and ships are predicted from the observations made upon scale models in *wind tunnels and water tanks. *See also* dimensional formula; dimensions.

dynamic stability Of a floating body. The amount of work performed in tilting the body over to a given angle from its position of equilibrium.

dynamic viscosity *See* kinematic viscosity; viscosity.

dynamo A machine that maintains an electric current when it is driven by mechanical means, such as a motor. *See* generator.

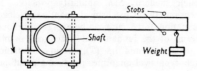

Prony brake dynamometer

dynamometer 1. An apparatus for measuring the torque exerted by a prime mover (*e.g.* petrol engine) or electric motor under dynamic conditions. There are two types: (i) Absorption dynamometers, which absorb the energy delivered by the machine under test and either convert it into internal energy by friction or work electrically by means of a generator. The best-known form is the *Prony brake*, which consists of two stout beams of wood clamped onto the shaft of the machine. The shaft (*see* diagram) rotates against the friction of the wooden beams, and the weight needed to keep the beam between the stops is determined. The rate of working, called the *brake horsepower* is calculable when the speed of rotation of the shaft has been measured with a *tachometer. (ii) Transmission dynamometers, in which the power developed is transmitted after measurement to a machine where it is usefully employed. One instrument, called a *torsionmeter* measures the angle of twist in a known length of the shaft

driven by the prime mover and the torque is deduced from this. If the speed of rotation is known, the power transmitted can be calculated. The *traction meter* is a transmission dynamometer used for measuring the tractive effort (i.e. the pull) of locomotives, etc. Basically, it consists of a spring balance between the locomotive and its load. (*See also* torquemeter.) **2.** *See* electrodynamometer.

dynamotor An electrical machine having a single magnetic field system and a single armature, the latter carrying two independent windings connected to two independent *commutators. The machine operates as a motor with one of the windings and simultaneously as a generator with the other. The armature windings are usually different so that the voltage on the generator side is different from the voltage on the motor side and the machine acts as a rotary transformer.

dynatron A thermionic vacuum tube with the grid placed very close to the anode and the anode treated to be a good emitter of secondary electrons. It has a negative resistance characteristic and has value as an amplifier of current or voltage under suitable conditions, but is generally used as a generator of oscillations.

dyne Symbol: dyn. A *CGS unit of force. 1 dyne = 10^{-5} newton.

dynode An electrode in an electron tube, whose primary function is to provide *secondary emission of electrons. *See* photomultiplier.

dystetic mixture A particular mixture of two or more substances that has the maximum melting point of all possible mixtures of those substances. If the composition of a dystetic mixture is changed, the melting point decreases. *Compare* eutectic.

E

e The charge on an *electron; the natural unit of electric charge. It is equal to $1.602\,177\,33 \times 10^{-19}$ coulomb.

early universe The universe at a time very soon after the big bang (*see* big-bang theory). Theoretical studies of the conditions in this period have led to a mutually beneficial interaction between cosmology and particle physics, especially *grand unified theories.

Because of the high temperatures shortly after the big bang, it is thought that the broken symmetries in *gauge theories may have become unbroken. As the universe cooled, phase transitions led to broken symmetry states.

One problem in cosmology that could be explained by studies of the early universe is the fact that the universe appears not to contain any *antimatter. Effectively, this means that the baryon number of the universe is not zero, which can be explained by nonequilibrium conditions in the early universe and the existence of processes accounted for by grand unified theories, in which the baryon number is not conserved.

One important idea in the theory of the early universe is *inflation*; i.e. that the nature of the vacuum state after the big bang gave rise to a rapid exponential expansion (the *inflationary universe*). This could explain some long-standing problems in cosmology, such as the flatness and homogeneity of the universe. *See also* phase transition.

earth *Syn.* ground. **1.** A large conductor, such as the earth, that is taken as the arbitrary zero in the scale of electric potential.
2. A connection, which may be accidental, between a conductor and the earth. A point on a body that is in good conducting contact with the earth is said to be *earthed*, or at *earth potential*. A wire soldered to a cold water pipe makes an effective earth, or connection can be made to an *earth electrode*, i.e. a large copper plate buried in moist soil.
3. The point or portion in an electric circuit or device that is at zero potential with respect to earth.

earth current 1. A current that flows to earth, particularly as a result of a fault in a system.
2. A current that flows in the earth and is associated with disturbances in the *ionosphere. **3.** A direct current in the earth that can cause corrosion to the lead sheaths of buried cables.

earth detector *See* leakage indicator.

earthed system A system of distribution of electrical current in which one conductor or the *neutral point is connected to earth. In this system, the normal maximum voltage to earth of the conductors, which are not earthed, is fixed automatically by the earth connection. *Compare* insulated system.

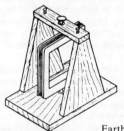
Earth inductor

earth inductor A coil of a few hundred turns of insulated wire wound on a suitable frame and connected to a *ballistic galvanometer. If the plane of the coil is initially perpendicular to the magnetic meridian, and is then rotated through 180° about a vertical axis, the linkages between the coil and the horizontal component H of the earth's magnetic field are reversed, and the quantity of electricity Q

flowing round the circuit is given by $Q = 2HA/R$, where A is the sum of the areas of the windings and R is the resistance of the circuit. If the coil is placed with its plane horizontal and rotated through 180° about a horizontal axis, the charge Q will equal $2VA/R$, where V is the vertical component of the earth's magnetic field. The value of V/H ($=$ tan of angle of dip) can be determined by comparing the throws of the ballistic galvanometer. The coil is often mounted on a suitable frame (*see* diagram) to facilitate the setting and rotation, and a commutator is usually fitted to rectify the output.

earthing reactor A reactor used sometimes in an a.c. *earthed system and connected between one conductor or the *neutral point and earth. Its purpose is to limit the *earth current when a fault occurs between a conductor and earth.

earthing resistor or resistance A resistor used in the same manner and for the same purpose as an *earthing reactor.

earth plane *Syn.* ground plane. A sheet of conducting material that is adjacent to an electrical circuit and is at earth potential. It provides a low-impedance earth at any point in the circuit. For example, one side of a double-sided printed circuit board may be used as the earth plane for the circuit on the other side, contacts being made through the board at any desired point in the circuit.

earth-return system A system of distribution of electrical energy in which a single insulated conductor is provided, the earth forming the return circuit.

earth's magnetic field *See* geomagnetism.

earth's mean density According to Newton's law of gravitation, $F = Gm_1m_2/d^2$, where F is the force of attraction between two particles of masses m_1, m_2, d their distance apart, and G the Newtonian constant of *gravitation. The earth exercises on a mass m placed at its surface a force equal to (mg), where g is the acceleration of free fall corrected for the effects of rotation, and this equals GMm/R^2, where M is the mass of the earth and R is its radius. If ρ is the mean density of the earth, $M = 4\pi R^3\rho/3$, and $g = 4\pi GR\rho/3$; by determining G, g, and R, ρ can be determined. g can be found readily, e.g. from the time of swing of a pendulum, when $T = 2\pi\sqrt{l/g}$; and R can be measured. The gravitational constant G has been determined by various methods; e.g. by large-scale methods, such as Bouguer's measurements in Peru of the relative masses of a mountain (whose mass was estimated from its volume and density) and of the earth, from the deflection of a plumb line placed near the mountain; by laboratory methods, such as those of Cavendish and Boys, who used a delicate torsion balance to determine the force of attraction between two spheres. Modern measurements give
$$G = 6.672\,04 \times 10^{-11} \text{ N m}^2 \text{ kg}^{-2};$$

the mean density of the earth
$$= 5.515 \times 10^3 \text{ kg m}^{-3}.$$
Because the density of solid matter near the earth's surface averages about 2.7 times that of water, the relative density of the core must exceed 5.5. It is calculated to be between 10 and 12, i.e. approximately that of iron-nickel under great pressure.

earth termination *Syn.* earth termination networks. Parts of a lightning protective system from which it is intended that lightning discharges shall emanate into the earth or at which the discharges shall be collected from the earth. *Compare* air terminations.

east-west effect *See* cosmic rays.

ebonite *See* rubber.

ebullition The process of boiling.

eccentricity Of an ellipse. The ratio *e* of the focal distance of any point on the ellipse to its distance from a fixed line, the directrix (*see* conic sections) of the ellipse. It is given by

$$e = \sqrt{\frac{a^2 - b^2}{a^2}},$$

where *a* and *b* are the major and minor semidiameters respectively. *Compare* ellipticity.

ECG Abbreviation for *electrocardiograph.

echelette grating A *diffraction grating for use primarily with infrared radiation with large grating interval and flat grooves inclined at an angle to reflect the radiation in the direction of the diffracted order intended to be bright.

Such gratings are now generally called *blazed gratings.

echelon grating A diffraction grating capable of high resolution (100 000 to 1 000 000) for a small portion of a spectrum, e.g. for studying hyperfine structure of lines, Zeeman effect, etc. Some twenty to forty accurately parallel plates of thickness all equal (to within a small fraction of a wavelength), are mounted in optical contact and staggered to form a series of steps of width about 1 mm (Michelson, 1898). The grating is used either as a reflection or transmission instrument, the former giving higher resolution. *See* interference of light; resolving power.

echo 1. When a sound pulse is incident upon a surface of large area, some part of the sound energy is reflected. If the time interval between the emission of the sound and the return of the reflected wave is more than about one-tenth of a second, the reflected sound is heard after a silent interval and is called an echo. This minimum time for the appreciation of an echo as a separate note implies a path difference of about 30 metres. A reflector having a large surface area relative to the wavelength of the sound gives the best echoes. Thus a high-pitched sound usually gives

a better echo than one of low frequency. The distance of the reflector from the source can be calculated from the time taken between the emission of the sound and the return of the echo. This principle has been extensively used in *echo sounding to find the depth of the sea bed beneath a ship. Echoes are very troublesome when they occur in large buildings since the interference they cause prevents the original sound from being heard distinctly. This can be overcome by the use of sound-absorbing materials and by avoiding curved surfaces, which act as concave mirrors and thus focus the echoes. (*See* auditorium acoustics.)

2. In communications, a wave returned to the transmitter with sufficient magnitude and delay to be distinguished from the transmitted wave.

3. In *radar, the portion of the transmitted pulse that is reflected back to the receiver.

echo chamber *See* reverberation chamber.

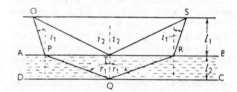

Echo prospecting

echo prospecting The *geophone* was developed originally in France during the First World War, for listening to the sounds sent through the earth as the result of military mining operations. It consists essentially of a heavy lead weight suspended through two light discs forming the case to which the heavy weight is connected circumferentially. The front of the case is pressed against the rock under test while the two ends of the listening tube are connected to the other side of the case. Any impulse impressed upon the system will not affect the heavy lead piece while the case will respond to it and so it will cause compression and rarefaction in the listening tube. By using two or more geophones the relative directions can be estimated by the binaural principle. (*See* interference of sound; binaural location.)

The geophone provides a useful instrument for echo prospecting, i.e. the exploring of the underground mineral strata. Taking as an example a surface of separation AB between a harder stratum above and softer earth below (*see* diagram), suppose the velocities in the two media are V_1 and V_2 respectively, and let a compressional sound be emitted at O just at the surface of the harder stratum, which joins it with the ground. At AB the waves emitted from O are partly reflected and partly refracted according to the angle of incidence. The critical angle in this case is given by $\sin^{-1} V_1/V_2$. If a ray is incident with an angle less than this angle, it will be refracted at P (say) and will be partly or entirely reflected at CD according to the layer beyond it. Let its path be represented by OPQRS. The rays that strike AB at an angle greater than the critical will be entirely reflected; let one of them be totally reflected at T and strike the earth at S, the same point as the ray OP, which traversed the lower medium. If the thickness of the strata OA and AD are l_1 and l_2 respectively, then the time taken for the path OTS

$$= 2l_1/V_1 \cos i_2$$

and that for the path OPQRS

$$= 2l_1/V_1 \cos i_1 + 2l_2/V_2 \cos r_1$$

while

$$\sin i_1/\sin r_1 = V_1/V_2.$$

If S is near O, the wave will traverse the path OTS quicker than the path OPQRS, but if S is far away from O, the wave *via* Q will have a shorter time (having a greater velocity) and so arrive first.

If many observation stations are set up on the surface OS, the reflected wave of OT will arrive in regularly increasing time as the observer recedes from O until there is a sudden drop or a kink in the time curve along OS. This means that there is a harder stratum below. From the measurement of these times and the direction i_1 and i_2 as obtained by the binaural geophones (as described previously), the quantities l_1, l_2; V_1 and V_2 can be calculated. These will give valuable information about the nature of the stratum ABCD. The method can also be used in the sea or a lake. Ultrasonic beams can be applied in a similar way to establish the existence of suspected discontinuities in a metal structure in the form of hidden cracks. *See* flaw detection.

echo sounding A sound beam can be used for estimation of the depth of the sea as well as the distance from the source to the nearest solid surface at which it is reflected. By measuring the time taken between the production of sound and the receipt of its echo, the distance can be evaluated if the speed of sound is known in the medium through which it is sent.

The earliest successful method was due to Behm, whose experiments began in 1911. A small charge is exploded in the water near one side of the ship. At the same time a *hydrophone near this place of explosion is actuated. On the other side of the ship there is another hydrophone, shielded from the effect of the original explosion, to record the echo. Behm arranged that the depth should be directly recorded on a dial by setting a disc in motion by the firing of the explosion and stopping it by the arrival of the echo; the angle through which the disc has revolved is calibrated to read depths directly.

The early system used by the British and United States navies for echo depth sounding was carried out by sending a train of undamped oscillations emitted by striking a diaphragm with a metal hammer. The sound is produced at certain definite instants. The current is supplied by a contact spindle rotating at a certain speed. On the same spindle there is another contact set in an ebonite disc, which periodically connects the telephones to the microphone receiving the echo. No sound is heard in the telephones unless the receiving circuit is closed at the instant the echo is received. The contact brush can be displaced round the spindle until the echo is caught; the distance moved by the brush is a measure of the depth of water.

The energy reflected from a pulse of audible sound is small unless the surface is of considerable extent. Furthermore, if the *specific acoustic *impedances of both media are nearly equal, e.g. sea water and ice, the reflecting power will be very small and therefore the detecting of icebergs is not possible. For these reasons Langevin employed piezoelectric quartz oscillators of sufficient power as the source of high-frequency sound. (*See* Langevin–Florissen sounder.) The British Admiralty used magnetostriction rods instead. (*See* Hughes's echo sounder.)

Distance is not the only information given by echo sounding, as something about the nature of the bed may be deduced from an examination of the record of the echo. If the latter is sharp and distinct a correspondingly distinct and hard bed is indicated. If, however, the record is fuzzy with the appearance of some of the radiation having penetrated before being reflected, the inference is that the bed is oozy and indistinct. Wrecks may also be detected by such echo-sounding methods.

The same principle is used in *radar.

ECL Abbreviation for *emitter-coupled logic.

eclipse Any of a number of astronomical phenomena resulting from the alignment of heavenly bodies. A planet, star, or satellite may pass behind the moon or a planet and so not be visible from earth. This is called an *occultation*. Venus and Mercury pass across the disc of the sun at irregular intervals, and satellites or the shadows of satellites pass across the disc of a planet. These phenomena are called *transits*.

The sun is eclipsed if the new moon passes directly between it and the earth. The full moon is eclipsed if the earth passes directly between it and the sun. The orbit of the moon is in a plane that makes an angle of about 5° with the ecliptic (*see* celestial sphere) hence usually the earth, moon, and sun are not in the same line at full and new moon. The orbit is, however, subject to *precession with a period about 19 years so at times, the moon passes through the ecliptic at such points as to cause eclipses.

A solar eclipse occurs when the shadow of the moon passes over the surface of the earth. A narrow strip of the surface may be in the *umbra so the eclipse is here *total*. Over a wider area there is a *partial* eclipse, in the *penumbra. The apparent sizes of sun and moon as seen from earth are very nearly equal but there is some variation because of the elliptical orbits, hence sometimes the moon does not completely cover the surface of the sun, causing an *annular* eclipse.

Lunar eclipses occur when the moon passes through the shadow of the earth.

Total eclipses have enabled astrophysicists to make valuable observations of the outer regions of the sun, the *chromosphere and *corona. The study of the eclipses of the satellites of Jupiter enabled *Roemer to make the first measurement of the speed of light. When this quantity was measured accurately in the laboratory, these observations permitted a more accurate determination of the distances of the planets. The earth lies nearly in the plane of the orbits of certain double stars, the light from which changes periodically as one star eclipses the other. Certain astronomical information has been obtained by the study of these *eclipsing binaries*.

ecliptic *See* celestial sphere.

Eddington, Sir Arthur Stanley (1882–1944) Brit. astronomer who made notable contributions to the theory of the internal constitution of the stars, to relativity, and to theories of the fundamental structure of the universe. He was also noted as a writer of semi-popular accounts of some of this work.

eddy current A current induced in a conductor when subject to a varying magnetic field. Such currents are a source of energy dissipation (*eddy-current loss*) in alternating-current machinery. The reaction between the eddy currents in a moving conductor and the magnetic field in which it is moving is such as to retard the motion, and can be used to produce *electromagnetic damping*. Although eddy currents were discovered by Joule, they are sometimes called Foucault currents.

eddy-current loss *See* eddy current.

eddy-making resistance The portion of the resistance to motion, of a body in a fluid flow, which is due to the formation of eddies in a turbulent wake.

eddy viscosity In the turbulent flow (*see* turbulence) of an incompressible fluid, the formation of eddies has the effect of increasing the rate of change of momentum of any portion of the fluid. This may be considered as an increased resistance or as an apparent viscosity of the fluid greater than that pertinent to nonturbulent motion. This apparent viscosity is called eddy viscosity.

In turbulent motion, using Cartesian coordinates, any point of the fluid can be considered as having three components of velocity:

$$u = \bar{u} + u', \quad v = \bar{v} + v', \quad w = \bar{w} + w',$$

where $\bar{u}$, $\bar{v}$, $\bar{w}$, are the temporal means of the three components u, v, w, and u', v', w', represent the component velocities of fluctuation about the mean values. The extra components of stress force on any given area in the fluid, due to turbulence, are proportional to the temporal means of the various products of u', v', w' (i.e. u'^2, v'^2, w'^2, $u'v'$, etc.).

The viscosity due to the eddying motion is considered to have a coefficient of kinematic viscosity (ε), which is called the coefficient of eddy viscosity or coefficient of turbulence. The coefficient of eddy

viscosity is not necessarily constant throughout the fluid but an average value is usually assumed.

edge connector On a *printed-circuit board, tracks are taken to one edge of the board and form the edge connector. This may be plugged into a suitable socket allowing input and output to the circuit on the board, and interconnections to other circuits to be made.

edge dislocation *See* defect.

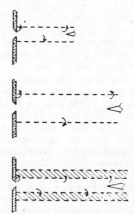

Edge tones

edge tones The sound made when a blade-shaped sheet of gas issues from a linear slit and meets an edge that may or may not be sharp. The distance between the slit and the edge apparently acts as a resonator and stabilizes the *jet tones produced without an edge. The frequency of the sound emitted is related to the number of vortices arriving at the edge per second. The frequency of the note, n, is approximately related to the velocity of efflux, V, and the distance between the slit and the edge, f_0, by the formula $V/nf_0 = k$. König and Brown independently have given empirical formulae that approach the experimental results more accurately. If V, the velocity of the air stream, is kept constant, there is a minimum value of f below which no edge tone is produced. If f is increased from this minimum value (f_0), the frequency emitted decreases to the suboctave at $f_1 = 2f_0$. The system then becomes unstable and the frequency and vortex spacing revert to their values at f_0. Concurrently, the spacing of the two rows increases to twice that at f_0 and having reached instability, reverts to the original spacing (*see* diagram for the three stages). Displacing the edge from the bisection of the two streams to the shaded portion shown on the figure gives no discontinuity in the decrease of frequency with increasing f. An obstacle deflecting the stream such that the edge will be in the shaded portion has a similar effect. In certain types of organ pipes, both these methods are used to stabilize the sound. If f, the spacing between the slit and the edge, is kept constant, there is a minimum V below which vorticity does not take place, and no sound is heard. The frequency of the edge tone increases with V, jumping to the octave above when the vortices rearrange themselves. Displacing the edge, or suitably positioning an obstacle in the stream, gives the lower octave again. Hence, also their use with organ pipes. A greater wind pressure means more energy for sound output, as well as a brighter tone quality. A circular orifice and suitable edge produces similar results. The action of the edge in the production of the edge tone is obscure. Smoke photographs show secondary vortices forming in the boundary layers of the edge as the fluid passes along. These are formed such that two vortex avenues of the appropriate spacing are formed. In some recent work the edge has been replaced by a wire without any material change in the edge tone.

If the edge tone is associated with a resonator to form a coupled system, the edge-tone conditions are modified by the less-damped resonator and a more stable tone, or quality depending mostly on the resonator, is produced. The open and closed flue pipes of the organ, the orchestral flute, the recorder, the bell whistle, the bird whistle – with a possible resonator between the two orifices – and the *Hartmann generator are examples of such coupled systems. For optimum sound production the edge tone frequency when uncoupled to the resonator, should be the same as one of the resonant frequencies of the resonator.

Edison, Thomas Alva (1847–1931) Amer. inventor. His very numerous inventions include the incandescent filament lamp (in conjunction with the Brit. inventor Sir Joseph Swan, whence "Ediswan" as a trade name) and the gramophone. *See also* Edison accumulator; thermionic emission.

Edison accumulator *Syn.* nickel-iron or Ni-Fe accumulator. A storage battery having steel grid plates: the positive plate is filled with a mixture of metallic nickel and nickel hydrate, and the negative plate is filled with iron oxide paste. Modern forms use cadmium or cadmium-iron alloy for the negative state. The electrolyte is a solution of potassium hydroxide of relative density about 1.2. The cells are strong, will deliver heavy currents and even withstand short circuits for a limited time, and they do not deteriorate on standing in the discharged state. They are lighter than the lead accumulator, but the voltage is lower – 1.3 to 1.4 volts per cell.

Edison–Butler bands Dark bands that appear in a continuous emission spectrum when a thin transparent plate is placed in the path of the light. They are caused by interference between waves reflected at the two surfaces, and are used in calibrating a spectrometer.

Edison effect *See* thermionic emission.

EEG Abbreviation for *electroencephalograph.

effective energy Of *heterogeneous radiation. The quantum energy of the beam of *homogeneous radiation that, under the same conditions, is absorbed or scattered to the same extent as the given beam of *heterogeneous radiation.

effective mass A parameter used in the theory of conductivity of solids to describe the behaviour of the charge carriers. When a potential difference is applied to a conductor, electrons are accelerated by the field produced. They have a mobility that depends on their position in the *energy band. This is described by an effective mass, which is a function of energy and can differ from the true mass.

effective photon theory At normal intensities electromagnetic radiation interacts with matter in *quanta of energy $h\nu$, where h is the *Planck constant and ν the frequency of the waves. It is said that a *photon is absorbed. It has been found, however, that at the enormous intensities that can be obtained by focusing the beam of a pulsed *laser, the simple quantum rule can be violated. For example, helium with *ionization potential 24.6 eV can be ionized by radiation in the far infrared with $h\nu \simeq 0.1$ eV. Various theories have been proposed to explain how the highly concentrated radiation acts as an *effective photon* of much higher frequency. *See also* photoelectric effect.

effective resistance The resistance of a conductor or other element of an electric circuit when used with alternating current. It is measured in ohms and is the power in watts dissipated divided by the square of the current in amperes. It may differ from the normal value of resistance as measured with direct current, since it includes the effects of *eddy currents within the conducting material, *skin effect, etc.

effective temperature *See* luminosity.

effective value *See* root-mean-square value.

effective wavelength Of *heterogeneous radiation. The wavelength of the beam of homogeneous radiation that, under the same conditions, is absorbed to the same extent as the given beam of heterogeneous radiation.

efficiency 1. For a *machine the efficiency η is the work done by the machine divided by the work done on it. For steady operation this is equal to the output power divided by the input power. It is usually expressed as a percentage.
2. For a *heat engine the efficiency is the work done by the engine divided by the heat input. For an ideal *reversible engine in which all heat input is at temperature T_1 and all waste heat is discharged at a lower temperature T_2, the efficiency is $(T_1 - T_2)/T_1$, the temperatures being *thermodynamic temperatures.
See Carnot cycle; Carnot's theorem.

effusion

effusion The leakage of gas through a fine orifice. *Graham's law applies at ordinary pressures when the mean free path is small compared with the dimensions of the orifice, the flow being governed by the laws of hydrodynamics and being analogous to a fluid jet forced out by pressure. At low pressures, when the mean free path is large compared with the dimensions of the orifice, the phenomenon has been termed *molecular effusion* and although the volume of gas escaping per second is also inversely proportional to the square root of the density, the mechanism is quite different. On the kinetic theory the volume diffusing per second into a vacuum is given by $s\sqrt{(kT/2\pi m)}$ where s is the area of the orifice, m the mass of a molecule, and k is the Boltzmann constant. The dependence of the velocity of effusion on molecular weight is used for detecting the presence of methane in mines. The methane effuses into a vessel containing air fitted with a manometer to register the resulting rise in pressure. The phenomenon plays an important part in high-vacuum techniques (*see* molecular flow) and is used in *isotope separation, as well as for the measurement of vapour pressures. *See* Egerton's effusion method.

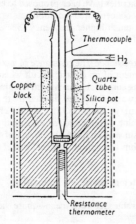

Egerton's effusion method of measuring vapour pressures of metals

Egerton's effusion method A method of measuring the vapour pressures of metals at their freezing points by which vapour pressures as low as 10^{-4} mm Hg can be accurately measured. (*See also* Knudsen absolute manometer.) The metal is placed in a silica pot with a tightly fitting lid in which a small hole of diameter 2 mm is drilled. The pot rests at the bottom of a quartz tube and is surrounded by a large copper block that may be maintained at a constant high temperature by heating coils wound round it. This temperature is measured by a resistance thermometer embedded in the block and the temperature of the specimen is measured by a thermocouple, whose leads emerge from the top of the quartz tube. The pot and contents are first weighed and hydrogen admitted to the quartz tube until all

temperatures are steady. On quickly evacuating the tube to a pressure of 10^{-5} mm Hg, molecules of the metal effuse through the hole in the lid of the pot and condense on the cool portions of the tube. The experiment is stopped after several hours by admitting hydrogen and the loss in weight of the pot enables the mass of vapour escaping per second (G) to be calculated. Since the mean free path of the molecules is large compared with the dimensions of the hole and tube,
$$G = \sqrt{p_1(p' - p'')/(W_1 + W_2)},$$
where p_1 is the density of the vapour at unit pressure and at the temperature of the system and $(p' - p'')$ is the pressure difference between which the flow resistance is W_2 in the tube itself and W_1 in the hole. The resistance of the hole is given by
$$W_1 = 2\sqrt{2\pi}/A,$$
where A is the area of the hole and the resistance of the tube, length L and radius R, is given by:
$$W_2 = \tfrac{3}{8}\sqrt{(2/\pi)}(L/R^3)$$
(*see* molecular flow). p' is the required vapour pressure and p'' the pressure at which the deposit occurs and may be neglected.

EHF Abbreviation for extremely high frequency. *See* frequency band.

Ehrenfest's rule The principle that if a system is described by quantized variables and subjected to an *adiabatic process, then the *quantum numbers of the system must either change suddenly to new values or remain the same. Furthermore, if the change occurs very slowly, then the quantum numbers must remain constant; the variables are then said to be adiabatically invariant. This implies that the converse may be true, that only quantities that are adiabatically invariant can be quantized.

EHT Abbreviation for extra high tension. It usually refers to the high-voltage supply for a cathode-ray tube or television picture tube.

Eigen, Manfred (*b.* 1927) German physical chemist; director of the Max Planck Institute (since 1964). He shared a Nobel prize (1967) with George Porter and Ronald Norrish for his work on the hydration of a proton using the relaxation technique. He later applied this technique to biochemical reactions and proposed a mechanism for the formation of proteins and nucleic acids.

eigenfunction and -value *See* wave function.

Einstein, Albert (1879–1956) German-born Swiss (later Amer.) mathematical physicist. He played a notable part in developing the quantum theory, applying it to photoelectricity (1905), and in the theory of specific heats (1907) (*see* Einstein's theory of specific heat capacities). His special theory of *relativity (1905) had a profound effect on the whole structure of science and philosophy. It was followed in 1915 by the general theory. The celebrated mass-energy relation (*see* Einstein's law) ($E = mc^2$, c = speed of light) is incorporated in the special theory. He carried out experiments demonstrating the gyromagnetic effect (with W. de Haas), 1915 and introduced Bose's statistics to the western world in 1924, applying the method to the perfect gas later in the same year.

Einstein and de Haas effect The reverse of the *Barnett effect. When an iron cylinder that is free to move is suddenly magnetized, it rotates slightly.

Einstein coefficients Coefficients representing the probability of radiative transitions between electronic states of atoms or molecules. If atoms in a level n are subjected to a beam of electromagnetic radiation of frequency ν, they may make a transition to a level of higher energy m by absorbing a photon of energy $h\nu$. The number of atoms making this transition is given by $B_{nm}N_n u(\nu)$, where $u(\nu)$ is the energy density of radiation of frequency ν and N_n the number of atoms in level n. B_{nm} is the Einstein coefficient for absorption, giving the transition probability for this process. Similarly, atoms in level m can interact with the radiation and undergo *stimulated emission of photons in changing to level n. The number of atoms making this change is given by $B_{mn}N_m u(\nu)$. Atoms in level m can also undergo spontaneous emission to level n with emission of a photon, the number of atoms making this transition being given by $A_{nm}N_m$. The Einstein coefficients are given by the equations:
$$B_{nm}/B_{mn} = g_m/g_n,$$
where g_m and g_n are the *statistical weights of level m and n respectively, and:
$$A_{nm} = 8\pi h\nu^3/c^3,$$
where h is the Planck constant.

For any system in equilibrium the probability of occupation of a state decreases with increasing energy. Thus if the equilibrium is disturbed by admitting a beam of radiation of frequency ν into the system, the probability of absorption is greater than the probability of stimulated emission so the beam is attenuated. It is possible to produce nonequilibrium conditions in which states of higher energy are more densely populated. In such cases the beam is intensified as the probability of stimulated emission exceeds that of absorption. This is the basis of the action of the *laser and *maser and of experiments to study the *Lamb shift. It is seen that, other things being equal, the coefficient of spontaneous emission is proportional to the cube of the transition energy. For this reason it is difficult to operate a laser for short wavelengths. Also the lifetime of highly excited states are generally very short.

Einstein shift *Syn.* gravitational redshift. A small *redshift in the lines of a *stellar spectrum caused by the gravitational *potential at the level in the star at which the radiation is emitted (for a bright line) or absorbed (for a dark line). This shift can be explained in terms of either the special or general theory of *relativity. In the simplest terms, a quantum of energy $h\nu$ has mass $h\nu/c^2$. On moving between two points with gravitational potential difference ϕ, the work done is $\phi h\nu/c^2$ so the change of frequency $\delta\nu$ is $\phi\nu/c^2$.

The shift has been measured in the laboratory using the *Mössbauer effect. A radioactive source emitting gamma radiation with very low quantum energy (usually Co-57) is placed at the top of a building at a height H, typically 20 m above the detector. ϕ is then Hg and $\delta\nu/\nu$ is about 2×10^{-15}. This minute change of frequency is detected by the use of a Mössbauer analyser. Very precise temperature control of source and detector is necessary as the atomic vibrations otherwise cause errors by the second-order *Doppler effect.

Einstein's law The law of equivalence of mass and energy. Mass m and energy E are related by the equation $E = mc^2$, where c is the speed of light in vacuum. Thus, a quantity of energy E has a mass m, and a mass m has intrinsic energy E. The law has been experimentally verified in the production and annihilation of electron-pairs and also from a study of nuclear transformations. *See also* relativity.

Einstein's photoelectric equation *See* photoelectric effect.

Einstein's theory of specific heat capacities Applying the quantum theory to the vibrations of the individual atoms, supposed to have a single frequency ν, Einstein obtained an expression for the specific heat capacity, c_ν of a solid:
$$c_v = 3R \frac{x^2 e^x}{(e^x - 1)^2},$$
where $x = \theta/T$ and $\theta = h\nu/k$, h being the Planck constant and k the Boltzmann constant. As a general theory, Einstein's theory has been superseded by that of *Debye.

Einthoven, Willem (1860–1927) Dutch physicist and physician who devised the sensitive *Einthoven galvanometer and was the first to use it to record the electrical activity of the heart muscle. *See* electrocardiograph.

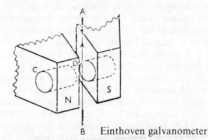

Einthoven galvanometer

Einthoven galvanometer *Syn.* string galvanometer. A type of *galvanometer consisting of a single conducting thread AB (often of silvered quartz) strung tightly between the poles of a powerful electromagnet NS. On passing a current through it, it is deflected at right angles to the direction of the magnetic field, and is viewed by a high-power microscope let into the pole piece at CD. Owing to its small inertia and high tension, it has a high natural frequency, is dead beat, and can be used for investigating rapidly fluctuating currents. Its sensitivity, which is high, is due to the very strong magnetic field employed, and the high magnification of the microscope. A current of 10^{-11} ampere can be detected.

Elasser, Walter Maurice (*b.* 1904) German-born Amer. geophysicist, who held professorships at various US universities. His main work concerned the origin of the earth's magnetic field. He also predicted electron diffraction (1925) and neutron diffraction (1936).

elastance The reciprocal of *capacitance. It is measured in farad^{-1}, sometimes called a *daraf* although this term is deprecated.

elastic collision *See* collision.

elastic constants Constants, such as the *Young modulus (E) and the *Poisson ratio (μ), relating *stress to *strain in a homogeneous medium. For an isotropic material, two constants are required to specify the behaviour and these are related by linear equations. The components of normal stress (σ) and linear strain (e) in the x direction are related by the equation:
$$e_x = (1/E)[\sigma_x - \mu(\sigma_y + \sigma_z)],$$
where σ_y and σ_z are the components of stress in the y and z directions. Similar equations apply to the components of strain in the y and z directions. In general, an anisotropic solid is described by 21 elastic constants.

elastic deformation A change in the relative positions of points in a solid body that disappears when the deforming stress is removed.

elastic hysteresis A phenomenon occurring when a stress is applied in steps to an elastic body, and then removed in equal steps: it is found that the strain on unloading is greater than at the corresponding stress when loading. This hysteresis is very small for substances such as steel but is very large for imperfectly elastic materials such as rubbers. On a graph of stress against strain, the area within the *hysteresis loop* represents the energy dissipation per unit volume in a cycle of loading and unloading.

A soft rubber ball dropped onto a hard surface will rebound to only about one half of the initial height, thus about a half of the energy is dissipated, mostly by hysteresis. For harder rubbers and slower processes the dissipation is less than this, but elastic hysteresis plays a significant part in the behaviour of tyres. High hysteresis is desirable in the tread to increase limiting friction, but is undesirable in the body of the tyre because of rolling friction (*see* friction).

Elastic hysteresis is important in the damping of oscillations, and in animal movements.

elasticity 1. The property of a body or substance by which it tends to resume its original size and shape after being subject to deforming stresses.
2. Short for *modulus of elasticity.

elastic limit The smallest stress that leaves a detectable permanent strain after removal (a necessarily vague term). *See* limit of proportionality; yield point.

elastic modulus 1. Of an elastic material. The ratio of stress to strain, within the *limit of proportionality. There are several moduli but these are not all independent. (*See* elastic constants; modulus of elasticity.) **2.** Of a cord (applied mathematics). The ratio of total load to fractional elongation (strain).

elastic scattering *See* scattering.

elastoresistance The change in electrical resistance of materials when they are stressed within their elastic limits. *See* magnetoresistance; resistance strain gauges.

E-layer or region *Syn.* Heaviside layer; Kennelly–Heaviside layer. *See* ionosphere.

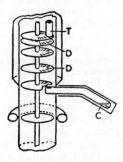

Eldridge's apparatus to prove Maxwell's law

Eldridge's experiment to verify Maxwell's law In an experiment to verify directly Maxwell's law of *distribution of velocities, a series of slotted coaxial discs D (*see* diagram) is rotated at high speed by an induction motor in an evacuated space. Molecules from heated cadmium C on passing through these slots are filtered out, those with different speeds being deposited at different places on the cooled target. A deposit of varying density is thus obtained on the target T and the distribution of speeds among the molecules leaving the furnace is obtained by examining the deposit with a photometer.

The rate at which molecules pass through the small hole from the source tube into the molecular beam is proportional to their speed. Thus the number of molecules in the beam with speed in the range v to $v + dv$ is proportional to v^3 (exp $mv^2/2kT)dv$. Verification of this formula for the beam confirms Maxwell's law for the molecules moving randomly in a gas.

electret A permanently electrified substance exhibiting electrical charges of opposite sign at its extremities. The first electret was made by Eguchi, who allowed a mixture of wax and resin to solidify between the plates of a capacitor in a strong electric field. Initially, the face that had been in contact with the anode showed a strong negative charge and that at the cathode a positive charge. These charges decayed in a few days, to be replaced by the building-up of charges of opposite sign that showed no measurable decay over a period of years. Electrets in many ways resemble permanent magnets. If, for example, they are cut, they separate into two complete electrets, each with positive and negative charges like the poles of a magnet. Electrets are best preserved with their faces short-circuited – analogous to the keeper of a permanent magnet. Electrets have been used in electrometers, and also in the construction of capacitor microphones.

electrical degrees In an electrical machine having two magnetic poles, in one revolution (i.e. 360 mechanical degrees), the alternating e.m.f. in any one armature conductor completes one cycle (i.e. 360 electrical degrees), and one mechanical degree corresponds to one electrical degree. In a 4-pole machine, in one revolution the alternating e.m.f. in any one conductor completes two cycles (i.e. 720 electrical degrees) and one mechanical degree corresponds to two electrical degrees, and so on. In all cases, physical rotation through an angle corresponding to two pole pitches is equivalent to 360 electrical degrees.

electrical discharge *See* conduction in gases.

electrical double layer *See* Helmholtz electric double layer.

electrical images A method, due to Kelvin, for solving electrostatic problems arising when a point charge is in the neighbourhood of a conducting surface. It can be shown that, in certain cases, the electrical effects due to the induced charges on the plane are identical with those that would be produced by a point charge situated at some particular point relative to the surface. This imaginary charge is known as the electrical image of the first charge in the surface. In the case of an infinite conducting plane the electrical image is equal and opposite in sign to the actual charge, and situated as far behind the plane as the original charge is in front of it.

electrical screening Shielding of apparatus from interference due to unwanted electrical disturbances, by surrounding it with an electrical conductor. It is a consequence of the inverse square law of force that no electric field is produced within a hollow closed conducting surface by any external electric charge.

electric axis The direction in a crystal of maximum electrical conductivity. It is the X-axis of a piezoelectric crystal.

electric braking A method of braking an electric motor by causing it to act as a generator, its output as a generator being either dissipated in a rheostat (*see* rheostat braking) or returned to the supply system (*see* regenerative braking). It has particular application in electric traction. *Compare* magnetic braking.

electric charge *See* charge.

electric constant Symbol: ε_0. The *permittivity of *free space, with the formally defined value:
$$\varepsilon_0 = 10^7/4\pi c^2 \text{ F m}^{-1}$$
$$= 8.854\,187\,817 \times 10^{-12} \text{ F m}^{-1},$$
where c is the speed of light in vacuum.

electric current *See* current.

electric displacement Symbol: D. If an electric field exists in a vacuum with magnitude E and a dielectric is introduced into the field, the electric flux per unit area (*electric flux density*) in the medium is D, the electric displacement. The permittivity of the medium (ε) is given by D/E. The divergence of electric displacement equals the surface density of charge. Displacement is measured in coulombs per metre squared.

electric endosmose *See* electrosmosis.

electric energy Any form of energy associated with electrical phenomena.
(1) *Electric potential energy*. If a charge Q moving between two points has work W done on it by an *electric field then the difference of potential energy is $W + QV$, where V is defined as the *potential difference between the points.
(2) *Electric field energy*. The energy per unit volume in an electric field E in a medium of permittivity ε is $\frac{1}{2}\varepsilon E^2$. For example, the energy of a charged capacitor is located in the field between the plates.
(3) *Magnetic field energy*. The energy per unit volume in a magnetic field H in a medium of permeability μ is $\frac{1}{2}\mu H^2 = \frac{1}{2}B^2/\mu$. The space around a conductor carrying a current contains a magnetic field that stores the energy of the current. Thus, when a current is switched off this energy is normally dissipated by generating a spark. The magnetic field energy of a circuit with *self inductance L carrying a current I is $\frac{1}{2}LI^2$.

electric field The space surrounding an electric charge within which it is capable of exerting a perceptible force on another electric charge.

electric field strength Formerly called *electric intensity*. Symbol: E. The strength of an electric *field at a given point in terms of the force exerted by the field on an infinitesimal charge at that point divided by the magnitude of the charge. The SI unit is the newton coulomb^{-1} or the volt metre^{-1}.

electric flux Symbol: Ψ. The quantity of electricity displaced across a given area in a dielectric. It is defined as the scalar product of the *electric displacement and the area; it is measured in coulombs. A line drawn in the field so that its direction at any point is the direction of the electric flux at that point is known as a *line of flux*. A space bounded by lines of flux forms a tube of electric flux. If the tube is so drawn that the flux across any cross section of it is unity, it is known as a unit tube of flux, or *Faraday tube. The flux density at any point in a dielectric is equal to the number of unit tubes of flux crossing a unit of area drawn at right angles to the direction of the flux at that point. *See* electric displacement.

electric flux density *See* electric displacement.

electric hygrometer (Dunmore, 1939) A type of *hygrometer that consists of a thin layer of lithium chloride on an insulator on which two palladium coils are wound side by side. The lithium chloride film is hygroscopic and an increase in the relative humidity leads to a change in the resistance of the film between the two coils. The action of the instrument is very rapid.

electric induction *See* electric displacement.

electric intensity *See* electric field strength.

electric line A conductor used for the transmission of an electric current from one place to another.

electric polarization *Syn.* dielectric polarization. Symbol: P. The *electric displacement D minus the product of the electric field strength E and the permittivity of free space (*electric constant), ε_0, i.e. $P = D - \varepsilon_0 E$. It is measured in C m^{-2}.

electric potential Symbol: V. The electric potential at a point in an electric field is the work that must be done against electric fields to bring an infinitesimal positive charge from infinity to that point, divided by the charge. The SI unit is the volt (joule coulomb^{-1}). *See also* potential.

electric wind A stream of air flowing outwards from a sharp point or projection on a highly charged electrical conductor. Gaseous molecules become ionized by collision in the intense field near the point, and those of the same sign as the conductor are repelled from it, dragging a stream of uncharged molecules with them.

electrization The equation defining *electric polarization P,
$$D = \varepsilon_0 E + P,$$
where D is the *electric displacement, ε_0 the *electric constant, and E the field strength, may be put in the form:
$$D = \varepsilon_0 (E + E_i),$$
where $E_i = P/\varepsilon_0$. E_i is known as the electrization.

electrocapillary phenomena Changes produced in the surface tension of a liquid by the existence of an electric field at the surface. *See* capillary electrometer.

electrocardiograph (ECG) A sensitive instrument that records the voltage and current waveforms associated with the action of the heart. The trace obtained is known as an electrocardiogram.

electrochemical equivalent The mass of any ion deposited from solution by a current of 1 ampere flowing for 1 second.

electrochemical series *See* electromotive series.

electrochemistry The study of electrolytes, including investigations of *electrolysis and cells.

electrode In general, a device for emitting, collecting, or deflecting electric charge carriers, especially a solid plate, grid, or wire for leading current into or out of an electrolyte, gas, vacuum, dielectric, or semiconductor. In certain electrolytic cells a liquid-mercury electrode is used. *See also* anode; cathode.

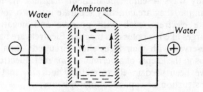

Electrodecantation

electrodecantation A modification, due to Pauli, of *electrodialysis in which a sol is concentrated at the sides and bottom of a cell arranged as in the diagram. The action is a combination of electrophoresis and gravitation. The supernatant liquid is removed by decantation.

electrode current The current flowing through a specified electrode; e.g. collector current, grid current, drain current, etc.

electrode efficiency The ratio of the actual yield of metal deposited in an electrolytic cell to the theoretical yield.

electrode potential The difference of potential between an electrode and the electrolyte with which it is in contact. *See* drop-electrode.

electrodialysis Dialysis assisted by applying the potential difference between electrodes situated on each side of the semipermeable membrane. The ions of any salts present are attracted to the electrodes and diffusion is accelerated.

electrodisintegration The disintegration of a nucleus by electron bombardment.

electrodynamic instrument An instrument in which the operating torque is produced by interaction of currents in a system of movable and fixed coils. The moving system consists of one or more coils that are pivoted so as to move in the magnetic field of the fixed coils and the magnetic circuit is devoid of ferromagnetic material. The currents in all the coils are obtained from a common source. Instruments of this type will operate with either direct or alternating current and for both uses a single calibration can be made to suffice. As a result of this, electrodynamic instruments are used sometimes as "transfer" instruments enabling, for example, an alternating current to be adjusted so that its *root-mean-square value is equal to a specified value of direct current.

electrodynamics The branch of science that studies the mechanical forces generated between neighbouring circuits when carrying electric currents.

electrodynamometer A measuring instrument actuated by the mechanical couple between a moving coil and one or more fixed coils when an electrical current passes through them. The moving coil may

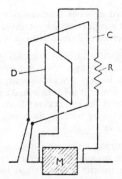

Electrodynamometer

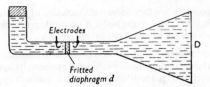

Electrokinetic transducer

be controlled either by a torsional or by a bifilar suspension, and is mounted so that its plane is at right angles to that of the fixed coil when no current is flowing. If the coils are in series, so that the same current passes through all of them, the couple is proportional to the square of the current, and the instrument can be used to measure either a.c. or d.c. With a high resistance in series, it acts as an a.c. voltmeter. The power consumption of a piece of electrical machinery M (*see* diagram) can be measured by passing the main current through the fixed coils C and shunting the moving coil D, through a high resistance R, across the terminals of the machine. The couple is then proportional to the power. *See* wattmeter; Siemens' electrodynamometer.

electroencephalograph (EEG) A sensitive instrument that records the voltage *waveforms associated with the brain. The trace obtained is known as an electroencephalogram.

electroendosmosis *See* electrosmosis.

electrography A method of studying the surface of metals in which a layer of absorbent paper, moistened with electrolyte, is sandwiched between a metal plate and the surface to be tested. When a current is passed with the specimen as anode, its ions move into the electrolyte and there react with the ions of the electrolyte, or with an added reagent, to form a coloured product. The distribution pattern formed gives an indication of surface state or structure, or can be used to identify the metal of the specimen.

electrokinetic phenomena Phenomena in which electrically charged particles move in a liquid under the influence of an electric field and in which, if the particles are restricted, the liquid moves instead. If the particles are ions moving in a liquid, the motion is called *electrolytic migration* and is studied under *electrochemistry. The other phenomena are dependent on a *Helmholtz electric double layer, which is set up at almost every phase boundary. *See* cataphoresis; migration potential; electrosmosis; streaming potential.

electrokinetic potential *See* zeta potential.

electrokinetic transducer A *transducer that generates an alternating current by using the electrokinet-

ic potential developed when a fluid streams through a porous body. Sound waves falling on the diaphragm D, set up oscillatory movement of liquid through a fretted diaphragm *d*, generating an alternating current of the same frequency at electrodes arranged one on each side of the diaphragm.

electroluminescence *Syn.* Destriau effect. The emission of light by certain phosphorescent substances when subjected to a fluctuating electric field. The phenomenon can be used for illumination by applying voltages of 400–500 volts across a dielectric coating, about 2.5 μm thick, in which the phosphor is dispersed. The illuminant has a flat plate support, and the voltage is applied between a metal foil backing and a transparent conducting layer of vaporized aluminium, which forms the front surface over the dielectric and allows the light to pass out through it.

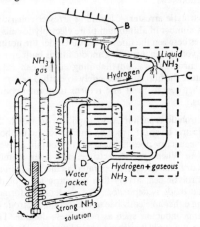

Electrolux refrigerator

Electrolux refrigerator an appliance without a mechanical compressor and with no moving parts. The refrigerant is ammonia, which is heated in the boiler A, to provide ammonia gas, which liquefies in the condenser B giving up its latent heat of vaporization to the surroundings with the help of the cooling fins shown. The liquid ammonia evaporates in C in a hydrogen atmosphere of partial pressure about 9 atmospheres, taking its heat from the space to be refrigerated. The mixture of hydrogen and ammonia passes into the cooled absorber D where the ammonia gas dissolves in the stream of dilute ammonia solution from the boiler A flowing through the baffle plates. The hydrogen is insoluble and passes back

173

into C, while the concentrated solution of ammonia is forced back to the boiler A via a side pipe wound round the heater. *See also* refrigerator.

electrolysis The production of chemical changes in a chemical compound or solution by causing its oppositely charged constituents or ions to move in opposite direction under a potential difference. *See* Faraday's laws of electrolysis.

electrolyte A substance that conducts electricity in solution or in the molten state because of the presence of ions.

Electrolyte solutions are often divided into two classes – strong electrolytes and weak electrolytes.

Strong electrolytes, such as mineral acids, are compounds that are totally dissociated into ions by the solvent (*see* electrolytic dissociation). Salts, such as sodium chloride, are strong electrolytes because their crystals are composed of positive and negative ions. Consequently, the salts are also totally ionic in solution. The *molar conductivity of strong electrolytes does not vary with the dilution.

Weak electrolytes are compounds that are only partially dissociated into ions in solutions. In general, the molar conductivity of such solutions will tend to increase with dilution because the degree of dissociation into ions is increased (*see* Ostwald's dilution law).

electrolytic arrester A lightning arrester consisting of a number of aluminium-plate electrolytic cells in series between the line and earth. Under the normal line voltage, an insulating anodic film is built up and no current flows, but lightning surges can easily rupture the film and pass safely to earth. The arrester then reseals itself and returns to normal conditions.

electrolytic capacitor Any *capacitor in which the dielectric layer is formed by an electrolytic method. The capacitor does not necessarily contain an electrolyte. When a metal electrode, such as an aluminium or tantalum one, is operated as the anode in an electrolytic cell, a very thin dielectric layer of the metal oxide is deposited. The capacitor is formed using either an electrolyte as the second electrode or a semiconductor, such as manganese dioxide. The electrolyte used is either in liquid form or in the form of a paste, which saturates a paper or gauze. Electrolytic capacitors have a high capacitance per unit volume but suffer from high *leakage currents.

electrolytic conduction *See* electrolytic dissociation.

electrolytic conductivity Symbol: κ. *See* conductivity.

electrolytic dissociation The reversible separation of certain substances into oppositely charged ions as a result of solution. For example, sulphuric acid (H_2SO_4) molecules break down in water to give hydrogen ions and sulphate ions:
$$H_2SO_4 \rightarrow 2H^+ + SO_4{}^{2-}.$$
Sulphuric acid is in fact totally dissociated into ions. Certain compounds, such as acetic acid, are

only partially dissociated in water. The theory of electrolytic dissociation was developed by Arrhenius to explain the properties of *electrolytes.

electrolytic ion When a salt, e.g. silver nitrate, is dissolved, free silver ions carrying a positive charge (Ag^+) and nitrate ions having a negative charge ($NO_3{}^-$) are formed. The former is the *cation and an electrolysis moves towards the cathode, while the latter is the *anion and moves towards the anode. *See* electrolysis; ionization.

electrolytic meter An instrument that measures electric charge in terms of the amount of gas liberated or metal deposited by electrolytic action. It can be calibrated directly in coulombs or ampere-hours. Although fundamentally it is a quantity meter, it can be calibrated in joules or kilowatt-hours for use on an electric supply of constant voltage, and hence can be used as a work meter.

electrolytic migration *See* electrokinetic phenomena.

electrolytic photocell A type of cell consisting of a metal electrode coated with metallic selenium, together with an electrode of platinum (or alternatively a similar electrode of selenium-coated metal), immersed in an aqueous solution of selenium dioxide. When a small external d.c. voltage is applied, the cell immediately becomes sensitive to light falling on the selenium electrode, and has instantaneous linear response. With an applied e.m.f. the short-circuit sensitivity is about 1 milliampere per lumen. The cell has high impedance.

electrolytic polarization The tendency for the products of electrolysis to recombine. It is measured by the minimum potential difference required to cause a permanent current to pass through the electrolyte. In a primary cell, electrolytic polarization causes a decrease in the effective e.m.f. of the cell, and is counteracted by various depolarizing agents. *See* depolarizer.

electrolytic rectifier A *rectifier consisting of two electrodes of dissimilar metals immersed in an electrolyte. With certain combinations of metals and electrolytes, the current passes very much more readily in one direction than in the other. An example is the rectifier employing electrodes of aluminium and lead placed in a solution of ammonium phosphate: the current passes readily when the aluminium is the cathode but when it is the anode, the cell is almost nonconducting.

electrolytic separation The separation of isotopes by *electrolysis. The method depends on the differing rates of discharge at an electrode of ions of different isotopes. Normally hydrogen, for example, contains one atoms of deuterium to six thousand of the light isotope. Thus one water molecule in six thousand is HDO, where D represents a deterium atom. On electrolysis, hydrogen ions are discharged at the cathode, producing hydrogen molecules. Molecules of HD are produced less rapidly in proportion to ones of H_2. Hence, the concentration of molecules of HDO in the remaining water is increased. When the concentration of deuterium becomes very large, a significant amount of heavy water (D_2O) is formed.

electrolytic solution pressure Every metal is assumed to have an electrolytic solution pressure (which can be likened to the vapour pressure of a liquid) tending to drive it into solution as ions when it is immersed in a liquid in which its ions can exist freely. This pressure operates against the *osmotic pressure of any similar ions already present in the liquid, the osmotic pressure tending to deposit them on the metal. Equilibrium is established between these two pressures, and the electrolytic potential difference established depends on the state of the equilibrium reached.

electrolytic tank A device for the solution of certain problems in electrostatics (or other branches of physics) by analogy, using measurements made on a suitably devised model immersed in a tank of conducting liquid (electrolyte).

If the electrolyte were replaced by an insulator, the equipotentials would be unaltered, while the lines of electric induction would follow the course of the former lines of current flow. The capacitance between two electrodes can be shown to be $\varepsilon/4\pi\kappa$ times the conductance (reciprocal of the resistance) when the electrolyte, of conductivity κ, replaces the insulator of permittivity ε.

To simulate the boundary conditions at an electrode in the electrostatic analogue, the bath electrodes are of high conductivity (cópper) while the electrolyte is of low conductivity (water or very dilute copper sulphate); a.c. operation is usually preferred to avoid polarization effects. Equipotentials are traced by using a probe connected through a null instrument to a selected point on a potentiometer joined in parallel across the bath.

Electrolytic tanks have been much used to aid design of electrode systems. They may also be used for other problems in which an important physical quantity, corresponding to the potential in the tank, satisfies the *Laplace equation, for example, steady-state problems on heat flow by conduction in isotropic media.

electromagnet An electrical circuit wound in a helix or solenoid so that the passage of an electric current through the circuit produces a magnetic field. The space within the windings is almost invariably filled with a core of ferromagnetic substance, to enhance the magnetic effect of the current. The strength of the magnet depends greatly on the design and continuity of the core. The saturation of the magnetization of iron sets an upper limit of about 2 tesla on the practical flux density. For higher values, superconducting coils can be used without a magnetic core.

electromagnetically maintained vibrations The vibrations of a tuning fork can be maintained by an *electromagnet, the current in which is made and broken intermittently by the motion of the prongs.

The contacts may be of the platinum-point type or may consist of a short spring fixed to one of the prongs and dipping into mercury. (*See* tuning fork.)

It has been shown by Hartmann-Kempf that the frequency of the fork is slightly altered from that of free vibration of the bar itself by this electromagnetic drive. This was proved by varying the frequency of the magnetic impulses on either side of the natural frequency, keeping the current strength constant.

The frequency of such vibrating systems also varies with the amplitude. It decreases firstly and then increases with the increase of amplitude, so that there is a minimum frequency corresponding to a certain amplitude at which the fork should be run.

It is obvious that an electrically maintained fork offers possibilities for uses as a frequency standard. To meet such a requirement, it must be designed with the utmost care. The supports of the fork must be massive. Any change of the constants of the electric circuit affects the period and consequently the frequency. The length of the gap between the contact springs and the contact points has a very marked effect on the period.

The effect of temperature change may be expressed as a temperature coefficient of frequency and varies from 1.04×10^{-4} at -25 °C to 1.43×10^{-4} at 56 °C for a steel fork. It is claimed that a well-designed fork gives a frequency constant to one part in 50 000. These forks can be made with frequencies as high as 200 hertz, but the difficulties of making and breaking the mechanical contacts carrying currents become very great at higher frequencies for which electronic maintenance is preferable.

electromagnetic braking *See* magnetic braking.

electromagnetic damping *See* eddy current.

electromagnetic deflection A method of deflecting an *electron beam using *electromagnets. It is most often applied to the beam in a *cathode-ray tube using two pairs of *deflector coils. *Compare* electrostatic deflection.

electromagnetic focusing *See* focusing.

electromagnetic induction As the result of a series of experiments, Faraday came to the following conclusions about the phenomenon now known as electromagnetic induction:

(a) When a conductor is moved so as to cut the flux of a *magnetic field an *electromotive force (e.m.f.) is induced in the conductor. The original experiment was carried out in the field due to a permanent magnet, but later, the same effects were noted in the magnetic field associated with a current-carrying solenoid.

(b) The size of the induced e.m.f. depends on the size of the relative motion, reverting to zero when the motion ceases.

(c) The direction of the induced e.m.f. depends on the orientation of the magnetic field.

Further experiments showed that an e.m.f. is induced in a conductor when placed in a region of varying flux, and that it is particularly noticeable in the flux due to an applied current at the instants of switching the current on and off.

Conclusion (b) was put in a quantitative form by Neumann, and is known as the *Faraday–Neumann law*: when a conductor cuts a magnetic flux Φ the induced e.m.f. E is proportional to the rate at which the flux is changing.

Conclusion (c) is generally expressed in the form of *Lenz's law*: The induced e.m.f. is in such a direction as to oppose the change that produces it. Lenz's law is in effect a particular case of the principle of conservation of energy in so far as it tells us that the magnetic field associated with the induced e.m.f. is in the reverse direction to that associated with the inducing source.

The two laws may be put in the form of an equation:
$$E = -d\Phi/dt,$$
the minus sign indicating the significance of Lenz's law. E is measured in volts when $d\Phi/dt$ is in webers per second. When the conductor is made part of a circuit, Ohm's law may be applied to show that the charge Q flowing when the flux changes by $\Delta\Phi$ is given by:
$$Q = \Delta\Phi/R,$$
R being the total resistance of the circuit.

When the current I in the circuit varies, the associated flux varies in proportion:
$$\Phi = LI.$$
Applying the Faraday–Neumann law,
$$E = -LdI/dt.$$
The back e.m.f. induced in a circuit when the current in that circuit varies is described as *self inductance* (formerly, self induction). L is the coefficient of self inductance, and is defined as numerically equal to the e.m.f. induced when the current changes at unit rate. L is measured in *henrys* when E is in volts and I is in amperes. Similarly, the change of current in one circuit can cause an e.m.f. to be induced in a neighbouring circuit due to the flux linkage. The relationships are identical:
$$\Phi_1 = MI_2,$$
$$E_1 = -MdI_2/dt.$$
M is called the coefficient of *mutual inductance*; it is measured in henrys. For an ideal mutual inductance:
$$M^2 = L_1L_2,$$
where L_1, L_2 are the self inductances of the component inductors. Mutual inductance is the principle behind the action of the *transformer. *Tuned circuits rely on the appropriate values of self inductances.

In order to establish a current I in a circuit with self inductance L, work must be done. The rate of doing work is EI, so the work done is given by:
$$W = \int EIdt = \tfrac{1}{2}IL^2.$$
This amount of energy is stored in the magnetic field around the circuit and must be dissipated on switching off, usually by generating a spark.

electromagnetic interaction The interaction between *elementary particles arising as a consequence of their associated electric and magnetic fields. The

electrostatic force between charged particles is an example. This force may be described in terms of the exchange of virtual photons (*see* virtual particle). Because its strength lies between *strong interactions and *weak interactions, particles decaying by electromagnetic interaction do so with a lifetime shorter than those decaying by weak interaction but longer than those decaying by strong interaction. An example of electromagnetic decay is:
$$\pi^0 \rightarrow \gamma + \gamma.$$
This decay process (mean lifetime 8.4×10^{-15} seconds) may be understood as the *annihilation of the *quark and the antiquark making up the π^0 into a pair of photons.

The following *quantum numbers have to be conserved in electromagnetic interactions: angular momentum, charge, baryon number, isospin quantum number I_3, strangeness, charm, parity, and charge conjugation parity. *See also* electroweak theory; quantum electrodynamics.

electromagnetic lens *See* magnetic lens.

electromagnetic mass That part of the total inertia of a charged body that arises from its electric charge.

electromagnetic moment *See* magnetic moment.

electromagnetic pump A pump with no moving parts for use with conducting liquids, such as liquid metals. The liquid is contained in a flattened pipe between two poles of an *electromagnet, a strong magnetic field being applied across the pipe. If an electric current is also passed through the liquid perpendicular to the magnetic lines of force, the liquid will experience a force along the axis of the pipe. This type of pump is sometimes used to move the liquid-sodium *coolant in *fast breeder reactors.

electromagnetic radiation A disturbance propagated both through a vacuum and through many materials, consisting of oscillating electric and magnetic fields with directions at right angles to each other and to the direction of propagation. It transfers energy, mass, momentum, and angular momentum. Although the values of these quantities depend equally upon the electric and magnetic fields, most interactions with matter depend primarily upon the electric field strength. Radiation is caused by the acceleration of charged paticles, the rate of emission of energy being proportional to the square of the charge times the square of the acceleration.

The speed of electromagnetic radiation in *free space, c, is exactly $2.997\,924\,58 \times 10^8$ m s^{-1} by definition of the metre, the *phase and *group speeds being identical. In a medium of refractive index n the phase speed $v = c/n$. The phase speed is usually less than c in a material medium, but can be greater. The group speed, which is the speed at which energy propagates, cannot exceed c.

The characteristics of electromagnetic radiations depend upon the frequence v which does not change on entering another medium, while the wavelength, which is the phase speed divided by v, does change.

The range of frequencies over which electromagnetic radiation has been studied is called the *electromagnetic spectrum* (*see* Appendix, Table 7). The methods of generating radiations and their interactions depend very much upon frequency. It can be shown that the rate of radiation of energy caused by the acceleration of a given charge is proportional to the square of the acceleration. In most cases it is apparent that electromagnetic radiation is generated in this way, but in the decay of certain elementary particles (for example, neutral *pions) intermediate states of the system involving oscillations of hypothetical virtual particles must be assumed. Interactions can usually be seen to be caused by the electric field of the radiation. Although half of the energy of the wave is associated with the magnetic field, interaction with matter is rarely caused directly by this.

The lowest frequencies (radio waves) are generated artificially by electrical oscillations in circuits. Some astronomical bodies emit radiations of this type (*see* radio astronomy). Detection uses an *aerial in which the incident wave generates electrical oscillations. Hot bodies, electric discharges, and luminescent bodies give *infrared radiation, *light, and *ultraviolet. These radiations may be affected or caused by molecular oscillations or rotations (especially for long-wavelength infrared) but usually both emission and absorption involve electrons moving between states of different energy in the outer structure of atoms, molecules, or ions. Characteristic *X-rays are generated by electrons entering states in the inner shells of atoms from which electrons have been ejected by high-energy electrons or other means. *Bremsstrahlung is caused by the acceleration of high-energy electrons in collision with atoms. *Gamma radiation is emitted by atomic nuclei. *See also* annihilation.

Electromagnetic radiations interact with matter as quanta. A quantum of radiation of frequency ν transfers energy $h\nu$, mass $h\nu/c^2$, and momentum $h\nu/c$, where h is the *Planck constant. The transfer of momentum causes *radiation pressure. Changes of angular momentum follow complicated rules, but in the simplest cases emission or absorption changes a component of angular momentum by an amount $h/2\pi$.

By whatever mechanism electromagnetic radiation is absorbed, the energy is usually degraded, ultimately raising the internal energy and hence the temperature. Hence the intensity of such radiations can often be measured absolutely using suitably calibrated instruments such as the *thermopile.

electromagnetic separation The separation of isotopes using an electromagnetic-deflection-type of mass spectrometer. *See* calutron.

electromagnetic spectrum *See* electromagnetic radiation.

electromagnetic theory of light *See* electromagnetic radiation; light.

electromagnetic units (e.m.u.) *See* CGS system of units.

electrometer A device for measuring potential difference. Originally electrometers were electrostatic instruments incorporating an *electroscope (*see* Dolezalek, Lindemann, quadrant electrometers). However, these are now becoming obsolete, and modern instruments consist of a very high input impedance amplifier known as an electrometer amplifier, which draws a negligible amount of current. Electrometer amplifiers may use valves in the circuit, or more commonly, *solid-state devices. *Field-effect transistors are widely used in the input stage of such an amplifier because of their very high input impedance characteristics.

electromotive force (e.m.f.) Symbol: E. The rate at which work is done electrically upon a circuit (power) divided by the current. It is measured in volts. Work may be done by *electromagnetic induction, *thermoelectric effects, or chemical reactions (*see* cell). If the process is driven in reverse, the circuit does work upon the source of e.m.f. Electromotive force (which is not a force in the normal sense) must be distinguished from potential difference. In a circuit, the latter is rate of energy dissipation divided by current, and is inherently irreversible.

If a battery of e.m.f. E and resistance b maintains a current I in an external resistance R, the rate of doing work is IE, which equals the rate of dissipation $(R + b)I^2$. The p.d. between the terminals is $V = RI$. If R tends to infinity, the value of E tends to that of V, so the e.m.f. is equal to the p.d. between the terminals on open circuit, although its nature is different. (In practice both E and b may vary with I, but this does not affect the results for very small currents.)

electromotive series *Syn.* electrochemical series. The chemical elements arranged in order of their *electrode potentials, beginning with the highest positive.

electromotor A machine for converting electrical into mechanical energy. *See* motor, electric.

electron A stable *elementary particle having a negative charge, e, equal to:
$$1.602\ 177\ 33 \times 10^{-19}\ C,$$
and a rest mass m_0 equal to:
$$9.109\ 534\ 5 \times 10^{-31}\ kg$$
(equivalent to 0.511 0034 MeV/c²).
It has a *spin of ½ and obeys *Fermi–Dirac statistics. As it does not have *strong interactions, it is classified as a *lepton.

The discovery of the electron was reported in 1897 by Sir J. J. *Thomson. He measured approximately the specific charge e/m_0 of the rays from the *cold cathode of a *gas-discharge tube and obtained a value about 1300 times that of a hydrogen atomic ion in electrolysis. Since the penetrating power of the rays (*see* Lenard rays) indicated only a small charge, he concluded that the mass was much less than that of an atom. Within a few years much more accurate measurements became possible and it was

electron affinity

established that particles with the same charge and mass were obtained from numerous substances by the *photoelectric effect, *thermionic emission, and *beta decay. Thus the electron was found to be part of all atoms, molecules, and crystals.

Free electrons are studied in a vacuum or a gas at low pressure. Beams are emitted from hot filaments or cold cathodes and are subject to *focusing by electric or magnetic fields. The force F_E on an electron in an electric field of strength E is given by $F_E = Ee$ and is in the direction of the field. On moving through a *potential difference V, the electron acquires a kinetic energy eV, hence it is possible to obtain beams of electrons of accurately known kinetic energy, provided that corrections are made for *contact potential difference and any initial kinetic energy. Since these corrections are of the order of volts, they can be omitted in experiments with very large values of V. In a magnetic field of *magnetic flux density B an electron with speed v is subject to a force $F_B = Bev \sin \theta$, where θ is the angle between B and v. This force acts at right angles to the plane containing B and v in the direction given by the Fleming left-hand rule (remembering that the charge is negative; see Fleming's rules).

The mass of any particle increases with speed according to the Einstein formula:

$$m = \frac{m_0}{\sqrt{1 - v^2/c^2}},$$

where m is the mass at speed v, m_0 is the rest mass, and c the speed of light. The kinetic energy $T = (m - m_0)c^2$, which gives the classical value $\frac{1}{2}m_0 v^2$ when $v \ll c$. In the case of the electron, $m_0 c^2 \simeq 0.5$ MeV hence if the particle is accelerated from rest through 5 kV, the mass is 1% greater than at rest. Thus account must be taken of relativity for calculations on electrons with quite moderate energies.

According to *wave mechanics a particle with momentum mv exhibits diffraction and interference phenonema similar to a wave with wavelength $\lambda = h/mv$, where h is the Planck constant. For electrons accelerated through a few hundred volts, this gives wavelengths rather less than typical interatomic spacing in crystals. Hence a crystal can act as a diffraction grating for electron beams. (See Davisson–Germer experiment; de Broglie waves; electron diffraction.)

At kinetic energies less than a few *electronvolt, electrons undergo *elastic collision with atoms and molecules. Because of the large ratio of the masses and the conservation of momentum, only an extremely small transfer of kinetic energy occurs, thus the electrons are deflected but not slowed down appreciably. At slightly higher energies collisions are inelastic. Molecules may be dissociated, atoms and molecules may be excited or ionized (see ionization potential). The excited particles or recombining ions emit *electromagnetic radiation mostly in the visible or ultraviolet. For electron energies of the order of several keV upwards, *X-rays are generated. Electrons of high kinetic energy travel considera-

ble distances through matter, leaving a trail of positive ions and free electrons. The energy is mostly lost in small increments (about 30 eV) with only an occasional major interaction causing X-ray emission. The range increases at higher energies and at a few MeV, electrons will penetrate matter of the order of 10 kg m^{-2}.

The relativistic wave mechanics of the electron was developed by *Dirac who introduced the concept of the *antiparticle. The antiparticle of the electron is the *positron, which has the same mass as the electron and equal charge of opposite sign. The positron is occasionally called the *positive electron*. See atom.

electron affinity Symbol: A or E_a. Many atoms, molecules, and free radicals form stable negative ions by capturing electrons (see electron attachment). The electron affinity is the least amount of work that must be done to separate the electron from the ion. It is usually expressed in electronvolts.

electron attachment The attachment of a free electron to an atom or molecule to form a negative ion. The process is sometimes called *electron capture* but this term is more usually applied to nuclear processes. See also capture; electron affinity.

electron beam A beam of *electrons, usually emitted from a single source such as the *thermionic cathode in an *electron gun.

electron biprism An arrangement of fields used to separate a beam of electrons or other charged particles so that interference can occur in an analogous way to an optical biprism.

electron capture 1. See capture. **2.** See electron attachment.

electron density 1. The number of electrons per unit mass of a given material. Most light elements (except hydrogen) have an electron density of about 3 $\times 10^{26}$ electrons per kilogram. **2.** The number of electrons per unit volume.

electron diffraction Owing to the fact that electrons are associated with a wavelength λ given by $\lambda = h/mv$, where h is the Planck constant and (mv) the momentum of the electron, a beam of electrons suffers diffraction in its passage through crystalline material, similar to that experienced by a beam of X-rays. The diffraction pattern depends on the spacing of the crystal planes, and the phenomenon can be employed to investigate the structure of surface and other films. See de Broglie waves; electron.

electronegative See electromotive series.

electronegativity A measure of the ability of an atom, in a compound, to attract electrons to itself. Pauling devised a scale of electronegativities based on *bond energies. If the bond energy of a molecule AB is E_{AB}, and E_{AA} and E_{BB} are the bond energies of molecules AA and BB respectively, then

$$X_A - X_B = E_{AB} - E_{AA}E_{BB},$$

where X_A and X_B are the electronegativities of A and B respectively. Using this relationship he constructed a scale for a large number of elements based on the value 4 for fluorine – the most electronegative element.

A different approach was used by Mulliken in expressing the electronegativity of an element as the arithmetic mean of its *ionization potential and its *electron affinity.

electron gas The concept of the free electrons in the solid or liquid state as a gas whose state may be compared with that of an actual gas dissolved in a solid or liquid. This model has found application in theories of electric and thermal conduction, thermionic emission, etc. Applying Fermi–Dirac statistics (*see* quantum statistics) to the electron gas, it is shown to be completely degenerate at ordinary temperatures, i.e. it obeys a totally different distribution law from that of an ideal gas. Furthermore it is shown that the energy content of the electron gas is to a first approximation independent of temperature, which means that its specific heat is vanishingly small to a first approximation.

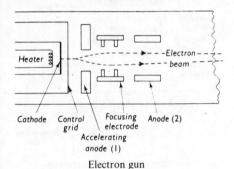

Cathode   Control   Focusing   Anode (2)
grid   electrode
Accelerating
anode (1)

Electron gun

electron gun A device, consisting of a series of electrodes, that produces an *electron beam. The beam produced is usually a narrow beam of high-velocity electrons whose intensity is controlled by electrodes in the gun. An electron gun is an essential part of many instruments needing electron beams, such as *cathode-ray tubes, *electron microscopes, *linear accelerators, etc.

Electrons are released from the indirectly heated cathode (*see* diagram). The control grid is a cylinder surrounding the cathode with a hole in front to allow passage of the electron beam. It controls the electron beam when its negative potential is varied. The electron beam is accelerated by the positively charged accelerating anode and passes through the focusing electrode before being further accelerated by the second anode.

electron–hole pair *See* hole.

electronic device A device in which conduction is mainly by the movement of *electrons in a vacuum, gas, or *semiconductor.

electronic flash A source of very brief but bright illumination, provided by a high-voltage discharge between electrodes in a gas-discharge tube containing xenon or neon. Electronic flash is used, for example, in photography and stroboscopy. In photography it takes the form either of battery-operated *flashguns* that may be mounted on a camera or mains-powered *studio flash*.

electronic instrument A musical instrument with the sound created and amplified by electrical or electronic means. The various techniques involved are best described in connection with particular examples.

The Hammond organ employs tone wheels, approximately circular but with irregular edges, that revolve at various fixed speeds in magnetic fields. The frequencies of the induced currents can be combined and amplified into a wide range of notes.

The Compton organ employs fixed insulating discs coated with charged layers, except along certain carefully calculated wavy paths. Each disc is scanned by a second metallic disc, rotating so as to pick up electric potentials from different parts of the first disc at appropriate frequencies. The frequencies are combined and amplified as in the Hammond organ.

Other electronic organs consist of a series of audiofrequency generators, the resulting frequencies again being combined and amplified.

The *Moog synthesizer* is an instrument designed to enable the player/operator to produce almost any desired type of sound that can be electronically generated. Many sine- and square-waves are available, either singly or in more-or-less complex groups, and a large number of filtering circuits are employed. From the outside it appears as a combination of keyboard and telephone exchange, with plug-in facilities for individual potentiometers. The synthesizer can either be set and played (in the conventional sense), or each sound can be prepared separately and at length, the sounds being recorded for reproduction when the complete work has been prepared.

Other methods of frequency generation that have been used for particular instruments include the modulation of a beam of light in a system similar to that of the sound-film technique; reed or string oscillators maintained either electrically or by air pressure, using electromagnetic or electrostatic pick-ups; and cathode-ray oscillators.

Electronic imitators of proper musical instruments are never completely successful for the following reasons:

(a) The tuning is generally too perfect (it is normally exactly equal temperament) to produce smooth blending and unlimited key modulation.

(b) The number and amplitudes of the harmonics generated for a given note rarely agree with the results obtained in the analysis of normal sounds.

(c) The "steady state" of most musical instruments gives a nonsteady state for individual harmonics.

electronics

(d) Transients are not reproduced, both at starting and (more particularly) at stopping.

(e) The use of loudspeakers, even if they are stereophonic, gives an unsatisfactory spatial effect.

Electrophonic instruments are those, such as the electric guitar and electric piano, in which the sounds are produced in a conventional manner, but are amplified by electronic means.

electronics The study, design, and use of devices based on the conduction of electricity in a vacuum, a gas, or a semiconductor. Modern electronics is principally concerned with semiconductor devices; vacuum and gas-filled devices are rapidly becoming obsolete, apart from a few specialized uses. *See* Appendix, Table 6 for symbols used in electronics.

electronic spectrum *See* spectrum.

electronic switch An *electronic device that can be used as a switch. Usually these devices operate as high-speed switches.

electron lens A device for *focusing an electron beam by using either a magnetic or an electrostatic field in a way that is analogous to the focusing of a light beam by an optical lens. These lenses can be combined, as in optics, in instruments such as the *electron microscope. *See* magnetic lens; electrostatic lens.

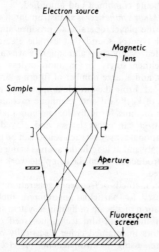

a Transmission electron microscope

electron microscope 1. *Transmission electron microscope.* An instrument, closely resembling the optical microscope but using, instead of light, a beam of energetic electrons. The *resolution, and consequently the magnification, is about one thousand times that of the optical microscope. The beam is focused (*see* focusing) by a *magnetic lens or sometimes an *electrostatic lens and has an energy of 50–100 keV. (Fig. *a*.) It is directed onto the entire area of the sample under investigation and the electrons emerging are focused by a second magnet-

ic lens onto a fluorescent screen, thus producing a visible image. A sharply focused image in one plane can only be obtained by using electrons of a single energy. To avoid energy losses in the incident beam, the sample must be extremely thin (usually no greater than 50 nanometres) so that the scattered electrons that form the image are not changed in energy. The thinness of the sample greatly limits the *depth of field and the image is therefore two-dimensional. Electrons accelerated to an energy of 100 keV have a wavelength of about 0.04 nm (*see* de Broglie equation) so that a resolution of between 0.2–0.5 nm is possible. The maximum magnification is approximately one million diameters, the image becoming blurred at greater values.

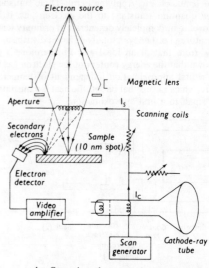

b Scanning electron microscope

2. The *scanning electron microscope*, developed after the transmission type, has lower resolution and magnification but produces a three-dimensional image from a sample of any convenient size or thickness rather than a thin slice. The beam of energetic electrons originates from a heated tungsten cathode with a diameter of between 20–50 μm. (Fig. *b*.) It is accelerated and focused by electric and magnetic fields forming a spot, with a diameter of about 10 nm, on the surface of an electrically conducting sample. (If the sample is nonconducting, a thin metallic layer, 5–50 nm thick, is evaporated onto the surface.) The electron beam is deflected by scanning coils so that the sample is scanned, point by point, in a *raster scan. Electrons striking the sample give rise to *secondary electrons whose number is dependent on the geometry and other properties of the sample. As no transmission process is evolved and no focusing of the secondary electrons is necessary, the sample can be of any shape.

The secondary electrons are collected by a positively charged electron detector in which they are accelerated to about 10 keV before striking a *scintillator and so producing a large number of photons. The photons, falling upon a *photomultiplier are converted into a highly amplified electric signal. This signal modulates the intensity of an electron beam inside a *cathode-ray tube whose screen is scanned in step with the electron beam that falls on the sample. Back-scattered electrons and photons emitted by the sample are also used to produce an image. Other types of image are produced by electrons transmitted through the sample and by currents induced in the sample.

The magnification of the microscope is determined by the ratio of the variable current I_s in the scanning coils to the current I_c in the deflection coil of the cathode-ray tube. It can vary continuously from 15 diameters to 100 000 diameters. The resolution is about 10–20 nm. The visual image on the screen has great depth of field due to the factor of 2–5000 between the diameters of the original electron source and the spot size on the sample. The visual image therefore apears three-dimensional and more lifelike.

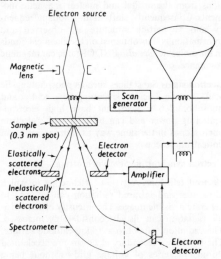

c Scanning-transmission electron microscope

3. The *scanning-transmission electron microscope* (STEM) combines the high resolution of the transmission instrument with the three-dimensional image of the scanning type. The electron source, a cold-cathode field emitter (*see* field emission) in the form of a tungsten tip, has a tiny diameter of about 10 nm. (Fig. *c*.) The electrons are accelerated by electric fields and focused magnetically to a spot, about 0.3–0.5 nm in diameter, which is made to scan the sample. The electrons transmitted through the sample fall on a circular electron detector with a small hole in the centre through which pass electrons either unscattered or involved in *inelastic collisions in the sample. Electrons involved in *elastic collisions, having a much greater deflection, strike the circular detector and produce an electrical signal.

The electrons passing through the hole enter a spectrometer and are separated according to energy. The unscattered electrons are removed and the remaining inelastically scattered electrons strike another detector, producing a second signal. Dividing the elastic signal by the inelastic signal gives an output proportional to *atomic number. This output is fed into a cathode-ray tube whose electron beam scans in synchrony with the primary electron beam. The output, controlling the brightness of the final visual image, responds to changes in atomic number rather than thickness of the sample. The contrast can be enhanced electronically. Other signals can be collected from low- and high-energy electrons and from X-rays and photons emitted by the sample. Resolution varies depending on the instrument but the highest resolution will be in the region of 0.3 nm.

See also atomic force microscope; field-emission microscope; field-ion microscope; scanning-tunnelling microscope.

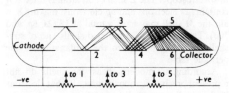

Electron multiplier

electron multiplier An electron tube in which current amplification is secured by *secondary emission of electrons. Primary electrons (released by photoelectric effect, or otherwise) are accelerated by application of a high potential and made to strike a good secondary emitter where they produce a greater number of electrons by impact (*see* diagram). These are then accelerated on to a further secondary emitter, the process being repeated several times within the same envelope, the set of anodes being called a dynode chain. The voltage of each anode must progressively increase above that of the preceding anode, so that the final plate has to be operated at a high potential, typically one or two kV above the cathode. *See* photomultiplier.

electron nuclear double resonance *See* ENDOR.

electronographic camera *See* image intensifier.

electron optics The study of the behaviour and control of an electron beam in magnetic and electrostatic fields. An analogy is drawn with the passage of a beam of light through refracting media. A limited magnetic or electric field is regarded as forming an *electron lens for *focusing or defocusing an electron beam.

electron paramagnetic resonance *See* electron spin resonance.

electron-probe microanalysis A technique for analysing small quantities of solids by examining the *X-ray spectrum emitted when the sample is bombarded by a fine beam of electrons. The elements in the sample are detected by the *characteristic X-radiation that they emit and the intensity of this radiation depends on the amount of substance present. The electron beam can be focused to a spot with a diameter of about 10^{-6} m and quantities as small as 10^{-16} kg can be detected. It is a particularly useful method for determining variations in composition over a solid surface. The technique is not very useful for elements with low atomic numbers as the intensity of X-ray emission falls with decreasing atomic number.

electron shell A group of electrons in an atom having a given total quantum number n. The innermost or K shell, with $n = 1$, cannot contain more than 2 electrons. The other shells are denoted in sequence by the letters L ($n = 2$), M (3), N (4), ... The electrons in these shells may be grouped into subshells, for which the letters s, p, d, f, g, h, ..., are used, of which the first four have a historical origin in spectroscopy. The L shell of an atom is filled if it contains $2s$ electrons and $6p$ electrons, making 8 in all; the M shell may have a total of 18 ($2s$, $6p$, and $10d$). The stability of completed shells and of subshells provides the key to the explanation of the periodic table.

According to modern ideas, the shells do not represent precise locations in space in that electrons "belonging" to one shell may well interpenetrate the orbits or tracks of electrons from another shell and even these orbits are no longer regarded as definite paths. *See* atomic orbital.

electron spectroscopy The measurement of the distribution of electron kinetic energies in a flux of electrons and the use of this information to determine energy levels in atoms, molecules, solids, nuclei, etc. The technique depends on accurate measurement of electron energies by some form of *electron spectrometer*. There are several rough subdivisions of electron spectroscopy:

(i) The study of electrons ejected from nuclei as a result of *beta decay. This is traditionally known as *beta-ray spectroscopy*.

(ii) The determination of the energy distribution of an electron beam after it has been inelastically scattered by a solid or gas (electron-energy loss).

(iii) The measurement of energies of electrons emitted from molecules as a result of some applied stimulus such as irradiation with photons, ions, metastable atoms, or other electrons. Techniques of this type are sometimes called *induced electron-emission spectroscopy*. They include *photoelectron spectroscopy and techniques based on the *Auger effect and *Penning ionization.

electron spin resonance (ESR) *Syn.* electron paramagnetic resonance (EPR). A phenomenon observed when paramagnetic substances containing unpaired electrons are subjected to high magnetic fields and microwave radiation. An electron has *spin and an associated magnetic moment. For each set of *quantum numbers describing spatial location there are two states, corresponding to the two possi-

ble spin orientations (*see* space quantization). In the absence of any magnetic field, these states have equal energy. When a field is applied, the energies became different. The difference in energy between these states is given by $gm_B\boldsymbol{B}$, where g is the *Landé factor of the electron, m_B is the Bohr *magneton, and $\boldsymbol{B}$ the magnetic flux density.

If a large number of atoms are considered, each having an unpaired electron, then at normal temperatures there is a statistical probability that slightly more atoms will be in the lower energy state than the higher one. If electromagnetic radiation is applied, they can be raised in energy to the higher state by absorption of a photon of frequency v, where $hv = gm_B\boldsymbol{B}$. For the fields normally used, v lies in the microwave region of the spectrum. In electron-spin resonance spectroscopy the sample is exposed to microwave radiation of variable frequency and at the frequency v, resonance occurs and the radiation is absorbed.

Free electrons have a Landé factor of about 2 but in most compounds this is modified by contributions from the orbital and nuclear magnetic moments. Consequently, shifts in the resonant frequency and *hyperfine structure are observed and information is thus obtained on the chemical bonds in the molecule. *See also* ENDOR; nuclear magnetic resonance.

electron stains Substances such as phosphotungstic acid, osmic acid, silicotungstic acid, and phosphomolybdic acid that have high electron-scattering power and can be used with *electron microscopes in the same way as staining media in optical microscopy.

electron synchrotron *See* synchrotron.

electron telescope A type of *telescope in which ultraviolet and infrared radiation can be used to generate a visible image. The intensity of faint visible radiation can also be considerably increased. The radiation falls onto a *photocathode surface. The resulting *secondary electrons are accelerated through a *series of voltage grids without being deviated and finally fall on a fluorescent screen. An accurate visible image is therefore obtained.

electron temperature Electrons in the plasma of a discharge tube have a *distribution of velocities that is approximately Maxwellian. The electron temperature of the plasma is that temperature at which gas molecules would have the same average kinetic energy as the electrons of the plasma.

electron torch A flame produced by high-frequency (of the order of 1000 megahertz) gas discharge at the end of an h.f. conductor, the discharge being stabilized by convection current or by gas blast. The flame produced by polyatomic gases can reach 3000 K.

electron tube Any electronic device in which the movement of electrons between two or more electrodes, through a gas or a vacuum, takes place in a sealed or continuously exhausted envelope. Examples include *cathode-ray tubes, *gas-discharge tubes, and *thermionic valves (now obsolete).

electronvolt Symbol: eV. A unit of energy extensively employed in atomic, nuclear, and particle physics. It is the work done on an electron that is displaced through a potential difference of one volt. It is equal to

$$1.602\ 189\ 25 \times 10^{-19}\ \text{joule}.$$

electron voltaic effect Photovoltaic cells (*see* photovoltaic effect) have been found to be sensitive to electron bombardment. The gain rises rapidly with voltage at low energies, reaches a maximum, and then decreases.

electro-optics The study of the changes in the optical properties of a dielectric produced by the application of an electric field. *See* Kerr effects.

electrophoresis The migration of fine particles of solid suspended in liquid to the anode (*anaphoresis) or cathode (*cataphoresis) when an electric field is applied to the suspension. *See* micelle.

electrophorus An early form of appliance for producing electrical charge by electrostatic induction. It consists of a flat plate of dielectric, known as the "cake", and a varnished metal plate with an insulating handle. The cake is charged by friction; the plate is placed on it and momentarily earthed, leaving it with an induced charge of opposite sign to that on the cake. The process can be repeated indefinitely, until the original charge on the cake leaks away.

electroplating A practical application of *electrolysis, in which the surface of one metal is covered with another, either for protection, or for decoration, or both.

electropolishing The production of a specularly reflecting surface by anodic etching of an initially rough metal.

electropositive *See* electromotive series.

electrorheological fluid *Syn.* smart fluid. A fluid that sets to a jelly-like solid when a high voltage is applied across it (about 3 MV m^{-1}). The stiffening is proportional to the electric field strength and is reversible. These fluids are essentially a suspension of particles in a nonconducting liquid.

electroscope An electrostatic instrument for the detection of electrical potential differences. The commonest form consists of a pair of gold leaves hanging side by side from an insulated metal support, and enclosed in a draught-proof case. If the support is given a charge, the leaves separate, owing to their mutual repulsion. One of the leaves may be replaced by a vertical metal plate. An instrument capable of accurate quantitative indication is usually called an *electrometer.

electrosmosis *Syn.* electric endosmose; electroendosmosis; electrosmose. The passage of an

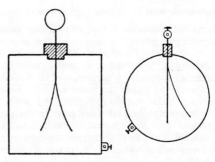

Electroscope

electrolyte through a membrane or porous partition under the influence of an electric field. If the pressure is kept the same on both sides of the partition, the volume of liquid transferred is proportional to the total electrical transfer, and is independent of the area and thickness of the partition. It depends on the nature of the electrolyte, and for any one electrolyte is approximately proportional to the *resistivity of the solution. The effect has been ascribed to the existence of a contact difference of potential between the material of the membrane and the electrolytic solution.

electrostatic deflection A method of deflecting an *electron beam using the electrostatic fields produced between two metal electrodes – most often applied to the beam in *cathode-ray tubes using two pairs of *deflector plates. *Compare* electromagnetic deflection.

electrostatic focusing *See* focusing.

electrostatic generator A machine for producing electric charge by electrostatic action, e.g. friction or (more usually) electrostatic induction. *See* Wimshurst machine; Van de Graaff generator.

electrostatic induction The production of electric charge on a conductor under the influence of an *electric field. Thus, if an uncharged conductor is placed near a positively charged body, the portion of the conductor nearest to the body becomes negatively charged, the more remote portions being positively charged. If the conductor is insulated, the induced charges are equal in magnitude. If the conductor is earthed and then insulated again, it will be left negatively charged. If the conductor completely surrounds the charged body, each of the induced charges is numerically equal to the inducing charge on the body. *See* ice-pail experiment.

The separation of charges in a dielectric by an electric field is also described as electrostatic induction.

electrostatic lens An *electron lens that uses electrostatic *focusing to focus the *electron beam. *Compare* electromagnetic lens.

electrostatics The study of the phenomena associated with electric charge at rest.

electrostatic units (e.s.u.) *See* CGS system of units.

electrostatic voltmeter A type of voltmeter based upon the principle of the quadrant or other form of electrometer. The Kelvin multicellular voltmeter is essentially a *quadrant electrometer, having a number of alternating quadrants, used idiostatically; the moving sectors are suspended by a metal strip, which also supplies the control and carries a pointer moving over a calibrated scale. In all types, the resistance of the instrument must be high, so that it takes only a very small current and hence produces no appreciable disturbance in the circuit to which it is connected.

electrostriction The stress of extension or compression that any body in a medium of relative permittivity different from its own experiences in an electric field. If the field is not homogeneous, a body of higher relative permittivity than its surroundings will experience a force towards the area of greater field strength and vice versa.

electrovalent bond *Syn.* ionic bond. A type of chemical bond caused by electrostatic attraction between ions of opposite charge. A simple example is the bond between sodium and chlorine in sodium chloride. Sodium atoms have one valence electron and can lose this to give a positive ion (Na^+). This has the stable electron configuration of the inert gas, neon. On the other hand, the chlorine atom has 7 valence electrons and can gain one to give a negative ion (Cl^-). Again this has the electron configuration of an inert gas, in this case argon. The bond in sodium chloride is simply electrostatic attraction between sodium and chloride ions. Solid sodium chloride is a regular array of these ions and no single NaCl molecule can be distinguished. Another example is sodium sulphate, which contains sodium ions and negative sulphate ions $SO_4{}^{2-}$. Each sulphate ion has a charge of 2 and requires two sodium ions to maintain neutrality. Within the sulphate ion the bonds between the sulphur atom and the oxygen atoms are *covalent bonds.

electroviscosity An effect in which the electrical resistance to shear in certain ionic liquids, due to the effect of an electrical potential gradient of the surface layers, becomes large in relation to the mechanical resistance to shear.

electroweak theory A *gauge theory (also called *quantum flavourdynamics*) that provides a unified description of both the *electromagnetic and *weak interactions. In the *Glashow–Weinberg–Salam* (*GWS*) theory, also known as the *standard model*, electroweak interactions arise from the exchange of *photons and of massive charged *$W^\pm$ particles and neutral *Z^0 particles of spin 1 between *quarks and *leptons. The interaction strengths of the gauge bosons to quarks and leptons and the masses of the W and Z bosons themselves are predicted by the theory in terms of a single new parameter, the Weinberg angle θ_W, which must be determined by experiment. The GWS theory successfully describes all existing data from a wide variety of electroweak processes, such as neutrino–nucleon, neutrino–electron and electron–nucleon scattering. A major success of the model was the direct observation in 1983–84 of the $W^\pm$ and Z^0 bosons with the predicted masses of 80 and 91 GeV/c^2 in high energy proton–antiproton interactions. The decay modes of the $W^\pm$ and Z^0 bosons have been studied in very high energy $p\bar{p}$ and e^+e^- interactions and found to be in good agreement with the standard model.

The six known types (or flavours) of quarks and the six known leptons are grouped into three separate *generations* of particles as follows:

1st gen.:	e^-	ν_e	u	d
2nd gen.:	μ^-	ν_μ	c	s
3rd gen.:	τ^-	ν_τ	t	b

The second and third generations are essentially copies of the first generation (which contains the electron and the up and down quarks making up the proton and neutron) but involve particles of higher mass. Communication between the different generations occurs only in the quark sector and only for interactions involving $W^\pm$ bosons. Studies of Z^0 boson production in very high energy electron–positron interactions have shown that no further generations of quarks and leptons can exist in nature (an arbitrary number of generations is *a priori* possible within the standard model) provided only that any new neutrinos are approximately massless.

The GWS model also predicts the existence of heavy spin 0 particle, not yet observed experimentally, known as the *Higgs boson*. This particle results from the so-called *spontaneous symmetry breaking* mechanism used to generate nonzero masses for the $W^\pm$ and Z^0 bosons and is presumably too massive to have been produced in existing particle *accelerators.

element One of the basic substances from which all others are built up by chemical combination. A table of elements is given in the Appendix, Table 9. An all-embracing definition is that of a substance consisting wholly of atoms having the same *atomic number. The word was used by the ancients in a similar context, being applied to what were believed to be the four fundamental entities of the universe, namely earth, fire, air, and water.

elementary particle A particle that, as far as is known, is not composed of other simpler particles. Elementary particles represent the most basic constituents of matter and are also the carriers of the fundamental forces between particles, namely the electromagnetic, weak, strong, and gravitational forces. The known elementary particles can be grouped into three classes: *leptons, *quarks, and *gauge bosons. *Hadrons, such strongly interacting particles as the *proton and *neutron, which are bound states of quarks and/or antiquarks, are also sometimes called elementary particles.

Leptons undergo electromagnetic and weak interactions but not strong interactions. Six leptons are known, the negatively charged *electron, *muon, and *tauon plus three associated *neutrinos: ν_e, ν_μ, and ν_τ. The electron is a stable particle but the muon and tauon decay through the weak interactions with lifetimes of about 10^{-6} and 10^{-13} seconds. Neutrinos are stable neutral leptons, which interact only through the *weak interaction.

Corresponding to the leptons are six quarks, namely the up (u), charm (c), and top (t) quarks with electric charge equal to $+\frac{2}{3}$ that of the proton and the down (d), strange (s), and bottom (b) quarks of charge $-\frac{1}{3}$ the proton charge. Quarks have not been observed experimentally as free particles but reveal their existence only indirectly in high-energy scattering experiments and through patterns observed in the properties of hadrons. They are believed to be permanently confined within hadrons, either in *baryons, half integer spin hadrons containing three quarks, or in *mesons, integer spin hadrons containing a quark and an antiquark. The proton, for example, is a baryon containing two up quarks and a down quark, while the π^+ is a positively charged meson containing an up quark and an antidown ($\bar{d}$) antiquark. The only hadron that is stable as a free particle is the proton. The neutron is unstable when free. Within a nucleus, protons and neutrons are generally both stable but either particle may transform into the other by *beta decay or *capture.

Interactions between quarks and leptons are mediated by the exchange of particles known as gauge bosons, specifically the *photon for electromagnetic interactions, the $W^\pm$ and Z^0 bosons for the *weak interaction, and eight massless *gluons in the case of the *strong interactions. The long-lived particles are given in the Appendix, Table 8.

elinvar An iron-nickel-chromium alloy the *Young modulus of which varies hardly at all with temperature. It is used for the hairsprings of watches and usually contains about 36% of nickel and 12% of chromium.

ellipse The closed oval curve obtained by cutting a right circular cone by a plane more nearly perpendicular to the axis of the cone than are the generators of the cone. The ellipse may be defined as the locus of a point moving in a plane so that its distance from a fixed point (the focus) is in a fixed ratio e (the *eccentricity), less than unity, to its distance from a fixed line (the directrix). There are two symmetrically placed foci and two corresponding directrices.

The foci S, S' lie on the *major axis* of the ellipse; the *minor axis* is perpendicular to this. The sum of the focal distances ($SP + S'P$) of any point P on the ellipse is constant and equal to the major axis ($2a$). The standard form of equation of an ellipse is
$$x^2/a^2 + y^2/b^2 = 1.$$
See also conic sections.

ellipsoid A closed surface with an equation of the type
$$x^2/a^2 + y^2/b^2 + z^2/c^2 = 1.$$

An *ellipsoid of revolution* is obtained by rotating an ellipse about its major axis (*prolate spheroid*) or its minor axis (*oblate spheroid*). The earth is closely approximated by an oblate spheroid.

ellipsometer An instrument for studying thin films on solid surfaces. It depends for its action on the fact that if *plane-polarized light is incident on a surface, it is reflected as elliptically polarized light. (*See* polarization.) The degree of ellipticity in the reflected beam depends on the thickness of the film.

elliptic functions A group of functions defined by certain integrals. See elliptic integrals.

elliptic integrals The elliptic integral of the first kind is:
$$F(k,\phi) = \int_0^\theta [d\phi/\sqrt{(1 - k^2\sin^2\phi)}];$$
that of the second kind is:
$$E(k,\phi) = \int_0^\theta \sqrt{(1 - k^2\sin^2\phi)}.d\phi.$$
These integrals are said to be *complete* if the upper limit is $\frac{1}{2}\pi$. The quantity k is the *modulus* of the integral. These integrals arise in certain problems on motion of particles: e.g., the integral of the second kind is proportional to a length of arc measured along an ellipse; the first kind occurs in the theory of oscillation of a pendulum bob in a finite arc. Values of the integrals are given in books of mathematical tables, often listed as *elliptic functions*. $K(k)$ denotes the complete elliptic function $F(k,\frac{1}{2}\pi)$. The modulus k may be replaced by $\theta = \sin^{-1}k$.

ellipticity *Syn.* oblateness; compression. The ratio $(a - b)/a$, where a and b are the major and minor semi-diameters of an *ellipse. For the earth, a section through the poles has an ellipticity of about 1/297. *Compare* eccentricity.

elongation 1. An increase in length. **2.** Tensional strain, i.e. increase in length per unit of original length. **3.** Of the moon or a planet: the difference between its geocentric celestial longitude and that of the sun.

Elster, Julius (1854–1920) German physicist. With Geitel, he worked extensively on conduction in gases and in the atmosphere and they were pioneers in the development of the photoelectric cell.

e/m The ratio of the charge to the mass of an electron; sometimes called the *specific charge* of an electron. Its value decreases (owing to increase in mass) as the velocity of the electron approaches that of light. For slow-moving electrons its value is $1.758\ 8047 \times 10^{11}\ C\,kg^{-1}$. *See* electron.

emanating power The rate of emission of radioactive inert-gas atoms (radon and thoron) from a given material expressed as a fraction of the rate of their production within the solid. The measurement of emanating power is used to study physical changes in solids (*Hahan technique*).

emanation An obsolete name for the radioactive gases given off by actinium, radium, and thorium. These are now known as actinon, radon, and thoron respectively, or as the isotopes 219, 222, and 220 of the element radon (z = 86).

embedded temperature-detector A resistance element or a *thermocouple forming part of temperature-measuring equipment and built into an electrical machine to give the temperature of an inaccessible point in the machine. Embedded temperature-detectors are commonly employed to indicate the temperature in the *rotor and *stator slots of turbo-alternators.

emergent-stem correction A correction to be applied to the reading of a thermometer whose stem is not wholly immersed in the substance whose temperature is being measured, in order to obtain the correct temperature of the substance.

emery A hard substance used as an abrasive. *See* corundum.

e.m.f. Abbreviation for *electromotive force.

emission 1. The liberation of electrons from the surface of a solid or liquid. The types are: (i) *Thermionic emission – emission resulting from the temperature of the substance. (ii) Photoelectric emission – emission resulting from the irradiation of the substance. (*See* photoelectric effect.) (iii) *Secondary emission – emission resulting from bombardment of the substance by electrons or ions. (iv) *Field emission – emission resulting from intense electric fields at the surface of the substance. (v) Emission resulting from the disintegration of radioactive substances.
2. The release of electromagnetic radiation from an excited atom, molecule, etc. *See also* emission spectrum.

emission spectrum The *spectrum of radiation coming directly from a source as distinct from the absorption spectrum when some absorbing medium has been interposed in the path of the radiation from a source emitting a continuous spectrum. Emission spectra may be continuous, line, or band.

emissivity Symbol: ε The ratio of the power per unit area radiated from a surface to that radiated from a black body at the same temperature. Alternatively it can be defined as the ratio of *radiant exitances, $\varepsilon = M_e / M'_e$, where M_e is the radiant exitance of the body and M'_e that of the black body. The emissivity is restricted to radiation produced by the thermal agitation of atoms, molecules, etc. *Compare* absorptivity. *See also* heat-transfer coefficient.

emittance Former name for *luminous exitance or *radiant exitance.

emitter The region in a bipolar junction *transistor from which *carriers flow, through the emitter junction, into the *base. The electrode attached to this region is the *emitter electrode* (usually shortened to *emitter*).

emitter-coupled logic (ECL) A family of integrated *logic circuits, so called because a pair of emitter-

186

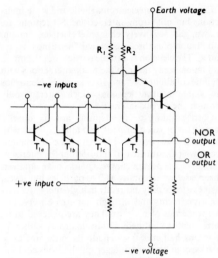

OR/NOR circuit

coupled transistors forms a fundamental part of the circuits. The input stage consists of the emitter-coupled transistor pair (known as a *long-tailed pair), which forms an excellent differential amplifier. The output is via an *emitter-follower buffer. ECL circuits are inherently the fastest logic circuits as the transistors are operated in nonsaturated mode and the *delay time is therefore exceedingly short (approximately 1 ns). The basic ECL gate has simultaneously both the function required and its complement. A simple OR/NOR circuit is shown in the diagram.

Input is via the three transistors $T_{1a,b,c}$. A fixed bias is applied to the base of the transistor T_2, the magnitude of which is half-way between a logical 1 and a logical 0. If a logical 0 is applied to all three input transistors, current flows through the transistor T_2 causing a voltage drop across R_2. This in turn produces a logical 0 at the OR output and a logical 1 at the NOR output. If any one or more of the input transistors $T_{1a,b,c}$ has a logical 1 applied, current flows through that transistor and a voltage drop is developed across R_1, causing the outputs to be reversed, i.e. a logical 1 occurs at the OR output. Typical values of applied voltages are: logical 0 –1.55 V, logical 1 –0.75 V, fixed bias –1.15 V d.c.

emitter follower An amplifier consisting of a bipolar junction *transistor with *common collector connection, the output being taken from the *emitter.

A signal is applied to the *base of the transistor, which is suitably biased so that it is nonsaturated and conducting. Since the transistor is conducting, the emitter will be one *diode forward voltage from the base at all times, and the emitter follows the signal applied to the base. Since the emitter voltage has a constant value relative to the base voltage, the voltage gain of the amplifier is almost unity, but the current gain is high. The amplifier is characterized

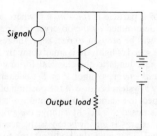

a Simple emitter follower

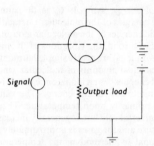

b Simple cathode follower

by high input impedance and low output impedance and is often used as a *buffer.

The analogue in *valve circuitry is the *cathode follower and in *FET circuitry the source follower, although neither is as efficient as a unity-gain buffer amplifier as the emitter-follower. The voltage gain (particularly of the source follower) is further from unity.

emmetropia The condition of the normal eye, which focuses parallel rays on the retina when at rest. Its far point is at infinity. Correction of refractive errors of the eye for distance consists of applying lenses to create artificial emmetropia. *See* ametropia; refraction of eye.

e.m.u. Abbreviation for electromagnetic units. *See* CGS system of units.

emulsifier *See* emulsion.

emulsion 1. A colloidal system in which the disperse phase and dispersion medium are both liquids. The latter are often not stable unless a small quantity of a third substance called an *emulsifier* is added.

2. A photographically sensitive coating applied to glass, plastic, or paper in the preparation of photographic materials. Usually this consists of grains of silver bromide with dimensions less than 10^{-6} m, dispersed in gelatine. Certain dyes are commonly added to give sensitivity to certain regions of the spectrum or for the production of coloured images. *See* photography.

emulsoid sol *See* sol.

enable To activate a particular electronic circuit or device, selected from a group, in order to effect its operation. An *enable pulse* is often used to select the desired circuit or device: this pulse must be present to allow other signals to be effective.

enantiomorphy A relationship that exists between a left and a right hand, or between any two bodies that can only be brought into coincidence by means of reflection across a plane.

encastré Built-in; e.g. an encastré beam is one in which the ends are firmly built into a wall or bolted on to a pier, so that they cannot tilt.

enclosed arc An *arc between electrodes of carbon situated in a transparent or translucent enclosure that is not airtight but that is designed to restrict the flow of air so that the arc burns in an atmosphere containing the products of combustion. *Compare* open arc.

end correction In the elementary theory of the vibrations in a *column of air, it is supposed that the open end of the pipe is a true displacement antinode. This is not true, for some sound energy escapes at each reflection from the open end and is radiated to the atmosphere in the form of spherical waves. The air beyond the open end of the pipe is in vibration and the effective length of the pipe is greater than the actual length.

The calculation of this end correction (x_0) was a matter of conflict from the theoretical point of view. Helmholtz and Rayleigh, considering an infinite flange flush with the end of the tube, gave the values $x_0 = \pi r/4 = 0.786r$ and $x_0 = 0.824r$ respectively, where r is the radius of the tube.

It can be simply determined experimentally by means of a resonance tube. This consists of a cylindrical tube whose distant end is closed by a movable piston or by a water surface whose level in the tube can be varied. The tube is used as a resonator to be tuned to a fork. By finding the first two successive lengths of the tube (say l_1 and l_2), which resounded to the fork, x_0 can be calculated. Since l_2 should be $3l_1$, in the absence of end correction (*see* column of air), we get accurately:
$$l_2 + x_0 = 3(l_1 + x_0).$$
By solving this equation, x_0 can be found. Recent experiments give $0.58r$ for a flangeless tube. The correction is approximately independent of the wavelength of the sound.

The magnitude of the correction for other than cylindrical ends depends upon the degree of openness or conductivity of the end; e.g. an uncovered hole in a flute may be considered as an open end, but the correction may be two or three times the radius. The pitch will be markedly below that calculated for a purely cylindrical pipe of the same length.

The change from plane to spherical waves at the mouth of the pipe gives rise to radiation impedance (*see* radiation resistance) and the effect is bound up with this end correction.

endoergic process *See* endothermic process.

end-on position of a magnet *See* Gauss positions.

ENDOR

ENDOR (*Electron–Nuclear Double Resonance*) A technique for studying molecules by a combination of *electron spin resonance and *nuclear magnetic resonance. The sample is irradiated at an electron resonant frequency and, simultaneously, with a second radiation whose frequency is swept over the range of nuclear resonant frequencies. The electron resonance signal is monitored as a function of this second variable frequency. The result is a much simpler spectrum than those obtained in electron-spin resonance, with much sharper lines.

endoscope A slender instrument used to view inaccessible locations, such as body cavities. *See* fibre-optics system.

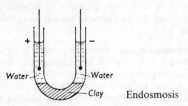

Endosmosis

endosmosis The movement of a liquid through a porous diaphragm when a potential difference is applied across it. In the example shown the two water surfaces will normally remain on the same level, but when a field is applied, the water level will rise in the anode tube.

endothermic process A process during which heat is absorbed by the system from outside. (*Compare* exothermic process.) When a nuclear process results in the absorption of energy, it is often called an *endoergic process*.

energy Symbol: E. The quantity that is the measure of the capacity of a body or a system for doing *work. When a body does work, W, its energy decreases by an amount equal to W. The energy of the body upon which it does work increases by exactly the same amount so the total energy of the system does not change. This is the *principle of conservation of energy*. The interaction of two bodies causes the transfer of energy between them. The kind of energy may be unchanged, or it may be partially or wholly changed. If two bodies are at different temperatures, it is possible for energy, Q, to be transferred from that at higher temperatures to that at lower without any apparent forces and displacements by which work could be done. This is the process *heat, which involves work on the molecular scale.

Kinetic energy, symbol: T or E_k, is the energy possessed because of motion and is equal to the work that a body would do if brought to rest with respect to a certain observer. In classical physics it can be shown that a particle of mass m with speed v has translational kinetic energy $T = \frac{1}{2}mv^2$, while a rotating body with moment of inertia I about its axis of rotation and angular velocity ω has rotational kinetic energy $T = \frac{1}{2}I\omega^2$. In *relativity the kinetic energy of a particle is $(m - m_0)c^2$, where m is the mass as determined by the observer with respect to whom it will be brought to rest, m_0 is the rest mass, and c the speed of light. If v is small compared with c, this formula tends to the classical form.

Potential energy, symbol V or E_p, is energy possessed by a system because of the position of a body with respect to a standard. For example, if a body of mass m is raised to a height h above the ground, the potential energy is mgh, where g is the acceleration of free fall. That is, if the body returns to ground level, the gravitational field of the earth will do this amount of work on it.

Internal energy, symbol: U, is the sum of the potential energies of the molecular interactions and the kinetic energies of the molecular motions within a body. For a solid, the vibrations of the molecules (or atoms, or ions) are nearly simple harmonic and U comprises equal amounts of molecular kinetic and potential energies. For a monatomic gas, U is almost entirely molecular kinetic energy. The value of U depends primarily upon temperature. There is no simple universal rule but U generally increases with temperature and for gases is approximately directly proportional to thermodynamic temperature. The value is usually affected slightly by pressure.

The internal energy of a body may be changed by both of the processes of work or heat:

$$\delta U = W + Q.$$

Thus it is misleading to use the popular term "heat energy" for U, since it is not uniquely related to heat. Moreover, it is possible to transfer internal energy from a colder to a hotter body by doing work.

Chemical energy Energy is required to break the chemical bonds in a substance. In many reactions new stronger bonds are formed so on balance, energy is released. For example, one kilogram of methane will react with 4 kg of oxygen according to the formula

$$CH_4 + 2O_2 \rightarrow CO_2 + 2H_2O.$$

Because of the very strong bonds in CO_2 and H_2O, about 56 MJ of energy is released in this reaction. The chemical energy is a property of the system as a whole; it is not a property of either constituent by itself. Strictly, the energy change depends on the change of *Gibbs function.

Nuclear energy In various nuclear reactions new bonds are formed that are stronger than those that must be broken so energy is released.

See also electric energy.

energy bands (i) *The band structure of crystals*. According to *quantum theory a single atom has a number of *quantum states, each defined by a set of *quantum numbers. If the atom is isolated, each state has a characteristic energy. Normally the electrons occupy the states of lowest energy, not more than one electron going into each state according to the *Pauli exclusion principle. The energy associated with any state can be changed by external fields.

If a large number of atoms are combined in a condensed substance, the energies become spread over certain bands by the interactions with

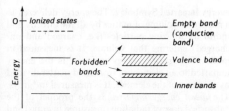

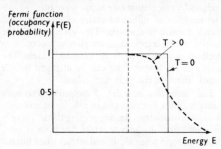

a Energies of water   Energy bands in a solid
   in a free atom   at absolute zero

neighbouring atoms. For pure crystals the system is as shown in Fig. *a*. Each state of each atom becomes a state of the crystal. Because of the very large number of states and the effects of the *uncertainty principle, the energies form a continuous distribution in the *allowed bands*. The number of states within a band is a small integral multiple of the number of atoms. Between these bands are ranges of energy called *forbidden bands* in which there are no quantum states in a pure substance.

The energies of the electrons in the inner states of the atoms are affected very little by the neighbouring atoms. These electrons remain tightly bound to their nuclei and play no part in electrical conduction.

The theory of these energy bands in solids depends on *quantum mechanics. In general, the *Schrödinger wave equation is solved for an electron moving in a varying electric potential, the periodicity of which is created by the spacing of the ions in the crystal lattice. The allowed solutions give the allowed bands of energy and the energies for which there are no solutions are the forbidden bands.

Fermi function
(occupancy
probability)

b   Energy distribution in a metal

(ii) *Energy distribution.* For a continuum the number of particles with energy between E and $E + dE$ is $N_E dE$, where $N_E = f_E\, g_E$. The quantity $g_E\, dE$ is the number of states with energy between E and dE. The *Fermi function* f_E is given by:

$$f_E = 1/\exp[(E-E_F)/kT + 1].$$

f_E is the probability that a state at energy E is occupied (Fig. *b*). At the *Fermi level* (or *Fermi energy*) E_F the value of f_E is exactly one half. Thus for a system in equilibrium one half of the states with energies very nearly equal to E (if any) will be occupied. The value of E_F varies very slowly with temperature, tending to E_0 as T tends to absolute zero. The function then becomes discontinuous, f_E

being *one* up to E_0 and *zero* above it.

The quantity g_E is zero in the forbidden bands. Fig. *c* shows schematically the form of g_E for a typical pure nonmetal. The quantum states of the valence electrons in the separate atoms form the *valence band*, which is separated by an *energy gap E_G* from the *conduction band*. If the value of E_G is greater than about 2 electronvolts, the substance is called an *insulator, if it is less, the substance is called a *semiconductor.

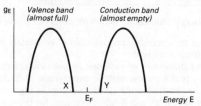

c Bands for a pure non-metal

At absolute zero the valence band would be full while the conduction band would be empty. From the form of the Fermi function it can be seen that at all attainable temperatures a small number of the states near the top of the valence band (region X in Fig. *c*) will be vacant, while a few of those near the bottom of the conduction band (region Y in Fig. *c*) will be occupied. The numbers increase rapidly with temperature, especially if E_G is small.

Metals do not have separate valence and conduction bands but a single band containing many more states than electrons to occupy them.

(iii) *Conduction of electricity.* If a potential difference is applied between the ends of a wire, the energies of the electron states are changed as illustrated in Figs. *d* and *e*. In the case of a metal there are, near the level of the Fermi energy, very many occupied states and empty states in the same locality. Thus electrons can move freely from occupied states into empty states of the same energy, giving a flow of negative charge towards the positive end. This process is initially conservative as there is a gain of kinetic energy equal to the loss of potential energy. As they move further, however, electrons undergo collisions with irregularities in the crystal structure (*see* defect; phonon) and give up energy to the vibrating ions. Thus the equilibrium energy distribution is maintained (very nearly) at all points along the wire.

In the case of a nonmetal there are relatively few electrons in the conduction band (*free electrons*). There are also a few empty states in the valence band (*holes*). The latter contribute to the conduction, as electrons can move into these states from adjacent occupied states at the same energy. The great majority of the electrons in the valence band cannot take part in conduction since there are no empty states of the same energy near them. The number of free electrons and holes in most pure nonmetals is incomparably less than the number of electrons and empty states near to the Fermi level of a metal,

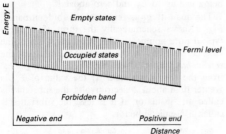

d Energy bands in a metal with potential gradient

hence the electrical conductivities of nonmetals are relatively very low.

(iv) *Absorption of radiation by nonmetals.* From Fig. *c* it is apparent that electromagnetic radiation can be absorbed if the quantum energy $h\nu$ is greater than the energy gap E_G. For example, for the semiconductor silicon $E_G \simeq 1.1$ eV hence pure silicon is transparent to infrared for frequencies up to about 2.7×10^{14} Hz ($\lambda \simeq 1.1$ μm) and absorbs at higher frequencies. The maximum quantum energy in the visible spectrum is about 3.0 eV hence substances with E_G greater than this will be transparent at all visible wavelengths. When an electron is raised into the conduction band by absorption of electromagnetic radiation, it may in certain circumstances remain there for a significant time. This causes *photoconductivity. *See also* semiconductor.

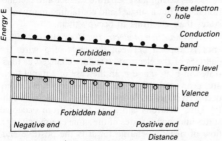

e Energy bands in non-metal with potential gradient

energy density The amount of (e.g. radiant) energy in unit volume.

energy ellipsoid *Syn.* Poinsot ellipsoid. *See* Poinsot motion.

energy equipartition *See* equipartition of energy.

energy fluence Symbol: Ψ. A quantity associated with nuclear reactions and ionizing radiation. It is the sum of the energies (excluding rest energies) of all particles that, within a time interval, are incident on a small sphere at a given point, divided by the cross-sectional area of the sphere. The *energy fluence rate* (or *energy flux density*), symbol ψ, is equal to $d\Psi/dt$.

energy flux density *See* energy fluence.

energy gap *See* energy bands.

energy imparted Symbol: ε. The energy delivered to a particular volume of matter by all the directly and indirectly ionizing particles (i.e. charged and uncharged) entering that volume. It is measured in joules. The *specific energy imparted* is the energy imparted to an element of irradiated matter divided by the mass of the element. It is measured in *grays. The *mean energy imparted* is the product of the *absorbed dose of radiation (measured in grays) and the mass of the irradiated element.

energy level The energy associated with a *quantum state under defined conditions. The term is often used to mean the state itself, which is incorrect because: (1) the energy of a given state may be changed by externally applied fields; (2) there may be a number of states of equal energy in the system. There is thus often confusion between changing the energy of a particle in a particular state and moving the particle into a different state with equal or different energy.

The electrons in an *atom can occupy any of an infinite number of bound states with discrete energies. For an isolated atom the energy for a given state is exactly determinate except for the effects of the *uncertainty principle. For typical lifetimes the fractional indeterminacy in the energy is of the order of 10^{-6} or 10^{-7}. The *ground state with lowest energy has infinite lifetime hence the energy is in principle exactly determinate. The energies of these states are most accurately measured by finding the wavelengths of the radiation emitted or absorbed in transitions between them (*see* line spectra). Theories of the atom have been developed to predict these energies by calculation (*see* atom; wave mechanics). The energies of unbound states of positive total energy form a continuum. This gives rise to the continuous background to an atomic spectrum as electrons are captured from unbound states. The energy of an atomic state can be changed by the *Stark effect or the *Zeeman effect.

The vibrational energies of molecules also have discrete values. For example, in a diatomic molecule the atoms oscillate in the line joining them. There is an equilibrium distance at which the force is zero, the atoms repel when closer and attract when further apart. The restraining force is very nearly proportional to the displacement hence the oscillations are simple harmonic. Solution of the *Schrödinger wave equation gives the energies of a harmonic oscillation as:

$$E_n = (n + \tfrac{1}{2})hf,$$

where h is the *Planck constant, f is the frequency, and n is the *vibrational quantum number*, which can be zero or any positive integer. The potential energy of interaction of atoms is described more exactly by the *Morse equation, which predicts departure from direct proportionality at large separations. Hence the oscillations are slightly *anharmonic (*see also* phonon). The vibration of molecules are investigated by the study of *band spectra. The equation shows that the lowest possible vibrational energy of an oscillator is not zero but $\frac{1}{2}hf$. This is the cause of *zero-point energy.

The rotational energy of a molecule is quantized also. According to the Schrödinger equation a body with moment of inertia I about the axis of rotation has energies given by:

$$E_J = h^2 J(J + 1)/8\pi^2 I,$$

where J is the *rotational quantum number*, which can be zero or a positive integer. Rotational energies are found from band spectra.

The energies of the states of the *nucleus can be determined from the spectra of *gamma radiations and from various *nuclear reactions. Theory has been less successful in predicting these energies than those of electrons in atoms because the interactions of nucleons are very complicated. The energies are very little affected by external influences but the *Mössbauer effect has permitted the observation of some minute changes.

See also energy bands; X-ray spectrum.

engineer's unit of mass *See* slug.

enhancement mode An operating mode of *field-effect transistors in which increasing the magnitude of the *gate bias increases the current.

enrich To increase the *abundance of a particular isotope in a mixture of the isotopes of an element. Applied to a nuclear fuel it means, specifically, to increase the abundance of *fissile isotopes.

The *enrichment* is the proportion of atoms of a specified isotope present in a mixture of isotopes of the same element, where this proportion is greater than that in the natural mixture. This is often expressed as a percentage. Applied to uranium, the amount of ^{235}U present is often expressed as a multiple of the natural abundance (C_0) of 0.71%. For example, $2C_0$ material contains 1.42% of fissile atoms and is said to be twice enriched. In isotope separation, enrichment is used to mean either the *enrichment factor* or the enrichment factor minus one (sometimes called the *degree of enrichment*).

The enrichment factor (E) of a material is the ratio of the abundance of a particular isotope in a product to its abundance in the starting material.

enthalpy Symbol: H. The former names *heat content* and *total heat* are not consistent with modern thermodynamic nomenclature. A thermodynamic function of a system equal to the sum of its energy (U) and the product of its pressure (p) and volume (V), i.e.

$$H = U + pV.$$

For a *reversible change at constant pressure the work done by the system is equal to the product of pressure times the change of volume. The heat absorbed in such a process is thus equal to the increase of enthalpy of the system. For example, many chemical processes are carried out in the atmosphere and occur at constant pressure. The heat change involved is thus an enthalpy of reaction – for *exothermic processes the enthalpy change is taken to be negative. The *heat capacity of a system at constant pressure is given by:

$$C_p = (\partial H/\partial T)_p.$$

The *Joule–Kelvin effect is a nonreversible flow process in which the initial and final states have the same enthalpy.

entrance port The image of the field *stop by the parts of an optical system lying in front of it. The entrance port is a stop image that subtends the smallest angle at the centre of the *entrance pupil and serves to indicate the extent of an object that can be seen through the instrument. For vision through a simple lens, the rim of the lens itself will be the entrance port. Dependent on the size of the entrance pupil and the distance between the latter and the entrance port is the *vignetting at the edges of the field. Roughly the illumination of the field, in accordance with the above definition, has dropped to half value. *See* apertures and stops in optical systems.

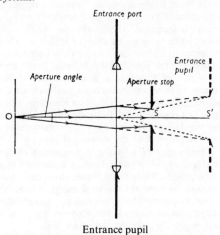

Entrance pupil

entrance pupil The image of the *aperture stop* (*see* apertures and stops in optical instruments) by the parts of an optical system lying in front of it. Rays filling the entrance pupil from a point in the object will just fill the aperture stop as they pass through the instrument. Its position may depend on the position of the object. It is that image (of a stop by the preceding parts of the system) that subtends the smallest angle at the object. Commonly it is the rim of the objective or the actual stop lying in front of a lens. The brightness of the image is determined by the size of the entrance pupil. It serves a similar purpose for the incident light that its conjugate through the whole instrument (*exit pupil) serves when delivering light to the image. For vision through a simple lens, the entrance pupil is the image of the pupil of the eye (more precisely the image of the entrance pupil of the eye) by the lens.

entropy Symbol: S. A property of a system that changes, when the system undergoes a *reversible change, by an amount equal to the energy absorbed by the system (dq) divided by the thermodynamic temperature T, i.e. $dS = dq/T$.

Entropy, like other thermodynamic properties such as temperature and pressure, depends only on the state of the system and not on the path by which that state is reached. It is a quantity with an arbitrary zero, only changes in its value being of significance. (*See also* third law of *thermodynamics.)

The concept of entropy follows from the application of the second law of *thermodynamics to the *Carnot cycle. The *thermodynamic temperature is so defined that the entropy given out at the lower temperature is equal to that taken in at the higher value. As for any completed cycle the entropy of the working substance is unchanged, hence there is no overall change in the entropy of the complete system. In any real cycle there will be some degree of irreversibility and the *efficiency will be less than for the Carnot cycle. Hence more heat is discharged at the lower temperature and the entropy output is greater than the input. Thus the entropy of the whole system is increased. All real changes are to some extent irreversible so all changes within a closed system cause an increase in entropy (*see* heat death). In applying this principle it is essential to consider every part of a system. It is quite usual for the entropy of some body to decrease, but this is always accompanied by a greater increase elsewhere. Contrary to a popular belief, living organisms do not in any way violate the second law of thermodynamics and the principle of the increase of entropy.

The entropy of a system is a measure of the unavailability of its internal energy to do work in a cyclic process. Thus if two bodies at unequal temperatures have the same internal energy, that at the higher temperature has lower entropy. More of this body's internal energy is available to do work in a heat engine than that of the cooler body.

The *Boltzmann entropy theory* relates entropy to the thermodynamic "probability", W, where W is the number of microscopically distinct states of a system that give the same macroscopically determinable state. In quantum terms W is the number of solutions of the *Schrödinger wave equation for the system giving the same distribution of energy. The relationship is:
$$S - S_0 = k \ln(W/W_0),$$
where k is the *Boltzmann constant and S_0 and W_0 are the entropy and probability for a standard condition. According to the third law of *thermodynamics the entropy of a perfect crystal at *absolute zero is zero. This corresponds to the idea that there is only one way in which this condition of lowest energy can be realized, i.e. taking this as a standard, W_0 is unity and as $\ln 1 = 0$ so $S_0 = 0$. A solid that is not perfect has a certain amount of disorder that gives it a *configurational entropy*. In general, entropy can be thought of as a measure of the molecular disorder of a system. This must not be confused with disorder on a macroscopic scale, a statue has the

same entropy as an irregular block of stone of the same mass.

Consider N molecules arranged in N places to form a condensed substance. If the body is solid, the placing of the first molecule leaves only $N-1$ places for the second, then there are $N-2$ available for the next and so on. Thus the solid can be assembled in $N!$ ways, so $W_S = N!$ For the liquid, W_L N^N as placing one molecule does not restrict the placing of subsequent ones. The entropy increase on melting is thus:
$$k(\ln W_L - \ln W_S) = k(N \ln N - \ln N!).$$
Since $\ln N! = N(\ln N - 1)$ for very large numbers, this change is kN. If N is the *Avogadro constant, the entropy of melting is equal to R, the *molar gas constant. In practice melting involves other changes (e.g. expansion) and experimental values of the *molar latent heat of fusion divided by the thermodynamic temperature of the melting point are typically 30% above this value.

Suppose an ideal gas expands to double its volume adiabatically and without external work (*see* Joule experiment). The number of quantum states available to each molecule is doubled, so the entropy change per molecule is $k \ln 2$. Hence for a mole $\Delta S = R \ln 2$. The molecular disorder has increased as there is more space in which each molecule can move. If two vessels containing different gases at the same temperature and pressure are connected, the gases diffuse into each other, the *entropy of mixing* being calculated as above, each molecule being free to move in the whole space. The entropy of the mixture is thus larger than that of the system in which the gases are separate, which is more ordered. This illustrates the principle of increase of entropy; gases will mix thereby increasing entropy, a mixture cannot decrease its entropy by separating the constituents. Separation can only be brought about by some interaction that increases the entropy of another body.

See also Carnot–Clausius equation; statistical mechanics.

Eötvös, Roland, Baron von (1848–1919) Hungarian physicist. He invented the torsion balance named after him, and also carried out investigations on capillarity, gravitation, and terrestrial magnetism.

Eötvös law (1886) An equation relating the surface tension and temperature of a liquid that experiment has shown to be of little value. Surface tension γ and temperature T are related by the equation:
$$\gamma = \gamma_0 (1 - bT)^n,$$
where for unassociated liquids $n = 1.2$, b is the reciprocal of the critical temperature in kelvin.

Eötvös torsion balance An instrument devised by Eötvös in 1888. It is used to determine the rate of change of g with horizontal distance. It is also used in *geophysical prospecting (*see also* g), when the density of the mineral deposit sought differs considerably from that of the surrounding strata.

It consists of a beam supported by a fine torsion wire carrying at its extremities two weights at different vertical heights, the lower of which is attached to the beam by a second torsion wire. The whole is enclosed in a metal case that can be rotated about a vertical axis.

epicadmium neutron A neutron with an energy just above the limit below which it would be absorbed by cadmium, i.e. about 0.5 eV (8×10^{-20} joule).

epicentre *See* seismology.

Epicurus (342–271 BC) Greek philosopher who revived and extended the atomic theory of *Democritus. Aristotle and the Christian theologists denied this theory because it was thought to have atheistic implications, and it did not come to light again until the seventeenth century.

epidiascope A projector that can be changed by a simple operation from one projecting slide transparencies to one that can project images of flat opaque objects (e.g. pictures and printing on paper). *See* episcope.

episcope A projector for throwing images of opaque objects onto a screen (generally flat objects, e.g. drawings). It is necessary to illuminate the object with high intensity over its area, and so requires sources of high luminous intensity, efficient condensers (generally reflectors), and the use of well-corrected and very large aperture projecting lenses.

epistemology *See* metaphysics.

epitaxial layer *See* epitaxy.

epitaxial transistor *See* transistor; diffused junction.

epitaxy A method of growing a thin layer of material upon a single-crystal substrate so that the lattice structure is identical to that of the substrate. The technique is extensively used in the manufacture of *semiconductors when a layer (known as the *epitaxial layer*) of different conductivity is required on the substrate.

epithermal neutron A neutron with an energy just above thermal energies, often taken to be in the range from 10^{-2} to 10^2 eV (1.6×10^{-21} to 1.6×10^{-17} joule). Neutrons in the middle of this range (or a logarithmic scale) have energies of the same order of magnitude as the energies of chemical bonds.

epoch 1. The precise time to which a set of astronomical data are referred or from which a new period or era is measured. The epoch for the coordinates right ascension and declination (*see* celestial sphere) is given by a date: α_{1970}, δ_{1970}. This indicates the astronomical conditions under which the measurements were made.
2. *See* simple harmonic motion.

epoxy resins Synthetic polymers that are widely used as structural plastics, surface coatings, and adhesives and for encapsulating and embedding electronic components. They are characterized by low shrinkage on polymerization, high strength, and good adhesion and chemical resistance.

EPR Abbreviation for electron paramagnetic resonance. *See* electron spin resonance.

EPROM Abbreviation for erasable programmable read-only memory, i.e. erasable *PROM. A type of semiconductor computer memory that is fabricated in a similar way to *ROM (read-only memory). The contents, however, are added after rather than during manufacture and then if necessary can be erased and rewritten, possibly several times. The contents are normally erased (i.e. reset to their nonprogrammed state) by exposure to ultraviolet radiation. The EPROM can then be reprogrammed using an electronic device known as a PROM programmer.

equalization In electronics, the introduction of networks that compensate for a particular type of *distortion over the frequency band required and hence reduce distortion in a system.

equalizer 1. *Syn.* equipotential connection. A low-resistance connection made between points on the winding of an electrical machine to ensure that such points are always at the same potential. In d.c. machines, if the e.m.f.s generated in all the parallel circuits in the armature are not exactly equal, a circulating current flows between the circuits via the brushes, and as a result the brushes may be overloaded. The fitting of equalizer rings to the winding provides a low-resistance connection for the circulating currents, which are thus diverted from the brushes.
2. A network designed to provide *equalization.

equation of condition *See* conditioned observations.

equation of continuity *See* continuity principle.

equation of energy *See* relativity.

equation of time *See* time; sundial.

equations of state *Syn.* characteristic equations. Equations showing the relationship between the pressure (p), volume (V), and thermodynamic temperature (T) of a substance.

(1) For a homogeneous fluid, the most familiar equation is:
$$pV = nRT,$$
where n is amount of substance and R the molar gas constant. This equation only holds good for an ideal gas, i.e. a gas that is made up of massive point particles that exert no forces on each other. This is shown by the fact that if the equation holds good over all ranges of pressure, when the pressure becomes infinitely great the volume is zero. To allow for the finite volume occupied by the particles, the equation should be written:
$$p(V - b) = nRT,$$
where b is the least volume into which the particles can be forced by an indefinitely large pressure. Further, the attractive forces between the particles result in a decreased pressure exerted on the walls of the containing vessel and the equation must be altered to:
$$(p + k)(V - b) = nRT.$$

Many characteristic equations have been proposed, the most famous of them being the *van der Waals equation of state in which $k = a/V^2$, so that the equation becomes:
$$(p + a/V^2)(V - b) = nRT.$$
The *Dieterici equation is:*
$$p(V - b) = nRT \exp(-a/RTV).$$
There are certain tests that may be applied to these characteristic equations. For example, the critical specific volume v_c, is equal to about four times the liquid specific volume (or volume of unit mass). Now the constant, b, is approximately equal to the liquid specific volume, so that $v_c = 4b$. Again, the quantity $RT_c/p_c v_c$ is roughly constant for unassociated fluids and is equal to $15/4$. Hence if the critical constants of the fluids under test are known, they should satisfy the values $v_c = 4b$ and $RT_c/p_c v_c = 15/4$. The van der Waals equation can be rearranged in the form:

$$v^3 - v^2\left(b + \frac{RT}{p}\right) + \frac{a}{p}v - \frac{ab}{p} = 0.$$

Let the three roots of this equation be V_1, V_2, V_3. Then we know that:
$$V_1 + V_2 + V_3 = b + RT/p,$$
$$V_1 V_2 + V_2 V_3 + V_3 V_1 = a/p,$$
$$\text{and } V_1 V_2 V_3 = ab/p.$$
If we draw the isothermals, that is, the curves connecting p and V, for different temperatures, T_1, T_2, T_3, etc. at a certain temperature (called the *critical temperature), the isothermals show a point of inflection, and at this temperature the three roots are all equal and the value of these is V_c. Hence we have
$$3v_c = b + RT_c/p_c;$$
$$3V_c{}^2 = a/p_c;$$
$$v_c{}^3 = ab/p_c,$$
which gives $v_c = 3b$; $p_c = a/27b^2$; $T_c = 8a/27Rb$. It will be seen that instead of $v_c = 4b$, van der Waals gives as a result $v_c = 3b$, and his value for $RT_c/p_c v_c = 8/3$. If we introduce the notion of reduced values, that is express the pressure as a fraction of the critical pressure and similarly for the volumes and temperatures, we obtain the reduced pressure $l = p/p_c$, reduced volume $m = V/v_c$ and reduced temperature $n = T/T_c$, which substituted for p, V, and T in van der Waals equation gives us the *reduced equation of state:
$$(l + 3/m^2)(3m - 1) = 8n.$$
In this equation, from which all terms expressing the individuality of the fluid have disappeared, it follows that one diagram can express the behaviour of all fluids and in particular if two fluids have the same values for the reduced pressure and volume, they will have the same values for the reduced temperature and they are said to be in corresponding states. The normal boiling points of unassociated liquids are approximately corresponding temperatures and it is for this reason that the comparison of physical properties of the liquids at their boiling points is most likely to yield valuable results.

(2) For a solid, using the *virial law of Clausius, the equation of state may be written:

$$pV + G(U) = -E[d(\log v)/d(\log V)],$$
where $G(U) = V(d/dV)W(U)$, and $W(U)$ is the potential energy per mole of the crystal when the atoms are at rest in their mean positions; v is the frequency of oscillation of an atom about its mean position; and E is the total energy of the oscillations given by
$$E = \int_0^T C_m \, dT,$$
where C_m is the molar heat capacity at constant volume. Debye deduced an equation of state for solids based on thermodynamics and statistical mechanics modified to include the quantum theory:

$$pV + V\frac{d\Phi}{dV}$$
$$= -\frac{d \log \theta}{d \log V} 9RT\left(\frac{T}{\theta}\right)^3 \int_0^{\theta/T} \frac{\xi^3}{e^\xi - 1} \, d\xi,$$

where θ is the characteristic temperature of the body and Φ is the increase in free energy when the body is compressed at absolute zero from a volume V_0 to a volume V. ξ is defined as hv/kT, where h is the Planck constant, k is the Boltzmann constant, and v is the frequency of the vibrations. This equation is essentially the same as that of Clausius. *Grüneisen's law follows from this equation as a first approximation.

equatorial mounting *See* astronomical telescope.

equatorial (planes, rays, section) *See* sagittal.

equilibrant A single force (if one exists) capable of balancing a given system of forces. Thus, if both system and equilibrant are applied to a body, it will be in equilibrium. (If the system of forces is equivalent to a couple together with a single force, there is no equilibrant.) *Compare* resultant.

equilibrium 1. The condition existing in a system of coplanar forces when the algebraic sums of the resolved parts of the forces in any two directions are both zero and the algebraic sum of the moments of the forces about any point in their plane is zero. If the system of forces is not coplanar, then the same results must hold between the components of the forces lying in any plane and also for the components lying in two other different planes.

As any system of forces can be reduced to a single force and a single couple, the condition of equilibrium is also that these shall both vanish.

2. A body is in stable, unstable, or neutral equilibrium according to whether the forces brought into play following a slight displacement tend to decrease, increase, or not affect the displacement respectively. Neutral equilibrium is sometimes called *indifferent* equilibrium. These concepts of equilibrium and of stability can be generalized to apply to other physical systems, e.g. a system of electric charges in a potential field; a soap bubble on the end of a tube connected to a reservoir of air.

In general, the potential energy of a system is a minimum or a maximum if the equilibrium is stable or unstable respectively. Neutral equilibrium may turn out to be either stable or unstable equilibrium if a large enough displacement be applied. (*See* least-energy principle.)

3. *See* buoyancy, for equilibrium of floating bodies, and centrifugal moment, for balance of rotating bodies.

equilibrium constant Symbol: K. *See* mass action, law of; Ostwald's dilution law.

equimomental Two distributions of matter are equimomental if they have the same total mass and the same principal moments of inertia and mass centre; e.g. a hoop of mass m and radius $a/\sqrt{2}$ is equimomental with a circular plate of mass m and radius a. Equimomental bodies, acted on by identical force systems behave in the same way.

equinoxes 1. The two points at which the ecliptic intersects the celestial equator (*see* celestial sphere).

2. The two days of the year when the sun is at these points, day and night being of equal length. The *vernal equinox* occurs about March 21 in the northern hemisphere (about September 23 in the southern hemisphere). The *autumnal equinox* occurs about September 23 in the northern hemisphere (about March 21 in the southern hemisphere). *Compare* solstice.

equipartition of energy The principle of equipartition of energy, based on classical statistical mechanics and enunciated by Boltzmann, states that the mean energy of the molecules of a gas is equally divided among the various *degrees of freedom of the molecules. The average energy of each degree of freedom is equal to $\frac{1}{2}kT$, where k is the Boltzmann constant and T is the thermodynamic temperature.

In the late nineteenth century the principle was extended to the vibrations of atoms in crystals and to electromagnetic radiation in a cavity (*see* blackbody radiation). Some of the results were consistent with experiment within certain conditions; for example, the principle predicts *Dulong and Petit's law for the specific heat capacities of solids, which was verified for most substances at the temperatures that were then attainable. In the case of radiation the principle led to difficulties and *Planck proposed the *quantum theory (1900) to overcome these. This led to extensive research, for example, the case of the *Nernst and Lindemann vacuum calorimeter to measure specific heat capacities at low temperatures. At the time of the first *Solvay Conference (1911) leading scientists agreed that the equipartition principle was untenable in general, although it is an admissible approximation in certain cases, especially at high temperatures.

equipollent Two systems of forces are said to be equipollent when (*a*) the vector sum of the forces of one system equals the vector sum of the forces of the other system, and, at the same time, (*b*) the sum of the moments of all the forces of one system about an arbitrary line is equal to the sum of the moments of all the forces of the other system about that line.

equipotential Having the same electric or gravitational *potential. An equipotential surface is a surface drawn so that all points on it are at the same potential. Hence no work is done in moving a small charge or mass in any direction in the surface.

equipotential connections *See* equalizer.

equitempered scale The major scale of equal temperament is a scale of eight notes to the octave, having frequencies proportional to 1, $2^{2/12}$, $2^{4/12}$, $2^{5/12}$, $2^{7/12}$, $2^{9/12}$, $2^{11/12}$, and 2, within limits detectable by the ear. The minor scales of equal temperament differ from this scale in that the ratio of the third to the keynote is $2^{3/12}:1$. The ratios of the 6th and 7th to the keynote are variable depending upon the particular minor scale in use. The chromatic scale of equal temperament has thirteen notes to the octave, the pitch interval between any two being $2^{1/12}$. *See* temperament; musical scales.

equivalent circuit An arrangement of simple circuit elements that has the same electrical characteristics as a more complicated circuit or device under specified conditions. *See also* equivalent network.

equivalent focal length The distance from a principal point (*see* principal planes and points) to its corresponding principal *focal point. With a zoom lens or variable focal length lens, the equivalent focal length is a variable. It can be considered as the ratio of the size of an image of a small distant object near the axis to the angular distance of the object in radians.

equivalent length of a magnet The poles of a magnet are not at its ends, and it is incorrect to use half the actual length in calculations. An "equivalent length" $2l$ can be chosen so that $2l$ times the pole strength is equal to the moment. This length is commonly about $5/6$ of the geometrical length for simple bar magnets.

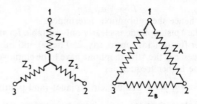

Equivalent networks

equivalent network An electrical network that may replace another network without materially affecting the conditions obtaining in the other parts of the system, but usually only at one particular frequency. An important practical example is the equivalence that may exist between *star and *delta connections of impedances provided certain relations between the impedances exist.

These two networks (*see* diagram), are equivalent if: (i) star impedances in terms of delta impedances,

$$Z_1 = \frac{Z_C Z_A}{Z_A + Z_B + Z_C},$$

$$Z_2 = \frac{Z_A Z_B}{Z_A + Z_B + Z_C},$$

$$Z_3 = \frac{Z_B Z_C}{Z_A + Z_B + Z_C}.$$

or, (ii) Delta impedances in terms of star impedances,

$$Z_A = Z_1 + Z_2 + \frac{Z_1 Z_2}{Z_3},$$

$$Z_B = Z_2 + Z_3 + \frac{Z_2 Z_3}{Z_1},$$

$$Z_C = Z_3 + Z_1 + \frac{Z_3 Z_1}{Z_2}.$$

It should be noted that, in general, the networks are equivalent only at one particular frequency. Also, in the above expressions the Z-terms are vector impedances.

equivalent piston Of a vibrating diaphragm for frequencies below its fundamental resonant frequency. A piston vibrating with the same speed amplitude as a given point on the diaphragm (usually the centre) and having an area such that the piston as a source of sound has the same strength as the diaphragm (the strength of a source of sound is the product of its area and its mean speed). The piston can also be given an equivalent mass and stiffness, and thus may represent the diaphragm completely as a vibrating body.

The piston, driven by an external alternating force, sets the air in front of the microphone in vibration and the correspondence voltage across the microphone terminals is observed. Knowing the amplitude of the piston (determined optically), the intensity of the sound I may be calculated from the formula:

$$I = \tfrac{1}{2} p_0 c a^2 n^2,$$

and hence the calibration determined.

The *pistonphone is clearly only suitable for the lower frequencies (up to, say, 250 hertz) and other devices, e.g. the *thermophone have to be employed at the higher frequencies.

equivalent plane *See* principal planes (and points).

equivalent positions (or **points**) A complete set of points in any given space-group that is obtained by performing the symmetry operations of the *space-group on a single point (x, y, z).

equivalent resistance The value of total *resistance that, if concentrated at a point in an electrical circuit, would dissipate the same power as the total of various smaller resistances at different points in the circuit.

equivalent sine wave A sine wave having the same *root-mean-square value as the given wave and also the same fundamental frequency.

erecting (lens, prism) The *Kepler telescope yields an inverted image but by using a terrestrial eyepiece can be made to provide an erect image, a more convenient arrangement for viewing terrestrial objects. The erecting lens was first used by Scheiner (1615). Such eyepieces increase the length of the instrument as they consist of four lenses; the two lenses nearer the objective constitute the erecting system proper, the remaining two form a *Huygens's eyepiece. Frequently the erecting system has unit magnification, with parallel rays between its two component lenses of equal focal length. Kepler telescopes can also yield erect images by the use of prism elements, which erect the image top to bottom and sideways. (*See* prism; binoculars; Porro prism.) Advantage is taken of such prisms to shorten the length of the tube, to increase the stereoscopic effect with binoculars, or for other specific purposes (e.g. rangefinders).

Projectors can be fitted with an erecting prism placed in front of the projector lens to erect the image of natural transparent objects placed in position between the condenser and projector lenses. A similar type of erecting prism may be arranged to rotate in periscope and gun-sight systems, to correct the tilt of the image produced by the rotation of the top prism.

erg The unit of energy in the *CGS system. 1 erg = 10^{-7} joule.

ergon A quantum of energy of an oscillator equal to the product of the frequency of oscillation and the *Planck constant.

eriometer A diffraction device for measuring the average diameters of samples of small particles or fibres (Young, 1817). The particles are thinly dusted upon a glass plate that is placed between the observer and a source provided by a small hole in a metal plate in front of a lamp. In modern work *monochromatic radiation, usually from a sodium-vapour discharge lamp, is employed. The diffraction pattern is a bright central spot surrounded by a series of light and dark rings. The radii of these rings in the plane of the source can be measured by adjusting the distance until the darker or lighter part of a ring coincides with very small holes in the metal plate at a known distance from the centre of the source. The dimensions of the particles are calculated from the radii of the rings, the distance between the source and sample, and the wavelength of the light. *See* diffraction of light.

error equation The equation
$$y = h\pi^{-1/2} \exp(-h^2(x - a)^2).$$
See frequency distribution.

errors of measurement 1. Accidental errors. In all physical measurements small errors, due to instrumental imperfections and inaccurate human judgments, always occur. It is often possible and always desirable to estimate the magnitudes of the errors in each part of the experiment and to combine these to find the likely error in the final result. The actual estimated errors in quantities that are added or subtracted should be added in finding the error in the result; the percentage errors should be added if the quantities are combined by multiplication or division. Graphical or arithmetical methods are used to combine observations of a similar type, on the grounds that a better result can be obtained than from one single observation. (*See* probable error.)

Heisenberg's *uncertainty principle shows that there is an irremovable minimum uncertainty in all physical measurements no matter how perfect the instruments or how accurate the observer.

2. Systematic errors. The foregoing comments do not apply to these errors, which must be removed by suitable design of apparatus and technique. *See* personal equation.

Esaki, Leo (*b.* 1925) Japanese physicist who discovered the *tunnel diode (*Esaki diode*, 1958). His main work has been on semiconductors.

Esaki diode *See* tunnel diode.

ESCA Abbreviation for electron spectroscopy for chemical analysis. *See* photoelectron spectroscopy.

escape speed The speed that a projectile, space probe, etc., must reach in order to escape the *gravitational field of a planet or satellite. It depends on the mass and diameter of the planet or satellite. For the earth, the escape speed is about 11 200 m s^{-1}; to escape the moon's gravitational field a speed of 2370 m s^{-1} must be attained.

The escape speed is equal to the lowest speed with which a body from remote space (such as a meteorite or returning space vehicle) approaches the planet.

ESR Abbreviation for *electron-spin resonance.

e.s.u. Abbreviation for electrostatic units. *See* CGS system of units.

etalon *See* Fabry–Perot interferometer.

eta-meson (η) An *elementary particle having zero spin, isospin, and charge, negative parity, and positive G-parity. It has a mass of 549 MeV/c^2.

etched figures Minute pits, bounded by small faces, that are formed on crystal surfaces by treatment with solvents. These are extremely useful in helping to determine the symmetry of crystals.

ether (or **aether**) A now-discarded hypothetical medium once thought to fill all space and to be responsible for carrying light waves and other electromagnetic waves. In order to facilitate description and to provide a physical explanation of various phenomena involving action at a distance, electricity, magnetism, transmission of light, and other radiations, a medium was postulated with mechanical properties adjusted to provide a consistent theory. For the transmission of electromagnetic radiation, it was assumed to pervade all space and matter, to be extremely elastic yet extremely light, to transmit transverse waves with the speed of light, to have a greater density in matter than in free space. Al-

though the existence of an ether is now regarded as an unnecessary assumption, the concept was partially responsible for the major advances in physical optics including the theory of relativity.

In 1887, Michelson and Morley attempted to measure the motions of the earth through the ether (ether drift) by an optical experiment. (*See also* Michelson–Morley experiment.) No such motion was detected and the result is the basis of the special theory of *relativity.

Ether thermoscope

ether thermoscope A simple instrument for the detection of radiant heat consisting of two bulbs, one of which (A) is blackened, connected by a tube. Air is removed from the apparatus and ether fills the tube CD. Radiation is more readily absorbed by A than B thus increasing the vapour pressure in A and forcing the liquid meniscus up the tube D and down the tube C.

Ettinghausen effect The establishment of a difference of temperature between the edges of a plate along which an electric current is flowing when a magnetic field is applied at right angles to the plane of the plate. The effect is very small, and for copper, platinum, and silver is unappreciable.

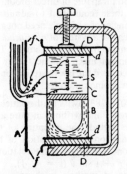

Thermal conductivity of crystals

Eucken's experiments (1911) **1.** He determined the specific heat capacity at constant volume (c_v) for hydrogen at low temperatures by a method similar to that used by Nernst (*see* Nernst and Lindemann vacuum calorimeter) for solids. The gas is contained in a thin-walled steel vessel wound with a constantan coil upon which a measured amount of work is done, the resulting rise of temperature of the gas

Euclid

being measured by a lead resistance thermometer. The method is only feasible at low temperatures where the thermal capacity of the metal vessel falls off so that it is of the same order as that of the gas. His values showed that at temperatures below 60 K hydrogen behaves as a monatomic gas, since the molar heat capacity at constant volume is about 12 joules per mole per °C. Moreover, experiments performed below 20 K suggest that in this region the molar heat capacities of hydrogen and helium may fall below the minimum value $3R/2$ possible on the classical theory. This phenomenon is known as degeneration of gases (see degenerate gas) and is explained on the quantum theory by Dirac.

2. He also determined the thermal conductivity of simple crystals at low temperatures by a modification of *Lees' method and showed that the thermal conductivity of such bodies is inversely proportional to the thermodynamic temperature. The apparatus is contained in a copper vessel V shown on its side in the diagram, S being the specimen and C a copper block enclosing a heating coil, resting on a wooden block B. Copper blocks D soldered to the vessel make contact with plates d, all surfaces being flat so that good contact is achieved. The copper lid A is soldered to V at f and the whole apparatus immersed in a low-temperature bath. Thermocouples measure the temperatures of C and d and the junctions of a thermocouple inserted in two holes in the specimen give the temperature gradient in the specimen when the steady state is reached. The corrections for the temperature drop at the contact faces and for the loss of heat by conduction from the sides of the specimen are obtained from experiments carried out with the apparatus filled in turn with air, hydrogen, and carbon dioxide.

Euclid (*c.* 330–275 B.C.) Greek mathematician. *Euclidian geometry* (as expounded in Euclid's "Elements of Geometry") is founded on axioms, regarded as self-evident and incapable of direct proof. By abandoning one of these (equivalent to admitting that the sum of the angles of a triangle need not add to 180°), one may achieve a system of *nonEuclidean geometry*, as did Riemann in 1854, for example. The "Elements" also contain much on the theory of numbers. Euclid discovered the laws of reflection in optics and wrote on geometric optics.

Euler, Leonhard (1707–1783) Swiss mathematician. His extensive researches and writings in pure and applied mathematics include the invention of the beta and gamma functions, the invention of the *calculus of variations, the proof of the relation
$$\exp(i\theta) = \cos\theta + i\sin\theta \ (i = \sqrt{-1}),$$
and the introduction of symbols, such as e, for certain important numbers. His work in astronomy introduced the concept of moving axes (see Eulerian angles). *Euler's constant* (γ) is the limit as $n \to \infty$ of
$$\left\{1 + \tfrac{1}{2} + \tfrac{1}{3} + \cdots + \frac{1}{n-1}\right\} - \log_e n,$$
and has the value 0.577....

198

Euler equations The three differential equations of motion of a rigid body (*a*) relative to the centre of mass, using the principal axes of the body through the centre of mass as coordinate axes, or (*b*) about a fixed point using the principal axes through this point as coordinate axes. They are:
$$A(d\omega_x/dt) - \omega_y\omega_z(B - C) = G_x;$$
$$B(d\omega_y/dt) - \omega_z\omega_x(C - A) = G_y;$$
$$C(d\omega_z/dt) - \omega_x\omega_y(A - B) = G_z.$$
A, B, and C are the principal moments of inertia, ω_x, ω_y, and ω_z are the components of angular velocity, and G_x, G_y, and G_z are the components of the applied torque about the principal axes OXYZ.

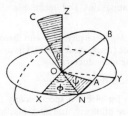

Eulerian angles

Eulerian angles A set of three angles (ω, ϕ, ψ) particularly useful in describing the position of a body moving about a fixed point O (*see* diagram). Cartesian axes, OABC, are fixed in the body (OC usually being an axis of symmetry, e.g. the axis of a top) and the motion is described relative to fixed Cartesian axes OXYZ (OZ is usually vertical). θ is the angle between the axis of the body OC and the axis OZ. The plane OAB in the body intersects the plane XOY (usually horizontal) in the *nodal line* ON. The angle $\phi = X\hat{O}N$ measures the *precession* of the axis OC around the vertical *precession axis* OZ. $\psi = A\hat{O}N$ measures the rotation of the body about its own axis OC. Variations in θ are referred to as *nutation*.

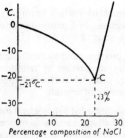

Cryohydric point

eutectic A mixture of two substances that solidifies as a whole when cooled, without change in composition. The *eutectic point* is the temperature at which the eutectic mixture solidifies. For aqueous salt solutions this point is termed the *cryohydric point* and is represented by the point C on the diagram. *See* alloy. *Compare* dystetic mixture.

eutectic composition *See* alloy.

eutectic point *See* eutectic; alloy.

evaporation 1. The conversion of a liquid to a vapour at a temperature below the boiling point. The process involves cooling of the liquid because molecules in the liquid state have negative potential energies resulting from intermolecular interactions.

2. The conversion of a substance, usually a metal, into a vapour at high temperatures, either from the liquid state or by sublimation from the solid metal. It is used for producing thin films of metal, used in *transistors and in studies of surface properties.

evaporator The part of a refrigerating plant in which the liquid refrigerant is evaporated, taking its latent heat from its surroundings.

even-even nucleus A nucleus that contains an even number of protons and an even number of neutrons. Well over a half of all stable nuclides have even-even nuclei.

even-odd nucleus A nucleus that contains an even number of protons and an odd number of neutrons. About a fifth of stable nuclides have even-odd nuclei.

even parity See wave function.

event A point in a *four-dimensional continuum, defined by three coordinates of space and one coordinate proportional to time.

Eve's constant The number of ions per cubic centimetre formed each second at STP 1 cm from a source of radium in equilibrium with 1 curie of its emanation.

Ewing, Sir James A. (1855–1935) Brit. physicist and engineer, who was responsible for many advances in experimental magnetism and for the theory of molecular magnets. See Ewing's theory of magnetism; hysteresis tester.

Ewing's hysteresis tester See hysteresis tester, Ewing's.

Ewing's theory of magnetism The theory that the individual atoms or molecules of ferromagnetic substances act as small magnets. In the unmagnetized state of the substance, these elementary magnets arrange themselves in closed chains so that the net effect of their poles externally is zero. Magnetization is produced by a realignment of the elementary magnets with their magnetic axes in the direction of magnetization, and saturation is reached when all are so aligned. The substance is prevented from following the changes in the magnetizing field owing to the force necessary to break up the molecular chains, thus explaining the phenomenon of magnetic hysteresis. This theory has been partially confirmed by modern investigations. See ferromagnetism.

exa- Symbol: E. A prefix denoting 10^{18}; for example, one exasecond (1 Es) equals 10^{18} seconds.

excess conduction In a *semiconductor, conduction due to electrons that are not required to complete the chemical bonding of the semiconductor and are therefore available to conduct charge. These electrons usually come from a donor impurity.

excess voltage See overvoltage.

exchange force A force acting between particles due to the exchange of some property. In *quantum mechanics such forces can arise when two interacting particles can share some property: for example, the *covalent bond responsible for the binding of the molecular hydrogen ion results from the two protons sharing an orbital electron. Alternatively, this electron may be regarded as being continually exchanged between the protons. In particle physics an exchange of particles, known as gauge bosons (see gauge theory), is now considered responsible for the four fundamental interactions – *strong, *electromagnetic, *weak, and *gravitational interactions. See also ferromagnetism.

exchange relation The statement
$$(pq - qp) = (h/2\pi i),$$
in which p and q are matrices replacing momentum and positional coordinates in *matrix mechanics. It replaces the old Wilson–Sommerfeld type of quantum condition. See quantum theory; quantum mechanics.

excitation 1. The addition of sufficient energy to an atom, molecule, etc., to change it to a state of higher energy. See excitation energy; collective excitation; quasiparticle.

2. The production of magnetic flux in an electromagnet by means of a current in a winding. The current is referred to as the *exciting current*.

3. The application of an electrical signal to drive a device such as an amplifier, tuned circuit, or piezoelectric oscillator.

excitation energy Syn. critical potential. The energy required to change an atom or molecule from one quantum state to another of higher energy. It is equal to the difference in energy of the states and is usually the difference in energy between the *ground state of the atom and a specified *excited state.

excitation loss The power loss in an electrical machine due to the flow of direct current in the exciting circuit. In addition to the power loss in the exciting winding itself, the term usually includes the loss in any control rheostat provided in the circuit.

excitation purity See chromaticity.

excited state The state of a system, such as an atom or molecule, when it has a higher energy than its *ground state. See also excitation energy.

exciter A small d.c. generator that supplies the current for the field coils of a larger generator. See excitation.

exciton An electron in combination with a *hole in a crystalline solid. The electron has gained sufficient energy to be in an *excited state and is bound by electrostatic attraction to the positive hole. The

199

exciton may migrate through the solid and eventually the hole and electron recombine with emission of a photon.

exclusion principle *See* Pauli exclusion principle.

exclusive OR gate *See* logic circuit.

exitance 1. *See* luminous exitance. **2.** *See* radiant exitance.

exit port Of an instrument. The image of the field stop by the portions of the instrument lying in front of it. *See* apertures and stops in optical systems.

exit pupil Of an instrument. The image of the aperture stop (*see* apertures and stops in optical systems) by the portions of the instrument lying behind it. It is the image of the *entrance pupil by the whole instrument. Rays filling the entrance pupil finally emerge from the instrument filling the exit pupil, which may be real or virtual. In the *Kepler telescope the exit pupil is real (it is the image of the objective mount by the eyepiece, i.e. the *eye ring*, or *Ramsden circle*). The observing eye should be placed with its entrance pupil coincident with the exit pupil of the instrument in order to avoid *vignetting and reduction of field. The ratio of entrance-pupil diameter to exit-pupil diameter is equal to the magnifying power of the instrument. Note that the exit pupil of a Galilean telescope (*see* refracting telescope) is virtual and lies inside the tube. For maximum illumination of the image through the Kepler telescope, the eye's entrance pupil should at least equal the diameter of the instrument's exit pupil.

The blur circles of inaccurate focusing for extra axial points are determined by projection from the exit pupil through the image point lying on the corresponding chief ray.

Exner's electrometer A form of gold-leaf electrometer in which the leaves open between parallel plates, which make the field more symmetrical. A scale is attached for direct reading of the deflections. In transporting the instrument the plates can be slid together to support the gold leaves.

exoergic process *See* exothermic process.

exosphere The outermost *atmospheric layer of the earth, lying above the *thermosphere and extending from about 400 km. The upper limit of the exosphere is unknown as the density in this region is almost negligible.

exothermic process A process during which heat is evolved from the system. (*Compare* endothermic process.) When a nuclear process results in the production of heat, it is often called an *exoergic process*.

exotic atom An unstable atom in which an electron has been replaced artificially by another negatively charged particle, such as a muon, pion, or kaon. Following capture the particle drops through the atomic energy states, causing X-ray photons to be emitted, before colliding with the nucleus. Exotic atoms are studied by means of these X-rays.

expanded sweep A technique whereby the electron beam in a *cathode-ray oscilloscope is made to move at greater speed during part of its horizontal traverse across the screen.

expander *See* volume compressors (and expanders).

expanding universe Lines in the spectrum of the light from remote galaxies are shifted towards the long wavelength end by an amount that is greatest for those galaxies known to be farthest away. If this *redshift is interpreted as due to a velocity away from the earth in the line of sight, then all galaxies (beyond the Local Group) are receding from us and those galaxies that are farthest away are moving fastest. This leads to the conclusion that the distance between clusters of galaxies is continuously increasing. Thus the universe is expanding. *See* big-bang theory.

expansion *See* coefficient of expansion; adiabatic process; isothermal process.

expansion ellipsoid A figure in which the length of axis in any direction is proportional to the coefficient of thermal expansion in the corresponding direction in the crystal.

exploring coil *Syn.* search coil. A coil used for measuring magnetic flux. It is commonly used in conjunction with a *ballistic galvanometer or a *fluxmeter.

explosion vent *See* relief vent.

exponential decay The decrease of some physical quantity, usually with time, according to a negative exponential law, represented by an equation of the type $y = y_0 e^{-at}$. Examples occur in many diverse branches of physics, e.g. the fall of amplitude in damped harmonic oscillations (*see* damped vibrations), the fall in voltage of a charged capacitor leaking through a high resistance, and the fall in activity of a pure radioactive substance with an inactive daughter product.

exponential functions Any expression of the form $y = A e^{ax}$ is an exponential function but *the* exponential function is e^x, where $e = 2.718\ 281\ 8\ldots$, the sum of the infinite series:

$$1 + 1/1! + 1/2! + 1/3! + \ldots + 1/n! \ldots$$

The *negative exponential function* refers to e^{-x}. For convenience of printing, the exponential functions may be written: $\exp(ax)$, especially when the index or exponent (whence the name) is a complicated expression. These functions are solutions of the differential equation $dy/dx = ay$, where a is a constant.

exponential horn An acoustic *horn whose cross-sectional area increases from throat to mouth according to the relation

$$S_x/S_0 = \exp mx,$$

where S_x and S_0 are the cross-sectional areas distance x from the throat and at the throat, m is a constant determining the rate of flare and the theoretical cut-off frequency.

exponential law The growth or decay of a physical quantity in accordance with a formula of the type $y = A\,e^{ax}$, in which A and a are constants and the variable x may be a distance, or a time, or other convenient coordinate. There is growth if a is positive and decay if it is negative.

exponential series The series:
$$e^x = 1 + x/1! + x^2/2! + x^3/3! + \ldots + x^n/n! + \ldots$$
to infinity. *See also* exponential functions.

exposure 1. Symbol: H. The product of the *illuminance, or the *irradiance, and the time, Δt, for which the material in question is illuminated or irradiated. (*See* light exposure; radiant exposure.) The quantity Δt is the *exposure time* and should not be confused with the term exposure.
2. *See* dose.

exposure meter *Syn.* light meter. A photographic instrument that measures light intensity by means of a *cadmium sulphide cell or *selenium cell. For a particular type of film it indicates the *f-number required for a given shutter speed, or vice versa, to give the correct *exposure. A similar instrument giving a measurement in terms of the actual light intensity rather than as the required f-number for a given shutter speed is called a *light meter*.

extensive air shower *Syn.* Auger shower. *See* cosmic rays.

extensometer A device for measuring the small change in length of an arbitrary length of a sample undergoing strain. The instrument utilizes a micrometer screw, mechanical lever, optical lever, or a rack and pinion, for magnification and measurement of the very tiny total displacements (e.g. 10 μm) involved.

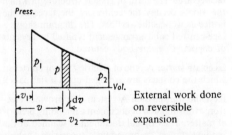

External work done on reversible expansion

external work Work done by a substance expanding against an external resistance. For a *reversible change the work done is:
$$\int_{v_1}^{v_2} p\,dv,$$
where v_1 and v_2 are the initial and final volumes and p is the pressure (*see* diagram). For a cyclic reversible process (*see* cycle) the external work done per cycle is given by the area in the diagram representing the cycle.

For real processes the system departs from equilibrium and pressure is no longer uniform in the substance or the immediate surroundings. The work done cannot then always be calculated accurately from the formula.

extinction coefficient *See* linear attenuation coefficient.

extraordinary ray *See* double refraction.

extrapolation The estimation of the value of a function for a value of the variable lying outside the range of those for which the function is known. This may be done graphically, by extending the graph of the function beyond the actual plotted points, or by calculation as for *interpolation. Extrapolation is of necessity less accurate than interpolation.

extremely high frequency (EHF) *See* frequency band.

extrinsic semiconductor A *semiconductor in which the charge *carrier concentration is dependent upon impurities or other imperfections. *Compare* intrinsic semiconductor.

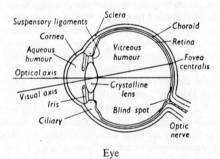

Eye

eye (human) *Anatomical and Physiological.* A roughly spherical body shown in the diagram by a nearly horizontal section. The outer protective coats are *sclera* (white), and the anterior transparent protuberance, the *cornea. The sclera is continued in the outer sheath of the optic nerve. Lining the sclera is the *choroid* (a layer of blood vessels and pigment). The choroid merges into the ciliary body (consisting of blood vessels and the *ciliary muscle), which in turn merges into the *iris (also vascular and muscle tissue). The iris forms a hole (the pupil), whose size varies. Lining the choroid from a region just adjacent to the ciliary body is the *retina, consisting of photosensitive cells (*rods* and *cones*) connected to nerve cells. Nerve fibres from these nerve cells pass over the inner surface of the retina to the optic nerve. When light strikes the retina, photochemical reactions occur in the rods and cones and nerve impulses pass along the optic nerve to the brain and give the sensation of sight.

The spherical shape of the eye is maintained by a small pressure exerted throughout the highly transparent *vitreous humour* (gelatinous) and *aqueous humour* (watery and saline). The *crystalline lens (a

201

highly transparent complicated cellular structure with a refractive index differing in different parts, and its elasticity decreasing with age), is suspended by ligaments attached to the ciliary body and lies between the vitreous and aqueous humours. The ciliary muscle, through the medium of the suspensory ligament, can relax the tension on the lens, which by virtue of its elasticity can become more convex. This describes *accommodation, enabling the eye to focus nearer objects. The normal unaccommodated eye focuses parallel rays on the retina. There is a small region of the latter – the *fovea centralis* (which is packed with cones) – for which vision is keenest (especially to recognize form and colour).

External muscles outside the eye cause it to turn so that the fovea receives the image of an object on which attention is fixed (the point of fixation). Other parts of the retina, where rods are more prevalent, are more susceptible to faint lights and movements. At the optic-nerve entrance is the blind spot. *See also* resolving power; colour vision.

eye (optics) The optical constants of the average eye are generally grouped under the title of *schematic eye*. A simplified form is the *Donders reduced eye consisting of a single spherical surface of radius 5 mm of water (index 4/3). A further simplification is the reduced equivalent lens in air with assumed thin-lens power of 60.0 D (focal length 16.67 mm), a figure that more closely agrees with latest values than Donders' focal length 15 mm.

The eye can be regarded as a variable-focus camera. When the eye is not accommodated, it may be too long for its focal power (myopic), too short (hypermetropic), or normal (emmetropic). Correction consists in throwing the focus backwards with diverging lenses in myopia, or forwards with converging lenses in hypermetropia, when accommodation is at rest. In *astigmatism, the focal power is different in different meridians of the eye. Due to hardening of the lens and consequent loss of elasticity with age, convex lenses are necessary to help the accommodation in near vision. *See* presbyopia; refraction of the eye.

eye lens The lens nearer to the eye in an eyepiece of an instrument as distinct from the more remote lens (the field lens).

eyepiece *Syn.* ocular. The single lens, doublet, or combination of lenses, acting virtually as a magnifying lens to examine the image formed by an objective. (*See* Ramsden, Huygens, Kellner, Fraunhofer eyepieces.) It is usual to arrange that the image from the objective lies in the focal plane of the eyepiece, which thus delivers parallel rays out of the instrument (infinity adjustment). *See* microscope; refracting telescope; reflecting telescope.

eye ring *See* exit pupil.

F

Fabry–Perot interferometer An interferometer that is a spectroscopic device of extremely high resolution and also serves as a laser resonant cavity. In its simplest form it consists of two parallel optically flat semisilvered or semialuminized glass plates. The plates are separated by an air gap of a few millimetres or centimetres, or of much greater length when serving as a resonant cavity. If the gap can be varied, the device is called an interferometer. If the separation is fixed, and the plates are only adjustable for parallelism, it is called an *etalon*.

A ray of light from a particular point on the source enters through the first partially reflecting plate and is multiply reflected within the gap. The rays transmitted through the second plate are focused on a screen, where they interfere to form either a bright or dark spot. All rays incident on the gap at a given angle will produce a single circular fringe. With a broad diffuse source, the interference pattern will be narrow concentric rings.

On account of its sharp fringes and high resolving power the device is used for accurate comparison of wavelengths and the study of the *hyperfine structure of spectral lines.

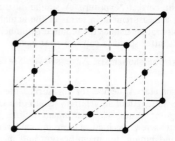

Face-centred unit cell

face-centred The form of crystal structure in which the atoms occupy the centres of the faces of the lattice as well as the vertices. The diagram shows a face-centred cubic arrangement typified by crystals of copper. *Compare* body-centred.

faceplate starter A type of electric *motor starter in which the contacts are arranged upon a flat panel.

facsimile transmission A system that provides electronic transmission of documents, including pictorial matter. The original document is scanned at the sending station, so providing a successive analysis from which an electrical representation (either analogue or digital) is produced. These electrical signals are sent over a communications channel to the receiving station, which produces a duplicate image on paper; this image is called a *facsimile*. The commercial system known as *Fax* uses the telephone network to transmit the information.

factor of merit *See* figure of merit.

factor of safety *Syn.* safety factor. The greatest estimated stress in any part of a machine or structure is called the working stress on that part. The ratio of the ultimate strength of material to the working stress is the factor of safety; it usually lies between 3 and 12.

fading In communications, variations in the signal strength at the receiver caused by variations in the transmission medium. It is usually caused by destructive interference between two waves travelling to the receiver by two different paths. If all frequencies in the transmitted signal are attenuated approximately equally, the fading is known as *amplitude fading* and results in a smaller received signal. If different frequencies are attenuated unequally, the fading is known as *selective fading* and results in a distorted signal at the receiver.

Fahrenheit, Gabriel Daniel (1686–1736) German physicist and instrument maker. He introduced mercury into the thermometer (1720) and played a notable part in standardization of temperature scales, his original scale running from 0° in a freezing mixture to 96° at blood temperature. He demonstrated the dependence of the boiling-point of a liquid on the pressure. *See* Fahrenheit scale.

Fahrenheit scale The temperature scale on which the ice point is defined as 32 °F and the steam point as 212 °F. It is no longer in use for scientific purposes.

Fahrenheit's hydrometer *See* hydrometer.

fail-safe device An automatic device that causes a system to cease operation when a failure occurs in the supply or control of power, or the overall structure is found defective. It is particularly important in nuclear equipment. The electric *fuse is a simple example of a fail-safe device.

Determination of density of saturated vapour

Fairbairn and Tate's measurements of vapour density (1860) Fairbairn and Tate devised an accurate method for the direct determination of the density of saturated vapour at various temperatures. After removing all air from the apparatus, a known small amount of the liquid is introduced above the mercury in A and a larger amount of the same liquid above the mercury in B. The whole apparatus is slowly heated, the mercury level in A being slightly higher because of the greater amount of the liquid in B. As soon as all the liquid in A has evaporated the level of the mercury in A begins to rise rapidly, since the saturation vapour pressure increases more rapidly with temperature than the pressure exerted by an unsaturated vapour. The temperature and volume of the space cut off by the mercury in A when the level begins to rise rapidly are noted and so the density of the saturated vapour is known at this temperature.

Fajans, Kasimir (1887–1975) Polish-born Amer. physical chemist who, independently of *Soddy, showed that elements consist of different isotopes. He worked in the field of radioactivity. *See* Soddy and Fajans' rule.

Fajans' rules The bond between atoms is more likely to be electrovalent than covalent if (a) the anion is small; (b) the cation is large; (c) the ionic charge is small.

fallout 1. Radioactive materials that fall to earth following a nuclear-bomb explosion. *Local fallout* is observed down-wind of the explosion after a few hours, no more than about 500 km from the source; it consists of large particles. During the month or so that follows, a *tropospheric fallout* of fine particles is observed in various locations at roughly the same latitude as the explosion. The particles that are drawn up to high altitudes often take many years before being deposited all over the surface of the earth, and are referred to as *stratospheric fallout*.
2. A substance that enters the atmosphere from a source on the earth's surface (e.g. a volcano, nuclear reactor, car exhaust, etc.) that is later deposited as particles either in the vicinity of the source or elsewhere.
3. Fallout is used with the same colloquial meaning as spin-off, i.e. technical or commercially valuable information derived directly or indirectly from an investigation, but in an incidental way (not being part of the general line of the investigation).

fall time A measure of the rate of decay of a periodic quantity. The time required for the ratio of the amplitude at a particular instant and its peak value to fall from 0.9 to 0.1.

false body *See* thixotropy.

fan-in The maximum number of inputs to a *logic circuit.

fan-out The maximum number of inputs to other circuits that can be driven by the output of a given *logic circuit.

farad Symbol: F. The *SI unit of *capacitance, defined as the capacitance of a capacitor that acquires a charge of one coulomb when a potential difference of one volt is applied. The farad is far too large a unit for ordinary use and the submultiples microfarad (μF), nanofarad (nF), and picofarad (pF) are generally employed.

Faraday, Michael (1791–1867) Brit. physicist and chemist. At one time an assistant to Sir Humphry Davy and then his successor at the Royal Institution, he was responsible for a succession of brilliant researches, notably in the field of electromagnetic induction, and in the study of dielectrics. In 1831 he made what was, in principle, the first electric generator. He carried out a great amount of work on electrolysis, and introduced the terms anode, cathode, cation, anion, ion, and ionization and formulated the two basic laws (Faraday's laws). The practical unit of capacity, the *farad, was named in his honour. *See also* Faraday constant.

Faraday cage *See* Faraday cylinder.

Faraday constant Symbol: F. The quantity of electricity equivalent to one mole of electrons, i.e. the product of the *Avogadro constant and the charge on an electron in coulombs. It is, therefore, the quantity of electricity required to liberate or deposit 1 mole of a univalent ion. Its value is
$$9.648\ 453\ 1 \times 10^4\ C\ mol^{-1}.$$

Faraday cylinder (or **cage**) **1.** A closed or nearly closed hollow conductor, usually earthed, placed round electrical apparatus to shield it from the external electric fields. (*See* electrical screening.)
2. A similar structure for the collection of a stream of charged particles (electrons or gaseous ions), usually shielded by an earthed cylinder. The inner conductor is insulated and connected to suitable detecting apparatus.

Faraday dark space *See* gas-discharge tube.

Faraday effect The rotation of the plane of polarization experienced by a beam of plane-polarized light when it passes in the direction of the magnetic lines of force, through certain substances exposed to a strong magnetic field. The effect was discovered by Faraday (1845) in a specimen of heavy flint glass (in which it is particularly marked). It also occurs in quartz and water. The Faraday effect differs from the rotation produced by crystalline media in the fact that the direction of rotation is independent of the sense in which the beam traverses the magnetic lines; thus if the beam is reflected back along its course, the rotation is doubled. The angle of rotation θ is directly proportional to the strength H of the magnetic field and to the length l of path in the substance. The quantity θ/lH, i.e. the rotation produced by traversing unit path in a unit field is called *Verdet's constant*. *See also* Kerr effects.

Faraday–Neumann law *See* electromagnetic induction.

Faraday's disc An early model of an electromagnetic *homopolar generator. It consists of a copper disc that can be rotated (usually by hand) about a horizontal axis through its centre at right angles to the plane of the disc, between the poles of a permanent horseshoe magnet. When the disc is in rotation, an e.m.f. is induced between the axis and the circumference, and can be tapped off by sliding contacts or brushes.

Faraday's laws of electrolysis The mass of any substance liberated from an electrolyte by the passage of current is proportional to the product of the current and the time for which it flows. If the same current passes for the same time through a series of different electrolytes, the masses of the different substances liberated are directly proportional to their relative atomic masses (atomic weights) divided by the charge of the ion carrying the current.

Faraday's laws of induction 1. Whenever the number of lines of magnetic induction linked with a conducting circuit is changing, an induced current flows in the circuit, which continues only so long as the change is actually taking place.
2. The direction of the induced current in the circuit is such that its magnetic field tends to keep the number of lines linked with the circuit constant.
3. The total quantity of electricity passing round the circuit is directly proportional to the total change in the lines of induction divided by the resistance of the circuit. *See* Lenz's law; Neumann's law.

Faraday's liquefaction method Faraday liquefied chlorine by generating it chemically within an L-shaped vessel with heat applied at one end to promote the reaction. The other end was cooled and the high pressure developed in the vessel was sufficient to ensure the liquefaction of much of the chlorine. He also successfully applied the method to other substances with relatively high *critical temperatures.

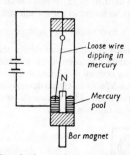

Faraday's rotation experiment

Faraday's rotation experiment A method of demonstrating the motion in a magnetic field of a current-carrying conductor. The apparatus is shown in the illustration. The wire rotates continuously around the pole of the magnet when current passes.

Faraday tube A tube of unit *electric displacement. If any small closed curve is drawn in the electric field so that the displacement across it is equal to unity, and lines of force are drawn at each point on the curve, the tubular space so obtained will be a tube of unit displacement, or unit Faraday tube. If the tube begins or ends on a conductor, the charge on the surface enclosed by the tube is equal to unit.

faradmeter An instrument for the direct measurement of *capacitance. The most common form employs a milliammeter to measure the current through the capacitance from a mains alternating-voltage supply.

far infrared or ultraviolet *See* near infrared or ultraviolet.

Farmer substandard dosemeter *See* ionization chamber.

far point A point conjugated with the retina when the *accommodation is relaxed. In *myopia, it lies in front of the eye (real), in *hypermetropia it lies behind (virtual). The focal point of the spectacle lens correcting for distance vision is made to coincide with the far point (also called punctum remotum; P.R.).

fast axis In negative crystals (e.g. calcite), the electric vibrations of the extraordinary ray (which travels faster) are parallel to the optic axis, which is then referred to as the fast axis. In positive crystals (e.g. quartz), the fast axis is at right angles to the optic axis.

fast breeder reactor A fast *nuclear reactor that breeds more *fissile material than it consumes (*see also* breeder reactor). These reactors are much more economical in the use of fuel than thermal reactors, being able to utilize some 75% of the uranium ore as it comes from the earth, compared to less than 1% in thermal reactors. After the first fuelling of a fast breeder reactor, which requires some 3000 kilograms of plutonium per 1000 megawatts of electricity produced, the net fuel requirement is a very small quantity of natural uranium.

Fuel must be arranged in a compact *core, resulting in a large heat flow from a small surface, a liquid metal (usually sodium) is used as the *coolant. Usually two sodium circuits are used: in the first the sodium passes through narrow channels round the fuel elements, becoming radioactive in the process. Heat exchangers are used to transfer heat from this radioactive sodium to another sodium circuit from which the heat is in turn extracted by a second heat exchanger in which steam is raised. The extra complexity of fast breeder reactors compared to *thermal reactors means that they are unlikely to be economic until the early decades of the next century.

fast fission Nuclear fission that is induced by *fast neutrons.

fast neutron A neutron with a kinetic energy greater than some specified value. The term is usually applied to neutrons with an energy above 0.1 MeV (1.6 × 10^{-14} joule). However, is is also applied to neutrons that have an energy greater than the fission threshold in ^{238}U, which is about 1.5 MeV. Neutrons of this energy are capable of initiating fast fission.

fast reactor *See* nuclear reactors; fast breeder reactor.

fast vibration direction The direction of the electric vector of the ray of light that travels with maximum velocity in a crystal and therefore corresponds to the least refractive index.

fathometer An instrument for determining depths under water. A sound or ultrasonic wave is emitted, and the time required for it to return after reflection from the sea bed (or equivalent) is measured. It is also called a sonic depth finder.

fatigue The progressive decrease of a property due to repeated stress, e.g. the elasticity of a metal under continuous vibration.

fault A defect in any apparatus that interferes with or prevents normal operation.

fault current Any electric current that flows from one conductor to another or from a conductor to earth as a result of a defect in the insulation or any other fault.

Faure cell A lead *accumulator cell, the plates of which consist of grids into which a mixture of lead oxide is forced under pressure. This type of accumulator plate takes a much smaller time to "form" than the Planté type in which the layer of active material is gradually built up on a solid lead plate by repeated charge and discharge of the cell. It is therefore cheaper to construct, but also less durable.

Fax *See* facsimile transmission.

F.B.R. Abbreviation for fast breeder reactor.

Fechner, Gustav Theodor (1801–1881) German philosopher and physicist.

Fechner fraction The smallest difference of luminance that can be detected by the eye when two fields of illumination are presented closely alongside, as in a photometer field; expressed as a fraction of the total brightness. It is only roughly constant for moderate degrees of brightness. *Fechner's law* (1860) is a deduction that assumes that the Fechner fraction is constant and states: the sensation of brightness varies as the logarithm of the stimulus.

Fechner's law *See* Fechner fraction.

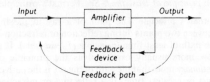

feedback The process of returning a fraction of the output signal of a signal device to the input. It usually applies to *amplifiers, the gain of the amplifier being either increased or reduced according to the relative phase of the returned signal (*see* diagram).

Feedback may occur through one electrical path (single loop) or through several paths (multiple loop). Capacitive feedback employs a *capacitor as the feedback device, and inductive feedback employs an *inductor or *inductive coupling. If the phase of the feedback is such that the input signal is increased, the feedback is known as *positive feedback*. If sufficient positive feedback is applied, the

feeder

amplifier will oscillate. If the phase is such that the input signal is decreased, the feedback is known as *negative feedback*. This is the type of feedback most commonly employed as it tends to stabilize the amplifier or reduce noise and distortion in the circuit: feedback used for this purpose may also be called stabilized feedback.

The feedback may be a function of the output current (current feedback), or voltage (voltage feedback), or both. An amplifier employing feedback is known as a feedback amplifier.

feeder 1. A system of wires or waveguides that conveys radio-frequency power between a radio aerial and a transmitter or receiver, with minimum loss.

2. An electric line that conveys electric power from a generating station to a point of a distributing network. It is not tapped at any intermediate points. *See* transmission line.

feeding point *Syn.* distributing point. A joining point of a *feeder and a *distributor.

feedthrough A contact between one layer of interconnections on a *printed circuit board and the next layer, passing through the insulating materials separating them. Usually only a double-sided board is used, but up to 12 layers have been mounted on a single board. Likewise, in an *integrated circuit with multilayer interconnections, feedthroughs can be used to make contact between one layer of interconnections and the next.

Felici balance A method of determining the mututal *inductance between the windings of an inductor by means of an alternating-current bridge circuit.

femto- Symbol: f. A prefix denoting 10^{-15}. For example, one femtometre (1 fm) equals 10^{-15} metre.

Fermat, Pierre (1601–1665) French mathematician. The founder of the modern theory of numbers, he also worked on maxima and minima and formulated the *principle of least time. See* Fermat's principle.

Fermat's principle The path of a ray in passing between two points during reflection or refraction is the path of least time (*principle of least time*). It is now more usually expressed as the principle of *stationary* time: that the path of the ray is the path of least *or* greatest time. If a reflecting or refracting surface has smaller curvature than the aplanatic surface tangential to it at the point of incidence, the path is a minimum; if its curvature is greater than the aplanatic surface, its path is a maximum.

fermi A unit of length used in nuclear physics, equal to 10^{-15} metre.

Fermi, Enrico (1901–1954) Italian-born Amer. physicist. In quantum theory, study of the application of the *Pauli exclusion principle to the statistical mechanics of systems of particles led to the *Fermi statistics* or *Fermi–Dirac statistics*, independently derived by Dirac (*see* quantum statistics). He worked on radiation, spectra, etc. and was responsi-

ble for recognition of the neutrino hypothesis in *beta decay (1934). He carried out most important studies of neutrons and reactions involving them, culminating in the demonstration of the possibilities of chain reactions and the construction of the first atomic pile (*see* Fermi pile). Nobel prize for physics 1938.

Fermi age theory An approximate method of calculating the *slow-down density of neutrons in a *nuclear reactor, based on the assumption that they lose energy continuously rather than in discrete amounts. The *age equation* is:
$$\nabla^2 q - dq/d\tau = 0,$$
relates the slowing-down density q to *neutron age, τ. Because of the assumptions made, the theory is least applicable to media containing light elements.

Fermi–Dirac distribution function Symbol: f_E. For any system of identical *fermions in equilibrium, the probability that a quantum state of energy E is occupied. It is given by:
$$f_E = 1/[\exp(\alpha + E/kT) + 1],$$
where k is the *Boltzmann constant, T is the *thermodynamic temperature, and α depends on the temperature and the concentration of particles. Since the exponential function cannot be less than zero, f_E cannot be greater than one, in agreement with the *Pauli exclusion principle.

For very low concentrations of particles, α has such values that the exponential function is much larger than one. The above equation can then be written:
$$f_E = e^{-\alpha}.e^{-E/kT} = Ae^{-E/kT}.$$
The distribution function is then indistinguishable from that of *Boltzmann's formula of classical physics. In the opposite extreme of very large concentrations, α can be written as $\alpha = -E_F/kT$, where E_F is the *Fermi level. The equation then becomes:
$$f_E = 1/[\exp(E - E_F)/kT + 1].$$
For $E = E_F$ this equation shows that $f_E = \frac{1}{2}$.

This function is of great importance for valence electrons in solids (*see* energy bands) and for matter in some exceptionally dense stellar interiors. *See* quantum statistics.

Fermi–Dirac statistics *Syn.* Fermi statistics. *See* quantum statistics.

Fermi function *See* energy bands.

Fermi gas model A model of the nucleus in which the neutrons and protons are regarded as independent particles obeying Fermi–Dirac statistics (*see* quantum statistics) but confined within a cube having a volume equal to that of the nucleus. The model is similar to the theory of electrons in solids. It is useful in describing collisions in high-energy nuclear processes.

Fermi level The value of particle energy at which the *Fermi–Dirac distribution function has the value one-half. *See also* energy bands.

fermion An *elementary particle with half-integer *spin. Fermions obey Fermi–Dirac statistics (*see* quantum statistics). All particles are either fermions or *bosons; *leptons, *quarks, and *baryons are fermions.

Fermi pile An arrangement of lumps of uranium embedded in blocks of graphite used for the liberation of energy by nuclear fission, and for the production of fissile material. This device was the first *nuclear reactor.

Ferranti effect An effect occurring in *transmission lines when the load is suddenly reduced. The charging current through the line inductance causes a sharp rise in the voltage at the end of the line.

ferrimagnetism The property of certain solid substances, such as ferrites, that show both ferromagnetic and antiferromagnetic properties. (*See* ferromagnetism; antiferromagnetism.) It is characterized by a small positive magnetic *susceptibility that increases with temperature. It is caused by the presence of two types of ion in the crystal with unequal electron *spins – arranged so that the magnetic moments of adjacent ions are antiparallel. Thus the situation is similar to that in antiferromagnetic materials with the difference that the magnetic moments are unequal.

ferrite A low-density ceramic oxide of iron to which another oxide has been added. The formula for a typical ferrite is $Fe_2O_3.XO$, where X is a divalent metal such as cobalt, nickel, zinc, or manganese. Ferrites possess insulating properties and exhibit *ferrimagnetism or *ferromagnetism according to the nature of X.

A typical example is the grade-6 nickel-zinc ferrite. A magnetic-field intensity of 15 000 amperes per metre produces a flux density of 0.12 tesla; the relative permeability is 7. The *loss factor at 10^8 hertz is less than 6000, and the ferrite can be used over the range 60 to 100 MHz. Its resistivity is about 1000 ohm metres, and the Curie point is 500 °C.

ferroelectric materials Dielectric materials, usually ceramics such as barium titanate, that in an alternating electric field develop very large values of the dielectric constant, generally in one particular direction, within a certain temperature range. They exhibit dielectric hysteresis and in most cases the piezoelectric effect. These properties are in many ways analogous to *ferromagnetism.

ferromagnetic materials Substances showing magnetic properties similar to those of iron; e.g. high magnetic susceptibility, permanent magnetism, etc. They include among the pure elements nickel and cobalt, and, in addition, many alloys. *See* ferromagnetism.

ferromagnetism A property of certain solid substances that, having a large positive magnetic *susceptibility, are capable of being magnetized by weak magnetic fields. The chief ferromagnetic elements are iron, cobalt, and nickel, and many ferromagnetic alloys based on these metals also exist.

Ferromagnetic materials display a characteristic variation of magnetization with increasing strength of magnetizing field, i.e. they exhibit magnetic *hysteresis. Their relative *permeability is much greater than unity and they achieve a maximum magnetization (magnetic saturation) at fairly low external magnetic field strengths.

The variation of magnetization with temperature is complicated. At a certain temperature, the Curie temperature, there is a change from ferromagnetism to *paramagnetism. The magnetic *susceptibility then varies according to the *Curie–Weiss law.

Ferromagnetics are able to retain a certain amount of magnetization when the magnetizing field is removed. Those materials that retain a high percentage of their magnetization are said to be hard, and those that lose most of their magnetization are said to be soft. Typical examples of hard ferromagnetics are cobalt steel and various alloys of nickel, aluminium, and cobalt. Typical soft magnetic materials are silicon steel and soft iron. The state of hardness or softness is governed by the *coercive force of the material, i.e. the reverse magnetizing field strength required to reduce the magnetization to zero after being magnetized to saturation. Hard ferromagnetics have a large coercive force and soft ferromagnetics a small coercive force.

The characteristic features of ferromagnetism are explained by the presence of *domains. A ferromagnetic domain is a region of crystalline matter, whose volume may be between 10^{-12} and 10^{-8} m^3, which contains atoms whose magnetic moments are aligned in the same direction. The domain is thus magnetically saturated and behaves like a magnet with its own magnetic axis and moment. The magnetic moment of a ferromagnetic atom results from the spin of the electrons in an unfilled inner shell of the atom. The formation of a domain depends upon the strong interatomic forces that are effective in a crystal lattice containing ferromagnetic atoms. These forces, discovered by Heisenberg, are known as *exchange forces*.

In an unmagnetized volume of a specimen, the domains are arranged in a random fashion with their magnetic axes pointing in all directions so that the specimen has no resultant magnetic moment. Under the influence of a weak magnetic field, those domains whose magnetic axes have directions near to that of the field grow at the expense of their neighbours. In this process the atoms of neighbouring domains tend to be aligned in the direction of the field but the strong influence of the growing domain causes their axes to align parallel to its magnetic axis. The growth of these domains leads to a resultant magnetic moment and hence magnetization of the specimen in the direction of the field. With increasing field strength the growth of domains proceeds until there is, effectively, only one domain whose magnetic axis approximates to the field direction. The specimen now exhibits strong magnetization. Further increases in field strength cause the final alignment and magnetic saturation in the field

ferroresonance

direction. This explains the characteristic variation of magnetization with applied field strength.

The presence of domains in ferromagnetic materials can be demonstrated by the use of *Bitter patterns or by the *Barkhausen effect.

The properties and uses of ferromagnetic materials can be illustrated by two examples:

*Mumetal is an alloy of 18% iron, 75% nickel, and 7% copper and chromium. It has a very high relative permeability – up to 8×10^5 – and a low coercive force of about 4 amperes per metre. Its *hysteresis loss is minute and it is easily magnetized. These properties make it suitable for uses involving changing magnetic flux, as in electric motors, generators, and transformers. It is also used in shielding equipment from stray magnetic fields.

Cobalt steel is an alloy of 45% iron, 35% cobalt, 9% carbon, and 11% chromium and tungsten. It has a low relative permeability, typically about 30, and a very high coercive force of about 2×10^4 amperes per metre. It has a high hysteresis loss and is very difficult to magnetize. These properties make it suitable for the construction of permanent magnets, such as those used in loudspeakers. See also antiferromagetism; ferrimagnetism.

ferroresonance Circuits containing resonating saturated iron-cored inductances may, under certain conditions, not behave according to Ohm's law, and they may fall into nonsinusoidal oscillation of frequency below that of the sinusoidal excitation. This condition, called ferroresonance, sometimes occurs in armoured cables.

Ferroxcube Proprietary magnetic materials that are cubic ferrites of composition $Me^{2+}Fe^{3+}O_4$, where Me^{2+} is a divalent metal. They have low eddy-current losses at high frequencies, high electrical resistivity, and are magnetically soft. See Ferroxdure.

Ferroxdure Proprietary magnetic materials consisting of oxidic ceramics not containing nickel or cobalt. They are magnetically hard and can be made into permanent magnets. They consist mainly of the oxide $BaFe_{12}^{III}O_{19}$ with hexagonal crystal structure and one axis of easy magnetization parallel to the hexagonal axis. See Ferroxcube.

fertile Of a nuclide. Capable of being transformed into a *fissile material in a *nuclear reactor. Uranium-238 is an example of a fertile nuclide.

Féry total radiation pyrometer A *pyrometer used for the direct measurement of temperature up to 1400 °C by measuring the total energy of radiation of all wavelengths from the source. The radiation falls on the concave mirror M (see diagram), and is focused onto a small blackened receiver R to which one junction of a thermocouple is attached. D is a polished diaphragm made in two halves inclined at a small angle to each other and is viewed by the eyepiece E through a small central hole in the mirror M. If the image of the furnace is formed on the surface of D, the two halves of the image seen in E

208

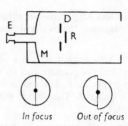

In focus   Out of focus

Féry total radiation pyrometer

appear undisplaced relative to each other, and M is moved in and out until this is achieved. Since the reading of the thermocouple depends only on the intensity of the image of the furnace, the millivoltmeter reading is independent of the distance of the instrument from the source, provided the whole of the aperture in D is filled by the image. For if the distance from the source be doubled, the amount of radiation received by M is reduced to a quarter but the area of the image is also reduced to a quarter so that the intensity of the image remains unchanged. If T and T_0 are the temperatures of the furnace and receiver respectively, the millivoltmeter readings cannot be assumed proportional to $(T^4 - T_0^4)$, chiefly because, when the receiver is some 80 °C above its surroundings, the e.m.f. of the thermocouple is not proportional to the temperature difference between its junctions. The instrument is therefore callibrated by sighting on a black-body furnace whose temperature is known either by a standard thermocouple or for temperatures above 1064.4 °C (the melting point of gold) by an *optical pyrometer. The range of the instrument may be extended beyond the calibration limit by the use of a rotating sector in combination with the instrument.

Fessenden oscillator An efficient form of electromagnetic or electrodynamic underwater sound generator and receiver. It is commonly used for signalling through sea when a large range is required.

The electromagnetic type consists of a steel diaphragm, which is in contact with water. The base of this diaphragm supports two copper rods fixed at their upper ends to two copper tubes that touch another cast-iron plate over and just clearing the first diaphragm. If a suitable alternating current with the requisite frequency is led through two coils wrapped around the two plates, it will throw them into violent and periodic attraction. The rods, together with the tubes and the diaphragm, will thus be set in vibration, sending sound waves into the water.

For receiving sound, this process is reversed, i.e. the vibrations impinging on the diaphragm will induce an alternating current through the coils. These are connected through an amplifier circuit to a pair of telephones or an indicating device.

In the electrodynamic type, there is an electrical transformer, the secondary of which is free to move in a very strong radial magnetic field. The primary winding is wound on the inner pole face of the electromagnet. The secondary winding is in the form of a copper tube, which has a very low electrical resistance. An alternating current passing in the primary will produce a corresponding alternating force in the secondary. This in turn is applied to a steel diaphragm in contact with the water.

As a transmitter this apparatus has a sound output of several kilowatts at a high efficiency. From the electrical point of view it is superior to the electromagnetic type on account of absence of troubles due to hysteresis and eddy-current losses in the iron. Furthermore, it is easily manipulated and all mechanical and electrical quantities involved in its design can be controlled. *See* transducer.

FET Abbreviation for *field-effect transistor.

fetron A junction *field-effect transistor, mounted in a package so that it can be directly plugged into a circuit as a replacement for a valve, with no special modifications of the circuit.

Feynman, Richard Phillips (1918–1988) Amer. physicist known for his contribution to *quantum electrodynamics, for which he shared a Nobel prize (1965) with Julian Schwinger and Shinitiro Tomonaga.

Feynman diagram *See* quantum electrodynamics.

fibre-optics system An optical system in which a single glass or plastic fibre or an array of fibres transmits light between two points. Such systems have many uses, including the direct transmission of images and illumination and medium- and long-distance communications. An optical fibre consists of a single flexible rod of high refractive index, less than 1 mm in diameter, having polished surfaces coated with transparent material of lower refractive index. This coating, known as *cladding*, prevents light from leaking between fibres in close proximity. In a *stepped-index fibre*, the indexes of both cladding and core are constant throughout. Light falling on one end within a certain solid angle will undergo *total internal reflection at the cylindrical surface of the glass core. The light is trapped within the core and travels in zig-zag paths down the length of the fibre with little or no absorption. The fibre can continue to reflect light when it is considerably curved, as long as the reflection angle remains greater than the critical angle.

In a *graded-index fibre*, the refractive index of the core decreases radially outwards. Light rays then spiral smoothly around the central axis rather than zig-zagging. This reduces the difference in time delays of the transmitted light rays. There may be many hundreds of ray paths, or modes, by which energy can propagate down the core. In a *single-mode fibre*, the core is very narrow relative to the cladding and rays travel parallel to the central axis; it may be stepped- or graded-index. These are the most efficient fibres for communications. A laser is used as the light source and information is carried by modulating the light. In general, all optical fibres have a much greater data-handling capacity than,

say, telephone systems and electric cables as the frequency of the transmitted signal is so much greater.

Images may be transmitted by using a bundle of optical fibres, between 0.01–0.5 mm in diameter, in a fixed array. If a pattern of light is displayed at one end, each fibre will transmit light from a small area of the pattern to the other end, and the image will be reassembled on that surface. If an *objective is placed at one end, and an *eyepiece or camera at the other end to view the image, the fibre bundle forms an *endoscope used in both industry and medicine to view inaccessible locations, an external layer of fibres transmitting light into the location. Resolving power is in the region of 100 lines per mm.

Screens on *cathode-ray tubes, *image intensifiers, etc., can be made from parallel glass fibres, fused together. There is therefore no loss in definition, which would normally occur when light from the inner phosphor layer passes through the glass envelope.

fibre photography A technique used in X-ray crystallography in which the specimen is a fibre or bundle of fibres placed with its length normal to the direction of incident X-radiation.

Fick's law A law expressing the process of *diffusion of liquids and solids in mathematical form. The mass of dissolved substance crossing unit area of a plane of equal concentration in unit time is proportional to the concentration gradient. The constant of proportionality is called the *coefficient of diffusion*, symbol: D.

fidelity Generally, the extent to which a system or part reproduces accurately at its output the characteristics of the input. There is no scale of fidelity, but relative and somewhat arbitrary terms such as good-fidelity or high-fidelity (hi-fi) apparatus are used. True fidelity is perfect reproduction of the original. Applied to radio receivers, it is the extent to which a receiver reproduces without attenuation or distortion all the modulation frequencies that may be present at its input.

field 1. A region under the influence of some physical agency. Typical examples are the *electric, *magnetic, and *gravitational fields that result from the presence of charge, magnetic dipole, and mass respectively; these are *vector fields. A field can be pictorially represented by a set of curves, often referred to as *lines of flux* (or *force*); the density of these lines at any given point represents the strength of the field at that point, and their direction represents the direction conventionally associated with the agency. Thus, electric fields run from positive to negative, magnetic fields from north-seeking to south-seeking, and gravitational fields from lighter to heavier.

A field is also used to describe the region inhabited by *nucleons, in which *exchange forces are set up. In addition, it has been used in connection with scalar quantities to describe distributions of temperature, electric potential, etc. *See also* quantum field theory.

2. *See* field of view.

field coil A coil that, when carrying a current, magnetizes a *field magnet of an electrical machine (dynamo or motor).

field-discharge switch A special form of switch connected in series with the field winding of an electrical machine. It is so designed that, when it is operated, a resistor (known as a *field-discharge resistor*) is connected in parallel with the field winding before the main-supply contacts are opened. By this means the e.m.f. of self-induction in the field winding is limited to a safe value.

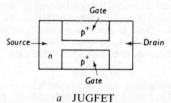

a JUGFET

field-effect transistor (FET) A *semiconductor device, a *transistor, in which current flow depends on the movement of *majority carriers only; it is thus a unipolar rather than a bipolar device. Current flows through a narrow conducting *channel* between two regions, the *source* and *drain*, to which electrodes are attached. Application of a suitable bias across the transistor causes the charge carriers to flow from source to drain. The current is modulated by an electric field applied to a third electrode attached to the *gate* region. Devices with n-type source and drain regions and hence an n-type channel are called *n-channel devices*; those with p-type regions are called *p-channel devices*. (*See* n-type conductivity; p-type conductivity.)

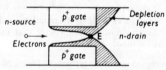

b   Drain voltage just at pinch-off

There are two main types of FET. In the *junction FET* (JFET), the conducting channel forms part of the structure of the device. In the *insulated-gate FET* (IGFET), the channel is formed in use by the action of the gate voltage.

The junction FET consists of a wafer of semiconductor material flanked by two highly doped layers of the opposite type (n^+ or p^+) (Fig. *a*.). If a positive voltage is applied to the drain, electrons in the source are attracted into flowing through the channel. As the drain voltage increases, *depletion layers associated with the p^+-n junctions increase in size and reduce the cross-sectional area of the channel, thus increasing the resistance of the device. When the depletion layers first meet, the drain voltage is

called the *pinch-off voltage*, V_p. At drain voltages above pinch-off, the depletion layers extend further into the drain region, but remain essentially constant in the channel (Fig. *c*). The current therefore remains substantially constant, until the voltage reaches such a value that *breakdown occurs. Electrons arriving at the pinch-off point, E, are swept across the depletion layer into the drain region. Applying a negative voltage to the gates will also increase the size of the depletion layers, and hence the pinch-off condition will be reached at a lower drain voltage. A family of characteristics will thus be generated for different values of gate bias (Fig. *d*). The gate voltage is therefore used to modulate the channel conductivity.

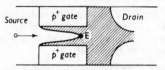

c   Drain voltage above pinch-off

The basic structure of an IGFET is shown in Fig. *e*. A wafer of semiconductor material has an insulating layer formed on the surface between two highly doped regions of opposite polarity to the substrate, which form the source and drain regions. A conductor is deposited on top of the insulating layer to form the gate electrode. Several alternative names and abbreviations are in common use: metal insulator silicon field-effect transistor (MISFET, MIST); metal oxide silicon field-effect transistor (MOSFET, MOST). MOS transistors have an oxide layer as the insulator and are a special case of MIS transistors. Most IGFETs are MOS devices.

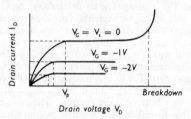

d   Characteristic curves of a JUGFET

If a positive voltage is applied to the gate electrode of an n-channel device, a depletion layer is formed in the substrate below the insulating layer. As the bias is increased, the depletion layer spreads into the semiconductor until a point is reached at which the semiconductor becomes inverted (i.e. of opposite conductivity type) at the surface. Beyond this point the depletion region ceases to grow, the effect of increasing the bias further being to increase the number of carriers in the inversion layer. The inverted layer constitutes a narrow channel connecting source and drain through which current can flow. The gate voltage at which inversion takes place is the *threshold voltage* and in practice is defined as the voltage at which a small current (say 10 μA) will flow (Figs. *f*, *g*).

Applying a small positive voltage to the drain attracts electrons from the source and current then

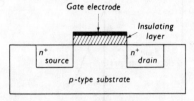

e IGFET

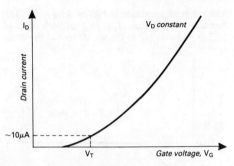

g IGFET transfer characteristic

flows through the conducting channel. The channel acts as a resistor and the drain current I_D is approximately proportional to the drain voltage V_D. (See Figs. *h*, *i*, and equation (2) below.) As the drain voltage is increased the channel depth is reduced near the drain until it eventually reaches a pinch-off point at which the channel depth is just zero. If the magnitude of the drain voltage is further increased, the current remains essentially constant since the channel remains constant, the depletion region near the drain increasing in size. The drain-current characteristics are similar to those of junction FETs. A p-channel device operates in a similar manner to an n-channel device, but negative gate and drain voltages are applied.

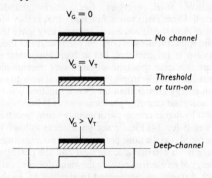

f Effect of increasing gate voltage

Field-effect transistors fall into two categories depending on the transfer characteristics of the devices (Fig. *j*). *Depletion-mode devices are those in which conduction takes place with zero gate bias, and *enhancement-mode devices are those in which a voltage must be applied to the gate before conduction can occur. All junction FETs are depletion-mode devices. Ideally, all insulated-gate FETs, as described above, are enhancement-mode devices, but when an oxide is formed on the surface of a semiconductor, a quantity of positive charge inevitably exists in the interface. In n-channel devices the presence of this positive charge can result in the presence of a spontaneous inversion layer even with zero gate bias. In p-channel devices the effect is to increase the magnitude of the threshold voltage.

Practical IGFETs can therefore be either enhancement or depletion-mode devices. Enhancement-mode devices are simpler to use in logic circuits, since they are "off" at zero gate bias and hence simpler to use as switches. n-channel devices are preferred to p-channel devices, however, due to the greater mobility of electrons compared to holes, which leads to a higher gain.

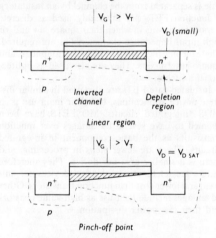

h Effect of increasing drain voltage for a given gate voltage

Field-effect transistors may be broadly described as square-law devices, i.e. the output current I_{DS} varies with the square of the input voltage V_{GS} (*compare* junction transistor, where there is an exponential dependence). The basic device equations are as follows.

Junction FET (beyond pinch-off):
$$I_{DS} = I_{DSS}[1 - (V_{GS}/V_P)]^2, \quad (1)$$
where I_{DSS} = saturation drain current at $V_G = 0$ and V_P = pinch-off voltage. The pinch-off condition is $V_{DS} = V_{GS} - V_P$.

Insulated gate FET (below pinch-off):
$$I_{DS} = k[2(V_{GS} - V_T) V_{DS} - V_{DS}^2], \quad (2)$$
where k = gain factor, V_T = threshold voltage.

IGFET (beyond pinch-off):
$$I_{DS} = k(V_{GS} - V_T)^2. \quad (3)$$

211

field emission

The pinch-off condition is $V_{DS} = V_{GS} - V_T$. The *mutual conductance, $\partial I_{DS}/\partial V_G$, of both types of FET is proportional to the square root of the drain current.

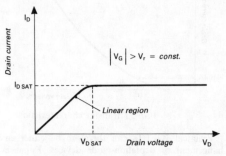

i Drain-current characteristic

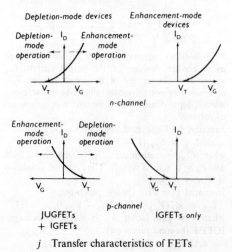

The input impedance of an FET is always very high. In the junction devices the input is through a reverse-biased diode, and in the insulated-gate version the input impedance is purely capacitive (the gate is separated from the channel by an insulator).

Junction FETs are invariably used as discrete devices in circuits in which their square-law and/or high input impedance characteristics are required. These include high input impedance *amplifiers and square-law *mixers. They are also used as bidirectional switches.

Insulated-gate FETs are also used in similar discrete device applications, but their main use is in MOS *integrated circuits. MOSFETS have been claimed to have several advantages over junction transistors as the basic component in integrated circuits. There are fewer steps in processing, and there is a smaller area per device. They therefore have a greater functional density (devices per unit area) and lower cost than junction transistors. Other advantages include their use as high-value resistors and their low power dissipation.

j Transfer characteristics of FETs

field emission Emission of electrons from a solid that is subjected to a high electric field. In a metal, the outer electrons of the atoms move through the metal as an *electron gas and their energies lie within allowed bands. The diagram shows one such band in an idealized metal; the electron energy is plotted against distance from the metal surface, both within the metal and outside it.

There are various mechanisms by which electrons may escape from the surface. (1) A very small proportion of the electrons in the distribution have positive total energies (*see* energy bands; Fermi–Dirac distribution function), that is they are above A in the figure. Such electrons may leave the solid spontaneously. At ordinary temperatures their number is infinitesimal, but at high temperatures there is a large spontaneous emission (*thermionic emission). (2) Electromagnetic radiation with quantum energy greater than the *work function ϕ may be absorbed causing *photoemission. (3) Bombardment by higher energy particles may cause *secondary emission. (4) Field emission arises as follows. If the metal is at the same potential as its surroundings, there is no external electric field and the potential energy of an electron outside the metal does not vary with distance, as represented by the line AC in the diagram. However, if the metal has a high negative potential with respect to an external electrode, there is a corresponding high electric field at its surface; the potential energy therefore falls off with distance, as represented by the line AD. If the field is very high, there is a barrier (represented by the area BAX) keeping the electrons inside the solid. Nevertheless, electrons can escape without having to surmount the barrier as a consequence of the *tunnel effect, and it is possible for electrons to appear at point X outside the metal. The probability of this process increases as the barrier width BX decreases. Consequently, the field-emission electron flux goes up if the electric field strength is increased and is greater for solids of low work function. High electric fields of the order of 10^{10} volts per metre are necessary for the effect to be observed and these are

Energy band at the surface of a metal

usually obtained by subjecting very sharp points to high potentials. Field emission is one of the factors causing breakdown in *gas-discharge tubes, is the basis of the *field-emission microscope, and is used in the scanning-transmission *electron microscope.

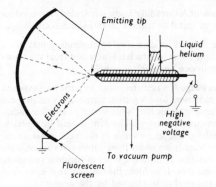

Field-emission microscope

field-emission microscope An instrument for observing the surface structure of a solid by causing it to undergo *field emission. The diagram shows a simple field-emission microscope. A metal tip of radius about 10^{-7} metres is maintained at a negative potential of about 10 kV with respect to a conducting fluorescent screen. The instrument is evacuated to a very low pressure to prevent gas discharge. Ideally the tip has a regular shape and is composed of a single crystal of material. Field emission occurs under the influence of the high local electric field and the resulting electrons are accelerated to the screen where they cause fluorescence. Field-emitted electrons leave a surface at right angles and if r_t is the tip radius and r_s its distance from the screen, an image of the surface of the tip is projected on the screen with a linear magnification of r_s/r_t. The resolution is limited by vibrations of the metal atoms and the tip is therefore usually cooled to liquid-helium or hydrogen temperatures. Individual atoms cannot be resolved (*compare* field-ion microscope) but a regular pattern of light and dark patches is observed corresponding to areas of different *work function on the tip. These can be interpreted in terms of different crystal planes on the metal surface. The technique is used to study the structure of alloys, the behaviour of impurities in metals, and the interaction of gas molecules with the surfaces of solids.

field index *See* accelerator.

field ionization Ionization of gaseous atoms and molecules by a high electric field at a solid surface. Electrons can normally only escape from an atom if they gain energy equal to the *ionization potential of the atom. If the atom is close to, say, a metal and there is a high electric field near the surface, an electron can tunnel (*see* tunnel effect) through the *potential barrier from the atom into the metal. The process is similar to *field emission, with the difference that electrons tunnel from atoms or molecules into a metal rather than out of a metal. The fields required for field ionization are of the order of 10^7 volts per metre and are produced by subjecting very sharp points to very high positive potentials ($\sim$10–20 kV). Ions formed at the metal surface are accelerated away by the field. Field ionization is the basis of the *field-ion microscope.

field-ion microscope An instrument for observing the surface structure of a metal using *field ionization. It is identical in form to the *field-emission microscope with the difference that a positive voltage is applied to the tip rather than a negative one and the image is formed by positive ions from gas atoms rather than by electrons from the metal itself. Helium is allowed into the microscope at low pressure and helium ions form at the surface of the tip and are accelerated to the screen where they cause fluorescence. The field-ion current from a point depends on the magnitude of the electric field, and on the atomic scale there is a local intensification of the field in the region of a surface atom. Consequently, a magnified image of the atomic structure of the surface is projected on the screen. The tip is cooled to liquid hydrogen or helium temperatures to reduce the vibrations of the metal atoms and improve the resolution of the instrument. Indeed, individual metal atoms can be "seen" on the surface of the metal. The technique is used in studies of alloy structure, the structure and behaviour of surfaces, and the interaction of low-pressure gases with metals.

field lens Of an eyepiece. The front lens of a two-lens eyepiece that serves to bend the chief rays towards the optical centre of the eye lens. It serves to bring the *exit pupil of the instrument nearer to the eye lens than it would otherwise have been, and at the same time reduces the necessary aperture of the eye lens. In addition, in combination with the eye lens, it commonly reduces lateral chromatic *aberration.

field magnet The magnet that provides the magnetic field in an electrical machine. Usually it is an electromagnet but may be a permanent magnet in small machines.

field of view *Syn*. field. The angular extent of object space that can be observed or embraced by an optical instrument. By drawing rays through the edges of the *entrance pupil and entrance port, the fields of full illumination, of half value, and the extreme field can be determined (*see* diagram). A similar delineation obtains on the image side field (*exit pupil and exit port). The field *stop is generally inserted in the focal plane of the eyepiece. Commonly the term field is applied to describe the angular extent over which the definition of the image is tolerable. Telescope fields extend to the order of 6°, camera lenses 30°–60° (wide angles 70°–90°). *See* apertures and stops in optical systems.

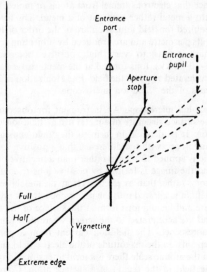

Field of view

field rheostat *Syn.* field regulator. A *rheostat that enables the current in the field winding of an electrical machine to be readily adjusted.

field-sequential colour television A sytem of *colour television in which the camera sees red, blue, and green images in turn, through a rotating disc containing colour filters, in the correct sequence. The receiver contains a similar disc rotating in synchronism, so that a sequence of colour images is produced and is synthesized into the complete coloured picture by the viewer's eye.

field spool A bobbin on which a *field coil of an electrical machine is wound. It may be constructed entirely of insulating material or may be built up from sheet steel and covered with insulating material. In small machines, field spools are sometimes not used, the field coils being former-wound.

field stop *See* stop.

field theory *See* field; quantum field theory.

field tube *See* tube of flux.

figure of merit Of a galvanometer. Another name for *sensitivity. It is sometimes taken as the deflection in scale divisions produced by a current, usually of 1 microamp. This usage is deprecated, as it neglects such factors as the resistance of the instrument and the periodic time of its suspended system, which are of considerable practical importance in estimating the behaviour of the instrument. The figure of merit, or factor of merit, is the sensitivity of the instrument expressed in millimetres deflection produced on a scale at a distance of 1 m by a current of 1 μA, after the deflection has been corrected for a coil-resistance of 1 ohm and a period of 10 seconds. A

formula employed by one firm of instrument makers gives the figure of merit as:

$$100 \ D/T^2R^{2/5},$$

where D is the sensitivity, R the coil resistance, and T the periodic time.

filament A threadlike body, particularly the conductor of metal or carbon in an incandescent lamp.

file A block of data used in computing that has a unique name by which it is accessed from the *storage device on which it is held.

film badge A dosemeter, worn by personnel exposed to radiation, for protection purposes. A small piece of photographic film is blackened by incident radiation (*see* dosimetry). The film is held in a plastic holder containing a series of different metal filters. The blackening under these filters gives an estimate of the dose of different types of incident radiation. Where there is no filter, the blackening is equivalent to the total dose received by the wearer.

film dosimetry *See* dosimetry.

film resistor *See* resistor.

film theory A theory of convection, based on the conduction of heat through a thin film of fluid adhering to the hot surface. In free convection this film is stationary while for forced convection, it is continually being removed and renewed.

Type	Pass band(s)	Attenuation band(s)
1. Low-pass	0 to f_c	f_c to ∞
2. High-pass	f_c to ∞	0 to f_c
3. High-pass	f_1 to f_2	0 to f_1, f_2 to ∞
4. Band-stop	0 to f_1, f_2 to ∞	f_1 to f_2

filter 1. A device for removing solid matter suspended in a liquid by forcing the suspension through a material (e.g. sand, filter paper) that retains the solid matter while allowing the liquid to pass.

2. An electrical network designed to transmit signals with frequencies that lie within one or more designated ranges (*pass bands*) and to suppress signals of other frequencies (in one or more *attenuation bands*). The cut-off frequencies are those that separate the several pass and attenuation bands (symbols f_1, f_2, etc. or f_c if there is only one). The four main types of filter are *low-pass, high-pass, band-pass,* and *band-stop filters.* Their frequency limits are given in the table.

3. A device for transmitting light (and also infrared and ultraviolet) with restricted ranges of wavelength. Commonly, transparent substances that absorb selectively are used (coloured glasses or films; *see also* Polaroid). *Interference filters* use a different principle and can be produced to yield a narrower band of wavelength 1–10 nm for half transmission. Filters for eye protection absorb harmful infrared or ultraviolet radiation. Filters find extensive application in various branches of photography, the study of colour and colour vision, isolation of spectral lines, etc. In one sense, the isolation by means of a slit inserted in a pure spectrum can be regarded as an elaborate form of filter. *See* focal isolation.

4. A sheet of a substance that transmits X-rays of a certain range of wavelengths and absorbs more strongly those of higher and lower wavelengths. The substance has an atomic number such that the required range is just on the long-wavelength side of an *absorption edge.

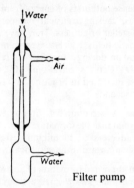

Filter pump

filter pump *Syn.* water aspirator. A fast-working vacuum pump in which a jet of water is used to trap and remove air (*see* diagram). The lowest pressure attainable is approximately the vapour pressure of water at the prevailing temperature.

finder *See* collimator.

fine structure Structure in a spectral line or band when it is viewed at high resolution. For instance, a line in an emission spectrum may consist of a number of closely spaced lines corresponding to different energy levels caused by interaction between the spin of the electron and that of the nucleus. Similarly, the existence of different nuclides in the sample may be a source of fine structure. *Hyperfine structure* is a similar effect, but requires higher resolution and is caused by more subtle effects, e.g. the *Lamb shift in quantum electrodynamics.

fine-structure constant Symbol: α. A dimensionless quantity formed from the four basic physical constants, electronic charge e, speed of light in vacuum c, the *Planck constant h, and the *permittivity of free space ε_0 (electric constant):

$$\alpha = e^2/2hc\varepsilon_0 = 7.297\ 3531 \times 10^{-3}$$
$$\simeq 1/137.$$

This constant was given its name because it first appeared in formulae specifying certain fine details in the allowed states of hydrogen. It has fundamental significance in that it is a convenient measure of the strength of the *electromagnetic interaction. The equivalent pure number measuring the strength of the *strong interaction is about 1. On this scale the *weak interaction has a strength of 10^{-13} and the gravitational interaction of 10^{-38}.

fissile Of a nuclide. Capable of undergoing *fission by interaction with *slow neutrons. *Compare* fissionable.

fission The splitting of a heavy nucleus of an atom into two or more fragments of comparable size usually as the result of the impact of a neutron on the nucleus. It is normally accompanied by the emission of neutrons or gamma rays. Plutonium, uranium, and thorium are the principal fissionable elements. *See* nuclear reactors.

fissionable Of a nuclide. Capable of undergoing fission by any process. *Compare* fissile.

Fitch, Val Logsdon (*b.* 1923) Amer. particle physicist. He became professor at Princeton (1960) and in 1964 collaborated with James Cronin in studies of *kaons and experiments to disprove *CP invariance, for which they shared a Nobel prize (1980).

FitzGerald, George Francis (1851–1901) Irish physicist. He suggested (1893) that the failure of the *Michelson–Morley experiment (1881) to detect motion relative to the *ether could be due to a shortening of all moving bodies (*FitzGerald contraction*) in the direction of motion, a hypothesis independently advanced by Lorentz in 1883. (*See* Lorentz–FitzGerald contraction.) FitzGerald also worked on Maxwell's electromagnetic theory and made attempts to detect electromagnetic waves; he suggested the pressure of solar radiation as the cause of the formation of the tail of a comet.

five-fourths power law The law of cooling applicable to free convection. The rate of loss of heat is proportional to the five-fourths power of the excess temperature of the body over the temperature of its surroundings. It was derived theoretically by Lorentz and verified experimentally by Langmuir.

fixation line (or **axis**) Of the eye. The line joining the centre of rotation of the *eye to the object in direct vision, i.e. when the fovea of the eye receives the image.

fixed end An end of a structure, e.g. a beam, that is clamped in a given position. In the case of a horizontal beam, at the end under consideration both the displacement, y, and the slope, dy/dx (where x is measured along the beam) are zero. When a column of air is contained in a pipe, a stopper in the pipe constitutes a fixed end.

fixed points Reproducible invariant temperatures used to define a temperature scale. In very early work the temperature of a healthy person was regarded as fixed. This was replaced by the melting point and boiling point of water, which were used to define nearly all temperature scales for two centuries. The *thermodynamic temperature scale now has one fixed point, the *triple point of pure air-free water. The *International Practical Temperature Scale of 1968 uses eleven fixed points all defined in terms of the equilibrium between phases of pure substances under specified conditions.

Fizeau, Armand H. L.

Fizeau, Armand H. L. (1819–1896) French physicist, a wealthy amateur whose best-known work is connected with light and efforts to detect motion of the *ether. He made the first determination of the speed of light by purely terrestrial observations (*see* Fizeau's method) and carried out experiments designed to test Fresnel's expression for the extent to which the ether was carried along with moving matter. He also applied interference phenomena in thin films (in fact, Newton's rings) to the measurement of thermal expansion (1864; *see* Fizeau's determination of expansion of small crystals).

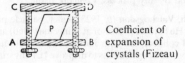

Coefficient of expansion of crystals (Fizeau)

Fizeau's determination of expansion of small crystals For the determination of the coefficient of linear expansion of small crystals. The crystal is cut into a plate P, a few millimetres thick, with one face resting on a metal disc AB and the other parallel face brought near to the glass plate CD. A beam of light falls normally on the system, and interference fringes are formed between light reflected from the lower surface of CD and that reflected from the upper surface of P. When the temperature rises, the thickness of the air film changes by an amount equal to the difference between the expansion of the plate and the frame supports, causing the fringes to shift. If a fringe moves a distance equal to the distance between two such fringes, the change in thickness of the air film is equal to half the wavelength of the monochromatic light used so that changes in thickness as small as 30 nm may be measured.

Fizeau's method (for determining the speed of light) A method in which a cogwheel was rotated at high speed to send a succession of flashes to a distant mirror that reflected the light back to the cogwheel, where an eye observed the rates or rotation to produce an eclipse of the returning light (1849). Improvements of the method have been employed by Cornu, Young, and Forbes.

flame collector An instrument for determining the potential at a point in the atmosphere. A small hydrogen or coal-gas glame burning at the end of an insulated metal tube furnishes a copious supply of gaseous ions of both signs. Ions of the appropriate sign will be attracted to the metal until its potential becomes equal to that of its surroundings, when, as the field has now been neutralized, further charging of the metal ceases. The potential reached by the tube can be measured on an electrostatic *electrometer. A metal wire, carrying a speck of radioactive material at its end, acts in a similar manner, and is more convenient to use. (*See* drop-electrode).

flame spectrum 1. The spectrum as revealed through a spectroscope when salts are introduced into Bunsen flames, e.g. by inserting a platinum loop holding the salt or by spraying salt solution into the gas supply.

2. The spectrum produced by any type of flame. This can be used to identify the molecules, atoms, ions, or fragments present in the flame, or various parts of the flame, and is a valuable tool in combustion research.

flare spots *See* interference of light.

flare stars Faint cool red main-sequence stars that undergo intense outbursts of energy from localized surface areas, causing transient but considerable increases in stellar brightness. They have unusually strong magnetic fields, which suggests a mechanism similar to that producing *solar flares. A remarkable example is UV Ceti (*spectral type M6), which on one occasion increased in brightness by six magnitudes in 45 seconds.

flash barrier A structure of fireproof material designed to minimize the formation of an electric arc between conductors or to minimize the damage caused by such an arc in an electrical machine.

flashover An abnormal formation of an arc or spark (called *arcover* and *sparkover* respectively) between two electrical conductors or between a conductor and earth.

flashover voltage The dry flashover voltage is the voltage at which the air surrounding a clean dry insulator (especially one supporting electric lines) breaks down completely and *flashover between the conductors occurs. The wet flashover voltage is the voltage at which flashover occurs when the clean insulator is wet (to simulate rain).

flash photolysis A method of studying the reactions of *free radicals, excited atoms and molecules, etc., produced in a solution or gas by a flash of light or ultraviolet radiation from a high-intensity source placed close to the sample. The absorption of radiation is proportional to the concentration of the absorbing species. The rates of reaction and lifetimes of the transient species produced by the flash can therefore be determined by studying the change in their *absorption spectrum with time. The spectra are usually obtained by using a second light source that emits a brief intense flash of light at any required time after the first decomposing flash. The light is passed through the sample and is then recorded on a photographic plate inside a spectrograph. Alternatively, a beam of light can be passed through the sample and into a *monochromator, set at the wavelength for maximum absorption of the free radical under observation. The emergent beam falls onto a *photomultiplier and the amplified signal appears as a trace on an *oscilloscope. As the absorption of the radical decreases during its reactions with other radicals, atoms, or molecules, the signal (representing transmitted light) increases until the free-radical concentration reaches a minimum. *Pulse radiolysis* is analogous to flash photolysis except that a pulse of *ionizing radiation is used to produce the free radicals.

flash point The lowest temperature at which a substance will provide sufficient inflammable vapour (under special conditions) to ignite upon the application of a small flame. In the US it is also referred to as the flashing point.

flash spectrum The spectrum of the sun at the instant preceding total eclipse: the absorption lines of the solar spectrum give place to the emission spectrum of the incandescent gases of the solar atmosphere.

flavour *See* quark.

flaw detection For the detection of flaws in the form of hairline cracks in metals, a system has been developed using ultrasonic waves (*see* ultrasonics); a short train of waves is sent out and its waveform, as recorded by the receiver, is noted on a *cathode-ray oscillograph. In one form of this detector, a *piezoelectric oscillator of quartz is placed at each end of an ingot (transmission method) or sometimes installed on two facets of a prismatic specimen (reflection or echo-sounding method). The presence of hairline cracks is shown in the latter method by a number of secondary echoes that reach the second quartz before the main reflection from the base of the prism. Whereas large fissures may be more easily detected by X-rays, the ultrasonic method is better for tiny cracks, since it needs but the smallest break in the continuity of the metal to give rise to a secondary reflection. *See* magnetic crack detection.

F-layer (or **region**), **F$_1$-layer, F$_2$-layer** *Syn.* Appleton layer. *See* ionosphere.

Fleming, Sir John Ambrose (1849–1945) Brit. physicist. The inventor of the thermionic valve, he made great contributions to the development of the electric lamp. He also studied electrical resistance at low temperatures.

Fleming–Kennelly law At a point near magnetic saturation, the *reluctivity of ferromagnetic substances varies linearly with magnetic field strength.

Fleming's rules Mnemonics, much used by practical electricians, for the relation between current, motion, and field in the dynamo and electromotor. *Right-hand rule* (dynamo principle). Hold the thumb, first finger, and middle finger of the right hand at right angles to one another. Point the thumb in the direction of motion, the first finger along the lines of the field; the middle finger will then point in the direction of the induced current. *Left-hand rule* (motor principle). Hold the thumb, first finger, and middle finger of the left hand at right angles to each other. Point the first finger in the direction of the field, the middle finger in the direction of the current; then the thumb points in the direction in which the force acts on the conductor.

Fleming valve The original diode thermionic valve invented by Fleming. It consisted of a heated cathode and simple anode contained in an evacuated glass envelope.

Flerov, Georgii Nikolaevich (*b.* 1913) Soviet physicist whose most important work was carried out at the Kurchatov Institute of Atomic Energy and the Joint Institute of Nuclear Research. His work centred on the synthesis of transuranic elements; he claimed to have been the first to identify elements 104 (1964) and 107 (1968).

flexural rigidity Of a beam. The product of Young's modulus and the geometrical moment of inertia about the natural axis of the cross section of the beam. It measures the resistance of the beam to bending.

flicker photometer A photometer that presents alternately to the eye two surfaces illuminated by the sources to be compared. If both sources are white and the frequency of alternation is not too high, disappearance of flicker signifies equality of brightness. Flicker photometry is useful when there are colour differences between the lights to be compared, as there occurs a speed of alternation at which brightness flicker exists only while colour flicker is absent. Direct comparison of steady luminosity of two different colours presents difficulties, the flicker method being preferred. *See* photometry.

Flinder's bar *See* semicircular deviation.

F-line A green-blue line in the *emission spectrum of hydrogen, wavelength 486.133 nm. It is used as a reference line for specifying the refractive index and dispersion of optical glass, etc.

flint glass *See* optical glass.

flip-chip A semiconductor *chip with thickened and extended *bonding pads, enabling it to be flipped over and mounted upside down on a suitable substrate, such as a *thin-film or *thick-film circuit.

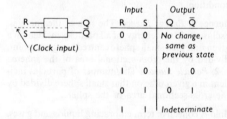

		Input		Output	
		R	S	Q	Q̄
		0	0	No change, same as previous state	
		1	0	1	0
		0	1	0	1
		1	1	Indeterminate	

Clocked R-S flip-flop

flip-flop *Syn.* bistable. An electronic circuit element that is capable of exhibiting either of two stable states and of switching between these states in a reproducible manner. Flip-flops are widely used in computers, especially in *logic circuits: the two states are made to correspond to logic 1 and logic 0, so that the flip-flop is a one-bit memory element. Various types of flip-flop have been developed, the simplest being the R-S flip-flop and the most useful being the J-K and D flip-flops. With a *clocked flip-*

flop, the device cannot change state until triggered by the application of a clock pulse.

(i) D-type flip-flop. ("D" stands for delay.) A clocked flip-flop whose output is a function of the input that appeared one clock pulse earlier. That is, if the input is high the output a pulse later will be high.

(ii) R-S flip-flop. A flip-flop with two inputs designated R and S. R-S flip-flops may be clocked or unclocked. The outputs corresponding to the various input combinations are shown in the table. It is assumed that a logical 1 never appears on both inputs together.

(iii) J-K flip-flop. A flip-flop with two inputs designated J and K. The outputs are the same as the R-S flip-flop except that when a logical 1 appears on both inputs, the flip-flop changes state regardless of the previous state. J-K flip-flops are clocked flip-flops.

floating Of a circuit or device. Not connected to a source of potential.

floating-carrier modulation *Syn.* controlled-carrier modulation. An *amplitude modulation system in which the amplitude of the *carrier wave is varied automatically in a manner dependent upon the amplitude (averaged over a short time interval) of the modulating wave so that the modulation factor remains approximately constant.

floppy disk *See* disk.

flotation (or **floatation**) The retention of small particles at the surface of a liquid or at the interface of two liquids, as when a needle floats on a water surface. The process is used industrially in separating minerals from soil, etc. To increase the area of the interface (usually water-air) a froth is formed with air bubbles (*froth flotation*). Some kind of oil may be added to modify the surface of the ore particles so as to fulfil the nonzero contact-angle condition.

fluence 1. *Energy fluence*. The sum of the energies, exclusive of *rest energy, of all particles incident in a given time on a small sphere centred at a given point, divided by the cross-sectional area of the sphere.

2. *Particle fluence*. The number of particles incident in a given time on that small sphere divided by the cross-sectional area of the sphere.

fluid A collective term embracing liquids and gases. A "perfect fluid" offers no resistance to change of shape (i.e. has zero viscosity).

fluid coefficient The reciprocal of the *viscosity coefficient of a fluid.

fluidics The study and use of jets of fluid in specially designed circuits to perform tasks usually carried out by electronic circuits. Fluidic systems are some million times slower than electronic systems, which enables them to be used to advantage in *delay lines, etc. They can generally be used at higher temperatures than electronics and they are unaffected by

*ionizing radiations; they have therefore been used in *nuclear reactors and spacecraft.

fluidity Symbol: ϕ. The reciprocal of dynamic *viscosity.

fluorescence A type of *luminescence in which the emission of electromagnetic radiation ceases as soon as excitation ceases (*compare* phosphorescence). The radiation emitted is usually, but not necessarily, light. Excitation is commonly by *ionizing radiation or by electromagnetic radiation of different wavelength from that which is emitted. Normally the emitted radiation is of longer wavelength than the incident electromagnetic radiation (Stokes's law), but it can be shorter if the substance is not initially in the ground state. Einstein (1905) cited Stokes's law as evidence that electromagnetic radiation is emitted and absorbed in quanta.

Fluorescent materials excited by ultraviolet may emit light of characteristic colours (fluorescene – yellow green; quinine sulphate – blue; chlorophyll – red). *See* fluorescent lamp.

X-ray fluorescence is the emission of the characteristic X-rays by a substance on exposure to X-radiation of sufficiently high quantum energy to eject electrons from the inner *electron shells.

Gamma rays can be absorbed by nuclei causing fluorescent gamma rays from the excited states.

Fluorescence is used for the detection of ultraviolet, X- and gamma-radiation, and high-energy particles. The *scintillation counter uses this effect.

fluorescent lamp A lamp in which light is generated by *fluorescence. The common forms of fluorescent lamp consist of a *gas-discharge tube containing a gas, such as mercury vapour, at a low pressure. The inner surface of the lamp is coated with a *phosphor. When an electric current is passed through the vapour, ultraviolet radiation is produced and this, in striking the phosphor, produces visible radiation. The phosphors used are such that the combination of colours emitted by the gas and by the phosphor gives white. Lamps of this type are more efficient than filament lamps because less of the radiation is in the infrared.

The common sodium-vapour and mercury-vapour street lamps do not have a fluorescent coating. The light is emitted directly from atoms of vapour that have been raised to an *excited state by electrons in the discharge.

fluorescent screen A surface coated with a luminescent material that fluoresces when excited by electrons, X-rays, etc., and hence displays visual information.

fluorite *Syn.* fluorspar. Natural calcium fluoride having low refractive index and dispersion and transparent to radiation over a very wide range 120–9000 nm. ($n_D = 1.4339$; $\omega = 0.010\ 48$.) It is used for the manufacture of prisms, etc., transparent to infrared radiation of wavelengths up to about 10 μm.

fluorspar *See* fluorite.

flutter An undesirable form of *frequency modulation heard in the reproduction of high-fidelity sound and characterized by variations in pitch above about 10 Hz. *See also* wow.

flutter echoes Echoes produced when a transient sound is emitted between two parallel reflecting walls. The sound is reflected first at one wall and then at the other, so that an observer between the walls hears multiple echoes that take the form either of a series of rapid pulses or of a note whose frequency is inversely proportional to the distance apart of the walls. Flutter echoes are only produced by transient sounds, since continuous notes excite the normal modes of vibration of the air between the walls and separate echoes are not formed. Fluttering can be produced by any impulsive sound such as footsteps, handclapping, or speech. When a person walks between the walls, the whole image system due to the sound of his footsteps moves with him so that he hears a note of constant frequency, but when another person remains still and listens to this note, the distance of successive images varies and the pitch becomes indefinite. If a person stands near one wall, the echoes will always appear to him to be coming from the opposite wall. Maa has shown that this is a psychological effect due to the echoes in one direction always arriving at the observer slightly before those in the opposite direction. Flutter echoes may prevent the logarithmic decay of sound in a reverberation chamber (*see* reverberation). They can be avoided by using walls that are not parallel.

flux 1. A chemical that will combine with a substance of high melting point (generally an oxide) forming a new, readily fusible substance. It is used in the smelting, soldering, brazing, and welding of metals.
2. A measure of the strength of a *field of force through a specified area. (*See* electric flux; magnetic flux.)
3. A measure of the rate of flow of a scalar quantity. (*See* luminous flux; sound-energy flux.)
4. A measure of a flow of particles, usually per unit area. *See* neutron flux.

fluxball Even though a magnetic field is nonuniform, the resultant flux through the surface of a sphere (provided that it does not contain any of the magnetic elements contributing to the field) is proportional to the field at the exact centre of the sphere. The fluxball is a test coil wound into the form of a solid spherical winding by combining a series of coaxial cylindrical windings of different lengths, but of equal and constant axial turn-spacing. It gives accurate values for the flux at its centre.

flux density 1. *See* magnetic flux density.
2. *See* electric displacement.
3. *See* neutron flux.

fluxmeter An instrument for measuring changes in magnetic flux. The *Grassot fluxmeter* is essentially a

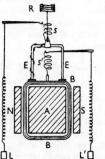
Grassot fluxmeter

moving-coil galvanometer in which the restoring couple on the moving coil BB is negligibly small (a single silk fibre attached to the spring R being employed as the suspension) and the electromagnetic damping is large. The ends of the coil are connected to terminals by fine silver spirals ss'. The terminals LL' are connected to an exploring coil of known area. Any change in the magnetic flux through the exploring coil causes an induced charge to circulate round the suspended coil, and the latter is deflected ballistically through an angle that is directly proportional to the change in flux through the exploring coil. The instrument is calibrated empirically, using a standard magnetic flux, such as the Hibbert magnetic standard, which consists of a block of hard steel, with a circular coil, wound on a brass cylinder that can be dropped into a cylindrical slot in the block. The latter is magnetized so that the magnetic flux cuts radially across the slot.

flux refraction An abrupt change of direction of lines or tubes of magnetic flux when they pass from one region to another of different *permeability. The ratio of the tangents of the angles of incidence and refraction are constant for any pair of media. The same effect is observed when lines of electric flux pass across a boundary between media of different *permittivity.

flyback *See* time base.

flying-spot microscope A microscope in which the lens system is used to produce a minute spot of light that, after passing through the object, falls on a photocell for subsequent amplification and display. The spot is made to scan the object and the image is produced on a *cathode-ray tube scanned in synchronization. This combination of electronic and optical techniques gives a brilliant projection display, the contrast of which is partly determined by the amplifier gain and can thus be varied. Further advantages are the greater *quantum efficiency of the photocell over the photographic plate and increased resolution.

flying-spot scanner A means of producing video signals from films or transparencies, using a high-intensity *cathode-ray tube as a light source. The flying spot of light generated on the screen of the cathode-ray tube by magnetic deflection coils, is used to scan the photographic image and the transmitted beam is modulated according to the density of the film. The modulated beam is focused on a *photocell resulting in electrical signals that can be amplified and transmitted.

FM (or **f.m.**) Abbreviation for *frequency modulation.

f-number *Syn.* relative aperture; stop number. The number, used especially in photography, equal to the ratio of the focal length of a particular lens to the effective diameter of the lens aperture for a parallel beam of incident light. For a ratio of, say, 4 it is written f/4, f4, f:4, etc. For a given shutter speed, it always corresponds to the same *exposure. The variable diaphragm used to change the effective diameter of a camera lens can be set to different f-numbers. The steps are chosen so that changing from one number to the next higher halves the image brightness and twice the exposure is necessary to produce the original brightness. British lens manufacturers use the series: 1.4, 2, 2.8, 4, 5.6, 8, 11, 16.

focal isolation A method devised by Rubens and Wood for isolating long-wave infrared radiation (about 107 μm). Quartz is transparent and has a high refractive index for this wavelength, and a lower index for shorter waves. Thus, a quartz lens can be constructed to focus the long waves into the plane of an aperture in a screen, which stops the shorter wavelengths.

focal length The distance from the *pole of a curved mirror, from the centre of a thin lens, or from the principal point of a system to the principal *focal point. In general, there are two focal lengths, anterior or first (f), and posterior or second (f'), and

$$n/f = -n'/f',$$

where n and n' are the refractive indexes of the medium on the two sides of the system. When the focal length is measured from the last vertex of a lens, it is referred to as the *back focal length. *See also* centred optical system.

focal lines In astigmatism, instead of a single point focus, two focal lines at right angles are produced in positions determined by the powers and direction of the principal meridians of power.

focal plane *See* focal point.

focal point *Syn.* focus. A point to which parallel light rays, incident on a lens or curved mirror, converge or from which they appear to diverge. The plane perpendicular to the axis through a focal point is the *focal plane*. The *principal focal point* is the focus for a light beam parallel to the principal axis of the system. (*See also* centred optical system.)

The concept of focal point (and *focal length) has been extended to electron lenses, acoustic lenses, and lenses and mirrors designed for use with infrared and ultraviolet radiation and radio waves. The focal point is then the point on the axis of the system to which a parallel beam of incident radiation converges.

focus 1. *See* conic sections.
2. *See* focal point.
3. *See* seismology.

focusing 1. Of charged particles in an accelerator. *See* accelerator.
2. Of an electron beam in, for example, a *cathode-ray tube. The principal methods are: (i) *Electrostatic focusing*. The beam is made to converge by the action of electrostatic fields between two or more electrodes at different potentials. The electrodes are commonly cylinders coaxial with the electron tube, and the whole assembly forms an electrostatic *electron lens. The focusing effect is usually controlled by varying the potential of one of the electrodes (called the focusing electrode). (ii) *Electromagnetic focusing*. The beam is made to converge by the action of a magnetic field that is produced by the passage of direct current through a focusing coil. The latter is commonly a coil of short axial length mounted so as to surround the electron tube and to be coaxial with it.

folded dipole *See* dipole aerial.

follow current Current at normal *power frequency that flows through a *surge diverter or other discharge path after discharge has been initiated by a high-voltage surge.

foot A unit of length now defined to be exactly 0.3048 metre.

foot candle An obsolete unit of illumination in the f.p.s. system, viz. a flux density of one lumen per square foot; or the intensity of illumination produced on the surface of a sphere of radius one foot by a uniform point source of one candle at its centre. 1 foot candle = 10.764 lux.

foot-lambert *See* lambert.

foot-pound *See* work.

foot-poundal *See* work.

foot-pound-second system of units (f.p.s. system) The system of units formerly used in English-speaking countries, based on the foot, pound, and second. It has now been replaced for scientific and technical purposes by *SI units. Appendix, Table 1, gives interconversions between SI, c.g.s., and f.p.s. units.

Forbes, George (1849–1936) Brit. physicist. He collaborated with James Young in a determination of the speed of light using *Fizeau's method (1880–1881), the results indicating that blue light had a greater velocity than red, but the measurements are now known to have been in error.

Forbes, James David (1809–1868) Brit. physicist. Noted chiefly for researches in heat, particularly on thermal conductivity (e.g. *Forbes's bar) and heating effects in the spectrum. He demonstrated for the first time the dependence on temperature of thermal conductivity. He also made observations on colour mixing.

Forbes's bar For determining the thermal conductivity of metals in the form of bars several feet long.

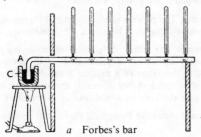

a Forbes's bar

One end A (Fig. *a*) is immersed in a crucible C of molten meltal, thermometers being placed in holes drilled at intervals along the bar. When the steady state is reached, the temperature distribution along the bar is plotted in a θ/x curve for the static experiment. A dynamic experiment is carried out to determine the heat flowing across any cross section by measuring the heat emitted by the portion of the bar lying beyond this section towards the cold end. A short piece of the bar is heated and allowed to cool under the same conditions as the static bar, a cooling curve of θ/t being plotted. Since

$$\frac{K}{\rho c}\left(\frac{\partial \theta}{\partial x}\right)_{x_1} = \int_{x_1}^{l} \frac{\partial \theta}{\partial t}\, dx,$$

where K is the thermal conductivity, ρ the density, and c the specific heat capacity of the bar of length l, by combining the two graphs obtained from the static and dynamic experiments to give a third graph of $\partial \theta/\partial t$ against x, the right-hand side of the equation is the shaded area F (Fig. *b*). The left-hand side is $(K/\rho c)$ times the slope of the tangent to the θ/x curve at x_1, i.e. $\tan \phi$.

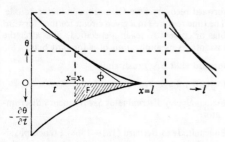

b Forbes's experiment to determine thermal conductivity

forbidden band *See* energy bands.

forbidden transition A transition between two states of a system that violates certain *selection rules. Such transitions are not necessarily impossible but have much lower probability than *allowed transitions of similar energy. In certain cases (e.g. beta emission) transitions can be classified according to the degree of probability as *first forbidden*, *second forbidden*, etc.

force Symbol: F. Unit: newton (N). In classical physics real forces are defined by a set of axioms, *Newton's laws of motion, with reference to an *inertial reference frame. By Newton's second law the resultant force F acting on a body of constant mass m is equal to ma, where a is the *acceleration of the body, and has the direction of a. Force is a *vector quantity.

Forces are either *long range* or *short range*. Long-range forces, such as *gravitation and the *Coulomb force, fall off less rapidly than the inverse fourth power of the distance. Short-range forces, such as those between molecules and forces within the atomic nucleus, fall off more rapidly than the inverse fourth power. It can be shown that in a condensed body, short-range forces are small at distances not much greater than those of near neighbours.

Some workers find it convenient to use fictitious forces in analyses. There are two kinds – *inertia forces* and *inertial forces*.

An inertia force is a fictitious force that is supposed to act on a body, being equal and opposite to the resultant of the real forces. Since it does not represent any actual interaction, an inertia force does not obey Newton's third law. According to the *principle of d'Alembert* (1742) any accelerated body can be treated as if it were in equilibrium under the action of the real forces and the fictitious one.

When a problem is considered from the point of view of an observer who is accelerated with respect to an inertial reference frame, Newton's laws are not applicable to real interactions. It is possible to apply these laws in such a case by introducing a fictitious force, in this case called an inertial force. In particular, an observer on a rotating body may use an inertial force called a Coriolis force, which is supposed to act at right angles to the path of a body that moves towards or away from the axis of rotation (*see* Coriolis theorem). Problems on projectiles and movements of the atmosphere and oceans are often treated in this way.

When a particle of mass m moves in a circular arc of radius r with uniform angular velocity ω, there is an acceleration $r\omega^2$ towards the centre, so by Newton's second law the resultant force on the particle, called the *centripetal force*, is $mr\omega^2$, acting radially inwards. In the system of d'Alembert there is a fictitious inertia force equal to this, supposedly acting radially outwards. Also an observer orbiting with the particle could introduce an inertial force that would be, in this case, equal and opposite to the centripetal force. Both these fictitious forces may be called *centrifugal forces* since they are directed away from the centre. Now the centripetal force is real, so there must be an equal and opposite real force acting on another body, by Newton's third law. This real force is also often called a centrifugal force.

Generalized force. This is the quotient of the work done by all the forces acting in a system, if one of the generalized coordinates alters by an infinitesimal amount while the others remain constant, divided by the change in that generalized coordinate.

forced convection Ventilated cooling in a strong draught. For a body cooling by this means, *Newton's law of cooling applies. *See also* convection (of heat); film theory.

force de cheval *See* cheval-vapeur.

forced oscillations *Oscillations produced in an electric circuit when it is acted on by an external driving force, as when a resonant circuit is coupled to a fixed-frequency oscillator. The resultant oscillations have two components: a transient component whose frequency is determined by the natural frequency of the circuit and decays rapidly, and a steady component whose frequency equals that of the external driving force. *See also* resonance; forced vibrations.

forced vibrations (or **oscillations**) Motion produced when a system, capable of vibrating, is acted upon by an external driving force. The resulting vibrations consist of two components, namely a transient component of frequency given by the natural frequency of the system and a steady component of frequency equal to that of the driving force.

The force equation for such a motion is

$$m\ddot{x} = -\kappa x - \mu\dot{x} + F \sin pt,$$

where x is the displacement, m, κ, and μ are the inertia, elastic, and resistive terms of the vibrating system, and F and p are the amplitude and pulsatance respectively of the driving force.

The solution of this equation is

$$x = A \sin (pt - \gamma) + Be^{-\alpha t} \cos (wt - \varepsilon),$$

where the amplitude of the steady vibrations,

$$A = F/\sqrt{[\mu^2 p^2 + (\kappa - mp^2)^2]},$$

and the phase angle,

$$\gamma = \tan^{-1} \frac{\mu p}{\kappa - mp^2}.$$

The amplitude B and phase angle ε of the transient vibration depend on the initial conditions; the damping factor of the system, $\alpha = \mu/2m$.

A may be written in the form:

$$\frac{F/m}{\sqrt{\frac{\mu^2 p^2}{m^2} + \left(\frac{\kappa}{m} - p^2\right)^2}},$$

and from this expression it can be seen that A is a maximum when $\kappa/m = p^2$, i.e. when the frequency of the driving force is equal to the natural frequency of the system without the effect of the damping. This condition is called *resonance, and the system is said to resonate with the driving force. Then $\gamma = \pi/2$ and the forced vibrations lag 90° behind the driving force.

When the frequency of the driving force is slightly different from that of the undamped system, the expression $(\kappa/m - p^2)$ is then either positive or negative and is called the *mistuning*. Examination of the expression for A shows that the mistuning has

the greatest effect on A when μ is small. *Compare* free vibrations.

force ratio *Syn.* mechanical advantage. *See* machine.

fore-pump *See* pumps, vacuum.

forging The process of shaping metal – previously made hot enough to be plastic – by hammering, etc.

formant The group of *partials of unvarying frequency associated simultaneously with *any* note within the range of a musical instrument. On the absolute-pitch theory of quality the formant is said to give the note its sound quality. *See* quality.

formed plates *See* Faure cell; Planté cell; accumulator.

form factor Of a periodic function (e.g. an alternating current or e.m.f.). The ratio of the *root-mean-square value of the function to the mean value taken over a half period beginning at a zero point. For a simple sine wave, the form factor is $\pi/2\sqrt{2}$, i.e. 1.111.

Fortin, Nicolas (1750–1831) French instrument-maker who designed many precision instruments for *Lavoisier and is remembered for his barometer.

Fortin's barometer *See* barometer.

Fortran A scientifically oriented high-level *programming language designed to facilitate the solving of mathematical problems. The current international standard, Fortran 77, is due to be replaced in the early 1990s.

forward bias *Syn.* forward voltage. A voltage applied to a circuit or device in such a direction as to produce the larger current (known as the *forward current*). The term commonly refers to *semiconductor devices.

forward lead *See* brush shift.

forward recovery time Of a semiconductor *diode. The time required in a given circuit for the *forward bias or current to reach a specified value after the instantaneous application of a forward bias.

forward shift *See* brush shift.

forward voltage *See* forward bias.

Foster–Seeley discriminator *See* frequency discriminator.

Foucault, Jean Bernard (1819–1868) French physicist. For a time a collaborator of Fizeau, he also worked mainly on the speed of light and on the *ether. Following a suggestion by Arago (1842), he used a rotating mirror to show (1850) that the speed of light was less in water than in air, thus establishing the validity of the wave theory (the corpuscular theory having predicated a contrary result) and then (1862) made an accurate determination of the absolute speed (*see* Foucault's method). He devised the *Foucault pendulum (1851) and the *gyroscope (1852), and he anticipated Bunsen and Kirchhoff in discoveries relating to spectra, notably in observing the reversal of the *D-lines by sodium vapour (1849).

Foucault currents *See* eddy current.

Foucault pendulum A simple pendulum that demonstrates the earth's rotation. The original, set up by Foucault (1851), consisted of a lead ball weighing about 28 kg suspended by a fine steel wire 67 m long. The plane of swing is invariable but, owing to the rotation of the earth, it appears to rotate through $360°$ in T hours, where T is equal to $24/\sin \lambda$. (λ is the latitude of the place of experiment. For Paris, this gives $T = 36$ hours.) The earth's rotation may be similarly demonstrated by an experiment with a *gyroscope.

Foucault's method Of measuring the speed of light (1850). A terrestrial method in which light is reflected by a rotating mirror to a distant stationary mirror that reflects it back to the rotating mirror. This in turn returns it displaced from the original source by an amount dependent on the speed of rotation and the speed of light. Improvements were made by Cornu, Newcomb, and Michelson.

four-dimensional continuum *Syn.* space-time continuum. In certain formulations of the theory of *relativity, use is made of a four-dimensional coordinate system in which three dimensions represent the space coordinates x, y, z and the *fourth dimension* is ict, where t is time, c is the speed of light, and i is $\sqrt{-1}$. Points in this space are called *events*. The equivalent to the distance between two points is the *interval* between two events. The distance between two points is not invariant under *Lorentz transformation equations, because the measurements of the positions of the points that are simultaneous according to one observer are not simultaneous according to an observer in uniform motion with respect to the first. By contrast, the interval between two events is invariant.

The equivalent to a vector in the four-dimensional space is a *four vector*, which has three space components and one time component. For example: the four-vector momentum has a time component proportional to the energy of a particle; the four-vector potential has the space coordinates of the magnetic vector potential, while the time coordinate corresponds to the electric potential.

Fourier, Jean Baptiste Joseph (1763–1830) French mathematician and physicist. A pioneer in experimental work on the conduction of heat, he also made a great contribution to theoretical physics in his book *La Théorie Analytique de la Chaleur* (1822), which expounded the method of representation of periodic functions by Fourier series (*see* Fourier analysis) and also introduced the concept of the dimensions of a derived physical quantity in terms of fundamental quantities. His work on conduction of heat was not only of importance in itself but it provided the basic conceptions for Ohm's work on conduction of electricity (1827).

Fourier analysis It is possible to express any single-valued periodic function as a summation of sinusoidal components, of frequencies that are multiples of the frequency of the function. Such a summation is called a *Fourier series*, and the analysis of a periodic function into its simple harmonic components is a Fourier analysis.

A function of time, $x = f(t)$, may thus be expressed as follows:

$$x = a_0 + a_1 \cos \omega t + a_2 \cos 2\omega t$$
$$+ a_3 \cos 3\omega t + \dots$$
$$+ b_1 \sin \omega t + b_2 \sin 2\omega t$$
$$+ b_3 \sin 3\omega t + \dots \quad (1)$$

The values of the coefficients $a_0, a_1, a_2, a_3, \dots, b_1, b_2, b_3, \dots$, may be obtained by integration as shown in the following analysis. Integrals that appear in the analysis, and whose values may be written down immediately are:

$$\int_0^{2\pi/\omega} \sin n\omega t \, dt$$
$$= \int_0^{2\pi/\omega} \cos n\omega t \, dt$$
$$= \int_0^{2\pi/\omega} \cos m\omega t.\cos n\omega t \, dt$$
$$= \int_0^{2\pi/\omega} \sin m\omega t.\sin n\omega t \, dt = 0,$$

and

$$\int_0^{2\pi/\omega} \sin^2 n\omega t \, dt$$
$$= \int_0^{2\pi/\omega} \cos^2 n\omega t \, dt = \pi/\omega;$$

where m and n are different integers.

Taking equation (1), and integrating with respect to t over a complete cycle, i.e. $t = 0$ to $t = 2\pi/\omega$, gives:

$$\int_0^{2\pi/\omega} x \, dt = a_0(2\pi/\omega), \quad (2)$$

therefore

$$a_0 = (\omega/\pi)\int_0^{2\pi/\omega} x \, dt.$$

Multiplying equation (1) by cos $n\omega t$ and integrating with respect to t over a complete cycle, gives:

$$\int_0^{2\pi/\omega} x \cos n\omega t \, dt = a_n(\pi/\omega),$$

therefore

$$a_n = (\omega/\pi)\int_0^{2\pi/\omega} x \cos n\omega t \, dt. \quad (3)$$

This gives the values for $a_1, a_2, a_3, \dots$, in turn by putting in the respective values for n.

Similarly, multiplying equation (1) by sin $n\omega t$ and integrating with respect to t over a complete cycle gives the values for b_n, namely:

$$b_n = \omega/\pi \int_0^{2\pi/\omega} x \sin n\omega t \, dt. \quad (4)$$

It may be possible to evaluate the integrals in equations (2), (3), and (4) algebraically if the expression for x is given, but otherwise the graphical method has to be employed. The latter method is usually undertaken by listing a table of values of x at small equal intervals of time throughout the cycle, and summating the appropriate terms arithmetically.

Fourier integral The representation of a function $f(t)$ as

$$(1/\pi)\int_0^\infty \int_{-\infty}^\infty f(u) \cos \omega(t - u)du.d\omega.$$

This is the limiting form of the Fourier series (*see* Fourier analysis), when the period is made indefinitely great. This representation is very useful for pulses and for limited trains of waves.

Fourier number

Fourier number Symbol: *Fo*. A dimensionless quantity used in the study of heat transfer. It is defined by the function $\lambda t / c_p \rho l^2$, where λ = thermal conductivity, t = time, c_p = specific heat capacity at constant pressure, ρ = density, and l = a characteristic length.

Fourier pair *See* Fourier transform.

Fourier series *See* Fourier analysis.

Fourier transform A mathematical operation by which a function expressed in terms of one variable, x, may be related to a function of a different variable, s, in a manner that finds wide application in physics. The Fourier transform, $F(s)$, of the function $f(x)$ is given by

$$F(s) = \int_{-\infty}^{\infty} f(x) \exp(-2\pi i x s) \, dx.$$

An analogous formula gives $f(x)$ in terms of $F(s)$. The variables x and s are called *Fourier pairs*. Many such pairs are useful, for example time and frequency.

four-layer diode *See* silicon controlled rectifier.

four-terminal transmission network *See* quadripole.

fourth dimension *See* four-dimensional continuum.

fourth power law *See* Stefan–Boltzmann law.

four vector *See* four-dimensional continuum.

fovea centralis *See* retina; eye.

Fowler, William Alfred (*b.* 1911) Amer. physicist who became professor at the California Institute of Technology (1946). His main interest has been stellar nuclear reactions and, with Fred Hoyle, he published *Nucleosynthesis in Massive Stars and Supernovae* (1965). He shared a Nobel prize (1983) with *Chandrasekhar.

f.p.s. system *See* foot-pound-second system of units.

fractal A type of geometric object that is produced by a repeated process. For example, taking an equilateral triangle, removing the middle third of each side, and replacing each of these by two sides of another equilateral triangle, gives a star-shaped figure with 12 sides. Repeating the process on each of these sides, and so on, produces a *snowflake curve*, which develops at each stage. Such figures are called fractals because the limit of the process is a curve of fractional dimension. The snowflake curve has a dimension of 1.26, lying between that of a curve (1) and a surface (2). The mathematical study of fractals (*fractal geometry*) has applications in certain physical theories, e.g. in crystallization and in *chaos theory.

fractional distillation A method of separating the constituents of a liquid mixture when the boiling points of all intermediate mixtures lie between the boiling points of the two pure liquids.

fractionating column A column made up of a series of compartments at different temperatures but each maintained at a steady temperature. A liquid mixture progressing down the column becomes richer in one component and the vapour progressing upwards becomes richer in the other component and in this way the two components may be separated.

fragmentation *See* quantum chromodynamics.

frame 1. In *television, the total area of the picture. (*See* scanning.)
2. In communications, one cycle of a regularly recurring number of pulses in a pulse train.
3. In *facsimile transmission, a rectangular area whose width is determined by the available line, and length by the service requirements.

frame aerial *See* loop aerial.

frame-frequency In *television, the frequency at which the frame is scanned. *See* scanning.

frame of reference 1. A rigid framework relative to which positions and movements may be measured; e.g. latitude and longitude define position on the earth's surface, the earth being used as a frame of reference. Astronomers prefer a frame of reference in which the sun is fixed and that does not rotate with the earth. (*See* coordinate.)
2. A Galilean frame of reference is a rigid framework isotropic with respect to mechanical and optical experiments (used in the special theory of *relativity). *See* inertial reference frame; Newtonian system.

framework structure A crystalline arrangement that is three-dimensionally openwork, because the atoms are linked by strongly directed bonds that are not confined to a single plane. *See* layer structure.

Franck, James (1882–1964) Amer. physicist. He associated with G. Hertz in the discovery of the retardation of electrons by inelastic collisions with atoms (1913) and was a pioneer worker in the elucidation of the fundamental processes in *photochemical effects.

Frank, Ilya Mikhaiboich (*b.* 1908) Soviet physicist, who became a professor at Moscow University. He and Igar Tamm were the first to explain *Cerenkov radiation, for which they shared a Nobel prize (1958) with Pavel Cerenkov.

Franklin, Benjamin (1706–1790) Amer. scientist and statesman. He advocated the one-fluid theory of electricity, introducing "positive" and "negative" electricity properties attributed to excess or deficit of normal quota. His researches on lightning, including discharges obtained with a kite, helped to establish the electrical nature of the discharge and he was a pioneer in advocating lightning conductors as protective devices.

Franklin, Rosalind (1920–1958) Brit. physicist. Working at King's College and Birkbeck College, London, on X-ray crystallography, she assisted James Watson and Francis Crick in establishing the helical structure of DNA. She died prematurely of cancer.

Franklin's pane Early form of capacitor, consisting of two equal sheets of tinfoil pasted opposite to each other on opposite sides of a much larger sheet of glass. The glass was usually shellacked to improve the insulation.

Fraunhofer, Joseph (1787–1826) Bavarian instrument maker, famous for the quality of his optical work and pioneering efforts in telescope objective design, spectroscopy, and diffraction. Although Wollaston had observed absorption lines in the solar spectrum (1802), Fraunhofer rediscovered them (1814), labelled eight of the more prominent lines by letters A, B, C, etc., their present-day designation, and suggested their use for standard reference. The method of measuring refractive index, using the minimum deviation position, is sometimes called the Fraunhofer method. His pioneering work with wire gratings gave an early standard wavelength (1821) and has led to his name being associated with the group of diffraction phenomena in which a plane wave passes an obstacle or through slits. The Fraunhofer design of telescope objective (double convex crown, plane concave flint) that corrects chromatic aberration and spherical aberration for one zone is very commonly used.

Fraunhofer diffraction *See* diffraction of light.

Fraunhofer eyepiece A terrestrial *eyepiece that has a lenticular erecting system in addition to the optical system of *Huygens's eyepiece or the *Ramsden eyepiece.

A	B	C,$H\alpha$	D_1
766·1	686·7	656·3	589·6
0*	0*	H	Na
D_2	D_3	E	b_1
589·0	587·6	526·9	517·8
Na	He	Fe	Mg
b_2	$F,H\beta$	$G',H\gamma$	G
517·3	486·1	434·0	430·8
Mg	H	H	Fe
g	$h,H\delta$	H	K, H'
422·7	410·2	396·8	393·4
Ca	H	Ca	Ca

Fraunhofer lines Absorption lines that occur in the spectrum of the photosphere (the sun's visible surface layer) and arise mainly as a result of absorption in the higher levels of the photosphere. A few lines are due to absorption in the terrestrial atmosphere.

History: W. H. Wollaston observed seven of these lines (1802) but considered most of them to be dividing lines between the colours of the spectrum. J. Fraunhofer (1814) observed many more, listing about 600 and naming the most prominent lines by letters. Although he knew that the *D-lines coincided with the emission lines of sodium, it was left to G. R. Kirchhoff (1859) to give the modern interpretation.

Nomenclature: The table gives the name (alternatives in some cases), wavelength (in nanometres), and origin of the main lines. * denotes lines of terrestrial origin. The line D_3 occurs as an emission line (*see* emission spectrum) from the chromosphere.

free axes *See* centrifugal moment.

free convection *Syn*. natural convection. Loss of heat vertically that occurs in the absence of draughts when the surrounding fluid circulates freely. The rate of cooling under these conditions obeys the *five-fourths power law although for a small temperature difference between the body and its surroundings *Newton's law of cooling may be applied. *See* convection (of heat).

free convection number *See* Grashof number; convection (of heat).

free electron An electron that is not permanently attached to a specific atom or molecule and is free to move under the influence of an applied electric field. *See also* energy bands; semiconductor.

free-electron paramagnetism *See* paramagnetism.

free end An end of a structure, e.g. a beam, that is not restrained in any way. In the case of a beam, at the end under consideration both the displacement, y, and the slope, dy/dx (where x is measured along the beam), may not be zero. When a column of air is contained in a pipe, the open end of the pipe is a free end.

free energy A thermodynamic function that gives the amount of work available when a system undergoes some specified change. *See* Gibbs function; Helmholtz function.

free fall The acceleration of a body under the action of a *gravitational field only, there being no air resistance or buoyancy. Near the surface of the earth the *acceleration of free fall*, g, is measured with respect to a nearby point on the surface. Because of the axial rotation the reference point is accelerated to the centre of the circle of its latitude, hence g is not quite equal in magnitude or direction to the acceleration towards the centre of the earth given by the theory of *gravitation. It varies slightly in magnitude with position on the surface because of the greater centripetal acceleration at lower latitudes, variation of distance from the centre, and local variations of density or level. g decreases on going upwards from the surface and increases slightly on going down a mine to any attainable depth. For some purposes the standard value, g_n = 9.806 65 m s^{-2}, is used.

The value of g can be measured by electronically timing the fall of a small sphere *in vacuo* between two levels determined by laser beams. Formerly, the most accurate methods for absolute measurements used the *pendulum. Variations of g are found using a *gravity meter* (*syn*. gravimeter). Moderately accurate measurements of variations due to local abnormalities are made with *gravity balances*. These are, in effect, sensitive spring balances in which the change in weight of a fixed mass is measured. Gradients of g

can be determined with a *gradiometer* such as the *Eötvös torsion balance.

A body in a state of free fall is said to be *weightless*. All living organisms are normally subject to stress as the force of gravity acts throughout the volume while support forces act on the surface. In the case of man the stress is large because the support usually acts over a small area. An astronaut in a freely falling space vehicle is not supported, so there is no stress (although there is still a gravitational force). Physiological changes occur, involving weakening of the bones.

free-field calibration Of a microphone. The determination of the open-circuit voltage produced by a certain sound pressure and the value of that pressure measured in the free progressive wave that existed before the introduction of the microphone. There are difficulties in the determination of this pressure since the free field is upset by the presence of the microphone. At low frequencies when the dimensions of the microphone are small compared with the wavelength, the pressure against the diaphragm is approximately that existing in the free field. At higher frequencies, however, reflections occur and the pressure on the diaphragm may reach twice that of the free sound wave. Resonance of the air just in front of the microphone also causes variations in the pressure on the diaphragm. As a result, the pressure calibration of a microphone in which the actual pressure on the diaphragm is measured varies considerably from the field calibration except at low frequencies. Ballantine has devised a method of finding the free pressure using a pressure-calibrated capacitor microphone housed in the spherical container. The relation between the free and actual pressures in this case is obtained from Rayleigh's mathematical treatment. In another method a very small capacitor microphone is used, which does not appreciably disturb the sound field, or a Rayleigh disc can be used to measure the velocity and the pressure calculated from this. Satisfactory results have been obtained with the disc between 10 000 and 300 hertz. To ensure that the sound wave used is a free progressive wave, these experiments are done in a heavily lagged room in which there are no echoes.

free magnetism An imaginary magnetic fluid to which the magnetic effects of a magnet are conventionally ascribed. In a bar magnet, the free magnetism is often regarded as being concentrated in two poles, one north-seeking, or positive, the other south-seeking or negative, but the actual distribution of free magnetism along the bar can be studied. The algebraical sum of the free magnetism on any specimen is always zero. *See* pole.

free oscillations *See* free vibrations.

free-piston gauge An absolute device for measuring high fluid pressures. The pressure is applied to one side of a small piston working in a cylinder and the force necessary to keep the piston stationary is a measure of the pressure.

free radical An atom or group of atoms with one or more unpaired electrons that readily enter into chemical bond-formation. It usually has a transient existence. Normally, the electrons in molecules exist, in a stable state, as pairs with *spins $\pm \frac{1}{2}$. A free radical can therefore increase its stability by its unpaired electron linking with another unpaired electron in a second free radical, thus leading to the formation of a molecule.

Free radicals occur in many chemical processes, including combustion, and are produced by the action of *ionizing radiation, ultraviolet radiation, or an intense flash of light.

free-running speed *Syn.* balancing speed. The constant speed at which the tractive effort of a traction vehicle, when running on a straight and level track, is exactly balanced by the forces opposing the motion of the vehicle.

free space A space that contains no particles (such as gas molecules) and no *fields of force. Formally it is distinguished from a *vacuum*, which contains no particles but may contain fields. For many purposes the two terms may be regarded as equivalent. Strictly a free space is only conceivable as a limit; for example, one cannot exactly describe the passage of electromagnetic radiation through free space, since the presence of the radiation means that the space is not free. The values of the properties possessed by free space fall into one of the following classes:

(a) zero (e.g. temperature);

(b) unity (e.g. *refractive index);

(c) the maximum possible (e.g. the *speed of light);

(d) a particular, formally defined value (e.g. *permeability and *permittivity).

free surface energy *See* surface tension.

free vibrations (or **oscillations**) A vibrating system when displaced from its neutral position oscillates about this position with a frequency characteristic of the system – the *natural frequency* of the system. The amplitude decays gradually, depending on the resistance of the medium to the motion and on the inertia of the system, until the energy supplied by the initial displacement has been expended into the medium. These vibrations are called free vibrations.

The expression for the displacement, x, in a free undamped vibration is given by:
$$x = a \sin (\kappa/m)^{1/2} t,$$
where a is the amplitude, κ and m are the elastic and inertia factors respectively, and t is the time.

The corresponding expression for free vibrations with damping is:
$$x = ae^{-\alpha t} \sin (\omega t - \delta),$$
where the damping factor α is equal to $\mu/2m$ (μ being the resistive term), and the angular frequency ω is given by:
$$\sqrt{[(\kappa/m) - \alpha^2]}.$$
δ is the angular displacement at $t = 0$ and is called the *epoch*.

Free oscillations occur not only in mechanical systems. They can also arise in electric circuits, as when a capacitor discharges through a resistance and inductance. The oscillations decay gradually, the frequency depending on the circuit parameters. *Compare* forced vibrations.

mixture	temperature
sodium chloride, water	$-20°$ C
ammonium nitrate, anhydrous sodium carbonate, water	$-10°$ to $-30°$ C
calcium chloride, ice	$-40°$ to $-70°$ C
sulphuric acid (dilute), ice	$-40°$ to $-90°$ C
solid carbon dioxide, ether	$-80°$ C

freezing mixture A mixture of two or more substances that absorb internal energy when they mix and thus produce a lower temperature than that of the original constituents. They are used in producing low temperatures for chemical reactions, freezing-point determinations, cold traps, etc. Some of the more common mixtures are shown in the table.

freezing point The temperature at which the solid and liquid phases of a substance can exist in equilibrium together at a defined pressure, normally standard pressure of 101 325 Pa. It is the same as the true *melting point*; however, a pure substance has a single reproducible melting point at a given pressure, and this may not be true for the freezing point. *See also* Blagden's law.

F-region *See* ionosphere.

Frenkel defect *See* defect.

frequency 1. The number of values of a statistical variable lying in a given range. *See* frequency distribution.

2. Symbol: ν or f. The number of complete oscillations or cycles in unit time of a vibrating system. It is measured in *hertz. The frequency, ν, is related to the *angular frequency, ω, by the formula $\omega = 2\pi\nu$.

3. Of an alternating current. The number of times the current passes through its zero value in the same direction in unit time; it is measured in hertz.

4. Of a wave. Ideally this has the same meaning as in (2), the oscillations being propagated through a medium. The frequency ν is then related to the wavelength λ and *phase speed v by: $v = \nu\lambda$. In practice ν and λ do not have exactly determinate values unless the wave has infinite duration. For an electromagnetic wave, emission and absorption occur in quanta of energy $E = h\nu$ where h is the *Planck constant. Suppose an atom is raised to an excited state of energy E and mean life T the process of emission can be regarded as the production of a train of waves of duration about T. The *uncertainty principle states that the energy has an uncertainty ΔE such that $\Delta E \simeq h/2\pi T$. Assuming typical values $E \simeq 5 \times 10^{-19}$ J, $T \simeq 10^{-9}$ s giving $\Delta E \simeq 10^{-25}$ J.

Thus $\Delta E/E \simeq 2 \times 10^{-7}$, so the frequency is indeterminate to about one part in five million in this case.

frequency analyser *See* wave analyser.

Wavelength	Band	Frequency
1 mm–1 cm	Extremely high frequency; EHF	300–30 GHz
1 cm–10 cm	Super-high frequency; SHF	30–3 GHz
10 cm–1 m	Ultra-high frequency; UHF	3–0.3 GHz
1 m–10 m	Very high frequency; VHF	300–30 MHz
10 m–100 m	High frequency; HF	30–3 MHz
100 m–1000 m	Medium frequency; MF	3–0.3 MHz
1 km–10 km	Low frequency; LF	300–30 kHz
10 km–100 km	Very low frequency; VLF	30–3 kHz

frequency band A range of *frequencies forming part of a larger continuous series of frequencies. The internationally agreed radio-frequency bands are given in the table.

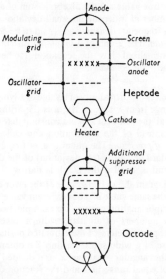

Heptode and octode valves as frequency changers

frequency changer 1. Generally, an electrical machine or circuit for converting alternating current at one frequency to alternating current at another frequency.

2. *See* mixer.

frequency deviation *See* frequency modulation.

frequency discriminator A device that selects input signals of constant amplitude and produces an output voltage proportional to the amount by which the input frequency differs from a fixed frequency. It is commonly employed in *automatic frequency-control systems (the output used to correct the frequency) and in *frequency-modulation systems (the frequency-modulated signals being converted to amplitude-modulated signals). The most common type is the *Foster–Seeley discriminator*.

frequency distortion *See* distortion.

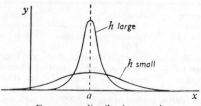

Frequency distribution graph

frequency distribution A table, graph, or equation describing how a particular attribute is distributed among the members of a group, e.g. the distribution of a set of measured quantities about their mean value.

When the *deviations of the members of the set from the true value are the algebraic sum of a very large number of independent small deviations, the resulting frequency distribution is said to be *normal* or *Gaussian*. The graph of a Gaussian distribution when normalized has the equation (sometimes called the *error equation*):

$$y = h\pi^{-1/2} \exp [-h^2(x - a)^2],$$

where $y.dx$ is the probability of a value of x lying in a small range from x to $x + dx$; a is the arithmetic mean of all the values of x; h is a constant determining the spread of the distribution and called the *modulus of precision*. The quantity $s = (h\sqrt{2})^{-1}$ is the standard deviation (*see* deviation) of the distribution and is small if the graph is narrow.

The diagram shows two graphs of the error equation for the same value of a but different values of h. These graphs might represent, for example, two sets of equal numbers of angular readings, taken on different spectrometers, of the estimated position of the centre of a wide spectrum line. To obtain the graph, the angular readings x are divided into groups of equal range in x and the fraction of the total number of readings y, in each group, is plotted against the mean value of x in that group. (As the area under the graph of the error equation is unity, the x-scale of the experimental graph needs adjusting by a factor to make the correspondence exact. This process is called *normalization.) Birge (1932) made 500 settings of a cross-wire on the centre of a wide spectrum line and found that the distribution was practically Gaussian. Bessel, using 470 readings of a particular astronomical angle, found good agreement with a Gaussian type of graph except that when $(x - a)$ was very large, the proportion of readings was greater than expected. Frequency distributions that are not normal are said to be *skew*. *See also* histogram.

frequency divider An electronic device, the output frequency of which is an exact integral submultiple of the frequency of the input.

frequency-division multiplexing A method of *multiplex operation in which a different frequency band is used for each of the input signals. The transmitted signal consists of a series of carrier waves of different frequencies, each modulated with a different input signal.

frequency doubler A particular type of *frequency multiplier in which the frequency of the output is twice the frequency of the input.

frequency function *Syn.* probability density function. *See* probability.

frequency meter An instrument for measuring the frequency of a wave or an alternating current. A *cavity resonator can be used to measure the frequency of an electromagnetic wave.

frequency-modulated cyclotron *See* synchrocyclotron.

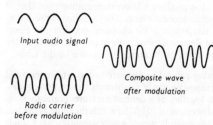

Input audio signal

Composite wave after modulation

Radio carrier before modulation

Frequency modulation

frequency modulation (FM, f.m.) A type of *modulation in which the frequency of the *carrier wave is varied above and below its unmodulated value by an amount that is proportional to the amplitude of the modulating signal and at a frequency equal to that of the modulating signal, the amplitude of the carrier wave remaining constant. (*See* diagram.) If the modulating signal is sinusoidal, then the instantaneous amplitude, e, of the frequency-modulated wave may be given as:

$$e = E_m \sin [2\pi Ft + (\Delta F/f) \sin 2\pi ft],$$

where E_m = amplitude of the carrier wave, F = frequency of the unmodulated carrier wave, ΔF = the peak variation of the carrier-wave frequency away from the frequency F, caused by the modulation, f = frequency of the modulating signal. Relevant definitions are:

Frequency swing. The variation (ΔF) in the frequency of the carrier wave.

Frequency deviation. The maximum value (ΔF_{max}) of the frequency swing for which the system has been designed.

Deviation ratio. The ratio of the frequency deviation to the maximum modulation frequency f_{max}.

Modulation index. The ratio of the frequency swing to the modulation frequency, i.e. $\Delta F/f$.

Modulation factor. The ratio of the frequency swing to the frequency of the unmodulated carrier wave, i.e. $\Delta F/F$. This factor is little used in practice.

Compared with *amplitude modulation, frequency modulation has several advantages, the most important of which is improved *signal-to-noise ratio. *Compare* phase modulation.

frequency multiplier An electronic device that produces an output signal whose frequency is an exact integral multiple of the frequency of the input. One type uses a nonlinear amplifier (usually *class B or *class C), so that the output is rich in harmonics of the input. The desired harmonic is selected by means of a *filter. Another type consists of a *multivibrator triggered by the input signal to produce oscillations with a frequency that is an exact integral multiple of the frequency of the input.

frequency swing *See* frequency modulation.

fresnel A unit of frequency equal to 10^{12} hertz.

Fresnel, Augustin Jean (1788–1827) French physicist. A key figure in establishing the wave theory of light, his earlier work on *interference was carried out in ignorance of that of Thomas Young but later they corresponded and were allies. Fresnel discounted earlier explanations of diffraction and interference phenomena as due to actions of edges, etc., by experiments with Fresnel mirrors, biprism, etc. (*see* interference of light), and he established the conditions necessary for two beams to interfere. His early work on wave theory assumed light to involve longitudinal displacements but experiments on polarization convinced him that the vibrations must be transverse. Assuming the wave motion to take place in a medium or *ether, he showed it must have properties of an elastic solid and he derived expressions (Fresnel's formulae, 1821) for the intensity of the reflected light at any angle of incidence on a bounding surface. In treating interference he introduced the method of dividing a surface into *zones*. He explained many of the phenomena of double refraction and he devised and constructed a glass rhomb (*Fresnel rhomb). (*See also* Fresnel lens.) Fresnel suggested (to explain observations by Arago) that a moving medium of refractive index μ communicates the fraction $(1 - 1/\mu^2)$ (Fresnel's convection coefficient) of its motion to the ether, a result that agreed with later experiments (1851) by *Fizeau.

Fresnel diffraction *See* diffraction of light.

Fresnel ellipsoid A figure in which the three rectangular axes are proportional in length to the ray velocities, that is, to the reciprocals of the principal refractive indices, $1/\alpha$, $1/\beta$, $1/\gamma$, and which lie parallel to the vibration directions (electric vectors) corresponding to α, β, γ.

Fresnel lens A lens consisting of a large number of steps, each one having a convex surface of the same curvature as the corresponding section of a normally shaped convex lens (*see* illustration). It was originally designed for use in lighthouses, to reduce the thickness and weight of the large lenses required. It is now also used as a *field lens in spotlights, camera viewfinders, etc., producing a large increase in image brightness. When close to a light source it is made of

Fresnel lens

glass; if no heat source is near it, as in a viewfinder, it is moulded in plastic (usually perspex), in which steps of a few tens of micrometres are used, the total thickness being about 0.5 mm.

Fresnel rhomb A glass rhombohedron that, by two internal reflections, changes *plane-polarized light into circularly polarized light.

Fresnel zone *See* diffraction of light.

Fricke dosemeter *See* dosimetry.

friction 1. Forces opposing the sliding of one surface over another. For a given value of the normal forces of interaction, N, between the surfaces, no sliding occurs for applied forces up to a limiting value F_l; this is the *limiting* or *static friction* at which sliding begins. For any pair of surfaces the *coefficient of limiting* (or *static*) *friction*, μ_l, is defined by:

$$\mu_l = F_l/N.$$

When sliding takes place at constant speed, the reuniting force F_k is the *kinetic* or *dynamic friction*. The *coefficient of kinetic* (or *dynamic*) *friction*, μ_k, is defined by:

$$\mu_k = F_k/N.$$

μ_k is lower than μ_l for low speeds and is roughly independent of speed, but it may rise considerably at high speeds.

By *Coulomb's laws of friction*: (1) μ_l and μ_k are constants, i.e. F_{lk} increases in proportion to N; (2) μ_l and μ_k are independent of the apparent area of contact for a given value of N. These laws are not quite exact but are good approximations over a wide range of values.

For typical metal-metal surfaces μ_l may have values 0.5 to 2.0, but much larger values may be given for very clean smooth surfaces. Friction between hard substances such as metals and oxides is caused by the interaction of *asperities*, i.e. minute irregularities that occur even on highly polished surfaces. Electrical resistance measurements show that even for large values of N the real area of contact is much less than the apparent area as the asperities interact, even welding together under sufficient loads. For softer materials, friction is caused by imperfect elasticity of layers extending some distance from the surface, in particular *elastic hysteresis.

Sliding friction is greatly reduced by *lubrication.

frictional electricity

(2) *Rolling friction*. The force F_R resisting the rolling of a circular body over a plane surface. The *coefficient of rolling friction* μ_R is defined by:
$$\mu_R = F_R / N.$$
μ_R has values typically 0.1 for rubber tyres on a hard road, and 10^{-2} for steel wheels on steel rails. It is caused mostly by elastic hysteresis and is unaffected by lubrication. The values increase with increasing load and decrease with increasing radius.

(3) *Journal friction*. The resistance to rotation of an axle in its bearing. For sliding bearings lubrication is necessary to prevent metal-metal contact. For most purposes ball or roller bearings are used, reducing the friction to the low values for rolling friction of hard bodies.

frictional electricity *Syn.* triboelectricity. The electric charge produced by rubbing together two dissimilar substances, e.g. ebonite and paper or glass and silk. The charges produced are equal and opposite, one of the substances becoming positively charged, the other negatively charged. Machines designed for the production of electricity by friction were known as *frictional machines*; they are now obsolete. *See* Wimshurst machine.

frictional machines *See* frictional electricity; Wimshurst machine.

friction loss The power loss in electrical machines due to friction in bearings and between sliding surfaces such as between *brushes and *commutator or *slip rings. It is usually expressed in watts.

Friedel's law X-ray diffraction measurements cannot distinguish the presence or absence of *centrosymmetry and therefore the complete process of *space-group determination cannot, in general, be carried out by X-ray analysis alone.

Friedmann, Alexandr Alexandrovich (1888–1925) Russian cosmologist, who applied Einstein's theory of general relativity to cosmology to give a series of models of the universe of constant average mass density in which space has a constant curvature. The different models, depending on whether space has negative, zero, or positive curvature, are called *Friedmann universes*.

fringes Bands, rings, or other patterns of alternate light and dark or of colour, produced by *interference or *diffraction of light.

Frisch, Otto Robert (1904–1979) Austrian-born Brit. physicist who with his aunt, Lise *Meitner, introduced the term nuclear fission and later worked on the development of the A-bomb.

front *See* air mass.

front-layer photocell *See* photovoltaic cell.

froth flotation *See* flotation.

Froude number In the relative motion of a floating body (e.g. a ship) and a fluid, the resistance force (drag) is dependent on the density of the fluid (ρ), the relative velocity (V), a characteristic length of the body (l), and the acceleration of free fall (g), the last term being due to the action of the gravitational attraction is producing waves. By the method of dimensions the resistance force D is given by:
$$D \propto \rho l^2 V^2 \phi(gl/V^2),$$
where $\phi(gl/V^2)$ represents a function of gl/V^2. This drag is the eddy-making and wave-making resistance but excludes the skin friction (due to viscosity of the fluid). The dimensionless number V^2/gl is called the Froude number. The resistance to motion per unit volume of a body, neglecting skin friction, is the same as the resistance to a scale model, providing the Froude numbers for each system are the same. For instance, if the scale of the model is $1/100$ then the equality of the Froude number requires that the velocity of the model is $1/10$ of the velocity of the prototype, and the two resistances per unit volume are equal providing that viscosity is not taken into account.

An analogous quantity is used in studies of animals walking or running. In this case l represents the height of the hip joint above the ground. The quotient of the length of stride divided by l is expressed as a function of this form of the Froude number.

fuel calorimetry *See* calorimetry.

fuel cell A device for the direct use of energy from an oxidation/reduction chemical process to maintain a flow of electricity. The requisite reagents are introduced continuously from outside the cell and react together with the aid of a catalyst. A typical simple fuel cell utilizes hydrogen and oxygen, which are fed to separate porous nickel plates in an electrolyte of weak potassium hydroxide solution; the plate fed by oxygen becomes the anode. The water that is formed from the gases so dilutes the electrolyte that its concentration must be increased from time to time. Fuel cells are able to deliver currents of twenty or more amperes for long periods, but are bulky and have efficiencies of 60% (as opposed to *accumulators, with efficiencies of 75%). The essential difference between a fuel cell and an accumulator is that the former feeds on chemicals and needs no charging, whereas the latter is recharged electrically and its chemicals do not need replenishing.

fuel element The smallest unit of a *fuel assembly* containing *fissile nuclides for powering a *nuclear reactor. The assembly (consisting of fuel elements and their supporting mechanism) together with the *moderator (if any) form the *core of the reactor.

fugacity Symbol: f. A corrected pressure used in the thermodynamic equations of real gases to give them the same form as the equations of ideal gases. It is related to *activity, λ, by:
$$f_A = \lambda_A \operatorname*{Limit}_{p \to 0} \left(\frac{y_A p}{\lambda_A} \right),$$
where p is the pressure and y_A is the number of moles of A divided by the total number of moles in the mixture.

fulchronograph An instrument for recording the characteristics of lightning currents. It consists of a set of magnet-steel laminations held in slots in the edge of a rotating aluminium wheel. During rotation each unit in turn passes between coils, which transfer the lightning discharge to earth, the laminations being magnetized in relation to the instantaneous current in the coil. A graph of current against time can be obtained by measuring the residual magnetic flux induced in each lamination.

full load The maximum output of an electrical machine or *transformer under certain specified conditions, e.g. of temperature rise.

full radiator See black body.

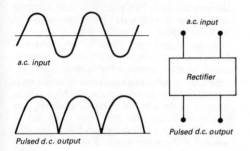

Full-wave rectifier circuit

full-wave rectifier circuit A rectifier circuit in which the positive and negative half-waves of the single-phase a.c. input wave are both effective in delivering unidirectional current to the load (see diagram). Compare half-wave rectifier circuit.

function generator A *signal generator producing specific waveforms, which may be used for test purposes, over a wide range of frequencies.

fundamental Generally that component of a complex vibration constituting a note by which the pitch of a note is described. In a given note it is generally the tone having the lowest frequency. However, if the fundamental and some of the lower *partials are removed from a note, the pitch of the note remains unchanged. See bell sounds.

fundamental constants See Appendix, Table 5.

fundamental interaction See interaction.

fundamental interval An arbitrary temperature difference assigned to the interval between two fixed points (the ice and steam points) in order to define a temperature scale. Celsius defined this interval as 100 °C, Fahrenheit as 180 °F, and Réaumur as 80 °R. The subdivision of the fundamental interval into the required number of equal parts (degrees) is carried out by using some property of the working substance of a thermometer, e.g. the value of the resistance of a platinum coil on the platinum-resistance thermometer scale. See International Practical Temperature Scale.

fundamental particles See elementary particle; particle physics.

funicular polygon A figure formed by a light flexible string suspended between two supports, to which weights are attached at various points. The name is also applied (in graphical statistics) to a diagram, complementary to a force diagram, drawn so as to display the lines of action of the various forces.

fuse 1. To melt or to cast. Care must be taken in interpreting this word. Some writers say that a substance is fused when it is in the liquid phase; thus an experiment is said to be done with a fused salt, meaning a molten salt. Alternatively, a substance may be said to be fused when it has been melted and then resolidified (see fused quartz).
2. A short length of easily fusible wire put into an electrical circuit for protective purposes. It is arranged to melt ("blow") at a definite current. The term includes all the parts of the complete device.

fused quartz Syn fused silica, silica glass, vitreous silica. A glass compound of pure silica (SiO_2) made by melting quartz, which cools down to form a supercooled liquid with all the normal properties of a hard solid other than crystal structure. It has a very small coefficient of expansion and is therefore used for weight thermometers, heat-resisting glassware, and *quartz-iodine lamps. It is transparent in the range 200 nm to 4000 nm so can be used for optical components for use in the ultraviolet and infrared as well as the visible. Unlike the crystalline material, it does not exhibit double refraction or optical activity (see quartz).

fused silica See fused quartz.

fusion 1. Syn. melting. The change of state of a substance from solid to liquid, which occurs at a definite temperature (melting point) at a given applied pressure. See Lindemann's theory of fusion.
2. See nuclear fusion.

fusion reactor Syn. thermonuclear reactor. A device in which *nuclear fusion takes place and in which there is a net evolution of usable energy. Intense research into the problems of designing such a device has occupied laboratories in many countries. The two central problems are (a) containing the *plasma in such conditions that it will yield more energy than is required to raise its temperature and confine it, and (b) extracting the energy in a usable form.

Three parameters control the first problem: temperature, *containment time, and plasma density. When fusion occurs the plasma temperature has to be high enough for the fusion energy released to exceed the energy lost by *bremsstrahlung radiation. The temperature above which this occurs is called the *ignition temperature*. The deuterium-tritium reaction:

$$^2H + {}^3H \rightarrow {}^4He + n \qquad + 17.6 \text{ MeV}$$

fusion reactor

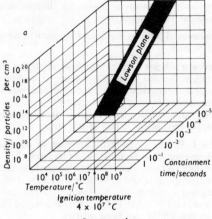

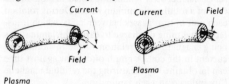

b Zeta-pinch device c Theta-pinch device

a Lawson plane

has the lowest known ignition temperature of 40 × 10^6 °C. This temperature was first achieved (for 3 microseconds) in a laboratory in 1963.

The problem of confining a plasma for long enough to release fusion energy has proved more difficult. Plasma instabilities have been the main cause of plasma leakage, but a workable plasma containment has now been achieved, though not at the same time as ignition temperature or adequate particle density. The basic criterion for determining the containment time at a given density and temperature was worked out in 1957 by J. D. Lawson. The *Lawson criterion is still in general use and for the deuterium-tritium reaction above its ignition temperature to reach break-even energy requires that the product of the plasma density (particles per cm^3) and the containment time in seconds must exceed 10^{14}. This condition defines a plane on a three-dimensional graph (Fig. a) that a given reactor using a particular reaction must achieve if it is to break even.

Several types of device have been used for fusion experiments; in most of them a strong pulse of current is passed through the gas to create the plasma. At the same time this current pulse creates a strong magnetic field that makes the charged particles in the plasma travel along helical paths around the lines of force of the field. This causes a contraction of the plasma away from the walls of the tube. This *pinch effect* partially solves the containment problem, but the confined plasma is not stable and tends to develop kinks. In *zeta pinch* devices, the current is passed axially through the plasma and the magnetic field forms round it (Fig. b). In the *theta pinch* (Fig. c), current-carrying coils run round the plasma and the magnetic field is axial. Both devices are often torus- (doughnut-) shaped. Since the mid-1970s most research into torus-shaped plasma confinement systems has been concentrated on the *tokamak configuration. This originated in the Soviet Union but research in many countries has confirmed that it has the greatest potential at pres-

ent for achieving the Lawson criterion. The Joint European Torus (JET) experiment at Culham in England uses this principle.

Linear devices are often called *magnetic bottles, their ends being "stoppered" with *magnetic mirrors. Greater stability in these linear devices is achieved by using extra current carriers; Fig. d illustrates the use of *Ioffe bars in a linear magnetic bottle. *Baseball bars create a *magnetic well within which the plasma is contained (Fig. e).

Another experimental device for the creation of plasma uses a pellet of fuel that is ionized instantaneously by a pulse from a high-power *laser.

Methods of extracting the energy of fusion reactions fall into two classes; those in which most of the energy is in the form of energetic neutrons, and those in which most of the energy is carried by charged particles. In the deuterium-tritium reaction, some 80% of the energy is carried by the neutrons. In this type of reactor the neutron energy could be absorbed by a liquid lithium *coolant surrounding the reactor tube. The heat so absorbed would be transferred to a conventional water, steam, and turbogenerator cycle. The neutron flux could also be used to breed *fissile fuel for fission reactors (*see* nuclear reactors).

In the deuterium-deuterium reaction a much larger proportion of the energy release appears as kinetic energy of charged particles. It has been proposed that some of this energy might be used directly to drive an electric current, without using a steam cycle. This would avoid the limitations on *efficiency imposed on heat engines by the laws of thermodynamics, but it is unlikely that more than a part of the energy release could be so used. Placing less reliance upon neutrons as the medium of energy transfer would also serve to reduce radioactive waste arising from neutron-induced radioactivity in the reactor structure. *See also* cold fusion.

G

g Symbol for the acceleration of free fall.

Gabor, Dennis (1900–1979) Hungarian-born Brit. physicist, who invented the technique of *holography (1948), for which he was awarded a Nobel prize (1971). His original work, using a mercury lamp, was designed to improve the resolution of electron microscopes. The use of lasers in holography was a later development (1961).

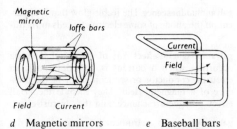

d   Magnetic mirrors
     with Ioffe bars

e   Baseball bars

Gaede molecular air pump See air pumps.

gain A measure of the advantage of using an electronic system. For an amplifier the gain is measured by the ratio of the power or voltage delivered by the amplifier to that of the input signal. For a directive aerial the gain is measured by the ratio of the voltage produced by a signal entering along the path of greatest sensitivity to that produced by the same signal entering an omnidirectional aerial. The gain is measured in *decibels, or sometimes in *nepers.

gain control A device that varies the *gain (i.e. amplification) of an *amplifier.

galaxy A giant assembly of stars, gas, and dust held together and organized largely by the gravitational interactions between its components. Galaxies contain most of the observable matter in the universe (see also dark matter). Few exist in isolation. The majority occur in groups, known as *clusters of galaxies*, which may contain up to a few thousand members.

Galaxies can be divided into three broad categories: *elliptical, spiral,* and *irregular galaxies.* Elliptical galaxies are dense spheroidal systems with no clearly defined internal structure. The stars are mainly cool and old and there is very little interstellar gas and dust. Spiral galaxies are disc-shaped systems with conspicuous spiral arms winding out from a dense central nucleus, which is sometimes bar-shaped. The arms contain mainly bright young stars and interstellar gas and dust, with older stars occurring in the nucleus. Irregular galaxies have no discernable shape or structure. They are small systems with a large amount of gas and dust. In many there is intense activity observed as emission of large amounts of radiation.

Although galaxies are of many sizes, an indication of the general properties might be:

major dimension	10^4 parsecs
absolute *magnitude	-20
velocity	3×10^5 m/s
mass	10^{10} solar masses
number of stars	10^{11}

The local group of galaxies, including our own and the Andromeda Galaxy, lie within a radius of about 6×10^5 parsecs. Beyond the local group the brightest galaxies lie in the distance range 10^6 to 10^7 parsecs. The space distribution of galaxies is of the order of 3 per 10^{20} cubic parsec.

The Galaxy refers to our own Milky Way System. It is a spiral system, roughly 2.5×10^4 parsecs by 4000 parsecs, with the sun situated about 10 kiloparsecs from the centre and only 10 parsecs north of the median plane, on the inner edge of a spiral arm. It has a mass of 2×10^{11} solar masses. In the region of the sun the Galaxy is rotating with a velocity of 2.2 $\times 10^6$ m s^{-1}. The whole system is thought to be 1.5 $\times 10^{10}$ years old.

The formation of galaxies is thought to have occurred several hundred thousand years after the big bang (see big-bang theory), and to have resulted from slight fluctuations in the density of the primordial gas. Details of the formation and subsequent evolution of galaxies are highly uncertain.

Galilean telescope See refracting telescope.

Galilean transformation equations The set of equations:

$$x' = x - vt,$$
$$y' = y,$$
$$z' = z,$$
$$t' = t.$$

They are used for transforming the parameters of position and motion from an observer at the point O with coordinates (x,y,z) to an observer at O' with coordinates (x',y',z'). The x axis is chosen to pass through O and O'. The times of an event are t and t' in the *frames of reference of observers at O and O' respectively. The zeros of the time scales are the instant that O and O' coincided. v is the relative velocity of separation of O and O'. The equations conform to Newtonian mechanics (see Newton's laws of motion). Compare Lorentz transformation equations.

Galileo (Galileo Galilei) (1564–1642) Italian mathematician, physicist, and astronomer. Usually acknowledged as the founder of modern physical science and scientific method. He discovered the constancy of period of the pendulum and applied it to make a *pulsilogium* for estimating the pulse-rates of patients. (He was at first a student of medicine.) The experiment (or "thought experiment") of dropping weights from the leaning tower of Pisa (1591) was to destroy the Aristotelian doctrine according to which heavy bodies fell fastest. Experiments on inclined planes led him to formulate the law of falling bodies (equivalent to $s = \frac{1}{2}gt^2$) and he demonstrated theoretically that the path of a projectile would be parabolic in the absence of air resistance. He invented perhaps the first thermometer, an air-thermometer (Padua, about 1593) and one of the earliest telescopes (1609), devised and constructed after hearing a report of the invention of a telescope in Holland. With his Galilean telescope (see refracting telescope) he observed mountains on the moon and four of the satellites of Jupiter. His support of the views of Copernicus concerning the solar system led to his persecution by the Church. His *Dialogues on the Ptolemaic and Copernican Systems* (1630) brought him to the Inquisition and he was forced to recant his views. Galileo is often also credited with inventing the microscope.

Galitzin pendulum *See* pendulum.

gallium arsenide (GaAs) devices Semiconductor devices based on the 3-5 *semiconductor gallium arsenide. The semiconducting properties of GaAs give it several advantages over silicon for certain applications. For example, it has a high *drift mobility, allowing it to be used for high-speed applications such as high-speed *logic circuits. It can be operated at microwave frequencies at which silicon devices cannot function: gallium arsenide is used for microwave devices, such as *Gunn diodes and *IMPATT diodes, and in microwave integrated circuits used, for example, in *direct broadcast by satellite and in phased-array *radar. In addition GaAs is a type of semiconductor known as a *direct-gap semiconductor*, allowing it to be used for optical components, such as *light-emitting diodes and *semiconductor lasers, and optically coupled devices. It therefore offers the potential for fabricating integrated optoelectronic circuits. GaAs devices are also more tolerant to ionizing radiation than silicon devices. It is, however, much more difficult to fabricate GaAs devices.

gallon A unit of volume now defined in the UK as being equal to exactly 4.546 09 cubic decimetres. The U.S. gallon = 0.832 674 UK gallons. The unit is not used for scientific purposes.

Galton whistle In order to determine the threshold of audibility of sound with regard to pitch (*see* audibility), Galton devised a miniature organ pipe in the form of a whistle to produce high frequencies. It consists essentially of a short cylindrical pipe blown from an annular nozzle of which the height of the mouth (distance of the nozzle from the edge of the pipe) can be varied by turning the lower micrometer screw. By suitable adjustment of this distance and the pressure of the air blast, the pipe is set into resonant vibration at a frequency corresponding to its length and diameter. (*See* edge tones.) Frequencies above the human audible limit (normally above 20 000 hertz) can be produced by means of this whistle; such sounds can be easily detected by a sensitive flame, a hot-wire microphone, or a *Rayleigh disc. Hartmann and Trolle have produced a powerful source of sound on the same principle as the Galton whistle. *See* Hartmann generator.

Galvani, Luigi Aloisio (1737–1798) Italian physiologist. He observed (1790) muscular action due to contact with dissimilar metals. Hence the terms *galvanic cell, galvanized iron (although the zinc coating is normally applied by dipping and not electrically), and *galvanometer.

galvanic cell Obsolescent name for voltaic cell.

galvanism 1. The study of the passage of electricity through electrolytes and cognate phenomena (obsolete).

2. The treatment of diseases by the application of a direct current.

galvanoluminescence The feeble glow that is apparent on the anode of some electrolytic cells or rectifiers.

galvanomagnetic effect Any of various phenomena occurring when a current is passed through an electrical conductor or semiconductor in the presence of a magnetic field. They include the *Hall effect, *magnetoresistance, and the *Nernst effect.

galvanometer An instrument for measuring or detecting small currents, usually by the mechanical reaction between the magnetic field of the current and that of a magnet. In the moving-magnet galvanometer (of which the *tangent galvanometer is an example), the magnetic field produced by the current flowing in a fixed coil of wires is used to deflect a small magnet suspended in a uniform magnetic field (e.g. that of the earth). In the moving-coil galvanometer, the coil carrying the current is suspended in the field of a permanent horseshoe magnet, and tends to set with its place at right angles to the lines of the magnetic field. The suspension supplies the necessary torsional control. For high-frequency currents, use may be made of a *thermogalvanometer* in which the temperature rise in a resistance wire through which the current is passing is employed to measure the current, the temperature rise being measured by means of a thermocouple. *See also* astatic system; Broca; Helmholtz coils; Einthoven galvanometer.

galvanometer constant The number by which a galvanometer's scale reading must be multiplied to give the current in any convenient units selected.

gamma (γ) **1.** The symbol used to denote the ratio of the principal *heat capacities C_p / C_V of a substance, where C_p is the heat capacity at constant pressure and C_V that measured at constant volume. The ratio may be determined experimentally for a gas by the methods due to (*a*) *Clément and Desourmes, (*b*) *Lummer and Pringsheim, (*c*) *Partington, (*d*) *Ruchardt, and (*e*) from the determination of the speed of sound in a gas, which is given by:
$$C = \sqrt{(\gamma p/\rho)} \text{ or } \sqrt{(\gamma RT)},$$
where R is the gas constant per unit mass of gas. Using the law of *equipartition of energy, the classical theory predicts that the ratio of the heat capacities should be given by:
$$\gamma = 1 + 2/F,$$
where F is the number of *degrees of freedom of the molecule. For a monatomic gas for which $F = 3$, $\gamma = 1.667$, in agreement with experimental determinations on the rare gases, but, for polyatomic molecules the experimental values are sometimes greater than those predicted by the formula and to give agreement F must sometimes be fractional. These difficulties are removed by application of the quantum theory to the problem but in any case, the fact that for complicated molecules the value of γ tends to unity, as indicated by making F large in the formula, is borne out in practice. (*See also* specific heat capacity.)

2. In photographic work. The gradient of the linear part of the graph of *density of exposed and processed photographic material against the logarithm of the *exposure. The gamma of a plate or film is a measure of the contrast the material will give in standard conditions.

3. A former unit of magnetic flux density, used in *geomagnetism, equal to 10^{-9} *tesla.

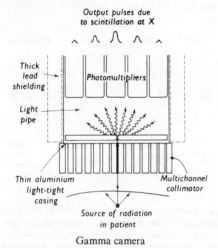

Output pulses due to scintillation at X

Thick lead shielding

Photomultipliers

Light pipe

Thin aluminium light-tight casing

Multichannel collimator

Source of radiation in patient

Gamma camera

gamma camera A device for visualizing the distribution of *radioactive compounds in the human body during diagnosis using radioisotopes. The gamma camera consists of a large thin *scintillation crystal, with an array of *photomultiplier tubes mounted above the crystal and connected with it by a section of transparent material. Usually the crystal is about 25 cm diameter by 1 cm thick, with 19 photomultiplier tubes mounted in circles above it. (*See* diagram.) A collimator is placed in front of the crystal to prevent scattered radiation interfering with the picture and to increase the resolution of the system. Radiation causing a scintillation at point X will create pulses in the photomultiplier tubes, the size of the pulses being dependent on the relative position of the tubes and X. The output of the photomultiplier tubes is fed into a circuit that analyses the pulses produced. The total sum of the pulses gives the intensity of radiation from X, i.e. the energy of the incident radiation; the relative sizes of the pulses gives positional information. The output, after amplification, controls the position of the spot on a *cathode-ray tube. If a camera is positioned to photograph the screen of the cathode-ray tube, after a suitable time interval a picture of the area below the crystal is built up. More sophisticated gamma cameras may have the output fed into and analysed by a computer, allowing corrections to be made for nonuniformities introduced by the camera itself. Because the crystal is thin, gamma cameras are more suitable for use with low-energy γ-ray emitters, such as technetium-99m (*metastable), which has a single γ-ray energy of 140 keV. Gamma cameras are useful

diagnostic aids as even a simple camera can produce a useful picture of the liver, for example, in about 5 minutes.

gamma rays (γ-rays) Electromagnetic radiations emitted by nuclei. A nucleus may be raised to an excited state by a nuclear interaction, such as the inelastic *scattering of a neutron or the absorption of radiation. A nucleus may be produced in an excited state by the emission of α- or β-rays, *electron capture, or the capture of a neutron. Excited states usually have lifetimes less than picoseconds and decay directly or indirectly to the ground state with the emission of one or more quanta of gamma radiation. Some states are much longer-lived because of certain *selection rules. In such cases decay by *internal conversion may compete with gamma-ray emission.

The quantum energies of gamma rays are mostly in the range 10^4 eV to 5×10^6 eV, giving wavelengths 10^{-10} m to 2×10^{-13} m. Some gamma-ray photons of cosmic origin have been detected with wavelengths down to 10^{-15} m. Those of long wavelength are almost totally absorbed by thin metal foils, while short-wave radiations are detectable through many centimetres of lead. The radiations usually form a line spectrum with a few wavelengths, but in several cases just one wavelength is observed. In many cases of α- or β-decay or electron capture, nuclei are produced in the ground state directly and therefore there is no gamma emission. X-rays caused by beta rays interacting with surrounding materials show the continuous spectrum of *bremsstrahlung and should not be confused with gamma rays.

Gamma rays ionize matter by the electrons they eject by the *photoelectric effect, or *Compton effect, or by *pair production. They can be detected by all forms of *counter, by *ionization chambers, and photographic emulsions. The rays that are most studied are those of the higher quantum energies, which are very penetrating; hence to obtain high detection efficiency counters must be used that have high stopping power, namely sodium iodide *scintillation counters or germanium *crystal counters.

Gamma rays are used for *radiography and in *nuclear medicine. Sources with low quantum energies are used in studies involving the *Mössbauer effect.

Electromagnetic radiations from sources other than nuclei are sometimes loosely called gamma rays when the quantum energy is more than a few hundred keV. Cases include *annihilation and the decay of some *elementary particles.

gamma-ray spectrum A series of wavelengths in the gamma-ray region emitted by a given gamma-ray source. *See also* spectroscopy.

gamma-ray transformation A radioactive disintegration accompanied by the emission of *gamma rays.

Gamow, George (1904–1968) Russian-born Amer. physicist. Educated in the Soviet Union, he worked in Göttingen and Cambridge before settling in

Gamow barrier

America in 1934, becoming professor at George Washington University and later at the University of Colorado. He was, with Ralph Alpher and Hans Bethe, responsible for the modern version of the big-bang theory, first presented (1948) in the so-called *alpha-beta-gamma* paper and later expanded in Gamow's *Creation of the Universe* (1952). In 1956 he postulated that elements with proton number in excess of 5 were formed later than the big bang in the interior of stars. His prediction of a uniform background radiation that originated with the big bang was confirmed by its discovery in 1968. His later work concerned the construction of amino acids from the genetic code contained in DNA.

Gamow barrier *See* nuclear barrier.

ganged circuits Two or more circuits having variable elements mechanically coupled by a single control so that the circuits can be adjusted simultaneously.

gap *See* air gap; horn gap; protective gap; spark gap; sphere gap.

gas A fluid that expands to fill any container, however large, without any change of phase. If the substance is below the *critical temperature it is called a *vapour*. Usually the term gas is understood to apply in this case also, that is, the vapour is a special case of a gas and not a distinct form.

In any substance molecules have translational kinetic energy (which is positive) and potential energy of intermolecular interaction (which is negative on the average). In a gas the total energy of the molecules is positive, whereas in condensed phases (liquid and solid) the total energy is negative.

gas amplification *See* gas multiplication.

gas breakdown A type of *breakdown that occurs when the voltage across a *gas-filled tube reaches a given value. Electrons in the gas are accelerated by the field to such energies that further ion-electron pairs are produced by collision but little recombination of ions occurs due to the high kinetic energies. Positive ions eject electrons from the cathode by *secondary emission. A multiplication effect is present causing breakdown of the gas. The process is analogous to *avalanche breakdown in a *semiconductor.

gas-bubble protective device *See* Buckholz protective device.

gas constant *See* molar gas constant.

gas-cooled reactor A type of thermal *nuclear reactor in which a gaseous *coolant is used. In the *Magnox reactors the coolant is carbon dioxide and the outlet temperature is about 350 °C: natural uranium metal fuel is used with a graphite *moderator, the fuel elements being encased in Magnox alloy. In the *advanced gas-cooled reactors* the fuel is ceramic uranium dioxide encased in stainless steel. The same coolant and moderator as in the Magnox type are used, but the outlet temperature is considerably higher – usually about 600 °C. As a result of

236

these higher temperatures, special precautions have to be taken to cool the graphite to avoid chemical attack.

In the high-temperature gas-cooled reactor (HTR), the reactor core is composed entirely of ceramic materials, with helium as coolant. A variant of this type is the pebble-bed reactor, in which ceramic pebbles (incorporating both fuel and moderator) are loaded into a vessel to form the reactor core.

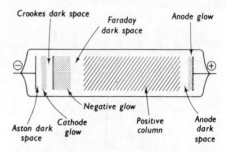

Gas-discharge tube

gas-discharge tube An *electron tube in which the presence of gaseous molecules contributes significantly to the characteristics of the tube. Normally a gas is a poor electrical conductor but if a sufficiently high electric field is applied, conduction can occur. If two plane electrodes are sealed in a tube and a potential difference applied between them, the gas can conduct as a result of an external ionizing agent, such as *ultraviolet radiation. If the ionization agent is removed, the current ceases. Under certain conditions the *discharge can be self-sustaining and independent of the external agent.

In self-sustaining discharges the ions and electrons initially formed in the tube are accelerated to the electrodes and the electrons cause further ionization along their path. Electrons are also produced by *secondary emission at the electrodes. Electrons and ions are removed at the *anode and *cathode respectively and by recombination. A stable state can be reached when the rate of production of ions and electrons is equal to the rate at which they are removed. The characteristics of the discharge depend on the gas, the pressure, the electric field, and the shapes and materials of the electrodes.

The most common type of discharge is the *glow discharge*, characterized by several luminous regions in the tube (*see* diagram). In the cathode region of the tube the electrons are emitted from the cathode as a result of ion bombardment (*see* secondary emission). They are accelerated towards the anode and for a short distance they have not enough kinetic energy to ionize the atoms of gas or to excite them. The positive ions moving towards the cathode have, in this region, a high velocity and a low probability of recombining with electrons. Any excited ions produced further down the tube have returned to their *ground state by the time they reach this region. Consequently, it emits no radiation and is called the *Aston dark space*. The *cathode glow* is a luminous region near the cathode where positive ions that have been excited by electrons return to their ground state with emission of luminous radiation.

In the *Crookes dark space* electrons moving from the cathode have gained enough kinetic energy to ionize atoms but the electrons thus produced do not have sufficient energy to excite atoms. Consequently, the region produces little radiation. In the *negative glow* the electrons have gained sufficient energy to cause excitation and the excited atoms return to their ground state with emission of radiation. A small amount of the radiation is also produced by recombination of ions and electrons in this region. In passing the region of the negative glow the electrons lose much of their energy and in the *Faraday dark space* they again have insufficient energy to excite or ionize the gas. Further along is a large luminous region (the *positive column*) in which the gas is excited and emits radiation. The relative sizes of the negative glow and positive column depend on the gas pressure, which determines the *mean free path of charged particles. At pressures below about 15 pascals the positive column often displays *striations*, i.e. alternate dark and light regions caused by the electrons alternately gaining and losing kinetic energy in their journey to the anode.

In a glow discharge the potential drop across the tube is independent of the current and does not vary uniformly down the length of a glow discharge. Most of the potential drop occurs between the cathode and the negative glow. This drop is called the *cathode fall* and depends on the material of the cathode as well as the nature of the gas. The current increases with gas pressure. When the current increases to a certain point the glow covers the whole of the cathode and beyond this point the voltage drop increases with current. This is called an *abnormal glow discharge*.

Two other types of discharge are distinguished. In an *arc discharge* the current density is very high and thermionic emission from the cathode can occur as well as secondary emission. In an arc discharge the voltage drop falls as the current increases. Arc discharges can occur over a very wide range of pressures and the term is applied to a large number of discharges that are not very well understood.

The *Townsend discharge* occurs at lower currents than the glow discharge and the voltage increases with increasing current. At low current densities the potential falls uniformly down the tube and there is a uniform luminous region between the two electrodes.

When discharges occur at low pressures, the ions and electrons rarely lose energy in collision with gas molecules. Electrons emitted from the cathode by the impact of positive ions are called *cathode rays*. They gain sufficient kinetic energy to produce *X-rays in their collision with the anode. If holes are bored in the cathode, positive ions can pass through

and cause the glass to fluoresce. These are known as *canal rays*.

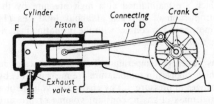

Gas engine

gas engine An internal combustion engine using air mixed with about one-eighth of its volume of coal or natural gas, admitted through a valve to the cylinder on the left of the piston B, connected by the rod D to the crank C. The ignition is usually by a spark, water circulating in the jacket F to prevent the cylinder becoming too hot. The exhaust valve E of the mushroom type is operated by a cam on a side shaft driven by the crankshaft C.

gaseous ions Positively or negatively charged systems formed in gases by the action of *ionizing radiation (e.g. X-rays); when an electric field is applied across the gas, the motion of the gaseous ions under the action of the field conveys an ionization current across the gas. They differ from electrolytic ions in the fact that they are not permanent, but recombine to form neutral molecules within a short time after the ionizing radiation has been cut off. In moist gases, and in some dry gases, the ion consists of an aggregation of molecules clustered around the original charged molecule or atom. *See* conduction in gases.

gas-filled relay *See* thyratron.

gas-filled tube An *electron tube containing a gas (or vapour, e.g. mercury vapour) in sufficient quantity to ensure that, once ionization of the gas has taken place, the electrical characteristics of the tube are determined entirely by the gas. *See also* conduction in gases.

gas focusing It is possible to focus the beam in a cathode-ray or other electron-beam tube by allowing a trace of gas to remain in the tube. This becomes ionized by collision with electrons and forms positive ions along the beam in such a manner that a concentrating field is set up. *See also* focusing.

gas laws Laws governing the variation of physical conditions (temperature, pressure, etc.) of a gas. *See* equations of state; ideal gas.

gas multiplication *Syn.* gas amplification. **1.** The process by which, in a sufficiently strong electric field, ions produced in a gas by *ionizing radiation can produce additional ions.

2. The factor by which the initial ionization is multiplied as a result of this process.

gas-pressure cable A pressure cable employing a gas (chemically unreactive) under pressure. It is a dry or

gas scales of temperature

oil-impregnated paper-insulated cable with a mechanically reinforced lead sheath, the contents of which are maintained at a high pressure by the introduction of nitrogen at about 1400 kPa (14 atmospheres) in contact with the insulating material. *Compare* compression cable.

gas scales of temperature Temperature scales in which changes in temperature are measured (1) by changes in pressure, (2) by changes in volume of a fixed mass of gas at constant volume or pressure respectively.

1. The Celsius temperature t_v measured on the constant-volume scale with a thermometer employing a particular gas is given by:

$$t_v = \frac{p_t - p_0}{p_{100} - p_0} \times 100,$$

where p_t, p_{100}, and p_0 are the pressures exerted by the given mass of gas when in equilibrium at the temperature t_v, the steam point and ice point respectively, at the constant volume V_0.

2. The Celsius temperature t_p measured on the constant-pressure scale with a thermometer employing a particular gas is given by the equation

$$t_p = \frac{v_t - v_0}{v_{100} - v_0} \times 100,$$

where v_t, v_{100}, and v_0 are the volumes occupied by the given mass of gas when in equilibrium at the temperature t_p, the steam point, and ice point respectively, at a constant pressure p_0.

Different gases give slightly different temperature scales when the initial pressure is finite, the same gas giving a different temperature on the two scales. When, however, the initial pressure tends to zero the scales given by different gases and different thermometers become identical and this extrapolated gas scale is chosen as the standard temperature scale. It is, of course, an ideal-gas scale, for an ideal gas obeys Boyle's law, and for such a gas it may be seen that the constant-pressure and constant-volume scales are identical, and each agrees with the scale defined by

$$t' = \frac{(pV)_t - (pV)_0}{(pV)_{100} - (pV)_0} \times 100.$$

It is of course in the condition of the initial pressure tending to zero that an actual gas behaves as an ideal gas. Thermometers employing actual gases have to be corrected to give the temperature t' on the ideal-gas scale. *See* constant volume gas thermometer; constant pressure gas thermometer.

Gassendi, Pierre (1592–1655) French philosopher and astronomer who revived the atomic theory of *Epicurus and suggested that atoms combine by a hook-and-eye mechanism.

gassing The evolution of gas in the form of small bubbles that occurs in an accumulator towards the end of the charging period.

gas thermometer *See* constant-pressure gas thermometer; constant-volume gas thermometer.

gas turbine *See* turbine.

gate 1. An electrode or electrodes in any of a number of devices. In a *field-effect transistor it is the electrode(s) to which a bias is applied for the purpose of modulating the conductivity of the channel.
2. *Digital gate.* A digital electronic circuit, with one or more inputs but only one output, frequently used in *logic circuits. The output is switched between two or more discrete voltage levels, depending on the input conditions.
3. *Analogue gate.* A *linear circuit or device, frequently used in radar or electronic control systems, that passes signals only for a specified fraction of the input signal. The output is a continuous function of the input signal for the period that the circuit is switched on.

gate array An integrated logic circuit comprising a two-dimensional array of digital logic *gates that can be interconnected in an arbitrary manner during manufacture. The interconnections determine the performance of the chip, according to the required application. It is thus a programmable device.

gauge boson *See* gauge theory.

gauge theory A quantum field theory for which all measurable transformations remain unchanged under a *gauge transformation*, in which the phases of the fields are altered by an amount that is a function of space and time.

There is, for a particular gauge theory, a symmetry group for the fields (the *gauge group*). Gauge theories are now believed to provide the basis for a description of all elementary particle interactions. *Quantum electrodynamics is a relatively simple form of gauge theory (the *group is Abelian), whereas *quantum chromodynamics and *electroweak theory are more complicated (non-Abelian) gauge theories.

In these theories, the interactions between particles are mediated by other elementary particles known as *gauge bosons*. The photon is the gauge boson of the electromagnetic interaction, the gluon is the gauge boson of the strong interaction, and the W and Z^0 bosons are gauge bosons of the electromagnetic-weak interaction. In the gravitational interaction, the gauge boson would be the *graviton, although there are considerable problems in developing *quantum field theory to include gravity.

gauge transformation The addition of a gradient of a function of space and time to the *magnetic vector potential and the subtraction of the partial derivative of this function with respect to time, divided by the speed of light, from the electric potential. Such transformations change the potentials but do not change the electric and magnetic fields.

gauss Symbol: G. The CGS-electromagnetic unit of *magnetic flux density. $1 \text{ G} = 10^{-4}$ tesla.

Gauss, Karl Friedrich (1777–1855) German mathematician, physicist, and astronomer. He investigated the occurrence of errors, the normal distribution being called *Gaussian* (*see* frequency distribution), and he established the method of *least squares for fitting curves to experimental results. His calculation of the elements of the minor planet Ceres, discovered in 1801, gave him fame as an astronomer. From about 1830 he worked especially on electricity and magnetism, inventing (with W. E. Weber) *magnetometers of various forms and other instruments and making systematic magnetic observations. In 1839 he published his general treatment of inverse-square forces, including the flux theorem known now as *Gauss's theorem. (*See also* Gauss positions.)

In optics, Gauss developed simplified formulae for the formation of images by thick lenses and combinations (1841). By finding the positions of six *cardinal (Gaussian) points, calculations of image formation can be simplified. The Gauss telescope objective aims at correcting chromatic aberration for two wavelengths by using crown and flint glass menisci. *See also* Gaussian eyepiece.

Gaussian curvature *Syn.* total curvature. *See* principal radii.

Gaussian distribution *Syn.* normal distribution. *See* frequency distribution.

Gaussian eyepiece A telescope eyepiece provided with a side window to illuminate the cross-wires, and a reflector to transmit the light through the objective for the purpose of *autocollimation.

Gaussian optics *Syn.* paraxial theory; first-order theory. A simplified theory of geometric optics concerned only with light rays close to the optic axis (i.e. paraxial rays). *See* centred optical system; Seidel aberrations.

Gaussian points (or **constants**) *See* centred optical system.

Gaussian system of units *See* CGS system of units.

gaussmeter *See* fluxmeter.

Gauss positions Two arrangements of magnets employed by Gauss in his verification of the inverse square law of force. A point on the axis, produced, of a bar magnet is said to be in the Gauss A (or end-on) position with respect to the magnet; a point on the magnetic equator of the magnet is said to be in the Gauss B (or broadside-on) position with respect to the magnet.

Gauss's theorem For any closed surface drawn in an electric field the integral $\int D.dS$ of the normal component of the *electric displacement, D, over the surface is equal to the total charge within the surface. If the surface encloses no charge the electric field strength within the space is equal to zero.

Gauss's theorem applies also to surfaces drawn in a magnetic field. Analogous statements of the theorem may be made for gravitational, magnetostatic, and fluid-velocity fields. The general mathematical statement is that the total flux of a vector field through a closed surface is equal to the volume integral of the divergence of the vector taken over the enclosed volume.

gauze tones *See* howling tube.

Gay-Lussac, Joseph Louis (1778–1850) French chemist and physicist. The discoverer of the law that chemical combination between gases involves simple proportions between the volumes of the separate reacting gases and of the product. This fact helped greatly to establish the atomic theory and led to *Avogadro's hypothesis. Gay-Lussac also experimented on the expansion of gases with rise of temperature and published his conclusions in 1802. (*See* Charles; Charles's law.) Gay-Lussac was, like Charles, a pioneer in balloon ascents and ascertained that the composition of the atmosphere did not appreciably change up to 6000 metres altitude.

Gay-Lussac's law 1. Of volume. The volumes in which gases combine chemically bear a simple relation to one another and to that of the resulting product if this is also gaseous. The volumes must all be measured under the same conditions of temperature and pressure. *See* ideal gas.

2. *See* Charles's law.

Gay-Lussac's method of measuring vapour density *See* density.

geepound *See* slug.

Geiger, Hans (1882–1945) German physicist. The inventor of the *Geiger counter, he is well known for his work on atomic theory and cosmic rays. *See also* Geiger–Nuttall relation.

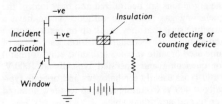

Geiger-counter circuit

Geiger counter *Syn.* Geiger–Müller counter; Geiger tube; Geiger–Müller tube. An instrument for counting ionizing particles and photons. A wire anode lies on the axis of a cylindrical cathode (*see* diagram). The electrodes are contained within a glass tube, or the cathode itself serves as a container. In order to detect particles of low penetration such as alpha rays and low-energy beta rays some tubes have thin windows of mica (2–4 mg per cm^2). Light alloy windows of rather greater absorption are suitable for medium- or high-energy beta rays. The tube contains argon at a pressure of a few tenths of an

Geiger–Nuttall relation

atmosphere, together with a small quantity of *quenching gas* (see below). A potential difference slightly lower than that required to produce a *discharge through the gas is maintained between the electrodes. Any charged particle that passes through the gas (except through small insensitive volumes near the ends) will initiate a discharge. The detection efficiency is very low for electromagnetic radiation, which must eject an electron into the gas by *photoelectric effect or Compton scattering (*see* Compton effect) in order to cause a response.

Whenever any free electrons, however few, appear in the sensitive volume of the tube, a discharge is initiated. The electrons move to the anode causing further ionization by collision. Ultraviolet radiation from excited atoms ejects photoelectrons from the cathode causing the discharge to grow until an *avalanche occurs. The surge of current to the anode causes a voltage pulse, which operates an electronic counter, usually through a *scaler (*see* dekatron). The size of this pulse is independent of the original amount of ionization so the instrument only counts particles or photons without distinguishing them or measuring their energies. (*See also* ratemeter.) As the discharge develops a large number of positive ions are produced, and until most of these are collected by the cathode the counter tube is insensitive to further incident radiation. This gives a *dead time* of the order of 10^{-4} s. To permit accurate correction for rays, which may be undetected because of this, it is usual for the electronic circuits to provide an accurately determined dead time larger than that of any tubes that are likely to be used.

When a positive ion reaches the cathode, it may eject further electrons by *secondary emission. If this occurs just after the end of the dead time, the discharge will be renewed. This is prevented by the quenching gas. This has molecules with *ionization potential less than that of argon so that on collision the argon ions are neutralized and only ions of the quenching gas reach the cathode. These have a structure such that they break up on neutralization and do not cause secondary emission. Early types of tube used organic vapours (ethanol; ethyl-formate) for quenching and operated typically at 1000 V. Later types usually use halogens and operate typically at 400 V. Counting rates up to a few hundred per second are practicable.

Compare proportional counter; scintillation counter.

Geiger–Nuttall relation (1911) The relation, discovered empirically, between the range, R, of an α-particle emitted by a given radioactive substance and the *decay constant, λ, of the substance:
$$\log \lambda = A + B \log R,$$
where B has the same value for all four radioactive series, while the constant A has a different value for each series. The law is of only approximate validity. Later researches have shown that similar laws, in which the energy of the α-particle replaces the range, can be applied with greater accuracy to certain groups of alpha emitters, for example the isotopes of

a given element. This is in agreement with theoretical analyses of the *tunnel effect. The most striking feature of the relation is the critical dependence of the decay constant upon the α-particle energy. A factor of two in energy causes a factor 10^{24} in λ.

Geissler, Henry (1814–1879) German scientific instrument maker.

Geissler pump *See* Toepler pump.

Geissler tube A *gas-discharge tube specially designed to demonstrate the luminous effects of an electrical discharge through a rarefied gas. It is used in *spectroscopy for investigating the spectra of gases.

Geitel, Hans Friedrich (1855–1923) German physicist who collaborated with Elster in extensive work on conduction in gases and in the atmosphere; they were pioneers in the development of the photoelectric cell.

gel The jelly produced when an emulsoid *sol coagulates.

Gell-Mann, Murray (*b.* 1929) Amer. physicist who has made several contributions to particle physics, including the concepts of *unitary symmetry, *strangeness, and the *quark model. He was awarded the Nobel prize for physics in 1969.

generalized coordinates Symbol: q_i or $\boldsymbol{q}$. Coordinates describing the motion of a mechanical system without specifying its exact nature. The *generalized momentum*, symbol: p_i or $\boldsymbol{p}$, is related to q_i by the relation:
$$p_i = \partial L/\partial q_i,$$
where L is the *Lagrangian function. *See also* Lagrange's equations; degrees of freedom.

generalized force *See* force.

generalized momentum *See* generalized coordinates.

general relativity *See* relativity.

generating set A set comprising one or more *generators complete with the prime mover that drives them.

generating station *See* power station.

generation time The mean period of time from the creation of a neutron in a fission process to a *fission brought about by the neutron itself.

generator A machine that drives an electric current when it itself is driven mechanically. In the electromagnetic generator (dynamo), a coil is moved so as to cut the lines of induction in a magnetic field. In the electrostatic generator (*see* Van de Graaff generator; Wimshurst machine), work is done in separating equal and opposite electrical charges produced by electrostatic induction or by friction. *See also* alternating-current generator; direct-current generator; induction generator.

geodesic 1. The path with minimum (or maximum) length between two points in a mathematically defined space. In three dimensions it is a straight line. On the surface of a sphere it is a *great circle.

2. The equivalent to (1) in the *four-dimensional continuum in the general theory of *relativity. It is the path of electromagnetic radiation, or of a particle that is subject to no nongravitational force.

geoid A surface defined by the term sea level, the ocean being imagined taken to any desired point by means of canals and to be free from tides. The geoid is thus a surface in the earth's gravitational field and is approximately an oblate spheroid in shape. The following data represent the shape of the geoid:

equatorial radius	$= 6.378\ 169 \times 10^6$ m
polar radius	$= 6.356\ 775 \times 10^6$ m
mean radius	$= 6.371\ 030 \times 10^6$ m
equatorial quadrant	$= 1.001\ 880 \times 10^7$ m
meridional quadrant	$= 1.001\ 202 \times 10^7$ m
ellipticity	$= 3.3543 \times 10^{-3}$
eccentricity	$= 0.081\ 837$
surface area	$= 5.100\ 68 \times 10^{14}$ m^2
volume	$= 1.083\ 22 \times 10^{21}$ m^3.

geomagnetism The study of the earth's magnetic field and its variations. At any point on the earth's surface three magnetic elements are defined:
(a) B_0 is the horizontal component of the magnetic flux density at the location;
(b) δ is the angle of *dip (often called the *inclination*), being the angle between the vector B_0 and the resultant magnetic flux density at the location;
(c) α is the angle of *declination (often called the *variation*), being the angle between the vector B_0 and the geographic true north.
B_0 may be determined by one of many standard methods, e.g. a method based on the vibration of a freely suspended magnet in the earth's field. δ may be determined using an *earth inductor or a *dip circle. α is determined by reference to the position of the sun and other astronomical bodies. The vertical component of the earth's magnetic flux density B_v is given by:
$$B_v = B_0 \tan \delta.$$
Two main kinds of variation in these elements are observed. Secular variations take place slowly and are associated with periodic and semiperiodic terrestrial and solar phenomena. Abrupt changes, termed *magnetic storms*, are the result of solar phenomena, e.g. *solar flares.

The approximate values of the magnetic elements in the UK are:
$$B_0 = 1.88 \times 10^{-5} \text{ T}$$
$$\alpha = 9.8° \text{ W}$$
$$\delta = 66.7° \text{ N}$$
$$B_v = 4.35 \times 10^{-5} \text{ T}.$$
The present locations of the earth's magnetic poles (i.e. the points on the earth's surface where $\delta = 90°$) are

N magnetic pole
= latitude 76° N, longitude 102° W;
S magnetic pole

= latitude 68° S, longitude 145° E.
Due to the slow change with time of the angle of declination, the position of the magnetic poles is changing with time. At times of great magnetic disturbance (usually linked with increased solar activity) the poles can be displaced by 150 km in short periods. It is also known that roughly every 200 000 to 300 000 years the magnetic poles reverse: north becomes south and vice versa.

The geomagnetic field is similar in shape to the field of a bar magnet. It is thought, however, to result from the presence of an internal dynamo, maintained in some way by the flow of matter within the inner layers of the earth.

geometric image *See* image.

geometric mean *See* mean.

geometric moment of inertia *See* moment of inertia.

geometric optics (Gauss) The study of reflection and refraction of rays of light without reference to the wave or physical nature of light. The elementary studies of light are essentially geometric. Conversely, although geometric optics has its restrictions, it is still a valuable tool in the more advanced studies of technical optics.

geometric series A mathematical series in which the ratio of a term to its predecessor is the same for every term. The general form is:
$$a + ar + ar^2 + \ldots + ar^{n-1}.$$
The sum of the terms is:
$$a[(1 - r^n)/(1 - r)].$$

geophone *See* echo prospecting.

geophysical prospecting The locating of mineral deposits by making measurements of gravitational, electrical, magnetic, and seismatic magnitudes that have been modified from their normal values by the nature and position of the deposit.

geophysics The physics of the *earth. Studies include the following: the evolution and constitution of the planet itself, its atmosphere, and oceans; movements within the earth, such as those associated with continental drift, mountain building, and earthquakes (*see* plate tectonics; seismology); *geomagnetism; *geophysical prospecting of mineral deposits; the circulation of the atmosphere and oceans.

geostationary orbit A circular orbit around the earth that lies in the plane of the equator and has a period equal to the period of the earth's rotation on its axis (nearly 24 hours). The altitude of the orbit is approximately 35 780 km. A *satellite in geostationary orbit will appear from earth to be very nearly stationary. An earth orbit with the same period but inclined to the equatorial plane is called a *geosynchronous orbit*.

geosynchronous orbit *See* geostationary orbit.

geothermal energy *See* renewable energy sources.

German silver

German silver *Syn.* nickel silver. Alloys of copper, nickel, and zinc of various compositions, e.g. 60% Cu, 25% Zn, 15% Ni; resistivity about 3×10^{-3} ohm metre. *See* alloy.

getter A material with a strong chemical affinity for other materials. Such materials may be used to remove unwanted atoms or molecules from an environment; for example, barium may be used in a sealed vacuum system to remove residual gases or phosphorus may be introduced into oxide layers on silicon to remove mobile impurities such as sodium. This latter is particularly important in the stabilization of MOS *field-effect transistors.

g-factor *See* Landé factor.

G-factor *Syn.* G-value. *See* dosimetry.

Ghiorso, Albert (*b.* 1915) Amer. physicist and chemist, who worked with Glenn T. Seaborg on the isolation of transuranic elements. He also invented an advanced accelerator.

Giaever, Ivar (*b.* 1929) Norwegian-born Amer. physicist, who worked on electron tunnelling through thin insulating films. His results gave information about the electron density in superconducting materials (*see* superconductivity).

giant star A highly luminous star of large dimensions in comparison with average stars like the sun. A typical giant star will have a diameter ten times that of the sun (i.e. 1.4×10^9 m), although supergiant stars may be as much as 500 times the sun's diameter. Giants have a dense core but a very tenuous atmosphere. They are in a late stage of stellar evolution and are situated above the main sequence in the *Hertzsprung–Russell diagram. *See also* red giant.

Giauque, William Francis (1895–1982) Canadian-born Amer. scientist. He became professor of chemistry at the University of California, where he used statistical mechanics to measure entropies at low temperatures. This led (1925) to his development of *adiabatic demagnetization as a method of achieving low temperatures (reaching 0.1 K). For this work he received the Nobel prize for Chemistry in 1949.

Gibbs, Josiah Willard (1839–1903) Amer. physicist who played a great part in the foundation of analytical thermodynamics, developing the use of thermodynamic potential functions and the representation of systems by graphical means. The concept of free energy had already been advanced by Helmholtz but Gibbs was responsible for the *Gibbs function. (*See also* Gibbs–Helmholtz equation.) The *phase rule is due to Gibbs and he was also largely responsible for modern *vector analysis and his notation, including the dot for scalar product and cross for vector product, is widely used.

Gibbs function *Syn.* Gibbs free energy; thermodynamic potential. Symbol: G. A thermodynamic function of a system given by its *enthalpy (H) minus the product of its *entropy (S) and its thermodynamic temperature (T), i.e. $G = H - TS$. In a

*reversible change occurring at constant temperature and pressure the change in the Gibbs function of a system is equal to the work done on it. If a system is considered at constant pressure and temperature and the only work done is that caused by changes in volume, it can be shown that the system is in equilibrium when G has a minimum value. In a chemical reaction the change in G (ΔG) is zero when equilibrium has been attained. If ΔG is negative for a particular reaction, it can proceed spontaneously to equilibrium, whereas if it is positive the reaction cannot occur without energy being supplied. The Gibbs function is often used in chemical reactions because these take place at constant pressure. Reactions at constant volume are much rarer. In these cases the *Helmholtz function would be used.

Gibbs–Helmholtz equation The thermodynamic expression for the internal energy (U) in terms of the free energy (A) and its variation with thermodynamic temperature:
$$U = A - T(\partial A / \partial T)_V.$$

Gibbs's adsorption formula Either solute or solvent will predominate in the surface layer of a solution so as to produce a surface with the least free energy. In general the osmotic pressure will oppose a completion of this state and a dynamic equilibrium will be attained. Gibbs' formula gives the difference in concentration between the surface and the bulk solution.

Application: the addition of a tiny quantity of alcohol (surface tension = 23 millinewtons/metre) to water (surface tension = 72 millinewtons/metre) markedly lowers the surface tension since the alcohol tends to concentrate in the surface layer. This effect is very marked with modern soapless detergents.

giga- Symbol: G. **1.** A prefix meaning 10^9; for example, one gigahertz (1 GHz) = 10^9 hertz.
2. A prefix meaning 2^{30} (i.e. 1 073 741 824), used in computing and other fields in which the binary-number system is used; for example, one gigabyte is 2^{30} bytes.

gilbert Symbol: Gb. The CGS-electromagnetic unit of magnetomotive force or magnetic potential. A point has a magnetic potential of one gilbert if the work done in bringing a unit positive pole up to that point is one erg. One turn of wire carrying a current of 1 ampere produces a magnetomotive force of $4\pi/10$ gilberts.

Gilbert, William (1544–1603) Brit. physician and scientist. One of the first to adopt experimental method in the modern sense, Gilbert made great advances in the subjects of magnetism and electrostatics. He correctly described the main features of the magnetic field of the earth and discovered the actions (attraction and repulsion) between magnetic poles. He introduced some of the modern terminology (*electricity, electric force, magnetic pole*). His great work *De Magnete* (full title: *De magnete magneticisque corporibus, et de magno magnete tel-*

lure: Physiologia nova) was published in London in 1600.

Giorgi units Units based on the metre, kilogram, and second as fundamental mechanical units, together with one electrical unit of practical size. When first proposed (in 1900) the ohm was the fourth unit chosen, although in 1950 this was replaced by the ampere. In 1954 Giorgi's system was superseded by *SI units. As well as unifying mechanical, thermal, and electrical units, Giorgi recognized and recommended the principle of *rationalization of electric and magnetic quantities.

Gladstone, John (1827–1902) Brit. chemist. In his study of refraction of light by gases, with T. P. Dale, he discovered the *Gladstone–Dale law.

Gladstone–Dale law If the density, ρ, of a substance is altered by compression or by increasing its temperature, there is a corresponding rise in the refractive index, n, given by:
$$(n - 1)/\rho = k,$$
where k is a constant.

Glaisher, James (1809–1903) Brit. meteorologist whose dew-point tables are still in use. *See* humidity.

glancing angle The complement of the angle of incidence, i, i.e. the angle $(90° - i)$.

Glan–Foucault polarizer *See* Nicol prism.

Glaser, Donald A. (*b.* 1926) Amer. physicist who invented the *bubble chamber. Nobel prize for physics, 1960.

Glashow, Sheldon Lee (*b.* 1932) Amer. physicist, who became professor at Harvard (1967). He introduced the concept of *charm into particle physics, which enabled Salam and Weinberg's theory of weak interactions to be extended into the standard model of the *electroweak theory, also known as the GWS theory (Glashow–Weinberg–Salam theory). For this work the three authors shared the 1979 Nobel prize.

Glashow–Weinberg–Salam theory (GWS theory) *See* electroweak theory.

glass *See* optical glass.

glass electrode F. Haber found that the potential difference between a glass surface and a solution varied with the pH of the solution. This property is used in the glass electrode, which consists of a thin-walled glass bulb filled with a buffer solution and carrying a platinum contact. The bulb is immersed in a solution of which the pH is to be found and the p.d. between the platinum wire and the external solution is measured. The instrument can be made to read directly in pH divisions.

glide The movement of one atomic plane over another in a crystal. It is the process by which a solid undergoes plastic deformation.

glide plane In metal physics, a plane upon which glide can take place upon application of a suitable shearing stress. Sometimes the glide is in a particular direction (the glide direction) in the plane.

glove box An enclosure with gloves fitted to holes in the walls, enabling a substance to be manipulated in an environment quite distinct from that of the operator. It is most commonly employed for working with sources of alpha particles and thus, because of their low power of penetration, the walls of the box are largely glass. It may also be used for working with beta sources and for work in environments with special properties (e.g. controlled humidity, sterilized, or inert). It is usual for the pressure to be maintained slightly above atmospheric to reduce the possibility of contamination from without.

glow discharge An electric discharge through a gas, usually at a relatively low pressure, in which the gas becomes luminous (*see* gas-discharge tube). A *glow lamp* (or *glow tube*) is a gas-discharge tube operated under conditions producing a glow discharge throughout the tube. The colour is characteristic of the gas: a *neon tube emits a red glow. Glow lamps are often used as voltage regulators.

gluon Symbol: g. An elementary particle that mediates the *strong interaction between *quarks (and antiquarks). *See* quantum chromodynamics.

GMT Abbreviation for Greenwich Mean Time. *See* time.

gnomonic projection From a point within a crystal (the pole of projection) lines are drawn normal to the crystal faces (or sets of planes in the crystal) and these produced will meet any plane in a pattern of points, which is the gnomonic projection of the crystal on that plane.

Goddard, Robert Hutchings (1882–1945) Amer. physicist who made the first rocket capable of flying in space, using a liquid fuel and liquid oxygen.

Golay cell A small transparent device containing gas, used to detect infrared radiation. A very thin film within the cell absorbs the incident radiation, causing the gas temperature and consequently the gas pressure to increase (for a fixed volume of gas). The amount of incident radiation can therefore be indicated by recording the changes in pressure.

gold-leaf electroscope *See* electroscope.

gold point The melting point of pure gold taken as a fixed point (1064.43 °C) on the *International Practical Temperature Scale.

Goldstein, Eugen (1850–1931) German physicist. In 1886 he discovered canal rays, visible behind a perforated cathode in a *gas-discharge tube and so called from their apparent origin in these channels. He had earlier investigated and named cathode rays (1876) and he claimed to have deflected them by an electrostatic field.

gon *Syn.* grade. A unit of angle (Germany) equal to one-hundredth of a right angle.

goniometry The measurement of angles. In crystallography it is the measurement of interfacial angles for the comparison of crystals of different development. The *contact goniometer* consists of two flat bars pivoted together like a pair of scissors and capable of being clamped in any position by means of the screw pivot. The angle between the bars is read off from a graduated semicircle (Carangcot, 1780). The crystal is fitted between the two bars and the angle between them is read off on the scale. In the case of small crystals, some type of reflecting goniometer is used (Wollaston, 1809). A fixed mirror is illuminated from a collimator so that part of the parallel beam falls on the crystal, which is fixed on an axis parallel to the mirror and a short distance above it, and is so adjusted that the edge of which the facial angle is to be measured, is parallel to the axis. An image of a horizontal slit is seen reflected in the mirror and the crystal face; the latter is rotated until the two images coincide, when the reading on a graduated scale attached to the axis is taken. The crystal is then rotated on the axis until coincidence is similarly obtained for the image reflected in the second face, i.e. when the second face is parallel to that originally occupied by the first face. The difference between the readings on the graduated scale gives the normal crystallographic angle between the two faces. Similarly, all the interfacial angles in a given zone can be found by further rotation of the crystal. The axis may be horizontal or vertical.

goniophotometer *See* photometry.

Goudsmit, Samuel Abraham (1902–1978) Dutch-born Amer. physicist. He became a professor at Michigan and North Western, moving to Brookhaven National Laboratory after retiring (1975). He was the first to suggest (1925) that electrons spin about an axis, producing a magnetic field, the theory of which was later worked out by Dirac. During World War II he worked on radar and was responsible for monitoring German progress towards an atom bomb.

G-parity A quantum number associated with *elementary particles that have zero *baryon number and *strangeness. It is conserved in *strong interactions only.

grade *See* gon.

graded-base transistor *See* drift transistor.

graded-index device *See* fibre-optics system.

gradient 1. Of a *graph at any point. The slope of the tangent to the graph at that point as measured by the increase of the ordinate divided by the increase of the abscissa. (*See also* derivative.)

2. (grad) Of a scalar field $f(x,y,z)$ at a point. The *vector pointing in the direction of the greatest increase in the scalar with distance (i.e. perpendicular to the level surface at the point in question). It has components along the coordinate axes that are the partial derivatives, f_x, f_y, f_z of the function with respect to each variable:

$$\text{grad } f = \nabla f = if_x + jf_y + kf_z,$$

where ∇ is the differential operator *del and i, j, and k are unit vectors along the x-, y-, and z-axes. Electric field is the negative gradient of electrical potential. *See* potential gradient.

gradient-index lens (GRIN lens) An optical lens composed of an inhomogeneous medium in which the refractive index varies in a prescribed fashion. It normally varies radially, decreasing parabolically from the central axis. GRIN lenses can take the form of small-diameter parallel flat-faced rods. These are usually grouped into large arrays.

grading shield *Syn.* arcing-shield; grading ring. An *arcing ring designed to improve the voltage distribution across the units of a string of insulators when used with an a.c. supply.

gradiometer *See* free fall.

Graetz number Symbol: Gz. A dimensionless coefficient of importance in the study of hydrodynamics:

$$Gz = q_m c_p / \lambda l,$$

where q_m = mass flow rate, c_p = specific heat capacity at constant pressure, λ = thermal conductivity, and l = a characteristic length.

Graham, Thomas (1805–1869) Brit. chemist. He is known for his work on the absorption of gases, osmosis, and diffusion (*see* Graham's law of diffusion). He introduced the term *colloid.

Graham's law of diffusion (1846) The rates of efflux of different gases through a fine hole at the same temperature and pressure are inversely proportional to the square roots of their densities. Knudsen showed that the law is only true when the mean free path in the issuing gas is at least ten times the diameter of the hole. *See* effusion.

Graham's pendulum (G. Graham, 1675–1751). *See* compensated pendulum.

grain In metallurgy, one of the small crystalline regions in a polycrystalline material, visible (usually with lens or microscopy) after suitable preparation of the surface by polishing and etching.

gram One-thousandth of a *kilogram.

gram-atom or -molecule The former name for a *mole.

Gramme ring An electromagnet having an iron ring as its core – sometimes used in the field of armature of an electric motor or generator.

Gramme winding *See* ring winding.

gramophone An instrument for using the undulating grooves in a gramophone record to generate sound waves. As the record, on a horizontal turntable, is rotated by a motor the stylus moves radially across the surface of the record, following the spiral groove in which it rests. This groove is cut into the surface of the record, and in a monophonic record the groove is modulated by its position being slightly varied radially, relative to the spiral the groove would trace if unmodulated. The stylus vibrates laterally in sympathy with the groove modulations, and these vibrations generate electrical signals by a transducer, usually called the *pick-up, and usually of electromagnetic or piezoelectric type. In the modulation technique used, the speed of the stylus is proportional to the amplitude of the recorded signal.

In a stereophonic record (*see* stereophonic reproduction) the groove is modulated both laterally and also by variations in groove depth. This depth variation corresponds to the difference between the left and right channels of the stereo recording, and the lateral modulation is the sum of the two channels. To obtain electrical signals proportional to the left and right channels, two transducers are required in the pick-up cartridge, detecting motion of the stylus in two directions, 45° either side of horizontal. This system ensures compatibility since a monophonic pick-up cartridge, sensitive only to lateral motion of the stylus, playing a stereo record will produce an electrical signal corresponding to the sum of the two channels, i.e. a mono version of the original stereo signal. A stereo pick-up playing a monophonic record, with only lateral groove modulations, will produce equal signals from the two transducers, again giving monophonic reproduction.

The pick-up cartridge is usually mounted on a pivoted counterbalanced arm that enables it to traverse the surface of the record as the groove spirals inwards. The counterbalancing is arranged to give the downward force necessary to locate positively the stylus in the groove. However, excessive force causes undue wear of both stylus and record.

Styluses are usually made of diamond or sapphire, and are conical in shape with a hemispherical or ellipsoidal tip. Modern records are made of vinyl, and may rotate at one of three standard speeds. 16 r.p.m., is used for speech recording, but is not common. 33 r.p.m. is used most frequently for "long playing" records, 12 in. in diameter, playing for about 25 minutes per side. 45 r.p.m. is commonly used for "single play" records of 7 in. diameter, playing about 5 minutes per side, and "extended play" records of the same diameter but twice the playing time. 78 r.p.m. was the standard speed for early records made of shellac, played with a fibre or steel needle, and originally reproduced mechanically rather than electrically.

Electrical signals from the pick-up cartridge are amplified and used to drive a loudspeaker. (Two loudspeakers in the case of stereo.) The pick-up cartridge and amplifier combination must have a response that falls with increasing frequency. This is necessary to compensate for the pre-emphasis of high frequencies introduced during the recording process. This technique has the effect of reducing high-frequency noise introduced on reproducing the recording by irregularities in the record groove. High-quality gramophones are capable of reproducing from a record all frequencies in the audible region, the ratio of loudest to quietest sounds obtainable being up to 55 decibels.

Quadraphonic records are intended to be reproduced by four loudspeakers to give directional sources of sound surrounding the listener. To ensure compatibility with mono and stereo equipment, it is necessary to carry the four signals required in coded form on the records.

Discrete systems record left (front plus rear) and right (front plus rear) signals as in normal stereo recording. This gives compatibility. Superimposed on these signals are supersonic tones modulated with left (front minus rear) and right (front minus rear) signals. These are not reproduced by mono or stereo equipment. A special pick-up cartridge and stylus is required to respond to these supersonic frequencies, the modulations of which are extracted by the decoder, which then combines them with the front plus rear signals by addition or subtraction to give the original four signals.

Matrix systems combine the four signals to be recorded, with differing degrees (depending on the system) of amplitude and/or phase change, into two composite signals recorded by normal stereo techniques. On replaying the record the two signals from the cartridge are combined, again with different degrees of amplitude and/or phase change, to give four signals to feed loudspeakers.

grand unified theory (GUT) A *quantum field theory that combines the strong, weak, and electromagnetic interactions in a single *gauge theory with one symmetry group. In most GUTs, the known interactions are considered to be a low-energy manifestation of a single unified interaction (thus, the *standard model is a result of *broken symmetry). The unification would take place at high energies (typically 10^{15} GeV); i.e. energies much higher than those currently accessible in particle accelerators. The most important feature of GUTs is that baryon number and lepton number would no longer be absolutely conserved, with the consequence that such processes as *proton decay would be possible. Searches for proton decay are being undertaken by many groups using large underground detectors, so far without success. Some GUTs also predict that the neutrino should have nonzero mass; again, the evidence for this is inconclusive.

See also Kaluza–Klein theory; string theory.

graph A type of diagram for exhibiting the relation between two quantities. It consists of two *axes* OX and OY at right angles. The point P is located by the distances marked *ordinate* (y) and *abscissa* (x) (Fig. a). These distances, measured on the scales shown on the axes, are collectively known as the *coordinates* of P, which is called the point (x, y). O, the point from which the scales start, is the *origin*. If the graph of the relation between two measured quantities x and y is the straight line shown (Fig. b), then

$$y = (a/b)x + c$$

is the mathematical relation between them, and is said to be *linear*. The quantity a/b is the *gradient* of the line and c is the *intercept* on the y-axis.

graphical methods

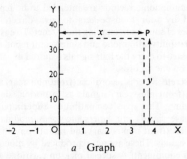

a Graph

Often the experimental data can be plotted in a form that will give a straight-line graph; e.g. suppose it is known that the periodic time of a simple pendulum is

$$T = 2\pi\sqrt{(l/g)}$$

and, from measurements of the length l and T, it is required to find the constant g. Then:

$$T^2 = (4\pi^2/g)l,$$

so that if a graph of T^2 (as ordinate) is plotted against l, the result will be a straight line passing through the origin, whose gradient (a/b) can be measured. As $a/b = 4\pi^2/g$, g can then be calculated. A more complex example is provided by the formula:

$$T = 2\pi\sqrt{(k^2 + h^2)/gh}.$$

Writing this in the form

$$hT^2 = \frac{4\pi^2k^2}{g} + \frac{4\pi^2h^2}{g},$$

we see that a plot of hT^2 against h^2 will give a straight line from whose slope and intercept k^2 and g may be deduced.

Many phenomena in physics involve power laws of the form $y = Ax^n$. In such cases it is useful to plot logarithmic graphs, since

$$\log y = \log A + n\log x.$$

A straight line is then obtained and the power n is found from the slope. Graph paper is printed with line-spacing following a logarithmic law so that points can be plotted directly in this way. Log-log paper has logarithmic spacing for both coordinates while log-linear paper has one logarithmic scale and one ordinary one.

It is usual to use the y-scale for the dependent variable and the x-scale for the independent one (e.g. in a graph of the temperature of a cooling body against time, the temperature is naturally considered to be the result of the time so that temperature is the dependent variable). In deducing the slope of a graph it is essential to observe the scales on the axes. It is usually incorrect to assume that the slope is the tangent of the angle made with the axis of abscissa. It is also important to recognize that a straight-line graph is only proof of direct proportionality between the variables if it passes through the origin. Note that there is no origin on a logarithmic scale.

It is now recommended that the quantities represented on graphs should be pure numbers, i.e. the physical quantity should be divided by the unit.

246

Thus a graph of current against time will have ordinate representing current/ampere and abscissa representing time/second.

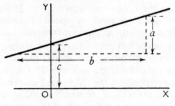

b Graph of $y = \dfrac{a}{b}x + c$

graphical methods *Geometric optics, as its name implies, lends itself to geometric constructions, scale diagrams, etc., for refraction and ray tracing, image formation, regulation of beams, etc. Much depends on the order of accuracy required when considering the advisability of using graphical methods. Figs. *a* to *d* and the explanatory notes give examples.

Graphical construction (Fig. *a*). AP is incident ray in medium of refractive index n at incidence angle APN.

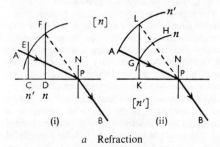

a Refraction

Construction (i). Erect perpendiculars at C and D, where PC:PD $= n':n$. With centre P and radius PE (E is the intersection of CE with AP), construct arc of circle intersecting perpendicular DF at F. Join FP and produce. Then PB is the refracted ray.

Construction (ii) With centre P draw two circles with radii in proportion to n and n'. Erect perpendicular KG, where G is the intersection of AP with circle (radius n). Join LP and produce. PB is the refracted ray.

Graphical construction (Fig. *b*). On the axis OP, mark to the same scale the position of object O (PO $= u$), centre of curvature C (PC $= r$), and the focal point F (PF $= f = r/2$). Erect the perpendicular PD at the pole of the mirror, and OA (drawn to a larger scale) for the (real) object. Draw AD parallel to the axis and join DF. Join AC. Then the intersection of DF and AC gives the image B of A. Drop the perpendicular JB; then J is the image of O.

In the illustrative case above for concave mirrors, C and F always lie in front and JB in this case is real. For convex mirrors, F and C will always be behind, while JB here is virtual.

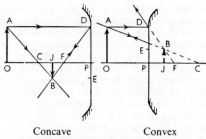

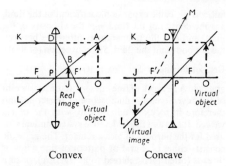

Concave   Convex

b Reflection by curved mirrors

d Lens with virtual object

Notice in addition that ray AF would be reflected parallel to the axis at E and that angle APC = angle BPC (concave mirror).

Graphical construction (Fig. *c*). Draw the optical axis OP, erect a perpendicular PD to represent the

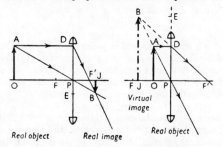

c Convex lens with real object

lens, and mark off the focal lengths PF and PF' (= *f*) and the object position PO (= *u*) to a certain scale. On a larger scale, erect OA to represent the object. Draw AD parallel to the axis and join DF'. Join AP and produce to intersect DF' at B. B is the image of A and by dropping the perpendicular BJ, J is the image of O.

Notice also that the ray AF would be refracted parallel to the axis at E.

Graphical construction (Fig. *d*). Exactly as previously, the rays KD and LP are chosen to pass virtually through A, KD being parallel to the axis. Notice that F' for the concave lens lies in front so that KD is refracted along DM virtually from F' (and B).

graphical symbols Symbols that represent the various types of components and devices used in electronics, electrical engineering, telecommunications, and allied subjects. *See* Appendix, Table 6.

graphic equalizer An electronic device in a radio, tape recorder, etc., that controls the *tone, i.e. alters the relative frequency response of the audiofrequency amplifier. The frequency range of the amplifier is divided into bands. The power of the output signal in each frequency band is adjusted by sliding contacts, the positions of which therefore indicate the frequency response in each band.

graphic instrument *Syn.* recording instrument; recorder; grapher; chart-recording instrument. An instrument that produces a record in the form of a graph of the quantity being measured. The graph is usually drawn in ink upon a suitable paper chart. A *cathode-ray oscilloscope can be used to display the graph, a permanent record being obtained photographically, if required.

graphics *Syn.* computer graphics. A mode of computer processing and output in which a large proportion of the output is in pictorial form. The information presented may be in the form of graphs, engineering or architectural drawings, maps, models, etc. It may be in one or more colours and may be labelled. The output may be displayed on the screen of a *VDU or may be recorded by a plotter. The information is fed into the computer by various means, such as *light pen or *mouse. The computer can be made to manipulate the information, for example by straightening lines, moving or removing specified areas, or expanding or contracting details. Apparently three-dimensional images can be produced, which can sometimes be observed from different viewpoints.

graphite *Syn.* plumbago; black lead. An allotropic form of carbon that occurs in crystalline forms, and more commonly in apparently amorphous masses of granules, which really consist of minute hexagonal crystals. It has a high thermal and electrical conductivity, namely:
electrical conductivity: $3.33 \times 10^4 \, \Omega^{-1} \text{m}^{-1}$ at 20 °C;
thermal conductivity: $126 \, \text{W m}^{-1} \text{K}^{-1}$.

It has a very high melting point (3500 °C) and is fairly resistant to chemical attack. It is used as a lubricant, and in lead pencils, electrodes, crucibles, etc. Pure graphite does not absorb neutrons strongly and it has therefore been used in block form as a *moderator in *nuclear reactors.

Grashof number *Syn.* free convection number. Symbol: *Gr*. The dimensionless parameter:
$$l^3 g \gamma \rho^2 \theta / \eta^2,$$
used in the dimensional analysis of convection in a fluid due to the presence of a hot body; *l* is a typical dimension of the body, *g* is the acceleration of free

fall, γ is the cubic expansion coefficient of the fluid, ρ is the density of the fluid, η is the viscosity of the fluid, and θ is the temperature difference between the hot body and the fluid. *See* convection (of heat).

Grassot fluxmeter *See* fluxmeter.

graticule A network of fine lines set at the focal point of the *eyepiece of a microscope or telescope, and therefore in focus simultaneously with the object viewed. It acts as a field reference system and may be used for the purpose of measurement. The graticule consists either of a grid or pattern of fine wires or threads (often then referred to as a *reticle*), or of a transparent glass disc with the lines engraved on it.

grating *See* diffraction grating.

graveyard *See* burial.

gravimeter *See* free fall.

gravitation The mutual attraction of all bodies, independent of electromagnetic, strong, or weak interactions.
Early ideas. Until 300 years ago gravitation, or gravity, was usually regarded as the "natural" tendency of dense bodies to move towards the earth if free to do so. *Galileo studied falling bodies and introduced the concept of *acceleration. He argued that in a vacuum all bodies would have the same acceleration, that of *free fall. Most scholars, however, accepted the views of *Descartes that gravitation was caused by movements of a subtle fluid, and that in a vacuum motion would be infinitely fast.
Newton's law and the gravitational constant. In 1687 Newton presented his *law of universal gravitation*, according to which every particle attracts every other particle with a force F given by:
$$F = Gm_1m_2/x^2,$$
where m_1, m_2 are the masses of two particles a distance x apart. G is the *gravitational constant*, which, according to modern measurements, has the value
$$6.672\ 59 \times 10^{-11}\ \text{m}^3\ \text{kg}^{-1}\ \text{s}^{-2}.$$
For extended bodies the forces are found by integration. Newton showed that the external effect of a spherically symmetric body is the same as if the whole mass were concentrated at the centre. Astronomical bodies are roughly spherically symmetrical so can be treated as point particles to a very good approximation. On this assumption Newton showed that his law was consistent with *Kepler's laws. Artificial satellites in orbits close to the earth have deviations from Kepler's laws, which permit precise measurement of the shape of the earth (*see also* mascon).
Early methods of determining G, and hence the *earth's mean density, include the deflection of a plumbline by a mountain (*Maskelyne (1774)), and measurement of the increase of g on going down a mine (*Airy (1854)), but most methods depend upon measuring the force acting on a delicately balanced body carried by a nearby mass. The first such experiment was that of *Cavendish (1798), which

248

was improved by *Boys (1895). Most such measurements have been made at distances of a few centimetres but Hirakawa (1980) extended the distance to more than 4 m by observing resonance in the oscillations of a suspended mass carried by the nonuniform field of a bar, which is rotated about its centre of mass. Until recently, all experiments have confirmed the accuracy of the inverse square law and the independence of the law upon the nature of the substances, but in the past few years evidence has been found against both.
Gravitational field and potential. The size of a gravitational field at any point is given by the force exerted on unit mass at that point. The field intensity at a distance x from a point mass m is therefore Gm/x^2, and acts towards m. Gravitational field strength is measured in newtons per kilogram. The gravitational potential V at that point is the work done in moving a unit mass from infinity to the point against the field. Due to a point mass
$$V = Gm\int_\infty^x \mathrm{d}x/x^2$$
$$= -Gm/x.$$
V is a scalar measured in joules per kilogram. The following special cases are also important: (a) Potential at a point distance x from the centre of a hollow homogeneous spherical shell of mass m and outside the shell:
$$V = -Gm/x.$$
The potential is the same as if the mass of the shell is assumed concentrated at the centre. (b) At any point inside the spherical shell the potential is equal to its value at the surface:
$$V = -Gm/r,$$
where r is the radius of the shell. Thus there is no resultant force acting at any point inside the shell (since no potential difference acts between any two points). (c) Potential at a point distance x from the centre of a homogeneous solid sphere and outside the sphere is the same as that for a shell:
$$V = -Gm/x.$$
(d) At a point inside the sphere, of radius r:
$$V = -Gm(3r^2 - x^2)/2r^3.$$
The essential property of gravitation is that it causes a change in motion, in particular the acceleration of free fall (g) in the earth's gravitational field. According to the general theory of *relativity, gravitational fields change the geometry of space-time, causing it to become curved. It is this curvature of space-time, produced by the presence of matter, that controls the natural motions of bodies. General relativity may thus be considered as a theory of gravitation, differences between it and Newtonian gravitation only appearing when the gravitational fields become very strong, as with *black holes and *neutron stars, or when very accurate measurements can be made. *See also* interaction.

gravitational collapse The contraction of an astronomical body resulting from the mutual gravitational pull of all its constituents. The term is most commonly applied to the sudden collapse of the core of a star when energy can no longer be produced by nuclear-fusion reactions. There is then no outwardly directed gas pressure or radiation pressure to counterbalance the inwardly directed gravitational force, and the hydrostatic equilibrium is destroyed. The three most likely end-products of such a collapse are (in order of mass) *white dwarfs, *neutron stars, and *black holes. In the first two the collapse can be halted by a quantum mechanical effect: a *degeneracy pressure* exerted by tightly packed electrons stripped from the atomic nuclei (in white dwarfs) or by tightly packed neutrons (in neutron stars).

gravitational constant Symbol: G. The universal constant that appears in Newton's law of *gravitation. It has the value:
$$6.672\,59 \times 10^{-11} \text{ N m}^2 \text{ kg}^{-2}.$$

gravitational field The space surrounding a massive body in which another massive body experiences a force of attraction. *See* gravitation.

gravitational lens An astronomical body (usually a galaxy or cluster of galaxies) whose gravitational field bends light and other radiation from a more distant source (usually a *quasar), so that multiple images of the latter are produced. This effect, known as *lensing*, can be explained by general *relativity and was first observed in 1979.

gravitational mass *See* mass.

gravitational potential *See* gravitation; potential.

gravitational redshift *See* Einstein shift.

gravitational unit A unit of force, pressure, work, power, etc., involving g, the acceleration of *free fall. The *slug is an example.

gravitational waves *Syn.* gravitational radiation. The propagation of a changing gravitational field at the speed of light, caused by the displacement of masses. Gravitational waves are predicted by the general theory of *relativity but have not yet been detected experimentally. Astronomical observations of *supernova explosions or the orbital motions of binary *pulsars could present indirect evidence.

graviton A hypothetical *elementary particle responsible for the effects of *gravitation; it is the quantum of the *gravitational field and is thus a gauge boson (*see* gauge theory). It is postulated to be its own *antiparticle, to have zero charge and rest mass, and a spin of 2. It is firmly predicted by theory but its direct observation is at present unlikely.

gravity 1. An alternative name for *gravitation.
2. For a body at or near the surface of a planet, the apparent force of gravitation. If this is combined vectorially with the centripetal force of axial motion it gives the real force of gravitation. *See* free fall; weight.

gravity balance *See* free fall.

gravity cell A primary electric cell in which two electrolytes are kept apart by their different densities.

gravity meter *See* free fall.

gravity wave A wave in the surface layers of a liquid, being controlled by gravity and not by surface tension. For example, if the depth of the liquid is large compared with the wavelength λ, the speed is given by:
$$v = \sqrt{(g\lambda/2\pi)},$$
where g is the acceleration of free fall. The amplitude falls off exponentially with depth, decreasing by a factor e in a distance $\lambda/2\pi$ (where e = 2.718). *See also* water wave.

gray Symbol: Gy. The derived SI unit of absorbed *dose of ionizing radiation, and of specific *energy imparted. It is equal to an absorption or delivery of one joule per kilogram of irradiated material. It replaces the *rad. 1 gray = 100 rad.

grease-spot photometer A design of photometer head (*see* photometry) due to Bunsen (1843), consisting of a thin white opaque paper with a translucent spot at the centre. Lights illuminate both sides. The intensity of illumination (i.e. the *illuminance) is assumed to vary inversely as the square of the distance. Three methods of use are as follows:
(i) *Substitution method.* Any convenient auxiliary source is fixed on one side. The two sources to be compared (of illuminance C_1 and C_2) are moved along a bench until the spot disappears for each in turn, at which point:
$$C_1/C_2 = (d_1/d_2)^2.$$
(ii) *Spot-disappearance method.* The two sources C_1 and C_2 are on opposite sides and the distances d_1 and d_2 when disappearance occurs on one side, and d'_1 and d'_2 for disappearance on the other side, are measured. Then,
$$C_1/C_2 = d_1 d'_1/d_2 d'_2.$$
(iii) *Equality of contrast.* Between the spot and surround on both sides. This requires a reflection device so that both sides can be viewed simultaneously. The lights to be compared are on the opposite sides. In this case, the condition is:
$$C_1/C_2 = (d_1/d_2)^2.$$

great circle A circle in which a plane passing through the centre of a sphere intersects the surface. The shortest distance between two points on the surface of a sphere is along the great circle joining them.

Green, George (1793–1841) Brit. mathematician. He developed many theories in electrostatics and magnetism and first introduced the concept of potential into electricity (1828). Several of his results were obtained independently by *Gauss somewhat later (1839) and are often attributed to the latter.

greenhouse effect A process in which an environment is heated by the trapping of *infrared radiation. It operates on several bodies in the solar system, including Venus (where it is responsible for the high surface temperature) and to a lesser (but increasing) extent on earth. In the case of earth, ultraviolet radiation in sunlight is absorbed by the earth's surface and is reradiated at longer (infrared) wavelengths. Although most of the infrared escapes from the atmosphere, some is absorbed by atmos-

Green's theorem

pheric gases, notably carbon dioxide (CO_2) and water vapour. The amount of CO_2 in the atmosphere is known to have risen recently, both as a result of increased combustion of fossil fuels (1 tonne of carbon burnt produces 3.7 tonnes of CO_2) and because vast areas of rain forest (which absorb CO_2 in photosynthesis) have been destroyed. With more CO_2 in the atmosphere, more of the reradiated infrared is absorbed by the atmosphere with a consequent rise in its temperature. This effect is enhanced by the depletion of the ozone layer by CFCs. With less ozone to absorb the sun's ultraviolet radiation, more reaches the earth and more is reradiated as infrared. The joint result of these processes is believed to be a rise in temperature of the earth's oceans of some 1 °C per decade.

Green's theorem A vector form of *Gauss's theorem.

Greenwich Mean Time (GMT) *See* time.

Gregorian calendar *See* time.

Gregorian telescope *See* reflecting telescope.

Gregory and Archer's method For determining the thermal conductivity of a gas. A temperature gradient is maintained between the walls and axis of a cylinder containing the gas. The rate at which energy is supplied in the steady state to a wire running along the axis is measured electrically. They investigated the losses due to convection in the gas and conduction along the wire and corrected for the nonuniformity of temperature along the wire. *See also* Schleiermacher's method to determine thermal conductivity of a gas.

grenz rays X-rays of long wavelength produced when electrons are accelerated by voltages of 25 kV or less. They are generated in many types of electronic equipment using electron beams but have a very low penetrating power.

grey body A body that emits radiation of all wavelengths in constant proportion to the *black-body radiation of the same wavelengths at the same temperature.

grid 1. *See* control grid; triode; modulator electrode.
2. The high-voltage transmission-line system that interconnects many large *power stations. Voltages of 275 kV, or in some cases 400 kV, are commonly used, though in some countries voltages as high as 735 kV are used.

grid bias A polarizing potential difference applied between the cathode and control grid of a *thermionic valve to cause it to operate on any desired part of its characteristic curve, or to modify its cut-off values.

grid control ratio *See* thyratron.

grid leak A high resistance between the grid of a thermionic tube and its cathode to prevent a charge accumulating on the grid.

grid modulation *Amplitude modulation carried out by means of a valve modulator in which the modulating signal is applied to the grid of the valve. *Compare* anode modulation.

grid stopper *See* parasitic oscillations.

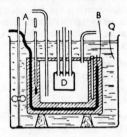

Determination of mechanical equivalent of heat (Griffiths)

Griffiths's apparatus An apparatus for determining *specific heat capacities using an electrically heated calorimeter. The heat gained by the calorimeter and contents D is equated to the electrical work $E^2 t / R$ done on a heater wire. The calorimeter is suspended in a partially evacuated steel vessel B maintained at a constant temperature by immersion in a large water bath Q whose temperature is controlled by a mercury regulator A. The potential difference across the terminals of the heater was measured by comparison with Clark cells on a potentiometer, experiments being carried out with different energy supplies so that the heat capacity of the calorimeter disappeared from the final equation. The difficulty is to measure R since the resistance of the wire varies with the temperature of the water and the water and wire are not at the same temperature. Griffiths carried out subsidiary experiments to give this temperature difference for various potential differences applied across the wire. Griffiths originally devised the method to evaluate the *mechanical equivalent of heat and showed that a comparison of the values obtained electrically and mechanically provided an independent check on the accuracy of the electrical standards which were then not very precisely established. He pointed out that a knowledge of the absolute values of the electrical standards is unnecessary in determining the variation of the specific heat capacity of water with temperature by this method.

Grimaldi, Francesco (1618–1663) Italian mathematician and physicist. He discovered the diffraction of light and described very accurately many of the phenomena; published posthumously (1666).

Grimsehl's electrometer An *electrometer in which an aluminium leaf is repelled from a fixed vertical plate and also attracted to a sloping fixed plate connected to the case of the instrument. The good distribution of the lines of force increases the sensitivity.

GRIN lens *See* gradient-index lens.

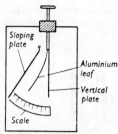

Grimsehl's electrometer

Grosse, Aristid (b. 1905) Amer. chemist. In 1927 he isolated protactinium, and in 1940 with Nier, Booth, and Dinning achieved the fission of uranium-235 by slow neutrons.

Grotthus' hypothesis Early form of the theory of electrolysis in which the molecules of the electrolyte were supposed to arrange themselves, under the action of an electric field, in a number of chains linking the anode and cathode. The negative ion of one molecule was supposed to be adjacent to the positive ion in the next molecule, and electrolysis was supposed to result from an interchange of partners, all along the line. The hypothesis has been superseded by the *dissociation theory.

ground *See* earth.

ground state The state of a system with the lowest energy. This was called the *permanent state* by *Bohr as an isolated body will remain indefinitely in it. It is possible for a system to have two or more ground states, of equal energy but with different sets of *quantum numbers. In the case of atomic hydrogen there are two states for which the quantum numbers n, l, and m are 1, 0, and 0 respectively, while the *spin may be $+\frac{1}{2}$ or $-\frac{1}{2}$ with respect to a defined direction (*see* atomic orbital). There is in fact an extremely small difference of energy according to whether the electron spin is parallel or antiparallel to the proton spin. For nearly all purposes one can assume that there is just one ground state of a unique energy, but transitions between these states do occur in interstellar atomic hydrogen, giving rise to radiation of wavelength 21 cm. *See also* excited state; zero-point energy.

ground wave An electromagnetic radio wave that is radiated from a transmitting aerial on the surface of the earth, and that travels along the surface of the earth. *Compare* ionospheric wave.

group (mathematics) A set of elements or operations $a, b, c, \ldots$ for which a law of "combination" may be defined so that the "product" ab of any two elements is well defined and satisfies the following conditions:

(1) If a and b belong to the set, so does ab.
(2) "Combination" is associative; that is $a(bc) = (ab)c$.
(3) The set contains an element e, called the *identity*, such that $ae = ea = a$ for all elements a of the set.

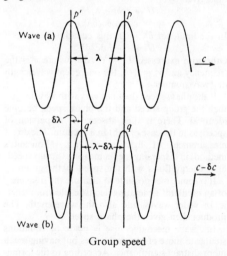

Group speed

(4) For every element a in the set there is an element b such that $ab = ba = e$. We denote b by a^{-1}, which is not necessarily the reciprocal of a, but depends on the combination (see below).

The law of combination referred to above is not necessarily ordinary multiplication. The set of integers

$$\ldots, -2, -1, 0, 1, 2, \ldots$$

form a group with addition as the law of combination. For example, take $a = 2$, $b = 3$ then ab means we take $2 + 3 = 5$ which is also an integer. The identity is 0 since $0 + n = n$ where n is any integer. The inverse of $a = 2$ is -2 since $aa^{-1} = 2 + (-2) = 0$. Some of the most important groups have *matrices as elements. For example, the set of all $n \times n$ matrices that have nonzero *determinants form a group called $GL(n)$ having matrix multiplication as its law of combination.

Two elements a,b of a group are said to commute if $ab = ba$. If all the elements of a group commute with each other the group is said to be commutative or an *Abelian group*. Non-commutative groups are *non-Abelian*; this distinction is important in *gauge theories. Group theory is also important in physics in the analysis of such symmetries as the rotations and reflections of molecules, which underlie the quantum theory of angular momentum. More abstract and generalized aspects of group theory are used to describe the fundamental *interactions by gauge theories.

group speed In certain forms of wave motion the *phase speed (i.e. the speed with which the phase of an oscillation is propagated) varies with the wavelength. As a result of this, a nonsinusoidal wave appears to travel with a speed distinctly different from the phase speed.

The phenomenon is most readily seen with waves on the surface of water. Considering the group of waves resulting from a stone dropped into water, it may be observed that the waves within the group travel faster than the group itself, fresh waves ap-

Grove, Sir William Robert

pearing at the rear of the group as the existing waves vanish at the leading edge. The speed of the group is called the group speed, while that of the waves within the group is the phase speed. Since wave speeds are usually measured by the arrival of the disturbance caused by the wave at different points in its path, it is seen that measurements usually give the group speed of the wave and not the phase speed.

An expression for group speed may be obtained by considering the propagation of two sinusoidal waves of slightly different wavelengths, λ and $\lambda - \delta\lambda$, say, the corresponding phase speeds being c and $c - \delta c$. The superposition of these waves produces *beats that travel with the group speed of the waves. Consider an instant at which a certain peak, p, of wave (a) in the diagram is coincident with a peak, q, of wave (b). Let a time δt elapse after which the preceding peak, p', of (a) then coincides with the preceding peak, q', of (b). In time δt, the crest of the resultant wave has travelled from the position occupied by p and q originally to the position occupied by p' and q' after time δt. This distance is $c\delta t - \lambda$.

For peaks p' and q' to coincide after time δt, p' must have travelled a distance $\delta\lambda$ more than q'. Hence

$$\delta\lambda = c\delta t - (c - \delta c)\delta t,$$

i.e.

$$\delta t = \delta\lambda/\delta c.$$

The speed of the beats, i.e. the group speed, U, is thus given by:

$$U = (c\delta t - \lambda)/\delta t$$
$$= c - \lambda(\delta c/\delta\lambda).$$

In the limit, for $\delta\lambda = 0$, this equation becomes

$$U = c - \lambda dc/d\lambda.$$

Otherwise expressed $U = d\nu/d\nu'$, where ν is the frequency and $\nu' = 1/\lambda$, the reciprocal wavelength or *wavenumber.

If the phase speed does not vary with wavelength, then the group speed and the phase speed become identical. There is then absence of dispersion of speed, as in the case of light in a vacuum. The direct measurement of light speed (e.g. *Foucault's method) in dispersing media measures group speed; light energy flows with the speed of the group.

It is sometimes possible to measure the frequency of an oscillator generating acoustic, electromagnetic, or water waves, and also the wavelength. The product then gives the phase speed.

In *wave mechanics use is made of equations similar to those of physical waves but having much more abstract significance. According to the formulation quite different values may be given to the phase speed associated with a given particle, but in each case a dispersion formula must be used so that the group speed is identical with the observed speed of the particle.

Grove, Sir William Robert (1811–1896) Brit. physicist who invented the *Grove cell (1839).

Grove cell A two-fluid primary *cell in which the negative element consists of a zinc rod in dilute sulphuric acid, and the positive element, which is separated from the negative by a porous partition, consists of a platinum plate immersed in fuming nitric acid. The e.m.f. is 1.93 volts.

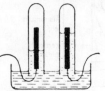

Grove gas cell

Grove gas cell A primary cell with two platinum electrodes immersed in dilute H_2SO_4, one being in contact with an atmosphere of oxygen and the other with an atmosphere of hydrogen. It is usually demonstrated by first electrolysing the acidified water, after which a current in the reverse direction can be drawn from the electrodes.

grown junction A *p-n junction formed in a single crystal of semiconductor by varying, in a precise manner, the types and amounts of impurities added to the semiconductor, while the crystal is being grown from the melt.

Grüneisen's law A law derived from the *equation of state for solids that states that the ratio of the coefficient of linear expansion of a metal to its specific heat capacity is a constant independent of the temperature at which the measurements are made.

guard band In allocating the bands of frequencies within the frequency spectrum to various communication channels, it is desirable that the frequency bands of adjacent channels be separated by narrow bands of frequencies (called guard bands) to reduce the possibility of mutual interference between the bands.

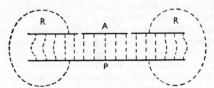

Principle of guard ring

guard ring 1. (electrical) A large metal plate surrounding and coplanar with a small metal plate from which it is separated by a narrow air gap. The device was invented by Kelvin, and used by him in his absolute *electrometer and standard capacitor to ensure a uniform and calculable field over the area of the smaller plate, which can be treated as an infinite plane – the variations in the field that occur as the edge of the plane is approached affect only the guard ring. The diagram shows the lines of force between the plate A with guard ring RR and a parallel earthed plate P. An extra electrode, equivalent to a guard ring, is commonly used in semiconductor devices and vacuum tubes.

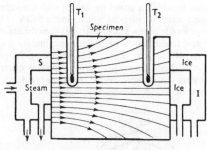

Guard ring (heat)

2. (heat) A device, used in experiments on heat flow, whose function is similar to the guard ring of an electrical capacitor. It produces a temperature gradient in the region all round the specimen identical with that down the specimen so that heat losses from the latter are eliminated (*see* diagram). The central part of the steam chest S and of the ice chamber I is isolated, the remainder forming a guard ring. The temperature gradient in the specimen is measured across the centre position by thermometers T_1, T_2.

guard wires 1. Earthed conductors placed beneath overhead-line conductors so that if the latter break, they will be earthed before they reach the ground. A series of guard wires arranged to form a net is known as a *cradle guard* and is used where a high-voltage line crosses a telephone wire or a thoroughfare.

2. *See* Price's guard wire.

Gudden–Pohl effect A form of *electroluminescence from a phosphor following exposure to ultraviolet radiation, which excites the phosphor into a metastable state.

Guericke, Otto von (1602–1686) German physicist who invented the vacuum pump and demonstrated the pressure of the atmosphere by showing that sixteen horses were unable to separate two joined and evacuated hemispheres.

Guillaume, Charles-Edouard (1861–1938) Swiss physicist noted for his work on standards of measurement. He began (1890) a search for an inexpensive substitute for the platinum-iridium alloy used for constructing standard metre bars. This led to the discovery of the nickel-steel alloy *Invar and the nickel-chromium-steel alloy *elinvar. Guillaume became director of the International Bureau of Weights and Measures and was a forceful advocate of the metric system.

Guillemin effect A type of *magnetostriction in which a bent bar of *ferromagnetic material tends to straighten out under the influence of a magnetic field applied along its length.

Guillemin line An electrical network designed to produce pulses with very sharp rise and fall times so that they are almost square.

gunmetal A bronze consisting of 88% copper, 8–8.5% tin, 3.5% zinc, with sometimes small amounts of nickel up to 0.5%. The name is also sometimes applied to a simple 90 Cu, 10 Sn bronze. Gunmetal shows a high resistance to corrosion and is easy to cast, being particularly suitable in hydraulic castings, which must be pressure-tight. It has good bearing properties and hence is much used in this capacity also.

Gunn diode A two-terminal device consisting of a sample of n-type gallium arsenide operated under conditions such that microwave oscillations are produced due to the *Gunn effect.

Gunn effect An effect observed by J. B. Gunn in 1963. If a d.c. electric field is applied across short n-type samples of gallium arsenide, at values above a threshold value of several thousand volts per cm, coherent microwave output is generated. The effect is caused by charge *carriers of different mobilities forming bunches, known as *domains*, under the influence of the electric field. Some of the conduction electrons move from a low-energy, high-mobility state into a higher-energy, low-mobility state causing domains of low mobility to be set up. It is these domains that produce the microwave output.

GUT *See* grand unified theory.

G-value *See* dosimetry.

GWS theory Glashow–Weinberg–Salam theory. *See* electroweak theory.

gyrator A component, usually used at microwave frequencies, that reverses the phase of signals transmitted in one direction but has no effect on the phase of signals transmitted in the opposite direction. The gyrator may be entirely passive, or contain active components.

gyrocompass A nonmagnetic compass using a *gyroscope fitted with a pendulous weight or some equivalent to induce precession due to gravity. The subsequent damped motion of the gyro aligns this with the true N-S direction. On a large ship, the master gyrocompass, which is usually electrically driven, is connected to repeater dials.

gyrodynamics The study of rotating bodies, particularly when subject to *precession.

gyromagnetic effects The relationships between the magnetization of a body and its rotation. *See* Barnett effect; Einstein and de Haas effect.

gyromagnetic ratio Symbol: γ. The ratio of the *magnetic moment of a system to its *angular momentum. An orbiting electron has a value of $e/2m$, where e is the electron charge and m its mass. The gyromagnetic ratio of an electron due to its *spin is twice this value.

gyroscope A device in which a suitably mounted flywheel or rotor is spun at high speed. The mounting, usually of gimbal type, allows the axis of rotation to be in any direction in space.

If a couple is applied to the frame of the gyroscope, the resulting motion of *precession tends to align the gyro with the axis of the couple. The rate of turning or precession is proportional to the moment of the applied couple and inversely proportional to the angular momentum of the gyro.

In the absence of disturbing couples, the direction in space of the spin axis stays constant; hence, gyroscopic devices are useful for guidance of aircraft during turns, etc., and they are widely used in automatic-guidance devices. Large gyros are used on some ships to achieve stability against rolling, either directly or indirectly. See gyrostat; gyrocompass.

gyrostat A *gyroscope, especially a version intended primarily to indicate or to use the constancy of direction of axis of a fast-running gyroscope.

H

h Symbol used for the Planck constant.

habit The set of natural faces that appear on a crystal.

Hadley, John (1682–1744) Brit. instrument-maker who invented the *sextant and one of the first *reflecting telescopes.

hadron An *elementary particle composed of *quarks and/or antiquarks that can take part in strong interactions. Hadrons with zero or integer spin are known as *mesons (consisting of a quark–antiquark pair), and those with half-integer spins as *baryons (consisting of three quarks). See also Appendix, Table 8.

hadronization See quantum chromodynamics.

Hagen–Poiseuille law See Poiseuille flow.

Hahan technique See emanating power.

Hahn, Otto (1879–1968) German chemist and physicist. He was the co-discoverer, with Lise *Meitner, of the radioactive element protactinium (1917) and he demonstrated, with Strassmann (1939), the fission of uranium by neutrons.

Haidinger brushes Faint yellow brushes that appear when the eye looks at a bright surface through a rotating *Nicol prism. They are believed to be due to double refraction by fibres at the *fovea centralis of the eye.

Haidinger fringes Interference fringes formed by rays reflected practically normally from two plane and parallel surfaces relatively widely separated. The observing eye or telescope must be focused for infinity (1849).

hail Roughly spherical ice particles, usually a few millimetres in radius, produced in very turbulent clouds. A hailstone may be carried upwards by an eddy several times, each time falling through a region in which it grows by impact with the super-cooled water droplets of which clouds below 0 °C are generally composed. On hitting the hailstone the droplet freezes onto the surface and minute fragments of ice are thrown off. According to one theory, these fragments have opposite electric charge to that of the hailstone and cause the electrification of thunderclouds. See snow.

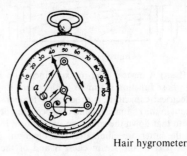

Hair hygrometer

hair hygrometer A *hygrometer that depends for its action on the increase in length of a hair occurring when the relative *humidity of the surrounding air increases. The hair is anchored at *a* and passes round three smooth pegs and a light pulley to which is attached a pointer. The spring *b* maintains a tension in the hair and the position of the pointer on the calibrated scale gives the relative humidity directly for temperature both above and below 0 °C.

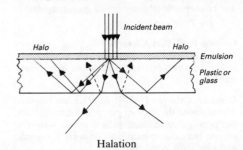

Halation

halation 1. An exposed ring surrounding a strongly illuminated spot on a photographic emulsion. It is caused by light scattered so as to strike the opposite face of the plastic or glass base at angles greater than the *critical angle, giving total reflection. It is prevented by an *antihalation backing* to the base, consisting of a film of refracting material containing a light-absorbing dye. (See diagram.)

2. A similar phenomenon when a fluorescent screen is coated on a sheet of transparent material, for example, the screen of a *cathode-ray tube. In this case the incident radiation is fully absorbed but the light is emitted at all angles, giving an illuminated ring around a bright spot.

half-cell One electrode of an electrolytic cell and the electrolyte with which it is in contact.

half-duplex operation See duplex operation.

half-life *Syn.* half-value period. Symbol: $T_{1/2}$, $t_{1/2}$. The time in which the amount of a radioactive nuclide decays to half its original value. It is given by:

$$T_{1/2} = (\log_e 2)/\lambda = 0.693\,15/\lambda,$$

where λ is the *decay constant, or by:

$$T_{1/2} = \tau \times 0.693\,15,$$

where τ is the *mean life.

half-period strips and zones Fresnel's method of subdividing a wavefront (in strips for cylindrical, in zones for spherical waves), so that the summation of the effects of secondary wavelets at a forward lying point is facilitated. Each successive zone in the spherical wavefront case is farther away from the point by a half wavelength, while all zones have equal areas.

half-power point The value on a response *characteristic that represents half the power of the value corresponding to the maximum power.

half-value period *See* half-life.

half-value thickness The thickness of a uniform sheet of material that, when interposed in a beam of radiation, will reduce the intensity or some other specified property of the radiation passing through it to one half. The half-value thickness is often used as a means of defining the quality of the radiation.

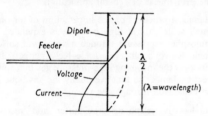

Voltage and current distribution in half-wave dipole

half-wave dipole An *aerial consisting of a straight conductor that is approximately half a wavelength the dipole is excited, it has a voltage node and current antinode at its centre, and a voltage antinode and current node at each end. The feeder is commonly, but not always, connected across a small gap in the centre of the dipole. *See also* dipole aerial.

half-wave plate A thin double-refracting optical element, often of quartz or mica, that can be used to change the *polarization of an incident wave. It is cut parallel to the optic axis of such thickness as to introduce a half-wavelength path difference, i.e. a relative phase difference of 180° between the ordinary and extraordinary rays. *Plane-polarized light incident normally on the plate has its plane of polarization rotated through twice the angle between the axis and the incident vibrations (Laurent, 1874). When inserted so as to halve the field of the polarizer, the analyser can be rotated until the two half fields are equally bright or dark, a setting useful

in saccharimetry or in determining rotation of plane of polarization (Laurent saccharimeter).

half-wave rectifier circuit A rectifier circuit in which only alternate half waves of the single-phase a.c. input wave are effective in delivering unidirectional current to the load. *Compare* full-wave rectifier circuit.

half-width Half the width of a spectrum line measured at half its height. In some branches of spectroscopy the half-width is used for the full width of the line at half its height.

Hall coefficient *See* Hall effect.

Hall effect An effect occurring when a current-carrying conductor is placed in a magnetic field and orientated so that the field is at right angles to the direction of the current: an electric field is produced in the conductor at right angles to both the current and the magnetic field. The field produced is related to the vector product of the current density j and magnetic flux density B by the relation

$$E_H = -R_H(j \times B).$$

The constant R_H (the *Hall coefficient*) is given by:

$$R_H = V_H A/aIB,$$

where V_H is the Hall e.m.f., I is the current, A is the cross-sectional area of the conductor, and a is the thickness in the direction perpendicular to I and B. R_H is given in cubic metres per coulomb. In general R_H is a function of B and the Hall voltage is nonlinear. The variation of R_H with B depends on the impurity density and crystal orientation.

In metals and *degenerate semiconductors, R_H is independent of B and is given by $1/ne$, where n = carrier density and e = electronic charge. In nondegenerate semiconductors additional factors are introduced due to the energy distribution of the current carriers. The energy bands are not in general spherically symmetrical and two or more bands may be effective in conduction so the situation is much more complicated.

The Hall effect is a consequence of the *Lorentz force acting on the charge-carrying electrons. A sideways drift is imposed on the motion of the electrons by their passage through the magnetic field. For materials in which the current is carried by positive charge carriers (*holes), the direction of the Hall field, E_H, is reversed.

Under certain conditions the *quantum Hall effect* is observed. The motion of the electrons must be constrained so that they can only move in a two-dimensional "flatland" – one degree of freedom has then been removed by quantization of motion; this can be achieved by confining the electrons to an extremely thin layer of semiconductor. In addition, the temperature must be very low (around 4.2 K or below) and a very strong magnetic flux density (of the order of 10 T) must be used. The magnetic field, applied normal to the semiconductor layer, produces the transverse Hall voltage as in the ordinary Hall effect. The ratio of the Hall voltage to the current is the *Hall resistance*. At certain values of flux density, both the conductivity and the resistivity of the solid become zero, rather like in a superconductor. A graph of Hall resistance against flux density shows steplike regions, which correspond to the values at which the conductivity is zero. At these points, then, the Hall resistance is quantized; calculations show that:

$$(V_H/I)n = h/e^2,$$

where n is an integer, h is the Planck constant, and e is the electron charge. The Hall resistance can be measured very accurately: it is equal to 25.8128 kΩ. Hence the quantum Hall effect can be used to calibrate a conventional resistance standard, and can also be used in the determination of h and e.

Halley, Edmund (1656–1742) Brit. astronomer. He played a great part in stimulating Newton's work in gravitation and in arranging the publication of Newton's *Principia* (1687). He constructed and published charts of magnetic declination and the earliest wind map (1688). In 1705 he identified a comet (*Halley's comet*) by records as having a period of 76 years. It is due for its next return in 2062.

Hall mobility Of a *semiconductor or conductor. Symbol: μ_H. The product of the Hall coefficient R_H (*see* Hall effect) and the electrical conductivity κ.

Hall resistance *See* Hall effect.

Hall voltage *See* Hall effect.

Hallwachs effect The emission of electricity by a metal illuminated with ultraviolet light. *See* photoelectric effect.

halos Coloured circles apparently surrounding light sources caused by small particles of matter diffracting the light. Another group of halos encircling the sun or moon with a much larger radius are due to reflection and refraction by ice particles floating in the air.

Hamilton, Sir William Rowan (1805–1865) Brit. mathematician and astronomer. He was the originator of Hamiltonian mechanics (*see* Hamilton's equations), which reformulated classical mechanics in a manner later found adaptable to quantum theory. He predicted the phenomenon of conical refraction. His method of *quaternions*, at one time in high repute, was a precursor of modern *vector analysis. It attempted to extend the methods of the *Argand diagram, representing complex quantities geometrically in a plane, to three dimensions, but required *four* components to each vector, whence the name.

Hamiltonian function *Syn.* Hamiltonian. Symbol: H. A function that expresses the energy of a system in terms of *generalized momenta, p, and positional coordinates, q; for example:

$$(p^2/2m + \mu q^2/2)$$

expresses the energy of a body in *simple harmonic motion. The Hamiltonian function may also involve the time. It is much used in *wave mechanics. *See* Hamilton's equations; Hamilton's principle.

Hamiltonian mechanics *See* Hamilton's equations.

Hamilton's equations A restatement of *Lagrange's equations with emphasis on momenta rather than forces. Much used in advanced mechanics including *quantum mechanics. There are twice as many Hamiltonian equations as Lagrangian equations but they are only first-order instead of second-order differential equations. They involve the *Hamiltonian function H, which in ordinary cases is the total energy expressed as a function of the *generalized coordinates q_i and momenta p_i:

$$dq_i/dt = \partial H/\partial p_i,$$
$$dp_i/dt = -(\partial H/\partial q_i).$$

Hamilton's principle If the configuration of a system is given at two instants, t_0 and t_1, then the value of the time-integral of the *Langrangian function, $L = T - V$, is stationary (maximum or minimum) for the path described in the motion compared with any other infinitely near paths that might be described (for instance under constraints) in the same time between the same configurations. That is,

$$\delta \int_{t_0}^{t_1} (T - V)dt = 0,$$

where T = total kinetic energy, V = total potential energy. It has been more freely stated as "Nature tends to equalize the mean potential and kinetic energies during a motion". The principle is important for it contains in itself all the $3n$ equations of motion of the n particles comprising the system. The form given here is that for a conservative system, but the principle is of general application.

Hampson process For the liquefaction of air on a small scale. Air at high pressure is expanded to atmospheric pressure through a valve and suffers Joule–Kelvin cooling (*see* Joule–Kelvin effect). The air cools the incoming high-pressure gas in a spiral heat exchanger so that eventually the gas liquefies on expansion through the valve.

handset A telephone transmitter and receiver mounted in a single holder so that they may easily be held to the mouth and ear, respectively, of the user.

Harcourt pentane lamp (Vernon-Harcourt, 1877). An obsolete standard lamp with specified dimensions and methods of manipulation that burnt a mixture of pentane vapour and air: originally with 1 candle power but later with 10 candle (international) power.

hard disk *See* disk.

hardness This is a quantity that is difficult to define, so as to cover all cases. In the case of a crystal, the hardness is the resistance that a face of the crystal offers to scratching, which may differ in different directions. For many substances, hardness is an inverse measure of plasticity, i.e. very plastic materials are soft, not hard, and the hardness may be measured by the *Brinell test*, which consists in determining the load necessary to produce an indent of measured dimensions on the material under test. The load is applied (*a*) by adjusting the force on a steel ball placed in contact with the specimen under test, or (*b*) by allowing the steel ball to drop from a known height onto a slab of the material to be tested. The test is sometimes applied by substituting for the steel ball a steel cone, point downards. *Mohs' scale* of comparative hardness uses ten selected solids arranged in such an order that a substance can scratch all substances below it in the scale, and cannot scratch those above it: (1) talc, (2) rock salt, (3) calcite, (4) fluorite, (5) apatite, (6) felspar, (7) quartz, (8) topaz, (9) corundum, (10) diamond. The scale is not quantitative, e.g. when we say that the blade of a penknife has hardness 6.5, we mean no more than that the penknife blade, as far as scratchability goes, lies between felspar and quartz.

hard radiation *Ionizing radiation with a high degree of penetration. The adjective "hard" is most commonly applied to X-rays of relatively short wavelength. *Compare* soft radiation.

hard-vacuum tube *Syn.* high-vacuum tube. A *vacuum tube in which the degree of the vacuum is such that ionization of the residual gas has a negligible effect upon the electrical characteristics. *Compare* soft-vacuum tube; gas-filled tube.

hardware The physical components of a *computer system, such as VDUs, disk drives, printers, and the electronic circuitry making up semiconductor memory and logic circuits. *See also* software.

Hare, Robert (1781–1858) Amer. chemist who invented the oxyhydrogen blowpipe and the *hydrometer that bears his name.

Hare hydrometer *See* hydrometer.

harmonic 1. An oscillation of a periodic quantity whose frequency is an integral multiple of the fundamental frequency.
2. A tone of a series constituting a *note, and having a frequency that is an integral multiple of the fundamental frequency of the note. Brass instruments use harmonics of the tube to give the musical notes required. String players sometimes lightly touch a vibrating string at the appropriate antinode to leave only the required harmonics sounding. A similar technique may be used with harps.

harmonic analyser A device that evaluates the coefficients of the Fourier series corresponding to a particular function. *See* Fourier analysis.

harmonic analysis *See* Fourier analysis.

harmonic distortion *See* distortion.

harmonic echo An echo is not always a faithful reproduction of the original note and under some conditions the echo of a complex note appears to be raised in pitch above that of the original sound. It is then called a harmonic echo. This effect has been examined mathematically by Rayleigh who showed that the intensity of the reflected sound varies inversely as the fourth power of the wavelength. Thus the relative intensity of the octave of a note will be sixteen times more in the reflected wave than in the original sound. In the case of a complex note the

higher harmonics will be reflected better than the fundamental and so the pitch of the echo will appear to be raised. This is often noticed in the echo from a group of trees. As would be expected, the sound transmitted through the trees is complementary to that which is reflected, and so the higher harmonics are considerably attenuated within the trees.

harmonic generator A *signal generator that produces a large number of odd and even *harmonics of the fundamental frequency of the input.

harmonic mean *See* mean.

harmonic series A mathematical series in which the reciprocals of the terms form an *arithmetic series. A typical example is:
$$1 + \tfrac{1}{2} + \tfrac{1}{3} + \tfrac{1}{4} + \ldots$$

Harrison, John (1693–1776) Brit. clockmaker and inventor who made the first accurate marine chronometers so winning a British Government prize of £20,000.

Harrison's grid-iron pendulum *See* compensated pendulum.

Hartley oscillator *See* piezoelectric oscillator.

Hartmann formula A formula giving the variation of refractive index n of a medium with the wavelength of light:
$$n = n_\infty + c/(\lambda + \lambda_0)^a,$$
where n_∞, λ_0, and a are constants. For common forms of glass it is usual to take $a = 1$. It is rather less useful than the *Cauchy dispersion formula.

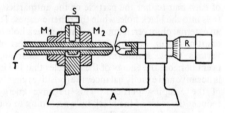

Hartmann generator

Hartmann generator An apparatus for producing ultrasonic edge tones on the principle of the *Galton whistle. It differs from this mainly in the greater blast velocity employed whereby the energy of the output is markedly increased.

The diagram shows a longitudinal section of Hartmann's generator. The nozzle T is fixed by the two nuts M_1 and M_2 to the base A. It is adjusted to be exactly opposite to the opening of the generator O by means of screws S. The position of the opening O is adjusted by the micrometer screw R.

If the velocity of the gas jet is increased above that of sound in air, a series of shock waves originate at the edges of the nozzle and are reflected to and fro on the confines of the jet. The compressions of the air or gas along these wave fronts are so intense that they can cast shadows on a photographic plate such as those taken by the *schlieren method. When these pulses fall on the opening of the little cylindrical pipe at the correct distance (from jet to pipe mouth), resonance occurs and the pipe emits powerful ultrasonic vibrations.

Frequencies up to 100 000 hertz can be produced. It has, however, considerable limitations: apart from the fact that it can only operate in a gas, its frequency is not very stable and there are many overtones sounding with the fundamental at the same time.

hartree *See* atomic unit of energy.

Harvard classification *See* stellar spectra.

Hawking, Stephen William (*b.* 1942) Brit. physicist, who became a professor at Cambridge, in spite of a progressive nervous disease that has confined him to a wheelchair, prevented him from writing, and impaired his speech.

Critical of Einstein's general theory of relativity for not being quantized, he has led a so-far unsuccessful search for a quantized theory of gravity. He later advanced several well-attested theories regarding black holes (*see* Hawking radiation). His book *A Brief History of Time* (1988) has been a popular best seller.

Hawking radiation The emission of particles by a *black hole as a result of quantum-mechanical effects. The gravitational field of the black hole causes production of particle–antiparticle pairs in the vicinity of the *event horizon (which is analogous to *pair production in an electric field). One member of each pair (either the particle or the antiparticle) falls into the black hole, while the other escapes. To an external observer, it appears that the black hole is emitting a flux of particles (and antiparticles) of various types. Furthermore, measured near the event horizon, the energy of the particles that fall in is negative and exactly balances the (positive) energy of the escaping particles. This negative energy reduces the mass of the black hole according to the relationship $E = mc^2$. The net result is that the emitted particle flux appears to carry off the black-hole mass. It can be shown that the black hole radiates like a *black body, with the energy distribution of the particles obeying *Planck's formula for a temperature that is inversely proportional to the mass of the hole. For a black hole of the mass of the sun this temperature turns out to be only about 10^{-7} K, so the process is negligible. However, for a 'mini' black hole, such as might be formed in the *early universe, with a mass of order 10^{12} kg (and a radius of order 10^{-15} m), the temperature would be of order 10^{11} K and the hole would radiate copiously (at a rate of about 6×10^9 W) a flux of gamma rays, neutrinos, and electron–positron pairs. (The observed levels of cosmic gamma rays put strong constraints on the number of such 'mini' black holes, suggesting that there are too few of them to solve the *missing-mass problem.)

H-bomb *See* nuclear weapons.

H–D curve *See* Hurter–Driffield curve.

head 1. A device that records, reads, or erases signals or data on a medium such as a magnetic tape or disk.
2. *See* pressure head.

health physics A branch of *medical physics concerned with the health and safety of personnel in medical, scientific, and industrial work. It is most particularly concerned with protection from *ionizing radiation and from neutrons.

Problems involved in radiation protection include the detection and measurement of ionizing radiation, cleaning of both personnel and surfaces contaminated by radioactive substances, disposal of *radioactive waste, design of laboratories and the *shielding* of equipment for radiation work, and the supervision of tolerance doses received by personnel in the course of their duties.

Common monitors used to detect radiation and measure the *dose received are the *ionization chamber, *Geiger counter, and film badge (*see* dosimetry). The harmful effects of radiation on biological tissue derive from ionization, and can follow from unintended ingestion of a radionuclide or exposure to radiation. Tolerance doses for given types of radiation acting either internally or externally have been formulated and no worker should receive greater than the maximum permissible *dose in a given period of time. Surface decontamination is generally achieved by washing, scraping, or the removal of a surface layer.

The shielding of equipment and personnel is usually achieved by using concrete (*see* loaded concrete) or lead. The conditions and the type of radiation under consideration determine the position and thickness of the shielding.

hearing aid A complete sound-reproducing system designed to increase the sound intensity at the ear. Hearing aids range from the now-obsolete ear trumpet to microminiature circuits that can be hidden in the frame of a pair of spectacles or even implanted. Modern hearing aids use a very small crystal microphone, battery-powered amplifier, and earpiece. Extremely good sound quality is obtained with sufficient power output for most cases. The amplifier can be designed to compensate for hearing loss over a specific band of frequencies only. Controls are provided to allow the wearer to vary the volume and tone. In certain types of deafness, in which the inner ear is normal, the output from the amplifier can be used to produce vibrations in a diaphragm pressed against the mastoid bone behind the ear. The vibrations are conducted through the bones directly to the cochlea, bypassing the defective outer or middle ear. Normally a modern hearing aid is in two parts connected by a thin wire. The battery and controls are housed in a small unit that can be hidden in a pocket and the microphone, amplifier, and earpiece are housed in a small unit that is worn in the ear. Often the earpiece is shaped to fit just inside the auditory canal. The development of *integrated circuits has assisted the design of modern hearing aids, allowing them to be extremely small and light and therefore much more comfortable.

hearing loss The degree of deafness of a person can be conveniently specified in terms of the hearing loss. This is defined as the threshold shift in decibels from the normal threshold intensity. If I is the threshold intensity for the person being tested and I_0 the threshold intensity for the normal ear, hearing loss = $10 \log_{10} I/I_0$. The hearing loss of a person at one frequency may not be the same as at another. For a proper indication of deafness, each ear should be tested over the whole of the audible range and an *audiogram should be plotted showing threshold intensity gainst pitch. By this method it is often found that deafness is confined to a region of frequencies only. The best method of testing for deafness is with an electronic audiometer, which allows a note of known frequency and intensity to be applied to the ear. Many rough methods of testing are often used, however, such as finding the person's ability to interpret numbers called in an average whisper at a certain distance. Similar tests with the click of coins or the tick of a watch are used. The amount of room noise and reflections of the sound from the walls make these tests unsatisfactory. Under certain conditions a deaf person appears to hear better than a normal one and this is used to trace the part of the ear that is defective. For example, if a tuning fork is pressed against the skull of a person he hears it by the medium of *bone conduction. Thus if a person has a defective middle ear he will hear such a tuning fork for longer than a normal person since room noise masks cochlea perception in the latter but not in the former case. Defects of the cochlea can be shown by placing a fork on the centre line of the skull. If a person hears it better with one ear than the other, there is a cochlear defect of the ear that hears less.

heat Symbol: Q. *Syn.* quantity of heat. The energy transferred from a body at a higher temperature to one at a lower temperature because of the difference of the temperature only. Before the principles of *thermodynamics were clearly established the word heat was used with various meanings, including *temperature and *internal energy. It is important to distinguish such quantities and to avoid any implication that heat now refers to any property or condition of a body, or anything but a process of transfer. The ambiguous term "heat energy" should be avoided. The processes are *conduction of heat, *convection, and *electromagnetic radiation.

Radiation is regarded as heat when the spontaneous emission from a hotter body is absorbed by one at lower temperature. All wavelengths emitted (ultraviolet, visible, infrared) are heat. Radiation can also transfer energy by doing work, for example a transmitter does work on a radio receiver, the temperatures being irrelevant.

The unit is the *joule (J); former units include the *calorie.

heat capacity Symbol: C. The quantity of heat required to raise the temperature of a body through one degree. It is measured in joules per kelvin. *See also* specific heat capacity.

heat death The condition of any isolated system when its *entropy is a maximum. The matter present is then completely disordered and at a uniform temperature, and there is therefore no internal energy available for doing work. If the universe is a closed system, it should eventually reach this state. This is called the *heat death of the universe*. It was first predicted in the 19th century by R. Clausius (the inventor of entropy) who said: "Die Energie der Welt ist constant, Die Entropie strebt einem maximum zu". (The energy of the universe is constant, its entropy tends to a maximum.)

heat engine A device that takes in heat from a hot source and does work, waste heat being discharged to a colder body, which is usually the atmosphere. In most cases work is done mechanically but, for example, a *thermocouple is a heat engine that does work electrically.

Ideally heat engines work cyclically, a *working substance* being taken through a sequence of operations and being returned to the initial state; in practice a fresh supply of working substance may be taken in for each cycle. A heat engine may be driven in reverse as a *refrigerator or *heat pump. *See* Carnot cycle; Diesel cycle; Otto cycle; Rankine cycle; efficiency.

heater In general, any *resistor used to provide a source of heat when carrying an electric current. Heaters have many applications as *heating elements, in particular for *indirectly heated cathodes. The term is also used to indicate a complete heating device, e.g. convector heater.

heat exchanger A device for transferring heat from one fluid to another without the fluids coming in contact. Its purpose is either to regulate the temperatures of the fluids for optimum efficiency of some process, or to make use of heat that would otherwise be wasted. The simplest form of heat exchanger consists of two coaxial pipes, the inner one finned on the outside to maximize the contact area, with the fluids moving through the pipes in opposite directions (counter-current flow). The radiator of a car is a typical example of a heat exchanger.

heat flow rate Symbol: Φ. The rate of heat flow across a surface; it is measured in watts. The *density of heat flow rate*, symbol: ϕ or u, is the heat flow rate per unit area. The *thermal conductivity is the density of heat flow rate divided by temperature gradient.

heating effect of a current *Syn.* Joule effect. When a current I is maintained in a resistor of resistance R by a potential difference V for a time t, the work done electrically is:

$$W = IVt = I^2Rt = V^2t/R.$$

By the first law of *thermodynamics the internal energy of the resistor increases by δU while heat Q is given out to the surroundings, such that $W = Q + \delta U$. In the steady state δU is zero and $Q = W$, hence the resistor supplies heat at the rate given by the equation above.

The law in the form $Q = I^2 R t$ was originally proposed by *Joule (1840) on the basis of experiment, before the electrical quantities were fully defined or the concept of energy clearly formulated. The law as stated here arises from the modern definitions of the electrical quantities and the laws of *thermodynamics.

heating element A complete and detachable assembly comprising a *heater with its former or other support as used in electric ovens, electric fires, etc.

heat pump A device for heating buildings, in the form of a *heat engine driven in reverse. The internal energy of some part of the environment (the atmosphere, soil, a river, etc.) is used as an energy source, giving heat Q_1 to the working substance. An electric motor does work W in taking the substance round a cycle in which the temperature is raised to slightly above that of the building. The heat given out Q_O is equal to $Q_1 + W$. Thus the internal energy of the colder body is decreased and that of the hotter body is increased. This does not violate the laws of thermodynamics since the process involves work being done upon the system and is not overall a process of heat transfer. Since the heat given to the building is greater than W, the cost of electricity supply is lower than for a simple *heater, but the capital costs of installation, maintenance, and replacement are greater. *See also* refrigerator.

heat shield A structure set round the nose cone of a spacecraft that is designed to return to earth in a recoverable state, protecting the craft from heat damage on re-entry into the earth's atmosphere. For the Apollo spacecraft the heat shield had a maximum thickness of 70 millimetres and consisted of several layers, the most important one being a reinforced synthetic material that glowed, charred, and melted, removing the heat by an ablative process. The speed reached by the spacecraft caused the temperature of the nose cone to rise from $-100\,°C$ to $3000\,°C$ in the space of a few minutes.

heat sink 1. A device employed (especially in association with *transistors and other electronic components) when it is essential to dispose of unwanted heat and prevent an unwelcome or damaging rise in temperature. It generally consists of a set of metal plates in a finlike formation that conducts and radiates the heat away.

2. A system that is considered to absorb heat at a constant temperature. The concept is useful in *thermodynamics, as in the operation of a heat engine.

heats of combustion and formation *See* calorimetry.

heats of reaction and solution *See* calorimetry.

heat-transfer coefficient The heat flow per unit time through unit area divided by the temperature difference. When applied to conduction of heat through a body, it is called the *thermal conductance* and has the symbol K. When applied to the emission of heat from a surface, it has the symbol E or α. It is measured in watt metre^{-2} kelvin^{-1}.

Heaviside, Oliver (1850–1925) Brit. physicist. He predicted the existence in the upper atmosphere of a layer capable of reflecting electromagnetic (radio) waves (the Heaviside layer – *see* ionosphere) and made many advances in the theory of wave propagation and in the practice of electrical communications. He developed the operational calculus and introduced complex numbers into problems involving alternating currents, especially a.c. networks. He played a considerable part in the development of modern *vector analysis.

Heaviside layer *See* ionosphere.

Heaviside–Lorentz units A *CGS system of electrostatic and electromagnetic units. It is a rationalized form of *Gaussian units, in which the magnetic constant has the value 4π and the electric constant $1/4\pi$. Like Gaussian units, Heaviside–Lorentz units are still used in particle physics and in relativity.

heavy-fermion system A substance in which some electrons have a very high effective mass (several hundred times the bare mass of the electron). Heavy-fermion compounds are compounds of actinide or lanthanide elements; an example is the cerium compound $CeCuSi_2$. The high-effective-mass electrons are f electrons in narrow energy bands associated with strong many-body effects.

These systems have unusual thermodynamic, magnetic, and superconducting properties, which are not well understood. It is thought that *superconductivity in this type of compound has a more complicated mechanism than the BCS theory, with the Cooper pairs being formed from high-effective-mass quasiparticles rather than electrons. In some substances superconductivity can occur at a much higher temperature than that for metals.

heavy water *Syn.* deuterium oxide (D_2O). The replacement of ordinary hydrogen by deuterium raises the molecular weight of water from 18 to 20, and there is an appreciable change in the physical properties (*see* isotope). When mixed with a compound of hydrogen, an exchange often takes place between atoms of deuterium and hydrogen and consequently heavy water has been used to trace chemical actions, especially those taking place in the bodies of animals. The use of heavy water in some *nuclear reactors is based on the fact that the deuterium nucleus has a much smaller capture cross section for neutrons and consequently makes an efficient *moderator.

heavy-water reactor (HWR) A type of thermal *nuclear reactor in which *heavy water is used as the *moderator, and sometimes also as the *coolant.

hecto- Symbol: h. A prefix denoting 100; for example, one hectometre (1 hm) = 100 metres.

hectometric waves Radio waves having wavelengths between 100 m and 1000 m.

Heisenberg, Werner Karl (1901–1976) German physicist. He introduced (1925) a system of quantum mechanics in which observable quantities alone enter (frequencies and amplitudes of radiation emitted). He advanced the principle of indeterminacy (now more usually called the *uncertainty principle) in 1927; he gave an explanation of *ferromagnetism in terms of *exchange forces between electrons in 1928. He was one of the first to suggest the modern concept that atomic nuclei consist of protons and neutrons, 1932. Nobel prize for physics, 1932.

Heisenberg force See exchange force.

Heisenberg uncertainty principle See uncertainty principle.

helix A curve lying on the surface of a cylinder and resembling a screw thread. The pitch is the distance parallel to the axis of the cylinder, between successive spirals.

Helmert's formula A formula, based on empirical measurements, that provides a value for the gravitational acceleration $g_{\phi, H}$ at any latitude ϕ and altitude H (in metres) above mean sea level. In metres per second squared,

$$g_{\phi, H} = 9.806\ 16 - 0.025\ 928 \times \cos 2\phi + 6.9 \times 10^{-5} \cos^2 2\phi - 3.086 \times 10^{-6} H.$$

helmholtz A unit of dipole moment per unit area proposed by E. A. Guggenheim. It is 1 *debye per square angstrom or, in SI units, 3.335×10^{-10} $C\,m^{-1}$.

Helmholtz, Hermann Ludwig Ferdinand von (1821–1894) German scientist – in turn physician, physiologist, physicist, and mathematician. An early supporter of the mechanical theory of heat and the conservation of energy, he played a considerable part in the development of *thermodynamics (see Gibbs–Helmholtz equation; Helmholtz function). His great work on vision (*Physiologische Optik*, 1856) included the development of Thomas Young's three-colour theory of vision (hence known as the *Young–Helmholtz theory*, see colour vision), the elucidation of the mechanism of accommodation of the eye, and an extensive study of defects of vision. In acoustics (*Lehre von den Tonempfindungen*, 1863; translated as *Sensations of Tone*, London, 1885) he analysed musical sounds, vowel sounds, etc., studied beats and combination tones, and produced a theory of harmony. The Helmholtz theory of *accommodation of the eye is the accepted theory (with additions) at the present day. See also tangent law.

Helmholtz coils A pair of identical cylindrical coils of wire mounted coaxially and separated by a distance equal to the radius of the coils. When a current is passed through the coils, connected in series, a uniform magnetic field is produced over a consider-able volume on either side of the midpoint between the coils. The arrangement was first used in the now-obsolete *Helmholtz galvanometer*.

Helmholtz electric double layer When a body is brought into contact with another body composed of a different material, the two bodies become oppositely charged, the substance with the higher relative *permittivity, ε_r, becoming positive. In order to explain this and other phenomena of electrification Helmholtz postulated that a film one molecule thick forming a double layer of positive and negative charges is set up and maintained by the inherent electrical forces of matter (see boundary conditions). Lenard extended the theory by suggesting that at the surface of any solid or liquid the molecules show orientation of the dipoles, negative charge outwards, forming an electric double layer. In materials of high ε_r, the attraction between the opposite charges is smaller, and a substance of small ε_r can thus remove free negative charges from one of greater ε_r. Contact needs to be close for this to happen (see frictional electricity). On separation of the charges, which initially were at molecular diameters apart, the lines of force are considerable extended, and the potential difference thus produced may be made very large.

Helmholtz function Syn. Helmholtz free energy. Symbol: A, F. A thermodynamic function of a system given by its internal energy (U) minus the product of its entropy (S), and its thermodynamic temperature (T), i.e.

$$A = U - TS.$$

If a system undergoes a *reversible change at a constant temperature, the Helmholtz function increases by an amount equal to the work done on it. The change in A (ΔA) between any two states of a system gives the maximum work that could be obtained from the system during this change if the optimum pathway were to be followed. If this change is negative, work is obtained from the system. See also Gibbs function.

Helmholtz galvanometer See Helmholtz coils.

Helmholtz–Kirchhoff correction For tube diameter. See Kundt's tube.

Helmholtz resonator In the analysis of sound, Helmholtz made use of air resonators. These are in the form of an air cavity, contained either in a spherical bulb connecting with the atmosphere through a neck where the length of the neck is negligible compared with the diameter or in a cylindrical bulb with a neck. The length of the cylinder is comparable with or greater than the diameter of the cylinder. In both cases the internal capacity of the resonator is larger than that of the orifice or neck by which the communication with the atmosphere is made.

These resonators in general have more selective resonance than the columns of air in pipes, since a very small proportion of energy is radiated into the atmosphere and therefore the damping is very small. This is true provided that the pressure amplitude

hemihedry

throughout the air cavity is practically uniform and negligible in the neck.

Rayleigh, in developing the theory of such resonators, regarded the air in the neighbourhood of the neck as acting like a piston, alternately compressing and rarefying the air in the cavity. He assumed also that the wavelength of the vibrations in the free air is large compared with the dimensions of the cavity and therefore at any instant the condensation will be uniform throughout the cavity.

The system in its simplest form is equivalent to a mass attached to a spring, the air piston in the neck being regarded as the mass and the air in the cavity as the spring. In acoustic terms the neck is considered as the inertance and the cavity as the capacitance (*see* acoustic impedance). The dissipation is mainly due to the energy radiated and this is equivalent to the acoustic resistance.

The resonance frequency v is that corresponding to the value of angular frequency for which the reactance term disappears, i.e. for which:

$$2\pi v = c\sqrt{(S/lV)},$$

where c is the speed of sound, l is the length of the neck, S its cross-sectional area, and V is the volume of the cavity.

The ratio S/l is of the dimension of a length and is called the conductivity k of the orifice or neck. For a circular aperture the conductivity is equal to the diameter.

Cylindrical Helmholtz resonators can be adjusted in volume by sliding one part over the other and it can be proved experimentally that the square of the resonant frequency is then inversely proportional to V, a relation that was found much earlier by Sondhauss. The *end correction must be taken into account and therefore l must be increased to $(l + 0.6r)$, where r is the radius of neck.

Although the pipe and the Helmholtz device both act as resonators, yet they differ in many aspects. In the pipe the length is comparable with λ, the wavelength of the sound, and the displacements vary sinusoidally along the column of gas due to stationary waves. In the Helmholtz resonator the dimension of the cavity is assumed to be smaller than λ and the particle displacement is negligible except in the neck. The fundamental frequency of a pipe of length L is given by $nL = $ a constant while that of a cylindrical resonator, whose volume V is proportional to L, is given by n^2L is equal to a constant.

Helmholtz resonators are used as extremely sensitive detectors of sound at a particular frequency. In order to obtain further increase of sensitivity, Boys suggested the use of a double resonator consisting of a closed pipe at the end of which was fitted a Helmholtz resonator in tune with it. Paris has substituted the pipe by another Helmholtz resonator or a conical horn and deduced the resonance frequencies. The double resonator has an important application when used in conjunction with the *hot-wire microphone or the *Rayleigh disc in the measurement of sound intensities.

hemihedry The possession by a crystal of only some of the symmetry elements that make up the full possible symmetry of the system to which it belongs.

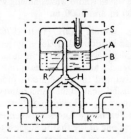

Determination of specific latent heat of vaporization (Henning)

Henning's heat experiments (1906) **1.** The determination of the specific latent heat of vaporization of steam at temperature from 30 °C to 180 °C by a constant-flow method in which the rate of electrical heating necessary to evaporate water at a known rate is measured. Water contained in a copper vessel A (*see* diagram) with an airtight lid S is heated by a constantan heating coil in the case B. The temperature of the steam is measured by thermocouples T, and the steam passes down the central tube R and may be condensed in either K' or K'' by turning the tap H. The pressure over the boiling water is varied by connecting a tube at H to a source of high or low pressure. Condensation occurs first in K' until the steady state is reached, when H is turned so that the water collected in a known time in K'' may be found. The tube R has an umbrella-like arrangement to prevent drops of water splashing down it and the boiler A is surrounded by an air space with an oil bath around it maintained at the same temperature as A. The heat loss from A is thus made small and is eliminated by carrying out two experiments of the same duration but with different rates of heating by B.

$$E_1 C_1 t + q = m_1 L$$
$$E_2 C_2 t + q = m_2 L$$

whence

$$L = [(E_1 C_1 - E_2 C_2)t]/(m_1 - m_2),$$

where q is the gain of heat by A in time t, m is the mass of water condensed in this time, and L its specific latent heat of vaporization at the boiling point.

2. The measurement of the variation of the vapour pressure of nitrogen with temperature over the range 60 to 80 K and of hydrogen between 14 and 20 K using constant temperature baths in which liquid nitrogen and hydrogen were boiled under reduced pressure maintained by powerful vacuum pumps. The results enabled temperatures in these two ranges to be measured by vapour-pressure thermometers of nitrogen and hydrogen respectively, calibration being achieved by the use of the *triple point of oxygen 54.24 K and of nitrogen 63.09 K as fixed points.

3. From a consideration of the experimental data for oxygen, he arrived at the value 22.414 litre atmospheres for the limiting value at zero pressure of the product (pV) at the ice point. This gave a value of 8.314 J K^{-1} mol^{-1} for the *molar gas constant R.

henry Symbol: H. The *SI unit of self- and mutual inductance, defined as the inductance of a closed loop that gives rise to a magnetic flux of one weber for each ampere of current that flows. *See* electromagnetic induction.

Henry, Joseph (1797–1878) Amer. physicist. He made considerable improvements in the design of electromagnets and in experiments with these discovered self-induction (1829, published 1832), anticipating Faraday (1834–1835), and discovered mutual induction about the same time. Henry was one of the early observers of the oscillatory nature of the discharge of a Leyden jar (1842). He also made a primitive telegraph; and he developed methods for weather forecasting. The unit of inductance is named after him.

Henry's law of solubility (W. Henry, 1775–1836, Brit. chemist: 1803). At constant temperature, the solubility of a gas in a liquid with which it does not react chemically is proportional to the pressure.

heptode A *thermionic valve having five grids between the cathode and anode (i.e. a total of seven electrodes).

Heraclides of Pontus (c. 350 B.C.) Greek philosopher who was the first to suggest that the earth rotates on its axis, rather than the heavens rotating about the earth.

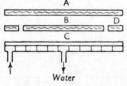

Determination of thermal conductivity of gases (Hercus and Laby)

Hercus and Laby's determination of thermal conductivity of gases (1919) The thermal conductivity of gases was found by heating them electrically from above to eliminate convection. B is a copper plate with an embedded heater and D is a guard ring at the same temperature as B to make the heat flow from B to C (which was water-cooled) linear. The plate A is maintained at the same temperature as B to prevent loss of heat from the top surface of B, all surfaces being plane and polished. Constantan wires formed thermocouples with the copper plates to which they were attached, enabling the temperatures of the plates to be found. The apparatus was made gastight by a rubber band clamped to A and C by steel bands. The plate B is supported from D by three ivory studs and the electrical work done on it is measured. About 5% of the whole transfer of heat

from B to C is by radiation, the allowance for this being found by a separate experiment on the loss of heat from a silvered Dewar flask.

Hering theory of colour vision In 1876, Hering suggested that there are three substances in the retina that were broken down and reformed (red-green, yellow-blue, white-black). The theory gives a reasonable theory of colour vision but fails to interpret colour blindness. *See* colour vision.

Hermitian matrix So named after the French mathematician Hermite. It is a matrix in which the elements A_{mn} and A_{nm} are equal when real, and conjugate to one another when complex. For example,
$$A_{mn} = \alpha + i\beta \text{ and } A_{nm} = \alpha - i\beta,$$
α and β being real. *See* matrix.

Hermitian operator An operator represented by a *Hermitian matrix.

Hero of Alexandria (c. A.D. 60) Alexandrian mathematician who first discovered that the area of a triangle can be calculated from its sides (Hero's formula). His *Mechanics* introduced the parallelogram of velocities and the principle of the lever. His *Catoptrics* proved the equality of the angles of incidence and reflection for mirrors, and his *Pneumatics* described many mechanical contrivances, such as the fountain and a primitive form of steam engine.

herpolhode cone *Syn.* space cone. *See* instantaneous axis.

Herschel, Sir John Frederick William (1792–1871) Brit. astronomer, son of Sir William Herschel. He made contributions to pure mathematics as well as to astronomy and optics. He made the first telescope survey of the southern skies. In 1823 he suggested that emission lines of metals could be used for their identification.

Herschel, Sir William (1738–1822) German-born Brit. astronomer (orig. Friedrich Wilhelm Herschel). An indefatigable worker, he constructed a great number of *reflecting telescopes and made very valuable observations, particularly on the fixed stars, aided by his sister *Caroline Herschel* (1750–1848). He discovered 5000 star clusters and many nebulae; established that the solar system is moving relative to the fixed stars; and discovered the planet Uranus (1781). He was the first to study the heating effects of different regions of the visible solar spectrum, using a very small mercury thermometer as a detector (1800). In this experiment he discovered that, in addition to the visible heat rays, the sun emits invisible heat radiation (infrared). His instrument was not sufficiently sensitive to detect the ultraviolet.

Herschel–Quincke tube An apparatus suggested by J. Herschel and later constructed by Quincke to demonstrate the *interference of sound. It consists of a tube that divides into two tubes of different lengths, the ends of which join together to form one tube again. One of the tubes can usually be varied in length by means of an arrangement similar to the slide of a trombone. A source of sound is placed at one end of the apparatus and the sound travels along the two different paths, the resultant being heard at the other end. When the path difference between the sound going through one tube and that going through the other is a whole number of wavelengths, they arrive at the ear in phase and reinforce each other. When the path difference is an odd number of half-wavelengths, however, they are out of phase and no sound is heard at all. This simple picture of the action of the tube has been extended by Stewart, whose theory allowed for the small amount of sound that passes through one tube and then travels back through the other tube and along the original one again, before it emerges from the apparatus. Stewart verified his theory by observation.

hertz Symbol: Hz. The *SI unit of *frequency, defined as the frequency of a periodic phenomenon that has a period of one second. This unit replaced the cycle per second.

Hertz, Gustav (1887–1950) German physicist, nephew of Heinrich Hertz. With J. *Franck he carried out experiments on inelastic collisions of electrons with atoms, establishing the existence of quantum states with discrete energies in the atoms (1913). He obtained a nearly complete separation of neon isotopes by a continuous many-stage effusion process (1932). (*See* isotope separation.) After the 1939–45 war, he was taken to the USSR to continue his research on nuclear *fission.

Hertz, Heinrich Rudolf (1857–1894) German physicist. He worked on the properties of cathode rays and noted in 1887 the effect of ultraviolet light in facilitating discharge across a spark gap, an observation that led to the discovery of photoelectricity. He added to Maxwell's electromagnetic theory and then (1888) demonstrated the existence and properties of electromagnetic waves generated from oscillatory circuits containing spark gaps. The unit of frequency is named after him.

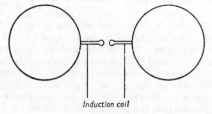

Induction coil

Hertzian oscillator

Hertzian oscillator An electrical system for the production of electromagnetic waves. It consists of two capacitors, e.g. two plates or spheres joined by a conducting rod in which there is a small spark gap. If the two halves of the oscillator are raised to a sufficiently high potential difference, a spark passes across the gap, rendering it temporarily a conductor; an oscillatory discharge takes place, the period

of the oscillations being equal to $2\pi\sqrt{LC}$, where L is the self inductance and C the capacitance of the system. Electromagnetic waves of the same period are given off during the discharge. The oscillator is usually activated by a small induction coil, and a group of waves is emitted at each discharge. Owing to the resistance of the spark gap, the waves are highly damped. Their wavelength is usually of the order of a few metres.

Hertzian waves Electromagnetic waves, such as those produced by a *Hertzian oscillator. The term is sometimes used generally for electromagnetic radiations produced directly by oscillations in an electric circuit, but is more usually restricted to radiations of the order of wavelength of a few metres, such as those studied by Hertz.

Hertzsprung, Ejnar (1873–1967) Danish astronomer who worked in Germany and discovered the existence of *giant and *dwarf stars. With H. N. Russell he devised the *Hertzsprung–Russell diagram.

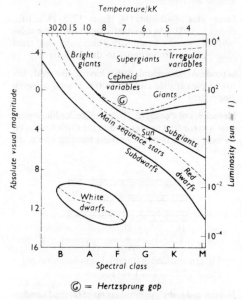

Ⓖ = Hertzsprung gap

Hertzsprung–Russell diagram

Hertzsprung–Russell diagram (H–R diagram) A diagram, first drawn up in 1913 by H. N. Russell from his own work and that of E. Hertzsprung, showing the variation of absolute *magnitude (i.e. intrinsic brightness) in stars against their spectral type, and hence temperature (*see* stellar spectra). Instead of a uniform distribution, the stars occupy well-defined regions on the diagram (*see* diagram). Some 90% lie along a diagonal band, known as the *main sequence*. The brighter *giant stars form another group, as do the brightest and relatively rare supergiants. Other groups can also be distinguished.

The H–R diagram is of great importance in studying *stellar evolution*. Diagrams determined from a theoretical basis can be tested against those obtained by astronomical observations. Diagrams can be drawn, for example, for the brightest stars, for stars in a particular locality (e.g. in the sun's neighbourhood), for pulsating *variable stars, and for globular clusters of stars (which contain some of the oldest stars in our Galaxy). These diagrams show different distributions of stars. For example, globular-cluster stars appear mainly in the giant branches of the H–R diagram.

A star spends most of its life on the main sequence, appearing there once nuclear-fusion reactions have begun in its core. Main-sequence stars obtain their energy from these reactions, in which hydrogen is converted to helium (see stellar energy). As a star burns up the last of its hydrogen, its radius increases and the star "moves" across from the main sequence to the giant region. This movement occupies only a relatively short length of time, and thus there is a space (known as the *Hertzsprung gap*) between the two regions. After eventually becoming *red giants, the stars tend to develop a variation in magnitude (see variable star), and they occupy the lower left-hand part of the giant region. In due course, after a sequence of as yet imperfectly understood events, the stars become *white dwarfs, *neutron stars, or possibly *black holes, depending on the final stellar mass. See also gravitational collapse.

Hess, Germain Henri (1802–1850) Swiss-born Russian physician, physicist, and geologist. He was a pioneer in the study of thermochemistry and is remembered for the law of heat summation that bears his name. See Hess's law.

Hess, Victor F. (1883–1964) Austrian-born Amer. physicist. He was one of the discoverers of cosmic rays (1910).

Hess's law The algebraic sum of the heats evolved or absorbed by each stage of a chemical reaction is equal to the total amount of heat evolved or absorbed when the reaction occurs directly; i.e. the net heat of reaction is independent of the path and depends only on the initial reactants and the final products.

heterochromatic photometry The science of comparing the illuminating powers of sources of light of different colours. Direct comparison is difficult and there is a wide range of uncertainty, not only as to the time of balancing but also as results differ between observers and between the results obtained by one observer at different times. The *Purkinje effect also introduces a complication, i.e. equal reduction of two colour sources from a condition of equal brightness results in a change of brightness – blues appear relatively brighter as intensities are reduced equally. Various methods have been adopted to reduce the region of uncertainty. The *cascade* or *step-by-step* method consists in comparing lights of nearly equal colour and so gradually bridging the colour gap by means of a succession of comparisons. The flicker method uses the principle that at certain

frequencies, brightness flicker exists as distinct from colour flicker. Physical photometry using photocells or photography appears to have a future for this type of measurement. See photometry.

heterodyne See beats.

heterodyne interference *Syn.* whistle. A spurious high audio note occurring in *superheterodyne receivers, due to beats between two *carrier waves with frequencies close together.

heterodyne reception See beat reception.

heterogeneous The opposite of *homogeneous.

heterogeneous radiation A particular type of radiation, such as X-rays or gamma rays, having a variety of wavelengths or quantum energies.

heterogeneous reactor A type of *nuclear reactor in which the fuel is separated from the *moderator. *Compare* homogeneous reactor.

heterogeneous strain See homogeneous strain.

heterojunction A junction between two dissimilar *semiconductors of opposite polarity (i.e. p-type and n-type) and with different energy gaps between the valence and conduction bands (see energy bands). Such junctions have several advantages over the usual *homojunction. They are used for example, in *semiconductor lasers, *light-emitting diodes, and in the *heterojunction bipolar transistor* (HJBT). See also heterostructure.

heteropolar generator A *generator in which the active conductors pass through magnetic fields of opposite sense in succession. Most modern generators are of this type. An alternating e.m.f. is induced in each conductor so that a *direct-current generator of this type must be fitted with a *commutator. *Compare* homopolar generator.

heterostructure The composite structure resulting when a layer of one semiconductor is deposited on a layer of a different semiconductor. There may be several layers of the two dissimilar materials. The two semiconductors are selected to have different energy gaps between the valence and conduction bands (see energy bands). Heterostructure semiconductors, involving ultrathin layers, can now be produced by advanced crystal-growing techniques. The choice of materials, and the thickness and number of layers, may be varied by design over wide limits. The electronic properties of such a device can therefore be designed for a particular application.

heuristic solution A solution to a mathematical problem for which no *algorithm exists. It employs inductive reasoning based on experience of similar problems and their solutions.

Heuse and Odo's experiments on expansion of gases 1. They measured the coefficient of expansion at constant pressure and the coefficient of increase of pressure at constant volume for helium, nitrogen, and hydrogen at various initial pressures and between the ice and steam points. Their values when extrapolated to those at zero pressure gave both coefficients equal to 0.003 6608 per °C giving absolute zero as –273.16 °C.

2. They tested the closeness of the agreement of the International Practical Temperature Scale with the gas scale. The international temperature was determined with a standard platinum thermometer alongside which was placed a helium-gas thermometer, maintained at the hydrogen point, the carbon dioxide point, and finally at the melting point of mercury. The readings of the helium thermometer were reduced to the ideal-gas scale by extrapolating to zero pressure and the greatest difference between the thermometer readings was 0.046 °C found at the carbon dioxide point.

Heusler alloys Alloys of manganese, aluminium, and zinc (or copper) that, although containing no ferromagnetic metal, exhibit ferromagnetic properties.

Hevesy, George von (1885–1966) Hungarian chemist and physicist. The discoverer (with D. Coster) of the element hafnium (1922). He introduced the use of *radioactive tracers into chemistry and biology.

Hewish, Antony (b. 1924) Brit. physicist and radio astronomer. He became professor of radio astronomy (1971) at Cambridge, where he was responsible for investigating the newly discovered *pulsars. For this work he shared a Nobel prize (1974) with Martin Ryle.

hexadecimal notation A number system, used in computing, having numbers to base 16, with the letters A to F normally used to represent decimal numbers 10 to 15. These numbers are often used to represent binary numbers in a convenient form, as each group of 4 *bits may be represented by 1 hexadecimal digit. For example,

1010 1101 0011 is AD3 in hexadecimal.

Compare octal notation; binary notation.

hexagonal system *See* crystal systems.

hexagon voltage *See* voltage between lines.

hexode A *thermionic valve having four grids between the cathode and anode (i.e. a total of six electrodes).

Heylandt's liquefier Heylandt devised a liquefier in which the compressed gas was first cooled by performing external work as in the *Claude process and then allowed to suffer Joule–Kelvin expansion as in the *Linde process. This principle was used by Kapitza for the liquefaction of hydrogen and helium.

Heyrovsky, Jaroslav (1890–1967) Czechoslovakian physicist who invented *polarographic analysis.

HF Abbreviation for high frequency. *See* frequency bands.

Hibbert's magnetic standard *See* fluxmeter.

hidden matter *See* missing mass.

Higgs boson A particle of zero spin and nonzero mass predicted to exist by certain *gauge theories (*see* electroweak theory). The Higgs boson has not been found but it is thought likely that it will be discovered in the next few years as higher-energy particle accelerators are built.

high elasticity A property that enables some substances to obey *Hooke's law with fair exactitude up to enormously greater strains than are normally met (e.g. cellulose hydrate and some other organic polymers).

high frequency (HF) *See* frequency bands.

high-level language *See* programming language.

high-pass filter *See* filter.

high-speed steel A steel alloy used for making high-speed cutting tools that maintain the sharpness of their edge even when red hot. The alloy contains tungsten or molybdenum and small amounts of other metals.

high-temperature gas-cooled reactor (HTR) *See* gas-cooled reactor.

high-temperature superconductivity *See* heavy-fermion system; superconductivity.

high tension (H.T.) High voltage, especially when applied to the anode supply of thermionic valves; usually in range 60 to 250 volts.

high voltage In electrical-power transmission and distribution, a voltage in excess of 650 volts.

high-voltage test A test applied to an electrical machine, transformer, capacitor, etc. to ascertain whether or not the insulation is adequate. A voltage in excess of normal value is applied between the parts intended to be insulated from one another.

Hilbert space A multidimensional space in which the proper (eigen) functions of *wave mechanics are represented by orthogonal unit vectors. *See* proper function and value.

Hillier, James (b. 1915) Canadian-born Amer. physicist who did most of his work at the Radio Corporation of America (RCA). He constructed the first high-resolution electron microscope (1940). *See also* Zworykin.

Hipersil A silicon-iron alloy (3–4% Si) with low impurities, large grain size, and crystal orientation produced by hot- and cold-rolling. The core loss in transformers made of this alloy is very low.

Hipparchus (190–125 B.C.) Greek astronomer who measured the eccentricity of the earth's orbit and calculated the length of the year to within six minutes. He invented trigonometry.

histogram A graphical representation of a frequency distribution in which rectangular areas standing on each interval into which the observations are grouped, show the frequency of observations in that interval.

Hittorf, Johann Wilhelm (1824–1914) German physical chemist remembered for his work on electrolysis. *See* Hittorf transport ratio.

Hittorf transport ratio For an electrolytic ion during electrolysis, the ratio of the velocity of the ion to the sum of the velocities of the two ions taking part in the action. If the two ions travel at different velocities, the loss of concentration in the neighbourhood of the electrode from which the faster ion is travelling will be greater than that at the other electrode. The transport ratio is measured by the ratio of loss of concentration at the appropriate electrode to total loss through electrochemical decomposition.

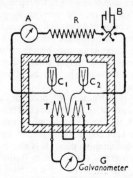

Determination of Stefan–Boltzmann constant (Hoare)

Hoare's determination of the Stefan–Boltzmann constant The value of the *Stefan–Boltzmann constant was found using both high- and low-temperature sources. F. E. Hoare modified Coblentz's method by using a Callendar *radio balance as receiver. Two copper cups C_1 and C_2, 8 mm long and 3 mm diameter are mounted in the same enclosure to allow for temperature variation of the surroundings, the temperature difference between the cups being obtained from the galvanometer deflection in the circuit containing thermocouple junctions T attached to the cups. Radiation from a black body is allowed to fall on C_1 and the *Peltier effect varied by adjusting R so that the temperature difference between C_1 and C_2 is small. C_1 is then shielded and the radiation allowed to fall on C_2, the current being reversed by the commutator but unaltered in magnitude, and the change in galvanometer deflection θ noted. Then $W = 2PI + \theta s$ microwatts, where W is the radiant energy incident per second on either cup, P is the value of the Peltier coefficient in millivolts, I the current in milliamps registered by A, and s is the sensitivity of the galvanometer obtained from ($s\theta_0 = 2PI_0$) by shielding both cups and measuring the deflection θ_0 obtained when a current I_0 is reversed through the Peltier junctions. The value P is obtained by having identical manganin heating coils in each cup and supplying power W_0 microwatts to one coil and passing a current I' through the Peltier junctions so that a small galvanometer deflection is obtained. If the same current is now passed through

the other heating coil and the main current reversed to give a change of galvanometer deflection θ', then $W' = 2PI' + \theta's$, whence P is obtained. The results obtained with the source at 100, 700, and 1000 °C all agreed, the mean value being

$$\sigma = 5.736 \times 10^{-8} \text{ W m}^{-2} \text{ K}^{-4}.$$

hoar frost A layer of ice deposited on surfaces whose temperature is below 0 °C by the condensation (strictly speaking, sublimation) of moisture from the atmosphere.

hoar-frost line *Syn.* sublimation curve. A curve expressing the relation between the saturation vapour pressure of ice and temperature. *See* triple point.

hodograph A device to facilitate visualizing the acceleration of a particle P moving along some curve, by reference to a fictitious particle P', which moves so that the *position vector of P' relative to some chosen origin O' is equal to the velocity of P. The path described by P' is the hodograph of the motion of P. The velocity of P' on the hodograph is equal to the acceleration of P at the corresponding point of its path.

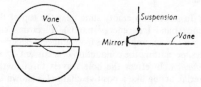

Hoffmann electrometer

Hoffmann electrometer A sensitive *electrometer using only a half-vane (*see* diagram) moving in two half-segments of a closed metal box, but otherwise working on the same principle as the *quadrant electrometer. Very heavy copper shields are used to minimize thermal variations, which might affect the needle, and the instrument is usually operated with the pressure reduced to a few hundred pascals.

Hofstadter, Robert (*b.* 1915) Amer. physicist, who became a professor (1954) at Stanford. His early work on infrared spectroscopy and photoconductivity was followed by scattering experiments using Stanford's linear accelerator. The various predictions he made about the nucleus led to a Nobel prize (1961), which he shared with Rudolph Mössbauer.

Holborn and Henning's specific heat capacity measurements (1911) The specific heat capacity at constant pressure was measured for gases at temperatures up to 1400 °C. The gas passes through heated spirals in X and through a heated tube RR into a double-walled platinum tube kept hot by passing an electric current through it via the leads Z, Z_1. The temperature of the gas before entering the calorimeter M is measured by a thermocouple L placed over the baffle-plates at D. The calorimeter M contains paraffin oil, the gas passing through three silver cylinders, $L_1 L_2 L_3$, packed with silver filings, which are immersed in the oil. The calorimeter is surround-

hole

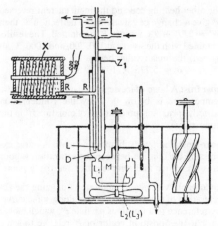

Specific heat capacity of gas at constant pressure

ed by a large oil bath maintained at constant temperature so that a radiation correction may be applied.

hole In a solid, an empty state near the top of the valence band (*see* energy bands). Electrons can move into such empty states from adjacent occupied states, permitting the conduction of a current – *hole conduction*. In effect the hole travels through the material, acting like a positive charge (as shown by the *Hall effect) with a positive mass of the same order of magnitude as the mass of an electron but not identical with it (*see* effective mass). Generally, the *mobility of holes is less than that of the electrons in the conduction band.

A hole – an empty state in the valence band – may result from an electron being thermally excited from the valence to the conduction band, generating an *electron-hole pair*, or from an electron being trapped by an acceptor impurity (*see* semiconductor). In an intrinsic semiconductor the numbers of holes and free electrons are equal, with concentration n_i per unit volume. In an extrinsic semiconductor the concentration of holes n_+ and electrons n_- is given by $n_+ \times n_- = n_i{}^2$. A p-type semiconductor is one in which n_+ greatly exceeds n_- so electrical conduction is mostly by the holes, which are the *majority carriers.

hole conduction *See* hole.

hole current In a *semiconductor the current associated with the transfer of *holes.

hologram *See* holography.

holographic interferometry A technique that involves the superposition on the same photographic plate of two or more holograms (*see* holography) of an object under study. Any movement of the object between the holographic exposures shows up as *interference fringes distributed across the reconstructed image. The movement may result from

vibration, heating, strain, etc. Analysis of the fringes provides information about the properties of the object.

holography A technique for the reproduction of a stereoscopic image without cameras or lenses. A monochromatic, coherent, and highly collimated beam of light from a *laser is separated into two beams, one of which is directed to a photographic plate coated with a film of high resolution. The second beam hits the subject whose image is to be reproduced, and is diffracted to the plate where a *hologram* is formed, consisting of an interference pattern (rather than a collection of light and dark areas as with a conventional photographic negative). The original subject may be recreated by placing the hologram in a beam of coherent light, generally from the same laser; the hologram behaves as a *diffraction grating, producing two beams of diffracted radiation, one giving a real image that may be recorded on a photograph, and the other a stereoscopic virtual image.

It is also possible now to produce full-colour holograms: three laser beams are used (instead of a single beam), the three wavelengths corresponding to three *primary colours, and a thick emulsion is used on the holographic plate. It is also possible to view such holograms with ordinary (reflected) sunlight or tungsten light and see a stereoscopic full-colour image.

holohedry The possession by a crystal of the full possible symmetry of the system to which it belongs.

Holtz machine An early type of machine for the continuous production of electrical charges at high potential by electrostatic induction. It was superseded by the *Wimhurst machine.

homocentric Converging to or diverging from a common point.

homogeneity principle *See* dynamic similarity.

homogeneous 1. A body is homogeneous if all parts of the body possess similar properties.

    2. (mathematics) Of the same kind so as to be commensurable. Of the same degree or dimensions.

homogeneous deformation That which is uniform throughout the crystal in any particular direction.

homogeneous radiation Radiation that has only one constant wavelength or quantum energy.

homogeneous reactor A type of *nuclear reactor in which the fuel and *moderator present a uniform medium to the neutrons; for example, the fuel, in the form of a salt of uranium, may be dissolved in the moderator. *Compare* heterogeneous reactor.

homogeneous solids Those in which the physical and chemical properties are the same about every point; they may be amorphous or crystalline.

homogeneous strain (1) When a body is strained (*see* strain), a particle whose Cartesian coordinates with respect to axes fixed outside the body are (x, y, z) is

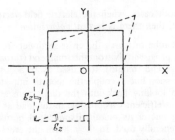

Shear strain

displaced to a new position (x', y', z'). If the following relations (in which the as, bs, and cs are nine constants) exist, the strain is said to be homogeneous and uniform:

$$x' = a_1x + a_2y + a_3z$$
$$y' = b_1x + b_2y + b_3z$$
$$z' = c_1x + c_2y + c_3z.$$

(No constant terms are included on the right-hand side as these would merely indicate a superimposed translation of the body.)

In a uniform homogeneous strain, a plane in the body remains a plane but changes its position; a parallelogram becomes a parallelogram in a different plane and with different angles. A sphere becomes an ellipsoid whose three mutually perpendicular axes are derived from three mutually perpendicular diameters of the sphere by their elongation and rotation.

If the relation between (x', y', z') and (x, y, z) are not linear, the strain is *heterogeneous*; if the as, bs, and cs vary with (x, y, z), the strain is not uniform.

(2) Strains may be expressed in terms of displacements, or shifts, of the particles; these are the quantities:

$$u = x' - x, \quad v = y'-y, \quad \text{and} \quad w = z'-z.$$

(3) If the squares and products of the displacements of a homogeneous strain are negligible in comparison with the displacements themselves, it can be shown that such strains can be combined by adding corresponding displacements. For example, two small homogeneous strains whose displacements are u_1, v_1, w_1, and u_2, v_2, w_2 are equivalent to a single strain with displacements $u_1 + u_2, v_1 + v_2, w_1 + w_2$. Using this theorem, it is found that any small uniform homogeneous strain can be resolved into two parts:

(a) a pure uniform homogeneous strain in which $c_2 = b_3, a_3 = c_1$, and $b_1 = a_2$. This may be written in terms of displacements as:

$$u = s_xx + g_zy + g_yz$$
$$v = g_zx + s_yy + g_xz$$
$$w = g_yx + g_xy + s_zz$$

(there being now only six different constants $s_x = (a_1 - 1), g_z = a_2, g_y = a_3$, etc.);

(b) a set of displacements representing a rotation of the body as a whole. (This brings the three mutually perpendicular diameters of the sphere into the same directions as the axes of the ellipsoid derived from them.)

Thus, a pure uniform homogeneous strain, alone, changes a sphere into an ellipsoid and the three mutually perpendicular diameters of the sphere become the axes of the ellipsoid by changing their lengths without rotating. This ellipsoid is known as the *strain ellipsoid* and its axes are called the *principal axes of strain*. If the axes OX, OY, OZ are parallel to the principal axes of strain, the constants g_x, g_y, and g_z vanish, and s_x, s_y, and s_z are known as the *principal strains*.

(4) A small pure uniform homogeneous strain may be resolved, by the theorem quoted, into six strains whose displacements are as follows:

$u_1 = s_xx$	$u_2 = 0$	$u_3 = 0$
$v_1 = 0$	$v_2 = s_yy$	$v_3 = 0$
$w_1 = 0$	$w_2 = 0$	$w_3 = s_zz$
$u_4 = g_zy$	$u_5 = 0$	$u_6 = g_yz$
$v_4 = g_zx$	$v_5 = g_xz$	$v_6 = 0$
$w_4 = 0$	$w_5 = g_xy$	$w_6 = g_yx$

The first three represent extensional strains without change of direction (in the first one, s_x is a change in length per unit length). The last three represent simple sheer strains: the 4th relation alone indicates that a square in the XY-plane is deformed into a rhombus in the same plane (*see* diagram); the shear strain is the total angular deformation at one corner $= 2g_z$.

The extensional strains s_x, s_y, and s_z together with the shear strains $2g_x, 2g_y$, and $2g_z$ are called the *components of the strain*. Any small uniform homogeneous strain is describable in terms of these six components together with a rotation of the body without change of shape.

homojunction A *p-n junction between two regions of opposite polarity (p-type and n-type) within a semiconductor. *Compare* heterostructure.

homopolar bond A *covalent bond, especially one with an equal distribution of charge.

homopolar crystals Those in which the interatomic bonds are all of the kind (i.e. homopolar or *covalent bonds) in which electrons are shared between neighbouring atoms.

homopolar generator A direct-current generator in which the active conductors pass through a magnetic field of constant magnitude and sense. This type of generator is best suited for relatively low-voltage, very heavy current machines, but even in this application the *heteropolar generator is almost universal.

Hooke, Robert (1635–1703) Brit. scientist. A very able but somewhat quarrelsome individual, he invented the conical pendulum, was a pioneer in the study of springs and elastic action of all kinds, and was in advance of most of his contemporaries in his ideas of optics, the structure of matter (especially crystals and gases), and the nature of combustion and of heat. *Hooke's law in elasticity was discovered about 1660 and published in 1676; the properties of coiled springs and their suitability for governing watches were recognized in about 1658 and a

269

Hooke's law

watch with a hair-spring of this type controlling a balance wheel was made in 1675. He improved the air pump for Boyle; and he gave a substantially correct explanation of the interference rings usually known as Newton's rings. He also made use of the microscope and made a number of biological experiments and observations.

Hooke's law (1676) The law that forms the basis of the theory of elasticity and in its most general form states that, for a certain range of stresses, the strain produced is proportional to the stress applied, is independent of the time, and disappears completely on removal of the stress. (*See* strain; stress.) A solid obeying this law accurately for all values of stress, however large, would be called a Hookean solid.

The point on the graph of stress versus strain for a real material, where it ceases to be linear, is known as the *limit of proportionality*. *See* yield point; elastic limit.

hook-up A temporary connection between electrical or electronic circuits or a temporary communications channel.

Hope, Thomas Charles (1766–1844) Scottish chemist who is remembered for his researches on the *maximum density of water and the apparatus (*see* Hope's apparatus) designed to measure it.

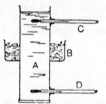

Hope's apparatus

Hope's apparatus An apparatus designed to show that water has a maximum density in the neighbourhood of 4 °C. A freezing mixture is placed in the tray B, round the metal vessel A containing water, and the thermometers C and D are read at regular intervals. At first the reading of C remains constant with time while that of D decreases, due to the cold water near B becoming more dense and falling to the bottom of A, until its temperature becomes steady at about 4 °C. At this stage the cooling below 4 °C of the water near B lowers the density so that the reading of C falls steadily to 0 °C. The temperature of maximum density is thus found at the point of intersection of the temperature/time curves for C and D.

horizontal polarization 1. *Polarization of electromagnetic radiation in which the electric-field vector is horizontal, and hence the magnetic-field vector is vertical.

2. The transmission of radio waves so that their electric-field vectors are horizontal. *Dipole aerials are arranged horizontally for the transmission and reception of horizontally polarized signals.

270

In both cases, when the electric-field vector is vertical, then there is *vertical polarization*.

horn A tube of which the cross-sectional area increases progressively from the small end (throat) to the large end (mouth). It is used as an acoustical transmission line to couple the acoustic impedance as seen looking back into the diaphragm at the throat as efficiently as possible to the load as seen looking out of its mouth. Exponential horns are most generally used. In these horns the cross-sectional area increases progressively from throat to mouth according to the relation

$$S_x = S_0 \exp(mx),$$

where S_x and S_0 are the cross-sectional areas at distance x from the throat and at the throat respectively, m is a constant that determines the rate of flare and theoretical cut-off frequency. In the exponential horn the impedance at the open end is a pure resistance at much lower frequencies than that of the conical horn of the same initial and final cross-sectional areas. It therefore radiates low frequencies more efficiently than the conical horn. Horns with other relations between S and x have been used in certain applications.

A horn has two main properties: (*a*) intensifying and (*b*) directive. The intensifying property is dependent upon the fact that by matching a sound source such as the moving coil and diaphragm of a loudspeaker system to the acoustic impedance of the air, there is a higher efficiency of energy transfer than would hold for a system without a means of matching impedances. A further element is generally introduced to match the mechanical impedance of the diaphragm to the acoustic impedance of the air. This usually consists of a small air cavity, one end of which is closed by the diaphragm, the other being coupled through a narrow orifice to the throat of the horn. If the dimensions of the mouth are small compared to the wavelength in use, a large proportion of the sound energy is reflected at the mouth with the formation of stationary waves in the horn and consequent reduction in the sound energy radiated. The horn then ceases to be a linear device and the natural resonances affect the quality of the sound transmitted. Increasing the dimensions of the mouth to values large compared to the wavelength in use reduces the reflection to a negligible quantity. (*See* radiation resistance.)

The directive property of a horn is also dependent upon the size of the mouth. Compared with the wavelength, the smaller the mouth, which acts as the effective sound source, the more pronounced is the diffraction effect, and hence the poorer the directive property. Recent work, however, has shown that the directive property of a horn is small compared with the amplifying property. The cross-sectional area must increase with distance along the horn at a mimimum rate. Below this minimum the horn must be considered as a cylinder or slightly conical pipe, viz. especially a resonator, with a large amount of reflection at the open end. The orchestral instruments are generally radiators of very low efficiency, even when terminated by a flare, for the latter is generally too sudden to be of use acoustically. The gradual flare of a loudspeaker system has very few resonances and is a good radiator.

horn antenna A microwave *aerial consisting of a metal device that flares out from the end of a *waveguide to a large circular, square, or rectangular aperture. The central axis may be straight, curved, folded, or bifurcated. Maximum radiation is along the centre of a straight horn.

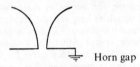
Horn gap

horn gap A spark gap horn-shaped as shown in the diagram to provide temporary earthing to protect circuits that may be subject to excessive transient potentials. An arc strikes across the narrowest part of the gap and is forced upwards by a combination of the heating and electromagnetic effects, making the arc take an increasingly longer path until it becomes broken. Such a device can provide lightning protection on high-voltage transmission lines.

horsepower (h.p.) A unit of power in the f.p.s. system equal to 745.7 watts. The *indicated* h.p. of an engine is the theoretical power developed in the cylinder by the steam, gas, or oil (according to the type of engine). The *effective* or *brake* h.p. is the rate of doing external work and is equal to the indicated h.p. minus the rate of working against friction in the engine itself. The brake h.p. is measured by some form of *dynamometer.

horseshoe magnet A permanent magnet or electromagnet shaped so that the two poles are close together.

hot Highly radioactive. A *hot atom* is one in an excited state or having kinetic energy above the thermal level of the surroundings, usually as a result of nuclear processes.

hot cathode A cathode, used for example in *electron guns, that is operated at high temperatures in order to provide a source of electrons by *thermionic emission. It may be directly heated by the passage of a current through the cathode circuit itself, or may be indirectly heated.

hotness A popular term for the property of a body that governs the direction of flow of heat when the body is placed in thermal contact with another body. A body A is hotter than a body B, if, when connected thermally, heat flows from A to B. The degrees of hotness is measured quantitatively by the *temperature of the body.

hot-wire ammeter An instrument in which the thermal expansion of a wire or strip due to the temperature rise caused by a current passing through it is employed to measure the current. Some mechanical

Hot-wire ammeter

device is used to magnify the actual increase in length of the wire, and to cause it to rotate a pointer over a circular scale. The scale is not uniform and the instrument must be calibrated empirically; but since the heating effect varies as the square of the current, the hot-wire ammeter can be used both for direct and for alternating current. It can be used as a voltmeter by adding suitable resistances in series, as in the *Cardew voltmeter*.

hot-wire anemometer An instrument that measures the speed of a fluid in motion by virtue of the convective cooling it experiences when exposed to the fluid. It has the advantage over other *anemometers that it can be made to occupy a very small space. It may, for example, take the form of a few centimetres of very fine nickel wire mounted on a fork and heated to 50 °C above its surroundings (in still air or water). The change of temperature of the wire when exposed to the flow is measured in terms of the change in its electrical resistance, recorded on a *Wheatstone bridge. The instrument is calibrated by measuring this change when the wire is exposed to known steady speeds in a wind tunnel or water channel.

A similar apparatus can be used to measure a fluctuating speed by replacing the galvanometer of the Wheatstone bridge with a cathode-ray oscillograph. Records of turbulent flow can be made in this way if adequate correction is made for the fact that the response of the hot wire diminishes as the frequency of the fluctuations in speed to which it is exposed increases.

hot-wire gauge A pressure gauge depending on the cooling by the gas of a hot filament. *See* Pirani gauge.

hot-wire microphone A device first applied by Tucker for the reception of sound waves from gunfire and used in sound ranging, but later it was modified for detection and measurements of continuous sounds.

The basic principle of this microphone is the change of the steady resistance of an electrically heated fine wire when subjected to sound waves. The resistance drop is the same as would result from a steady draught of speed equal to the maximum speed of the SHM. So if the resistance–speed curve of the wire in a steady draught has been obtained, this can be used as a calibration curve to obtain from the measured drop of resistance the particle speed of the sound.

The instrument consists of a fine wire grid of platinum or nickel mounted at the entry of the neck H of a *Helmholtz resonator. Connection is made

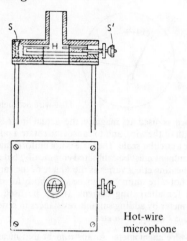

Hot-wire
microphone

from the two ends of the wire to two annular discs of silver foil, one above and one below a piece of mica, which forms the frame of the grid, as shown in the diagram.

The resonator is inserted to increase the sensitivity of the hot wire. The natural frequency of the resonator depends upon the volume of the container and the form of the neck. For experimental purposes it is desirable to use a microphone whose frequency can be varied by arranging the container to be of two tubular parts sliding over each other.

The change in the resistance experienced by the wire when exposed to alternating draught can be regarded as partly steady and partly oscillatory:

(a) The steady drop δR_1 is proportional to U^2 where U is the particle speed in the neck of the resonator and therefore is a measure of the intensity of sound.

(b) The oscillatory change δR_2 is proportional to $U \sin 2\pi\nu t$, where ν is the fundamental frequency of the imposed sound and therefore is a measure of the amplitude.

The value of δR_1 is measured directly by a Wheatstone's-bridge method while δR_2 is obtained by coupling the oscillatory change to a low-frequency amplifier. By using the direct-flow calibration curve the absolute value of U can be determined.

The sensitivity of the hot-wire microphone is affected by the fineness of the wire as well as by its temperature above its surroundings, or in other words, by the heating current passing through it. This is shown by the equation given by Tucker and Paris:

$$\delta R = 0.2(R - R_0),$$

where R_0 is the resistance of the wire when cold, R is the resistance when heated with a current of a certain value, δR is the change of resistance when it is exposed to a sound of certain intensity. The sensitivity, which is measured by $\delta R/R$, is, however, limited by the convection current set up.

Paris has adapted the principle of the hot-wire microphone to the double resonator consisting of

two Helmholtz resonators joined by a neck in which the grid is placed and protected from casual draughts. By this method, greater sensitivity can be obtained. Tucker subsequently used the double hot-wire microphone to determine the speed of sound in gases, while Richardson has applied the hot-wire microphone in many researches on sound problems in both the audible and ultrasonic frequencies.

In general, the instrument has wide applications in comparative measurements of sound intensity and distribution of sound in acoustical buildings. The microphone also serves as a sensitive tuned detector of signals at long ranges. The damping of the microphone is fairly small, and from this point of view the resonant curve of that with a single resonator is selective. The resonance curve for a double resonator has two resonant tones that are separated by a frequency interval depending on the relative dimensions of the various parts of the two resonators.

hour-angle *See* celestial sphere.

howl A high-pitched audiofrequency tone heard in receivers and generally caused by acoustic feedback: the sound output of the loudspeaker can be detected and amplified by electronic circuits in the sound-reproduction system; above a critical level, oscillations occur and produce the howl in the loudspeaker. The process is similar to electrical *feedback, which can also produce howl. An oscillator designed to produce such a tone is called a *howler*.

howling tones *See* howling tube.

howling tube In the *singing flame the tones are rather pure and free from overtones. It is, however, possible to maintain thermal vibrations in another way with much power and correspondingly with many overtones. That is the reason these tones are called *howling tones* and sometimes called *gauze tones*. They were first discovered by Rijke in 1859, who introduced a piece of metal gauze in the lower half of the vertical tube (best about a quarter of the way up) and heated it to red heat by a flame. The tube sounds loudly for a few minutes after the withdrawal of the flame until the temperature of the air and the gauze are equalized. The sound can be maintained if the gauze is heated by an electric current. The explanation of this maintenance is that the intermittance in this case is not supplied by the source of heat; it is supplied only by the motion of air in the tube. The air flow near the gauze consists of an alternating flow due to the vibrations in the tube superimposed on a steady flow due to convection. That is to say, every upward movement of the air in vibration brings cold air on to the heated gauze, whereas the downward movement brings the already hot air back on to the gauze and therefore the compression occurs as the air goes up because of the big difference of temperature. When the gauze is in the lower half, this upward phase occurs just before the maximum compression (because the air is moving then towards the central displacement node) and therefore the vibration is assisted. When the gauze is in the upper half, this phase precedes a rarefaction and so the motion is damped.

The converse case was discovered by Bosscha. In this case the gauze is cold and placed in the upper half of the tube while the current of air is hot. To make Bosscha's gauze and tube sound continuously, the gauze can be cooled by water in a circulating pipe wound round it while the hot air rises from a candle near the lower end of the tube.

Hoyle, Sir Fred (*b.* 1915) Brit. mathematician and cosmologist who, with Hermann Bondi and Thomas Gold, postulated the *steady-state theory. With J. V. Narlikar he proposed a theory of gravitation in which the mass of a particle is regarded as a consequence of the presence of all the other particles in the universe.

h.p. Abbreviation for *horsepower.

h-parameter *Syn.* hybrid parameter. *See* transistor parameters.

H–R diagram *See* Hertzsprung–Russell diagram.

HI region *See* interstellar matter.

HII region *See* interstellar matter.

H.T. Abbreviation for *high tension.

HTR Abbreviation for high-temperature *gas-cooled reactor.

Hubble, Edwin Powell (1889–1953) Amer. astronomer who was the first to postulate that stellar red shifts are due to recession of the galaxies. *See* Hubble constant.

Hubble constant *Syn.* Hubble parameter. Symbol: H_0. According to the theory of the *expanding universe, the *redshifts observed in the spectra of galaxies represent recessional velocities. The Hubble constant is defined as the ratio of recessional velocity to distance, and is commonly measured in kilometres per second per megaparsec. Galaxy redshifts, and hence recessional velocities, can be accurately determined. Distances are known with less certainty, especially for the more distant galaxies, and this leads to uncertainty in the value of H_0. Current estimates place H_0 as either about 55 or 80–100 km s^{-1} Mpc^{-1}.

The reciprocal of H_0 has the dimensions of time. It is a measure of the age of the universe, but only if the rate of expansion has always been constant. Since gravitation tends to diminish the expansion rate, $1/H_0$ can only give an upper limit: using the value 55 km s^{-1} Mpc^{-1} gives an age of 18×10^9 years.

hue *See* colour.

Huggins, Sir William (1824–1910) Brit. astronomer, a pioneer in the use of the spectroscope in astronomy. He observed the Doppler shift in stellar spectra (1868) and found a series of lines in the spectrum of hydrogen in the ultraviolet (1880).

Hughes, David Edward (1831–1900) Brit. inventor who was responsible for the first microphone, *Hughes's echo sounder, and *Hughes's induction balance.

Hughes's echo sounder In this system the high-frequency sound signal is produced and the echo received by *magnetostriction oscillators. It consists of a pile of annealed nickel stampings with central holes mounted coaxially in a conical air-filled reflector about 25 cm in diameter. This pile has coils wrapped round it so that the nickel stampings expand and contract round their circumference in unison. It is immaterial whether they are piled loosely or cemented together but they should be electrically insulated. Both transmitter and receiver are of the same type and have the same frequency, which is equal approximately to the speed of sound in nickel divided by the mean circumference of the ring. The transmitting oscillator is excited by the discharge of a capacitor of suitable capacitance and voltage through the winding of the coil. Therefore large currents are produced in the oscillator, which is set into resonance and emits a damped train of waves. These are directed by the reflector towards the sea bed when determining depth. The echo reflected to the receiver is collected by its directional reflector. The coil of the receiver oscillator is initially magnetized by a suitable direct current and so the high-frequency impulse of the echo produces radial stresses in the nickel, which results in corresponding magnetic alternations. The latter induce currents in the coil winding that are amplified, rectified, and recorded. The motor that drives the recorder also controls the instant of transmission of the sound signal. Two marks are made at each traverse, one corresponding to instant of transmission and the other to instant of arrival of echo. The magnetostriction transmitter and receiver are fitted in the ship quite close together. The system is tuned mechanically and electrically to a high frequency of about 16 kHz and so it is insensitive to the noises of low frequency. The apparatus records the depth to an accuracy of 30 cm in water of any depth up to 60 m.

Hughes's induction balance A balance consisting of two mutual inductances placed at some distance apart. The two primary coils are connected in series with a source of interrupted or alternating current, and the two secondaries are connected in series with a telephone receiver, so that the induced e.m.f.s. in the secondaries act in opposite directions through the telephone. The positions of the secondaries are adjusted until there is no sound in the telephone, and the two mutual inductances are then equal. If a piece of metal of the same size and shape is introduced into each secondary, the balance will be disturbed unless the conductivities of the two are identical.

hum An extraneous low-pitched droning noise heard in sound-reproduction systems. It originates in an associated or nearby electric power circuit. The most

humidity

common hum is that caused by mains 50 hertz alternating current.

humidity A measure of the degree of wetness of the atmosphere. *Absolute humidity* is the mass of water vapour present in unit volume of moist air and is measured in kilograms (or, in practice, grams) per metre cubed. In meteorological studies this is often called the *vapour concentration*. The absolute humidity is proportional to the ratio of the actual *vapour pressure to the thermodynamic temperature. Because the absolute humidity of a volume of air is strongly dependent on temperature, a more useful measure is the *relative humidity* (symbol: U), defined as the ratio e/e' of the actual vapour pressure to the *saturation vapour pressure over a plane liquid water surface at the same temperature, expressed as a percentage. This is the quantity normally referred to by the single word "humidity". As the temperature of the air is reduced, the water-vapour concentration increases until, at a particular temperature known as the dew point, θ_d, the air becomes saturated. Because only temperature has been altered, the actual vapour pressure at a temperature θ must equal the saturation vapour pressure at the dew point. Hence:

$$U = 100e'(\theta_d)/e'(\theta).$$

Because values of e' for water vapour are well known, this formula for U offers a particularly simple way of evaluating the relative humidity (*see* Regnault's hygrometer). The definition of U is not identical with the ratio of the actual absolute humidity to the absolute humidity at saturation, but differs inappreciably from it at temperatures below about 95 °C, and thus it may be used in connection with the *chemical hygrometer. The *specific humidity* is defined as the mass of water vapour divided by unit mass of moist air. The *mixing ratio* (symbol: r) is defined as the ratio of the mass of water vapour to the mass of dry air that contains it. If p is the total atmospheric pressure (in pascals),

$$r = 0.622 \, e/(p - e).$$

Humidity, either relative or absolute, is measured with a *hygrometer. A rough indication of the humidity is given by a *hygroscope.

hunting Variation of a controlled quantity such as the temperature of a thermostat above and below the desired value. Time-lags or bad positioning of regulating devices may lead to violent hunting. The term is applied particularly to cyclic variation in the angular velocity of the rotor of an electrical machine. It is usually undesirable and is eliminated by fitting a *damper winding.

Hurter–Driffield curve (H–D curve) The *characteristic curve of a photographic material, showing the relation between the transmission density and the exposure. *See also* reciprocity law.

Huygens, Christian (1629–1695) Dutch physicist, mathematician, and astronomer. (The form Huyghens is sometimes used to prevent softening the -g- sound.) Huygens developed mechanics from the stage to which it had been brought by Galileo, producing solutions of problems in rotation, introducing the idea of centrifugal action, and giving an exhaustive treatment of the compound pendulum. He brought the pendulum clock to a high stage of perfection and showed how to make the pendulum isochronous, even for large amplitudes, by the use of cycloidal checks. He made his own lenses, telescopes, and microscopes, inventing an eyepiece (*see* below) still much used. He was the first to appreciate the true form of the rings of Saturn.

Huygens made a careful study of the phenomena of double refraction and showed how to explain them in terms of a wave theory of light. This was based on the concept of wavelets spreading from each point affected by a disturbance and combining to give a fresh wavefront that envelops the wavelets that create it. This *Huygens's principle* was used by Fresnel in his development of the wave theory. Huygens conceived the waves as longitudinal in nature and could not explain polarization. (Fresnel did likewise at first but eventually realized that transverse waves would have the properties he needed.)

Huygens made many other contributions to physics and introduced the use of the bell jar and plate into vacuum-pump work.

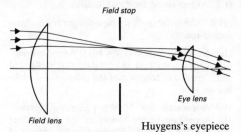

Huygens's eyepiece

Huygens's eyepiece An *eyepiece that is very commonly used in telescopes and microscopes, particularly those of rather simple design. It consists of two planoconvex lenses, having the convex faces towards the incident light and a field stop between them to reduce the somewhat large *spherical aberration. The focal length of the field lens is usually from two to three times that of the eye lens. The distance between the lenses is half the sum of the focal lengths, which minimizes *chromatic aberration.

Light from the objective converges towards a virtual focus inside the eyepiece (*see* diagram) so an external *graticule or cross-wires cannot be used. Because the eye lens by itself is uncorrected for aberrations, a graticule placed in the field stop would be unsuitable. Thus the Huygens eyepiece is not used for measuring instruments, although it is satisfactory for those used purely for observation. *See also* Kellner eyepiece; Ramsden eyepiece.

Huygens's principle *See* Huygens, Christian.

HWR Abbreviation for *heavy-water reactor.

hybrid *See* integrated circuit.

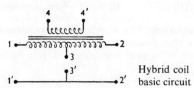

Hybrid coil
basic circuit

hybrid coil *Syn.* hybrid transformer. Telephone circuits are normally capable of transmitting signals (speech) in both directions. To compensate for losses in long lines, amplifiers are introduced into the circuit. This presents difficulties since an amplifier is essentially a one-way device and to prevent the setting up of continous oscillations (called singing) its output and input circuits must be isolated from one another. This separation of circuits is usually effected by means of a hybrid coil, which is a transformer having, fundamentally, two windings and four pairs of terminals. Provided that certain relationships between the impedances connected across the pairs of terminals, 1, 1' and 2, 2' (*see* diagram) are maintained, the application of a voltage to terminals 4, 4' produces no voltage between terminals 3, 3'.

hybrid integrated circuit *See* integrated circuit.

hybrid parameter *Syn.* h-parameter. *See* transistor parameters.

hybrid transformer *See* hybrid coil.

Hycomax *See* Alnico.

hydrated electron *Syn.* aqueous electron. When an aqueous solution is irradiated with *ionizing radiation, the water molecules become ionized and secondary electrons are released. Such an electron rapidly loses its energy by ionizing and exciting neighbouring water molecules. After about 10^{-11} seconds from its formation, the water molecules have sufficient time to orientate themselves around the electron without it escaping, and a region of radial polarization results, trapping the electron in the centre. (The water molecule is polarized such that the hydrogen atom is slightly more positive than the oxygen atoms.) This species is called the hydrated electron, the electron existing in a *potential well. It is an extremely reactive species.

hydraulic accumulator A device consisting of a vertical cylinder A, fitted with a ram B, loaded with large weights C. Water is pumped in through the pipe D and raises the piston E against the weights so storing energy that can be used to work hydraulic machinery such as a lift or crane. When the falling weights release the lever F, the pumps start working again to refill the accumulator.

hydraulic press *Syn.* Bramah's press. A device for producing large forces, consisting basically of two cylinders, one much wider than the other, fitted with pistons A and B (Fig. *a*) and connected by a pipe C,

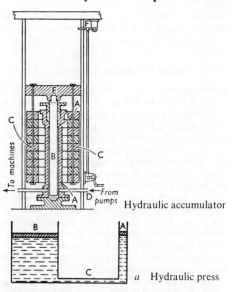

Hydraulic accumulator

a Hydraulic press

the whole being filled with water. A force f applied downwards at A develops a pressure equal to f/A where A is the cross-sectional area of A. This pressure is transmitted throughout the fluid (*see* Pascal's principle) and is thus applied at B. The total force upwards on B equals the pressure times the area of B $= fB/A$. Hence a much larger force appears at B if B is much larger than A. On the other hand, if A moves downwards, B moves upwards by a much smaller distance and the work done on B never exceeds that done on A. (*See* intensifier.)

Bramah (1796) invented a leather collar L that fitted in a groove round the large cylinder wall D and prevented water leaking past the piston B at high pressure (Fig. *b*).

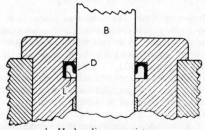

b Hydraulic press piston

hydrodynamics A branch of the science of deformable bodies, being a study of the motion of fluids (liquid and gaseous). The classical theory of hydrodynamics is concerned with the mathematical treatment of perfect fluids. This theory is subject to modification to include the effect of the viscosity of a real fluid. *Aerodynamics is essentially a specialized branch of hydrodynamics.

hydroelectric power station An electricity-generating station powered by water falling under gravity

275

hydrogen bomb

through a water turbine. Usually a dam is built across a river to provide a large reservoir of water, which permits continued operation during a dry season.

hydrogen bomb *See* nuclear weapons.

hydrogen bond A type of chemical bond occurring between the hydrogen atom in molecule and an electronegative atom in a neighbouring molecule. The bond is electrostatic in origin. For example, in a water molecule the electronegative character of the oxygen atom causes distortion of the electrons in the covalent bond between the oxygen and the hydrogen atoms. This results in some negative charge on the oxygen and positive charge on the hydrogen atoms. The hydrogen bonding is caused by intermolecular electrostatic attractions between hydrogen and oxygen atoms. Similar bonds can be formed between molecules containing fluorine and nitrogen atoms.

Although hydrogen bonds are weaker than electrovalent and covalent chemical bonds, they are of immense importance in biological systems (especially in the structure of proteins and nucleic acids). They also account for the properties of certain liquids, for example, the high boiling point and low vapour pressure of water (compared with hydrogen sulphide and selenide, which are gases at room temperature).

hydrogen electrode An electrode system in which hydrogen is in contact with a solution of hydrogen ions. It consists of a *half-cell in which a platinum foil is immersed in a dilute acid solution. Hydrogen gas is bubbled over the foil, which is usually coated with finely divided platinum to increase absorption of the hydrogen. Hydrogen electrodes are used in cells to measure standard *electrode potentials.

hydrogen-ion concentration *See* pH value.

hydrogen overvoltage *See* overvoltage.

hydrogen spectrum The *emission spectrum of a *gas-discharge tube containing hydrogen has a faint continuum caused by the recombination of ions with electrons, *band spectra of hydrogen molecules (H_2) and molecular ions (H_2^+), and lines in the visible, ultraviolet, and infrared attributed to separated hydrogen atoms. Both emission and absorption lines of H atoms are observed in stellar spectra.

The study of the spectrum of atomic hydrogen has been of outstanding importance in the understanding of spectra, atomic structure, and quantum theory. The first discovery of a regularity in spectra was made by Balmer (1885) who showed that the wavelengths λ of the visible lines of hydrogen were represented accurately by the empirical equation:
$$1/\lambda = R(\frac{1}{4} - 1/n^2)$$
where R is the *Rydberg constant and n is an integer greater than or equal to 3. The first members of the *Balmer series* are given below in nanometres.

C (H_α)	656.3	(strong, red)
F (H_β)	486.1	(medium, blue-green)
f (H_λ)	434.0	(weak, violet)
h (H_δ)	410.2	(weak, extreme violet)

The series limit, for $n = \infty$, has $\lambda = 364.6$ nm and is in the near ultraviolet.

Later, other series were found, following the general formula:
$$1/\lambda = R(1/n_1{}^2 - 1/n_2{}^2),$$
where n_1 and n_2 are integers. The value $n_1 = 1$ gives the *Lyman series* in the *vacuum ultraviolet; $n_1 = 3$ gives the *Paschen series*; $n_1 = 4$ gives the *Brackett series* and $n_1 = 5$ gives the *Pfund series*. The last three are all in the infrared.

The formation of the Balmer series was discussed by Bohr in his theory of the *atom (1913). He obtained the formula for the energy E_n of a quantum state of the H atom:
$$E_n = -me^4/8h^2\varepsilon_0{}^2n^2,$$
where m is the *reduced mass of the electron of charge e, h is the *Planck constant, ε_0 the *electric constant, and n is a positive integer. When the atom goes to a state of lower energy, electromagnetic radiation is emitted with a *quantum of energy (*photon) $h\nu$, where ν is the frequency of the waves. The formula gave values agreeing with experiment to within a few percent, but later values of the quantities involved show that it is very much more accurate than this. The solution of the *Schrödinger wave equation (1926) gave the same formula with a more satisfactory theoretical basis.

On further investigation it was found that the spectrum lines of hydrogen have *fine structure, which is most clearly shown by the Balmer red line caused by transitions from states with $n_2 = 3$ to those with $n_1 = 2$. This was at first believed to be a doublet and in 1915 *Sommerfeld modified the theory of Bohr by considering the variation of mass with speed given by *relativity, giving approximate agreement with experiment. After many years of theoretical research *Dirac (1928) gave a form of the *wave equation of the electron consistent with the principles of relativity. This predicted five components for the Balmer red line, which agreed with measurements using very high resolution. Later still it was shown that there are seven components, which was explained by *quantum electrodynamics (*see* Lamb shift).

The spectrum of *deuterium is very similar to that of ordinary hydrogen but all the frequencies are changed in the ratio of the reduced masses of the two atoms. There are also minute differences caused by the electric *quadrupole moment of the deuterium nucleus and the difference in the nuclear magnetic moments (*see* hyperfine structure).

The theory of the one-electron atom can be extended to singly ionized helium, double ionized lithium, etc. The energies of the states, and hence the frequencies of the quanta, are increased in the ratio of the squares of the atomic numbers and the reduced mass has to be calculated in each case. (*See also* positronium.) An atomic nucleus can capture a negative *muon into an orbit that is much closer to the nucleus than are the electrons, permitting treatment analogous to that of the H-atom. The very high binding energies and the significant probability that the muon is at any time inside the nucleus forbid an analytical solution, but valuable results have been obtained by computation using Dirac's wave equation. A similar system is formed when a nucleus captures a negative *pion, but as this particle is a *boson computation is based on the *Klein–Gordon equation. The study of the spectra of the radiations from such systems has given useful information about atomic nuclei.

hydrometer *Syn.* aerometer. An instrument for determining relative densities. Under this definition, such an instrument as the hydrostatic balance should be included, but it is customary to restrict the definition to the determination of the relative density of a liquid. It is usual to include the *Hare hydrometer*, which consists of two vertical glass tubes, one standing in a vessel of water, the other in the liquid under test. The tubes are connected at their upper ends by a glass T-piece, and by applying suction, the liquids may be raised in the tubes; the relative densities are inversely proportional to the height to which the liquids are raised in the tubes.

Hydrometers may be divided into two classes – hydrometers of constant immersion, and hydrometers of variable immersion. In the latter class, the hydrometer consists of a glass tube blown out to two bulbs at its lower end. The lower bulb contains mercury, so that the hydrometer may float vertically in whatever it is immersed, and the graduated scale, which is fixed to the tube, will indicate the relative density of the liquid in which it floats. The hydrometer scale may be graduated in different ways. In the common hydrometer, the relative density is obtained by dividing the reading by 1000. In the hydrometer of constant immersion, the instrument carries a pan at its upper end and is adjusted by means of weights put into the pan until the hydrometer is sunk to a fixed mark on its neck; the relative density of the liquid may be obtained from a knowledge of the weights required, first in water, and then in the liquid under test. In this form the hydrometer is known as *Fahrenheit's hydrometer*. In *Nicholson's hydrometer*, the lower end of the instrument carries a scale pan and this addition permits the determination of the relative density of a solid. *Sike's hydrometer*, which is much used in alcoholometry, consists of a graduated brass rod carrying a weighted bulb at its lower end. The range of the hydrometer may be increased by adding brass weights to a platform below the bulb.

hydrophilic *See* sol.

hydrophobic *See* sol.

hydrophone An instrument for detecting sounds under water. It usually consists of a carbon microphone or electromagnetic detector fixed to a diaphragm in contact with the water. In a common form of hydrophone a carbon-button microphone is fixed to the centre of a metal diaphragm that is rigidly clamped at its edges to a massive case and that covers a cylindrical cavity in the case. The direction from which an underwater sound originates may be found with a bidirectional hydrophone. In this instrument two diaphragms are used, one on either side of the heavy annular casing of the instrument. In the cavity between the diaphragms a small button microphone is fixed so that it picks up sounds from either side. Thus when one diaphragm faces the incident sound, there is a response depending on the pressure difference between the two faces. When the plane of the diaphragm is in the direction of the incident sound, however, there is no pressure difference and so no response. This hydrophone may be made unidirectional by placing a circular *baffle plate a few inches away from one of the diaphragms. It is usual to find the line of the sound from zero position with a bidirectional hydrophone and the actual direction with a unidirectional instrument. These pressure receivers are generally more sensitive than "displacement" receivers in view of the high acoustic resistance of water, but an instrument of the displacement type – the light-body hydrophone – has been constructed by Young and Wood. This employs a button microphone mounted in a spherical container that floats on water. It is a directional instrument. For maximum sensitivity a hydrophone should be resonant and tuned to the frequency of the source, but for actual recognition of sound, e.g. that of a ship's propellers, a nonresonant instrument is used.

hydrosphere *See* meteorology.

hydrostatic analogy A comparison of various relations between potential difference, resistance, and current in an electric circuit by analogy with the flow of water under a hydrostatic head.

hydrostatic balance *See* balance.

hydrostatic equation An equation giving the relation between atmospheric pressure, p, and altitude z:
$$dp/dz = -g\rho,$$
where g is the acceleration of free fall and ρ is the density. The rate of fall of pressure with altitude is usually sufficiently regular to allow pressure readings to be used to determine altitude. Strictly, the hydrostatic equation is only valid when the atmosphere has no accelerations in the vertical; these are usually very small compared with g.

hydrostatics *See* statics.

hygristor An electronic component with an electrical *resistance that varies with *humidity. It is used as the basis of some recording *hygrometers.

hygrograph An instrument for measuring and recording the relative *humidity of the air.

hygrometer An instrument for the measurement of the *humidity of air. The main types are (1) the *chemical hygrometer, in which the mass of water vapour actually present is measured and can be

hygroscope

compared with the mass of vapour to saturate the same volume at the same temperature; (2) the dew-point instruments (*see* Daniell; Crova; Dines; Regnault), in which the *dew point is measured and the relative humidity obtained from the ratio of the saturation pressure of water vapour at the dew point to that at the temperature of the air; (3) the *wet and dry bulb hygrometers, in which the temperature of two thermometers placed side by side, one in air and the other with its bulb covered with muslin dipping into water, are noted, the relative humidity being then obtained from special tables. The *Mason hygrometer of this type operates in a light air draught as does the *psychrometer of Assmann; (4) the recording instruments, in which an indicator gives the relative humidity directly, the instrument having been previously calibrated. The *electric and *hair hygrometers are both of this type.

hygroscope A device for giving a rough guide to the *humidity of the atmosphere. A very simple form of hygroscope is a card impregnated with chemicals that change colour when the moisture content increases above a certain level, e.g. cobalt chloride (changing from blue to pink).

hyperbola *See* conic sections.

hyperboloid mirror *See* mirror.

hypercharge Symbol: Y. A *quantum number associated with an elementary particle. It is the sum of its *baryon number and *strangeness and is conserved in strong and electromagnetic interactions.

hyperfine structure *See* fine structure.

hypermetropia *Syn.* hyperopia. A refractive defect of the eye when it is too short for its unaccommodated focal length. The far point is virtual and behind the eye. The correction is a converging lens that focuses parallel rays to the far point. It is called long sight because an uncorrected person by using accommodation (provided the defect is not too great) can focus objects at distance. His near point, however, is farther away than is normal for his age. *See* eye; refraction of eye.

hyperon The collective name given to long-lived *baryons other than the proton and neutron. Long-lived in this context is taken to mean not decaying by *strong interaction, i.e. particles with lifetimes much greater than 10^{-24} seconds. The lamda, sigma, xi, and omega-minus particles are hyperons. *See* elementary particle; strangeness; charm.

hyperopia *See* hypermetropia.

hypersonic speed A speed not less than five times the speed of sound through a medium at the same level and under the same physical conditions, i.e. not less than *Mach 5.

hypothesis A provisional supposition that, if true, would account for known facts and serves as a starting point for further investigation by which it may be proved or disproved.

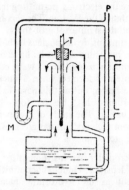

Hypsometer

hypsometer An apparatus for the calibration of a thermometer at the steam point. The thermometer T is placed in the steam above the water boiling under a known pressure, usually atmospheric, applied at P, the manometer M being used to ensure that the water is not being boiled too vigorously. The central space is enclosed by a steam jacket from which any condensed liquid flows back to the boiler. The boiling point under the measured pressure is deduced from tables and the reading of the thermometer T corrected.

hysteresis A delay in the change of an observed effect in response to a change in the mechanism producing the effect.

1. (magnetic) A phenomenon shown by *ferromagnetic materials, whereby the magnetic flux through the medium depends not only on the existing magnetizing field, but also on the previous state or states of the substance. The existence of permanent magnets is due to hysteresis. The phenomenon necessitates a dissipation of energy when the substance is subjected to a cycle of magnetic changes. This is known as the magnetic *hysteresis loss. *See* hysteresis loop.

2. (dielectric) *See* dielectric hysteresis.

3. (elastic) *See* elastic hysteresis.

4. (torsional) When a wire is subject to many twists and untwists between the limits $+\theta$ and $-\theta$, it settles down to a condition in which the couples called into play by the twists $+\theta$ and $-\theta$ have definite values, and the couple for an intermediate twist θ' has two values, one corresponding to the passage from $+\theta$ to $-\theta$ and the other to the passage from $-\theta$ to $+\theta$. The wire is in a cyclic state and the twist lags behind the couple: this has been called torsional hysteresis by Ewing and corresponds to magnetic and dielectric hysteresis.

hysteresis loop (magnetic) A closed curve obtained by plotting the magnetic flux density, B, of a *ferromagnetic material against the corresponding value of the magnetizing field H. The area enclosed by the loop is equal to the *hysteresis loss per unit volume in taking the specimen through the prescribed magnetizing cycle. The general form of the hysteresis loop for a symmetrical cycle between H and $-H$ is

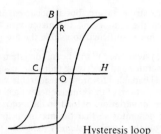

Hysteresis loop

shown in the diagram, but any complete magnetizing cycle, say between the limits $H + h$ and $H - h$, will give rise to a hysteresis loop. OC is the *coercive force and OR the *remanence. The area enclosed by the loop varies with the nature and heat treatment of the magnetic substance, being a minimum for electrolytic iron and reaching a value some twenty times greater for tungsten steel. *See* ferromagnetism.

hysteresis loss 1. (magnetic) The dissipation of energy that occurs, due to magnetic *hysteresis, when the magnetic material is subjected to changes (particularly cyclic changes) of magnetization. *See* hysteresis loop.
2. (dielectric) The dissipation of energy that occurs, due to *dielectric hysteresis, when the dielectric is subjected to a varying (in particular, an alternating) electric field.
3. (elastic) The dissipation of energy through *elastic hysteresis.

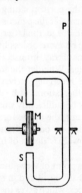

Ewing's hysteresis tester

hysteresis tester, Ewing's An instrument for the rapid determination of the *hysteresis loss of a given specimen of magnetic material. The specimen, M, built up out of strips of a prescribed shape and size, is rotated by hand between the poles of a horseshoe magnet NS, which is balanced so as to be free to turn on an axis in the same straight line as the axis of rotation of the specimen. The latter is rotated sufficiently fast to produce a steady deflection of the horseshoe magnet, which is read on a scale by means of a long pointer P. The hysteresis loss in the specimen is proportional to the sine of the angle of deflection. The instrument is callibrated by using a specimen, the magnetic properties of which are known.

I

IAT Abbreviation for *International Atomic Time.

IC Abbreviation for *integrated circuit.

ice The solid form of water, the transition point at the *standard atmosphere being defined as 0 °C (*see* ice point). The specific latent heat of fusion is 0.3337 MJ kg^{-1}. Its density at 0 °C is 916.0 kg m^{-3}, compared to water at 0 °C with a density of 999.8 kg m^{-3}. There are several allotropic forms of ice, mostly stable only under high pressure.

ice calorimeter *See* Bunsen ice calorimeter.

ice line *Syn.* solidification curve. A curve expressing the relation between the melting point of ice and the applied pressure. It may be calculated by using the *Clausius–Clapeyron equation.

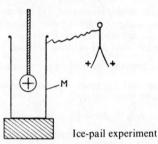

Ice-pail experiment

ice-pail experiment An experiment performed by Faraday, to demonstrate the equality of the charges produced by electrostatic induction. A tall metal can M (an ice pail was used in the original experiment) is insulated from earth, and connected to an uncharged gold-leaf electroscope. An insulated positively charged metal ball is introduced, without touching the sides, into the can, and the electroscope registers a positive charge. The can is momentarily earthed, and the charged ball withdrawn. The electroscope then registers the same deflection as before, but the charge is now negative. The charged conductor is again introduced into the can and brought into contact with it. The electroscope then shows zero charge; thus showing that the two induced charges and the inducing charge are all equal in magnitude.

ice point The temperature of equilibrium of ice and water at standard pressure, the water being saturated with dissolved air. Its former importance was as the lower fixed point on the Celsius scale of temperature. Now, however, *thermodynamic temperature and the kelvin are based on the *triple point of water, and its value (273.16 K) has been chosen so as to make the ice point equal to 0 °C within the limits of experimental measurement. *Compare* steam point.

iconoscope *See* television camera.

ideal crystal The crystal structure considered as perfect and infinite, that is, ignoring all questions of *crystal texture.

ideal gas *Syn.* perfect gas. A gas defined for the purposes of *thermodynamics as one that obeys *Boyle's law and that, in addition, has an internal energy independent of the volume occupied, i.e. it obeys *Joule's law of internal energy. These two requirements are, from the point of view of the kinetic theory, both equivalent to saying that the intermolecular attractions are to be negligible, but the first requires also that the molecules shall be of negligible volume. An ideal gas in fact obeys (*a*) Boyle's law, (*b*) Joule's law of internal energy, (*c*) Dalton's law of partial pressures, (*d*) Gay-Lussac's law, and (*e*) Avogadro's hypothesis exactly, whereas real gases obey them only as their pressure tends to zero.

The equation of state for 1 mole of an ideal gas is given by:
$$pV = RT,$$
R being the molar gas constant. The isothermals of a perfect gas on a p/V graph therefore form a family of rectangular hyperbolas. For a treatment of gases that are not ideal, *see* equation of state.

ideal-gas constant *See* molar gas constant.

idle component *See* reactive current; reactive voltage; reactive volt-amperes.

iff If and only if. *See* logic circuit.

IF strip The part of a *superheterodyne receiver that is concerned with intermediate-frequency amplification.

IGFET Abbreviation for insulated-gate *field-effect transistor.

ignition temperature 1. The temperature to which a substance must be heated before it will burn in air (or some other specified oxidant).
2. The temperature to which a *plasma has to be raised in order for the fusion energy generated to exceed the energy lost in *bremsstrahlung radiation. *See* fusion reactor.

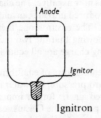

Ignitron

ignitron A type of mercury-arc rectifying tube in which the discharge is initiated by a subsidiary *ignitor electrode*. The tube has an anode and a mercury-pool cathode in which the ignitor is immersed (*see* diagram). The ignitor usually consists of a rod of semiconductor, such as silicon or boron carbide. The voltage applied to the anode is insuffi-

cient to strike an arc, but if a current is passed between the ignitor and the mercury, a hot spot forms that is sufficient to enable the arc to strike.

Ilgner system The *Ward–Leonard system in which the generator and its driving motor are coupled to a flywheel capable of storing considerable kinetic energy. Variations of this kinetic energy partially compensate for variations of load on the motor supplied by the generator so that the maximum power demand on the supply system is reduced.

illuminance *Syn.* illumination; intensity of illumination. Symbol: E_v, E. The *luminous flux, Φ_v, incident on a given surface per unit area. At a point on a surface, of area dS, the illuminance is given by:
$$\Phi_v = \int E_v\,dS.$$
It is measured in lux. *See also* cosine law. *Compare* irradiance.

illumination 1. *See* illuminance.
2. The extent to which a surface is illuminated or the application of visible radiation to a surface. The brightness of an object depends on its *illuminance and its *reflectance. Many properties of vision, such as *visual acuity, depend on illumination.

image 1. (light) From the geometric optics point of view an image point is the point to which rays are converged (*real image*) or from which they appear to diverge (*virtual image*) after reflection or refraction. The real object point from which rays have diverged (and the virtual object point to which incident rays may be converging), and the corresponding image point, are said to be *conjugate. If the pencils of rays do not reunite to foci in the image, the latter suffers from *aberration and may be more or less blurred; the greatest concentration of ray intersections is taken as the *geometric image*. Whether the image is blurred by virtue of the receiving screen being placed out of focus, or by aberrations, or by diffraction, should the blur circles be sufficiently small (say 0.1 mm for viewing at 25 cm), the image formation may be practically regarded as being sharp. This means that there is an allowable *depth of focus of the image.

From the physical optics point of view, the distribution of light in the image is considered in relation to phase and path differences, which gives the image a focal depth, throughout which there is little deterioration of quality. The *image space* is a convenient mathematical conception to describe where images may lie; it may be real or virtual, i.e. beyond or in front of the second principal point. A convenient device to represent image-side quantities is to use the accented (dash) letters, e.g. l' the image distance, y' the size of the image. The lateral magnification of the image is y'/y where y is the size of the object. The aberration of oblique astigmatism leads to the reunion of rays into two curved surfaces (tangential and sagittal) referred to as the *image surfaces*. If the astigmatism is corrected, the image lies on a curved surface whose curvature is called the *curvature of image*. When there is lack of geometrical similarity between the object and image, the latter is said to be distorted.

2. (electrical) *See* electrical images.

image-attenuation coefficient *See* image-transfer coefficient.

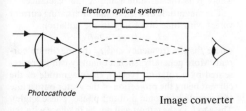

Electron optical system

Photocathode

Image converter

image converter An evacuated electron tube that is similar to the *image intensifier but operates with infrared, ultraviolet, X-ray, or electron images rather than faint optical images; such images occur, for example, in astronomy, microscopy, and medical diagnosis. The focused image falls on a suitable surface from which electrons can be liberated by, for instance, the *photoelectric or *photovoltaic effect. The electrons are accelerated and focused on a detector or recorder, such as a positively charged fluorescent screen, so that a visible image is produced.

image dissector A type of *television camera in which scanning is effected by *focusing electrons from each portion of the sensitive photoelectric plate in turn onto a collector electrode, rather than using an electron beam to scan a static charge pattern as in the iconoscope or image orthicon. This type of camera is no longer widely used.

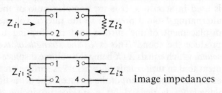

Image impedances

image impedances Of a *quadripole. The two impedances Z_{i1} and Z_{i2} that satisfy both the following conditions: (i) When Z_{i2} is connected across one pair of terminals, the impedance between the other pair is Z_{i1}. (ii) When Z_{i1} is connected across the other pair of terminals, the impedance between the first pair is Z_{i2}. *Compare* iterative impedance.

image intensifier An evacuated electron tube used to intensify a faint optical image. The image falls on a *photocathode so that electrons are emitted by the *photoelectric effect. The electrons are accelerated by an electric field and may be detected and recorded by a variety of methods. They can be focused (by an *electron-lens system) on a positively charged *fluorescent screen; the resulting optical image is many times brighter than the original image, and can be photographed.

In some devices the image on the screen can be made to fall on a second photocathode so that the intensification process is repeated; several image intensifiers can be linked in this way to form a multistage device. The image produced can be photographed or it can be recorded by a special TV camera. The TV signal is fed to and stored in a computer. The image of an extremely faint object can hence be slowly built up. The stored image (with the noise removed) can be displayed, analysed, and/or manipulated electronically.

In the *electronographic camera*, the liberated electrons are focused on a very sensitive high-resolution photographic emulsion in which the electrons are recorded directly. The density at every point on the developed image is proportional to the intensity at the corresponding point in the optical image over almost the entire intensity range (unlike a photographic image).

image orthicon *See* camera tube.

image phase-change coefficient *See* image-transfer coefficient.

image potential If a charged particle (electron or ion) is a distance r from a metal surface, it experiences an electrostatic force. The interaction is equivalent to the interaction between the particle and an image of the particle a distance r below the surface. The potential energy of the particle is then $e^2/16\pi\varepsilon_0 r^2$, where e is its charge and ε_0 the *electric constant. *See* electrical images.

image processing The analysis and manipulation, by computer, of information contained in images, such as those obtained from satellites and spacecraft, medical diagnostic equipment, or electron microscopes. The original may be an actual object or scene, or may be a photograph, drawing, etc. This is converted into a digitized form by spatial *sampling, i.e. it is converted into a two-dimensional array of tiny elements, and a set of numbers produced corresponding to the brightness and possibly the colour of each element. The sampling is often done by a form of TV camera. The numerical version of the image is stored in a computer, and manipulated in various ways to highlight different aspects of the original, compare or superimpose slightly different images, correct over- or underexposure or blur, etc.

image space *See* image.

image surfaces *See* image.

image-transfer coefficient *Syn.* image-transfer constant. Of a *quadripole. The quantity:
$$\theta = \tfrac{1}{2}\log_e[(E_1 I_1)/(E_2 I_2)],$$
where E_1 and I_1 are the voltage and current, respectively, at the input terminals, and E_2 and I_2 are the corresponding quantities at the output terminals under steady-state conditions when the network is terminated in its *image impedance. The voltages and currents are to be expressed in vector (e.g. complex) form and θ is in general complex. Its real

part is the *image-attenuation coefficient* and its imaginary part is the *image phase-change coefficient*. *Compare* propagation coefficient.

image tube An *image intensifier or *image converter.

imbibition (Literally "drinking in".) A phenomenon peculiar to gels; gelatine, for example, is capable of absorbing water and swelling takes place even against considerable opposing pressure. It is a specific property of the gel since it may imbibe some liquids and not others. The reverse of imbibition is sometimes observed when a dilute gel, protected from evaporation, exudes a liquid that is a dilute solution of the gel in water. This is called *syneresis*.

immersion heater A device containing a *heater for raising the temperature of a liquid in which it is immersed. Immersion heaters are usually controlled by a *thermostat.

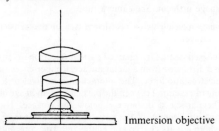

Immersion objective

immersion objective Microscope objectives use the principle of *aplanatic refraction, and to reduce the refraction at the front lens of the objective for higher powers, cedar-wood oil (refractive index 1.517) is placed between the cover glass (index 1.51, say) and the plane surface of the front lens of higher index (*see* diagram). Besides aiding aplanatism the *numerical aperture is increased by this process, which therefore increases resolving power.

immittance A term used to include both the terms *impedance and *admittance.

IMPATT diode Abbreviation for *imp*act ionization *a*valanche *t*ransit *t*ime. A diode that provides a very powerful source of microwave power. It consists essentially of a p-n junction that is reverse biased into *avalanche breakdown. It then exhibits negative resistance at microwave frequencies and may be used as an *oscillator. The current is delayed, usually by half a cycle, with respect to the voltage. This delay is due to (and characteristic of) the avalanche and to the transit time during which the charge carriers are collected by the electrodes.

impedance In general the ratio of one sinusoidally varying quantity (e.g. force or e.m.f.) to a second quantity (e.g. acceleration or current) that measures the response of the system to the first quantity. Usually the term is used for electrical impedance:

  1. (electrical) Symbol: Z. If an alternating e.m.f. is applied to an electric circuit, the *alternating current produced is affected by the capacitance and

inductance of the circuit as well as its resistance. The consequent opposition is the *reactance of the circuit and the total opposition to current flow is the impedance. Impedance is measured in ohms. It is the ratio of *root-mean-square voltage to root-mean-square current and is equal to $\sqrt{(R^2 + X^2)}$, where R is the resistance and X the reactance.

The magnitude of a sinusoidal alternating current varies with time according to the equation:

$$I = I_0 \cos (2\pi ft),$$

where f is the frequency and I_0 the maximum current. More generally, such a quantity can be represented by a rotating *vector, the magnitude of the current being the projection of the vector onto a line (*see* simple harmonic motion; phase; phase angle). The e.m.f. in the circuit will not be in phase with the current if reactance is present. It follows an equation of the form:

$$V = V_0 \cos (2\pi ft + \phi),$$

where ϕ is the phase angle. It is often convenient to represent such quantities by complex numbers on an *Argand diagram. Thus

$$I = I_0 e^{i\omega t}$$
$$V = V_0 e^{i(\omega t + \phi)},$$

where ω is $2\pi f$. The real parts of these are the instantaneous current and voltage. The impedance is then the complex voltage divided by the complex current, i.e. $Z = V/I$, and is thus equal to $|Z| e^{i\phi}$, where $|Z|$ is $\sqrt{(R^2 + X^2)}$. This quantity is sometimes called the *complex impedance*. It is given by:

$$Z = R + iX,$$

the real part being the resistance and the imaginary part the reactance.

  2. (acoustic) Symbol: Z_a. Just as the electric impedance is the ratio between an alternating e.m.f. and current, a similar term, the acoustic impedance, is used in acoustics. It is the complex ratio of the alternating *sound pressure to the rate of volume displacement of the surface that is vibrating to produce the sound. This is U, the *strength of the sound*, which equals X (X being the volume displacement that in turn equals $S\zeta$, where S is surface area and ζ the displacement of the surface). The acoustic impedance is defined for a surface producing a simple sinusoidal source of sound: It is related to *acoustic resistance* (R_a) and *reactance* (X_a) by:

$$Z_a = R_a + iX_a.$$

The *specific acoustic impedance* (Symbol: Z_s) is the product of acoustic impedance and the area, i.e. $Z_s = Z_a A$. It is the complex ratio of sound pressure to the velocity of particles in the medium, i.e. to the *particle velocity. Acoustic impedance has units of pascal second per metre squared. Specific acoustic impedance has units of pascal second per metre.

  3. (mechanical) Symbol: Z_m; ω. The complex ratio of the force acting in the direction of motion to the velocity. It is related to mechanical *resistance (R_m) and *reactance (X_m) by:

$$Z_m = R_m + iX_m,$$

and is measured in newton second per metre.

impedance drop (or **rise**) *See* voltage drop.

impedance magnetometer An instrument for measuring local variations of the magnetic field of the earth (e.g. in a building) by measurement of the change in impedance of a nickel-iron wire of high permeability caused by the axial component of the field in which the wire is placed. The wire is of mumetal, 0.45 mm diameter, and about 25 cm long.

impedance matching (electrical) In a system in which power is transferred, the matching of the *impedances of parts of the system to ensure optimum conditions for transfer of power. If power is transferred from an *amplifier to a *load, the load impedance is made the *conjugate of the amplifier output impedance to effect the transfer of maximum power. (*See also* emitter follower.) In *transmission lines, the line impedance is made equal to the generator output impedance, and also to the load impedance, to ensure no reflection of wave power in the transmission line. If transmission lines of differing line impedance are joined in a system, a section of line, one quarter of a wavelength long, is used to couple the lines and effect matching. Such a section is termed a *quarter-wavelength transformer* or a *quarter-wavelength line*.

impedance matching in sound The function of the horn is not primarily for confining the emitted sound to one direction, although it has some directive effect. Nor is it a resonator; the effect of resonance in the horn is fatal for good reproduction (*see* horn). Its essential purpose is to load the diaphragm at the narrow end by increasing the pressure against which it has to work and deliver the energy it receives to the atmosphere over an area sufficiently large to avoid reflection back into the horn with consequent stationary vibration and resonance. For these purposes the throat of the horn must be small and it is usually designed so that it opens out of a small chamber of which the vibrating diaphragm constitutes the opposite end. The impedance of the air in this chamber must be adjusted to that of the diaphragm. This process is known as impedance matching. It is carried out in practice by adjusting the inertance and capacitance of the throat by calculation (*see* Helmholtz resonator) to be equal to the mass and the stiffness respectively of the diaphragm. On this basic principle the throat acts as an acoustic transformer, the product of the pressure and velocity on either side of the constriction being approximately the same. To prevent reflection at the open end, the specific impedance at the open end must be adjusted to be equal to that of the open air. The waves start from this constriction as spherical waves and so they emerge into the atmosphere with considerable radius. If the wavelength is not too large, the acoustic impedance is nearly the same as for plane waves. Therefore radiation takes place efficiently and there is little reflection back into the horn except at low frequency.

This fact can be shown mathematically by treating the mouth of the horn as part of a spherical source whose radius r_0 bears the following relation to R, radius of mouth:
$$S = \pi R^2 = 1.39\pi r_0{}^2,$$
(S being the surface area). In this case the impedance is $\rho CS(X' + iY')$, where:
$$X' = k^2 r_0{}^2/(k^2 r_0{}^2 + 1),$$
and
$$Y' = k r_0/(k^2 r_0{}^2 + 1),$$
where $k = 2\pi/\lambda$ (λ being the wavelength of the applied source of sound). The first term $\rho CS X'$ is the resistance term. As k increases (on decreasing λ), X' approaches 1 and Y', the reactance term, will diminish. When $X' = 1$, Y' is negligible.

Therefore the impedance is a pure resistance ρCS. Hence the specific impedance of the mouth is ρC, which is equal to that of the plane waves in the medium into which the waves are being radiated.

Acoustical matching plays a very big role, especially in transmission problems. The reason for most of the difference between the theoretical and actual performance of a filter is to be found in the improper matching of the impedances of the line and the filter. *See* acoustic filters.

2. The mammalian ear exhibits a very important example of impedance matching. If sound in air were to impinge directly on a membrane enclosing the liquid in the cochlea, only about one part in ten thousand could be transmitted. The system by which the sound agitates the ear drum, linked by a lever system to the cochlea, provides impedance matching that greatly increases the sensitivity.

impedance voltage *See* voltage drop.

impregnated cable A paper-insulated cable in which the paper tapes are impregnated with an insulating compound, usually oil.

impregnated carbon Arc-lamp carbon consisting of carbon mixed with other materials selected to colour the light emitted by the arc.

impulse Of a constant force F, the product, Ft, of the force and the time t for which it acts. If the force varies with time, the impulse is the integral of the force with respect to the time during which the force acts. In either case, impulse of force equals the change of momentum produced by it. An *impulsive force* is one that is very large but acts only for a very short time; it can be represented by a *Dirac function.

impulse current *See* impulse voltage (or current).

impulse flashover voltage *See* impulse voltage (or current).

impulse generator *Syn.* surge generator. An electronic circuit for producing single pulses, usually by charging and discharging a capacitor.

impulse noise *See* noise.

impulse puncture voltage *See* impulse voltage (or current).

impulse ratio *See* impulse voltage (or current).

impulse voltage

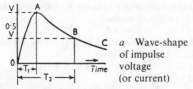

a Wave-shape of impulse voltage (or current)

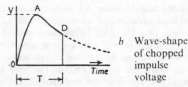

b Wave-shape of chopped impulse voltage

impulse voltage (or **current**) A unidirectional voltage (or current) that rises rapidly to a maximum value without appreciable superimposed oscillations and then falls to zero more or less rapidly. Fig. *a* shows a typical waveshape, which is described as a T_1/T_2 wave.

Relevant terms are: *Peak value*: maximum value, *V*. *Wavefront*: rising portion, OA. *Wavetail*: falling portion, ABC etc. *Duration of the wavefront*: time interval (T_1) for the voltage (or current) to rise from zero to its peak value (usually measured in microseconds). *Time to half value of the wavetail*: time interval (T_2) for the voltage (or current) to rise from zero, pass through its peak value (V), and then fall to half its peak value $(0.5\ V)$ on the wavetail (usually measured in microseconds). *Steepness of the wavefront*: average rate of rise of voltage (or current) on the wavefront.

An impulse voltage having $T_1 = 1\ \mu s$ and $T_2 = 50$ μs (described as a 1/50 microsecond wave) has been found to be representative of those produced by surges that are propagated along a transmission line as a *travelling wave as a result of lightning. Accordingly, the standard testing wave produced in a *surge generator has these proportions.

Additional terms used in connection with an impulse voltage are: *Full impulse voltage*: impulse voltage of the type shown in Fig. *a*, which does not cause flashover or puncture. *Chopped impulse voltage*: impulse voltage that collapses rapidly owing to flashover or puncture. This is illustrated in Fig. *b*. *Impulse flashover* (or *alternating puncture*) *voltage*: (i) For flashover (or puncture) on the wavefront. Actual value of the impulse voltage at the instant of flashover (or puncture). (ii) For flashover (or puncture) on the wavetail (as in Fig. *b*). Peak value of the impulse voltage, which causes the flashover (or puncture).

Impulse ratio for flashover: Of an insulator used in an a.c. power transmission system. Ratio of the impulse-flashover voltage (*see* above) to the peak value of the power-frequency flashover voltage. Common values for normal types of high-voltage insulators are between about 1.2 and 1.5. *Impulse ratio for puncture*: Definition similar to that of impulse ratio for flashover (*see* above) with "puncture" substituted for "flashover". *Time to flashover* (or, alternatively, *puncture*): time interval T (*see* Fig. *b*) between the beginning of the impulse voltage and the instant (point D) at which the wave is chopped by flashover (or puncture).

impulsive force *See* impulse.

impulsive sound A sharp sound that is completed in a small period of time and is extended throughout the aural spectrum. A hammer blow can be typically described as impulsive.

impurities In a semiconductor. Foreign atoms, either naturally occurring or deliberately introduced into the semiconductor. They have a fundamental effect on the amount and type of conductivity. *See* semiconductor.

incandescence The emission of visible radiation from a substance at a high temperature. The term also refers to the radiation itself. *Compare* luminescence.

incandescent lamp An electric lamp in which light is produced by the heating effect of a filament of carbon, osmium, tantalum, or (more usually) tungsten. Inert-gas fillings are often used to suppress disintegration of the filament at the high temperature (>2600 °C), and efficiency is often increased thermally by winding the filament into a close spiral, and then this into a second close spiral (coiled coil) to reduce heat loss by conduction through the gas. *See also* quartz-iodine lamp.

incidence (angle of) The angle between the ray striking a reflecting or refracting surface (i.e. the incident ray) and the normal to the surface at the point of incidence.

inclination, magnetic *See* dip.

inclined plane A rigid plane, inclined at an angle to the horizon and used to facilitate the raising of heavy bodies. The lifting force may be applied along the plane or at an angle to it.

inclinometer An instrument for measuring the magnetic inclination or dip. *See* dip circle.

incoherent Denoting radiation that is not *coherent radiation. If two sources of optical radiation are independent, there is normally no regular relationship between the phases of waves from them arriving at any point. Although *interference has been observed between waves from independent sources in exceptional cases, it is usually necessary to employ coherent radiation to demonstrate this phenomenon. If the light from a *gas-discharge tube is split into two beams to form a pair of virtual sources, interference fringes can be produced on recombining the waves, provided the difference of paths is not too great (typically less than 10^{-1} m). Much greater path differences are possible without incoherence using a *laser as the primary source.

incremental permeability A useful concept in cases where direct and alternating magnetizations are superimposed. It is defined as the ratio of ΔB to ΔH (where B is a magnetic flux density and H is magnetic field strength) for any position on a magnetization curve, where ΔB and ΔH may be of any magnitude but ΔH must be in reverse direction from the immediately preceding change. The incremental permeability at any point is the slope of the small *hysteresis loop corresponding to the increment in H starting at that point.

incremental plotter A device for plotting graphs from instructions generated by a *computer. It normally approximates to curved lines by means of small increments in the x-, y-, and diagonal directions.

On-line plotting is controlled directly by the computer, the instructions being carried out immediately they are generated. This is inefficient, because the computer is inherently faster than the plotter.

Off-line plotting uses instructions from a previously prepared *magnetic tape or *disk. A *microfilm plotter* uses film instead of paper as its output medium. Animation may be achieved by plotting a sequence of frames for cine projection.

incremental recorder An *off-line device for recording data as it is generated, usually on *magnetic tape or *paper tape, in a form that can be fed straight into a *computer.

indeterminacy principle *See* uncertainty principle.

index error *Syn.* zero error. A scale error on a measuring instrument such that the instrument shows a reading *x* when it should show zero reading. Provided there is no other error, all readings on the instrument require a correction of –*x*.

index of refraction *See* refractive index.

indicating instrument An instrument for indicating or measuring an electrical quantity, e.g. *ammeter, *voltmeter, *wattmeter.

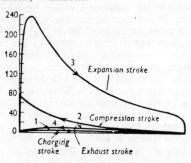

Indicator diagram

indicator diagram The cycle traced out during the motion of a piston in the cylinder of an engine. Vertical displacements represent the pressure (*P*) and horizontal displacements (*D*) the volume of the working substance. The area enclosed by the diagram gives the work done per cycle and is used in estimating the efficiency of the engine.

indicator tube A minute *cathode-ray tube, often with a screen diameter measured in millimetres, in which the shape or size of the image on the screen varies with the input voltage *V* in such a manner that it is used to measure *V*, and hence can indicate the value of a varying signal.

indifferent equilibrium *Syn.* neutral equilibrium. *See* equilibrium.

indirect effect *See* ionizing radiation.

indirect lighting A system of lighting in which most of the luminous flux reaches the area to be illuminated after diffuse reflection from a ceiling or other object external to the light sources.

indirect lightning surge *See* lightning surge.

indirectly heated cathode *See* hot cathode.

indirect stroke *See* lightning stroke.

induced current *See* electromagnetic induction.

induced electron emission spectroscopy *See* electron spectroscopy.

induced e.m.f. *See* electromagnetic induction.

induced lightning surge *See* lightning surge.

induced radioactivity *Syn.* artifical radioactivity. *See* radioactivity.

inductance A property of an electric circuit that results from the magnetic field set up when a current flows. Inductance relates the *magnetic flux through the circuit to the current flowing in that circuit – self inductance – or in a nearby circuit – mutual inductance. *See* electromagnetic induction.

induction 1. *See* electromagnetic induction.
2. (magnetic) *See* magnetic flux density.
3. *See* electrostatic induction.

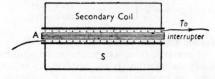

a   Induction coil

induction coil A device for producing a series of pulses of high potential and approximately unidirectional current by *electromagnetic induction. It consists of a primary circuit of a few turns of wire, wound on an iron core A, and insulated from a secondary coil S of many turns, which surrounds it coaxially (Fig. *a*). The primary is supplied with an interrupted current, from an interrupter. Owing to the high resistance introduced into the primary circuit by each break of the circuit, the *time constant of the primary is much smaller at break than when the contact is remade, and the induced e.m.f.

in the secondary is consequently much higher. The efficiency of the coil thus depends on the sharpness of the break, and various types of interrupter have been designed to improve the performance. For small coils, a vibrating hammer break (working on the principle of the electric bell) is sufficient. For larger coils, mercury breaks, in which contact is made and broken by a rapidly rotating jet of mercury, are employed. The output from the secondary consists of a succession of sharp pulses, corresponding to the breaks in the primary circuit, with much smaller inverse pulses produced when the current is remade (Fig. *b*).

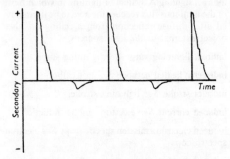

b   Secondary current in an induction coil

induction flowmeter A device whereby the rate of flow of a conducting liquid passing through a tube T (see diagram) in a magnetic field can be measured by the e.m.f. induced across a diameter, between electrodes E. The relationship is $e = BLv$, where e is the e.m.f. in volts, B is the flux density in tesla, L is the tube diameter in metres, v is the velocity in m s^{-1}, and B, L, and v are mutually perpendicular.

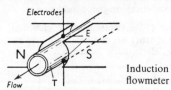

Induction
flowmeter

induction generator Fundamentally, an *induction motor operated as an a.c. generator. Its primary (usually stator) windings are connected to an a.c. source that provides the necessary a.c. excitation, and when the rotor is driven mechanically above the *synchronous speed corresponding to the frequency of the a.c. excitation, a.c. energy is delivered to the source at the frequency of the excitation. An induction generator will operate only if connected in parallel with one or more *synchronous alternating-current generators, since it cannot provide its own excitation.

induction heating The heating effect of induced *eddy currents in a conducting material subjected to a varying magnetic field. The field can be produced by an alternating current flowing in a coil surrounding the material. Induction heating can be employed for

metal melting. The advantage is that the heat is generated in the metal itself, and after melting the eddy currents set up circulatory movements that stir the melt.

induction instrument An instrument in which the deflecting force or torque is produced by the interaction of *eddy currents induced in a movable conducting mass (usually in the form of a disc or cylinder) and the magnetic field of an a.c. electromagnet.

induction interaction *See* van der Waals forces.

induction machine A machine for the production of electrical charges by electrostatic induction. *See* Van de Graaff generator; Wimshurst machine.

induction meter A type of motor meter whose action is an *induction motor in one form or another.

induction motor An a.c. motor consisting of a *stator and a *rotor in which the current in one member (usually the rotor) is generated by *electromagnetic induction when alternating current is supplied to a winding on the other member (usually the stator). The torque is produced by interaction between the rotor current and the magnetic field produced by the current in the stator. For motors used industrially, there are two main types of rotor: (i) *cage rotor* (originally called *squirrel-cage rotor*) in which all the rotor conductors are permanently short-circuited at both ends of the rotor by means of end-rings; (ii) *slip-ring rotor* (or *wound rotor*), which carries a *polyphase winding connected to *slip rings. The object of the slip rings is to enable resistances to be connected temporarily in the rotor circuits so that the torque and current at starting may be controlled. The brushes on the slip rings are short-circuited when the motor is running normally. A very small induction motor of the type used, for example, in an *induction instrument may have a rotor consisting of a solid conducting mass (disc or cylinder) in which case the rotor currents are *eddy currents.

induction voltage regulator A voltage regulator for varying the voltage of an a.c. supply, or for maintaining a constant voltage under varying load conditions. In its simplest form for use on a single-phase supply it has two windings (primary and secondary), having a common magnetic circuit. The primary winding is connected in *shunt with the supply and the secondary is connected in *series with the load circuit so that the voltage applied to the load is the vector sum of the supply voltage and the voltage at the terminals of the secondary windings. The latter voltage is variable at will by adjusting the relative positions of the series and shunt windings. The constructional features closely resemble those of an *induction motor fitted with a slip-ring rotor, especially in the case of a three-phase regulator.

inductive Of an electric circuit or winding. Having an appreciable *inductance. Since, however, it is difficult to obtain a circuit that is absolutely devoid of inductance, the term is usually applied to a circuit in which, for the purpose in view, the effect of inductance cannot be neglected. *Compare* noninductive.

inductive coupling *See* coupling.

inductive load *See* lagging load.

inductive-output device An electronic device in which electromagnetic induction between the stream of electrons and the output electrode enables energy to be transferred from the former to the latter. The output electrode usually surrounds the electron stream but it does not collect the electrons.

inductive reactance *See* reactance.

inductive tuning *See* tuned circuit.

inductor *Syn.* reactance coil. A coil, or other piece of apparatus, possessing *inductance and selected for use because of that property. *See also* choke.

inductor generator An a.c. generator in which the armature (a.c.) winding and the *field coils are fixed in position relative to one another (usually both are wound on the stator of the machine). The movement (usually rotation) of masses of ferromagnetic material (usually soft iron) is arranged to produce cyclic variations in the *reluctance of the magnetic circuit, which in turn produces cyclic variations in the magnetic flux, thereby inducing alternating e.m.f.s in the armature (a.c.) winding.

inelastic collision *See* collision.

inelastic scattering *See* scattering.

inertance *See* acoustic inertance.

inert cell A primary *cell that is inert until water is added to produce an electrolyte. It contains the chemicals and other necessary ingredients in solid form.

inertia 1. The property of a body by virtue of which it tends to persist in a state of rest or uniform motion in a straight line. *See* Newton's laws of motion.
  2. In photographic work, the *optical density of an exposed and processed negative varies linearly with the logarithm of the *exposure over the usual working portion of the range. The inertia is the exposure that would just begin to affect the density of the linearity held at such small exposures.

inertia ellipsoid *See* product of inertia.

inertia force *See* force.

inertial force *See* force.

inertial observer 1. In classical physics, an observer who finds that *Newton's laws of motion are valid.
  2. In *relativity, an observer who finds that the special theory of relativity is valid.
  See inertial reference frame.

inertial reference frame *Syn.* inertial coordinate system. A frame of reference used by an *inertial observer in which any body that is not subject to a

resultant force has constant *velocity, i.e. the speed and direction of motion are unchanging.

inflation *See* inflationary universe.

inflationary universe A possible phase in the very *early universe when its size increased by an immense factor. It is postulated (Guth et al) that at an age of 10^{-35} second the state of the universe changed in such a way that additional energy was released, allowing the existing expansion of the universe to accelerate rather than decelerate (*inflation*). The phenomena occurring during this rapid expansion are very speculative. The existence of such an inflationary phase can lead, for example, to an explanation of the origin of galaxies and stars. After a period of time the inflationary expansion changed to that of the standard *big-bang theory.

information technology (IT) Any form of technology, primarily electronic equipment and techniques, used by people to handle and distribute information. It incorporates the technology of both computing and of telephony, television, and other forms of telecommunication.

information theory An analytical technique for determining the optimum (generally minimum but sufficient) amount of information required to solve a specified problem in communication or control.

infrared astronomy The study of astronomical sources emitting *infrared radiation. These include the dust clouds around many stars and all newly forming stars, and extragalactic objects such as *quasars and *Seyfert galaxies. Observations can be made from the ground through several *atmospheric windows up to a wavelength of about 20 μm. Satellites, rockets, and balloons are required for longer-wavelength studies. The equipment used includes reflecting telescopes to collect the radiation plus detectors such as *photovoltaic and *photoconductive cells and *bolometers. The detectors, and sometimes the telescope optics, must be cooled to very low temperatures to minimize thermal emission and hence noise. Siting ground-based telescopes at high altitudes minimizes atmospheric absorption of the infrared radiation, which is due mainly to water vapour and carbon dioxide.

infrared radiation (IR) Long-wave radiation emitted by hot bodies, with wavelengths ranging from the limit of the red end of the spectrum, about 730 nm, to about 1 mm. The longest wavelengths are adjacent to the microwave and radio-wave regions of the electromagnetic spectrum, where the radiation is produced electronically. The shorter wave or near infrared (the original region detected by Herschel in 1800 by its heating effect just beyond the visible spectrum), is examined by a spectroscopic method using fluorite or other material prisms in place of glass, and concave reflectors in place of lenses. This is because most glasses are absorbent at wavelength 2 μm. The quartz limit is 4 μm, fluorite 10 μm, rocksalt 15 μm, sylvin 23 μm. The near infrared

infrared spectroscopy

radiation is copiously emitted by molten metal surfaces in furnaces and is responsible for the prevalence of cataract among furnace workers (Crookes' sage-green glass is recommended as protection). Detectors operating in the near-infrared region, and in some cases at longer wavelengths, include the *bolometer, *thermopile, and semiconductor devices such as *photoconductive and *photovoltaic cells. Photographic methods can be used only to about 1 μm. To examine the far infrared (to about 75 μm) the method of *selective reflection, using the residual rays, is employed. Focal isolation methods are used for the extremely long radiations.

infrared spectroscopy *See* spectroscopy.

infrared windows *See* atmospheric windows.

infrasound Vibrations of the air below a frequency of about 16 hertz, recognized by the ear as separate pulses rather than as sound. The sensation of detonation caused by these waves is due to the rapid pressure change at the ear, the intensity depending on the suddenness of the initial compression of the air. Infrasonic waves are generated at the muzzles of guns and by some mechanical devices.

Ingen-Hausz, Jan (1730–1799) Dutch physician. His chief discovery was that green plants absorb carbon dioxide by day and exude it by night, but he also devised the well-known method of comparing the thermal conductivities of materials in the form of rods (*see* Ingen-Hausz's apparatus).

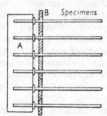

Ingen-Hausz's apparatus

Ingen-Hausz's apparatus To compare the thermal conductivities of good conductors. The specimens are in the form of long rods of equal length and radius of cross section, and are coated with wax. These rods are placed as shown in the diagram, A being a tank containing boiling water and B a radiation shield. When conditions become steady the lengths l_1, l_2, l_3, etc., along the rods from which the wax has melted are measured and then

$$\lambda_1 : \lambda_2 : \lambda_3, \text{ etc.} = l_1{}^2 : l_2{}^2 : l_3{}^2, \text{ etc.},$$

where λ is the thermal conductivity of the material of a rod.

inherent regulation 1. Of an *alternator. A rise in terminal voltage above normal when full load (i.e. rated output at rated power factor and rated voltage) is thrown off, the excitation and speed being maintained constant at their full-load values. It is expressed as a percentage of the full-load voltage.
2. Of a d.c. generator. A change in terminal voltage when full load (i.e. rated output at rated

voltage) is thrown off, the speed being maintained constant at its full-load value and no external adjustment being made to the exciting circuit. (N.B. The excitation is not necessarily constant.) It is usually expressed as a percentage of the full-load voltage.
3. Of an a.c. transformer with constant primary voltage. The difference betwen the no-load and full-load secondary terminal voltages, for constant supply voltage and frequency, at any stated power factor of the load. It is usually expressed as a percentage of the no-load secondary terminal voltage. *See* regulation; voltage drop.

inhibiting input An input to a *gate that prevents any output that might otherwise occur.

injection 1. In general, the application of a signal to an electronic circuit or device.
2. In a *semiconductor, the process of introducing carriers (electrons or holes) into the semiconductor so that the total number of carriers exceeds the number present at thermal equilibrium. The carriers may be introduced in various ways, e.g. across a forward-biased junction or by irradiation.
When the number of excess carriers is small, the injection is called *low-level injection*. If it is comparable to the thermal-equilibrium numbers, the injection is *high-level injection*.

injection efficiency The efficiency of a *p-n junction when forward bias is applied, defined as the ratio of the injected minority *carrier current to the total current across the junction.

injection laser *See* semiconductor laser.

in parallel *See* parallel.

in phase *See* phase.

in-phase component *See* active current; active voltage; active volt-amperes.

input The signal or driving force applied to a circuit, device, machine, or other plant. Also the terminals to which this is applied.

input impedance The *impedance presented by a circuit or device at its input terminals.

input/output (I/O) Operations and devices concerned with the passage of information into or out of a *computer.

in series *See* series.

insertion loss (or **gain**) The reduction (or increase) of power in a load that occurs when a network is interposed between the load and the generator supplying the load. It is usually expressed in *nepers or *decibels and, in general, it is a function, not only of the network parameters, but also of the load and generator impedances.

installed load The sum of the rated inputs of the electrical apparatus installed at a customer's premises for connection to the electricity-supply system. A knowledge of the installed load is important for tariff purposes, but in this connection, apparatus that serves entirely as standby is not usually included in the assessment of the installed load.

instantaneous axis That straight line in a rigid body about which it may be regarded as rotating at any instant. If a rigid body is constrained to rotate about a fixed point O (*compare* Poinsot motion), the instantaneous axis will occupy different positions in the body but will always pass through O. Its locus in the body is a cone with vertex O, called the *body cone* and its locus in space is the *space cone*. The body cone touches the space cone along the instantaneous axis.

instantaneous carrying-current The peak value of the current, at rated voltage, that a switch, circuit-breaker, or similar apparatus is capable of carrying instantaneously under specified conditions.

instantaneous centre The point at which the *instantaneous axis of a plane rigid body moving in its own plane meets that plane. The curve traced by the instantaneous centre relative to axes of reference outside the moving body is known as the *space centrode* and the path of the instantaneous centre in the moving body is the *body centrode*. It can be shown that the body centrode rolls on the space centrode.

instantaneous frequency The rate of change of phase of an electric oscillation in radians per second, divided by 2π. It has particular applications in connection with *frequency and *phase modulation.

instantaneous value The value of any varying quantity at a particular instant of time, or (strictly) the average value over an infinitesimal period of time. The symbol for an instantaneous value is usually the lower-case form of the capital letter used for the quantity itself; e.g. p, i, and v (or u) are the symbols for instantaneous power, current, and potential difference.

instrument transformer An a.c. *transformer utilizing the property of voltage transformation (*see* voltage transformer) or current transformation (*see* current transformer) and used in conjunction with an instrument, relay, or other similar apparatus. The design is such that the proportionality between the primary and secondary quantities (voltages or currents as the case may be) and the phase displacement between them (usually 180°) are maintained with precision under prescribed conditions. *See* phase difference; ratio.

insulate To surround or support an electrical (or thermal) conductor by insulating material so that the flow of electricity (or heat) is confined to the desired path.

insulated-gate field-effect transistor (IGFET) *See* field-effect transistor.

insulated-return system A system for the distribution of electric power to trains and vehicles, in which the outgoing and return conductors are both insulated. *Compare* track-return systems.

insulated system A system of distribution of electrical power in which, under normal conditions, no point is connected to earth. The system has the advantage that a fault to earth does not prevent the system from being used but the *earthed system is employed to a greater extent in practice.

insulating barrier A screen of insulating material fitted to electrical apparatus (such as switch gear and fuses) to prevent damage either to the operator or to the apparatus by the spreading of an arc.

insulating material *See* dielectric.

insulating resistance The resistance between two electrical conductors or systems of conductors that are normally separated by an insulating material. It is usually expressed in megaohms or, in the case of cables, in megaohms per mile (or kilometre).

insulation 1. Material that insulates an electrical conductor.

2. Material that reduces the transmission of heat, sound, etc., from a body or region.

insulation-testing set An instrument for measuring insulation resistance. It usually incorporates a small hand-driven magneto-electric generator that generates an e.m.f. of about 500 volts (common value) so that the test is carried out with this voltage applied to the insulation. The instrument gives direct indications on a scale that is divided into multiples and submultiples of a megaohm. *Compare* ohmmeter.

insulator A substance that provides very high resistance to the passage of an electric current; an appliance made of insulating material used to prevent the loss of electric charge or current from a conductor. *See also* energy bands; dielectric.

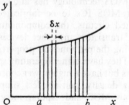

Integral

integral The area between the x-axis, the lines $x = a$, $x = b$, and the curve $y = f(x)$ is the integral of $f(x)$ with respect to x between the limits a and b. It is written as

$$\text{Area} = \int_a^b f(x)\mathrm{d}x.$$

It can also be regarded as the limit of a sum, e.g. if ordinates are erected at small distances δx apart between $x = a$ and $x = b$, then the limit of the sum

$$\sum_{x=a}^b f(x)\delta x \text{ as } \delta x \to \text{zero}$$

is equal to the integral. (The rigorous mathematical definition of an integral is an arithmetic one.)

integral transform

Integration is the converse of differentiation in that if the *derivative of the integral between the limits 0 and x is taken, the original function results.

integral transform A function $f(p)$ of a parameter p, defined as

$$\int_a^b f(x)K(p, x)\mathrm{d}x,$$

is an integral transform of $f(x)$. If the kernel $K(p, x)$ has the form e^{-px} and $a = 0, b = \infty$, this is a Laplace transform; if $K(p, x) = \sin px$ or $\cos px$, again with $a = 0, b = \infty$, it is a Fourier sine or cosine transform.

Such transforms arise in the solution of differential equations.

integraph A device for drawing a graph of the integral of a function when presented with a graph of the function.

integrated circuit (IC) A complete electronic circuit manufactured in a single package and implementing a particular function. Both *digital and *linear circuits are produced. Normally, all the circuit components – transistors, etc. – are manufactured in or on top of a single *chip of *semiconductor, usually silicon. Interconnections between the various parts of the circuit are made by means of a pattern of conducting material on the surface of the IC. The individual parts are not separable from the complete circuit. This type of circuit is often called a *monolithic integrated circuit*.

In contrast, a *hybrid integrated circuit* has the individual circuit components attached to an insulating substrate and interconnected by conducting tracks laid down on the substrate. The individual devices are unencapsulated, and may be diodes, transistors, monolithic ICs, or thick-film resistors and capacitors; the complete circuit is very small.

In *MOS integrated circuits, the active devices are MOS *field-effect transistors, which operate at low currents and high frequencies. They have a high *packing density and consume very little power. The development of MOS technology has allowed extremely complex MOS ICs to be fabricated. In *bipolar integrated circuits, the components are bipolar junction *transistors and other devices that are fabricated using the p-n junction properties of semiconductors. They have higher operating speeds than MOS circuits but have a high power consumption, low packing density, and are less simple to manufacture. Both MOS and bipolar integrated circuits are monolithic.

The complexity of digital circuits that may be produced on a single chip is usually described in terms of the number of transistors or the number of logic gates involved. This leads to the following groupings, in order of complexity:

SSI	small-scale integration
MSI	medium-scale integration
LSI	large-scale integration
VLSI	very large-scale integration. LSI and VLSI technologies produce at least 10 000 and 100 000 transistors respectively on a single chip.

integrating frequency meter *Syn*. master frequency meter. An instrument that integrates the number of cycles through which an alternating-supply voltage has passed in a given period of time, thus enabling a comparison to be made with the number of cycles through which it would have been passed in the same period of time if the prescribed frequency had been maintained.

integrating galvanometer A *galvanometer in which inertia and control of the moving system are so chosen that the change in flux produced in an exploring coil can be measured even though the changes last over a period of several minutes.

integrating meter A measuring instrument that integrates the measured quantity with reference to time. Examples are: ampere-hour meter; *electrolytic meter; *induction meter; *integrating frequency meter; *mercury motor meter; *motor meter; volt-ampere hour meter; watt-hour meter.

integrating photometer In order to measure the mean spherical *luminous intensity of a source, a large sphere is painted white inside with a matt or opal window whose brightness can be matched with a standard lamp on an outside photometer bench. The lamp under test is completely surrounded by the sphere; the internal brightness and that of the window are proportional to the total flux emitted by the lamp. For best results the interior should be coated by magnesium oxide, deposited from a burning magnesium ribbon.

integration The evaluation of an *integral.

integration time The period over which a noisy signal is averaged in order to improve the *signal-to-noise ratio in an electronic system.

integrator A mechanical or electrical device for performing the mathematical operation of integration, e.g. (1) a shaft A drives a second shaft B at a gear ratio of $n{:}1$ (i.e. B makes n turns for one turn of A); then for a rotation $\mathrm{d}x$ of the shaft A, B rotates through $n\mathrm{d}x$. If the gear ratio n changes while the shaft is rotating, the total rotation of B is the sum of the contributions $n\mathrm{d}x$, i.e. the integral required. (2) the direct current i flowing into the capacitor C gradually builds up a voltage on the capacitor equal to $(1/C)\int i\mathrm{d}t$, i.e. an integration of i with respect to time is performed.

integrometer *Syn*. moment planimeter. An instrument for evaluating the moment and moment of inertia about a given line of the area enclosed by a curve.

intensifier 1. (photographic) A substance used to strengthen the image on a negative or positive medium. Chemical intensifiers, such as chromium or mercury salts, increase the optical density to an extent depending on the original image density.

2. (hydraulic) A device for increasing a hydraulic pressure. Two pistons rigidly connected together work in two cylinders of different diameters placed in line. The pressure on the larger piston multiplied by its area is the force applied to the other piston. The pressure in the smaller cylinder is this force divided by the smaller piston area and this is greater than the other pressure. *See* hydraulic press.

intensifying screen A screen coated with a fluorescent material, such as calcium tungstate crystals, that emits light under the action of X-rays; each X-ray photon produces several hundred light photons. Such screens are used in medical X-ray diagnosis, the emitted light being recorded on photographic film adjacent to the screen. They reduce the exposures needed in radiography by a factor of about 60. There is, however, some reduction of definition as the screen impresses its own grain on the film.

intensity A measure of the concentration of some factor, such as sound or light, usually over a given area or volume. (*See* sound intensity; luminous intensity; radiant intensity.) The term *illuminance* is replacing *intensity of illumination*, and the terms *magnetic* and *electric field strength* have replaced *magnetic* and *electric intensity*.

intensity modulation *Syn.* z-modulation. The variation of the brilliance of the spot on the screen of a *cathode-ray tube in accordance with the magnitude of a signal.

intensity of magnetization For a uniformly magnetized body, the *magnetic dipole moment per unit volume.

interaction A process in which bodies exert forces on each other, or in which electromagnetic radiation and particles exert mutual forces. Usually one is concerned with those processes in which one or more bodies undergo some change of structure.

Every particle of matter in the universe feels the influence of the force of *gravitation, which derives from the existence of mass. *Electromagnetic interactions bind matter on a smaller scale, holding atoms and molecules together. In nuclear science *strong interactions occur between *quarks, and between *hadrons and systems of hadrons (nuclei). Such interactions are attributed to *exchange forces involving virtual bosons. The forces are short-range (*see* force) and do not operate significantly at separations much greater than 10^{-15} m, taking place in times typically 10^{-23} s. At a deeper level, strong interactions can be explained in terms of the exchange of virtual *gluons between quarks or antiquarks (*see also* quantum electrodynamics). *See also* electroweak theory; gauge theory. *Weak interactions are typically 10^{-12} times as powerful as strong ones. They are most commonly observed in decay processes (such as *beta decay) when there is some principle (e.g. a conservation law) that prevents the operation of a strong or electromagnetic interaction.

interactive Allowing continuous two-way transfer of information between user and *computer. An *interactive *program interrogates the user during its execution, and can modify its subsequent action as a

result of the information so obtained. An *interactive *terminal allows the user to send or receive information from a computer. *See* time sharing.

interconnected star connection *See* zig-zag connection.

interconnecting feeder *See* trunk feeder.

interconnector *See* trunk feeder.

interelectrode capacitance The capacitance between specified electrodes of an electronic device, in which the electrodes form a small capacitor, e.g. between emitter and base of a *transistor. These capacitances may have a significant effect on the operation of such devices.

interface A common boundary between two parts, devices, or systems. It may be a surface separating two fluids of different densities or velocities. Another example is the electronic circuitry plus associated software that allows communication between two computer systems or two components of a computer system.

interfacial angles Angles between the normals to crystal faces.

interfacial surface tension Surface tension of the interface between two immiscible liquids. Antonow (1907) stated that this is the difference between the surface tension of the two liquids separately. This is true if by "the two liquids separately" is meant the actual liquids in contact, each of which will have dissolved a little of the other. *See* surface tension.

interference In a communications system. A disturbance to a signal caused by undesired signals, which may be man-made or may arise from natural causes, such as changes in the atmosphere. In radio reception, electrical machinery and apparatus (especially commutating machines and apparatus employing gas-discharge tubes) commonly give rise to interference. Motor-car ignition systems sometimes produce serious interference with the reception of television signals. When the interference is man-made it can usually be prevented by fitting special devices (*suppressors*) to the offending apparatus. *See* crosstalk; hum.

interference figure A pattern of coloured rings and black bands given by a crystal in convergent polarized light between crossed *Nicol prisms. *See* interference of light; isogyres.

interference filter *See* filter.

interference fringes *See* interference of light.

interference microscope *See* interference of light; microscope.

interference of light (1) *Discovery.* By Thomas Young in 1801. Fig. *a* shows the apparatus schematically, the distance between the pinholes A and B being actually about a thousandth of their distance from the primary pinhole S. The latter was illuminated by white light, causing coloured fringes in the

interference of light

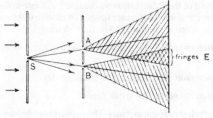

a Young's fringes

region E. The beams going through A and B *interfere*, since each alone gives a patch of light on the screen without fringes. Later experiments have used *monochromatic radiation giving light and dark fringes, and slits (perpendicular to the plane of the figure) have been used instead of pinholes.

(2) *Theory of interference fringes.* These and similar fringes are readily explicable on wave theory and were used by Fresnel and Young as evidence to establish wave theory. A bright fringe will be observed at P (Fig. *b*) if the path difference BP–AP is one wavelength λ or any integral multiple $n\lambda$. The separation of successive fringes comes to $\lambda D/2b$, for slits $2b$ apart, with the screen at distance D. Incident white light gives coloured fringes since red fringes (of greater λ) are further apart than green, etc.

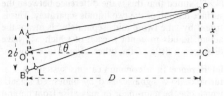

b Formation of interference fringes

(3) *Interference systems.* (*a*) A typical system uses two images of a source for A and B, or a source and its own image. Thus *Lloyd's mirror* has a slit parallel to a mirror or face of a glass slab and close to the reflecting plane. Light from the slit is reflected from the surface at nearly grazing incidence, travelling as if from the slit image in the surface, and interferes with light coming direct. The fringes may be viewed through a short-focus eyepiece. *Fresnel's biprism* is a glass plate prepared to be equivalent to two thin prisms (refracting angle about 1°) base to base. The two halves of this device deviate light from a slit in opposite senses so that two virtual images of the slit are produced to serve as A and B in the above account. An observing eyepiece traversed by a micrometer arrangement enables spacing to be measured and λ to be calculated (assuming monochromatic light). Other systems include *Fresnel's mirrors*, a pair of plane mirrors at very slight inclination to one another; and the *Billet split lens, with two separated halves of a convex lens forming images of the primary slit. All the above have a pair of effective sources (A, B) and light emerging from each is *coherent radiation, derived from sources that necessarily are in fixed-phase relationship.

(*b*) Thin films (e.g. soap bubbles; oil on water) often display brilliant colorations when reflecting white light and show fringes when in monochromatic light. Here, light reflected from the front and back surfaces may be out of step by various amounts, *destructive interference* and consequent darkness occurring in some directions while *constructive interference* or reinforcement occurs in others. The irregular appearance of the coloured bands results from the uneven thickness of the film. Parallel-sided layers give *fringes of equal inclination*, visible by eye or telescope focused on infinity; thin films give fringes of equal thickness, effectively contours, located in or near the film. *Newton's rings*, formed between a convex-lens face and a plane glass slab, are of the latter type. Fringes become much sharper when formed by multiple reflections between surfaces of high reflecting power, e.g. if the film is between surfaces thinly silvered so that light can enter and leave the interspace.

(4) *Interferometers.* There are several important interferometers of precision. The *Fabry–Perot etalon has parallel thinly silvered surfaces enclosing an airspace, the separation being either accurately fixed by distance pieces, or variable with an accurately calibrated screw. The *Michelson interferometer uses a thinly silvered inclined surface to divide the incident light into beams of comparable strength, travelling at right angles until returned by mirrors. It is possible to obtain fringes similar to those of equal inclination or of equal thickness and to count them while one mirror is displaced by known amounts. Both the above interferometers have been used to make highly accurate wavelength measurements. Adaptations of the Michelson interferometer have been made for testing lenses, mirrors, and prisms (*Twyman–Green apparatus*).

The *Jamin interferometer or refractometer and the Rayleigh refractometer (*see* refractive-index measurement) are both useful for measuring refractive indices of gases.

The *Lummer–Gehrcke plate is an accurately prepared glass or quartz slab in which multiple reflections give rise eventually to sets of emergent beams with large path-differences. The high resolving power of this device is best realized in conjunction with a prism spectrograph, especially a constant-deviation instrument. It is used in the parallel beam between the collimator and the prism. Each line in the prism-spectrum is split into repeated patterns of horizontal lines, each group being a high-resolution representation of the "spectrum-line". This arrangement is favourable to demonstration of the *Zeeman effects.

The *echelon grating (Michelson, 1898) consists of a pile of glass plates in steplike formation. Used in transmission, some light leaves each stepface giving beams separated by quite considerable path-differences. There may be as many as 30 plates. The resolving power is very high, of the order of 5×10^5.

(5) *Conditions for interference.* (*a*) For interference to occur between two beams of light, they must be

coherent, i.e. any phase changes occurring in one beam must be matched by equal changes in the other. Formerly, this meant that the beams had to be derived from a common source. Two *lasers can now be used. (b) The beams must intersect at an angle that is not too great. (c) They must *not* be polarized in planes at right angles. Condition (c) led to the recognition that light waves must be transverse and not longitudinal, as Huygens and, at first, Fresnel had thought. (d) The beams must have approximately equal intensity, i.e. the waves must be of equal amplitude. (e) For complete destructive interference the phase difference must be any odd integral multiple of half a wavelength; for complete constructive interference the phase difference must be an integral multiple of one wavelength.

It is possible to obtain interference fringes from a highly monochromatic source with a Michelson interferometer with a path-difference of half a meter, containing some million wavelengths. This implies that wave trains of considerable length are emitted from the radiating atoms in the source.

(6) *Applications.* (a) Control of reflection is possible by a thin coating on a surface. A layer of thickness $\frac{1}{4}\lambda$ (where λ is the wavelength in the medium) and of refractive index the geometric mean of the media on either side would give no reflection at all at near-normal incidence, owing to destructive interference between the light reflected from the two interfaces, but the condition would hold only for one wavelength. In practice, a *coated or bloomed lens or other optical component has a layer, or a composite of layers, adjusted to give minimum reflection in the most effective part of the spectrum (yellow for vision, blue for some photographic work). Flare spots due to images formed by multiple reflections between the various surfaces, and general fogging are much reduced in intensity and transmission is improved.

(b) Testing of optical surfaces of high-precision work is often achieved by observation of thin film fringes formed between the surface and a complementary test surface, or between the surface and an image of a test surface. Surfaces can be made true to 1/10 or even 1/50 of a wavelength by interference tests, the last corresponding to 10^{-8} metre.

(c) The *interference microscope*, in Dyson's form, uses a thinly silvered inclined surface to divide the light into two beams, much as in the Michelson interferometer, and one beam is returned by the surface of the specimen to interfere with the other, reflected from a silver pot. Differences of level as small as 3 nm are revealed. The multiple-beam interferometric technique of Tolansky uses interference in the space between a thinly silvered glass plate and the face of the specimen and permits even smaller vertical displacements to be recognized. (*See* microscopes.)

(d) Interference fringes are used in the measurements of very small movements as in the elastic strain of short specimens and in the study of thermal expansion. (*See* Fizeau.)

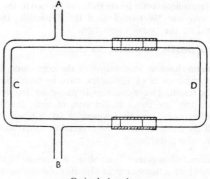

Quincke's tubes

interference of sound The superposition of two or more waves originating from a common source, but traversing different paths, results in regions in the transmitting medium at which there is a minimum intensity and in other regions at which there is a maximum intensity. This phenomenon is called interference, and the resulting pattern in the field of radiation is called an *interference pattern*. Interference can also be produced by waves originating from two or more different sources.

A ready demonstration of interference in sound is provided by the *Herschel–Quinke tube. Sound from a source at A (*see* diagram) reaches a receiver at B by way of the tubes C and D. If the difference of path between tube C and D is exactly an odd number of half-wavelengths, then the two components of the note arriving at B will be out of phase, thereby cancelling each other out and producing a minimum of intensity. If now tube D is extended, by the telescopic means provided, until the path difference between C and D is exactly an even number of half-wavelengths, then the two components reinforce each other to give a maximum intensity for the particular note sounded. Noting the consecutive positions of extension of D enables the wavelength of the note produced by the source to be determined.

This clearly indicates a method of determining the wavelength of a given note, and the method in a refined form is employed in all interferometers.

Since, among other ways, interference is produced by the interaction of a wave reflected from a wall and the corresponding incident wave, the phenomenon is frequently experienced in everyday life. It contributes, in particular, to the poor audibility at certain positions in some theatres, and in the wavering of a high-pitched note to an observer walking in the vicinity of a reflecting surface.

interference pattern *See* interference of sound.

interferometer 1. *See* interference of light.

    2. *See* radio telescope.

interlaced scanning *See* television.

intermediate frequency *See* superheterodyne receiver.

intermediate vector boson Either of the two forms of boson, the *W particle and the Z particle, that mediate the *weak interaction.

intermetallic compounds See alloy.

intermodulation Modulation of the component sinusoidal waves of a complex wave by each other. The resulting wave contains, in particular, frequencies that are equal to the sum of and also the difference between the frequencies, taken in pairs, of all the components of the original complex wave. See also distortion.

internal absorptance Symbol: α_i. A measure of the ability of a substance to absorb radiation as expressed by the ratio of flux absorbed between the entry and exit surfaces of the substance to the flux leaving the entry surface. Internal absorptance does not apply to loss of intensity by scattering or to reflection of radiation at the surface of the substance. (Compare absorptance.) It is related to the *internal transmittance (τ_i) by:
$$\alpha_i + \tau_i = 1.$$

internal-combustion engine An engine in which liquid or gaseous fuel is injected into a cylinder containing air where it is ignited. See gas engine; petrol engine.

internal conversion The process in which a nucleus in an *excited state decays to a lower state and gives up energy to one of its orbital electrons, usually a K-electron. If this energy is large enough to overcome the *binding energy of the electron, then it is ejected from the atom as a *conversion electron. The process is independent of *gamma-ray emission; it is not the production of a gamma-ray photon, which then knocks the electron out by the photoelectric effect. ^{99}Tc nuclei decay by internal conversion at a rate that depends on the chemical composition of the compound involved and thus on the state of the surrounding electrons. Usually excited states of nuclei have very short lifetimes and decay by gamma emission within picoseconds so internal conversion is not observed. Sometimes, however, the *selection rules prevent rapid decay by gamma rays and the excited state is relatively long-lived. In these circumstances internal conversion becomes significant and may be the principal mode of decay. The conversion electrons have a line spectrum so are easily distinguished from beta rays. Electrons from outer shells fall into the vacant states caused by conversion, hence the substance emits its characteristic X-rays.

internal energy Syn. thermodynamic energy. Symbol: U or E. A thermodynamic function of a system that changes by an amount equal to the algebraic sum of the heat received by the system (δq) and the work done on it (δw):
$$dU = \delta q + \delta w.$$
It is necessary to suppose that a system possesses internal energy in order to apply the principle of conservation of energy to thermal systems (see thermodynamics). The above equation is a symbolic representation of the first law. If a system changes from one state to another, the heat change and work change both depend on the path taken but the change in internal energy is only dependent on the initial and final states, not on the way the change is made. The internal energy is never absolutely determined, only changes in its value are important. Sometimes a conventional standard state is considered to have U equal to zero and other states have a value of U equal to the change in internal energy in moving to this state from the standard state. The internal energy of a system is equal to the sum of all the kinetic and potential energies of the molecules. See also Joule's law of internal energy.

internal friction The effect that causes a damping of elastic vibrations in a solid and similar effects. It is analogous to viscosity in liquids and results from the *anelasticity of the material.

internal photoelectric effect See photoconductivity.

internal resistance Of a cell, accumulator, or dynamo. The resistance obtained by dividing the difference between the generated e.m.f. and the potential difference between the terminals of the device by the current.

internal standard line The spectral line due to a known amount of a known element added to a material that is undergoing *spectrographic analysis. The intensity of this line is compared with that of the *analysis line to determine the concentration of the particular element under investigation.

internal transmission density Syn. absorbance. Symbol: D_i. A measure of the ability of a body to absorb radiation as expressed by the logarithm to base ten of the reciprocal of the *internal transmittance:
$$D_i = \log_{10}(1/\tau_i).$$

internal transmittance Symbol: τ_i. A measure of the ability of a material to transmit radiation as expressed by the ratio of the flux reaching the exit surface of the body to the flux leaving the entry surface. The internal transmittance only applies to regular transmission and not to substances that scatter light or to reflection at the surfaces of the body. (Compare transmittance.) The internal transmittance is related to the *internal absorptance (α_i) by:
$$\tau_i + \alpha_i = 1.$$

internal work The work done in separating the molecules of a system against their forces of attraction. Its value is zero for an *ideal gas.

international ampere See ampere; international system.

International Atomic Time (IAT, TAI in France) The most precisely determined timescale now available, set up by the Bureau Internationale de l'Heure in Paris and adopted in 1972. Atomic time is measured by means of atomic *clocks, the fundamental unit being the SI *second. Civil timekeeping is based on IAT.

international candle A unit of luminous intensity agreed upon between the standardizing laboratories of France, Great Britain, and the United States (1909) and maintained by electric incandescent lamps. It was superseded (in 1948) by the *candela.

international ohm See ohm; international system.

International Practical Temperature Scale (IPTS) A temperature scale based on *thermodynamic temperature, consisting of certain fixed points (physical properties of pure substances) at which temperatures are defined absolutely, together with experimental procedures for measuring temperature between these points. The original scale was introduced in 1927, and there have been a number of changes since. The 1968 version (known as IPTS–68) had eleven fixed points defined in both Celsius and thermodynamic temperatures. The most recent version was introduced in 1990 (IPTS–90) and has sixteen fixed points with temperatures assigned in kelvins as follows:

Triple point of hydrogen	13.8033
Boiling point of hydrogen (33 321.3 Pa)	17.035
Boiling point of hydrogen (101 292 Pa)	20.27
Triple point of neon	24.5561
Triple point of oxygen	54.3584
Triple point of argon	83.8058
Triple point of mercury	234.3156
Triple point of water	273.16 (0.01 °C)
Melting point of gallium	302.9146
Freezing point of indium	429.7485
Freezing point of tin	505.078
Freezing point of zinc	692.677
Freezing point of aluminium	933.473
Freezing point of silver	1234.93
Freezing point of gold	1337.33
Freezing point of copper	1357.77

At low temperatures (0–5 K), intermediate temperatures between fixed points are measured by vapour-pressure determinations of ^{3}He and ^{4}He. In the range 3–24.5561 K, a constant-volume gas thermometer is used. A platinum-resistance thermometer is used for temperatures above 13.8033 K and, at high temperatures ($>$1234.93), radiation pyrometry is used. For particular temperature ranges, specified fixed points and equations are defined.

international steam-table calorie (IT calorie) A standardized heat unit now replaced by the joule. 1 cal$_{IT}$ = 4.1868 joules. See calorie.

international system A former system of units for measuring electrical quantities, based on the international ohm and the international ampere. It has now been replaced by SI units.

interocular distance The distance between the centres of rotation of the eyes (adults average 63 mm).

interpolation Estimation of the value of a function, $f(x)$, for a value of the variable, x, which lies between those for which the function is known. This may be done by graphing $f(x)$ against x using the known values and reading off the value of $f(x)$ for the x required. Alternatively, various interpolation formulae (due to Newton, Bessel, etc.) exist by which, in principle, a polynomial,

$$y = ax^n + bx^{n-1} + \ldots + lx + m,$$

of degree n, which is one less than the number of known values of y, is fitted to the data so that the value of y corresponding to any x can then be determined.

Linear interpolation consists of taking two pairs of corresponding values (x_1, y_1) and (x_2, y_2) and using simple proportion to determine the value of y corresponding to any other x between x_1 and x_2. This is the simplest case of the general method in that a linear equation $y = ax + b$ has, in principle, been fitted to the data. See extrapolation.

interpole See compole.

interrupter See induction coil.

intersecting storage ring (ISR) See accelerator.

interstage coupling In a multistage *amplifier, employing several amplifying stages in *cascade, the system that effects the transfer from the output of one stage to the input of the next. Common types of *coupling are direct, resistive, capacitive, etc.

interstellar matter Syn. interstellar medium (ISM). The matter – both gas and dust – that occurs in the regions between the stars of our Galaxy and tends to be concentrated in the spiral arms. The gas is mainly hydrogen, gathered into immense clouds. Roughly spherical clouds of predominantly ionized hydrogen (*HII regions*) are known to exist, usually less than 200 parsecs across. There are also smaller, more diffuse, and relatively cool (about 70 K) clouds of neutral predominantly atomic hydrogen (*HI regions*). Between the HI regions there is more tenuous neutral hydrogen gas at temperatures of several thousand kelvin. In addition, very cool (10–20 K) very dense *molecular clouds* exist, consisting principally of molecular hydrogen but with a large variety of other molecules; these are major sites of star formation. These different regions have been detected and studied through their radio, X-ray, ultraviolet, and infrared emissions.

Interstellar dust is found throughout the interstellar region. It causes dimming and reddening of starlight by absorption and *scattering; the effect is greatest for observations directed towards the centre of the Galaxy, where the extent and density of the dust is greatest. The dust also produces partial *polarization of starlight. The dust consists of solid grains, mainly of carbon, between about 0.01 to 0.1 μm in size.

interstitial structures Crystalline arrangements in which small atoms occupy some of the interstices between large atoms, which themselves form a regular crystalline pattern. They are of considerable importance in connection with the structure of steels

interval

and other alloys as well as *semiconductors. *See* defect.

Table 1   **Intervals of Just Temperament**

unison	1/1
major second	9/8
major third	5/4
perfect fourth	4/3
perfect fifth	3/2
major sixth	5/3
major seventh	15/8
octave	2/1

interval 1. The relationship in frequency between two notes of a scale, the relationship being expressed either as a ratio or logarithmically. If the ratio form is used, it is conventionally written as a value greater than one. The most common units of interval expressed in a logarithmic form are the *millioctave* and the *cent*. An interval I in millioctaves between frequencies f_1, f_2 is:

$$I = (10^3/\log_{10}2)\log_{10}(f_1/f_2)$$
$$= 3322 \log_{10}(f_1/f_2).$$

In cents the interval is:

$$I = (1200/\log_{10}2)\log_{10}(f_1/f_2)$$
$$= 3986 \log_{10}(f_1/f_2).$$

(The octave is an interval of size $2/1 = 1000$ millioctaves = 1200 cents.) Although the first-named unit is associated with metric measurements, the latter is in more common use because of the convenient sizes of intervals in the scales of equal *temperament. The advantage of expressing intervals logarithmically is that they are combined by addition. Intervals expressed as ratios must be combined by multiplication.

Since musical scales have been developed gradually, and sometimes without any mathematical approach, the actual intervals used in musical history show a wide variation of size. Modern scales, being associated among other things with musical instruments of fixed design and operation, have intervals of designated sizes, although the ear is capable of accepting – and expects – variations from the standard values. Diatonic scales of *just temperament* contain seven intervals, of three different kinds. The largest intervals are tones: the larger whole tone (T_1) is an interval of $9/8 = 170$ millioctaves = 204 cents, and the lesser whole tone (T_2) is an interval of $10/9 = 152$ millioctaves = 182 cents. The smallest intervals are semitones (S) of size $16/15 = 93$ millioctaves = 112 cents. In the major scale the order of interval is:

$$T_1 \; T_2 \; S \; T_1 \; T_2 \; T_1 \; S.$$

Chromatic scales of just temperament contain twelve intervals, and are derived from the corresponding diatonic scales by dividing the whole tones into two semitones. Each larger whole tone is divided into a diatonic semitone and a larger chromatic semitone of size $135/128 = 77$ millioctaves = 92 cents. Similarly, each lesser whole tone is divided into a diatonic semitone and a smaller chromatic semitone of size $25/24 = 59$ millioctaves = 71 cents. In scales of *equal temperament* all the semitones have the same size, $2^{1/12} = 83$ millioctaves = 100 cents (exactly). Each whole tone equals two semitones. The sizes of various named intervals as ratios are given in Tables 1 and 2. Perfect intervals become diminished by reduction of the ratio by a chromatic semitone of appropriate size. Major intervals become minor by reduction of the ratio by a chromatic semitone of appropriate size; they become diminished by reduction of ratio by 1125/1024 (i.e. the sum of both chromatic semitones). All intervals become augmented by an increase of the ratio by a chromatic semitone of appropriate size. Perfect intervals become diminished by reduction of the ratio by $2^{1/12}$. Major intervals become minor by reduction of the ratio by $2^{1/12}$; they become diminished by reduction of ratio by $2^{2/12}$.

2. The separation of two events in a *four-dimensional continuum.

Table 2   **Intervals of Equal Temperament**

unison	2^0	=	1.00
major second	$2^{2/12}$	=	1.12
major third	$2^{4/12}$	=	1.26
perfect fourth	$2^{5/12}$	=	1.33
perfect fifth	$2^{7/12}$	=	1.50
major sixth	$2^{9/12}$	=	1.68
major seventh	$2^{11/12}$	=	1.89
octave	$2^{12/12}$	=	2.00

interval of Sturm *See* astigmatic interval.

intrinsic conductivity The *conductivity of a *semiconductor that is associated with the semiconductor itself and is not contributed by impurities. At any given temperature equal numbers of charge carriers – electrons and holes – are thermally generated, and it is these that give rise to the intrinsic conductivity.

intrinsic equation of a curve The relation between the length of the arc of a curve as measured from a reference point to any other point on the curve and the inclination of the tangent at the latter point.

intrinsic mobility The mobility of *carriers in an *intrinsic semiconductor. Electrons are approximately three times as mobile as *holes.

intrinsic pressure A term in the *equation of state of a liquid resulting from intermolecular attractions. It has the form a/V^2 in the van der Waals equation of state.

intrinsic semiconductor *Syn.* i-type semiconductor. A pure *semiconductor in which the *electron and *hole densities are equal under conditions of thermal equilibrium. In practice absolute purity is unattainable and the term is applied to nearly pure materials.

Invar An alloy of iron with 36% of nickel, which has a very small coefficient of thermal expansion. It is used in instruments, pendulums, and accurate standards of length.

invariable line *See* Poinsot motion.

invariable plane *See* Poinsot motion.

invariant (optical) A quantity that remains constant during an optical action. *Snell's law,

$$n \sin i = n' \sin i',$$

expresses an invariant relation. For the small angles encountered during paraxial refractions several invariant relations hold, many developed from the refraction law, $ni = n'i'$.

inverse Compton effect *See* Compton effect.

inverse gain The *gain of a bipolar junction *transistor when it is connected in reverse, i.e. with the *emitter acting as the *collector, and the collector as the emitter. It is usually less than the gain normally observed, as the emitter has a higher *doping level than the collector and therefore a higher *injection efficiency into the *base than the collector.

inverse-speed motor *See* series-characteristic motor.

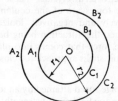

Inverse square law

inverse-square law A law relating the intensity of an effect to the reciprocal of the square of the distance from the cause. The law of *gravitation is an inverse-square law, as is *Coulomb's law relating the force associated with static electric charges. Other important cases include:

(1) (sound) Waves carry both momentum and energy, so if we assume a point origin of sound and no dissipation of energy due to friction of the medium, we can prove that the intensity of sound falls off as the inverse square of the distance from the source.

If O (*see* diagram) is the point source, and the spherical wave $A_1 B_1 C_1$ of the emitted sound has covered a sphere of radius r_1, a receiver at A_1 will gain energy I_1 per unit area and therefore the total energy over all the surface of this spherical wave equals $4\pi r_1{}^2 I_1$. Similarly at a later instant the spherical wave will cover a large sphere $A_2 B_2 C_2$ with radius r_2; another receiver at A_2 will gain energy I_2 per unit area and the total energy over all this sphere equals $4\pi r_2{}^2 I_2$.

By the principle of conservation of energy

$$4\pi r_1{}^2 I_1 = 4\pi r_2{}^2 I_2 \text{ or}$$
$$I_1 / I_2 = r_2{}^2 / r_1{}^2.$$

It follows that the energy crossing any unit area at right angles to the direction of propagation must vary inversely as the square of the distance of the area from the source. Since energy is proportional to the square of the amplitude, the amplitude is therefore inversely proportional to the first power of the distance from the source.

On account of absorption due to viscosity of the medium, the intensity of sound spreading in free air falls more rapidly than the inverse-square law would indicate, especially if the sound is of large amplitude where serious loss due to heat conduction and radiation will occur. (*See* absorption of sound.)

(2) (light) Light radiation from a point source produces an intensity of illumination of a screen normal to the light direction varying inversely as the square of distance.

inverse thixotropy *See* dilatancy; thixotropy.

inverse time lag *See* time lag.

inverse Zeeman effect *See* Zeeman effects.

inversion 1. A reversal in the usual direction of a process or variation, as in the change of density of water at 4 °C or, in meteorology, an increase in temperature with altitude as opposed to the normal decrease. The term is similarly employed in connection with the Joule–Thomson porous-plug experiment.

2. The production of a layer of opposite type in the surface of a *semiconductor, usually under the influence of an applied electric field. The presence of mobile minority carriers is necessary for inversion to take place, otherwise a *depletion layer forms. The phenomenon is utilized in the formation of the channel in an insulated-gate *field-effect transistor. A spontaneous inversion layer is often found in the surface of p-type semiconductor material in contact with an insulating layer even when no external electric field is applied.

3. (mathematics) The process of inversion with respect to a circle (or a sphere) consists in replacing each point of a diagram by another point on the same radius such that the product of the distances of the two points from the centre is equal to the square of the radius. Some geometrical results yield fresh information when the whole figure is inverted.

inversion axis A symmetry axis involves rotation through an angle of $2\pi/n$ ($n = 1, 2, 3, 4, 6$) followed by reflection across a plane normal to the axis.

inversion temperature 1. If one junction of a *thermocouple is kept at a constant low temperature, the temperature to which the other junction must be raised in order that the thermoelectric e.m.f. in the whole circuit shall be zero is known as the inversion temperature. For the same thermocouple, the sum of the temperatures of the two junctions is a constant, at the inversion point. Thus, for a copper-iron junction with the colder junction at 0 °C the inversion temperature is 550 °C; with the colder junction at 100 °C, the inversion temperature is 450 °C. If the inversion temperature is exceeded, the direction of the e.m.f. in the thermocouple is reversed.

2. The temperature at which the *Joule–Kelvin effect of a gas changes sign, given by:

$$T(\partial V/\partial T)_p = V.$$

The initial temperature of a gas must be below this temperature if the gas is to be liquefied by expansion through a porous plug or throttle.

inverter 1. Any device that converts d.c. into a.c., particularly a rotating machine designed for the purpose.
   2. *Linear inverter.* An amplifier that inverts the polarity of a signal, i.e. introduces a 180° phase shift.
   3. *Digital inverter.* A *logic circuit whose output is low when the input is high and vice versa.

I/O *See* input/output.

Ioffe bars Heavy current-carrying bars, named after the Russian physicist M. S. Ioffe, used in experimental fusion devices to increase *plasma stability. *See* illustration under fusion reactor.

ion An electrically charged atom, molecule, or group of atoms or molecules. A negative ion (or *anion) contains more electrons than are necessary for the atom or group to be neutral, a positive ion (or *cation) contains less. *See* electrolytic ion; gaseous ions.

ion-beam analysis The analysis of materials on microscopic and macroscopic scales using a beam of positively charged light ions – protons, deuterons, or alpha particles. The ions generally have energies of a few MeV, the beams being produced in particle accelerators. Various analytical techniques have been developed, including *Rutherford back scattering* (RBS), *nuclear reactions analysis* (NRA), and *particle-induced X-ray emission* (PIXE). The ions are allowed to impinge on the surface of the sample being analysed, and the radiations subsequently emitted from the sample – particles from RBS and NRA, gamma rays from NRA, and X-rays from PIXE – are detected and processed using conventional devices. Information is obtained on where sample atoms are sited within the crystal lattice.

ion burn A defect of the fluorescent coating on cathode-ray screens that may appear as a rash of darker areas when they have been subjected for a time to ion bombardment.

ion engine A proposed form of propulsion engine for use in space vehicles. A stream of ions is accelerated to a very high speed by electrostatic propulsion and expelled from the vehicle. By *Newton's third law of motion the ions exert an equal and opposite force on the vehicle.

ion exchange (IX) A reversible process in which a liquid runs through or over a suitable solid and in so doing exchanges cations or anions. The solid employed is often an ion-exchange resin (a synthetic or natural polymer), zeolite (a synthetic or natural alumino-silicate of sodium, calcium, etc.), or a specially prepared carbonaceous mineral. The process is employed for water softening, desalination of brine, isotope separation, and the extraction of metals from their ores.

ionic atmosphere In an electrolyte, the accumulation of anions around cations, and vice versa. When an electric field is applied, the ions migrate in the reverse direction to their ionic atmosphere, and the symmetry of the atmosphere with respect to the ion is disturbed in such a manner that the ion is retarded. If, however, rapidly alternating (or short-duration direct) current is applied, there is insufficient time to disturb the symmetry and a higher value of conductivity is found. Further, as the rate of build-up of asymmetry is dependent on the field strength the conductivity depends, under these conditions, on the applied voltage gradient. *See* Wien effect.

ionic bond *See* electrovalent bond.

ionic conduction In a *semiconductor, the movement of charges within the semiconductor due to the displacement of ions within the crystal lattice. An external contribution of energy is required to maintain such movement.

ionic crystals *Syn.* electrovalent crystals. Those in which the interatomic forces are of the coulomb type, in which positively and negatively ionized atoms attract each other, the attraction being balanced by the repulsive force that comes into play when the outer electronic shells approach too closely.

ionic mobility The average speed attained by an ion when acted on by an electric field of unit strength. It is usually measured in m/s per V/m, i.e. $m^2 V^{-1} s^{-1}$.

ionic semiconductor A solid in which the electrical conductivity due to the flow of *ions predominates over that due to the movement of *electrons and *holes.

ion implantation A technique used in the manufacture of *integrated circuits and *transistors in which the *semiconductor material is bombarded by high-velocity ions under controlled conditions. The ions penetrate the surface of the semiconductor and can be made to assume lattice positions within the semiconductor crystal. The technique may be used in conjunction with diffusion or as an alternative to it.

ionization The process of forming *ions. Ionization occurs spontaneously when an electrolyte dissolves in a suitable solvent. Ionization in gases requires the action of some *ionizing radiation, e.g. X-rays, α-, β-, or γ-rays. *See* conduction in gases.

ionization chamber A chamber containing two oppositely charged electrodes so arranged that when the gas in the chamber is ionized, e.g. by X-rays, the ions formed are drawn to the electrodes, creating an ionization current. This is used as a measure of the intensity of the *ionizing radiation. (*Compare* cloud chamber.)
   The sensitivity of an ionization chamber is dependent on the mass of gas enclosed in the sensitive volume. Extremely large ionization chambers have been developed for measuring *background radia-

tion levels, whereas extremely small chambers are used for calibrating high-output beams of X-rays or electrons. The most versatile type of ionization chamber is the *Farmer substandard dosemeter*. This is a small ionization chamber with an *air wall* (i.e. the wall of the chamber is made with a material that has the same effective atomic number as air). The sensitivity is checked periodically against a national standard, and the chamber is used for routine calibrations and measurements.

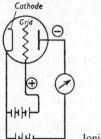

Ionization gauge

ionization gauge A vacuum pressure gauge consisting basically of a three-electrode thermionic valve and used for measuring small gas pressures of the order of micropascals. The tube is fused to the gas system to be measured, and is connected up as shown. Electrons are accelerated between the cathode and grid, but cannot reach the plate since it is at negative potential. Some electrons, however, pass through the grid and collide with gas molecules and ionize them, leaving them positively charged. The positively charged gas molecules then go to the plate and the plate current produced provides a measure of the number of molecules present.

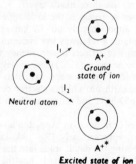

Ionization of lithium atom

ionization potential Symbol: I. The minimum energy necessary to remove an electron from a given atom or molecule to infinity. It is thus the least energy that causes an ionization:
$$A \rightarrow A^+ + e^-,$$
where the *ion and the electron are far enough apart for their electrostatic interaction to be negligible and no extra kinetic energy is produced by the ionization. The electron removed is that in the outermost

orbit, i.e. the least strongly bound electron. It is also possible to consider removal of electrons from inner orbits in which their *binding energy is greater. The minimum energy required to remove the second least strongly bound electron from a neutral atom is called its *second ionization potential*. This process is shown in the diagram for ionization of the lithium atom. The first ionization potential (I_1) corresponds to formation of the *ground state of the singly charged ion. The second ionization potential (I_2) corresponds to formation of the ion in its first *excited state. Similarly, there are third, fourth, fifth, etc., ionization potentials for atoms or molecules with larger numbers of electrons. These higher ionization potentials should not be confused with those used in some branches of chemistry where, for example, the second ionization potential is the energy required to form the doubly charged positive ion:
$$A \rightarrow A^{2+} + 2e.$$

Ionization potentials were originally defined as the minimum potential through which an electron would have to fall in order to ionize the atom. In this sense they were measured in volts. In the present definition the ionization potential is an energy and is conveniently measured in *electronvolts. Ionization potentials are determined by Rydberg spectroscopy (*see* Rydberg spectrum), *appearance potential measurements, and *photoelectron spectroscopy. *Compare* electron affinity.

ionizing agent Any physical agent that produces ionization.

ionizing radiation Any radiation that causes *ionization or *excitation of the medium through which it passes. It may consist of streams of energetic charged particles, such as electrons, protons, alpha particles, etc., or energetic ultraviolet, X-rays, or gamma-rays. A large number of ions, secondary electrons, and excited molecules are produced in the medium by particles; electromagnetic radiation produces a lower number by processes such as the *photoelectric effect, *Compton effect, and *pair production.

Ionizing radiation occurs naturally as *cosmic rays and the *solar wind, and is emitted by *radionuclides. It is produced artificially by X-ray machines and particle *accelerators. Its effects can be observed visually by using such apparatus as the *bubble chamber or *spark chamber, or by examination of the tracks made in photographic emulsion. More quantitative measurements are made with *counters.

This radiation can cause extensive damage in molecules. Energetic charged particles have either a *direct* or an *indirect effect* on a medium. A direct effect occurs when energy is transferred directly to the atom or molecule, usually in a solid or gaseous state. With a large molecule, the secondary electrons released by ionization at many different sites, can migrate along molecular bonds to some common reactive site, thus accentuating the radiation damage. Direct effects can take place in solutions of high concentration but in dilute solutions the energy is transferred to the more prolific solvent molecules and the chemical effects are brought about indirectly by the interaction, with solute molecules, of the highly reactive transient species resulting from reaction of the radiation with the solvent. The ionization and excitation of molecules along the track of a charged particle occur in clusters or *spurs*. In a single spur there is one initial ionization (or excitation) product, plus secondary ionizations and excitations produced by the energetic electrons released from the ion. The resulting transient species will, in general, be the same in a particular medium regardless of the type of energy of the radiation. It is the rate of energy loss that depends on the form of radiation and its energy.

Linear energy transfer (LET) is the average energy, dE, locally imparted to the medium by a charged particle, of specified energy, in traversing a small distance, dl: (LET) $= dE/dl$. A beam of energetic electrons has a low LET on account of the small electronic mass and the spurs are therefore widely separated initially.

As the electron is slowed down, its LET increases and the spurs eventually overlap. For high-LET radiation, such as alpha particles, the spurs cannot be separated into individual ionization events and a column of ions and excited species are formed along the track of the particle. If low-LET radiation is delivered at a very high dose rate, the spurs from different particles will overlap.

In biological tissue, the indirect effect of ionizing radiation is extremely important, since there is an 80% water content. In dilute aqueous solutions, the ejection of the secondary electron from a water molecule leaves the resulting ion in a highly unstable state and the most probable reaction is the formation of a hydroxyl *free radical, $\cdot$OH (the dot representing an unpaired electron), within 10^{-14} seconds. The excited water molecule may also dissociate into a hydroxyl and a hydrogen free radical:

$$H_2O \xrightarrow{\text{ionization}} H_2O^+ + e^-$$
$$\searrow H_2O^*$$
$$H_2O^+ + H_2O \rightarrow \cdot OH + H_3O^+$$
$$H_2O^* \rightarrow \cdot OH + \cdot H$$

The secondary electron rapidly loses energy by further ionization and excitation, and becomes a *hydrated electron, e_{aq}^-, trapped by the water molecules. If the solution is acidic, the hydrated electron combines with a hydrogen ion to form a hydrogen radical, $\cdot$H:

$$e_{aq}^- + H^+ \rightarrow \cdot H$$

These transient species are all exceptionally reactive because of their unpaired electron, the $\cdot$OH radical being a powerful oxidizing agent and the e_{aq}^- and $\cdot$H radical having an equally strong reductive capacity.

Radical-radical reactions proceed in the spur followed by diffusion of these products and the original transients, if the LET is low enough, into the bulk of the liquid. If the spurs overlap, the considerably higher concentration of transients results in the

majority reacting together and much less diffusion will occur. The subsequent attack on solute molecules by the free radicals and the e_{aq}^-, usually within a microsecond, results in the solute molecule becoming a free radical or radical ion. Although these are not as reactive as the original radicals, they will either combine with one another, decompose, or remain as a fairly stable free radical.

Radicals resulting from irradiation of biological tissue attack proteins, nucleic acids, and other vital molecules occurring in the tissue cells. The sequence of reactions following this attack is extremely complicated but in some cases the molecular structure is altered and the function of the molecule is interfered with.

The uses of this radiation include both diagnosis and therapy in medicine, sterilization of perishable food, cloth, etc. *See also* dosimetry.

ionosphere A spherical shell of ionized air surrounding the earth, extending from about 50 km (the top of the *stratosphere) to over 1000 km. Nitrogen and oxygen molecules are split into atoms, ions, and free electrons by *ionizing radiation from space, especially ultraviolet radiation and X-rays from the sun.

Following the first radio transmission, it was postulated by Heaviside and Kennelly in 1902 that transmission was achieved by reflection of radio waves from a layer of charged particles in the atmosphere. The particles were detected by Appleton in 1924. Long-distance radio transmission between any two points on the earth's surface is still obtained sometimes by successive reflections from the ionosphere.

The ionosphere can be divided into distinct layers or regions whose degree of ionization varies with time of day, season, latitude, and state of solar activity. There are three major layers.

The *D-layer* or *region* is the lowest ionospheric layer, lying approximately 60–90 km above the earth; it contains a relatively low concentration of free electrons and reflects low-frequency waves.

The *E-layer* or *region* (*Syn*. Heaviside layer; Kennelly–Heaviside layer) lies approximately 90–150 km above the earth, has a higher electron concentration than the D-layer and reflects medium-frequency waves.

The *F-layer* or *region* (*Syn*. Appleton layer) is the highest layer, approximately 150–1000 km above the earth. During the day it splits into the F_1-*layer* (lower) and F_2-*layer* (higher). It has the highest fractional concentration of free electrons and is the most useful region for long-range radio transmissions at frequencies up to about 30 GHz.

At night the electron concentrations in the D- and E-layers fall owing to the absence of sunlight and the consequent recombination of electrons and ions. In the higher F-layer the density is lower and collisions between electrons and ions are less frequent. The F-layer can therefore be used for radio transmission at all times.

Radio waves deflected by the electrically conducting ionospheric layers are called *ionospheric waves* (or *sky waves*). Some wavelengths, lying in the *radio window between about a millimetre and 30 m, are not reflected but transmitted through the ionosphere; long-distance television, broadcast at high frequencies, must therefore be reflected by means of artificial *satellites, usually in *geostationary orbits. *Radio astronomy is restricted to using these transmitted frequencies.

ionospheric wave *Syn.* sky wave. *See* ionosphere.

ion pair A pair of positively and negatively charged ions generated when an electron transfers from one atom or molecule to another.

ion pump A type of vacuum pump in which the gas is ionized by a beam of electrons and the positive ions attracted to a cathode and thereby removed from the system. It is only operated at very low pressures (less than about 10^{-6} Pa) and the gas is not completely removed from the system but simply trapped on the cathode. The pump thus saturates after a certain time. The capacity of ion pumps can be increased by continuously evaporating a film of metal onto the cathode during its operation. One way of doing this is to maintain a discharge in the gas by the application of a high electric field. This causes ionization. The fresh film of metal is produced by *sputtering of the cathode. Pumps of this type depend for their action on a combination of ion-pumping and getter-ing (*see* getter). *See also* pumps, vacuum.

ion source A device that provides ions, especially for use in a particle *accelerator. A minute jet of gas, such as hydrogen or helium, is ionized by bombard-ment with an electron beam and the resulting pro-tons, alpha particles, etc., are ejected into the accel-erator.

ion trap In a *cathode-ray tube, a device to prevent the *ions present in the tube impinging on the phosphor coating of the screen and so causing blem-ishes.

IPTS Abbreviation for *International Practical Temperature Scale.

IR Abbreviation for *infrared.

iridescence A display of colours on a surface, com-monly as a result of the *interference of light of the various wavelengths reflected from superficial layers in the surface.

iris The structure lying in front of the crystalline lens of the *eye, pigmented behind and consisting largely of muscular tissue, that controls the circular opening (pupil) through which light enters the eye. The variable diaphragm used as an aperture stop in the photographic lens is called the *iris diaphragm.*

iris diaphragm *See* iris.

I²R loss *Syn.* copper loss. The power loss due to the flow of electric current in the windings of a machine or transformer. It is calculated by multiplying the square of the current by the resistance of the wind-ing.

iron loss *See* core loss.

irradiance Symbol: E_e, E. The *radiant flux of electromagnetic radiation, Φ_e, incident on a given surface per unit area. At a point on the surface, of area dS, the irradiance is given by:
$$\Phi_e = \int E_e \, \mathrm{d}S.$$
It is measured in joules per square metre. *Compare* illuminance.

irradiation The exposure of a body or substance to *ionizing radiation, either electromagnetic (X-rays and gamma rays) or corpuscular (alpha particles, electrons, etc.).

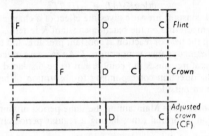
Irrationality of dispersion

irrationality of dispersion When prismatic spectra have been produced by different glasses so that lines at the extreme red and blue ends are in coincidence, the intermediate lines will not be in coincidence – the spectra are not geometrically similar (*see* dia-gram). In consequence, if achromatism is effected for two lines of the spectrum, there will be residual chromatic aberration. *See* secondary spectrum.

irreversible change *See* reversible change.

irreversible colloid *See* reversible colloids.

irrotational motion Of a fluid. Motion such that the equation of relative motion of any element of a finite portion of the fluid does not include rotational terms. The mathematical condition of irrotational motion is:
$$\text{curl } V = \nabla \times V = 0,$$
where V is the vector velocity of an element of the fluid. When the motion is irrotational, there exists a *velocity potential and conversely when a velocity potential exists, the motion is irrotational. If once irrotational, then the motion of a fluid under con-servative forces is always irrotational. Any motion of a fluid, such that the component angular veloci-ties of rotation do not vanish together, is called *rotational* and a velocity potential does not then exist.

isenthalpic process A process that takes place with-out any change of *enthalpy, i.e. so that the total heat energy (internal plus external) remains con-stant.

isentropic process A process that occurs with no change in entropy. *Compare* adiabatic process.

isobar 1. A line on a map passing through places of the same atmospheric pressure.

2. One of two or more nuclides that have the same *mass number but different *atomic numbers. Isobars are different elements and have different properties. Examples are $^{234}_{90}$Th and $^{234}_{91}$Pa.

isobaric Taking place without change of pressure.

isochore A curve representing two variables involved in an isometric (constant volume) thermodynamic change, e.g. pressure/temperature, temperature/entropy.

isochore of reaction The van't Hoff equation
$$\partial(\log K)/\partial T = -H/RT^2$$
at constant pressure gives the effect of a change of temperature on the reaction constant K in terms of H, the heat of reaction at constant pressure, and the thermodynamic temperature T. An increase in temperature causes the equilibrium to be displaced in the direction corresponding to absorption of heat from outside.

isochronous Maintaining the same period of vibration or orbital time; having a regular periodicity.

isoclinal A curve drawn in such a manner that all places on the curve have the same magnetic *dip. See aclinic line.

isodiapheres Two or more *nuclides that have the same difference in the number of neutrons and protons. For example, the commonest isotope of radium ^{226}Ra possesses 88 protons and 138 neutrons; the difference in the number of *nucleons is therefore 50. ^{226}Ra decays naturally into an isotope of radon, ^{222}Rn, with 86 protons and 136 neutrons, the nucleon difference again being 50. ^{226}Ra and ^{222}Rn are isodiapheres.

isodisperse Dispersible in solutions having the same *pH value.

isodispersion A sol of a natural substance in which the colloidal particles are all of the same size.

isodynamic A curve drawn in such a manner that the total magnetic field strength of the earth's magnetic field is identical at all points on the curve.

isoelectric point A critical condition of a colloidal suspension in an electrolytic medium for which the *cataphoresis of the suspended particles is zero. The condition is reached as the concentration of the electrolyte is increased; if the concentration is further increased, the cataphoresis may reverse sign. A sol is usually most liable to coagulate at the isoelectric point when electrical repulsion between particles is reduced.

isogam A line on a map joining points at which the acceleration of free fall is constant. Used in geophysical prospecting.

isogonal A curve drawn in such a manner that the magnetic *declination (or variation of the compass) is the same at all places on the curve. Maps showing the isogonal curves over the surface of the earth are important in navigation. See agonic line.

isogyres Black bands or brushes to be seen on *interference figures.

isolating The act of disconnecting a circuit or piece of apparatus from an electric supply system: it usually implies the opening of a circuit which, at the time, carries no current. See isolating switch.

isolating switch A switch for making and breaking, nonautomatically, an electric circuit when it is not on load. See isolating.

isolating transformer A *transformer used to isolate any circuit or device from its power supply.

isolation diode A *diode used in a circuit to allow signals to pass in one direction but not in the other, thus preventing damage from surges in the reverse direction.

isolator A device that allows microwave radiation to pass in one direction, while absorbing it in the reverse direction.

isomagnetic lines Lines joining points at which the magnetic field is equal. They need not be concerned with terrestrial magnetism.

isomers 1. Compounds of the same relative molecular mass and percentage composition differing in some or all of their chemical and physical properties.

2. See nuclear isomerism.

isometric change A change in a gas that takes place at constant volume.

isomorphism Similarity of crystalline form or of structure in substances that are chemically related.

iso-osmotic Having equal osmotic pressures. See isotonic.

isophote A line on a diagram joining points of equal flux density or intensity.

isoporic charts Charts of equal annual rate of change for magnetic declination, inclination, and the strengths of the various components of the earth's magnetic field. See geomagnetism.

	I	I_3		I	I_3		I	I_3
n	$\frac{1}{2}$	$-\frac{1}{2}$	π^-	1	-1	Σ^-	1	-1
p	$\frac{1}{2}$	$\frac{1}{2}$	π^0	1	0	Σ^0	1	0
			π^+	1	1	Σ^+	1	1

isospin Syn. isotopic spin; i-spin. Symbol: I. A *quantum number associated with *elementary particles. It is found experimentally that the *strong interaction between two protons and between two neutrons is the same. This suggests that the proton and neutron may be regarded as two states of the same "particle" as far as strong interactions are concerned. Similarly, the three pions, π^+, π^0, and π^-, may be regarded as three states of a single "particle" when only strong interactions are considered. When *electromagnetic interactions are taken into account, there will be differences between interactions involving π^+ and π^0 because only the π^+ has a charge. However, as electromagnetic interactions are about 100 times weaker than strong interactions they can often be ignored.

*Hadrons with very similar masses and differing only in their charge can thus be combined into groups (called *multiplets*) that can be regarded as different states of the same object: the mathematical treatment of this characteristic is identical to that for *spin (angular momentum). It is found that to each hadron two quantum numbers I and I_3 may be assigned. The quantum number I is the isospin. It can take values

$$0, \tfrac{1}{2}, 1, \tfrac{3}{2}, 2, \ldots$$

and is the same for all particles in a multiplet. The *isospin quantum number* I_3 can have values

$$-I, -I + 1, \ldots, I - 1, I$$

and labels the particles in a multiplet. Examples of isospin multiplets are the nucleon doublet and the pion and sigma triplets (*see* table). In general, the charge Q of any elementary particle is related to its *hypercharge Y and the quantum number I_3 by the equation:

$$Q = I_3 + \tfrac{1}{2}Y.$$

For systems of strongly interacting particles a total isospin may be defined. Two particles having isospin quantum numbers (I, I_3) and (I', I'_3) have a total I_3 quantum number given by $I_3^{TOT} = I_3 + I'_3$. The quantum number I^{TOT} of the combined system can have a number of different values:

$$I^{TOT} = I + I' \text{ or } I + I' - 1,$$

down to the larger of $|I_3^{TOT}|$ and $|I - I'|$. Strong interactions only depend on the total quantum number I^{TOT} of the system and are independent of I_3^{TOT}. Both I^{TOT} and I_3^{TOT} are conserved in strong interactions. For electromagnetic interactions there is a dependence on the charge of the particles and I^{TOT} is no longer conserved although I_3^{TOT} is.

isotherm (or **isothermal**) A line joining all points on a graph that correspond to the same temperature, as in *Andrews's curves or the *Langmuir isotherm.

isothermal 1. Occurring at constant temperature.
2. *See* isotherm.

isothermal process A process that occurs at a constant temperature. For example, if a gas is expanded in a cylinder by a piston, its temperature can be kept constant by supplying heat from a thermostatically controlled source during the expansion. In such a process the wall separating the gas from the source has to allow them to remain in thermal equilibrium with each other. It is then called a *diathermic* wall. *Compare* adiabatic process.

isotones Nuclides having the same *neutron number.

isotonic Solutions apparently exhibiting equal osmotic pressures with respect to a particular semipermeable membrane are isotonic. This is not the same

as iso-osmotic since a particular membrane may allow the passage of dissolved substances to some extent.

isotopes Two or more *nuclides that have an identical nuclear charge (i.e. the same atomic number) but differ in nuclear mass; the nuclides are said to be *isotopic*. Such substances have almost identical chemical properties but differing physical properties, and each is said to be an isotope of the element of given atomic number. The difference in mass is accounted for by the differing number of *neutrons in the nucleus. For most elements several different naturally occurring isotopes have been discovered, and artificial radioactive isotopes can generally be prepared by bombardment of suitable materials by high-speed particles, or by slow neutrons.

In the case of hydrogen three isotopes are known. Ordinary hydrogen (^1H), with no neutrons, makes up 99.985% of the total and deuterium (^2H), with one neutron, makes up the remaining 0.015%. The artificial isotope tritium (^3H), with two neutrons, has a *half-life of 12.26 years and decays through *beta decay. The difference in physical properties between isotopes is nicely illustrated by comparing two forms of water (see table).

	^1H_2O	^2H_2O	
density (25° C)	997·1	1104·7	kg/m^3
relative permittivity	81·5	80·7	
surface tension	0·0728	0·0678	N/m
viscosity	1·310	1·685	mPa s
melting point	0	3·80	°C
boiling point	100	101·42	°C
refractive index (D line)	1·3330	1·3283	

isotope separation Methods of separating isotopes use either physical or chemical processes taking place at a rate that depends on the mass of the atoms or molecules. The most important methods are:

(1) *Gaseous diffusion*. This method is based on the fact that light molecules diffuse through a porous barrier faster than heavier molecules (*see* Graham's law of diffusion). It can only be applied to gaseous or volatile substances and has been used to separate ^{235}U from ^{238}U by using the volatile hexafluoride of uranium (UF_6).

(2) *Thermal diffusion*. This depends on the fact that if a temperature gradient is maintained through a gas, the lighter atoms preferentially diffuse into the warmer region. A *Clusius column* is a long vertical column (about 30 m high) with a radial temperature gradient produced by an electrically heated wire along its axis. The lighter isotope tends to concentrate around the wire and the heavier isotope concentrates near the cool walls of the column. Convection currents carry the lighter isotope to the top of the tube.

(3) *Centrifuge separation*. This is a simple method depending on the fact that the centrifugal force on

atoms is proportional to their mass. Thus, in a centrifuge radial concentration gradient is produced and enriched material can be removed from a point near the axis. The method is applied to gaseous and liquid samples.

(4) *Distillation*. Isotopes of different mass evaporate from a liquid at slightly different weights. At equilibrium the relative concentrations of isotopes in a liquid is different from those in its vapour and the isotopes can be separated by distillation. Separation can also be effected by pumping away the vapour so that molecules leaving the liquid do not return.

(5) *Electrolytic separation*. If a compound is subjected to electrolysis (*see* electrolytic dissociation), the rate at which an ion is discharged at an electrode depends on its mass. For example, in the electrolysis of water hydrogen ions are discharged at the cathode: $H^+ + e \rightarrow H$. The hydrogen atoms form molecules and hydrogen gas is evolved. This reaction is faster than the corresponding reaction for deuterium ions and prolonged electrolysis of water results in an increase in its concentration of heavy water.

Another electrolytic method of separation depends on the fact that the mobility of isotopic ions in a liquid under the influence of an electric field depends on their mass.

(6) *Chemical exchange*. This method depends on exchange reactions of chemical compounds. For example, in the reaction

$$^{13}CO + {}^{12}CO_2 \rightarrow {}^{12}CO + {}^{13}CO_2$$

the molecules exchange carbon atoms. There are slight differences in reactivity between ^{13}CO and ^{12}CO and $^{13}CO_2$ and $^{12}CO_2$. Consequently, the equilibrium constant of the reaction differs slightly from unity ($K = 1.086$).

(7) *Electromagnetic separation*. In this technique the material is ionized and the beam of ions is deflected by a magnetic field. The amount of deflection depends on the mass of the ion; the heavier isotopic ions are deflected less than the lighter ones and each type of action is collected at a different position. The device is, in fact, a large *mass spectrometer. It can only be used for small samples.

Most of the methods described have a low efficiency and only give a small enrichment of one isotopic species over the other. In practice a large number of stages are used in series. This arrangement is called a *cascade*.

isotopic number *Syn.* neutron excess. The difference between the number of neutrons and the number of protons in a nuclide.

isotropic Possessing a property or properties, such as *permittivity, *susceptibility, or *elastic constants, that do not vary with direction.

ISR Abbreviation for intersecting storage ring. *See* accelerator.

IT Abbreviation for *information technology.

iterative impedance Of a *quadripole. The impedance presented by the quadripole at one pair of terminals when the other pair is connected to an impedance of the same value. In general, a quadripole has two iterative impedances, one for each pair of terminals. Sometimes the two are equal and in this case their common value is called the *characteristic impedance* of the network. *Compare* image impedance.

i-type semiconductor *See* intrinsic semiconductor.

IX Abbreviation for *ion exchange.

J

Jaeger and Diesselhorst's determination of thermal conductivity of metals (1900) A cylindrical rod of the metal under experiment was heated to 18 °C and to 100 °C by means of an electric current passed through it until the steady state was reached. Kohlrausch had shown that for such a case:

$$\int_{T'}^{T} \frac{K}{\sigma} \, dT + \tfrac{1}{2}v^2 + \alpha v + \beta = 0,$$

where T and v are respectively the temperature and electric potential in a plane section at right angles to the length of the rod and α and β are integration constants. These constants were determined by measurements of the temperatures and potentials at two points on the bar, and by measuring the same quantities at another point the ratio K/σ of the thermal to the electrical conductivity at that temperature could be calculated. They used bars of metal 20 or 30 cm long and 1 or 2 cm in diameter with the ends in constant-temperature baths. In order to obtain a difference in temperature between the middle and the ends of the bars of a few degrees, a current of about 350 A had to be used. Allowance was made for the heat lost from the surface by surrounding the bar by a double-walled copper jacket through which water or steam was circulated.

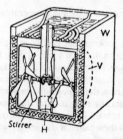

Mechanical equivalent of heat (Jaeger and Steinwehr)

Jaeger and Steinwehr's determination of J (1921) The value of the *mechanical equivalent of heat was found electrically by an accurate calorimetric method in which a large mass (50 kg) of water was used in a cylindrical vessel V, surrounded by an air jacket, and a water bath W maintained at a constant temperature by water circulating through copper pipes. The potential difference across the heater H was measured by a potentiometer and the current through H deduced from the potential drop across a standard 0.1 ohm coil arranged in series. The rise in temperature was made small (1.4 °C) so that Newton's law of cooling could be assumed in making the cooling correction and was measured by a platinum resistance thermometer. The value obtained for J was 4.186×10^7 ergs per 15° calorie, accurate to within a few parts in ten thousand.

Jamin, Jules C. (1818–1886) French physicist. As a result of a suggestion by Brewster he invented the Jamin interferometer or refractometer.

Jamin refractometer *Syn.* Jamin interferometer. An instrument for measuring the refractive index of gases. By reflection at the front and back surfaces of a thick plate inclined at 45° to the incident light, two parallel beams pass towards a similar glass plate placed parallel to the first, which causes recombination of the beams and interference. Two similar tubes are placed one in each parallel beam and the displacement of interference fringes is noted as gas is introduced in one tube. *See* interference of light; refractive-index measurement.

jamming Deliberate *interference in communications and radar caused by an undesired signal that is so strong that the desired signal cannot be understood.

jansky Symbol: Jy. A unit of *radiant flux density that is used in astronomy throughout the spectral range but especially for radio and infrared measurements. It refers to a particular frequency. One jansky is equal to 10^{-26} W m^{-2} Hz^{-1}.

Jansky, Karl (1905–1950) Amer. radio engineer who, by discovering in 1931 that some static radio interference is due to radio waves from space, was the instigator of *radio astronomy.

Jansky noise High-frequency static disturbance of cosmic origin. *See* radio noise.

Janssen, Pierre Jules César (1824–1907) French astronomer who worked on spectroscopy. He discovered (1868) lines in the solar spectrum that Normal Lockyer assigned to a new element, helium. William Ramsey discovered helium on earth in 1895, identifying it by the lines in its spectrum.

Jeans, Sir James Hopwood (1877–1946) Brit. physicist, mathematician, and astronomer who made important contributions to the kinetic theory of gases and to the stability of aggregations of matter. He advocated a theory of the origin of the solar system in which a star passing near our sun raised a tide that developed into a cigar-shaped filament from which the planets condensed. He also gave a theory of formation of spiral galaxies. He was jointly responsible with Rayleigh for a radiation formula (*see* Rayleigh–Jeans formula). Jeans made an important report on the *quantum theory in its early days

(1914); he is also noted for his treatises on the mathematical theory of electricity and magnetism and on the dynamical theory of gases and for numerous more popular works.

Jena glass O. Schott in collaboration with E. Abbe developed a great variety of optical glasses with wide variation of refractive index and dispersion. Whereas previously higher dispersion went hand in hand with higher index, the Jena glass list included glasses with high dispersion associated with low index and vice versa. This helped the firm of Zeiss to construct microscopic and photographic objectives with better corrections of chromatism and spherical aberration.

Jensen, Johannes Hans Daniel (1907–1973) German theoretical physicist. A professor at Heidelberg, he jointly received the 1963 Nobel prize for physics with Maria Goeppert Mayer for their (independent) work on the *shell model of the nucleus.

JET Abbreviation for Joint European Torus. *See* fusion reactor.

jet propulsion Propulsion of aircraft or other vehicles in which one or more jets of hot gases are ejected at high speed from backwardly directed nozzles. The ejected gases exert forces in the forward direction upon the system. Jet engines are devoid of reciprocating parts; air is drawn through an intake into a compressor, whence it passes to a combustion chamber, where it mixes with an oil fuel. The products of the combustion are expanded into the jet, driving the compressor on their way by means of a turbine. *See also* ramjet.

jet tones The rather unsteady tones produced when a stream of air is projected into still air from an orifice. If a stream of fluid issues from a linear slit in an infinite plate, into stationary fluid – it is generally assumed that it is a homogeneous jet – surfaces of discontinuity arise between the moving and stationary fluids. The moving fluid tends to curl outwards into the stationary fluid forming alternate vortices on each side of the jet. Instability in jets is, however, very high and where the velocity of efflux and the fluid are suitably chosen to give sufficient vortices per second for an audible sound, the tones produced are weak, uncertain, and fluctuating. The sound, generally of high frequency, is more suitably described as a variable hiss. Experiments show that every vortex pattern in a free jet is equally unstable and that it is only when some other mechanism is used that the vortices may be guided into periodically repeated and stable patterns. (*See* edge tones.) A circular orifice may be used; the pertinent patterns may be sometimes observed when smoke is emitted from a factory chimney. The general frequency for a circular orifice is approximately proportional to the velocity of efflux. The relation between frequency and diameter is not simple.

JFET Abbreviation for junction field-effect transistor. *See* field-effect transistor.

jitter

jitter A short-term instability in either the amplitude or phase of a signal, particularly the signal on a *cathode-ray tube. It has the effect of causing momentary displacements of the image on the screen.

Johnson–Lark–Harowitz effect The change in resistivity of a metal or *degenerate semiconductor due to scattering of the charge carriers by impurity atoms.

Johnson noise Thermal *noise.

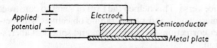

Johnson–Rahbeck effect

Johnson–Rahbeck effect If a semiconducting plate of material such as slate or agate is placed against a metal plate, the two hold strongly together during the application of a potential of about 200 volts. The mechanism of this action depends on the fact that the plate and stone are only in actual contact at a few points through which a very small current flows to equalize a high potential difference. This potential difference is therefore applied across a very small distance and the forces of attraction are correspondingly great.

Joint European Torus (JET) *See* fusion reactor.

Joliot-Curie, Irène (1897–1956) French nuclear physicist, daughter of Pierre and Marie Curie and wife of J. F. Joliot. Shared Nobel prize for physics with her husband (1935).

Joliot-Curie, Jean Frédéric (1900–1958) French nuclear physicist. Noted for his work on nuclear physics, he produced, in collaboration with his wife Irène, the first artificial radioactive nuclide by bombarding boron with fast alpha particles. Shared Nobel prize for physics with his wife (1935).

Jolly, Philipp Gustav (1810–1884) German physicist who opened at Heidelberg (1846) the first physical laboratory for students in a German university.

Jolly's balance *See* balance.

Joly, John (1857–1933) Irish physicist and geologist. His most notable work was in the proof that *pleochroic haloes are a radioactive phenomenon and can be used to estimate the age of rocks. He invented a photometer, and two forms of steam calorimeter. He was a pioneer in the radium treatment of cancer.

Joly photometer (1884) A photometer consisting of two paraffin wax blocks separated by a thin sheet of opaque foil. Each side of the block is illuminated by different sources of light, which are adjusted to produce equality of brightness on both sides, whence a comparison of *luminous intensities can be effected.

Joly's steam calorimeter A calorimeter for the determination of the specific heat capacities of solids. A pan A is suspended from one arm of a chemical

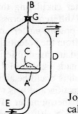

Joly's steam calorimeter

balance by a fine wire, B, passing through a plaster plug, G, in an enclosure, D, through which steam can be passed via the pipes E and F. The balance is counterpoised both with and without the specimen C, giving m_1 the mass of the specimen at temperature t_1. Steam at temperature t_2 is then passed through the chamber, condensation at G being prevented by a small heating coil encircling the wire B. On reweighing, the mass of steam m_2 condensed on the pan (of *water equivalent W) and specimen is found and so
$$m_1 c(t_2 - t_1) = m_2 L - W(t_2 - t_1),$$
where c is the specific heat capacity of the specimen and L the specific latent heat of steam at temperature t_2. W is determined by a similar experiment carried out with the pan empty. All weighings must be corrected for the buoyancy of the air and steam.

There is also a differential form (*see* differential steam calorimeter), used for determining specific heat capacities of gases at constant volume.

Jones, Sir Harold Spencer (1890–1960) Brit. astronomer. (Astronomer Royal 1933–1955). He was associated especially with the accurate estimation of the dimensions of the earth's orbit round the sun.

Jordan, Ernst Pascual (b. 1902) German theoretical physicist who was a professor at Berlin and Hamburg. He was one of the founders of modern quantum theory and, independently of Carl Brans and Robert Dicke, also developed a variant of Einstein's theory of general relativity in which the gravitational constant varies with time. The theory is sometimes known as the *Brans–Dicke–Jordan* theory.

Josephson, Brian David (b. 1940) Brit. physicist who became professor at Cambridge. While a research student (1962) he predicted the *Josephson effect in semiconductors. For this work he shared the 1973 Nobel prize with Leo Esaki and Ivar Giaever.

Josephson constant *See* Josephson effect.

Josephson effect Any of the phenomena that occur at sufficiently low temperatures when a current flows through a thin insulating layer between two superconducting substances (*see* superconductivity). The narrow insulating gap between the superconductors is known as a *Josephson junction*, and is usually in the form of a very thin film. The electrons forming the current are able to leak across the junction as a result of the *tunnel effect.

The current can flow across the junction in the absence of an applied voltage: this is the *d.c. Joseph-

son effect. In certain circuit configurations of Josephson junctions, the superconducting current is highly sensitive to a magnetic field. This allows it to be used as an extremely fast electronic switch with very low power dissipation. (*See also* squid.)

If a voltage is applied across a Josephson junction, then an alternating current flows through the junction: this is the *a.c. Josephson effect*. The current varies at a microwave frequency, v, which is related to the voltage V:

$$v = (2e/h)V,$$

where h is the Planck constant and e the electron charge; the quantity $2e/h$ is called the *Josephson constant*. Conversely, if microwave radiation (frequency 10–100 GHz) impinges on a Josephson junction, the microwave frequency can be related to increments in the voltage developed across the junction when a superconducting current flows. The voltage increments are very precise, being equal to multiples of $(h/2e)$ times the frequency. It is now possible to connect many thousands of Josephson junctions in a long line to obtain a measurable voltage. These voltages can be used, for example, to compare laboratory voltage standards, and to standardize the volt to within a few parts in 10^8.

Josephson junction *See* Josephson effect.

joule Symbol: J. The *SI unit of all forms of *energy (mechanical, thermal, and electrical), defined as the energy equivalent to the work performed as the point of application of a force of one newton moves through one metre distance in the direction of the force. In electrical theory the relationship 1 J = 1 W s (watt second) is most useful. Since 1948 the joule has replaced the calorie as a unit of heat. As a formally defined conversion factor 1 calorie = 4.1868 joules.

Joule, James Prescott (1818–1889) Brit. physicist. He carried out a very thorough investigation into the relationship between heat and mechanical work over the period 1839–1878, the best known of his many experiments in this field being the water-stirring one described in a separate article (*see* Joule experiment (1847)). His first paper establishing the existence of a *mechanical equivalent of heat, presented in 1843, was not published until 1846 and he was thus anticipated by J. R. von Mayer, who published a short essay in 1842, backed by a calculation (based on the difference of the principal specific heat capacities of a gas) in 1845. As part of this research, Joule investigated the *heating effect of an electric current, discovering Joule's law, and the expansion experiment. (*See* Joule experiment (1845).) With William Thomson, he carried out a series of experiments (1852–1862) on temperature changes in a gas on passing through a porous plug. The resulting change is known variously as the Joule–Thomson effect and the *Joule–Kelvin effect as Thomson later became Lord Kelvin.

The energy unit the *joule is named in honour of Joule and the mechanical equivalent of heat is given the symbol J.

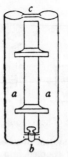

Maximum density of water (Joule and Playfair)

Joule and Playfair's experiment on maximum density of water An experiment to determine the temperature for the maximum density of water by finding two temperatures, one slightly above and one slightly below 4 °C at which the density is the same and taking the mean. Two long columns of water a, a at these temperatures are connected at the bottom by a tube b and a small glass float is placed in the connected trough at c. The temperature of one limb is adjusted until this float is stationary indicating the absence of convection currents when the density of water is the same in both limbs.

Joule calorimeter An electrically heated calorimeter such as that used in *Griffiths's apparatus.

Joule effect The liberation of heat by the passage of a current through an electric conductor, due to its resistance. *See* heating effect of a current.

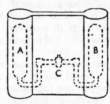

Joule experiment; (expansion of gas)

Joule experiment (1845) An experiment devised to see if a gas does any internal work when expanding into a vacuum from a high pressure. In this case no external work is done by the gas in expanding. Two vessels, A and B, were placed in a common water bath, B being exhausted and A filled with air at a pressure of 22 atmospheres. On opening the stopcock, C, the gas expanded to fill both vessels but no change in temperature of the water bath was observed implying that no internal work was done in separating the molecules against the molecular attractions. Joule concluded that the internal energy of a gas is independent of its volume; i.e. $(\partial U/\partial v)_T$ = 0. Later, it was shown that this law is true only for an ideal gas. The Joule experiment was not sufficiently sensitive to detect the small temperature change because the heat capacity of the calorimeter was very large. *See* Joule–Kelvin effect.

Joule experiment (1847) An experiment to determine the amount of heat equivalent to a given amount of mechanical work. Slow-falling large

Joule heating

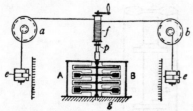

Mechanical equivalence of heat (Joule)

weights *e*, *e* operating on a system of wheels and axles *a*, *b*, rotate the shaft *f* coupled to a paddle system in the stationary calorimeter A, B, which rests on an insulating base *g*. On reaching the floor the weights are raised by winding *f* with the pin *p* removed, and after replacing the pin the weights are allowed once more to stir the paddles in the water whose temperature rise is measured by a thermometer. Allowance is made for friction at the pulleys, elasticity in the cords and for the kinetic energy of the weights on reaching the floor. The heat lost by the calorimeter to its surroundings during the experiment is allowed for by plotting a cooling curve.

Joule heating *See* heating effect of a current.

Joule–Kelvin effect *Syn.* Joule–Thomson effect. A change of temperature observed when a gas undergoes an irreversible adiabatic expansion on being pumped continuously through a porous plug or a very fine orifice (*throttle*). Provided that the change of kinetic energy of the flowing gas is small, the *enthalpy is unchanged in the process. According to the form of the *equation of state, the work done by part of the gas upon the gas in front of it on leaving the system may be greater or less than the work done on it by the pump, so the net external work may be of either sign. Generally, it is found that for each gas there is an *inversion temperature (dependent on the pressure) above which there is a rise of temperature on expansion and below which there is a fall.

The change of temperature with pressure at constant enthalpy, *H*, is called the *Joule–Thomson* (or *Joule–Kelvin*) *coefficient*, symbol: μ, and is given by:

$$\mu = \left(\frac{\partial T}{\partial P}\right)_H = \frac{T\left(\dfrac{\partial v}{\partial T}\right)_p - v}{c_p},$$

where *v* is the specific volume, *p* the pressure, *T* the thermodynamic temperature, and c_p the specific heat capacity at constant pressure.

The Joule–Kelvin effect is used in *refrigerators and in the *liquefaction of gases. In the important cases of hydrogen and helium the inversion temperatures are far below room temperature so these gases must first be cooled to below their inversion temperatures before further cooling can be produced by this effect.

From the equation it is seen that μ would be zero for an *ideal gas.

308

Joule magnetostriction Positive *magnetostriction.

joulemeter An instrument for measuring the energy supplied from the mains to an electric circuit. It is more commonly known as a *watt-hour meter or electricity meter.

Joule's equivalent The *mechanical equivalent of heat.

Joule's law 1. The principle that the heat produced by an electric current, *I*, flowing through a resistance, *R*, for a fixed time, *t*, is given by the product of the square of the current, the resistance, and the time, i.e. $q = I^2Rt$. If the current is expressed in amperes, the resistance in ohms, and the time in seconds then the heat produced is in joules.

2. The principle that the *internal energy of a gas is independent of its volume. It only applies to *ideal gases, i.e. when there are no intermolecular forces. *See* Joule's experiment (1845).

Joule–Thomson effect, coefficient *See* Joule–Kelvin effect.

journal friction *See* friction.

J/psi particle A massive unstable *elementary particle (3097 MeV), more precisely a meson resonance (*see* resonances). When detected (1974) by two independent US groups (hence its dual name), the width of the resonance peak was found to imply a lifetime of 10^{-20} second. This is considerably longer than the 10^{-23} second characteristic of resonance decay, and led to the concept of *charm: the J/psi is composed of a charm quark and a charm antiquark ($c\bar{c}$) but the particle itself has zero charm. The decay inhibition results because decays into final states containing charmed hadrons are kinematically forbidden, the rest masses of these hadrons being too large for these decays to occur.

JUGFET Abbreviation for junction field-effect transistor. *See* field-effect transistor.

junction 1. A contact between two different conducting materials, e.g. two metals, as found in a *rectifier or *thermocouple.

2. In a *semiconductor device. A transition region between semiconducting regions of differing electrical properties. *See* p-n junction.

3. A connection between two or more conductors or sections of transmission lines.

4. *See* Josephson effect.

junction field-effect transistor *See* field-effect transistor.

junction transistor *See* transistor.

just intonation The major scale of just intonation is a scale with eight notes to the octave having frequencies proportional to 24, 27, 30, 32, 36, 40, 45, 48, within the limits detectable by the ear. The minor scale of just intonation is a similar scale except that the interval between the keynote and third note of the scale is a minor third. There are several minor scales for each keynote, depending on the intervals between the 6th and 7th notes and the keynote. The chromatic scale of just intonation has thirteen notes to the octave. Three pitch intervals are observed – 16/15, 25/24, 135/128. *See* musical scale.

Juvin's rule When a capillary tube of internal radius r stands vertical in a liquid of density ρ and surface tension γ the liquid rises a distance h up the tube, given by:
$$h = (2\gamma/rg\rho)\cos\alpha,$$
where α is the angle of contact between the liquid and the walls of the tube, and g is the acceleration of free fall. For a liquid that does not wet glass, α exceeds $90°$ and h will be negative: the liquid is depressed in the bore below the general level.

K

k The symbol for the *Boltzmann constant.

kaleidoscope A toy consisting of two mirror strips inclined at $60°$ suitably mounted so that a regularly repeated pattern produced by multiple reflection can be seen when looking through a peephole at one end, while the objects (coloured beads, etc.) lie at the opposite end and are suitably illuminated. The objects can be shaken to vary the pattern.

Kaluza–Klein theory A type of *unified-field theory in which the theory of *relativity is extended to more than four space–time dimensions. In five dimensions this accounts for electromagnetic interactions. In higher dimensions, Kaluza–Klein theories give general relativity and more general *gauge theories. A combination of Kaluza–Klein theory and *supersymmetry gives rise to *supergravity in eleven space–time dimensions. In such theories it is proposed that the higher dimensions are "rolled up" to become microscopically small (a process known as *spontaneous compactification*), with four macroscopic space–time dimensions remaining. Like *grand unified theories, Kaluza–Klein theories predict the existence of magnetic monopoles. There are, however, difficulties with such theories, in particular the problem of *renormalization.

Kamerlingh-Onnes, Heike *See* Onnes, Heike Kamerlingh.

kaon *Syn.* K meson. *See* meson.

Kapitza, Peter (1894–1984) Soviet physicist. While working at Cambridge (England), he produced high transient magnetic fields and succeeded in measuring magnetic properties of substances by measuring the mechanical force on them. He liquefied hydrogen and helium by the adiabatic expansion method, overcoming the difficulty of lubrication by the use of a loosely fitting piston and such a rapid expansion that the amount of gas escaping is small. He was awarded the Nobel prize in 1978.

Kaplan, Joseph (*b.* 1902) Amer. physicist born in Hungary who made extensive contributions to spec-

troscopy and produced the first *aurora spectrum in the laboratory.

Kastler, Alfred (*b.* 1902) French physicist; professor at Bordeaux and Paris. He received the 1966 Nobel prize for physics for his work on *double-resonance techniques in spectroscopy using visible and radio-frequency radiation.

Kater, (Captain) Henry (1777–1835) Brit. physicist, drawn into the field of scientific instruments through his survey work for the army. He designed a form of reversible *pendulum, which he used for an accurate determination of the acceleration of free fall.

Kater's pendulum *See* pendulum.

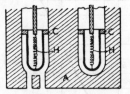
Katharometer

katharometer An instrument for detecting the presence of small quantities of an impurity (say hydrogen) in air. It depends for its action on the fact that the thermal conductivity varies greatly from gas to gas so that a heated filament loses heat to a different extent when the surrounding gas is contaminated. The platinum heaters H, H, are mounted in frames consisting of copper loops soldered to copper rings C, C. Both cells are contained in a copper block A, one being completely closed and the other communicating with the atmosphere. The coils H, H are mounted in a resistance bridge balanced when both coils are surrounded by pure air so that the rate of loss of heat from both is the same. If the apparatus is now placed in air contaminated with hydrogen, the open cell loses heat at a greater rate, resulting in a lowering of the temperature and a decrease in resistance of the coil H. The deflection of the galvanometer in the bridge circuit is a measure of the impurity of the gas and a proportion of one part in 500 000 parts of air is detectable. The principle is related to that of the *Pirani gauge.

K-capture *See* capture.

Keeler, James Edward (1857–1900) Amer. astronomer and physicist who was one of the first to use the spectroscope in astronomy. He studied the structure of Saturn's rings and obtained photographs of many galaxies.

keeper *Syn.* armature. A piece of iron or steel that is placed across the extremities of a permanent horse-shoe magnet, or across pairs of extremities of permanent bar magnets, when the magnets are not in use, thereby completing the magnetic circuit. The regions near the ends of a magnet produce an induced flux within it opposing the original magnetizing flux, the effect being greatest for the shortest magnets.

Keesom, Wilhemus H.

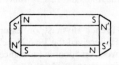

Keepers for bar magnets

Keeper for horse-shoe magnet

The magnetization of the keeper(s) neutralizes the demagnetizing effect.

Keesom, Wilhemus H. (1876–1956) Dutch physicist who first solidified helium by working at high pressures.

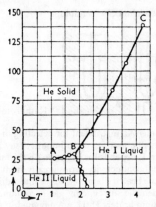

Solidification curve for helium

Keesom's solidification of helium Liquid helium was compressed in a narrow brass tube joined to two German-silver tubes, these being cooled in a liquid-helium bath. The system became blocked at a pressure of 130 atmospheres due to the formation of solid and the solidification curve ABC, was traced from 1.1 K to 4.2 K. At the lower temperatures the curve BA shows a tendency to become parallel to the temperature axis so that $\mathrm{d}p/\mathrm{d}T \rightarrow 0$ as $T \rightarrow 0$. But since

$$\mathrm{d}p/\mathrm{d}T = (s_2 - s_1)/(v_2 - v_1),$$

where s_1 and v_1 are the entropy and volume of unit mass of solid and s_2 and v_2 are the corresponding values for the liquid, all at temperature T, if $\mathrm{d}p/\mathrm{d}T = 0$ then $s_1 = s_2$ at $T = 0$ K. This indicates that the *Nernst heat theorem applies to liquid helium since the entropy of unit mass of liquid helium is equal to that of unit mass of solid helium at the absolute zero.

Kellner eyepiece A type of *Ramsden eyepiece with an *achromatic eye lens that corrects chromatic aberration and distortion inherent in the original design. It is commonly used as an eyepiece in *prism binoculars.

kelvin Symbol: K. The *SI unit of *thermodynamic temperature, defined as 1/273.16 of the thermodynamic temperature of the *triple point of water. The kelvin is also used as a unit of temperature differ-

ence on the thermodynamic and Celsius scales, where 1 K = 1 °C. *See also* degree Celsius.

Kelvin, Lord (William Thomson; 1824–1907) Brit. physicist. Converted to the doctrine of the conservation of energy by Joule, Kelvin made great contributions to the new science of *thermodynamics and showed how to construct a temperature scale independent of the properties of any particular substance. He joined with Joule in a series of experiments (1852 to 1862) to determine the extent to which actual gases deviate from the ideal (*see* Joule–Kelvin effect). From the results of such experiments, it is possible to correct the readings of actual gas thermometers to give readings on the Kelvin scale. (*See* constant-pressure gas thermometer; constant-volume gas thermometer; gas scales of temperature.) He discussed the behaviour of refrigerators and showed that the thermodynamically most efficient way of producing heat is to run a refrigerator in reverse. The Kelvin "warming engine" (1852) is now realized in practice as the *heat pump.

Kelvin made many investigations in electricity and some of his inventions are separately treated (*see* Kelvin balance; Kelvin double bridge; Kelvin replenisher). He also invented the *quadrant electrometer and some of its variants, and the *attracted-disc electrometer. Kelvin made great contributions to telegraphic signalling. (He was knighted in 1866 for his part in laying the first two transatlantic cables.) He first showed that the delay in simple submarine cables was due to high capacity and he showed how to reduce the lag. He also improved the nautical magnetic compass and the method of depth sounding.

He applied thermodynamic reasoning to the thermoelectric circuit of two metals and showed that the maintenance of a current by keeping the two junctions at different temperatures could not be explained by the *Peltier effect at the junctions alone, so leading to the discovery of the Thomson (or Kelvin) effect, and to the so-called "specific heat of electricity" (the *Thomson coefficient). He also made important contributions to the theory of damped electromagnetic oscillations. (*See* Kelvin's formula.)

Kelvin made a celebrated estimate of the age of the earth, assuming it to have cooled to its present condition from a wholly molten state, obtaining as limits 2×10^7 and 4×10^7 years (1862).

Other contributions to science include a tide calculator. He engaged in the manufacture of scientific instruments.

Kelvin balance A type of *current-balance instrument that consists of six coils, four fixed and two that move between them on a balanced rod. The suspension consists of two flexible multiple copper ribbons that serve to carry current to the coils in the manner shown in the diagram, so that each fixed coil tends to displace the balanced arm in the same direction when current flows. A rider, moving along an arm graduated in amperes, is used to rebalance

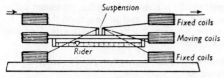

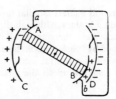

Kelvin balance

Kelvin replenisher

the coil system – the scale divisions being uneven as the displacement of the weight is proportional to the square of the current. If the instrument is connected in the opposite direction, the current is changed in direction in all the coils so that the deflection of the arm is the same. The instrument is thus suitable for measuring alternating currents. Balances are made with ten ranges up to 2500 amperes.

The instrument can also be adapted to measure wattage. The current flows through thick windings in the four fixed coils and the two movable coils consist of a considerable number of turns of fine wire with consequent high resistance. These movable coils are connected in parallel with the part of the circuit in which the power is to be measured and thus the current in them is proportional to potential difference.

Kelvin contacts A means for testing or making measurements on electronic circuits or components. Two sets of leads are used to each test point, one set carrying the test signal and the other going to the measuring instrument. This removes the effect of the resistance of the leads on the measurement.

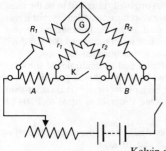

Kelvin double bridge

Kelvin double bridge A special development of the d.c. *Wheatstone bridge for precision measurement of low resistances. A is the low resistance to be measured, and B is a known low resistance of the same order. They are placed in series and a current passed. A bridge R_1, r_1, R_2, r_2 and galvanometer G are set up as shown in the diagram. The bridge is balanced by varying R_1 and r_1 with the connection K open and closed until the balance is exact for both conditions. Then it can be shown that:
$$A/B = R_1/R_2 = r_1/r_2.$$
The method eliminates possible errors due to contact resistance and the resistance of leads. A and B are usually four-terminal resistors, as shown.

Kelvin effects See thermoelectric effects.

Kelvin replenisher A simple *electrostatic generator. Curved metal plates A and B, mounted on an insulating arm, are arranged to rotate between larger curved plates C and D. Metal wipers are fixed as shown at a, b, and to the plates C and D.

The action is equivalent to that of the *water dropper, energy in the replenisher being supplied mechanically by the rotation of the arm in the clockwise direction. At the position shown A and B are connected by the wire, and A being nearer to the high potential of C, a current flows to leave A negative and B positive. These charges are removed on contact with the wipers fixed to D and C respectively to increase the negative and positive potentials of D and C, so that on further rotation stronger charges are produced on the rotor by these increased charges.

Kelvin's formula The approximate formula, $T = 2\pi\sqrt{LC}$, relating the period to the inductance (L) and capacitance (C) of an electric circuit with negligible resistance.

Kennelly, Arthur Edwin (1861–1939) Amer. electrical engineer who contributed to the theory of alternating currents and, independently of *Heaviside, discovered the reflecting layer of the *ionosphere that enables long-distance radio communication to be achieved.

Kennelly–Heaviside layer See ionosphere.

Kepler, Johann (1571–1630) German astrologer and astronomer. He developed the two cardinal principles of modern astronomy (laws of elliptical orbits and of equal areas – see Kepler's laws) and catalogued 1005 stars. He also made contributions to the science of optics, the law of refraction, and the astronomical telescope, studied gravitational attraction and suggested that tides are due to the influence of lunar attraction. He developed the theory of planetary revolutions and the use of logarithms.

Kepler's laws Three fundamental laws of planetary motion.

1. Every planet moves in an ellipse, the sun occupying one focus of the ellipse.

2. The radius vector drawn from the sun to the planet sweeps out equal areas in equal times (i.e. the areal velocity is constant).

3. The squares of the times taken to describe their orbits by two planets are proportional to the cubes of the major semiaxes of the orbits.

The first two laws were published in 1609, the third in 1619.

These laws were later shown to apply also to the orbits of comets around the sun, and to natural and artificial satellites around planets. Similar laws apply to the orbits of double stars. Although the laws are followed very accurately, there are small deviations that are explained by *perturbation theory, the imperfect symmetry of the central body, or the theory of *relativity, according to the case. See gravitation.

Kepler telescope Syn. astronomical telescope. See refracting telescope.

keratometer A clinical instrument for measuring the astigmatism of the front surface of the cornea as well as its power – a development of the *ophthalmometer.

kerma (kinetic energy released in matter) Symbol: K. The sum of the initial kinetic energies of all charged particles produced by the indirect effect of *ionizing radiation in a small volume of a given substance divided by the mass of substance in that volume. The SI unit is the *gray.

Kerr, John (1824–1907) Brit. physicist, discoverer of the electro-optical and magneto-optical effects known by his name. See Kerr effects.

Kerr cell See Kerr effects.

Kerr effects Two effects concerned with the optical properties of matter in electric and magnetic fields.

The *electro-optical effect* is the effect in which certain liquids and gases become double-refracting when placed in an electric field at right angles to the direction of the light. The substance acts as a *uniaxial crystal with optic axis parallel to the field. If n_1 and n_2 are the refractive indexes of light with planes of polarization respectively parallel and perpendicular to the field, then:

$$n_1 - n_2 = k\lambda E^2,$$

where k is the *Kerr constant*, λ the wavelength of the light, and E the electric field strength. The *Kerr cell* consists of two parallel plate electrodes immersed in a liquid that shows a marked electro-optical effect. Polarized light passes through the cell and can be interrupted by the application of an electric field. The device is also called an *electro-optical shutter*. (*See also* Pockel effect.)

The *magneto-optical effect* refers to the production of a slight elliptic polarization, produced when plane-polarized light is reflected from the polished pole face of an electromagent. The incident light is plane-polarized in, or normal to, the plane of incidence. See also Faraday effect.

Kerst, Donald William (b. 1911) Amer. physicist who became a professor at the University of Illinois and later at Wisconsin. He developed (1939) the idea of the *betatron accelerator and was responsible for the building of the largest such machine (310 MeV) at Illinois in 1950.

Kew magnetometer A type of *magnetometer used to make accurate measurements of the earth's magnetic field and the magnetic declination. The mag-

netic needle is a steel tube with a graduated transparent scale on one end and a lens on the other. Its precise position can be observed with a coaxial telescope.

Kew-pattern barometer Syn. Adie barometer. See barometer.

keyboard A manually operated device by means of which people can communicate with a computer. It consists of an array of labelled keys, operated by finger pressure as in a typewriter. Operation of a particular key (or combination of keys) produces a coded digital signal that can be fed directly into a computer. A keyboard usually has a standard QWERTY layout plus some additional keys. These can include a control key, function keys, cursor keys, and a numerical keypad.

kilo- Symbol: k. **1.** A prefix meaning 10^3, i.e. one thousand; for example one kilometre is equal to 1000 metres.
2. In computing, etc., where the binary rather than the decimal system of numbers is used, a prefix meaning 2^{10}, i.e. 1024; for example, one kilobyte is equal to 1024 bytes. The symbol K is not recommended.

kilogram Symbol: kg. The *SI unit of mass represented by the international prototype kilogram at the International Bureau of Weights and Measures at Sèvres in France. It consists of a cylinder whose height is equal to its diameter and is made from an alloy consisting of 90% platinum and 10% iridium. The kilogram was originally intended to be the mass of 1 dm³ of water at its maximum density, but it was subsequently found, using the international prototype, that this mass of water occupies 1.000 028 dm³. This volume was formerly called the *litre.

Decimal multiples and submultiples of the kilogram are formed by adding the SI prefixes to the word gram, for example milligram (mg) rather than microkilogram. One kilogram is equal to 2.204 62 pounds.

kilometric waves Radio waves having wavelengths between 1000 m and 10 000 m.

kilowatt-hour Symbol: kWh. A unit of energy equivalent to the work done when power of 1 kilowatt is expended for 1 hour, or equal to one thousand watthours. It is used for electric work.

kinematics The branch of *mechanics dealing with the motion of bodies without reference to mass or force.

kinematic viscosity (coefficient of) Symbol: ν. The ratio of the coefficient of *viscosity (η) to the fluid density (ρ). It is used in modifying the equations of motion of a perfect fluid to include the terms due to a real fluid. The units of kinematic viscosity are metres squared per second. At room temperature water has a kinematic viscosity of 10^{-6} m² s⁻¹. The ordinary viscosity coefficient is often called the coefficient of *dynamic viscosity* to avoid confusion.

kinetic energy *See* energy.

kinetic equilibrium *See* dynamic equilibrium.

kinetic friction *See* friction.

kinetic potential *See* Lagrangian function. *See also* Hamilton's principle.

kinetics *See* dynamics.

kinetic theory The work of Rumford, Joule, and others, led to the establishment of the concept of heat as a process of *energy transfer. The kinetic theory combines this conclusion with the molecular theory of chemistry and interprets the internal energy of a body as being the energy of the motions and positions of the molecules of which the body is made up. The basis of the theory is in fact that of the kinetic theory of matter as a whole, namely, that the particles of matter in all states of aggregation are in a violent state of agitation.

In gases the molecules move rapidly in all directions, being so small that they are mostly removed from one another at distances large compared with their own dimensions, being virtually free from the influence of other molecules. In a liquid, on the other hand, the molecules are very close to one another and their mutual influence is significant. The molecules are in continuous motion, but there seems to be some semblance of a patterned structure, which is not present in a gas. In the solid state matter may be crystalline, exhibiting a very definite patterned structure, or amorphous without such a pattern. Each fundamental unit of the structure is vibrating about a mean position, these vibratory motions constituting thermal agitation, becoming more energetic with increase of temperature. The energy of these motions accounts for the greater part of the specific heat capacity of a solid substance. Evidence of molecular agitation is afforded by such phenomena as *diffusion, while the *Brownian movement gives an outline of matter in the gaseous state, namely that the molecules are in a state of incessant irregular motion, frequently colliding with one another.

During their motion gas molecules will collide with the walls of the vessel in which they are contained, thereby delivering momentum to them and giving rise to an exertion of pressure. The kinetic interpretation of an *ideal gas is one in which the molecules of such a gas occupy a space entirely negligible compared with the total volume of the gas, and these molecules exert no influence on one another except when they actually collide, collisions being on the average perfectly elastic in equilibrium. The pressure exerted by an ideal gas may be shown to be given by the expression:
$$p = \tfrac{1}{3}mvC^2,$$
where m is the mass of the molecules, v is the number of molecules per m^3, and C is the *root-mean-square velocity of the molecules.

Maxwell's law of the *distribution of speeds gives the actual distribution of speed among the molecules in equilibrium state, and is based on the main concepts of classical statistics. These concepts lead also to the principle of *equipartition of energy, namely that the total energy of a system is equally divided between the different degrees of freedom, and that each degree of freedom possesses a mean energy $\tfrac{1}{2}kT$, where k is the Boltzmann constant and T is the thermodynamic temperature.

The application of kinetic theory leads to the relation for gases:
$$\gamma = 1 + 2/n,$$
where γ is the ratio of the principal *specific heat capacities and n is the number of *degrees of freedom for each molecule. In the case of solids it leads to *Dulong and Petit's law that the *molar heat capacity of a solid is constant and equal to $3R$. The simple kinetic theory of specific heat capacities is quite unable to account for the variation of specific heat capacity with temperature both in solids and in gases, and the classical concepts of equipartition had to be replaced by those of the *quantum theory.

Kirchhoff, Gustav Robert (1824–1887) German physicist, who made important contributions to the study of thermal radiation (*see* Kirchhoff's law) and to spectrum analysis in which he collaborated with Bunsen, discovering with him the elements rubidium and caesium (1861). In the course of this work he gave the first satisfactory explanation of dark lines in spectra such as the *Fraunhofer lines, obtaining in the laboratory a reversal of spectra by passing the light from a white source (oxy-hydrogen limelight) through a flame fed with appropriate salts (1859).

Kirchhoff developed the theory of diffraction of light giving a mathematical formulation of Huygens' principle. He also gave rules for solving steady current *networks. *See* Kirchhoff's laws.

Kirchhoff formula A formula for the variation of *vapour pressure with temperature:
$$\log p = A - B/T - C \log T,$$
where A, B, and C are constants. It is valid over limited temperature ranges.

Kirchhoff's law (for radiation) The principle that at a given temperature the spectral *emissivity of a point on the surface of a thermal radiator in a given direction is equal to the spectral *absorptance for incident radiation coming from that direction. A *thermal radiator* describes a body emitting radiation as a result of thermal vibration of the atoms or molecules. The adjective *spectral* implies that the emissivity or absorptance is considered for monochromatic radiation.

Kirchhoff's laws (for an electric circuit) **1.** The algebraic sum of the electric currents that meet at any point in a network is zero.

2. In any closed electric circuit the algebraic sum of the products of current and resistance in each part of the network is equal to the algebraic sum of the electromotive forces in the circuit.

Kittell, Charles (*b.* 1916) Amer. physicist who worked at Bell Telephone Laboratories and became professor at the University of California (Berkeley).

Klein–Gordon equation

A leading authority on solid-state physics, he worked on magnetic and conduction properties, and is noted for his textbook *Introduction to Solid State Physics* (first published 1953).

Klein–Gordon equation An equation used in relativistic quantum mechanics to describe the fields for spin-zero particles such as mesons:

$$(\partial^2/\partial t^2 - \nabla^2 + m^2)\phi(x,t) = 0,$$

where t is time, ∇^2 is the Laplace operator, m is the mass, x the space coordinate, and ϕ is the scalar field wavefunction. Originally it was developed as an equation for single particles but is more successful if used for processes involving particles associated with spin-zero fields. The field interpretation of the equation also shows that antiparticles exist for bosons; i.e. the existence of boson antiparticles is a consequence of combining quantum mechanics with relativity, just as fermion antiparticles are a consequence of the *Dirac equation.

klydonograph An instrument for recording the characteristics of a lightning discharge or other electric surge. It consists essentially of a photographic plate (or film) placed between electrodes, one of which is connected to earth and the other to the conductor under investigation. A potential difference between the electrodes affects the emulsion of the plate and a figure is obtained upon development. The plate or film may be fixed relatively to the electrodes or may run past them in order to provide a continuous record.

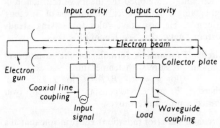

Two-cavity klystron amplifier

klystron An *electron tube that employs *velocity modulation of an electron beam, and is usually used for either the amplification or generation of *microwaves. Several varieties of the basic klystron exist.

In the simple two-cavity klystron (*see* diagram), a beam of high-energy electrons from an electron gun is passed through a *cavity resonator excited by high-frequency radio waves. The interaction between the high-frequency waves and the electron beam produces velocity modulation of the beam. After leaving the cavity, bunching of the electrons will occur; the current density of the beam thus varies and has the same frequency as the exciting radio waves.

The modulated beam then passes through a second cavity resonator, where its current-density variations produce a voltage wave. This is tuned to the exciting radio frequency or a harmonic of it. Voltage amplification is obtained by conversion of the energy of the original beam into r.f. energy in the output cavity, power being taken from the beam. If positive feedback to the input cavity is employed, the device can be made to oscillate. A *reflex klystron* employs only one cavity, the electron beam being reflected after velocity modulation has occurred. Reflex klystrons are most commonly used as low-power oscillators. *Multicavity klystrons* employ more than two cavities in the beam, and a higher overall gain may be achieved. The intermediate cavities are excited by the velocity modulation from the input cavity, and in turn modulate the beam further. They are used when extremely high-power pulses or continuous waves of moderate power are required.

K meson *Syn.* kaon. *See* meson.

knife switch A switch in which the moving part consists of one or more current-carrying blades each hinged so that it moves in its own plane and enters the fixed contact or contacts.

Knudsen, M. H. C. (1871–1949) Danish physicist, noted for work on kinetic theory and phenomena occurring at low pressures.

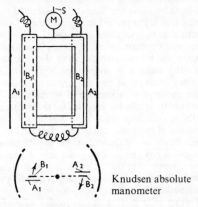

Knudsen absolute manometer

Knudsen absolute manometer A device for the absolute measurement of very low pressures where the mean free path of the molecules is large compared with the dimensions of the apparatus. Two cold plates B_1, B_2, at a temperature T_2 are free to rotate in the evacuated vessel about a vertical quartz suspension S. Stationary plates A_1, A_2, are electrically heated to a temperature T_1. The gas molecules striking B from the side A have greater momentum than those striking the other side of B and so the vanes B_1, B_2 experience a force per unit area equal to F in the direction shown. The deflection of the suspended system is obtained from the deflection of a beam of light reflected from the mirror M and together with the torsion constant of the fibre enables F to be calculated. The pressure p of the gas is given by:

$$p = \frac{2F}{\sqrt{\dfrac{T_1}{T_2} - 1}}.$$

Knudsen flow *See* molecular flow.

Knudsen gauge A gauge of the same general form as the *Knudsen absolute manometer but without the constants being necessarily known. The total pressure of all gases and vapours present is measured. (*Compare* McLeod gauge.) These gauges are sometimes misleadingly called *radiation* or *radiometer* gauges. *Compare* Crookes radiometer.

Knudsen number *See* molecular flow.

Knudsen's equation *See* molecular flow.

Kodachrome *See* colour photography.

Kohlrausch, Friedrich (1840–1910) German physicist. An early supporter of the theory of *electrolytic dissociation, he recognized the independence of ionic mobilities in 1875. He used alternating currents to avoid polarization in measuring the resistance of electrolytes (*Kohlrausch bridge*, 1879). His *Lehrbuch der praktische Physik* is an authoritative manual on advanced laboratory work.

Korolev, Sergei Pavlovich (1906–1966) Soviet engineer and physicist who designed the satellite in which Gargarin made the first space flight in 1961.

Krigar–Menzel law *See* Young-Helmholtz laws.

Kundt, August A. E. E. (1839–1894) German physicist, best known for his work in acoustics and in light.

Kundt's rule The principle that the refractive index of a medium does not vary continuously with wavelength in the region of absorption bands. *See* anomalous dispersion.

Kundt's tube An apparatus devised by Kundt in 1866 to measure the speed of sound in gases under different controllable conditions of temperature, density, and humidity. It also gives valuable information on the molecular aggregation of the constituents of these gases, which may be supplied in a small quantity.

A column of gas in a tube D is closed by a reflector piston R at one end and has a source of sound at the other; in between there is a dry powder such as lycopodium or cork or pith dust for detecting resonance (there are some other methods that are mentioned below). If the piston is adjusted so that the length of the gas column gives an exact number of stationary waves, the dust will be violently disturbed at the displacement antinodes and will form a series of striations. The source of sound may be a rod clamped at its centre C, and having at one end a diaphragm nearly fitting the cross section of the tube. Other sources may be a telephone diaphragm actuated with an alternating current supplied by a

suitable oscillator, a supersonic quartz crystal, or a rod vibrated by *magnetostriction.

Other less primitive methods exist for detecting resonance. (*See* interference of sound; Rayleigh disc; manometric flame; hot-wire microphone.)

The phenomena occurring in Kundt's tube are much more complex than the early observers supposed. Between the displacement nodes the powder arranges itself in striae. Using a powerful source of sound and smoke particles observed in scattered light, Andrade was able to measure the amplitude of the vibrations in the tube and establish the existence of the circulation that was predicted by Rayleigh. This circulation takes place from antinode to node in the neighbourhood of the walls and from node to antinode along the centre. In fact, all the phenomena of Kundt's tube were explained in terms of vortex motion and circulation by Andrade.

Kundt's tube has many important applications:

The absolute determination of speed of sound in a gas from the relation $c = f\lambda$, where c is the speed of sound, f is the frequency, and λ is the wavelength, equal to $2d$ (d being the distance between two nodes or antinodes in the tube).

The comparison of speed of sound in a gas and in a solid rod, from the relation
$$c_{gas}/c_{rod} = d/l,$$
where l is the length of the rod the longitudinal vibration of which throws the gas in the tube into resonance.

The comparison of speeds of sound in two different gases (double-tube method). The same rod is used to excite resonant vibrations in two tubes. If d_1 and d_2 are the respective nodal separations, c_1 and c_2 the corresponding speeds,
$$d_1/d_2 = c_1/c_2.$$

The determination of the ratio of the specific heat capacities of a gas (γ), especially those supplied in small quantities as rare gases, from the relation $c = \sqrt{(\gamma p/\rho)}$, where p and ρ are the pressure and density of the gas.

The measurement of the variation of speeds of sound with different conditions of temperature, pressure, humidity, or with the nature of the gas.

The value of the speed as determined by Kundt's tube is subject to the Helmholtz–Kirchhoff correction for tube diameter. The speed of sound in a tube of radius r for a frequency n is given by the formula

$$c' = c\left\{1 - \frac{k}{2r\sqrt{\pi n}}\right\},$$

in which c is the speed of the free gas and c' is the speed in the tube; k is a constant depending on the viscosity and heat conduction.

Kundt's tube

Kurchatov, Igor Vasilievitch (1903–1960) Soviet physicist who studied neutron reactions and was largely responsible for the USSR's first nuclear weapons.

Kusch, Polycarp (*b.* 1911) German-born Amer. physicist who did most of his work at Columbia University, New York City. He investigated hyperfine splitting in electron-spin resonance for molecular beams and showed (1947) a discrepancy in observed energy states significant in *quantum electrodynamics. He shared the 1955 Nobel prize for physics with Willis Lamb (*see* Lamb shift).

L

labelled *See* radioactive tracer.

Laby and Hercus' method for measuring J (1927) Laby and Hercus used a continuous-flow method for the direct determination of the *mechanical equivalent of heat, using a form of electromagnetic brake or induction dynamometer. An external magnet is rotated about a vertical axis round a cylindrical stator consisting of stalloy stampings slotted longitudinally to receive fourteen copper tubes T through which water flows (*see* diagram). The water enters at the top through a brass end-ring with an annular channel communicating with each tube, and flows out through a similar arrangement at the bottom. The stator is connected by a glass tube G to the inner sleeve of a large diameter ballbearing carefully constructed to avoid any motion parallel to the axis of rotation. A horizontal torsion wheel is attached to this sleeve, the whole system being suspended by a torsion wire attached to a rigid plate. When the stator is held stationary, the rotating magnetic field induces currents in the copper tubes thus heating the water. When the steady state is attained, the work done is found from the couple applied to the torsion wheel and wire and the number of rotations per second of the electromagnet. The heat produced is obtained by measuring the difference between the inlet and outlet temperatures of the water by means of two differential platinum resistance thermometers H and the mass of water flowing through the stator per second. Heat losses are made small (0.04%): (*a*) by surrounding the stator with a vacuum flask and measuring the temperature difference between the inside and outside walls of the flask by means of a thermocouple F; (*b*) by surrounding the resistance thermometer tubes with vacuum jackets; (*c*) by adjusting the temperature of the surroundings to equality with that of the top of the stator as measured by a thermocouple K so that heat losses from the top of the stator are eliminated. In order that the applied couple might be measured to 1 part in 10 000, pulleys mounted on agate knife edges were used, a correction for the slight displacement of the knife edges from the centre of the pulleys being made. The axis of rotation of the stator was made parallel to that of the rotating magnet, which was also mounted on

ballbearings. They obtained a value for the mechanical equivalent of heat:

$$J = 4.186 \times 10^7 \text{ ergs per } 15° \text{ calorie.}$$

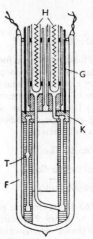

Mechanical equivalent of heat (Laby and Hercus)

ladder filter A network consisting of a succession of series and shunt impedances, usually acting as a transmission line with a known attenuation or delay. *See* filter.

laevorotatory *Syn.* laevorotary. Capable of rotating the plane of polarization of polarized light in an anticlockwise direction as viewed against the direction of motion of the light. *See* optical activity.

lag 1. Of a periodically varying quantity. The interval of time or the angle in *electrical degrees by which a particular phase in one wave is delayed with respect to the similar phase in another wave. (*Compare* lead.)
2. The time elapsing between the transmission and reception of a signal.
3. In a control system. The delay between a correcting signal and the response to it.
4. In *camera tubes. The persistance in the electrical image which may last for several frames. *Compare* delay time.

lagging current An alternating current that, with respect to the applied electromotive force producing it, has a *lag. *Compare* leading current.

lagging load *Syn.* inductive load. A *reactive load in which the inductive *reactance exceeds the capacitive reactance and therefore carries a *lagging current. *Compare* leading load.

Lagrange, Joseph Louis (1736–1813) Italian-born French mathematician. The creator (at the age of nineteen) of the *calculus of variations, he also made great advances in the treatment of differential equations and applied his mathematical techniques to problems of mechanics, especially those arising in astronomy. In his great work, *Mécanique Analytique* (1788), he used the conservation of mv^2 as the foundation of his mechanics and applied the principle of virtual velocities and the *least-action principle. Lagrange gave (1758) the complete solution of the problem of transverse vibrations of a string, and discussed beats and other acoustical phenomena. His generalized equations of motion were formulated in 1780. He had already introduced the idea of potential (for the gravitational case), 1777. Among other contributions to science, he developed the theory and practice of interpolation. He was also influential in establishing the metric system in France after the Revolution.

Lagrange law *Syn.* Smith–Helmholtz equation. If, from a small object y perpendicular to the axis in a medium of refractive index n, a ray from the axial point of the object makes an angle α with the axis, then using the accented letters for the image conjugates,

$$ny \tan \alpha = n'y' \tan \alpha'.$$

Lagrange's equations A set of second-order differential equations for a system of particles that relate the kinetic energy T of the system to the generalized coordinates q_i, the generalized forces Q_i, and the time t. There is one equation for each of the n *degrees of freedom possessed by the system:

$$\frac{d}{dt}\left(\frac{\partial T}{\partial \dot{q}_i}\right) - \frac{\partial T}{\partial q_i} = Q_i, \quad (i = 1, 2 \ldots n).$$

Here $\dot{q}_i$ denotes dq_i/dt.

Lagrange's equations provide a uniform method of approach for all dynamical problems.

Lagrangian function *Syn.* Lagrangian; kinetic potential. Symbol: L. An expression for the kinetic minus the potential energy in a conservative system:

$$L = T(q_i, \dot{q}_i) - V(q_i, \dot{q}_i),$$

where q_i are the *generalized coordinates. The Lagrangian formulation of dynamics has an advantage over Newton's formulation in that, instead of dealing with many vector quantities, one has only to deal with two scalar functions, T and V. *See also* Lagrange's equations; Hamilton's principle.

Lagrangian points Five points in space, associated with a system of two bodies orbiting around a common centre of mass, at which a small body can maintain a stable orbit despite the gravitational influence of the two much more massive bodies. Groups of minor planets are found at two Lagrangian points in the sun-Jupiter gravitational field (60° ahead and behind Jupiter in its orbit). The other three points, along the line joining the centres of mass, are in unstable equilibrium; this is the case in any system.

Lalande cell A primary *cell with zinc and iron electrodes in caustic soda solution as electrolyte and with copper oxide as depolarizer.

Lamb, Willis Eugene (*b.* 1913) Amer. physicist who was awarded a Nobel prize in 1955 for his work on the *Lamb shift.

lambda particle Symbol: λ. An uncharged elementary particle, a *hyperon, with spin $\frac{1}{2}$ and a mass about 1.1 times that of the proton. It can replace a neutron in a nucleus to form an extremely unstable *hypernucleus.*

lambda point (λ) **1.** The temperature (2.186 K at the equilibrium vapour pressure) at which the two forms of liquid helium can exist together. *See* superfluid.

2. *See* Simon and Bergmann's specific-heat-capacity experiments.

lambert A former unit of *luminance equal to 1 lumen of flux emitted per square centimetre of surface assumed to be perfectly diffusing. In terms of the *candela:

$$1 \text{ lambert} = 1/\pi \text{ cd cm}^{-2}.$$

Lambert, Johann H. (1728–1777) German mathematician, whose contributions to physics were mainly concerned with heat and light, especially photometry and colour.

Lambert's law 1. *Syn.* cosine law of emission. The *luminous intensity of a small element of a perfectly diffusing surface in any direction is proportional to the cosine of the angle between the direction and the normal. The law is used to define the perfect diffuser, and since brightness is defined as the luminous intensity in a prescribed direction per unit area projected perpendicular to the direction, such a surface will have the same photometric luminance in different directions. *See* photometry.

2. *See* linear absorption coefficient.

Lamb shift A small difference in energy between the energy levels of the $^2S_{1/2}$ and $^2P_{1/2}$ states of hydrogen. These levels would have the same energy according to the *wave mechanics of *Dirac. The shift can be explained by a correction to the energies on the basis of the theory of the interaction of electromagnetic fields with matter (*see* quantum electrodynamics) in which the fields themselves are quantized.

The effect was verified by experiments upon a beam of atomic hydrogen entering an evacuated container from a small vessel at high temperatures in which molecules were dissociated. Atoms in the beam were excited to the $^2S_{1/2}$ state by electron impact. This state is metastable as a *selection rule forbids decay to the $^1S_{1/2}$ ground state and the very small energy excess above the $^2P_{1/2}$ state gives a very low probability of spontaneous decay to this state. The metastable excited atoms were detected by their ejecting electrons from a fine wire. The beam passed through a cavity into which microwave radiation was fed. When the frequency ν was such that the quantum energy $h\nu$ was equal to the difference of energy between the states, atoms went from the $^2S_{1/2}$ to the $^2P_{1/2}$ state by stimulated emission (*see* Einstein coefficients) and decayed immediately to the ground state, thus the current at the detector falls. As the frequency of the microwaves cannot be varied easily, the energy difference between the states was varied in a known way by applying a magnetic field.

lamellar crystal A *mosaic crystal composed of thin sheets.

lamellar field A *field in which the *vector associated with the field is derivable from a scalar potential (by taking its *gradient). The scalar potential field may be mapped by level surfaces (or laminae), hence the name. Such a vector field has no *curl.

laminar flow Steady flow in which the fluid moves in parallel layers or laminae, the velocities of the fluid particles within each lamina not being necessarily equal. In the motion of a fluid through a straight horizontal pipe the velocities of the particles within each lamina are the same until the *critical velocity is attained and the motion changes from laminar to *turbulence. When a velocity potential exists the flow is called *potential flow* and is essentially laminar flow, although the laminae are not necessarily plane.

lamination A thin steel or iron stamping, oxidized or lightly varnished on the surface, a number of which can be built up to form the *core of a transformer, transductor, relay, choke, or similar apparatus. The laminations reduce losses by preventing *eddy currents circulating in the core.

Lamy's theorem (B. Lamy, 1679) If a particle is in equilibrium under the action of three forces P, Q, and R, then:
$$P/\sin \alpha = Q/\sin \beta = R/\sin \gamma,$$
where α is the angle between Q and R, β the angle between R and P, and γ the angle between P and Q.

LAN Abbreviation for *local area network.

Lanchester's rule *See* precession.

Land, Edwin H. (1909–1968) Amer. inventor who devised a polarizer used as a camera filter and invented the Polaroid camera which takes, develops, and prints a photograph in less than one minute.

Landau, Lev Davidovich (1908–1968) Soviet physicist who contributed to the modern form of the *quantum theory and carried out research into the theory of condensed matter.

Landau damping The damping of a space charge oscillation by a stream of particles moving at a speed slightly less than the phase speed of the associated wave.

Landé factor *Syn*. g-factor. Symbol: g. A constant factor used in expressions for changes in energy level in a magnetic field. It is a correction for the fact that there is not a simple relationship between the total *magnetic moment of an atom, nucleus, or particle and its angular momentum. The Landé factor is necessary to explain fine structure in spectral lines due to coupling between orbital and spin angular momenta. It is also used in the magnetic moments of particles resulting from their *spin. For example, a nucleus with a spin quantum number I has a magnetic moment given by:
$$g\sqrt{[I(I + 1)]} \cdot \mu_N,$$
where μ_N is the nuclear *magneton and g is a constant for a particular nucleus.

landscape lens A simple *meniscus lens or achromatic doublet with front or rear stop rarely greater than $f/11$ used as a cheap photographic lens; oblique astigmatism restricts the useful field to about $40°$.

Langevin, Paul (1872–1946) French physicist. He developed the theory of magnetic susceptibility of a paramagnetic gas (1905). He obtained ultrasonic waves in water by using piezoelectric vibrations of quartz, developing the method in World War I for detection of submarines. He also carried out important work on Brownian motion and on secondary X-rays.

Langevin–Florissen sounder A device based on a suggestion made at the time of the *Titanic* disaster in 1912 by L. E. Richardson, who proposed the use of an ultrasonic beam to detect submerged objects such as icebergs or wrecks by *echo sounding.

The Langevin–Florissen system for ultrasonic echo sounding consists of a matrix of piezoelectric quartz crystals in an oscillator of high power. Electric sparks generate oscillations in a circuit containing the crystals. The circuit is tuned until its frequency equals the natural frequency of the crystal. Therefore each spark generates an electrical damped train of waves that are transferred into mechanical oscillations in the crystals and so into a train of compressional waves in the water that is in contact with the crystal. To ensure that plane waves are used, the crystals are inserted between two large metal plates forming the electrodes and having a diameter of 20 to 30 cm. The semiangle of the primary conical sound beam emitted or received is given by:
$$\sin^{-1} 1.22\lambda/D,$$
where λ is the wavelength (*see* diffraction of sound). For a frequency of 30 kilohertz, λ will be equal to about 5 cm in water and if D, the diameter of the plate, is 30 cm then the semiangle is $12°$. Thus the waves can be considered as effectively plane and directed in a beam just like a beam of light when concentrated through an aperture. These plates are cemented to the opposite faces of the quartz. The reflected wave train from the sea bed or the wrecked vessel that is being detected is received by the transmitter and the resulting oscillation of the quartz generates an alternating e.m.f. that is amplified electronically. The amplifier will pass the fluctuations in the circuit due to departure and arrival of ultrasonic waves to a sensitive string galvanometer contained in an analyser or a cathode-ray oscilloscope. This analyser sets in motion at constant speed a spot of light moving along a scale and exhibiting kicks in its motion at the passage and return of the waves. The apparatus can be adjusted so that the first kick registers the depth of the transmitter below the surface of the water and the second kick indicates on the scale the depth of the surface from which reflection occurs.

The ultrasonic beam may also be directed horizontally and the reflection received from a submarine or from the hull of another ship.

langley A former unit of energy density used for solar radiation; it is equal to one calorie per cm^2 or 4.1868×10^4 J m^{-2}.

Langley, Samuel Pierpont (1834–1906) Amer. physicist. He investigated the infrared spectrum, designing a sensitive bolometer for the purpose. He followed the solar spectrum beyond 5000 nm, studied fluctuations in the *solar constant, and investigated atmospheric absorption. He also made aeronautical studies and built a pilotless aircraft, which achieved half a mile (1896).

Langmuir, Irving (1881–1957) Amer. chemist and physicist. In his well-known work on surface chemistry, Langmuir demonstrated the occurrence of unimolecular surface films. His extensive work on gases and vapours and discharges in gases produced the *Langmuir vacuum pump. Langmuir also carried out extensive convection studies and verified the *five-fourths power law proposed by Lorentz; he also invented the gas-filled tungsten lamp. Following G. N. Lewis, he developed a theory of isomorphism based on similarity of outer electronic structures of the ions concerned (1919, 1921). The Lewis–Langmuir atomic mode, with its tendency of electrons to form octets, was a static counterpart of the electron-shell model that supplanted it. *See also* Langmuir effect.

Langmuir–Blotchett film (LB film) An ordered layer of organic molecules that is formed on a solid surface as it is passed through the surface of a liquid on which is spread a quasi-solid monomolecular layer of the film material. Perfect films can now be achieved. Multilayer structures are also being produced. Large areas can be fabricated. LB films have many (potential) applications, for example as insulating coatings and layers in transistors; multilayer films have even greater potential.

Langmuir effect An ionization that occurs when atoms of low *ionization potential come into contact with a hot metal of high *work function. It has been used in the production of intense beams of ions of such elements as the alkali metals.

Langmuir isotherm An *isotherm with the equation:
$$\theta = bp(1 + bp),$$
relating the fractional coverage (θ) of a solid surface by adsorbed gas to the equilibrium gas pressure (p) where b is a constant. The equation was derived on the assumption that adsorption occurs onto discrete surface sites and that each adsorbed molecule or atom occupies one such site.

Langmuir probe A small electrode that is inserted into a *gas-discharge tube at a point along its length. The current drawn by the electrode can be used to determine the potential at the point at which the electrode is inserted.

Langmuir vacuum pump A condensation pump in which mercury, issuing from an annular orifice, impinges on the water-cooled walls of the pump where it rapidly condenses.

lanthanide contraction The steady decrease in the size of atoms and ions of the elements in the lanthanide series with increasing *atomic number. The lanthanide series is composed of the elements between lanthanum and lutetium. Each element in the series has one extra nuclear charge and one more $4f$ electron than the preceding element. The contraction results because f electrons are inefficient at shielding other f electrons from the nuclear charge and as the atomic number increases, the $4f$ shell experiences a greater effective nuclear charge.

Laplace, Pierre Simon, Marquis de (1749–1827) French mathematician and astronomer. An extremely able scientist, he borrowed freely from others, often without proper acknowledgment. His work on the solar system was published in his *Mécanique céleste* (5 vols., from 1799 to 1805). Laplace's hypothesis that the solar system evolved from a rotating nebula was published in 1796. He also introduced the *Laplace equation for the potential in free space.

Laplace made important contributions to the theory of probability and to the method of least squares, and to the solution of certain differential equations by transformation. He gave an extensive treatment of capillarity on the basis of molecular short-range forces. He corrected (1816) Newton's formula for the speed of sound. (*See* Laplace equation (for speed of sound).)

Laplace carried out some experimental work in physics in collaboration with Lavoisier, chiefly the determination of specific heat capacities (approx. 1780–1784) by use of an elaborate form of ice calorimeter.

Laplace coefficients Spherical harmonics that first appeared in Laplace's treatment of the gravitational attraction of a spheroid on an external point (1784). They are the analogues in three dimensions of Legendre polynomials (*see* Legendre equation).

Laplace equation 1. A linear differential equation of the second order:
$$\frac{\partial^2 V}{\partial x^2} + \frac{\partial^2 V}{\partial y^2} + \frac{\partial^2 V}{\partial z^2} = 0.$$

V may, for example, be the potential at any point in an electric field where there is no free charge.

2. For speed of sound. An equation relating the speed of sound (c) in a gas to the density (ρ), pressure (p), and ratio of heat capacities (γ) of the gas. It has the form:
$$c = \sqrt{(\gamma p / \rho)}.$$
See dispersion of sound in a gas.

Laplace operator The differential operator

$$\left(\frac{\partial^2}{\partial x^2} + \frac{\partial^2}{\partial y^2} + \frac{\partial^2}{\partial z^2}\right),$$

often represented by the symbol ∇^2. *See* del.

lapse rate The rate of decrease of a quantity, usually temperature, with height in the atmosphere. The *dry adiabatic lapse rate* (DALR) is the rate of cooling for dry air that is subjected to an adiabatic ascent. Its value is equal to 9.76 °C per kilometre. This is also the lapse rate for moist air that remains unsaturated.

large-scale integration (LSI) *See* integrated circuit.

Larmor, Sir Joseph (1857–1942) Brit. physicist, noted for contributions to atomic physics and electron theory.

Larmor precession A uniform magnetic field applied to a plane electron orbit of an atom causes the plane to precess about the direction of the field in such a way that the normal to the plane traces a cone with its axis in the field direction. The frequency of precession is given by:
$$\nu = eB/4\pi m,$$
e being the electron charge and m its mass, and B the magnetic flux density.

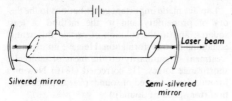

Silvered mirror   Semi-silvered mirror

Gas laser operating as electrical oscillator

laser (*l*ight *a*mplification by *s*timulated *e*mission of *r*adiation) A source of near monochromatic radiation in the visible, ultraviolet, and infrared regions of the spectrum. (X-ray lasers are under development.) It operates by the production of a large population of electrons in certain high-energy states in a medium. These electrons are then stimulated to emit a narrow beam of *coherent radiation.

At normal temperatures, the majority of electrons in a system are in the *ground state. Following *excitation, electrons raised to higher energy states will return to lower levels or to the ground state with the emission of *photons. This emission is usually spontaneous and cannot be controlled. *Stimulated emission* is a process whereby an incoming photon of energy $h\nu$ (where h is the *Planck constant and ν the frequency) can stimulate an electron in a high-energy state E_1 to jump to a lower energy state E_2, where:
$$E_1 - E_2 = h\nu.$$
The photon resulting from this process has the same frequency,
$$\nu = (E_1 - E_2)/h,$$
as the stimulating photon and travels in the same direction. If there are sufficient electrons in the high-energy level, both stimulating and stimulated pho-

320

tons can cause further stimulated emission and a narrow beam of monochromatic radiation results, the intensity of which increases exponentially. The beam is coherent (i.e. spatially and temporally in phase) and can have a very high energy density.

A laser beam is produced by stimulated emission but can only operate efficiently if a large number of electrons are in a particular high-energy level. This condition, called *population inversion*, is a nonequilibrium condition and power must be fed into the system to maintain the inversion.

Laser action has been achieved in gaseous, solid, and liquid media. The beams may be pulsed or continuous wave (CW), and vary widely in the power generated and the overall efficiency. The first laser (1958) was the pulsed ruby laser, output wavelength 694.3 nm; it is still in use. Chromium ions in the ruby lattice are excited by an intense flash of light. Electrons, raised to a high-energy state, decay immediately to a slightly lower *metastable state, and thus population inversion occurs. The level slowly depopulates and this spontaneous emission triggers stimulated emission of the same frequency. The photons are reflected back and forth along the cylindrical ruby crystal between two parallel flat external mirrors, one only partially silvered. During this process stimulated emission continuously builds up the number of photons, so increasing the power; a small percentage of photons emerge from the semisilvered end, forming an intense pulse of light lasting about a millisecond. The laser plus mirrors act as an optical *cavity resonator.

There are many other solid-state lasers, their outputs varying in wavelength and power. For example, trivalent rare earths undergo laser action in a variety of media, including yttrium aluminium garnet (YAG) and glass (which may both be doped with neodymium). *Semiconductor lasers are especially important, being small, robust, and cheap to produce.

Gas lasers form another large group, operating across the spectrum from far infrared to far ultraviolet. Examples include the helium-neon, argon, krypton, carbon dioxide, and hydrogen fluoride lasers. The helium-neon laser usually consists of a *gas-discharge tube containing a mixture of helium and neon at low pressure. Helium atoms, excited by a continuous electrical discharge, collide inelastically with neon atoms and an energy transfer occurs. The helium atom relaxes to its ground state and the neon atom becomes excited, a large population inversion occurring in certain high-energy levels. Once stimulated emission has been triggered, a continuous laser beam is obtained at wavelengths including 1.152 μm, 3.391 μm (both infrared), and 632.8 nm (visible). The continuous output is usually a few tens of milliwatts.

The carbon dioxide (CO_2) laser has a much greater efficiency. The usual medium is a mixture of CO_2 and nitrogen in a gas-discharge tube. Vibrationally excited N_2 atoms collide with and excite CO_2 atoms, and laser action occurs on a rotation-vibration transition of the electronic ground state of CO_2. The output, at 10.6 μm, can be between 0.5–2 kW for a continuous beam and up to 50 kW for a pulsed output.

As with the solid-state lasers, two reflecting surfaces are used with the gas-laser tube to form a cavity resonator. One example, with the reflectors at the *Brewster angle, is shown in the diagram.

Liquid lasers are frequently liquid organic dyes. These dye lasers have been made to lase at frequencies from the IR to the UV. They have the advantage that they can be tuned continuously over a range of wavelengths.

Chemical lasers are compact inexpensive lasers that produce more than 1 kW of power without the necessity of a large power supply. Their efficiency, over 12%, is higher than electric lasers. In one such device, sulphur hexafluoride (SF_6) is dissociated into fluorine and sulphur atoms. The fluorine expands through a nozzle forming a supersonic jet into which hydrogen is diffused. The reaction between hydrogen and fluorine produces vibrationally excited hydrogen fluoride (HF*), in two vibrational states and several rotational states. Photons arising spontaneously from thermal noise in the cavity interact with the HF*. Several changes in energy level can occur accompanied by the emission of laser light in the wavelength region 2.6 to 2.9 μm.

The laser is not truly monochromatic but has a linewidth of the order of 10^4 hertz. This is narrower than other radiation sources, the best "monochromatic" beams having a linewidth between 10^8–10^9 Hz. An important feature of the laser is the fact that the beam emerges almost perfectly collimated, apart from the very small diffraction associated with the large aperture. In consequence, laser beams can be brought to a very fine focus in which the intensity (power/area) can attain extremely high values. Many of the applications depend upon this. *See also* holography; Einstein coefficients.

latch An electronic device in which a single bit of data is stored temporarily. The storage is under the control of a clock signal, a given transition of which fixes the contents of the latch at the current value of its input. The contents remain fixed until the next transition. The latch is an extension of a simple *flip-flop.

latent heat Symbol: L. The quantity of heat absorbed or released in an isothermal transformation of phase. The quantity of heat released or absorbed per unit mass is called the specific latent heat. (*See* specific latent heat of fusion; specific latent heat of vaporization; specific latent heat of sublimation.) The quantity of heat absorbed or released per unit amount of substance (per mole) is called the *molar latent heat*.

latent heat equations Thermodynamic equations involving the latent heats of a substance. *See* Clausius–Clapeyron equation; Clausius's equation.

latent image *See* photography.

latent magnetization The property possessed by certain metals, such as manganese and chromium, that are weakly magnetic in themselves but form strongly magnetic alloys or compounds.

lateral aberration As applied to chromatic *aberrations, lateral aberrations are chromatic differences of magnification, or sometimes the actual difference in sizes of the images of an object for two colours, e.g. F and C lines. When applied to the geometric theory of spherical aberration, it is the distance from the axis at which a ray from an outer zone intersects a plane normal to the principal axis through the paraxial focus. The term has also been applied to the difference between the *reciprocals* of the distances from the lens or surface to the paraxial focus, and the intersection of a zonal ray with the axis.

lateral magnification The inverse ratio of the size (y) of an object perpendicular to the axis of a reflecting or refracting system, to the size (y') of the image, i.e. $m = y'/y$. According to the sign convention adopted, the algebraic sign of m will determine whether the image is erect or inverted.

latitude 1. The geographical latitude of a place is the angle between the normal to the earth's surface and the plane of the equator. It is the latitude found from astronomical observations.

2. The geocentric latitude is the angle between a line joining the place to the earth's centre and the plane of the equator. It differs from the geographical latitude because the earth is an oblate spheroid and the maximum difference (at latitude 45°) is equal to 11'36".

3. Celestial latitude. *See* celestial sphere.

lattice 1. A regular repeated three-dimensional array of points that specify the positions of atoms, molecules, or ions in a crystal. *See* Bravais lattice; crystal systems.

2. The internal structure of the *core of a *nuclear reactor, consisting of a regular array of *fissile material and nonfissile material, especially a *moderator.

lattice constant The length of edge or the angle between the axes of the *unit cell of a crystal. It is usually the edge length of a cubic unit cell.

lattice dynamics The study of the excitations that a crystal *lattice can experience and their effects on the thermal, optical, and electrical properties of solids.

Laue, Max von (1879–1960) German physicist. He suggested (1911) that X-rays might be diffracted from the atoms of a crystal, if these were regularly arranged in a three-dimensional lattice. This was confirmed experimentally by W. Friedrich and P. Knipping (1912). If the incident X-rays form a narrow pencil and contain a continuous range of wavelengths, a photographic plate receiving the diffracted radiations records an array of Laue spots, the whole being a *Laue diagram. For this work he was awarded a Nobel prize (1912). Laue's theory

Laue diagram

was comprehensive and somewhat elaborate. The Bragg treatment (*see* Bragg's law), which is much simpler, was produced soon afterwards.

M. von Laue also made important contributions to the special theory of *relativity.

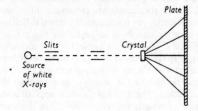

Laue diagram

Laue diagram A symmetrical pattern of spots produced on a photographic plate by a collimated beam of X-rays that have passed through a stationary single crystal (*see* diagram). There is a range of wavelengths present in the X-ray beam, which usually comes from a tungsten source operated at a voltage too low to excite characteristic K radiation (*see* X-rays). A heterogeneous beam of electrons or neutrons could also be used. The different atomic planes diffract the beam and give rise to the series of symmetrically arranged spots on the Laue diagram. From the diagram it is possible to determine the type of crystal and calculate its crystal structure.

Laurent half-wave plate *See* half-wave plate.

Laurent saccharimeter *See* half-wave plate.

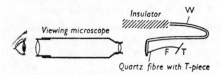

Lauritsen electroscope

Lauritsen electroscope An electroscope in which a metal wire W, as shown, carries a metal-coated quartz fibre, F. It is attracted to the wire when under charge. A T-piece, T, on the fibre is viewed by a microscope with an eyepiece scale.

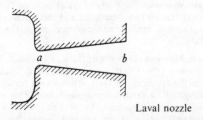

Laval nozzle

Laval nozzle Invented by De Laval (patented 1889) and designed originally to increase the efficiency of steam turbines by increasing the velocity of discharge of the steam. The nozzle has uses with regard to general fluids when it is necessary to increase the

velocity of flow. The cross section is illustrated in the diagram, the nozzle being essentially of a convergent to divergent nature. The important feature of the Laval nozzle is the straight-walled divergent pipe from *a* to *b*. The angle of divergence is approximately 10°; and the length from *a* to *b*, and the areas of cross sections, are so determined that the expansion of fluid between *a* and *b* is adiabatic and the flow is streamline. The complete expansion of the fluid increases the flow velocity at *b*.

A *wind tunnel for supersonic flow often takes the form of a Laval nozzle, with the velocity of discharge in the contraction exceeding that of sound.

Lavoisier, Antoine Laurent (1743–1794) French chemist. He established the nature of combustion and the composition of water (naming oxygen and hydrogen) and the conservation of mass in chemical change. One of the founders of modern chemistry, he nevertheless was executed by guillotine in the French Revolution. For his specific heat capacity measurements, *see* Laplace.

law A theoretical principle deduced from particular facts expressed by the statement that a particular phenomenon always occurs if certain conditions are present.

Lawrence, Ernest Orlando (1901–1958) Amer. physicist, a pioneer in the construction of particle *accelerators capable of delivering charged particles at high energies. The *linear accelerator had to await wartime development of high-power oscillators of very high frequency before its potentialities could be realized but the *cyclotron was at once successful and the forerunner of many varieties of cyclic accelerator. Nobel prize for physics, 1939. The artificially prepared element *lawrencium* (atomic number 103) is named in his honour.

Lawson criterion The product of the particle density of a *plasma (in particles per cm^3) and the *containment time (in seconds) at or above its *ignition temperature, such that the fusion energy released equals the energy required to produce and confine the plasma. For the deuterium-tritium reaction the value of the Lawson criterion is 10^{14} s/cm^3. *See* fusion reactor.

layer structure *Syn.* layer lattice. A crystalline arrangement in which the forces between atoms along one general direction are weaker than those at right angles or nearly at right angles to this direction, so that the atoms tend to form layers.

lay ratio Of the core of a cable or the wire of a stranded conductor. The ratio of the axial length of one complete turn of the helix to the mean diameter of the helix.

LC A prefix to an electronic device, circuit, or effect (as with LC filter, LC network, LC coupling) whose action is based on the properties and arrangement of one or more inductors and capacitors.

LCC Abbreviation for *leadless chip carrier.

LCD Abbreviation for *liquid-crystal display.

lead 1. An electrical conductor.

2. Of a periodically varying quantity. The interval of time or the angle in *electrical degrees by which a particular phase in one wave is in advance of the similar phase in another wave. *Compare* lag.

lead accumulator *See* accumulator.

lead-covered cable A cable having a sheath of commercially pure lead or of lead alloy for the purpose of excluding moisture from the insulation and the conductors. The lead-covered cable is plain if it has no exterior layer of protecting material, and served (*see* serving) if it is provided with an exterior layer of protecting material, such as jute, yarn, or tape. For many purposes lead-covered cables have now been replaced by plastic-covered cables.

lead equivalent A measure of the absorbing power of a radiation screen expressed as the thickness of metallic lead (usually in millimetres) that could give the same protection as the given material under the same conditions.

leader stroke The initial discharge that establishes the track of a *lightning flash. It is described as downward or upward according to whether it develops from the cloud towards earth or from the earth towards the cloud, respectively. If its development is continuous, it is called a *dart leader stroke* but if it develops in a series of definite steps of relatively short length, it is called a *stepped leader stroke*. A high-current discharge that flows upwards through a lightning path as soon as a downward leader stroke has made contact with earth is called a *return stroke*.

leading (and **trailing**) *See* brush; pole horn; pole tip.

leading current An alternating current that, with respect to the applied electromotive force producing it, has a *lead. *Compare* lagging current.

leading load *Syn.* capacitive load. A *reactive load that takes a *leading current. *Compare* lagging load.

leadless chip carrier (LCC) A form of package commonly used for integrated circuits in which connections to the device are made by means of small metallic contacts arranged around the outer periphery of the package, flush with the edges. This allows the package to be inserted in, for example, printed circuit boards.

leakage 1. The flow of an electric current, due to imperfect insulation, in a path other than that intended.

2. *See* magnetic leakage.

3. A net loss of particles from a region or across a boundary in a nuclear reactor.

leakage current A *fault current due to *leakage. It is small in magnitude compared with that of a short circuit.

leakage flux In any electrical machine or transformer in which there is a magnetic circuit. The flux that is outside the useful portion of the flux circuit.

leakage indicator An instrument that measures or indicates the value of a *leakage current to earth in an electrical system.

leakage protective system *See* protective system.

leakage reactance Reactance caused in a transformer by the leakage inductance associated with losses due to some of the magnetic flux cutting one coil but not the other.

leap second *See* time.

least-action principle A principle, first propounded by Maupertius (1740), which is applicable to conservative dynamical systems. It is less useful than *Hamilton's principle and states that the action has a stationary value for the actual path, as compared with various paths between the same points for which the total energy has the same constant value. That is,

$$\delta \int_{t_0}^{t_1} 2T dt = 0,$$

where T is the kinetic energy of the system at any time t.

least-energy principle A dynamic system is in stable equilibrium only if the potential energy of the system considered as a whole is a minimum. *See* Le Chatelier's rule.

least squares, method of A technique for finding the equation that gives the best fit for a set of experimental data. A simple example is in fitting a linear equation $y = ax$ to a set of measurements of the dependent variable y in terms of the independent variable x. A value of a is chosen and the *deviation of experimental values can be obtained. The best fit is considered to occur for a value of a at which the sum of the squares of these deviations is a minimum. The method is due to Gauss and Legendre (1806).

least-time principle *See* Fermat's principle.

leaving edge *See* brush.

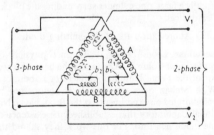

Leblanc connection

Leblanc connection A method of interconnecting transformer windings to provide *three-phase to *two-phase transformation or vice versa. In the diagram shown, A, B, and C are the primary windings wound on three separate cores, or on the limbs of an ordinary three-limb core as used for three-

Le Chatelier, Henri Louis

phase transformers. If on the two-phase side, $V_1 = V_2 = V$ then the voltages of the secondary windings must be:

Winding:	a_1	b_1	b_2	c_1	c_2
Voltage:	$\frac{2}{3}V$	$\frac{1}{3}V$	$V/\sqrt{3}$	$\frac{1}{3}V$	$V/\sqrt{3}$

Compare Scott connection.

Le Chatelier, Henri Louis (1850–1936) French chemist, a pioneer in physical chemistry, particularly chemical thermodynamics. He enunciated the principle of mobile equilibrium in 1884. *See* Le Chatelier's rule.

Le Chatelier's rule (or **principle**) *Syn.* Le Chatelier–Braun principle. When a constraint is applied to a dynamic system in equilibrium, a change takes place within the system, opposing the constraint and tending to restore equilibrium.

lecher wires If oscillations of very high frequency ($>10^8$ hertz) are produced in two parallel wires, a *Geissler tube placed across them will glow or remain dark at various equally spaced positions as it is arranged to bridge the wires and is moved along them. There are in fact nodes and antinodes of potential in a manner similar to those acoustically formed in an organ pipe or in *Kundt's tube.

Leclanché, Georges (1839–1882) French chemist who invented (1867) the *Leclanché cell.

Leclanché cell A primary *cell in which the anode is a rod of carbon, and the cathode a zinc rod, which may be amalgamated. The electrolyte is 10–20% NH_4Cl solution. The depolarizer consists of manganese dioxide mixed with graphite or crushed carbon, contained in a fabric bag or porous pot. The e.m.f. is about 1.5 volts, but falls off fairly rapidly on closed circuit as the depolarizer is slow in action. The cell is particularly useful for intermittent applications. In another form, the *agglomerate cell*, an attempt is made to reduce the internal resistance by having the depolarizer made into solid blocks held to a carbon plate by rubber bands. In this variant the cathode is usually a large zinc cylinder surrounding the blocks. *See* dry cell.

LED Abbreviation for *light-emitting diode.

Lederman, Leon Max (*b.* 1922) Amer. physicist. A professor at Columbia University, he became director of the Fermi National Accelerator Laboratory. With Melvin Schwartz and Jack Steinberger he collaborated on experiments at the Brookhaven National Laboratory that established the existence of the muon neutrino. For this work they shared the 1988 Nobel prize for physics.

Leduc effect *Syn.* Righi effect. If heat is flowing through a metal strip, a magnetic field set up perpendicular to the plane of the strip causes a temperature difference to appear across the strip. The disposition of the higher and lower temperature regions depends on the metal of which the strip is composed. The effect is related to the *Nernst effect.

Lee, Tsung Dao (*b.* 1926) Chinese-born Amer. particle physicist who, together with *Yang, first showed that *parity is not conserved in *weak interactions.

Lees, C. H. (1869–1952) Brit. physicist, best known for his extensive work on heat transmission.

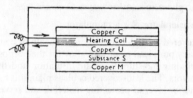

Determination of thermal conductivity of bad conductors (Lees)

Lees' disc For the determination of the *thermal conductivity of bad conductors. The specimen in the form of a thin cylindrical slice S (*see* diagram) is sandwiched between copper blocks, U, M, and heated electrically by a coil contained between the copper discs C, U. The temperatures of all copper blocks are determined by thermocouples. Thermal contact is established between all contacting surfaces by smearing them with glycerine. The apparatus is varnished all over so that all surfaces have the same heat transfer coefficient E and is suspended in a constant temperature enclosure. Then, if x is the thickness of a disc of radius r and θ denotes the excess of temperature over that of the enclosure when the steady state is reached, the thermal conductivity W is given by:

$$W\left(\frac{\theta_U - \theta_M}{x_S}\right) = E\left[\theta_M + \frac{2}{r}\left\{\left(x_M + \frac{x_S}{4}\right)\theta_M + \frac{x_S}{4}\theta_U\right\}\right],$$

assuming that the heat flowing through the specimen S is the mean of that flowing into it and the heat flowing out of it. The value of E is obtained in terms of H, the rate at which heat is supplied by the heater in the steady state by equating H to the total rate of emission of heat from the various surfaces:

$$H = \pi r^2 E\left[\theta_C + \theta_M + \frac{2}{r}\left(x_M\theta_M + x_S\frac{\theta_M + \theta_U}{2}\right.\right.$$

$$\left.\left. + x_U\theta_U + x_C\theta_C\right)\right].$$

Lees' method for the thermal conductivity of metals at low temperatures The specimen is in the form of an accurately turned cylinder R (*see* diagram), 8 cm long and 0.585 cm in diameter, placed vertically inside a Dewar vessel. The lower end fits into a copper disc D accurately turned to fit the copper tube T wound with platinoid wire p and carrying a separate heating coil P. Thin brass sleeves A, B, C, can be moved easily along R. A and B carry platinum thermometer coils and C a heating coil of resistance equal to that of the windings p round the tube T. The amount of heat supplied to the apparatus is made constant by arranging for the heating current to pass through p whenever it is switched off the coil C.

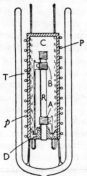

Determination of thermal conductivity of metals (Lees)

For determinations at low temperatures the Dewar vessel is filled with liquid air and after the apparatus has cooled down the remaining liquid is poured out. The heating coil P is used to raise the temperature of the tube quickly if desired. To allow for changes in temperature of the tube T during a determination, the resistances of the coils A and B are determined, together with the rate of doing work electrically, when the conditions are steady with current passing (1) through the tube coil p, (2) through the coil C, (3) through p again. The difference of resistance δ that would have existed between A and B had the tube remained at constant temperature may be shown to be given by:

$$\delta = \delta_2 - \tfrac{1}{2}(\delta_1 + \delta_3),$$

(where δ_1 and δ_3 are the differences found when the current flowed round the tube and δ_2 the difference when the current flowed round the rod), provided the rate of rise of temperature of the apparatus is constant, thus enabling the temperature gradient along the rod, and hence the thermal conductivity, to be measured.

Lees' rule For moments of inertia. Lees' rule is more elegant than *Routh's rule and has the great advantage that it includes the expression for the moment of inertia (I) of a cylinder about an axis through its centre, perpendicular to the cylinder axis:

$$I = \text{mass} \times \{a^2/(3 + n) + b^2/(3 + n')\},$$

where n and n' are the numbers of principal curvatures of the surface that terminates the semiaxes in question and a and b are the lengths of the semiaxes. Thus, if the body is a rectangular parallelepiped, $n = n' = 0$, and

$$I = \text{mass} \times (a^2/3 + b^2/3).$$

If the body is a cylinder as mentioned above, then $n = 0$ and $n' = 1$ and

$$I = \text{mass} \times (a^2/3 + b^2/4).$$

If I is desired about the axis of the cylinder, then $n = n' = 1$ and $a = b = r$ (the cylinder radius) and

$$I = \text{mass} \times (r^2/2).$$

Leeuwenhoek (pron. Layvenhook), **Anthony van** (1632–1723) Dutch microscopist. He made simple (i.e. one-lens) microscopes of considerable power and succeeded in observing bacteria, capillary blood vessels, etc.

left-hand rule For electric motor. *See* Fleming's rule.

Legendre equation The differential equation of the form

$$\frac{d}{dx}\left((1 - x^2)\frac{dy}{dx}\right) + ay = 0.$$

The solutions of this equation are known as *Legendre polynomials*.

Leibniz (or **Leibnitz**), **Gottfried Wilhelm von** (1646–1716) German mathematician, philosopher, and scientist. Inventor of the modern notation of differential calculus. He claimed to have priority over Newton in the invention of infinitesimal calculus but there is some evidence that his work was prompted by suggestions obtained from one of Newton's manuscripts. He developed the calculus, carrying out some integrations, and applying the methods to geometrical and mechanical problems.

Leibniz' theorem The nth differential coefficient of the product uv of functions u, v is given by the expansion:

$$\mathbf{D}^n(uv) = \mathbf{D}^n u \cdot v + n\mathbf{D}^{n-1}u \cdot \mathbf{D}v$$

$$+ \frac{n(n-1)}{1.2} \mathbf{D}^{n-2}u \cdot \mathbf{D}^2 v + \cdots$$

$$+ n\mathbf{D}u \cdot \mathbf{D}^{n-1}v + u \cdot \mathbf{D}^n v,$$

where D denotes the operator d/dx. The coefficients occurring in the expansion are the same as those in the *binomial expansion.

Lenard, Philipp (1862–1947) German physicist, chiefly noted for his work on cathode rays. He also made notable contributions to the theory of phosphorescence. The penetration of solid fuel by electrons (*see* Lenard window) led Lenard to postulate a dynamic model of the atom (about 1903), which had some of the attributes of the nuclear atom introduced later by Rutherford. Lenard discovered (1902) the dependence of the energy of electrons liberated by photoelectric action on the frequency of the radiation but not on its intensity.

Lenard rays Cathode rays, but the term is applied to cathode rays that have passed, through a metal foil window in their generating electronic tube, into the air.

Lenard's mass absorption law The absorption of electrons moving with a speed at least one-fifth of that of light is determined only by the mass of absorbing matter traversed and is unaffected by its chemical nature.

Lenard spiral The electrical resistance of bismuth increases considerably under the influence of a strong magnetic field at right angles to the direction of the current. The Lenard spiral consists of a small noninductively wound spiral of bismuth wire, mounted between mica plates. It is connected to one arm of a *Wheatstone bridge and is used to measure the strengths of magnetic fields by measurement of the change in its resistance.

Lenard window A very thin foil window first devised in 1893, through which fast electrons can pass out of an evacuated *gas-discharge tube.

length of lay Of the core of a cable or the wire of a stranded conductor. The axial length of one complete turn of the helix.

Lennard–Jones potential See van der Waals forces.

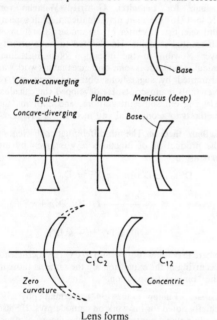

Lens forms

lens 1. A piece of transparent substance (commonly glass, plastic, quartz, etc.) bounded by two surfaces of regular curvature. Most commonly the surfaces are parts of spheres but they may be cylindrical, parabolic, toroidal, plane, etc. The general function of a lens is to change the curvature of wavefronts so that light may be focused to a desired position. Lenses are classed as *convergent* or *divergent*. Most lenses are intended to be *stigmatic*, uniting rays to point foci; others (see cylindrical lens; toric lens) have a different converging effect in meridians at right angles, producing two focal lines instead of one point focus – such lenses are *astigmatic*.

Lens shape should not be confused with *lens form* – the shape refers to that of the periphery while form

refers to the relative allocation of surface curvature (*see* diagram). *Lens transposition* and *bending* is a process involving change of form without change of central power, and is necessitated by the *gauge* to remove aberrations. A *lens desire* or *lens measure* is a mechanical device for reading directly the power of a surface of a lens depending on the spherometer principle, but its calibration is only correct for one refractive index of glass (generally spectacle glass of index 1.523). A *thin lens* is one whose thickness is small compared with the focal lengths of its surfaces or of the focal length of the lens: the *powers of the individual surfaces add up to the power of the lens. This is not true for thick lenses. (*See* lens formula.)

Designations, specifications, and lens terminology cover a wide range according to purpose, design, form, sign convention, etc. (*see* objective; eyepiece; Fresnel, concentric, spectacle, achromatic lenses, etc.). A *lens of zero curvature* is one whose thickness is equal to the distance between the centres of curvature of the surfaces (it is a converging meniscus). *See also* blooming of lenses; aberrations of optical systems; crystalline lens (of the eye).

2. See electron lens.

lens antenna A microwave *aerial with an electronic focusing arrangement placed in front of the radiator in order to produce a desired shape and direction in the radiated beam. This is achieved by introducing selected phase shifts over different paths through the lens, which could, for example, be a system of metal slats or shaped insulator segments.

lens formula The relationship between object distance, u, and image distance v, from a thin lens:
$$1/v + 1/u = 1/f,$$
where f is the *focal length. For a *real object or image, v and u are positive; for a virtual object or image, v and u are negative. For a convex lens, f is positive; it is negative for a concave lens. For thick lenses, refraction must be taken into account. The relation for refraction at a single curved surface is:
$$n_2/v + n_1/u = (n_2 - n_1)/2f,$$
where n_2 and n_1 are the refractive indexes of the lens material and the medium.

Lenz, Heinrich Friedrich Emil (1804–1865) German physicist, who did much of his work in Russia. His main work was on electrical conductivity, and he has a claim to have anticipated Joule in discovering the factors determining the heating produced by a current. His law governing the direction of electromagnetically induced currents (*see* Lenz's law) was published in 1834.

Lenz's law The current induced by any change in an electrical circuit or the surrounding magnetic field is in such a direction that it tends to oppose the change. *See* electromagnetic induction.

Leonardo da Vinci (1452–1519) Italian painter, scientist, and engineer. He studied optics, especially the action of the eye; mechanics, including the lever; hydrostatics and wave motion, as well as attaining eminence as an engineer, an architect, a physiologist, and an artist.

lepton The class of *elementary particles that do not take part in *strong interactions. They have no substructure of *quarks and are considered indivisible. They are all *fermions. There are six distinct types: the *electron, *muon, and *tauon (which all carry an identical charge but differ in mass) and the three *neutrinos (which are all neutral and thought to be massless or nearly so). In their interactions the leptons appear to observe boundaries that define three families, each composed of a charged lepton and its neutrino.

The families are distinguished mathematically by three *quantum numbers, l_e, l_μ, and l_τ, called *lepton numbers*:

Particle	l_e	l_μ	l_τ
e^-, ν_e	1	0	0
$e^+, \bar{\nu}_e$	−1	0	0
μ^-, ν_μ	0	1	0
$\mu^+, \bar{\nu}_\mu$	0	−1	0
τ^-, ν_τ	0	0	1
$\tau^+, \bar{\nu}_\tau$	0	0	−1
others	0	0	0

In *weak interactions the total lepton numbers l_e^{TOT}, l_μ^{TOT}, and l_τ^{TOT} (obtained by adding up the values of l_e, l_μ, and l_τ for the individual particles) are conserved.

Leslie, Sir John (1766–1832) Brit. physicist. Inventor of the differential air thermometer and *Leslie cube.

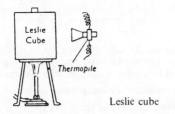

Leslie cube

Leslie cube A simple cubic metal box whose vertical faces have different surface finishes. Water may be boiled in the cube, and using a thermopile it may be shown that a dull surface is a better emitter than (say) a polished surface. See Ritchie's experiment.

LET Abbreviation for linear energy transfer. See ionizing radiation.

lethargy Symbol: u. The negative natural logarithm of the ratio of the energy (E) of a neutron to a specified reference energy (E_0), i.e.

$$u = -\log_e(E/E_0).$$

level The ratio of the magnitude of a quantity to a specified reference magnitude. A level generally implies measurement on a logarithmic scale and in acoustic terms the *decibel is usually employed. See sound-pressure level; band pressure level; loudness level.

level-compounded See compound-wound machine.

lever A simple *machine that can be regarded as a rigid bar that can turn about a pivot. The relative positions of load, effort, and pivot determine the type of lever.

levigation The reduction of a substance to powder by grinding in water followed by fractional sedimentation to separate the particles of different sizes.

levitation Any process by which a body is supported in a vacuum or in a gas by means of radiation or a field of force.

Laser levitation can be produced in a vacuum by projecting a continuous laser beam upwards and converging it with a lens. Spheres with radii of order 10^{-6} m to 10^{-4} m can be supported just above the focus using a beam of power about one watt. The suspended body is stable against vertical displacements because the support force falls off with height. To give stability against lateral displacements the intensity of the beam must vary across its area of cross section. To support reflecting spheres the centre of the beam must have lower intensity than the outer regions. The reverse is necessary for spheres supported more by refraction than by reflection. In each case the support depends upon the momentum of radiation, which is equal to the energy transfer divided by the speed of light.

Acoustic levitation is analogous to the above, being produced by focusing an ultrasonic beam in a gas.

Magnetic levitation (or *maglev*) is the repulsion of a magnetized body by a suitably magnetized surface. One form of this levitates a small magnet in the space over a dish-shaped piece of superconducting material, using the Meissner effect (see superconductivity.) A considerable amount of work has been done on transport systems in which vehicles are raised by magnetic levitation, so as to provide frictionless motion. These have generally proved to be expensive and uneconomic. Those maglev systems that have been constructed use a *linear motor for forward propulsion.

Lewis number Symbol: Le. A dimensionless number used in problems involving both heat and mass transfer. It is equal to $\lambda/\rho Dc$, where λ is the thermal conductivity, ρ is the density, D is the *diffusion coefficient, and c is the *specific heat capacity.

Leyden jar A historic form of *capacitor consisting of a glass jar, with metal foil on the outside and inside surfaces. The connector to the inside foil usually terminated in a small brass knob. (See illustration.)

LF Abbreviation for low frequency. See frequency bands.

Lichtenberg figures

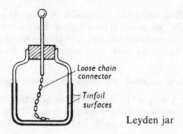

Loose chain
connector

Tinfoil
surfaces

Leyden jar

Lichtenberg figures The surface of a solid dielectric is affected by proximity of a potential high enough to produce ionization of the surrounding air. If fine powder is dusted over the surface and then blown by a current of air, some still remains adherent in a symmetrical star-shaped pattern, often of great complexity. This is called a Lichtenberg figure and is of use in the study of the properties of insulators under high fields. The figures can also be obtained on photographic emulsions and can be developed by ordinary photographic chemicals. In some dielectrics they can be "developed" directly on the dielectric by subsequent heating.

lidar (*li*ght *d*etection *a*nd *r*anging) A radar-like technique employing pulsed or continuous-wave *laser beams for remote sensing. It is used, for example, in atmospheric physics to study the distribution of clouds, dust particles, and pollutants. The incident laser beam is backscattered by a target of interest and detected by a receiver. The photons in the beam interact with the target species by elastic or inelastic *scattering.

lifetime 1. In a *semiconductor. The mean time interval between generation and recombination of a charge *carrier.

2. *See* mean life.

lift coefficient For a body and a fluid in relative motion. The ratio of the component of resistance force perpendicular to the direction of relative motion (lift) to the quantity $\rho l^2 V^2$, where ρ is the fluid density, V the relative velocity, and l is some characteristic length of the body. The quantity $\frac{1}{2}\rho V^2$ is sometimes used in place of ρV^2. The coefficient is a function of the *Reynolds number and is dependent on the circulation round the body. For an aerofoil the quantity l^2 is replaced by S, the area of projection of the aerofoil on the chord. *Compare* drag coefficient.

light The agency that when it strikes the retina of the eye serves as the adequate stimulus to excite the sensation of the same name. Theories of light include the obsolete *corpuscular theory; the elastic solid *ether theory involving transverse waves (Fresnel); the *electromagnetic theory (Maxwell), in which there are transverse oscillations of an electric field associated with similar transverse oscillations of a magnetic field at right angles to the electric field; the *quantum theory (Einstein), in which light is absorbed in packets of light quanta or *photons.

The electromagnetic-wave theory and the quantum theory are regarded as complementary; phenomena involving the propagation of light can be interpreted adequately on the wave basis, but when interactions of light (e.g. the *photoelectric effect) are under consideration, the quantum theory must be employed.

The wavelength range of light in nanometres for normal persons is approximately 400 (violet) to 730 (red). When light frequencies are multiplied by the *Planck constant (h) 6.636×10^{-34} J s, the energy of the photon for the frequency concerned is obtained. *See also* colour; speed of light.

light-emitting diode (LED) A small cheap p-n junction *diode formed from certain *semiconductor materials, e.g. gallium arsenide, in which direct radiative recombination of excess electron-hole pairs is possible: a photon of radiation is emitted as an electron in the conduction band recombines with a hole in the valence band (*see* energy bands). When the p-n junction is *forward biased, light emission can take place; the intensity is proportional to the bias current, i.e. to the numbers of excess minority carriers. The useful light obtained will be dependent on the optical quality of the crystal surfaces, and the colour will depend on the particular material used. Such diodes are used as small indicator or warning lights and in self-luminous display devices.

light exposure Symbol: H_v, H. **1.** The surface density of the total quantity of light received by a material.

2. A measure of the total amount of light energy incident on a surface per unit area, expressed as the product of the *illuminance and the time for which it is illuminated. When the intensity of the light varies over the period of illumination the exposure is given by the integral $\int E_v \, dt$, where E_v is illuminance and t is time. It is measured in lux second. *Compare* radiant exposure.

light guide An informal name for an optical fibre. Large-diameter (about 5 mm) fibres are generally called *light pipes*. *See* fibre-optics system.

light meter *See* exposure meter.

lightning arrester *See* surge diverter.

lightning conductor A *lightning protective system having an *air termination and a single conductor connected to earth.

lightning flash A complete lightning discharge along a single discharge path. It may be made up of more than one *lightning stroke, in which case it is described as a multiple stroke. *See also* leader stroke.

lightning protective system A complete system of conductors intended to protect a building from damage by lightning.

lightning stroke A discharge that is a component of a complete *lightning flash. It is the discharge of one of the charge regions of a thunder cloud. The polarity of a lightning stroke is the polarity of the electric charge that is brought to earth. A lightning stroke to any part of a power or communication system is described as a *direct stroke*. It is an *indirect stroke* if it induces a voltage in that system without actually striking it. *See* lightning surge.

lightning surge 1. Direct lightning surge. A *surge in a conductor of a power or communication system by a direct stroke (*see* lightning stroke) or by a *flash-over to that conductor.

2. Indirect lightning surge (or induced lightning surge). A surge induced in a conductor of a power or communication system either by an indirect stroke (*see* lightning stroke) or by a direct stroke that terminates on an earthed part of the system without causing back flashover.

light pen A penlike device attached to an *on line *VDU by which information may be input to a *computer. This is usually achieved by pointing the pen at small areas of the screen, for example indicating a selection from a displayed list, or sometimes by drawing shapes on the screen. The pen is connected by cable to the computer, which is able to identify the position of the pen when light from the screen is detected by an electronic photodetector in the pen.

light pipe *See* light guide.

light-year (l.y.) A unit of distance employed particularly in popular works on astronomy and equal to the distance travelled by light through a vacuum in one year. It equals 9.460 528 × 10¹⁵ metres or 0.306 5949 *parsec.

limit cycle *See* attractor.

limiter In general, an electronic or electrical device that automatically sets a boundary value or values on some output characteristic. In particular, a device that for inputs below a specified instantaneous value gives an output proportional to the input, but for inputs above that value gives a constant peak output. The output of a *base limiter* comprises that part of an input signal that exceeds a predetermined value.

limiting current *Syn.* diffusion current. In electrolysis the rate of diffusion of ions towards the electrodes may not keep pace with their deposition. There is thus a limiting value of current that can be passed under the particular conditions of ionic concentration and characteristics of the electrolyte.

limiting friction *See* friction.

limit of proportionality *See* Hooke's law.

limit switch A switch fitted to a mechanical device to limit the travel of a moving part. It is operated by the mechanism that it controls.

linac *See* linear accelerator.

Linde, Carl von (1842–1934) German engineer who is remembered for his work on the liquefaction of gases. *See* Linde process.

Lindemann, F. A. *See* Cherwell.

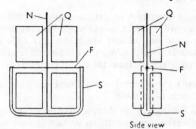
Lindemann electrometer

Lindemann electrometer A light needle N is supported by a torsional fibre F in an S-shaped stirrup, as shown. Metal plate quadrants Q surround the needle on all sides and apply the deflecting electrostatic field. The movement of the tip of the needle is measured by a microscope.

Lindemann's theory of fusion Lindemann worked on the assumption that the breakdown of a crystal lattice occurs when the amplitude of oscillation of the atoms becomes comparable with the separation between atoms, and the crystal melts. He gave a formula connecting the atomic frequency (ν) with the melting point T:
$$\nu = cV^{-1/3}\sqrt{(T/M)},$$
where V is the atomic volume, M the relative molecular mass, and c is a constant.

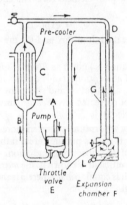
Linde process for liquefying air

Linde process For the liquefaction of air. This process is similar to the *Hampson process, the chief difference being that the air is expanded through the valve to a pressure of about 40 atmospheres instead of to atmospheric pressure, resulting in a more efficient process suitable for large plants. Air is compressed by the pump A, and passes through B to the cooler C where the heat of compression is removed (*see* diagram). The compressed air passes through D and expands through a throttle valve E into the vessel F. The cooled air then returns through the heat exchanger G to the pump chamber A, cooling the oncoming air in DE, until finally liquid air is formed in F and may be withdrawn at L.

lineage structure A crystal that begins as a dendritic (branched) skeleton, but in which the branches are not quite parallel and which therefore ends as a *mosaic crystal in which there is a single centre but irregular growth from the centre outwards.

linear 1. Having components arranged in a line, as in a *linear accelerator.

2. Having only one dimension or (when applied to mathematical relationships) containing only terms of the first *degree. (*See also* graph; interpolation.)

3. Having an output that is directly proportional to its input (said of a component, circuit, or device, e.g. a *linear amplifier).

linear absorption coefficient *Syn.* absorption coefficient. Symbol: a; unit: m^{-1}. During its passage through a medium radiation is absorbed to an extent that depends on the wavelength of the radiation and the thickness and nature of the medium. If $d\Phi$ is the change in *radiant flux or *luminous flux of a parallel beam of monochromatic radiation on passing through a small thickness dl of an absorbing medium, the linear absorption coefficient is defined by the equation:

$$a = \Phi^{-1}(d\Phi/dl).$$

Thus a is the *internal absorptance of a path element of the material minus $(d\Phi/\Phi)$ divided by the length of the path element (dl). The equation integrates to:

$$\Phi_x/\Phi_0 = \exp(-ax),$$

where Φ_0 is the initial flux and Φ_x the flux after a distance x.

This equation is known as *Bouguer's law* or *Lambert's law* of absorption. It only applies in practice if factors such as reflection and scattering are negligible or can be corrected for. (*See also* linear attenuation coefficient.)

For X-ray absorption it is often more convenient to consider the mass per unit area, rather than the thickness of the absorbing radiation. The corresponding coefficient is known as the *mass absorption coefficient*. It is equal to a/ρ, where ρ is the density of the material.

linear accelerator (linac) A particle *accelerator in which electrons or protons are accelerated along a straight evacuated chamber by an electric field of radio frequency produced by a *klystron or *magnetron. In older low-energy machines, cylindrical electrodes (or *drift tubes*) of the RF supply are aligned coaxially with the chamber (Fig. *a*). In an electron accelerator, electrons are accelerated towards a positively charged drift tube; they enter and travel through the tube during the phase of the radio wave that would produce negative acceleration, i.e. when the electrode is negatively charged. As they emerge from the tube, the phase has changed again and the electrons are accelerated towards the next electrode. They are only accelerated in the gaps between the electrodes. As the electron velocity increases, the length of the drift tube must be increased according-ly to keep the electrons in phase with the electric field; the distance between the drift tube centres is also increased. The final beam, travelling at close to the speed of light with an energy in the MeV range, is pulsed, the pulse length being of the order of microseconds. The initial electrons can be produced by *thermionic emission from a heated filament. In high-energy linacs energetic electrons or protons are injected into the system in phase with the RF field, from an external source such as a *Van de Graaff accelerator.

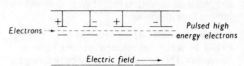

Electrons → Pulsed high energy electrons

Electric field ──→

a Linear accelerator with drift tubes

Modern high-energy linacs are usually travelling-wave accelerators in which particles are accelerated by the electric component of a *travelling wave set up in a *waveguide. No drift tubes are used, the RF being boosted at regular intervals along the chamber length by means of klystrons (Fig. *b*). Only a small magnetic field, supplied by *magnetic lenses between the RF cavities, is required to focus the particles and maintain them in a straight line. Typical rates of energy gain in a linac are 7 MeV per metre (electrons) and 1.5 MeV m^{-1} (protons). The most energetic electrons at present produced are 50 GeV. Electrons are accelerated on a travelling wave down a 10 cm diameter copper tube about 3.5 km long. The engineering difficulties and the expense involved in building such long accelerators have meant that higher energies are difficult to achieve in linear accelerators; *proton synchrotrons are therefore being designed and constructed for purposes of extra high-energy research.

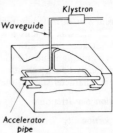

Klystron

Waveguide

Accelerator pipe

b Cavity of a travelling-wave linear accelerator

linear amplifier An *amplifier in which there is a *linear relation between instantaneous output currents and input voltage.

linear attenuation coefficient *Syn.* linear extinction coefficient; extinction coefficient. Symbol: μ; unit: m^{-1}. A measure of the ability of a medium to diffuse and absorb *radiation. If a collimated beam of radiation is passing through the medium it loses intensity due to absorption and scattering. The linear attenuation coefficient is defined by the equation:

$$\mu = -\Phi^{-1}(\mathrm{d}\Phi/\mathrm{d}l),$$

where $\mathrm{d}\Phi$ is the decrease in *luminous flux or *radiant flux Φ passing through a section $\mathrm{d}l$ of the material perpendicular to its face. The linear attenuation coefficient is more general than the *linear absorption coefficient, which only applies to an absorbing medium. The part of the linear attenuation coefficient not due to absorption is sometimes called the *scattering coefficient*.

linear circuit *Syn.* analogue circuit. A circuit in which the output varies continuously as a given function of the input. *Compare* digital circuit.

linear distortion *See* distortion.

linear energy transfer (LET) *See* ionizing radiation.

linear expansion coefficient *See* coefficient of expansion.

linear extinction coefficient *See* linear attenuation coefficient.

linear inverter *See* inverter.

linearly graded junction A *p-n junction in which the concentration of impurities does not change suddenly from *acceptors to *donors, but varies linearly across the junction.

linear momentum *See* momentum.

linear motor A type of *induction motor in which stator and rotor are linear and parallel rather than cylindrical and coaxial.

linear network *See* network.

linear stopping power *See* stopping power.

linear strain *See* strain.

line broadening *See* line profile.

line choking coil *Syn.* screening reactor. An inductive reactor for connection in series with electrical plant such as power transformers to provide protection against the effects of steep-fronted (*see* wavefront) or high-frequency *surges that may occur on the line supplying the plant. The line choking coil usually causes partial absorption of the surge due to I^2R losses and *core losses and partial reflection of the surge. *See* travelling wave.

line communication A physical path, such as a wire or waveguide, between the terminals of a communications system.

line defect *See* defect.

line frequency *See* television.

line of force *Syn.* line of flux. An imaginary line whose direction at all points along its length is that of the electric, gravitational, or magnetic field at those points. *See* field.

line pair A particular line in a *spectrum together with the *internal standard line with which it is compared in a *spectrographic analysis.

line printer A computer output device that prints a line at a time. Typical speeds are in the range 200 to 3000 lines per minute.

line profile A plot of intensity against wavelength or frequency for a spectral line, showing the fine structure of the line. The natural width of a spectral line is determined by quantum-mechanical uncertainty. Other factors, however, can lead to additional *line broadening*. These include the *Doppler effect, *Zeeman effect, and high density and hence high pressure of the emitting or absorbing material (*pressure broadening*). Analysis of the line profile yields information about the physical conditions at the source.

line-sequential colour television A system of *colour television in which each of the red, blue, and green video signals is transmitted in turn for the duration of one scanning line.

line spectrum A *spectrum consisting of bright lines on a dark ground when light has been admitted through a slit from an incandescent gas (metallic arc, electric discharge through a vacuum tube), and has been dispersed by prism refraction or by a diffraction grating. The lines are due to individual atoms (changes of electronic orbits). They occupy characteristic positions that can, in a few cases, be calculated by reference to certain series formulae.

line-turn *Syn.* maxwell-turn. A unit of linkage of magnetic flux. *See* linkage.

line voltage *See* voltage between lines.

linkage Of magnetic flux. A measure of the flux and the number of turns of the coil or circuit with which it links. Quantitatively, it is the product of the number of lines of magnetic flux and the number of turns of the coil or circuit through which they pass. This is sometimes called a line-turn (or maxwell-turn).

Linke–Fuessner actinometer *See* pyrheliometer.

lin-log receiver A radio receiver that has a linear response for small input signals and a logarithmic response for large signals.

Lippershey, Hans (died *c.* 1619) Dutch spectacle lens worker accredited with the first invention (1608) of the telescope (Dutch or Galilean type).

Lippmann, Gabriel (1845–1921) French physicist. The inventor of the *capillary electrometer and of a process of colour photography with which he obtained the first coloured photograph of the solar spectrum. He was awarded the Nobel prize for physics, 1908.

Lippmann effect Surface tension changes produced by a potential difference across the boundary between two immiscible liquid conductors.

Lippmann electrometer *See* capillary electrometer.

Lippmann process of colour photography

Lippmann process of colour photography A process depending on the production of standing waves by reflection at a coating on the back of the photographic emulsion. After development, silver layers exist at the electrical antinodes. In suitable illumination, the original colours are faithfully reproduced but the process is difficult to carry out. The phenomenon had been used by O. Wiener to determine whether the electric or the magnetic vector is the photographically effective component of electromagnetic waves. The process was invented in 1891.

liquefaction of gases The following methods for the liquefaction of gas are available:

(1) The application of high pressure after cooling the gas below its critical temperature, as in *cascade liquefaction.

(2) The Joule–Kelvin effect in which gas at high pressure is cooled by expansion through a porous plug or throttle valve, as in the *Linde process for air and the *Hampson process.

(3) Adiabatic expansion in which a compressed gas is cooled by performing external work as in the *Claude process. Heylandt and Kapitza developed liquefiers using this principle in conjunction with the Joule–Kelvin effect.

(4) Adiabatic *desorption in which the gas is adsorbed in cooled charcoal and further cooling produced when the gas is removed adiabatically as in *Simon's method of liquefying helium.

liquid A phase of matter, intermediate between a gas and a solid, that is characterized by ease of flow and near incompressibility. It takes the shape of its container but, unlike a gas, does not expand to fill the container. The intermolecular forces are intermediate in strength between those in gases and solids. Liquids possess a short-range structural regularity and these bundles of atoms, molecules, or ions can move relative to each other.

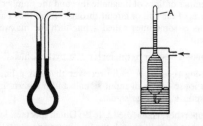

a Open liquid-column   *b* Closed liquid-column
manometer   manometer

liquid-column manometers 1. Open liquid-column *manometers are based on the U-tube of liquid, which measures the pressure difference between the two sides. If the difference in vertical level of the surfaces of the liquid is h (Fig. *a*) and its density is ρ, the applied pressure difference is $h\rho g$ (g being the acceleration of free fall).

The mercury *barometer is a special case in which one pressure is zero. When measuring very high pressures (e.g. 100 atmospheres), which would need

a prohibitively great difference in height, several U-tubes of mercury are used in tandem with either compressed air or a light liquid in between.

2. Closed liquid column (or compressed gas) manometer. The compressed gas A (Fig. *b*) may be hydrogen and the liquid mercury.

liquid controller A *motor controller for an electric motor, employing a liquid as the resistor.

liquid crystal An arrangement of certain kinds of long molecules in a liquid with the result that the liquid does not suffer an entire loss of fluidity. In *nematic* liquid crystals, the molecules are aligned in the same direction but randomly so. In *cholesteric* and *smectic* liquid crystals, the molecules are still aligned but in distinct layers; their long axes are parallel to (cholesteric) or usually perpendicular to (smectic) the plane of the layers. Liquid crystals have characteristic optical properties. Cholesteric and nematic types can have very large rotatory powers (*see* optical activity). Smectic and nematic structures both exhibit *double refraction and are optically positive. Cholesteric types exhibit *iridescence as a result of *dichroism, but often only in a certain temperature range.

liquid-crystal display (LCD) A low-power device used to display numbers, letters, and other characters (black on white) in, for example, digital watches, calculators, measuring instruments, and some computer displays. The LCD depends for its action on the change produced by an electric field on the optical properties of *liquid crystals.

In one form, a thin film of liquid crystal is sandwiched between thin transparent electrodes. The upper electrode is etched into a pattern of segments making up the characters of the display. Two crossed *polarizers (i.e. with their vibration planes at right angles) lie on either side of the film. Light falling on an unenergized liquid crystal segment is largely reflected back and it appears transparent: the plane-polarized light entering the segment has its plane of polarization rotated through 90° so that it can then pass through the second polarizer, and is reflected back. When an electric field is applied across selected segments, they appear dark: the rotatory power of the liquid-crystal segments is affected by the field and the light cannot pass through the second polarizer.

liquid-drop model A model of the nucleus in which the nuclear matter is regarded as being continuous. The interactions between *nucleons are thought of as being analogous to those between molecules in a liquid. As in a liquid drop, the net effect of the interaction of particles near the surface of the nucleus is interpreted as a *surface tension, which maintains the shape of the nucleus. The liquid-drop model is most applicable to heavy nuclei and is used in the theory of *nuclear fission.

liquid starter A *motor starter for an electric motor, employing a liquid as the resistor.

Lissajous, Jules Antoine (1822–1880) French physicist who worked on acoustics and optics. *See* Lissajous' figure.

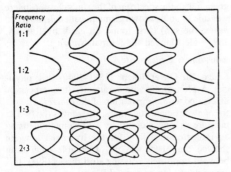

Lissajous' figures

Lissajous' figure The displacement pattern traced out by the superposition of two vibrations in directions at right angles to each other. These figures can be constructed graphically; or they may be obtained practically using either a mechanical device or a cathode-ray oscilloscope.

The simplest pattern is a straight line, being obtained from two vibrations of equal frequency in phase with each other. The patterns can become very involved if the ratio of the frequencies is not a simple one. Examples are given in the illustration for various frequency ratios and phase differences between 0 and π. The figures are useful, in particular, in identifying the phase relationship of two vibrations of the same frequency, and in verifying that two given vibrations are of the same frequency.

lithium fluoride dosimetry *See* dosimetry.

lithography The techniques, including *photolithography, used in the manufacture of *integrated circuits, *thin-film circuits, *printed circuits, etc. Advantages, especially increased resolution, are obtained if the light or ultraviolet radiation used in photolithography to expose the resist is replaced by a beam of X-rays or high-energy electrons or ions.

lithosphere *See* meteorology.

litre Symbol: l or L. A unit of volume formerly defined as the volume occupied by a mass of 1 kg of pure water at its maximum density and under standard atmospheric pressure. This volume is equal to 1.000 028 decimetres cubed. Subsequently the litre was defined as a special name for the decimetre cubed, but owing to confusion between the two definitions the unit is not recommended for scientific purposes, although the ml (millilitre) is used synonymously with the cc where great accuracy is not implied.

littrow prism A 30°, 60°, 90° prism silvered on the longer perpendicular face, used as a dispersing prism with *autocollimation.

litzendraht wire (litz) A multistranded wire composed of many fine conducting filaments, used, for example, in low-loss coils for filters and radios to reduce their high-frequency resistance. *See* skin effect.

live 1. Connected to a voltage supply; not at earth potential.

2. Reverberant; having a reverberation time that is normal or above normal.

3. Involving the direct transmission of a TV or radio signal without prior recording.

live-end, dead-end principle A principle used in the design of broadcasting and sound-recording studies to give the right acoustic conditions both for the artistes and for sound reproduction. From the point of view of the listener the reverberation time of the studio is superimposed on that of the listening room. Since the latter cannot usually be controlled, it appears desirable to use a studio in which the reverberation time is very low. This was attempted in the early days of broadcasting but considerable difficulties were experienced by the performers. The atmosphere seemed oppressive and the unnaturally dead effect made it a strain for singers and orchestra to keep in tune. As a result, the "live-end, dead-end" studio was developed in which the artistes performed at one end, which was quite reverberant, and the microphone was placed at the other end, which was treated with sound-absorbing material. Modern studios, however, are designed to have the correct reverberation time–frequency characteristics, and directional microphones are used to pick up a large proportion of direct sound and very little reflected sound. *See* reverberation.

Lloyd, Humphrey (1800–1881) Irish physicist. He was the first to observe (1833) internal conical refraction, using aragonite, verifying the theory of Sir William Hamilton. *See* Lloyd's mirror.

Lloyd's mirror (1834) A glass or metal mirror for producing interference fringes in overlapping beams – one direct and the other after grazing reflection. It satisfies the condition to produce achromatic fringes. *See* interference of light.

load 1. A device or material in which electrical signal power is dissipated or received, i.e. a device that absorbs power from a source of electrical signals. Examples include loudspeakers, TV and radio receivers, logic circuits, and the material to be heated by dielectric and induction heating.

2. The power delivered by a machine, generator, transducer, or electronic circuit or device.

3. The mechanical force applied to a body.

4. The weight supported by a structure.

load curve A graph in which the power supplied by an electrical transmission and distribution system is plotted against time.

loaded concrete Concrete containing material of high atomic number or capture cross section, such as barium, iron, or lead, to increase its effectiveness as a radiation shield in *nuclear reactors.

load factor The ratio of the number of units of electrical work done (kilowatt-hour) during a given period of time to the number of units that would have been done if the maximum demand had been maintained during the same period. It is usually expressed as a percentage. The *plant load factor* relates specifically to a generator or a group of generators, and is the ratio of the number of units actually done in a given period to the number of units that would have been done if the generator or generators had operated at their maximum continuous rating during the same period. This also is usually expressed as a percentage.

load impedance The *impedance presented by a *load to the driver circuit that supplies power to it.

loading A process that can improve the transmission properties of a line used for telecommunication purposes by artificially increasing its inductance. There are two common methods of effecting this: (i) Continuous loading. A tape or wire of iron or other magnetic materials is wound continuously round the conductor to be loaded from end to end. (ii) Coil loading (or lumped loading). Inductance coils are connected in series with the line and are spaced at uniform intervals along the line. This method of loading is effective only throughout a limited range of frequencies.

load line A line drawn on the graph of a family of characteristics of an electronic device, showing the graphical relationship between voltage and current for the particular *load of the circuit under consideration.

lobe A region on a *radiation pattern representing enhanced response of an aerial. The *main* or *major lobe* corresponds to the direction of best transmission and reception. All the other lobes are called *side lobes* and are usually unwanted.

local action A chemical reaction that takes place on the surface of a zinc or other electrode immersed in an electrolyte, owing to the presence of impurities in the electrode. The impurities form with the metal of the electrode small voltaic cells, as a result of which the electrode corrodes even when no current is being taken from the cell. It is prevented by amalgamating the electrode. *See* amalgamation.

local area network (LAN) A simple communications system linking a number of computers within a small and defined locality, such as an office building, industrial site, or university. This allows the computers (usually microcomputers) to share single and/or expensive resources, such as a line printer or hard disks, and to share data files and databases. Messages can be sent by electronic mail. The transmission medium is usually electric cables or optical fibres.

local oscillator *See* superheterodyne receiver.

Lockyer, Sir Norman (1836–1920) Brit. astronomer and physicist. He was first to suggest (1873–1874) that *line spectra are characteristic of atoms and *band spectra of molecules. He argued (1878) that a green line in the spectrum of the sun's chromosphere was due to an unknown element, called *helium* (Gk. *helios*, the sun) (first found on earth by Ramsay, 1895). He founded the journal *Nature*.

lodestone *See* magnetite.

Lodge, Sir Oliver Joseph (1851–1940) Brit. physicist. A pioneer of radio communication, especially associated with the invention of the *coherer*, a detector of electromagnetic waves. Lodge was the first to measure an ionic mobility by a direct method, using an indicator in a jelly for the hydrogen ion. He showed that there was no observable ether drag between oppositely rotating steel discs (1893), this experiment playing a notable part in the development of the theory of *relativity.

logarithmic decrement If a system, such as a simple pendulum, be set to execute oscillations in a vertical plane, then, if frictional forces are absent, the pendulum will continue to oscillate with undiminished amplitude and the equation of motion will be:
$$\mathrm{d}^2x/\mathrm{d}t^2 = -qx.$$
If, however, frictional forces are present, due to, say, the viscosity of the medium in which the pendulum is oscillating, the amplitude will diminish according to a certain law. The equation of motion will now contain a term proportional to the velocity of the pendulum bob at any point and the viscous force will act in opposition to the direction of motion; that is, the equation of motion will now become:
$$\mathrm{d}^2x/\mathrm{d}t^2 = -p(\mathrm{d}x/\mathrm{d}t) - qx.$$
The solution of this equation is, if the viscous forces are not too great, given by:
$$x = A \exp\left(-\tfrac{1}{2}pt\right) \sin\left\{\frac{2\pi t}{\sqrt{(q - \tfrac{1}{4}p^2)}} - \alpha\right\},$$
where A and α are constants of integration, and the motion is oscillatory so long as q is greater than $p^2/4$. The system then oscillates with diminishing amplitudes. If the successive amplitudes are $a_1, a_2, \ldots a_n$, then it follows from a consideration of the equation of motion that:
$$a_1/a_2 = a_2/a_3 \ldots = a_{n-1}/a_n = \exp(pT/4) \\ \exp(\lambda),$$
where $\lambda = pT/4$ and is called the logarithmic decrement of the system:
$$\lambda = \log_e a_1 - \log_e a_2 = \log_e a_2 - \log_e a_3,$$
and therefore:
$$\lambda = (\log_e a_1 - \log_e a_n)/(n - 1).$$
If $q = p^2/4$ the damping is said to be critical. If $q < p^2/4$, then the motion ceases to be oscillatory, the moving body swings out once and then returns slowly and asymptotically to its equilibrium position. The motion is then said to be aperiodic.

Commonly the friction experienced by a pendulum is caused by turbulence and is proportional to the square of the speed. For the very small frictional forces normally encountered the theory given above is sufficiently accurate for most purposes. It is inapplicable to heavily damped systems such as rolling ships.

logarithmic resistor A form of variable resistor designed so that the fractional change in resistance is directly or inversely proportional to the movement of the contact.

logarithmic series *See* Taylor series.

logic circuit A circuit designed to perform a particular logical function based on the concepts of "and", "either-or", "neither-nor", etc. Normally these circuits operate between two discrete voltage levels, i.e. high and low logic levels, and are described as binary logic circuits. Logic using three or more logic levels is possible but not common.

The devices used to implement the elementary logic functions are called *logic gates*. The basic gates are:

(*a*) *AND gate*. A circuit with two or more inputs and one output in which the output signal is high if and only if (sometimes written iff) all the inputs are high simultaneously.

(*b*) **Inverter (NOT gate)*. A circuit with one input whose output is high if the input is low and vice versa.

(*c*) *NAND gate*. A circuit with two or more inputs and one output, whose output is high if any one or more of the inputs is low, and low if all the inputs are high.

(*d*) *NOR gate*. A circuit with two or more inputs and one output, whose output is high if and only if all the inputs are low.

(*e*) *OR gate*. A circuit with two or more inputs and one output whose output is high if any one or more of the inputs are high.

(*f*) *Exclusive OR gate*. A circuit with two or more inputs and one output whose output is high if any one of the inputs is high.

These circuits are for use with *positive logic*: that is the high voltage level represents a logical 1 and low a logical 0. *Negative logic* has high level representing a logical 0 and low a logical 1. The same circuits may be used in negative logic, but become the complements of the positive logic circuits, i.e. a positive OR circuit becomes a negative AND circuit.

Any logical procedure may be effected by using a suitable combination of the basic gates. Binary logic circuits are extensively used in *computers to carry out instructions and arithmetical processes. They may be formed from discrete components or, more commonly, from *integrated circuits. Families of integrated logic circuits exist based on bipolar transistors. (*See* diode transistor logic (DTL); emitter-coupled logic (ECL); resistor-transistor logic (RTL); transistor-transistor logic (TTL).) *MOS logic circuits are based on *field-effect transistors.

logic element In a computer or data processing system. The smallest parts of the system, usually a logic gate, that may be represented by mathematical operators in symbolic logic. *See also* logic circuit.

logic gate *See* logic circuit.

London, Fritz (1900–1954) German-born Amer. theoretical physicist who worked in Oxford, Paris, and at Duke University, North Carolina. With Walter Heitler (1927) he gave a wave-mechanical description of the covalent bond in the hydrogen molecule. He also worked with his younger brother, Heinz London, on superconductivity. They gave a theoretical account of the *Meissner effect. London's later research was on the *superfluid state in liquid helium.

London, Heinz (1907–1970) German-born Brit. theoretical physicist. He collaborated at Oxford with his elder brother, Fritz London, in theoretical work on the *Meissner effect. Later he worked at the Atomic Energy Research Establishment, Harwell, on the *superfluid state in liquid helium.

longitude 1. The terrestrial longitude of a place is the angle between the terrestrial meridian through that place and the meridian through Greenwich; it is measured eastwards or westwards from Greenwich from 0–180°.

2. The celestial longitude. *See* celestial sphere.

longitudinal aberrations Aberration distances measured along the principal axis. In *chromatic aberration it is the distance between the foci for the two standard colours, e.g. F and C. In *spherical aberration, it is the distance from the paraxial focus to the intersection of a zonal ray with the axis.

longitudinal mass In special *relativity theory. The ratio of force to acceleration in the direction of the existing velocity of a particle. It is given by:
$$m_1 = m_0 / \sqrt{(1 - \beta^2)^3},$$
where m_0 is the rest mass of the particle and $\beta = v/c$, i.e. its velocity expressed as a fraction of the speed of light. *Compare* transverse mass.

longitudinal strain *See* strain.

longitudinal vibrations Vibrations in which the displacement is along the main axis or direction of the vibrating body or system. A typical example is the note produced by a metal rod when stroked with a resinated cloth.

longitudinal waves Waves in which the particles of the transmitting medium are displaced along the direction of propagation. The speed, c, of longitudinal waves in a bar is given by $c = \sqrt{(E/\rho)}$, where E is the Young modulus and ρ is the density; the corresponding equation for longitudinal waves in a fluid is $c = \sqrt{(k/\rho)}$, where k is the bulk modulus. Sound waves in a gas form the chief example of longitudinal wave motion. *See* speed of sound.

long-range force *See* force.

long-range order Crystalline order that extends over a distance covering several *unit cells, and is measured by the extent to which positions of atoms in different cells of the lattice are correlated. This is given by ordinary X-ray diffracting data.

long sight *See* hypermetropia; eye.

long-tailed pair Two matched bipolar *transistors that have their *emitters coupled together, with a common-emitter bias resistor acting as a constant-current source. The name is derived from the physical resemblance of the bias resistor to a tail: the larger (i.e. longer) the bias resistor, the more nearly it resembles a constant-current source (because of the relatively large voltage developed across it). Originally the name was given to matched thermionic valves with the cathodes connected to a common-cathode bias resistor. The long-tailed pair forms the basis of most *differential amplifiers.

long-wave Designating radio waves with wavelengths exceeding 1000 metres, i.e. with frequencies below 300 kHz.

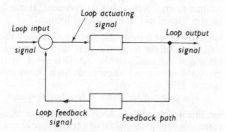

loop For feedback and control. *Syn.* feedback control loop. A means of control used in many types of control system, in which part of the output derived from the control system is fed back to the input circuit in order to control the output in a desired manner. The external signal applied to the loop is the loop input signal, and the loop output signal is the controlled signal extracted from the loop (*see* diagram).

The *loop error* is the difference between the actual output value and the desired output value. The signal derived from mixing the loop input signal and the feedback signal gives the loop actuating signal, which is used to produce the controlled output. In some types of control system the loop error is the loop actuating signal, and it is then called the loop error signal.

loop aerial *Syn.* frame aerial. An *aerial that is essentially a coil having one or more turns of wire wound on a frame and having an axial length that is usually small compared with its other linear dimensions. The plane of the coil is the direction of maximum sensitivity and transmission. It is commonly employed in radio direction finders and in small portable radio receivers.

loop test A test carried out to locate a fault in the insulation of a cable (e.g. short circuit or fault to earth). The faulty cable is arranged to form part of a closed circuit or loop. According to the type of test, one or more sound cables, running along with the faulty cable, are necessary. The resistance of the fault, if comparatively low, does not affect the result of the test, but if it is very high it may adversely affect the sensitivity.

Lorentz, Hendrik Antoon (1853–1928) Dutch theoretical physicist. He was one of the principal founders of the electron theory of matter, providing explanations of normal and anomalous dispersion and of the Zeeman effect. Lorentz obtained expressions to transform coordinates of an event as observed in one frame of reference into coordinates in another frame. *See* Lorentz–FitzGerald contraction; Lorentz transformation equations.

Lorentz–FitzGerald contraction The theory that a material body moving through the *ether with a velocity v, contracts by a factor of $\sqrt{(1 - v^2/c^2)}$ in the direction of motion, where c is the speed of light in vacuum. The theory was advanced to account for the failure of the *Michelson–Morley experiment to detect the earth's motion through the ether, but has been superseded by the theory of *relativity. According to relativity theory the length of a body as measured by an observer in uniform relative motion is less than that measured by an observer at rest with respect to the body by the factor given above. This is not a physical change in the body so should not be confused with the hypothetical change postulated in the older theory. *See* FitzGerald, G. F.

Lorentz force The force acting on a moving charge q in magnetic and electric fields. It is given by:
$$F = q(E + v \times B),$$
where F is the force, E the electric field, and $v \times B$ the *vector product of the particle's velocity and the *magnetic flux density.

The magnetic contribution to this force is often called the Lorentz force and this, in nonvector notation, is given by:
$$F = qvB \sin \theta,$$
where θ is the angle that the direction of motion of the particle makes with the magnetic field. The force acts in a direction that is perpendicular to both the direction of motion and the magnetic field.

Lorentz–Lorenz law The molecular refraction r of a substance of relative molecular mass M and density ρ is given by:
$$r = M(n^2 - 1)/\rho(n^2 + 2),$$
where n = refractive index.

Lorentz transformation equations A set of equations for transforming the position-motion parameter from an observer at a point $O(x, y, z)$ to an observer at $O'(x', y', z')$, moving relative to one another. The equations replace the *Galilean transformation equations of Newtonian mechanics in *relativity problems. If the x-axis is chosen to pass through OO' and the time of an event is t and t' in the *frame of reference of the observers at O and O' respectively (where the zeros of their time scales were the instant that O and O' coincided) the equations are:
$$x' = \beta(x - vt)$$
$$y' = y$$
$$z' = z$$
$$t' = \beta(t - vx/c^2),$$
where v is the relative velocity of separation of O, O', c is the speed of light, and β is the function $(1 - v^2/c^2)^{-1/2}$.

Loschmidt, Joseph (1821–1895) Austrian physicist. He estimated (1865) the sizes of molecules from their mean free path and the volume occupied in the liquid state, thus obtaining an estimate of the number of molecules per unit volume (*see* Loschmidt constant).

Loschmidt constant Symbol: N_L, n_0. A constant that gives the number of molecules in one cubic metre of an ideal gas at STP. It is equal to
$$2.686\ 763 \times 10^{25}\ \text{m}^{-3}.$$
It is the ratio of the *Avogadro constant to the molar volume of an ideal gas.

loss *See* core, copper, dielectric, eddy-current, stray load, and windage losses.

loss angle Of a capacitor or dielectric when subjected to alternating electric stress. The angle by which the angle of *lead of the current is less than 90° when the applied voltage is sinusoidal. It is due mainly to dielectric *hysteresis loss.

loss factor 1. The ratio of the average power dissipation to the power dissipation at peak load in a transmission line, circuit, or device.
2. The product of the *power factor and the relative *permittivity of a *dielectric. It is proportional to the heat generated in a material in a given alternating field.

lossy Denoting an insulator that dissipates more energy than is considered normal for the class of material.

lossy line A *transmission line designated to have a high degree of *attenuation.

loudness The magnitude of the sensation produced when a sound reaches the ear. Although loudness is related to *sound intensity, there is no simple connection between the two and it is difficult to produce a scale of loudness that correctly represents the physiological sensation caused by the sound. The basis of loudness scales is the *Weber–Fechner law, which states that the sensation is proportional to the logarithm of the stimulus. In the *decibel scale of intensity level, the sound intensity is logarithmically related to the threshold intensity at the same frequency. This suffers from the disadvantage that the sensitivity of the ear to changes of intensity varies with frequency. The *phon scale of equivalent loudness overcomes this by relating the intensity of a sound to a fixed reference tone of defined intensity and frequency. The phon scale is widely used since it places all sounds in order of their loudness. However, it assumes the Weber–Fechner law in expressing the ratio of the actual sensations produced by different sounds. Experimental work on a purely psychological scale of loudness has been done by Fletcher and other workers. These experiments rely on the ability of an observer to judge a doubling in the sensation produced by a sound. Several methods were used, such as adjusting a sound heard with one ear until it produced the same sensation as another of the same frequency heard with both ears. Consis-

tent results were obtained by many different methods and the logarithmic scale was justified over the main audible range.

loudness level A measure of the strength of a sound as expressed by the *sound pressure level of a pure tone of specified frequency that is judged, by a normal listener, to be equally as loud as the sound.

loudspeaker *Syn.* speaker. A device in which an electrical signal is converted into sound. It is the final unit in any broadcast receiver or sound reproducer. In the most common types of loudspeaker the current is passed through a small coil fixed to the centre of a diaphragm and moving in an annular gap across which is a strong magnetic field. Alternating current in the coil causes the diaphragm to vibrate at the same frequency and emit sound waves. For high efficiency a small diaphragm is used at the mouth of a large exponential *horn. Although the horn gives suitable loading to the diaphragm, it is impractical for most indoor work on account of its size. Instead, a speaker is used having a large conical or elliptical diaphragm with the coil at its apex. The cone is made of stiff paper and is supported round its edge by a metal frame. The magnetic field is produced either by a permanent magnet or an electromagnet and the coil is held in position in the centre of the gap by a flexible mounting. The cone should be set in a large *baffle to prevent direct passage of sound from front to back and so improve the low-frequency response. In most commercial sound reproducers the cabinet forms the baffle. This type of speaker gives a good response over a moderate range of frequencies but careful design is necessary for good reproduction at high or low frequencies. Unless the cone is large it has a small *radiation resistance at low frequencies and its output falls. It is usual to make the resonant frequency of the cone very low to improve the output at these frequencies, but unless the cone is considerably damped there are strong resonant peaks giving a boomy sound. At high frequencies the output falls owing to the mass of the vibrating system. To overcome this, two speakers are sometimes used: a large cone speaker to handle the low notes and a small one for the high notes. In some speakers multiple coils and cones enable the effective mass to be reduced at high frequencies. A modern cone speaker gives a uniform response between about 80 and 10 000 hertz. Its efficiency of energy conversion is about 5% compared with up to 50% for a horn speaker.

Lovell, Sir Alfred Charles Bernard (*b.* 1913) Brit. radio astronomer who was primarily responsible for the design and construction of the 250-ft. dish telescope at Jodrell Bank in Cheshire.

low-angle scattering A halo of diffracted radiation immediately surrounding the incident beam, which is dependent only on the size and shape of the scattering particles and is independent of their internal character.

lower sideband *See* sideband.

low frequency (LF) *See* frequency band.

low-level language *See* programming language.

low-pass filter *See* filter.

low voltage In electrical power transmission and distribution. A voltage that does not exceed 250 volts.

LSI Abbreviation for large-scale integration. *See* integrated circuit.

lubrication The reduction of *friction between two solid surfaces sliding over each other by interposing a layer of liquid, or a solid with much lower coefficient of friction. *Hydrodynamic lubrication* uses a thin layer of liquid, usually oil, which prevents contact between the solid surfaces. Resistance to sliding is thus reduced to the effect of viscosity in the lubricant (the thickness is normally too small to permit turbulence even at high speeds). In machines such as gears the surfaces may be driven forcibly against each other, squeezing the fluid layer. As the viscosity of most lubricants increases with pressure, the liquid is not normally forced out from between the solid surfaces. Under large stresses, however, asperities on the surfaces may penetrate the liquid layer and for this reason *barrier lubrication* may be used. The oil receives certain additives that react with metal surfaces coating them with layers of soft material giving much lower friction if the liquid layer is broken, than that for the untreated metal.

Certain substances containing both hard and soft components may be self-lubricating. Thus the extremely hard wood lignum vitae contains a wax that gives low-friction bearings with steel.

Some solids in the form of powders may act as lubricants, in particular some forms of graphite.

lumberg The luminous energy radiated per second in unit solid angle by a standard point source of 1 candela, all the components of the light having their mechanical values weighted according to their luminosity values.

lumen Symbol: lm. The *SI unit of *luminous flux, defined as the luminous flux emitted by a uniform point source, of intensity one candela (cd), in a cone of solid angle of one steradian. Thus 1 lm = $(1/4\pi)$ cd.

lumen-hour A unit quantity of light, i.e. a flux of 1 lumen continued for 1 hour.

lumen-second A unit quantity of light, i.e. a flux of 1 lumen continued for 1 second; it is used, for example, in measurement of light flashes.

Lumière, Auguste Marie Louis Nicholas (1862–1948) French chemist, inventor of a system of colour photography.

luminance Symbol: L_v, L. The brightness, for a specified direction, of a point source of light or a point on a surface that is receiving light. For sources of light it is defined as the luminous intensity, I_v, per unit projected area, i.e.

$$L_v = \mathrm{d}I_v/\mathrm{d}A.\cos\theta,$$

where A is the area and θ is the angle between the surface and the specified direction. For illuminated surfaces it is defined as the illuminance (E_v) per unit solid angle (Ω).

$$L_v = \mathrm{d}E_v/\mathrm{d}\Omega.$$

The *illuminance is taken over an area perpendicular to the direction of the incident radiation. The general equation of luminous intensity, applying to both a point source and a point receptor, is:

$$L_v = \mathrm{d}^2\Phi_v/(\mathrm{d}\Omega.\mathrm{d}A.\cos\theta),$$

where Φ_v is the luminous flux. Luminance is measured in candela per square metre. *Compare* radiance.

luminance signal *See* colour television.

luminescence The emission of electromagnetic radiation from a substance as a result of any nonthermal process. The term is also applied to the radiation itself and is usually used for visible radiation. Luminescence is produced when atoms are excited, as by other radiation, electrons, etc., and then decay to their *ground state. If the luminescence ceases as soon as the source of energy is removed, the phenomenon is *fluorescence*. If it persists, the phenomenon is *phosphorescence*. More precisely, in fluorescence the persistence is less than about 10^{-8} seconds and in phosphorescence it is greater.

If certain solids are subjected to ionizing radiation, electrons may be released within the solid and trapped at *defects. These electrons may be released when the solid is heated and the energy produced is emitted as visible radiation. This is known as *thermoluminescence* (TL). The number of electrons is proportional to the intensity of the incident radiation (i.e. the number of incident photons), hence TL *dosimetry has been developed to monitor radiation levels wherever radioactive sources are used – hospitals, factories, etc. TL can also be used for archaeological and geological *dating, the radiation sources being, for example, the common radionuclides ^{40}K and the ^{235}U and ^{232}Th decay series.

Luminescence can also be produced by the friction of solids (*triboluminescence*) and chemical reaction (*chemiluminescence*). *Compare* incandescence.

luminosity 1. The attribute of a source of light that gives the visual sensation of brightness. The luminosity depends on the power emitted by the source, i.e. on the *radiant flux, but also on the fact that the sensitivity of the *eye varies for different wavelengths. *Radiant quantities are pure physical quantities based on absolute energy measurements, whereas *luminous quantities depend on some judgment of brightness by an observer and thus on the spectral sensitivity of the eye. By using the methods of heterochromatic *photometry, it is possible to compare the radiant flux of light at one wavelength with that at a different wavelength. (*See* spectral luminous efficiency.)

2. Symbol: L. The intrinsic or absolute brightness of a star or other celestial body, equal to the total energy radiated per second from the body. It is related to the body's surface area and *effective tem-*

perature, T_e (i.e. the surface temperature expressed as the temperature of a *black body having the same radius and radiating the same total energy per unit area per second as the body), by a form of the Stefan–Boltzmann law:
$$L = 4\pi R^2 \sigma T_e^{\ 4},$$
where σ is the Stefan–Boltzmann constant and R the radius of the body. Hence stars with similar T_e but greatly different luminosities must differ in size: they belong to different *luminosity classes*. Luminosity is also related to the absolute *magnitude of a celestial body.

luminous A qualifying adjective denoting physical quantities used in *photometry in which energies of light are evaluated by an observer (*see* luminosity). They are distinguished from their corresponding *radiant quantities by adding a subscript v (for visual) to their symbols.

luminous efficacy A property relating *luminous flux to *radiant flux for radiation or for a source. It was formerly called luminous efficiency but this term is now more properly applied to a dimensionless ratio.
1. Symbol: K. The ratio of the luminous flux, Φ_v, of a radiation to its radiant flux, Φ_e:
$$K = \Phi_v / \Phi_e.$$
If monochromatic radiation is considered, the property is called *spectral luminous efficacy*, symbol: $K(\lambda)$, given by the ratio:
$$\Phi_{v,\lambda} / \Phi_{e,\lambda}.$$
2. Symbol: η_v, η. The ratio of the luminous flux emitted by a source to the power it consumes. (*Compare* radiant efficiency.)
Luminous efficacy is measured in lumen per watt.

luminous efficiency Symbol: V. A dimensionless quantity defined by the ratio K/K_m, where K is the *luminous efficacy and K_m the maximum spectral luminous efficacy. If monochromatic radiation is considered, the property is called the *spectral luminous efficiency*, symbol: $V(\lambda)$, given by $K(\lambda)/K_m$, where $K(\lambda)$ is the spectral luminous efficacy.
The term was formerly applied to what is now called luminous efficacy.

luminous emittance Former name for luminous exitance.

luminous energy Symbol: Q_v, Q. A measure of a quantity of light expressed as the product of luminous flux and its duration:
$$Q_v = \int \Phi_v \, dt.$$
It is measured in lumen seconds.

luminous exitance Symbol: M_v, M. At a point on a surface, the *luminous flux leaving the surface per unit area. It is measured in lumen per square metre. *Compare* radiant exitance.

luminous flux Symbol: Φ_v, Φ. The rate of flow of radiant energy as evaluated by the luminous sensation that it produces. The luminous flux is obtained from the *radiant flux of the source corrected according to the effect it has on the observer, i.e.

according to the spectral sensitivity of the receptor. Consider, for example, a source of monochromatic radiation of wavelength λ and with radiant flux Φ_e. The luminous flux is then proportional to $\Phi_e V(\lambda)$, where $V(\lambda)$ is the spectral *luminous efficiency. This factor weights the radiant flux according to the sensitivity of a standard observer (in *photopic vision) to radiation of wavelength λ. Specifically the luminous flux is given by:
$$\Phi_v = K_m \Phi_e V(\lambda),$$
where K_m is a constant relating the units of luminous flux to those of radiant flux. For *polychromatic radiation the radiant flux will generally vary with wavelength, and luminous flux can be defined by:
$$\Phi_v = K_m \int (d\Phi_e / d\lambda) V(\lambda) \, d\lambda,$$
where $(d\Phi_e / d\lambda) \, d\lambda$ is the *radiant flux of light with wavelengths in the range $\lambda \to \lambda + d\lambda$.

The constant K_m can be obtained by applying the above formula to a black body at the temperature of freezing platinum and has the value of 680 lumens per watt (for photopic vision). It is the maximum spectral *luminous efficacy. Luminous flux is measured in lumens.

luminous intensity Symbol: I_v, I. The *luminous flux emitted per unit solid angle by a point source in a given direction. It is measured in candela. A source may radiate unequally in different directions and the direction has to be specified. If the luminous intensity is averaged over all directions, it is called the *mean spherical intensity*. For extended sources the luminous intensity per unit area, or *luminance, is used. *Compare* radiant intensity.

Lummer, Otto (1860–1925) German physicist. With Pringsheim, he investigated experimentally the distribution of energy in the black-body spectrum at various temperatures (1893 onwards). Also with Pringsheim (1894), he devised a method for measurement of the ratio of the principal specific heat capacities of a gas (γ), recording the temperature change in an adiabatic expansion, by using a sensitive *bolometer. This method has been improved and developed by *Partington. With Brodhun, Lummer devised two accurate forms of photometer which may be regarded as refinements of the Bunsen grease spot instrument. *See* Lummer–Gehrcke plate.

Lummer and Pringsheim's determination of γ The ratio of the principal specific heat capacities of a gas was found by a modification of *Clement and Desormes' method in which temperatures were measured instead of pressures. The use of a sensitive *bolometer for this purpose was criticized by *Partington, who modified the method and followed the temperature oscillations occurring on expansion of the gas by means of an *Einthoven galvanometer in the bolometer bridge circuit.

Lummer–Brodhun photometer *See* photometry.

Lummer–Gehrcke plate An interferometer using an accurately parallel-sided glass or quartz plate of considerable thickness in which multiple reflections occur, giving rise to interference effects. It gives a resolving power of the order of 10^6 but it is difficult to manufacture to the degree of accuracy necessary. *See* interference of light; resolving power.

lumped parameter Any parameter of a circuit, such as inductance, capacitance, or resistance, that, for the purposes of circuit analysis, can be treated as a single localized parameter throughout the frequency range under consideration.

lux Symbol: lx. The *SI unit of *illumination, defined as the illumination of one lumen uniformly over an area of one metre squared.`

l.y. Abbreviation for *light-year.

Lyman, Theodore (1874–1954) Amer. physicist who pioneered ultraviolet spectroscopy. He discovered the series of lines, in the ultraviolet region of the *hydrogen spectrum, that bears his name.

Lyman series Of hydrogen. *See* hydrogen spectrum.

lyophilic *See* sol.

lyophobic *See* sol.

M

Mach, Ernst (1838–1916) Austrian physicist, psychologist, and philosopher. Mach's chief importance lies in his influence on scientific thought but he is becoming best known for the association of his name with the phenomena of high-speed motion (*see* Mach angle; Mach number), which he studied. He was a pioneer of spark photography for ballistic phenomena (1887).

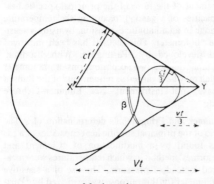

Mach angle

Mach angle If a body moves, with a supersonic velocity V, through a fluid from a point X to a point Y in time t, then when the body is at Y the spherical pressure wave originating from X will have a radius ct where c is the local speed of sound in the fluid. Similarly, the pressure waves from the other points between X and Y will have corresponding radii such

that all the spherical pressure waves combine to form a right-conical wavefront with its vertex at Y. The semiangle of the cone (β) is called the Mach angle:

$$\beta = \sin^{-1} ct/Vt = \sin^{-1} c/V$$
$$= \sin^{-1} 1/Ma,$$

where Ma is the *Mach number.

machine A device for doing work. In a machine a comparatively small force called the *effort* is used to overcome a larger force (e.g. the weight lifted by a system of pulleys) called the *load*.

The ratio

$$\frac{\text{distance moved by effort}}{\text{distance moved by load}}$$

is the *velocity ratio* of the machine.

The ratio

$$\frac{\text{load}}{\text{effort}}$$

is the *mechanical advantage* or *force ratio* and usually varies with the load.

The principle of work states that: work done by effort = work done on load + work lost in friction.

The fraction

$$\frac{\text{work done on load}}{\text{work done by effort}}$$

is the *efficiency* and is necessarily less than 1. It is usually multiplied by 100 and expressed as a percentage.

Simple machines include levers, pulleys, gears, gear trains, the inclined plane, and the screw.

machine code, instructions *See* program.

Mach number Symbol: Ma. A dimensionless number being the ratio of the relative velocity of a body and fluid to the local speed of sound in the fluid. A Mach number in excess of 1 indicates a supersonic velocity; in excess of 5 it is said to be hypersonic. The Mach number appears in all problems of flow in which compressibility is of importance. The resistance to motion of a body moving at high speed in a fluid of small viscosity is, in general, a function of both the Mach and Reynolds numbers. The latter is of high value and the compressibility is of more importance than viscosity. *See also* Mach angle.

Maclaurin series *See* Taylor series.

macro- (prefix) Of everyday size; large. *Compare* micro-.

macroscopic state The state of matter characterized by the statistical properties of its components. *Kinetic theory is an analysis of the macroscopic state. *Compare* microscopic state.

Madelung constant The potential energy of any ion in a rock-salt-type of structure is proportional to the square of the electronic charge times the reciprocal of the distance of nuclear separation of neighbouring unlike ions. The constant of proportionality is the Madelung constant, and this is independent of the lattice dimensions, but has different values for other types of ionic structure.

magic numbers Certain values of the number of protons (Z) in a nucleus or the number of neutrons (N) that produce unusual stability in that nucleus. These values of Z or N are 2, 8, 20, 28, 50, 82, and 126. There is therefore, a tendency for nuclei to prefer magic Z or N: for example, there are six stable nuclides with $Z = 20$, whereas the average number of stable nuclides for a given Z in this part of the periodic table is about two. It is also found that the energy required to remove a nucleon from a nucleus with magic N or Z is higher than for neighbouring nuclei with nonmagic values of N or Z. *See also* shell model.

maglev Magnetic levitation. *See* levitation.

magnalium A group of aluminium-magnesium alloys, with 70–95% aluminium and 30–5% magnesium, that combine strength with lightness. Special varieties have been developed containing small amounts of nickel, copper, lead, and tin.

magnet A body possessing the property of *magnetism. Magnets are either temporary or permanent (*see* permanent magnet). A magnet consisting of a single bar of steel is called a simple magnet; compound magnets comprise several bars fastened together and previously magnetized. *See* electromagnet; magnetite.

magnetic amplifier An amplifier in which *transductors are used to produce power amplification of the input signal.

magnetic analysis The determination of the magnetic characteristics, under either direct or alternating fields, of ferromagnetic alloys. This can throw light on their phase structure, and has been used in identifying components in alloy systems, and the effects of heat treatment and other physical and mechanical variations.

magnetic axis *See* axis of permanent magnet.

magnetic balance 1. A device used for the direct determination of attraction or repulsion between magnetic poles. One magnet is suspended on a knife-edge system so that it takes up a horizontal position, and the force is applied to one end by bringing a magnet pole near to it, the magnet being restored to horizontal by direct addition of weights or the action of a movable rider. The magnets should be long to reduce interference by interaction between their second poles.
2. A type of *fluxmeter in which the force required to prevent movement of a current-carrying coil in a magnetic field is measured.

magnetic blow-out A device fitted to a *circuit-breaker to produce a magnetic field in the neighbourhood of the arc so that the length of the arc path increases, thereby rapidly extinguishing the arc.

magnetic bottle Any configuration of magnetic fields used to confine a *plasma, especially a linear device in which the ends are stoppered with *magnetic mirrors. *See* fusion reactor.

magnetic braking *Syn.* electromagnetic braking. A method of braking used in electric traction in which brake shoes are applied to the track rails by magnetic attraction, the exciting current for the electromagnets being obtained either from the motors acting as generators or from a separate supply. *Compare* electric braking.

magnetic circuit The completely closed path described by a given set of lines of *magnetic flux.

magnetic constant Symbol: μ_0. The *permeability of free space, with the formally defined value:
$$\mu_0 = 4\pi \times 10^{-7} \text{ H m}^{-1}$$
$$= 1.256\,637\,0614 \times 10^{-6} \text{ H m}^{-1}.$$
For the derivation of the magnitude of μ_0 *see* rationalization of electric and magnetic quantities.

magnetic crack detection If a magnetizing field is applied to a ferromagnetic body there is often leakage of lines of force, or uneven distribution of magnetization, at points where discontinuities occur at or near the surface. These discontinuities become evident when the surface is painted with a magnetic fluid consisting of very finely divided particles of iron or of magnetic oxide of iron dispersed in oil. The particles concentrate above the discontinuities. *See also* flaw detection.

magnetic declination *See* declination.

magnetic dipole moment *See* magnetic moment.

magnetic disk *See* disk.

magnetic element *See* geomagnetism.

magnetic equator *See* aclinic line.

magnetic field The *field of force, surrounding a magnetic pole or a current flowing through a conductor, in which there is a *magnetic flux.

magnetic field strength Symbol: H. Formerly called magnetic intensity. A *vector quantity given by the ratio B/μ, where B is the *magnetic flux density and μ the *permeability of the medium. Its magnitude gives the strength of the magnetic field at a point in the direction of the *line of force at that point. The integral of the magnetic field strength along a closed line is equal to the *magnetomotive force. Magnetic field strength is measured in amperes per metre.

magnetic flux Symbol: Φ. The product of a particular area under consideration and the component, normal to the area, of the average magnetic flux density over it. For an element of area dA the flux

magnetic flux density

dΦ is the scalar product $B . dA$. It is measured in *webers.

magnetic flux density *Syn.* magnetic induction. Symbol: B. The *magnetic flux passing through unit area of a magnetic field in a direction at right angles to the magnetic force. It is a *vector quantity whose magnitude at a point is proportional to the *magnetic field strength and whose direction at that point is that of the magnetic field. It indicates the strength of a magnetic field, often in terms of the effects of the field. For example, the vector product of the magnetic flux density and the current in a conductor gives the force per unit length of conductor. (*See also* Lorentz force.) Magnetic flux density is measured in teslas.

magnetic hysteresis *See* hysteresis.

magnetic induction *See* magnetic flux density.

magnetic intensity Former name for magnetic field strength.

magnetic leakage The loss of *magnetic flux from the core of a transformer or transductor, which reduces the overall efficiency of operation. The leakage is the portion of the total magnetic flux that follows a path such that it is ineffective for the desired purpose. Thus, for example, in a d.c. machine the effective magnetic flux is that which passes from a pole face, across the air gap, and enters the armature core: the leakage flux passes directly from one pole to another without entering the armature core. *See* magnetic-leakage coefficient.

magnetic-leakage coefficient Symbol: σ. A coefficient defined by

$$\sigma = \frac{\text{total magnetic flux}}{\text{effective (or useful) magnetic flux}}.$$

σ usually exceeds unity on account of *magnetic leakage. It follows that:

$$\sigma = \frac{\text{useful flux + leakage flux}}{\text{useful flux}}$$

$$= 1 + \frac{\text{leakage flux}}{\text{useful flux}}.$$

A typical value for σ in electrical machines is 1.2.

magnetic lens *Syn.* electromagnetic lens. An *electron lens that focuses an electron beam by means of the magnetic field of one or more electromagnets. Only electrons of one energy can be focused at a particular point. *See* focusing; electron microscope.

magnetic levitation *See* levitation.

magnetic link *See* surge current indicator.

magnetic meridians Imaginary lines drawn along the earth's surface in the direction of the horizontal component of the earth's field at all points along

their length. They converge on the magnetic poles of the earth. *See* geomagnetism.

magnetic mirror A region of high magnetic field strength that reflects ions from a plasma back into a *magnetic bottle. *See* fusion reactor.

magnetic moment 1. Symbol: m. A property possessed by a *permanent magnet or a current-carrying coil, used as a measure of the magnetic strength. It is the torque experienced when the magnet or coil is set with its axis at right angles to a magnetic field of unit size. The torque can thus be expressed as the *vector product $m \times H$ of magnetic moment and magnetic field strength; m is then often called the *magnetic dipole moment*. It is measured in weber metres. The torque is also expressed as the vector product $m \times B$, where B is the magnetic flux density; m is then often called the *electromagnetic moment*, and is measured in amperes per metre squared.
  2. Of a particle. Symbol: μ. A property of a particle arising from its *spin. It is measured in $A\,m^2$ or $J\,T^{-1}$. The electron magnetic moment, symbol: μ_e, is very nearly equal to μ_B, where μ_B is the Bohr *magneton. It has the value
$$9.284\,770 \times 10^{-24}\,J\,T^{-1}.$$
In a system of particles, such as an atom, a particle also has a magnetic moment associated with its orbital motion in the system. The magnetic moment of an orbital electron is equal to $l\mu_B$, where l is the orbital quantum number.

magnetic monopoles Hypothetical magnetic particles (analogous to the electrical particles, the electron and proton) with a magnetic charge of either north or south. They have been postulated on conservation and symmetry principles: an electric particle gives rise to an electric field, and when set into motion gives rise to a magnetic field; a magnetic particle should give rise to a magnetic field and, in motion, produce an electric field. Monopoles would emit and absorb electromagnetic radiation, as do electrons, and could be produced as a pair of monopoles by energetic photons (*see* pair production). Neither *quantum theory nor classical electromagnetic theory bars the existence of the magnetic monopole. *Maxwell's equation would prove completely symmetrical if such particles did exist. They are thought to be more massive than *nucleons. They could be created by the interaction of very high energy particles. In particular, certain *gauge theories predict their existence with Higgs bosons (*see* electroweak theory). Some *grand unified theories also predict monopoles with masses of 10^{16} GeV. Despite these reasons for their existence, and intensive searches for them, no individual monopoles have yet been detected.

magnetic pole strength A measure no longer generally employed for the intensity of a magnet. Magnetic pole strength was originally explained by applying the inverse square law of forces to the poles possessed by a magnet. It can best be understood as the ratio of *magnetic (dipole) moment to magnetic length, and is measured in webers. Modern practice favours the use of magnetic moment in place of this quantity, since the precise location of the poles (regions of concentrated magnetism) is indefinite, and hence the exact magnetic length is unknown.

magnetic potential Former name for magnetomotive force.

magnetic potential difference Symbol: U_m or U. The difference between the magnetic states of two points in a magnetic field. It equals the line integral of the magnetic field strength between the two points. In general, when electric currents are present, this line integral is many-valued and the concept of magnetic potential difference is invalid. It is, however, applicable to regions having boundaries that make it impossible for any closed path to link an electric current.

magnetic quantum number *See* atomic orbital; spin.

magnetic recording In magnetic sound recording, a continuously moving steel wire or tape, or iron-oxide-impregnated plastic tape is longitudinally magnetized so that variations in magnetization represent variations that occur in the audiofrequency currents. If the tape or wire is fed past suitable electromagnets, currents are induced in the coils corresponding to the original magnetizing currents. In practice, microphone currents are amplified electronically and fed to coils surrounding magnetic poles shaped so that a very small length of the recording medium completes the *magnetic circuit. The recording medium is moved at a uniform speed past the recording head. A reproducing head of similar design to the recording head is used to transform the magnetic flux variations into small current variations, which are then amplified and fed to a loudspeaker system. The reproduction may take place immediately after recording. The record is reasonably permanent, but may be easily erased by passing the record medium close to an electromagnet carrying a large direct current that magnetizes the material uniformly. Various forms of recording medium have been used: specially heat-treated steel tape and wire, the latter of 0.004 in. diameter, and a cellulose acetate tape ¼ inch wide and coated on one side with finely divided ferric oxide in the γ form. Tapes are now made with several separate tracks on them. A coated disc, which is traversed spirally by recording and reply heads, has also been used. In a well-designed magnetic-recording and reproducing apparatus it is difficult to detect any difference in tone quality between the reproduction and the original. The ratio of maximum sound to the minimum that may be recorded and reproduced is about 60 decibels. The length of time of a single playing is dependent upon the length of material used.

Magnetic recording of information is used extensively in *computer technology. Data can be stored on and retrieved from *magnetic tape and *disk.

magnetic resistance *See* reluctance.

magnetic resonance imaging (MRI) A technique that is based on *nuclear magnetic resonance of protons, and is used in diagnostic medicine to produce images (proton-density maps) of the body.

magnetic saturation *See* saturation.

magnetic screening A process whereby an area may be screened from magnetic effects by enclosing it with material of high *permeability.

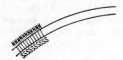

Magnetic shell

magnetic shell A thin iron sheet magnetized across its thickness. It can be considered as an infinite number of small bar magnets. The strength of the shell is expressed as *magnetic moment per unit area.

magnetic shunt A piece of magnetic material mounted near a magnet in an electrical measuring instrument, and having means whereby its position relative to the magnet can be adjusted so that the useful *magnetic flux of the magnet can be varied.

magnetic storms *See* geomagnetism.

magnetic susceptibility *See* susceptibility.

magnetic tape A plastic strip coated on one side with iron oxide, used for *magnetic recording of sound or as a storage medium for information in computing. In the former case, the tape is ¼″ wide with one, two, or four separate recording tracks.

Tape used in computing is essentially the same as that used in domestic audio and video cassettes. Binary information is stored in the form of rows of magnetized dots, typically 7 or 9 across the tape, and up to tens of thousands per inch along it. The tape is wound on a plastic or metal former, commonly in 2400 ft lengths. Items of data are recorded and retrieved from magnetic tape by read and write heads in a device called a (*magnetic*) *tape unit*, the heads operating in accordance with signals from the computer.

magnetic vector potential *Syn.* vector potential. Symbol: A. A quantity defined so that curl $A = B$, where B is the *magnetic flux density.

magnetic viscosity In most ferromagnetic substances (*see* ferromagnetism), there is a time lag between application of a magnetic field and the resulting magnetization, which is accounted for by the eddy currents induced in the substance. In some materials, however, the persistence and magnitude of the change of magnetization are much too great to be accounted for in this way, and this phenomenon is called magnetic viscosity.

magnetic well A configuration of magnetic fields for containing a *plasma in experimental fusion devices. *See* fusion reactor.

magnetism

magnetism The property of magnetism was first discovered in the natural oxide of iron, magnetite. All materials have some magnetic properties caused by the motion of their electrons. *Diamagnetism is a weak effect common to all substances and results from the orbital motion of electrons. In certain substances this is masked by a stronger effect due to electron *spin, *paramagnetism. Some paramagnetic materials, such as iron, also display *ferromagnetism (see also ferrimagnetism; antiferromagnetism).

A *magnetic field can be produced by an electric current (see electromagnet) or by a permanent *magnet. The earth also has a magnetic field (see geomagnetism).

magnetite Syn. lodestone. A naturally occurring form of ferrous-ferric oxide ($FeO.Fe_2O_3$) that shows magnetic properties.

magnetization Symbol: M. The difference between the ratio of the *magnetic flux density (B) to the *magnetic constant (μ_0) and the *magnetic field strength (H):
$$M = B/\mu_0 - H.$$
It is measured in amperes per metre.

magnetization curves The magnetic properties of ferromagnetic substances are usually studied by drawing curves relating the magnetization of the material to the strength and variations of the magnetizing field. See ferromagnetism; hysteresis.

magneto An electrical generator, usually one in which the *magnetic field is provided by a permanent magnet. It can produce periodic high-voltage pulses, and can thus be used to provide the spark in internal-combustion engines.

magnetobremsstrahlung See synchrotron radiation.

magnetocaloric effect Syn. thermomagnetic effect. A fall in temperature occurring when a paramagnetic substance suffers *adiabatic demagnetization. It increases as the initial temperature of the substance is lowered so that the effect has been used for the production of temperatures approaching absolute zero. At very low temperatures paramagnetic substances become antiferromagnetic, restricting further cooling.

magnetochemistry The application of the study of magnetic properties (such as susceptibility) to the solution of chemical problems.

magnetodamping An increase in internal damping of acoustic vibrations in a metal such as nickel when it is subjected to a strong magnetic field.

magnetohydrodynamics (MHD) Syn. magnetofluid dynamics. The study of the behaviour of a conducting fluid (e.g. an ionized gas, *plasma, or collection of charged particles) under the influence of a *magnetic flux. The motion of the fluid gives rise to an induced electric field that interacts with the applied magnetic field, causing a change in the motion itself.

A *magnetohydrodynamic (MHD) generator* or *magnetoplasmadynamic (MPD) generator* is a source of electrical power in which a high-speed flame or plasma flows between the poles of a magnet. The free electrons in the flame constitute a current when they flow, under the influence of the magnetic field, between electrodes in the flame. The concentration of free electrons in the flame is increased by adding elements of low *ionization potential, such as sodium or potassium.

magnetometer Any of a variety of instruments for comparing *magnetic field strengths (H) at different places, or for comparing *magnetic moments (m). One form consists of a short magnet, freely suspended by a jewelled pivot as in a compass needle, and carrying a long pointer, which moves over a graduated circle. The pivoted needle is deflected from its N–S direction by a magnet placed near to it. In the broadside position (with the magnet in an E–W direction, due N or S of the needle):
$$m/H = (d^2 + l^2)^{1/2} \tan\theta,$$
where d is the distance of the centre of the magnet from the needle, $2l$ the length of the magnet between its poles, θ the angle of deflection of the needle. If l is small compared with d then:
$$m/H = d^3 \tan\theta.$$
In the end-on position (with the magnet in an E–W direction and due E or W of the needle):
$$m/H = \{(d^2 - l^2)^2/2d\} \tan\theta$$
or, if l is small compared with d, then:
$$m/H = (d^3 \tan\theta)/2.$$
By measuring θ for various positions of the magnetometer, using the same magnet, or by using two or more magnets in the same positions with respect to the needle, the measurements mentioned above can be made by use of the equations given.

magnetomotive force (mmf) Symbol: F_m. The circular integral of the *magnetic field strength, H, round a closed path:
$$F_m = \oint H\,dx.$$
It is measured in amperes.

magneton A fundamental constant, first calculated by Bohr, for the intrinsic *magnetic moment of an electron. The circulatory current created by the angular momentum p of an electron moving in its orbit produces a magnetic moment $\mu = ep/2m$ where e and m are the charge and mass of the electron. By substituting the quantized relation $p = jh/2\pi$ (h = the Planck constant; j = magnetic quantum number), $\mu = jeh/4\pi m$. When j is taken as unity the quantity $eh/4\pi m$ is called the *Bohr magneton*, symbol: μ_B; its value is
$$9.274\,0780 \times 10^{-24}\ \text{A m}^2.$$
According to the *wave mechanics of Dirac, the magnetic moment associated with the *spin of the electron would be exactly one Bohr magneton, but *quantum electrodynamics shows that there is a small difference.

The *nuclear magneton*, μ_N, is equal to $(m_e/m_p)\mu_B$, where m_p is the mass of the proton. The value of μ_N is

5.050 8240 × 10⁻²⁷ A m².

The magnetic moment of a proton is, in fact, 2.792 85 nuclear magnetons. The magneton is also often given a symbol m_B or m_N.

magneto-optical effects Optical phenomena resulting from the presence of a magnetic field. They include the *Zeeman, *Kerr, *Faraday, *Voigt, and *Cotton–Mouton effects.

magnetopause See magnetosphere.

magnetoplasmadynamic generator See magnetohydrodynamics.

magnetoresistance The change in electrical resistance that ferromagnetic substances (see ferromagnetism) undergo when magnetized. It is closely associated with the change in resistivity (*elastoresistance) caused by tension within the elastic limit of the materials, and is also associated with their *magnetostriction. In materials with negative magnetostriction, the effects of tension and longitudinal magnetic field are in opposite directions, resistivity increasing under an applied field.

magnetosphere A region surrounding the earth and most of the other planets in which ionized particles are controlled by the magnetic field of the planet. It is bounded by the *magnetopause* and includes any radiation belts, such as the *Van Allen belts of the earth. The earth's magnetosphere extends to some 60 000 km on the sunward side but is drawn out to many times this distance on the side away from the sun by the *solar wind.

magnetostatics The study of magnetic forces between poles not in motion.

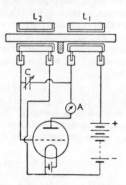

Magnetostriction oscillator

magnetostriction Stresses of compression or extension experienced by a body in a magnetic field and that in ferromagnetic materials are sufficiently great to cause mechanical deformation. Conversely, when such a ferromagnetic material is subjected to mechanical stress, a change in its *permeability occurs. An increase in length of a ferromagnetic rod on application of an axial magnetic field is described as *positive* (or *Joule*) *magnetostriction*; *negative magnetostriction* occurs in materials that decrease in

length as the field density increases. (See also Guillemin effect; Wiedemann effect.)

Magnetostriction is best demonstrated when a bar of nickel or iron is magnetized by a coil carrying a direct current on which is superimposed an alternating current. If the a.c. frequency coincides with the natural frequency of the rod, the mechanical vibrations have large amplitudes. This is used to generate acoustic waves with frequencies ranging from audible to ultrasonic, depending on the dimensions and mode of vibration of the rod. Such vibrations have many applications, especially at ultrasonic frequencies when they can be used, for example, to hasten certain chemical reactions, break the oxide film on aluminium for soldering, or for cleaning purposes.

Magnetostriction oscillators use this principle (rod plus a.c. coil) to produce frequency-controlled oscillations at frequencies from 25 000 hertz downwards. A *tuned circuit is incorporated so that when its frequency is tuned to correspond to the natural frequency of the ferromagnetic rod, the oscillations set up can be maintained. A magnetostrictive rod can also be used to control the frequency of an oscillator (in a similar manner to a crystal-controlled *piezoelectric oscillator). The frequency of the oscillator is adjusted to be close to the natural frequency of the rod; the induced vibrations in the rod are used to pull the oscillator frequency to the rod's natural frequency and maintain it at a substantially constant value.

There are many other devices whose operation is based on magnetostriction; these include *magnetostriction transducers, loudspeakers, microphones,* and *filters.*

magnetostriction oscillator See magnetostriction.

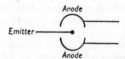

a Magnetron structure (cross section)

magnetron An *electron tube that produces high-frequency microwave oscillations. The typical structure of a magnetron is shown in Figs. *a, b*. The cylindrical cathode is surrounded by a coaxial cylindrical anode with *cavity resonators in its inner surface. The whole tube is placed in a uniform magnetic field parallel to the cylindrical axis. Electrons are emitted from the heated cathode and in the absence of the magnetic field would travel radially to the anode under the influence of the electric field of the anode. The presence of the magnetic field, however, causes them to head into a cycloidal path (see Fig. *c*). The maximum distance an electron can travel towards the anode is determined by the magnetic field. The critical field is reached when electrons just fail to reach the anode. A sufficiently large magnetic field turns most of the electrons back towards the cathode, resulting in a sheath of electrons rotating about the cathode.

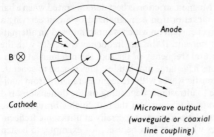

b   Magnetron structure

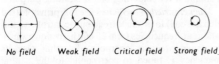

No field   Weak field   Critical field   Strong field

c   Effect of magnetic field on electrons in magnetron

Due to the structure of the anode, the fields associated with this electron cloud induce radiofrequency (r.f.) fields in the cavity resonators, and this field further interacts with the electrons. Depending on the point of interaction with the r.f. field, electrons either travel towards the anode in "spokes" and give up kinetic energy to the field, or they are turned back towards the cathode (*see* Fig. *d*). The kinetic energy received from electrons by the r.f. field is greater than the power required to turn electrons back to the cathode and the result is a net power gain by the r.f. fields. The closed nature of the circuit provides built-in positive feedback and oscillations can occur. The frequency of the oscillations depends critically on the geometrical structure of the anode and the magnitudes of the electric and magnetic fields. The electrons returned to the cathode cause *back heating* of the cathode reducing the *heater current required when the tube is running, and also stimulate *secondary emission of electrons, which forms a significant portion of the total electron emission.

magnification (optics) Symbol: M. When the word is unqualified it refers (dependent on the context) either to the *magnifying power* of an instrument, or to the *lateral magnification* of the image. In the former case, it is the ratio of the angular subtense (ω) of the image seen through the instrument to the angular subtense (ω_o) of the object (*a*) placed 25 cm away for microscopes; (*b*) *in situ* for telescopes, i.e. $M = \omega/\omega_o$. *Lateral magnification* is the ratio y'/y, where y is the height of the object perpendicular to the axis, and y' the corresponding height of the image. *See also* angular magnification.

magnifying glass *Syn.* magnifier. *See* microscope.

magnifying power (MP) *Syn.* instrument magnification. The magnifying power of an instrument is

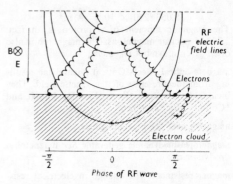

d   Electron paths in the presence of the RF field

defined as the ratio of the size of the retinal image of an object seen with the instrument, to the size of the retinal image of the object seen with the unaided eye. In the case of the microscope, to find the retinal image for the unaided eye, the object is placed at the conventional distance of most distinct vision, i.e. 25 cm. The above definition leads to the magnification as the ratio of the angle subtended at the eye by the image of the object as seen through the instrument, to the angle subtended by the object (*a*) *in situ* for *telescopes, (*b*) when placed 25 cm from the eye for *microscopes; this latter ratio is often called the *angular magnification*.

magnitude A means of expressing the brightness of astronomical bodies, the brighter the body, the lower its magnitude (which may have a negative value). It is based on a logarithmic scale.

The *apparent magnitude*, symbol: m, is the magnitude as observed, correcting for atmospheric absorption. Its value depends on the body's *luminosity, its distance, and the amount of light absorption by intervening *interstellar matter. Two bodies of luminous intensities I_1, I_2 will have magnitudes m_1, m_2 related by:

$$m_1 - m_2 = 2.5 \log_{10} (I_2/I_1).$$

Thus a difference of five magnitudes means a hundred times the luminous intensity. The reference point of the scale has been fixed as $m = 0$ (in the visible region of the spectrum) when $I = 2.65 \times 10^{-6}$ lux.

The *absolute magnitude*, M, is the apparent magnitude the body would have at a distance of ten parsecs from the observer. It can be shown that:

$$M = (m - 5) + 5 \log_{10} x,$$

where x is the distance of the body in parsecs.

Magnox A group of proprietary magnesium alloys used to encase the *fuel elements in certain types of *nuclear reactors. (*See* gas-cooled reactor.) Magnox A consists of magnesium containing 0.8% aluminium with 0.01% beryllium.

Magnus, H. G. (1802–1870) German physicist and chemist. *See* Magnus effect.

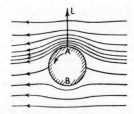

Magnus effect

Magnus effect If a cylinder or sphere rotates about its axis while at the same time it is in relative motion with a fluid, there is a resultant force on the cylinder or sphere perpendicular to the direction of relative motion. The presence of this force was first demonstrated by H. G. Magnus (1852) and is therefore known as the Magnus effect. The resultant flow pattern for a rotating cylinder and a streaming fluid is illustrated in the diagram (the motion is two-dimensional). At A the velocity is greater than the velocity of the undisturbed stream and at B the velocity is less than that of the undisturbed stream. By the application of *Bernouilli's theorem the pressure at B is greater than the pressure at A giving a resultant lift force (L) perpendicular to the direction of streaming. It may be shown that the lift per unit length of the cylinder is given by $\rho\Gamma V$ where ρ is the density of the fluid, V is the relative flow velocity, and Γ is the *circulation.

It is in virtue of the Magnus effect that a spinning shell or golf ball is diverted from its direction of propagation.

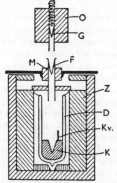

Specific heat capacity of metals (Magnus)

Magnus' method of measuring the specific heat capacity of metals (1915) The specific heat capacity is determined at high temperatures by the method of mixtures. The tapered specimen G (*see* diagram) is allowed to fall into the conical hole in a copper block K after being heated to a known temperature in a furnace O. The rise in temperature of the copper is measured, the block being in a Dewar vessel D surrounded by a water jacket Z. The falling specimen first fits into a thin copper sheath F lightly held at M, the sheath protecting the specimen from the air during the last part of the fall. The cover Kv automatically closes as the sheathed specimen enters the conical hole in K.

Maiman, Theodore Harold (*b.* 1927) Amer. physicist who, working at the Hughes Research Laboratories, Miami, designed and operated the first optical *laser (1960). He later founded his own company to develop and manufacture high-power lasers.

main (electrical) A conductor or group of conductors used for the transmission and/or distribution of electrical power. *See also* ring main; distributor; trunk feeder.

mainframe Any large general-purpose *computer system.

main-sequence star A star, such as the sun, situated on the main sequence of the *Hertzsprung–Russell diagram.

main store *See* memory.

Majorana force *See* exchange force.

majority carrier In a *semiconductor. The type of *carrier constituting more than half of the total charge-carrier concentration.

make-and-break A type of switch that is automatically activated by the circuit in which it is incorporated, and that repetitively makes and breaks the circuit, i.e. rapidly closes and opens the circuit. It is used, for example, in an electric-bell circuit.

making current Applied to a switch, *circuit-breaker, or similar apparatus. The *peak value of the current (including any d.c. component) during the first cycle after the circuit is closed on a short-circuit. *Compare* breaking current.

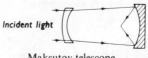

Incident light

Maksutov telescope

Maksutov telescope A telescope consisting of a concave spherical mirror, the spherical *aberration of which is reduced by a convexo-concave *meniscus lens. (*See* diagram.) The image is either focused onto a curved photographic plate or is formed outside the telescope by an additional optical system such as that used in the Cassegrain telescope (*see* reflecting telescope).

malleability A property of a metal whereby it can be shaped when cold by hammering or rolling. Gold is the most malleable metal.

Malter effect If a layer of semiconductor of high *secondary emission ratio (e.g. caesium oxide) is separated by a thin film of insulator (e.g. aluminium oxide) from a metal plate, it can become strongly positively charged on electron bombardment. The potential may be up to 100 volts with an insulating layer of about 0.1 μm thick.

Malus, Etienne Louis (1775–1812) French physicist. He discovered polarization by reflection at a glass surface (1808). The law of Malus states that the transmission through a polarizer placed at an angle with the analyser varies as the square of the cosine of the angle. (*See* plane-polarized light.) The theorem of Malus states that the optical path between any two wave fronts is constant.

manganin An alloy of 15–25% Mn, 70–86% Cu, and 2–5% Ni, that has a high electrical resistivity (about 38 ohm metre) and low temperature coefficient of resistance. It is used for electrical resistances.

A circuit containing resistors and copper connecting wire will normally act as a *thermocouple, and differences of temperature may generate an e.m.f., which can cause errors in precise direct-current measurements. Manganin has the advantage over cheaper resistor alloys (*see* constantan) that it gives a very small thermoelectric e.m.f. with copper.

Mangin mirror (1876) A diverging *meniscus lens, silvered on the outer convex surface, used with signalling lamps and searchlights to throw out parallel light. A combined refraction and reflection results in a system corrected for spherical aberration and coma. Paraboloid mirrors are used for larger apertures.

manometer A device for measuring a fluid pressure. A *differential manometer* is a device for measuring fluid-pressure differences. Some writers restrict the word manometer to instruments measuring fluid-pressure differences, although the etymology does not justify this restriction. However, the word *micromanometer* is generally restricted in this way since the term *vacuum gauge* is available for instruments measuring single pressures. *See* pressure gauges; liquid-column manometers.

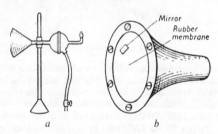

a   *b*

Manometric capsule

manometric flame One of the earliest methods to detect the vibration of air inside an organ pipe was to lower a very light membrane of parchment or rubber stretched on a ring and covered with sand. Maximum agitation of the sand particles indicated a displacement antinode and minimum agitation a node.

König used the manometric flame to locate the maximum pressure changes along the column of air or resonator. The method involves boring a small hole in the walls of the pipe where a thin rubber membrane is stretched. This membrane is held in position by a metal capsule with two apertures; one is connected to the gas supply and the other to a fine

gas jet (Fig. *a*). As the sound falls on the membrane, it vibrates. This will vary the volume of the capsule and set up unsteadiness in the gas supply. The flame, which serves as indicator, rises and falls with the vibration of the membrane. The motion of the flame can be examined stroboscopically (*see* stroboscope) or by means of a revolving mirror. If a number of small manometric capsules are mounted at intervals in the walls of a pipe, there will be rhythmic variations in the flame height as the capsule is moved from a pressure node to an antinode.

A simplified and more useful form of König's manometric capsule has been devised by E. G. Richardson (Fig. *b*). It consists of a short cylindrical portion that widens conically to a flange over which the membrane is stretched. The flame is now replaced by a small mirror attached by a rubber cement to a point of the rubber membrane mid-way between the centre and the edge. The vibration of the membrane gives a slight motion to the mirror and consequently to a beam of light reflected from it to a distant scale. The capsule is calibrated by applying various small pressures measured on a water manometer and noting the corresponding deflection of the image of the light on the scale. The apparatus is mounted on the sounding pipe, a hole being made into which the cylindrical portion of the capsule is sealed. The pressure amplitude can be read from the calibration curve when the amplitude of the deflection has been observed. An investigation by this method has the disadvantage that the pipe or the resonator must be perforated at all the points to be tested. *See* sensitive flame.

many-body problem The problem that there is no general analytical solution to the equations describing the motion of several mutually interacting bodies or particles. If two bodies are involved, e.g. the sun and the earth, the equations of motion can be solved exactly using Newton's laws of motion and Newton's law of gravitation. If, however, three bodies are involved – e.g. the sun, the earth, and the moon – the problem is much more complex, and there is no general analytical solution to the equations of motion. This – the *three-body problem* – was intensively investigated in the eighteenth century because of its importance in celestial mechanics. It was also investigated in the twentieth century because of its importance in atomic theory. Thus, the hydrogen atom, having one positive nucleus and one electron, interacting by electrical forces, is a 'two-body problem', and there is an exact solution to the equations describing the system (in terms of probabilities). A helium atom, however, has a nucleus and two electrons. Not only do the electrons individually interact with the nucleus, they also interact with each other. This means that the description of the helium atom is a three-body problem, and there is no exact solution.

In general, the situation is even worse – there is a many-body problem (or *n-body problem*), involving a number of interacting bodies or particles. For example, the solar system, atoms with more electrons than helium, and molecules are all many-body systems. These are commonly treated by approximate methods, in particular, *perturbation theory. For instance, it is possible to calculate the energy levels of the helium atom by assuming that the two orbiting electrons independently interact with the helium nucleus, and then applying a correction to account for the 'perturbation' caused by the effect of one electron on another.

Marconi, Guglielmo (1874–1937) Italian inventor of wireless telegraphy.

Marconi aerial Fundamentally, an aerial one quarter of a wavelength long, earthed at its lower end. In practice, the aerial length is of the order of a quarter wavelength and the aerial is connected to earth by an inductive or capacitative reactance of such a value that it is electrically equivalent to a quarter-wavelength aerial. If the series reactance is variable (this is usual) the aerial may be tuned to resonate at any frequency within certain limits.

Mariotte, Edmé (*c.* 1620–1684) French physicist. A pioneer of experimental physics in France, he established by experiment the relation between pressure and volume of an imprisoned mass of air, publishing the result in 1676, 14 years after Boyle's publication of the same result. Thus in France, the law known elsewhere as Boyle's law is often called *Mariotte's law*.

mark-space ratio In a pulse waveform, the ratio of the duration of the pulse to the time between pulses. In a perfect *square wave the mark-space ratio is unity.

Marx effect The stopping potential required to prevent the escape of photoelectrons from an illuminated potassium or sodium surface is decreased when the surface is at the same time illuminated by light of lower frequency.

mascon A region of unexpectedly high gravitational potential that is found when lines of equal gravitational potential are plotted over the moon's surface. Mascons may be the result of subsurface convective processes (rather like continental drift on the earth).

maser (*m*icrowave *a*mplification by *s*timulated *e*mission of *r*adiation) Any of a class of microwave amplifiers and oscillators that operate by the same principles as the *laser, but with the beam occurring in the *microwave region of the spectrum. The first maser (1951) predated the construction of a laser by several years. Masers generate less *noise than other types of oscillator or amplifier, producing monochromatic *coherent radiation in a narrow beam, which can have very high energy density. A variety of gas masers and solid-state masers exist.

mask A means of shielding selected areas of a *semiconductor chip during the manufacture of semiconductor components and *integrated circuits. The circuit layout is described on a set of photographic masks, which may be either emulsion on glass or etched in a thin film of chromium on glass. These masks are used during the *photolithography process to define the patterns of openings in the oxide layer through which the various diffusions are made, the windows through which the metal contacts are formed, and the pattern in which the desired metal interconnections are formed.

In the manufacture of *thin-film circuits, metal-foil masks are used to define the pattern of material deposited as a thin film, by vacuum evaporation, onto a substrate.

Maskelyne, Nevil (1732–1811) Brit. astronomer. Founder of the *Nautical almanac* (first published 1767). He estimated the *gravitation constant *G* by observing the deflection of a plumb-line in the vicinity of the mountain Schiehallion (Perthshire, Scotland) in 1774.

masking of sound If a person with a normal ear listens to a pure note and at the same time the intensity of another pure note is gradually raised, a point is reached when the first note becomes inaudible. It is then said to be masked by the second note. Wegel and Lane have conducted experiments on the masking of one pure tone by another. They defined the masking as the logarithm of the ratio of the threshold intensity of a note with masking to that without. The magnitude of the note was taken as the logarithm of the ratio of its pressure to the threshold pressure. A linear relationship between masking and the magnitude of the note was obtained. Experiments by Meyer indicated that a low-frequency note easily masks one of higher frequency but the reverse is much more difficult. Further work by Fletcher shows that a high-intensity note is required to mask one of much higher frequency and also that one note can mask another of slightly lower frequency. Masking is greatest when the notes are nearly alike, provided beats are not produced. Masking is of great importance in telephony and communications work where intelligence must be picked out from background noises.

Mason's hygrometer A *hygrometer of the wet and dry bulb type used as the standard British instrument. The instrument must be exposed to an air draught of 1 to 1.5 metres/second on which the Meteorological Office Hygrometric Tables are based.

mass 1. Newton (1687) defined mass as the quantity of matter of a body, expressed as the product of volume times density. For example, a ball of wool is assumed to have the same amount of matter, whether it is closely compressed or allowed to spread out into a large volume. *Newton's laws of motion assume that mass so defined represents the *inertia* of a body, i.e. its resistance to *acceleration. In principle the masses of two bodies can be compared by comparing their inertias: when subject to equal forces (for example, when interacting with each other, according to the 3rd law of motion), the ratio of the masses is equal to the inverse ratio of their acceleration. Newton's theory of *gravitation assumes that all free bodies have the same acceleration in a gravitational field, hence the *gravitational mass* can be identified with the inertia. This identity is assumed also in the general theory of *relativity. Recent discoveries concerning gravitation may require a reconsideration of this point.

In practice masses are compared (by weighing or by their inertial properties) indirectly with the international prototype *kilogram.

Einstein (1905) showed by his theory of relativity that the mass of a body m as measured by an observer moving with a speed v with respect to it is given by:

$$m = m_0 / \sqrt{(1 - v^2/c^2)},$$

where m_0 is the mass measured by an observer at rest with respect to the body, and c is the speed of light. From this result, and assuming that energy is rigorously conserved in all processes, he showed that the transfer of energy E entails the transfer of mass m where $E = mc^2$. This leads to the conclusion that mass also is conserved.

Mass must be distinguished from *amount of substance, which is defined in terms of the number of constituent particles in a body. Although different observers with different relative motions determine different values for a mass, they find the same value for the amount of substance, since this in principle is purely a question of counting.

2. In particle physics the rest energy of a particle $m_0 c^2$ is sometimes called the "mass". This is contrary to the international system and leads to unresolved ambiguities when considering systems of particles.

mass absorption coefficient See linear absorption coefficient.

mass absorption law See Lenard's mass absorption law.

mass action, law of The principle that the speed of a chemical reaction between substances is proportional to the product of their concentrations. If a chemical reaction is reversible, it reaches a state of dynamic equilibrium in which the concentrations of all the substances present become constant with time. For a particular reaction,

$$aA + bB \rightarrow cC + dD,$$

the concentrations [A], [B], [C], and [D] at equilibrium are related by the equation:

$$\frac{[C]^c [D]^d}{[A]^a [B]^b} = K,$$

where K is the *equilibrium constant* for that reaction at a given temperature. If the reaction occurs in the gas phase, partial pressures may be used instead of concentrations. The application of the law to *electrolytic dissociation leads to *Ostwald's dilution law.

mass defect The difference between the sum of the rest masses of the constituent *nucleons of a particular nucleus and the mass of the nucleus itself. This difference is due to the emission of energy when the nucleus is formed. Energy must therefore be supplied to the nucleus to break it up into its constituents. This energy, the energy equivalent of the mass defect, is the *binding energy of the nucleus. If a particular *fission process is to be used as a source of energy, the sum of the mass defects of the fragments must be greater than that of the original nucleus. For *nuclear fusion to be of practical use, the total mass defect of the combining nuclei must be less than that of the resulting nuclei.

mass–energy equation Einstein's equation, $E = mc^2$. See mass; relativity.

mass excess See mass number.

Massieu function Symbol: J. The negative of the *Helmholtz function (A) divided by the thermodynamic temperature (T): $J = -A/T$. Compare Planck function.

mass–luminosity law A theoretical law relating the mass m and total outflow of radiation, or luminosity, L of normal stars. The law may be represented by the approximation

$$\log L = 3.3 \log m,$$

where m and L are in *solar units.

mass moments See centre of mass.

mass number Syn. nucleon number. Symbol: A. The number of *nucleons in the nucleus of a particular atom. It is the number nearest to the atomic mass, m_a, of a nuclide. The difference ($m_a - Am_u$) is the *mass excess*, where m_u is the *unified atomic mass unit.

mass reactance See reactance (acoustic).

mass resistivity The product of the mass and electrical resistance of a conductor, divided by the square of its length. The units are kilogram ohms per metre squared.

mass spectrograph See mass spectrum.

mass spectrometer See mass spectrum.

mass spectrum The separation of a beam of gaseous ions into components with different values of mass divided by charge. Occasionally negative ions are studied, but most usually positive. Ions which have lost two or more electrons may be observed, but most measurements are made with singly charged ions so the spectrum is divided simply according to mass.

The apparatus is normally highly evacuated except for the ion source, where ionization is caused by an electric discharge through a gas at low pressure, or ions are emitted from a solid surface, for example, by high temperature. Ions are separated by various combinations of electric and magnetic fields and are focused onto a detector. A common system is that in which ions are accelerated from rest through a p.d.

V, giving kinetic energy $\frac{1}{2}mv^2 = eV$. The ion beam then passes between the poles of a magnet and is deflected into a circular arc of radius R given by:

$$R = mv/Be,$$

where B is the magnetic flux density. Thus ions with different values of m/e are deviated by different amounts.

When the ion beams are detected with a photographic plate the instrument is called a *mass spectrograph*. This is suitable for the precise measurement of the relative masses of the ions, and in particular is used for the determination of the masses of *isotopes.

When the ion beams are detected with an electrometer the instrument is called a *mass spectrometer*. This is suitable for the precise measurement of the relative abundances of the ions. It may be used to find the relative abundances of the isotopes of an element, for chemical analysis, or for the measurement of *appearance potentials.

The study of mass spectra originated with the work of J. J. *Thomson (1913) and F. W. *Aston (1919) who confirmed the existence of isotopes, which had been proposed by F. *Soddy (1912) from studies in radioactivity. The techniques have been developed greatly by later workers. By using alternating electric fields, it is possible to select ions of different masses according to their times of flight, thus avoiding the need for a heavy magnet and thereby making possible the use of small portable instruments. The identification of substances by their mass spectra is used in detecting leaks in vacuum apparatus and in studying chemical reactions (*see* molecular beam). Measurement of isotopic abundances permits the use of rare stable isotopes (e.g. oxygen 18) as *tracers.

mass stopping power *See* stopping power.

master frequency meter *See* integrating frequency meter.

master oscillator The frequency of an oscillator depends to a certain extent upon the load on the oscillator. When a high degree of frequency stability is required, it is usual to employ an oscillator of high inherent frequency stability to drive a power amplifier, the latter supplying the power to the load. An oscillator used in this manner is called a master oscillator. When the power amplifier consists of several stages, that which follows the master oscillator is usually designed to operate as a *buffer.

matched load *See* matched termination.

matched termination In a network or transmission line. A termination at which no reflected waves are produced. A load that absorbs all the power incident from a transmission line and forms a matched termination is called a *matched load. See also* impedance matching.

matched waveguide A *waveguide that has no reflected waves at any of the transverse sections.

mathematical series There are certain well-known and important mathematical series:

(a)
$$e = \lim_{n \to \infty} \left(1 + \frac{1}{n}\right)^n$$

$$= 1 + 1 + \frac{1}{2!} + \frac{1}{3!} + \dots \text{ ad inf.}$$

$$\simeq 2 \cdot 718\ 281\ 828\ 5. \dots$$

(b) Exponential series.

$$e^x = 1 + x + \frac{x^2}{2!} + \frac{x^3}{3!} + \dots \text{ ad inf.}$$

(c)
$$a^x = 1 + x \log_e a + \frac{x^2}{2!} (\log_e a)^2$$

$$+ \frac{x^3}{3!} (\log_e a)^3 + \dots \text{ ad inf.}$$

(d) Logarithmic series.
$$\log_e(1 + y) = y - \frac{1}{2}y^2 + \frac{1}{3}y^3 - \frac{1}{4}y^4 + \dots$$
(y being numerically less than 1).

(e) Binomial theorem.

$$(1 + x)^n = 1 + nx + \frac{n(n - 1)}{1 \cdot 2} x^2$$

$$+ \frac{n(n - 1)(n - 2)}{1 \cdot 2 \cdot 3} x^3 + \dots \text{ ad inf.}$$

(x being numerically less than 1).

(f)
$$\cos \alpha = 1 - \frac{\alpha^2}{2!}$$

$$+ \frac{\alpha^4}{4!} - \frac{\alpha^6}{6!} + \dots \text{ ad inf.}$$

(g)
$$\sin \alpha = \alpha - \frac{\alpha^3}{3!}$$

$$+ \frac{\alpha^5}{5!} - \frac{\alpha^7}{7!} + \dots \text{ ad inf.}$$

(h) Gregory's series
$$\theta = \tan \theta - \frac{1}{3} \tan^3 \theta + \frac{1}{5} \tan^5 \theta - \dots$$
$$(-\pi/4 < \theta < +\pi/4)$$
See also Taylor series.

matrix A mathematical concept introduced originally to abbreviate the expression of simultaneous linear equations. It is like a determinant but differs from it in not having a numerical value in the ordinary sense of the term. It obeys the same rules of multiplication, addition, etc. An array of mn numbers set out in m rows and n columns is a matrix of order $m \times n$. The separate numbers are called *elements* (or sometimes constituents or coordinates). Such arrays of numbers, treated as single entities

matrix mechanics

and manipulated by the rules of matrix algebra, are of use wherever simultaneous equations are found (e.g. changing from one set of Cartesian axes to another set inclined to the first; quantum theory; electrical networks). Matrices are very prominent in the mathematical expression of quantum mechanics.

matrix mechanics A mathematical form of *quantum mechanics that was developed by Born and Heisenberg and originated simultaneously with but independently of *wave mechanics. It is equivalent to wave mechanics, but in it the *wave functions of wave mechanics are replaced by *vectors in a suitable space (Hilbert space) and the observable things of the physical world, e.g. energy, momenta, coordinates, etc., are represented by *matrices.

The theory involves the idea that a measurement on a system disturbs, to some extent, the system itself. With large systems this is of no consequence, and the system obeys the rules of classical mechanics. On the atomic scale, however, the result depends on the order in which the observations are made. Thus if p denotes an observation of a component of momentum and q an observation of the corresponding coordinate, then $pq = qp$. Here p and q are not physical quantities but operators. In matrix mechanics they are matrices and obey the relationship

$$pq - qp = ih/2\pi,$$

where h is the Planck constant and $i = \sqrt{-1}$. This leads to the quantum conditions for the system. The matrix elements are connected with the transition probabilities between various states of the system. *See also* uncertainty principle.

matter waves *See* de Broglie waves.

Matthiessen's rule The product of the *resistivity and temperature coefficient of resistance of a metal is the same whether the metal be pure or impure. Normally impurities and alloying elements increase the resistance of a metal markedly, but this effect is accompanied by a corresponding decrease in change of resistance with temperature. The rule is not exact but is often a useful approximation.

Matthiessen's sinker method For the determination of the coefficient of absolute expansion (α) of a liquid. A sinker of weight w in air, is suspended from one arm of a chemical balance and is completely immersed in liquid at a temperature θ_1, its apparent weight w_1 being found by counterpoising the balance. The experiment is repeated with the liquid at a different temperature θ_2, giving an apparent weight w_2 for the sinker, whence

$$\alpha = \frac{w_2 - w_1}{(w - w_2)(\theta_2 - \theta_1)} + \frac{w - w_1}{w - w_2}\gamma,$$

where γ is the coefficient of cubic expansion of the material of the sinker.

Maupertuis, Pierre Louis Moreau de (1698–1759) French mathematician, physicist, and astronomer.

In 1736 he led an expedition to Lapland to measure a degree of longitude, thereby verifying Newton's hypothesis that the earth is not perfectly spherical. He is best known for his formulation (1744) of the *least-action principle (sometimes called *Maupertuis's principle*).

Maupertuis's principle *See* least-action principle.

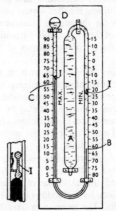

Maximum and minimum thermometer

maximum and minimum thermometer An alcohol thermometer due to Six that records both the highest and the lowest temperatures reached since setting the thermometer. The bulb A is filled with alcohol or spirit while the U-tube between B and C contains mercury. The small bulb, D, is partially filled with alcohol, the rest of the space containing alcohol vapour and air. Movement of the mercury in the U-tube, due to the expansion or contraction of the alcohol in A, causes the mercury to push tiny steel indicators I along the tubes. These indicators remain in position if the mercury meniscus recedes, being held against the walls of the tube by tiny springs. The lower ends of the indicators on the left- and right-hand limbs indicate maximum and minimum temperatures respectively. The thermometer is reset by using an external magnet to draw the indicators into contact with the mercury.

maximum demand The greatest value of the power, kilovolt-amperes, or current taken by a consumer of electricity during a demand-assessment period. It is not, however, an instantaneous value but is the greatest of the average values over each of successive time intervals (called demand-integration periods, which are commonly of 15 min or 30 min duration). The averaging process is carried out either by integration or by using an instrument fitted with a *time lag. A knowledge of this quantity is necessary for tariff purposes. *See* maximum-demand indicator.

maximum-demand indicator An instrument that determines the *maximum demand of a consumer of electricity during a demand assessment period of, for example, three months. In order that the consumer shall not be penalized on account of momentary heavy demands, the instrument either has a *time lag or alternatively is of the type that indicates the maximum average value of the demand over successive equal periods of time (e.g. 15 or 30 minutes).

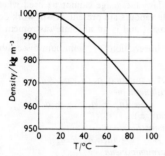

Maximum density of water

maximum density of water Water at 0 °C when heated, contracts until the temperature is 4 °C after which it expands normally (*see* diagram). Owing to the hydrogen bonds between water molecules, ice crystals have a very open three-dimensional tetrahedral structure. When ice melts, this structure collapses and the water molecules become closer packed; small aggregates of molecules can, however, continue to survive up to 4 °C. Thus water has a maximum density at 4 °C. The following workers have devised methods for determining the exact temperature corresponding to the maximum density – *Hope; *Despretz; *Joule and Playfair.

maximum permissible dose *See* dose.

maxwell Symbol: Mx. The electromagnetic unit of magnetic flux in the CGS system, now replaced by the *weber. $1 \, Mx = 10^{-8}$ weber.

Maxwell, James Clerk (1831–1879) Brit. physicist who in 1864 published four fundamental equations (*see* Maxwell's equations) and a theory that linked light with electromagnetic phenomena (*see* electromagnetic theory of light). His researches into colour mixing and colour equations were pioneering. The *Maxwellian view* refers to the method of making a lens apparently flooded with a uniform brightness: a real image of a source of light is formed by a lens in the pupil of the eye. An extended area of bright white light or coloured light is produced and has wide application in photometry and colorimetry.

Maxwell also made important contributions to thermodynamics and to kinetic theory.

Maxwell–Boltzmann law *See* distribution of velocities.

Maxwell distribution *See* distribution of velocities.

Maxwellian view *See* Maxwell, James Clerk.

Maxwell's bridge An early form of inductance bridge (*see* diagram). If the battery key K_1, is closed first, followed by the ballistic galvanometer key K_2, no ballistic throw is observed for the balance condition: $R_1 \dot{R}_3 = R_2 R_4$.

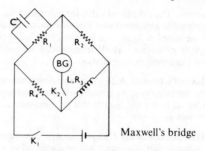

Maxwell's bridge

If, now, K_2 is closed before K_1, and there is still no ballistic throw, the balance condition

$$L = R_2 R_4 C$$

has been achieved, and the value of L may be deduced. If the ballistic balance is not perfect, the ratio R_1 / R_2 must be altered and the double balancing repeated. The circuit may be set up using an alternating current source of electricity, and headphones for detection.

Maxwell's cyclic currents *See* network.

Maxwell's demon An imaginary creature to whom Maxwell assigned the task of operating a door in a partition dividing a volume containing gas at uniform temperature. The door was opened to enable fast molecules to move (say) from left to right through the partition. In this way, without expenditure of external work, the gas on the right could be made hotter than before and that on the left made cooler.

The concept was presented to illustrate how (in principle) it might be possible to violate the second law of *thermodynamics. According to modern ideas the demon would have to interact with the molecules by means of radiation in order to determine their speeds, and would therefore cause other changes, which would invalidate the argument.

Maxwell's equations A series of classical equations that govern the behaviour of electromagnetic waves in all practical situations. They connect vector quantities applying to any point in a varying electric or magnetic field. The equations are:

curl H $= \partial D / \partial t + j$
div B $= 0$
curl E $= -\partial B / \partial t$
div D $= \rho$

H is the *magnetic field strength, D is the *electric displacement, t is time, j is the *current density, B is the *magnetic flux density, E is the *electric field strength, and ρ is volume density of charge.

From these equations, Maxwell demonstrated that each field vector obeys a wave equation: he showed that where a varying electric field exists, it is accompanied by a varying magnetic field induced at right angles, and vice versa, and the two form an electromagnetic field that could propagate as a transverse wave. He calculated that in a vacuum, the speed of the wave was given by $1/\sqrt{(\varepsilon_0 \mu_0)}$, where ε_0 and μ_0 are the *permittivity and *permeability of vacuum. The calculated value for this speed was in remarkable agreement with Fizeau's measured value of the speed of light, and Maxwell concluded that light is propagated as electromagnetic waves. *See also* electromagnetic radiation.

Maxwell's formula A formula connecting the relative permittivity ε_r of a medium with its refractive index n. If the medium is not ferromagnetic, the formula is $\varepsilon_r = n^2$.

Maxwell's rule or law Unless otherwise constrained, a movable part of a circuit will be displaced so as to give the maximum possible magnetic flux linkage with the circuit.

Maxwell's thermodynamic relations The equations relating the four thermodynamic variables S, p, T, and V, referring to a given mass of a homogeneous system, namely

$$\left(\frac{\partial T}{\partial V}\right)_S = -\left(\frac{\partial p}{\partial S}\right)_V, \quad \left(\frac{\partial T}{\partial p}\right)_S = \left(\frac{\partial V}{\partial S}\right)_p,$$

$$\left(\frac{\partial V}{\partial T}\right)_p = -\left(\frac{\partial S}{\partial p}\right)_T, \quad \left(\frac{\partial S}{\partial V}\right)_T = \left(\frac{\partial p}{\partial T}\right)_V,$$

where S is the entropy, V is the volume, p is the pressure, T is the thermodynamic temperature.

maxwell-turn *Syn.* linkage-turn. *See* linkage.

Mayer, Julius Robert van (1814–1878) German physiologist and physicist. His most important work was in thermodynamics (*see* Mayer's formula). He argued that the quantity mv^2 (*vis viva*) was conserved in all processes, including inelastic ones, and considered a quantity analogous to internal energy except that it was regarded as wholly kinetic. This work is often claimed to be a discovery of the principle of the conservation of energy and of the first law of *thermodynamics, but there are obvious difficulties with this interpretation.

Mayer, Maria Goeppert (1906–1972) German-born Amer. physicist who became professor of physics at the University of California (La Jolla). She shared the 1963 Nobel prize for physics with Hans Jensen for their (independent) formulation of the *shell model of the nucleus.

Mayer's formula A theoretical formula, derived from a consideration of the work done by an expanding gas, for the difference between the principal *specific heat capacities of a gas:

$$c_p - c_v = R/M_r,$$

where R is the *molar gas constant and M_r is the molecular weight of the gas. The derivation of the formula depends on the assumption that the internal energy of a gas is independent of its volume, i.e. that the gas behaves as an *ideal gas.

Mayer's original form of the formula was

$$(c_p - c_v) = R/J,$$

which he used to determine the *mechanical equivalent of heat, J.

McLeod gauge A mercury-in-glass vacuum pressure gauge in which a known large volume of gas is compressed into a small volume at which the pressure, now much larger, is measured. Being based on Boyle's law, the gauge cannot be used when condensable vapours are present. It will work down to 10^{-3} pascal and is an absolute instrument.

m.d.s. Abbreviation for *minimum discernible signal.

mean Of n numbers a_1, a_2, a_3, ... a_n:
1. *Arithmetic mean*
$$= (a_1 + a_2 \ldots + a_n)/n$$
(*See* average.)
2. *Geometric mean*
$$= (a_1 a_2 a_3 \ldots a_n)^{1/n}$$
3. *Harmonic mean*
$$= n\bigg/\left(\frac{1}{a_1} + \frac{1}{a_2} + \cdots + \frac{1}{a_n}\right);$$
4. *Quadratic mean* (or root-mean-square value)
$$= \sqrt{\{(a_1^2 + a_2^2 + \ldots + a_n^2)/n\}}$$
5. *See* weighted mean.

mean current density *See* current density.

mean curvature Of a surface. *See* principal radii.

mean density of matter In the universe. The factor that determines the dynamical behaviour of the universe, i.e. whether it is a continuously expanding open system or a closed system that must eventually stop expanding and contract. It is a function of the *Hubble constant and the gravitational constant. The critical density, which if exceeded will lead to an eventual halt to expansion, is about 5×10^{-27} kg m^{-3} (for a Hubble constant of 55 km s^{-1} Mpc^{-1}).

mean deviation *See* deviation.

mean free path Symbol: λ, l. 1. In *kinetic theory, the mean distance that a molecule moves between two successive collisions with other molecules. It is related to the molecular cross section $\pi\sigma^2$ by the relationship:
$$\lambda = 1/\sqrt{2}\pi n\sigma^2,$$
where n is the number of molecules per unit volume. The most important means of determining λ is through its connection with viscosity η. According to kinetic theory:
$$\lambda = k\eta/\rho u,$$
where ρ = density and u = mean molecular velocity. The value of k lies between $\frac{1}{3}$ and $\frac{1}{2}$ according to the degree of approximation introduced into the theory.

2. In atomic, nuclear, and particle physics, the mean distance that a particle travels in a medium before undergoing a particular type of interaction. For example, there are mean free paths for absorption, elastic scattering, inelastic scattering, fission, etc., for various types of particle and medium. If the number of target particles per unit volume is N and

the *cross section for the particular process is σ then $\lambda = 1/N\sigma$.

mean lethal dose *See* median lethal dose.

mean life *Syn.* average life or lifetime. Symbol: τ. **1.** The average time for which the unstable nuclei of a radionuclide exist before decaying. It is the reciprocal of the *decay constant and is equal to $T_{1/2}/0.693\,15$, where $T_{1/2}$ is the *half-life.
2. The average time of survival for an elementary particle, ion, etc., in a given medium or a charge carrier in a *semiconductor.

mean solar time *See* time.

mean spherical intensity *See* luminous intensity.

mean square velocity The average value of the square of all the velocities of a system of particles, given by the relation:
$$C^2 = (n_1c_1{}^2 + n_2c_2{}^2 + n_3c_3{}^2 + \ldots n_rc_r{}^2)/n,$$
where n_1 particles have velocity c_1, n_2 particles have velocity c_2, etc., and
$$n = \Sigma_r^1\, n_r$$
is the total number of molecules.
Its value for a gas may be calculated on the *kinetic theory from the expression $p = \frac{1}{3}\rho C^2$, where p and ρ are the pressure and density respectively of the gas. For an ideal gas $C^2 = 3rT$, where r is the gas constant for unit mass of the gas. This expression shows that the mean velocity is dependent only on the temperature of a given gas. By the Maxwell *distribution of velocities, for the molecules of a gas in a steady state, the mean square velocity has the value $C^2 = 3kT/m$. *Compare* mean velocity.

mean sun *See* time.

mean tone scale For use with keyboard instruments with 13 notes to the octave, to permit tolerable approach to the just intonation scales starting on the different keynotes. The mean tone system of *temperament is based on major thirds being accurate, the other intervals being adapted. The system gives 6 major and 3 minor scales reasonably accurately, but the other common scales produce impossible intervals. The out-of-tuneness is sometimes called the "wolf". Mean tone scales are not used today.

mean velocity The average value of the velocities of a system of particles, given by the relation
$$\bar{C} = (n_1c_1 + n_2c_2 + \ldots n_rc_r)/n,$$
where n_1 particles have velocity c_1, n_2 particles have velocity c_2, etc., and
$$n = \Sigma_r^1\, n_r$$
is the total number of particles.
The Maxwell *distribution of velocities for the molecules of a gas in a steady state yields a value:
$$C = 2/\sqrt{(\pi hm)},$$
where $h = 1/(2kT)$.

mechanical advantage *Syn.* force ratio. *See* machine.

mechanical equivalent of heat *Syn.* Joule's equivalent. Symbol: J. Before 1948 the unit generally employed for the measurement of heat and internal energy, in some cases even those involving electrical heating, was the calorie, defined as the quantity of heat required to raise the temperature of one gram of water through a specified one degree Celsius; in particular the "fifteen-degree calorie" was used, defined over the range from 14.5 °C to 15.5 °C. The usual procedures of calorimetry enabled experimental determinations of such quantities as specific and latent heats to be made in terms of the calorie, but other experiments were required to relate the calorie to the mechanical units of energy, the *joule and the *erg. The mechanical equivalent of heat was defined as the ratio of an amount of work in ergs to the amount of heat in calories to which it is equivalent.
Since 1948 it has been recommended that *all* kinds of energy be measured in "mechanical units" and the mechanical equivalent of heat has been formally defined values (for the different 1 °C ranges) that represent the experimentally determined values very closely. For the fifteen-degree calorie
$$J = (4.1855 \pm 0.0005) \times 10^7 \text{ erg cal}_{15}{}^{-1}$$
$$= 4.1855 \pm 0.0005 \text{ J cal}_{15}{}^{-1}.$$
The recommendation concerning the unit of heat and internal energy has only been followed generally since the late 1960s. Until this time the calorie employed had been that defined by the Fifth International Conference on Properties of Steam, 1956, and called the international table calorie. The definition is derived from the formally adopted value for the mechanical equivalent of heat:
$$J = 4.1868 \text{ J cal}_{IT}{}^{-1}.$$
Experiments formerly used for evaluating J are now employed to determine the specific heat capacities of the substances used. Where the substance was water, the value was assumed to be $1 \text{ cal}_g{}^{-1}\,°C^{-1}$. *See* the experiments of Joule (in which internal energy was increased by doing work on a body); *also* those of Griffiths, and Callendar and Barnes (in which work was done electrically).

mechanical equivalent of light *Radiant flux expressed in mechanical units, which are equivalent to the unit of *luminous flux, at the wavelength of maximum visibility. It is 0.0015 watts per *lumen at 555 nm. Its reciprocal is also quoted with the same title (660 lumens per watt).

mechanical impedance Symbol: Z_m. The mechanical impedance of a vibrating system is the complex ratio of the force acting in the direction of motion at a point or surface to the velocity at that point or surface. It is measured in newton seconds per metre. This quantity is meant to act as the mechanical analogue of electric *impedance and *acoustic impedance. The concept has been further extended by writing
$$Z_m = R_m + iX_m.$$
R_m is the mechanical resistance and X_m the mechanical reactance.

mechanical rectifier A rectifier that consists essentially of a rotating or oscillating commutator, operated synchronously so that alternate half waves of the input alternate current are inverted thereby producing a unidirectional output current.

mechanics The branch of science, divided into dynamics, statics, and kinematics, that is concerned with the motion and equilibrium of bodies in a particular frame of reference. *See also* wave mechanics; quantum mechanics; statistical mechanics.

mechanomotive force In any machine that develops an alternating force, the *root-mean-square value of the force developed.

median 1. (general) The central term of a sequence of values arranged in order of magnitude.
2. (geometrical) The line joining the apex of a triangle to the midpoint of the opposite side.

median lethal dose (MLD) The absorbed *dose of ionizing radiation that will kill, in a prescribed time, half of a large population of a particular species.

medical physics The application of physics to medicine. An extremely wide field that is constantly expanding with the development of more and more sophisticated medical techniques. Most medical physicists at present are found in the fields of radiotherapy, *nuclear medicine, diagnostic physics, dosimetry, and medical electronics. Their duties include the measurement and responsibility for all sources of *ionizing radiation, calculation of dose distributions in radiation therapy, development of electronic equipment for many uses, and the provision of a consulting service for doctors who wish to apply a particular physical technique (e.g. ultrasonics) to a particular situation. A computing service is often provided by the medical physicist, who will advise on the application of computers for analysis of medical problems and often provides the programs.

medium frequency (MF) *See* frequency bands.

medium voltage In electrical power transmission and distribution, a voltage in excess of 250 volts but not exceeding 650 volts.

medium-wave Designating radio waves with wavelengths in the range $0.1-1$ km, i.e. with frequencies of $3-0.3$ MHz.

mega- Symbol: M. **1.** A prefix meaning 10^6 (i.e. one million); for example, one megahertz (1 MHz) is 10^6 hertz.
2. In situations where the binary number system is used, such as computing, a prefix meaning 2^{20} (i.e. 1 048 576); for example, one megabyte is 2^{20} bytes.

megaphone An instrument for amplifying and directing sound. It consists of a conical or rectangular horn about 30 cm long, the small end of which is held near the mouth of the speaker. The horn increases the efficiency of the voice by providing a suitable loading for it. Provided the solid angle of a conical horn is not large, the wave front of the sound emerging from the open end is almost plane (*see*

horn). However, the diameter of the open end of the horn should be large compared with the wavelength of the sound or reflections are produced, causing a loss in radiation efficiency. The megaphone horn also has directive properties depending on its dimensions. To enable a narrow beam of sound to be radiated the open end of the horn should be large compared with the wavelength of the sound, otherwise diffraction effects cause a spreading of the beam. Owing to the practical limitations on the size of a megaphone, these theoretical requirements are not usually fulfilled. As a result the low-frequency components of speech are not radiated efficiently and there is also very little directive effect at these frequencies. Megaphones may be designed to spread the sound in one direction but to give a narrow beam at right angles to this. Thus, a megaphone with a long narrow rectangular opening will spread the sound in the plane of the smaller side and confine it in the plane of the larger side.

megger A portable insulation tester. Two coils are fixed at right angles to each other and move in a strong magnetic field. One is the pressure coil and is connected, in series with a resistance, across the terminals of a hand generator, which gives about 500 volts. The other is the current coil and is connected, in series with the resistance to be measured, to the generator. The coils oppose each other and reach a balance position where the turning moments are equal and opposite. The position of balance depends on the external resistance, and the scale of the instrument is calibrated directly in megaohms.

Meissner effect *See* superconductivity.

Meitner, Lise (1879–1968) Austrian physicist. With Otto Hahn, she discovered the radioactive element protactinium (1918). She proved experimentally (1926) that a photon is emitted from a nucleus *after* undergoing a transmutation and not simultaneously. With Hahn and Strassmann, she caused the first *fission of uranium by neutron bombardment (1934), but the effects were at the time wrongly attributed to the creation of elements of higher atomic number than uranium.

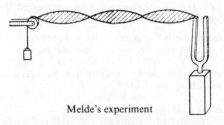

Melde's experiment

Melde's experiment An example of the maintenance of transverse vibrations of a string when coupled to that of a rod such as a tuning fork or a reed. It was first accomplished by Melde who gave the apparatus the form shown. A long horizontal thread is attached at one end to the prong of a massive tuning fork while the other end passes over a frictionless pulley and carries a scale pan containing weights. If the fork is excited by striking or bowing or by electrical maintenance, the thread is forced to vibrate with the same frequency as that of the fork. As the thread has a definite tension and mass per unit length, the only variable is the vibrating length. At certain values of the length, the thread divides itself by a series of displacement nodes into vibrating segments of equal length. As the tension or the load in the scale pan is increased, the vibrating lengths become longer and the number of segments into which the thread divides itself becomes smaller. If the load X giving an exact number of segments p is adjusted, then it will be found that $Xp^2 =$ constant. This result is derived from the formula:

$$f = (p/2l)\sqrt{X/m}$$

(see stretched string), where f is the frequency, l is the total length of the thread, and m is the mass per unit length. When adjustment for resonance is not exact, pseudo-nodes and antinodes will be observed due to the vibrations in the string forced by the fork. These coupled vibrations are possible not only with the fork vibrating in a direction at right angles to the string, but also with the vibrations in the direction of the length of the string. In the latter case, Melde showed that the frequency of the tone in the string is half that of the fork. When the fork is at an oblique angle to the string, *Lissajous' figures result, due to vibrations corresponding to the horizontal and vertical positions.

melting See fusion.

melting point See freezing point.

membrane in vibration A theoretical membrane is a very thin, completely flexible, and uniform lamina stretched outwards in all directions by a constant tension. The equation of motion for transverse vibration of a member in the x, y plane is

$$m \cdot \frac{d^2z}{dt^2} = T\left(\frac{d^2z}{dx^2} + \frac{d^2z}{dy^2}\right)$$

where m is the mass per unit area and T is the tension. For a rectangular membrane there is a series of *partials having frequencies given by:

$$n = \sqrt{\frac{T}{4m}\left(\frac{p^2}{a^2} + \frac{q^2}{b^2}\right)}$$

where a and b are the lengths of the sides and p and q are integers. $(p - 1)$ nodal lines of displacement are produced parallel to the side of length b and $(q - 1)$ lines parallel to a. Some of the partials are harmonic but the lower tones lie close together and the general effect is that of a noise with a predominant note corresponding to $p = q = 1$. In the case of a circular membrane the solution is more complex, the fundamental mode having a frequency given by:

$$n = \frac{0{\cdot}765}{2a}\sqrt{\frac{T}{m}},$$

a being the radius. The partials are inharmonic and the nodal lines consist of a series of concentric circles. The theoretical membrane cannot be realized in practice owing to the natural stiffness of any actual membrane. However, membranes of paper, skin or metal can be employed to demonstrate the nodal lines predicted by theory. The nodal lines can be seen by sprinkling fine dry sand on the membrane. A common use of membranes is in drums. In the case of the kettledrum, a resonant cavity of air is used to give it a definite fundamental tone. The drum is struck with a soft hammer about a quarter of the way along a diameter. At this point fewest inharmonic partials are produced. In the *capacitor microphone, a very thin steel membrane is stretched almost to its elastic limit and has a natural frequency of the order of 5000 hertz.

memory Syn. store; storage. A device or medium in which data and *programs can be held for subsequent use by a *computer. It may be either *backing store or main store (also called main memory or simply memory). Main store now consists of *semiconductor memory – either *RAM (random-access memory) or *ROM (read-only memory). Main store is closely associated with the central processor of the computer: program instructions and associated data are stored there temporarily, awaiting use by the processor.

The memory is divided into storage locations, each of which can be uniquely identified by its address; each location holds the same number of *bits, usually 8, 16, or 32 (see word; byte). The processor can retrieve information from a particular location extremely rapidly.

Programs can only be executed when they are in main store. They and their associated data are not held permanently in main store, however, but are kept in larger-capacity backing store (usually magnetic *disks or tapes) until required by the processor.

Mendeleev, Dmitri Ivanovitch (1834–1907) Russian chemist. He arranged the elements in the *periodic table (1869), and predicted (1871) the existence and properties of elements as then unknown, required to fill gaps in the table.

Mendenhall and Forsythe's verification of radiation laws (1914) Mendenhall and Forsythe compared the scales of a total *radiation pyrometer and a *disappearing filament pyrometer by measuring with both the temperature of a black-body furnace outside the range in which standard thermocouples can be used. The agreement was within experimental error, confirming the radiation laws upon which the scales are based, viz. the *Stefan–Boltzmann law and the *Wien radiation law.

meniscus A concave or convex upper surface of a liquid column that is due to capillary action.

meniscus lens

meniscus lens A convexo-concave or concavo-convex lens. Such types in spectacle lenses are usually called *deep* meniscus lenses and in general provide a wider field of good definition. *See* landscape lens; lens.

meniscus correction When reading the height of a liquid in a glass tube it is usual to note the position of the highest point of the meniscus for mercury and the lowest point for liquids such as water. A correction must often be made to this height, before it is used in calculations, to allow for the pressure difference between the two sides of the curved surface.

mercurial air pump *See* Sprengel air pump.

mercury barometer *See* barometer.

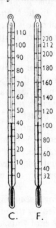

Celsius and Fahrenheit
C.   F.   thermometers

mercury-in-glass thermometer A type of thermometer in which mercury acts as the thermometric fluid in a glass bulb attached to a graduated fine capillary tube. During manufacture all air is excluded from the capillary tube, a small bulb being left at the top of the tube as a safeguard against breakage should the thermometer be raised to a temperature beyond the highest value on the graduated scale. The thermometer is calibrated by immersion first in melting ice, then in steam in a *hypsometer, the positions of the mercury meniscus being marked. In the Celsius (centigrade) thermometer, the distance between the marks is divided into 100 equal parts, each part corresponding approximately to a degree Celsius on the mercury in glass scale. In the Fahrenheit thermometer, freezing point is 32 °F and boiling point is 212 °F.

Although it has the advantage of giving a direct reading that is easily read, for accurate work so many corrections have to be applied that this type of thermometer has been replaced by the platinum *resistance thermometer. Since the coefficient of expansion of mercury is not independent of temperature and since the expansion of the glass is not negligible, the thermometer readings can only be corrected to the gas scale by a direct comparison with a *gas thermometer. The chief errors for which

correction is necessary are (1) nonuniformity of the bore of the capillary tube; (2) errors in marking the ice and steam points; (3) hysteresis of the glass causing a change in the ice point; (4) the effect of external pressure on the bulb; (5) the emergent stem correction. *See also* International Practical Temperature Scale.

mercury-in-steel thermometer A type of thermometer used industrially for recording temperatures some distance away from the hot body. The expansion of the mercury operates a form of *Bourdon tube, the temperature being obtained from the position of a pointer on a scale.

mercury-jet magnetometer An instrument that allows the measurement of *magnetic field strength to be made without introducing a ferromagnetic body to distort the field. A stream of mercury flowing through a narrow nonconducting pipe has two electrodes at opposite ends of a diameter; in a magnetic field an e.m.f. is set up in the flowing mercury, directly proportional to the field strength. By this means a rapidly fluctuating field can be measured over a very small area of the field.

mercury motor meter A *motor meter in which some part of the moving sytem is immersed in mercury. The mercury provides electrical contact with the moving part and since the latter tends to float in the mercury, bearing friction is very small. Instruments of this type are used on d.c. electrical supply systems to indicate the quantity (e.g. ampere-hours) of electricity consumed or, with special design, the (e.g. kilowatt-hours) work done.

mercury spectrum Mercury arcs and glow discharges produce intense lines in the visible part of the spectrum; the green mercury line (546.07 nm) is frequently used, the other lines being absorbed by filters. The lines in the ultraviolet are transmitted by using quartz envelopes.

mercury switch A switch in which contact is established between two mercury surfaces usually enclosed in a glass tube. Arcing is often suppressed by filling the tube with an inert gas and sometimes a porcelain tube is fused in at the point of contact to eliminate breakage from heat shock. There are many types. Usually they are operated by tilting, and may have delayed make, or break, or both, by the mercury having to flow through a constriction in a side tube. The mercury may flow over a series of contact pools in turn. In one form an external solenoid pulls a soft iron core into a reservoir of mercury and displaces it over the contacts.

mercury-vapour lamp An incandescent arc of mercury vapour between mercury electrodes in an enclosed tube. The light is rich in ultraviolet radiation and in *fluorescent lamps some of this is converted to visible wavelengths by fluorescent powders coated onto the interior of the tube.

mercury-vapour rectifier An electron tube in which a discharge passes in one direction only from a hot-wire cathode to an anode via an atmosphere of ionized mercury vapour. After the initial ionization is achieved, the voltage drop across the tube is only 10 to 15 volts and is almost independent of the current. With a mercury-pool cathode, the voltage drop is 20 to 25 volts.

meridian 1. A *great circle passing through a point on the surface of a body such as the earth (or the *celestial sphere), and through the north and south poles, and crossing the equator at right angles.
   2. *See* magnetic meridians.
   3. (optics) *See* meridian plane.

meridian circle *Syn.* transit circle. *See* astronomical telescope.

meridian plane *Syn.* tangential plane. In an optical system. A plane that contains both the optic axis and the *chief ray (i.e. the one passing through the centre of an aperture). It is perpendicular to the *sagittal plane. Rays in a meridian plane, e.g. in oblique *astigmatism, converge to the meridian image point; the focal line at this point is perpendicular to the meridian plane, and is referred to as the *meridian focal line* or the *tangential focal line*. The corresponding focal surface is the tangential surface. (*See* tangential foci.)

Mersenne, Marin (1588–1648) French natural philosopher who was the first to measure the speed of sound. By timing echoes he obtained a value of 316 metres per second (modern value 332 m s^{-1} at sea level at 0°C). He also discovered the law of vibrating strings (*see* monochord). For many years he maintained a regular correspondence with scholars in various countries, notably *Descartes, and hence played a major part in the development of science in a period preceding the establishment of scientific societies and journals.

Mersenne's law *See* monochord.

Merton grating A *diffraction grating produced by the method due to Sir Thomas Merton (1948). In this, a fine helical thread is first cut on a cylinder and errors in this are smoothed by cutting a second helix further along the same cylinder, the linkage including a *Merton nut*. This has a resilient lining, e.g. of pith, which moulds itself onto a wide portion of the first helix and averages out the errors. A plastic replica is then made, the plastic film applied to the helix being slit and peeled off, and applied to mould a sheet of moist gelatin into a plane grating.

Merton nut *See* Merton grating.

mesa transistor A type of bipolar junction transistor in which the *base region is first diffused into the substrate, the areas surrounding the base region then being etched to leave a plateau above the substrate. The substrate below the base region forms the *collector. The *emitter may be either alloyed into the base region or diffused into it. The latter is a double-diffused transistor.

mesh *See* network.

mesh connection A method of connection used in *polyphase systems in a.c. working in which the windings of a transformer, a.c. machine, etc., are all connected in series to form a closed circuit so that a polygon may be used to represent them diagrammatically. A special form is the *delta connection. *Compare* star connection.

mesh contour *See* network.

mesh voltage *See* voltage between lines.

meson A collective name given to *elementary particles that can take part in *strong interactions and that have zero or integral spin. By definition, mesons are both *hadrons and *bosons. *Pions* and *kaons* are mesons. Mesons have a substructure composed of a *quark and an antiquark bound together by *gluons. For example, the π^+ meson contains an up (u) quark and an antidown ($\bar{d}$) quark and the K$^-$ meson contains a strange (s) quark and an antiup ($\bar{u}$) antiquark. *See also* Appendix, Table 8.

mesosphere An *atmospheric layer of the earth, lying above the *stratosphere, extending from an altitude of 50–60 km up to 80–90 km. The temperature decreases rapidly with height reaching a minimum of between −70 to −90 °C at the *mesopause*. This is the upper boundary of the mesosphere at which the air pressure is approximately 10^{-6} bar.

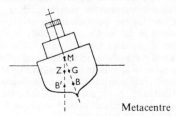

Metacentre

metacentre The point at which a vertical line through the centre of *buoyancy B′ of a tilted ship (or other floating body) intersects the line joining the centre of mass G and the centre of buoyancy B of the upright ship. There are two metacentres according to whether rotation is about a horizontal axis parallel or perpendicular to the keel. If G is below M, the force of buoyancy (upwards through B′) together with the weight of the ship (downwards through G) tends to rotate the ship back to the upright position.
   The horizontal distance (GZ) of G from B′M is known as the *righting lever*. The distance GM is the *metacentric height* and although the ship is safe from capsizing if this is large, it is liable to roll heavily in a rough sea.
   The surface described by B as the ship rotates about various horizontal axes, without changing the volume displaced, is the *surface of buoyancy*. The two principal normal sections of this surface are known as the *curves of buoyancy* and the centres of curvature of these curves are the metacentres.

metacentric height *See* metacentre.

metadyne

metadyne 1. Metadyne generator. A commutator-type d.c. generator having, in addition to the usual brushes supplying the load, a set of brushes that are either short-circuited or connected together through stator windings of low resistance. The extra brushes have a position intermediate between the others on the commutator. The current in the circuit containing the additional brushes produces the whole or a substantial part of the magnetic flux that gives rise to the output voltage; *armature reaction plays an important part in the operation. The armature is driven mechanically as in a normal d.c. generator. Fundamentally, the output current is constant, i.e. is independent of the resistance of the load but the output may be controlled by means of a separately excited control winding arranged to produce a voltage in the circuit containing the additional brushes. One form of this machine is known as a Rosenberg generator.

2. Metadyne converter. A machine that is similar in construction to a metadyne generator. The additional brushes are not, however, connected together but are connected to an independent d.c. supply (usually of constant voltage). Fundamentally, this machine converts the input power from the d.c. supply at constant voltage into output power at constant current (i.e. output current is independent of load resistance), the mechanical power required to drive the armature being very small. Additional windings are provided for control purposes. This type of machine has applications in connection with the control of d.c. motors, particularly in electric traction. It provides smooth acceleration of the motors and enables *regenerative braking to be used even at very low speeds. In the metadyne (generator or converter) the pole pieces are usually divided into two so that in any particular machine there are twice as many polar projections as there would be poles in a normal d.c. generator using the same armature. *Compoles may also be provided. *Compare* amplidyne.

metal-arc welding Arc-welding in which the additional metal is provided by the melting of a metal electrode.

metallic crystals Crystals in which a regular arrangement of positive metallic ions is held together by an "atmosphere" of free electrons.

metallizing The covering of an insulating material with a film of metal or other substance to render it electrically conducting. The technique is widely used in solid-state electronics. The conducting film is etched and forms interconnections on *integrated circuits. It is also used in forming *bonding pads for integrated circuits and discrete components.

metallurgy The study and practice of producing, alloying, fabricating, and heat-treating metals.

metal rectifier A *rectifier that depends for its action upon the fact that when a metal is placed in contact with a suitable solid (such as a semiconductor or an oxide of the metal), the resistance offered to the passage of an electric current is very much less when the current flow is in one direction (e.g. from the solid to the metal) than it is in the other direction. Typical materials used are cuprous oxide on copper, and selenium on copper. The potential applied must not exceed a few volts so that for high voltages a series of elements is required. The current passed depends on the area of contact.

metal V-ring or collar Of a commutator. *See* V-ring.

metaphysics The branch of speculative enquiry that deals with first principles. It is usually considered to include *ontology* (the study of being) and *epistemology* (the study of the nature of knowledge).

metastable state 1. (chemistry) A state of pseudo-equilibrium in which a system, such as a *supersaturated vapour or supercooled liquid (*see* supercooling), has acquired more energy than that normally required for its most stable state, and yet is not unstable. It is often achieved by attaining the state very slowly. A slight disturbance, such as the addition of a small amount of the stable *phase (*see* phase rule), will produce the stable state. Water can be cooled very slowly to a temperature below 0 °C. The addition of a piece of ice will cause the water to freeze rapidly.

2. (physics) Symbol: m. A comparatively stable *excited state of a radionuclide that decays into a more stable lower-energy state with the emission of gamma-rays. The nuclide technetium-99m decays into technetium-99, the half-life being 6 hours. It is often an excited state from which all possible transitions to lower states are *forbidden transitions according to the relevant *selection rules.

3. A comparatively stable electronically *excited state of an atom or molecule.

meteor A lump of matter from space that enters the earth's atmosphere and is detected either optically by virtue of its luminosity as a result of friction with air particles, or by radio means by virtue of the trail of ionized gas left in its wake. Isolated meteors are described as *sporadic*. More obvious are the meteor *showers*; between five and a hundred meteors per hour can be observed, which appear to radiate from a small area of space called the *radiant*. The paths of the meteors are in fact parallel and the apparent convergence is due to *parallax. The showers are identified by the constellation that contains the radiant.

Typical of meteor showers is the Perseids (radiant in Perseus), visible between about August 8 to August 16 each year. At maximum some 40 per hour are observed, travelling with velocities of about 6×10^4 metres per second.

Most showers can be associated with former comets, and represent the debris resulting from a break-up. The Perseids have an orbit about the sun that relates them to the great comet of 1862.

Since the surface of a typical meteor reaches a temperature of the order of 3000K, very little of the estimated 10^6 kg of meteoric material captured by the earth reaches its surface, and only a minute proportion has been recovered. Such specimens are called *meteorites*. 45% of meteorites have a stony constitution, 50% are made up largely of iron and nickel alloys, and the remaining 5% have a mixture of metals and minerals. These figures do not represent the true proportions among meteorites in general, but reflect the relative difficulty of discovering stone meteorites (since they are easily confused with earth-originated stones). The densities of the three classes are approximately 3500, 8000, and 6000 kilograms per cubic metre respectively. Meteorites tend to produce craters with a volume about 5×10^4 times that of the specimens themselves, and one important theory of the formation of lunar craters is that they result from unimpeded meteoric bombardment onto an atmosphere-free body.

meteorite *See* meteor.

meteorograph A device for recording some or all of the following: temperature, relative humidity, pressure, and wind speed. It is usually carried into the upper air by a balloon or kite.

meteorology The science of the atmosphere, especially with relation to the weather and climate.

meter 1. Any measuring instrument. In electrical engineering, the term, when unqualified, usually implies an *integrating meter.
    2. American spelling of *metre.

method of mixtures A method of calorimetry in which a substance is added to a calorimeter at a different temperature, the mixture being stirred to reach equilibrium at an intermediate temperature. The unknown heat capacity may be calculated by equating the heat lost by one part of the system to the heat gained by the remainder of the system since the law of conservation of energy applies if there is allowance made for heat exchange with the surroundings. *See* radiation correction.

metre Symbol: m. The *SI unit of length, defined (since 1983) as the length of the path travelled by light in vacuum during a time 1/299 792 458 second.
    The metre was originally intended to be one ten-millionth of the quadrant from the equator to the north pole through Dunkirk, but difficulties in measurement led to the adoption of the length of a prototype bar instead. Developments in the precision of the measurements of optical wavelengths led to the adoption of a definition in terms of the wavelength of a spectrum line of krypton. The great precision and reliability of atomic *clocks has permitted the current definition in terms of the second.

metre bridge A form of the *Wheatstone bridge in which a uniform resistance wire 1 metre long, which can be tapped at any point along its length, takes the place of two of the four resistors.

metre-kilogram-second electromagnetic system of units (MKS units) A system of absolute units (due to Giorgi (1901), following a suggestion of Maxwell),

in which the fundamental units of length, mass, and time are respectively the metre, the kilogram, and the second, and in which the permeability of free space is 10^{-7} henrys per metre. In many electromagnetic equations of practical importance, a factor 4π appears. This factor can be transferred from these equations to others less commonly used by taking the permeability of free space as $4\pi \times 10^{-7}$ henrys per metre and this gives the *rationalized* MKS system of units. *SI units are based on the MKS system and have replaced this system.

metre slug *See* slug.

metric waves Radio waves having wavelengths between 1 m and 10 m.

metrology The branch of science concerned with the accurate measurements of the three fundamental quantities: mass, length, and time. It is often extended to mean the systematic study of weights and measurements.

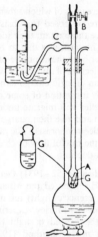

Victor Meyer's apparatus

Meyer, Victor (1848–1897) German chemist. He devised a method for measuring the vapour density of an unsaturated vapour, in which a known mass of the liquid was vaporized by dropping it into a bulb maintained at a constant temperature above the boiling point of the liquid. Air, which is less dense than the vapour, is displaced from the bulb and is collected over water or mercury. The liquid of mass m contained in a small stoppered phial, G (*see* diagram), is dropped into the bulb A, surrounded by boiling water, and the clip at B is then closed. The vapour pressure of the liquid forces out the stopper and all the liquid evaporates forcing air out through C into the jar D (volume V). The density of the unsaturated vapour at 100 °C and pressure B is then given by

$$d_{100} = \frac{m}{V} \frac{B}{B-h} \text{ g cm}^{-3}$$

where h is the saturated vapour pressure of water vapour at the temperature of the collecting jar.

MF Abbreviation for medium frequency. *See* frequency band.

MHD Abbreviation for *magnetohydrodynamics.

mho The reciprocal ohm, formerly used as a unit of conductance. This unit is now replaced by the *siemens.

mica A mineral consisting of complex silicates, characterized by a perfect basal cleavage enabling the crystals to be split into very thin plates. It has a low thermal conductivity and high dielectric strength, being widely used for electrical insulation.

mica capacitor A *capacitor in which *mica is used as the dielectric. Mica capacitors are characterized by low loss and a near-constant capacitance over a wide frequency and temperature range.

mica V-ring *See* V-ring.

micelle 1. Molecular aggregates of which substances like gelatin and agar are supposed to consist. On solution of the substance these retain their identity but become solvated. The term was due to Nageli (1858) and arose in describing the structure of gels.
  2. An assembly of molecules of up to microscopic size that has acquired an electric charge either by adsorption of ions or by ionization of some of the surface molecules. Micelles will migrate under the action of an electric field and give their charge to an anode or cathode. (*See* electrophoresis.) The ratio of rate of flow of mass to the rate of flow of charge is high in this type of conduction.

Michelson, Albert Abraham (1852–1931) German-born Amer. physicist. He invented the Michelson interferometer (*see* interference of light) and used it in the famous *Michelson–Morley experiment. Michelson's work on the *speed of light (1883) employed the rotating mirror method (1926 at Mount Wilson, 1929 with evacuated pipe line). The Michelson *stellar interferometer* (1920) is an attachment to a telescope to increase the effective resolving power so that stellar diameters may be measured.

Michelson interferometer *See* interference of light; Michelson–Morley experiment.

Michelson–Morley experiment An experiment (1887) that attempted to measure the velocity of the earth through the *ether. Using a Michelson interferometer (*see* diagram) Michelson and Morley attempted to show that there is a difference in the speed of light as measured in the direction of the earth's rotation compared to the speed at right angles to this direction. If there was such a difference the interference fringes observed in the interferometer would be shifted when the instrument was turned through 90°. This shift would correspond to a change of optical pathlength of approximately $2dv^2/c^2$, where v is the velocity of the earth with

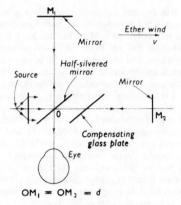

Michelson's interferometer

respect to the ether in the direction OM_2. No shift was observed, indicating the absence of an ether wind.

  This fact was of considerable importance and was responsible for the downfall of the ether concept. Attempts to reconcile this concept with the null results of the experiment led to the postulate of the *Lorentz–FitzGerald contraction. The special theory of *relativity rejects the concept of an ether with respect to which there can be determinate motion.

micro- 1. Symbol: μ. A prefix meaning 10^{-6} (i.e. one-millionth); for example, one microsecond (1 μs) is 10^{-6} second.
  2. A prefix meaning very small or concerned with very small quantities or objects.

microbalance A *balance capable of weighing very small masses (e.g. down to 10^{-5} mg). It is not practical to use such balances with standard weights (except perhaps for calibrating). Instead, the beam is made to balance by varying the air pressure in the balance case so altering the upward buoyant force on a bulb fixed on one end of the beam. A manometer is provided for measuring the pressure when the beam is balanced; if the temperature is constant, the density of the gas (and thus the upward buoyant force) is proportional to the pressure.

  Microbalances have been used for finding the relative density of a gas by measuring the pressure needed in the case to balance the beam first with the gas and then with oxygen. The density ratio is the inverse ratio of these pressures if the temperature is the same during both measurements. Aston (1913) found the density of neon to within 1 part in 1000 using less than 1 cm³ of gas.

microcalorimeter A differential calorimeter used for the measurement of very small quantities of heat. Wertenstein used such a calorimeter to measure the heat evolved by a small quantity of radioactive substance. Two identical Dewar vessels D, are embedded in a thermostatically controlled iron block A, mercury (shaded black) being used to obtain good thermal contact. Two identical vessels a, b, are

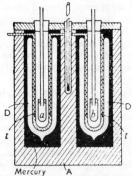

Mercury   A   Microcalorimeter

separated from the Dewar vessels by paraffin; the source of heat is contained in *a* and a small heating coil in *b*, the current through the coil being varied until there is no difference in temperature between *a* and *b* as indicated by no deflection of a galvanometer in the circuit of a large number of copper-constantan thermocouples *t, t*, arranged differentially.

microchip A *chip of semiconductor containing complex microcircuits, usually integrated circuits.

microcomputer A compact *computer system in which the central processor is fabricated on a single *chip (or a small number of chips) of semiconductor; this processing unit is called a *microprocessor*. In addition, a microcomputer contains storage and input/output facilities for data and programs on different chips, or possibly on the same chip as the microprocessor. The capability of the system depends not only on the characteristics of the microprocessor but also on the amount of storage provided, the types of *peripheral devices that can be used, the possibility of expanding the system, etc.

microdensitometer A device for automatically measuring and recording small changes in *transmission density across a sample, such as a photographic plate. It can be used to determine the change in absorption with wavelength (in the visible and near ultraviolet) of a substance whose *absorption spectrum is recorded on a photographic plate in a *spectrometer. As the absorption of the substance increases, the optical density on the plate increases.

microdiffusiometer *See* diffusiometer.

microelectronics The branch of *electronics concerned with the design, production, and application of electronic components, circuits, and devices of extremely small dimensions. Increased miniaturization not only reduces size and weight but is also cost-effective, and is extremely desirable particularly in the field of *computers. *Integrated circuits are widely used in microelectronics.

microfilm plotter *See* incremental plotter.

microgravity The condition of near *weightlessness induced by free fall or unpowered space flight.

micromanometer A device for the measurement of very small pressure differences. They include:

(1) The U-tube manometer with one arm of the U nearly horizontal so that pressure changes causing only a small difference in vertical height produce easily visible movements of liquid in this arm.

(2) Diaphragm gauges. The two pressures are applied on either side of the diaphragm and optical methods are used to measure the tiny displacement.

(3) The *Chattock gauge.

micrometer eyepiece An eyepiece, generally a *Ramsden eyepiece, provided with cross-wires that can be displaced by means of a *micrometer screw. It is used for the measurement of small objects or small separations of objects, lines, etc.

micrometer screw A device for use when measuring small and/or accurate distances, e.g. with micrometer calipers or a depth micrometer. Such instruments are fitted with a drum that when rotated advances a screw of known pitch; the drum is calibrated in fractions of a revolution, which can be interpreted in terms of the distance advanced by the screw.

micron Symbol: μ. A former name for the micrometre; 10^{-6} m.

microphone A device for converting an acoustic signal into an electric signal. It forms the first element of the telephone, the broadcast transmitter, and all forms of electrical sound recorders. The types of microphone most generally used are the *carbon, *crystal, *moving-coil, *capacitor, and *ribbon microphones. Many other types exist for specialized purposes however, such as the magnetostriction, induction-coil, and hot-wire microphones. Most of these use a thin diaphragm that vibrates under the influence of the sound waves. The diaphragm is mechanically coupled to some device, the motion of which changes the properties of a component of an electric circuit or induces an e.m.f. in it. The force exerted by the sound against the diaphragm is usually proportional to the sound pressure, but in the case of the ribbon microphone it is proportional to the particle velocity. For good quality reproduction, resonance in the mechanical system of the microphone should be avoided. This is done by making the resonant frequency of the moving parts either much higher or much lower than the frequency of the sound to be reproduced. Lack of sensitivity is not a great disadvantage since it is usual to amplify the output from the microphone. In all cases a battery or other power supply is needed. Nearly all microphones have directional properties and in many these vary with frequency as a result of diffraction.

microprocessor *See* microcomputer.

microradiography *See* microscope.

microscope 1. An optical instrument for producing an enlarged image of small objects. The *simple microscope* (or *magnifying glass*) consists of a strong converging lens system, corrected for chromatic and

microscope

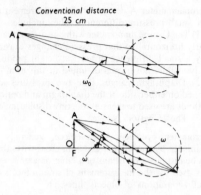

Conventional distance 25 cm

a Simple microscope

spherical *aberrations, and used for low-power work. The object is usually placed at the focus of the lens system, producing an image at infinity. The magnification obtained is the ratio of the angle ω, subtended at the eye by the image produced by the microscope to the angle ω_0, subtended by the object at the unaided eye (Fig. *a*). The normal unaided eye sees an object most clearly at the near point (25 cm). The magnification is therefore $\omega/\omega_0 = 25/f$ (where f is in cm).

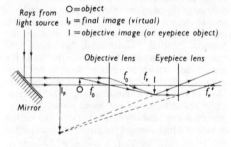

Rays from light source
O = object
I_F = final image (virtual)
I = objective image (or eyepiece object)

Objective lens   Eyepiece lens

Mirror

b Compound microscope

The *compound microscope* (Fig. *b*) consists essentially of two lens systems and gives a much greater magnification, up to about 1500 diameters. A very short focal length *objective forms a magnified real image of the object, which is further magnified by the *eyepiece acting as a simple microscope. The total magnification is the product of the objective and eyepiece magnification. There is usually a choice of objectives on a compound microscope giving low, medium, and high magnification. Greater magnification can be obtained with oil immersion (*see* resolving power). Several types of eyepiece are in common use, including *Huygens's eyepiece, the *Ramsden eyepiece (used with measuring microscopes), and the *compensating eyepiece (used with objectives having *apochromatic lenses). The maximum *resolving power of the optical microscope is between 200 and 300 nm (i.e. half the wavelength of blue or orange light). The *binocular microscope* has two eyepieces. Light from the objective is split into two beams by using prisms. It

has the advantage of depth perception and greater eye comfort. There can also be a third eyepiece to which a camera can be attached. The *stereoscopic microscope* has two eyepieces and two objectives, the object being viewed by reflected rather than transmitted light. The magnification is usually of the order of 100 diameters.

Illumination for low-power work with transparent objects is achieved by using a mirror mounted below the object to reflect light from the light source onto the object. For higher magnifications, a substage *condenser, such as the *Abbe condenser or a modified Abbe condenser (variable focus) is necessary. This concentrates the light within a cone of larger angle than that achieved by the mirror, being positioned between the reflecting mirror and the object. Examination of opaque objects requires illumination from above and the objective becomes, in effect, the condenser. The optical microscope can be focused sharply only in one plane so that a two-dimensional image is obtained. If the object is fairly transparent, different depths can be brought into focus, but since material above and below the plane of focus can interact with the light, the image may be blurred. The microscope therefore works best with thin samples viewed by transmitted light, or with flat samples viewed by reflected light. The shape of the object can only be obtained at low magnifications of about 200 diameters.

Details in objects are seen because of varying density regions. With a strongly lit background, a small transparent object is very difficult to observe. *Dark-field illumination* increases the visibility of small objects. An opaque disc is placed over the centre of the condenser so that the borders of the object are illuminated by marginal rays, which do not enter the objective. Some light is refracted by the specimen, some of which enters the objective. A bright image is thus obtained against a dark background; however, little detail can be seen. (The same effect can be obtained by illuminating the object obliquely so that no direct light enters the objective.)

The refractive index of a transparent specimen varies slightly from point to point. These variations give rise to *diffraction patterns in the focal plane of the objective. The diffracted light passing through the object is one quarter of a wavelength out of *phase with undiffracted light, which has not been transmitted through the object. In *phase-contrast microscopy* a transparent *phase plate* is used to produce a quarter-wavelength shift in the undiffracted light. The phase shift is produced by shaping the surface of the phase plate, which may either have a shallow circular indentation around the centre or a slightly raised disc at the centre. The plate is placed at the focal plane of the objective (Fig. *c*) onto which the image of a substage annular diaphragm falls. The final image has higher contrast due to interference between the diffracted and undiffracted beams.

In *interference microscopy* a transparent object is placed between two semi-silvered surfaces. Light passing through the object interferes with light that

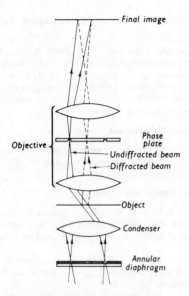

c Phase contrast microscope

has not passed through, and interference patterns can be observed. It is possible to view opaque objects in a similar way.

Reflecting microscopes use a reflecting objective rather than the conventional lens system and can focus wavelengths ranging from infrared to ultraviolet at the same point. In *ultraviolet microscopy* (*see* resolving power) the resolving power of the microscope is increased to 100 nm by using shorter-wavelength ultraviolet radiation. The image is made visible by using a photographic plate to record it. *X-ray microscopy* (or *microradiography*) further increases the resolution, the image being recorded on film or on a fluorescent screen.

2. *See* electron microscope.

microscopic state The state of matter characterized by the actual properties of each individual elemental component. *Quantum theory is typically an analysis of the microscopic state. *Compare* macroscopic state.

microstrip A transmission line used in ultrahigh frequency applications consisting of a strip of conducting material in close proximity to an *earth plane.

microtome An apparatus for cutting thin sections of material – either frozen or embedded in a supporting medium – for microscopic examination.

microwave An electromagnetic wave with a wavelength in the range 1 mm to 0.1 m (or sometimes 0.3 m), i.e. with a frequency in the range 300 to 3 GHz (or sometimes 1 GHz). The microwave region of the electromagnetic spectrum thus lies between the infrared and radio regions, overlapping the radio region (*see* frequency band). Microwaves are generated by devices such as the *maser, *klystron, and

*magnetron. They are used in *radar, high-frequency communications, and also for cooking (in microwave ovens).

microwave background radiation *See* cosmic background radiation.

microwave spectroscopy *See* spectroscopy.

microwave tube An *electron tube that is suitable for use as an amplifier or oscillator at *microwave frequencies. They usually employ *velocity modulation of the electron beam. *Klystrons, *magnetrons, and *travelling-wave tubes are examples.

middle wire *See* three-wire system.

Mie scattering Scattering of light by spherical particles of diameters comparable with the wavelength; an extension of *Rayleigh scattering, applicable to particles small compared with the wavelength.

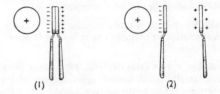

(1)   (2)

Mie's double plate

Mie's double plate Two small discs in contact, supported by insulating handles, are put into an electrostatic field and then separated. The charge on one disc is then measured by a cup electrometer, to determine the electric displacement at that point in the field.

migration area The area required for a neutron to slow down from fission energy to thermal energy plus that required to diffuse the energy. The former area is defined formally as one-sixth of the mean square distance between the source and the point where the neutrons reach thermal energy. The diffusion area is formally one-sixth of the mean square distance between the point where the neutron is in thermal equilibrium with the surroundings and the point where it is captured. The *migration length* is the square root of the migration area. *See also* neutron age.

migration of ions The ions of an electrolyte migrate when a current is passed through it, and play a part in the transport of electricity. (*See* electrolytic dissociation.) The cations and anions do not always move at the same velocity, and thus transport different fractions of the current. Progressive changes take place in the concentration of electrolyte around the electrodes. The fraction of the current carried by either ion is known as its *transport number* (or *transference number*), symbol: *t*.

migration potential *Syn.* sedimentation potential; Dorn effect. The potential difference due to the settling or centrifuging of charged colloidal particles. It may be regarded as the reverse of *cataphoresis.

mil One thousandth of an inch.

mild steel Steel with not more than about 0.25% of carbon and without special alloying constituents. It is comparatively soft.

milking generator A low-voltage d.c. generator used for charging one or more cells in an accumulator battery independently of the others.

Miller effect In an electronic device, the phenomenon whereby the *interelectrode capacitance provides a feedback path between the input and output circuits, which can affect the total input admittance of the device. The total dynamic input capacitance of the device will always be equal to or greater than the sum of the static electrode capacitances because of this effect.

Miller indices *See* rational intercepts, law of.

milli- Symbol: m. A prefix meaning 10^{-3} (i.e. one-thousandth); for example, one millimetre (1 mm) is 10^{-3} metre.

Millikan, Robert Andrews (1868–1953) Amer. physicist, chiefly noted for his measurements of the electronic charge by the oil-drop method (1910), for his determination of the *Planck constant from photoelectric measurements (1916), and for his work on *cosmic rays. In measuring the electronic charge, he observed the motion of small charged droplets subjected to an adjustable electric field and showed that all observed charges were multiples of one elementary charge. The very large number of measurements and the great care with which they were made and presented caused his value for the electronic charge to be credited with a much higher precision than was justified, and it was assumed in calculations until about 1949, despite considerable evidence against it. In particular, the value assumed for the viscosity of air was inaccurate. He explained the great energy of the primary cosmic rays, believed for a time to be photons, in terms of the annihilation of particles somewhere in the universe but it is now known that the main incoming radiation is corpuscular.

mimetic twinning The symmetrical growing together of two or more individual crystals in such a way as to simulate higher symmetry (pseudosymmetry).

minicomputer Loosely, a medium-sized *computer, usually less capable in terms of performance (and hence cheaper) than a *mainframe computer. There is no clear boundary now between minicomputers and the more sophisticated types of *microcomputers.

minimal surfaces *See* principal radii.

minimum deviation *See* deviation.

minimum discernible signal (m.d.s.) The smallest value of input signal to an electronic circuit or device that just produces a discernible change in the output.

Minkowski, Hermann (1864–1909) Russian-born German mathematician, who became professor of mathematics at Göttingen. He is chiefly remembered in physics for the four-dimensional geometry that he used (1907) to give an elegant formulation of Einstein's general *relativity theory.

Minkowski space-time *See* relativity.

minority carrier In a *semiconductor, the type of *carrier constituting less than half of the total charge-carrier concentration.

minus tapping *See* tapping.

minute 1. Symbol: min. A unit of time equal to 60 seconds. Although not an *SI unit, the minute is of practical importance and may be used with the SI units.
2. *Syn.* minute of arc; arc minute. Symbol: ′. A unit of angle equal to $1/60$ of a degree, i.e. 0.291 milliradian.
3. *Syn.* centesimal minute. A unit of angle = 0.01 grade.

mirror An optical device for producing reflection, generally having a surface that is plane or a portion of a sphere, paraboloid, or ellipsoid. *Concave mirrors* are hollowed out, *convex mirrors* are dome-shaped. The *mirror formulas* generally describe the conjugate focus relations for spherical mirrors. The commonest form is:
$$1/v + 1/u = 2/r = 1/f,$$
in which u is the object distance, v the image distance, r the radius of curvature, f the focal length. Objects and images lying in front of the mirror are real and the distances are taken as positive. For a virtual *image, v is negative. For a concave mirror, f is positive; for a convex mirror, it is negative. The magnification, M, is equal to v/u and is positive for inverted images.

A *thick mirror* is a lens with the back surface silvered or possibly a lens in combination with a curved mirror with or without separation.

Although mirrors are free from *chromatic aberration, in general they suffer from *spherical aberration. The paraboloid form (*see* parabolic reflector) focuses parallel rays accurately and is used with reflecting telescopes, searchlight mirrors, etc.; the *ellipsoid mirror* focuses light from one focus to the other focus (both foci are real); the *hyperboloid mirror* reflects light directed to one focus (virtual) to the opposite focus (real). The concave meniscus *Mangin mirror corrects spherical aberration reasonably. The *Schmidt telescope uses a Schmidt corrector to correct spherical aberration of spherical mirrors.

Mirrors are used not only as optical devices but also in the infrared, ultraviolet, and X-ray regions of the spectrum. A mirror may be a finely polished metal surface or a piece of glass with a reflective coating on the front or back surface, often of silver.

Silver is an efficient reflector of infrared and ultraviolet. Vacuum-evaporated coatings of aluminium on highly polished substrates are now used in quality mirrors, often with protective coatings of silicon monoxide or magnesium fluoride. The aluminium forms a harder more stable surface than silver and can reflect shorter wavelengths. Mirrors can also be formed of multilayered dielectric films. *See also* magnetic mirror.

mirror nuclides Two nuclides having the same number of nucleons each, but where the number of protons (or neutrons) in one is equal to the number of neutrons (or protons) in the other. The nuclides will have the general form, $_n^m X$ and $_{m-n}^m Y$, and will be the source and product nuclides in a beta-capture or *beta decay. In the special case of the pair $_n^{2n+1} X$ and $_{n+1}^{2n+1} Y$, they are known as *Wigner nuclides*.

MISFET, MIST *See* field-effect transistor.

mismatch The condition arising when the impedance of a *load is not equal to the output impedance of the source to which it is connected.

missing mass The mass of the universe's hypothetical invisible matter. The existence of this invisible matter (also known as *dark matter* and *hidden matter*) is postulated on a number of grounds, including: the unexplained velocity distribution of stars perpendicular to the plane of the Galaxy; the view that the outer regions of galaxies contain matter that contributes mass but not radiation; studies of galactic dynamics and some cosmological theories suggest that the mean density of the universe exceeds that due to visible matter.

A number of solutions to the missing mass problem have been suggested; these include the existence of many planet-sized bodies and rocks as well as various exotic massive particles, such as massive *neutrinos, *axions, and so-called *weakly interacting massive particles* (*WIMPs*).

mistuning *See* forced vibrations.

mixed crystal A solid solution of two chemical substances with isomorphous or closely related structures.

mixer *Syn.* frequency changer. A device used in conjunction with a beat-frequency oscillator to produce an output having a different frequency from the input. The amplitude of the output bears a fixed relationship to the amplitude of the input (usually approximately linear) and the device is used in a *superheterodyne receiver for changing the frequency of an amplitude-modulated *carrier wave while retaining the modulation characteristics. If the frequency of the input is f_1 and the desired output frequency f_2, the *conversion conductance* is obtained by dividing the current output at frequency f_2 by the voltage input at frequency f_1 when the output load impedance at frequency f_2 is zero, i.e. the conversion conductance is the short-circuit output current at

frequency f_2 per volt input at frequency f_1; it is usually expressed in milliamperes per volt.

mixing length For turbulent motion in which the mean flow is the x-direction (Cartesian coordinates), the temporal mean value of velocity being $\bar{u}$ (a function of y alone), and the components of fluctuation of velocity being u', v', w' (in the respective x, y, and z directions); then from momentum considerations it is known that the stress or apparent friction per unit area, due to turbulence, is in the y-direction and given by:

$$\tau = \rho\varepsilon \, d\bar{u}/dy = -\overline{\rho u'v'}$$

where ρ is the density of fluid, ε is the coefficient of *eddy viscosity, and where $<u'v'>$ indicates the temporal mean value of the product $u'v'$. The dimensions of ε are those of [length $\times$ velocity]. If the velocity
$$V = \sqrt{<u'v'>}$$
is kept constant then the variation of ε, indicating various degrees of turbulence of the fluid, is achieved by varying some length l, such that $\varepsilon = Vl$ and

$$\tau = \rho l^2 \left|\frac{d\bar{u}}{dy}\right|\frac{d\bar{u}}{dy}.$$

This quantity of length l is called (after Prandtl) the mixing length.

mixing ratio *See* humidity.

MKS system *See* metre-kilogram-second electromagnetic system of units.

MLD Abbreviation for *median lethal dose.

mmf Abbreviation for *magnetomotive force.

mmHg Abbreviation of millimetres of mercury, a former unit of pressure equal to 133.32 pascals. The standard atmosphere was 760 mmHg.

mobility *See* plasticity; mobility coefficient; Hall mobility; drift mobility; intrinsic mobility.

mobility coefficient *Syn.* mobility. The average speed of diffusion, in the direction of the concentration gradient, of the molecules in a solution, at unit concentration and unit osmotic pressure gradient.

mode 1. The value of the abscissa corresponding to the maximum ordinate of a *frequency distribution curve.

2. *Syn.* transmission mode. Any of the several different states of oscillation of an electromagnetic wave, of given frequency, in a waveguide, cavity resonator, etc. *See* waveguide.

model scale *See* dynamic similarity.

modem Abbreviation of modulator-demodulator. A device that converts the signals from one particular type of equipment into a form suitable for use in another. For example a modem can convert the digital signals from a computer into an analogue form for use over an (analogue) telephone system.

moderator In *nuclear reactors. Material used to slow down the fast neutrons created in a *fission process to the lower velocities appropriate to the type of reactor in use by scattering without appreciable capture.

modulated amplifier The *amplifier stage in a transmitter in which the modulating signal is introduced for the purpose of *modulation of the *carrier wave.

modulation In general, the alteration or modification of an electronic or acoustic parameter by another. In particular, the process of varying one electronic signal, called the carrier, according to the pattern provided by another signal, as when an audiofrequency signal is impressed onto a higher-frequency *carrier wave for radio transmission. See amplitude, frequency, phase, and pulse modulation. See also velocity modulation.

modulation factor See amplitude modulation; frequency modulation.

modulation index See frequency modulation.

modulator 1. Any device that effects the process of *modulation.
2. A device used in radar for generating a succession of short pulses to act as a trigger for the *oscillator.

modulator electrode An electrode used for modulating the flow of current in a device. In a *cathode-ray tube it is the electrode controlling the intensity of the electron beam. In a *field-effect transistor it is the gate electrode(s), which control(s) the conductivity of the channel.

modulus 1. Of a real quantity. Syn. absolute value. The magnitude of a quantity without regard to its sign.
2. Of a vector. The magnitude of the vector.
3. Of a complex quantity expressed as $re^{i\theta}$ or $(x + iy)$: the modulus is r or $\sqrt{(x^2 + y^2)}$.
4. See elliptic integrals.

modulus of compression Syn. bulk modulus. See modulus of elasticity.

modulus of decay In a system exhibiting damped oscillations of the form $a = a_0 e^{-\alpha t}$, where a_0 is the initial amplitude and a its value after time t, then the modulus is equal to the time t_1 at which the amplitude has fallen to $1/e$ of its initial value. Taking natural logarithms:
$$\log (a_0/a) = \alpha t_1 = 1.$$
Thus the modulus of decay is the reciprocal of the damping factor α. See damped vibrations.

modulus of elasticity The ratio of *stress to *strain for a body obeying *Hooke's law. There are several moduli corresponding to various types of strain:
1. The Young modulus $(E) =$

$$\frac{\text{applied load per unit area of cross section}}{\text{increase in length per unit length}},$$

It applies to tensional stress when the sides of the rod or bar concerned are not constrained. (See Poisson ratio.)
2. Bulk modulus (or volume elasticity) $(K) =$

$$\frac{\text{force per unit area}}{\text{change in volume per unit volume}};$$

It applies to compression or dilation, e.g. when a body is subject to changes in hydrostatic pressure. Fluids, as well as solids, have bulk moduli.
3. Shear (or rigidity) modulus $(G) =$

$$\frac{\text{tangential force per unit area}}{\text{angular deformation}}.$$

4. Axial modulus, or modulus of simple longitudinal extension is defined in the same way as the Young modulus with the proviso that the sides of the specimen are restricted so that there is no lateral change.

Since strain is a ratio and so dimensionless, the moduli have the dimensions of stress, i.e. force/area. The various moduli and the Poisson ratio for an isotropic solid are interrelated. The moduli given in Physical Tables, and most often used, are measured under isothermal conditions; the adiabatic values are always greater.

If stress is not proportional to strain (as in cast metals, marble, concrete, wood), the moduli have to be defined as the ratio of a small change in stress to a small change in strain at a particular value of the stress.

modulus of precision See frequency distribution.

modulus of torsion The couple required to give a wire a twist of one radian per metre of length.

moho See Mohorovicic discontinuity.

Mohorovicic, Andrija (1857–1936) Serbian geologist who, while studying the seismic waves produced by a Balkan earthquake in 1909, discovered the boundary of the earth's crust now called the *Mohorovicic discontinuity.

Mohorovicic discontinuity Syn. moho. A discontinuity lying between the earth's crust and mantle at a depth of about 35 km under continents but only 6–10 km under the oceans. It affects the velocity at which seismic waves travel. See also seismology.

Mohs, Friedrich (1773–1839) German mineralogist. He worked on crystal structure and introduced, in 1820, the scale of hardness known by his name. See hardness.

moiré pattern The pattern produced by overlying sets of parallel threads or lines, the sets being slightly inclined to one another. The overlaps produce the appearance of dark bands running athwart the individual lines. If the two sets are perfectly regular, these bands are straight, but deviations in either or both give wavy lines as in the characteristic appearance of moiré silk. Transparent diffraction gratings and replicas from them may be compared by superimposing them and examining the resulting moiré pattern.

Moissan, Henri (1852–1907) French chemist. He developed the electric furnace as a laboratory tool and claimed to have made diamonds artificially.

molality *See* concentration.

molar A term now restricted in its meaning to divided by *amount of substance. In practice this means "per mole"; e.g. the molar heat capacity is the heat capacity per mole, written C_m.

molar conductivity Symbol: j_m. The *conductivity of an *electrolyte solution divided by the concentration of electrolyte present. If the conductivity is measured in siemens per metre and the concentration is moles per cubic metre, the molar conductivity has units of $S\ m^2\ mol^{-1}$. Note that the word molar in this case means "divided by concentration" and not "divided by amount of substance".

molar gas constant Symbol: R. The constant occurring in the *equation of state for 1 mole of an *ideal gas, namely:
$$pV = RT.$$
It may be shown to be a universal constant for all gases. Actual gases obey this equation of state only in the limit as their pressure tends to zero.

The pressure exerted by a gas according to *kinetic theory is shown to be:
$$p = \tfrac{1}{3}mvC^2,$$
where m is the mass of the molecules, v is the number of molecules per cubic metre, and C is the root-mean-square velocity of the molecules. Considering one mole of the gas, L being the number of molecules present (i.e. the Avogadro constant), and V the *molar volume then:
$$pV = \tfrac{1}{3}mLC^2,$$
and V is independent of the nature of the gas at given values of p and T. Hence,
$$RT = \tfrac{2}{3}(\tfrac{1}{2}mLC^2).$$
This expression shows that R is equal to two-thirds of the total translational energy of the molecules in 1 mole of a gas at a temperature of 1 kelvin.

The Boltzmann constant k is given by the ratio R/L and R has the value
$$8.314\ 510\ J\ K^{-1}\ mol^{-1}.$$

molar heat capacity Symbol: C_m. The *heat capacity of unit *amount of substance of an element, compound, or material. Molar heat capacities are measured in joules per kelvin per mole.

molarity *See* concentration.

molar latent heat *See* latent heat.

molar volume The volume occupied by 1 mole of a substance. According to *Avogadro's hypothesis all ideal gases have the same molar volume at the same pressure and temperature. The value at STP is
$$2.241\ 3837 \times 10^{-2}\ m^3\ mol^{-1}.$$

mole Symbol: mol. The *SI unit of *amount of substance, defined as the amount of substance of a system that contains as many elementary entities as there are atoms in 0.012 kilograms of carbon-12. The elementary entities must be specified, and may be atoms, molecules, ions, electrons, other particles, or specified groups of particles.

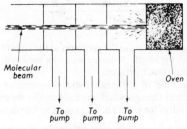

Molecular beam

molecular beam A collimated beam of atoms or molecules at low pressure, in which all the particles are travelling in the same direction and few collisions occur between them.

Such beams are produced in an apparatus connected to several fast vacuum *pumps. Beams of metal atoms are formed by heating the metal in an oven and allowing the vapour to escape through a small hole. This enclosure is called the "oven", even when the molecular beam is formed from a permanent gas escaping from an unheated enclosure. The vacuum system is usually made of several sections, each connected to a pump and separated by partitions with collimated apertures. Each section is at a lower pressure than the preceeding section. Molecules that do not pass through the apertures are pumped away. (*See* diagram.)

Molecular beams of this type are used in the study of the adsorption of gases on surfaces. In studies of chemical reactions, two molecular beams can be crossed at an angle and the angular distribution and concentration of products determined with a mass spectrometer (*see* mass spectrum). They are also used in experiments similar to the *Stern–Gerlach experiment and *Eldridge's experiment.

molecular cloud *See* interstellar matter.

molecular crystal A crystalline arrangement in which the atoms are first formed into molecules, which are then held together by weaker residual forces.

molecular effusion *See* effusion.

molecular flow *Syn*. Knudsen flow. A type of gas flow occurring at low pressures, when the *mean free path of the gas molecules is large compared with the dimensions of the pipe through which the gas is flowing. The rate of gas flow is determined by collisions between the molecules and the wall of the tube, rather than by collisions between molecules. Thus the flow does not depend on the viscosity of the gas.

molecular gauges

Knudsen assumed that the direction of a molecule rebounding from a wall is independent of the incident direction (*see* cosine law). In this way he derived the equation (known as *Knudsen's equation*) for the volume of gas measured at unit pressure flowing through a tube:

$$Q = \frac{4\sqrt{2\pi}}{3} \frac{P_1 - P_2}{\sqrt{\rho}} \frac{R^3}{L},$$

where P_1 and P_2 are the pressures at the ends of the tube, L the length, R the radius, and ρ the density of the gas at unit pressure.

The ratio of a characteristic dimension of the apparatus through which the gas flows to the mean free path of the gas is known as the *Knudsen number*.

molecular gauges *Syn.* viscosity gauges or manometers. Devices used for measuring low gas pressures and whose action is based on the dependence of gas viscosity on pressure at low pressures. In one gauge, a disc is turned rapidly at a uniform speed and a second disc parallel to the first tends to follow its rotation due to the viscous drag of the air. The couple acting on the second disc is a measure of the pressure. The instrument works from about 10^{-1} to 10^{-5} Pa, and is usually calibrated with reference to a *McLeod gauge.

In another form, a flat quartz fibre, fixed at one end, vibrates in the gas and the damping, which is observed, depends on the viscosity. This instrument, called a *decrement gauge* or *quartz-fibre manometer*, is most useful from $1-0.01$ Pa.

molecular orbital In an atom the electrons moving around the nucleus have *atomic orbitals that are often represented as a region around the nucleus in which there is a high probability of finding the electron. When molecules are formed, the valence electrons move under the influence of two or more nuclei and their *wave functions are known as molecular orbitals. These can also be represented by regions in space. It is usual to think of molecular orbitals as formed by a combination of atomic orbitals. Two atomic orbitals combine to give two molecular orbitals of different energies and forms. In the lower-energy orbital there is a concentration of charge between the nuclei, which serves to hold them together and thus form the chemical bond; this is called a *bonding orbital*. The orbital of higher energy does not have an internuclear charge concentration and the nuclei tend to repel one another; this is called an *antibonding orbital*. Each molecular orbital can be occupied by two electrons with opposite spins in accordance with the *Pauli exclusion principle.

In hydrogen for example, each hydrogen atom has one electron. When a molecule is formed the pair of electrons occupy the lower-energy bonding orbital forming a stable molecule. The higher-energy antibonding orbital is unoccupied.

370

molecular polarization When a molecule is subjected to an electric field, there is a small displacement of electrical centres that induces a dipole in the molecule. If $m = \alpha E$, where m is the electric *dipole moment induced by a field strength E, then the constant α is called the *polarizability* of the molecule.

molecular pump *See* pumps, vacuum.

molecular sieves *See* zeolites.

molecular spectra *See* band spectrum.

molecular weight Former name for *relative molecular mass.

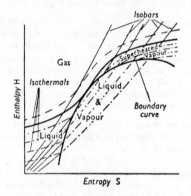

Mollier diagram

Mollier diagram An *enthalpy–*entropy diagram on which isothermals, isobars, and lines of equal dryness are plotted from steam table data. It is used for the graphical computation of the efficiency of various steam-engine cycles.

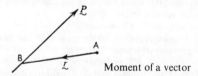

Moment of a vector

moment 1. *Syn.* torque. The moment of a force about an axis is the product of the perpendicular distance of the axis from the line of action of the force, and the component of the force in the plane perpendicular to the axis. The moment of a system of coplanar forces about an axis perpendicular to the plane containing them is the algebraic sum of the moments of the separate forces about that axis (anticlockwise moments are taken conventionally to be positive and clockwise ones negative).

2. Moment of momentum about an axis. *Syn.* angular momentum. (i) Of a particle. The product of the component of momentum of the particle in a plane perpendicular to the axis and the perpendicular distance of the axis from a line through the position of the particle in the direction of the velocity. (*See* conservation of momentum.) (ii) Of a rigid body or system of particles. The algebraic sum of the moments of momentum of the individual particles of the body about the same axis.

The concept is particularly useful (*a*) when a particle is moving in a plane under the influence of a central force, in which case the axis taken passes through the centre of force and is perpendicular to the plane; the angular momentum is then constant throughout the motion; (*b*) when dealing with the motion of a rigid body about a fixed axis, in which case the laws of motion have the same form as those of rectilinear motion, e.g. in the equations of the latter we replace mass by moment of inertia, linear velocity by angular velocity, force by torque, and linear momentum by angular momentum.

3. Similar definitions of moment about an axis (as in defs. **1** and **2**) apply to any vector quantity; the moment so defined is a scalar and is given a positive or negative sign as in def. **1**. When dealing with systems in which forces and motions do not all lie in one plane, the concept of moment about a point is needed.

4. The moment of a vector *P* (e.g. force or momentum) about a point A (*see* diagram) is a *pseudovector *M* equal to the *vector product of *r* and *P*, where *r* is any line joining A to any point B on the line of action of *P*. (The vector product $M = r \times P$ is independent of the position of B.)

The relation between the scalar moment about an axis (**3**) and the vector moment about a point on that axis (**4**) is that the scalar is the component of the vector in the direction of the axis.

5. *Varignon's theorem*: the sum of the moments of a number of vectors with a common origin, about a line, is equal to the moment of their resultant, with the same origin, about that line. Also, the vector sum of the vector moments about point A, of a number of vectors with a common origin B, is equal to the vector moment about A of their vector sum with origin B.

momental ellipsoid *See* product of inertia.

moment of inertia 1. Of a body about an axis. Symbol: *I*. The sum of the products of the mass of each particle of the body and the square of its perpendicular distance from the axis. (This addition is replaced by an integration in the case of a continuous body.) The *kinetic energy of the body (of moment of inertia *I*) rotating about that axis with angular velocity ω is $\frac{1}{2}I\omega^2$, which corresponds to $\frac{1}{2}mv^2$ for the kinetic energy of a body of mass *m* translated with velocity *v*. (*See* moment for further details of the analogy between rectilinear motion and rotation about a fixed axis. *See* product of inertia, for further definitions. *See also* Routh's rule; theorem of parallel axes.)

2. Of a surface about an axis. The moment of inertia of an imaginary sheet of matter whose mass/unit area is unity and which coincides with and has the same boundaries as the surface considered. This is known as the *geometric moment of inertia* of the surface.

moment of momentum *See* moment.

momentum 1. *Syn.* linear momentum. Of a particle. Symbol: *p*. The product of the mass and the velocity of the particle. It is a *vector quantity directed through the particle in the direction of motion. The linear momentum of a body or of a system of particles is the vector sum of the linear momenta of the individual particles. If a body of mass *M* is translated (*see* translation) with a velocity *V*, its momentum is *MV*, which is the momentum of a particle of mass *M* at the centre of gravity of the body. (*See* Newton's laws of motion, II; conservation of momentum.)

2. Angular momentum. *Syn.* moment of momentum. Symbol: *L*. *See* moment.

monel A corrosion-resisting alloy containing about 60–70% nickel; 25–35% copper; 1–4% iron; 0–2% manganese; and a little silicon and carbon. (Relative density 8.80; M.P. 1330 °C.)

monochord *Syn.* sonometer. The transverse vibrations of a stretched string form one of the most ancient methods of producing musical sound. It was familiar to the ancient Egyptians as well as the Greeks. Pythagoras was the first to use the monochord, a sounding board and a box with a scale on which one string or more may be stretched. Mersenne of Paris in 1636 gave the relations connecting the frequency, the length, the diameter, the density, and the tension of the string in his *Harmonic Universelle*, and proved them experimentally. Galileo probably deduced these relations quantitatively before Mersenne, but it seems that his publication was delayed for some time. *Mersenne's law*, as it is usually called, can be put in the form

$$n_p = (p/2l)\sqrt{(T/m)}$$

(*see* stretched string), where *l* is the length of the string, *T* the tension, *m* the mass per unit length, *p* may be 1, 2, 3, etc. according to the number of loops into which the string divides itself and n_p is the frequency of the harmonic mode (*see* harmonic).

The monochord is used to verify such relations or to compare the pitch of tones. It consists in its modern form of a thin metallic wire (steel piano wire is usually employed) stretched either horizontally or vertically over two bridges by means of a weight hanging over a pulley or by a spring tensioning device. A movable bridge provides a convenient means of varying the vibrating length of the wire. In order to increase the volume of the sound as in most stringed instruments, the string and the bridges are mounted on a *soundboard consisting of a hollow box, the air in which is forced to vibrate to a certain extent by the vibration of the string. The vibration may be excited in any convenient manner, e.g. by plucking, striking, bowing, or by electromagnetic means. Sonometers are often provided with a second wire of fixed length and tension to provide a standard frequency as a basis for comparison. Two methods are suitable for such tuning: either the two wires may be sounded together and adjustment made so that the beats between them vanish, or the resonance principle may be employed. In this case, the standard wire alone is excited and the second, if in tune, will vibrate so strongly as to throw off light paper riders hung on it.

monochromatic radiation Radiation restricted to a very narrow band of wavelengths: ideally one wavelength. (*Compare* polychromatic radiation.) Even spectral lines embrace a range of a few ten-thousandths of a nanometre (they are more homogeneous the lower the temperature of the vacuum-discharge tube). Monochromatic light can be isolated with varying degrees of purity by using filters (interference filters enable narrower limits of wavelength to be isolated than do ordinary glass filters when used with white light); by incandescent solids and gases with or without monochromatic filters; by isolating a portion of the spectrum by means of a slit; by the use of *lasers. Sodium-yellow light is the most commonly used for the specification of the refractive indexes of substances. Monochromatic aberrations refer to the spherical aberrations when, as is usual with spherical-aberration theory, light of one colour is under consideration.

monochromator *See* spectrometer.

monoclinic system *See* crystal systems.

monodisperse system A colloidal system in which the particles in the disperse phase are of approximately the same size. *Compare* polydisperse system.

monolayer *See* surface films.

monolithic integrated circuit *See* integrated circuit.

monomolecular film *See* surface films.

monostable A type of circuit having only one stable state, but which can be triggered into a second quasi-stable state by the application of a *trigger pulse. One form consists of a *multivibrator with resistive-capacitive coupling. Monostables are used to provide a fixed duration pulse and can be utilized for pulse stretching, or shortening, or as a delay element.

Moog synthesizer *See* electronic instruments.

Morley, Edward Williams (1838–1923) Amer. chemist who collaborated with Michelson in the classic *ether drift experiment. *See* Michelson–Morley experiment.

Morse, Samuel Finlay Breese (1791–1872) Amer. artist and inventor. He devised a telegraphic apparatus (1837) in which a moving armature produced dots and dashes on a moving paper strip and invented the *morse code* in which each important symbol (letter, figure, etc.) is represented by a group of dots and/or dashes.

Morse equation An empirical equation giving the potential energy of two atoms in a molecule as a function of their separation. It has the form:
$$V = D_e\{1 - \exp[-\beta(r - r_0)]\}^2,$$
where V is the potential energy, β is a constant, r is the distance between the atoms, and r_0 is the equilib-

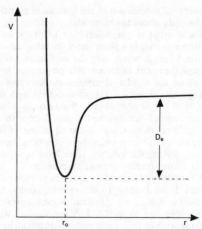
Potential energy curve for diatomic molecule

rium distance, i.e. the bond length. A typical potential energy curve for a diatomic molecule is shown in the diagram. At small separations the energy is very high because of repulsion between the nuclei. At large separations the energy is constant with separation because the molecule has dissociated into two atoms. D_e, the energy from the minimum to the dissociation level, is called the spectroscopic heat of dissociation. Note that the energy required to dissociate the molecule is the chemical heat of dissociation where $D_e = D_0 + \frac{1}{2}h\nu$, $\frac{1}{2}h\nu$ being the *zero-point energy of the molecule.

MOS Abbreviation for metal oxide semiconductor. *See* field-effect transistor; MOS integrated circuit; MOS logic circuits.

mosaic crystal An imperfect crystal: the majority of crystals are composed of smaller crystals or crystallites that grow together so as to be nearly or exactly parallel, but with discontinuity at their mutual surfaces. *See* crystal texture.

mosaic electrode The light-sensitive surface of a *camera tube.

Moseley, Henry Gwyn-Jeffreys (1887–1915) Brit. physicist. A pioneer of X-ray spectroscopy, he established the connection between the frequency of characteristic X-ray emission and the atomic number of the element (*see* Moseley's law), so providing a physical verification of the significance of atomic number and ensuring that there were no missing columns in the *periodic table.

Moseley's law The *X-ray spectrum of a particular element can be split into several distinct line series: K, L, M, and N. Moseley's law states that the square root of the frequency f, of the characteristic *X-rays of one of these series, for certain elements, is linearly related to the atomic number Z. A graph of Z against $\sqrt{f}$ is called a *Moseley diagram*. Moseley found that two lines corresponding to the K series could be drawn on such a diagram for atomic numbers between 13 (aluminium) and 30 (zinc), excluding argon. Five lines corresponding to the L series could be drawn for elements between zirconium and gold. For each series:

$$f = a(Z - b)^2,$$

where a and b are constants for the series in question.

The law, which is only approximate, can be explained in terms of the energies of the electrons in the various inner shells of the atom. Thus the energy E_K of a K electron is given in electronvolts approximately by:

$$E_K = -13.6 (Z - 1)^2.$$

That of an L electron is given more roughly by:

$$E_L = -(13.6/4)(Z - 5)^2,$$

with similar formulae for higher energy states.

MOSFET, MOST *See* field-effect transistor.

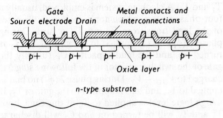

a Cross section of *p*-channel aluminium gate circuit

MOS integrated circuit A type of *integrated circuit based on insulated-gate *field-effect transistors. MOS circuits have several advantages and account for a substantial proportion of all semiconductor devices produced. They usually have a higher functional *packing density than *bipolar integrated circuits as MOS transistors are self-isolating and no area-consuming isolation diffusions are required. MOS transistors may be used as active load devices (*see* field-effect transistor); thus no separate process is required to form resistors, as in bipolar integrated circuits. When used as load devices *pulse operation of the circuit is easily obtainable using the gate electrodes to activate the device: power dissipation is greatly reduced, involving less complicated heating problems. A characteristic of MOS transistors is their exceptionally high input impedances, allowing the gate electrodes to be used as temporary storage capacitors (enabling the circuits to be relatively simple). This is called *dynamic operation*, and usually these circuits operate above a minimum specified frequency.

The relatively few processing steps required in manufacture, compared to bipolar integrated circuits, enables large chips to be made thus further increasing the functional compactness, and reducing the costs.

MOS circuits tend to be slower than their bipolar counterparts due to their inherently lower *mutual conductance, and to the fact that their speed is extremely dependent on the load capacitance.

Two main types are used, having slightly different structures: aluminium gate circuits, and self-aligned gate circuits.

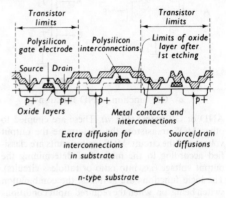

b Cross section of *p*-channel self-aligned circuit

(i) *Aluminium gate circuits.* A typical section is shown in Fig. *a*. *Source/*drain diffusions are made through a thick oxide layer, into the substrate, and a new oxide layer grown over the diffused regions. These layers may be additionally thickened if required using a low-temperature vapour-phase reaction. Openings are then made in the oxide layer between source and drain regions and a thin oxide layer grown to form the *gate insulators. Contact windows to the source and drain regions are opened, and aluminium deposited to form the gate electrodes and the circuit interconnection pattern.

(ii) *Self-aligned gate circuit.* In this type of circuit the gate electrodes are formed before the source/drain diffusions are made; the most widely used method is known as the *silicon gate technology* (Fig. *b*). Openings are made in the initial thick oxide layer and the thin layer of gate oxide grown. This is immediately covered with a layer of polycrystalline silicon (polysilicon) using a vapour-phase reaction. The polysilicon is then etched to form the gate electrodes, together with some interconnections. The gate oxide is then removed from the regions on which it is not covered with polysilicon and the source/drain diffusions are made. The edges of these diffused regions are defined by the previously etched gate regions. The polysilicon regions are simultaneously doped with the diffusing material and this has the desirable effect of reducing their resistivity. The whole slice is covered with a further oxide layer, contact windows etched and the final metal layer is deposited and etched to form the interconnections. Two layers of interconnections are possible using this technique, plus some interconnections in the substrate itself by diffusing extra regions when the source/drain diffusions are made.

MOS logic circuits *Logic circuits constructed in *MOS integrated circuits. They consist of combinations of MOS *field-effect transistors in series or in parallel that perform the logic functions (e.g. act as

MOS logic circuits

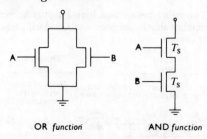

OR function   AND function

a   OR function; AND function

AND or OR gates) (Fig. *a*). These are coupled to other MOS transistors that determine the output voltages of the circuit. MOS logic circuits are classified according to the method of determining the output voltage (i.e. into ratio or ratioless circuits). The logic functions are switches, the combination switch being 'on' when the required input conditions are fulfilled, e.g. both A and B are high logic levels to produce a high-level output for the AND gate in Fig. *a*. The high logic level is chosen to be greater than the threshold voltage V_T; the low level is lower. Here the function is represented by a single switch T_S.

(*a*) *Ratio circuits*. The switch transistor T_S is connected in series with a load transistor T_L. The drain of the load transistor is connected to the power supply, and the source of the switch transistor to earth (Fig. *b*). The output is taken from the node between the transistors. The circuit will usually be driving similar MOS logic gates, which are represented by a capacitor C_L, due to the very high input impedance.

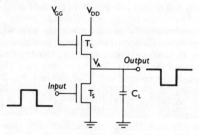

b   Ratio circuit

A voltage of magnitude equal to or greater than the drain voltage is applied to the gate of the load transistor either continuously (static operation) or on the application of a clock pulse (dynamic operation) to reduce dissipation.

If a low logic level is applied to the gate of T_S, the switch transistor will be off and C_L will be charged by T_L until the output voltage V_A rises sufficiently to turn off T_L (i.e. to $V_{GG} - V_T$, or V_{DD}, whichever is lower).

When the switch transistor is on, due to the application of a high logic level to the gate, C_L will discharge through T_S and V_A will fall to a value determined by the relative impedances of the two transistors.

It can be shown that the voltage V_A at the node depends on the ratio of the aspects ratios of the devices, and these are manufactured to ensure an output voltage suitable for a low logic level, i.e. less than the threshold voltage of the following state.

The circuit provides inversion of the logic function; thus an AND function in T_S provides a NAND output, etc.

If the dynamic version of the circuit, with the gate voltage of the load transistor clocked, is used, a minimum rate of clocking must be specified to prevent loss of information at the output due to leakage paths causing the charge on the load capacitor to decay.

(*b*) *Ratioless circuit*. In this type of circuit the load transistor is replaced by two transistors T_1 and T_2, connected in series with the switch transistor, and the output voltage V_B taken from the node between T_1 and T_2. A clocking system is employed, usually a four-phase system, which applies a bias to the gates of the load transistors T_1 and T_2 in turn. During phase 1 (ϕ_1) bias is applied to the gate of T_1, T_1 is turned on, and the load capacitor C_L (usually the gate of the switch transistor of the following stage) is charged to $V_{GG} - V_T$. During phase 2 (ϕ_2) no bias is applied to T_1, and bias is applied to the gate of T_2. If a high logic level is applied to T_S at this time, both T_2 and T_S will be turned on and C_2 will discharge through them and the output voltage V_B will fall to the low logic level. If T_S is not turned on, i.e. a low logic level is applied to the gate of T_S during ϕ_2, C_L will not discharge as no conducting path exists, and V_B will remain at the high logic level. The output of the circuit is sampled by the following circuit during phases ϕ_3 and ϕ_4, thus information may only be supplied to T_S once in every four clock phases.

Operation of this circuit does not depend on the impedances of the devices and it is therefore termed ratioless. Power dissipation is very low since no conducting path ever exists directly between the power supply and earth; the circuits depend solely on charge storage in the load capacitance.

The circuit is inverting and two gates are frequently combined to provide a noninverting circuit. If used in a dynamic *shift register, for example, six transistors are needed for each *bit of information.

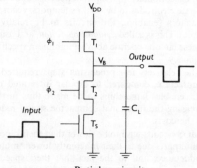

c   Ratioless circuit

Mössbauer, Rudolf (*b.* 1929) German-born Amer. physicist noted for his researches into gamma rays and for the *Mössbauer effect. For this work he shared a Nobel prize (1961) with *Hofstadter.

Mössbauer effect An effect observed when certain nuclei emit or absorb gamma radiation of low quantum energy. A photon of energy $h\nu$ transfers momentum $h\nu/c$, hence when a stationary nucleus emits or absorbs a quantum, it must recoil to conserve momentum. The kinetic energy of recoil of a free nucleus of mass M is given by:

$$E = (1/2M)(h\nu/c)^2.$$

If a gamma ray is emitted by a nucleus on falling from a given excited state to the ground state, the quantum energy is therefore not sufficient to raise a similar nucleus from the ground to the excited state. Mössbauer showed that for low quantum energies an atom bound in a crystal may remain so bound on emitting or absorbing radiation. In this case the recoil momentum is shared between as many atoms as an acoustic wave will reach during the process of emission or absorption. Hence the mass M in the equation must be replaced by the relatively enormous mass of this large number of atoms, so that E becomes infinitesimal. Such processes are called *recoilless*.

An atom bound in a crystal vibrates at a very high frequency, typically 10^{12} Hz, so during the fairly long lifetime of the relevant excited states, it completes a very large number of cycles of operation. The average velocity of a body in a complete cycle is zero, so the frequency of the radiation is not affected by the first-order *Doppler effect. The mean square velocity is nonzero, hence the frequency of the radiation is affected by the second-order Doppler effect. As the value of the mean square velocity depends on temperature, there is a temperature coefficient of the quantum energy, of the recoilless gamma rays. Although this is extremely small, the very high precision possible in the measurements requires careful control of temperature.

Very many experiments have been done using the nuclide $^{57}_{26}$Fe. Electron capture in $^{57}_{27}$Co produces $^{57}_{26}$Fe in an excited state of energy 14.4 keV and lifetime 0.1 μs. In iron foils at room temperature about two-thirds of the emissions and absorptions of the 14.4 keV radiation are recoilless. In a *Mössbauer analyser* the source is mounted so that it can be moved at steady speeds (of the order 10^{-5} m s^{-1}) towards or away from the detector. The latter consists of an iron foil, usually highly enriched in the rare isotope ^{57}Fe, behind which is a proportional counter or similar instrument. The movement of the source changes the quantum energy by the Doppler effect. Over a very limited range of speeds corresponding to a line width of the order of 10^{-13} there is a decrease in the count rate as nuclear absorption is added to the normal absorption by the *photoelectric effect. Similar experiments have been done using other gamma emitters, some of which have to be cooled to very low temperatures to give a large enough proportion of recoilless interactions.

The technique can detect fractional changes in energy of the order of one in 10^{16}. It has been used to study: the *Einstein shift in the earth's gravitational field; the quantum states of nuclei; magnetic fields inside ions; and some chemical compounds. The energies of the quantum states of the nucleus are affected by the orbital electrons and thus the quantum energy of the radiation depends slightly upon the compound in which the nucleus occurs. The difference between the energy of a state in the pure element and that in a compound is called the *chemical shift*.

most probable speed The speed corresponding to the maximum in the curve of the Maxwell *distribution of velocities. It has the value $\sqrt{2kT/m}$.

MOS transistor *See* field-effect transistor.

motion *See* Newton's laws of motion.

motional impedance *See* driving point impedance.

motor Electric motor. A machine that does work mechanically when it is driven by an electric current. *See* induction motor; synchronous-induction motor; synchronous motor; universal motor.

motor controller A device for adjusting the speed of an electric motor. It is usually designed to be suitable for continuous operation at any of the speed settings. (*Compare* motor starter.) The term does not include shunt-field rheostats.

motor converter *See* converter.

motor generator A complete assembly of one or more generators coupled directly to one or more driving motors.

motor meter An *integrating meter that incorporates some form of electric motor. Particular examples are the *induction meter and the *mercury motor meter.

motor starter A device for starting an electric motor and for accelerating it to normal speed. It is not designed for continuous operation in any other than the full-on position. *Compare* motor controller. *See* number of steps.

Mott, Sir Nevill Francis (*b.* 1905) Brit. physicist. Professor at Bristol and later Cambridge, his early work centred on the quantum theory of atomic collisions. He later worked on solid-state problems and shared a Nobel prize for physics (1977) with Philip Anderson and John Van Vleck for their work on the electronic structure of magnetic and disordered systems.

Mottelson, Benjamin Roy (*b.* 1926) Amer.-born Danish physicist. Working in the Institute of Theoretical Physics in Copenhagen (now the Niels Bohr Institute) with Aage Bohr he was awarded a share of a Nobel prize (1975) with Bohr and L.J. Rainwater for their work (1950–53) on the collective (or unified) model of the nucleus. His later work concerned pair vibrations and nuclear-field theory.

mouse A computer device that is used as a pointer, i.e. it passes two-dimensional spatial information to a computer. It is moved by hand around a flat surface. These movements are communicated to the computer and lead to corresponding movements of the cursor on a display screen. A mouse has one or more buttons to indicate to the computer that the cursor has reached the desired position.

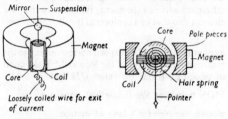

| *a* Moving-coil galvanometer | *b* Plan of another type of moving-coil galvanometer |

moving-coil galvanometer A type of *galvanometer in which a small, light coil of many turns is delicately suspended in the space between the curved poles of a strong permanent magnet (Figs. *a*, *b*). A soft-iron core serves to concentrate the lines of force and to make the field more uniform. The instrument is independent of the earth's field since this is so weak in comparison with that of the permanent magnet. The controlling torque may be the torsion in the suspending wire, which is often a strip of phosphor-bronze, or the effect of a light hair spring. Usually the coil is wound on a metal former, induced *eddy currents in which make the instrument *deadbeat. The current is led into the coil by the suspension or hair spring.

The galvanometer coil may have a resistance of from a few ohms to a few thousand ohms, while full-scale deflection may be given by a few milliamps for a robust instrument (Fig. b) or a few nanoamps for a very sensitive instrument (Fig. a). *See* ammeter; voltmeter.

moving-coil instrument *See* permanent-magnet moving-coil instrument.

moving-coil microphone A type of *microphone in which the diaphragm is connected to a coil and moves it backwards and forwards in a stationary magnetic field, thus inducing an e.m.f.

moving-iron instruments Instruments for measuring electric currents or voltages that depend on the attraction of a soft-iron armature into a current-carrying coil (Fig. *a*), or on the mutual repulsion between the poles induced in two soft-iron rods, both of which being within the coil have similar poles induced in the ends near to each other (Fig. *b*). Moving-iron instruments give the same deflection whatever the direction of the current, and can thus be used with alternating currents provided the fre-

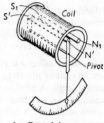

| *a* Attraction type of moving-iron instrument | *b* Repulsion type of moving-iron instrument |

quency is not too great for the iron to follow the changes.

MP Abbreviation for *magnifying power

MRI Abbreviation for *magnetic resonance imaging.

MSI Abbreviation for medium-scale integration. *See* integrated circuits.

Mueller, Erwin Wilhelm (1911–1977) German-born Amer. physicist who worked at the Fritz Huber Institute in Germany and, later, at Pennsylvania State University. He is noted for his investigations of solid surfaces, in particular, his inventions of the *field-emission microscope (1936) and the *field-ion microscope (1951).

MUGA scan *See* nuclear medicine.

Muller, Karl Alexander (*b.* 1927) Swiss physicist. Working at IBM's Zürich research laboratory, he headed the solid-state department. He shared a Nobel prize with J.G. Bednorz for their work on the superconductivity of oxides at relatively high temperatures, which stimulated a world-wide search for substances that become superconducting at temperatures high enough for the practical transmission of electrical power.

Mulliken, Robert Sanderson (1896–1986) Amer. physicist and chemist; professor at Chicago and at Florida State University. Mulliken first suggested (1922) the separation of isotopes by evaporative centrifuging. Much of his research was concerned with the quantum mechanics of molecules. For this work he received the 1966 Nobel prize for chemistry.

multichannel analyser A test instrument that splits an input waveform into a number of channels in respect of a particular parameter of the input. A circuit that sorts a number of pulses into selected ranges of amplitude is known as a *pulse-height analyser*. A circuit that splits an input waveform into its frequency components is known as a *spectrum analyser. In general, a multichannel analyser will have facilities for carrying out both these operations.

multielectrode valve A *thermionic valve that contains two or more sets of electrodes within a single envelope, each set of electrodes having its own independent stream of electrons. The sets of electrodes may have one or more common electrodes (e.g. a common cathode). A typical example is the double-diode-triode, which contains the electrode assemblies of two diodes and one triode.

multiple reflection The reflection of light and the formation of a number of images when two or more mirrors reflect light several times in succession.

multiple scattering *See* scattering.

multiple stroke *See* lightning flash.

multiplet 1. A group of spectrum states specified by the values of the quantum numbers L (vector sum of orbit and angular momenta of individual electrons) and S (vector sum of spin momenta of individual electrons). The group of states gives rise to a set of spectrum lines.
2. A set of quantum-mechanical states of *elementary particles having the same value of certain *quantum number(s). Individual members of the set are distinguished by having different values of other quantum numbers. The word multiplet is most commonly used in connection with sets of states that are transformed into each other by operations that form the elements of a *group. For example, *unitary symmetry multiplets are sets of particles having the same values of the quantum numbers J (*spin), P (*parity), and B (*baryon number). The particles forming these sets are distinguished by different values of *hypercharge and the *isospin quantum number I_3. Mathematically these sets are associated with the group SU_3 (*see* unitary symmetry).

multiplexer *See* multiplex operation.

multiplex operation The use of a single path for the simultaneous transmission of several signals without any loss of identity of an individual signal. The various signals are fed to a *multiplexer*, which allocates the transmission path to the input according to some parameter (e.g. *frequency-division multiplexing or *time-division multiplexing). At the receiving end a *demultiplexer*, operating in sympathy with the multiplexer, reconstructs the original signals at the outputs. The transmission path may be in any of the available media, e.g. wire, waveguide, or radio waves.

multiplication The process by which additional neutrons are produced by a *chain reaction as a result of *fission in a *nuclear reactor.

multiplication constant or factor (effective) Symbol: k_{eff}. In a *nuclear reactor. The ratio of the number of neutrons liberated in one generation to the number liberated in the previous generation.

multipolar machine An electrical machine having a *field magnet with more than two poles. *Compare* bipolar machine.

multivibrator A form of *oscillator consisting of two linear *inverters coupled so that the input of one is derived from the output of the other. The action of

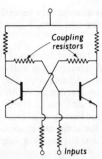

Multivibrator using resistive coupling

the various types of multivibrator is determined by the coupling used.

Capacitive coupling gives an *astable multivibrator* with two quasistable states; once the oscillations are established the device is free-running, i.e. it can generate a continuous waveform without any *trigger.

Capacitive-resistive coupling gives a *monostable multivibrator.

Resistive coupling (*see* diagram) gives a *bistable multivibrator that has two stable states and can change state on the application of a trigger pulse. *See* flip-flop.

mu-meson Former (and incorrect) name for muon.

Mumetal *See* ferromagnetism.

muon A negatively charged *lepton similar to the electron except for its mass, which is 206.7683 times greater than that of the electron. It has a mean life of $2.197\,09\,\mu s$, and decays into an electron, a *neutrino, and an antineutrino:
$$\mu^- \rightarrow e^- + \nu_\mu + \bar{\nu}_e.$$
The decay process of the muon's antiparticle, the *antimuon*, is:
$$\mu^+ \rightarrow e^+ + \bar{\nu}_\mu + \nu_e.$$
It was originally thought to be a meson. *See* elementary particles.

muon-catalysed fusion *See* cold fusion.

mush (or **random**) **winding** A type of winding used in transformers and electrical machines, having multilayer coils, each of which is wound in a random manner, i.e. the layers are indefinite.

musical echo An echo is heard when a short sound (e.g. a handclap) is reflected from a stepped structure (e.g. palings). The impulse is reflected by the elements of the structure and the observer hears a regular succession of impulses. A musical note is heard if the pulses are received in sufficient number per second. The frequency (f) of the note obviously depends on the spacing (d) of the elements of the stepped structure, the direction (θ) from which reflections proceed and the speed (c) of the sound in the air:
$$f = c/(2d \cos \theta).$$

musical sand

The first received sounds are generally masked by the initial impulse and hence θ may be considered small.

musical sand Sand in certain parts of the earth when struck by a hard object (or even walked upon), emits a musical sound of definite pitch. This phenomenon has so far been observed in the Arabian deserts, and in several coastal districts of North America and Britain. It is found that the sand grains in such cases are very nearly the same size, uncontaminated with impurities, and dry. Impurities or water added to the sand immediately damp out the sound. The elastic vibration of spheres has been suggested as a possible explanation but the fundamental frequency of such small spheres would be very high. Another suggested explanation is that the sound is due to simple friction between highly polished grains of quartz. This does not explain why such a definite pitch is heard.

musical scale A series of notes progressing from any given note to its octave by prescribed *intervals chosen for musical effect. Each scale covers all the chief notes in a particular musical composition of a period or people. Of the very large number of scales used throughout history, only a few survive, and it is upon these that series Western music is composed. In the musical world there has been a number of attempts to introduce new scales, although many of these are based on the scales in common current use. With the possible exception of the octave, not one interval is common to all the scales. The octave, perfect fifth, and perfect fourth (*see* intervals) are common to many scales. The musical scales of the East are mainly for melodic purposes and generally contain more notes to the octave than those used in the West. Frequency observations on the Western current scales show ratios which are relatively simple–within limits detectable by the ear. This result links with the ideas of concord and *beats.

The method of development of the musical scale is unknown, but has been attributed to speech inflexions and, in later years, to the needs of harmonic progression. The pentatonic scale is thought to be associated with the early development. All scales have equal authority if they are the basis of providing music that is aesthetically satisfying. Attempts to prove that certain scales are more basic than others, as for example, those built on the arithmetic harmonic series, are fundamentally unsound. Again, frequency ratios quoted for intervals are the average of a large number of readings – or, in certain cases the theoretical values, which assume accuracies beyond those of the ear. In musical reproduction the notes produced and perceived by musicians may differ considerably from the stated ratios, depending upon the context of the note. In keyboard instruments the tuning depends upon the tuner, and the intervals will only approximate to the mathematically expressed ratios. Pythagoras did much to stabilize the intervals of scales from which the present *diatonic scale* has developed. This latter scale consists of eight notes to the octave, some of which are separated by *tones and some by semitones. Later all the tones were divided into semitones, first, for the use in embellishment of the melodic line, and secondly, for harmonic colouring. The extra notes provided are used much less frequently than the others. All notes of the scale bear a relation to the keynote or starting note of the scale. Microtonal scales have been introduced by some composers. Quarter tones or less are known, e.g. the Hindu octave is divided into twenty-two notes.

The *chromatic scale* has thirteen notes to the octave, the notes being the same as those found on the diatonic scale with each tone divided. All the notes of the chromatic scale, however, are used with equal authority, but are still related to the keynote. The *duodecuple scale* uses the same notes as the chromatic scale but considers all notes of equal value, viz. there is no keynote. The *wholetone scale*, which has been used by some recent composers, consists of six notes to the octave, there being no semitones. Only two series of notes are possible. The *pentatonic scale* with five notes to the octave having the spacing tone, $1\frac{1}{2}$ tones, tone, tone, $1\frac{1}{2}$ tones, has been used for a considerable time in countries as far apart as Scotland, China, India (hill tribes), Japan, Africa, and America. The *bagpipe scale* of Scotland covers from G (392 hertz) to A octave, and corresponds approximately to the white notes on the keyboard except that C and F are about a quarter tone sharp.

The *equitempered scale and the *just intonation scale, which have intervals differing only slightly from the chromatic scale, have been introduced to permit the adequate use of the chromatic scales based on all or most of the different keynotes, with only twelve notes between a note and its octave. Such limitation is imposed by musical instruments with the traditional keyboard. Various keyboards have been proposed to permit a more near approach to the intervals of the diatonic scale in all keys, e.g. the Janko keyboard. Comparison of intervals between some of these scales and the diatonic scales of C and D are given under *temperament.

Musschenbroek, Pieter van (1692–1761) Dutch physicist who reported (1746) his invention of the *Leyden jar (independently discovered in 1745 by the German Georg von Kleist).

mutual branch *See* network.

mutual capacitance A measure of the extent to which two *capacitors can affect each other, expressed in terms of the ratio of the amount of charge transferred to one, to the corresponding potential difference of the other.

mutual conductance *Syn.* transconductance. Symbol: g_m. Of an amplifying device or circuit. The ratio of the incremental change in output current, I_{out}, to the incremental change in input voltage, V_{in}, causing it, the output voltage remaining constant:

$$g_m = \partial I_{out} / \partial V_{in} \qquad (V_{out} \text{ constant}).$$

mutual inductance *See* electromagnetic induction.

myopia A refractive defect of the eye in which the length of the eye is too great relative to its unaccommodated (longest) focal length. The furthest point of distinct vision (conjugate with the retina when the eye is unaccommodated) lies in front. The correcting spectacle lens must have its second focal point coincident with this far point: the lens is concave. The nearest point of distinct vision is closer than the normal for the age, giving rise to the description of the defect as *short sight*. *See* eye; refraction of eye.

myria- A prefix meaning 10 000. It is not approved for use with *SI units.

myriametric waves Radio waves having wavelengths greater than 10 km.

N

NA Abbreviation for *numerical aperture.

NAA Abbreviation for neutron *activation analysis.

nabla *See* del.

nadir *See* celestial sphere.

NAND gate *See* logic circuit.

nano- Symbol: n. A prefix meaning 10^{-9}; for example, one nanometre (1 nm) is 10^{-9} metre.

Napier, John (1550–1617) Brit. (Scots) mathematician, the inventor of logarithms. The first table of logarithms, published in 1614, was of *natural logarithms. Napier also invented mechanical devices for carrying out arithmetic operations (*Napier's rods*) and he discovered formulae used in spherical trigonometry (*Napier's analogies*).

Napierian logarithms *See* natural logarithms.

natural abundance *See* abundance.

natural convection *See* convection (of heat).

natural frequency *See* free vibrations.

natural logarithms *Syn.* Napierian logarithms; hyperbolic logarithms. Logarithms to the base e.

natural units A system of units based on *Gaussian or *Heaviside–Lorentz units for electromagnetic quantities. These units are often used in particle physics in place of the generally more widely used *SI units. In natural units, quantities having dimensions of length, mass, and time are given the dimensions of power or energy (usually expressed in *electronvolts), which effectively makes the rationalized *Planck constant and the *speed of light both equal to unity.

n-channel device *See* field-effect transistor.

near infrared or ultraviolet Parts of the infrared or ultraviolet regions of the spectrum of *electromag-

netic radiation that are close to the visible region. The regions that are far from the visible region and close to the X-ray and microwave regions, are called the *far ultraviolet* and *far infrared* respectively.

The terms are used rather loosely and it is not possible to give any definite ranges of wavelength. In the case of infrared radiation the near infrared is usually the region in which molecules absorb radiation by making transitions between vibrational *energy levels. The far infrared is usually the region in which absorption is due to changes in rotational energy levels.

near point Of the eye. The nearest point for which, with accommodation fully excited, clear vision is obtained. Normally since the crystalline lens becomes harder with age, the accommodation amplitude decreases and the near point tends to recede with age. When it has receded beyond a comfortable reading distance, the condition is described as *presbyopia* or old age sight. It should not be confused with the conventional distance of distinct vision (25 cm or 10 in.). The difference of the reciprocals of the distances (in metres) of the far point (R) and the near point (P) measures the amplitude of accommodation.

nebula A cloud of interstellar gas and dust that can be observed either as a luminous patch – a *bright nebula* – or as a dark region against a brighter background – a *dark nebula*.

Various processes can cause a nebula to become visible. Ultraviolet radiation from a nearby source, usually a hot young star, can ionize the interstellar gas atoms, and light (and other radiation) is emitted when the ions interact with free electrons in the nebula. This is called an *emission nebula*, and it has an emission spectrum. *Planetary nebulae* are emission nebulae in which the gas is ionized by ultraviolet radiation from the remnant of a dying star lying within the nebula; *supernova remnants are another example. With a *reflection nebula*, light from a nearby star or stellar group is scattered by the dust grains. It has essentially the same spectrum as the illuminating star(s). In contrast, dark nebulae have no nearby stars to illuminate them.

nebular hypothesis The hypothesis, due to Laplace, that the solar system has evolved from a giant mass of gas that, mainly as a result of rotation and cooling, condensed into the sun and planets. The hypothesis is untenable, raising many problems in connection with the known momenta and energies in the solar system.

Néel, Louis Eugène Félix (*b.* 1904) French physicist who became professor of physics at Strasbourg and, later, at Grenoble. He was awarded the 1970 Nobel prize for physics for his work on magnetism, which included his predictions of antiferromagnetism (1930) and ferrimagnetism (1947).

Néel temperature *See* antiferromagnetism.

Ne'emen, Yuval (*b.* 1925) Israeli physicist who contributed, independently of *Gell-Mann, to the theory of *unitary symmetry.

negative 1. A photographically produced image in which the dark and light parts of the subject appear as light and dark, respectively. (*See* photography.) In a colour negative the colours of the object appear as *complementary colours, being converted to the original colours by the colour-printing process. *See* colour photography.
2. Of an electric charge. Having the same polarity as the charge of an electron.
3. Of a body or system. Having a negative charge; having an excess of electrons.

negative bias A potential that is applied to an electrode of an electronic device and is negative with respect to earth potential (or some other fixed reference potential).

negative booster *See* booster.

negative crystal *See* optically negative crystal.

negative electron *Syn.* negatron; negaton. An electron with a negative charge as opposed to the positively charged electron or *positron.

negative feedback *See* feedback.

negative feeder *Syn.* return feeder. A *feeder in an electric traction system that connects the negative conductor rail or the track rails to the negative *busbar at a generating station or at a substation.

negative feeder-booster A *booster employed in an earthed-return system to reduce the potential difference between any two points of the earth return. It is used extensively in electric traction systems to ensure compliance with statutory regulations that demand that the potential difference between any two points of the earthed return must be less than 7 volts, and the potential of any point must not exceed 4.2 volts above earth.

negative glow *See* gas-discharge tube.

negative logic *See* logic circuit.

negative magnetostriction *See* magnetostriction.

negative phase sequence *See* phase sequence.

negative principal and nodal points Negative principal points are those conjugate points of a system for which the lateral magnification is -1, i.e. at $2f$ in both object and image spaces from the (usual) principal points. The negative nodal points are points for which the angular *magnification is unity and negative. They lie as far from the focal points as the ordinary *nodal points but on opposite sides. Also known respectively as antiprincipal and antinodal points. *See* centred optical system.

negative resistance A property of certain electronic devices whereby a portion of the voltage–current characteristic has a negative slope, i.e. the current decreases as the applied voltage increases. Such devices include the thyristor (*see* silicon controlled rectifier), the *magnetron, and the *tunnel diode.

negative specific heat capacity A property of certain substances that under stated conditions need heat to be extracted from them if their temperature is to be raised. The most familiar example is a *saturated vapour, for which:

$$c = (c_p)_1 + \mathrm{d}L/\mathrm{d}T - L/T,$$

where $(c_p)_1$ is the specific heat capacity of the liquid at the equilibrium temperature T, L is the specific latent heat of vaporization. When the vapour rises in temperature, it must simultaneously be compressed to keep it saturated since the density of saturated vapour rises with increasing temperature. For steam, the heat of compression is so great that the vapour becomes superheated so that heat must be extracted from it. This occurs through the evaporation of more water, the specific latent heat for this evaporation coming from the supersaturated steam. If the heat of compression is small, the vapour becomes superheated, condensation of liquid occurring with evolution of heat to the saturated vapour so that the specific heat capacity is positive. If the heat of compression is just sufficient to keep the vapour saturated, the specific heat capacity is zero and the curve of saturated vapour pressure coincides with the curve of adiabatic compression. In the case of steam an adiabatic expansion causes it to become supersaturated so that it condenses forming a fog.

negative transmission *See* television.

negatron or negaton *See* negative electron.

nematic structure A state intermediate between the crystalline and the liquid in which the lengths of the molecules are parallel, but in which no further regularities exist. *Compare* smectic structure.

neon tube A *gas-discharge tube containing neon at low pressure, the colour of the glow discharge being red. The striking voltage is between 130 and 170 volts and within a range of current the voltage across the tube remains constant so that for small currents (up to about 100 milliamp) it can be used as a voltage stabilizer. Electrodeless neon tubes will easily glow in the presence of high-frequency currents of high voltage.

neper Symbol: Np. A dimensionless unit used for comparing two currents almost exclusively in telecommunication engineering, the two currents being usually those entering and leaving a transmission line or other transmission network. Two currents I_1 and I_2 are said to differ by N nepers when:

$$N = \log_e |I_1/I_2|.$$

If the input and output impedances Z_1 and Z_2 have equal magnitudes, i.e. $|Z_1| = |Z_2|$, then:

$$N = \log_e |E_1/E_2| \text{ nepers},$$

where E_1 and E_2 are the input and output voltages respectively. If the two impedances have equal resistances,

$$N = \tfrac{1}{2} \log_e (P_1/P_2) \text{ nepers},$$

where P_1 and P_2 are the input and output powers respectively. One neper equals 8.686 decibels. *See also* bel.

nephoscope An instrument for studying the motion of clouds.

neptunium series *See* radioactive series.

Nernst, Walter (1864–1941) German physical chemist. He formulated a theorem (*see* Nernst heat theorem) on the entropy of substances and deduced the vanishing of specific heat capacities and of coefficients of expansion as the absolute zero of temperature is approached. With Lindemann, he devised a calorimeter (*see* Nernst and Lindemann vacuum calorimeter) for specific heat capacity measurements at low temperatures and with it confirmed the *Debye theory of specific heat capacities. Nernst was the first to give a convincing explanation of the origin of the electromotive force in a voltaic cell, which he did (1889) in terms of the osmotic pressures of ions.

Nernst and Lindemann's experiment They showed that experimentally the constant A in the formula
$$c_p - c_v = A c_p^2 T$$
for a solid is equal to $0.0214/T_M$, where T_M is the melting point of the solid.

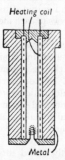

Heating coil

Metal

Vacuum calorimeter
(Nernst and Lindemann)

Nernst and Lindemann calorimeter A calorimeter designed for the measurement of specific heat capacities at low temperatures. If the substance is a metal, the calorimeter has the form shown and is made of the metal itself. If not, the specimen is placed inside a silver vessel with a platinum heating coil wound on the outside. The whole is suspended in a closed vessel, which is connected to a vacuum pump and a tube containing charcoal cooled in liquid hydrogen. When the calorimeter has been cooled to the required temperature, the vessel is exhausted. A known amount of electrical work is done on the heating coil and the small rise in temperature deduced from the measured resistance of the platinum forming the heating coil.

Nernst effect When heat flows through a strip of metal in a magnetic field, the direction of flow being across the lines of force, an e.m.f. is developed perpendicular to both the flow and the lines. The direction of the current the e.m.f. produces depends on the nature of the metal of which the strip is composed. *See* Leduc effect.

Nernst filament An element, formerly used as a source of infrared radiation, consisting of a rod of rare-earth oxides. It only conducts electricity at a high temperature, and has to be provided with a preliminary heater, and a resistance must be added in series to prevent over-running because of the negative temperature coefficient of resistance.

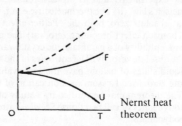

Nernst heat theorem

Nernst heat theorem Also known as the third law of *thermodynamics. If a chemical change occurs between pure crystalline solids at absolute zero, there is no change in *entropy, i.e. the entropy of the final substance equals that of the initial substances. Planck extended this by stating that the value of the entropy for each condensed phase is zero at absolute zero. In the *Gibbs–Helmholtz equation:
$$U = F - T(\partial F/\partial T),$$
the internal energy U may be written as
$$U_0 + \int_0^T \Sigma \, n C_p \; dT,$$
where the suffix 0 denotes the value at absolute zero. n is the number of moles considered, and C_p the molar heat capacity. Differentiating, the equation gives
$$\frac{\partial U}{\partial T} = -T \frac{\partial^2 F}{\partial T^2}, = \Sigma \, n C_p,$$
so that,
$$\frac{\partial F}{\partial T} = -\int_0^T \frac{\Sigma \, n C_p \, \partial T}{T} + \left(\frac{\partial F}{\partial T} \right)_0,$$
in which the first term on the R.H.S. may be evaluated from the known variation of specific heat capacity with temperature. (*See* Debye theory of specific heat capacities.) Since $(\partial F/\partial T)_0$ is the change in the entropy value of the system at absolute zero occurring during the reaction, according to the heat theorem $(\partial F/\partial T)_0 = 0$ and hence $U_0 = F_0$ and $(\partial U/\partial T)_0 = 0$. If, therefore, the free energy F and internal energy U are plotted against the thermodynamic temperature the two curves have the same value at absolute zero and both become horizontal at that point (*see* diagram).

The following deductions may be made from the Nernst heat theorem. (i) The coefficient of expansion of all condensed phases vanishes at absolute zero. This follows since by *Maxwell's (thermody-namic) relations:

$$\left(\frac{\partial V}{\partial T}\right)_p = -\left(\frac{\partial S}{\partial p}\right)_T$$

and at absolute zero $S = 0$ and S becomes independent of p and T. This is verified by experiments on crystals at low temperature (*see* Nernst and Lindemann's experiment) and is in accordance with *Grüneisen's law. (ii) The thermoelectric e.m.f. vanishes at absolute zero, both the *Peltier effect and the *Thomson effect becoming zero. (iii) The magnetic susceptibility of a paramagnetic crystal vanishes at the absolute zero. (iv) The entropies of the solid and liquid states of helium have the same value at absolute zero. (*See* Keesom's solidification of helium.) The theorem may be verified by a study of the thermodynamics of cell reactions. *See also* entropy; thermodynamics.

Nernst–Thomson rule The electrolytic dissociation and the dielectric constant of a solvent are related in such a way that dissociation of substances is higher in media of high dielectric constant. This is not invariably true since in media of low dielectric constant a minimum sometimes occurs in the equivalent conductance. According to Walden, the cube of the dielectric constant divided by the minimum conductance is a constant in these cases.

net radiometer An instrument for measuring the difference in intensity between radiation entering and leaving the earth's surface. The radiation can be direct, diffuse, and reflected solar radiation or infrared radiation from the sky, clouds, and ground. A similar thermopile system is used to that found in *solarimeters, except that both sides of the thermopile are exposed to radiation and the resulting e.m.f. is proportional to the difference in intensity of the incoming and outgoing radiation. This system is used in the *Funke net radiometer* and the *Gier and Dunkle net radiometer*. *See also* pyrheliometer.

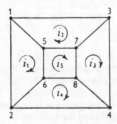

Electrical network

network 1. In electronics. A number of conductors connected together to form a system consisting of a set of interrelated circuits that performs one or more specific functions. The conductors are resistors, capacitors, and inductors, in all forms, i.e. they possess impedance. The behaviour of the network depends on the *network parameters* (or *network constants*),

which are the values of the impedances – resistances, capacitances, etc. – of the components of the network, and the manner in which the components are interconnected.

Networks are described as *linear* or *nonlinear*, depending on whether or not there is a linear relation between the voltages and currents. They are *bilateral* or *unilateral* depending on whether they pass currents in both directions or only one direction. A network is described as *passive* if it contains no source or sink of energy (the latter does not include energy dissipated in the resistance elements of the network); otherwise it is described as *active*.

A point in the network at which more than two conductors meet is called a *branch point* or a *node* (e.g. points 1 to 8 in the diagram), and a conducting path between two branch points is called a *branch* (e.g. 1 to 2). A *mesh* is the portion of the network included in any closed conducting loop in the network (e.g. 1-3-7-5-1) and its boundary is the *mesh contour*. The *mesh currents* are the currents that may be considered to circulate round the meshes (Maxwell's cyclic currents). Any branch that is common to two or more meshes is a *common branch* or *mutual branch* (e.g. 5 to 6).

Analysis of linear networks can be achieved by considering the network as a *quadripole, and deriving sets of equations relating the currents, voltages, and impedances at the input and output. It is also possible to apply *Kirchhoff's laws to each mesh in the network in turn.

2. In computing. A number of computer systems that are often widely separated but are interconnected in such a way that they can exchange information by following agreed procedures. The computer systems must be able to transmit information onto and receive information from the connected system. The information is sent as an encoded digital signal, and is transmitted along telephone lines, satellite channels, electric cables, etc.

Each computer is not directly linked to every other computer on the network. Direct connections are only made at certain points in the network, called *nodes*. Computing facilities are attached to some or all of the nodes. Nodes may be at a junction of two or more communication lines or at an endpoint of a line. A particular piece of information has to be routed along a set of lines to reach its specified destination, being *switched* at the nodes from one line to another.

network parameter *Syn.* network constant. *See* network.

Neumann, Franz Ernst (1798–1895) German mathematician, minerologist, and physicist. He made contributions to heat, optics, and electricity. Neumann's law of *electromagnetic induction is named after him.

Neumann, John von (1903–1957) Hungarian-born Amer. mathematician who contributed to the theory of quantum mechanics, devising a rigorous proof to show that macroscopic concepts of cause and effect do not apply to subatomic phenomena. His contribution to operator theory helped in the construction of computers.

Neumann's law *Syn.* Faraday–Neumann law. *See* electromagnetic induction.

Neumann's triangle *See* spreading coefficient.

neutral 1. Devoid of either positive or negative electric charge.
2. At earth potential. *See* neutral point.
3. Lacking hue; black, white, or grey.

neutral axis *See* neutral surface.

neutral equilibrium *Syn.* indifferent equilibrium. *See* equilibrium.

neutral filter A light filter that absorbs equally all wavelengths: it reduces light intensity without change of relative spectral distribution.

neutralization (electrical) The provision of negative *feedback in an amplifier to a degree sufficient to neutralize any inherent positive feedback. Positive feedback in an amplifier is usually undesirable since it may give rise to the production of oscillations. Neutralization is commonly employed with radio-frequency amplifiers to counteract the *Miller effect, and also with *push-pull operation to avoid *parasitic oscillations.

neutralization of lenses The process of finding the power of an unknown lens by using an equal and opposite power in contact with it and observing the neutralization of movement as the pair are displaced laterally. The neutralizing movements of lenses consist in observing the apparent displacement of a line as the lens is displaced at right angles to the line across the line of view. A concave lens makes the line appear to move in the same direction as the movement of the lens (with movement). Provided a real image is not formed between the lens and the eye, convex lenses produce an apparent (against) movement.

neutral point 1. Of an electrical system. A point having the same potential as that of the star point of a group of equal resistances arranged to form a *star connection with their free ends connected to the main terminals or lines of the system. The number of these resistances is 2 for a single-phase system, 4 for a two-phase 4-wire system, and 3 for three-phase, six-phase, or twelve-phase systems.
2. Of a symmetrical *polyphase system. A point in the system with respect to which the potential of the conductors is symmetrical. Usually it is connected to earth either directly or through a resistor (earthing resistor) or reactor (earthing reactor).
3. In a complex magnetic field. Those points at which opposing magnetic forces are balanced. A compass needle put at a neutral point will maintain any position in which it is set.

neutral position Of an electrical commutating machine. The position of the brushes (*see* brush shift)

that gives the same speed at the same load for either direction of rotation. It agrees closely with the position for which the mutual induction between the armature and field windings is zero.

 Neutral surface of bent beam

neutral surface If a beam is bent so that its axis becomes a curve lying in one plane, the longitudinal fibres of the beam closest to the axis of bending will be compressed and those farther away will be extended. There will, between these, be a layer of fibres whose length is unchanged and these define the position of the neutral surface. The line of intersection of any cross section with the neutral surface is its *neutral axis*. (Some writers give the name neutral axis to the central longitudinal fibre of the neutral surface.)

neutral temperature For a *thermocouple with one junction maintained at 0 °C, the temperature θ of the hot junction causes the e.m.f. E to vary according to the formula:

$$E = \alpha\theta^2 + \beta\theta,$$

where α, β are constants. The maximum value of E occurs when $\theta = -\beta/2\alpha$; this is called the neutral temperature. It is usual to restrict the use of a thermocouple to the range between 0 °C and its neutral temperature.

neutral zone Of an electrical commutating machine. The zone of the commutator in which the voltage between adjacent commutator bars is substantially zero when the machine is running at no-load.

neutrino A neutral *elementary particle with *spin $\frac{1}{2}$ that only takes part in *weak interactions. Neutrinos are *fermions and are classified as *leptons. They are generally regarded as stable, but it has been suggested that a neutrino might change into a system of neutrinos and antineutrinos. The rest mass is believed to be zero, so according to the theory of *relativity the particle must move with the speed of light with respect to any observer. A neutrino that transfers mass m also transfers energy mc^2 and momentum mc. Three kinds of neutrino are known, one (ν_e) is associated with the *electron, another (ν_μ) is associated with the *muon, and the third (ν_τ) is associated with the massive *tauon. The antiparticle of the neutrino is the *antineutrino* $(\bar{\nu})$.

The neutrino was first postulated (*Pauli, 1930) to explain the continous spectrum of *beta rays. It is assumed that there is the same amount of energy available for each *beta decay of a particular nuclide and that this energy is shared according to a statistical law between the electron and a light neutral particle (now classified as the antineutrino, $\bar{\nu}_e$). Later it was shown that the postulated particle would also conserve angular momentum and linear momentum in beta decays.

In addition to beta decay, the electron neutrino is also associated with, for example, *positron decay and *electron capture:

$$^{22}Na \rightarrow \ ^{22}Ne + e^+ + \nu_e,$$
$$^{55}Fe + e^- \rightarrow \ ^{55}Mn + \nu_e.$$

The absorption of antineutrinos in matter by the process

$$^1H + \bar{\nu}_e \rightarrow \ ^1n + e^+$$

was first demonstrated by Reines and Cowan. The muon neutrino is generated in such processes as:

$$\pi^+ \rightarrow \mu^+ + \nu_\mu,$$
$$\mu^- \rightarrow \bar{e} + \nu_\mu + \bar{\nu}_e.$$

Although the interactions of neutrinos are extremely weak the cross sections increase with energy and reactions can be studied at the enormous energies available with modern *accelerators. In some forms of *grand unified theories, neutrinos are predicted to have a nonzero mass, although no evidence has been found to support this prediction.

neutron An elementary particle with zero charge and a rest mass equal to:

$$1.674\,929 \ \times \ 10^{-27} \ kg$$
$$\text{i.e. } 939.200\,3 \ MeV/c^2.$$

It is a constituent of every atomic *nucleus except that of ordinary hydrogen. Free neutrons decay by *beta decay with a mean life of 914 s. The neutron has *spin ½, *isospin ½, and positive *parity. It is a *fermion and is classified as a *hadron as it has *strong interactions. *See* quark.

Neutrons can be ejected from nuclei by high-energy particles or photons; the energy required is usually about 8 MeV, although sometimes it is less, and it is lowest in the cases of ^9_4Be (1.6 MeV) and ^2_1H (2.2 MeV). *Fission is the most productive source. They are detected using all normal detectors of ionizing radiation as a result of the production of secondary particles in *nuclear reactions, for example $^{10}_5B(n, \alpha) \ ^7_3Li$ and $^6_3Li(n, \alpha) \ ^3_1H$. The discovery of the neutron (1932) by *Chadwick involved the detection of the tracks of protons ejected by neutrons by elastic collisions in hydrogenous materials.

Unlike other nuclear particles, neutrons are not repelled by the electric charge of a nucleus so they are very effective in causing nuclear reactions. When there is no *threshold energy, the interaction *cross sections become very large at low neutron energies, and the *thermal neutrons produced in great numbers by *nuclear reactors cause nuclear reactions on a large scale. The capture of neutrons by the (n, γ) process produces large quantities of radioactive materials, both useful nuclides such as ^{60}Co for cancer therapy and undesirable by-products.

neutron activation analysis (NAA) *See* activation analysis.

neutron age One-sixth of the mean square displacement of a neutron as it slows down, through a specified energy range, in an infinite homogeneous medium. *See* Fermi age theory.

neutron diffraction A technique for determining the crystal structure of solids by diffracting of a beam of neutrons. It is similar in principle to *electron diffraction and can be used in place of *X-ray crystallography. The wavelength of a neutron is related to its velocity by the *de Broglie equation, a neutron with a velocity of about $4 \times 10^3 \ m\,s^{-1}$ having a wavelength of about 10^{-10} m. The technique of neutron diffraction is more difficult than that of X-ray crystallography because of the problems of producing beams of neutrons and measuring their intensity. However, it does have certain advantages.

One use of neutron diffraction is in the study of antiferromagnetic materials. The interaction between magnetic moments of atoms in the crystal and the moments of the neutrons allows the magnetic unit cell to be detected (*see* antiferromagnetism).

Another use of neutron diffraction is in the study of crystals containing hydrogen atoms. Hydrogen atoms scatter X-rays weakly and the contribution that they make to an X-ray diffraction pattern is often masked by the effects of heavier atoms. Neutrons, on the other hand, are strongly scattered by hydrogen atoms and can be used to determine their position.

neutron excess *See* isotopic number.

neutron flux *Syn.* neutron flux density. The product of the number of free neutrons per unit volume and their mean speed. The neutron flux in a *power reactor lies in the range $10^{16}-10^{18}$ per square metre per second.

neutron number Symbol: N. The number of neutrons present in the nucleus of an atom. The neutron number is obtained by subtracting the *atomic number from the *mass number.

neutron star A star that, having exhausted its nuclear sources of energy, has undergone *gravitational collapse. It reaches a stage of electron *degeneracy, but has sufficient mass (> 1.4 solar masses) for further contraction to occur. When the density exceeds 10^7 kg m^{-3} equilibrium between protons, electrons, and neutrons shifts in favour of the neutrons until at densities of 5×10^{10} kg m^{-3} 90% of the protons and electrons have interacted to form neutrons. If the mass of the star is less than 2.0 solar masses, strong repulsive forces between neutrons are set up causing a rapid rise in pressure. Contraction is halted and a stable neutron star is formed. It has a diameter of only 20 to 30 km. *Pulsars are almost certainly rotating neutron stars.

neutron temperature A temperature T defined by the equation $E = ^3/_2 kT$. E is the energy possessed by neutrons in thermal equilibrium with their surroundings, and k is the *Boltzmann constant. The formula is analogous to the *kinetic theory energy-temperature relationship for gases. *See also* thermal neutrons.

new achromats *See* achromatic lens; optical glass.

new candela *See* candela.

newton Symbol: N. The *SI unit of *force, defined as the force that provides a mass of one kilogram with an acceleration of one metre per second per second.

Newton, Sir Isaac (1642–1727) Brit. mathematician and physicist. He consolidated the ideas on mechanics developed by Galileo and others, and, by taking the bold step of extending the idea of gravitational attraction to cover the whole universe, was able to include celestial mechanics. This great systemization of knowledge was published in the *Philosophiae Naturalis Principia Mathematica* (usually called simply the *Principia*) in 1687, thanks to Edmund Halley. He invented the infinitesimal calculus but his notation (e.g. $\dot{x}$ for dx/dt) was generally inferior to that introduced by Leibnitz and the unfortunate controversy on priority of discovery of the calculus, which clouded much of Newton's life, delayed adoption in England of the more practical notation.

In optics, Newton made a celebrated investigation of the formation of a spectrum by a prism and he investigated the mixing of colours. His false conclusion that dispersion would be proportional to refraction in all media, which would have rendered impossible the construction of achromatic lenses, led him to make a *reflecting telescope (1672). He was familiar with phenomena of interference and diffraction but attributed them to the behaviour of corpuscles of light, which he believed to have a kind of *polarity* (whence the *polarization* of light) and to have fits of easy transmission and of easy reflection when encountering a bounding surface.

Newton's rings are circular interference fringes formed between a lens and a glass plate with which the lens is in contact. Their discovery and study is closely associated with Hooke, who worked with Newton for some time (*see* interference of light).

Newton's *Optiks* appeared in 1704.

Newton also derived a formula for the *speed of sound, which was later corrected by Laplace.

Newtonian fluid A fluid in which the amount of strain is proportional to the product of the stress and the time (Newton, 1685). The constant of proportionality is known as the coefficient of viscosity. *See* viscosity; anomalous viscosity.

Newtonian force *Syn.* Coulomb force. A force between points that falls off as the inverse square of the distance between them. *Compare* Coulomb's law.

Newtonian frame of reference *See* Newtonian system.

Newtonian mechanics *See* Newton's laws of motion.

Newtonian potential A potential that varies with distance in the same way as gravitational potential, i.e. any potential associated with an inverse square law of force (e.g. electrostatic and magnetostatic potentials).

Newtonian system *Syn.* Newtonian frame of reference. Any frame of reference relative to which a particle of mass m, subject to a force F, moves in accordance with the equation $F = kma$ where a is the acceleration of the particle, m its mass, and k a universal positive constant equal to unity in *SI units.

Such a frame of reference is one in which the centre of mass of the solar system is fixed, and does not rotate relative to the fixed stars. Any other frame of reference that moves relative to this with a uniform velocity is also a Newtonian system; this is the classical principle of *relativity. For non-Newtonian systems, e.g. a frame of reference fixed to the earth's surface, Newton's equation $F = kma$ can be made to hold if the fictitious inertial *forces are added to F.

Newtonian telescope *See* reflecting telescope.

Newton's formula (for a lens) The distances p and q between two conjugate points and their respective foci are related by $pq = f^2$ (with a suitable sign convention); f is the focal length of the lens. For a mirror, the foci coincide but the relationship is unaltered (except for sign in some conventions).

Newton's law of cooling The rate of loss of heat from a body is proportional to the excess temperature of the body over the temperature of its surroundings. Strictly the law applies only if there is *forced convection, but provided the temperature excess is small, the law is fairly well obeyed, even in the case of free or natural convection.

Newton's law of gravitation *See* gravitation.

Newton's laws of fluid friction 1. The force of resistance, D, opposing the relative motion of a body and a fluid is given by $k_0 A V^2 \rho$, where V is the relative velocity, ρ the density of fluid, A some projected area of the body, and k_0 a constant of proportionality (*compare* drag coefficient). The law was formulated from considerations of the change of momentum in the direction of relative motion.

2. The shearing force between two infinitesimal layers of viscous fluid is proportional to the rate of shear in a direction perpendicular to the direction of motion of the layers. This force is expressed:
$$F = \eta \partial u / \partial y,$$
where u is the velocity in the direction of motion and y is perpendicular to the direction of u. η is called the coefficient of *viscosity, and in classical hydrodynamics is a factor peculiar to the molecular nature of the fluid alone.

In the limited range of conditions studied by Newton, law **1** was found to apply to motion in very fluid media such as air and water, while law **2** applied to very viscous liquids. Later research showed that for low values of the *Reynolds number, law **1** applies for all fluids, while in all cases law **2** applies for high values.

Newton's laws of motion In his *Principia* Newton (1687) stated the three fundamental laws of motion, which are the basis of *Newtonian mechanics*.

Law I. Every body perseveres in its state of rest, or uniform motion in a straight line, except in so far as

385

it is compelled to change that state by forces impressed on it. This may be regarded as a definition of force.

Law II. The rate of change of linear momentum is proportional to the force applied, and takes place in the straight line in which that force acts. This definition can be regarded as formulating a suitable way by which forces may be measured, that is, by the acceleration they produce,

$$F = \mathrm{d}(mv)/\mathrm{d}t$$
i.e. $$F = ma + v(\mathrm{d}m/\mathrm{d}t),$$

where F = force, m = mass, v = velocity, t = time, and a = acceleration. In the majority of nonrelativistic cases, $\mathrm{d}m/\mathrm{d}t = 0$ (i.e. the mass remains constant), and then:

$$F = ma.$$

Law III. Forces are caused by the interactions of pairs of bodies. The force exerted by A upon B and the force exerted by B upon A are: simultaneous; equal in magnitude; opposite in direction; in the same straight line; caused by the same mechanism.

Note: the popular statement of this law in terms of "action and reaction" leads to much misunderstanding. In particular: any two forces that happen to be equal and opposite are supposed to be related by this law, even if they act on the same body; one force, arbitrarily called "reaction", is supposed to be a consequence of the other and to happen subsequently; the two forces are supposed to oppose each other, causing equilibrium; certain forces such as forces exerted by supports or propellants are conventionally called "reactions", causing considerable confusion.

The third law may be illustrated by the following examples. The gravitational force exerted by a body on the earth is equal and opposite to the gravitational force exerted by the earth on the body. The intermolecular repulsive force exerted on the ground by a body resting on it, or hitting it, is equal and opposite to the intermolecular repulsive force exerted on the body by the ground. (Note: this law tells us nothing about the relationship between the gravitational force and the intermolecular repulsions, contrary to popular belief.)

It is sometimes noted that the force supposed to be exerted magnetically by one current element upon another is not equal and opposite to that supposed to be exerted by the second upon the first. There is in fact no violation of Newton's law in the interaction of two circuits considered as whole bodies; the paradox arises because an isolated current element cannot exist.

The unit of force as defined by *Law II* is that which acting on unit mass, produces unit acceleration. In *SI units this force is called the *newton.

The weight W of a body is given by $W = mg$, where m is its mass and g is the acceleration of *free fall.

The laws of motion applied to a system of particles (and in this connection a rigid body is a system of particles in which internal forces maintain the relative positions of the particles) show that: (*a*) The rate of change of momentum in any direction equals the resolved part of the external forces in that direction. (*b*) The mass centre of a system moves like a particle of mass equal to the total mass of the system acted on by a force equal to the *vector sum of all the external forces acting on the system. (*c*) The rate of change of moment of momentum of the system, about a point that is either fixed, or moving with the mass centre, equals the total moment of the external forces about that point. (*d*) The motion of the mass centre and the motions of the system relative to the mass centre may be treated independently, e.g. the kinetic energy of the system is $\frac{1}{2}Mv^2$ + kinetic energy of motion about the mass centre (where M = total mass, and v = velocity of the centre of mass).

A more general system of mechanics has been given by Einstein in his theory of *relativity. This reduces to Newtonian mechanics when all speed relative to the observer are small compared with those of light.

Because of certain difficulties with the concept of simultaneity in relativity theory, Newton's third law does not always apply exactly. Nevertheless, the law of conservation of momentum does still apply in relativity.

Newton's rings *See* interference of light.

Nicholson, William (1753–1815) Brit. physicist. He constructed, with Sir Anthony Carlisle, the first voltaic pile in England and with it decomposed water by electrolysis (1800). He invented the Nicholson *hydrometer. He was editor of *Nicholson's Journal of Natural Philosophy, Chemistry and the Arts*, at one time an important periodical.

Nicholson hydrometer *See* hydrometer.

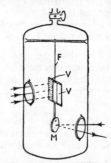
Nichols's vane radiometer

Nichols's vane radiometer A sensitive modification of the *Crookes radiometer. Two mica vanes V with one side blackened are suspended from a torsion fibre F in a glass envelope containing air at low pressure. Radiation entering a fluorite window falls on one vane causing an angular displacement of the suspended system that is measured by the deflection of a beam of light reflected from a small mirror M carried by the system.

Nichrome A heat-resistant alloy with high resistivity that is used in electrical heating elements and resistors. The composition varies but is approximately 62% Ni, 15% Cr, and 23% Fe.

nickel-iron accumulator *See* Edison accumulator; accumulator.

Nicol, William (1768–1851) Brit. physicist. He made contributions to the techniques associated with the microscope and invented the Nicol prism.

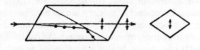

Nicol prism

Nicol prism A prism made of calcite once widely used for polarizing light and analysing *plane-polarized light. The crystal is cut in a special direction, sliced, and recemented together with Canada balsam. Light entering one face undergoes *double refraction; the extraordinary ray passes straight forward through the balsam, whereas the ordinary ray is reflected to the lower face where it is absorbed by a black coating (*see* diagram). The emerging extraordinary ray is plane-polarized with vibrations parallel to the short diagonal of the rhomb-shaped section as viewed from the emergent face.

The Nicol prism has been largely superseded by more effective polarizers. One example is the *Wollaston prism. Two others are the *Glan–Foucault* (or *Glan-air*) *prism* and the *Glan–Thompson prism*. Both are composed of two calcite halves but the former has a film of air between the sections while the latter is cemented. The beam strikes the surface normally, passes undeviated through the first half and is split at the interface. The Glan–Thompson has the wider angular aperture of the two, but the Glan-air, being uncemented, can handle high-power laser radiation and transmits the broad spectral range of calcite (about 5000 to 230 nm).

NiFe accumulator *See* Edison accumulator.

night glasses *See* refracting telescope.

nile A unit used to measure the departure of a *nuclear reactor from its critical condition. 1 nile corresponds to an *activity of 0.01.

nit Symbol: nt. A unit of *luminance, defined as one candela per metre squared. The unit is identical with the *SI unit of luminance, but the name has not received international recognition.

NMR Abbreviation for *nuclear magnetic resonance.

Nobel, Alfred Bernhard (1833–1896) Swedish chemist and engineer. The inventor of dynamite, he became very wealthy and bequeathed a fortune to found the *Nobel prizes*, normally awarded annually for contributions to the cause of peace, and in the fields of literature, of medicine or physiology, for chemistry, and for physics. A prizewinner is a *Nobel Laureate*. The first awards were made in 1901. See list of prizewinners for physics in Appendix, Table 14.

Nobili, Leopoldo (1784–1835) Italian physicist who invented the thermopile and the astatic galvanometer.

nodal line *See* Eulerian angles.

nodal points Two axial and conjugate points of a system: a ray from the object space passing through the first or anterior nodal point emerges in the image space in a parallel direction through the second or posterior nodal point, i.e. they are conjugate points with unit angular magnification. When the object and image spaces are the same the nodal and principal points coincide. For a single surface, the two nodal points coincide at the centre of curvature. It will be found that in general $FH = N'F'$ and $FN = H'F'$. (*See* centred optical system.)

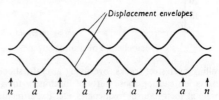

Nodes and antinodes in a vibrating string

node 1. A point or region in a *standing wave at which some characteristic of the wave motion, such as particle displacement, particle velocity, or pressure amplitude, has a minimum (or zero) value. An illustration of displacement nodes is readily given by a vibrating string plucked at a particular point along its length. Positions marked n in the illustration are positions at which the string vibrates with minimum amplitude and are therefore nodes; at those marked a the string has maximum amplitude and these are *antinodes*. For standing waves in an air column, displacement nodes are pressure antinodes and conversely. Generally, any kind of wave has two types of disturbance and the nodes for one type are the antinodes for the other.

2. In electricity or electronics. A point at which the current or voltage has a minimum value.

3. *See* network (defs. 1, 2).

4. In astronomy. Two points at which the orbit of a celestial body intersects the ecliptic (*see* celestial sphere).

node voltage Of a *network. The voltage of some point in the network with respect to a *node.

noise 1. (general) Sound that is undesired by the recipient. Although noise is usually a discordant sound that is neither music nor speech, this definition includes any sound that, though desired by one person, constitutes a noise to another. A common source of noise is machinery in which some of the rotating parts are slightly out of balance. Other everyday sources are explosive noises from the exhausts of petrol engines, noises caused by trains

noise abatement

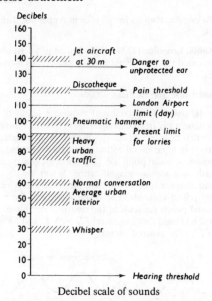

Decibels

160	
150	
140	Jet aircraft at 30 m — Danger to unprotected ear
130	
120	Discotheque — Pain threshold
110	London Airport limit (day)
100	Pneumatic hammer
90	Heavy — Present limit for lorries
80	urban traffic
70	
60	Normal conversation
50	Average urban interior
40	
30	Whisper
20	
10	
0	Hearing threshold

Decibel scale of sounds

rolling over steel tracks, overflying aircraft, shock excitation causing loose parts of machinery to vibrate at their natural frequencies, etc. Experiments have been carried out to determine whether noise has any effect on the working capacity of a person subjected to it. It is generally found that the worker soon becomes accustomed to noise and ceases to be aware of it. For routine tasks it has little effect on working capacity but the worker's oxygen intake is increased. When considerable mental effort is required, it has an adverse effect on output and in some cases on the worker's resistance to disease also. Physical injury to the ear is rarely caused unless a person is exposed to a very intense noise level for considerable periods. Noise impairs the value of sleep even though it is not loud enough to wake the sleeper. *Masking of sound is caused by noise and this is a problem in radio and telephone communication. The frequency components of a noise can be obtained with a sound analyser but it is usually only necessary to know its loudness. This is expressed either in decibels or phons and is measured with an *audiometer or noise-level meter. The diagram shows various sounds and noises on a decibel scale.

2. In telecommunication engineering. Spurious unwanted energy (or the associated voltage) in an electronic or communications system. *Interference often produces noise, but not always (see crosstalk; hum). There are two main types of noise, *white noise* and *impulse noise*.

(i) White noise has a wide frequency spectrum. It is caused by various sources, the most common being *thermal noise* and *random noise*. Thermal noise is due to the thermodynamic interchange of energy in a material or between a material and its surroundings. Random noise is due to any random transient disturbances. White noise in a communications sys-

tem gives rise to loudspeaker hiss or television-screen snow. (ii) Impulse noise is due to a single momentary disturbance or a number of such disturbances when they are separated from one another in time. In audiofrequency amplifiers this type of noise gives rise to clicks in the loudspeaker. *See also* Schottky noise; Jansky noise; signal-to-noise ratio; noise factor.

noise abatement Noise (or acoustic pollution) is a natural byproduct of modern life with its increasing use of machinery, and many attempts have been made to reduce unnecessary noise. Whilst the exact effects of noise on health are not quite agreed, it is definitely known that it wears down the nervous system and reduces the beneficial effects of sleep. The necessity for its reduction is greatest in large cities, where sounds must be magnified perhaps a million times above the threshold of hearing before they become audible. Several legal steps have been taken to reduce noise. Motor vehicles must have silencers fitted to their exhausts and must comply with a specified noise level; excessive noise due to a vehicle itself or its load is prohibited. Motor horns must not be sounded at night in built-up areas, and the output from radios, televisions, and record players must be kept within reasonable limits. Special attention has been paid to the problem of noise abatement in underground tube trains. In the confined space the sound reverberates and is much louder than in the open air. Some improvement has been achieved by the use of longer rails, lagging of the walls and roof, and by putting ballast between the sleepers.

In passenger aircraft it has been found necessary to soundproof the cabins by lining them with special sound-absorbing material. However, the problem of noise from overflying aircraft (especially at take-off and landing) is one of the most prominent in noise abatement. This problem has been accentuated by the advent of supersonic transport.

By careful design and the use of sound insulators, noise can be considerably reduced in buildings. In offices it is best cut down at the source by employing quiet office machinery, but if this is not possible, sound insulators can be used or the noise absorbed by special materials. Problems of noise abatement are now receiving serious attention in all the main cities of the world.

noise factor A measure of the *noise introduced into a circuit or device. It is defined as the ratio of the *signal-to-noise ratio of the power at the source to the corresponding ratio at the output, i.e. the ratio of the actual noise at the output to the noise at the output due only to the source.

noise level A measure of the loudness of a noise. The units in which it is expressed are either *decibels, which indicate the intensity level, or *phons, which measure equivalent loudness. These two units are approximately equal over a large part of the audible range. Some typical noise levels are shown in the diagram under noise (general).

Instruments for measuring noise intensity can be divided into two classes – subjective and objective. *Subjective audiometers* produce a source of sound of known intensity having the desired frequency or band of frequencies. The observer then has to judge equality of loudness between the noise and the sound from the audiometer. This is often difficult if there is much difference of frequency between the two, but with care a reliable indication of equivalent loudness can be obtained. Another method of using this instrument is to vary the sound in intensity until it just masks the noise. (*See* masking.) *Objective noise meters* eliminate the individual judgment and measure the noise level directly. These usually consist of a microphone, amplifier, and output meter (*see* audiometer). Such a system must be carefully designed to give reliable readings. The frequency response is either made linear or similar to that of the ear. In the first case, intensity level is measured and in the latter, equivalent loudness can be read direct. It is usually possible to measure the noise level either at one particular frequency or within selected bands of frequencies. Resonance in any part of the system should be corrected for, and the instrument is calibrated against a standard.

noise meter *See* audiometer; noise level.

no-load Operation of any electrical or electronic circuit, device, machine, etc., under rated operating conditions (*see* rating) of voltage, speed, etc., but in the absence of a *load.

nomography A graphical method, based on collinear points, whose object is to solve all equations of a given type by means of one diagram called a nomogram. It was developed by d'Ocagne (1884) and is widely used for exhibiting the relation between three variables. Three scales, not necessarily straight or uniformly spaced, one for each variable, are drawn on a plane in such a way that a straight edge passing through given values of two of the variables intersects the third scale at the appropriate value.

non-Abelian group *See* group.

nondegeneracy The normal state of matter, i.e. matter not cooled to a very low temperature nor subject to excessive stress such that the density is abnormally high. A nondegenerate gas is characterized by Maxwell's law of the *distribution of velocities.

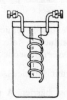

Non-inductive
resistance

noninductive An electric circuit or winding in which, for the purpose in view, the effect of its inductance is negligible. Coils are often wound of bifilar wire or wire doubled back on itself so that the return wire neutralizes the inductive effect of the forward wire (*see* diagram). A circuit completely devoid of inductance is very difficult to obtain. *Compare* inductive.

nonlinear distortion *See* distortion (electrical).

nonlinear network *See* network.

nonlinear optics The study of the effects produced by the electric and magnetic fields of extremely intense beams of light, such as focused *laser beams. The usual classical treatment of the propagation of light – reflection, refraction, superposition, etc. – assumes a linear relationship between the electromagnetic light field and the response of the atomic system constituting the medium. The electric field associated with a focused beam of a high-power laser can be 10^8 volts per metre or more, which is sufficient to generate appreciable nonlinear optical effects.

One effect is the *self-focusing* of laser light. The passage of an intense beam through glass induces local variations in the refractive index of the glass, causing it to act as a converging lens. The beam therefore contracts, its intensity increases, and the contraction process continues. The effect can be sustained until the beam diameter is about 5 μm, when it becomes totally internally reflected.

Frequency mixing is another effect. If two intense laser beams of different frequencies are sent through a suitable dielectric crystal, the nonlinear effects can result in the production of radiation whose frequency is equal to the sum or the difference of the original frequencies.

non-Newtonian fluid *See* anomalous viscosity.

non-perturbative phenomenon *See* perturbation theory.

nonreactive An electric circuit or winding in which, for the purpose in view, the *reactance is negligible.

nonreactive load A *load in which the alternating current is in phase with the terminal voltage. *Compare* reactive load.

nonstoichiometric crystal *See* defect.

nonthermal radiation Electromagnetic radiation, such as *synchrotron radiation, that is produced by the acceleration of electrons or other particles but is nonthermal in origin, i.e. it is not *black-body radiation.

nonvortical field *See* curl.

NOR gate *See* logic circuit.

normal 1. To a surface. The normal to a plane surface at a point is a line perpendicular to the plane and passing through the point. The normal to any surface at a point is the line perpendicular to the tangent plane at that point.

2. To a curve in space. There is an infinite number of normals (i.e. lines perpendicular to the tangent) to a curve at a point. The *principal normal* to a curve lies in the *osculating plane* at the point, and the *binormal*

normal distribution

is perpendicular to this plane. The osculating plane is that plane containing three consecutive points on the curve that are infinitely close to one another.

normal distribution *Syn.* Gaussian distribution. *See* frequency distribution.

normal equations A set of linear equations derived from the *observational equations* and *equations of condition* to be solved in connection with the method of *least squares.

normalization The process of introducing a numerical factor into an equation $y = f(x)$ (in which $y \to 0$ as $x \to \pm \infty$) so that the area under the corresponding graph (if finite) shall be made equal to unity. The process is of importance (*a*) in quantum mechanics (where an extended definition is applicable) and (*b*) in statistics, where the total area under the error equation graph represents the probability of a value of x lying between $+ \infty$ and $- \infty$ and must be 1. (*See* frequency distribution, where $h\pi^{-1/2}$ is the *normalizing factor*.) *See also* renormalization.

normal stress *See* stress.

north polar distance *See* celestial sphere.

nose suspension A method of mounting an electric traction motor. One side of the motor frame is axle hung, i.e. is supported by special bearings on the axle. The other side is supported by a nose or lug that projects from the motor case and rests upon the framework of the truck. As in the case of *bar suspension, this method is suitable for use with geared drive.

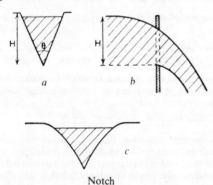

Notch

notch A device for measuring the flow in a channel, usually being of an overflow structure. Rectangular sharp-crested weirs and a variety of nonrectangular weirs are called notches. The most important type of notch is the V-notch (Figs. *a* and *b*), its principal feature being that whatever the level of the reservoir, the stream sections through the notch are similar. The discharge (or quantity of flow per second) is given by $Q = CH^{5/2}$, where H is the head of fluid and C is a quantity depending on the angle of the notch, the stream velocity, and on the head. It has been shown that for a 90° notch $Q = 2.52 \, H^{2.47}$.

The formula $Q = CH^n$ can be used for various values of n greater than one to provide various shapes of notches. For example, if $n = 3/2$ the shape of notch is rectangular; if $n = 2$, the shape is parabolic with axis vertical; if $n > 5/2$, the notch is convex upwards (Figs. *a* and *c*).

note 1. A musical sound of specified pitch (frequency) produced by a musical instrument, voice, etc.

2. A sign in musical score representing pitch and time value of a musical sound.

3. A key of a pianoforte, or other keyboard instrument.

American usage restricts the term "note" to the second definition. *See* tone.

note frequency *Syn.* beat frequency. *See* beats.

NOT gate *See* logic circuit.

nova A faint *variable star that can undergo a considerable explosion during which the *luminosity increases by up to 100 000 times. The peak luminosity is attained in a few hours or days, then a slow decrease occurs over months or years until the original luminosity is reached.

A nova occurs in a *binary star system where the two components are very close. One component is a *white dwarf. The other star is expanding and is losing mass (hydrogen) to the white dwarf, around which a disc of gas forms. The gas can spiral down to the surface of the white dwarf, and after some 10 000 to 100 000 years enough has accumulated to cause a thermonuclear explosion – the nova outburst. The explosion does not disrupt the binary system, however, and the flow of gas from the other star can continue. *Compare* supernova.

n-p-n transistor *See* transistor.

NQR Abbreviation for *nuclear quadrupole resonance.

NTP Abbreviation for normal temperature and pressure (no longer used). *See* STP.

NTSC National Television Standards Committee; the name for the *colour television system used in America.

n-type conductivity Conductivity in a *semiconductor caused by a flow of *electrons, as opposed to *p-type conductivity, which is caused by a flow of *holes.

n-type semiconductor An *extrinsic semiconductor in which the density of conduction *electrons exceeds that of mobile *holes. *See also* semiconductor.

nuclear Relating to an atomic *nucleus.

nuclear barrier *Syn.* Gamow barrier. A region of high potential energy that a charged particle must pass through in order to enter or leave an atomic nucleus.

nuclear cross section *See* cross section.

nuclear energy Energy released by nuclear reactions, in particular *nuclear fusion and *fission. *See* binding energy; fusion reactor; nuclear reactors.

nuclear energy change *See* Q-value.

nuclear fission *See* fission.

nuclear fusion A *nuclear reaction between light atomic nuclei with the release of energy (*see* binding energy). Such reactions can be caused in the laboratory using an *accelerator to produce beams of deuterons or other light nuclei to bombard suitable targets. In such processes the energy required is incomparably greater than that released. To give a net energy output it is necessary to use *thermonuclear reactions*, i.e. those reactions that occur in a *plasma at very high temperatures.

The greater part of *stellar energy is released by thermonuclear reactions. The processes involve *weak interactions and are only possible because of the great quantity of matter involved. The development of fusion energy on earth requires *strong interactions using different substances from those primarily involved in stars. This was first achieved in 1952 in the *hydrogen bomb.

Since the 1950s considerable efforts have been made to produce a controlled thermonuclear reaction. The central problem is to contain a high-temperature ionized gas (or *plasma, as it is called) as no solid material can exist at these temperatures. For this purpose most current research uses magnetic fields to confine the plasma and various configurations of *magnetic bottles have been tried (*see* fusion reactors).

The principal fusion reactions that are likely to be used in any future fusion reactors are:

^{2}H + ^{2}H → ^{3}He + n $\quad$ + 3.2 MeV
^{2}H + ^{2}H → ^{3}H + ^{1}H $\quad$ + 4.0 MeV
^{2}H + ^{3}H → ^{4}He + n $\quad$ + 17.6 MeV
^{2}H + ^{3}He → ^{4}He + ^{1}H $\quad$ + 18.3 MeV
^{6}Li + ^{1}H → ^{3}He + ^{4}He $\quad$ + 4.0 MeV
^{6}Li + ^{3}He → ^{4}He + ^{4}He
$\qquad\qquad$ + ^{1}H $\quad$ + 16.9 MeV
^{6}Li + ^{2}H → ^{7}Li + ^{1}H $\quad$ + 5.0 MeV
^{6}Li + ^{2}H → ^{3}He + ^{4}He
$\qquad\qquad$ + n $\quad$ + 2.6 MeV
^{6}Li + ^{2}H → 2^4He $\quad$ + 22.4 MeV
^{7}Li + ^{1}H → 2^4He $\quad$ + 17.5 MeV

In all of these processes in which neutrons are produced, there will be extra energy released by the interactions of these particles in surrounding substances. Typically 8 MeV is released, usually by means of gamma radiation, on capturing a neutron.

Although deuterium is only present as one atom in 6000 in hydrogen the total quantity available is very great. Tritium can be produced from lithium by the reaction

$$^6Li(n, \alpha)^3H$$

in a *nuclear reactor.

nuclear heat of reaction *See* Q-value.

nuclear isomerism When nuclei exist with the same mass number and atomic number (*see* isobar; iso-

topes) but show different radioactive properties, they are said to be nuclear isomers. They represent different energy states of the nucleus.

nuclear magnetic resonance (NMR) An effect observed when radio-frequency radiation is absorbed by matter. A nucleus with a *spin has a nuclear *magnetic moment. In the presence of an external magnetic field this magnetic moment precesses (*see* precession) about the field direction. Only certain orientations of the magnetic moment are allowed and each of these has a slightly different energy.

If the nucleus has a spin I there are $2I + 1$ different quantum states due to this quantization; each is characterized by a different value of the magnetic quantum number m, which can have values:

$$I, I - 1, \ldots -(I - 1), -I.$$

The difference in energy depends on the strength of the applied magnetic field. The nucleus can make a transition from one state to another with the emission or absorption of electromagnetic radiation according to a selection rule that $\Delta m = \pm 1$.

The technique used is to apply a strong magnetic field (~ 2 tesla) to the sample, which is usually a liquid or solid. This field can be controlled by a current through two small coils fixed to the pole pieces of the magnet. A radio-frequency field (1–100 MHz) is imposed at right angles and a small detector coil is wound around the sample. As the magnetic field is varied, the spacing of the energies changes and at a certain value of the magnetic field, this spacing is such that radio-frequency radiation is strongly absorbed. This *resonance produces a signal in the detector coil. A plot of the detected signal against the magnetic field gives an NMR spectrum that can be used for determining nuclear magnetic moments. The energies of the nuclei depend to some extent on the surrounding orbital electrons and this is useful in studying chemical compounds. For example, the NMR spectrum of the hydrogen nuclei in methanol, CH_3OH, shows two peaks, one corresponding to the changes in energy of the hydrogen nuclei (protons) attached to the carbon and the other to the proton attached to the oxygen. The difference between the position of a peak and a standard position is a *chemical shift*. The technique is widely used in chemistry. It is also now used in medicine to construct images showing the variation of proton density in organs such as the heart, breasts, and lungs.

See also electron spin resonance.

nuclear magneton *See* magneton.

nuclear medicine The branch of medicine concerned with the application of radioactive nuclei in diagnosis and therapy. The storage of the radioactive materials, instruments for the measurement and visualization of distributions of activity in a patient, interpretation of the results, and the production of suitable compounds for use are interrelated problems calling for cooperation between clinicians and physicists.

nuclear quadrupole resonance

Radioisotopes are used both for diagnosis and treatment of disease, especially cancer. A radioisotope will follow the same path inside the body as a nonradioactive, normally ingested isotope of the same element, and will accumulate in the same areas. Measurement of the radioactivity at these areas will indicate any abnormal activity in the body. A high level of radioactivity means that there are overactive cancer cells present. The most common uses of radioisotopes are the diagnosis and treatment of thyroid disorders, kidney studies, and liver studies.

Equipment commonly used includes *scintillation counters, *Geiger counters, *scanners, and *gamma cameras, often with computer analysis of the outputs of this equipment.

Radiotherapy is concerned with the use of beams of *ionizing radiation, such as X-rays, energetic electrons, and the stream of gamma rays from the radionuclide cobalt-60, in the treatment of cancer. A sufficient amount of radiation must be given to kill the cancer cells without harming intervening tissue. This is achieved by irradiating from several different directions with a narrow beam so that the tumour receives the maximum dose. The direction, dose, and length of irradiation are determined from knowledge of the site and size of the tumour.

In *nuclear cardiology*, radionuclides are used in the study and diagnosis of heart disease. An intravenous injection of a radionuclide that is a gamma emitter enables a gamma camera and computer to construct an image of the heart. For example, in a *MUGA scan* (multiple-gated arteriography), the patient's red cells are injected with technetium-99 to form an image of the blood pool within the heart at specified points in the cardiac cycle.

nuclear quadrupole resonance (NQR) An effect similar to *nuclear magnetic resonance, but applied to the quadrupole moments of nonspherical atomic nuclei (e.g. ^{35}Cl). Nuclear quadrupole resonance techniques have been applied particularly to the study of molecular crystals.

nuclear reaction A reaction between an atomic *nucleus and a bombarding particle or photon leading to the creation of a new nucleus and the possible ejection of one or more particles. Nuclear reactions are often represented by enclosing within brackets the symbols for the incoming and outgoing particles or quanta, the initial and final nuclides being shown outside the brackets. For example:

$$^{14}N(\alpha, p)^{17}O$$

represents the reaction:

$$^{14}_{7}N + {}^{4}_{2}He = {}^{17}_{8}O + {}^{1}_{1}H.$$

nuclear reactors 1. *Energy from nuclear fission*. On the whole, the nuclei of atoms of moderate size are more tightly held together than the largest nuclei, so that if the nucleus of a heavy atom can be induced to split into two nuclei of moderate mass, there should be a considerable release of energy (*see* binding energy). The uranium isotope ^{235}U will readily accept a neutron but the nucleus ^{236}U so formed is

very unstable and one-seventh of the nuclei stabilize by gamma emission while six-sevenths split into two parts (*see* fission). Most of the energy released (about 170 MeV) is in the form of the kinetic energy of these fission fragments. In addition an average of 2.5 neutrons of average energy 2 MeV and some gamma radiation is produced. Further energy is released later by radioactivity of the fission fragments. The total energy released is about 3×10^{-11} joule per atom fissioned, i.e. 6.5×10^{13} joule per kg conserved (by comparison one kg of coal interacting with about three times its mass of oxygen releases about 4×10^{7} joule).

Natural uranium consists mainly of atoms of a heavier isotope, ^{238}U, and this will accept neutrons to form ^{239}U but without fission unless the incident neutron is a *fast neutron, when the excess energy that it brings to the nucleus may lead to its fission.

To extract energy in a controlled manner from fissionable nuclei, arrangements must be made for a sufficient proportion of the neutrons released in the fissions to cause further fissions in their turn, so that the process is continuous (*see* chain reaction). At present, the energy released is transferred by heat and is used in the same way as ordinary fuel in order to raise steam, etc.

2. *Types of reactor*. A reactor with a large proportion of ^{235}U or plutonium ^{239}Pu in the fuel uses the fast neutrons as they are liberated from the fission; such a reactor is called a *fast reactor*. Natural uranium contains 0.7% of ^{235}U and if the liberated neutrons can be slowed down before they have much chance of meeting ^{238}U atoms, the latter are not likely to absorb them. A high proportion of the neutrons will then travel on until they meet a ^{235}U atom and then cause another fission. To slow the neutrons, a *moderator* is used containing light atoms to which the neutrons will give kinetic energy by collision. As the neutrons eventually acquire energies appropriate to gas molecules at the temperature of the moderator, they are then said to be *thermal neutrons and the reactor is a *thermal reactor*.

3. *Thermal reactors*. In a typical thermal reactor, the fuel elements are rods embedded as a regular array in the bulk of the moderator. The typical neutron from a fission process has a good chance of escaping from the relatively thin fuel rod and making many collisions with nuclei in the moderator before again entering a fuel element. Suitable moderators are pure graphite (C atoms of mass number 12), heavy water (D_2O, deuterium or heavy hydrogen atoms of mass number 2 and oxygen atoms of mass number 16), and ordinary water (H_2O). Very pure materials are essential as some unwanted nuclei capture neutrons readily. Using graphite or D_2O as a moderator, a reactor will work with natural uranium. Because of the absorption of neutrons by protons, reactors using H_2O as a moderator require enriched fuel. The reactor *core is surrounded by a *reflector* made of suitable material to reduce the escape of neutrons from the surface. Each *fuel element is encased (e.g. in magnesium alloy or

stainless steel can) to prevent escape of radioactive fission products. The *coolant, which may be gaseous or liquid, flows along the channels over the canned fuel elements. There is an emission of gamma rays inherent in the fission process and also many of the fission products are intensely radioactive. To protect personnel from gamma rays and from escaping neutrons, the assembly is surrounded by a massive *biological shield, of concrete, with an inner iron *thermal shield* to protect the concrete from high temperature caused by absorption of radiation.

A *homogeneous reactor* is one in which fuel and moderator are in the same medium (e.g. a solution of fuel, such as uranium salt, in a liquid moderator). In a *heterogeneous reactor* the fuel and moderator are separated in a regular pattern called a lattice. To keep the power production steady, *control rods are moved in or out of the assembly. These contain material that captures neutrons readily (e.g. cadmium or boron). The power production is held steady by allowing the currents in suitably placed *ionization chambers automatically to modify the settings of the rods. Further absorbent rods, the shut-down rods, are driven into the core to stop the reaction, as in an emergency if the control mechanism fails. To attain high thermodynamic efficiency so that a large proportion of the liberated energy can be used, the heat should be extracted from the reactor core at a high temperature. This presents many problems, particularly as materials that happen to absorb neutrons heavily must be excluded, however desirable their other properties may be. Various coolants and canning materials are used in thermal reactors. (*See* gas-cooled reactors, boiling-water reactors, pressurized-water reactors, and heavy-water reactors.)

Fuel rods become in time depleted of fissile material and accumulate quantities of fission products. Some of these are strong absorbers of neutrons and are *poisons* and therefore the control rods have to be withdrawn as the charge of fuel ages, in order to maintain the chain reaction. In time, the fuel element has to be withdrawn for chemical processing to remove the poisons (and also any plutonium, etc., that may have been formed).

The construction materials of the reactor suffer from the bombardment by neutrons and progressive change in their properties must be watched.

4. *Fast reactors*. In fast reactors no moderator is used, the frequency of collisions between neutrons and fissile atoms being increased by enriching the natural uranium fuel with ^{239}Pu or additional ^{235}U atoms that are fissioned by fast neutrons. The fast neutrons thus build up a self-sustaining chain reaction. In these reactors the core is usually surrounded by a blanket of natural uranium into which some of the neutrons are allowed to escape. Under suitable conditions some of these neutrons will be captured by ^{238}U atoms forming $^{239}_{92}$U atoms, which by two *beta-decay stages are converted to $^{239}_{94}$Pu. As more plutonium can be produced than is required to

enrich the fuel in the core, these are called *fast breeder reactors. *See also* fusion reactors.

nuclear recoil The mechanical recoil suffered by the residual nucleus of an atom on radioactive or other disintegration. It can lead to physical effects such as abnormally high volatilities, or to chemical effects such as initiation of polymerization or molecular rupture.

For a typical alpha emitter the energy release is about 5 MeV and the mass of the nucleus is about fifty times that of the alpha particle. Hence the kinetic energy of the recoiling nucleus is about 10^5 eV, i.e. enough to break tens of thousands of chemical bands. By contrast the emission of beta or gamma rays may give recoil energies of from tens of eV down to a small fraction of an electronvolt. (*See also* Mössbauer effect.)

nuclear weapons The first nuclear weapon was the *atom bomb* or *A-bomb*, exploded by the Americans over the city of Hiroshima in Japan in 1945. It consisted of two small masses of *fissile material each of which was below the *critical mass. When the bomb was detonated the two subcritical masses were brought rapidly together to form a supercritical mass within which a single *fission set off an uncontrollable *chain reaction. This first bomb consisted of only a few kilograms of uranium-235, but it had an explosive effect of 20 000 tons (20 kilotons) of TNT. Later models of this fission bomb used plutonium to even greater effect, but these weapons are small compared with the *hydrogen* or *fusion* or *H-bomb*, the first of which was detonated by the Americans in 1952. The hydrogen bomb consists of a fission bomb surrounded by a layer of hydrogenous material, such as lithium deuteride. The fission bomb elevates the temperature of the hydrogen to its *ignition temperature so that a *nuclear fusion reaction takes place with the evolution of enormous quantities of energy – equivalent to tens of megatons of TNT (1 kg of TNT releases about 4 MJ). *See also* fall out.

nucleon The collective term for a *proton or *neutron, i.e. for a constituent of an atomic nucleus. *See* atom; nucleus; isospin.

nucleonics The practical applications of nuclear science and the techniques associated with these applications.

nucleon number *See* mass number.

nucleor It has been postulated that *nucleons (neutrons and protons) have cores called nucleors. A nucleon is supposed to be a nucleor surrounded by a cloud of *pions.

nucleosynthesis The creation of the chemical elements by nuclear reactions. Hydrogen and helium were created in the very early universe (*see* big-bang theory). After 100 seconds, protons and neutrons could combine to form deuterium nuclei, which could then combine to form helium nuclei. Most of the helium in the universe today was formed at this time. Once the temperature had dropped sufficiently, electrons and protons could form hydrogen atoms. The heavier elements are synthesized in nuclear reactions occurring largely in stars and in *supernova explosions.

nucleus The most massive part of the atom; it was shown by Rutherford and others to have a positive charge given by Ze, where Z is the *atomic number of the element and e the charge on an electron. The size of nuclei has been estimated from scattering experiments to be about 10^{-15} m. More recently the radius has been shown to be related to the mass number A of an atom by the formula: $r = c.A^{1/3}$, where c is a constant equal to 1.2×10^{-15} m.

Nuclei consist of protons and neutrons collectively called *nucleons. The number of protons in nuclei of the same element is equal to the atomic number Z. The number of neutrons N associated with the Z protons varies within limits, the different numbers of neutrons giving rise to the various isotopes of that element. The total number of nucleons in a given isotopic nucleus is called the mass number A: $A = N + Z$.

A nucleus is completely defined by the value of its atomic number Z and mass number A. This allows an abbreviated form of nomenclature in which a given nucleus is represented by its chemical symbol with Z and A as subscript and superscript respectively, e.g. the common uranium isotopes are $^{235}_{92}U$ and $^{238}_{92}U$.

The nucleons are maintained within a roughly spherical volume by forces inside the nucleus (*see* strong interaction; weak interaction). These attractive binding forces act between pairs of nucleons, being operative over a distance that is less than the nuclear radius. Various theories have been put forward to explain the structure of the nucleus. The *liquid-drop model of the nucleus has been developed by Bohr and Wheeler to give an explanation of several nuclear phenomena, in particular, the fission of heavy nuclei like uranium. (*See also* shell model.) In addition the presence of Coulomb repulsive forces between the protons limits the numbers of neutrons that can combine with a given number of protons to give stable nuclei. (*See* magic numbers.)

The mass of a given nucleus is always less than the sum of the rest masses of the constituent nucleons. This is due to the emission of energy on creation of the nucleus. The energy equivalent of the mass lost (*mass defect) is indicative of the degree of cohesion of the nucleons and is known as the *binding energy of the nucleus.

Most naturally occurring atoms have stable nuclei. The naturally occurring radioactive atoms have unstable nuclei giving rise to nuclear transmutations in which the atomic number is altered and the product atom is chemically different. Artificial nuclei are produced by bombarding stable nuclei with high-energy charged particles such as protons, deuterons, etc., or with neutrons. The collision process that occurs is called a *nuclear reaction. *See also* radioactivity; elementary particles; particle physics.

nuclide An atom as characterized by its *atomic number, *mass number, and nuclear energy state. *Compare* isotope.

null method *Syn.* balance method. A method of measurement in which the quantity being measured is balanced by another of a similar kind so that the indicating instrument reading is adjusted to zero (as with a *Wheatstone bridge).

number density Symbol: n. The number of particles, atoms, molecules, etc., per unit volume.

number of poles Of a switch, circuit-breaker, or similar apparatus. The number of different electrical conducting paths that the device closes or opens simultaneously. The device is described as single-pole, double-pole, triple-pole, or multipole if it is suitable for making or breaking an electrical circuit on one pole, two poles, three poles, or more than one pole, respectively.

number of steps Of a starter (*see* motor starter) for an electric motor. The number of steps between the off and full-on positions. An N-step starter has $N-1$ accelerating positions between the off and full-on positions.

numeric 1. A pure number.
2. A physical magnitude having no dimensions.

numerical aperture (NA) A parameter used to express the angle of view of *objectives used in *microscopes and also the light-gathering power of these lenses. It is the product of the refractive index of the medium in which the objective is situated (air, oil, etc.) and the sine of half the angle of view of the objective. As the numerical aperture is increased, the *resolving power of the microscope improves.

Nusselt number Symbol: Nu. The dimensionless group ($hl/K\theta$), where h is the rate of loss of heat per unit area of a hot body immersed in a fluid, l is a typical dimension of the body, θ is the temperature difference between the body and the fluid, K is the thermal conductivity of the fluid. *See* convection (of heat).

Nusselt's equation A dimensional equation on the basis of the film theory of convection giving the constant E in the formula for the rate of loss of heat:
$$-dQ/dt = EA(\theta - \theta_0),$$
where A is the area and θ and θ_0 are temperatures.

nutation *See* Eulerian angles.

Nyquist noise theorem The law that relates the power P due to thermal *noise in a resistor to the frequency f of the signal. At ordinary temperatures T,
$$dP = kTdf,$$
where k is the *Boltzmann constant.

O

OASM system A system of fundamental units based on the ohm, ampere, second, and metre.

object Extended natural objects consist of points, self-luminous or otherwise, that deliver diverging pencils of light. They are classed as *real objects* when they are delivering rays to some optical system under consideration. Commonly an optical system follows some preceding system that may focus an *image, real or virtual, in front of the second system – this image becomes a real object for the second system. If the second system lies in such a position as to intercept the rays before they have converged to a focus, the real image from the first system becomes the *virtual object* for the second system. Incident rays are convergent when a virtual object is under consideration.

The *object space* is a mathematical conception covering both the region lying in front of the system (real) and behind the system (virtual) in which real or virtual objects may lie, and possessing the same refractive index throughout – that of the preceding region. It completely coexists with the similarly conceived image space.

object contrast *See* contrast.

objective The lens (generally compound), in an optical instrument, that lies nearest to the object viewed. The term is sometimes applied to a mirror used for the same purpose.

oblateness *See* ellipticity.

oblique astigmatism *See* radial astigmatism.

observable The name used for the measurable things of physical science. In *quantum mechanics they are represented by matrices (matrix form of quantum mechanics) or, alternatively, by operators (wave mechanics).

occlusion The absorption of gases by solids.

occultation The passage of an astronomical body in front of another, especially the moon in front of a star or planet, thus obscuring its light, radio emission, etc. Planets can sometimes occult stars and also their satellites. The precise timing of an occultation provides information about the obscured body and also the obscuring body. *See* eclipse.

OCR Abbreviation for *optical character reader.

octal notation A number system, used in computing, having numbers to base 8. These numbers are often used to represent binary numbers in a convenient form, as each group of 3 *bits may be represented by 1 octal digit. For example,
101 010 111 in binary is 527 in octal.
Compare hexadecimal notation; binary notation.

octave 1. An *interval having the frequency ratio 2:1.

2. An interval of an octave together with the notes included in that interval. Men and women singing unison songs are singing generally at an interval of an octave. The octave is one of the earliest and most common intervals in all scales.

octode A *thermionic valve having five grids between the cathode and the main anode and an additional anode between the two innermost grids (i.e. a total of eight electrodes).

ocular *See* eyepiece.

odd-even nucleus A nucleus that contains an odd number of *protons and an even number of *neutrons. There are more than fifty stable nuclides with odd-even nuclei.

odd-odd nucleus A nucleus that contains an odd number of both *protons and *neutrons. Most of these nuclei are unstable but $^2_1$H, $^6_3$Li, $^{10}_5$B, and $^{14}_7$N are stable.

odd parity *See* wave function.

Odén's balance *See* sedimentation.

oersted Symbol: Oe. The *CGS electromagnetic unit of magnetic field strength.
$$1 \text{ Oe} = 10^3/4\pi \text{ A m}^{-1}.$$

Oersted (or **Orsted**), **Hans Christian** (1777–1851) Danish physicist. He discovered the magnetic action of a current (1820), correctly deducing, from the deflections of a compass needle near a wire carrying a current, that the lines of force form circles round the wire (supposing this to be straight). A former unit of magnetic field strength was named in his honour.

Oersted's experiment The earliest discovery (1820) of the magnetic effect of electricity was made when Oersted noticed that a compass needle was deflected when brought near to a wire carrying an electric currrent. The direction of deflection is given by *Ampère's rule.

off-circuit tap changer *See* tap changer.

off line 1. Not connected to a *computer, switched off, or disconnected.
2. Controlled by instructions generated earlier and held on a computer *storage device. *Compare* on line.

ohm Symbol: Ω. The *SI unit of electric *resistance, defined as the resistance between two points on a conductor through which a current of one ampere flows as a result of a potential difference of one volt applied between the points, the conductor not being the source of electromagnetic force. This unit, at one time called the absolute ohm, replaced the *international ohm* (Ω_{int}) in 1948. The latter was defined as the resistance of a column of mercury of mass 14.4521 grams, length 106.300 cm at 0 °C, and uniform cross section. $1 \Omega_{int} = 1.000 49 \Omega$.

Ohm, Georg Simon (1787–1854) German physicist. He investigated by experiment the relative lengths of wires of different materials that had equal effects in a circuit; then showed that the current in a circuit could be ascribed to the joint action of the exciting force (electromotive force) and the resistance (1826). His main experiments used a copper-bismuth thermocouple (with junctions in ice and steam) as generator as this was steadier than the voltaic cells then available, but he also established rules for cells in series and in parallel. In 1827 he published a theoretical deduction of his law (see Ohm's law), using ideas probably derived from those of Fourier on the flow of heat (published in 1822). He later (around 1833) turned his attention to acoustics. (See Ohm's law of hearing.)

ohmic contact An electrical contact in which the potential difference across it is proportional to the current flowing through it.

ohmic loss Power dissipation in an electrical circuit arising from its resistance, rather than from other causes such as magnetic hysteresis.

ohmmeter An instrument for measuring the electrical resistance of conductors or of insulators. It gives a direct indication on a scale graduated in ohms or in suitable multiples, or submultiples of that unit. Compare insulation-testing set.

Ohm's law The electric current I in a conductor is directly proportional to the potential difference V between its ends, other quantities (especially temperature) remaining constant. As the resistance R is defined by $R = V/I$ the law can be stated as R = constant. (The equation defining R is true even when the resistance is not constant, for example, for a hot filament, so the equation does not express the law.)

Ohm's law is very accurately obeyed by pure metals and alloys over a very large range of currents. It is less reliable for nonmetals.

Ohm's law of hearing In 1843, G. S. Ohm stated his law of hearing in the following terms: "Every simple harmonic motion of the air is perceived by the ear as a simple tone, all others are resolved by the ear into a series of simple tones of different periods." In other words, the ear performs a *Fourier analysis on any complex note, breaking it up into its simple harmonic constituents. The *quality of a note from a musical instrument depends on the number, amplitude, and distribution of its associated harmonics. The ear is not a perfect Fourier analyser since this would involve recognition of the phase relationships of the partials of a complex note. Experiments show that variation of the phase difference between two notes has no effect on the ear.

oil-break Of an electrical switch, circuit breaker, or similar apparatus. Having contacts that separate in oil.

oil-filled cable An impregnated cable in which the impregnating medium can flow freely at all working temperatures. See oilostatic cable.

oil-immersed Of electrical apparatus. Having the main conducting parts submerged in oil. The oil used is specially prepared. It must be a good insulator and must possess other properties such as low flash point, freedom from sludge formation, etc.

oil-immersion (microscope) See resolving power; immersion objective.

oilostatic cable An oil-filled cable in which the impregnating medium (oil) is maintained at a pressure in excess of atmospheric. Cables of this type are employed at voltages in excess of about 66 kV and can be made for working voltages of the order of 250 kV.

Olbers, Heinrich Wilhelm Malthaus (1758–1840) German astronomer who made substantial contributions to astrophysics, including the observation that if the universe is finite, light must be dimmed by its passage through space (see Olbers' paradox).

Olbers' paradox A paradox first suggested by the German astronomer Heinrich Wilhelm Olbers stating that the generally accepted idea of an infinite number of stars, uniformly distributed in space, would mean that the night sky should glow with uniform brightness. The assumption of a finite number of stars is unnecessary to explain the dark night sky as it is now thought that the recession of galaxies, as indicated by their *redshifts, causes the brightness of distant stars to be greatly diminished.

old achromats See achromatic lens; optical glass.

Oliphant, Sir Mark Laurence Elwin (b. 1901) Australian physicist who built one of the first cyclotrons and was the first to produce *tritium.

omega-minus particle Symbol: Ω^-. An *elementary particle classified as a *hyperon. See unitary symmetry.

omni-aerial Syn. omnidirectional aerial. An aerial of any type that is essentially nondirectional, i.e. is equally effective as an emitter (or collector) of radiation in all directions having the same angle of elevation. Compare directive aerial.

onde de choc The first sound heard when a high-speed projectile travels overhead. The projectile is the source of sound waves that travel out from it. If the projectile itself has a speed less than sound, these waves gain upon it and reach the listener before the projectile passes over. If its speed is supersonic, it carries its shock waves with it. In military sound ranging, the onde de choc is, accordingly, the first sound registered at the listening post. See also sonic boom.

ondograph An instrument that produces a graph of an alternating voltage.

on line Of a computer *peripheral device. Directly controlled by a *computer; executing instructions as they are generated. Compare off line.

on-load tap changer See tap changer.

Onnes, Heike Kamerlingh (1853–1926) Dutch physicist. A pioneer in low-temperature physics, the first to liquefy helium, reaching a temperature below 1 K. He was the discoverer (1911) of the low-temperature phenomenon of *superconductivity. Onnes first introduced the idea of virial coefficients (*see* virial expansion) A, B, C, for representing the behaviour of real gases by means of the equation
$$pv = A + Bp + Cp^2 + \dots.$$
He also measured the compressibility of various gases at different temperatures by measuring the pressure required to compress a known mass of the gas into a known volume. In 1908 at Leiden he liquefied helium by pre-cooling it below its inversion temperature with hydrogen boiling under reduced pressure and then expanding it through a valve, in a small-scale model of the apparatus used for liquefying hydrogen.

ontology *See* metaphysics.

opacity The ratio of the *radiant flux incident on an object (e.g. part of an exposed and processed photographic plate) to the flux transmitted. It is the reciprocal of *transmittance. It is a measure of the ability of a solid, liquid, or gaseous body to absorb radiation.

op-amp *See* operational amplifier.

open arc An arc between electrodes of carbon arranged to permit free access of air. The arc may, however, be shielded from draughts. *Compare* enclosed arc.

open circuit *See* circuit.

open-circuit voltage The voltage at the output terminals of a transformer, transmission line, d.c. or a.c. generating machines, etc., at *no-load under specified conditions of, for example, input voltage in the case of transformers and transmission lines, excitation and speed for d.c. and a.c. machines.

opening time of a circuit breaker The time interval (excluding any intentional delay time caused by a *time lag) between the instant at which power is applied to the tripping device and the instant at which the arcing contacts separate.

opera glass *See* refracting telescope.

operating point The point on the family of *characteristic curves of an active electronic device, such as a transistor, that represents the magnitudes of voltage and current for the particular operating conditions under consideration.

operational amplifier *Syn.* op-amp. A very high *gain voltage *amplifier, with very high input impedance, usually having a differential input (i.e. its output voltage is proportional to (and very much greater than) the voltage difference between its two inputs). It is invariably used with considerable *feedback, which determines its *transfer characteristics. The feedback circuits make op-amps operate as voltage amplifiers with a gain precisely defined by the values

of resistance and/or capacitance used in the feedback circuits, or else enable them to perform mathematical operations such as integration or signal-conditioning functions such as filtering. They are thus used in a very wide range of instrumentation and control applications.

An operational amplifier is usually a multistage device designed for insertion into other equipment. It is supplied as a complete packaged unit, commonly as a single monolithic *integrated circuit.

operator A mathematical symbol, such as a plus sign, integral sign, or differential operator, indicating a specific operation to be carried out on a number, variable, function, etc.

ophthalmometer An instrument (Helmholtz, 1854) for measuring the radius of curvature of the cornea of the eye. It depends on the measurement of the size of the reflected image of an object of known size at a known distance. *See* keratometer.

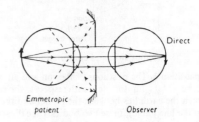

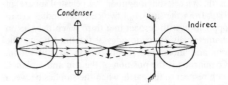

Ophthalmoscope

ophthalmoscope An optical instrument for illuminating the back of the eye and for observing the reflex from the choroid coat and retina (the fundus), (Babbage, 1847; Helmholtz, 1851).

There are two systems:

1. *Direct method. Illumination stage.* Light is converged on the pupillary area to illuminate an area at the back of the eye (fundus). *Observation stage.* Light reflected back from the fundus emerges parallel (*emmetropia), convergent (*myopia), or divergent (*hypermetropia) and the observing eye looks through the peephole situated at the centre of the illuminating mirror. The ophthalmoscope is held close to the eye under observation. Instruments may be nonluminous or luminous.

2. *Indirect method. Illumination stage.* Light is reflected by a mirror onto a condensing lens that converges the light through the pupil to illuminate the fundus. *Observation stage.* Light is reflected back from the fundus as above, but a real inverted image is formed by retransmission, through the condenser lens. The observer views this inverted image from a distance of about an arm's length away.

Oppenheimer, J. Robert (1904–1967) Amer. nuclear physicist. He is noted for his contributions to quantum mechanics and his leadership of the team that developed the first atomic bomb. Opposed to the hydrogen bomb, he was regarded as a security risk by Senator McCarthy's committee.

Oppenheimer–Phillips (O–P) process *See* stripping.

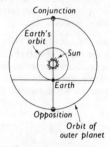

Opposition and conjunction

opposition 1. The positions of the sun and an outer planet (Mars, Jupiter, etc.), with respect to the earth, when the earth lies between the planet and the sun on the line joining these two bodies. (*See* diagram.) The planet is then, approximately, at its closest to earth, at a celestial longitude (*see* celestial sphere) of 180° from the sun. Any two celestial bodies are in opposition with respect to a third when they lie on diametrically opposite sides of the third body. The sun and moon are in opposition at full moon. *Conjunction* is the position of the sun and a planet, with respect to the earth, when the sun lies between the planet and earth, all three lying on the same line. The sun and planet have the same longitude. Any two celestial bodies are in conjunction when a third reference body lies on an extension of the line joining the two bodies. The sun and moon are in conjunction at new moon.

2. Two periodic quantities with the same frequency and waveform are in opposition when they differ in *phase by 180°.

optical activity The ability of certain solutions and crystals to rotate the plane of polarization of *plane-polarized light in proportion to the length of substance traversed and to the concentration in the case of solutions. The angle through which the plane is rotated is the *angle of optical rotation*, symbol: α, and is measured in radians. When looking towards the oncoming light, if the rotation is clockwise the optical activity is called right-handed (dextrorotatory); if the rotation is anticlockwise, it is called left-handed (laevorotatory).

optical axial angle The acute angle between the two optical axes of a *biaxial crystal.

optical axial plane The plane defined by the two optical axes of a *biaxial crystal.

optical axis *See* optic axis.

optical bench A rigid table, wooden beam, steel girder, etc., that permits the easy slide of mounts for lenses, mirrors, photometer heads, etc., along a straight course and with attached scales to enable the position (or travel) of the optical element to be accurately determined. Designs differ according to the purpose and accuracy required. It is used for focal-length determination, testing of aberrations, photometer slide, interference and diffraction experiments, etc.

optical centre From the practical point of view a spot marked on the surface of a lens where the optic axis intersects the surface. More precisely for thicker lenses, it is defined as that axial point through which all undeviated rays pass (incident and emergent rays are parallel with or without lateral displacement). The position of the optical centre depends only on the surface radii and thickness, and is independent of wavelength. *See* centration.

optical character reader (OCR) A device used in computing systems for transferring information in the form of letters, numbers, or other characters printed on paper, to a *computer or *storage device. A specially designed typeface is no longer necessary. *Optical character recognition* is the process involved in identifying the information as valid characters.

optical density *See* transmission density.

optical disk A light, cheap, and physically compact computer storage device. The most common form uses audio *compact-disc technology, and cannot therefore be written to by a computer. Writeable and erasable forms are also available.

optical distance *See* optical path.

optical fibres *See* fibre-optics system.

optical flat A surface that is so flat that there are no irregularities of surface flatness greater than a fraction of the wavelength of light. The surface is tested by observing interference fringes when a known flat is placed in contact (*optical contact*).

optical glass Glass with which special precautions are taken during manufacture to avoid mechanical and optical defects (density, strain, heterogeneity, colour, refractive index, etc.). The lens designer requires a selection of glasses with prescribed refractive indexes and dispersions measured to a high degree of accuracy, in order to provide lenses corrected to various degrees of accuracy for chromatic and spherical aberration. The glass manufacturer therefore provides catalogues of glass varieties (melts) with numbers and descriptions, such as hard crown, dense barium crown, telescope flint, etc. These catalogues show:
the *refractive index for yellow D light (n_0),
mean *dispersion ($n_F - n_C$),

Optical Glass   Illustrative Table

	n_D	$n_F - n_C$	v
Fluor Crown	1·4785	0·00682	70·2
Baryta Crown	1·5881	0·00962	61·1
Dense Flint	1·6182	0·01697	36·4
Heaviest Flint	1·9044	0·04174	21·7

	$n_D - n_C$	$n_F - n_D$	$n_{G'} - n_F$
Fluor Crown	0·00202	0·00480	0·00363
Baryta Crown	0·00284	0·00678	0·00544
Dense Flint	0·00484	0·01213	0·01031
Heaviest Flint	—	0·03023	0·02726

*Abbe number $V = (n_D - 1)/(n_F - n_C)$,

partial dispersions $(n_D - n_C)$, $(n_F - n_D)$, $(n_G - n_F)$, and relative partial dispersions as fractions of the mean dispersion.

Reference to the table shows n_D varying between about 1.48 and 1.90 and Abbe numbers from about 70.0 to 22.0. A system of specification has also been used in which the first three digits (of six-digit numbers) are the first three decimal places of the refractive index and the last three, the three significant figures of constringence, e.g. light flint 576410; dense flint 618364.

Prior to the Jena works (1880), glasses were classified as crowns and flints (the latter with a lead content), in which refractive index and dispersive power, i.e. reciprocal constringence, varied roughly together – high index with high dispersion. They are called "old" glasses and lenses made under such restriction are called *old achromats*. Abbe, having found the necessity of having available a wider selection (e.g. high index with low dispersion), collaborated with Schott in the production of "new" (Jena) glasses and rendered possible new achromats with better correction of spherical aberration, anastigmats, etc.

optically negative crystal A crystal in which *double refraction occurs and in which the refractive index (ω) for the ordinary ray is greater than the index (ε) for the extraordinary ray. In an *optically positive crystal*, $\varepsilon > \omega$. This is the case for *uniaxial crystals. In *biaxial crystals, β is nearer to γ than to α (optically negative) or β is nearer to α than to γ (optically positive). Here, α, β, γ are the principal refractive indices, in ascending order ($\alpha < \beta < \gamma$).

optically positive crystal *See* optically negative crystal.

optical maser Former name for laser.

optical path *Syn.* optical pathlength; optical distance. The distance traversed by light multiplied by the refractive index of the medium (nd). When light passes through different media the total pathlength is:

$$n_1 d_1 + n_2 d_2 + \ldots = \Sigma (nd).$$

It is the distance in a vacuum that would contain the same number of waves as occur in the actual path in the medium, so that the optical paths between two wavefronts are all the same.

optical pyrometer A *pyrometer in which the luminous radiation from the hot body is compared with that from a known source. The instrument therefore measures the temperature of a luminous source without thermal contact. The two chief types are the *disappearing filament pyrometer and the *polarizing pyrometer.

optical rotation *See* optical activity.

optical switch A device whose optical properties (e.g. refractive index and polarizing properties) can be varied by an externally applied field or other influence; electric, magnetic, and surface acoustic wave techniques are all used for this purpose. Light (often a laser beam) can thus be deflected from a detector, so switching the beam.

optical window *See* atmospheric windows.

optic axis 1. The path of rays passing through the centres of the *entrance pupil and *exit pupil of an optical system. The *cardinal points lie on this line.
2. The direction (not a single line) in a doubly refracting crystal in which the ordinary and extraordinary rays apparently do not exhibit *double refraction, while their velocities are equal, i.e. the direction in a crystal along which the polarized components of a ray of light will be transmitted with a single velocity.

optics A branch of physics concerned with the study of light, its production, propagation, measurement, and properties. On account of its intimate connection with sight and with optical instruments, it has deep roots in physiology and ramifications in engineering. Since to a first degree of approximation, light travels in straight lines, the ray treatment of light is called *geometric optics* as distinct from *physical optics*, which attempts to explain the objective phenomena of light. *Physiological optics* is concerned with light and vision, i.e. with the interactions of light with the eye. *Ophthalmic optics* (optometry) specializes in the application of optical principles to measure and correct by optical means visual defects of persons. There is an interweaving of optical interest in illuminating engineering, optical engineering, optical working, meteorology, astronomy, etc. *See also* nonlinear optics; optoelectronics.

optoelectronics The convergence of optical and electronic technology in gathering, processing, storing, and displaying information: it is concerned with the generation, processing, and detection of optical signals that represent electrical quantities. Major areas of application include communications and computing. *See also* fibre-optics system; light-emitting diode; semiconductor laser; photodiode; phototransistor.

optoisolator An optoelectronic device whereby two unconnected electric circuits can exchange signals by means of an optical link yet remain electrically isolated.

optometer A class of instrument for measuring refractive errors of the eye. They are commonly direct reading and may use subjective or objective methods. *Optometry* is the term applied to the general measurement and optical treatment of refractive errors in America (ophthalmic optics in the United Kingdom).

optometry *See* optometer.

orbit A curved path, such as that described by a planet or a comet in the field of force of the sun, or by a particle in a field of force. The term is especially used for the locus of an extra-nuclear electron in an *atom in the Rutherford–Bohr atomic model.

orbital 1. *See* atomic orbital.
2. *See* molecular orbital.

orbital quantum number *Syn.* orbital angular-momentum quantum number; azimuthal quantum number. A *quantum number that governs the orbital angular momentum of a particle, atom, nucleus, etc. The symbol is l for a single entity or L for a whole system. *See* atomic orbital.

orbital velocity The velocity required by a *satellite or spacecraft to enter and maintain a particular orbit around the earth or some other celestial body. The orbital velocity needed for a 24-hour orbit (*see* geostationary orbit) around the earth is approximately 3.2 kilometres per second, at an altitude of about 36 000 km.

order of interference or diffraction A whole number that characterizes a position of an interference fringe according to whether there is interference arising from one, two, three, etc., wavelength difference of path, or according to the direction of the maxima of illumination produced by diffraction.

order of magnitude The value of a number or of a physical quantity given roughly, usually within a power of 10. Thus, 2.3×10^5 and 6.9×10^5 are of the same order of magnitude and 5×10^8 is 3 orders of magnitude greater than either.

ordinary ray Double refracting crystals give rise to two refracted beams of light called the ordinary ray and the *extraordinary ray. Only the ordinary ray obeys the ordinary laws of refraction. The plane-polarized vibrations in the ordinary ray are perpendicular to the principal plane. *See* double refraction.

ordinate *See* graph.

organ pipes *See* column of air.

OR gate *See* logic circuit.

orientation interaction *See* van der Waals forces.

O-ring A circular ring, usually made of neoprene, that is used as an oil or air seal.

orthicon A type of television *camera tube in which the image to be scanned is produced on a photomosaic.

orthochromatic film Photographic film that is sensitive to green as well as to the blue end of the spectrum.

orthogonal 1. Mutually perpendicular, as in orthogonal axes.
2. Having or involving a set of mutually perpendicular axes as in orthogonal crystals.

orthogonal projection A projection from a pole at infinity, along a direction parallel to a crystal edge, onto a plane normal to that direction.

orthohelium Those helium atoms in which the *spins of the two electrons are parallel, in contrast to *parahelium* in which the spins are antiparallel. By the *Pauli exclusion principle the ground state is a state of parahelium. By electron impact or other processes, helium can be excited into ortho- or para- states, which usually decay rapidly to states of the same kind, as transitions between ortho- and para- require changes of spin orientation that are not easily produced. Thus the emission spectrum of helium shows two different sets of strong lines, together with very faint lines caused by transitions between the two kinds of state.

orthohydrogen One of two forms of ordinary hydrogen depending on the *spins of the nuclei. In orthohydrogen both nuclei spin in the same direction whereas in the other form, *parahydrogen*, they spin in opposite directions. Under normal conditions hydrogen is an equilibrium mixture of the two containing about 75% of orthohydrogen and 25% of parahydrogen. The equilibrium concentration of parahydrogen decreases with temperature; at $-253\ °C$ it is 99.8%, and almost pure parahydrogen can be obtained at low temperatures by using a catalyst to increase the rate at which equilibrium is attained. Once formed, pure parahydrogen can be kept for long periods at room temperature. Transitions between the two forms are very rare as they depend upon interactions with the extremely small magnetic moments associated with nuclear spins (*see* magneton). The two forms of hydrogen are chemically identical but differ slightly in physical properties – such as melting point, boiling point, and thermal conductivity.

The phenomenon of existing in ortho- and para-forms is not peculiar to hydrogen. It occurs for all similar diatomic molecules in which the nuclei have spins. However, hydrogen is peculiar in that there is such a (relatively) high difference in physical properties between the forms. This is a consequence of the low moment of inertia of the hydrogen molecule.

orthorhombic (rhombic) system *See* crystal systems.

orthoscopic Free from distortion. Photographic objectives should be orthoscopic as should good-quality magnifiers and eyepieces.

oscillating current (or **voltage**) A current (or voltage) waveform whose amplitude periodically increases and decreases with time according to some mathematical function. Oscillating waveforms may be sinusoidal, sawtooth, square, etc.

oscillation 1. A vibration.

2. A periodic variation of an electrical quantity, such as current or voltage.

3. A phenomenon that occurs in an electrical circuit if the values of self inductance and capacitance in the circuit are such that an *oscillating current arises from a disturbance of the electrical equilibrium in the circuit. A circuit in which oscillations can freely take place is called an *oscillatory circuit*. Oscillations that result from the application of a direct voltage input to the circuit and continue until the direct voltage is removed are called *self-sustaining oscillations* (*see* oscillator). *See also* free vibrations; forced oscillations; parasitic oscillations.

oscillation photography A crystal-diffraction method in which a single crystal is allowed to oscillate uniformly through a small angle (usually < 15°) about an axis normal to a beam of monochromatic X-radiation (or electrons, neutrons, etc.). *See* X-ray analysis.

oscillator An electric circuit designed specifically to convert direct-current power into alternating-current power, usually at relatively high frequencies. Application of the direct-voltage supply to the circuit is usually sufficient to cause it to oscillate, and for the electrical *oscillations to be maintained until the direct voltage is switched off.

A simple oscillator consists essentially of a frequency-determining device, such as a *resonant circuit, and an active element that supplies power to the resonant circuit and also compensates for damping due to resistive losses. The active element can be considered as supplying a *negative resistance of sufficient value to counterbalance the positive resistance of the resonant circuit. Once started, then, the oscillations will continue.

The effective negative resistance can be provided by the use of positive *feedback to overcome the damping, or by means of an electronic device that exhibits negative resistance on a portion of its characteristic curve. *See also* piezoelectric oscillator; relaxation oscillator.

oscillatory circuit *See* oscillation.

oscillogram The record produced by a recording *oscillograph, or the reading from an *oscilloscope.

oscillograph An *oscilloscope equipped to make a permanent record of the parameter being measured. *See* cathode-ray oscilloscope; Duddell oscillograph.

oscilloscope An instrument used to produce a visual image of one or more rapidly varying electrical quantities. The *cathode-ray oscilloscope is the most usual type.

osculating plane *See* normal.

osmometer An instrument for measuring osmotic pressure. *See* osmosis.

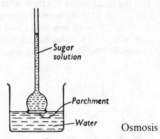

Osmosis

osmosis A process in which certain kinds of molecules in a liquid are preferentially transmitted by a *semipermeable membrane. There is diffusion of a solvent through the membrane into a more concentrated solution. For example, parchment will allow water molecules to pass but will hinder sugar molecules in a solution. In the diagram, water molecules are incident on the underside of the parchment and water and sugar molecules on the upper. The sugar does not get through and probably hinders the water molecules. The net effect is that more water passes upwards than downwards and a hydrostatic pressure builds up in the tube increasing the diffusion of water downwards through the membrane until ultimately a state of dynamic equilibrium is reached. The hydrostatic pressure then balancing osmosis is called the *osmotic pressure*, symbol: Π.

Quantitative measurements were first made by Pfeffer in 1877 and van't Hoff showed theoretically that, at great dilution and if the solute molecules do not dissociate, the osmotic pressure of a solution is equal to the pressure that the solute molecules would exert if they were a gas of the same volume. Thus the osmotic pressure of a solution is given by $\Pi V = RT$, where V is the volume of solution containing unit amount of solute. *See also* van't Hoff factor.

osmotic pressure *See* osmosis.

Ostwald, Wilhelm (1853–1932) German physical chemist. He made important contributions to the theory of electrolytic dissociation. Applying the law of mass action to ionization gave *Ostwald's dilution law (1888).

Ostwald was an important figure in the foundation of physical chemistry as a subject and he has importance also as an editor and scientific historian (e.g. *Ostwald's Klassiker der exakten Wissenschaften*, containing (in German translation) many of the most important papers in science).

Ostwald's dilution law A law obtained by applying the law of *mass action to *electrolytic dissociation.

Ostwald viscometer

If unit amount of acid is dissolved in a volume (V) of water, the acid dissociates into ions according to the equation:

$$HA \rightleftharpoons H^+ + A^-.$$

At equilibrium the concentrations of acid and ions are related by:

$$[H^+][A^-]/[HA] = \alpha^2/(1 - \alpha)V = K,$$

where α is the fraction of acid dissociated and K a constant. K is the equilibrium constant of the reaction and is often called the *dissociation constant*. For weak *electrolytes α is much less than 1 and the law is often stated as $\alpha = \sqrt{KV}$. As the dilution of the electrolyte increases, the degree of dissociation also increases.

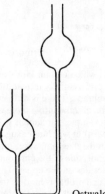

Ostwald viscometer

Ostwald viscometer If $\partial p/\partial x$ is the gradient of pressure along a cylinder of liquid of length δx and radius r, the force due to it is $\pi r^2(\partial p/\partial x)\delta x$ and the frictional force round its periphery is $-2\pi r\delta x\eta\, \partial u/\partial r$ where u is the flow induced by the pressure gradient at radius r, and η is the coefficient of *viscosity. By equating these expressions we get the gradient of velocity across the flow:

$$\frac{\partial u}{\partial r} = -\frac{r}{2\eta}\cdot\frac{\partial p}{\partial x}.$$

By integration we find the velocity to be proportional to $(a^2 - r^2)$, if a is the radius of the wall of the tube. A second integration gives the overall outflow in unit time as

$$\frac{Q}{t} = \frac{\pi a^4}{8\eta}\cdot\frac{\partial p}{\partial x}.$$

This equation is the basis of a number of instruments for measuring or comparing the viscosities of liquids, of which the Ostwald form is the commonest. The liquid is sucked into the upper bulb that it fills. It is then released and the time (t) taken for all the liquid to flow out of the bulb measured. The density ρ of the liquid must be known and the instrument calibrated by timing the flow of a standard liquid, say, water, of which the viscosity at a given temperature is accurately known and for which the corresponding quantities in the instrument are t_0 and ρ_0. The pressure differences that each liquid suffers, in succession, are the same, except that at equal differences of level between the free surfaces, the respective falls of pressure are proportional to the respective densities. All other factors in the above equations are the same for the two liquids except η and t. Finally,

$$\eta/\eta_0 = \rho t_0/\rho_0 t.$$

Otto, Nicholas (1832–1891) German engineer. The inventor of the four-stroke cycle known as the *Otto cycle.

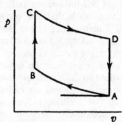

v Otto cycle

Otto cycle A four-stroke cycle, two strokes being charging and exhausting processes. After the explosive mixture of air and petrol has been drawn in, it is compressed adiabatically, AB, and fired at B after which the pressure and temperature are increased rapidly at constant volume, BC. The piston then moves out causing adiabatic expansion, CD, after which the exhaust valve opens reducing the pressure to atmospheric, DA. The next inward motion of the piston sweeps out the exhaust gases to complete the cycle.

outers *See* three-wire system.

outgassing 1. In any vacuum system, the removal by heating of some of the air adsorbed on the inside surfaces of the system.
   2. The slow deterioration of the vacuum due to the release of adsorbed gases from the interior surfaces of the vacuum system.

out of phase *See* phase.

output 1. The power, voltage, or current delivered by any circuit, device, or plant.
   2. The terminals or other place where the signal is delivered. *See also* input/output.

output impedance The *impedance presented to the *load by a circuit or device.

output transformer A transformer used to couple an output circuit (usually of an *amplifier) to the *load.

overcompounded *See* compound-wound machine.

overcurrent *See* overcurrent release.

overcurrent release *Syn.* overload release. A tripping device that operates when the current exceeds a predetermined value (usually adjustable). A current that causes the release to operate is called an *overcur-*

rent (*compare* undercurrent release). If the device is arranged so that the tripping action is delayed for a definite time, which, although being adjustable, is independent of the magnitude of the overcurrent, it is called a definite time-lag overcurrent release. If the tripping action is delayed for a time inversely dependent upon the magnitude of the overcurrent, it is an inverse time-lag overcurrent release. If the tripping action is delayed for a time inversely dependent upon the magnitude of the overcurrent and the time delay approaches a definite minimum value with increase of overcurrent, it is an inverse and definite minimum time-lag overcurrent release.

overdamped *See* damped.

overdriven amplifier An *amplifier operating at an input voltage exceeding that for which the circuit was designed, resulting in distortion being introduced.

Overhauser, Albert Warner (*b.* 1925) Amer. physicist who became professor at Purdue University. He is noted for his work in solid-state physics and his prediction (1953) of the *Overhauser effect. Overhauser has worked on a number of other topics including the prediction (1960) of the existence of spin-density and charge-density waves in metals, the properties of alkali metals, and many-body theory.

Overhauser effect An effect in which polarization of nuclear spins occurs in a solid as a result of excitation of electron spins (*see* electron spin resonance). It occurs because of coupling between the electron and nuclear magnetic moments and is used in producing polarized targets for particle-scattering experiments and in such double-resonance techniques as *ENDOR.

overhead crossing A device used in electric traction systems at the crossing of two contact wires to permit current collectors to pass along either wire.

overhead line An electric *line that is situated above the ground. The conductors are usually supported on separate insulators.

overlap span *See* section gap.

overload (electrical) Any *load that exceeds the rated output of a machine, transformer, or other apparatus. It is expressed numerically as the amount of the excess or may be given as a percentage of the rated output.

overload release *See* overcurrent release.

overscanning Scanning of an electron beam beyond the phosphor of a television or oscilloscope tube.

overshoot *See* pulse.

overtone A constituent of a musical note other than the fundamental or lowest tone. The first overtone is the second *harmonic. Overtone and upper *partial are synonymous terms.

overvoltage *Syn.* excess voltage. **1.** Voltage that exceeds the normal value between two conductors or between a conductor and earth.

2. The potential of a gas or metal liberated from an electrolyte, relative to the electrolyte, depends on the nature of the electrode. The amount by which the e.m.f. necessary to liberate hydrogen on an electrode exceeds the e.m.f. necessary to liberate hydrogen on platinized platinum is known as the hydrogen overvoltage of the electrode. *See also* oxygen overvoltage.

overvoltage protective device A device that affords protection from damage that might be caused by an overvoltage.

overvoltage release A *tripping device that operates when the voltage exceeds a predetermined value. *Compare* undervoltage release.

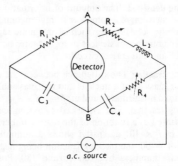

Owen bridge circuit

Owen bridge A four-arm a.c. bridge, designed by David Owen in 1914, used to measure the self inductance, L, of an element in terms of the capacitance, C, and resistance, R, of other elements in the arms (*see* diagram). The current in each arm is balanced such that there is no potential difference between points A and B. Then
$$R_1 R_4 C_3 = L_2; \qquad R_1 C_3 = R_2 C_4.$$
These relationships are thus independent of frequency. R_1, C_3, and C_4 are known values. R_2 is varied until a minimal current flows through the detector (a galvanometer, etc.); R_4 is then varied until the current is zero.

oxygen overvoltage As with hydrogen overvoltage (*see* overvoltage) at a cathode, an anode can show oxygen overvoltage effects since the potential of the anode must be raised to an appreciably higher value than the theoretical one for the gas to be evolved.

oxygen point The temperature of equilibrium between liquid and gaseous oxygen at a pressure of one standard atmosphere, taken as a fixed point in the *International Practical Temperature Scale at 90.188 kelvin.

ozone layer *Syn.* ozonosphere. *See* stratosphere.

ozonizer Ozone is produced when oxygen is subjected to an electrostatic glow discharge, the gas being passed through an annular space between two electrodes. (*See* gas-discharge tube.) This production of ozone accounts for the unpleasant smell caused when an electrical machine is being operated.

P

pachimeter An instrument for measuring the elastic shear limit of a solid material.

package A set of computer programs that is directed at some application in general, such as computer graphics, computer-aided design, word processing, statistics, mathematics, or mapping, and that can be tailored to the needs of a particular instance of that applicationn.

packing density 1. The amount of information in a given dimension of a computer storage medium, e.g. the number of *bits per inch of magnetic tape.
   2. The number of devices or logic *gates per unit area of an *integrated circuit.

packing fraction The difference between the exact nuclear mass M of a nucleus and its *mass number A, divided by the mass number; $f = (M - A)/A$. The packing fraction, first expressed by Aston, can be represented as a function of the mass number. The curve of $f \times 10^4$ against A shows a minimum at about $A = 50$, the packing fraction being negative for mass numbers between 16 and 180. Positive values indicate a tendency to instability, and nuclides with these mass numbers ($16 > A > 180$) can be used in *nuclear fusion and *fission processes. *See* binding energy.

pad 1. A fixed-value *attenuator, placed in a circuit to reduce the amplitude of electrical signals without introducing distortion.
   2. *See* bonding pad.

paired cable *Syn.* twin cable. A cable composed of bundles of *twisted pairs encased in an outer protective sheaf. There may be several thousand pairs in a large cable.

pair production The simultaneous formation of a *positron and an *electron from a *photon. It occurs when a high-energy gamma-ray photon (> 1.02 MeV) passes close to an atomic nucleus. *See also* annihilation.

PAL Abbreviation for phase alternation line, the *colour television system adopted by most European countries. In this system the colour information is carried on a chrominance subcarrier with two colour-difference modulating signals in *quadrature, the relative phase of the signals being reversed at regular intervals. This reduces errors in the colour decoder.

palaeomagnetism The study of the residual magnetization of certain rocks in order to determine the direction of polarization of the earth's magnetic field at the time of the rock's formation. The age of the rock can be found by radiometric *dating. A graph of polarity versus time shows that the earth's field has reversed many times during its history (i.e. north and south poles have interchanged) and that there is a variable period of time, a *magnetic interval*, between reversals.

panchratic eyepiece An *eyepiece of a telescope in which the magnifying power can be varied by displacement of the *erecting lens while maintaining the focus at infinity.

panchromatic film A photographic film that is sensitive to all colours of the visible spectrum. *Compare* orthochromatic film.

panoramic Of a lens or an instrument. Taking in a wide angle of view. In the panoramic camera, the film is curved and the lens is rotated about its second nodal point, which is coincident with the centre of curvature of the film.

pantagraph A sliding current-collector employed in the overhead-contact system of electric traction. A bow-shaped contact strip is mounted on a diamond-shaped framework that is hinged to permit the contact strip to move vertically.

pantograph An instrument used in copying drawings on the same or on a reduced or enlarged scale. It consists of a series of pivoted levers fitted with a tracing point (which is moved round the original outline) and a pencil (which produces the new drawing).

paper tape *Syn.* punched tape. A strip of paper upon which information is coded in the form of holes punched across it, each combination of holes representing a specific character, number, or symbol. There are usually 5 to 8 data hole positions across the tape, with a continuous row of smaller sprocket holes. A *paper-tape reader* is a device for decoding information on the tape and displaying it, storing it, or transmitting it to a *computer. A *paper-tape punch* is a device for preparing paper tape and is operated either manually from a keyboard, from a source of experimental data, or automatically as the output of a computer.

Papin, Denis (1647–1712) French physicist. He made important contributions to the development of the air pump, also introducing the bell jar and plate (1674), an idea he obtained from Huygens (1661). He also attempted to design water-pumping machinery; and he invented the *pressure cooker and the safety valve.

parabola *See* conic sections.

parabolic reflector *Syn.* paraboloid reflector. A concave paraboloid reflecting surface, i.e. with a shape that results from rotating a parabola about its axis of symmetry. A beam of radiation striking the surface parallel to the axis is reflected to a single point (the focus) no matter how wide the aperture. The surface is thus free of *spherical aberration (but not of *coma). Such surfaces are used in reflecting telescopes and radio and microwave dishes. Conversely, a small source at the focus delivers parallel rays with little divergence. This is used in microscope illuminators, searchlight condensers, directive aerials, etc.

parabolic rule *See* Simpson's rule.

paraboloid The surface traced out by the rotation of a parabola about its axis of symmetry. The volume between the apex and a circular cross section of radius r perpendicular to the axis, at a distance h from the apex, is $\frac{1}{2}\pi r^2 h$. *See also* parabolic reflector.

parahelium *See* orthohelium.

parahydrogen *See* orthohydrogen.

parallax 1. If a remote object is viewed from two points at the end of a baseline, the angle between lines drawn from the object to each end of the base is the parallax. In astronomy, the parallax of remote celestial bodies is measured with respect to various baselines. For example, the *annual parallax* of a star is the maximum angle subtended at the star by the mean radius of the earth's orbit around the sun, the angle earth-sun-star being 90°.
2. The apparent change in the separation between two objects when viewed from different positions. Consider two objects in line with an eye; if the latter moves to the right, the more distant object will appear to have moved to the right of the nearer object. Objects farther away show the same parallactic displacement as the eye displacement.

Parallax errors may occur in reading a scale if the pointer and scale are separated. Such errors may be eliminated (i) by putting a mirror behind the pointer, parallel to the scale, and placing the eye so that the pointer and its reflection are seen superimposed; (ii) by using a *telecentric optical system.

a Resistors in parallel   *b* Capacitors in parallel

parallel Involving the simultaneous transfer (or in computing the simultaneous processing) of the individual parts of a whole. Circuit elements connected so that the current divides between them and later reunites, are said to be *in parallel*. For resistors of resistances $r_1, r_2, r_3, \ldots r_n$ in parallel (Fig. *a*), the total resistance R is given by:
$$1/R = 1/r_1 + 1/r_2 + 1/r_3 + \ldots 1/r_n.$$
The current in any branch is: $i_n = i(R/r_n)$ where i is the total current. For capacitors of capacitances $c_1, c_2, c_3, \ldots c_n$ in parallel (Fig. *b*), the total capacitance C is given by:
$$C = c_1 + c_2 + c_3 + \ldots c_n.$$
They behave as a large capacitor of the total plate area.

The greatest current a cell can produce (i.e. with zero external resistance) is limited by its internal resistance. Consequently, several cells may be used in parallel to reduce the total internal resistance of the cell circuit so that a large current can be drawn. *Compare* series. *See* shunt.

parallelogram of vectors A geometrical construction expressing the law of addition of two vectors. *See* vector.

parallel system of distribution *See* shunt system of distribution.

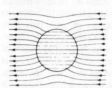

Paramagnetic substance in a magnetic field

paramagnetism The property of substances that have a positive magnetic *susceptibility. It is caused by the *spins of electrons, paramagnetic substances having molecules or atoms in which there are unpaired electrons and thus a resulting *magnetic moment. There is also a contribution to the magnetic properties from the orbital motion of the electron. The relative *permeability of a paramagnetic substance is thus greater than that of a vacuum, i.e. it is slightly greater than unity. For platinum its value is 1.000 02 at 20 °C.

Lines representing the flux in a uniform magnetic field (*see* diagram) become closer together when passing through a paramagnetic material; similarly if a paramagnetic substance is placed in a nonuniform field, a force acts on it in the direction from the weaker to the stronger part of the field. If a bar of paramagnetic substances is freely suspended in a field, it rotates so that its longer axis is aligned with the field.

A paramagnetic substance is regarded as an assembly of magnetic dipoles that have random orientation. In the presence of a field the magnetization is determined by competition between the effect of the field, in tending to align the magnetic dipoles, and the random thermal agitation. For small fields and high temperatures, the magnetization produced is proportional to the field strength. (At low temperatures or high field strengths, a state of saturation is approached.) As the temperature rises the susceptibility falls according to *Curie's law or the *Curie–Weiss law.

Solids, liquids, and gases can exhibit paramagnetism. Some paramagnetic substances are ferromagnetic below their Curie temperature. (*See* ferromagnetism; diamagnetism; antiferromagnetism; ferrimagnetism.)

Certain metals, such as sodium and potassium, also exhibit a type of paramagnetism resulting from the magnetic moments of free, or nearly free, electrons in their conduction bands (*see* energy bands). This is characterized by a very small positive suscep-

parameter

tibility and a very slight temperature dependence. It is known as *free-electron paramagnetism* or *Pauli paramagnetism*.

Another effect, known as *Van Vleck paramagnetism*, is characterized by a very small positive susceptibility and no temperature dependence at all. It occurs for certain compounds, such as compounds containing Co^{2+} ions, where there is a large difference in energy between two states of the atom.

parameter A quantity that is constant in a given case but takes a particular value for each case considered. Some examples are: (*a*) A constant in the equation to a curve or surface by whose variation the equation is made to represent a family of such curves or surfaces. (*b*) (astron.) The data necessary to determine the orbit of a celestial body. (*c*) The values of the resistances, capacitances, inductances, etc., that determine the properties of an electrical *network. (*d*) Crystal parameters. The distance from the origin of the axis of a crystal, to the intersection of any axis with a face in terms of the arbitrary unit selected for measurement along that axis. (*See* rational intercepts, law of.)

parametric amplifier A low-noise microwave amplifer in which gain is achieved by periodically varying the *reactance of the device, usually by applying an external voltage to a *varactor. Energy can then be transferred from the external signal so that amplification is achieved.

parasitic capture *Capture of a neutron by an atomic nucleus without any consequent nuclear *fission occurring.

parasitic oscillations Unwanted oscillations that may occur in the circuit of an amplifier or oscillator. Such oscillations usually have a frequency very much higher than the frequencies for which the circuit has been designed, since it is mainly determined by stray inductances and capacitances (e.g. in connecting leads), and interelectrode capacitances. Devices incorporated in the circuit to prevent the generation of parasitic oscillations are called *parasitic stoppers*. These are commonly resistors used in the input and output circuits of the device.

parasitic stoppers *See* parasitic oscillations.

paraxial rays Rays of light close to the optic axis of a system. Under these conditions angles of incidence, etc., are small so that sines of angles can be replaced by angles expressed in radians and *spherical aberration does not have to be considered. The ordinary *conjugate focus relations only apply strictly to paraxial rays.

parent Any *nuclide from which a given nuclide, the daughter product, is formed by radioactive *decay. Thus if a nucleus A decays to give a nucleus B, A is the parent and B the *daughter*.

parity *Syn.* space-reflection symmetry. Symbol: *P*. The principle of space-reflection symmetry, or parity invariance, states that no fundamental distinction

can be made between left and right; that the laws of physics are the same in a right-handed system of coordinates as they are in left-handed system. This is true for all the phenomena described by classical physics.

A *wave function, $\psi(x, y, z)$, describing a quantum mechanical system is said to have parity $+1$ if it remains unchanged on reflection through the origin, i.e. $\psi(x, y, z) = \psi(-x, -y, -z)$. If $\psi(x, y, z) = -\psi(-x, -y, -z)$ the wave function is said to have parity -1. In general, wave functions do not have a definite parity since $\psi(-x, -y, -z)$ will not usually be proportional to $\psi(x, y, z)$. It is found that the wave functions describing individual elementary particles have a definite symmetry under reflection. This means that an intrinsic parity can be associated with elementary particles.

The parity of the total wave function describing a system of elementary particles is conserved in *strong interactions and *electromagnetic interactions. However, *weak interactions do not exhibit parity invariance and parity is not conserved in these interactions. *Beta decay is an example of a process in which parity is not conserved. Parity invariance requires that, if in some interaction a particle is produced with left-polarization (i.e. the particle spins in an opposite sense to the direction of motion), then it must also be possible for a right-polarized particle to be produced in a similar interaction and that, on average, equal numbers of each will occur. In beta decay it is found that the electron is always left-polarized.

parsec Symbol: pc. An astronomical unit of length equal to the distance at which a baseline of one *astronomical unit subtends an angle of one second of arc. 1 parsec $= 3.085\,677 \times 10^{16}$ metres or about 3.26 light-years.

Parsons, Sir Charles Algernon (1845–1931) Brit. engineer. The creator of the steam turbine. He made considerable improvements in optical engineering, especially in the production and working of large glass elements (telescope mirrors, etc.).

partial A musical note consists generally of the simultaneous sounding of a group of tones, the frequency of each tone usually being related to the generating tone (namely the fundamental) by the equation $A = nF$, where n is an integer, and A and F are frequencies of a tone of the series and of the fundamental respectively. Certain musical sounds contain tones that are not part of the harmonic series, e.g. *bell sounds. Tones above the fundamental are known as overtones or upper partials, these two terms being synonymous. Harmonics are partials, but, since partials may be inharmonic, the converse may not be true. The fundamental is known as the first harmonic or first partial; the octave, second note of the harmonic series, as the second harmonic or second partial or first overtone. The quality of a note depends upon the intensity and number of the upper partials in relation to the funadmental.

partial ionic character The character of chemical bonds that are neither wholly ionic nor wholly covalent. The partial ionic character of a bond between two atoms depends on their abilities to attract electrons, i.e. on their *electronegativities. Hannay and Smyth have shown that there is an empirical relationship equating the percentage ionic character of a bond between two atoms A and B to $16(X_A - X_B) + 35(X_A - X_B)^2$, where X_A and X_B are the electronegativities of A and B.

partial pressure The partial pressure of a gas in a mixture of gases occupying a fixed volume is the pressure that the gas would exert if it alone occupied the total volume. *See also* Dalton's law of partial pressures.

particle A body of infinitely small dimensions; or, a mathematical point endowed with mass. Any extended body may be regarded as composed of particles. The concept of a particle is a convenient fiction in mechanics. *See also* elementary particle.

particle physics The study of the structure and properties of *elementary particles and *resonances and their interactions (*see* electromagnetic, strong, and weak interactions; electroweak theory; gauge theory; grand unified theory; quantum chromodynamics; quantum electrodynamics). The interactions of particles are responsible for their scattering and transformations (decays and reactions). During the first forty years of the twentieth century only the proton and neutron (nucleons) and the electron and positron had been detected; however, the number of particles with a claim to elementarity now approaches 150 (excluding resonances). The task of the particle physicist has been to try to correlate systematically the properties of these particles, as the chemists succeeded earlier in correlating the properties of an approximately equal number of chemical elements. Only partial success has been achieved (*see* unitary symmetry), and it has relied more on a study of conservation laws (*see* conservation of energy, mass, momentum, and charge; quantum number) than on the detailed elucidation of the mechanics of nuclear interactions.

The problem of elementarity has also yet to be resolved, the concepts of theorists (*see* quarks) not having been supported by experimental evidence from the high-energy physicists. (*See* accelerator.)

particle velocity Symbol: u. The alternating component of the velocity of a medium that is transmitting sound. It is thus the total velocity of the medium minus the velocity that is not due to sound propagation. The velocity is changing regularly with time and the term *particle velocity* is usually taken to mean its *root-mean-square value. Otherwise the terms *instantaneous particle velocity* and *maximum particle velocity* are used.

Partington's method for determining γ Partington found the ratio of the principal specific heat capacities of gases using a modification of *Lummer and Pringsheim's determination, the rapid temperature

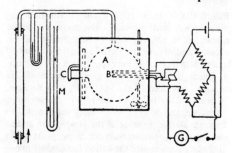

Ratio of specific heat capacities of gases

changes of a bolometer being recorded by means of an *Einthoven galvanometer. Pure dry gas is admitted to the copper vessel A, of 130 litres capacity, and its temperature measured by a bolometer of fine platinum wire with compensating leads to eliminate the error due to conduction along the leads. The size of the expansion valve C is altered by inserting brass diaphragms so that no overshooting occurs, the galvanometer deflection being instantaneous and steady. In the experiment, the resistance in the arm of the bridge opposite to the bolometer was lowered by an arbitrary amount and the initial excess pressure $(p - p_0)$ registered by the oil manometer M was varied, until the galvanometer G showed no deflection if its circuit was closed immediately after the expansion. The temperature change corresponding to the fixed lowering of resistance was found afterwards by adding iced water to the bath containing A until the galvanometer again showed no deflection. This temperature change $(T - T_0)$ and the temperature prior to the expansion (T) were measured on a standardized thermometer. Assuming *Berthelot's equation of state to hold for the gas, then

$$\left\{\frac{p}{p_0}\right\}^{(\gamma - 1)/\gamma} = \frac{T}{T_0}\left\{1 + \frac{27}{32}\frac{\pi}{\theta^3}\right\},$$

then

$$\pi = \frac{p + p_0}{2p_c}; \quad \theta = \frac{T + T_0}{2T_c},$$

where the suffix c refers to a value at the critical point.

partition coefficient and law See distribution law.

partition function See statistical mechanics.

parton A hypothetical pointlike particle postulated to exist inside nucleons (associated with quarks). The *parton model* was introduced to explain the results of high-energy experiments on nuclei, which seemed to show that nucleons had an internal structure. The parton model is now understood in terms of asymptotic freedom in quantum chromodynamics.

pascal Symbol: Pa The *SI unit of *pressure, defined as the pressure that results from a source of one newton acting uniformly over an area of one square metre.

Pascal, Blaise

Pascal, Blaise (1623–1662) French mathematician, philosopher, and physicist. His most important researches in physics were on the subject of the vacuum and of pressure in fluids. In 1646 he heard incomplete reports on the experiments of *Torricelli, which he repeated and greatly extended. He became convinced of the possibility of a vacuum, contrary to the belief of nearly all scholars since *Aristotle, and he conducted many experiments to demonstrate the pressure of the atmosphere and to show that it could account for those phenomena (including suction and the action of siphons) previously attributed to nature's "abhorrence" of a vacuum. The decisive experiment in 1648, showing that the barometric height was much less at the top of a mountain than at the base, was conducted according to his instructions. His later work, which systematized hydrostatics, was completed by 1653 but not fully published until 1663 (*see* Pascal's principle). The principle that the pressure exerted by a liquid depended only on the height and not on the shape of the vessel was demonstrated using *Pascal's vases*, which were variously shaped vessels fitted with pistons such that the force required to support the liquid could be measured. He distinguished between the uniform compression caused by fluid pressure and the distortion caused by stresses in solids, thereby explaining why animals are not harmed by the former.

Pascal constructed the first calculating machine (1641, improved 1649). With Fermat he founded the theory of probabilities.

Pascal's principle Pressure applied at any point of a fluid at rest is transmitted without loss to all other parts of the fluid. *See* hydraulic press.

```
            1
          1   1
        1   2   1
      1   3   3   1
    1   4   6   4   1
    . . . . . . . .
```

Pascal's triangle A triangle of numbers in which each number is obtained by summing the two adjacent to it in the preceding row. It contains, in successive lines, the coefficients in *binomial expansions.

Paschen, Friedrich (1865–1947) German physicist who investigated the infrared *hydrogen spectrum.

Paschen–Back effect An effect similar to the *Zeeman effect, but applicable to magnetic fields so strong that the vectors due to orbital and spin angular momentum of electrons each separately take up their possible orientations relative to the field direction. The split pattern of spectral lines produced is quite different from those of the Zeeman effect, the lines being due to transitions between the quantum states of the electron orbits.

Paschen series *See* hydrogen spectrum.

Paschen's law The *breakdown voltage for a discharge between electrodes in gases is a function of the product of pressure and distance; e.g. if the distance between the electrodes is doubled, the pressure must be halved for the same breakdown potential difference.

passband *See* filter.

passivation Protection of the junctions and surfaces of electronic components, including integrated circuits, from harmful environments. With silicon chips, a protective layer of silicon dioxide is usually formed on the surface.

passive aerial *See* directive aerial.

passive circuit A circuit or *network that contains only *passive components. Such circuits are capable only of attenuation, and are often designed for use as passive *filters.

passive component An electronic component that is not capable of amplifying or control function, e.g. resistors, capacitors, and inductors.

Pasteur, Louis (1822–1895) French scientist. His earlier work was on optical activity (rotation of the plane of polarization) and he showed that some crystalline substances exist in two forms, related as object and image in a plane mirror and this external form is related to the right-handed and left-handed rotations exhibited by solutions of these forms. (*See* dextrorotatory; laevorotatory.) His main work (e.g. on fermentation, bacterial origin of disease) lies in other fields (biochemistry, bacteriology, etc.). Pasteurization is the sterilization of a liquid (wine, milk, etc.) by controlled heating in the absence of air.

path integral *See* quantum mechanics.

Paul, Wolfgang (*b.* 1913) German physicist working at the University of Bonn. He shared the 1989 Nobel prize with Hans Dehmelt for their work on the *ion-trap technique*, which makes it possible to isolate a single ion or electron for accurate measurement of energy levels.

Pauli, Wolfgang (1900–1958) Austrian-born Swiss physicist. His contributions to quantum theory include space quantization of electron orbits in a magnetic field (1920), the very important *Pauli exclusion principle (1925) and its applications to electron theory of metals, to spectra, and to explanation of the periodic table. He suggested (1927) the *neutrino to explain the continuous beta-ray spectrum obtained in *beta decay.

Pauli exclusion principle The principle that no two identical *fermions in any system can be in the same quantum state, that is have the same set of *quantum numbers. The principle was first proposed (1925) in the form that not more than two electrons in an atom could have the same set of quantum numbers. This hypothesis accounted for the main features of the structure of the *atom and for the *periodic table. With the introduction of the fourth quantum number (*see* spin) it was seen that only one electron could be in a given state. An electron in an atom is characterized by four quantum numbers, *n*, *l*, *m*, and *s*. A particular *atomic orbital, which has fixed values of *n*, *l*, and *m*, can thus contain a maximum of two electrons, since the *spin quantum number *s* can only be $+\frac{1}{2}$ or $-\frac{1}{2}$. Two electrons with opposing spins in an atomic orbital are said to be spin-paired.

The concept of exclusion is necessary to explain the nonobservation of electron states of negative energy in the relativistic wave mechanics of *Dirac (*see* annihilation; antiparticle; pair production). It has been shown to be a general principle applicable to every type of fermion including nucleons, and is a consequence of Pauli's *spin-statistics theorem of relativistic quantum field theory.

Pauli paramagnetism *See* paramagnetism.

Pauthenier generator An electrostatic generator in which a blast of air at high pressure carries powdered glass close to a heated filament from which it gains an electric charge. The powdered glass then strikes a metal collector on which the charge is given up. Air currents of about 50 metres per second are used and the potential achieved is limited only by the *brush discharge from the collector.

PCB Abbreviation for *printed circuit board.

p-channel device *See* field-effect transistor.

PCM Abbreviation for pulse-code modulation. *See* pulse modulation.

p.d. Abbreviation for potential difference.

peak factor The ratio of the *peak value of an alternating or pulsating quantity to its *root-mean-square value. For a *sinusoidal quantity, the peak factor is $\sqrt{2}$.

peak forward voltage The maximum instantaneous voltage applied to a device in the forward direction, i.e. the direction of minimum resistance of the device.

peak inverse voltage The maximum instantaneous voltage applied to a device in the reverse direction, i.e. the direction of maximum resistance of the device. The peak inverse voltage applied to a rectifying device must be less than the *breakdown voltage of the device to prevent avalanche breakdown in a *semiconductor or arc formation in a valve. A rated value is often used to specify the maximum voltage that a device can withstand.

peak value *Syn.* amplitude. **1.** Of an alternating quantity. The maximum positive or negative value. The positive and negative values need not necessarily be equal in magnitude.
2. Of an *impulse voltage (or current). The maximum value of the voltage (or current).

Peierls, Sir Rudolf (*b.* 1907) German-born physicist who has worked in England since 1929. Together

with Otto *Frisch he calculated the critical mass of U-235 required to cause a nuclear explosion.

Peltier, Jean Charles Athanase (1785–1845) French watchmaker and scientific amateur who discovered the *Peltier effect in 1834.

Peltier constants Constants that measure the heat developed (or absorbed) at the junction of a crystal bar and an isotropic metal, when the bar is cut with different crystallographic directions normal to the junction plane, and an electric current is then sent across the junction. *See* thermoelectric effects.

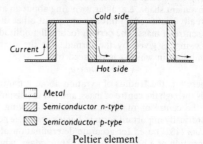

□ Metal
▨ Semiconductor n-type
▧ Semiconductor p-type

Peltier element

Peltier effect The liberation of heat at one junction and the absorption of heat at the other junction when a current is passed round a circuit consisting of two different metals. The Peltier effect, the converse of the Seebeck effect (*see* thermoelectric effects), is reversible; if the current is reversed, the hot junction cools and the cold one heats up. The temperature difference obtained using two metals is not high. However, *semiconductors can be used to produce larger temperature differences. A junction made of metal and a *p-type semiconductor produces a voltage of opposite sign to that occurring in a metal–*n-type semiconductor junction. (*See* diagram.) A *Peltier element* contains a number of these metal-semiconductor junctions and is used as a heating or cooling element.

pencil of rays A slender cone or cylinder of rays that traverses an optical system, the pencil being limited by the *aperture stop. Pencils may be diverging (from real object points or from virtual image points), parallel, or converging (to virtual objects or real images). The central ray is the *chief ray* of the pencil. An incident pencil fills the *entrance pupil through the centre of which the chief ray passes. A collection of pencils constitutes a beam.

pendulum A device consisting of a mass, suspended from a fixed point, that oscillates with a known period, *T*. The various types are as follows.

(1) *Simple pendulum.* A small weight suspended from a point by a light thread. The period of oscillation for small amplitudes of swing is determined by the formula:
$$T = 2\pi\sqrt{(l/g)},$$
where *l* is the length of the thread and *g* is the acceleration of *free fall.

Cycloidal pendulum

(2) *Compound pendulum.* A rigid body of any convenient shape, e.g. a bar, swinging about an axis (usually a knife edge) through any point other than its centre of mass. The period, for small amplitudes of swing, is given by the formula for the simple pendulum in which l is replaced by:

$$\sqrt{(k^2 + h^2)/h},$$

where k is the *radius of gyration about a parallel axis through the centre of mass, and h is the distance of the centre of mass from the axis of swing. A historically important form is *Kater's reversible pendulum* (1817) used for accurate determination of g, consisting of a bar carrying two knife edges, which are set facing one another, one on each side of the centre of mass. They serve as axes about which the body can oscillate under the action of gravity. On the bar are also mounted two sliding masses, one of which is much larger than the other. The position of the larger mass is adjusted to make the times of oscillation about the two knife edges equal. The period of oscillation is that of a simple pendulum whose length is the distance between the two knife edges, which can be measured very accurately. There are various small corrections. The method was greatly improved by *Bessel.

(3) *Horizontal pendulum* A compound pendulum whose axis of rotation is nearly vertical. Used for finding the alteration in the direction of the force of gravity with time. A massive horizontal pendulum is the basis of a seismograph (known as the *Galitzin pendulum*).

(4) *Cycloidal pendulum.* The period of a simple pendulum is not quite independent of the magnitude of the arc, but it can be made so (Huygens, 1657) by suspending the pendulum by means of a pair of cords slung between two pieces of metal M-shaped in the form of a cycloid as indicated in the diagram.

(5) *Conical pendulum* A simple pendulum in which the bob swings in a horizontal circle. The period, for a very small radius of swing only, is the same as for the simple pendulum.

(6) *Equivalent simple pendulum* A simple pendulum that is capable of free oscillations in unison with a given freely oscillating pendulum.

(7) *Ballistic pendulum* A large mass hanging by a rod from a horizontal axle. The mass is set into motion by a horizontal impact and the vertical height to which it then rises is used to measure the impulse given to it. It is used to measure the speeds of projectiles.

(8) *See* compensated pendulum.

penetrating shower *See* cosmic rays.

Penney, William George, Baron (*b*. 1909) Brit. nuclear physicist and former director of the Atomic Energy Commission.

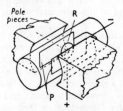

Penning gauge

Penning gauge *Syn.* Philips' gauge. A high-vacuum gauge consisting of a wire ring electrode, R, between two plates, P (*see* diagram). 1000–2000 volts is applied between the ring and the plates. Electrons emitted from the plates move to the ring, but are compelled to do so in long spiral paths by an applied magnetic field. Ions are formed by collisions with remaining gas molecules and add to the normal current between the electrodes.

Penning ionization The ionization of gas atoms or molecules by collision with atoms in a *metastable state. Metastable helium atoms, for example, in one excited energy state have an energy 21 eV greater than the ground state and can transfer this energy to other atoms and molecules, thus causing them to ionize:

$$M + He^* \rightarrow M^+ + He + e^-.$$

This process occurs if the *ionization potential of the atom or molecule is less than the energy released when the metastable atom reverts to the ground state. The excess energy is carried away as kinetic energy of the electron. If sufficient energy is available, it is possible to produce ions in excited states. Penning ionization is one of the reactions occurring in gas discharges and is also used in *electron spectroscopy.

Penrose, Roger (*b*. 1931) Brit. mathematician and theoretical physicist, who became professor of mathematics at Oxford. Much of his work has been on the properties of *black holes and on attempts to unite general relativity with quantum mechanics. Penrose showed (1965) that a body undergoing gravitational collapse must eventually form a singularity (this work was extended in collaboration with Stephen *Hawking). Penrose also put forward the hypothesis of 'cosmic censorship' – that a singularity must be surrounded by an event horizon.

pentagonal prism A glass prism with two silvered surfaces inclined at 45° for producing 90° deviation, being independent of the angle of incidence on the prism (used at the ends of the baseline of self-contained range-finders, etc.).

pentagrid *See* heptode.

pentode A *thermionic valve with five electrodes. It is equivalent to a tetrode with an additional electrode between the screen grid and anode. This electrode, which is of open-mesh design, is at a negative potential with respect to both anode and screen so that it can prevent low-velocity secondary electrons from the anode returning to the screen.

penumbra See shadows.

Penzias, Arno Allan (b. 1933) Amer. physicist who worked at Bell Telephone Laboratories. He shared the 1978 Nobel prize for physics with Robert Wilson for their discovery (1964) of *microwave background radiation.

Pepper's ghost A stage illusion in which a ghost appears to move through solid objects. A large sheet of glass is used to reflect the image of an actor moving out of view of the audience (e.g. in the wings or the orchestra pit). The glass both reflects and transmits light, and the actor seems to be superimposed on furniture on the stage.

Peregrinus, Petrus (13th century), French scientist who discovered the laws of attraction between magnets and invented the compass with a graduated scale. (1269)

perfect crystal A single crystal in which the arrangement of atoms is uniform throughout. Compare mosaic crystal.

perfect fluid See fluid.

perfect gas See ideal gas.

pericynthion The point at which a satellite launched from earth into lunar orbit is nearest to the surface of the moon. Compare apocynthion; perilune.

perigee The point in the orbit of the moon or an artificial earth satellite that is nearest the earth. The most distant point in an earth orbit is called the apogee.

perihelion The point in a solar orbit that is nearest the sun. The orbiting body could be a planet, comet, or artificial satellite. The earth is at perihelion on Jan. 3. The most distant point in a solar orbit is called the aphelion. The earth is at aphelion on July 4.

perilune The point at which a satellite launched from the moon into lunar orbit is nearest to the surface of the moon. Compare apolune; pericynthion.

period 1. Syn. periodic time. Symbol: T. The time occupied in one complete to and fro movement of a given vibration or oscillation. The period is related to the frequency v, and the angular frequency ω, by:
$$T = 1/v = 2\pi/\omega.$$
2. See periodic table.

periodic A varying quantity that repeats itself at regular intervals is said to be periodic, and the interval between two successive repetitions is known as the *period.

periodic law See periodic table.

periodic table The classification of chemical elements, introduced by *Mendeleev in 1869, which depends on his periodic law that the elements when arranged in the order of their atomic weights show a periodicity of properties, chemically similar elements recurring in a definite order. The sequence of elements is thus broken up into horizontal periods and vertical groups, the elements in each group showing close chemical analogies, e.g. in valency, chemical properties, etc. Mendeleev arranged the periodic table and another such table was published independently by Lothar Meyer. Certain anomalies exist if the elements are set out in order of their atomic weights, but these are largely removed when the elements are in order of their *atomic numbers. Since Mendeleev's first table, many forms have been proposed. The modern form is given in the Appendix, Table 10.

The elements are arranged in eight vertical columns called Groups 1 to 8, with a ninth column, '0', the inert gases; hydrogen is placed in the first column for convenience but its position there is anomalous. The number of a group corresponds to the valency of the elements it contains. There are seven horizontal periods, each of which ends in an inert gas; the first period consists of hydrogen and helium only, hydrogen being at one end and helium at the other end of the period. The ordinal numbers of the elements are their atomic numbers.

By fixing the elements in the Table, it was possible for Mendeleev to correct some of the atomic weights (e.g. that of beryllium) in use in 1869 and to predict the properties of a number of elements yet to be discovered.

Soddy showed that the loss of an α-particle reduces the nuclear charge by two and hence lowers the atomic number by two, and the position of the element in the periodic table by two groups. Fajans pointed out that the loss of a β-particle raises the nuclear positive charge by unit and thus moves the element up by one group in the table. (See also radioactive series.)

periodic time See period.

peripheral devices Syn. peripherals. Equipment that can be connected to and controlled by a *computer. Peripherals are external to the central processor of the computer. They may be input devices such as keyboards, output devices such as printers, or backing store such as disk or magnetic-tape units.

periscope 1. An optical instrument to provide a view over or around an obstacle or from a submarine. In its simplest form, it consists of two parallel mirrors at 45° to the direction of view; the top mirror receives light from the object and directs it down to the lower mirror close to the eye of the observer. More elaborate periscopes used a top rotating reflecting prism, a tube containing unit magnification transfer systems dependent on the length of the tube, together with a rotating erecting prism, deliv-

permalloy

ering the light to the eyepiece with which is a reflecting prism to provide for comfortable vision. The rotating erecting prism compensates for the tilting action produced by the rotating top prism.

2. A form of thin lens approaching *meniscus shape, having ±1.25 D base curve (minus base for convering lenses).

permalloy An alloy with a high magnetic *permeability at low magnetic flux density and a low *hysteresis loss. The term was originally applied to an alloy of 78.5% nickel and 21.5% iron but is now used for a variety of materials containing cobalt, manganese, etc.

permanent gas A gas whose *critical constants are such that it remains gaseous under very high pressure at normal temperatures. A gas that cannot be liquefied by pressure alone.

permanent magnet A magnetized mass of steel or other ferromagnetic substances, of high retentivity and stable against reasonable handling. It requires a definite demagnetizing field to destroy the residual magnetism. *See* ferromagnetism.

permanent magnet moving-coil instrument *Syn.* moving-coil instrument. An electrical instrument in which the action depends upon the mechanical torque exerted by a fixed permanent magnet on a movable coil carrying a current. Instruments of this type are suitable for use only with direct current but for such use they have two outstanding advantages over other types: (*a*) Relatively low power consumption on account of the high torque/weight ratio. (*b*) Uniformly divided scale. They can be adapted for alternating-current measurements by means of rectifiers. *See* rectifier instrument; moving-coil galvanometer.

permanent set The strain remaining in a material after the stresses have been removed.

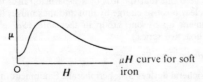

μH curve for soft iron

permeability Symbol: μ. The ratio of the *magnetic flux density in a body or medium to the external *magnetic field strength inducing it, i.e. $\mu = B/H$. It has the unit *henry per metre. The *permeability of free space*, μ_0, is sometimes called the *magnetic constant and has the value $4\pi \times 10^{-7}$ H m^{-1} in the SI system. The relative permeability, μ_r, is the ratio μ/μ_0. For most substances μ_r has a constant value. If it is less than unity the material is diamagnetic (*see* diamagnetism); if μ_r exceeds unity it is paramagnetic (*see* paramagnetism). *Ferromagnetic materials have high permeabilities, which are not constant but vary with the field strength. A typical μH curve for soft iron is shown in the illustration. *See also* permittivity.

permeability tuning *See* tuned circuit.

permeameter An instrument for measuring the magnetic characteristics, in particular the *permeability, of ferromagnetic materials.

permeance Symbol: $\wedge$. The reciprocal of *reluctance. It is measured in henrys.

Perminvar A variety of *permalloy, having a constant magnetic permeability for small magnetic field strengths. It is composed of 45–60% Ni, 30–25% Fe, 25–15% Co.

permittivity Symbol: ε. The ratio of the *electric displacement in a dielectric medium to the applied *electric field strength, i.e. $\varepsilon = D/E$. It indicates the degree to which the medium can resist the flow of electric charge. It is measured in farads per metre. The *permittivity of free space*, ε_0, is sometimes called the *electric constant. It is equal to $1/(c^2\mu_0)$, i.e.

$$8.854\ 187\ 817 \times 10^{-12}\ \mathrm{F\ m}^{-1},$$

where c is the speed of light and μ_0 is the *permeability of free space. The relative permittivity, ε_r, is the ratio of the permittivity of a medium to the permittivity of free space, i.e. $\varepsilon/\varepsilon_0$. The value varies from unity (for a vacuum) to over 4000 (for *ferroelectric materials) but normally does not exceed 10. The quantity ε_r is also called the *dielectric constant* when it is independent of electric field strength and refers to the dielectric medium of a capacitor. It is better defined under these conditions as the ratio of the capacitance of the capacitor to the capacitance it would possess if the dielectric were removed.

Perrin, Jean Baptiste (1870–1942) French physicist. He made a thorough study of the *Brownian movement and the vertical distribution of colloidal particles in gamboge. The results were presented to the first *Solvay Conference (1911) as evidence for the existence of molecules, the reality of which had been disputed, notably by *Mach. It was shown that the vertical distribution followed the same law as for molecules in an ideal gas and permitted the evaluation of the *Avogadro constant. Previous observations on Brownian motion had indicated kinetic energies incomparably less than predicted by kinetic theory, but Perrin was able to explain these results and to interpret his own measurements using *Einstein's diffusion equation. This gave values for the Avogadro constant in close agreement with the vertical distribution and also with values reported by other workers using unrelated methods. He demonstrated (1895) the negative charge carried by cathode rays.

persistence *Syn.* afterglow. **1.** The interval of time following excitation during which light is emitted from the screen of a *cathode-ray tube. The amount of persistence depends on the type of phosphor used for the screen and varies very widely.

2. The faint luminosity, observable in certain gases for a considerable period after the passage of an electric discharge.

personal equation A systematic error of measurement, made by an experienced observer, e.g. habitually estimating the passage of an object across the wires in the focal plane of an eyepiece slightly after the actual event. If the observer tires, or is inexperienced, he ceases to have a personal equation that can be corrected for, and contributes random errors.

perspex dosimetry *See* dosimetry.

perturbation theory An approximate method of solving a difficult problem if the equations to be solved depart only slightly from those of a problem already solved. For example, the orbit of a single planet round the sun is an ellipse; the perturbing effect of the other planets modifies the orbit. It is not possible to obtain an analytical solution of the equations describing the motions of all the planets (*see* many-body problem). In perturbation theory, the system is divided into two parts: an exactly calculable part (e.g. the elliptical orbit of the planet), and a correction term (the effect of other planets), which acts as a 'perturbation' of the exactly calculable part. In general, the technique enables the system to be described by an infinite series, terms of which are 'correction terms' for the calculable system.

Although originally developed in classical physics for use in celestial mechanics, perturbation theory is extensively used in quantum mechanics (e.g. for calculating the energy levels of molecules). It is also used in relativistic quantum-field theory, in which case the terms in the series describing the system can be represented by Feynman diagrams (*see* quantum electrodynamics). *Non-perturbative phenomena* are phenomena that can not be predicted or described by perturbation theory.

perveance The space-charge-limited characteristic between electrodes in an electron tube. It is equal to $j/V^{3/2}$, where j is the current density, and V the potential of the collector.

peta- Symbol: P. A prefix denoting 10^{15}; for example, one petajoule (1 PJ) $= 10^{15}$ joules.

Petersen coil *See* arc-suppression coil.

Petit, Alexis Thérèse (1791–1820) French physicist. He collaborated with P. L. Dulong in several important investigations, including studies of thermal expansion of gases and of liquids, the cooling of bodies and the heat capacities of substances. *See* Dulong and Petit's law.

petrol engine An engine using light fuel oil that has a high vapour pressure. Fuel and air are ignited electrically, operating a piston in a two-stroke or four-stroke cycle (*Otto cycle).

Petzval surface When the tangential and sagittal image surfaces due to oblique refraction by a compound lens have been made coincident, the image is stigmatic but lies on a curved surface, the Petzval surface, with curvature given by $\Sigma(F/n)$, where F is the power of the lens, n its corresponding refractive index (Petzval formula, 1843).

Pfeffer, William (1845–1920) German botanist who discovered that a substance in solution exerts a pressure (*see* osmosis) on the boundary of the solution and that this pressure is in many cases proportional to the concentration of the solute. *See* Pfeffer's pot.

Pfeffer's pot A *semipermeable membrane used in studying *osmotic pressure. Pfeffer formed such a membrane of copper ferrocyanide at the junction of solutions of copper sulphate and potassium ferrocyanide in the pores of a porous pot (1877).

Pfund series *See* hydrogen spectrum.

PGA Abbreviation for *pin grid array.

pH A logarithmic measure of the hydrogen ion (or hydroxonium ion, H_3O^+) concentration of a solution. It equals the logarithm to the base 10 of the reciprocal of the hydrogen (or hydroxonium) ion concentration in moles per dm^3. Thus, if there are 10^{-8} mol dm^{-3} of hydrogen ions present, the solution has a pH of 8. If the pH is greater than 7, the solution is alkaline, and if it is less, the solution is acid.

phantom circuit The two pairs of wires of two telephone circuits (transmission lines) may be used effectively as the two sides of a third circuit. This is a phantom circuit since it is superposed on the other two, the latter being called physical or side circuits. As far as the phantom circuit is concerned, the two wires in each pair are effectively in parallel. To avoid *crosstalk, the individual circuits must be carefully balanced.

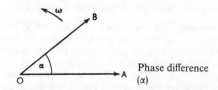

Phase difference (α)

phase 1. Any homogeneous and physically distinct part of a system that is separated by definite bounding surfaces from other parts of the system; e.g. the various crystalline forms of ice, water, and water vapour are the phases of the water system. (*See* phase rule; phase diagram; phase transition; alloy.)

2. As applied to an operation that repeats periodically: (i) the stage of development of the process; (ii) the fraction of the whole *period that has elapsed, measured from some fixed datum. A quantity that varies sinusoidally may be represented by a rotating vector OB (*see* diagram and simple harmonic motion), whose length is proportional to the *peak value of the quantity. OB rotates through 360° about O during the period, T, of the oscillation. The angular velocity, ω, of OB is related to the frequency, ν, of the oscillation by:

$$\nu = 1/T = \omega/2\pi.$$

The phase of the quantity with respect to another such quantity, OA, is given by the angle, α, between

OB and OA. This is called the *phase angle if the two quantities have the same frequency (*see also* phase difference).

Particles in periodic motion due to the passage of a wave are said to be in the same phase of vibration if they are moving in the same direction with the same relative displacement. Particles in a wavefront are in the same phase of vibration and the distance between two wavefronts in which the phases are the same is the wavelength. For the simple harmonic wave,
$$y = a \sin 2\pi(t/T - x/\lambda),$$
the phase difference of the two particles at x_1 and x_2 is:
$$2\pi(x_2 - x_1)/\lambda.$$
When light is reflected at the surface of a denser medium, there is a change of phase of π.

Periodic quantities having the same frequency and *waveform are said to be *in phase* if they reach corresponding values simultaneously; otherwise they are said to be *out of phase*. If the waveforms are not alike, these terms are used in connection with the fundamental components of the waves.

3. One of the separate circuits or windings of a *polyphase system, machine, or other apparatus.

4. One of the lines or terminals of a polyphase system.

phase advancer An electrical machine connected in series with the secondary (rotor) circuit of an *induction motor to inject into the rotor circuit an e.m.f. that leads the rotor current for the purpose of improving the *power factor of the motor. This is especially useful when the motor is operating with small loads since, under such conditions, the inherent power factor of an induction motor is low.

phase angle The angle between the two vectors that represent two sinusoidal alternating quantities having the same frequency. (*See* phase.) The term may be used in connection with periodic quantities that are not sinusoidal but that have the same fundamental frequency and it is then the angle between the vectors representing their fundamental components. *See* phase difference.

phase constant *Syn.* phase-change coefficient. *See* propagation coefficient.

phase-contrast microscope *See* microscope.

phase converter *Syn.* phase changer; rotary phase converter. An electrical machine for converting alternating current having one number of phases into alternating current having a different number of phases. For example, single phase to three phase.

phased-array radar *See* radar.

phase delay The ratio of the *phase shift undergone by a periodic quantity to its frequency.

phase diagram A graph combining two conditional parameters (e.g. temperature, pressure, entropy, volume) of a substance drawn so that a particular curve represents the boundary between two *phases of the

substance. For example, on a plot of pressure against volume, the *critical temperature isothermal separates the gas phase from the vapour or liquid phase; on a plot of pressure against temperature, the *triple point represents the intersection of the three curves demarcating the solid/liquid, liquid/gas, and solid/gas boundaries.

phase difference 1. Symbol: ϕ. The difference of *phase between two sinusoidal quantities that have the same frequency. It may be expressed as a time or as an angle (the *phase angle).
2. In an instrument transformer. The angle betweeen the reversed secondary vector (current in a current transformer, voltage in a voltage transformer) and the corresponding primary vector. The phase difference is positive or negative according to whether the reversed secondary vector leads or lags the primary vector respectively. In this application an alternative term is *phase error*, but this is deprecated. (*See* phase angle.)

phase discriminator A *detector circuit in which the amplitude of the output wave is a function of the *phase of the input wave.

phase error *See* phase difference.

phase meter *See* power-factor meter.

phase modulation A type of *modulation in which the phase of the *carrier wave is varied about its unmodulated value by an amount proportional to the amplitude of the modulating signal and at a frequency equal to that of the modulating signal, the amplitude of the carrier wave remaining constant. A phase-modulated wave in which the modulating signal is sinusoidal may be represented by:
$$e = E_M \sin [2\pi Ft + \beta \sin 2\pi ft],$$
where E_M = amplitude of carrier wave, F = frequency of the unmodulated carrier wave, β = the peak variation in the phase of the carrier wave, caused by modulation, f = frequency of the modulating signal. *Compare* frequency modulation.

phase plate *See* microscope (phase-contrast).

phase rule Substances are capable of existing in very different states of aggregation. These states are called *phases. Thus, water at ordinary temperatures and pressures can exist in a solid, liquid, or vapour phase. The term *component* is applied to the least number of chemically identifiable substances required to define completely the existing phases. Thus, in the three phases described above, there is only one component, namely water. The phase rule, enunciated in 1874 by Willard Gibbs, is defined by the equation:
$$F = C - P + 2,$$
where C is the number of components in the system, P the number of phases present, and F the number of *degrees of freedom* of the system (the least number of independent variables defining the state of the system). Thus, if there is only one phase present and one component, namely water, the number of degrees of freedom is two; that is, the system is not

completely defined until the pressure and the temperature are fixed. If two phases are present, the number of degrees of freedom is one; that is to say, to each pressure there corresponds a fixed temperature and on the pressure-temperature diagram (see phase diagram) there is a definite line separating the liquid from the vapour phases. Similar lines exist separating the solid-liquid phases and the solid-vapour phases. These three curves meet at the *triple point and this represents the only pressure and temperature at which the three phases can exist in equilibrium.

phase sequence The order in which the three *phases of a three-phase system (see polyphase system) reach a maximum potential of given sign. In a given system, the normal order is called the positive sequence and the reverse order is called the negative sequence.

phase-sequence indicator An instrument for indicating the *phase sequence of a *polyphase system.

phase shift Of a periodic quantity. Any change that occurs in the *phase of one quantity or in the *phase difference between two or more quantities.

phase space A multidimensional space in which the coordinates represent the variables required to specify the state of the system, in particular a six-dimensional space incorporating three dimensions of position and three of momentum.

phase speed The speed with which wave crests and troughs travel through a medium; in fact the speed with which the phase in a homogeneous train of waves is propagated. It is expressed by λ/T, λ being the wavelength and T the period of vibration. This is equivalent to ν/σ, ν being the frequency of vibration and σ the *wavenumber. Compare group speed.

phase splitter A circuit that has a single input signal and produces two separate outputs with a predetermined *phase difference. An example is the *driver for a push-pull amplifier (see push-pull operation).

phase transition A change from one state to another; for example, a change from solid to liquid, liquid to gas, etc. The term is also applied to other types of discontinuous physical change – for example, a change from ferromagnetic to paramagnetic behaviour. Such changes occur at a certain *transition temperature*. Phase transitions are frequently characterized by *broken symmetry and, in gauge theories, they are thought to be important in understanding the cosmology of the *early universe.

phase voltage The voltage in one *phase. of a polyphase system. It is sometimes used as a synonym for voltage to neutral but this is deprecated since such use gives rise to ambiguities.

phase waves See de Broglie waves.

Philips' gauge See Penning gauge.

phlogiston theory A theory of combustion suggested by *Becher in the 17th century and refuted by *Lavoisier in the 18th. All combustible materials were supposed to contain phlogiston, which was released during combustion to leave calx (ash).

phon The unit of equivalent loudness of a sound, judged subjectively. It is a measure of the intensity level relative to a reference tone of defined intensity and frequency. The accepted reference tone has a *root-mean-square sound pressure of 2×10^{-5} pascal and a frequency of 1000 hertz (this being the threshold intensity at this frequency). The experimental method of measuring equivalent loudness is as follows: a normal observer listens with both ears to the standard tone and the sound being measured alternately. The standard tone is varied in intensity until the observer judges it to be as loud as the sound under test. If then the standard tone is n *decibels above the reference intensity, the equivalent loudness of the sound being measured is defined as n phons. This should be compared with the decibel scale of intensity levels where the intensity of a note is referred to its threshold intensity at the same frequency. The decibel and phon scales are not the same since the sensitivity of the ear to changes of intensity varies with frequency. The two scales are nearly the same between 500 and 10 000 hertz, but below this range there is considerable variation. The phon scale is subjective, since it depends on the judgment of equal loudness by an observer. It should be noted, however, that only when the actual sensation of loudness is logarithmically related to the intensity does the phon scale give a true quantitative measure of loudness sensation, although it always places sounds in the order of their loudness. See also sone; loudness.

phonetics The science of phonetics seeks to describe and classify the various sounds produced in speech and to allocate a symbol(s) to each sound. Electrical analysis of speech sounds, examination of the mouth, etc. have materially helped in these studies. Certain language schools use phonetics to help students accurately to produce given sounds. Phonetic symbols are the written representations of sounds where the same symbol is always used for the same sound. Rationalized systems of spellings have been built on such symbols.

phonic wheel A device due to Rayleigh and La Cour that enables the speed of rotation of a motor to be kept constant as controlled by an electrically maintained tuning fork. It is a very useful principle for driving a stroboscopic disc. The device consists of a soft-iron wheel with teeth revolving between the poles of an electromagnet. The current feeding the electromagnet is made and broken by the vibration of the fork. If the soft-iron wheel is rotating at such a speed that every time the electromagnet is excited, a pair of teeth is exactly opposite to the poles, then the speed of the wheel is unaffected. If the rotation of the wheel tends to slow down or to move faster, forces of attraction will be brought into play to restore the steadiness of the speed. Such a phonic wheel is not self-starting, it must be run up to synchronous speed by hand or by an auxiliary motor. The phonic wheel provides a very convenient and accurate method of determining the frequency of electrically maintained forks.

If a *tuning fork of frequency n controls a phonic wheel having m teeth, the speed of the motor is consequently n/m revolutions per second. If the motor makes p revolutions in time t, the frequency must be equal to pm/t; p can be noted by attaching a counter to the shaft of the motor. The greater the time interval, the greater the accuracy of the method. An accuracy of 1 in 10 000 is easily obtainable at low frequencies.

phonon In a solid the atoms do not vibrate independently but the oscillations are transmitted through the substance as acoustic waves of extremely high frequency f (typically of the order of 10^{12} Hz). The energy transmitted by the waves is quantized; the quantum is called a phonon and has value hf, where h is the Planck constant.

For many purposes the phonons can be treated as if they were gas molecules moving within the space occupied by the solid, the *mean free path being limited by various scattering processes. *Defects and boundaries scatter generally. For large amplitudes the vibrations are anharmonic (*see* anharmonic motion) and the phonons are then scattered by free electrons and by other phonons. These effects are very important in limiting *thermal conductivity and increasing electrical *resistivity, especially at high temperatures.

In theories of *specific heat capacity a phonon can be identified with the quantum of a mode of the *standing wave system in the substance. (*See* Debye theory of specific heat capacities.) *See also* collective excitation.

phosphor A substance that exhibits *luminescence, emitting light at temperatures below the temperature at which it would exhibit *incandescence. Particular examples are fluorescent substances such as those used on the screens of cathode-ray tubes or in fluorescent lamps.

	Copper	Phosphor bronze
resistivity/ ohm metre (at 18° C)	$1{\cdot}8 \times 10^{-8}$	$5{-}10 \times 10^{-8}$
modulus of rigidity/N m^{-2}	$4{\cdot}0 \times 10^8$	$4{\cdot}36 \times 10^8$
Young's modulus/N m^{-2}	$12{\cdot}4 \times 10^8$	$12{\cdot}0 \times 10^8$

phosphor bronze Bronze containing not less than 0.1% of phosphorus, the addition of which greatly increases tensile strength, ductility, and shock resistance. A typical composition is 90% copper, 9.7% tin, and 0.3% phosphorus. Phosphor bronze strip has been extensively used for galvanometer suspen-

sions and similar purposes. Some properties are tabulated in the table in comparison with copper.

phosphorescence A type of *luminescence in which light emission may persist for some considerable time after the excitation has ceased.

phosphoroscope An instrument invented by A. H. Becquerel, to estimate the duration of the phosphorescent glow after a phosphorescent body has been illuminated.

phot A unit of intensity of illumination (now called illuminance) equal to 1 lumen per cm^2. 1 phot $= 10^4$ *lux.

photocathode A *cathode from which electrons are emitted as a result of the *photoelectric effect.

photocell Any light–electric *transducer. Originally the device was a *photoelectric cell* consisting of a valve diode with a *photocathode and anode in which a current flows when the photocathode is illuminated. The word is now commonly used to designate a *photoconductive cell*, consisting simply of a slab of *semiconductor with *ohmic contacts fixed at opposite ends. When illuminated with light or other radiation of a suitable wavelength, a marked increase in conductivity occurs due to the generation of charge *carriers (*see* photoconductivity). A photocurrent then flows in an external circuit. The gain is defined as $\Delta I/eG_{pair}$ where $\Delta I =$ photocurrent in amperes, $G_{pair} =$ total number of electron-hole pairs created per second per photon absorbed, $e =$ electron charge.

The term photocell is also sometimes applied to a *photodiode or a *photovoltaic cell.

photochemical effect Any effect in which a chemical reaction is caused of induced by electromagnetic radiation, usually visible or ultraviolet radiation.

photochromic substance A substance that changes colour when light falls on it. Many photochromic substances, e.g. dyes, revert to the original colour when the light is removed. The exposure of light generally causes darkening to take place. If the substance is transparent, the fraction of light transmitted is changed, and this property may be made use of for the storage of information in conjunction with a *laser.

photoconductive cell *See* photocell.

photoconductivity Enhanced conductivity of certain *semiconductors as a result of exposure to light or other *electromagnetic radiation. The absorption of a photon in the material increases the energy of an electron in the valence band (*see* energy bands) of the solid. If the photon energy exceeds the *work function of the solid, the electron is liberated by the *photoelectric effect. However, if the photon energy is not sufficient to liberate the electron, it may be enough to increase the electron's energy so that it is excited into the conduction band. This is sometimes called the *internal photoelectric effect*. The presence of extra electrons in the conduction band causes the photoconductivity.

Selenium, in the grey crystalline form, shows this photoconductivity to a marked degree and it is also shown to a lesser degree by copper oxide, molybdenite, and thallium oxysulphide as well as normally insulating crystals such as diamond, zinc blende, cinnebar, and stibnite. Some substances including lead sulphide and lead selenide are sensitive to the infrared region of the spectrum to wavelengths of several micrometres. *See also* photocell.

photodetachment The removal of an electron from a negative ion by a *photon of electromagnetic radiation to give a neutral atom or molecule. The process is thus:

$$M^- + h\nu \rightarrow M + e,$$

where $h\nu$ represents a photon of frequency ν. It is exactly the same as *photoionization of a neutral species and the *ionization potential of the negative ion is equal to the *electron affinity of the atom (or molecule).

photodetector Any electronic device activated by light. *See* photocell; photodiode; phototransistor.

photodiode A semiconductor *diode that produces a significant photocurrent when illuminated. There are various types. One form uses a p-n junction that is reverse biased but operated below the *breakdown voltage. When exposed to electromagnetic radiation of suitable frequency, excess charge carriers (electron-hole pairs) are generated as a result of *photoconductivity. The carriers normally recombine rapidly, but those produced in or near the *depletion layer present at the junction can cross the junction and produce a photocurrent. This is superimposed on the normally very small reverse saturation current. A common type of such a device is the *p-i-n photodiode*. This has a layer of intrinsic semiconductor between the p and n regions that wholly contains the depletion layer, allowing devices to be manufactured with a depletion width suitable for optimum sensitivity and frequency response.

photodisintegration *See* photonuclear reaction.

photoelasticity *Syn.* mechanical (or stress) birefringence. The effect whereby a normally transparent isotropic substance can become optically anisotropic under mechanical stress, and thus exhibit *double refraction. Marked effects are thus produced with polarized light. The phenomenon is used by engineers to study the stresses in structures by the examination of models made of transparent plastics.

photoelectric cell *See* photocell.

photoelectric constant The ratio of the *Planck constant to the charge of the electron. It is obtained by *photoemission experiments and has been used in determining the Planck constant.

photoelectric effect The liberation of electrons from matter by electromagnetic radiation of certain frequencies. The effect was first noticed by Hertz who found that sparks more readily jumped a gap when it was illuminated by ultraviolet radiation. For solids, electrons are only liberated when the wavelength of the radiation is shorter than a certain value (the *photoelectric threshold*). Most solids emit electrons when this value is in the *vacuum-ultraviolet region of the spectrum although some metals (e.g. Na, K, Cs, and Rb) and semiconductors emit for visible and near-ultraviolet radiation.

*Lenard (1902) showed that the maximum speed of the electrons was independent of the intensity and appeared to increase linearly with frequency. Their number was directly proportional to intensity for a given frequency. Einstein explained this behaviour by assuming that the energy of the incident radiation was transferred in discrete amounts (*photons), each of magnitude $h\nu$, where h is the *Planck constant and ν the frequency. Each photon absorbed will eject an electron provided that the photon energy ($h\nu$) exceeds a certain value Φ – the *work function. The maximum kinetic energy of the electrons E (by the *conservation of energy principle) is then given by $E = h\nu - \Phi$. This is known as the *Einstein photoelectric equation*. The electrons with this maximum kinetic enegy are the least strongly bound electrons in the solid. More strongly bound electrons will also be ejected with energies lower than E. Although proposed in 1905, Einstein's equation was not well verified until the work of *Millikan (1916).

The photoelectric effect does not only apply to solids. Gases and liquids can also emit electrons under the effect of light. For gases each electron is removed from a single atom or molecule and the work function in the Einstein equation is replaced by the *ionization potential. (*See* photoionization.) The photoelectric effect is used in *photocells, *photomultipliers, and is the basis of *photoelectron spectroscopy. *See also* internal photoelectric effect (*under* photoconductivity).

The enormous intensities obtainable by focusing radiation from a *laser can cause photoelectric emission at frequencies far below the normal threshold value, thus the Einstein equation does not apply in all circumstances.

Electromagnetic radiation of very short wavelength (X-rays and gamma rays) causes the photoelectric effect in all substances. The Einstein equation applies with Φ replaced by the binding energy of the electron in an inner shell (*see* atomic orbital). For a given frequency the probability of interaction is proportional to the fourth power of the *atomic number Z. Hence bone, which contains calcium ($Z = 20$) and phosphorous ($Z = 15$) absorbs more strongly than soft tissue in which the most important absorber is oxygen ($Z = 8$). Lead ($Z = 82$) is used for shielding.

photoelectric threshold *See* photoelectric effect.

photoelectron spectroscopy A form of *electron spectroscopy for measuring the *ionization potentials of atoms and molecules. In *ultraviolet photoelec-*

tron spectroscopy a sample gas is passed into the target region of an electron spectrometer (*see* electron spectroscopy) and irradiated with a beam of monochromatic ultraviolet radiation. A helium discharge tube is commonly used as a source of ultraviolet radiation with a quantum energy of 21.21 eV. The atoms or molecules in the sample undergo *photoionization and the energies of the electrons produced are determined in the spectrometer. In general, the spectrum obtained consists of a number of peaks due to groups of electrons of energies E_1, E_2, etc. Each group arises from photoionization of a particular energy level, i.e. from the removal of electrons from a particular atomic or molecular orbit. The electron kinetic energies are given by:

$$E_1 = h\nu - I_1, \quad E_2 = h\nu - I_2, \text{ etc.}$$

where I_1, I_2, etc. are the first, second, etc., ionization potentials of the substance studied. Since $h\nu$ is known (21.21 eV) these ionization potentials are thus determined. The importance of the technique lies in the fact that the ionization potential corresponding to removal of an electron from a particular orbital is approximately numerically equal to the energy of the atomic orbital. When molecules are studied the spectrum of electron energies usually consists of a number of bands because the ion may be formed in a number of vibrationally excited states.

In *X-ray photoelectron spectroscopy* the principle is exactly the same but a beam of monochromatic *X-rays is used instead of ultraviolet radiation. This allows the determination of the energies of very strongly bound electrons in the inner shells of atoms and molecules. The binding energy of the inner electrons in a particular element depends on the chemical compound of the element. The technique is used for qualitative and quantitative analysis and is sometimes called *ESCA* (*E*lectron *S*pectroscopy for *C*hemical *A*nalysis).

photoemission Emission of electrons as a result of bombardment by photons, as in *photoionization or the *photoelectric effect.

photofission *See* photonuclear reaction.

photofluorography Instead of taking an X-ray shadowgraph on a large photographic plate, the light image formed by the rays on a fluorescent screen is photographed on a small plate, thus saving material.

photographic objectives Lenses designed to form sharp real images of distant or near objects on a photographic film. Designs vary over a wide field owing to the large and diverse applications of photography, e.g. landscapes, high speed, portraiture, process work, cinematography, interiors, etc. Some lenses work with an *f-number as large as f/0.8, requiring this to reduce the time of exposure by increasing the illumination in the image, i.e. increasing the speed of the lens. Such large apertures increase the difficulties of design. The *Schmidt telescope works at an f-number as high as f/0.5. Rapid rectilinear lenses reduce distortion by using a

central stop between a pair of lenses symmetrically arranged. Modern optical glasses enable the development of the new *achromatic, *apochromatic, and anastigmatic lenses with flattening of the field in a large number of designs, some of which include convertible elements, i.e. either or both components of a lens can be used separately. Aerial reconnaissance lenses necessarily have long focal length. The *telephoto lens increases back focal length by virtue of its rear diverging lens. *See also* zoom lens.

photography The production of permanent images by use of sensitized emulsions. Light falling on a photographic emulsion sets up a series of photochemical reactions in which silver ions belonging to silver salts in the emulsion are converted to neutral silver atoms. In the development of this *latent image*, a chemical reducer liberates further silver atoms clustered round the original atoms giving opaque specks of metallic silver. Subsequent fixing (e.g with hypo, sodium thiosulphate) removes unaffected silver salts so giving a photographic *negative*, which is darkest where the original image was lightest. A *positive* (print) is produced by exposure of a further emulsion, e.g. one coated on a paper backing, behind and in contact with the negative (contact printing) or by projecting the image of the negative onto the paper using an enlarger and then chemically processing the paper.

The blackening of an emulsion depends mainly on *exposure*, defined as illumination × time (*see* reciprocity law); photographic methods are therefore useful in *photometry and especially in spectrophotometry (*see* spectrometer). High illumination enables short exposures to be made (normal flash photography). Electronic flash photography, using extremely short electronically produced flashes may be used for studying high-speed phenomena.

Photographic emulsions are inherently very sensitive to ultraviolet and blue light but they can be treated with dyes to sensitize them into the infrared. X-rays and all nuclear radiations are photographically active in varying degree. *See also* camera; photographic objectives; colour photography.

photoionization Ionization of an atom or molecule by electromagnetic radiation. The energy of a *photon of radiation is $h\nu$, where ν is its frequency and h is the *Planck constant. It can only remove an electron if its energy exceeds the first *ionization potential (I_1) of the atom. In accordance with the principle of *conservation of energy, the excess energy ($h\nu - I_1$) is taken up by the positive ion and the electron and distributed between their kinetic energies. Since the mass of the ion is always much greater than that of the electron, its extra energy is negligible and thus $E = h\nu - I_1$, where E is the kinetic energy of the electron. This is simply the Einstein photoelectric equation (*see* photoelectric effect) applied to a single atom or molecule. The radiation capable of photoionizing molecules and atoms has energy greater than the photoelectric threshold, which lies in the ultraviolet region of the spectrum. If the quantum energy of the radiation is high enough, more strongly bound electrons may be removed from the neutral species. For example, electrons may be ejected with energy E_2 ($<E_1$), where $E_2 = h\nu - I_2$ and I_2 is the second *ionization potential. The atom is then left in an electronically *excited state.

In the photoionization of molecules, part of the excess energy may be taken up by the ion to leave it in a vibrationally or rotationally excited state. *See also* photoelectron spectroscopy; photodetachment.

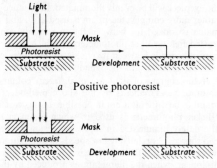

a Positive photoresist

b Negative photoresist

photolithography A technique used during the manufacture of *integrated circuits, *semiconductor components, *thin-film circuits, and *printed circuits. In this technique a desired pattern is transferred from a photographic *mask onto the substrate material ready for a processing step to be carried out. The clean substrate is first covered with a solution of *photoresist, which is allowed to dry and is then exposed to light or UV through the mask. The depolymerized portions of the photoresist are then washed away with a suitable solvent, the polymerized portion remaining and acting as a barrier to etching substances or as a mask for deposition processes. When the processing step is complete, the photoresist may be stripped off with a suitable solvent. If positive photoresist is used, the exposed portion is depolymerized and removed during development (Fig. *a*). If negative photoresist is used, the exposed portion is polymerized and remains after development (Fig. *b*).

photolysis The chemical decomposition or dissociation of molecules as a result of the absorption of light or other electromagnetic radiation. For example, if chlorine is irradiated with light of wavelength below 480 nm, some chlorine atoms are produced:
$$Cl_2 + h\nu \rightarrow 2Cl.$$
The first step is absorption of a photon followed by dissociation of the molecule as a result of its extra energy. Clearly the photon energy must exceed the strength of the chemical bond for photolysis to occur. *See also* flash photolysis.

photomagnetism Paramagnetism produced in a substance when it is in a phosphorescent state.

photometer *See* photometry.

photometry Photometry is concerned with measurements of light intensity and amounts of illumina-

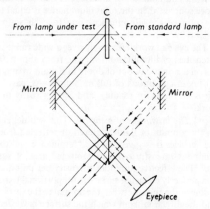

Lummer–Brodhun photometer

tion. Two types of measurement are possible. In one, the radiation is evaluated according to its visual effects, i.e. according to judgment by observers (*see* luminosity; spectral luminous efficiency). The physical quantities measured in this way are preceded by the adjective *luminous*. This distinguishes them from physical quantities measured in units of energy, for which the adjective *radiant* is used.

Visual photometry is the branch of photometry in which the eye is used to make comparisons. In *physical photometry* the measurements are made by physical receptors, such as the *photocell, *thermopile, and *bolometer.

(i) *The *luminous intensity of a source*. The intensity of light from a source can be obtained by comparing it with a *primary standard of light*, i.e. a standard light source used to establish a unit of measurement. For the present standard, *see* candela. A *secondary standard of light* is a constant reproducible light source, e.g. an electric lamp, whose luminous intensity is known by direct or indirect comparison with a primary standard.

(ii) The *luminous flux is the quantity of light crossing a surface in unit time. The unit is the *lumen, which is the flux emitted into unit solid angle by a uniform point source of one candela.

(iii) *Intensity of illumination of a surface*. The *illuminance of a surface is the quantity of light falling per second on unit area of the surface, i.e. the flux per unit area. A flux of 1 lumen per square metre corresponds to an illuminance of 1 *lux.

According to the *inverse square law*, the intensity of illumination of a surface is inversely proportional to the square of its distance from the light source. The law applies strictly only in the case of point sources and may be much in error for surfaces near an extended source.

The general expression for the intensity of illumination of a surface a distance r away from a point

photomicrography

source of luminous intensity J is $J \cos \theta / r^2$, where θ is the angle between the normal to the surface and the line from source to surface.

If the illumination of a surface is due to two or more sources, then the total illuminance is equal to the sum of the illuminances due to each source separately.

The eye can work effectively over a wide range of intensities of illumination, varying from about 10^5 lux on a bright day out of doors in Britain to about 0.1 lux in conditions of full moonlight. At least 30 lux is desirable for reading, and 500 for really fine work.

(iv) *The brightness of a surface*. This will depend on the amount of light scattered or reflected from the surface. It is given by the *luminance, i.e. the luminous intensity per unit projected area of surface. The effective luminous intensity per unit area of a perfectly diffusing surface is proportional to the cosine of the angle between the normal to the surface and the direction from which the surface is viewed. This result is known as *Lambert's law.

(v) *The brightness of optical images*. It can be shown that, provided the eye pupil is filled with light, the brightness of the image seen through an optical system of perfect transparency is equal to that of the object. In practice there is bound to be some absorption and scattering of light in the system so that the brightness of the image is less than that of the object. If the eye pupil is not filled with light, the brightness of the image is less than that of the object even when the absorption in the system is discounted.

In a telescope, for example, provided the *exit pupil is larger than the eye pupil, the brightness of the image would be equal to that of the object if no absorption losses occurred. If, however, the diameter of the exit pupil is made smaller than that of the eye pupil, the brightness of the image falls off in proportion to the ratio of the area of exit pupil to area of eye pupil.

There are two exceptions to this general rule. In the case of *star images*, the fact that the object subtends so small an angle means that the geometrical image remains a point even under conditions of high magnifying power. This leads to the result that the brightness of such an image may be greater than that of the object. Moreover, since a small exit pupil can now be used, the brightness of the background may be thereby reduced and enhanced contrast obtained. The second exception concerns the use of *night glasses*. Under conditions of poor illumination, *scotopic vision is employed by the eye. The result is that an increased magnification leads to images that may appear brighter than the object viewed directly with the naked eye.

(vi) *Photometers*. Practical measurements of luminous intensity are made by comparing the intensity of the lamp under test with that of a standard lamp. The general method is to adjust the position of a screen with respect to the two lamps until the intensities of illumination of the screen due to the two lamps are equal. The luminous intensities of the two sources are then proportional to the squares of their respective distances from the screen.

Measurements such as these are made with photometers, one type of which – the *Lummer–Brodhun* – is shown in the diagram.

In the (simplified) arrangement shown, the light reaching the eyepiece is seen to consist of a central pencil of rays from the left-hand side of the screen C and an outer bundle from the right-hand side of C. When the intensities of illumination of the two sides of the screen are equal, the field of view seen through the eyepiece will be evenly illuminated. In practice a rather more complicated prism is used at P, and a match is looked for between the two evenly contrasted fields. Other photometers in which a comparison is made between two fields are the *grease-spot photometer and the *shadow photometer.

If the sources of light are of different colours, accurate measurements by the above method are found to be difficult if not impossible. In this case a *flicker photometer is used in which a screen is alternately illuminated in rapid succession by the two sources. At relatively low speeds of alternation, it is found that a flicker effect is noticeable, which disappears when the illuminations produced by the two sources are equal. This appears to hold good even when the sources are of different colours. In the case of most lamps, the effective luminous intensity depends on the direction from which the lamp is viewed. It is usual, in such cases, to produce polar diagrams in which the source is taken as origin and the *radius vector in any particular direction is made equal to the luminous intensity in that direction. A number of such diagrams in different planes are required to give the complete picture. From such polar curves the value of the total luminous flux emitted by the lamp can be obtained by a construction known as the *Rousseau diagram*. A photometer for measuring the directional characteristics of a source is called a *goniophotometer*.

It is sometimes convenient with sources of this kind to refer to the average value of the luminous intensity in all directions. This is the mean spherical luminous intensity of the source and is equal to the total luminous flux emitted, divided by 4π steradians. The total flux emitted by the lamp can be measured directly in practice by placing the source inside a hollow sphere, which has an internal matt white coating (*see* integrating photometer). The flux received by a photometer pointed at a small window in the side of the sphere is proportional to the total light flux emitted by the lamp. The measurement is usually used to give a comparison between two lamps rather than to provide absolute results.

photomicrography The recording of microscope images on photographic media. The recorded image forms a *photomicrograph*.

photomultiplier An *electron multiplier in which the primary electrons causing the cascade are produced by the *photoelectric effect. The cathode of such a tube is a *photocathode, which is illuminated by some means. Photomultiplier tubes are frequently used with a suitable *scintillation crystal as radiation detectors or counters.

photon 1. The *quantum of electromagnetic radiation. It has an energy of $h\nu$ where h is the *Planck constant and ν the frequency of the radiation. For some purposes photons can be considered as *elementary particles travelling at the *speed of light (c) and having a momentum of $h\nu/c$ or $h\lambda$ (where λ is the wavelength). Photons can cause *excitation of atoms and molecules and more energetic ones can cause ionization (*see* photoionization). (*See also* Compton effect; photonuclear reaction.)
2. A unit of retinal illumination equal to the illumination produced by a surface having a luminance of 1 nit (1 candela per square metre) when viewed by an eye with a pupil of 1 square millimetre.

photoneutron *See* photonuclear reaction.

photonuclear reaction *Syn.* photodisintegration. A reaction occurring when a high-energy *photon, such as a gamma ray or X-ray photon, collides with an atomic nucleus. As a result the nucleus disintegrates. In certain cases the photon appears to knock out a neutron (*photoneutron*) or proton (*photoproton*). In other cases the nucleus appears to absorb the photon and then break up as a result of its higher energy – in some cases fission occurs (*photofission*).

photopic vision Vision by the normal *eye when the cones in the retina are the principal receptors of light. This occurs when the eye is adapted to high levels of *luminance (several candelas per square metre) and the sensation of colour is produced. The maximum *spectral luminous efficiency occurs at longer wavelengths than in *scotopic vision. *See* luminosity.

photoproton *See* photonuclear reaction.

photoresist An organic photosensitive material used during *photolithography. Negative photoresists are materials that polymerize due to the action of light; positive photoresists are polymeric materials that are depolymerized by the action of light. The polymerized material acts as a barrier during processing steps.

photosensitivity The property of responding to electromagnetic radiation (especially light) by the creation of an electric potential (*see* photoelectric effect), by changing colour (*see* photochromic substance), etc.

photosphere A layer within a star from which most of the observed radiation is emitted. Very little radiation comes directly from deeper layers because of absorption and refraction in the photosphere, and very little from further out because of the low density.
In the sun the photosphere has a thickness of several hundred kilometres and an average tempera-

ture about 6000 K. Although it is a gas of low density the total amount of matter is sufficient to give a continuous spectrum in the emitted radiation. It gives the illusion of a definite surface to the sun.

phototransistor A bipolar junction *transistor activated by light or UV. The *base electrode is left *floating and the base signal is supplied by excess *carriers generated by illumination of the base. Once equilibrium has been reached, the device behaves like a normal transistor with the base signal being a function of the intensity of the radiation.

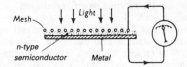

Photovoltaic cell

photovoltaic cell An electronic device that uses the *photovoltaic effect to produce an e.m.f. An example is the *solar cell, the basis of which is an unbiased p-n junction. Other types use a metal-semiconductor junction. These depend for their action on the formation of a potential barrier across the unbiased junction (*see* Schottky effect); they are often called *rectifier photocells* or *barrier-layer photocells*. A typical structure is shown in the diagram. The contact to the n-type semiconductor is in the form of a mesh to minimize reflection of the incident light.

photovoltaic effect An effect arising when a junction exists between two dissimilar materials and one of the materials is exposed to electromagnetic radiation, usually in the range near-ultraviolet to infrared. The two materials may, for example, be a metal and a semiconductor or two semiconductors of opposite polarity (the combination forming a *Schottky barrier and a *p-n junction respectively). A forward voltage appears across the illuminated junction and power can be delivered to an external circuit (*see* photovoltaic cell).
    The incident radiation imparts energy to electrons in the valence band (*see* energy bands), and electron-hole pairs are generated in the depletion region existing around the p-n junction and in the Schottky barrier. As the pairs are produced they are able to cross the junction (due to the inherent field) and produce the forward bias: a migration of electrons into the n-type semiconductor produces a negative bias while an excess of holes migrating into the p-type semiconductor or the metal produces a positive bias.

physical adsorption *Adsorption on a surface in which the adsorbate is held by weak physical forces rather than by chemical forces. The adsorbed layer is easily removed in a vacuum. *Compare* chemisorption.

physical optics *See* optics.

Piccard, Auguste

Piccard, Auguste (1884–1962) Swiss physicist, best known for his balloon ascents to investigate *cosmic rays (1931, 1932) and for his deep-sea exploration (bathyscaphe or bathysphere, 1947).

pick-up A *transducer that converts information (usually recorded) into electric signals. The term is particularly applied to the electromechanical devices that reproduce the signals recorded in grooves in gramophone records. Several types of pick-up are in common use.

Crystal pick-ups consist of a *piezoelectric crystal that is stressed by the mechanical vibrations in the grooves of the rotating record, producing a corresponding e.m.f.

Ceramic pick-ups are similar to crystal pick-ups in that their output is also due to the piezoelectric effect. The ceramic materials, e.g. barium titanate, are more reliable and stable under ambient conditions.

Magnetic pick-ups have a small inductance coil in the field of a magnet. Mechanical vibrations in the grooves of the record cause the coil to move and hence the magnetic flux through the coil changes. The induced current in the coil depends on the magnitude of the vibrations and provides the signal for the audio-system.

pico- Symbol: p. A prefix denoting 10^{-12}; for example, 1 picofarad (pF) $= 10^{-12}$ farad. The use of $\mu\mu$ as a symbol for this prefix is now deprecated.

Pictet, Raoul (1846–1929) Swiss physicist. The first to liquefy oxygen (1877) (achieved independently by Cailletet at about the same time), Pictet subsequently liquefied also nitrogen, carbon dioxide, and hydrogen, using the *method of cascades*. See cascade liquefaction.

picture frequency *See* television.

Pierce crystal oscillator *See diagram* under piezoelectric oscillator.

piezoelectric crystal A crystal exhibiting the *piezoelectric effect. All crystalline *ferroelectric materials are piezoelectric as well as certain nonferroelectric crystals and some ceramics. Examples include quartz crystal, Rochelle salt, and barium titanate (a ceramic).

piezoelectric effect An effect exhibited by *piezoelectric crystals whereby the surfaces become oppositely electrically charged when subject to stress; the sign of the charges changes when a compression of the crystal is changed to a tension. The converse effect, in which the crystal expands along one axis and contracts along the other when subjected to an electrical field, also occurs. The magnitude of the piezoelectric effect depends on the direction of the stress relative to the crystal axes. The maximum effect is obtained when the electrical and mechanical stresses are applied along the X-axis (the electric axis) and Y-axis (the mechanical axis) respectively. The third major axis in a piezoelectric crystal is the Z-axis (the optic axis). See piezoelectric oscillator.

piezoelectricity Electricity generated by the *piezoelectric effect.

piezoelectric oscillator A highly stable *oscillator in which a *piezoelectric crystal is used to determine

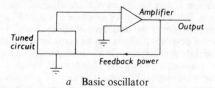

a   Basic oscillator

the frequency. If an alternating electric field is applied across the crystal, mechanical vibrations result (*see* piezoelectric effect). The crystal is usually cut with its major surface either perpendicular or parallel to the X-axis. It is then mounted between the plates of a capacitor in order to apply the alternating voltage. Normally a metallic film is formed on the crystal faces for this purpose. The crystal is supported by lightweight supports that touch it at mechanical nodes.

Piezoelectric crystals are most conveniently set in vibration by the aid of undamped electric oscillations, which may be generated electronically at any desired frequency and intensity. The connections between the crystal and the oscillator circuit may be made in various ways. In general, the circuits used can be divided into two main types.

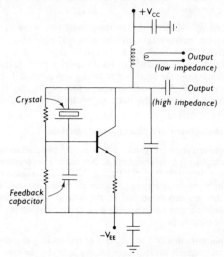

b   Pierce crystal oscillator with crystal between base and collector

In the *crystal oscillator*, the crystal replaces the *tuned circuit in the oscillator, thus providing the resonant frequency of the oscillator (Figs. *a*, *b*). In the *crystal-controlled oscillator*, the crystal is coupled to the oscillator circuit, which is tuned approximately to the crystal frequency. The crystal controls the oscillator frequency by *pulling the frequency to its own natural frequency and so preventing drift of the oscillator frequency (Fig. *c*).

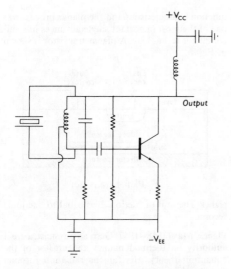

c Hartley crystal-controlled oscillator

piezometer A bulb fitted with a capillary tube, or a metal cylinder fitted with a piston, used in the determination of the *compressibility of a liquid. It is filled with a liquid and then is subjected to fluid pressure over its external surface and over the end of the capillary (there may be a small amount of mercury over the liquid in the capillary). The movements of the liquid in the capillary, or the piston in the cylinder, enable the compressibility of the liquid to be determined.

pile *See* nuclear reactor.

pilot *Syn.* pilot wire. In an electrical power transmission system. An auxiliary line for telecommunication purposes or for use with protective or measuring apparatus.

pilot balloon A small balloon used in studying wind directions and speeds in the earth's atmosphere. The rate of rise of the balloon relative to the air is known and its track is generally observed with only one theodolite.

pi-meson (π-meson) *Syn.* pion. *See* meson.

pinch An air-tight glass seal through which the leading-in wires pass into the bulb of an electric lamp, thermionic valve, etc.

pinch effect An effect whereby an electric current passing through a liquid or gaseous conductor tends to cause its cross section to contract as a result of the electromagnetic forces set up. This effect is important in thermonuclear plasmas, the pinch effect keeping the plasma from contact with the walls of the container. *See* fusion reactor.

pinch-off *See* field-effect transistor.

p-i-n diode A semiconductor *diode with a region of almost *intrinsic semiconductor between the p-type and n-type regions. *See* IMPATT diode; photodiode.

pin grid array (PGA) A form of package used for complex *integrated circuits, capable of providing up to several hundred connections to one chip. Connections are made by means of an array (i.e. several parallel rows) of output pins on the package periphery, either on two opposite edges or around all four edges. The pins are connected through the casing to the bonding pads of the chip.

pin insulator An insulator mounted rigidly on a pin, the latter serving to fix the insulator to its support, commonly employed in electric-power transmission systems at voltages up to about 20 kV.

pion *See* meson.

Pirani gauge A low-pressure gauge in which electrically heated wire loses heat by conduction through the gas. A constant potential difference is maintained across the wire and its resistance variation with pressure observed. Alternatively, the resistance is kept constant by varying the applied p.d., which may then be measured.

piston gauge *See* free-piston gauge.

pistonphone A device for calibrating a microphone consisting of a small piston and its associated cylinder. *See* equivalent piston.

pitch 1. A subjective quality of a sound that determines its position in a musical scale. It may be measured as the frequency of the pure tone of specified intensity that is judged by the average normal ear to occupy the same place in the musical scale. Although pitch is measured in terms of frequency, it is also dependent on the loudness and quality of the note. As the intensity is increased, the pitch of a low-frequency note is lowered, while that of a high note is raised. This pitch variation amounts to a fall of about 20% when a 200 hertz note has its loudness raised from 60 to 120 decibels. Fletcher has shown experimentally that the pitch of a complex note does not alter when some of its components are removed. He found that even when the fundamental and the first seven overtones were removed from a musical note the only effect was a change of quality, the pitch remaining the same. This is thought to be due to subjective *combination tones produced in the ear itself. (*See* quality.)
2. Of a screw thread, etc. The distance apart of successive threads (or of successive teeth of a gear wheel).

pitometer An instrument making a continuous record of the velocity of a liquid.

Pitot-static tube *See* Pitot tube.

Pitot tube An instrument used for measuring the total (static and dynamic) pressure of a fluid stream. It consists essentially of a tube of small bore connected at one end to a *manometer, the other end being open and pointing upstream. Fluid cannot

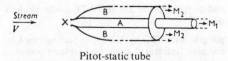

Pitot-static tube

flow through the tube; the pressure registered at the manometer is the stagnation pressure at the nose of the tube, which, by *Bernouilli's theorem, will be $p_0 + \frac{1}{2}\rho V^2$, where p_0 is the static pressure, ρ the density, and V the velocity of the undisturbed stream.

The term Pitot tube is commonly used for the true *Pitot-static tube* (*see* diagram). The tube A represents the Pitot tube connected to the manometer M_1 (which registers the total pressure), the point X being the stagnation point. The tube B is the static tube connected to manometer M_2 registering the static pressure. The difference pressure is the quantity $\frac{1}{2}\rho V^2$. The Pitot-static tube is used to measure the stream velocity and is used on aircraft to measure the relative wind speed. *See* static tube.

PL/1 (programming language one). A scientifically oriented high-level computer *programming language constructed to embody the best features of previous languages and to give greater flexibility.

plain steel conduit A conduit of light-gauge steel tubing that may be close-jointed, welded, brazed, or solid-drawn. The ends are not screwed. Adjacent lengths are connected together by means of a plain coupler.

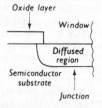

a   Planar process

planar process The most commonly used method of producing junctions in the manufacture of *semiconductor devices. A layer of silicon dioxide is thermally grown on the surface of a silicon substrate of the desired conductivity type. Holes are etched in the oxide layer, using *photolithography, and suitable impurities are diffused into the substrate through these holes to produce a region of opposite polarity. The oxide acts as a barrier to diffusion, except through the holes. The impurities will diffuse into the substrate in directions normal and parallel to the surface so that the junction meets the surface of the substrate below the oxide (Fig. *a*). Several diffusions can be carried out, one after another. Usually a final layer of oxide is grown over the entire chip (except for the contacts) to provide a stable surface for the silicon and keep surface-leakage currents to a minimum. In early junction *transistors, surface-leakage effects tended to dominate the

junction characteristics and the planar process was one of the most important single advances in semiconductor technology. A planar transistor is shown in Fig. *b*.

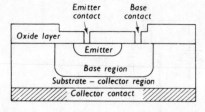

b   Planar transistor

planck The unit of *action. It is equal to one joule second.

Planck, Max (1858–1947) German physicist, a great authority on thermodynamics and creator of the *quantum theory. His famous radiation formula (*see* Planck's formula) was first obtained by interpolation between those of Wien and of Rayleigh and Jeans; to justify it, the idea of continuity in emission and absorption of radiant energy was abandoned (1900). In these processes, discrete amounts of energy or *quanta* are involved, each quantum being of magnitude hf, where f is the frequency of the radiation and h is a universal constant, the *Planck constant.

The development of the quantum theory was largely due to others (notably Einstein). Planck himself made many efforts to link quantum theory with classical (i.e. pre-quantum theory) physics. His contributions to thermodynamics were especially valuable in the field of chemical equilibria, in formulation of the laws of thermodynamics, especially the *Nernst heat theorem, and in linking the concept of entropy to the probabilities that occur in statistical mechanics, continuing Boltzmann's work. Planck also added (1908) to the Einstein principle of the inertia of energy and thus gave an interpretation of radiation pressure.

Planck constant Symbol: h. A universal constant having the value $6.626\,076 \times 10^{-34}$ J s. In quantum mechanics the constant $h/2\pi$ is often used $(1.054\,573 \times 10^{-34}$ J s). This has the symbol $\hbar$ (called "crossed h"). It is known as the *rationalized Planck constant* or the *Dirac constant. See* Planck's law.

Planck function Symbol: Y. The negative of the *Gibbs function (G) divided by the thermodynamic temperature (T): $Y = -G/T$.

Planck length The length $(Gh/2\pi c^3)^{1/2}$, where G is the *gravitational constant, h is the *Planck constant and c the speed of light; equal to $1.615\,99 \times 10^{-35}$ m. It arises in theories relating quantum theory to gravitation. *See* Planck mass.

Planck mass The mass $(hc/2\pi G)^{1/2}$, where h is the *Planck constant, c is the speed of light, and G is the *gravitational constant; equal to $2.176\ 84 \times 10^{-8}$ kg. See Planck length.

Planck's formula A formula showing how the energy radiated from a *black body varies with wavelength and temperature. The spectral concentration of *radiant exitance $M_{e\lambda}$ is given by the equation:

$$M_{e\lambda} = \frac{C_1}{\lambda^5} \frac{1}{e^{C_2/\lambda T} - 1},$$

where
$C_1 = 2\pi hc^2 = 3.741\ 50\ (\pm 0.000\ 09) \times 10^{-16}$ W m^2,
$C_2 = hc/k = 1.438\ 79\ (\pm 0.000\ 06) \times 10^{-2}$m K,
h is the Planck constant, k is the Boltzmann constant, c is the speed of light in vacuum, and T is the thermodynamic temperature. λ is measured in metres. The spectral concentration of radiant exitance is the radiant exitance over an infinitesimal range of wavelength divided by this infinitesimal wavelength, taken at the wavelength of interest, i.e.
$$M_{e\lambda} = dM_e/d\lambda.$$
See also radiation formula.

Planck's law The law that forms the basis of *quantum theory. The energy of *electromagnetic radiation is confined to small indivisible packets or *photons, each of which has an energy hf, where f is the frequency of the radiation and h is the *Planck constant.

Planck time The time $(Gh/2\pi c^5)^{1/2}$ taken for a *photon travelling at the speed of light to travel a distance equal to the *Planck length; equal to $1.708\ 63 \times 10^{-43}$ s.

Planck units A system of units used in quantum theories of gravity based on the *Planck length, *Planck mass, and *Planck time. The *gravitational constant, the *speed of light, and the rationalized *Planck constant are all assigned the value of unity, thus all quantities that normally have dimensions involving mass, length, and time become dimensionless in this system.

plane group A set of operations (rotation about an axis, reflection across a line, or translation) that when carried out on a periodic arrangement of points in a plane brings the system of points to self-coincidence.

plane of flotation See buoyancy.

plane of symmetry The plane across which reflection of each point in a lattice or other system of points will bring the system to self-coincidence.

plane-polarized light Syn. linearly polarized light. Light in which the vibrations are rectilinear, parallel to a plane, and transverse to the direction of travel. (In fact all electromagnetic radiation can be plane-polarized.) Light reflected from a surface of polished glass (of refractive index n) at an angle of incidence $\tan^{-1} n$ is plane-polarized with vibrations parallel to the surface (Brewster's law). It is said to be polarized in the plane of incidence, the plane of vibration (electric vector) being perpendicular to the plane of polarization. Plane-polarized light may be produced by reflection or transmission through a pile of plates, by *double refraction in dichroic substances, e.g. tourmaline, Polaroid, by *Nicol prisms, etc. Optically active substances rotate the plane of polarization. If the analyser of a *polariscope on rotation, produces extinction, the existence of plane-polarized light is established. If θ is the angle between the polarizer and analyser, and I_0 the intensity of the polarized light incident on the analyser, then the transmitted intensity is $I_0 \cos^2 \theta$ (Malus law). See optical activity; polarization.

planetary electron An electron orbiting around the nucleus of an *atom.

plane wave A wave in which the *wavefronts form a set of planes, each generally perpendicular to the direction of propagation. It is the simplest example of a three-dimensional wave. See also progressive wave.

planimeter A mechanical device for the rapid measurement of area on a plane surface. A pointer, attached to the instrument, is made to move round the boundary of the area required. See also integrometer.

planoconcave See concave.

planoconvex See convex.

plan position indicator (PPI) See radar.

Planté, Gaston (1834–1889) French physicist who constructed the first secondary cell. See Planté cell.

Planté cell The first primitive accumulator, consisting of rolled lead sheets (Planté plates) dipping into dilute sulphuric acid.

plant load factor See load factor.

plasma 1. A region of ionized gas in a *gas-discharge tube, containing approximately equal numbers of electrons and positive ions.
2. A highly ionized substance that can be formed at very high temperature (as in stars) or by *photoionization (as in interstellar gas). The atoms present are nearly all fully ionized and the substance consists of electrons and atomic nuclei moving freely. This has been described as a fourth state of matter. See fusion reactors.

plasma oscillations Under certain conditions oscillations of the ions and electrons (independent of the conditions of an external circuit) may be set up in the *plasma of a *gas-discharge tube, and they can cause a scattering of a stream of electrons greater than that explainable by ordinary gas collisions.

plasmatron A hot-cathode helium-filled diode in which large currents can be controlled with only several volts anode potential and that, unlike the *thyratron, operates continuously. An auxiliary circuit provides a plasma of electrons and positive ions, and the control circuit provides ionization to neu-

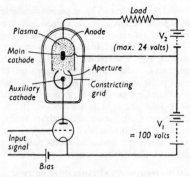

Plasmatron

tralize the space-charge and affect the cathode-anode current.

plasmon A collective excitation for quantized oscillations of the free electrons in a metal.

plastic deformation A permanent deformation of a solid subjected to a stress. It is sometimes produced in single crystals of metals even by vanishingly small forces.

plasticity The property that enables a material to be deformed continuously and permanently without rupture during the application of a stress that exceeds the yield value.

Bingham proposed a law of plastic flow as follows:

$$F - f = (A/\mu)(dv/dx).$$

F is the tangential force between two parallel layers of the substance of area A, dx apart, moving with relative velocity dv; f/A is the value of shear stress at the yield point. The equation does not hold if $F < f$ for then the material does not flow. The quantity μ (which has the dimensions of fluidity) is called the mobility.

If a plastic material flows down a capillary tube the equation predicts that there will be a rigid plug moving down the centre. The equation has had some success with clay suspensions, paints, greases, etc.

plastics A group of organic materials that, though stable in use at normal temperatures, are plastic at some stage in their manufacture, and can be shaped by the application of heat and pressure. There are two main types of plastic, namely *thermoplastic* and *thermosetting* compositions. The former are mouldable by heat and the shape impressed on them is preserved when they are cooled in the mould. The latter, on the other hand, initially soften on heating and then harden on further heating in the mould. Most, if not all, plastic materials are high polymers, i.e. materials composed of giant molecules.

plate 1. The anode in a thermionic valve. The term is more common in the USA than in the UK.

2. An electrode in a *capacitor or *accumulator.

Plateau, Joseph (1801–1883) Belgian physicist, noted for researches on physiological optics and on

properties of liquids, especially surface tension and the mechanism of drop formation. Some of his work was carried out under his direction after he was blind.

Plateau's sphere When a drop of liquid breaks away and falls it is followed by a smaller drop called Plateau's sphere.

plate in vibration The theory of a vibrating plate differs from that of a membrane in that its stiffness must be taken into account while the effect of tension is negligible. The plate can be considered as an extension of a transversely vibrating bar. The mathematical analysis is extremely complicated. In the case of a thin circular plate clamped at the rim, the fundamental frequency is given by $n = 0.47 Vt/r^2$, where t is the thickness, r the radius, and V the velocity of longitudinal waves in the material. For a plate:

$$V = \sqrt{E/(1 - \mu^2)\rho},$$

where E is the Young modulus (*see* modulus of elasticity), μ is the *Poisson ratio, and ρ the density. The upper *partials have different numerical factors and are inharmonic. The shape of the nodal lines on a vibrating rectangular plate can be considered to be due to the superposition of two sets of *standing waves, each having nodal lines parallel to one pair of sides. The two superimposed sets of perpendicular nodal lines give a set of nodal curves. The nodal lines in a vibrating plate may be demonstrated by Chladni's method of sprinkling find sand onto the plate. (*See* Chladni's plates.)

The theory of the bell can be approached as an extension of the plate but this method is only partly successful and most of the work on bells is empirical.

The diaphragm is part way between a membrane and a plate with its main features similar to the latter. It is used extensively in telephone earpieces and microphones as a source or a detector of sound.

plate tectonics The theory that the earth's surface is composed of a number of large but relatively thin slabs of rigid material that are moving relative to each other. These *plates* extend through the earth's crust into the upper mantle and are supported on the viscous layer known as the asthenosphere. Many of the earth's structures and processes, including oceanic trenches, midocean ridges, major faults, earthquake zones, volcanic belts, mountain building, and continental drift, are explained by the movements and consequent collisions of the plates.

platinum resistance thermometer An instrument in which changes in the electrical resistance of a platinum coil P immersed in a body are used to measure the temperature changes of the body. Inaccuracy due to the unknown temperature of the leads is eliminated by incorporating in the thermometer short-circuited compensating platinum leads L of the same dimensions as the coil leads C. These leads are mounted in close proximity to each other, the two sets being connected to opposite arms of a *Wheatstone bridge circuit. The slide wire of this

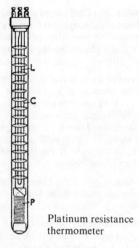

Platinum resistance
thermometer

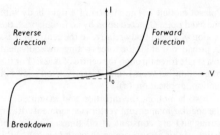

Voltage characteristic of p–n junction

bridge can be calibrated directly against temperature at certain fixed points.

pleochroic haloes Regions of darker colour found around small amounts of radioactive material present in certain crystals. The colour results from the radioactive decay and these zones have been used in studies of the age of rocks.

pleochroism That property of a crystalline substance whereby it appears to be differently coloured in different directions. The colours themselves result from the *polarization of light, and the differences result from the *anisotropic nature of the crystal.

plotter *See* incremental plotter.

Plucker, Julius (1801–1868) German physicist who investigated *diamagnetism. He discovered cathode rays in *gas-discharge tubes and was the first to suggest spectrum analysis.

plug-in A device that may be rapidly connected or disconnected from a complex electrical or electronic system. It usually consists of a standardized *printed circuit board or *integrated circuit package that may be inserted or removed rapidly from the main equipment. The use of such devices facilitates maintenance and repair of complex equipment.

Plumbicon *See* camera tube.

plural scattering *See* scattering.

plus tapping *See* tapping.

pneumatics The branch of physics dealing with the dynamic properties of gases.

p-n junction The region at which two *semiconductors of opposite polarity (p-type and n-type) meet. A p-n junction can perform various functions depending on the geometry, the bias conditions, and the doping level in each semiconductor region. Most *diodes, *transistors, etc., utilize the properties of

one or more p-n junctions. If the materials are dissimilar, e.g. silicon and germanium, the junction is a *heterojunction*. Normally the same material is used but doped so as to produce two different conductivity types; this is a simple *homojunction*. A typical voltage characteristic is shown in the diagram. Under reverse-bias conditions (i.e. negative bias applied to the p-type semiconductor) a depletion layer is produced at the junction as the *holes in the p-type material and the *electrons in the n-type region are attracted to the electrodes and very little current flows, until *breakdown occurs. Under forward-bias conditions, carriers are attracted across the junction into the region of opposite type (where they become *minority carriers) and a current flows in the external circuit. The forward current in a homojunction increases exponentially with the voltage, i.e.

$$I = I_0(e^{eV/kT} - 1),$$

where I_0 = reverse saturation current, e = electronic charge, V = applied voltage, k = Boltzmann constant, and T = thermodynamic temperature. Resistance in the material reduces the rate of rise of current through the device after a few tenths of a volt.

pnpn device A semiconductor device consisting of alternating layers of p-type and n-type semiconductor (almost always silicon), with at least three p-n junctions. Such devices are used for power-switching purposes, examples being the *silicon controlled rectifier and the *silicon controlled switch.

p-n-p transistor *See* transistor.

Pockel effect The *Kerr effect as observed in a *piezoelectric crystal.

Poggendorff, Johann C. (1796–1877) German physicist and scientific author and editor. He made a number of improvements in experimental technique, including the use of lamp and scale for measuring small angular deflections, and suggested some to others (e.g. the use of a thermoelectric generator to Ohm in place of the then unsteady voltaic cells). He was the author, editor, or initiator of several books and periodicals.

Poinsot ellipsoid *Syn.* energy ellipsoid. *See* Poinsot motion.

Poinsot motion

Poinsot motion 1. The motion of a rigid body with one fixed point O, acted upon by no forces, or, **2.** the motion of a rigid body relative to the centre of mass O, provided that all the forces acting are equivalent to a single force through the centre of mass. (For the connection between **1.** and **2.** *see* Newton's laws of motion, application (*d*)).

In such motion, the direction and magnitude of the angular momentum vector (*see* moment) drawn through O are constant at all times. The line of action of this vector, drawn through O is called the *invariable line* and a fixed plane perpendicular to it, called the *invariable plane*, contains the extremity of the angular velocity vector (drawn from O) at all times. The body moves in such a way that an ellipsoid, whose axes are along the principal axes of the body through O (*see* product of inertia) and whose shape (but not size) is that of the momental ellipsoid, rolls on the invariable plane, the point of contact being at the extremity of the angular velocity vector. The ellipsoid is known as the *Poinsot ellipsoid* (*syn.* energy ellipsoid). A line joining the point of contact of plane and ellipsoid to O traces out a cone which is the *space cone* (*see* instantaneous axis). An observer moving with the body, to whom the Poinsot ellipsoid appears stationary, sees the extremity of the angular velocity vector trace a curve on the surface of this ellipsoid. If the points on this curve are joined to O, the *body cone* is obtained.

point 1. A unit used for measuring wind direction. 1 point = $^{1}/_{32}(360°)$ = $11¼°$.

2. A unit of mass used for precious stones. 1 point = 0.01 metric carat = 2×10^{-6} kg.

3. A termination in an electric wiring installation intended for the attachment of a lighting fitting, or of a device, such as a socket or switch-socket, which enables an electrical appliance to be connected to the supply.

point-contact transistor *See* transistor.

point defect *See* defect.

point function A quantity whose value depends on the position of a point in space, e.g. magnetic field, temperature, density.

point group A set of symmetry operations (rotation about an axis, reflection across a plane, or combinations of these), not including translation, that when carried out on a periodic arrangement of points in space brings that system of points to self-coincidence.

point source A source of light of very small size, i.e. small compared with the distance away, so that the angular subtense is very small. The intensity of illumination varies inversely as the square of the distance from such sources (distances should be greater than, say, ten times the source size). Whilst geometrical optics would make the image of a point source a point, actually the image is a diffraction pattern of finite dimensions.

poise Symbol: P. The CGS unit of dynamic *viscosity. It is defined as the tangential force per unit area required to maintain unit difference in velocity between two parallel planes in a liquid that are separated by unit distance. 1 poise = 0.1 pascal second.

poiseuille Symbol: Pl. A unit of dynamic *viscosity, defined as the viscosity of a liquid that sets up a tangential stress of one newton per square metre across two planes separated by one metre when the velocity of streamlined flow is one metre per second. The unit is identical with the *SI unit of dynamic viscosity (the pascal second), but the name has not received international recognition.

Poiseuille, Jean (1799–1869) French physicist, remembered for his work on the flow of fluids. *See* Poiseuille flow.

Poiseuille flow The steady laminar flow of a viscous fluid through a pipe of circular cross section was first investigated by Poiseuille and Hagen. This type of fluid flow is named after Poiseuille and is such that the velocity distribution has the form of a paraboloid of revolution, the velocity being a maximum at the centre of the pipe, zero at the boundary walls, and the velocity being constant along any line parallel to the axis of the pipe. Assuming *Newton's laws of fluid friction for a viscous fluid, the quantity of fluid flowing per second is given by:

$$Q = \pi(p_1 - p_2)r^4/8\eta l,$$

where p_1, p_2 are the initial and final pressures on a cylinder of fluid, length *l*, and coefficient of viscosity η, the radius of the pipe being *r*. This is called the *Poiseuille equation* or the *Hagen–Poiseuille law*. The calculation of the coefficient of viscosity of a fluid can be accurately made from the results of observations and measurements taken with this type of flow in a pipe. (*See* Ostwald viscometer.)

Poisson, Siméon Denis (1781–1840) French mathematician. He developed the application of *Fourier series to physical problems and made major contributions to the theory of probability and to the calculus of variations. He also wrote on mechanics and mechanical properties of matter and made major advances in the theory of attractions, covering electrostatics, magnetism, and gravitational attraction. His generalization of the Laplace equation (Poisson equation) was published in 1813. His great knowledge of the theory of waves enabled him to show that Fresnel's wave theory of light led to the prediction of a bright spot in the centre of the shadow of an opaque object, a result confirmed experimentally by Arago (although it had in fact been observed in 1715 by Delisle).

Poisson distribution A *frequency distribution often applicable in practice to discontinuous variables. It can be applied to a radioactive decay process to predict the probability that a specific event (decay) will occur in a given period. It is the limit of the *binomial distribution. As the number of trials, *n*, increases, the probability, *p*, decreases and $np = m$ where *m* is the average number of times an event occurs in n trials. The probability of r successes in n trials is then: $m^r e^{-m}/r!$.

Poisson equation In SI units:

$$\frac{\partial^2 V}{\partial x^2} + \frac{\partial^2 V}{\partial y^2} + \frac{\partial^2 V}{\partial z^2} = -\frac{\rho}{\varepsilon_0},$$

or $\nabla^2 V = -\rho/\varepsilon$, when V is the electric potential at any point, ρ is the charge density, and ε the permittivity.

Poisson ratio Symbol: μ or ν. The ratio of lateral contracting strain to the elongation strain when a rod is stretched by in-line forces applied to its ends, the sides being free to contract. If the volume does not change under stretching, this ratio $= 0.5$, but the value is often less in practice being typically 0.3 for a metal. The reciprocal of Poisson ratio, denoted by m, is often used by engineers.

polar axis 1. A crystal axis of rotation that is not normal to a reflection plane and does not contain a centre of symmetry. Certain crystal properties will be dissimilar at opposite ends of such an axis.

2. *See* coordinate.

polar coordinates *See* coordinate.

polar diagram A graphical figure in which a physical quantity, such as the relative field strength radiated in any direction in a given plane by a transmitting aerial, is represented in polar *coordinates. Polar diagrams are usually drawn for planes that, relative to the surface of the earth, are horizontal and vertical (horizontal and vertical polar diagrams respectively). *See also* lobe.

polarimeter An accurate instrument for measurement of rotation of the plane of polarization (*see* plane-polarized light) by optically active liquids and solids. *See* saccharimeter.

polariscope An instrument for studying polarization phenomena and applied to study strain in glass, the distribution of stress in small-scale plastic models of engineering structures, etc. It consists of a *polarizer from which polarized light passes to a transparent substance under investigation and then to a rotatable analyser. The simple (Biot) polariscope, which depends on polarization by reflection, consists of two inclined glass plates, one to polarize the light and the other (rotatable) to analyse. Sometimes the analyser is a *Nicol prism or *Polaroid; sometimes both analyser and polarizer are Nicol prisms.

polarity 1. (general) The condition of a body or system in which there are opposing physical properties at different points.

2. (magnetic) The distinction between the north- and south-seeking *poles of a magnet.

3. (electrical) The distinction between the positive and negative parameters (e.g. voltage, charge, current, carrier type) in an electrical circuit or device.

polarizability *See* molecular polarization.

polarization 1. (electrical) In a simple cell consisting of two dissimilar plates in an electrolyte, such as Zn and Cu in dilute H_2SO_4, the current obtained soon falls considerably. This is due to a layer of hydrogen bubbles that collects on the copper plate and not only partially covers the plate and increases the internal resistance of the cell, but also sets up an e.m.f. of opposite direction to that of the cell. This phenomena is known as polarization. To make cells effective for a longer period, some means must be adopted to prevent gas deposition. In the *Daniell cell, for example, it is changed, by using two solutions, into a copper coating on a copper anode, which has no back e.m.f., and in the *Leclanché cell a chemical depolarizer is used, which reacts with the hydrogen produced.

2. (of radiation) Unpolarized (natural or ordinary) *electromagnetic radiation consists of waves in which the vibrations are transverse without any one-sidedness or preferential direction of vibration. Under various circumstances, the direction and characteristics of the vibration are more restricted. In *plane-polarized light or other radiation, the electric field resides entirely in one plane, called the *plane of vibration*. Sometimes the direction of the electric-field vector is not restricted to one plane but rotates with constant angular frequency (as viewed along the direction of propagation), although its magnitude stays constant; this is described as *circularly polarized radiation*. If the electric-field vector not only rotates but also changes its magnitude, then the radiation is *elliptically polarized*. A quarter-wave plate has sufficient thickness to produce a *phase difference of $\pi/2$ between the ordinary and extraordinary vibrations, and if plane-polarized light vibrating at 45° to the vibration plane strikes it, the light emerges circularly polarized; if vibrating in any other plane, it yields elliptically polarized light. Interference of polarized light requires that the vibrations are in the same plane and are obtained from the same beam.

3. *See* dielectric polarization.

4. *See* molecular polarization.

polarizer A crystal or conglomerate of crystals used to produce *plane-polarized light (pile of plates, *Polaroid, *Nicol prism, etc.).

polarizing angle *Syn.* Brewster angle. When light strikes a glass surface at an angle of incidence given by $\tan^{-1}(n)$, where n is the refractive index, the reflected light is plane-polarized. At this angle of incidence, the refracted ray makes an angle of 90° with the reflected ray (*Brewster's law*). *See* plane-polarized light; refractive index measurements.

polarizing pyrometer An instrument in which light from a hot source is polarized and its intensity compared with that from a fixed lamp, also polarized, whose filament is maintained at a fixed but unknown temperature. The *Wanner optical pyrometer* is such an instrument, used for measuring the temperature of a hot source (*see* diagram). A *Rochon prism is used, in conjunction with a bipr-

polarographic analysis

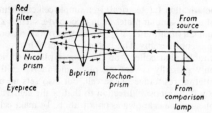

Wanner optical pyrometer

ism, to allow beams from the source and comparison lamp, which are polarized at right angles, to emerge from a diaphragm. These beams then pass through a *Nicol prism and a red filter, and an optical system producing two adjacent semicircular patches of light. In one position of the Nicol, the light from the lamp is completely extinguished and for a rotation of the Nicol through an angle ϕ both patches are made equally bright, whence:

$$E_\lambda / L_\lambda = \tan^2 \theta,$$

where E_λ and L_λ are the intensities of light of wavelength λ emitted by the source and lamp respectively. If E_λ is the intensity of the source when the temperature is changed from T to T' and ϕ' is the corresponding angular setting of the Nicol with the lamp unaltered:

$$\frac{E_\lambda}{E_\lambda{'}} = \frac{\tan^2 \phi}{\tan^2 \phi'}$$

and since

$$\log_e \frac{E_\lambda}{E_\lambda{'}} = K\left\{\frac{1}{T'} - \frac{1}{T}\right\}$$

by the *Wien radiation law:

$$2(\log_e \tan \phi - \log_e \tan \phi') = K\left\{\frac{1}{T'} - \frac{1}{T}\right\}.$$

The relation between $\log_e \tan \phi$, and $(1/T)$ is thus linear so that the scale may be calibrated using a source at a standard temperature T'. This type of pyrometer is useful since it dispenses with the need for rotating sectors or other devices to cut down the intensity when the source is at a very high temperature.

polarographic analysis A method of chemical analysis (invented by J. *Heyrovsky) by recording current-voltage curves of a solution electrolysed, at continuously increasing potential, with a dropping mercury electrode. Each constituent present is electrodeposited in turn when the potential reaches a certain value fundamental to the element, the current increasing at this point to give a curve consisting of a series of steps with horizontal sections between them. The potential at which each step starts is characteristic of the ion deposited, and the height of the current step is determined by the quantity of that ion in solution. The method is applicable even at considerable dilution.

Polaroid *Tradename* A thin transparent film containing ultramicroscopic polarizing crystals with their optic axes lined up parallel (Land, 1932). One component of polarization is absorbed and the other is transmitted with little loss. There are several different methods of production, the original form using synthetic dichroic crystals; others use stretched polyvinyl alcohol films impregnated with iodine or treated with hydrogen chloride to make the film strongly dichroic (*see* dichroism). The advantages of Polaroid are that it provides a large area of polarization with increased fields of operation, e.g. motor car headlights, visors, sunglasses, camera filters. By placing the axis of the Polaroid in an appropriate direction, stray plane-polarized light (produced, for example, by reflection) causing glare, can be removed.

Polaroid (Land) camera *Tradename* A camera that yields finished positive prints or transparencies within about 60 seconds (colour) or 10 seconds (black and white) of the exposure, the developing and processing of the film (*see* photography) taking place inside the camera. The negative and positive mediums of the *Polaroid film* are in separate rolls or stacks of connected sheets. Small closed capsules, incorporated into the negative emulsion, contain an activating reagent for the developer and also a jellifying agent. After exposure of the negative, the negative and positive materials are pulled through steel rollers, breaking the capsules, and allowing the activating chemicals to spread evenly between the two sheets in contact. The negative image is developed in contact with a suitable receiving layer such that unused silver salts diffuse into this layer, forming a positive image. This process is called *diffusion transfer*. The two layers are then stripped apart. Diffusion transfer is used in some types of photocopying machine, although the physical processes differ between machines. In colour-diffusion processes the negative material consists of three layers sensitized to blue, green, and red light (*see* colour photography) each of which is associated with dye molecules of the complementary colour (yellow, magenta, and cyan) coupled with developer molecules. Following exposure, an activating jelly is released and starts the processing by which the dye-developer molecules diffuse into the associated layer, develop the negative image and become fixed in the layer. Unused dye-developer molecules diffuse into the receiving layer of the positive material and a positive colour image results.

Polaroid cameras are used to record information on *cathode-ray oscilloscope screens, microscope images, etc., being attached directly to the apparatus.

polaron An electron coupled to an ion in the conduction band of a perfect ionic crystal.

polar vector *See* vector.

pole 1. Of electrical apparatus or a circuit. Each of the terminals or lines between which the main circuit voltage exists. (*See* number of poles; quadripole.)

2. The place toward which lines of magnetic flux converge, or from which they diverge. It usually exists near a surface of magnetic discontinuity, and in the material of higher permeability. A north pole of a magnet is that which is subject to a force towards the north (magnetic) pole of the earth. (*See also* polarity.)

3. The midpoint of a convex or concave mirror. The line joining the *centre of curvature and the pole is the *principal axis* of the mirror.

pole chamfer *See* pole-face bevel.

pole-changing control A method of obtaining two or more running speeds of a three-phase motor by changing the number of magnetic poles. This is done either by providing separate stator windings each designed to produce the required numbers of poles or by suitably reconnecting a single winding. In a two-speed machine, the speeds are usually in the ratio 2:1, but other ratios are possible.

pole core In an electrical machine. That part of a *pole piece on which is placed one or more *field coils.

pole face The end surface of the *core of a magnet through which surface the useful magnetic flux passes. In particular, in an electrical machine it is that surface of the core or pole piece of a *field magnet that directly faces the armature.

pole-face bevel The part of the *pole face in an electrical machine that has been bevelled, i.e. is not coaxial with the armature.

pole horn The part of a *pole piece or pole shoe that projects circumferentially beyond the *pole core in an electrical machine. It is described as leading or trailing in the same manner as a *pole tip.

pole piece Either of the pieces of ferromagnetic material attached to the ends of an electromagnet or permanent magnet and specially shaped to control the flux in various electrical devices.

pole shoe The separate part of a *pole piece that faces the armature in an electrical machine. It is usually built of *laminations riveted together and is attached to the pole body by means of countersunk screws.

poles of a magnet *See* pole.

pole strength *See* magnetic pole strength.

pole tip Either of the two edges of a *pole face that are usually parallel to the axis of rotation in an electrical machine. The pole tip that is encountered first, during revolution, by a fixed point on the armature or stator is called the leading pole tip, and the other is called the trailing pole tip.

polhode cone *Syn.* body cone. *See* instantaneous axis.

polychromatic radiation Electromagnetic radiation that is a mixture of radiation with more than one wavelength. *Compare* monochromatic radiation.

polydisperse system A colloidal system in which the particles in the disperse phase are of various sizes. *Compare* monodisperse system.

polygon of vectors A geometrical construction extending the law of addition of two *vectors to any number of vectors. If the vectors are represented in magnitude and direction by the sides of an unclosed polygon, not necessarily confined to the same plane, the directions of the vectors all leading in the same way round the figure, then the vector sum or resultant is given by the line joining the beginning of the first to the end of the last side of the polygon.

polymerization The chemical phenomenon in which a number of similar molecules combine to form a single molecule. The compound of higher molecular weight is called the *polymer* of the other.

polymorphism The existence of more than one crystalline modification of the same substance (usually more than three, the terms dimorphism, trimorphism being applied to two or three respectively).

polyphase system An electrical system or apparatus in which there are two or more alternating supply voltages displaced in *phase relative to each other. A symmetrical polyphase (n-phase) system has n sinusoidal voltages of equal magnitude and frequency, with mutual phase differences of $2\pi/n$ radians (or $360/n$ degrees). An exception to this is the two-phase system (quarter-phase) in which the two voltages have a phase difference of $90°$. For $n > 2$, the system requires a minimum of n line wires.

polyphase transformer A *transformer for use in a *polyphase system. The magnetic circuits corresponding to the various phase windings have parts in common.

Popov, Aleksandr Syepanovich (1859–1905) Russian physicist who was the first to use a suspended wire as a radio aerial. As a result of this, and other innovations in radiotelegraphy, he is accepted in Russia as the inventor of radio.

population inversion *See* laser.

Porro prism A total reflection prism used in the construction of prismatic telescopes and binoculars. The simplest form is a $45°$, $90°$ prism receiving light through the hypotenuse face, and reflecting it back parallel to the original direction after successive internal reflection at the other two faces (e.g. in *prismatic binoculars). This prism inverts in one direction only. The second prism of a binocular completes the inversion at right angles by placing its roof edge at right angles to the first. *See* prism.

port An access point in an electronic circuit, device, network, etc., where signals can be fed in or out or where the variables of the system can be observed or measured.

position vector In mechanics, a line joining a point on the path of a particle to some reference point.

positive booster *See* booster.

positive column *See* gas-discharge tube.

positive crystal A uniaxial crystal in which the refractive index for the extraordinary ray is greater than that of the ordinary ray (e.g. quartz). The ordinary (spherical) wave surface encloses the extraordinary (ellipsoidal) wave surface.

positive electron *See* positron.

positive feedback *See* feedback.

positive glow *Syn.* positive column. *See* gas-discharge tube.

positive logic *See* logic circuit.

positive magnetostriction *See* magnetostriction.

positive phase sequence *See* phase sequence.

positive transmission *See* television.

positron *Syn.* positive electron. The *antiparticle of the *electron, i.e. an elementary particle with electron mass and positive charge equal to that of the electron. According to the relativistic *wave mechanics of Dirac, space contains a continuum of electrons in states of negative energy. These states are normally unobservable, but if sufficient energy can be given, an electron may be raised into a state of positive energy and become observable. The vacant state of negative energy behaves as a positive particle of positive energy, which is observed as a positron. (*See* pair production; annihilation.)

positronium A short-lived association between a *positron and an *electron, similar to a hydrogen atom. There are two types: *orthopositronium* in which the *spins of the particles are parallel and *parapositronium* in which the spins are antiparallel. The orthopositronium decays with a mean life of about 10^{-7} s to give three photons. The parapositronium has a smaller mean life and produces two photons (*see* annihilation).

It is also possible that the positronium molecule, consisting of two positrons and two electrons, may have a brief existence.

post-office box A type of *Wheatstone bridge that consists of a number of resistance coils arranged in a special box. Each coil is connected between adjacent metal blocks so that it can be shorted out of circuit by inserting a metal plug between the blocks. The arrangement, and resistance values, are as shown in the diagram. Tapping keys are inserted in both cell and galvanometer circuits. AB and BC are known as the ratio, or proportional arms. If R_4, the unknown resistance, is small then R_1/R_2 should be made 1000/10, and if large 10/1000, to make fullest use of the arm R_3. Thus from 0.1 to 10^6 ohms can be measured. The post-office box can also be used for other measurements within the scope of the Wheatstone bridge, or can be used as a *potentiometer.

potassium-argon dating *See* dating.

potential 1. Electrostatic, magnetostatic, and gravitational potentials, at a point in the field: the work done in bringing unit positive charge, unit positive pole, or unit mass respectively from infinity (i.e. a place infinitely distant from the causes of the field) to the point. Gravitational potential is always negative but the electrostatic and magnetostatic potentials may be positive or negative. Since these are *conservative fields, the potential is a function only of the position of the point. The difference in potential between two points is the work done in taking the unit object from one point to the other. Potential is a scalar quantity.
   2. *See* kinetic potential.
   3. *See* thermodynamic potential.
   4. *See* velocity potential.

potential barrier The region in a field of force in which the potential is such that a particle, which is subject to the field, encounters opposition to its passage.

potential difference Symbol: V, U. The line integral of the electric field strength between two points. The work done when a charge moves from one to the other of two points (by any path) is equal to the product of the potential difference and the charge. *See* electromotive force.

potential divider A chain of resistors, inductors, or capacitors connected in series and tapped to allow a definite fraction of the voltage across the chain to be obtained across one or more of the individual components. *See* potentiometer.

potential energy Symbol: E_p, V, Φ. The work done in changing a system from some standard configuration to its present state. Thus, if a body of mass m is raised vertically through a height h, the work done $= mgh =$ increase in potential energy. If the work done is independent of the way in which the change is made, the system is said to be *conservative*. Systems with friction between their parts are nonconservative (*compare* conservative field). Potential energy is indeterminate to within an additive constant, since the standard configuration may be chosen arbitrarily. If a conservative system is in equilibrium, the change in potential energy in any infinitesimal displacement is zero.

potential flow *See* laminar flow.

potential function From the theory of functions of a complex variable, it is shown that a relation of the form:
$$w = \phi + i\psi = f(z),$$
represents a two-dimensional irrotational motion of a fluid in the xy-plane, where z is the complex variable $(x + iy)$, ϕ the *velocity potential, ψ the *stream function, i is $\sqrt{-1}$, and $f(z)$ means a function of z. The complex function $w = (\phi + i\psi)$ is called the potential function; equating the real and imaginary parts of the relation $w = f(z)$ gives the

lines of equivelocity potential and the stream lines of the irrotational motion.

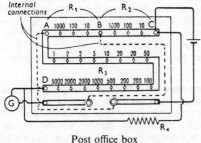

Post office box

$$AB = R_1; \ BC = R_2; \ AD = R_3$$

potential gradient The rate of change of electric potential, V, at a point with respect to distance x, measured in the direction in which the variation is a maximum. It is measured in volts per metre. The electric field strength, E, is numerically equal to the potential gradient but in the opposite sense:

$$E = -dV/dx.$$

potential scattering *See* scattering.

potential transformer *See* voltage transformer.

potential well A region in a field of force in which the potential decreases abruptly, and on either side of which the potential is greater.

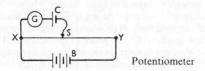

Potentiometer

potentiometer 1. A form of *potential divider using a uniform wire as the resistive chain. A moving sliding contact can tap off any potential difference less than that between the ends of the wire. A typical use is in measuring an unknown e.m.f., such as that of a cell, C (*see* diagram). A battery B supplies a steady current along the resistance wire XY. The slider S is moved until the galvanometer G shows no deflection. The distance XS = l_1 is noted, the cell is replaced by a standard cell, and a new point of balance l_2 is found. Then,

$$E_1/E_2 = l_1/l_2,$$

where E_1 and E_2 are the e.m.f.s of the unknown and standard cell respectively. It will be seen that the true e.m.f. is given, as no current flows through the cell. For precision work, more elaborate forms of potentiometer are available.

2. (pot) Any variable resistor with a third movable contact. The geometry of the device can be arranged so that the output voltage is a particular function (linear, logarithmic, sine or cosine) of the applied voltage.

pound A unit of mass formerly based on a platinum cylinder called the Imperial Standard Pound and kept by the Board of Trade. The pound is now defined as 0.453 592 37 kilograms.

poundal The f.p.s. unit of force, equal to the force required to give a mass of one pound an acceleration of 1 foot per second per second. 1 poundal = 0.138 255 newton.

powder diagrams *See* X-ray analysis.

powder photography A crystal-diffraction method in which the specimen is a randomly orientated crystalline powder (which is usually rotated), in a parallel beam of monochromatic X-radiation (or electrons, neutrons, etc.). *See* X-ray analysis.

Powell, Cecil Frank (1903–1969) Brit. physicist who was a professor at Bristol. He devised the photographic method of studying nuclear processes and using this method was the first to provide experimental evidence for the existence of the *pion. For this work he was awarded a Nobel prize (1950). He is also known for his investigations of the nuclear processes involved in *cosmic rays.

power 1. Symbol: P. The rate at which energy is expended or work is done. It is measured in *watts.

The power P developed in a direct-current electric circuit is given by the expression $P = VI$, where V is the potential difference in volts and I is the current in amperes. In an alternating-current circuit, the *active power*, P, is given by $VI \cos \phi$, where V and I are the rms values of the voltage and current and ϕ is the *phase angle between the current and the voltage. The product IV, called the *apparent power*, is measured in volt-amperes. Cos ϕ is called the *power factor. $IV \sin \phi$ is called the *reactive power* and is measured in *vars.

2. (optic) The power of a lens or mirror is the reciprocal of the focal length in metres, generally positive if converging and most commonly applied to the *dioptric power* of a lens. For mirrors, the term *catoptric power* is sometimes used.

According to the origin of measurement, the dioptric power may be qualified, e.g. *back vertex* power, *front vertex* power, *principal point* power. The dioptric power of a surface is the reciprocal of the reduced focal length, viz. $(n' - n)/f$. The *effective power* of a lens at a certain point is the reciprocal focal distance measured from that point. Narrow-angle prisms have a power expressed as an angle of deviation in either *centrads or *prism dioptres. *Magnifying power* is the ratio of the retinal image produced when an instrument is used, to the retinal image of the object (*in situ* for telescopes; at 25 cm for microscopes). *Dispersive power* (symbol ω) (*see* dispersion) is equal to:

$$[(n_F - n_C)/(n_D - 1)],$$

and is the reciprocal of *constringence.

power amplification 1. In an *amplifier. The ratio of the power level at the output terminals to that at the input terminals.

power amplifier

2. In a *magnetic amplifier. The product of voltage amplification and current amplification using a specified control circuit.

3. In a *transducer. The ratio of the power delivered to the load to the power absorbed by the input circuit, under specified operating conditions.

power amplifier An *amplifier producing an appreciable current flow into a relatively low impedance or a large increase in output power. It is usually used so that the output is not applied to the input of a further amplifying stage, but to an output *transducer, such as an *aerial or *loudspeaker.

power component The *active current, *active voltage, or *active volt-amperes.

power factor The ratio of the actual *power in watts developed by an a.c. system (as measured by a wattmeter) to the apparent power in volt-amperes (as indicated by voltmeter and ammeter readings). If the voltage and current are *sinusoidal, the power factor is equal to the cosine of the *phase angle between the voltage and current vectors.

power-factor meter *Syn.* power-factor indicator; phase meter. An instrument for measuring directly the *power factor of the circuit to which it is connected. Fundamentally it measures the *phase difference between two electrical quantities (e.g. voltage and current) having the same frequency. It operates on the electromagnetic principle and may be an *electrodynamic instrument or a *moving-iron instrument. Instruments are available for indicating the power factor of single-phase circuits or of *polyphase systems with balanced and unbalanced loads.

power frequency The frequency used in a.c. power systems. Standard values are: 50 hertz in the British Isles, 60 hertz in the USA. For a.c. traction, frequencies of 25 hertz and 16.67 hertz are sometimes used.

power-level difference Symbol: L_P. The quantity:
$$\tfrac{1}{2} \log_e (P_1/P_2),$$
where P_1 and P_2 are two powers. The term *sound-power level* is used when P_1 and P_2 are sound powers (*see* sound-energy flux), P_2 being a reference power. L_P is usually expressed in *nepers or *decibels.

power line *See* transmission line.

power pack A device that converts power from an a.c. or d.c. *power supply, usually the mains, into a form suitable for operating an electronic device.

power reactor A *nuclear reactor designed to produce useful quantities of electrical power.

power station *Syn.* generating station. A complete assemblage of plant, equipment, and the necessary buildings at a place where electric power is generated on a large scale. The main types are: *thermal power station, nuclear power station, and *hydro-electric power station.

power supply Any source of electrical power in a form suitable for operating electrical or electronic devices. Alternating-current power may be derived from the mains, either directly or by means of a suitable *transformer. Direct-current power may be supplied from batteries or suitable rectifier/filter circuits. A *bus is frequently used to supply power to several circuits or to several different points in one circuit. Suitable values of voltage are derived from the common supply by some form of *coupling.

power transistor A *transistor designed to operate at relatively high values of power or to produce a relatively high power gain. Power transistors are used for switching or amplification and because of the relatively high power dissipation, which ranges from 1 watt to 100 watt, they usually require some form of temperature control.

Poynting, John Henry (1852–1914) Brit. physicist. He made a considerable study of gravitation, including a determination of the gravitational constant with a balance of knife-edge type. To electromagnetic theory, he contributed the concept of a *Poynting vector (*see also* Poynting's theorem) and he studied radiation pressure, predicting, and with Simon Barlow in 1904, establishing experimentally the existence of a tangential force from an oblique beam of radiation. He showed that radiation pressure might exceed gravitational pull on small particles and explained thereby the observed behaviour of the tails of comets.

Poynting's theorem The space through which *electromagnetic radiation passes is filled with electric and magnetic fields at right angles to each other and to the direction of propagation of the radiation. The rate of energy transfer is proportional to the product of the electric and magnetic field strengths, i.e. to the surface integral of the *Poynting vector formed by the components of the field in the plane of the surface.

Poynting vector A *pseudovector giving the direction and magnitude of the rate of energy flow perpendicularly through unit area in an electromagnetic field. It is equal to the vector product of the electric and magnetic field strengths at any point.

PPI Abbreviation for plan position indicator. *See* radar.

Prandtl number The dimensionless group $(C\eta/K\rho)$ occurring in the dimensional analysis of convection in a fluid due to the presence of a hot body, where C is the heat capacity per unit volume of the fluid, η is the viscosity of the fluid, K is the thermal conductivity of the fluid, ρ is the density of the fluid.

Its reciprocal is known as the *Stanton number*. *See* convection (of heat).

preamplifier An *amplifier used as an earlier stage to the main amplifier. It is frequently placed near the signal source (aerial, pick-up, etc.), being connected by cable to the main amplifier. This improves the *signal-to-noise ratio as amplification of the initial signal occurs before it traverses the path to the main amplifier.

precession If a body is spinning about an axis of symmetry OC (where O is a fixed point) and C is rotating round an axis OZ fixed outside the body, the body is said to be precessing round OZ. OZ is the precession axis. A gyroscope precesses due to an applied torque (called the *precessional torque*).

If the moment of inertia of the body about OC is I and its angular velocity is ω, a torque K whose axis is perpendicular to the axis of rotation will produce an angular velocity of precession Ω about an axis perpendicular to both ω and the torque axis where $\Omega = K/I\omega$. The direction of precession is given by *Lanchester's rule*: View the gyroscope from a point in its plane of rotation and apply a torque to the axis of rotation, the axis of this torque and the line of sight being coincident. Precession of the rotating axis then occurs so that a point on the gyroscope's circumference appears to describe an ellipse whose sense is that of the applied torque. *See* Eulerian angles.

precessional torque *See* precession.

pre-emphasis (and **de-emphasis**) In a radiocommunication system employing *frequency modulation or *phase modulation, an improvement in the *signal-to-noise ratio can be effected by using pre-emphasis at the transmitter, i.e. by artificially increasing the *modulation factor for the higher modulation frequencies as compared with the lower modulation frequencies, and by using de-emphasis at the receiver, i.e. by reducing the relative strength of the higher audio frequencies to compensate for the effect of the pre-emphasis. Although pre-emphasis and de-emphasis can be applied to a system using amplitude modulation, this is rarely done since the resulting improvement in signal-to-noise ratio is only slight. The technique is also used for tape recordings and gramophone records.

preferential orientation *See* alloy.

preon A hypothetical entity postulated as the building blocks of *leptons and *quarks. There is no experimental evidence for their existence, nor is there likely to be with the energies of current *accelerators. They do, however, have certain theoretical attractions.

presbyopia With age, the crystalline lens of the human eye hardens so that the *accommodation for viewing nearer objects decreases. The *near point of the eye recedes beyond the convenient reading distance. Convex lenses are required to aid near vision (generally required after 50 years of age), sometimes in addition to a distance correction, i.e. two pairs of glasses or a pair of bifocal lenses are required.

pressure Symbol: p. At a point in a fluid, the force exerted per unit area on an infinitesimal plane situated at the point. In a fluid at rest the pressure at any point is the same in all directions. In a liquid it increases uniformly with depth, h, according to the formula:

$$p = \rho g h,$$

where ρ is the density of the fluid, and g is the acceleration of *free fall. In a gas under isothermal conditions it decreases exponentially with height h, according to the formula:

$$P_h = p_0 e^{-(\mu g/RT)h},$$

where μ is the relative molecular mass, R is the *molar gas constant, and T is the thermodynamic temperature.

The SI unit of pressure is the *pascal (i.e. newton per metre squared) but the *bar is also officially recognized. For example, in meteorology, pressure is measured in millibars (mb), where one pascal is equal to 0.01 mb. The atmosphere, millimetre of mercury (mmHg), and torr are now obsolete units of pressure and should only be used for rough comparisons. *See also* standard atmosphere; standard temperature and pressure.

pressure broadening An effect observed in *line spectra in which high density, and hence pressure, of the emitting or absorbing material causes a broadening of the spectral lines; the higher the density and pressure, the greater the width of the lines. The broadening is caused by collisions with other atoms while the atom is emitting or absorbing radiation. *See also* Doppler broadening; line profile.

pressure cable An electric cable in which the insulating material (dielectric) is maintained at a pressure in excess of that of the atmosphere. *See* gas-pressure cable; compression cable.

pressure coefficient Symbol: β. The quantity relating pressure p and thermodynamic temperature T at constant volume:

$$\beta = (\partial p/\partial T)_V.$$

The *relative pressure coefficient*, symbol: α, is equal to $\beta(1/p)$.

pressure cooker A cooking vessel with a steam-tight lid and a weighted valve that blows off at preset pressures. The boiling point of the liquid is raised due to the high pressure (up to 2 atmospheres) enabling quick high-temperature cooking to take place within the vessel.

pressure gauges 1. Primary gauges include the liquid-column *manometers and the *free-piston gauge.

2. Secondary gauges include the *Bourdon tube and *resistance gauge.

3. *See* micromanometers.

4. Vacuum gauges include the *McLeod and *Knudsen gauges, which are primary gauges (although the Knudsen gauge is often made as a secondary gauge), and the *Pirani, *ionization, and *molecular gauges.

pressure head The height of a column of liquid capable of exerting a given pressure, e.g. the head h of liquid, corresponding to a pressure p, is $h\rho g$ where

ρ is the density of the liquid and g the acceleration of *free fall.

pressurized-water reactor (PWR) A type of thermal *nuclear reactor in which water under pressure (to prevent boiling) is used as both coolant and moderator. The effectiveness of water as a moderator and the fact that water absorbs neutrons means that the core is compact and that the fuel must be slightly enriched in ^{235}U. The fuel rods of uranium dioxide are clad in a zirconium alloy. Water enters the core at about 280 °C and leaves at 310 °C under a pressure of 16 MPa; a substantial pressure vessel is therefore needed, about 200 mm thick. At this pressure and temperature the coolant does not boil; these conditions are maintained with an electrically heated device called a pressurizer. A strong containment building encloses the pressure vessel and pipework carrying the cooling water; this is designed to contain the coolant should a pipe break. Because depressurized water at these temperatures would rapidly turn to steam, reserve supplies of water are available.

A variant of the PWR is the *boiling-water reactor* (BWR). In this device water is allowed to boil within the core and the resulting steam is separated in the pressure vessel and passed directly to the turbine generators. This avoids the expense of the heat exchangers (steam generators) of the PWR but additional equipment is needed to contain any radioactive gases from the coolant, which might escape from the turbine generator.

preventive reactor A *reactor used in conjunction with an on-load *tap changer to limit the current that circulates between adjacent tappings when the latter are momentarily bridged by the selector switches during the operation of the tap changer.

preventive resistor A *resistor used for the same purpose as a *preventive reactor.

Prévost, Pierre (1751–1839) Swiss physician. He is noted for his theory of exchanges, which he advanced in 1791.

Prévost's theory of exchanges (1791) A body emits precisely the same radiant energy as it absorbs when it is in temperature equilibrium with its surroundings.

When a body is in thermal equilibrium with its surroundings, conduction and convection cease, but not necessarily thermal radiation. In fact, when the temperature of the body is constant, a state of dynamic equilibrium is set up, the rate of emission equalling the rate of absorption.

Price's guard wire A conductor (wire) employed mainly in measurements of insulation resistance (of cables, in particular). It is arranged so as to prevent the surface leakage current from flowing through the measuring instrument so that the latter indicates only the *leakage current that flows through the insulating material.

primary See primary winding.

primary cell See cell.

primary colours A set of three coloured lights (or pigments) that when mixed in equal proportions produce white light (or black pigment). One group consists of red, green, and blue; cyan (greenish-blue), magenta (reddish-blue), and yellow form another set. Red and cyan, green and magenta, and blue and yellow are pairs of *complementary colours. Three lights of primary colours can be mixed in suitable proportions to produce any other colour, excluding black, by an *additive process. Three pigments, paints, dyes, etc., of primary colours can be mixed to produce any colour, excluding white, by a *subtractive process. The proportions of the primaries required to produce a certain coloured light are not the same as those needed to produce the same coloured pigment, paint, etc. *See also* colour television; colour photography; chromaticity.

primary cosmic rays See cosmic rays.

primary electrons Electrons incident on a surface distinguished from the secondary electrons that they release. *See* secondary emission.

primary extinction An increase in absorption or decrease in diffraction in a perfect crystal resulting from interference between multiple-reflected beams.

primary standard A standard used nationally or internationally as the basis for a unit, e.g. the international prototype *kilogram. *Compare* secondary standard.

primary winding The winding on the supply (i.e. input) side of a *transformer, irrespective of whether the transformer is of the step-up or step-down type. *Compare* secondary winding.

principal axis of a body 1. *See* product of inertia. **2.** *See* pole.

principal directions The directions about which there is symmetry of crystal properties, such as refractive indexes, coefficients of thermal expansion, magnetic susceptibility, thermal conductivity, etc.

principal focus See focal point.

principal moments of inertia See product of inertia.

principal normal See normal.

principal planes (and **points**) Two conjugate planes (and axial points in these planes) for which the lateral magnification is unity. They are also called *unit planes*. A ray directed to a point in one principal plane emerges from a point in the other plane at an equal distance from the axis and on the same side. A ray directed towards one principal point (on the axis) emerges from the other (axially conjugate) in a direction as though it suffered a small angle refraction between the first and last media. Once the principal points and principal focal points are known, the calculations of conjugate image formation are reduced to the effect of a single system – the usual conjugate focus relations hold, in which distances are measured from the principal planes. The graphical solution is also simplified.

For a single surface the principal planes coincide with the tangential plane at the pole. For an infinitely thin lens, the planes coincide with the lens. For a thick lens in air, the principal planes are separate but each coincides with its corresponding nodal plane (they are then called *equivalent planes*). For any two surfaces, lenses, or principal systems:

$$F = F_1 + F_2 - \delta F_1 F_2;$$
$$h = -\delta F_2 / F;$$
$$h' = \delta F_1 / F;$$

where F is the reduced principal power of the combination, F_1 and F_2 are the reduced principal powers of the components, δ is the reduced distance between adjacent principal points, h and h' are the positions of the principal planes of the combination from the first principal point of the first system and the second principal point of the second system respectively. *See* centred optical systems; antiprincipal points.

principal quantum number 1. *See* atom.
2. *See* atomic orbital.

principal radii A section of a surface made by a plane passing through a *normal to the surface is called a *normal section*. A principal normal section is a normal section of maximum or minimum radius of curvature. There are two such sections and they are perpendicular to one another; their radii are called *principal radii of normal curvature*, or briefly, *principal radii*.

The curvature of a section is the reciprocal of its radius. The sum of the curvatures of any two perpendicular normal sections equals the sum of the two principal curvatures.

A *synclastic surface* is one for which the centres of curvature of the principle sections are on the same side of the surface. An *anticlastic surface* has these two centres of curvature on opposite sides (e.g. a saddle); in this case the two radii have opposite signs.

Half the sum of the principal curvatures is called the mean curvature of the surface. The product of the principal curvatures is called the *Gaussian* or *total curvature*.

Surfaces whose mean curvature is zero are called minimal and have the least surface area consistent with their fixed boundaries (e.g. soap films on wire frames when the pressure differences between the two sides of the film are equal).

Surfaces whose total curvature is everywhere zero are called *developable* (the converse is not true); they are obtainable from a plane sheet by deformation without stretching.

principal strains *See* homogeneous strain.

principal stresses *See* stress components.

principle A highly general or inclusive law, exemplified in a multitude of cases.

principle of coincidence *See* vernier scale.

principle of equivalence *See* relativity.

principle of exclusion *See* Pauli exclusion principle.

principle of indeterminacy *See* uncertainty principle.

principle of least time *See* Fermat's principle.

principle of stationary time *See* Fermat's principle.

principle of work *See* machine.

Pringsheim, Ernst (1859–1917) German physicist. He worked with Lummer on black-body radiation and other researches in heat (*see* Lummer), and also did much research on spectroscopy, in particular on the emission spectra of compounds.

printed circuit An electronic circuit together with conducting interconnections fabricated on a thin rigid insulating sheet, usually fibreglass. The supporting sheet plus circuit is called a *printed-circuit board* (PCB). The interconnections are produced by first applying a conducting film, usually copper, to the sheet, coating parts of the film with a protective material using *photolithography, then removing the unprotected metal by etching. This leaves the pattern of interconnections required by the circuit design. The components, usually *integrated circuits, are finally soldered into position between the conducting tracks to complete the circuit.

Doubled-sided PCBs are commonly produced in which both sides of the board have a circuit formed on them, with *feedthroughs to connect the two sides as required. Printed circuits have also been produced with several alternating layers of metal film and thin insulating film mounted on a single board. *See also* edge connector.

printer In computing, a device that converts the coded information from a computer into a readable form printed on paper. There are many types, varying in the method and speed of printing and the quality of the print. Some printers print a single character at a time. Line printers print a complete line of characters at a time. Page printers, such as the laser printer, produce a complete page at a time. The printed characters may have a solid form or may consist of patterns of closely spaced dots.

prism A refracting medium bounded by intersecting plane surfaces or a combination of such. The main uses of prisms are (*a*) for deflection or small-angle deviation – *narrow-angle* refracting prisms; (*b*) for large-angle deviation and dispersion – in *spectrometers, *refractometers; (*c*) for changing ray direction or inverting or erecting images using total internal reflection (sometimes silvered to permit reflection); (*d*) for miscellaneous purposes, polarization, doubling images, photometer heads, etc.

The prismatic effect of *narrow-angle* prisms is given by $P = (n - 1)A$, where n is the refractive index, A the angle of the prism, i.e. the angle made by the refracting faces in a principle section perpendicular to the refracting edge. P is the angle of

prismatic binoculars

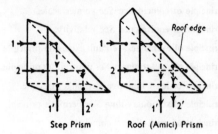

Step Prism   Roof (Amici) Prism

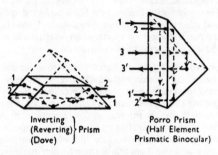

Inverting
(Reverting) } Prism
(Dove)

Porro Prism
(Half Element
Prismatic Binocular)

Reflecting prisms

deviation (usually expressed in degrees) or the power of the prism (expressed in *prism dioptres). The angle A is also called the *apical angle*; the region of widest section is the *base* of the prism. Rays are really deflected towards the base direction, while objects viewed through the prism are apparently displaced towards the apex. When the base apex lines are parallel, the resultant power is the algebraic sum of the separate prism powers ($P = P_1 + P_2 +$ etc.) but if the base apex lines are inclined, the parallelogram vector composition law must be used. *Variable power prisms* (*Risley rotary prisms) consists of two equal prisms rotating in opposite directions; such prisms are used in ophthalmic optics (photometers) and in various optical instruments (e.g. rangefinders). The *chromatic aberration C of a single prism is ωP, where ω is the dispersive power. For two prisms the chromatic aberration C is ($\omega_1 P_1 + \omega_2 P_2$) and total power $P = P_1 + P_2$. For a *direct-vision prism* $P = 0$ (Amici); and for an *achromatic prism* $C = 0$.

For large angle deviation and dispersion, the apical angle must be large. Such prisms have a minimum deviation (*see* deviation (angle of)), a property used for the determination of refractive index and of dispersion using spectrometers. Among the *dispersing* prisms are the *constant-deviation prism (*see also* Wadsworth prism). The Pulfrich prism is used in the *Pulfrich refractometer employing grazing incidence on one face.

The *reflecting* prisms include the step prism, *Porro prism, *erecting or Dove prism, *Amici or roof prism. Miscellaneous prisms include *Nicol prism, *biprisms, *Rochon prism, *Fresnel rhomb.

prismatic binoculars Binoculars in which each half consists of a *Kepler telescope, the prisms serving

the two-fold purpose of reducing the length of the instrument and also erecting the image. The prisms are two *Porro prisms; one prism inverts in one direction only, the second prism completes the inversion, since its roof edge is crossed at right angles to the roof edge of the former.

prism dioptre A unit of deviating power of a *prism based on a tangent measured in centimetres on a scale placed one metre away or a similar proportion. If θ is the angle of deviation then the power P, in prism dioptres, is equal to $100 \tan \theta$. The unit is used mainly for narrow-angle prisms. *See* centrad.

probability The numerical value quantifying the chance that one specified outcome out of several possible outcomes will occur as a consequence of an unpredictable event.

Independent events. If n is the total number of ways in which an event can occur, and m is the number of ways in which an event can occur in a specified way, then the ratio m/n is the *mathematical* (or *a priori*) *probability*. When a dice is tossed, the mathematical probability of obtaining the number 4 is 1/6. If there are 4 white counters and 5 black counters in a bag, the mathematical probability of drawing out a white is 4/9. In a random sequence of n trials of an event with m favourable outcomes, as n increases indefinitely the ratio m/n has the limit P, where P is the probability.

If in a number of trials an event has occurred n times and failed m times, the probability of success in the next trial is given by $n/(n + m)$. This is the *empirical* (or *a posteriori*) *probability*. The probability of a man not dying at a certain age, based upon past observations (recorded, say, in a mortality table), is an empirical probability.

The probability of a given number of successes in a certain number of trials, when the probability of success in a single trial does not vary, is given by the *binomial distribution. If p is the probability of success in one trial and q (equal to $(p - 1)$) the probability of failure, then the probability of r successes in n repeated trials is:

$$\frac{n!}{(n-r)!r!} p^r q^{n-r}.$$

Dependent events. If two or more events are so related that the outcome of one affects the outcome of the other or others, the notion of *conditional probability* must be used. The individual probability of each event is calculated in sequence and the product of these gives the conditional probability. If there are 4 white counters and 5 black counters in a bag, the probability of picking 3 white counters in the first 3 tries would be:

$$(4 \times 3 \times 2)/(9 \times 8 \times 7).$$

If a variable X can take on a set of discrete random values x_i, each of which has a probability p_i, the relative frequency of occurrence of any one of these values is $x_i p_i$; the cumulative frequency F, up to the value x_n is given by:

$$F(x_n) = \sum_{i=1}^{n} x_i p_i.$$

If X varies continuously, then the cumulative frequency up to a value x_n is given by

$$F(x)_n = \int_{-\infty}^{x} f(x)\mathrm{d}x,$$

where $f(x)$ represents the relative frequency of any specific value of x. The function $f(x)$ is called the *frequency function* or *probability density function*. The graph of $F(x)$ is a normal *frequency distribution.

probability density function *See* probability.

probable error That error such that the chances of the absolute magnitude of the *deviation being greater than or less than it are even. The arithmetic mean of n observations has a probable error that is $n^{-1/2}$ times the probable error of a single observation. (*See* frequency distribution.)

Experimental results are often quoted in scientific literature as $(x \pm \delta)$, δ being a small quantity that may be (*a*) the probable error; (*b*) the standard deviation; (*c*) the error intelligently guessed (*see* errors of measurement).

probe 1. A lead that contains or connects to a measuring or monitoring circuit and is used for testing purposes. The circuit may consist of either active or passive components.
2. A resonant conductor inserted into a *waveguide or *cavity resonator for the purpose of injecting or extracting energy.

processor *See* central processing unit; microcomputer.

product of inertia If Cartesian axes $OXYZ$ are fixed relative to a rigid body, the product of inertia of the body with respect to the axes OY, OZ is $\Sigma myz = F$, the summation being carried out for every particle of the body. ($x, y,$ and z are the coordinates of a particle of mass m.) There are three such products of inertia: F, $G = \Sigma mxz$, and $H = \Sigma mxy$ and also three moments of inertia A, B, C about the axes OX, OY, OZ respectively given by:
$$A = \Sigma m(y^2 + z^2), B = \Sigma m(z^2 + x^2),$$
$$C = \Sigma m(x^2 + y^2).$$
The moment of inertia, I, of the body about any axis L through the origin of coordinates is expressible in terms of A, B, C and F, G, H, and the *direction cosines α, β, γ of the axis L with respect to the coordinate axes OX, OY, OZ:
$$I = A\alpha^2 + B\beta^2 + C\gamma^2$$
$$- 2F\beta\gamma - 2G\gamma\alpha - 2H\alpha\beta).$$
If a line is drawn from O whose direction is that of the axis L and whose length from O is equal to $1/\sqrt{I}$, where I is the moment of inertia about the axis, then as L takes various directions in turn, the end point of this line will trace out the surface of the ellipsoid:
$$Ax^2 + By^2 + Cz^2 - 2Fyz - 2Gxz - 2Hxy = 1,$$
which is known as the *momental* or *inertia ellipsoid* at O.

If the set of axes is rotated about O, new positions can be found such that the equation of this ellipse with respect to these axes changes to the form:

$$A^*x^2 + B^*y^2 + C^*z^2 = 1.$$

These axes, about which the moments of inertia are A^*, B^*, and C^*, and the products of inertia are all zero, are known as the *principal axes* of the body with respect to O and the quantities A^*, B^*, and C^* are called the *principal moments of inertia*. An axis of symmetry of the body will be a principal axis. (*See* centrifugal moment.)

profile The shape of a wave, pulse, spectral line, etc., found by plotting the amplitude or intensity against time, distance, frequency, or some other temporal or spatial variable.

profile drag *See* profile resistance.

profile resistance Of a body in relative motion in a fluid. The sum of the *eddy-making resistance and the *skin friction. The term *profile drag* is used for the total resistance to motion of a two-dimensional aerofoil.

program A sequence of statements that can be submitted to a *computer system and used to direct the way in which the system behaves. The program must be expressed precisely and unambiguously, and must therefore be written in any of a large variety of programming languages. Before a program can be executed by a computer, it must be translated (automatically) into a sequence of *machine instructions*, expressed in the particular *machine code* appropriate to that computer.

programmable ROM *See* PROM.

programming language Any of a large variety of artificial languages designed for the precise description of computer *programs. A *high-level language* (such as Fortran, Pascal, C, or Basic) has features that reflect the requirements of the programmer (i.e. it is application-oriented). It is close to natural language or mathematical notation than a *low-level language*, which is nearer to machine code (*see* program).

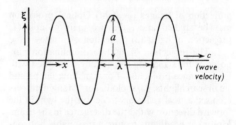

Plane progressive harmonic wave

progressive wave A wave propagated through an infinite homogeneous medium. For any type of wave motion a plane-progressive wave may be represented by the equation of wave motion:
$$\partial^2\xi/\partial t^2 = c^2(\partial^2\xi/\partial x^2),$$
where ξ is the particle displacement at distance x from a fixed point along the direction of propaga-

tion, c is the wave velocity, and t is time measured from a fixed instant.

The solution of this equation, for a wave travelling along the positive direction of x, may be written $\xi = f(x - ct)$, where $f(x - ct)$ depends on the wave shape. Thus, according to the latter equation, the instantaneous waveform existing along the x axis at time $t = t_1$ will have been transferred en bloc a distance $c(t_2 - t_1)$ from the instant $t = t_1$ to the instant $t = t_2$. For a wave travelling along the negative direction of x, the corresponding equation is $\xi = f(x + ct)$. Combining the two forms gives the general solution:

$$\xi = f_1(x - ct) + f_2(x + ct).$$

A plane progressive harmonic wave may be represented by the formula:

$$\xi = a \sin (2\pi/\lambda)(ct - x),$$

where ξ is the particle displacement at distance x from a fixed point along the direction of propagation, at time t from a fixed instant of time, a is the maximum particle displacement or amplitude, λ is the wavelength, and c is the wave velocity.

Such a wave is illustrated in the diagram. The wavelength, λ, is the separation between two points at which the displacements differ in phase by 2π radians (one period). For a given value of x, the displacement ξ varies through a complete cycle when $(2\pi/\lambda)(ct - x)$ changes by 2π radians; the corresponding change in t is T, the *period. Thus, $(2\pi/\lambda)cT = 2\pi$ and $T = \lambda/c$.

The frequency f (which is the number of waves passing a fixed point per unit time) is the reciprocal of T, and, therefore, $1/f = \lambda/c$, or $c = f\lambda$. The intensity of the wave, I, i.e. the energy passing per unit time per unit area normal to the direction of propagation, is proportional to the square of the product of the amplitude, a, and the frequency f, i.e. $I \propto a^2f^2$. For a sound wave the intensity is given by $I = \frac{1}{2}\rho_0 ca^2f^2$, where ρ_0 is the average density. (*See* intensity of sound.)

projected-scale instrument An indicating instrument in which a beam of light is used to project an image of the scale upon a screen.

projection of images (or **light**) Optical projection may be studied under the two main groups: (1) projection of light; (2) projection of images.

1. The projection of light for illumination purposes or for signalling is generally concerned with long-distance projection. For signalling lamps and small searchlights, paraboloidal or *Mangin mirrors are used as rear reflectors to deliver the light in the forward direction with little divergence in the beam. Motor car headlights require a reasonable horizontal divergence. Larger searchlights are provided with large-aperture paraboloidal reflectors. Lighthouse projectors consist of an elaborate system of dioptric (refractory lenses and prisms – stepped or *Fresnel lenses) and catadioptric (reflecting prisms) systems so mounted that light is collected from all round the source and delivered with little divergence in a single horizontal direction; ships' lanterns use stepped lenses to project all round in a horizontal direction.

2. The projection of images requires the twofold consideration: (i) of condensing the light on the object (*see* condenser) and (ii) the collection of this light by the *projector* proper whose function is to deliver a sharply focused image free from chromatic and spherical *aberrations with a uniform distribution of illumination over the field of the projected image (*see* epidiascope; episcope, etc.). For photographic negative projection, the condenser focuses the image of the source within the entrance pupil of the projector (this ensures uniform illumination and freedom from *vignetting when all are adequately centred). For cinematograph microprojection, where the object is small, the condenser images a uniform source in the plane of the object to be projected. The projecting lens in the case of microprojection is termed the projection eyepiece.

Prokhorov, Alexandr Mikhaylovich (*b*. 1916) Soviet physicist. His initial work on microwave spectroscopy led him to cooperate with Nikolai Basov on the development of the maser (1955) and later the semiconductor laser (1958). For the development of the maser they shared a Nobel prize (1964) with Charles *Townes, who had developed the maser independently in America.

PROM Abbreviation for programmable read-only memory, i.e. programmable ROM. A type of semiconductor memory that is fabricated in a similar way to *ROM. The contents required are, however, added after rather than during manufacture, and cannot be altered from that time. The memory contents are fixed electronically by means of a device called a *PROM programmer*. *See also* EPROM.

prompt neutron A neutron produced in a reactor by primary fission rather than by decay of a fission product. *Compare* delayed neutron.

Prony brake *See* dynamometer.

proof Proof spirit has a relative density of 0.919 76, i.e. 49.28% of alcohol by weight. If the over-proof strength of a spirit be added to 100, the sum is the number of volumes of proof spirit that can be made from a hundred volumes of the given spirit. (Similarly the underproof strength should be subtracted from 100.) Absolute alcohol is 75.35° over proof so that 100 volumes would make 175.35 volumes of proof spirit.

proofed tape A tape of cotton cloth coated with a rubber compound. The tape is applied to the insulation of rubber-insulated cables.

proof plane A small metal plate held by an insulating handle, used for transferring a small part of the electric charge on a body to an electrometer to investigate the distribution of the charge on the body. The size of the plane must be small compared with that of the body tested so that the distribution of electric charge on the body is not affected.

propagation coefficient *Syn.* propagation constant. Symbol: P or γ. A measure of the attenuation and phase change of a wave travelling along a *transmission line. It is defined for a uniform line of infinite length supplied at its sending end with a sinusoidal current having a specified frequency. At two points along the line separated by unit distance, let the currents under steady-state conditions be I_1 and I_2, where I_1 is the current at the point nearer the sending end of the line. Then

$$P = \log_e(I_1/I_2)$$

at the specified frequency; not that the vector ratio of the currents is used. P is a complex quantity and may be expressed as

$$P = \alpha + i\beta,$$

where $i = \sqrt{-1}$. The real part, α, is the the *attenuation constant and is measured in *nepers per unit length of line. It measures the transmission losses in the line. The imaginary part, β, is the *phase-change coefficient* or *phase constant* and is measured in radians per unit length of line. It is the *phase difference between I_1 and I_2 introduced by the line.

If at a given instant the displacement at a given point is a maximum and given by p_1, then the instantaneous displacement p_2 at the same instant distance x along the direction of propagation is given by:

$$p_2 = p_1 e^{-Px} = p_1 e^{-(\alpha + ib)x}$$
$$= p_1 e^{-\alpha x} \times e^{-i\beta x}.$$

The factor $e^{-\alpha x}$ is an attenuation factor, whereas the factor $e^{-i\beta x}$ is a factor that changes the phase by $-\beta x$ radians. A line of infinite length is not physically realizable but the conditions in a line of finite length, which is terminated in an impedance equal to its *characteristic impedance, simulate those of the infinite line.

propagation constant *See* propagation coefficient.

propagation loss The energy loss from a beam of *electromagnetic radiation as a result of absorption, scattering, and the spreading out of the beam.

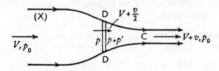

Flow pattern for ideal propeller

propeller, ideal A propeller for which the effects of fluid viscosity and compressibility are neglected and for which the propeller is considered alone in the fluid stream. The ideal propeller is treated as a thin disc (called the actuator disc) perpendicular to the direction of fluid flow, and responsible for the change in the velocity of the slipstream. The thrust of the ideal propeller is considered as the sum of thrust forces, having a constant value per unit area, acting over the whole disc. The idealized flow pattern (slipstream) is illustrated in the diagram. The actuator disc D is at rest and the undisturbed fluid velocity is V, the static pressure $p_0(X)$. At the disc the fluid velocity rises to $V + \tfrac{1}{2}v$, the pressure falls to p. Across the disc there is an increase in pressure to $p + p'$ after which the fluid experiences a further increase in velocity to $V + v$. The pressure meanwhile decreases to the static pressure p_0 and is accompanied by a reduction in the area of cross section of the slipstream. The point C is called the *vena contracta. By the application of *Bernouilli's theorem to the flow before and after the actuator and from a consideration of the momentum change, the thrust of the ideal propeller is

$$A\rho(V + \tfrac{1}{2}v)v,$$

where A is the area of the disc and ρ the density of the fluid.

proper function and value *Syn.* eigenfunction and -value. *See* wave function.

proper time Time measured on a clock at rest with respect to a particle or system under study, as distinct from time measured, for example, in a laboratory, by a clock that is in motion relative to it. *See* relativity.

proportional counter A gas-filled radiation counter that operates in the *proportional region. Such counters may be used to study the energy spectrum of ionizing radiations, or to distinguish between two types of radiation, one of which is much more strongly ionizing than the other. An important example of the latter kind is the boron trifluoride (BF_3) counter for detecting neutrons in the presence of gamma radiation. Neutrons interact with the isotope boron-10, which is present to about 19% in natural boron: $^{10}_{5}B(n,\alpha)^{7}_{3}Li$. The alpha particle and lithium nucleus fly apart, causing intense ionization and hence a large pulse in the counter circuit. A counter threshold voltage can be chosen so that only the large signals from neutron capture are recorded, the large number of smaller pulses caused by secondary electrons generated by gamma rays being excluded. Higher sensitivity can be obtained by using boron that is highly enriched in this isotope.

proportional region The operation voltage range for a radiation counter in which the *gas multiplication exceeds 1, its value being independent of primary ionization. A counter operating in this region is called a proportional counter.

The pulse size in the counter is proportional to the number of ions produced as a result of the initial ionizing event. *See* Geiger counter; ionization chamber.

protanopia *See* colour vision.

protective colloid Emulsoid sols, such as gelatin, which are themselves not coagulated by the addition of small concentrations of electrolytes, are able when present in small quantities to protect suspensoid sols from the coagulation effect of electrolytes. It is probable that the gelatin, or other protective colloid, is adsorbed on the surface of the suspensoid particles.

protective gap A *surge diverter consisting of a spark gap connected between a conductor of an electrical system and earth partially to divert to earth any *surge of excessively high voltage and thus to avoid protection to the system and any associated apparatus. The term applies only to the spark gap and does not include any series resistor that may be included in the earth connection.

protective horn See arcing horn.

protective relay A *relay that protects electrical apparatus against the damaging effects of abnormal conditions (e.g. overloads and internal faults). When such abnormal conditions occur, the relay causes a *circuit breaker to open so that the faulty apparatus is automatically disconnected from the power supply and from other associated equipment.

protective system An electrical system designed to protect electrical plant and equipment from the damaging effects of abnormal conditions such as surges, faults to earth, etc. The system is sensitive to these abnormal conditions and when they occur it causes the faulty section to become isolated. Special forms of protective system are described as follows for the conditions stated: (i) Discriminating. When the protective system operates only in the event of an abnormal occurrence in a particular part of the electrical plant and disconnects only the fault section. (ii) Reverse-power. When the protective system responds only to reversals in the direction of flow of electrical power. (This type is also discriminating.) (iii) Leakage. When the protective system responds to *leakage current only.

proton A positively charged *elementary particle that forms the nucleus of the hydrogen atom and is a constituent particle of all nuclei. It is about 1836 times heavier than the electron. It is a stable *baryon of mass 938.2796 MeV/c^2 (1.672 6231 × 10^{-27} kg) having charge $Q = 1$. It has *spin $J = \frac{1}{2}$, *isospin $I = \frac{1}{2}$, and positive *parity. It also has an intrinsic *magnetic moment of 2.793 nuclear *magnetons. Although there are many elementary particles having a lower mass, the proton cannot decay into these as it is the lowest mass particle with baryon number $B = 1$; baryon number is conserved in all interactions. See hadron; quark.

proton microscope A microscope similar to the *electron microscope but using a beam of protons rather than electrons. This allows a better resolving power and contrast.

proton number See atomic number.

proton–proton chain A series of thermonuclear reactions (see nuclear fusion) by which hydrogen nuclei (protons) are converted to helium nuclei with the release of a considerable amount of energy. The reactions require a temperature of 10^7 kelvin. They are almost certainly the major source of energy in the sun and in all stars of lower mass than the sun, and occur in the dense stellar core. There are various

sequences that can be followed in the proton-proton chain, but the principal sequence is as follows:

$$^1H + {}^1H \rightarrow {}^2H + \nu + e^+$$
$$^2H + {}^1H \rightarrow {}^3He + \gamma$$
$$^3He + {}^3He \rightarrow {}^4He + 2\,{}^1H,$$

where ν, e$^+$, and γ are a neutrino, positron, and γ-ray respectively. Compare carbon cycle.

proton resonance *Nuclear magnetic resonance of hydrogen nuclei.

proton synchrotron A cyclic *accelerator of very large radius that can accelerate protons to extremely high energies: 70 GeV at Serpukhov, USSR; 450 GeV at CERN, Geneva; and 900 GeV at FNAL, USA. It is basically similar to a *synchrotron. In a synchrotron a fixed orbit is maintained by increasing the magnetic field strength in proportion to the relativistic increase in mass. This leads to a constant angular frequency. This is possible because the electrons are travelling at close to the speed of light at an energy of a few MeV. The equivalent stable orbit at constant angular frequency is not achieved by protons until they have an energy of about 3 GeV. To maintain a fixed orbit up to this energy, the frequency of the accelerating electric field must be varied (it is constant in the synchrotron). The frequency of the field and the particle beam must remain in synchrony and also satisfy the relation $v = \omega r$ where v is the proton velocity, r is the radius of the orbit, and ω is the angular frequency of the protons; $\omega = 2\pi f$, where f is the electric-field frequency. The protons are accelerated by radio-frequency fields between the magnets, and can make several million revolutions in one second. The beam is both focused and maintained in a circular orbit by means of magnets. Strong focusing (see accelerator) is used. Usually each of the magnets surrounding the particle track performs both functions. At Batavia there are separate deflecting and focusing magnets; it is therefore called a separated-function synchrotron. A 0.2 GeV *linear accelerator feeds protons to a cyclic accelerator, which injects 8 GeV protons into the main 6300 m accelerator.

prototype The first or original of anything, from which copies are made (e.g. prototype kilogram).

Prout, William (1785–1850) Brit. physician and biochemist, remembered for his supposition that hydrogen is a common constituent of all elements. See Prout's hypothesis.

Prout's hypothesis Based on the observation that the atomic weights known to him appeared all to be integers. Prout suggested that hydrogen is a constituent of all elements. The idea was dropped as more accurate determinations of atomic weight revealed values undoubtedly not integers but interest was revived later by the discovery of isotopes. The masses of individual atoms are close to integral multiples of the mass of one nucleon (proton or neutron).

proximity effect An effect that occurs when conductors carrying alternating current are placed close to one another, the distribution of current across the cross section of any one being influenced by the magnetic fields set up by the others. The change in current distribution modifies the *effective resistance of the conductors. The effect is particularly important in coils used at high (e.g. radio) frequencies.

pseudoscalar A quantity defined by a single magnitude but distinguished from a *scalar as it is an odd function, i.e. its sign is reversed on reversing the signs of a set of coordinate axes. Examples are pressure and volume.

pseudoscope A device that reverses stereoscopic effects, e.g. reversing the pictures of a stereoscope. *See* Stratton pseudoscope.

pseudovector *Syn* axial vector. A quantity defined in terms of a magnitude and the direction of an axis. Pseudovectors obey the same rules of algebra and calculus as polar *vectors but are even functions, i.e. signs of the components are unchanged on reversing the directions of the coordinate axes. Examples include area, torque, and magnetic flux density. The vector product of two polar vectors is a pseudovector.

psi/J particle *See* J/psi particle.

psychrometer A *hygrometer in which a strong draught is obtained past the wet and dry bulbs either by whirling on a sling or by a fan. The best-known type is the *Assmann psychrometer* in which the necessary ventilation is provided by a clockwork-driven fan.

Ptolemaic system *See* Ptolemy.

Ptolemy (Claudius Ptolemaeus) (100–178) Alexandrian astronomer who produced a great work on astronomy (usually known by its Arabic name as the *Almagest*), in which the motions of the planets are described as taking place in epicycles, circular paths round centres themselves in circular motion, this system giving a complicated but moderately good approximation to their apparent motion round the earth. Although the ideas were mostly those of Hipparchus (active period in Alexandria 160–127 BC), they are usually referred to as the *Ptolemaic system*. Ptolemy also wrote on geography and he is credited with a book on optics that includes an approximate treatment of refraction.

p-type conductivity Conductivity in a *semiconductor caused by the effective movement of mobile *holes.

p-type semiconductor An *extrinsic semiconductor in which the density of mobile *holes exceeds that of conduction *electrons. *See also* semiconductor.

public address system A complete reproducing installation for rendering sounds such as speech or music audible to a large gathering of people. It is used extensively for large open-air meetings, on railway stations, and in large halls. The system usually consists of one or more microphones, a power amplifier, and several loudspeakers. The amplifier and speakers should be designed to handle the power required to give an adequate sound level over the whole area to be covered. This level should be sufficient to enable the sound to be heard above the noise level of the crowd. Unless the sound system is carefully designed, there is a tendency for the amplified sound to be fed back to the microphone, which results in howling due to self-oscillation. To prevent this, the amplifier should be made very stable by the use of negative *feedback and the nearest speaker should be placed some distance from the microphone. Care must be taken in the placing of loudspeakers where several of them are used to cover a large area, otherwise a person hears a sound from one speaker and then after a distinct interval the same sound is heard more faintly from the others in turn. This produces a very disturbing effect, similar to that of several echoes. One method of avoiding this in large open-air installations is to employ a time delay on each speaker, which increases with its distance from the source. Thus a sound is emitted by each speaker at the same time as the direct sound arrives there. Another method is to use two clusters of speakers on a pole, one group being a metre or more above the other. The two groups are fed in opposite phase so that there is a zone of almost complete silence at a certain distance from the speakers. In this way it is arranged that the sound from one group of speakers does not interfere with that from another group. For indoor work it is more usual to employ directional speakers, each feeding a small section of the auditorium. No speakers should be placed near the source of sound since the direct sound will usually be sufficient. Directional microphones are sometimes used to reduce the background noise picked up from the audience.

Pulfrich refractometer A *critical angle refractometer for the accurate measurement of refractive index and dispersion. The unknown material rests on the *Pulfrich prism* (*see* prism); light at grazing incidence is refracted at the critical angle in the Pulfrich prism, and is further refracted at the exit face. The field of view in the observing telescope is a half-field of colour, on the edge of which a cross-wire is set, and an angular reading made on an appropriate scale. Reference must be made to tables to convert angular readings to index and dispersion. *See* refractive-index measurement.

pulley Any of a number of simple *machines in which forces are transmitted by ropes, belts, etc., used in conjunction with pulley wheels and axles. The wheels usually have a grooved rim in which the rope, etc., can run. In a frictionless (i.e. theoretical) system, the force (or pull) in any part of a continuous rope has a constant value, F. Consider a system of vertically aligned equal-sized wheels, connected by a continuous rope, the topmost wheel fixed to a support. A weight w, attached to the lowest wheel, is to be raised. If the number of supporting ropes is n, then $nF = w$. The mechanical advantage, i.e. the ratio of load to effort, is then equal to n.

pulling An effect occurring when an electronic oscillator is coupled to a circuit in which there is another independent oscillation: the oscillator frequency tends to change towards that of the independent oscillation. The tendency is particularly strong if the two frequencies differ by only a small percentage. Complete synchronization can sometimes be achieved. Pulling is used to control the frequency of an oscillator, as in crystal-controlled oscillators.

pulsar A member of a class of astronomical objects that all possess extreme characteristics, including (*a*) their energy output varies with exceptional regularity at a fast rate; (*b*) they have small dimensions (20–30 km across) and are the densest observable form of matter; (*c*) their energy per unit area is very large. Pulsars are generally believed to be examples of *neutron stars.

Pulsars are rapidly spinning bodies emitting energy either in the radio or the X-ray regions of the spectrum. (Some radio pulsars have also been detected at optical and gamma-ray wavelengths.) The period of the pulsed output represents the rotation rate. In the case of radio pulsars, the radio waves (which are polarized) originate from the pulsar's extremely strong magnetic field. As the pulsar rotates, the radio beam may be observed as a pulsed signal as it sweeps past the earth. The period of most radio pulsars lies between 0.1 and 4 seconds, indicating that these pulsars are rotating in the range 10–0.25 times per second. All, however, are gradually slowing down as they lose rotational energy; the rate of slowdown varies considerably. A small group of radio pulsars rotate even more rapidly. Their pulse rate is a few milliseconds and they are thus called *millisecond pulsars*.

The millisecond pulsars and a few other radio pulsars are in close orbit with another star, i.e. they are binary stars. In contrast, all X-ray pulsars are in binary systems. The X-rays are generated when gas, transferred from the companion star, falls onto the pulsar. The gas flow affects the pulsar's spin, and as a result all X-ray pulsars are gradually speeding up. The periods of some X-ray pulsars are a few seconds but many have longer periods, amounting to several minutes.

It is believed that some but not all pulsars originate in *supernova explosions.

pulsatance *See* angular frequency.

pulsating current An electric current that varies in magnitude in a regularly recurring manner. The term implies that the current is unidirectional.

pulsating star *See* variable star.

pulse A single transient disturbance manifest as an isolated wave, or one of a series of transient disturbances recurring at regular intervals, or a short train of high-frequency waves, as used in *echo sounding and *radar. A single pulse consists of a voltage or a

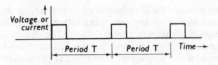

a Square pulse train

current that increases from zero to a maximum value and then decreases to zero in a comparatively short time. A pulse is described as square (Fig. *a*), triangular (Fig. *b*), etc., according to the geometrical shape of the pulse when its instantaneous value is plotted as a function of time.

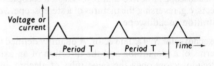

b Triangular pulse train

In practice a perfect geometrical shape is never achieved and a practical rectangular pulse is shown in Fig. *c*. The magnitude of the pulse normally has a constant value, ignoring any spikes or ripples, and is called the *pulse amplitude* or *pulse height*. A practical pulse has a finite *rise time*, usually occurring between 10% and 90% of the pulse height, and a finite *decay time*, occurring between the same limits. The *pulse duration* or *pulse width* is the time between the rise and decay time. A practical pulse frequently rises to a value above the pulse height and falls to the pulse height with damped oscillations. This is called *overshoot* and *ringing*. A similar phenomenon occurs as the pulse decays to the base level. *Droop* can occur in a rectangular pulse when the pulse height falls slightly below the nominal value. This is particularly associated with inductively coupled circuits (*see* inductive coupling).

A group of regularly recurring pulses of similar characteristics is called a *pulse train*, and is usually identified by the type of pulses in the train, e.g. square wave, sawtooth wave, etc. The *pulse-repetition frequency* (or *pulse rate*) is the average number of pulses per second, expressed in hertz.

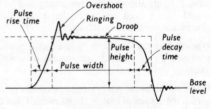

c Practical rectangular pulse

pulse-code modulation (PCM) *See* pulse modulation.

pulse decay time *See* pulse.

pulse droop *See* pulse.

pulse generator An electronic circuit or device that produces current or voltage pulses of a desired waveform.

pulse height *See* pulse.

pulse-height analyser *See* multichannel analyser.

pulse-height discriminator An electronic circuit that selects and passes *pulses whose amplitude lies between specified limits.

pulse-mode multiplexing A form of *time-division multiplexing in which the signals to be transmitted are used to modulate a *carrier pulse train before transmission.

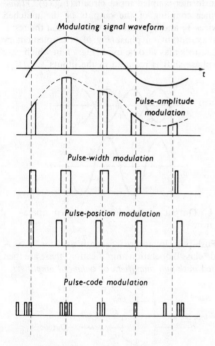

Forms of pulse modulation

pulse modulation A form of *modulation in which (most commonly) a pulse train is used as the carrier. Information is conveyed by modulating some parameter of the pulses with a set of discrete instantaneous samples of the message signal.

In *pulse-amplitude modulation*, the amplitude of the pulses is modulated by the corresponding samples of the modulating wave. In *pulse-time modulation*, the samples are used to vary the time of occurrence of some parameter of the pulses. Particular forms of pulse time modulation are *pulse-duration modulation* (*Syn. pulse-length* or *pulse-width modulation*) in which the time of occurrence of the leading edge or trailing edge is varied from its unmodulated position; *pulse-frequency modulation* in which the frequency of the carrier pulses is varied from its unmodulated value; and *pulse-position* (or *phase*)

modulation in which the time of occurrence of a pulse is modulated from its unmodulated time of occurrence. All these types of pulse modulation are examples of uncoded modulation.

In *pulse-code modulation*, the amplitude (usually) of the modulating signal is sampled and a digital code is used to represent the sampled value: sample amplitudes falling within specified ranges of values are assigned different discrete values, each of which is represented by a specific pattern of pulses. The signal is thus transmitted in the form of a digital code (i.e. as a stream of *bits), which is converted back to the analogue signal at the receiving point.

pulse operation Any means of operation of electronic circuits or devices in which the energy is transferred in the form of pulses.

pulse radiolysis *See* flash photolysis.

pulse regeneration In most forms of *pulse operation the pulses can get distorted by circuits or circuit elements. Pulse regeneration is the process of restoring the original form, timing, and magnitude to a *pulse or pulse train.

pulse-repetition frequency *Syn.* pulse rate. *See* pulse.

pulse rise time *See* pulse.

pulse shaper Any circuit or device that is used to change any of the characteristics of a *pulse.

pulse train *See* pulse.

pulse width *See* pulse.

pulsometer pump A pump that depends upon the pressure of steam forcing water out of one of two chambers alternately. While not economical in steam consumption, it is simple and very suitable for temporary installation.

pumps, vacuum Modern kinetic vacuum systems usually use two pumps in tandem. The backing pump or forepump, which works directly to the atmosphere, is one of the many forms of rotary oil pump. These pumps reduce the pressure to between 100 and 0.1 Pa according to the type. The second pump is usually a *diffusion* or *condensation pump*, which is rather similar in principle to the *filter pump but uses a jet of oil vapour instead of water. There may be several such jets acting in tandem.

Molecular pumps, developed by Gaede and Holweck, work on the principle that a gas molecule striking a rapidly moving surface may be given a high velocity in the direction of the exit pipe. These pumps need a backing-pump. *See also* air pumps; filter pump; ion pump; sorption pump; Sprengel pump; Toepler pump.

punched tape *See* paper tape.

punch through A type of *breakdown that can occur in both bipolar junction *transistors and *field-effect transistors. If the collector-base voltage applied to a bipolar transistor is increased, the *depletion layer associated with the collector-base junction spreads into the base region. At a sufficiently high voltage, the *punch-through voltage*, the depletion layer reaches the emitter region and a direct conducting path is formed from emitter to collector; carriers from the emitter "punch through" to the collector and breakdown occurs.

In an FET, a similar process occurs. When the drain voltage reaches a sufficiently high value, the depletion layer associated with the drain spreads across the substrate and meets the source junction; carriers can then punch through the substrate.

punctum remotum (P.R.) *See* far point.

puncture voltage The value of a gradually increasing voltage applied to an insulator, at the instant when the insulator is punctured electrically. *See also* impulse voltage.

pupil The aperture of the eye bounded by the iris, which acts as the aperture stop of the eye. The image of the pupil by the cornea is visible from the front and constitutes the *entrance pupil being 1.12 times the pupil itself. The diameter varies according to the brightness of the field of view from 1.5 mm in bright light to about 8 mm in extreme darkness. When the magnification of an instrument is such that its *exit pupil is equal to the entrance pupil of the eye, it is referred to as normal magnification – it is the maximum value to avoid restriction of brightness and the minimum for full resolution.

The size of the pupil in bright light is approximately the optimum for resolution; a smaller diameter would give more diffraction and a larger one would give more aberration. Variation of the aperture contributes far less to the range of sensitivity of the eye than do changes in the retina, but it is achieved very rapidly by muscular contractions in the *pupillary reflex*.

Purcell, Edward Mills (*b.* 1912) Amer. physicist known for his work on the magnetic moments of atomic particles.

Purkinje effect The sensitivity of the eye to light depends on the wavelength of the light and is usually a maximum in the green-yellow region of the spectrum. As the level of illumination falls to low levels this maximum shifts towards the blue end of the spectrum. This change in the position of maximum sensitivity is called the Purkinje effect. *See also* spectral luminous efficiency.

purple plague The formation of an alluminium-gold eutectic mixture at the bond between the gold connecting wire and the aluminium *bonding pads on silicon *integrated circuits. The compound is very brittle and results in a mechanically weak bond that is susceptible to failure. The name purple plague is due to the purple colour of the eutectic ($AuAl_2$).

push-pull operation The use in a circuit of two matched devices in such a way that they operate with a 180° *phase difference. The output circuits combine the separate outputs in phase. One means of achieving the desired phase shift in the inputs is a

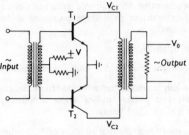

a Transformer-coupled class A push-pull operation

transformer-coupled input circuit (Fig. *a*). Transformer coupling of the outputs of the matched devices T_1 and T_2 combines them so that the resultant a.c. output is in phase (Fig. *b*). *Complementary transistors may also be used (Fig. *c*), in which case no phase shift is required in the inputs.

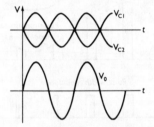

b Output of transformer-coupled operation

Push-pull circuits are frequently used for *class A and *class B operation amplification; these are then called *push-pull amplifiers* or *balanced amplifiers*.

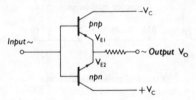

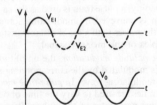

c Complementary transistor push-pull operation

PWR Abbreviation for *pressurized-water reactor.

pyknometer A form of *relative-density bottle consisting of a bulb AB joined to two capillary tubes,

ADC and BEF. It is filled to a reference mark G by sucking at F with the end C immersed in the liquid.

It is made for measuring the density of a liquid or of a solid by determining the mass of a known volume.

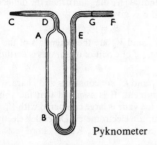

Pyknometer

pyranometer 1. An instrument for measuring the diffuse solar radiation (direct sun excluded) incident on a horizontal surface.

2. An instrument for measuring total solar radiation (sun plus sky) incident on a horizontal surface. *See* solarimeter. *Compare* pyrheliometer.

Pyrex The proprietary name of a heat-resisting glass containing a high percentage of silica with boron trioxide and smaller quantities of alkalis and alumina. It is tougher than, and softens at a higher temperature than, ordinary soda glass, and has a low expansion coefficient. It is widely used for the manufacture of heat-resistant glassware.

pyrheliometer An instrument for measuring the intensity of direct solar radiation at normal incidence, diffuse radiation being excluded. It can also measure radiation from a selected part of the sky. (*Compare* pyranometer.) These work on three main principles.

1. In the *Ångstrom pyrheliometer* two identical strips of blackened platinum are mounted so that one is exposed to the radiation at normal incidence, while the other is shielded. A difference in temperature between the two strips is determined by using a *thermocouple attached to the back of each strip and connected in series with a galvanometer. An electric current is passed through the shaded strip until the galvanometer registers no deflection. The two strips are then at the same temperature and solar radiation absorbed by one strip equals the work done electrically on the other. The current gives a measure of the intensity. Each strip is exposed in turn to the radiation and a mean value is found for the required current. It is a standard instrument requiring no external calibration.

2. In the *silver-disc pyrheliometer* a blackened silver disc is supported by fine steel wires inside a copper shell and has a thermometer embedded in it. It is exposed to a narrow angle of solar radiation at normal incidence. The rate of temperature increase determines the radiation intensity.

3. A thermocouple method is used in the third type. In the *Linke–Fuessner actinometer* (a pyrheliometer), the temperature difference between a blackened surface, exposed to a narrow angle of solar radiation, and a reference point is determined. The surface consists of a *thermopile of similar design to that used in a *solarimeter. The surface increases in temperature until its rate of loss of heat by all causes is equal to the rate of gain of heat from radiation. This increase in temperature should depend only on the intensity of the radiation, and must be independent of external conditions, such as ambient temperature, and wind speed. *See also* solarimeter; net radiometer; pyranometer.

pyroelectricity The development of opposite electric charges at the ends of polar axes in certain crystals when there is a change of temperature. Such crystals (e.g. tourmaline, lithium sulphate) do not possess a centre of symmetry.

pyrolysis Chemical decomposition as a result of the application of high temperatures.

pyrometer An instrument for measuring high temperatures. There are several types:

(1) The *optical pyrometer, which depends for its action on *Planck's formula. Instruments of this type are used for the measurement of high temperatures on the *International Practical Temperature Scale. Two examples of this class are the *disappearing filament pyrometer and the *polarizing pyrometer.

(2) The *total radiation pyrometer, which depends for its action on the Stefan–Boltzmann law $E = \sigma T^4$. The most convenient form of this type of pyrometer is the *Féry total radiation pyrometer.

(3) *See* resistance pyrometer.

(4) *See* thermocouple pyrometer.

Pythagoras (*c.* 582–500 BC) Greek philosopher and mathematician. He founded a school that became virtually a learned secret society and that outlasted his death by more than a century. Geometry and arithmetic were made logical and intellectual pursuits. A treatment of music and harmony was evolved, based on the simple relations between the lengths of strings corresponding to related notes. The Pythagoreans believed in a rotating earth, and they supposed that vision involved the passages of particles from the object seen into the eye in contrast to the Platonists and others who held that something emanated from the eye.

Q

QCD Abbreviation for *quantum chromodynamics.

QED Abbreviation for *quantum electrodynamics.

Q-factor *Syn.* quality factor. Symbol: Q. A measure of the quality of performance of a resonant system, especially a *resonant circuit. It indicates the ability of the system to produce a large output at the resonant frequency.

QFD

For a resonant circuit comprising resistance R, capacitance C, and inductance L,
$$Q = 1/R\sqrt{(L/C)}.$$
In the case of a simple series resonant circuit at the resonant frequency, $\omega_0 L = 1/\omega_0 C$, where ω_0 is 2π times the resonant frequency. Thus,
$$Q = \omega_0 L/R \text{ or } Q = 1/\omega_0 CR.$$
This is effectively the ratio of the total inductive or total capacitive *reactance to the total series resistance at resonance. In the case of a simple parallel resonant circuit,
$$\omega_0 = \sqrt{(1/LC)}\sqrt{(1 - 1/Q^2)}.$$
If Q is large, this reduces to the value for a series resonant circuit. The *selectivity of a resonant circuit is given by $1/Q$.

A single reactive component may be capable of resonance (i.e. if the self inductance of a coil or self capacitance of a capacitor is large enough). For a single component, Q is the ratio of reactance to effective series resistance, which for an inductance Q is equal to $\omega_0 L/R$ and for a capacitor to $1/\omega_0 CR$.

For a resonant system such as a *cavity resonator, for which values of R, C, and L cannot be specified,
$$Q = \omega_0 E_s/\eta,$$
where E_s is the stored energy and η is the rate of energy dissipation.

QFD Abbreviation for quantum flavourdynamics. *See* electroweak theory.

QSO Quasi-stellar object. *See* quasar.

quadrant 1. A quarter of a circle, especially of its circumference.

2. Any one of the four parts into which a plane is divided by rectangular Cartesian axes.

3. A right angle.

4. An instrument for angular measurements in astronomy and navigation, incorporating a 90° graduated arc.

quadrantal deviation Masses of soft iron disposed horizontally in a ship can affect its compass needle in a manner that changes its direction four times during one complete revolution of the ship. This is called quadrantal deviation and is corrected by placing two hollow soft-iron spheres (compensators) one on each side of the compass and level with its card. Varying poles are induced in the spheres by the earth's field, and these balance the deviation as the ship rotates.

Quadrant electrometer

quadrant electrometer A type of *electrometer in which a light foil-covered vane, supported by a quartz fibre, moves within hollow quadrantal segments of a cylindrical metal box. Opposite quad-

rants are connected together, but insulated from the case of the instrument. A mirror is carried to reflect a spot of light and measure deflections of the vane. If the vane hangs symmetrically within the quadrants when the needle and both pairs of quadrants are at zero potential (*see* diagram), then the deflection θ of the needle is given by:
$$\theta = k_1(V_A - V_B),$$
where V_A and V_B are the potentials of the pairs of quadrants; if one pair of quadrants is earthed, then:
$$\theta = k_2 V_A,$$
where k_1 and k_2 are constants and characteristic of the instrument. It is assumed that the voltage applied to the vane is large compared with V_A and V_B. *See* Compton electrometer; Dolezalek electrometer; electrostatic voltmeter; Kelvin, Lord.

quadraphonic recording *See* gramophone.

quadrature Two periodic quantities having the same *frequency and *waveform are said to be in quadrature when they differ in phase by 90°, i.e. one wave reaches its maximum value when the other passes through zero.

quadrature component *See* reactive current; reactive voltage; or reactive volt-amperes.

quadripole An electrical *network that has only four terminals – a pair of input terminals and a pair of output terminals. Its behaviour is usually described by the impedances presented at its terminals at specified frequencies. A common arrangement, used especially for *filters and *attenuators, consists of a number of impedances arranged in series and in parallel in a ladder-shaped network. This arrangement can be broken down for analysis into identical T-shaped and π-shaped sections, each with the same characteristic impedance.

quadrupole A distribution of charge or magnetization equivalent to two equal electric or magnetic *dipoles arranged very close together and set in opposite directions. The potential falls off as the inverse cube of the distance. For an arbitrary distribution of charges, the potential V_r at a distance r is given by

$$V_r = \frac{e}{4\pi\varepsilon_0 r} + \frac{p}{4\pi\varepsilon_0 r^2} + \frac{q}{4\pi\varepsilon_0 r^3} + \cdots$$

The first term gives the coulomb potential, e being the net charge. The second gives the dipole potential, p being the net dipole moment. Similarly, the third gives the quadrupole potential, when q is the *quadrupole moment*. ε_0 is the electric constant. A similar expression can be used for magnetic fields except that there is no known monopole. For a distribution of mass the terms with even powers are missing but there is the equivalent of a quadrupole term. Quadrupole sources give rise to quadrupole radiation, with *selection rules analogous to those related to dipoles.

quadrupole lens A device for focusing beams of charged particles. Four electrodes or magnetic poles with alternate polarities are arranged around the beam. The lens focuses in one plane but defocuses in the plane at right angles. Two lenses are used at a short distance apart with the polarities displaced by 90°, giving resultant focusing in both planes. (The system can be compared with a combination of two optical lenses of equal and opposite powers that form a convergent system when they are slightly separated.)

Magnetic quadrupole lenses are commonly used to focus beams of very high energy particles in *accelerators. Electric quadrupole lenses are used in types of mass spectrometer.

quadrupole resonance See nuclear quadrupole resonance.

quality of sound 1. The fidelity of reproduction of a sound. The highest quality occurs when the reproduced sound creates the same effect as that obtaining if the original sound were present.

2. The term is used in the same sense as the term *timbre, viz. the distinguishing quality other than pitch or intensity of a note produced by a musical instrument, voice, etc.

Different musical instruments may produce notes of the same pitch and intensity, but of varying quality, e.g. piano, trumpet, violin, and oboe. If a series of steel strings of suitable length are stretched on a sound-board and appropriately tensioned to give the same pitch, it will be found that the quality of the note emitted varies with such diverse factors as the physical properties of the strings, the method of excitation, e.g. bowing, striking or plucking, the position and area of excitation, and the design of the soundboard. Room acoustics will also give a general change in quality. Two theories of sound quality have been suggested.

(a) The absolute pitch theory assumes that whatever the pitch of the note, there are associated simultaneously with any fundamental, an unvarying group of partials known as the *formant and it is this formant that gives the note its quality. A similar theory occurs in connection with the production of vowel sounds in song and speech. (b) The relative pitch theory assumes that there is a group of partials–generally harmonics–associated with each fundamental. The relative intensity and number of the partials remain the same within limits and give a note its distinct quality. The intervals between the various upper partials and fundamental remain constant, and thus the frequencies of the partials change with a change of pitch as opposed to a group of fixed frequencies on the absolute pitch theory.

By means of *Fourier analysis, a complex recorded sound wave may be reduced to a number of sine wave components. If the sine wave components are generated electrically and mixed in their relative intensities given by the Fourier analysis, and reproduced by a loudspeaker system (see electronic instruments), the steady-state conditions of a note very similar to the one originally analysed, are obtained. The Fourier analysis of a note is often known as a *spectrum of sound*. It has been found that the relative phases of the various partials have negligible effect, if any, on the sound quality. The pitch of the note is not affected by removing some of the lower partials, including the fundamental. In such cases there will be a change in quality. This supports the relative pitch theory.

The relative pitch theory is believed to give the main explanation of the quality of a note, but there are considered to be components in a note, which may be thought of as a formant of unvarying frequency composition. These add a little to the quality. A simple example of this is the noise of the bow on the string when a violin is played. Since in music the steady state conditions of a note are rarely maintained for any length of time, the starting and ending transients are of some importance. The sound quality as well as the amplitude of these transients differs from that of the steady-state condition. Particularly the amplitude of the various partials varies with time. The relative phases of the partials is also of importance. Every instrument has its own characteristic *formant. *Ohm's law of hearing is pertinent in connection with the quality of musical notes. This states that the ear recognizes as pure tones only those due to simple harmonic vibrations and it resolves any other complex vibration into simple harmonic components, perceiving the note as a summation of pure tones.

quality factor 1. See Q-factor.
2. See dose.

quantity of electricity Symbol: Q. The time integral of electric current, i.e. $\int I dt$, equivalent to the electric charge.

quantity of heat Symbol: Q. See heat.

quantity of light Symbol: Q. The time integral of the *luminous flux, i.e. $\int \Phi dt$.

quantization The process, used in electronics and computing, of constructing a set of discrete values that represents a quantity that varies continuously. One example is the measurement of the amplitude of a signal at discrete intervals of time, when the signal itself varies continuously with time. Another is the measurement of the brightness of small picture elements (pixels) making up a picture, which may be regarded as space-continuous.

quantized See quantum theory; quantum number.

quantum The smallest amount of energy that a system can gain or lose. The change in energy corresponding to a quantum is very small and only noticeable on an atomic scale. See quantum theory.

quantum chromodynamics (QCD) A *quantum field theory that is a *gauge theory of the *strong interactions based on the exchange of massless gluons between quarks and antiquarks. QCD is similar to *quantum electrodynamics (QED), the quantum field theory of electromagnetic interactions, but with the gluon as the analogue of the photon and with a quantum number known as *colour* replacing that of electric charge. Each quark type (or flavour) comes in three colours (red, blue, and green, say), where colour is simply a convenient label and has no connection with ordinary colour. Unlike the photon in QED, which is electrically neutral, gluons in QCD carry colour and can therefore interact with themselves. Particles that carry colour are believed not to be able to exist as free particles. Instead, quarks and gluons are permanently confined inside *hadrons (strongly interacting particles, such as the proton and the neutron).

The gluon self-interaction leads to the property known as *asymptotic freedom*, in which the interaction strength for the strong interactions decreases as the momentum transfer involved in an interaction increases. This allows perturbation theory to be used and quantitative comparisons to be made with experiment, similar to (but less precise than) those possible in QED. QCD has been tested successfully in high-energy muon–nucleon scattering experiments and in proton–antiproton and electron–positron collisions at high energies. Strong evidence for the existence of colour comes from measurements of the interaction rates for $e^+e^- \rightarrow$ hadrons and $e^+e^- \rightarrow \mu^+\mu^-$. The relative rate for these two processes is found to be a factor of three larger than would be expected without colour; this factor measures directly the number of colours (i.e. three) for each quark flavour. Energetic quarks and gluons produced in very high energy particle collisions undergo a process known as *hadronization* or *fragmentation*, in which the quark or gluon becomes a collimated jet of hadrons (mostly pions) aligned with the original quark or gluon direction, and it is these jets of particles that are observed experimentally. In e^+e^- collisions, interactions are observed in which three separated jets or hadrons are produced. These are interpreted as being due to the underlying process $e^+e^- \rightarrow q\bar{q}g$, with the subsequent fragmentation of the quark, antiquark, and gluon into the observed jets of hadrons; such interactions provide direct evidence for the existence of the gluon.

quantum discontinuity The discontinuous emission or absorption of energy accompanying a quantum jump.

quantum electrodynamics (QED) A relativistic quantum-mechanical theory of *electromagnetic interactions. QED's descriptions of the photon-mediated electromagnetic interactions have been verified

a Feynman diagrams

over a great range of distances and have led to highly accurate predictions. QED is a *gauge theory in which the gauge group is non-Abelian. In QED, the

electromagnetic force can be derived by requiring that the equations describing the motion of a charged particle remain unchanged in the course of local symmetry operations. Specifically, if the phase of the *wave function by which a charged particle is described is altered independently at every point in space, QED requires that the electromagnetic interaction and its mediating photon exist in order to maintain symmetry.

In the *Feynman propagator* approach, the scattering of electrons and photons is described by a matrix (the scattering matrix), which is written as an infinite sum of terms corresponding to all the possible ways the particles can interact by the exchange of virtual electrons and photons (*see* virtual particle). Each term may be represented by a diagram (called a *Feynman diagram*). These diagrams are built up from vertices representing the emission of a (virtual) photon by an electron, and propagators that represent the exchange of virtual photons or electrons, as in Fig. *a*. Fig. *b* shows the first few diagrams for electron-electron scattering.

In these diagrams all lines joining two vertices are propagators. The lines having a vertex at only one of their ends and the other end free, represent the physical particles before and after the interaction. A set of simple rules enables the contribution to the scattering matrix to be calculated from each of these diagrams.

b Feynman diagrams for electron–electron scattering

quantum electronics An application of the principles of *quantum mechanics to the study of the production and amplification of power at microwave frequencies in solid crystals.

quantum field theory A quantum mechanical theory in which particles are represented by fields whose normal modes of oscillation are quantized. *Elementary particle interactions are described by relativistically invariant theories of quantized fields (i.e. by *relativistic quantum field theories*). In *quantum electrodynamics, for example, charged particles can emit or absorb a *photon, the quantum of the electromagnetic field. Quantum field theories naturally predict the existence of *antiparticles and both particles and antiparticles can be created or destroyed; a photon, for example, can be converted into an electron plus its antiparticle, the positron. Quantum field theories provide a proof of the connection between spin and statistics underlying the *Pauli exclusion principle. *See also* electroweak theory; gauge theory; quantum chromodynamics.

quantum flavourdynamics (QFD) *See* electroweak theory.

quantum Hall effect *See* Hall effect.

quantum mechanics A mathematical physical theory that grew out of Planck's *quantum theory and deals with the mechanics of atomic and related systems in terms of quantities that can be measured. The subject developed in several mathematical forms all of which are, in fact, equivalent:

Wave mechanics developed from de Broglie's theory (1924) that a particle can also be regarded as a wave (*see* de Broglie equation). This idea arose from the analogy between *Fermat's principle for light and the *least-action principle in mechanics. Wave mechanics is based on the *Schrödinger wave equation describing the wave properties of matter. It relates the energy of a system to a *wave function, and in general it is found that a system (such as an atom or molecule) can only have certain allowed wave functions (eigenfunctions) and certain allowed energies (eigenvalues). These are *stationary states* of the system. The physical interpretation of the wave function of a particle is that its square is proportional to the probability of finding the particle per unit volume at a particular point (*see also* atomic orbital). In wave mechanics the quantum conditions arise in a natural way from the basic postulates as solutions of the wave equation.

Matrix mechanics was developed by Born and Heisenberg at about the same time that Schrödinger introduced wave mechanics. In matrix mechanics observable physical quantities such as momentum, energy, and position are represented by matrices. The theory involves the idea that a measurement on a system disturbs, to some extent, the system itself. With large systems this is of no consequence and the system obeys the rules of classical mechanics. However on the atomic scale, the observations may be expected to have a disturbing effect. If two successive observations are made on a system, the result depends on the order in which the observations are made. Thus if p denotes an observation of a component of momentum and q an observation of the corresponding coordinate then $pq = qp$. Here p and q are not physical quantities but operators. In matrix mechanics they are matrices and obey the relationship $pq - qp = ih/2\pi$. This leads to the quantum conditions for the system. The matrix elements are connected with the transition probabilities between various states of the system. Schrödinger showed that matrix mechanics and wave mechanics are simply different mathematical formulations of the same fundamental principles. (*See also* uncertainty principle.)

An alternative and equivalent formulation is the *path-integral method* due to Feynman. In this the particle is assumed to take all possible paths between two points. Interference between adjacent paths gives the most probable state of the particle.

Relativistic quantum mechanics. Pauli in 1925 suggested that an electron in an atom has a fourth quantum number with a value $\pm\frac{1}{2}$ (*see* atomic orbital). This was interpreted, by Uhlenbeck and Goudsmit, as due to the *spin of the electron. Dirac, in 1928, extended the principles of quantum mechanics so that they also satisfied the principle of *relativity. This allowed the properties of spin to be obtained, in a natural way, from the relativistic Schrödinger equation. Dirac's theory also involves the idea that an electron may have states with negative energy. He suggested that free space may be filled with electrons occupying all possible states of negative energy. These cannot be detected when all the states are filled but the absence of an electron with negative energy (and charge) appears to be a particle with positive energy and charge and the same mass as the electron. The *positron is such a particle; according to Dirac's theory it is a *hole in a distribution of negative-energy electrons. *See also* quantum electrodynamics; Klein–Gordon equation; spin–statistics theorem.

quantum number In *quantum mechanics, it is often found that the properties of a physical system, such as its angular momentum and energy, can only take certain discrete values. Where this occurs the property is said to be *quantized* and its various possible values are labelled by a set of numbers called quantum numbers. For example, according to Bohr's theory of the *atom an electron moving in a circular orbit could not occupy any orbit at any distance from the nucleus but only an orbit for which its *angular momentum (*mvr*) was equal to $h/2\pi$, or $2h/2\pi$, or $3h/2\pi$, etc. Thus the only possible orbits of

Table of Conserved Quantum Numbers

Quantum Number / Interaction	Angular Momentum J, J_3	Charge Q	Baryon Number B	Isospin I	Isospin Q.N. I_3	Strangeness S	Parity P	C-Parity C	G-Parity G	Lepton Numbers l_e, l_μ, l_τ
Strong	√	√	√	√	√	√	√	√	√	√
Electromagnetic	√	√	√	×	√	√	√	√	×	√
Weak	√	√	√	×	×	×	×	×	×	√

the electron are given by a set of equations of the type $mvr = nh/2\pi$, where n is an integer (0, 1, 2, 3, etc.). Thus the property of angular momentum is quantized and n is a quantum number that gives its possible values. The Bohr theory has now been superseded by a more sophisticated theory in which the idea of orbits is replaced by that of regions in which the electron may move, characterized by quantum numbers n, l, and m (see atomic orbital).

Properties of *elementary particles are also described by quantum numbers. For example, an electron has the property known as *spin, which is often visualized as a rotation of the electron, although this must not be taken literally. It is found that an electron can exist in two possible energy states depending on whether this spin is set parallel or antiparallel to a certain direction. The two states are conveniently characterized by quantum numbers $+\frac{1}{2}$ and $-\frac{1}{2}$. Similarly properties such as *charge, *isospin, *strangeness, *parity, and *hypercharge are characterized by quantum numbers. In interactions between particles, a particular quantum number may be conserved, i.e. the sum of the quantum numbers of the particles before and after the interaction remains the same. It is the type of interaction – *strong, *electromagnetic, *weak – that determines whether the quantum number is conserved (see table). See also energy level.

quantum of action See quantum theory.

quantum state See stationary state.

quantum statistics Statistics concerned with the equilibrium distribution of elementary particles of a particular type among the various quantized energy states. It is assumed that these particles are indistinguishable.

In *Fermi–Dirac statistics*, the *Pauli exclusion principle is obeyed so that no two identical *fermions can be in the same quantum mechanical state. The exchange of two identical fermions does not affect the probability of distribution but it does involve a change in the sign of the *wave function.

The *Fermi–Dirac distribution law* gives f_E, the average number of identical fermions in a state of energy E:
$$f_E = 1/[e^{\alpha + E/kT} + 1],$$
where k is the *Boltzmann constant, T is the *thermodynamic temperature, and α is a quantity depending on temperature and the concentration of particles. For the valence electrons in a solid, α takes the form $-E_1/kT$ where E_1 is the *Fermi level.

In *Bose–Einstein statistics*, the Pauli exclusion principle is not obeyed so that any number of identical *bosons can be in the same state. The exchange of two bosons of the same type affects neither the probability of distribution nor the sign of the wave function.

The *Bose–Einstein distribution law* gives f_E, the average number of identical *bosons in a state of energy E:
$$f_E = 1/[e^{\alpha + E/kT} - 1].$$

The formula can be applied to photons, considered as quasiparticles, provided that the quantity α (which conserves the number of particles) is zero. Planck's formula for the energy distribution of *black-body radiation was derived from this law by *Bose.

At high temperatures and low concentrations both the quantum distribution laws tend to the classical distribution:
$$f_E = Ae^{-E/kT}.$$
For any distribution law the number of particles in the energy range E to $E + dE$ is found by multiplying the quantity f_E by $g_E dE$, the number of states in this range of energy.

quantum theory A departure from the classical mechanics of Newton involving the principle that certain physical quantities can only assume discrete values. It was introduced by Max *Planck in 1900. Until this time physics was based on the mechanical laws formulated by Newton and it was believed that these were capable of giving a complete description of the properties of matter. The *kinetic theory of gases is one example of the successful application of Newtonian mechanics. Several problems, however, remained unsolved, in particular, the explanation of the curves of energy against wavelength for radiation emitted from a *black body. It was found that classical theory was unable to explain the characteristic maximum found in these spectral distribution curves. These attempts were based on the idea that the enclosure producing the radiation contained a number of *standing waves and that the energy of an oscillator is kT, where k is the *Boltzmann constant and T the thermodynamic temperature. It is a consequence of classical theory that the energy does not depend on the frequency of the oscillator. This inability to explain the phenomenon has been called the *ultraviolet catastrophe*.

Planck tackled the problem by discarding the idea that an oscillator can gain or lose energy continuously. He suggested that it could only change by some discrete amount, which he called a *quantum*. This unit of energy is given by $h\nu$, where ν is the frequency and h is the *Planck constant. h has dimensions of energy × time, or *action, and was called the *quantum of action*. According to Planck an oscillator could only change its energy by an integral number of quanta, i.e. by $h\nu$, $2h\nu$, $3h\nu$, etc. This meant that the radiation in an enclosure had certain discrete energies and by considering the statistical distribution of oscillators with respect to their energies, he was able to derive the Planck *radiation formula.

Planck's hypothesis can be expressed in a different way. If the momentum (p) of a linear harmonic oscillator is plotted against the displacement (q), an ellipse is obtained with one area $\int p dq$. It can be shown that the product of this area and the frequency is the energy of the oscillator. Thus, $\int p dq = nh$, where n is an integer 0, 1, 2, etc.

The idea of quanta of energy was applied to other problems in physics. In 1905 Einstein explained

Flavour	Mass (GeV/c^2)	Charge Q	Isospin I_3	Strangeness S	Charm C	Bottomness B	Topness T
d	≈ 0.3	$-\frac{1}{3}$	$-\frac{1}{2}$	0	0	0	0
u	≈ 0.3	$+\frac{2}{3}$	$+\frac{2}{3}$	0	0	0	0
s	≈ 0.5	$-\frac{1}{3}$	0	-1	0	0	0
c	≈ 1.5	$+\frac{2}{3}$	0	0	$+1$	0	0
b	≈ 5.0	$-\frac{1}{3}$	0	0	0	-1	0
t	> 90	$+\frac{2}{3}$	0	0	0	0	$+1$

features of the *photoelectric effect by assuming that light was absorbed in quanta (*photons). In 1907 he used the idea to interpret the behaviour of the heat capacities of solids at low temperatures (*see* Einstein's theory of specific heat capacities). A more successful theory was developed in 1912 by Debye (*see* Debye theory of specific heat capacities).

A further advance was made by Bohr (1913) in his theory of atomic spectra (*see* atom; hydrogen spectrum) in which he assumed that the atom can only exist in certain energy states and that light is emitted or absorbed as a result of a change from one state to another. He used the idea that the angular momentum of an orbiting electron could only equal an integral number of units ($nh/2\pi$ where $n = 0, 1, 2,$ etc.). A refinement of Bohr's theory was introduced by Sommerfeld in an attempt to account for fine structure in spectra. Other successes of quantum theory were its explanations of the *Compton effect, and *Stark effect.

In quantum theory certain conditions are imposed on physical quantities restricting their values to a number of discrete values. They are then said to be *quantized*. The changes of energy (corresponding to a quantum) are small and only noticeable on the atomic scale. Large-scale systems are usually adequately described by classical mechanics. In spite of the success of quantum theory in certain applications, there were many problems in which it gave inadequate or misleading results. It involved the application of quantum conditions to classical mechanics and is sometimes referred to as the "old" quantum theory. Later developments involved the formulation of a new system of mechanics known as *quantum mechanics.

quantum yield *See* quantum efficiency.

quark A fundamental constituent of *hadrons, i.e. of particles that take part in *strong interactions. Quarks are never seen as free particles (*see* quark confinement) but their existence has been demonstrated in high-energy scattering experiments and by symmetries in the properties of observed hadrons. They are regarded as elementary *fermions, with *spin ½, *baryon number ⅓, strangeness 0 or –1 and *charm 0 or +1. They are classified in six *flavours* [up (u), charm (c), and top (t), each with charge ⅔

the proton charge; down (d), strange (s), and bottom (b), each with –⅓ the proton charge]. Each type has an antiquark with reversed signs of charge, baryon number, strangeness, and charm. The top quark has not been observed experimentally but there are strong theoretical arguments for its existence. The top quark mass is known to be greater than about 90 GeV/c^2, too heavy to be detected in current high-energy particle accelerators.

The fractional charges of quarks are never observed in hadrons, since the quarks form combinations in which the sum of their charges is zero or integral. Hadrons can be either *baryons or *mesons. Essentially baryons are composed of three quarks while mesons are composed of a quark–antiquark pair. These components are bound together within the hadron by the exchange of particles known as *gluons*. (Gluons are neutral massless gauge bosons; *see* quantum chromodynamics.)

The quarks and antiquarks with zero strangeness and zero charm are the u, d, ū, and d̄. They form the following combinations:

proton (uud), antiproton (ūūd̄);
neutron (uud), antineutron (ūd̄d̄);
pions: π^+ (ud̄), π^- (ūd), π^0 (dd̄, uū).

The charge and spin of these particles is the sum of the charge and spin of the component quarks and/or antiquarks.

In the strange baryons (e.g. the Λ and Σ particles), one or more of the quarks are the s flavour. In the strange mesons (e.g. the K mesons), either the quark or antiquark is strange. Similarly, the presence of one or more c quarks leads to the charmed baryons and a c or c̄ to the charmed mesons.

It has been found useful to introduce a further subdivision of quarks, each flavour coming in three *colours* (red, green, blue). Colour as used here serves simply as a convenient label and is unconnected with ordinary colour. A baryon comprises a red, a green, and a blue quark and a meson comprises a red and antired, a blue and antiblue, or a green and antigreen quark and antiquark. In analogy with combinations of the three *primary colours of light, hadrons carry no net colour, i.e. they are 'colourless' or 'white'. Only colourless objects can exist as free particles. The characteristics of the six quark flavours are shown in the table.

quark confinement The theory that *quarks can never exist in the free state, which is substantiated by lack of experimental evidence for isolated quarks. The explanation given for this phenomenon in the form of *gauge theory known as *quantum chromodynamics (QCD), by which quarks are described, is that quark interactions become weaker as they come closer together and fall to zero when the distance between them is zero. The converse of this proposition is that the attractive forces between quarks become stronger as they move apart; as this process has no limit, quarks can never separate from each other.

In some theories, it is postulated that at very high temperatures, as might have prevailed in the early universe, quarks can separate; the temperature at which this occurs is called the *deconfinement temperature*.

quarter-phase *See* two-phase; polyphase system.

quarter-wavelength line *Syn.* quarter-wavelength transformer. A *transmission line one quarter of a wavelength long, used as an *impedance-matching device (i.e. impedance transformer). It is used extensively in systems operating at the higher radio frequencies.

quarter-wavelength transformer *See* quarter-wavelength line.

quarter-wave plate A thin double-refracting optical element, often of quartz or mica, that can be used to change the *polarization of an incident wave. It is cut parallel to the optic axis and its thickness is such that it introduces a quarter-wavelength path difference, i.e. a relative phase shift of 90°, between the ordinary and extraordinary waves. When plane-polarized light, at 45° to either principal axis, is incident on the plate, it is converted to circularly polarized light (and vice versa).

quartz The most abundant mineral, consisting of crystalline silicon dioxide (silica, SiO_2) and having diverse physical properties and uses. It is a *piezoelectric crystal, much used as a *piezoelectric oscillator. It also produces *double refraction, being an optically positive uniaxial crystal (*see* optically negative crystal). It rotates the plane of polarization to the left or right according to the variety, and to different extents for different colours. In addition, it transmits wavelengths between 180 nm (in the UV) and 4000 nm (in the IR). A variety of quartz prisms, lenses, and other optical elements make use of these properties. Quartz fibres – extremely fine filaments of quartz – are used as torsion threads in delicate instruments (*see also* molecular gauges).

quartz-crystal clock *See* clocks.

quartz-crystal oscillator *See* piezoelectric oscillator.

quartz-fibre dosemeter A dosemeter for recording radiation dose consisting essentially of a portable electroscope having a charged quartz fibre as the indicator. *See* dosimetry.

quartz-fibre manometer *See* molecular gauges.

quartz fibres Filaments of quartz of extreme fineness and uniformity of diameter used as torsion threads in delicate instruments.

quartz-iodine lamp *Syn.* quartz-halogen lamp. A tungsten filament electric lamp with a fused quartz envelope containing iodine vapour. The filament operates at a higher temperature than that of conventional lamps giving higher *luminous efficiency. The lamp is much smaller than a conventional lamp of similar power so the envelope is at a higher temperature than ordinary glasses could withstand. Tungsten atoms separating from the filament react with the iodine forming molecules that are prevented from condensing on the envelope because of its high temperature. On hitting the filament the molecules decompose, restoring the tungsten to the filament and the iodine to the vapour. Thus the lamp does not deteriorate by the sublimation of the tungsten at very high temperatures as would otherwise occur.

quartz wedge A wedge of quartz cut so that the long direction of the wedge is parallel to the vibration direction of the ordinary rays in the quartz (or occasionally normal to this direction). It is used to determine directions of fast and slow rays in crystal plates, to determine the order of interference colours, and to make optical-sign determinations with or without *interference figures.

quasar *Syn.* quasi-stellar object (QSO). A member of a class of astronomical bodies that lie well beyond our Galaxy and emit an immense amount of energy from a compact region of space. They were first detected (1963) as the bright optical counterparts to some powerful radio sources, but in fact only about 1% of quasars are radio sources. The greater part of the energy output is in the infrared region of the spectrum. They are also strong X-ray sources.

Quasars all have large *redshifts. The first value determined was $z = 0.158$; some have recently been detected with exceptionally large redshifts, $z > 4$. The quasar redshifts are now generally interpreted as being due to the *Doppler effect arising from the expansion of the universe. This explanation makes quasars extremely distant: those with the largest redshifts are the most distant objects, and hence the youngest objects, observed in the universe. Why there are so few quasars at shorter distances is not yet known.

Quasar spectra are dominated by bright emission lines superimposed on a continuum. The quasar redshift is determined from the emission spectra. Many quasar spectra also show absorption lines, which sometimes have a wide range of redshifts up to the value of the emission redshift. Those close to the emission redshift possibly arise from matter close to the quasar.

To be visible at such great distances, quasars must be exceedingly luminous: many have absolute *magnitudes exceeding –27. Their light-producing region, however, has been found to be extremely small (in some cases less than a light-day). Quasars are now thought to be the energetic cores of galaxies. The only process known to be efficient enough to generate the quasar energy output is some form of accretion of matter onto a supermassive *black hole at the heart of the quasar.

quasiparticle A long-lived single-particle *excitation in the quantum theory of many-body systems, in which the excitations of the individual particle are modified by interaction with the surrounding medium.

quaternions A system of vector analysis invented by W. R. Hamilton in the 19th century. To change one vector OB ($= b$) into another vector OR ($= a$) it is necessary to rotate OB through an angle so that it coincides with OA and to increase or reduce its length to that of OA, so that B and A coincide. These two steps are performed by an operator or quaternion q and we can then write:

$$qp = a.$$

This is the basis of *quaternion analysis*. It has been superseded by other systems of vector analysis.

quench A capacitor, resistor, or combination of the two, placed across a contact to an inductive circuit to inhibit sparking when the current ceases. Typically, a quench is employed across the make-and-break contacts of an induction coil.

quenching 1. See alloy.
2. See Geiger counter.

quick-break switch A switch in which the contacts are arranged to break the circuit rapidly and at a speed independent of that at which the operating handle, lever, etc., is moved. The speed of contact separation is usually controlled by a spring. Compare slow-break switch.

quiescent component A component of an electronic circuit not, at the instance of being described as quiescent, in operation, but shortly to become so.

quiescent current In any circuit, the current flowing in the circuit under conditions of zero applied signal.

quiet automatic gain-control A biased *automatic gain-control system combined with a device that automatically suppresses signals insufficiently strong to operate the automatic gain-control.

quill drive A form of drive used in electric traction. The driving motor is either coupled directly or geared to a hollow cylindrical sleeve (the quill), which runs in bearings nominally concentric with the driving axle. The motor and quill bearings are mounted in a frame supported by springs from the locomotive and the drive is transmitted from the quill to the driving wheels by means of a flexible coupling incorporating springs. The flexible coupling takes up the play in distance and alignment between the axes of the quill and the driving axle. The particular advantage of this method is the

reduction in the uncushioned load on the driving axle. *Compare* bar suspension; nose suspension.

Quincke, George H. (1834–1924) German physicist, best known for his work in surface tension and in acoustics. He also studied many optical phenomena, such as the penetration of light into the second medium at total reflection; and his method of measurement of magnetic permeability of a liquid, using a U-tube with one limb in a strong magnetic field, is still of interest.

Quincke's tubes *See* Herschel–Quincke tube; interference of sound.

Q-value *Syn.* nuclear energy change; nuclear heat of reaction. The amount of energy produced in a *nuclear reaction, often expressed in megaelectronvolts (MeV).

R

R.A. Abbreviation for right ascension. *See* celestial sphere.

Rabi, Isidor Isaac (1898–1988) Austrian-born Amer. physicist who is noted for his contributions to *particle physics and *quantum mechanics, especially his invention (1937) of the atomic and molecular beam magnetic resonance method of observing atomic spectra, for which he won a Nobel prize (1944). He originated the concept of *CERN international laboratory and was a founder of the Brookhaven National Laboratory.

rad A former unit of absorbed *dose of radiation, equal to 0.01 joule per kilogram of material. It has been replaced by the *gray, equal to 100 rad.

radar (RAdio Direction And Ranging) A system for locating distant objects by means of reflected radio waves, usually of microwave frequencies. Modern systems are highly sophisticated and can produce precise and detailed information about stationary and moving objects. Radar is used for navigation and guidance of aircraft, ships, etc., and is also used in meteorology and astronomy and for military purposes.

A radar system consists of a source of microwave power, a transmitting aerial emitting a narrow beam together with a receiving aerial, a receiver to detect the echo, and a *cathode-ray tube – the *radar indicator* – to display the output in suitable form.

Pulse radar systems transmit short bursts of microwaves, and the reflected pulse is received in the interval between the transmitted pulses. The same aerial is generally used for transmission and reception. *Continuous wave* (CW) systems transmit energy continuously, a small proportion being reflected by the target and returned to the transmitter. This uses less bandwidth than the conventional pulse systems.

Doppler radar employs the *Doppler effect to distinguish between stationary and moving targets.

radial astigmatism

The change in frequency between transmitted and received waves is measured to give the velocity. *V-beam radar* can determine range, bearing, and height of an object. It uses two transmitters simultaneously, the fan-shaped beams rotating continuously. One beam is vertical, the other inclined to it at ground level.

In any of the above systems the direction and distance of the target is given by the direction of the receiving aerial, and the time between transmission of the signal and reception of the echo. The transmitting and receiving aerials are usually dish aerials, steered so as to scan an area. A common procedure is to rotate the aerials in a horizontal plane, and produce a synchronous circular scan on the radar indicator. A target is displayed as a luminous spot on a radial line. Such a presentation is called a *plan position indicator* (PPI).

Phased-array radar is a highly sophisticated system. It typically consists of a flat rectangular arrangement of small identical radiating elements. Each is fed a microwave signal of equal amplitude but the relative phase of the signals can be altered electronically across the face of the array. This allows the direction of the radar beam to be altered rapidly: there is no mechanical movement of the aerials, the beam being steered through the principle of wave interference. The same set of delays that steered the transmitted beam brings all the constituent signals of the echo back into phase at the array, ready for processing. Computer control enables several hundred targets to be tracked simultaneously.

radial astigmatism *Syn*. oblique astigmatism. *Astigmatism due to oblique incidence on a lens system.

radial heat flow Flow occurring between two coaxial cylinders maintained at different temperatures. This method was used for the determination of the thermal conductivity of gases by *Schleiermacher's method and by *Gregory and Archer's method.

radian An SI unit of angle. One radian encloses an arc equal to the radius of a concentric circle. 2π radians = $360°$; 1 radian = $57.296°$.

radiance Symbol: L_e, L. 1. For a point of radiant energy, the *radiant intensity, in a specified direction, per unit projected area:

$$L_e = \mathrm{d}I_e/\mathrm{d}A \cdot \cos\theta,$$

where A is the area and θ is the angle between the specified direction and the surface.

2. For a point on a surface that is receiving radiant energy, the *irradiance (E_e) per unit solid angle (Ω):

$$L_e = \mathrm{d}E_e/\mathrm{d}\Omega.$$

The irradiance is taken over an area perpendicular to the direction of incident radiation.

The general equation of luminous intensity applying to both a point source and a point receptor, is:

$$L_e = \frac{\mathrm{d}^2\Phi_e}{\mathrm{d}\Omega \cdot \mathrm{d}A \cos\theta},$$

where Φ_e is the *radiant flux. Radiance is measured in watts per steradian per square meter. *Compare* luminance.

radiant A qualifying adjective denoting pure physical quantities used in *photometry in which *electromagnetic radiation is evaluated in energy units. Radiant quantities are distinguished from their corresponding *luminous quantities by adding a subscript e (for energy) to their symbols.

radiant efficiency Symbol: η_e, η. The ratio of the *radiant flux emitted by a source of radiation to the power consumed. It is the radiant equivalent of the photometric quantity *luminous efficacy (def. 2).

radiant energy Energy in the form of *radiation. The total power emitted or received by a body in the form of radiation is the *radiant energy flux*, or simply *radiant flux.

radiant exitance Symbol: M_e, M. The *radiant flux leaving a surface per unit area. It was formerly called the *radiant emittance*. It is measured in watts per square metre. *Compare* luminous exitance.

radiant exposure Symbol: H_e, H. 1. The surface density of the total radiant energy received by a material.

2. A measure of the total energy of the radiation incident on a surface per unit area, expressed by the product of the *irradiance, E_e and the irradiation time: $H_e = E_e \int \mathrm{d}t$. It is measured in joules per square metre. *Compare* light exposure.

radiant flux *Syn*. radiant power. Symbol: Φ_e, Φ. The total power emitted or received by a body in the form of *radiation. The term is usually applied to the transfer of energy in the form of *electromagnetic radiation as opposed to particles, but it is not usually applied to radio waves. It is measured in watts. *Compare* luminous flux.

radiant flux density Symbol: ϕ. Either the *irradiance or *radiant exitance of a surface. For *radiant flux Φ, it is given by $\Phi = \int \phi \mathrm{d}S$ for area S.

radiant intensity Symbol: I_e, I. The *radiant flux (Φ_e) emitted per unit solid angle (Ω) by a point source in a given direction:

$$I_e = \mathrm{d}\Phi_e/\mathrm{d}\Omega.$$

It is measured in watts per steradian. *Compare* luminous intensity.

radiant power *See* radiant flux.

radiation Anything propagated as rays, waves, or a stream of particles but especially light and other electromagnetic waves, sound waves, and the emissions from radioactive substances.

radiation belts Regions within a planet's *magnetosphere in which energetic particles – mainly electrons and protons – are trapped by the planet's magnetic field. *See* Van Allen belts.

radiation chemistry The study of the chemical effects of *ionizing radiation on matter, both living and nonliving. *Compare* radiochemistry.

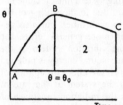

Time   Radiation correction

radiation correction A correction employed in *calorimetry to allow for heat transfer between the body and its surroundings. If the recorded temperature of the contents of a calorimeter follows the curve ABC (*see* graph), while the temperature of the surroundings is given by the line $\theta = \theta_0$, then the maximum θ_B will be too small and a radiation correction may be made assuming *Newton's law of cooling.

$$(\theta_B)_{true} = \theta_B + \frac{\text{Area 1}}{\text{Area 2}} (\theta_B - \theta_C).$$

The areas are obtained by counting squares, area 1 being counted positive if the corresponding temperatures are above the temperature of the surroundings, and negative if these lie below the line $\theta = \theta_0$. In Rumford's form of the correction, by having the calorimeter initially as much below room temperature as finally it is above, the area 1 is made up of two almost equal positive and negative parts and so is practically zero. The maximum temperature recorded is thus almost the true value.

radiation diagram *See* radiation pattern.

radiation formula The formula, devised by *Planck, to express the distribution of energy in the normal spectrum (spectrum of black-body radiation). Its usual form is:

$$8\pi ch d\lambda/\lambda^5(\exp[ch/k\lambda T] - 1),$$

which represents the amount of energy per unit volume in the range of wavelengths between λ and $\lambda + d\lambda$. The meanings of the other symbols are $c =$ speed of light in free space, $h =$ the *Planck constant, $k =$ the *Boltzmann constant, $T =$ thermodynamic temperature. *See also* Planck's formula.

radiation gauge *See* Knudsen gauge.

radiation impedance *See* radiation resistance.

radiation pattern *Syn.* radiation diagram. A graphical representation of the distribution in space of radiation from any source, especially an *aerial; for a particular aerial, the pattern for transmission is identical to that for reception. The graph is normally in polar coordinates (*see* polar diagram).

radiation physics The study of radiation, particularly *ionizing radiation, and the physical effects it can have on matter.

radiation pressure The pressure exerted upon a surface exposed to electromagnetic radiation. The transfer of energy E is associated with the transfer of momentum p where $p = E/c$, c being the speed of light; hence any body that absorbs, reflects, refracts, or scatters electromagnetic radiation experiences a force. If a parallel beam of intensity I (power/area) is totally absorbed by a surface normal to the beam, the radiation pressure is I/c. If the radiation is perfectly reflected straight back, the pressure is double this. For uniformly diffused radiation the pressure is $\rho/3$ where ρ is the energy density of the radiation. This result was used by *Boltzmann in his thermodynamic proof of *Stephan's law, and by *Wien in the derivation of his displacement law. *See* levitation; Nichols's vane radiometer; Poynting, John Henry.

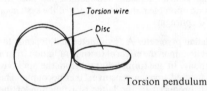

Torsion pendulum

radiation pressure of sound The steady pressure exerted on a surface by sound waves. This pressure is to be distinguished from the oscillatory change of pressure observed at the displacement node of the *standing waves. Rayleigh deduced the relation $\bar{P} = \frac{1}{2}(\gamma + 1)E \ldots (a)$ where E is the energy density and γ the ratio of the specific heat capacities. This relation was derived on the assumption that an adiabatic process takes place. If we assume Boyle's law, we find $\bar{E}$. This is in fact the sound pressure for a progressive wave; if the waves are reflected then we get $\bar{P} = 2E$. Lebedew also had verified the law ascribing the sound pressure to the energy arriving at the wall per second, i.e. as $P = I/c$ or $2I/c \ldots (b)$ if the waves are reflected, where I is the energy crossing unit area per second (corresponding to *sound intensity) and c is the speed of sound. On account of this radiation pressure, a resonator experiences a repulsion away from a tuning fork or source of sound placed near its mouth. Altberg applied this principle to the absolute measurement of the intensity of sound. The disc of a torsion pendulum was arranged so as to close a hole in a surface exposed to sound waves. The torsion head was then rotated so as to bring the disc into the plane of the surface. By knowing the angle of twist θ, the area of the disc S, the coefficient of torsion of the wire k, and the length of arm of the disc r, the pressure can be calculated from the formula $P = k\theta/Sr$. Using the previous equations (a) and (b), the intensity of sound:

$$[I = cP/(1 + \gamma)],$$

can be evaluated. This method has been extensively used for the measurements of the intensity of ultrasonic waves both in gases and liquids. The diagram shows a typical torsion pendulum for this purpose, in which the vertical disc receives the radiation and the horizontal disc acts as counterpoise.

Another direct method for measuring radiation pressure is due to Barus. A double resonator is arranged with a hole connecting the body of the resonator to a mercury surface. The displacement of the mercury is measured by an optical interferometer when the resonator is turned to the applied sound. A better method is due to Gerlach in which the forces on a diagram due to the impinging sound waves are compensated by applying to it an electric force with the same frequency as that of the sound. The diaphragm is thus kept stationary under the joint action of the opposing mechanical and electrical forces. The value of the latter forces in mechanical units gives the fluid forces acting on the diaphragm and hence the intensity of the impinging sound. It is an absolute null method and does not depend upon the natural frequency of the system or the diaphragm.

radiation pyrometer A *pyrometer that depends for its action upon the effect of thermal radiation from a hot body. In one form, heat rays from the hot body are focused upon a sensitive *thermocouple. The e.m.f. produced in the latter is a function of the temperature of the hot body and is either measured by means of a potentiometer or is utilized to produce a deflection of a galvanometer or a millivoltmeter.

radiation resistance 1. At a surface vibrating in a medium. The portion of the total *resistance (unit-area, acoustical, or mechanical) due to the radiation of sound energy into the medium.

The mean power radiated per unit area of a plane wavefront is:
$$\tfrac{1}{2}\rho c \xi^2_{max} = \tfrac{1}{2}\rho c f^2 a^2,$$
where ρ is the density of the medium, c the speed of sound in it, ξ_{max} is the maximum sound particle speed, which is equal to fa (f being the frequency and a the amplitude of the vibration). In the analogous electrical case, this equation represents the power of dissipation in a circuit of resistance ρc. The quantity ρc is usually known as the radiation resistance or impedance of a medium that is transmitting plane waves.

The characteristic impedance for a spherical wave of radius r is expressed as:
$$z = \rho c(X' + iY'),$$
where
$$X' = k^2 r^2/(k^2 r^2 + 1)$$
$$Y' = kr/(k^2 r^2 + 1)$$
$$k = 2\pi/\lambda = 2\pi f/c,$$
(λ being the wavelength).

The first term $\rho c X'$ is a resistance term and the second term is a reactance. If r is very large, z will reduce to ρc as for plane waves. For a place at a small distance (compared with λ) from the source, kr is small and the radiation impedance or resistance is approximately equal to:
$$\rho c k^2 r^2 = 4\pi^2 f^2 \rho r^2/c.$$
This impedance may be included whenever there is a change from plane to spherical waves as, for example, in the propagation of sound through pipes at the mouth. The effect is bound up with the *end correction of the tube. In general, the resistance or friction coefficient in any vibrating system must include not only the internal friction due to the sliding of the elements with respect to each other, but also air-resistance factors or, in more general terms, the effects of immersing the vibration system in the medium. In the first place, sound energy is radiated from the system and this represents a dissipation of the energy supplied to the system at the rate of $\tfrac{1}{2}R_1\xi^2_{max}$, where R_1 is the radiation resistance of the medium. Besides radiation there is a pure frictional term due to laminar motion in the surrounding air. The total work done by the system is against resistances, R_1 (radiation) + R_2 (internal) and the rate of expenditure of energy is
$$\tfrac{1}{2}(R_1 + R_2)\xi^2_{max}.$$
2. *See* aerial radiation resistance.

radiation sickness Illness resulting from exposure of body tissue to a large dose of *ionizing radiation. Mild short-term effects include headache, nausea, vomiting, and diarrhoea. Repeated or prolonged exposures leads to delayed effects, which can include sterility, cataract formation, disturbances in blood cell formation, and cancer. In severe cases (e.g. the effects of a nuclear explosion) the symptoms can include extensive haemorrhages, loss of hair and teeth, prolonged clotting and bleeding times, and changes to genetic material.

radiative capture A type of *nuclear reaction in which a particle (usually a neutron) is captured by a nucleus and gamma radiation is emitted. The radiative capture of a neutron produces an isotope of the original element with the *mass number increased by unity. The *cross sections for the process vary very greatly between nuclides, and for a given nuclide usually decrease as the neutron energy increases except for peaks of characteristic energies. The isotopes produced are very often radioactive, usually emitting beta rays, which are sometimes accompanied by gamma radiation. Useful radioactive substances are often produced by exposing suitable elements to the large flux of neutrons in *nuclear reactors.

radiative collision A collision that takes place between two charged particles and from which *electromagnetic radiation is emitted due to the conversion of part of the kinetic energy.

radio 1. The use of *electromagnetic radiation to transmit or receive electrical impulses or signals without connecting wires. Also the process of transmitting or receiving such signals. The term is usually confined to the communications system transmitting audio information (wireless).

2. A *radio receiver.

3. Denoting electromagnetic radiation in the frequency range 3 kHz to 300 GHz. (*See* radio frequency.)

4. A prefix denoting *radioactivity.

radioactive *Syn.* active. Possessing or relating to radioactivity.

radioactive constant *See* decay constant. The term radioactive constant is no longer in use.

radioactive series Most of the natural radionuclides have *atomic numbers (Z) in the range $Z = 81$ to $Z = 92$. These substances can be grouped into three radioactive series: the *uranium series, thorium series*, and *actinium series*. The *mass numbers of these radionuclides can be represented by a set of numbers: $4n$ (thorium series), $4n + 2$ (uranium series), and $4n + 3$ (actinium series), where n is an integer between 51 and 59. The *parent nuclides at the head of the series are long-lived with *half-lives in the region 10^9–10^{10} years. They are uranium-238 (uranium series), thorium-232 (thorium series), and uranium-235 (actinium series). The actinium series contains actinium-227 produced by decay of uranium-235 and palladium-231. The other members of the three series are formed mainly by *alpha decay or *beta decay of the preceding nuclide. The final substances are all stable isotopes of lead. The series are shown in the Appendix, Table 11; they are illustrated by plotting mass number against atomic number. An emission of α-particles is indicated by a downward displacement of 4 units and to the left by 2 units, β-decay by a displacement to the right by 1 unit; half-lives are shown for some of the decay processes. *Branching occurs in several places in all the series.

There is a fourth radioactive series, the *neptunium series* in which the half-lives of the three nuclides at the top of the series, plutonium-241, americium-241, and neptunium-237 are much shorter than the parents of the other three series. These substances have therefore either disappeared from the earth's surface or are present in negligible quantities. They can be synthesized as artificial radionuclides (*see* radioactivity). Neptunium-237 has the longest half-life, and has been found, with plutonium, in uranium deposits. The mass numbers of the members of this series are represented by $4n + 1$.

radioactive tracer A definite quantity of *radioisotope introduced into a biological or mechanical system so that its path through the system and its concentration in particular areas can be determined by measuring the radioactivity with a *Geiger counter, *gamma camera, or a similar device. Measurements are made after a period of time or at intervals of time, often after a chemical reaction has occurred. A compound containing a radioisotope and used as a tracer is said to be *labelled*.

radioactive waste Solid, liquid, and gaseous waste products from nuclear reactors, uranium processing plants, hospitals, etc., that are radioactive. Because the radioactivity of some materials will remain for thousands of years, their disposal must be undertaken with great care. High-level waste (spent nuclear fuel, etc.) needs artificial cooling and is therefore stored by its producers for several decades before disposal. Intermediate-level waste (reactor components, filters, sludges, etc., from processing plants) is solidified and stored mixed with concrete in steel drums in power stations prior to burial in deep mines or beneath the seabed in concrete chambers, where it cannot contaminate ground water. Low-level waste (solids and liquids contaminated by traces of radioactivity) presents fewer problems; in the UK, since 1988 it has been disposed of in steel drums in concrete-lined trenches at Driggs in Cumbria. This work has been undertaken by Nirex Ltd, a company set up jointly by the government and the nuclear industry. Other countries make similar arrangements. Until 1983, when it was suspended by international agreement, low- and intermediate-level wastes were disposed of in the deep Atlantic in steel drums cast in concrete. In addition some very dilute low-level gaseous and liquid wastes have been discharged into the air and the sea.

radioactivity The spontaneous disintegration of the nuclei of some nuclides (called radionuclides) with the emission of *alpha particles or *beta particles, sometimes accompanied by a *gamma ray. The processes involved in *alpha decay and *beta decay alter the chemical nature of the atom involved, because of the change in *atomic number, and usually result in a more stable nucleus. Specific energy changes take place in the nucleus during a disintegration and any excess energy possessed by the nucleus after the expulsion of an α- or β-particle is emitted by gamma radiation or *internal conversion. Another particle that can be emitted by the nucleus is the *positron (an antielectron), the disintegration process being analogous to β-decay. It is possible for a γ-ray alone to be emitted, when a *metastable state of a radionuclide decays to a lower energy state of the same nuclide. A radionuclide can disintegrate into two different energy states of the same nucleus, forming a pair of nuclear *isomers. This occurs in the uranium series (*see* radioactive series). Electron *capture is another disintegration process.

Natural radioactivity is the disintegration of naturally occurring radionuclides. It was discovered in 1896 by Becquerel who found that the radiation from uranium salts caused blackening of photographic film. Rutherford and Soddy showed that this emission consisted of two types of radiation that could ionize gas, thus producing a measurable current. One type was an easily absorbed radiation, which they called α-rays, the other, a more penetrating type, they named β-rays. It was later discovered that an α-particle is a helium nucleus and that a β-particle is a high-energy electron.

The Curies discovered (1898) the radioactive elements polonium and radium in pitchblende and demonstrated that their ionizing ability, which is a measure of the *activity of a nuclide, was much higher than that of uranium, the activity depending on the quantity of radionuclide present. Since the emission of radiation was unaffected by temperature, pressure, etc., it was inferred that the radiation originated in the nucleus after the discovery of the latter in 1912.

The fact that the emission of an α- or β-particle alters the chemical nature of an atom was confirmed by the identification of the chemical properties of the *daughter product nuclei produced by radioactive transformations. The isolation of the various radioactive products produced by uranium, thorium, and actinium enabled three of the *radioactive series to be determined, indicating how these three elements give rise to successive transformations, ending in stable isotopes of lead. Apart from the members of the radioactive series, some other naturally occurring elements are known to contain radioisotopes. These include carbon, lutecium, neodymium, potassium, rhenium, rubidium, samarium, and scandium. Following the discovery of *artificial radioactivity*, a fourth radioactive series (the neptunium series) was drawn up.

Artificial radioactivity was first demonstrated in 1934 by the Joliot-Curies. They showed that the bombardment of nuclei (boron and aluminium) by α-particles produced *artificial radionuclides*. These substances decay by the same processes as natural radionuclides. It has been possible to create many nuclides having an atomic number greater than that of uranium (92). These are the *transuranic elements*, produced by bombarding heavy stable atoms with high-energy protons, neutrons, deuterons, carbon atoms, etc. They are all radioactive and decay, principally by the emission of alpha particles or by electron capture, into a stable natural substance.

The activity of both natural and artificial radionuclides decreases exponentially with time. The time take for half a given number of atoms of a particular nuclide to be transformed is called the *half-life. It can vary from 1.5×10^{-8} seconds up to 10^{17} years. The fraction of atoms decaying in a certain time is not truly constant. Radioactive decay is a statistical phenomenon and the half-life is an average value of very many disintegrations.

The activity of a given specimen is measured by determining the ionizing ability of its characteristic radiation, using one of a number of *counters or *detectors. These include the *Geiger counter, *scintillation counter, and *ionization chamber. Counters can measure the number of disintegrations occurring in a specimen in a given time. The unit of activity is the *becquerel.

The radiation emitted by radionuclides, especially artificial radionuclides, has now many uses in medicine (*see* nuclear medicine), scientific research, and industry. Radionuclides are also used in *dating. (*See also* radioactive tracer.)

radio astronomy The study of astronomy through the radio signals emitted by some celestial bodies. The signals originate from sources of nonthermal radiation and are associated with bodies both within and beyond our Galaxy. A *radio telescope is used for observing. Radio sources within our Galaxy include *pulsars, *supernova remnants, and HI and HII regions – regions of neutral predominantly atomic hydrogen and predominantly ionized hydrogen respectively – in interstellar space. Extragalactic radio sources include *quasars and the highly active radio galaxies.

radiobalance An instrument due to Callendar in which the heating due to the absorption of radiation is neutralized by the cooling due to the *Peltier effect at one junction of a thermocouple, thus enabling the amount of incident radiation to be measured absolutely. Two copper cups $C_1 C_2$ 3.5 mm in diameter and 1 cm high are connected by a differential copper-constantan thermocouple with one junction soldered to the bottom of each cup (*see diagram under* Hoare's determination of the *Stefan–Boltzmann constant). Two other thermocouples T also are connected to the cups and are in circuit with a sensitive galvanometer G. Radiation is allowed to fall on one cup and a current is passed from the battery B in such a direction that Peltier cooling occurs at this cup. The current is adjusted by means of the rheostat R until the galvanometer G shows no deflection. The mean current I when each cup is exposed in turn to the radiation is obtained from the ammeter A. Then the heat H absorbed per unit area per unit time is given by $H = 2PI/A$ where P is the Peltier coefficient for the couple in volts and A is the mean aperture of the two cups. The apparatus has been used for the determination of the Stefan–Boltzmann constant by Hoare and the solar constant.

radiobiology The branch of science concerned with the effects of radiation on living matter.

radiocarbon dating *See* dating.

radiochemistry The production of radioactive nuclides and their compounds by chemically processing irradiated materials or naturally radioactive materials, their subsequent use in elucidating chemical problems, and the study of the special techniques involved.

radio direction-finding Determining the direction of radio waves received from a distant radio transmitter in order to find the direction of the latter with respect to the radio receiver. The radio receiver used for this purpose, complete with all its associated apparatus (e.g. aerials), is called a direction-finder.

radio frequency (r.f.) Any frequency electromagnetic radiation in the *frequency band 3 kilohertz to 300 gigahertz, or of alternating currents in this frequency range.

radio-frequency heating *Dielectric heating or *induction heating, when the alternating field has a frequency greater than about 25 kHz.

radio galaxy A *galaxy that emits radio waves, many of which have been detected by *radio telescopes. *See also* radio source.

radiogenic Resulting from radioactive *decay.

radiogoniometer An apparatus by means of which the bearing of radio waves incident upon a fixed (i.e. nonrotating) aerial system to which it is connected may be determined. It consists fundamentally of two fixed coils mounted with their axes at right angles and a third coil that can be rotated inside the other two.

radiography The production of shadow photographs (*radiographs*) of the internal structure of bodies opaque to visible light by the radiation from X-rays, or by gamma-rays from radioactive substances.

radio interferometer *See* radio telescope.

radioisotope An *isotope of an element that undergoes *disintegration.

radiology The study and application of X-rays, gamma rays, and other penetrating *ionizing radiation.

radiolucent *See* radiopaque.

radioluminescence The emission of visible electromagnetic radiation from a radioactive substance.

radiolysis The chemical decomposition of materials into ions, excited atoms and molecules, etc., by *ionizing radiation. *See* pulse radiolysis *under* flash photolysis.

Radiometal A proprietary alloy of about 50% iron, 45% nickel, and 5% copper. It has a high *permeability and low *hysteresis loss, and is therefore used in transformer and transducer cores.

radiometer An instrument for measuring the total energy or power received from a body in the form of radiation. It is used in physical *photometry, astronomy, and meteorology. The term is used in particular for instruments that detect and measure radiation in the infrared, visible, and near ultraviolet. *See* thermopile; bolometer; net radiometer; pyrheliometer; radiomicrometer.

radiometer gauge *See* Knudsen gauge.

radiometric age The age of a geological or archaeological specimen, determined by radiometric *dating.

radiomicrometer An extremely sensitive detector of radiation devised by C. V. Boys. A thermocouple of antimony and bismuth is joined in series with a single copper loop suspended by a quartz fibre between the two poles of a magnet. Radiation falling on one Sb-Bi junction causes a current to flow in the copper loop, which is therefore deflected, the deflection being obtained from the movement of a spot of light reflected from a mirror carried by the loop.

radio noise In radio communication, a certain amount of unwanted sound over a wide range of audio frequencies is always present. Some of this is due to electrical interference from discharges in the *troposphere or *ionosphere, or may have its origin in the stars in outer space (*Jansky noise). The remainder of the noise is an inherent property of all electronic circuits and is due to random electron flow. *See* noise.

radionuclide Any radioactive *nuclide.

radiopaque Opaque to radiation, especially X- and gamma rays. Radiopaque substances, such as bones, are visible on a radiograph (*see* radiography). *Radiotransparent* is the converse of radiopaque, i.e. transparent to radiation. Radiotransparent substances, such as skin, are not visible on a radiograph. If a medium is almost entirely transparent to radiation, it is *radiolucent*.

radio receiver *Syn.* radio; wireless. A device that converts radio signals into audible signals. A simple receiver consists of a receiving aerial, a tuner that can be adjusted to the desired *carrier frequency, preamplifier, detector, audiofrequency amplifier, and a loudspeaker. A common refinement of a simple receiver is the *superheterodyne receiver. Radios can detect frequency-modulated (FM) signals and/or amplitude-modulated (AM) signals. High-fidelity devices usually contain extra circuits associated with the audiofrequency amplifier to restore the bass and treble response of the output to that of the original audible source.

radiosonde system A compact apparatus comprising a *meteorograph and radio transmitter that is carried into the earth's atmosphere by a balloon and transmits radio signals indicative of the temperature, pressure, and humidity.

radio source An extraterrestrial source detected with a *radio telescope. Many sources are extended. *See* radio astronomy.

radiospectroscope An apparatus for displaying (usually on a cathode-ray tube) an analysis of the radio-frequency energy arriving at an aerial. The wavelengths in actual use for transmission at any particular time are shown, and by the height and spread of the trace on the tube face, some indication of the field strength and modulation is given.

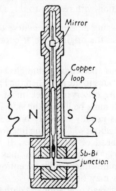

Boys' radiomicrometer

radio telescope A type of telescope used in *radio astronomy to record and measure the radio-frequency emissions from celestial radio sources. All radio telescopes consist of an *aerial or system of aerials, connected by *feeders to one or more *receivers. The aerials may be in the form of large metal *dishes or simple linear *dipoles. The receiver outputs may be displayed, or recorded for computer analysis.

The steerable dish consists of a wire mesh approximately shaped and mounted so that it can be directed to any part of the sky without serious distortion. A pencil beam of radiation is received from a small but not highly defined area of sky. A well-known example is the telescope at Jodrell Bank near Manchester. It has a diameter of 76 metres, an altazimuth mounting (see astronomical telescope), and is efficient at wavelengths above 0.1 metres.

The *radio interferometer* is a form of radio telescope that consists of two or more fixed or steerable radio aerials separated by a known distance and connected to the same radio receiver. *Interference occurs between waves from a radio source that are received by the aerials. The position of the source can thus be determined. If large numbers of aerials are used, they are generally arranged in parallel rows or in two rows at right angles to each other. Alternatively the method of *aperture synthesis may be employed. The interferometer is more sensitive than a single dish as it can detect radiation from sources of small angular diameter. As resolution increases, however, large-scale structure is lost.

radiotherapy See nuclear medicine.

radiotransparent See radiopaque.

radio waves *Electromagnetic radiation of *radio frequency, used in *radio and *television broadcasting and other communications systems. See aerial; modulation; ionosphere; radio astronomy.

radio window See atmospheric windows.

radius of curvature See centre of curvature.

radius of gyration The root mean square distance of the mass distribution of a rigid body from its axis of rotation. For a rigid body of *moment of inertia I about a given axis, $I = mk^2$, where m is the mass and k is the radius of gyration with respect to this axis.

radius vector A line joining a point on a curve to a reference point (such as the origin of polar coordinates). See coordinate.

rainbow The continuous spectrum of sunlight seen as one or more circular arcs in the sky when the light falls on raindrops. The observer's back must be towards the sun, which must be low in the sky. The primary bow, which is usually the only one seen, makes a mean angle of about 41° with the line from the sun through the head of the observer. If the background is very dark, the secondary bow may be seen at about 52° from this line. Under exceptional conditions other bows may be seen.

The primary bow is spread over about 2° with the red on the outside. As the disc of the sun subtends about 0.5°, the spectrum is impure. The bow is formed at minimum deviation by light, which undergoes one partial internal reflection in a raindrop and is deviated and dispersed on entering and leaving. The secondary bow is spread over more than 3° with the violet on the outside. The light undergoes two partial reflections in a drop so it is far less intense than the primary bow.

Rainwater (Leo) James (b. 1917) Amer. physicist. A professor at Columbia, he collaborated with Aage Bohr and his associate Benjamin Mottelson in formulating the concept of the collective (or unified) model of the nucleus. He was later director of the Nevis Cyclotron Laboratory.

RAM Abbreviation for random-access memory. A type of *semiconductor memory used in computers for which both recording and retrieval of data by the user is possible – i.e. the user can both write to and read from RAM (*compare* ROM). The basic storage elements – often called *cells* – are microscopic devices fabricated as an *integrated circuit. A single cell can store one *bit – either a binary 0 or a binary 1. Very large capacity memories can be produced. The cells are formed in a rectangular array so that each one can be uniquely identified by its row and column. Any cell can thus be accessed directly in any order (and extremely rapidly), i.e. there is *random acess to the cells. To preserve the cell contents, RAM requires its power supply to be maintained.

RAM devices can be classified as *static RAM* or *dynamic RAM* (DRAM). Static RAM is fabricated from either bipolar or MOS components (*see* integrated circuit); each cell is formed by an electronic *latch whose contents remain fixed until written to. Dynamic RAM is fabricated from MOS components, the cells utilizing the charge stored on a capacitor as a temporary store; leakage currents require the cell contents to be "refreshed" at regular intervals (typically every millisecond). Compared with static RAMs, DRAMs have larger cell densities, lower power consumption, but slower access times. There may be many thousands of cells on one RAM chip: a 64 K RAM chip stores a total of 64 kilobits (i.e. 65 536 bits) of data.

Raman, Sir Chandrasekhara (1888–1970) Indian physicist. His best-known work was on molecular scattering of light (*see* Raman effect), for which he was awarded a Nobel prize (1930). He also made contributions to the theory of musical instruments and worked on diffraction.

Raman effect A type of inelastic scattering of electromagnetic radiation (usually light) by atoms and molecules, in which the scattered radiation is changed in wavelength by discreet amounts characteristic of the scattering medium. The effect is similar to the inelastic scattering of X-rays in the *Compton effect. It was discovered by the Indian physicists Sir Chandrasekhara Raman and K. S. Krishnan and, independently, by the Soviet physicists G. Landsberg and L. Mandelstam (who called it *combination scattering*).

Raman scattering is a low-intensity effect; the intensity of the radiation scattered by the Raman effect is about one thousandth of the intensity of that in *Rayleigh scattering. For this reason, investigations of the Raman effect usually involve use of lasers. *Raman spectroscopy* is a technique for investigating molecular structure and energy levels. A sample is irradiated with a monochromatic source of light and the spectrum of the scattered light is taken. Generally, the Raman spectrum consists of a main line at the frequency of the incident radiation, with additional lines at lower frequency (*Stokes lines*) and higher frequency (*anti-Stokes lines*). The differences in frequency between the scattered radiation and the incident radiation correspond to changes in vibrational and rotational energy levels of the molecule. Raman spectroscopy gives similar information to infrared spectroscopy but, because different selection rules apply, it can be used to investigate molecules that are not suitable for infrared investigation.

Raman scattering *See* Raman effect.

Raman spectroscopy *See* Raman effect.

ramjet *Syn.* aerothermodynamic duct or athodyde. A propulsion engine in which a fuel burns in air that has been compressed by the forward motion of the engine only. It consists of a suitably shaped duct into which fuel is fed at a controlled rate, the combustion products being expanded in a nozzle. The shape of the duct depends on whether or not it is to operate at supersonic velocities.

Ramsauer effect When the energies of electrons incident on the inert gases fall below certain critical values, a rapid reduction of the scattering cross section of the atoms is observed.

Ramsay, Sir William (1852–1916) Brit. chemist. He explained the *Brownian movement in terms of molecular bombardment (1879). He identified (1895) a gas that could be obtained from the mineral cleveite as being helium, previously only known by its spectrum as an element in the sun. In association with Travers and with Rayleigh he discovered also the other inert gases (argon, 1894; neon, krypton, and xenon, 1898).

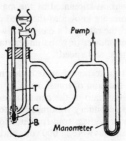

Measurement of saturated vapour pressure (Ramsay and Young)

Ramsay and Young's measurement of vapour pressure The saturation vapour pressure was determined by a dynamical method depending on the fact that a liquid boils when its S.V.P. equals the external pressure. The bulb of a thermometer T is surrounded by cotton wool C soaked in the liquid and the pressure in the system is reduced by pumping until the thermometer reading is steady (*see* diagram). A bath B at different temperatures surround the tube in which the thermometer is situated, the saturation vapour pressure being read on the manometer. A large reservoir vessel is placed between the pump and the actual apparatus to stabilize the system, and to condense the vapour.

Ramsden, Jesse (1735–1800) Brit. optical-instrument maker, he invented the equatorial mounting of *astronomical telescopes and the *Ramsden eyepiece.

Ramsden circle *See* exit pupil.

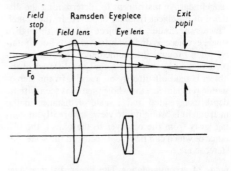

Ramsden and Kellner eyepieces

Ramsden eyepiece A type of *eyepiece that, in its most elementary form, consists of two planoconvex lenses of equal focal length, each equal to the separation of the curved surfaces facing one another. Commonly the separation is reduced to $2/3$ of the lens focal length. It is better than *Huygens's eyepiece for spherical aberration, distortion, and longitudinal chromatic aberration, but suffers from lateral chromatism. The Ramsden eyepiece is used in measuring instruments, such as microscopes and spectrometers, with cross-wires or a *graticule in the plane of the field stop. The achromatized Ramsden is called the *Kellner eyepiece. *See* achromatic lens.

Ramsey, Norman Foster (*b.* 1915) Amer. physicist who works at Harvard. He was awarded the 1989 Nobel prize for physics for his resonance method using separated oscillatory fields, which forms the basis of atomic *clocks.

Randall, Sir John Turton (1905–1984) Brit. physicist who invented the *magnetron, which enabled high-power *radar to be used in the second world war.

random access A method of retrieval or storage of data in which the individual storage locations can be accessed (read or written to) directly, in any order. There is random access with *disk storage. *See also* RAM; ROM.

random noise

random noise *See* noise.

random structure A crystalline arrangement in which *equivalent positions are not necessarily occupied by atoms of a single kind.

random winding *See* mush winding.

range 1. The distance from the starting point at which a projectile reaches a horizontal plane through that point.
   2. The penetration distance of a beam of ionizing radiation.
   3. The maximum distance over which reception is possible from a radio or television transmitter. The distance over which a radar set is effective.
   4. *See* transducer.

rangefinder An instrument for determining the distance of an object from an observer. In the coincidence rangefinder, the object, as seen through a single eyepiece, is divided into two parts horizontally; when brought into coincidence to form a complete picture by turning a screw, the distance of the object is read off directly on a scale. In the stereoscopic rangefinder, the whole picture is seen with its depth exaggerated and a scale of distance in the instrument is brought into view, apparently stretching away from the observer to the object, the distance of which can be read directly on the scale. *See* stereoscope.

range of accommodation The linear distance over which the eye can accommodate for clear vision (from far point to near point); it is not to be confused with *amplitude of accommodation.

Rankine, William John M. (1820–1872) Brit. (Scots) engineer. He played a considerable part in establishing the form of the second law of *thermodynamics and in applying thermodynamics to engineering problems.

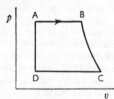

Rankine cycle

Rankine cycle An ideal steam-engine cycle since it is theoretically reversible. It differs from the *Carnot cycle in that a separate boiler and condenser are used. Beginning at A with the working substance as water in the boiler, AB represents isothermal expansion on boiling at constant pressure (*see* diagram); BC the adiabatic expansion of the steam in the cylinder or turbine during which it cools to the temperature of the condenser; and CD the isothermal compression on condensing at constant pressure. DA represents the transfer of the cold water to the boiler and at A the water is heated to the temperature of the boiler. The first three stages are identical with those of the Carnot cycle.

In later forms of this cycle the steam from the boiler enters a *superheater* in which the temperature is raised at constant pressure before entering the turbine. This increases the *efficiency and reduces the harmful effects of steam condensing in the turbine.

Rankine's formula An approximate formula, derived from experiment, for the compressive load F that causes a column of length l and cross-sectional area A to collapse:
$$F = \sigma A / [1 + \sigma l^2 / \pi^2 k^2 E].$$
σ is the safe compressive stress, k is the least radius of gyration, and E is Young's modulus.

Rankine temperature Symbol: °R. An obsolete thermodynamic temperature scale linked to the Fahrenheit degree. Absolute zero on this scale is –459.67 °F, therefore °R = °F + 459.67. The ice point is 491.7 °R, often taken as 492 °R. 1K = 1.8 °R.

Raoult's law When a solute is added to a solvent to form a dilute solution, the fractional drop in vapour pressure $(p - p')/p$ is equal to $N_1(N_1 + N_2)$, where N_1 and N_2 are respectively the total numbers of solute and solvent molecules present (solute undissociated). This enables molecular weights in solution to be deduced either by the elevation of the boiling point or by direct measurement of vapour pressure.

rarefaction The converse of *compression.

raster scan A method of producing pictorial images, the picture being built up line by line. It is used in *television and in most computer-graphics displays.

ratemeter *Syn.* counting-rate meter. An instrument that gives a continuous indication of the rate of arrival of pulses, averaged over a time of the order of a minute, from a *Geiger counter or similar radiation detector. It is quicker and easier to use than a *scale but is less accurate.

rating The limitations of performance of a machine, transformer, or other piece of apparatus under certain conditions, as stipulated by the manufacturer. The conditions are known as the rated conditions. For example, the rating of a d.c. motor would be its output in horsepower or watts under stated conditions of voltage and speed.

ratio Of a *transformer. For a single-phase power transformer the ratio of transformation or, shortly, the ratio, is the ratio of the e.m.f.s induced in the primary and secondary windings. For example, if these e.m.f.s in a given transformer under normal working conditions were 11 000 volts and 2200 volts respectively, the ratio would be 5/1. (This transformer may also be described as an 11 000/2200 volt transformer to indicate the normal working voltages in addition to the ratio.) The following definitions are used in practice since they make allowance for special methods of connecting the windings (*see* polyphase transformer) and for special applications:

(1) *Voltage ratio of a power transformer*. The ratio of the voltage between terminals on the higher-voltage side to the voltage between terminals on the lower-voltage side at no-load. This ratio may differ from the turns ratio (*see* below), since it is a ratio of terminal voltages, which may differ from the ratio of the phase voltages as determined by the method of connection. A transformer has more than one voltage ratio if it is provided with tappings.

(2) *Turns ratio of a transformer* (general). The ratio of the number of turns in the phase winding associated with the higher-voltage side to the number of turns in the corresponding phase winding associated with the lower-voltage side. For a single-phase transformer, the voltage ratio is substantially equal to the turns ratio but this is not generally the case for a polyphase transformer. It is possible for the latter type to have a turns ratio that is less than unity while the voltage ratio is greater than unity.

(3) *Of an *instrument transformer*. (i) *Voltage transformer. The ratio of the primary terminal voltage to the secondary terminal voltage under specified load conditions. (ii) *Current transformer. The ratio of the primary current to the secondary current under specified load conditions.

ratio adjuster *See* tap changer.

ratio circuit *See* MOS logic circuits.

ratio detector A detector used for frequency-modulated carrier waves. *See* frequency discriminator.

ratio error Of an *instrument transformer. The amount by which the magnitude of the secondary quantity (voltage or current as the case may be) differs from the nominal value due to incorrect *ratio. It is taken as positive if the secondary quantity is greater than the nominal value and it is usually expressed as a percentage of the true (actual) value of the secondary quantity.

ratioless circuit *See* MOS logic circuit.

ratiometer An electrical instrument in which the indications are determined by the ratio of the currents in two circuits or by the ratio of the currents in the two branches of a circuit.

rational indices *See* rational intercepts.

rational intercepts, law of If the edges formed by the intersections of three faces on a crystal are chosen as axes of reference OX, OY, OZ and a fourth face intersects these axes in A, B and C, then any other face on the crystal will intercept the axes in A', B', C' such that $OA/OA' = h$, $OB/OB' = k$, $OC/OC' = l$ where h, k, and l are rational whole numbers rarely exceeding 6, and where (h, k, l) are said to be the *Miller* or *rational indices* of the face, relative to the axes OA, OB, OC.

rationalization of electric and magnetic quantities A technique that has been used to modify electrical and magnetic equations to provide a more rational "common sense" approach. The technique is best explained by reference to three examples. The mag-

nitude of the *magnetic flux density B at a point distance r from an infinitely long straight wire carrying a current I and situated in a vacuum can be derived from Ampere's theorem as:

$$B = 2\mu_k I/r,$$

where μ_k is the *permeability of free space. This constant has unit value in the CGS-electromagnetic system (*see* CGS system of units) by definition. In the system of *SI units, it can be shown that μ_k is equal to 10^{-7} SI units. In practice this is written as $\mu_k = 10^{-7} K$ and hence:

$$B = 2\mu_k I/Kr. \qquad (1)$$

A similar discussion to determine the magnetic flux density at the centre of a flat circular coil of radius r and N turns yields:

$$B = 2\pi\mu_k NI/Kr. \qquad (2)$$

A completely different kind of analysis gives the capacitance C of an isolated evacuated sphere of radius r as:

$$C = K\varepsilon_k r, \qquad (3)$$

where ε_k is the *permittivity of free space.

On further consideration, each of these three formulae has an irrational appearance. The magnetic field around a point on a straight wire is well known to be circular, and yet the quantity 2π that characterizes the concept of circularity does not occur in equation (1). The magnetic field at the centre of a flat circular coil is quite uniform and parallel, and yet equation (2) contains the 2π that was expected in equation (1). The electric field emanating from a sphere is, of course, three-dimensionally symmetrical yet no 4π appears in equation (3). Each of the equations can be thrown into a rational form by putting $K = 4\pi$. Thus

for a straight wire $\qquad B = \mu_k I/2\pi r,$
for a circular coil $\qquad B = \mu_k NI/2r,$
for a spherical capacitor $\qquad C = 4\pi\varepsilon_k r.$

0π implies linearity, 2π implies circular symmetry, and 4π implies spherical symmetry. The permeability and permittivity of free space have magnitudes changed by a factor of 4π, and are termed the *magnetic and *electric constants.

Many formulae are unaffected by the process of rationalization, particularly those (such as Ohm's law) that are purely concerned with electric currents. *See also* Heaviside–Lorentz units.

rationalized Planck constant *See* Planck constant.

ray A mathematical concept to give a first-order representation of the rectilinear propagation of light and basic to geometric optics theory. In an isotropic medium, the rays are normal to the wavefront; in *double refraction, the ordinary rays are normal to the wavefront, while it is only in special cases that the extraordinary rays are so. In general, rays are the shortest optical paths between wavefronts.

Ray tracing is the technical term applied to the calculation of the paths of rays through a system generally to a high degree of accuracy (six to eight significant figures) by the successive application of *Snell's law, etc., to paraxial and marginal rays. From the systematic tabulation of calculations, etc.,

Rayleigh, Lord

a picture of the path of a ray is obtained so that the designer may, by modifying radii or indices, etc., produce an instrument or lens free from harmful *aberration.

A narrow cone of rays is a *pencil of rays*, the *central ray* of the pencil is the *chief ray*. An incident pencil of rays on a system would be determined by the entrance pupil of the system.

Rayleigh, Lord (John William Strutt; 1842–1919) Brit. physicist. An extremely versatile mathematical and experimental physicist, whose researches ranged over hydrodynamics, sound, physical optics, magnetism, and electricity, particularly the establishment of a system of absolute electrical units. In addition, in collaboration with Sir William Ramsay, he discovered the element argon (1894).

Rayleigh, Lord (Robert John Strutt; 1875–1947) Brit. physicist. His researches covered a wide range of topics, in particular he pioneered *dating of minerals by studying their content of radioactive substances and their decay products.

Rayleigh criterion *See* resolving power.

Rayleigh current balance *See* current balance.

Rayleigh disc A device due to Lord Rayleigh based upon the principle that a light disc tends to set itself at right angles to the direction of an air stream whether the stream is alternating or direct. A small disc is suspended by a torsion thread so as to lie at an angle to the opening of a cylindrical resonator when it is unexcited.

The original disc was, in fact, a small galvanometer mirror with small magnets attached to give a restoring couple when it is displaced. This disc was suspended by a fibre in a magnetic field. The deflection could be measured by a beam of light reflected from the disc itself through the side of a glass resonator onto a scale outside. The disc usually used is made of mica 1 cm in radius, suspended by a quartz fibre and set at an angle of 45° to the axis of the tube when it is undeflected. To indicate the movement of the disc, a galvanometer mirror is attached to the fibre outside the resonator. The hole by which the fibre enters the latter must be small consistent with free motion of the fibre. When the resonator is excited, the alternating air flow round the disc causes it to rotate. For small angles of deflection, the rotation is proportional to the sound intensity in the resonator tube and consequently to the intensity in the undisturbed field. To increase sensitivity, the disc is suspended in the connecting neck of a double resonator (*see* Helmholtz resonator). It is important to note that in any form of Rayleigh disc, its diameter should be small compared with the wavelength of the incident sound. The device has been used extensively as a means of comparing sound intensities and has often been used to check other methods.

The formula obtained by König connecting the couple G and the mean speed of the stream is given as:

$$G = (4/3)\rho r^3 V^2 \sin 2\theta,$$

where ρ is the density of gas, V is the speed of the stream if steady or the root mean square if alternating, r is the radius of the disc and θ is the angle between the normal to the disc and the direction of the stream. For maximum sensitivity the disc must be set so that $\theta = 45°$.

Against the König theory of the Rayleigh disc, there are two main criticisms to be considered:

(a) The drag of the medium: the disc, apart from its rotation, will have a displacement depending on the density of the medium, the mass and dimensions of the disc. King and A. B. Wood have independently examined this correction. The above equation for the couple must be multiplied by the inertia factor $(1 - \beta^2)$, where β is the ratio of the velocity amplitude taken up by the disc to that of the medium itself, which excites it into vibration.

(b) Production of vortices: Merrington and Oatley have investigated this correction experimentally by introduced smoke into the sounding resonator. Vortices rather like those in *Aeolian tones were discovered at each edge of the disc rotating in opposite directions. The critical velocities of the departure from streamline flow could, however, not be observed since the vortices were present down to the lowest frequency of vibration that could be obtained.

Rayleigh–Jeans formula (1900) The formula for the energy distribution of *black-body radiation given by classical *statistical mechanics assuming *equipartition of energy. In the far infrared it approximated to *Planck's formula, and can be written

$$M_{e\lambda} = \frac{C_1}{C_2 \lambda^4} = \frac{2\pi ckT}{\lambda^4},$$

where the symbols are those given for Planck's formula. On integrating over all wavelengths the Rayleigh–Jeans formula gives an infinite denisty of radiation. This result is called the *ultraviolet catastrophe* (*see* quantum theory).

Rayleigh limit To prevent detectable deterioration in the quality of an image, the optical path differences should not exceed $\lambda/4$.

Rayleigh refractometer *See* refractive index measurement; interference of light.

Rayleigh scattering The scattering of light by particles of dimensions small compared with the wavelength of light. For plane-polarized incident light of wavelength λ, the scattered intensity bears to the incident intensity the ratio

$$\frac{I}{I_0} = \frac{\pi^2 \sin^2 \theta}{r^2} (\varepsilon_r - 1)^2 \frac{V^2}{\lambda^4},$$

where θ is the angle between the electric vector of the incident beam and the direction of viewing, r is the distance from the particle to the point at which observations are made, and the particle has volume V and is of material of dielectric constant ε_r relative to the surroundings. For unpolarized light, $\sin^2\theta$ is replaced in the formula by

$$\tfrac{1}{2}(1 + \cos^2\phi),$$

where ϕ is the angle between the incident light and the direction of observation.

The fourth power dependence on wavelength means that blue light is much more strongly scattered than red light from a medium containing very fine particles. This is called the *Tyndall effect*. It accounts for the bluish appearance of smoke and of clear sky when the observation is not along the direction of illumination. The setting sun, seen through a considerable thickness of atmosphere, appears reddish because the light has been robbed of much of the blue end of the spectrum.

Rayleigh's law In magnetic materials subjected to magnetic fields low in comparison with the maximum coercive force, the *hysteresis loss in a cycle varies directly as the cube of the magnetic flux density. The law ceases to be valid at the field value at which the *Barkhausen effect takes place.

ray surface The figure representing the distances travelled by rays of light coming from a point of origin within a crystal, in various directions, during a given interval of time.

reactance 1. (electrical) Symbol: X. If an alternating e.m.f. is applied to a circuit the total opposition to the flow of an alternating current is called the *impedance. The part of the impedance that is not due to pure *resistance is called the reactance and is caused by the presence of *capacitance or *inductance. If the alternating e.m.f. is given by:

$$E = E_0 \cos 2\pi f t = E_0 \cos \omega t,$$

(ω = angular frequency), then the peak value of the current in a series circuit with resistance R and inductance L is I_0, where:

$$I_0 = E_0 / \sqrt{[R^2 + (\omega L^2]}.$$

$\sqrt{[R^2 + (\omega L)^2]}$ is the impedance and ωL is the reactance, in this case the *inductive reactance*. Similarly for a series circuit with resistance R and capacitance C,

$$I_0 = E_0 / \sqrt{[R^2 + (1/\omega^2 C^2)]}.$$

Here $1/\omega C$ is the *capacitive reactance*.

The reactance is the imaginary part of the complex impedance Z, i.e.

$$Z = R + iX.$$

It is measured in ohms.

2. (acoustic) Symbol: X_a. The magnitude of the imaginary part of the acoustic *impedance Z_a, i.e.

$$Z_a = R_a + iX_a,$$

where R_a is the acoustic resistance. It is measured in pascal second per metre cubed. The *specific acoustic reactance* (Symbol: X_s) is the imaginary part of the specific acoustic impedance and is measured in pascal second per metre. If the reactance is caused solely by inertia, it is called the *acoustic mass reactance*. If it is due to stiffness, it is called the *acoustic stiffness reactance*. The product of acoustic mass reactance and angular frequency is called the *acoustic mass* (Symbol: m_a). The product of acoustic stiffness reactance and angular frequency is called

the *acoustic stiffness* (Symbol: S_a). For an enclosure of volume V with dimensions that are small compared with the wavelength of sound, the acoustic stiffness is given by $\rho c^2 / V$, where ρ is the density and c the speed of sound in the medium.

3. (mechanical) Symbol: X_m. The magnitude of the imaginary part of the mechanical *impedance Z_m, i.e. $Z_m = R_m + iX_m$ where R_m is the mechanical resistance. It is measured in newton second per metre. If the reactance is caused by inertia, it is termed the *mechanical mass reactance*. If it is due to stiffness, it is called the *mechanical stiffness reactance*. The product of mechanical stiffness reactance and angular frequency is called the *mechanical stiffness* (Symbol: s).

reactance coil *See* inductor.

reactance drop (and **rise**) *See* voltage drop.

reactance transformer A device consisting of pure reactances arranged in a suitable circuit. It is commonly employed for matching impedances at radio frequencies.

reaction 1. An interaction between atoms or molecules (chemical reaction) or between nuclides (*nuclear reaction).

2. *See* Newton's laws of motion.

3. A condition that arises in electronic valve circuits when part of the amplified current is fed back positively to the *control grid. It can give additional *gain or can cause oscillation.

reaction alternating-current generator An alternating-current generator having a rotor with *salient poles not fitted with *field coils. It can generate only when connected in parallel with one or more *synchronous alternating-current generators since its a.c. exciting current must be supplied by an independent a.c. source. The rotor is driven at *synchronous speed and the machine is a form of synchronous alternating-current generator.

reactive current *Syn.* reactive component; wattless component; idle component; quadrature component, of the current. The component of an alternating current that is in quadrature with the voltage, the current and voltage being regarded as vector quantities. *Compare* active current.

reactive factor The ratio of the *reactive volt-amperes to the total *volt-amperes.

reactive load The *load in which the current and the voltage at the terminals are out of phase with each other. *Compare* nonreactive load.

reactive power *See* power.

reactive voltage *Syn.* reactive component; wattless component; idle component; quadrature component of the voltage. The component of an alternating voltage that is in quadrature with the current, the voltage and current being regarded as vector quantities. *Compare* active voltage.

reactive volt-amperes *Syn.* reactive component; wattless component; idle component; quadrature component of the volt-amperes. The product of the current and the *reactive voltage, or the product of the voltage and the *reactive current. *Compare* active volt-amperes.

reactivity An indication of the departure of a *nuclear reactor from the condition in which the reaction can just take place (the *critical* condition). The reactivity is defined by the expression $(1 - 1/k)$, where k is the ratio of the number of neutrons produced in a generation to the total number absorbed or lost; it is called the effective *multiplication constant. $k > 1$ implies a supercritical condition for the reactor. $k < 1$ implies a subcritical condition. Reactivity may be measured in *niles where 100 niles is the reactivity corresponding to $k = 1$.

reactor 1. An electrical device possessing *reactance and selected for use because of that property. Special types are described as: *current-limiting, *earthing, *preventive, *screening reactors. (*See also* choke.)
2. *See* nuclear reactor.

read Of a computer *storage device or *memory. The operation of retrieving information from a storage location. The read operation may be either destructive (DRO–destructive read operation) or nondestructive (NDRO), depending on the nature of the memory. If the read operation is destructive, it must be immediately followed by a *write operation, to restore the information if it is to be preserved.

reading glass A large aperture converging lens held at a distance from the eye before reading matter to magnify print, etc. (a form of simple microscope).

read-only memory *See* ROM.

read time Of a storage component or memory. The time taken to extract information from a storage location.

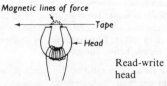

Magnetic lines of force
Tape
Head
Read-write head

read-write head An electromagnet used to retrieve or record data on a magnetic medium, such as magnetic tape or computer *disks. (*See* diagram.) A very small gap between the poles of the magnet gives high resolution since the greatest magnetic field only cuts a small area of the magnetic medium.

real crystal A crystal structure and texture considered as a whole.

real gas An actual gas whose molecules do not necessarily possess the properties assigned to those of an *ideal gas. In particular its molecules have a

nonzero size and exert forces on each other. *See* equations of state.

real object (and **image**) A natural *object that delivers pencils of diverging rays to an optical system. A real *image is one that could be received sharply on a screen – where the rays converge to form the image. Using the "real is positive" convention of signs, the conjugate focus relations for lenses and mirrors are the same:
$$1/v + 1/u = 1/f.$$

real-time processing A method of *computer operation in which the computer is *on line to a source of data, and processes the data as they are generated. The computer must be able to process the data at least as fast as they are generated.

Réaumur, René Antoine Ferchault, Seigneur de (1638–1757) French physicist and naturalist. He constructed thermometers using alcohol-water mixtures, observing the expansion between the freezing and boiling points of water to be from 1000 to 1080 units of volume for his favourite mixture. He therefore used a degree equal to 1/80 of the interval between these points.

Réaumur scale A temperature scale in which the ice point is taken as 0° R and the steam point as 80° R.

recalescence The sudden evolution of heat by a metal due to an exothermic structural change. *Compare* decalescence.

receiver The part of a communication system that converts the transmitted waves into perceptible signals of the desired form. *See* radio receiver; television; superheterodyne receiver.

reciprocal lattice A theoretical lattice associated with a crystal lattice. If a, b, and c are the sides of a *unit cell of the real lattice then a', b', and c' define the reciprocal lattice, where
$$a' = \frac{b \times c}{a \cdot (b \times c)}, \quad b' = \frac{c \times a}{a \cdot (b \times c)},$$
and
$$c' = \frac{a \times b}{a \cdot (b \times c)}.$$
Reciprocal lattices are used in crystallography and in theories of the solid state.

reciprocal theorem If a force F applied to one point in an elastic system produces a deflection d at another point, the same force F applied at the second point in the direction of the original deflection produces a deflection d at the first point in the direction in which F was first applied. The theorem is due to Clerk Maxwell.

reciprocal wavelength *See* wavenumber.

reciprocity failure *See* reciprocity law.

reciprocity law 1. (photography; Bunsen and Roscoe, 1857) The density of processed photographic material is a function of exposure (= *illuminance × time) only, for a standard procedure of processing. (More accurately, density depends on It^p where p varies from 0.8 to 1.1 according to the emulsion.) When the illuminance (or exposure time) varies greatly from that to which a given emulsion is most sensitive, the reciprocity law no longer holds (*reciprocity failure*). Exposures at very low or very high illuminences produce a less dense image than an exposure at an illuminance corresponding to the sensitivity.

2. (electrostatics; Gauss) If two alternative sets of charges Q_1, Q'_1 on a set of conductors produce potentials V_1, V'_1 respectively, on those conductors, then:
$$\Sigma Q_1 V_1 = \Sigma Q'_1 V_1$$
(summations over all conductors).

reciprocity relations (Onsager, 1931.) If two flows (of heat, electricity, matter, etc.) J_1, J_2, produced by gradients or forces X_1, X_2, so interact that:
$$J_1 = L_{11}X_1 + L_{12}X_2$$
$$J_2 = L_{21}X_1 + L_{22}X_2,$$
then, subject to a condition restricting magnitudes, $L_{12} = L_{21}$; and similarly for three or more interacting flows.

These relations can be used, for example, to establish relations between thermoelectric coefficients; to justify the use of a single mutual inductance between two circuits and to establish the symmetry of the *tensors applicable to flow of heat and of electricity in anisotropic crystals.

recombination rate The rate at which electrons and holes in a *semiconductor recombine, tending to restore the system to thermal equilibrium. An electron in the conduction band may recombine with a hole in the valence band. Alternatively, there may be electron capture or hole capture by a suitable acceptor or donor impurity in the semiconductor.

recording instrument *See* graphic instrument.

recording of sound The mechanical and/or electrical process of producing a permanent or semipermanent record of sounds, which may be realized as and when required by using the record in a replay or reproducing apparatus (*see* reproduction of sound). Four main methods of sound recording are used: (1) electromechanical sound recorders (*see* gramophone); (2) sound on film (*see* soundtrack); (3) *magnetic recording; (4) digital recording (*see also* digital audio tape; compact disc). Each method has its advantages, disadvantages, and applications, and some of these are mentioned under the appropriate headings. Records for gramophone reproduction are mainly used in the commercial provision of recording sound for home and broadcast entertainment. Many copies of an original recording may be made with comparative ease. This recording method is also much used for the recording of broadcast performances and in the initial stages of recording on sound film where normally speech, music, and sound effects are recorded separately and then combined at suitable amplitudes into one sound track. Sound on film is used almost entirely in the cinema world. With a much expanded time scale it is often used for the analysis of sounds. The method permits of records being made of much longer duration than those made on disc.

Magnetic recording provides a recording and replay system that is as good as, and possibly better than, the other two systems. It is used in place of gramophones, in broadcasting, in offices, and in law courts.

The requirements for good sound recording and reproduction might be summarized as follows: (*a*) The ability to produce sounds at one place that are a faithful copy of the original. The acoustics of the reproducing room generally militate against this (*see also* stereophonic reproduction; distortion). (*b*) The ability to reproduce the sounds at sufficient volume level to give a sound field at the hearer's position approximating in amplitude to that which would hold at a suitable position in the presence of the original sound. These requirements are rarely satisfied in practice.

recovery The disappearance (possible slowly) of deformation on removal of the deforming stress.

recovery voltage The voltage applied to a circuit that is opened under fault (e.g. short-circuit) conditions by a circuit breaker, switch, or fuse. It is the d.c. voltage or, in the case of an a.c. circuit, the *root-mean-square value of the normal frequency voltage that appears across the terminals of the circuit breaker, etc., after final arc extinction. It is less than the normal circuit voltage mainly owing to the effects of *armature reaction in the generator(s) supplying the circuit. After final arc extinction the voltage recovers and ultimately reaches its normal value. *Compare* restriking voltage.

rectifier 1. An electrical device that permits current to flow in only one direction and can thus make alternating into direct current. It operates either by suppressing or attenuating alternate half-cycles of the current waveform or by reversing them. The most common rectifiers are semiconductor *diodes. Other types include the *mercury-vapour rectifier and such mechanical devices as the *commutator.

2. A device for obtaining a pure sample of a substance such as an element from, say, liquid air, and depending for its action on the difference in the boiling points of constituents of liquid air. (*See* fractional distillation; Claude's process of rectification.)

rectifier instrument A d.c. instrument that can be made suitable for a.c. measurements by using a *rectifier to convert the alternating current to a unidirectional current. A common arrangement consists of four diodes and a moving-coil instrument, M, connected to form a bridge circuit (*see* diagram). Instruments of this type are usually cali-

rectifier photocell

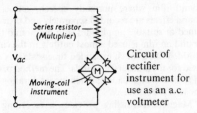

Series resistor (Multiplier)

V_{ac}

Moving-coil instrument

Circuit of rectifier instrument for use as an a.c. voltmeter

brated to read root-mean-square values on a.c. supplies of sinusoidal waveform, and when they are used with any other type of waveform, the readings are incorrect (i.e. rectifier instruments are subject to waveform error).

rectifier photocell *See* photovoltaic cell.

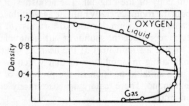

Density/temperature curve for oxygen (liquid and gas)

rectilinear diameters The law of rectilinear diameters, discovered by Cailletet and Mathias, states that since the equilibrium curves of the densities of liquid and vapour against temperature are roughly parabolic, meeting at the *critical point, the curve showing the variation with temperature of the mean of the liquid and vapour densities is a straight line passing through the critical point (*see* diagram). The law enables values for densities at the critical point to be extrapolated from determinations made at temperatures near to the critical point.

rectilinear propagation The progress of light in straight lines in an isotropic medium (to a sufficiently high degree of accuracy for everyday practical purposes); the *ray is the geometrical representation of it. On account of the wave character of light, it is only approximately true (*see* diffraction), so that the elementary theory of the pinhole camera, shadows, etc., generally given under rectilinear propagation is only superficially true. According to the general theory of *relativity, light rays travelling through free space are deflected towards any massive bodies they may pass.

rectilinear scanner *See* scanner.

recurrent-surge oscilloscope An instrument used for testing and for research in connection with electrical surges. It usually consists of a *surge generator and a *cathode-ray oscilloscope with other ancillary apparatus. The surge phenomenon being investigated is repeated at a rate such that a steady picture, suitable for visual or photographic purposes, is obtained in the oscilloscope.

red giant A type of cool *giant star emitting light in the red region of the spectrum. A normal star expands to a red giant as it exhausts its nuclear fuel. *See* stellar spectra; Hertzsprung–Russell diagram.

redshift A shift in the spectral lines of an astronomical body towards longer wavelength values relative to the wavelengths of these lines in the terrestrial spectrum; optical lines are shifted towards the red end of the visible spectrum. The *redshift parameter*, z, is given by the ratio

$$(\lambda' - \lambda)/\lambda',$$

where λ and λ' are the wavelengths of a spectral line from a terrestrial and astronomical source respectively.

Redshifts of objects within our Galaxy are due to the Doppler effect, arising from the movement of a star or other source away from the observer. The value of z is then v/c, where v is the velocity of recession and c the speed of light. Redshifts of extragalactic objects (e.g. other galaxies and *quasars), are also interpreted in terms of the Doppler effect, which for these objects arises from the expansion of the universe. Since the recessional velocities for these objects can be very great, the relativistic expression for redshift must be used:

$$z = [(c + v)/(c - v)]^{1/2} - 1.$$

See also Hubble constant.

reduced distance A distance in a medium divided by the refractive index of the medium. It may be regarded as an air-equivalent distance, and conjugate focus relations of refraction from one medium to another, using reduced distances, are the same as for a thin lens in air. The reciprocal of a reduced distance is *reduced vergence*. If l = the object distance in a medium of refractive index n, l/n is the reduced distance and $L = n/l$ is the reduced vergence. If F is the reduced power,

$$F = n/f = n'/f';$$

$L + F = L'$ is the conjugate focus relation using reduced vergences.

reduced equation of state An *equation of state in which the variables, p, V, and T are expressed as fractions of the *critical pressure, *critical volume, and *critical temperature respectively, these fractions being known as the *reduced pressure* (α), *volume* (β), and *temperature* (γ). The reduced form of the *van der Waals equation is:

$$[\alpha + (3/\beta^2)](3\beta - 1) = 8\gamma.$$

reduced mass Let a small particle of mass m be attracted by, and describe a closed orbit about, a heavier one of mass M. Then the equations describing the motion, assuming M to be fixed, may be transformed into those holding if M is not fixed (in which case both particles revolve around their common centre of gravity) by replacing m by μ, the reduced mass, where

$$1/\mu = 1/m + 1/M.$$

reduced pressure, temperature, volume *See* reduced equation of state.

reduced vergence *See* reduced distance.

reduction factor *See* tangent galvanometer.

redundancy 1. In a transmission system. The existence of information in excess of that required for the essential information to be transmitted. This is often deliberately included to allow for loss in the transmission system.

2. In electronic circuits or systems. The inclusion of extra components or circuits to increase the reliability of the system, i.e. if a fault should develop, the function of that part of the circuit may be taken over by the redundant circuits or components provided.

reed A thin bar of metal or cane clamped at one end and set into transverse vibration generally by a flow of air. In a musical box, the reeds are excited by plucking. Reeds may be considered as a special case of the vibrating bar. The vibrating portion of the reed is known as the tongue. The frequency of the note emitted depends on the material of the reed and its dimensions. In certain applications the tone quality of the reed is modified by coupling it to a resonator. Reeds are of two varieties: the beating reed, where the reed vibrates against an air slot completely stopping the flow of air at certain parts of its vibration, and the free reed, where the reed vibrates through a slot of dimensions slightly larger than the reed tongue.

Reeds may also be single or double; the latter consists of two reeds together with a slight orifice between them through which the wind passes. The reed tongue vibrations are almost entirely simple harmonic. The wind passing through the reed, however, is interpreted in a far from simple harmonic manner.

In organ reed pipes a beating reed is coupled to a pipe resonator, generally so that the free resonance frequencies of the pipe and the reed are very nearly the same. For optimum operation the wind container in which the reed is held (the boot), should also be correctly proportioned. By suitable variation of the coupling, material, and dimensions of the pipe resonator, reed tongue, and air opening, different sound qualities are produced.

The reed is the generating source in several orchestral wind instruments. In most of these the note generated by the wind passing through the reed is dependent to a great extent on the air-column resonance. In the clarinet a single beating reed vibrating against a slot in the mouthpiece is used. The player can control the length of the vibrating reed and the wind pressure to produce approximately a free tone of the same frequency as the resonating column. In the oboe a double reed is used–the reeds beating against each other. In the bagpipes, double reeds are used for the chanter and a single reed for each of the drones. In the mouth organ, concertina, accordion, harmonium, and reed organ, a separate reed is used for each note, sometimes with a small resonator.

In brass instruments of the orchestra the mouthpiece consists of a cup or cone-shaped orifice that fits over the player's lips, the latter forming a twin reed. For high frequencies the player must compress his lips more tightly and increase the blowing pressure. The vocal chords of the larynx are often considered to be a pair of free reeds, the pitch produced being altered by tension and wind pressure. However, the range of tension and thickness available seem too small for the vibrations covering two octaves over which a normal voice may operate. It is suggested that the sound is rather produced by a sort of *jet tone in which the sides of the vocal chords also take part. This may be the explanation of sound production by the players' lips in brass instruments.

reed relay *See* relay.

reference tone An accepted standard pure tone of known intensity and frequency. Some such tone is necessary as the basis of any scale of sound intensity or loudness. For the measurement of sensation units in *decibels, the reference tone used is a pure tone of normal threshold intensity and of the same frequency as the sound being measured. For the *phon, the unit of equivalent loudness, the reference tone is more carefully specified. It must be a plane sinusoidal wave train having a frequency of 1000 hertz. The intensity is that corresponding to a *root-mean-square sound pressure of 2×10^{-5} pascal and it must be measured in the free progressive wave. Great importance should be attached to this last condition, since the pressure exerted by the sound on any measuring instrument such as a *Rayleigh disc may vary from that of the undisturbed sound wave to twice this amount depending on whether the surface of the instrument is small or large compared with the wavelength of the sound. (*See* free-field calibration.)

reflectance Symbol: ρ. The ratio of the *radiant or *luminous flux reflected by a body to the incident flux. The term may be qualified by the adjectives *specular*, *diffuse*, and *total* according to the nature of the reflecting surface. *See also* reflectivity.

reflected current *See* reflection coefficient.

reflected wave *See* travelling wave.

reflecting microscopes *See* microscope.

reflecting telescope *Syn.* reflector. An optical (or infrared) *telescope with a large-aperture concave mirror, usually paraboloid, for gathering and focusing light from astronomical bodies. There is no chromatic *aberration and very little spherical aberration and coma, all of which occur in the *refracting telescope. The mirror is supported on its back surface, and has either an equatorial mounting, with one axis parallel to the earth's axis, or an altazimuth mounting, with one axis vertical (*see* astronomical telescope).

There are several types of reflecting telescopes that differ from each other by the additional optical system used to bring the image to a convenient

reflection

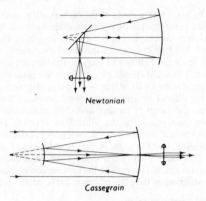

Newtonian

Cassegrain

Reflecting telescopes

point, where it can be viewed through an eyepiece or recorded photographically or electronically. The earliest reflecting telescope, the *Newtonian telescope* was built by Newton to overcome the apparent impossibility in his day of producing achromatic lenses for refracting telescopes. A small prism or flat mirror deviates the light from the concave mirror to the side of the telescope where it is viewed or recorded. In the *Cassegrain telescope* a small hyperboloid convex mirror has one focus coincident with that of the coaxial concave mirror. The other focus lies at the pole of the concave mirror. Light passing through a hole in the mirror at this point is viewed or recorded. The *Gregorian telescope* uses an additional small concave mirror to converge the light through a hole at the pole of the main mirror onto an eyepiece. In the *Herschelian telescope* the light is reflected from the concave mirror to a point off the axis. The concave mirror can be used without an additional optical system, a photographic plate being placed at the focus of the mirror. These four types are widely used in observatories as they have high-quality definition at the centre of the field and are very stable. The 200″ reflector at Mount Palomar Observatory is a Cassegrainian design.

The intensity of the image depends mainly on the amount of light collected (i.e. on the area of the mirror) and the time for which the image is viewed or recorded. In an observatory the high stability of the telescope structure, and the great precision with which it can track an astronomical object in its diurnal motion, permits long exposures of photographic emulsions and of highly sensitive photoelectric devices. As a result, very faint objects can now be detected. A spectrograph can be used in conjunction with the telescope to obtain *stellar spectra.

The *Schmidt telescope uses a combination of a specially curved lens (the *correcting plate) and a spherical concave mirror. In this way spherical aberration and coma are reduced to a minimum and a large field of view is possible. The *Maksutov telescope works on a similar principle.

reflection 1. (Of light) The process occurring when light strikes a surface of separation of two different

media such that some is thrown back into the original medium. If the surface is smooth, reflection is *regular*, otherwise it is *diffuse* and the light is scattered. The two *laws of reflection*, namely that incident ray, normal, and reflected ray lie in the same plane, and the angle of incidence (with the normal) is equal to the *angle of reflection* (with the normal), suffice to determine the position and attributes of the image, whether at plane, curved, or multiple mirrors, etc. *Total internal reflection* takes place when the incident ray in a medium strikes a surface of a less dense medium (lower refractive index) at an angle greater than the *critical angle. Reflection can be regarded mathematically as a special case of *refraction ($n' = -n$). *Selective reflection* is said to occur when certain wavelengths are reflected more strongly than others. When reflection occurs at a denser medium, there is a change of phase of π; when it occurs at a less dense medium, there is no change of phase.

Equations giving the fractions of light reflected and transmitted for different planes of polarization when light is incident on the interface between two refracting media at various angles of incidence were first obtained by *Fresnel. In the simple case of normal incidence with relative refractive index n the fraction reflected (independently of polarization) is equal to

$$(n - 1)^2/(n + 1)^2.$$

See also polarizing angle.

Other forms of electromagnetic radiation, such as infrared and radio waves, also undergo reflection. In the case of X-rays, reflection can only occur at *grazing incidence* i.e. for very small *glancing angles, up to a critical angle.

2. (Of sound) A similar process occurring for a sound wave, the geometrical laws of reflection of sound waves being the same as those of light waves. The apparent differences between light and sound in reflection are merely questions of scale. The typical wavelength of sound being 100 000 times the wavelength of light it requires the dimensions of the reflecting surface to be 100 000 times the corresponding one in light to produce diffuse reflection or scattering. A mirror or lens to produce concentration of sound must be enormous compared with mirrors and lenses used in optical work. The same remark applies to gratings in sound diffraction (*see* acoustic grating). Plane waves when reflected may produce *standing waves. If the wavelength of the sound is small compared with the dimensions of the reflector, the ordinary geometrical laws of optics are applicable. The reflection of sound waves on a large scale is familiar in the echoes from a cliff, the edge of a wood, or the walls of a house. In large buildings and halls, echo is one of the main defects in acoustical engineering (*see* acoustics of buildings).

Echoes from atmospheric discontinuities are common, such as those observed from clouds. The surface of water forms a good reflector for sound waves in air, only about one part in a thousand being transmitted at normal incidence. This is due to the fact that the characteristic impedances or resistances are widely different from each other (*see* specific acoustic impedance). For similar reasons *echo-sounding methods are very useful for measurement of depth of sea as well as the detection of wrecked vessels and submarines.

reflection coefficient 1. *Syn.* return-current coefficient. In telecommunication engineering a uniform *transmission line is said to be correctly terminated (or matched) when the terminating impedance is equal to the *characteristic impedance of the line (symbol Z_0). If the terminating impedance (Z_R) differs from Z_0, the actual current in the line at the termination under steady-state conditions may be regarded as the *vector sum of two currents, one being the current that would flow if Z_R is made equal to Z_0 (called the incident or initial current), the other being a current that is reflected from Z_R (called the return current or reflected current). The vector ratio of the return current to the incident current is called the reflection coefficient. In terms of the impedances Z_0 and Z_R, the reflection coefficient is:

$$(Z_0 - Z_R)/(Z_0 + Z_R).$$

In this expression, the sign convention is such that the actual current in the line at the termination is the vector sum of the incident and return currents. *Compare* reflection factor.

2. *See* acoustic absorption coefficient.

reflection density Symbol: D. The logarithm to base ten of the reciprocal of the *reflectance, i.e. $D = -\log_{10}\rho$.

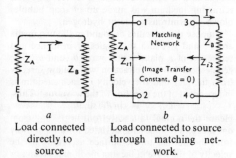

a
Load connected directly to source

b
Load connected to source through matching network.

reflection factor 1. *See* reflectance.
2. In telecommunications. Fig. *a* shows a source of e.m.f. E and internal impedance Z_A supplying a load impedance Z_B, the current in Z_B being I. In Fig. *b*, a network having image impedances Z_{i1} and Z_{i2} equal respectively to Z_A and Z_B at the source and load ends, has been inserted between the source and the load for the purpose of matching Z_B to Z_A and the *image-transfer constant of this network is assumed to be zero. [*Note.* Such a network may not be physically realizable.] The current in the load under these conditions is I'. The reflection factor of Z_A and Z_B is the vector ratio of I and I' and, in terms of Z_A and Z_B, is given by

$$\text{reflection factor} = \frac{I'}{I} = \frac{\sqrt{4Z_A Z_B}}{Z_A + Z_B}.$$

It is therefore the vector ratio of the current that actually flows in the load impedance to that which would flow if the load impedance were matched to that of the source by a suitable network having zero image-transfer constant. The ratio of the powers in Z_B for the two cases illustrated, expressed in decibels, is the *reflection loss* (or *reflection gain*, if negative) between Z_A and Z_B. Thus:

$$\text{reflection loss} = 10\log_{10}(I'/I).$$

Compare reflection coefficient.

reflection loss (and **gain**) *See* reflection factor.

reflectivity Symbol: ρ_∞. The *reflectance of a layer of material sufficiently thick that no change of reflectance would occur with increase in thickness.

reflector 1. *See* mirror.
2. *See* reflecting telescope.
3. A layer of material surrounding the *core of a *nuclear reactor, whose purpose is to scatter some of the escaping *neutrons back into the core.
4. *See* directive aerial.

reflex klystron *See* klystron.

refracting angle (and **edge**) The angle formed by the two refracting surfaces of a prism in the principal section, i.e. a section perpendicular to the *refracting edge*, which is the edge formed by the intersection of the two refracting surfaces. This region is also called the apex of the prism.

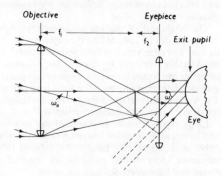

a   Kepler telescope

refracting telescope *Syn.* refractor. An optical telescope consisting essentially of two lens systems. The *objective is a convex lens of long focal length, f_1; the *eyepiece is a lens of short focal length, f_2.

In the *astronomical* or *Kepler telescope* (Fig. *a*), the eyepiece is convex and in normal adjustment the lenses are separated by the sum of the focal lengths producing a real but inverted image at infinity. The eye is therefore unaccommodated (fully relaxed) when viewing the image. The objective lens constitutes both the *aperture stop and the *entrance pupil. The *exit pupil is real and for maximum illumination the eye's entrance pupil should coincide with and have the same diameter as the exit pupil. When the image is formed at infinity the magnification is the ratio (f_1/f_2) of the focal length of the lenses. This is equivalent to the ratios of the angles ω/ω_0.

Since the image is inverted, this telescope is only suitable for work where there is not a problem such as in astronomy or in making physical measurements.

A *Huygens, *Ramsden, or *Kellner eyepiece can then be used. The *Fraunhofer eyepiece has an additional lenticular erecting system so that the instrument is suitable for terrestrial purposes.

*Prismatic binoculars and *rangefinders use two Kepler telescopes in conjunction with a prism system for making the image erect.

The *Galilean telescope* (Fig. *b*) consists of a concave eyepiece separated from the convex objective by a distance equal to the difference $(f_1 - f_2)$ in focal lengths. This produces an erect image, but the *exit port is virtual and lies inside the instrument. As the eye cannot be placed here, its best position is as near the eyepiece as possible thus restricting the field of view. The eye pupil acts as the exit pupil, different parts of the objective and eyepiece being viewed as the eye rotates. Under faint illumination, the pupil of the eye expands and the resulting increased exit pupil renders the telescope more efficient at night than the Kepler telescope and is therefore employed in *night glasses*. The magnification is the ratio (f_1/f_2) of the focal lengths and is rarely greater than six. As the length of the instrument is much shorter than Kepler telescopes adjusted for terrestrial use, Galilean telescopes of low magnification are used in *opera glasses*.

The brightness of the image depends on how much light can be collected by the objective, i.e. on its area. The observation of faint objects is therefore considerably improved by using a large-diameter objective and focusing the real image onto a photographic plate, exposing the plate for a long period. The Yerkes Observatory in the US has a refracting telescope with a 40″ diameter objective.

Large objective lenses are very difficult to grind and to mount. In addition, chromatic and spherical *aberration, coma, etc., have to be reduced to a minimum. Most of these problems are removed in *reflecting telescopes. *See also* telescope.

refraction 1. (Of light) The change of direction that a ray undergoes when it enters another transparent medium. The *laws of refraction* are:
(1) the incident ray, normal, and refracted ray all lie in the same plane;
(2) *Snell's law, $\sin i/\sin i' = n$ (a constant; *see* refractive index).
According to the wave theory, the direction of the wavefront is altered because of the change of velocity. The action of prisms, lenses, etc., is explained by refraction and the continued application of the laws.

2. (Of sound) The change in direction in sound waves on reaching the boundary between two media can be easily realized by applying *Huygens' princi-

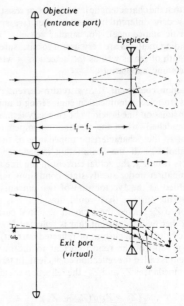

b Galilean telescope

ple as is usually done in optics. Early experiments were made on gases by Sondhauss who gave a qualitative explanation of the phenomena. The gases constituting the lenses and prisms were contained in thin membranes, which must vibrate as a whole; this might be a source of criticism in the method. Tyndall made similar experiments using a reed as a source and a sensitive flame as a receiver. The refracting medium was made up of soap bubbles blown with nitrous oxide or hydrogen.

Suppose the velocities of sound in the two media are C_1 and C_2; and the angles of incidence and refraction of a plane sound wave θ_1 and θ_2. The geometrical law of refraction will follow directly from the fact that the velocity in each medium is independent of the direction of the wavefront. Thus,
$$C_1/C_2 = \sin\theta_1/\sin\theta_2.$$
Hence the normal to a wavefront in passing from one medium to another is deviated towards or away from the normal to the surface according as the velocity of the wave in the first medium is greater or less than in the second medium. The critical angle, when $\theta_2 = 90°$, is given by:
$$\sin\theta_1 = C_1/C_2,$$
and total reflection may take place when the sound travels in a medium of low towards one of higher wave velocity.

Refraction of sound may take place whenever the wave reaches a point at which the wave velocity changes. This occurs not only by a complete change of medium, but also by the gradual change of the properties of the same medium, for example by wind or temperature gradients. Such temperature or wind refraction in the atmosphere is analogous to the optical phenomena of mirage and has a very important influence on the range of transmission of sound in the atmosphere. On the basis of such wind or temperature gradients, the silent zone observed in large or small areas may be explained.

3. (Of lines of force) Electrical lines of force are refracted on passing at an angle from one dielectric medium to another. The tangents of the angles of incidence and refraction are in the ratio of the relative *permittivities.

refraction, atmospheric Owing to refraction by the atmosphere, astronomical objects as observed are displaced towards the *zenith. If z is the true zenith distance and z_0 is the observed zenith distance,

$$\sin z = n \sin z_0$$

n is the refractive index of the air at the earth's surface. For small zenith distances this formula reduces to

$$z - z_0 = (n - 1) \tan z$$

The value of n depends on wavelength λ. With λ in μm,

$$n_\lambda = 1.000\ 287\ 6 + 1.629 \times 10^{-6}\ \lambda^{-2}$$
$$+ 1.36 \times 10^{-8}\ \lambda^{-4}$$

at $0\ °C$ and standard atmospheric pressure. n also varies with pressure, temperature, and humidity.

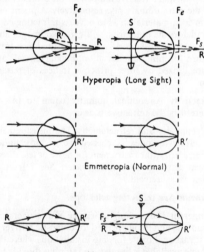

Refraction defects of the eye

refraction of the eye The refractive defects of the eye or the measurement of these defects. *See* illustration. *See also* astigmation of the eye.

refractionometer A direct reading instrument for measuring refractive defects of the eye.

refractive index Symbol: n. The ratio of the sine of the angle of incidence to the sine of the angle of refraction. If the first medium is a vacuum, the value is the *absolute refractive index*. The absolute refractive index is thus the ratio of the speed of light in a vacuum to the *phase speed of light in the medium, i.e. $n = c_0/c$. The value of the ratio for two media is

the *relative refractive index*. If n_{12} is the relative index from medium 1 to medium 2, and n_{23} from medium 2 to medium 3, etc., then:

$$n_{12} \times n_{21} = 1; \text{ i.e. } n_{12} = 1/n_{21}$$
$$n_{12} \times n_{23} \times n_{31} = 1; \text{ i.e. } n_{23} = n_{13}/n_{12}.$$

In general,

$$n_{12} \times n_{23} \times n_{34} \times \ldots n_{k1} = 1.$$

The relative refractive index is the ratio of the phase speeds of light in the two media, i.e. $n_{12} = c_1/c_2$, where c represents phase speeds. Commonly, the term refractive index refers to the value for sodium yellow ($\lambda = 589.3$ nm) relative to air whose absolute index is 1.000 29. *Dispersion of light arises on account of differences of refractive index for different colours, i.e. for different wavelengths. Refractive index is measured with a prism and spectrometer or by refractometer, etc. *See* refractive-index measurement.

refractive-index measurement 1. *General.* The measurement of refractive index is important in the optical industry to check the suitability of materials intended for components of lenses, etc.; it can be used for identification (gemstones, mineralogy) and as a method of analysis (e.g. to control a component in a gaseous mixture).

2. *Prism methods.* High accuracy can be attained with a solid that can be used as a prism with at least two polished faces using a *spectrometer. If A is the angle between the faces and D is the angle of minimum deviation, the refractive index $n = \sin \frac{1}{2}(A + D)/\sin \frac{1}{2}A$. A hollow glass prism is used when a liquid is under examination. The best spectrometers give angles to one second of arc and n to the fifth or even sixth place of decimals, but this is only physically significant if the temperature is closely controlled.

Even a gas may be used, with a prism of wide angle (e.g. 160°) but a micrometer device is necessary to record the telescope movement when air is replaced by the sample of gas.

3. *Total reflection.* Several methods rely on finding the critical angle of incidence on an interface between two media, when reflection becomes total. If one of the media is air, the ordinary refractive index of the other medium is then known ($n = \text{cosec } i_c$, where i_c is the critical angle of incidence); otherwise a relative index is found. *Wollaston's method* has a glass block that slides along a matt black surface, with diffuse illumination on the far side. A simple sighting device, using a slit and a mark on the block, enables the light/dark division of the field to be located and a simple calculation gives the refractive index. It is mainly used to give a rough value for a liquid placed under the base of the prism. Better values are obtainable by adapting the method to a prism on a spectrometer table; a liquid film is retained by a cover glass held by capillarity to a prism face.

The *Pulfrich refractometer* uses a glass block of high refractive index μ_g, with its top face horizontal and polished with the specimen on this face (as a drop, if a liquid, or in a small cell on the top face; as

refractivity

a block if solid, with air film excluded by a film of liquid of high refractive index). A special angled observing telescope moving round a vertical graduated circle helps the location of the limiting ray emerging through the vertical side face of the glass block and if the angle of emergence below the horizontal is α,

$$n = \sqrt{(n_G^2 - \sin^2 \alpha)}.$$

This instrument gives indices to about ± 0.0001 and differences between close indices to about $\pm 0.000\,02$, using a special micrometer adjustment. There are facilities for temperature control on the better versions of the instrument.

The *Abbe refractometer* is more convenient and rapid to use than the Pulfrich instrument. It has a telescope directed down on a pair of prisms that imprison a layer of liquid, the light coming up from below. The telescope is swung round a scale until the light/dark boundary is on the crosswires and the index is read directly on a scale to the third decimal place, the fourth being obtainable by interpolation. By ingenious use of *Amici prisms, dispersion in the specimen can be compensated (the boundary losing its colour and becoming sharp) and the dispersive power of the medium may be roughly estimated. For solids, one prism only is used and the procedure is much as for the Pulfrich refractometer.

4. *Immersion method.* Small specimens of solids may be immersed in each of a succession of liquids and viewed under a microscope. The changing appearance as the microscope is racked up depends on whether the index of the liquid is higher or lower. Sometimes a pair of liquids is mixed in various proportions until the specimen appears to vanish and a drop of the successful mixture is then tested in an Abbe refractometer.

5. *Interference methods.* Light of wavelength λ_0 in a vacuum will have wavelength $\lambda = \lambda_0/n$ when in a medium of refractive index n. A length l in this medium contains $l/\lambda = ln/\lambda_0$ wavelengths. If therefore light from a given source is divided into two channels of equal length, one in a vacuum and the other in the given medium, there is an effective path difference of $(n-1)l/\lambda_0$ and on reuniting the two beams, interference results. The *Jamin refractometer* divides the light by using reflections at the front and back surfaces of oblique thick glass blocks and uses two parallel tubes, one evacuated and the other slowly filled with gas while fringes are counted passing across the eyepiece.

The *Rayleigh refractometer* (originally described in 1896; modified since by others) uses a pair of parallel slits across a collimating lens to give two beams through the gas tubes. The interference fringes are compared with a fixed fringe system and a compensating device of inclined glass plates enables the effective path difference to be compensated and the achromatic (uncoloured) fringes of the two systems to be brought to coincidence. The instrument can be made so sensitive that small changes in composition of a gaseous mixture can be detected in suitable conditions.

6. *Miscellaneous methods.* (1) The *apparent depth* of a layer viewed normally is $1/n$ of its true depth. Use of a travelling microscope yields values of n to 1 per cent or better, for liquids and plane parallel slabs of transparent solids. (2) A thin prism of small refracting angle A deviates light through an angle $(n-1)A$ approx. A triangular glass or plastic trough of liquid deviates a well-defined pencil of light onto a scale calibrated directly in refractive index. This is quick and capable of moderate accuracy. (3) When an incident beam and the refracted beam are at right angles, the reflected beam is completely plane-polarized. The tangent of the angle of incidence (*Brewster angle* or *polarizing angle*) is then equal to the refractive index (*Brewster's law*, 1815). This position can be found by trial, using a *Nicol prism or a Polaroid screen. If the incident light is suitably polarized, the reflection will disappear entirely at this setting. This method is suitable for strongly absorbing media but contaminated or altered material in the surface layers will prevent the setting from being sharp. (4) *See* Boys' method for measuring the refractive index of glass.

7. *Birefringent materials.* Uniaxial crystals have two principal refractive indices, for the ordinary and for the extraordinary rays respectively. A prism, cut from such material so as to have its refracting edge parallel to the optic axis, will give two images and the two refractive indices can be measured by setting in turn for minimum deviation. Biaxial crystals are characterized by three principal indices of refraction.

refractivity An optical quantity equal to $(n-1)$ where n is the refractive index.

refractometer and refractometry The measurement of refractive index is classed as a study under *refractometry*; the instrument for more or less direct measurement is the *refractometer*. *See* refractive index measurement.

refractor *See* refracting telescope.

refrigerant The working substance of a refrigerator. Ammonia and freon are used for domestic refrigerators; carbon dioxide for ship installations where the volume available is limited; and sulphur dioxide for plants of large volume when very low temperatures are not required. Freons are a group of refrigerants consisting of chlorofluorocarbons (CFCs), some of which are now known to damage the ozone layer. Care therefore has to be taken in selecting suitable CFCs and in some cases they are being replaced by other refrigerants.

refrigerator A device for maintaining a chamber at a lower temperature than its surroundings. The ideal refrigerator is a *heat engine working backwards, work being done in taking heat from a condenser and transferring heat to the surroundings at a higher temperature. The three main types are (1) the *vapour-compression refrigerator, extensively used for big plants; (2) the *vapour-absorption refrigerator type, of low efficiency; and (3) the *Electrolux type, which is a type of vapour-absorption refrigerator having no moving parts.

regelation If two blocks of melting ice are pressed firmly together the increase in pressure lowers the melting point and so the ice at the contact faces melts, taking its latent heat from the neighbouring ice whose temperature thus falls below 0 °C. When the pressure is removed the film of water previously formed freezes, giving its latent heat to the neighbouring ice and thus joining the two blocks into a single block of ice. This process is known as regelation.

regenerative braking A method of *electric braking used with electric motors particularly in electric vehicles. During the braking period, each motor is used as a generator so that energy is returned to the supply system and a retarding torque is exerted by the armature of the machine. On account of the energy returned to the supply system, this method is more economical than, for example, *rheostat braking. In the development of battery-operated electric cars and other vehicles, regenerative braking is an important means of extending the range (by some 13–16%).

regenerative cooling A process used, e.g. in the *Linde process for liquefying air, in which compressed gas is cooled by expansion through a nozzle and the cool expanded gas is used to cool the oncoming compressed gas in a heat exchanger before it is cooled by expansion.

Regge, Tullio (*b.* 1931) Italian physicist who became professor at the University of Turin. He is noted for his work in introducing the idea of complex angular momenta into elementary particle physics (*see* Regge pole model), and has also contributed to relativity and group theory.

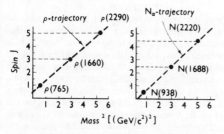

Regge trajectory of two Regge poles, mass being given in brackets

Regge pole model A theoretical model used to describe the scattering of *elementary particles at high energies. In general, it is found that the *strong interactions involved in such processes cannot be described in terms of the exchange of a single elementary particle. Although the contribution from the exchange of low-mass particles is usually the most important contribution to the *scattering amplitude, the contributions from the exchange of the higher-mass *resonances cannot be neglected.

Mathematically it is possible to describe the collective effect of exchanging all these particles in terms of the exchange of a few objects called Regge poles whose *spins increase with their effective masses. The path traced out by the spin of a Regge pole as its "mass" varies is called a *Regge trajectory*. On a graph of spin against the square of the mass (*see* diagram), Regge trajectories are found to be approximately straight lines. The individual particles represented by a Regge pole have all quantum numbers the same except for their spins, which differ by $\Delta J = 2n$, where n is an integer.

register A semiconductor device that acts as a storage location in the processing unit of a computer. A register usually stores a single *word or sometimes a *byte or *bit, the information being held only temporarily before it can be operated on. Storage and retrieval of the information must be extremely rapid. A register thus normally consists of a group of *flip-flops, each storing one bit.

Regnault, Henri Victor (1810–1878) French chemist and physicist who is remembered for his experimental work on heat. *See also* Regnault hygrometer.

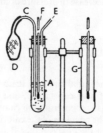

Regnault hygrometer

Regnault hygrometer Regnault devised a *hygrometer of the dew-point type consisting of two silver vessels A and G (*see* diagram), mounted side by side. Air may be blown from D through a tube C dipping into ethoxyethane (ether) contained in A. This causes the ether to evaporate through E thus cooling the tube A until eventually, at the dew point, moisture condenses on the outside of A giving it a dull appearance compared with the surface of G. This temperature, and that at which the dullness disappears on allowing the apparatus to stand, are noted on the thermometer F, the mean giving the dew point; this, with the room temperature, enables the relative humidity of the air to be calculated.

Regnault's heat experiments 1. Regnault determined the coefficient of absolute expansion of a liquid directly by balancing a hot and a cold column of the liquid. The vertical tube AB (Fig. *a*) is maintained at a high temperature t_2, while the tubes, CD, HG, and FE are all at a lower temperature t_1. AC is a narrow connecting tube with a small hole at L, and pressure is applied to the two inner tubes by a pump connected to K. Then the coefficient of absolute expansion, α, is given by:

$$\alpha = h/(H - h)(t_2 - t_1),$$

where h is the difference in level of the liquid in GH

regulation

and FE and H is the length of the column AB or CD.

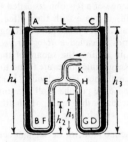

a Determination of coefficient of absolute expansion (Regnault)

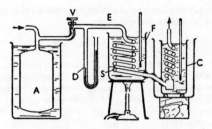

b Specific heat of gas at constant pressure (Regnault)

2. He determined the specific heat capacity of gases at constant pressure. Gas from a container A (Fig. b) passes through a regulator V and a capillary tube E into a spiral S in a hot bath F and thence through a second spiral immersed in a calorimeter C. The flow of gas is made steady by the use of V and the manometer D and the mass of gas m passing is calculated from the decrease in pressure of the gas in A.

Then

$$mc_p\left(t - \frac{t_1 + t_2}{2}\right) = C(t_2 - t_1),$$

where t_1 and t_2 are the initial and final temperatures of the calorimeter of heat capacity C, and t is the temperature of the gas on entering the calorimeter. After correcting for the heat loss of the calorimeter a correction is made for the heat conducted along the copper tube connecting the spiral in F to that in C by performing an experiment with no gas flowing. Experiments were carried out at pressures up to 10 atmospheres by allowing the gas to emerge from C through a fine capillary.

3. He also determined the saturation vapour pressure of water vapour at temperatures from 0 °C to 50 °C by arranging two barometer tubes side by side in the same trough of mercury, with a constant temperature bath surrounding the upper ends of the tubes. Water was introduced into one tube until a layer of liquid was obtained on the mercury meniscus. The difference in height of the two columns gave the saturation vapour pressure of water vapour at that temperature and was read by a cathetometer.

regulation Of electrical generators, transformers, and power transmission lines. The changes that take place in the available voltage due to internal resistance (for direct curent) or to internal impedance (for alternating current) when the load is changed under specified conditions. For generators (d.c. and a.c.) *see* inherent regulation. For power transmission lines, the regulation is defined as the increase in receiving-end voltage (or decrease in a.c. systems under certain loading conditions, in which case the regulation is negative), which occurs when a speci-

fied load (e.g. rated load) at a specified power factor (in a.c. systems) is thrown off, the sending-end voltage remaining constant. It is usually expressed as a percentage of the receiving-end voltage when on load. *See* inherent regulation; voltage drop.

Reines, Frederick (b. 1918) Amer. physicist who worked at the Los Alamos Scientific Laboratory and, later, became professor at the University of California (Irvine). With Clyde Cowan, he obtained (1953) tentative evidence of the existence of the neutrino in radiation from nuclear reactors; more definitive results were obtained at the Savannah River nuclear reactors in 1956. Subsequently, Reines investigated the existence of neutrinos in cosmic radiation using large underground detectors.

Reines' and Cowan's experiments A series of experiments to detect *antineutrinos emitted in *beta decay. During the operation of a *nuclear reactor there are produced some very short-lived beta-active nuclides, which have exceptionally high decay energies. Thus while a reactor is running it is the source of antineutrinos of sufficiently high energy to make their detection possible, using the reaction with free protons (hydrogen nuclei):

$$p + \bar{\nu} \rightarrow e^+ + n.$$

From the known spectrum of beta rays and the theory of beta decay the *cross section for this reaction was calculated to be 6×10^{-48} m^2, that is the *mean free path in hydrogen of the density of the sun is about 10^8 times the solar diameter. To detect such penetrating particles, it was necessary to use a large tank of water as absorber, placed underground and with a special shield to minimize background radiation. A liquid *scintillation counter sytem was used to detect the radiation caused by the absorption of antineutrinos. Firstly the positron is captured within 10^{-10} s and the two 0.51 MeV quanta characteristic of the *annihilation process are detected. The neutron is slowed down and captured in about 10^{-6} s by a cadmium salt in solution in the water. The radiation capture of neutrons by the isotope ^{113}Cd gives a total gamma-ray energy about 9 MeV, hence the counter system was designed to record only those events corresponding to the expected time interval and energies.

The first, inconclusive, results were published in 1953, and the completed research in 1956. A counting rate of 2.88 ± 0.22 counts/hour were recorded when the reactor was running, this being twenty times the background. When the water was diluted with 50% of D_2O the detection rate was halved, in agreement with the theory that the cross section is much less for protons bound in the nucleus.

Reisz microphone *See* carbon microphone.

rejector A parallel *resonant circuit. The *impedance of a circuit comprising inductance and capacitance in parallel has a maximum value at one particular frequency. The maximum impedance is called the *dynamic impedance*. Since the effective resistance of the parallel branches of the circuit cannot, in practice, be zero, the dynamic impedance cannot be infinite. *Compare* acceptor.

relative aperture *See* f-number.

relative atomic mass Symbol: A_r. The average mass per atom of a given specimen of an element, expressed in unified *atomic mass units. The value depends on the isotopes present in the specimen. The natural isotopic composition is assumed unless otherwise stated. Formerly called *atomic weight*.

relative density *See* density.

relative-density bottle A small flask with a perforated glass stopper that may be completely filled with a liquid. In order to determine the relative *density of the liquid, the bottle is weighed empty (m_1), full of liquid (m_2), and finally, full of water (m_3). The relative density of the liquid is then
$$(m_2 - m_1)/(m_3 - m_1).$$
Ingenious modifications of the procedure enable the relative density of powders, and of quantities of liquid insufficient to fill the bottle, to be found. *Compare* pyknometer.

relative humidity *See* humidity.

relative molecular mass Symbol: M_r. The average mass of a molecule or other molecular entity, expressed in unified *atomic mass units. It is equal to the sum of the *relative atomic masses of the constituent atoms. Formerly called *molecular weight*.

relative permeability *See* permeability.

relative permittivity *See* permittivity.

relative pressure coefficient *See* pressure coefficient.

relative velocity The velocity of A relative to B is the velocity that B, supposing himself at rest, assigns to A. If A and B are moving in the same direction the relative velocity of A to B is $v_A - v_B$; if moving in opposite directions, it is $v_A + v_B$. This only applies when v_A and v_B are very small compared to the speed of light. *See* relativity.

relativistic particle A particle the speed of which with respect to a particular observer is not small compared with the speed of light, such that the observer must use the theory of *relativity instead of classical physics. For example, a particle has a mass m_0 when measured by any observer at rest with respect to it,

but an observer with relative speed v obtains the value m given by:
$$m = m_0 (1 - v^2/c^2)^{-1/2},$$
where c is the speed of light. Thus, if the relative speed is one-tenth of that of light, m exceeds m_0 by $1/2\%$. An electron must be accelerated from rest with respect to an observer through 510 kV for the apparent mass to be doubled. The equivalent value for a proton is 938 MV.

It must be stressed that the particle is not itself changed. It is the large relative motion of the particle and observer that requires the more exact theory.

relativistic quantum field theory *See* quantum field theory.

relativity Two theories pioneered by Einstein, the *special theory of relativity* (1905) and the *general theory of relativity* (1915).

The special theory gives a unified account of the laws of mechanics and of electromagnetism (including optics). Before 1905 the purely relative nature of uniform motion had in part been recognized in mechanics, although Newton had considered time to be absolute and also postulated absolute space. In electromagnetism the *ether was supposed to provide an absolute basis with respect to which motion could be determined. (*See also* Galilean transformation equations; Michelson–Morley experiment; Newton's laws of motion.) Einstein rejected the concepts of absolute space and time, and made two postulates. (1) The laws of nature are the same for all observers in uniform relative motion. (2) The speed of light is the same for all such observers, independently of the relative motions of sources and detectors. He showed that these postulates were equivalent to the requirement that the coordinates of space and time used by different observers should be related by the *Lorentz transformation equations. The theory has several important consequences.

The transformation of time implies that two events that are simultaneous according to one observer will not necessarily be so according to another in uniform relative motion. This does not affect in any way the sequence of related events so does not violate any concepts of causation. It will appear to two observers in uniform relative motion that each other's clock runs slowly. This is the phenomenon of *time dilation*; for example, an observer moving with respect to a radioactive source finds a longer decay time than that found by an observer at rest with respect to it, according to:
$$T_v = T_0/(1 - v^2/c^2)^{1/2},$$
where T_v is the mean life measured by an observer at relative speed v, T_0 is the mean life measured by an observer relatively at rest, and c is the speed of light. This phenomenon is of great importance in studies of the decays of *elementary particles moving at very high speeds with respect to the laboratory. It was first convincingly demonstrated about 1941 by *Rossi and others in the case of the *muons in secondary *cosmic rays.

relaxation methods

Among the results of relativity optics is the deduction of the exact form of the *Doppler effect, which was verified in 1960 using the *Mössbauer effect.

In relativity mechanics, mass, momentum, and energy are all conserved. An observer with speed v with respect to a particle determines its mass to be m while an observer at rest with respect to the particle measures the *rest mass* m_0, such that:

$$m = m_0/(1 - v^2/c^2)^{1/2}.$$

This formula has been verified in innumerable experiments, of which the first of fairly high precision was by Bucherer (1909) using beta rays. An important special case is that for a system of zero rest mass, which must be propagated in vacuum and speed c. One consequence of this law is that no body can be accelerated from a speed below c with respect to any observer to one above c, since this would require infinite energy. Einstein deduced that the transfer of energy δE by any process entailed the transfer of mass δm where $\delta E = \delta mc^2$, hence he concluded that the total energy E of any system of mass m would be given by:

$$E = mc^2.$$

(*See* conservation of mass and energy.) The kinetic energy of a particle as determined by an observer with relative speed v is thus $(m-m_0)c^2$, which tends to the classical value $\frac{1}{2}mv^2$ if $v \ll c$.

Attempts to express *quantum theory in terms consistent with the requirements of relativity were begun by *Sommerfeld (1915). Eventually *Dirac (1928) gave a relativistic formulation of the *wave mechanics of conserved particles (*fermions). This explained the concepts of *spin and the associated magnetic moment, which had been postulated to account for certain details of spectra. The theory led to results of extremely great importance for the theory of elementary particles (*see* annihilation; antiparticle; pair production), the theory of *beta decay, and for *quantum statistics. The *Klein–Gordon equation is the relativistic wave equation for *bosons, which is used in the theory of the *strong interaction of *Yukawa.

A mathematical formulation of the special theory of relativity was given by Minkowski. It is based on the idea that an event is specified by four coordinates: three spatial coordinates and one time coordinate. These coordinates define a four-dimensional space and the motion of a particle can be described by a curve in this space, which is called *Minkowski space-time*. See four-dimensional continuum.

The special theory of relativity is concerned with relative motion between nonaccelerated frames of reference. The general theory deals with general relative motion between accelerated frames of reference. In accelerated systems of reference, certain fictitious forces are observed, such as the centrifugal and Coriolis forces found in rotating systems. These are known as fictitious forces because they disappear when the observer transforms to a nonaccelerated system. For example, to an observer in a car rounding a bend at constant velocity, objects in the car appear to suffer a force acting outwards. To an observer outside the car, this is simply their tendency to continue moving in a straight line. The inertia of the objects is seen to cause a fictitious force and the observer can distinguish between noninertial (accelerated) and inertial (nonaccelerated) frames of reference.

A further point is that, to the observer in the car, all the objects are given the same acceleration irrespective of their mass. This implies a connection between the fictitious forces arising from accelerated systems and forces due to gravity, where the acceleration produced is independent of the mass (*see* free fall). For example, a person in a sealed container could not easily determine whether he was being driven towards the floor by gravity or if the container was in space and being accelerated upwards by a rocket. Observations extended in space and time could distinguish between these alternatives, but otherwise they are indistinguishable. This leads to the *principle of equivalence* from which it follows that the inertial mass is the same as the gravitational mass.

A further principle used in the general theory is that the laws of mechanics are the same in inertial and noninertial frames of reference.

The equivalence between a gravitational field and the fictitious forces in noninertial systems can be expressed by using *Riemannian space-time*, which differs from the Minkowski space-time of the special theory. In special relativity the motion of a particle that is not acted on by any forces is represented by a straight line in Minkowski space-time. In general relativity, using Riemannian space-time, the motion is represented by a line that is no longer straight (in the Euclidean sense) but is the line giving the shortest distance. Such a line is called a *geodesic*. Thus space-time is said to be curved. The fact that gravitational effects occur near masses is introduced by the postulate that the presence of matter produces this curvature of space-time. This curvature of space-time controls the natural motions of bodies.

The predictions of general relativity only differ from Newton's theory by small amounts and most tests of the theory have been carried out through observations in astronomy. For example, it explains the shift in the perihelion of Mercury, the bending of light or other electromagnetic radiation in the presence of large bodies, and the *Einstein shift. Very close agreement between the predictions of general relativity and their accurately measured values have now been obtained. *See also* gravitational waves.

relaxation methods A mathematical technique developed by Sir R. V. Southwell and his collaborators for solving that wide range of problems in physics, which can be formulated in terms of simultaneous equations. Differential equations can be solved by first replacing them by their finite difference equivalents. The method is a numerical one and proceeds successively to more and more accurate solutions. Problems soluble by this method include: the displacements in loaded elastic frameworks, torsion of shafts of nonuniform diameter, characteristic frequencies of vibrating systems, the temperature distribution in the piston of an internal combustion engine, and the percolation of liquid through a porous material. Many of these, and other, problems are now better solved by *computers.

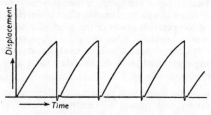

Saw-tooth waveform

relaxation oscillations 1. Oscillations characterized by a sawtooth type of waveform, the vibrating system apparently relaxing at each peak and returning quickly to the zero position from which the build-up recommences. Such oscillations can only be maintained by the existence of an effectively steady applied unidirectional force, and the motion may be represented by the equation:

$$m\ddot{x} - f(x)\dot{x} + \kappa x = E,$$

where m and κ are the inertia and elastic terms respectively, $f(x)$ is a function that is positive for low values of x but rapidly decreases to $-\infty$ after a certain value of x has been reached, and E is the steady applied unidirectional force. The presence of the E indicates the asymmetry that is always produced in such oscillations. The frequency of vibration depends almost entirely on the elasticity and resistive terms, and an expression for this has been derived by B. van der Pol.

2. Oscillations of a system to which an impulse is applied intermittently. In this sense, the waveform of a system in relaxation starts a succession of short trains of damped oscillations, for which the amplitude is renewed from time to time. Examples are the sounds of the *singing flame and certain vowel sounds in speech.

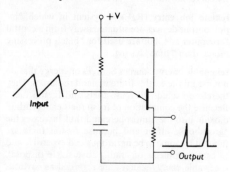

Relaxation oscillator

relaxation oscillator An oscillator in which one or more voltages or currents change suddenly at least once during each cycle. The circuit is arranged so that during each cycle energy is stored in and then discharged from a reactive element (e.g. a capacitor or inductor), the two processes occupying very different time intervals. An oscillator of this type has an asymmetrical output waveform that is far from being sinusoidal, a *sawtooth waveform being commonly generated. Square or triangular waveforms can be produced by means of a suitable circuit. The output waveform is very rich in harmonics and for some purposes this is particularly useful. Common types of relaxation oscillator include the *multivibrator and *unijunction transistor (see diagram), but many other circuit arrangements are possible. See relaxation oscillations.

relaxation time Symbol: τ. **1.** The time required for the *electric polarization of any point of a suitably charged dielectric to fall from its original value to $1/e$ of that value, due to the electric conductivity of the dielectric.

2. Generally, the time required for an exponential variable to decrease to $1/e$ of its initial value.

3. The time required for a gas, in which the Maxwellian distribution of speeds has been temporarily disturbed, to recover that state.

4. A measure of the time for the shear in a viscous liquid to disappear when no fresh shear is applied.

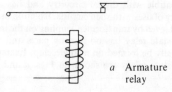

a Armature relay

relay An electrical device in which one electrical phenomenon (current, voltage, etc.) controls the switching on or off of an independent electrical phenomenon. There are many types of relay, most of which are either electromagnetic or solid-state relays.

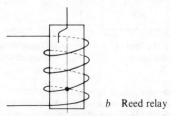

b Reed relay

1. *Electromagnetic relays.* (i) Armature type. In this type of relay a coil wound on a soft-iron core attracts a pivoted armature that operates contacts or tilts a mercury switch. (Fig. *a*). Several variants of the armature exist. Differential relays have two coils and only operate when the currents in the coils are additive not subtractive. The armature may be split, so that a small section of the metal operates with small currents independently of the main contacts, which require large currents to move the whole armature. Polarized relays have a central permanently magnetized core and operate differently with currents in different directions. (ii) Reed relay. These relays have a coil wound around a glass envelope containing fixed contacts and a centrally placed reed contact in the form of a thin flat metal strip (Fig. *b*). When the coil is energized the reed is deflected, either making or breaking contact. (iii) Diaphragm relay. This type of relay has a coil wound around a central core, with a thin metal diaphragm plate mounted close to its end (Fig. *c*). When the coil is energized the central portion of the diaphragm moves towards the core and makes contact with it.

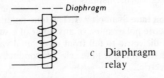

c Diaphragm relay

2. *Solid-state relays.* A true solid-state relay has all its components made from solid-state devices. Isolation between input and output terminals is provided using a *light-emitting diode (LED) in conjunction with a *photodetector. The switching is achieved using a *silicon controlled rectifier (SCR) or more commonly two SCRs (a triac). This type of relay is compatible with digital circuitry and has a wide variety of uses with such circuits. Isolation may also be achieved by transformer coupling on the input. A solid-state relay has no moving parts and cannot normally be formed on a single chip. Examples of solid-state relays are shown in Figs. *d* and *e*.

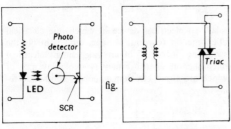

d LED-coupled solid-state relay

e Transformer-coupled solid-state relay

Solid-state relays have advantages over electromechanical relays because of increased lifetime, particularly at a high rate of switching, decreased electrical noise, compatibility with digital circuitry, ability to be used in explosive and corrosive environments since there are no contacts across which *arcs can form; the lack of physical contacts and moving elements also gives increased resistance to corrosion. No mechanical noise is associated with them. This is particularly important for certain applications where noise could be an annoyance, such as hospitals, etc. Disadvantages include the substantial amount of heat generated necessitating cooling for applications above a current of several amperes,

greatly increased production costs for multipole devices compared to single-pole devices, and in certain applications a physical disconnection may be required for safety purposes and this is not available in solid-state relays.

Other types of relay include thermionically operated relays in which the heating effect of a current is used to operate contacts or the effect of a heating coil on a bimetallic strip. Gas-filled relays such as the *thyratron have also been used, but these are being superseded by solid-state relays using thyristors.

release *See* overcurrent release; overvoltage release; reverse-power release; undercurrent release; undervoltage release.

relief vent (or **explosion vent**) Of a transformer. A device fitted to the transformer tank to relieve any sudden increase of internal pressure caused by an internal fault.

reluctance Symbol: R. The ratio of the *magnetomotive force, F_m, applied to a magnetic circuit or component, to the magnetic flux Φ, in that circuit, i.e. $R = F_m/\Phi$. It has the units henry^{-1}. Its reciprocal is the *permeance.

reluctivity The reciprocal of magnetic *permeability.

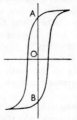

Hysteresis curve, showing remanence (OA or OB)

remanence *Syn.* retentivity; residual magnetism. The residual *magnetic flux density in a substance when the magnetizing field strength is returned to zero. It is represented by OA and OB in the hysteresis curve shown. *See* hysteresis.

remote job entry (RJE) A system in which *input/output devices are situated away from a central *computer site, but are used for *batch processing rather than *time sharing.

renewable energy sources Sources of energy that do not use up the earth's finite mineral resources. They therefore exclude all fossil fuels and fission fuels. Because the combustion of fossil fuels emits carbon dioxide into the atmosphere and thus increases the *greenhouse effect and because fission fuels are thought by some to be hazardous and expensive and to cause problems of *radioactive waste disposal, renewable energy sources are regarded as environmentally desirable. The renewable forms of energy currently being explored and developed are solar energy, wind energy, tidal energy, wave energy, geothermal energy, and biomass energy. Hydroelectric energy is already widely exploited and has limited scope for further expansion; fusion energy is a virtually renewable source that offers enormous reserves of energy once the problems of *fusion reactors have been solved.

Solar energy is already being used, especially in hot countries. With some 10^{17} joules per second of solar power falling on the earth there remains great scope for widening its use. Two methods are used. The commonest is to heat water flowing through special panels on a building's roof. The second method is to use *solar cells. If these are to have a commercial future there will have to be a considerable fall in cost (e.g. in the cost of silicon).

Wind energy is a pollution-free cheap source. It does, however, require a great deal of land and a method of storing electricity for use when the wind drops. About 12 wind turbines are already feeding electricity to the UK national grid, mostly making use of Atlantic winds on the west coast. If all the usable sites in the UK could be harnessed for this purpose, some 20% of the country's energy could probably be wind-generated.

Tidal energy makes use of the tides to collect water behind a barrage, which is later released to turn a turbogenerator. Tidal power stations are in use in the USSR and France (on the River Rance). One is proposed for the Severn estuary, which, with its tidal rise of 8.8 m could generate 7% of the UK electricity; another possibility is a Mersey barrage.

Wave energy uses the energy of the waves to make a string of floats (called *Salter ducks*) bob up and down. This bobbing motion is harnessed to turn a generator. Off the coast of the UK there are probably sufficient suitable sites to generate over 100 GW of electricity, once technical problems have been solved.

In geothermal energy, geysers and hot springs are tapped. Iceland, Italy, New Zealand, and the USA all make use of this form of energy. However, in the UK the very deep drilling required to reach usable sources involves considerable technical problems.

Biomass energy relies on the combustion of biofuels, such as methane generated by sewage or by farm, industrial, or household organic waste. Biofuels can also include specially cultivated organisms or crops (e.g. cane sugar) grown for their energy potential. It is estimated that by 2025 some 20% of the UK's energy could be met by renewable sources.

renormalization A procedure used in relativistic quantum field theory to deal with the fact that, in *perturbation theory, calculations give rise to infinites beyond the first term. Renormalization was first used in *quantum electrodynamics, where the infinities were removed by taking the observed mass and charge of the electron as "renormalized" parameters rather than the "base" mass and charge in the *Lagrangian for electromagnetic interactions. In QED, perturbation-theory calculations using renormalization lead to agreement with experiment to a very high degree of accuracy.

Theories for which finite results for all perturbation-theory calculations exist by taking a finite number of parameters from experiment and using renormalization are called *renormalizable*. Only certain types of *quantum field theories are renormalizable. Theories that need an infinite number of parameters are called *non-renormalizable* theories. The *gauge theories describing the *strong, *weak and *electromagnetic interactions are renormalizable but the quantum theory of *gravitational interactions is a non-renormalizable theory, which perhaps indicates that gravity needs to be unified with other fundamental interactions before a consistent quantum theory of gravity can be realized.

Similar techniques to renormalization theory are used in the many-body problem in quantum theory and *statistical mechanics, particularly the theory of *phase transitions.

repeater A device, used especially in telegraphic and telephonic circuits, that receives signals in one circuit and automatically delivers corresponding signals to one or more other circuits. A repeater usually amplifies the signal or may perform pulse regeneration on transmitted pulses.

reproduction of sound The process by which a sound is recreated as and when required by use of a permanent or semi-permanent record and suitable replay equipment. Four main methods of recording, and hence of suitable reproducing equipment, are in use – digital recording (*see* compact disc; digital audio tape), electromechanical recording (*see* gramophone), sound-film recording (*see* soundrack), and *magnetic recording. For general comments on these methods, *see* recording of sound. *See also* loudspeaker.

repulsion-induction motor A type of a.c. *single-phase motor similar to a *repulsion motor except that the *rotor carries two separate windings arranged in slots, one above the other. The outer winding at the top of the slots is the repulsion winding, which is connected to a commutator as in a repulsion motor. The inner winding at the bottom of the slots is a squirrel-cage winding (*see* induction motor). During the process of starting and running up to full speed, the relative effects of the two windings change considerably, that of the repulsion winding decreasing while that of the squirrel-cage winding is increasing, so that fundamentally the machine starts as a repulsion motor and runs as an induction motor, the transition being gradual. It is a *shunt-characteristic motor. *Compare* repulsion-start induction motor.

repulsion motor A type of a.c. *single-phase motor. The *stator carries the a.c. winding housed in slots on its inner periphery. The *rotor carries in slots on its outer periphery a winding similar to that of a d.c. armature and this winding is connected to a *commutator. One or more pairs of brushes on the commutator are short-circuited. Rotating the brushes round the commutator to different positions by means of the *brush rocker provides a means of speed control. It is a *series-characteristic motor.

repulsion-start induction motor

Compare repulsion-induction motor; repulsion-start induction motor.

repulsion-start induction motor Fundamentally, a *repulsion motor provided with a centrifugally operated device that short-circuits the *commutator bars when the rotor has reached a predetermined speed during the starting process. The machine starts as a repulsion motor and runs at normal speed as a plain *induction motor. To avoid unnecessary wear of the commutator and brushes, arrangements may be made for the latter to be raised clear of the former by the action of the centrifugal device. It is a *shunt-characteristic motor. Compare repulsion-induction motor.

residual See deviation.

residual charge In a capacitor there may be some viscous movement of the dielectric under charge and some of the charge may penetrate into the dielectric. On rapid discharge only that part of the charge near to the plate first passes, and further smaller charges can be drawn later, these being called residual charges.

residual magnetism See remanence.

residual rays Syn. reststrahlen. Almost monochromatic infrared rays selectively reflected by quartz, fluorite, etc., that, while perfectly reflecting these rays, absorb radiation of shorter and longer wavelength. For quartz these rays have a wavelength of 8.8 μm; for sylvite, 70 μm; for potassium iodide, 96 μm.

resilience The amount of potential energy stored in an elastic substance by means of elastic deformation. It is usually defined as the work required to deform an elastic body to the elastic limit divided by the volume of the body. It has the units $J\,m^{-3}$.

resistance 1. (electrical) Symbol: R. The ratio of the potential difference across a conductor to the current flowing through it. (See Ohm's law.) If the current is alternating, the resistance is the real part of the electrical *impedance Z, i.e.
$$Z = R + iX,$$
where X is the *reactance. Resistance characterizes a dissipation of energy as opposed to its storage. It is measured in ohms. See also resistor; temperature coefficient of resistance.

2. (acoustic) Symbol: R_a. The real part of the acoustic *impedance Z_a, i.e.
$$Z_a = R_a + iX_a,$$
where X_a is the acoustic reactance. It is measured in pascal second per metre cubed. The specific acoustic resistance (Symbol: R_s) is the real part of the specific acoustic impedance and is measured in pascal second per metre.

3. (mechanical) Symbol: R_m. The real part of the mechanical *impedance Z_m, i.e.
$$Z_m = R_m + iX_m,$$
where X_m is mechanical reactance. It is measured in newton second per metre.

4. (thermal) Symbol: R. (i) The reciprocal of the *thermal conductance. The units are $m^2\,K\,W^{-1}$. (ii) The ratio of the temperature difference between two points and the mean rate of flow of entropy between them. The units are $K^2\,W^{-1}$.

resistance-capacitance coupling (RC coupling) See interstage coupling.

resistance coupling See interstage coupling.

resistance drop See voltage drop.

resistance gauge A gauge used for measuring high fluid pressures by means of the change in electrical resistance produced in manganin or mercury by those pressures. These gauges are calibrated with reference to the *free-piston gauge.

resistance pyrometer See resistance thermometer.

resistance strain gauge An instrument for measuring structural strains by the increase in electrical resistance of a wire or grid of wires attached to, or supported by, the structure. The wires are often wound on a small paper former and the gauge is attached by adhesive to the structure. Static or dynamic strains may be measured according to the type of recording apparatus used, and the gauges can be adapted to allow measurement of weights or pressures by application to bodies that deform regularly under the stress imposed.

resistance thermometer A thermometer in which the change in electrical resistance of a wire is used as the thermometric property. A small coil of wire, usually platinum but of other metals or carbon for use at low temperatures, is wound on a mica former and enclosed in a silica or porcelain sheath. The change in resistance is determined by placing the coil in one arm of a *Wheatstone bridge. As the measuring galvanometer is often at a great distance, duplicate leads are added to the other balancing arm to compensate for temperature variations in the leads. The platinum instrument is useful over a very wide range (–200 °C to over 1200 °C). See platinum resistance thermometer. Carbon or certain alloys can be used at liquid-helium temperatures.

resistance voltage See voltage drop.

resistance welding A welding process in which the heat is produced by passing an electric current across the contact surface between the components to be welded. During the process the components are kept in contact by external pressure.

resistivity 1. Symbol: ρ. The electrical quantity defined by the equation $\rho = RA/l$, where R is the resistance of a wire of length l and cross-sectional area A. It is measured in ohm metres. For pure metals at ordinary temperatures ρ is of the order $10^{-8}\,\Omega\,m$ while for good insulators it may exceed $10^{14}\,\Omega\,m$. The reciprocal of resistivity is *conductivity. The product of the resistivity and the density is sometimes called the mass resistivity.

2. Thermal resistivity. Symbol: ϕ. The reciprocal of the *thermal conductivity. It is measured in m K W^{-1}.

resistor A piece of electrical apparatus possessing *resistance and selected for use because of that property. *Carbon resistors* are widely used in electronic circuits. They consist of finely ground carbon particles mixed with a ceramic material, encapsulated into insulated tubes. The casing has a set of coloured stripes denoting the value of the resistance. For more precise values of resistance, *wire-wound* and *film resistors* are used. In the former, a *constantan or *manganin wire of uniform cross section is wound into a suitable shape; in the latter, a thin uniform layer of resistive material is deposited in a continuous pattern on an insulating core. *See also* inductive, noninductive, and nonreactive; earthing resistor; heater; preventive resistor.

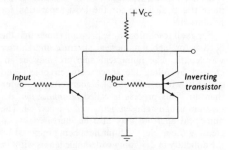

Two input NOR gate

resistor-transistor logic A family of integrated *logic circuits in which the input is via a resistor into the *base of an inverting *transistor. The basic NOR gate is shown. The output is high (corresponding to logical 1) only if both the inputs are low (corresponding to a logical 0). If either input is high the transistor conducts and saturates and the voltage at the output is low. RTL circuits tend to be slow and susceptible to noise and are now little used.

resolute *See* resolution.

resolution 1. The separation of a vector into its components. By reversing the process of vector addition (*see* polygon of vectors) it is possible to decompose (i.e. resolve) any vector into the sum of a number of component vectors. The most useful instance is the resolution of a vector into three components parallel to the axes of Cartesian coordinates. The component vectors are also known as *components*, *resolutes*, or *resolved parts*.

The expression *horizontal component* means the resolved part in a horizontal direction, it being implied that there is only one other component, which is vertical.

2. The amount of information or detail revealed in an image produced by a telescope, microscope, computer, etc. It can be expressed in numerical terms. *See also* resolving power.

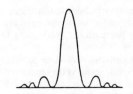

a Two patterns overlapping completely

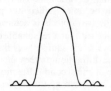

b Two patterns very close together

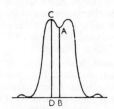

c Two patterns sufficiently far apart to be separately distinguished

resolving power 1. *General.* For a *telescope or a *microscope, the resolving power gives a measure of the ability of the instrument to produce detectably separate images of objects that are close together. A spectroscopic instrument is required to separate two wavelengths that are very nearly equal; the measure of its ability to do this is known as the *chromatic resolving power.*

2. *The resolving power of the eye.* The ability of the eye to see as separate two objects that are close together varies with the type of objects, the conditions of illumination, and so on. In round figures a normal eye is able to distinguish two points close together as separate points, provided they subtend an angle of not less than one minute of arc at the eye.

3. *The resolving power of telescopes.* In the case of optical telescopes, the resolving power is measured by the angular separation of two point sources that are just detectably separated by the instrument. The smaller this angle, the greater is the resolving power. The image of a point source formed by the primary mirror or lens will consist of a diffraction pattern (*see* diffraction of light(5)). The pattern has a central bright spot surrounded by alternate dark and light rings. Two point sources close together will give rise to two overlapping diffraction patterns. A cross section of the variation in intensity of the light in such a diffraction pattern is illustrated in Fig. *a*, while Figs. *b* and *c* show the resulting intensity produced by two patterns overlapping to different extents.

With two patterns very close together a broad central disc will be seen. As the separation increases, a decrease in intensity occurs at the centre of the pattern (AB < CD). It is at a separation such as this that the eye may be expected to observe that in fact two patterns are responsible for the illumination and not one; that is, when the central dip in the intensity graph becomes sufficiently pronounced, resolution of the two objects is achieved.

Rayleigh proposed that a reasonable criterion for resolution of two point sources was that the inner dark ring of one diffraction pattern should coincide with the centre of the second diffraction pattern, a condition that is known as the *Rayleigh criterion*. This leads to the condition that two point sources are resolved by a telescope mirror or lens of aperture *D*, provided their angular separation in radians is not less than $1.22\lambda/D$, where λ is the wavelength. The Rayleigh criterion, which is quite arbitrary, corresponds in the combined diffraction pattern to an intensity ratio saddle to peak (AB/CD) of 0.81, for sources of equal intensity.

From practical determinations the astronomer Dawes proposed the rule that two point sources are resolved if their angular separation is at least $11/D$ seconds of arc, *D* being the diameter of the objective measured in cm. This corresponds to a smaller separation for resolution than that proposed by Rayleigh.

Experiments by Abbe, mainly on microscopes, have suggested the Abbe criterion for resolution that the angular separation should not be less than λ/D. This corresponds to an intensity ratio saddle to peak of 0.98 for equal intensity sources, and is again less stringent than the Rayleigh criterion. Finally Sparrow suggested, as the result of an empirical study, that the condition for resolution is simply that there should be some central dip in the combined diffraction pattern.

The Rayleigh criterion is still largely used as giving ready comparisons between different instruments as regards the resolving powers obtainable.

The Rayleigh criterion can be applied not only to optical but also to infrared and radio telescopes, and indicates the need for a large aperture to give the necessary resolving power. Objects will only be resolved, however, if there is sufficient magnification to produce the necessary separation of the images.

By using two slits as widely separated as possible in front of the objective instead of the whole aperture of the objective, the ability to resolve two point sources subtending a small angle is almost doubled. This principle was used by Michelson in his *stellar interferometer* to measure the angular diameters of certain stars. The angles to be measured may be as small as one-hundredth of a second of arc. To obtain the necessary separation of the two slits, Michelson arranged a system of four mirrors fixed to a beam

attached in front of the telescope. The two outer mirrors (which may be 20 metres apart) reflect the light onto the two inner mirrors placed in front of the objective and hence down the telescope tube.

4. *The resolving power of microscopes.* The resolving power of a microscope is measured by the actual distance between two object points that can be detectably separated by the instrument. The greater the resolving power the smaller will this distance be. Application of the Rayleigh criterion shows that the least separation for resolution is:

$$0.61\ \lambda/n \sin i,$$

where λ is the wavelength of the light used, *n* the refractive index of the medium between object and objective, and *i* the semi-angle subtended at the object by the edges of the objective. Abbe called the quantity $(n \sin i)$ the *numerical aperture of the objective. Experiments by Abbe led him to the conclusion that a separation of $0.5\ \lambda/n \sin i$ gave a more practical figure for the least separation for resolution.

For good resolution the expression shows that the need is for higher numerical aperture and shorter wavelength. The numerical aperture may be increased by filling the space between object and objective with a medium of higher refractive index than air. This process is called *oil-immersion*, the medium usually chosen being cedar-wood oil. The numerical aperture may also be increased by increasing the angle, the limit here being imposed by the difficulty in designing wide-angle lenses without introducing objectionable aberrations in the images. A good, modern, high-power microscope objective will have a numerical aperture of perhaps 1.6, corresponding to about 200 nm as the least separation for resolution.

The second way of increasing the resolving power is to decrease the wavelength of the radiation used. Ultraviolet radiation is employed in *ultraviolet microscopy* using wavelengths roughly half those of visible light and so theoretically increasing the resolving power by a factor of 2. The resulting image is of course photographed. Again in the *electron microscope use is made of the fact that the wavelength associated with a swiftly moving electron is much smaller than the wavelength of light. By using suitably focused electron beams, photographic images are obtained that bring down the limit of resolution to as little as 0.2 nm.

5. *Chromatic resolving power.* In a spectroscopic instrument, the requirement is for a detectable separation of wavelengths that are very nearly equal. In such cases, the resolving power is measured as the ratio of the wavelength studied to the difference in wavelengths that can just be separated.

If a prism *spectrometer is used then, assuming the Rayleigh criterion, the resolving power is given by the expression $(t\ dn/d\lambda)$, where *t* is the maximum thickness of prism traversed by the beam and $dn/d\lambda$ is the ratio of change in refractive index to change in wavelength for the material of the prism. In a simple

Meson Resonances

	Mass (MeV/c^2)	Isospin, I	Spin, J	Parity, P	G-parity, G	Charge conj. parity, C
ρ	765	1	1	−1	+1	−1
ω	784	0	1	−1	−1	−1
K*	892	$\frac{1}{2}$	1	−1	—	—
$\eta(X^\circ)$	958	0	0 or 2	−1	+1	+1
ϕ	1019	0	1	−1	−1	−1
f	1260	0	2	+1	+1	+1
A_2	1310	1	2	+1	−1	+1
K**	1420	$\frac{1}{2}$	2	+1		

laboratory spectrometer of this type a resolving power of about 10^3 is possible.

If a *diffraction grating is used, the resolving power is the product of the total number of lines illuminated and the order number of the spectrum being used. For a 7.6 cm concave grating used in the second order a resolving power of about 10^5 is theoretically possible.

Various other devices of high resolving power are used. They include the *Fabry–Perot interferometer, the *Lummer–Gehrcke plate and both transmission and reflection echelons (see echelon grating). With such devices it is possible to obtain resolving powers of one million or more.

resonance 1. A condition in which a vibrating system responds with maximum amplitude to an alternating driving force. The condition exists when the frequency of the driving force coincides with the natural undamped oscillatory frequency of the system. (See forced vibrations; sharpness of resonance.)

2. A condition existing when an oscillatory electric circuit responds with maximum amplitude to an external signal of angular frequency ω. The *impedance of an a.c. circuit with inductance L and capacitance C in series is given by:

$$Z = \sqrt{\{R^2 + [(\omega L) - (1/\omega C)]^2\}}.$$

When $\omega L = 1/\omega C$ a condition is achieved where the impedance depends on the resistance alone, and as the resistance may be quite low for high values of L and C the current flowing will be high. This is the condition for resonance. Similar resonance can also be obtained when the capacitance and inductance are in parallel. (See resonant circuit; tuned circuit.)

3. See resonances.

resonance cross section See cross section.

resonances Extremely short-lived *elementary particles that decay by *strong interaction in about 10^{-24} second. They are thus hadrons. Resonances may be regarded as *excited states of the more stable particles. Over one hundred *meson and *baryon resonances are known; some of the more important of these are given in the tables.

resonance scattering See scattering.

Baryon Resonances

	Mass (MeV/c^2)	Isospin, I	Spin, J	Parity, P
N(1470)	1470	$\frac{1}{2}$	$\frac{1}{2}$	+1
N(1520)	1520	$\frac{1}{2}$	$\frac{3}{2}$	−1
N(1535)	1535	$\frac{1}{2}$	$\frac{1}{2}$	−1
Δ(1236)	1236	$\frac{3}{2}$	$\frac{3}{2}$	+1
Λ(1405)	1405	0	$\frac{1}{2}$	−1
Λ(1520)	1520	0	$\frac{3}{2}$	−1
Σ(1385)	1385	1	$\frac{3}{2}$	+1
Ξ(1530)	1530	$\frac{1}{2}$	$\frac{3}{2}$	+1

resonance theory of hearing A sound wave reaching the ear passes down the auditory canal and causes vibrations of the drum behind which is a mechanical coupling of three bones leading to a window in the cochlea. The latter is a spiral tube of bone gradually diminishing in size, its cross section being divided into two by the basilar membrane. This membrane has about 24 000 fibres, arranged in groups from which a separate nerve leads to the brain. The resonance theory of hearing, suggested by Cotugno and developed by Helmholtz, considers the fibres stretched radially throughout the basilar membrane to act as resonators. These fibres vary in length, tension, and loading throughout the membrane so that their natural frequencies cover the audible range. The incoming sound excites a small group of fibres having resonant frequencies near the frequency of the sound. The vibrations thus set up are communicated to the hair cells and the corresponding nerve transmits the sensation to the brain. The theory explains many of the known facts about hearing. In spite of the relatively small number of fibres in the cochlea, the pitch sensitivity of the ear is accounted for if each pure tone excites a region of fibres, the brain recognizing the resonator having maximum response. Thus, pitch sensitivity is dependent on the ability of the brain to distinguish between two nearby points of maximum stimulation. The sensitivity and damping at different frequencies are of the order to be expected from such a set of resonators. *Ohm's law of hearing is explained since any number of groups of resonators can be stimulated at the same time by a complex note, each group transmitting its own nervous stimulus to the brain. Subjective *combination tones are considered to be due to the nonlinear response of the ear drum owing to its one-sided loading. Experiments in which a small region of the ear has been destroyed show that the sensation of a particular note is produced by the motion of a small area of the basilar membrane as would be expected on this theory. *Masking of sound and binaural phase perception (*see* binaural location) have not yet been explained but no existing theory of hearing is completely satisfactory and the resonance theory explains most of the facts.

resonant cavity *See* cavity resonator.

resonant circuit A circuit that contains both inductance and capacitance so arranged that the circuit is capable of *resonance. The frequency at which resonance occurs – the *resonant frequency – depends on the values of the circuit elements and their arrangement.

A *series resonant circuit* contains the inductance and capacitance in series. Resonance occurs at the minimum *impedance of the circuit, and a very large current is produced at the resonant frequency. The circuit is said to accept that frequency. Although large voltages are developed across the individual elements, these are out of *phase so that the total voltage developed is relatively low.

A *parallel resonant circuit* contains the circuit elements arranged in parallel. Resonance occurs at or near the maximum impedance of the circuit. The currents in each branch can be very large but are out of phase. This results in a minimal overall current at the resonant frequency, a voltage maximum being produced. The circuit is said to reject that frequency. Parallel resonant circuits are thus also called *antiresonant circuits*. *See also* tuned circuit.

resonant frequency Of an oscillatory circuit. That frequency of the applied *sinuisoidal voltage at which the rate of change of a specified current or voltage with frequency becomes zero when changing sign (i.e. that supply frequency at which the specified current or voltage reaches a mathematical maximum or minimum value). Alternatively, it is that frequency at which a specified phase angle passes through zero. Several interpretations are possible and, for a given a.c. circuit, they may give different values for the resonant frequency. Probably the most common interpretation in connection with simple a.c. series and parallel circuits is the frequency at which the circuit has unit *power factor, i.e. at which the supply current is in phase with the supply voltage. *See* resonance.

rest energy The energy equivalent of the *rest mass of a body or particle, usually given in electronvolts.

restitution *See* coefficient of restitution.

rest mass The *mass of a body as determined by an observer who is at rest with respect to it. *See* relativistic particle; relativity.

restriking voltage Applied to an a.c. circuit that includes a circuit breaker, switch, or other device suitable for opening the circuit when on load or under fault (e.g. short-circuit) conditions. It is the transient voltage of high frequency that occurs at, or very near to, each zero current pause during the arcing period while the contacts of the circuit breaker, etc., are being separated. An important quantity in the design of circuit breakers, etc., for use with high-voltage systems is the rate of rise of restriking voltage since this influences the total time interval between the instant at which the contacts begin to separate and the instant of final arc extinction. *Compare* recovery voltage.

reststrahlen *See* residual rays.

resultant The sum of a number of *vectors (e.g. forces, velocities) found according to the *polygon of vectors. The resultant of a system of forces applied to a body is the single force that has the same translational effect as the system itself.

resultant tone *See* combination tones.

retentivity *See* remanence.

reticle (or **reticule**) *See* graticule.

retina The inner coat of the *eye, consisting of nerve fibres and endings sensitive to light (rods and cones). The innermost layer of nerve fibres transmits nerve impulses to the brain by way of the optic nerve where the blind spot exists. At an equal distance from the posterior pole to the blind spot and on the opposite side (i.e. towards the temple) is the important macular area with a central depression (*fovea centralis*). For keenest vision, an image of an object should be focused on this fovea, where only cones are present. A little distance away from the fovea is a region that is most sensitive to faint light. Rods are more numerous. Surrounding this, the rods become still more numerous and the eye more sensitive to movement. The sensitivity of the retina depends very much upon the level of illumination, increasing by a factor of many thousands on going from daylight into a dark room. *See* adaptation of the eye; pupil.

return-current coefficient *See* reflection coefficient.

return feeder *See* negative feeder.

return stroke Of lightning. *See* leader stroke.

reverberation The persistence of audible sound after the source has been cut off. If the time difference between reception of direct sound and its echo is less than about 1/15th of a second, true reverberation occurs and a number of echoes gives the sensation of a continuous sound of diminishing intensity. In the acoustics of large halls the *reverberation time* is of considerable importance. It is defined as the time for the energy density to fall to the threshold of audibility from a value 10^6 times as great, i.e. a fall of 60 *decibels. Franklin's theory shows that the reverberation time of a room of volume V and surface area S is given by

$$T = 0.05 \ V/\alpha S,$$

where α is the average absorption coefficient. This agrees with a more general expression obtained by Eyring except when α is large. The expression has been confirmed experimentally by Sabine who measured reverberation times by recording on a chronograph the time between the cutting off of an organ pipe of known frequency and intensity and the instant when it became inaudible.

The optimum reverberation time is a matter of taste depending on the size of a room and the type of sound used. If the time is too long, there is an overlapping of successive sounds, which makes their analysis by the ear very difficult. A very short reverberation time makes a room seem dead and very tiring to a speaker. Sabine found the optimum time under various conditions by two methods. He varied the reverberation time of a room until a committee of experts pronounced it to be the best, and he also calculated the reverberation times of rooms considered to be acoustically good. Optimum periods generally lie between 1 and 2.5 seconds. The value should be low for speech and light music but high for orchestral music. In general, the optimum period is proportional to the linear dimensions of the room. In the case of speech there is an exponential growth and decay of each syllable. For good articulation one syllable must die away to a considerable extent before the next one is heard. This would imply a very short reverberation time but in practice it is not preferred since there is a reduction in loudness and the effect is unnatural. When the reverberation time is long a speaker must talk very slowly to be understood. *See* auditorium acoustics.

reverberation chamber *Syn.* echo chamber. A room that has a very long *reverberation time and is carefully designed to allow a uniform energy distribution of sound to be produced. For a room to have a long reverberation time there must be very little absorption by the exposed surfaces. Thus the walls and ceiling are usually plastered and then painted to ensure uniformity of surface throughout the room. However, these highly reflecting surfaces tend to produce *standing waves, which disturb the measurement of the energy distribution. Standing waves may be avoided by using a room in which no two walls are parallel or by having a large steel reflector rotating silently in the room. It is also customary to use a revolving source that produces a *warble tone and to take measurements at various points in the room. All sound from outside the chamber must be excluded and elaborate soundproofing is necessary.

Reverberation rooms are used to measure the absorbing power of certain materials by the Sabine method. This involves finding the decay times for two known sound intensities, both in the empty room and with the material under test in the room. Test chambers should be large since their surface area must be great relatively to that of the material, which itself should be large in relation to the wave-

length of the sound. A reverberation chamber together with an adjacent room can also be used to measure sound transmission by Sabine's method if part of one of the walls between the two rooms is removed and replaced with the material under test.

reverberation time *See* reverberation.

reversal 1. (photographic) The transformation of a photographic negative into a positive (*see* photography). This involves chemical removal of reduced silver from the exposed negative (*bleaching*), followed by an all-over exposure and normal development. Silver is thus deposited in all places not originally exposed. A self-reversal (*solarization) occurs in badly overexposed negatives, e.g. in the centres of intense lines in a spectrum plate.

2. (spectroscopic) Bright emission lines in the spectrum of a discharge tube or flame may be *reversed*, i.e. apparently transformed into dark absorption lines, when intense white light traverses the source and enters the spectroscope, due to selective absorption by the gas or vapour at the same frequencies as it emits. Under certain conditions the lines produced by emission appear as broad lines with a narrow dark line down their centre. This is caused by emission from a high-temperature gas, which results in the broadened diffuse emission line. This light is then absorbed by cooler gas in a different region of the source. Since the gas is cooler the absorption line is narrower than the emission line and the line appears to be a doublet. This effect is called *self-reversal*.

reverse bias *See* reverse direction.

reverse direction *Syn.* inverse direction. The direction in which an electrical or electronic device has the larger resistance. Voltage applied in the reverse direction is the *reverse voltage* or *reverse bias* and the current flowing is the *reverse current*.

reverse-power release *Syn.* reverse-current release. A *tripping device that operates when the power in the circuit reverses its direction and reaches a predetermined value in that reverse direction.

reversibility principle (optics) If a reflected or refracted ray is reversed in direction, it will retrace its original path.

reversible change A change that is carried out so that the system is in *equilibrium at any instant and so that an infinitesimal change in the factor effecting the change causes every feature of the forward process to be completely reversed. This often means that the change must be carried out infinitely slowly since any kinetic energy could not change sign, and there must be no friction since this would always result in the evolution of heat. Such a process is never realizable in practice but close approximations to it can be attained. All practical processes involve *irreversible changes*, i.e. changes in which the system is not in equilibrium at all instants during the change.

reversible colloids Colloids are sometimes classified as reversible or irreversible according to whether the coagulated colloid can be redispersed or not under certain arbitrary conditions. The classification corresponds roughly to emulsoid and suspensoid sols but is best avoided as the result depends on the method of peptization given in the definition.

reversible engine A *heat engine that operates reversibly, in the sense described under reversible change. *See* Carnot cycle; Carnot's theorem.

revolution 1. Motion of a body about an axis or around an external point. In astronomy the word is restricted to orbital motion of a body about its centre of mass, the term rotation being used for axial motion.
2. One complete cycle of such motion.

Reynolds, Osborne (1842–1912) Brit. physicist and engineer. His main contributions to physics concerned fluid flow and *kinetic theory of gases (theory of radiometer action, 1874, 1876); but he also took part in a determination of the *mechanical equivalent of heat and he was a pioneer of the theory of the lubrication of bearings. He was the first to study the transition from laminar or streamline to turbulent flow, which is governed in cases of simple geometry by the attainment of a critical value by the dimensionless *Reynolds number.

Reynolds' law The pressure head h required to maintain a liquid flow at constant speed v through a pipe of length l and radius r is given by the equation:
$$h = klv^p/r^q.$$
The constants k, p, and q are known as the *Unwin coefficients*. $p \approx 1$; $q \approx 2$.

Reynolds number Symbol: Re. A dimensionless quantity equal to $\rho vl/\eta$, where ρ is the density of a fluid of viscosity η, in motion with speed v relative to some solid characterized by the linear dimension, l. For steady flow through a system with a given geometry, the flowlines take the same form at a given value of the Reynolds number. Thus flow of air through an orifice will be geometrically similar to that of water through a similar orifice if the dimensions and speed are chosen to give identical values of the Reynolds number. At the critical value, turbulence will arise in either system. For a straight pipe of diameter l, the critical value is of order 2000 but it is affected by roughness of the pipe. For a sphere of radius l moving through a fluid the critical value is about unity. Note that the linear dimension l can be chosen in different ways (e.g. diameter for radius), leading to ambiguity in some instances. *See* convection (of heat); dynamic similarity.

r.f. Abbreviation for *radio frequency.

rheology The study of the deformation and flow of matter. *See* plasticity.

rheometer An instrument designed to measure the flow properties of solid materials by investigating the relationship between stress, strain, and time.

Many studies have been made with constant-stress rheometers, i.e. the compressive or tensile forces applied to the specimen are varied as the cross-sectional area of the test piece changes so as to keep the mean stress constant.

rheostat A variable *resistor connected into a circuit, in series, to vary the current flowing in the circuit. It often consists of a linear or circular wire-wound resistor with a sliding contact, or it may have a number of small resistors that can each be brought into the circuit by a rotary switch. The word is usually applied to physically large devices. Small rheostats are usually called *potentiometers.

rheostat braking *Syn.* dynamic braking. A method of *electric braking used with electric motors particularly in electric traction. During the braking period, each motor is used as a generator, the energy being dissipated in one or more *rheostats, and a retarding torque is exerted by the armature of the machine. Since the energy is wasted, this method is less economical than *regenerative braking.

rheostatic control Of electric motors. A method of control provided solely by *rheostats connected in series with the armatures of the motors. It is used particularly in electric traction.

rhombic system *See* crystal systems.

rhombohedral system *See* crystal systems.

rhumbatron *See* cavity resonator.

ribbon microphone A type of *microphone that makes use of the simple dynamo principle that when a conductor moves perpendicular to a magnetic field, an e.m.f. is induced in it. The conductor in this case is a very thin strip of aluminium alloy a few millimetres wide, loosely fixed in a strong magnetic field parallel to the plane of the strip. The resulting force on the ribbon due to a sound wave is proportional to the difference in pressure between the front and back of the ribbon. When the acoustic path difference to the two sides is much smaller than a quarter of a wavelength, the resultant pressure on the ribbon is proportional to the product of the particle velocity and the frequency. If the resonant frequency of the ribbon is lower than the frequency of the sound, the resulting e.m.f. is then independent of frequency. Sound waves originating in the plane of the ribbon arrive at its front and back faces in phase so that no resultant force is produced. Thus the microphone has strong directional characteristics that can be used to reduce the pick-up of unwanted *noises. By a method due to Gerlach the microphone can be used to measure sound intensities. A current having the same frequency as the sound is passed through the ribbon and varied in amplitude and phase until the ribbon is brought to rest. The forces due to the current then balance the mechanical forces of the sound whose intensity can be calculated from the current amplitude and the magnetic field strength.

Richardson, Sir Owen Willans (1879–1959) Brit. physicist known for his work on the thermal emission of electrons by hot bodies. *See* Richardson–Dushman equation.

Richardson–Dushman equation *Syn.* Richardson's equation. The basic equation of *thermionic emission relating the temperature of a body to the number of electrons it emits. It has the form:
$$j = AT^2 e^{-b/T},$$
where j is the emitted current density, A and b are constants, and T the thermodynamic temperature. A depends on the nature of the metal surface and b can be put equal to ϕ/k, where ϕ is the *work function and k the *Boltzmann constant.

Richardson derived this equation by thermodynamic reasoning and it was later derived by Dushman using quantum mechanics.

Richter, Burton (*b.* 1931) Amer. physicist who worked at the high-energy physics laboratory at the University of California (Stanford). Here, in the 1960s, he was responsible for the design and development of the Stanford Positron–Electron Accelerating Ring (SPEAR), and he later (1974) used it to detect the *J/psi particle. He shared the 1976 Nobel prize with Samuel *Ting, who had independently made the same discovery.

Riemann, Georg Friedrich Bernhard (1826–1866) German mathematician who is remembered for his nonEuclidean geometry, which was used by Einstein in his formulation of the general theory of *relativity.

Riemannian space-time *See* relativity.

Righi effect *See* Leduc effect.

right ascension *See* celestial sphere.

right-handed axes *See* coordinate.

right-handed rotation Of the plane of polarization of light. Looking against the direction of the light, the plane is rotated clockwise. *See* rotation of plane of polarization.

right-hand rule For dynamo. *See* Fleming's rules.

righting lever *See* metacentre.

rigid body A body in which the distance between every pair of particles remains constant under the action of any forces. An abstract but useful concept in mechanics.

rigidity modulus *See* modulus of elasticity.

rime A layer of clear ice produced in certain frosty conditions by the deposition and subsequent rapid freezing of tiny supercooled water drops on surfaces.

ring current A strong electric current flowing westwards in the earth's upper atmosphere. It arises from the net flow of electrons (eastwards) and protons (westwards) trapped in the *Van Allen belts, and causes variations in the normal pattern of the earth's magnetic field.

ringing *See* pulse.

ring main 1. An electric main that is closed upon itself to form a ring. If the ring is supplied by a power station at one point only, then between that point and any other point in the ring to which a consumer may be connected there are two independent electrical paths. Hence, in the event of a fault in the ring, the latter can be broken and the faulty section disconnected without interrupting the supply to the consumer. Also in comparison with the simple system in which a single main is run between the power station and the consumer, the ring main has the advantage of better regulation.
2. A domestic wiring system in which individual outlets have their own fuses, a number of such outlets being connected in parallel to a ring circuit.

rings and brushes A pattern produced by interference between ordinary and extraordinary rays when the substance (uniaxial) is placed between *Nicol prisms in highly convergent light (dark brushes–Nicols crossed). In white light the rings are coloured.

ring winding *Syn.* toroidal winding; Gramme winding. A winding in an electrical machine in which the coils are wound on an annular magnetic core, one side of each turn being threaded through the magnetic ring.

ripple (electrical) An a.c. component superimposed on a d.c. component, resulting in variations in the instantaneous value of a unidirectional current or voltage. The term is used particularly in connection with the output of a *rectifier. The ratio of the *root-mean-square value of the ripple (i.e. a.c. component) to the mean value (i.e. d.c. component) is called the *ripple factor*. A circuit (usually a type of *filter) designed to reduce the magnitude of a ripple is called a *smoothing circuit*.

ripples On a fluid surface. Waves of small amplitude on the surface of a fluid for which the wavelength is small and the effects of the *surface tension of the fluid are important. The velocity of ripples of length λ on a fluid of density ρ and surface tension σ is given by:
$$\sqrt{(\lambda g/2\pi + 2\pi\sigma/\lambda\rho)},$$
g being the acceleration of *free fall.

ripple tank The similarity between a plane section of a three-dimensional sound wave and *ripples on a water surface has been used to study the motion of sound waves. Ripples are produced in a rectangular tank by rods that can be dipped just under the water surface. The action of a continuous note is represented by attaching the rod to an electrically maintained tuning fork, which causes it to dip into the water about fifty times per second. For demonstration purposes the ripple tank is usually fitted with a glass bottom through which light is projected from beneath and a 45° mirror above the tank to reflect the light horizontally onto a ground-glass screen. If the light is interrupted at the same frequency as that of the tuning fork, the ripples appear stationary. Reflection may be studied by placing suitable objects in the tank, reflection from the sides of the tank being damped out by the use of shelving beaches. The ripples travel more slowly in shallow water and so give the effect of a denser medium. Refraction can be studied by placing objects beneath the water surface, the effect of an acoustic lens being obtained by placing an optical lens just below the surface. Diffraction is shown by using suitable obstacles and apertures. If two dipping rods are used, the *superposition principle may be demonstrated. The acoustics of buildings can be investigated by using a two-dimensional wooden model projecting above the surface. Photographs taken by this method give similar results to the spark technique but the ripple tank is easier to use and has the advantage of a visual presentation if this is required. In the use of the tank the following points should be considered. The wave pattern is sometimes a combination of progressive and standing waves, which are not distinguished either by photography or interrupted illumination. The pattern seen is produced by lens effects of the liquid surface, neither the longitudinal nor the transverse displacements being observed directly, only the positions of maximum or minimum curvature being indicated. *See* sound-wave photography.

rise time *See* pulse.

Risley rotary prism A variable prism consisting of two equal narrow-angle prisms rotating in opposite directions. It is used in ophthalmic instruments, and as a separate unit for measuring defects of the external muscles of the eye.

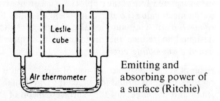

Emitting and absorbing power of a surface (Ritchie)

Ritchie's experiment (1833) Ritchie showed that the emitting power of a surface is proportional to its absorbing power by an experiment with a Leslie cube having one face blackened and the opposite face silvered, and a *differential air thermometer with one bulb blackened and the other polished (*see* diagram). When the blackened face of the cube faced the polished bulb and the polished face the blackened bulb, the liquid index remained stationary when hot water was placed in the cube.

Ritter, Johann Wilhelm (1776–1810) German physicist who discovered ultraviolet radiation.

RJE Abbreviation for *remote job entry.

rms value *See* root-mean-square value.

Rochelle salt Crystals of sodium potassium tartrate; they are of particular importance because of their ferroelectric properties. *See* ferroelectric materials.

Rochon prism A double-image prism consisting of two quartz prisms, the first to receive the light, cut parallel to the axis, the second with optic axis at right angles; their deviations are in opposition. The ordinary ray passes through undeviated and is achromatic; the extraordinary ray is deflected (doubling is produced). The prism can be used to produce plane polarization by placing a screen at a distance from the prism thereby intercepting the extraordinary ray.

rocker, brush *See* brush rocker.

rocket A missile or space vehicle powered by ejecting gas, that carries both its own fuel and oxidant (if required). Rockets are therefore independent of the earth's atmosphere and are the power systems used in space flights. Most space flights have been powered by chemical rockets in which the thrust is obtained by the expansion occurring when a solid or liquid fuel (e.g. alcohol) reacts chemically with the oxidant (e.g. liquid oxygen). Most rockets are multistage devices, the first, or booster stage, being jettisoned in the less dense region of the upper atmosphere. This has the double advantage of making the vehicle lighter and therefore easier to accelerate to its *escape speed, and it also reduces friction heating as the escape is not reached until the air has very low density.

rocksalt Naturally occurring crystalline mineral, sodium chloride, NaCl. It is transparent to infrared up to 14.5 μm and to the ultraviolet to 175 nm, it is used in the construction of prisms for infrared spectroscopy.

rods and cones *See* retina; eye; colour vision.

Roemer (or **Romer** or **Römer**), **Ole** (1644–1710) Danish astronomer. He made the first reasonable estimate (1676) of the speed of light, deduced from irregularities in the observed times of eclipse of one of the satellites of Jupiter. Roemer made several advances in the use of instruments in observational astronomy and introduced micrometers and reading microscopes into observatory practice.

roentgen (or **röntgen**) Symbol: R. A unit of exposure *dose of X- or gamma rays such that one roentgen produces in air a charge of 2.58×10^{-4} C on all the ions of one sign, when all the electrons released in a volume of air of mass 1 kg are completely stopped. $1 R = 2.58 \times 10^{-4}$ C kg^{-1}. The roentgen is rarely now used as exposure in SI units is measured in coulombs per kilogram.

Roentgen, Wilhelm Konrad (1845–1923) German physicist. (*Note*: Röntgen is the correct rendering, Roentgen being an accepted version of the modified vowel.) He is best known for his somewhat fortuitous discovery (1895) of X-rays (or *Roentgen rays* as they were formerly widely called). His thorough investigation of the properties of these rays is a model in scientific method. This discovery set Becquerel on a search for other such rays and led indirectly to the discovery of radioactivity.

Roentgen's other work included studies of the rotatory polarization of light and of thermal conduction in crystals. He demonstrated (1888) that magnetic effects similar to those of an ordinary current arise from motion of an electrically polarized medium (*Roentgen current*).

Roentgen rays *See* X-rays.

Roget, Peter Mark (1779–1869) Brit. scientist and philologist who discovered the persistence of vision.

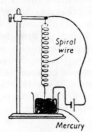

Spiral wire

Mercury   Roget's spiral

Roget's spiral An electric current is passed along a vertically arranged spiral wire (*see* diagram) dipping into a mercury pool for its lower contact. The mutual attraction of the turns of the spiral causes it to shorten, and the lower end lifts out of the mercury, breaking the circuit. The spiral lengthens again under its own weight, remakes the contact, and the operation is repeated, setting up vibrations in the spiral.

Rohrer, Heinrich (*b.* 1933) Swiss physicist. Working at the IBM Research Laboratory in Zurich with Gerd Binnig, he invented and built the first *scanning tunneling microscope. For this work they shared half a Nobel prize (1986), the other half being awarded to Ernst *Ruska.

rolling friction *See* friction.

ROM Abbreviation for read-only memory. A type of *semiconductor memory used, especially in computers, for the storage of information that does not require modification. ROM is fabricated in a similar way to *RAM, with a rectangular array of storage elements, but the contents of the storage elements are fixed during manufacture. They can only be read. Programmable ROM does exist (*see* PROM; EPROM), but the contents cannot be modified within a computer. As with RAM, any storage element in ROM can can be uniquely identified and directly accessed in any order, i.e. there is *random access to the storage elements.

roof conductors Conductors in a lightning protective system that interconnect various *air terminations and extend the zone of protection.

roof prism *Syn.* Amici prism. *See* direct-vision prism; prism.

root-mean-square (rms) value *Syn.* effective value; virtual value. The square root of the mean value of the squares of the instantaneous values of a current,

voltage, or other periodic quantity during one complete cycle; it is the effective value of current or voltage in an alternating current provided the resistance is constant. The rms value of a sine wave is the peak value divided by $\sqrt{2}$.

root-mean-square (rms) velocity The square root of the *mean square velocity defined as:

$$C = \left\{ \frac{n_1 c_1{}^2 + n_2 c_2{}^2 + \dots n_r c_r{}^2}{n} \right\}^{1/2},$$

where n_1 particles have velocity c_1, n_2 particles have velocity c_2, etc., to n_r particles having velocity c_r, and

$$n = \sum_1^r n_r$$

is the total number of particles.

Rosenberg generator *See* metadyne.

Rose's alloy *Syn.* Rose's metal. An alloy having a low melting point ($\sim 100\ °C$), consisting of 50% bismuth, 28–25% lead, 22–25% tin.

Rossi, Bruno (*b.* 1905) Italian-born Amer. physicist who has made substantial contributions to the theory of *meson decay as a result of his work on *cosmic rays.

rotameter An instrument that measures the rate of flow of fluids. It consists of a small float that, due to the motion of the fluid, moves vertically in a transparent calibrated tube. The height of the float gives a measure of the speed of the fluid.

rotary converter *See* converter.

rotary phase converter *See* phase converter.

rotary transformer *See* dynamotor.

rotating sector A device used in conjunction with a *pyrometer when measuring very high temperatures in order to cut by a known fraction the amount of radiation incident on the pyrometer. The true temperature T of a source is related to the temperature T_1 recorded by a total radiation pyrometer used with a sector subtending an angle θ at the centre, by the equation:
$$T_1{}^4 = [(2\pi - \theta)/2\pi]T^4.$$
The speed of rotation of the sector is immaterial provided it is not too low.

rotation 1. The turning or spinning motion of a body, geometrical configuration, etc., about an axis that passes through it. *See also* irrotational motion.
   2. One complete cycle of such motion.
   3. (rot) *See* curl.

rotational field *See* curl.

rotational quantum number *See* energy level.

rotation of plane of polarization Certain substances called optically active substances (such as quartz and sugar solution) rotate the plane of polarization of light, some to the right at the top when facing oncoming light (clockwise, dextrorotatory, or right-handed), others to the left (laevorotatory or left-handed). The rotation is proportional to the product of the length of substance traversed times the concentration, in the case of solutions. The *specific rotation* of a solution is that produced by 10 cm of solution with concentration of 1 g of active substances per cm^3. For solids it is expressed usually per mm. *Rotatory dispersion* arises from the fact that rotation of the plane of polarization is different for different wavelengths.

rotation photography A crystal-diffraction method in which a single crystal is allowed to rotate about an axis normal to an incident beam of monochromatic X-rays (or electrons, neutrons, etc.), the photographic film, plane or cylindrical, being stationary.

rotation spectrum *See* spectrum.

rotatory dispersion *See* rotation of plane of polarization.

rotor The rotating part of an electrical machine. The term is usually applied only to an a.c. (as distinct from a d.c.) machine. The term is also applied to the rotating part of a *motor meter. *Compare* stator. *See also* slip-ring rotor; induction motor.

Rousseau diagram *See* photometry.

Routh's rule The moment of inertia of a uniform solid body about an axis of symmetry is given by the product of the mass and the sum of squares of the other semiaxes, divided by 3, 4, or 5 according to whether the body is rectangular, elliptical, or ellipsoidal.

The circle is a special case of the ellipse. The rule works for a circular or elliptical cylinder about the central axis only, but for circular or elliptical discs it works for all three axes of symmetry. For example, for a circular disc of radius a and mass M, the moment of inertia about an axis through the centre of the disc and lying (a) perpendicular to the disc, (b) in the plane of the disc, is:

(a) $\frac{1}{4}M(a^2 + a^2) = \frac{1}{2}Ma^2$,
(b) $\frac{1}{4}Ma^2$.
See Lees' rule.

Rowland, Henry A. (1848–1901) Amer. physicst. Rowland's outstanding achievements were the production of high-grade optical *diffraction gratings (1882) and a careful redetermination (1879) of the *mechanical equivalent of heat by a development of the water-stirring technique of Joule. He also demonstrated that moving electrical charges produce magnetic fields (1878; *Rowland effect*).

Rowland grating A *diffraction grating ruled by H. A. Rowland, or on a machine to his design. Such machines have the grating blank mounted on a carriage advanced by a lead-screw. A pawl and ratchet device moves the carriage by short steps (equal to the desired grating spacing) and causes the ruling diamond to slide across the work. Defects in the wheels and screws lead to inaccuracies in spacing

and, when these recur with a periodic pattern, the diffraction pattern given by monochromatic radiation is not a simple line in each order but contains also faint *Rowland ghosts*, which may be mistaken for genuine faint lines. In 1883, Rowland constructed concave gratings of high resolving power on metal surfaces (up to 15 000 lines per in. for 5 to 6 in.). The radius of curvature of the grating was 21 ft. A circle whose diameter is equal to the radius of curvature of the grating surface is called a *Rowland circle* and possesses the property that if the slit, grating, and receiving screen lie on its circumference the spectrum will be in focus. The *Rowland mounting satisfies the requirements of the Rowland circle by mechanical means. Since the grating acts by reflection and does not require any lenses, it is suitable for use with ultraviolet radiation, including that of very short wavelength. By working at grazing incidence it has been used with soft X-rays of wavelength of the order of nanometres.

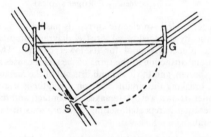

Rowland mounting

Rowland mounting A mounting for a concave *diffraction grating that has the plate-holder H and the concave grating G mounted on carriages running on two rails fixed at right angles, with the slit S mounted at their intersection. O is the centre of curvature of the grating (*see* diagram). It can be shown that all rays diffracted at a given angle from such a grating will be brought to a focus at a point on the broken circle, of diameter equal to the radius of curvature of the grating. As carrier G moves towards S, successively higher orders are brought into focus. The plate at H records only a small part of the spectrum at one time and it should be bent to lie along the circle. In practice a number of plates may be arranged along an arc of this circle.

RTL Abbreviation for resistor-transistor logic.

rubber Raw rubber is the solid material obtained by coagulating latex, a milky juice from the bark of certain trees: latex is an emulsion containing rubber particles of about 1–2 micrometres diameter. In the raw state rubber is weak, plastic, and inelastic, being unsuitable for most manufactured goods before it is vulcanized. This process converts the raw rubber into strong, elastic, nonplastic, and insoluble material by combination with sulphur. The rubber can be given a wide range of properties by the inclusion before vulcanization of certain compounding ingredients. Ebonite is a hard rubber, also known as vulcanite, being a black product of vulcanizing mixtures in which the rubber: sulphur ratio usually lies between 65:35 and 70:30. It has an electrical resistivity of 2×10^{13} Ω m. Rubber is extensively used for its electrical insulating properties, but it should be noted that conducting rubber is now available for use where dangerous electrostatic charges may develop, e.g. in belt drives to machinery. Many special-purpose synthetic rubbers are also available.

The elastic properties of different forms of rubber differ greatly, the Young modulus being of the order 10^5 N m^{-2} for very soft rubbers and many hundred times this for highly vulcanized rubbers. Extension may exceed 1000%, the elasticity changing considerably with strain. Generally the elasticity, especially for soft rubbers, is very imperfect with considerable *creep and *hysteresis. Special rubbers with either very high or relatively low hysteresis have been developed because of the effects on *friction of tyres.

Rubbia, Carlo (*b.* 1934) Italian physicist. Senior physicist at the European Organization for Nuclear Research (formerly CERN) and professor at Harvard, he proposed the colliding-beam experiments at CERN, which provided the evidence (1983) for the existence of W and Z particles. For this work he shared a Nobel prize (1984) with Simon *van der Meer.

rubidium-strontium dating *See* dating.

ruby The red variety of corundum crystals, hexagonal in form. Artificial rubies may be made by fusing together alumina, potassium carbonate, calcium fluoride, and a little potassium chromate to give the red colour. They may also be made by fusing alumina with a little chromic oxide.

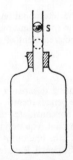

Determination of ratio of specific heat capacities of gases (Ruchardt)

Ruchardt's determination of γ (1929) Ruchardt found the ratio of the principal specific heat capacities of a gas by measurements of the slow oscillations of a small metal sphere S in a vertical glass tube of uniform circular section, fitting into a large vessel of volume V containing the gas (*see* diagram). The time period of the oscillations is given by:

$$T^2 = 4\pi^2 mV/\gamma pA^2,$$

where m is the mass of the ball, A the cross-sectional area of the tube, and p the pressure in the vessel when the sphere comes to rest. Broderson developed

a photographic method of determining the time period for the limited number of oscillations available.

Ruhmkorff's coil *See* induction coil.

Rumford, Sir Benjamin Thompson, Count (1753–1814) Amer.-born scientist, soldier, and statesman, worked in the UK and Bavaria. His main scientific work was in the subject of heat. He cast some doubt upon the caloric theory by his observation of the unlimited production of heat in the operation of cannon-boring when the borer was blunt (1789). He held advanced ideas on the economical and efficient use of heat and he founded the *Royal Institution of Great Britain* (1799), largely with the idea of making scientific knowledge of this kind available to artisans and others but he soon left the management. He was made a Count of the Holy Roman Empire in 1790 by the Elector of Bavaria, for whom he carried out many reforms. (*See* shadow photometer.)

Runge, Carl David Tolmé (1856–1927) German physicist who worked on the theory of functions and, in spectroscopy, on the theory of the *Zeeman effect.

running in of mechanical parts This practice produces a *Beilby layer to an appreciable depth on the surface and, in the case of the cast iron used for cylinders of internal combustion engines, graphite (a lubricant) is brought to the surface.

Ruska, Ernst August Friedrich (1906–1988) German electrical engineer. While working at the Technical University in Berlin (of which he later became a professor) he invented (1933) the electron microscope. He later became director of the Institute for Electron Microscopy (1955–1972) and was awarded half a Nobel prize (1986) for his invention.

Russell, Henry Norris (1877–1957) Amer. astronomer who independently of Hertzsprung worked on the theory of stellar evolution. *See* Hertzsprung–Russell diagram.

Russell–Saunders coupling *See* coupling (def. 2).

rutherford Symbol: rd. A unit of *activity. It is the quantity of a nuclide required to produce 10^6 disintegrations per second. 1 rutherford = 2.7×10^{-5} *curie = 10^6 *becquerel.

Rutherford, Lord (Ernest Rutherford; 1871–1937) Brit. physicist born in New Zealand, whose brilliant researches established the existence and nature of radioactive transformations, the electrical structure of matter, and the nuclear nature of the atom. He identified α-particles and β-particles and with Soddy advanced the theory of spontaneous atomic disintegration, which resulted in the discovery of *isotopes. He originated the concept of the nucleus of the atom and was successful in converting nitrogen into an isotope of oxygen by bombarding it with α-particles, this being the first transmutation by artificial means of one element with another. He was knighted in 1914 and created a baron in 1931. (*See also* atom; nucleus.)

rydberg *See* atomic unit of energy.

Rydberg, Johannes Robert (1854–1919), Swedish physicist who developed a formula for a series of spectral lines independently of Balmer (*see* hydrogen spectrum). The constant in this formula is named after him.

Rydberg constant Symbol: R. The quantity appearing in the equation that gives the *wavenumbers k of the lines in the spectra of atoms containing a single electron (hydrogen, deuterium, singly ionized helium, etc.):

$$k = R(1/n^2 - 1/m^2)Z^2.$$

Z is the atomic number, n and m are positive integers. R contains the *reduced mass of the electron as a factor so its value is slightly different for each type of atom. Assuming the actual rest mass of the electron (i.e. the mass of the nucleus is regarded as infinite), the value is

$$R = 1.097\ 373\ 153\ 4 \times 10^7\ \text{m}^{-1}.$$

Apart from the effect of finite nuclear mass there are certain very small corrections to this value to allow for nuclear magnetic moment, nuclear quadrupole electric moment, and relativistic effects. *See* atom; hydrogen spectrum; Rydberg energy.

Rydberg energy The quantity:

$$Rhc = me^4/8\varepsilon_0^2 h^2,$$

where R is the *Rydberg constant, h is the Planck constant, c the speed of light, ε_0 the electric constant, and m the mass of the electron. Subject to certain small corrections, this represents the binding energy of the electron in a hydrogen atom in the ground state, the nuclear mass being regarded as infinite. Value = 13.605 804 36 eV.

Rydberg spectrum An absorption *spectrum of a gas taken with ultraviolet radiation and used for determining *ionization potential. The spectrum contains a large number of lines, each corresponding to excitation of an electron from its normal orbit (in the *ground state) to an allowed orbit further from the nucleus (an *excited state). The lines form a series (*see* hydrogen spectrum) and become closer together as the energy increases. At one particular energy they merge into a continuum. This is the energy required to ionize the atom or molecule. At higher energies absorption of radiation is continuous because part of the energy of the absorbed photon is used to ionize the atom and the remainder is taken up by the kinetic energy of the electron.

Ryle, Sir Martin (1918–1984) Brit. radio astronomer who became professor of astronomy at Cambridge. A pioneer of radio telescopy, he developed a technique for studying distant radio sources using two radio telescopes to give a large effective aperture. His observations gave experimental support to the big-bang theory (*see* cosmology). He shared the 1974 Nobel prize for physics with Antony Hewish.

S

Sabattier effect *See* solarization.

sabin An f.p.s. unit of equivalent absorption area of a surface to sound. One sabin is equivalent to 1 square foot of a surface, with a reverberation absorption coefficient of unity, that wouls absorb sound energy at the same rate as the surface being investigated. This unit is sometimes called the open window unit.

saccharimeter An instrument for measuring the rotation of the plane of polarization of optically active solutions, especially sugars, the quality of which is tested for commercial and import duty reasons. There are numerous designs to secure greater accuracy of measurement generally using the half shadow principle (equality of brightness of two adjacent fields). *See* half wave plate.

safety factor *See* factor of safety.

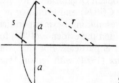

Sagitta of arc

sagitta (sag) Of an arc. The maximum elevation of a convex surface (or depression of a concave surface). If r is the radius, a the semiaperture and s the sagitta, then

$$s(2r - s) = a^2$$

and approximately for shallow curves

$$s = a^2/2r.$$

The sagitta relation is useful for the determination of lens thickness, theory of spherometer and lens measure, curvature of wavefront, and treatment of reflection and refraction.

sagittal coma *See* coma.

sagittal (plane, rays, section) During discussion of reflection or refraction of oblique pencils (centric or eccentric), two principal sections are considered, the meridian (tangential) and the sagittal (equatorial). The planes are at right angles and the line of intersection contains the chief ray of the pencil. Rays in a sagittal section are converged to a point on oblique refraction by a lens, and the other rays form a focal line with a slender figure-eight section (called the *sagittal focal line*), which lies practically in the meridian plane.

Saha, Meghnad N. (1894–1956) Indian astrophysicist who became professor of physics at Calcutta. His main work was on the spectra of stars; he was able to explain the lines in stellar spectra by the presence of ions formed by thermal ionization at high temperatures. *See* Saha equation.

Saha equation An equation that predicts the degree of thermal ionization in a gas at constant pressure. It is usually written in the form:

$$\log_{10} K_p = \frac{-5050}{T} I + \frac{5}{2} \log_{10} T - 6{\cdot}5,$$

where K_p is the equilibrium constant of the ionization reaction at constant pressure, T is the thermodynamic temperature, and I is the *ionization potential of the species being ionized.

Salam, Abdus (b. 1926) Pakistani physicist who has worked at Imperial College, London, and at the International Centre for Theoretical Physics, Trieste. Salam shared the 1979 Nobel prize for physics with Sheldon Glashow and Steven Weinberg for work done on the *electroweak theory of particle interactions.

salient pole A type of *pole piece that is attached to the *yoke and projects towards the armature in an electrical machine.

Salter duck See renewable energy sources.

sampling A technique of measuring only some portions of a signal, the resultant set of discrete values being taken as representative of the whole. For the information contained in a signal to be contained in the output sample value without significant loss, the rate of sampling of a periodic quantity must be at least twice the frequency. A sampling circuit is one in which the output is a set of discrete values representative of the input values at a series of different instants. The technique is widely used for conveying information in such devices as *analogue/digital converters, *digital voltmeters, *multiplex operation, *pulse modulation, etc.

sapphire The blue variety of crystals of *corundum.

satellite A natural or artificial body orbiting another body so large that the centre of mass of the system is well within the larger body. The name is not normally applied to the planets of the solar system, although they are strictly satellites of the sun.

Six of the major planets possess natural satellites, the earth (one – the moon), Mars (two), Jupiter (at least 16), Saturn (at least 22), Uranus (at least 15), and Neptune (at least six).

Many artificial satellites have been launched, either to orbit the earth itself, or to orbit other bodies of the solar system. They fall into two classes:

(i) *Information satellites* are designed to provide information concerning the earth, other celestial objects, or space itself, and to relay it back by radio. They carry a variety of measuring instruments and cameras, plus support equipment to control and power them and store data prior to transmission. Typical of the observations made are: depth, pressure variation, and resistance of the atmosphere; ionospheric structure; components and strength of *cosmic rays; presence and sizes of magnetic fields; studies of optical sources in the universe; studies of radio sources, infrared sources, and ultraviolet, X-ray, and γ-ray sources; ultraviolet and infrared radiation from the sun; atmospheric composition; meteorological conditions; navigational information; surveys of the earth's surface; hydrogen in space; stellar magnitudes; *solar wind.

(ii) *Communications satellites* are designed to provide high-capacity communications links between widely separated locations on the earth's surface. Telephone services and live TV broadcasting are achieved by transmission of radio signals from one point on earth to a satellite, where it is amplified before being relayed back (at a different frequency) to another or other locations on earth. The orbits of communications satellites lie above the earth's *atmosphere. High-frequency radio waves (microwaves), which can penetrate the *ionosphere, must therefore be used. Most satellites are in a *geostationary orbit in the plane of the earth's equator. An orbiting geostationary satellite appears stationary to observers on the ground and it can cover an extensive area of the earth's surface.

The orbits of satellites are approximately ellipses with the centre of the primary at one focus. (*See* Kepler's laws.) Departures from the ideal laws are caused by attractions by other bodies (*see* perturbation theory), imperfect spherical symmetry of the primary (*see* ellipticity; mascon), or friction in the upper atmosphere. For the limiting case of a circular orbit of radius r around a body of mass m the period T is given by:

$$T^2 = 4\pi^2 r^3 / Gm,$$

where G is the gravitational constant.

saturable reactor See transductor.

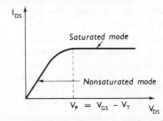

Saturated and nonsaturated modes of an FET

saturated mode The operation of a *field-effect transistor in the portion of the characteristic beyond pinch-off, i.e. $V_{DS} \geqslant V_p$, where V_p is the pinch-off voltage (*see* diagram). The drain current I_{DS} is independent of the drain voltage V_{DS} in this region. *Nonsaturated mode* is operation of the device in the portion of its characteristic below pinch-off.

saturated vapour A vapour in equilibrium with the liquid or solid phase. The pressure exerted by a saturated vapour is dependent upon temperature (*see* Clausius–Clapeyron equation; triple point) and upon the curvature of a liquid surface. Values of this *equilibrium vapour pressure*, or *saturated* or *saturation vapour pressure* (SVP), are given for flat liquid surfaces. The value over a surface of radius of curvature R is changed by

saturated vapour pressure

$$\delta p = \frac{2\gamma}{R}\left(\frac{\rho_v}{\rho_L - \rho_v}\right)$$

where γ is the surface tension, ρ_L and ρ_v are the densities of liquid and vapour respectively, and δP is positive for concave curvature.

If the actual vapour pressure is greater than that for equilibrium, the vapour is *supersaturated*. Under these conditions the vapour will condense on any suitable surfaces or nuclei that are present. Water vapour in the atmosphere usually condenses on minute crystals of sea salt, or if below 0 °C upon particles of clay (*see* snow). (*See also* cloud chamber.) If the pressure of the vapour is less than the equilibrium value, the liquid or solid will evaporate. *See* negative specific heat capacity; supersaturated vapour.

saturated vapour pressure (SVP) The pressure exerted by a *saturated vapour.

saturation 1. (magnetic) The degree of magnetization of a substance that cannot be exceeded however strong the applied magnetizing field. In this state all the domains (*see* ferromagnetism) are assumed to be fully orientated along the lines of force of the magnetizing field.
2. (electronic) A condition in which the output current of an electronic device is substantially constant and independent of voltage. In the case of a *field-effect transistor, saturation is an inherent function and produces the maximum current inherent to the device. With a bipolar junction *transistor, saturation occurs because the output from the collector electrode is limited by the elements of the external circuit; changing these alters the magnitude of the *saturation current* drawn from the transistor.
3. (light) *See* colour.

saturation current The portion of the static *characteristic of an electronic device in which further increases in the voltage do not lead to a corresponding increase in the current, until *breakdown is reached. The actual value of the saturation current is a function of the device and the external circuit.

saturation resistance The resistance beween the *collector and *emitter terminals of a bipolar *transistor under specified conditions of *base current, when the collector current is limited by the external circuit.

saturation vapour pressure (SVP) *See* saturated vapour.

saturation voltage The residual voltage between the collector and emitter of a bipolar junction *transistor, under specified conditions of base current, when the collector current is limited by an external circuit. *See* saturation.

Savart, Felix (1791–1841) French physicist who, with *Biot, discovered the law named after them. He also invented a wheel to measure the frequency of notes and investigated the polarization of light.

SAW devices *See* surface acoustic wave devices.

sawtooth waveform A periodic waveform whose amplitude varies approximately linearly between two values, the time taken in one direction being very much longer than the time taken in the other (*see* diagram). Sawtooth waveforms are usually produced by *relaxation oscillators and are frequently used to provide a *time base.

Sayers, James (*b.* 1912) Brit. physicist who helped to develop the *magnetron thus enabling *radar to be used in World War II.

scalar A quantity defined by a single magnitude (as distinct from a *vector, which has magnitude and direction and thus needs three numbers to define it). Examples are mass, time, and wavelength.

scalar product *See* vector.

scale of temperature *See* International Practical Temperature Scale.

scaler A device that produces an output pulse when a specified number of input pulses have been received. It is frequently used for counting purposes, particularly in conjunction with *Geiger counters and *scintillation counters.

scanner *Syn.* rectilinear scanner. A device used for visualizing the distribution of a *radioactive compound in a particular system, usually the human body. The scanner consists of a *scintillation crystal, *photomultiplier tube, and amplifying circuits. The crystal is collimated so that only the radiation from a small area is received by the crystal. The crystal is driven by a motor across the area to be investigated, reversing direction and stepping along by a specified amount at the end of each line. The output from the amplifier is used to drive either a light source or a mechanical printer, producing a film whose blackening depends on the intensity of radiation, or a printed picture composed of dots (usually coloured) the number of dots and/or the colour depending on the radiation intensity. The resolution and sensitivity of the scanner is dependent on the particular radioactive nuclide used, the speed at which the scanner is driven, the size of crystal used, and the efficiency of the collimator. Anomalies can occur when the radioactive source is deep in the patient's body, and to overcome this problem, double-headed scanners have been developed in which two crystals simultaneously scan the area under investigation, their outputs being summed.

scanning 1. The process of exploring an area or volume in a methodical manner, in order to produce a variable electrical output whose instantaneous value depends on the information contained in the small area examined at each instant. The information can then be reproduced by a suitable receiver. The technique is most often used in *television, *facsimile transmission, and *radar. (*See also* scanner.)
2. In a *cathode-ray tube, causing one complete horizontal or vertical traverse of the spot of light on

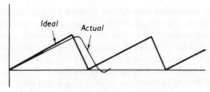

Sawtooth waveform

the screen in response to the voltage generated by a *time-base circuit.

scanning electron microscope (SEM) *See* electron microscope.

scanning-transmission electron microscope (STEM) *See* electron microscope.

scanning tunnelling microscope (STM) A type of electron microscope based on the *tunnel effect and used primarily for studies of surfaces. If two electrical conductors are brought very close together, it is possible for electrons to tunnel from one to the other. The sample to be analysed in the microscope forms one conductor and a very fine metal tip forms the other; if the sample is nonconducting it must be coated with a thin layer of conducting material. An electric potential is applied between the conductors and electrons tunnel across, producing a small current. The tip is scanned over the surface of the sample, and in the process is moved up and down so that the current remains constant. The tip therefore remains at a constant distance from the sample surface and its vertical movements are processed by computer to provide a topographical map of the surface. Horizontal and vertical resolution is approximately 0.2 nm and 0.01 nm respectively. *See also* atomic force microscope.

scattering 1. The deflection of light by fine particles of solid, liquid, or gaseous matter from the main direction of a beam. If the particles are relatively large, reflection and refraction as well as diffraction play a part; if the particles are small (smaller than a wavelength or as small as a molecule), the effect is diffractive. In *Rayleigh scattering the scattered intensity for small particles varies as $(1/\lambda^4)$ so that white light scattered by very small particles is bluish (*Tyndall effect). Chalk dust particles are whitish – they are larger and reflection effects are more noticeable. The blue of the sky is due to scattering by air molecules, and the red sun is due to the removal of the blue by scattering from the direct beam. Retardation of velocity in a medium of higher refractive index has been explained in terms of interference of scattered light with the primary wave, resulting in a change of phase – in effect a change of velocity.

2. The deflection of radiation resulting from the interaction of individual particles or photons with the nuclei or electrons in the material through which the radiation is passing or with the photons of another radiation field. *Single scattering* is the result of one interaction; *plural scattering* is the result of a few interactions; *multiple scattering* is the cumula-

tive effect of many interactions, each of which (in the case of charged particles) causes very little deflection.

Inelastic scattering occurs as a result of an inelastic *collision in which net changes occur in the internal energies of the participating systems and in the sum of their kinetic energies before and after the collision; there is no such energy change in *elastic scattering*, which occurs as a result of an elastic collision.

Thomson scattering is the scattering of electromagnetic radiation by free (or loosely bound) electrons, which can be explained by classical physics or nonrelativistic quantum theory in terms of the forced vibrations of the electrons of an atom that is absorbing radiation. These oscillating electric charges become the source of electromagnetic radiation of lower energy than the incident radiation and it is emitted in all directions. If I_0 is the intensity per unit area of the incident radiation, the total intensity, I, of the scattered radiation is given by:
$$(8\pi/3)(e^2/mc^2)^2 I_0,$$
where e and m are the electronic charge and mass and c is the speed of light. The quotient I/I_0 has the dimensions of an area and is the scattering *cross section of the electron.

The elastic scattering of photons by electrons, known as the *Compton effect, produces a reduction of energy of the photons. The *Raman effect involves scattering of photons by molecules. The Coulomb field of the nucleus produces *Coulomb scattering of particles due to electrostatic repulsion. (*See also* scattering amplitude.)

Resonance scattering occurs at high energies when the incident wave can penetrate the nucleus and interact with its interior. If the wave is reflected at the nuclear surface, *potential scattering* takes place. *Shadow scattering* results from the interference of the incident wave and scattered waves.

scattering amplitude A mathematical function specifying the *wave functions of elementary particles scattered in a collision. One of the most important ways in which the interactions between particles can be investigated experimentally is by allowing a beam of high-velocity particles to collide with other particles. As a result of the collisions that occur, some of the particles are deflected and in some cases the final particles may have different *quantum numbers from the initial ones (*see* strong interactions). Details of the scattering, such as the angular distribution of the final particles, will depend on the nature of the forces that act during the collision. Scattering amplitudes describe these scattering processes theoretically. From these amplitudes, the angular distribution of the final particles can be calculated. Scattering amplitudes are usually written in terms of relativistically invariant quantities.

scattering coefficient *See* linear attenuation coefficient.

scattering cross section Symbol: σ_s. A measure of the probability that a specified particle will be scattered

499

scattering of sound

by a specified nucleus or other entity through an angle greater than or equal to a specified angle, θ. (*See* cross section; scattering.) The *differential scattering cross section* is a measure of the probability of scattering through an angle lying between θ and $\theta + d\theta$.

scattering of sound In treating the reflection of sound waves, it is assumed that the wavelength is smaller than the reflecting surface so that by decreasing this wavelength the effectiveness of a small reflector improves. If the incident sound wave is complex, consisting of the fundamental tone together with the higher harmonics, the component tones will be reflected in increasing proportion towards the higher frequencies, i.e. towards the shorter wavelength. Rayleigh has dealt with this question mathematically referring to the reflection as secondary waves and the phenomenon as scattering of sound. He showed that the amplitude of these secondary waves varies directly as the volume V of the scattering element, inversely as the square of the wavelength λ of the incident sound, and inversely also as the distance apart from the element, i.e.

$$a_s/a_i \propto V/\lambda^2 r,$$

where a_s and a_i are amplitudes of the scattered and incident waves respectively. Therefore

$$I_s/I_i \propto V^2/\lambda^4 r^2,$$

where I_s and I_i are the intensities of the scattered and incident waves. In general, the intensity of the scattered wave varies inversely as the fourth power of the wavelength; this is a well-known law in optics. (*See* scattering.) In sound it has been shown that the octave of a tone reflected from a small scattering object is sixteen times more intense than in the incident sound. Echoes of such a character have been called *harmonic echoes. The shorter wavelength components have been scattered by the small volume obstacle in a much greater proportion than the fundamental or long wavelength component; so the sound behind the obstacle appears to be purified.

Schawlow, Arthur Leonard (*b.* 1921) Amer. physicist. A professor at Stanford, he worked with Charles Townes on the maser and is regarded as a coinventor of the laser, for which he shared a Nobel prize (1981) with Nicolaas Bloembergen and Kai Siegbahn.

Determination of specific heat capacity of gas at constant pressure

Scheel and Heuse's determination of specific heat capacity (1912) The specific heat capacity at constant pressure for gases down to liquid air temperatures was measured by a continuous flow method due to Swann. The glass calorimeter is sealed inside

an evacuated tube M (*see* diagram), to reduce heat losses, the whole being immersed in a constant temperature bath. The gas passes through a long spiral immersed in the bath, entering the calorimeter at the bottom where its temperature is recorded by the platinum resistance thermometer P_1. It then passes up between the walls G and B and then down between B and A, and finally up up past the constantan heater K. Any heat escaping from the heater is thus brought back by the oncoming gas. The gas is thoroughly mixed by a copper gauze cylinder surrounding K before its temperature is measured by the resistance thermometer P_2 connected differentially to P_1. The electrical power in the steady state is then equated to the heat gained by the amount of gas flowing through per second.

schematic eye *See* eye (optics).

Schering bridge A particular form of alternating-current *bridge extensively used for measuring the losses in cables and insulators at high working voltages and also for testing capacitors at high and low voltages.

Schleiermacher's method to determine thermal conductivity of a gas (1888) Schleiermacher used the hot-wire method in which the gas was contained in a cylinder radius r_1 along whose axis was a resistance wire of radius r_2. The rate of flow of heat, Q, between the heated wire and the coaxial cylinder in the steady state was determined from the electrical power supplied to the wire. If T_1 and T_2 are the temperatures of the cylinder and wire respectively:

$$k = \frac{1}{2\pi l} \frac{Q}{(T_2 - T_1)} \log_e \frac{r_1}{r_2},$$

where l is the length of the coaxial cylinder. The temperature of the wire can be determined from its electrical resistance. Weber made a study of the elimination of end effects and convection effects. The former are determined by using two vessels of different lengths and the latter by using the tube in both the horizontal and vertical positions and by varying the pressure of the gas.

schlieren method A method (due to A. Toepler, 1866) of exhibiting inhomogeneities in transparent media (e.g. flaws in glass, convection currents, shock waves, pulses of sound). It depends on the use of special illumination to make changes in refractive index apparent as the density of the medium changes. For one of the simplest of the possible arrangements, *see* schlieren photography of sound waves.

schlieren photography of sound waves To adapt the *schlieren method to photography of sound waves, light from a spark gap L is focused by a good-quality lens onto the edge of a screen. Another spark gap S is placed between the lens and the screen and a camera behind the screen is focused onto S. This spark gap is used to produce the sound that spreads out from it in a spherical pulse. The pulse is a thin layer of the

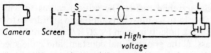

Photograph of sound waves

medium having a very high density and refractive index relative to the surrounding air. Without the sound wave, the field of the camera is dark when the illuminating gap sparks, since the light is stopped by the screen. When the light passes through a sound wave from S, however, it is deviated by the highly compressed shell of air and appears on the camera plate as a bright line on a dark ground. It is necessary for the sound to be produced a fraction of a second before the light pulse if a spherical wave is to be photographed. This is arranged by having the two spark gaps in the same circuit with a capacitor C in parallel with the illuminating gap. The light pulse is then delayed by an amount depending on the capacitance of the capacitor. A concave mirror can be used instead of the converging lens if desired. The method has been used to show reflection of a sound pulse. Photographs have also been obtained by this method of the shock waves due to a projectile travelling at supersonic speeds and the explosion wave from the mouth of a tube. See shock waves; supersonic flow.

Schmidt, Bernhard (1879–1935) Swedish-born German astronomer who developed a new method for dealing with spherical aberration in astronomical telescopes. See Schmidt telescope.

Schmidt, Maarten (b. 1929) Dutch-born Amer. astronomer. Schmidt worked at the Hale Observatories in California and is noted for his work (1963) in interpreting the spectra of quasars in terms of Doppler shifts.

Schmidt corrector See aberrations of optical systems.

Schmidt number Symbol: Sc. A dimensionless parameter equal to the ratio v/D, where v is the *kinematic viscosity and D the diffusion coefficient.

Schmidt telescope (or **camera**) A wide-field astronomical telescope that uses a thin figured transparent plate – a *Schmidt corrector* – at the centre of curvature of a short-focus spherical primary mirror to remove *spherical aberration; *coma and oblique *astigmatism are also negligible. A large field of view can be sharply focused on a curved surface, to which a photographic plate can be sprung. See aberrations of optical systems; reflecting telescope.

Schmitt trigger A type of *bistable circuit that gives a constant high-voltage output when the input signal exceeds a specific voltage, and falls to a constant low-voltage output below this value. A constant output is obtained irrespective of the input waveform, which may be sinusoidal, sawtooth, etc. This type of circuit is frequently used as a *trigger

circuit and in binary *logic circuits to maintain the logical 1 and logical 0 levels.

Schottky barrier See Schottky effect; Schottky diode.

Schottky defect See defect.

Schottky diode A *diode consisting of a metal-semiconductor junction (*Schottky barrier*), which has rectifying characteristics similar to a *p-n junction. When a forward bias is applied, *majority carriers with sufficient energy can cross the barrier and a current flows by the process of thermionic emission (see Schottky effect). A Schottky diode differs from a p-n junction diode in that the *diode forward voltage is different (lower for commonly used materials) and there is no charge stored when the diode is forward biased. The device can therefore be turned off very rapidly by the application of reverse bias, as the *storage time is negligible.

Schottky effect A reduction of the *work function of a solid due to the application of an external electric field, leading to a consequent increase in its *thermionic emission. The presence of an accelerating field lowers the potential energy of electrons outside the solid. This leads to a distortion of the potential barrier and consequent lowering of the work function. In the case of a metal there is also a contribution from *image potentials. The lowering of the work function increases the electron current due to thermionic emission. In the Schottky effect the electrons leaving the solid pass over the potential barrier, as opposed to tunnelling through it. (See tunnel effect; field emission.)

If the vacuum is replaced by a semiconductor the metal-semiconductor junction is called a *Schottky barrier*. Similar effects occur although generally the decrease in work function is less.

Schottky noise Syn. shot noise. Variations in the current output from an electronic device that arise due to the random manner in which electrons or holes are emitted from an electrode (e.g. the collector of a *transistor or the source of an FET).

Schottky TTL See transistor–transistor logic.

Schrage motor A form of a.c. *commutator motor in which control of speed is effected by shifting the position of sets of brushes on the commutator. It is a shunt-characteristic motor and the usual speed control range is from about 0.5 to about 1.5 times the *synchronous speed.

Schrieffer, John Robert (b. 1931) Amer. physicist who collaborated with John Bardeen and Leon Cooper in the formulation of the BCS theory of *superconductivity (1957).

Schrödinger, Erwin (1887–1961) Austrian physicist, whose work on *wave mechanics initiated this branch of physics. See Schrödinger wave equation.

Schrödinger wave equation (1926) The basic nonrelativistic equation of *wave mechanics expressing the behaviour of a particle in a field of force. The time-dependent equation describing *progressive waves, applicable to the motion of free particles, is

$$\nabla^2\psi - \frac{4\pi m}{ih}\left(\frac{\partial\psi}{\partial t}\right) - \frac{8\pi^2 mU}{h^2}\,\psi = \phi$$

where ∇^2 is the *Laplace operator, m is the mass of the particle, U is the potential energy, h is the Planck constant, and $i = \sqrt{-1}$. The *wave function ψ is a function of the coordinates and time.

A particle bound in a system, such as an electron in an atom, is analysed by the time-independent form of this equation describing *standing waves,

$$\nabla^2\psi + \frac{8\pi^2 m}{h^2}\,(E - U)\psi = \phi$$

where ψ is now a function of the coordinates only. The quantity ψ is usually a complex function. The most commonly accepted meaning is that proposed by *Born; that the value of $|\psi|^2 dV$ at a point represents the probability of finding the particle in the element of volume dV at this point.

Schrödinger's time-independent equation can be solved analytically for a number of simple systems but for most problems it is necessary to use *perturbation theory or other approximate methods. It may be noted that the time-dependent equation is of the first order in time but of the second order with respect to the coordinates, hence it is not consistent with *relativity. The solutions for bound systems give three *quantum numbers, corresponding to three coordinates, and an approximate relativistic correction is possible by including a fourth *spin quantum number. *See also* *Dirac equation; Klein–Gordon equation.

Schroteffekt German for *shot effect.

Schumann region *See* ultraviolet radiation.

Schuster, Sir Arthur (1851–1934) Brit. physicist, born in Germany, primarily concerned with astronomy and mathematical physics and a pioneer in work on spectra and discharge of electricity in gases. He determined e/m for cathode rays by the magnetic-deflection method; and he studied terrestrial magnetism. The Schuster–Smith magnetometer is commonly used for determining the horizontal component of the earth's field.

Schwartz, Melvin (b. 1932) Amer. physicist. A professor at Stanford University, he founded his own company, Digital Pathways Inc. in 1970. With Leon Lederman and Jack Steinberger, he devised an experiment at the Brookhaven National Laboratory that established the existence of the muon neutrino. For this work they shared a Nobel prize.

Schwarzschild, Karl (1873–1916) German mathematician and astronomer. He made considerable contributions to observational astronomy and is also noted for his solution of the equations of general relativity and the concept of the *Schwarzschild radius (*see also* black hole).

Schwarzschild radius The critical radius to which matter in space must be compressed in order to form a *black hole. It is given by $2GM/c^2$, where M is the mass, G the gravitational constant, and c the speed of light. It is true for a nonrotating black hole. The surface having such a radius is the *event horizon* of the black hole and defines the boundary from inside which neither mass nor radiation can escape. The enormous gravitational tidal forces inside the black hole draw matter towards the centre where it is destroyed in a region of infinite curvature, a space–time *singularity*, where the known laws of physics break down.

Schwinger, Julian Seymour (b. 1918) Amer. physicist who became a professor at Harvard University. He is noted for his theoretical work in developing *quantum electrodynamics, for which he shared the 1965 Nobel prize for physics with Richard Feynman and Sin-Itiro Tomonaga.

scintillation 1. The production of small flashes of light, near ultraviolet and near infrared, from certain materials (*scintillators*) as a result of the impact of radiation. Each incident particle or interacting quantum produces one flash.
2. Rapid irregular variations in the intensity of light or other radiation as it passes through an irregular medium. Starlight is deflected by mobile irregularities in the refractive index of the earth's atmosphere so that a star's image wanders rapidly about its mean position, i.e. the star *twinkles*.

scintillation counter A detector of ionizing radiation based on *scintillation. The scintillator may be an inorganic crystal (commonly sodium iodide with a small quantity of thallium iodide), an organic substance (for example, a mixture of naphthalene and anthracene), a plastic, or a liquid (for example, a mixture of triethylbenzene and terphenyl). On the passage of an ionizing particle or the interaction of a high-energy photon there is a scintillation that is detected by one or more *photomultipliers. The magnitude of the observed pulse is proportional to the energy given up to the scintillator by the particle or quantum, hence the spectrum of the radiation can be studied using a *multichannel analyser. Alternatively, the device can be used as a simple counter of highly ionizing particles, such as alpha particles, the counter circuit being designed so that weakly ionizing events are not recorded. For this purpose zinc sulphide detectors may be used, as in the historic *spinthariscope.

Organic solids, plastics, and liquids are generally used to study high-energy charged particles. The scintillation is extremely brief (a few nanoseconds) thus very short time intervals and very high counting rates can be studied.

Gamma rays of high quantum energy interact by the *photoelectric effect, the *Compton effect, or *pair production (above 1.02 MeV). Only in the case of photoelectric effect is the whole energy normally given up to the scintillator, thus a spectrum is confused by signals corresponding to lower-energy transfers by high-energy quanta, including radiation scattered into the scintillator by the surroundings. The probability of photoelectric interaction is proportional to the fourth power of the atomic number Z, hence organic materials for which Z is low are not suitable for studying gamma-ray spectra unless very large scintillators (such as tanks of liquid) can be used; in this case Compton-scattered radiation will almost all be absorbed by subsequent reactions. Usually inorganic crystals are more suitable, for example the iodine ($Z = 53$) in NaI has high-absorption cross section for gamma rays. The duration of the signal is about a hundred times that for organic materials, but this still permits very high counting rates.

Scintillation counters can be used as *scintillation spectrometers* to measure the energies of quanta in a spectrum of gamma radiation provided the scale can be calibrated using a source of known quantum energy, for example, the 0.511 MeV *annihilation radiation always present when a radionuclide produces *positrons.

scintillation crystal A material that emits light quanta of a characteristic wavelength when exposed to *ionizing radiation.

scintillation spectrometer *See* scintillation counter.

scintillator *See* scintillation.

scope Short for oscilloscope.

scotophor A material, such as potassium chloride, that darkens under electron bombardment and recovers on heating. It is used on screens of cathode-ray tubes (in place of the usual phosphor) when long persistence is required.

scotopic vision Vision by the normal *eye when the rods in the retina are the principal receptors of light. The rods are sensitive to low levels of *luminance (less than a few hundredths of a candela per square meter) but do not produce the sensation of colour. The maximum *spectral luminous efficiency occurs at shorter wavelengths than in *photopic vision. *See* luminosity; Purkinje effect.

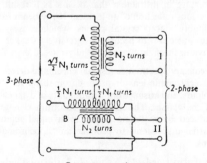

Scott connection

Scott connection A method of interconnecting two transformers to provide *three-phase to *two-phase transformation and *vice versa*. In the connection

diagram shown both transformers have similar magnetic circuits and similar secondary windings. The relative numbers of primary turns are indicated so that when connected to a symmetrical three-phase supply, the voltages of the two secondary phases are equal and have a phase difference of 90° with respect to each other. To distinguish between the two transformers A and B in the diagram, A is sometimes called the *teaser transformer* and B the main transformer.

SCR Abbreviation for *silicon controlled rectifier.

scrambler A circuit or device used in communication systems to make the transmitted signal unintelligible unless the appropriate circuit is used to unscramble it after reception.

screen 1. The front surface of a TV, VDU, or other *cathode-ray tube, suitably coated, on which the visible pattern is displayed.
2. *See* shielding.

screened cable A multicore cable (as used for electric power supply) in which each core with its external insulating material is separately enclosed in a conducting film, the latter being usually in the form of a metallized perforated paper tape. The conducting films of the individual cores are in electrical contact with one another and with the metallic cable sheath and they are usually earthed. Each conducting film has a circular cross section and is coaxial with its core so that the electric field between the film and the core is entirely radial. Three-core cables of this type are commonly used on 33 kV three-phase systems.

screen grid *See* thermionic valve.

screening *See* electrical screening; magnetic screening.

screening constant A number than when subtracted from the *atomic number, Z, gives an effective atomic number. It allows for the screening effect on the nuclear charge, Ze, of the inner electron shells.

screening reactor *See* line choking coil.

screw A cylinder of metal, having a helical groove cut in it, thus leaving a projecting screw-thread, which works in a similarly cut, fixed support. As the screw is made to revolve about its axis, it moves simultaneously in a direction parallel to its length; during one revolution, it moves longitudinally a distance equal to the pitch of the screw. The screw may be used to lift weights, as in the screw jack. In the differential screw, there is an outer screw, to which the force is applied and which works in a fixed block; the inside of the screw is hollow to admit a smaller screw carrying the weight. For one rotation of the outer screw, the weight moves a distance equal to the difference between the pitches of the outer and inner screws.

screw axis The axis of symmetry of a crystal where rotation about the axis, through an angle $2\pi/n$, is accompanied by translation along the axis through a distance xI/n, where I is the identity distance, and x = $1,2 \ldots (n-1)$.

screw dislocation *See* defect.

scribing Scoring a wafer of *semiconductor containing several circuits or components with a precision diamond tool for the purpose of separating individual chips for packaging.

scribing channel The gap left between the areas on a wafer of *semiconductor, containing circuits or components, to allow for *scribing into *chips.

SCS Abbreviation for *silicon controlled switch.

S-drop *See* strange matter.

sealing end A closed box fitted to the end of a cable where a connection with an external conductor is made. It protects the insulation of the cable from air or moisture.

search coil *See* exploring coil.

searchlights In order that light may be projected to great distances, a powerful source must be used. As much light as possible is collected by a large-aperture system and is delivered after reflection or refraction in a beam of little spread. A concave spherical mirror of wide aperture will show a large spread if the source (small) is placed at the focus, owing to the great spherical aberration; a paraboloid mirror wil produce parallel light if the source is at the focus, but there remains spread of light owing to the extension of the source, while the extra-axial pencils will also show some degree of aberration (sine condition not fulfilled). The inverse square law for calculating the intensity of illumination on the axis can be applied only to points beyond a certain distance (at least 20 metres), the apparent luminous intensity being AB where A is the area of the mirror aperture and B the brightness of the source (absorption neglected).

Searle's method For determining the thermal conductivity of a good conductor a well-lagged bar is used so that the temperature gradient is uniform. The polished bar is surrounded by dry felt and enclosed in a wooden box. One end of the bar (diameter about 5 cm) is steam heated, the other end being cooled by a steady stream of cold water passing through a copper spiral soldered to the bar. The temperature gradient is determined by the readings (θ_1 and θ_2) of two thermometers placed in small holes filled with mercury a known distance (d) apart in the bar. The rate of flow of heat in the steady state is deduced from the steady inlet and outlet temperatures (θ_3 and θ_4) of the water in the cooling spiral. If m is the mass of water flowing per second and A the cross-sectional area of the bar:

$$kA \frac{\theta_1 - \theta_2}{d} = m(\theta_4 - \theta_3).$$

The method is used mainly for teaching purposes.

SECAM (SEquential Couleur A Memoire) A *colour television system adopted by France and the Soviet Union. The colour difference signals are transmitted line-sequentially (i.e. one colour per line), the human eye integrating the result into the original shades. The system avoids the use of in *quadrature transmission of the chrominance information, but requires a line-interval state and a high degree of stability in the receivers.

Secchi, Angelo (1818–1876) Italian astronomer who made the first spectroscopic survey of the sky and was the first to classify stars according to spectral type (*see* stellar spectra).

second 1. Symbol: s. The *SI unit of time defined as the duration of 9 192 631 770 periods of the radiation corresponding to the transition between two hyperfine levels of the *ground state of the caesium-133 atom.
  2. *Syn.* arc second; second of arc. Symbol: ". A unit of angle equal to 1/3600 degree.
  3. *Syn.* centesimal second. A unit of angle equal to 0.0001 *grade.

secondary *See* secondary winding.

secondary cell *See* accumulator; cell.

secondary cosmic rays *See* cosmic rays.

secondary electron An electron emitted from a material as a result of *secondary emission.

secondary emission The process occurring when an electron moving at sufficiently high velocity strikes a metal surface, the impact causing other electrons to escape from the surface. The total energy of one incident electron is often sufficient to eject several secondary electrons. The principle is used in the *electron multiplier and in *storage tubes. Secondary emission is also produced by the impact of positive ions on surfaces. The *secondary emission ratio* δ, defined as the number of secondary electrons emitted per incident particle, is used as a measure of the effect.

secondary extinction An increase in absorption or decrease in diffraction that occurs as a result of previous reflection of an incident X-ray beam from suitably placed crystal planes, which causes the incident beam to become progressively weaker on passing through a large partially mosaic crystal.

secondary spectrum A residual chromatic defect of an achromatic lens or prism chromatized for two colours, and arising from *irrationality of dispersion of the glasses. The secondary spectrum is largely reduced by matching glasses with equal relative partial dispersions, to produce an *apochromatic lens.

secondary standard 1. A copy of a *primary standard for which the difference between the copy and the primary standard is known.
  2. A quantity accurately known in terms of the primary standard and used as a unit, e.g. the wave-

lengths of lines in the arc spectrum of iron, which are known in terms of the wavelength of the primary standard (the wavelength of the red cadmium line).

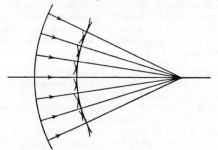

Secondary wavelets

secondary wavelets According to *Huygens's principle (1678), each point in a wavefront is a source for secondary waves, a useful concept for explaining refraction (regular and double) and diffraction. The new wavefront is the forward envelope of the wavelets (a back wave does not exist). The amplitude in a secondary wavelet falls off in proportion to:
$$(1 + \cos\theta),$$
where θ is the angle with the forward direction

secondary winding The winding on the load (i.e. output) side of a *transformer irrespective of whether the transformer is of the step-up or step-down type. *Compare* primary winding.

second moment of area A geometrical property of a lamina, of significance in studies relating to elasticity. The second moment of area I_a is related to the moment of inertia I of the lamina about an axis by:
$$I_a = (A/m)I,$$
where A is the area, and m the mass of the lamina. The units are metre to the power of four.

second sound A form of temperature wave propagated in *superfluid liquid helium II. It involves flow and counterflow of the normal fluid and superfluid components such that the density remains constant (unlike ordinary sound) while the temperature oscillates.

section gap *Syn.* air gap. An arrangement of the line work by which a contact wire in an electric traction system is divided mechanically and electrically into sections while providing a continuous path for the current collectors. The usual method is to overlap the ends of adjacent contact wires in a horizontal plane.

section insulator A device that enables a contact wire in an electric traction system to be divided electrically into sections without interfering with the mechanical continuity or with the continuous path for the current collectors.

sectored disc A rotating disc from which one or more sectors of total angle α have been removed. A beam of radiation of intensity I that falls on the disc is

diminished in intensity to $(\alpha/360)I$. By using such a disc the range of many measuring instruments (e.g. *pyrometers) can be extended.

sedimentation The free fall of particles in suspension in a liquid. If the particles are spherical, they attain terminal velocities dependent on the relative density of solid and liquid, the viscosity of the latter, and the size of the particles (*see* Stokes's law). This result is used to separate particles of different sizes or to analyse the particle-size distribution in a powder from this piling up of the particles towards the base of the tank containing the liquid. In *Odén's balance*, one balance pan is held near the bottom of the tank, the successive accumulation on it being counterbalanced by weights added from time to time on the other pan. Alternatively, the analysis may be done by recording the increase in opacity of the suspension near the bottom of the pan photoelectrically.

sedimentation equilibrium or velocity *See* ultracentrifuge.

sedimentation potential *See* migration potential.

Seebeck, Thomas Johann (1770–1831) German physicist who invented the *thermocouple, a device whose operation depends on the Seebeck effect (*see* thermoelectric effects).

Seebeck effect *See* thermoelectric effects.

seed crystal A small crystal on which crystallization can begin, e.g. in a supersaturated solution or a supercooled liquid. It is used in the manufacture of *semiconductor devices.

Seger cones Cones usually made of ceramic material that are used to estimate temperatures of about 1000 °C in kilns, from the amount of bending-over suffered by the vertex.

Segrè, Emilio Gino (1905–1989) Italian-born Amer. physicist who did most of his work at the University of California (Berkeley). Segrè was involved in the discovery of the elements technetium (1937), astatine (1940), and plutonium (1940). He shared the 1959 Nobel prize for physics for his part in the discovery (1955) of the antiproton.

Segrè chart *Syn.* chart of the nuclides. A graph in which the number of protons in nuclides is plotted against the number of neutrons. Stable examples will be found on, or in the vicinity of, a line with a gradient of 1: the gradient diminishes somewhat for nuclides of high atomic number.

Seidel aberrations Five aberrations due to sphericity of surface, namely spherical aberration, coma, astigmatism, curvature of field, and distortion. Seidel (1855) deduced correcting terms to the paraxial theory (first order), to include the next term in the expansion of sines (third power or third order), expressing the corrections as Seidel sums (S_1 to S_5) (*see* aberrations of optical systems (1)). To be free from aberration to this order all should be zero; e.g. $S_1 = 0$ would mean freedom from axial spherical aberration, $S_1 = S_2 = 0$ would mean coma removed, and so on, in cumulative order.

seismograph An instrument used to register the movement of the ground due to distant earthquakes, underground nuclear explosions, etc., consisting in principle of a massive pendulum set into motion by the force developed at its point of suspension.

seismology The study of the structure of the earth by means of the waves (*seismic waves*) produced by earthquakes, explosions, etc. *Seismographs at various points on the earth's surface record the arrival of the different kinds of waves that travel through the earth and along its surface. *P-waves* (longitudinal waves) and the slower *S-waves* (shear waves) travel deep into the earth, where they are refracted at the boundaries between layers of different density. The source of the earthquake is called its *focus* and the nearest point on the earth's surface to the focus is the *epicentre*.

selection rules Rules derived by *quantum mechanics specifying the transitions that may occur between different *quantum states of a system. For example, in a change between two quantum states of vibration of a molecule the selection rule is that the vibrational quantum number can only change by one unit, i.e. $\Delta VC = \pm 1$. Transitions that follow the selection rules are *allowed transitions*. *Forbidden transitions* are ones in which the rules are not followed and they are very unlikely, but sometimes not impossible.

selective absorption 1. The absorption of radiation of certain wavelengths in preference to others. Coloured glasses and pigments, etc., exhibit selective absorption, their colour being determined by the remaining light transmitted (glasses and filters) or reflected (pigments) after the absorption has occurred. When all wavelengths are equally absorbed, absorption is said to be neutral. All substances show strong absorption at some region or other of the radiation spectrum, which, for many transparent colourless optical media, occurs in the infrared and ultraviolet regions.
2. *See* anomalous absorption of sound.

selective fading *See* fading.

selective radiation Radiation from a body with a relative spectral energy distribution differing from that of a *black body, or of a *grey body in which the radiation for all wavelengths is a constant fraction of that given by a black body. Although all substances show selective temperature radiation over the visible portion, many give an approximately grey radiation. With such, it is possible to specify a temperature for a black body that will give roughly a similar distribution of spectral energy as the luminous selective radiator (*see* colour temperature). Incandescent gases and vapours show more or less extreme selectivity.

selective reflection The strong reflection of certain wavelengths in preference to others, which occurs

for all substances, although the wavelengths may be outside the visible spectrum; the wavelengths reflected generally correspond with those of the absorption bands. Metallic reflection, of a higher order, is responsible for surface colour as distinct from pigment or body colour, which arises from absorption of light penetrating the substance prior to its reflection.

selectivity The ability of a radio receiver to discriminate against signals having *carrier frequencies different from that to which the receiver is tuned (*see* tuned circuit). Selectivity is usually expressed by means of a graph in which the inverse of the overall receiver gain, relative to the gain at the frequency to which the receiver is tuned, is plotted against frequency. The graph shows the number of times by which the input must be increased to maintain constant receiver output as the carrier frequency is varied from the frequency to which the receiver is tuned. In obtaining selectivity curves, the *modulation factor and modulation frequency must be specified. If the receiver is fitted with *automatic gain control it is usual for this to be rendered inoperative.

selectron *See* supersymmetry.

selenium cell A *photovoltaic cell consisting of a thin layer of selenium coated onto a metal disc and covered with a film of gold or platinum that is sufficiently thin to allow light to pass through. The selenium is subsequently annealed at above 180 °C to produce the light-sensitive grey allotrope. Light falling on the gold surface creates a potential difference between the gold and the selenium. A sensitive galvanometer incorporated into the circuit will show a deflection, the current increasing with the square root of the intensity. It is used in *exposure meters and can be built into cameras. It has a much lower sensitivity than the *cadmium sulphide cell but does not require an external power supply.

selenium rectifier A type of semiconductor *rectifier using a selenium-iron junction as the rectifying element. Such rectifiers usually consist of a number of such junctions in series.

selenology The study of the geography and geology of the moon's surface.

self-aligned gate *See* MOS integrated circuit.

self capacitance The inherent capacitance between individual turns of an inductance coil or a resistor. To a first approximation it may be represented as a single capacitance connected in parallel with the coil or resistor.

self-conjugate particle *See* antiparticle.

self-consistent field method An approximate method, due to Hartree, of solving central field problems in atomic physics, e.g. finding the *wave function of an electron electrostatically bound to a nucleus in the presence of other bound electrons. The method is self-consistent in the sense that it includes a means of recognizing when the correct solution is reached.

self-excited 1. Of an electrical machine (generator). Having the *field magnets wholly or substantially excited by a magnetizing current generated in the machine itself. (*Compare* separately excited.)
2. Of an *oscillator. Being in a state in which the oscillations build up to a steady output value, following application of power to the circuit, without any separate input of the required output frequency.

self-field *See* accelerator.

self inductance *See* electromagnetic induction.

self-reversal *See* reversal.

self-starting synchronous motor A *synchronous motor having a winding housed in slots or tunnels in the pole faces so that the machine may be started as a squirrel-cage *induction motor. Ultimately the machine runs at *synchronous speed with d.c. excitation for the poles as a normal synchronous motor.

Sellmeier equation A mathematically deduced formula to give the variation of refractive index, n, with wavelength, λ, in the neighbourhood of an absorption band (λ_0):
$$n^2 = 1 + A_0 \lambda^2/(\lambda^2 - \lambda_0^2),$$
where A_0 is a constant for a given material. It fails within the region of the absorption band, agreeing better in the regions of wavelength for which the substance is transparent.

SEM Abbreviation for *scanning electron microscope.

semicircular deviation In an iron ship there may be deviation of the compass needle due to permanent magnetization of the ship or to the effect of masses of soft iron contained in it. This magnetization, and the effect caused by vertically disposed masses of soft iron, cause semicircular deviation that is east for half, and west for half of the complete circle turned by the ship. Permanent magnetization is corrected by fixing small permanent magnets, of opposite effect, near to the compass needle. Deviation caused by vertical soft-iron masses is corrected by putting a vertical soft-iron bar (*Flinder's bar*) just in front of, or just behind, the compass in a position found by trial and error.

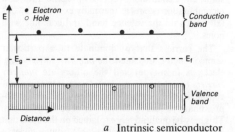

a Intrinsic semiconductor

semiconductor A material having a *resistivity between that of conductors and insulators and usually having a negative *temperature coefficient of resistance.

(1) *Intrinsic semiconductors.* The *energy bands of a perfect crystal semiconductor are shown in Fig. *a*. The energy gap between the conduction band and the valence band is from a few tenths of an electronvolt up to 2 eV. Such a semiconductor is called an *intrinsic semiconductor.* At a given temperature T_1 some electrons will be thermally excited into the conduction band. The number (n) of these electrons is given by:

$$n = \int_{E_c}^{E_2} \frac{g_c^{\cdot}(E)\,\mathrm{d}E}{\mathrm{e}^{(E-E_F)/kT} + 1},$$

where $g_c(E)$ is the number of quantum states per unit range of E in the conduction band (the density of states), E_c and E_2 are the lower and upper limits of the band respectively, the *band-edge energies*, E_F is the *Fermi level of the material, k is the Boltzmann constant, and T the thermodynamic temperature. Each electron that is thermally excited into the conduction band leaves behind a vacant state in the valence band and in a pure crystal the number of vacancies equals the number of electrons in the conduction band. The action of an applied field causes conduction both in the conduction band and in the valence band. The conduction electrons are accelerated by the applied field and carry charge through the semiconductor. The electrons in the valence band move to occupy the adjacent vacancy and the net effect is that the vacancy moves through the material as if it were a positive charge (Fig. *b*). The vacancies are known as *holes and are treated as positive charge carriers.

(i) ● ● ○ ● ●
(ii) ● ○ ● ●
(iii) ○ ● ● ●

● *Electron*
○ *Hole*　　　　*b* Hole conduction

(2) *Extrinsic semiconductors.* In practice, absolutely pure crystals do not exist. The presence of defects in the crystal lattice of a semiconductor, such as vacancies in the lattice or interstitial atoms, distort the energy bands near the defect position and cause states to be formed in the forbidden band. If these occur near the valence band, electrons without sufficient energy to cross the forbidden band may have sufficient energy to enter the states formed. These states act as traps for electrons and conduction can take place in the valence band, as empty states will be left. Conduction caused by defects is known as *defect conduction.* A similar type of conduction occurs in *insulators. If the states occur near the conduction band, electrons in the conduction band can be trapped and the overall conductivity is decreased.

The presence of impurities in a semiconductor affects the conductivity significantly. A semiconductor whose properties depend on the presence of impurities is an *extrinsic semiconductor*, and the properties depend on the type of impurity.

semiconductor

Donor impurities are atoms that, when present in the crystal lattice, have more valence electrons than are required to complete the bonds with neighbouring atoms. The presence of these atoms affects the distribution of quantum states in the immediate vicinity, and states are formed in the forbidden band, close to the conduction band (Fig. *c*). At absolute zero each of these *donor states* will contain an electron. At a temperature typically within a few tens of degrees from room temperature the Fermi level will be at the same energy as the donor states so

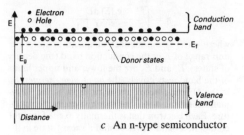

c An n-type semiconductor

half will be occupied. At the temperature shown in Fig. *c* less than half are occupied. The electrons fill most of the holes in the valence band and enter states in the conduction band. The number of electrons in the conduction band, n_e, exceeds the number of holes in the valence band, n_p, and the semiconductor is said to be *n-type*. The product $n_e n_p$ equals n_i^2, where n_i is the number of free electrons (or holes) in the intrinsic material. As the donor states are localized in space, no conduction can occur by electrons moving between them, but the electrons in the conduction band and holes in the valence band can move. Thus as the temperature rises the conductivity increases until nearly all the donor states are empty. There is then usually a range of temperature of a few tens of degrees over which the number of free charge carriers is roughly constant so the conductivity falls as the temperature rises and there are more collisions between electrons and *phonons. At higher temperatures the conductivity rises again as electrons are thermally excited across the energy gap.

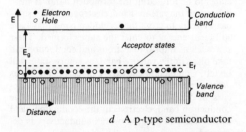

d A p-type semiconductor

Acceptor impurities are atoms that have fewer valence electrons than are required to complete the bonds with neighbouring atoms and they therefore accept electrons from any available source to complete the bonds. These extra electrons are almost as

tightly bound to the atom as the valence electrons and the presence of acceptor impurities results in states just above the valence band (Fig. *d*). The electrons in the valence band only need a small increment of energy to occupy the *acceptor states* and provide the source of electrons for the acceptor atom. Mobile holes are left in the valence band, but the electrons are bound to the acceptor atom, which is ionized by the electron capture. Mobile holes thus predominate and the semiconductor is known as *p-type*.

The effect of the impurity is to move the *Fermi level near to the conduction band in n-type, and near to the valence band in p-type as the distribution of available states has changed.

The conductivity of an extrinsic semiconductor will depend on the type and amount of impurities present and this may be controlled by adding impurities of a particular type to achieve the desired type of conductivity. This process is known as *doping* and the amount of impurity is the *doping level*. This is usually carried out by *diffusion or *ion implantation of the impurity into the crystal. *Doping compensation* is carried out to compensate for the effects of one type of impurity by diffusing the opposite type of impurity into the material. It is possible to produce a semiconductor with properties similar to an intrinsic semiconductor using doping compensation (Fig. *e*). This is sometimes known as a *compensated intrinsic* semiconductor.

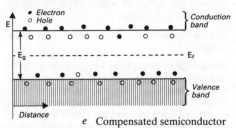

e Compensated semiconductor

(3) *Movement of charges in a semiconductor.* At conditions of thermal equilibrium a dynamic equilibrium exists in the semiconductor. Mobile charges move around the crystal in a random manner due to scattering by the nuclei in the crystal lattice. The crystal retains overall charge neutrality and the number of charge carriers remains essentially constant, but a continuous process of regeneration and recombination occurs as thermally excited electrons fall back into the valence band and combine with holes.

The carriers that predominate in a particular semiconductor are called the *majority carriers* (e.g. electrons in n-type) and the others are *minority carriers* (e.g. holes in n-type). In an extrinsic semiconductor the number of majority carriers is approximately equal to the number of impurity atoms. This may be qualitatively explained on the basis that electrons are more readily available from or given to the impurity states, than cross the forbidden gap and fewer thermally generated pairs will be present, as the levels they would normally occupy are filled by majority carriers due to the impurities. If extra charge carriers are generated by, for example, light energy, these will have a limited lifetime in the material. The lifetime of excess carriers is defined in terms of minority carriers, and in the bulk of a homogeneous semiconductor the *bulk lifetime* is the average time interval between generation and re-combination of minority carriers. Recombination also takes place near the surface of a semiconductor; the mechanism is rather different from the direct recombination occurring in the bulk of the material. This is defined by the *surface recombination velocity*, the ratio of the component normal to the surface of the electron (hole) current density and the excess electron (hole) volume charge density close to the surface.

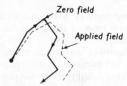

f Movement of a carrier in a semi-conductor

If an electric field is applied to the semiconductor, charge carriers move under the influence of the field, but still undergo scattering processes. The result of the field is to impose a drift in one direction onto the random motion of the carrier (Fig. *f*). The *drift mobility* of a charge carrier is the average drift velocity of carriers per unit electric field, and the average distance travelled by a carrier during its lifetime is the *diffusion length*. The mobility of electrons has been shown to be about three times that for holes.

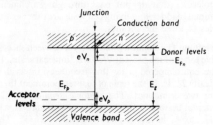

g Energy bands before equilibrium

(4) *Junctions between semiconductors.* Junctions between semiconductors of different types form a fundamental part of modern electronic components and circuits. If two semiconductors are brought into contact with no external applied field, thermal equilibrium will be established between the two, and the condition of zero net current requires the Fermi level to be constant throughout the sample. Fig. *g* shows the situation before equilibrium is established between samples of p- and n-type semiconductor. Electrons and holes will cross the junction until an equilibrium is set up by the establishment of two

*space-charge regions preventing further flow of carriers (Fig. *h*). The space-charge regions are due to ionized impurities on either side of the junction, and a voltage drop is formed across the junction, while maintaining the overall neutrality of the sample. The *diffusion potential* or *built-in potential* V_{bi} is given by:

$$eV_{bi} = E_g - e(V_n + V_p) = E_{Fn} - E_{Fp}$$

(Figs. *g* and *i*), where *e* is the charge of an electron.

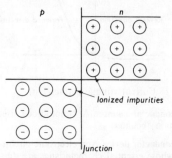

h Space charge regions at equilibrium

The space-charge region around the junction forms a *depletion layer, the width of which depends on the electric field in the interface and the numbers of acceptor and donor ions on either side of the junction.

If a voltage V is applied across the junction, the electrostatic potential is given by $(V_{bi} + V)$ for reverse bias (positive voltage on the n-region) and $(V_{bi} - V)$ for forward bias (positive voltage on the p-region). The depletion width increases with reverse bias as the holes in the p-region are attracted to the negative electrode and vice versa in the n-region, and very little current flows across the junction. The reverse-bias current flowing is due to minority holes in the n-region and electrons in the p-region crossing the junction.

Under conditions of forward bias the depletion width decreases as the built-in potential is reduced by the applied field, and the current through the sample increases exponentially with voltage for a few tenths of a volt (Fig. *j*).

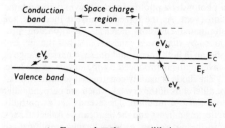

i Energy bands at equilibrium

(5) *Types of semiconductor material.* Commonly used materials are elements falling into group 4 of the periodic table, such as silicon or germanium. The donor and acceptor impurities are group 5 and group 3 elements, differing in valency by only one

electron. Certain compounds, such as gallium arsenide (which has a total of 8 valence electrons), also make excellent semiconductors. These materials are classified as 3–5 or 2–6 depending on their position in the periodic table. Suitable impurities in these cases would be 2, 4, and 6 or 3 and 5 respectively.

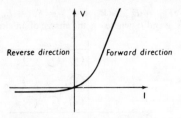

j   Current/voltage characteristic

semiconductor counter A *photodiode used as a radiation *counter.

semiconductor device Any electronic circuit or device whose essential characteristics are due to the flow of charge *carriers within a *semiconductor.

semiconductor diode *See* diode.

semiconductor laser *Syn.* diode laser; injection laser. A small robust cheap and flexible type of *laser fabricated from *semiconductor material. It consists essentially of a *p-n junction diode, being a refinement of the *light-emitting diode.

Laser action is nearly impossible to produce in silicon, and semiconductor materials such as gallium arsenide must be used. Under a forward bias, electrons flow across the junction from the n-side to the p-side, where they form an excess minority-carrier concentration; the process is called electron *injection. These electrons can recombine with *holes in the p-side, emitting a photon by spontaneous emission; the photon energy is approximately equal to the energy gap between conduction and valence bands, E_g. At sufficiently large values of applied voltage, great numbers of electrons can cross the junction. Above a threshold current, stimulated emission (*see* laser) can occur: an electron excited into the conduction band is induced to emit a photon by a photon from a previous recombination event. As the injection current increases, the stimulated emission increases. The photon produced by stimulated emission matches the incident photon in both energy and phase; a narrow range of wavelengths is produced by the device as a whole.

The diode is usually constructed to have two flat parallel ends of the crystal, which are perpendicular to a flat p-n junction. These ends act as partially reflecting mirrors and the light can be reflected back into the p-n region, causing further amplification. The diode acts as a resonant cavity, the light and its reflection being in phase. An intense laser beam emerges from the mirror ends of the crystal. A very high current density is required, and to prevent overheating at room temperature the beam has to be pulsed.

A continuous laser beam can be achieved by using a modified crystal. A region of pure GaAs is made adjacent to a region of aluminium gallium arsenide in which some of the gallium atoms in the GaAs crystals have been replaced by aluminium atoms. The junction between these two regions of similar crystal structure (a *heterojunction*) can be used to reduce the threshold current required to achieve laser action, and a continuous laser beam is possible. The output power is tens of milliwatts at wavelengths between 900–700 nm and the efficiency can be as high as 10%.

A range of different materials and modifications to the basic structure have now been produced to provide lasers of different wavelengths and to optimize the operation. Semiconductor lasers have an extensive range of applications.

semiconductor memory *Syn.* solid-state memory. Any of various types of cheap compact computer *memory composed of one or more *integrated circuits fabricated in *semiconductor material (usually silicon). The integrated circuit comprises a rectangular array of possibly thousands of microscopic electronic devices, each of which can store one bit of data – either a binary 1 or a binary 0. Data can be accessed extremely rapidly. The different types include *RAM, *ROM, *PROM, and *EPROM. There is *random access to all these types.

semipermeable membrane A membrane used in *dialysis and *osmosis that allows certain molecules in a fluid to pass through while stopping others. In general, the larger molecules are stopped but the process is not altogether a matter of the relative sizes of pores and molecules, e.g. the pore spaces in a copper ferrocyanide membrane are more than a hundred times as great as the diameter of sugar molecules that do not pass through. Parchment, animal bladder, and living cell walls are other examples of these membranes.

semitone The smallest pitch interval between successive tones of the present Western *musical scales. In the equitempered scale this frequency interval is ideally $2^{1/12}$. In the scale of just temperament there are two semitones. The *diatonic semitone* is the interval between notes of different designation, e.g. G♯ to A, having a pitch interval of 16/15. The *chromatic semitone* is the interval between notes of the same designation, e.g. G to G♯, E♭ to E, with a frequency interval of 25/24 or 135/128 depending whether it divides a lesser or larger whole tone.

semitransparent cathode A type of *photocathode in which electrons are emitted from the opposite surface to that on which the radiation falls.

sensation level A measure of the intensity I of a sound with reference to the minimum audible intensity I_0. If the sensation level L is measured in decibels,

$$L = 10 \log_{10}(I/I_0).$$

I_0 is usually taken as 2.5 picowatts per square metre.

sensing element *See* transducer.

sensitive flame Using suitable gas pressure and size of orifice, an ignited flame will give an indication of sound disturbances. If a gas or liquid jet is made visible by smoke or suitable colouring, a length of streamline flow followed by an area of turbulent motion will be observed. Increasing the velocity of efflux decreases the streamline portion; the two being related hyperbolically. The breakdown of streamline flow to turbulent motion is connected with the viscous drag of the stationary fluid on the jet. After the stream has travelled a certain distance, it is unable to withstand the large shearing forces brought into play because of the large differences in velocity across the stream. With low velocity of efflux and narrow bore orifice, it is possible to obtain a long streamline jet that will break down into a short turbulent jet if the velocity of efflux is slightly increased, or if an air disturbance, such as hand clapping or musical sounds, takes place near the jet. If ignited gas is used as the jet, such an arrangement is known as a sensitive flame or jet, and it may be used for the detection of sounds up to frequencies of 100 kilohertz. An ignited flame when adjusted to its sensitive point acts very similarly to an unignited one. A periodic disturbance causes a periodic change in length. For a given orifice, a jet is sensitive only in a very small range of gas pressure. Screening of the jet shows that the tip is the seat of the sensitivity.

Various types of sensitive flames have been used; the pin-hole gas burner used with fairly high gas pressure provides the simplest flame for high-frequency detection. Rayleigh used a thin membrane in the gas supply line, the sound vibrations on the membrane altering the gas pressure over a small interval. Fishtail burners have directional properties – sensitivity being zero when the direction of the sound is at right angles to the line of the jets. Instead of igniting the total jet the gas may be lit on the upper side of a wire gauze placed some suitable distance above the orifice of the jet. (*See* manometric flame.)

sensitivity 1. Generally, the response of a physical device to a unit change in the input.

2. Of a measuring instrument. The magnitude of the change of deflection or indicated value that is produced by a given change in the measured quantity.

3. Of a radio receiver. A measure of the ability of the receiver to respond to weak input signals. Quantitatively, it is the smallest input at the receiver that will produce a certain output under specified conditions, particularly a designated *signal-to-noise ratio.

sensor *See* transducer.

separated function synchrotron *See* proton synchrotron.

separately excited Of an electrical machine (generator). Having the *field magnets wholly, or substan-

tially, excited by a magnetizing current that is obtained from a source other than the machine itself. The advantages of this arrangement over that of a *self-excited machine are that the excitation current is entirely independent of the load current supplied by the machine and that reasonably satisfactory operation can be obtained over a range of voltage from zero to the maximum that the machine can generate. Separately excited generators are sometimes used in battery-charging equipment where the number of cells connected in series may vary considerably and they are also used as *boosters.

separation energy The energy needed to remove one proton or neutron from the nucleus of a particular nuclide.

separation factor A measure of the efficiency of any process for separating isotopes. It is the ratio of the concentration of isotopic forms in one phase to that in the new phase after the process is carried out. Usually the value is only a little above unity, but in the case of electrolytic separation of deuterium, the factor may be as high as 6.

separation of isotopes *See* isotope separation.

sequential access A method of data organization in *computers. In order to reach the nth record of a *file by this method, the previous $(n-1)$ records must first be scanned. Data on a *magnetic tape, for instance, must be retrieved by a sequential-access method. *Compare* direct access.

sequential scanning *See* television.

serial 1. Involving the sequential transfer or processing of the individual parts of a whole. For example, in *serial transmission*, the individual *bits making up a unit of data are transmitted one after the other along the same path.

2. *Syn.* sequential. Involving the occurrence of two or more computing events or activities such that one must finish before the next begins. *See* sequential access.

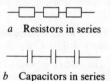

a   Resistors in series

b   Capacitors in series

series 1. Pieces of electrical apparatus are *in series* when they are connected so that one current flows in turn through each of them. For conductors of resistances $r_1, r_2, r_3, \ldots r_n$ in series, the total resistance, R, is given by:
$$R = r_1 + r_2 + r_3 + \ldots + r_n.$$
For capacitances $c_1, c_2, c_3, \ldots c_n$ in series:
$$1/C = 1/c_1 + 1/c_2 + 1/c_3 + \ldots + 1/c_n.$$
Cells in series add their e.m.f.s. *Compare* parallel.

2. *See* mathematical series.

series-characteristic motor *Syn.* inverse-speed motor. Any electric motor, the speed of which decreases substantially with increase of load, as with a *series-wound or heavily *compound-wound machine. *Compare* shunt-characteristic motor.

series circuit Of an electrical instrument. *See* current circuit.

series current *See* current circuit.

series-parallel connection 1. An arrangement in which electrical machines or electronic devices or circuits may be connected in series or in parallel as alternatives.
2. An arrangement in which electrical machines or electronic devices or circuits are connected so that some are in series and some are in parallel with one another.

series-parallel control A method of control commonly employed in electric traction in which d.c. motors or groups of d.c. motors initially (i.e. at starting) are connected in series and finally (i.e. normal running) are connected in parallel. *See* bridge transition; shunt transition.

series resonant circuit *See* resonant circuit.

series system of distribution A distribution system in which all the consuming devices are connected in *series so that the same current traverses all of them. *Compare* shunt (or parallel) system of distribution.

series winding *Syn.* series coil. A winding or coil connected in *series with a main circuit so as to carry the current in that circuit. *Compare* shunt winding.

series-wound machine A d.c. machine in which the *field magnets are wholly, or substantially, excited by a winding that is connected in *series with the armature winding, or alternatively is connected so as to carry a current proportional to that in the armature winding. *Compare* compound-wound machine; shunt-wound machine.

service cable An electric cable by which a consumer's installation is connected to the general supply system.

service line A line by which a consumer's electrical installation is connected to a *distributor.

service voltage The voltage at no-load between the lines of the circuit to which an electrical machine or other apparatus is to be connected for use, as specified by the purchaser. For a single-phase system, it is the voltage between the two principal conductors; for two-phase, the voltage between the two conductors of one phase; for three-phase, the voltage between any two of the principal conductors.

serving A layer or layers of fibrous material, such as hessian or jute, applied to the exterior of a metal-sheathed or armoured cable. It is usually dressed with a waterproof preservative compound.

servomechanism In general, a closed-sequence control system that is automatic and power-amplifying.

For example, in servo-assisted brakes in motor vehicles a small foot pressure in the brake pedal activates a servomechanism that applies a much greater pressure to the brake shoes or pads. The servosystem may include *feedback. Suppose a heavy shaft is required to rotate at the same speed as a light shaft when the latter is not capable of driving it directly. The light shaft is made to regulate the power supply driving the heavy one in such a way that the power is increased or decreased according to the error in the relative angular position of the two shafts.

sessile drop (and **bubble**) A large stationary drop resting on a plane surface that it does not wet: a bubble in a liquid resting under a plane or concave-downward surface. The dimensions of the principal section of one or the other are measured as a stage in the determination of the surface tension of the liquid.

set-up scale instrument *See* suppressed-zero instrument.

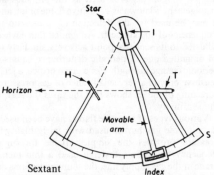

Sextant

sextant An instrument for measuring angles (up to 120°) between two objects, and particularly the angle between an astronomical body and the horizon (i.e. its altitude). The horizon is observed through the upper clear half of a fixed horizon glass H by applying the eye to the eyepiece of the telescope T. The index glass I is rotated until the image of a star is reflected from I into the lower silvered half of H, and thence to T. The required angle is read from the graduated scale S at the point indicated by an arm attached to I.

Seyfert galaxy Any of a group of *galaxies with an exceptionally bright central region. Most are otherwise normal nearby spiral galaxies. Although the central region is an optical source, most of the emission is in the infrared, with strong UV and X-ray emission. This emission is substantially nonthermal. It is thought that the central activity could be powered by a *black hole, and that Seyferts are a less powerful example of the *quasar phenomenon.

shadow mask A perforated metal sheet placed between the *electron guns and the screen of some *colour-television picture tubes to allow the correct selection of colours on the screen.

shadow photometer A simple photometer invented by Count Rumford. It uses a rod in front of a screen and two light sources to be compared are adjusted until the shadows thrown just touch and match in intensity. The *luminous intensities are then in the ratios of the squares of the distances of the sources from the shadows thrown by each.

shadows The shape and relative density of shadows can be explained for most practical purposes on a macroscopic scale in terms of geometric optics. A point source delivers rays to an obstacle, the extremities of which limit light that may pass; the shadow throughout is complete (umbra). With an extended source the shadow shows variation of density – umbra and penumbra. When the edges of geometric shadows are examined more carefully, *diffraction phenomena are evident and these are more marked as the size of the obstacle is decreased.

shadow scattering *See* scattering.

shank, pole *See* pole core.

shape factor Of a lens. The ratio:
$$(r_2 + r_1)/(r_2 - r_1),$$
where r_2 and r_1 are the radii of the surfaces, using the geometric convention of sign, light travelling in the positive x direction.

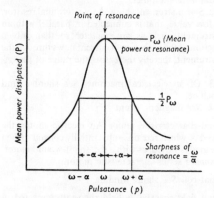

Sharpness of resonance

sharpness of resonance The reciprocal of that fractional change in frequency from the point of resonance of a system subject to forcing (*see* forced vibrations), which reduces the mean power dissipated by half. An expression for the mean power, P, is $\alpha A^2 p^2$, and this gives the condition as:
$$\alpha A_p p^2 / \alpha A \omega^2 = \tfrac{1}{2},$$
where ω is the angular frequency at resonance, p the forcing angular frequency, and A_p and A are the respective amplitudes. Inserting the appropriate expressions for A_p and A and solving for p in terms of α and ω gives the result $p = \omega \pm \alpha$, α^2 being assumed small compared with ω^2.

The reciprocal of the fractional change in frequency is therefore ω/α, and this may be called the sharpness of resonance.

shear *See* strain; stress.

shear modulus *See* modulus of elasticity.

shear stress *See* stress.

shear wave A *transverse wave that travels without compression of the medium. The S waves propagated through the solid body of the earth in earthquakes are an example (*see* seismology).

shell *See* atomic orbital; electron shell; shell model.

shellac A resinous secretion of the insect *Laccifer lacca*, originally used as a source of dyestuff, but now almost entirely for its content of resinous matter. It is used extensively for the preparation of French polish, printing inks, and insulators on account of its high gloss, adhesiveness, hardness, and toughness. The properties of shellac are improved by compounding with fillers and modifying resins to give thermoplastic compositions. Shellac plastics have poor heat resistance and high water absorption.

shell model *Syn.* quasi-atomic model. Any of various models of the nucleus in which the interactions between *nucleons are approximated by assuming the nucleons move in a single central potential (much as the electrons in the *atom move in the electrostatic field of the nucleus). By solving the *Schrödinger wave equation corresponding to this potential, a set of possible *quantum states are obtained. The sets of states corresponding to the same energy are called *shells*. Being *fermions, the nucleons must obey the *Pauli exclusion principle. Therefore no two nucleons can occupy the same quantum state. The nucleons will try to reach the lowest energy possible by filling the lowest energy states first. The more nucleons there are, the greater the number of shells that will be filled. This is exactly analogous to the filling of atomic shells by orbital electrons in heavy elements. The shell model is useful in explaining why nuclei with certain proton or neutron numbers (called *magic numbers) are more stable than others. These are thought to correspond to nuclei in which the number of nucleons is just sufficient to fill a given number of shells, by analogy with the inert gases (helium, argon, etc.), which owe their stability to filled atomic shells.

shell-type transformer A *transformer in which the *core encloses the greater part of the windings. The core is made of *laminations that are usually built up around the windings, the latter having been assembled beforehand. *Compare* core-type transformer.

Shenstone effect A considerable increase in photoelectric emission produced in some metals (e.g. bismuth) after an electric current has passed through.

SHF Abbreviation for super-high frequency. *See* frequency band.

shield 1. A mass of material (such as concrete, etc.) surrounding a *nuclear-reactor core, or other source

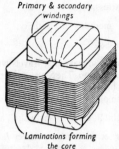

Primary & secondary windings

Laminations forming the core

Single-phase shell-type transformer

of radiation, to absorb neutrons or other dangerous radiations. A *biological shield* is specially designed to protect laboratory workers and plant operators from harmful radiations. *See also* health physics.
2. *See* shielding.

shielding *Syn.* screening. Removal of the influence of an external energy field by surrounding a region with a *shield* or *screen* of suitable material. In the case of an electric field, earthed metal walls are required. Stray magnetic fields are removed by using a shield of high *permeability. (*See also* grading shield.)

shift register A digital circuit, a *register, used to displace a set of information, stored in the form of pulses, either to the right or left. If the information is a pattern of *bits representing a binary number, a shift in position to the left (or right) is equivalent to multiplying (or dividing) by a power of two. Shift registers are extensively used in *computers and calculating machines as storage or delay elements.

SHM Abbreviation for *simple harmonic motion.

shock excitation The production of natural oscillations in an oscillatory system by suddenly introducing energy into the system from an external source.

Shockley, William Bradford (1910–1989) Amer. physicist who in 1948 developed the first transistor with John Bardeen and Walter Brattain.

shock waves Waves of compression and rarefaction that originate in the neighbourhood of sharp points or roughness on obstacles exposed to the flow of a compressible fluid at high speeds. The local compressions and rarefactions set up in this way are not instantly reversible and are propagated out into the fluid as, in effect, sound waves. They transport momentum from the vicinity of the obstacle to a distance or convert kinetic energy into internal energy. *See also* sonic boom.

shoe *See* pole shoe.

short circuit An electrical connection of relatively very low resistance made intentionally or otherwise between two points in a circuit.

short-circuited rotor A squirrel-cage rotor. *See* induction motor.

short-circuit ratio Of a *synchronous alternating-current generator. The ratio of the field current when the machine is generating its rated open-circuit voltage at rated frequency to the field current required for the machine to generate its rated current at rated frequency when on sustained symmetrical short circuit. Common values lie between 0.5 and 1.5. It follows that the steady short-circuit current with normal field current also lies between about 0.5 and 1.5 times the rated current.

short-circuit transition *See* shunt transition.

short-range force *See* force.

short sight *See* myopia; eye.

short-wave Designating radio waves with wavelengths in the range 10–100 metres, i.e. in the high-frequency band.

shot effect *Syn.* shot noise. *See* Schottky noise.

shot noise *See* Schottky noise.

shower A group of *elementary particles and *photons arising from the impact of a single particle of high energy. *See* cosmic rays.

shunt 1. (general) If two electrical devices or circuits are connected in *parallel, either one is said to be in shunt with the other.
2. *Syn.* instrument shunt. A four-terminal resistor, of low value, that is connected in parallel with an instrument such as an ammeter so that only a fraction of the main circuit current flows through the instrument, thereby increasing the range of the latter.
3. Of an electrical machine. *See* shunt-wound machine.
4. *See* magnetic shunt.

shunt-characteristic motor An electric motor the speed of which remains substantially constant when the load is varied, as with a d.c. shunt-wound motor. *Compare* series-characteristic motor.

shunt coil *See* shunt winding.

shunt field rheostat A field rheostat connected in series with the shunt winding of a *shunt-wound machine.

shunt (or parallel) system of distribution A distribution system in which all the consuming devices are connected in *parallel to the supply so that they all operate at the same nominal voltage. This is the system most commonly used. *Compare* series system of distribution.

shunt transition *Syn.* short-circuit transition. A method of transition from series to parallel connection used in the *series-parallel control of d.c. electric motors. During the transition from series to parallel the sequence is as follows: (*a*) Two motors A and B initially in series (*see* diagram). (*b*) Motor A short-circuited. (*c*) Motor A open-circuited (disconnected). (*d*) Motor A connected in parallel with motor B. *Compare* bridge transition.

(a)

A   B

(b)

A   B

(c)

A   B

(d)

A   B

Shunt transition
(N.B. Control
rheostat omitted)

shunt winding A winding or coil connected in *shunt with the whole or part of a main current. *Compare* series winding.

shunt-wound machine A d.c. machine in which the *field magnets are wholly, or substantially, excited by a winding that is connected in *shunt with the armature winding. *Compare* compound-wound machine; series-wound machine.

sideband In *modulation. A band of frequencies embracing either all the upper or all the lower *side frequencies, called the *upper* and *lower sideband* respectively.

side circuit *See* phantom circuit.

side frequency Any frequency produced as a result of *modulation. For example, in *amplitude modulation if a *carrier wave of frequency f_c is modulated by a sinusoidal signal of frequency f_s ($f_s \ll f_c$), the resulting wave has three components, the frequencies of which are f_c, $f_c + f_s$ (the upper side frequency), and $f_c - f_s$ (the lower side frequency).

sidereal day, year, and time *See* time.

Siegbahn, Kai (*b.* 1918) Swedish physicist and son of K. M. G. *Siegbahn. His work on the energy spectrum of electrons ejected from substances exposed to a narrow X-ray beam led to ESCA (electron spectroscopy for chemical analysis) and the related technique of UV photoelectron spectroscopy. For this work he shared a Nobel prize (1981) with Nicolaas Bloembergen and Arthur Schawlow.

Siegbahn, Karl Manne Georg Swedish physicist noted for his work in *X-ray spectroscopy. In 1914 he initiated a programme of research to develop highly accurate spectrometers for investigating the X-ray emission spectra of atoms of all 92 natural elements. His work established the idea that electrons in an atom are in shells (K, L, M, etc.) and extended the ideas of Henry Moseley. Siegbahn also showed (1924) that X-rays could be refracted by a prism. He was awarded the 1924 Nobel prize for physics; his son, Kai *Siegbahn, received the 1981 prize.

siemens Symbol: S. The *SI unit of electrical *conductance, defined as the conductance of an element that possesses a resistance of one ohm. The unit used to be called the mho or reciprocal ohm.

Siemens, Sir William (1823–1883) German-born Brit. engineer who invented a number of scientific instruments, including the *Siemens electrodynamometer. The unit of *conductance is named after him.

Siemens electrodynamometer An *electrodynamic instrument in which the torque produced by the electromagnetic forces is balanced against the torque of a spiral spring by adjustment of a calibrated torsion head attached to the spring. It may be an ammeter, voltmeter, or wattmeter and can be used with direct or alternating current. *See* electrodynamometer.

sievert Symbol: Sv. The *SI unit of *dose equivalent, used for protection purposes in the case of ionizing radiation. It has replaced the rem: 1 Sv = 100 rem. In terms of other SI units, 1 Sv = 1 J kg^{-1}.

sigma particle Symbol: Σ. An *elementary particle, charged or neutral, classified as a *hyperon.

sigma pile A neutron source plus *moderator, but without any fissile material, used to analyse the properties of the moderator.

signal The variable parameter, such as current or voltage, by means of which information is conveyed through an electronic circuit or system.

signal generator Any circuit or device producing an electrical parameter that is adjustable and controllable. Most commonly it supplies a specified voltage with variable amplitude, frequency, and waveform. The term is reserved for continuous-wave generators, especially sine-wave generators. *See also* pulse generators.

signal level The magnitude of a signal at any point in a transmission system, usually with reference to some arbitrarily chosen value.

signal-to-noise ratio (S/N ratio) The ratio of one parameter of a wanted signal to the same parameter of the *noise at any point in an electronic circuit, device, or transmission system. It is often measured in *decibels.

signature A collection of symbols that can be used for identification purposes or to select a particular interpretation. One example is a group of spectral lines, usually in an emission spectrum, that identifies an atom or molecule, and possibly its ionization state and isotopic form.

Sikes hydrometer *See* hydrometer.

silencing *See* sound insulation.

silent discharge An inaudible electrical discharge that takes place at high voltage and involves a relatively high dissipation of energy. Such a discharge readily takes place from a conductor that has a sharp point.

silica (SiO_2) One of the most important constituents of the earth's crust, known in the amorphous form, and in a number of crystalline forms, such as

515

silicon controlled rectifier

*quartz, tridymite, and cristobalite. Naturally oc-curring silica is frequently discoloured, yellow sand, for example, consisting of silica discoloured by ferric oxides. Common flint is amorphous silica, that has become discoloured by iron oxide.

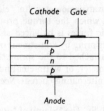

a Silicon controlled rectifier

silicon controlled rectifier (SCR) A *semiconductor *rectifier whose forward anode-cathode current is controlled by means of a signal applied to a third electrode, called the *gate*. The device is constructed as a four-layer *chip of semiconductor material forming three *p-n junctions, with *ohmic contacts to three of the layers (Fig. *a*).

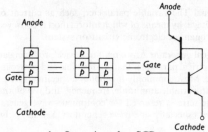

b Operation of an SCR

The operation of the device is most easily under-stood by representing it as a combination of two bipolar junction *transistors with two common elec-trodes (Fig. *b*). The gate electrode forms the collec-tor of the p-n-p transistor and the base of the n-p-n transistor. A voltage is applied across the device between the anode and the cathode. If no signal is applied to the gate electrode ($V_{GC} = 0$), the n-p-n transistor will not conduct as no current is supplied to its base. The p-n-p transitor will not conduct either as the base current that must flow to allow transistor action is supplied via the n-p-n transistor (which is off). If a positive current is supplied to the gate electrode, the n-p-n transistor turns on and draws current from the base of the p-n-p transistor, which can also conduct. When the transistors are both turned on, current flows through the device from anode to cathode provided that the anode is positive with respect to the cathode. The two transis-tors provide a positive current feedback loop for each other and current continues even after the cessation of applied signal to the gate. The current may only be cut off by reducing the anode voltage to near zero, when the emitter-base junction of the p-n-p transistor becomes reverse biased and current ceases to flow. Alternatively, it can be cut off by

reducing the current through the device to a low value, when the gains of each transistor become so low that insufficient current is supplied to the bases to allow conduction to continue. The minimum current for continuation of conduction is the *holding current*. The SCR is the solid-state equivalent of the *thyratron valve and was originally called a *thyris-tor*. The current-voltage characteristic (Fig. *c*) is similar to that of the thyratron.

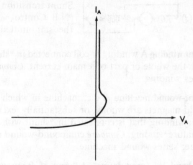

c Current/voltage characteristic of an SCR

The most important applications of SCRs are in a.c. control systems and solid-state *relays. If an a.c. signal is applied to the anode, the device may be switched on at any desired portion of the positive half-cycle using a *trigger pulse, or by illuminating the gate region causing electron-hole pairs to be formed. The device will automatically switch off at the end of the half-cycle when the anode voltage drops below the turn-off level. This type of opera-tion is known as *phase control*. If the gate turn-on current is supplied at the beginning of the positive half-cycle, a single SCR acts as a half-wave rectifier. Full-wave rectification can be achieved by using a bridge rectifier circuit in conjunction with an SCR (Fig. *d*) or by using two SCRs in antiparallel connec-tion (Fig. *e*). A version of the SCR for switching a.c. is known as the *triac*. It is almost equivalent to two antiparallel SCRs made in a single chip, but is able to operate with a single gate connection; its opera-tion relies on a complex pattern of current paths within the device.

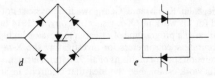

d   *e*

Full-wave rectification by means of SCR

silicon controlled switch (SCS) A switch fabricated in a silicon chip. Like the *silicon controlled rectifi-er, it is a *pnpn device but has connections to all four semiconductor layers of the device. Under conditions of forward bias, it can be turned on by a voltage pulse to the control gate electrode (connect-ed to the inner p-type layer – see Fig. *a* at silicon controlled rectifier) and can be turned off by a voltage pulse to a second gate electrode that is connected to the inner n-type layer.

silver-disc pyrheliometer *See* pyrheliometer.

silvering, etc. The deposition of thin films of metal (such as silver) on glass, etc., which is usually carried out by one of four processes: (i) Chemical deposition (especially silver). (ii) Burning-on process in which a chemical salt in an oily solution is heated on the surface to be coated (e.g. platinum, gold, silver, iridium). (iii) *Sputtering from the cathode of a glow discharge (aluminium does not sputter well). (iv) Evaporation of the metal in vacuum and subsequent condensation on the surface to be coated.

silver point The temperature of equilibrium of solid and liquid silver at standard atmospheric pressure taken as a fixed point (961.93 °C) on the *International Practical Temperature Scale.

similarity principle *See* dynamic similarity.

Simon, Sir Francis Eugene (1893–1956) German-born Brit. physicist who made many valuable contributions to low-temperature physics and developed the diffusion process for the separation of isotopes.

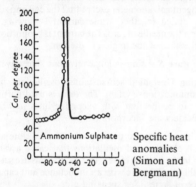

Ammonium Sulphate   Specific heat anomalies (Simon and Bergmann)

Simon and Bergmann's specific-heat-capacity experiments (1930) They discovered anomalies in the specific heat capacities of many solids within a narrow range of temperature. From the shape of the specific-heat-capacity temperature curves (*see* diagram), some of these are known as lambda (λ) points.

Simon and Lange's adiabatic vacuum calorimeter (1924) The apparatus was used for thermal measurements at low temperatures. Heat losses from the calorimeter are eliminated by surrounding it with an enclosure kept at the same temperature as the calorimeter itself. The substance being investigated is placed in a thin-walled copper calorimeter K (*see* diagram), containing a constantan heating coil, a lead resistance thermometer, and platinum leads L, which pass out through the tube T used for the pumping. The calorimeter is suspended freely inside the enclosure Th by the wire F. This enclosure is a brass cylinder wound with a heating coil. One junc-

tion of a copper-constantan thermocouple is attached to the outside of the calorimeter and the other to the enclosure, which is heated electrically so that it is always at the same temperature as the calorimeter as indicated by the thermocouple. The calorimeter K and its enclosure Th are enclosed in a vessel M that can be evacuated and immersed in liquid hydrogen contained in a closed Dewar vessel D and the temperature of the liquid may be varied by varying the pressure in this vessel. For the determination of the *specific latent heat of liquid hydrogen, the liquid hydrogen in the Dewar vessel is boiled under reduced pressure until the hydrogen in the calorimeter K is all liquefied. The enclosure is then evacuated and a current passed through the heater in the calorimeter. The liquid vaporizes and is collected in a reservoir attached to T, the mass of liquid vaporized being deduced from the pressure and volume of the reservoir and the *equation of state for hydrogen.

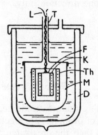

Adiabatic vacuum calorimeter (Simon and Lange)

Simon's method of liquefying helium Simon employed the principle of adiabatic *desorption for the liquefaction of small amounts of helium surrounded by a vessel containing charcoal cooled by a liquid hydrogen bath. Helium gas was adsorbed in the charcoal and the heat developed removed by the bath. On isolating the charcoal and pumping off the helium, cooling occurred, which liquefied the helium in the inner vessel.

simple harmonic motion (SHM) The motion of a body subjected to a restoring force directly proportional to the displacement from a fixed point in the line of motion. The force equation is:

$$m\ddot{x} = -kx,$$

where m is the inertia of the body, and k is the restoring force per unit displacement. Putting k/m equal to ω^2, the equation becomes $\ddot{x} = -\omega^2 x$, which in words may be stated as: acceleration is directly proportional to displacement.

The solution of the latter equation may be written:

$$x = A \cos(\omega t + \alpha),$$

where A is the maximum displacement, ω is the *angular frequency ($\omega = 2\pi f$), and α is the angle determining the displacement at $t = 0$ and is called the *epoch*. The period, T, is given by $T = 2\pi/\omega$.

SHM may be represented graphically as the projection onto a straight line of the path of a particle travelling in a circle with uniform angular velocity (*see* diagram). The amplitude A is the radius of the

simple microscope

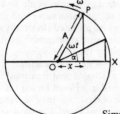

Simple harmonic motion

circle and the angle α is the angular displacement from the fixed line OX at $t = 0$. At time t the displacement, x, is given by:
$$x = A \cos(\omega t + \alpha).$$

simple microscope *See* microscope.

simplex operation The operation of a communications channel between two points in which signals or data can be carried in one direction only.

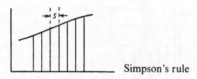

Simpson's rule

Simpson's rule For finding the area underneath a curve. Any odd number of ordinates are erected at equal distances s apart; then the area intercepted between the curve, the base and the first and last ordinates is given by: Area = $s/3 \times$ (sum of the first and last ordinate plus twice the sum of the other odd ordinates plus four times the sum of the even ordinates).

It is sometimes called the *parabolic rule* since it assumes each section of the curve of length $2s$ to be a parabola.

simulator A device, especially one under the control of a computer, that performs a *simulation*, i.e. that mimics the behaviour of an actual system but is made from components that are easier, cheaper, or more convenient to manufacture. Simulators are frequently used for solving complex problems such as weather forecasting, and as a design aid and training tool. Simulation is the major application of *analogue computers.

simultaneity In classical physics different observers in uniform relative motion are assumed to use the same scale of time (*see* Galilean transformation equations) hence any two events that are found to be simultaneous by one observer will also be found to be so by any other observer in uniform relative motion. According to the theory of *relativity this is not generally true. Two events occuring in different places may be found to be simultaneous by one observer, but will not be simultaneous according to an observer moving uniformly relative to the first one. This does not affect the temporal sequence of

related events nor imply any failure of determination.

sine condition If n and n' are refractive indexes of media in front of and beyond a surface, y and y' the

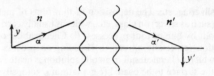

Sine condition

object and image sizes, α and α' the angles made between the conjugate rays and the axis passing through the axial feet of the object and image, then:
$$ny \sin \alpha = n'y' \sin \alpha',$$
is the sine condition (Abbe, 1873). It had already been enunciated by Seidel and others. The sine condition is also referred to as the condition of aplanatism (viz. that a sharp image of a plane object element may be formed). The condition is necessary for freedom from *coma and is equivalent to the condition imposed by Seidel's second sum. *See* Seidel aberrations.

sine galvanometer An instrument similar to the *tangent galvanometer except that the coil and scale are rotated together, while current is flowing, to return the needle to zero. The current is proportional to the sine of the angle of rotation.

sine wave *See* sinusoidal; equivalent sine wave.

singing Unwanted self-sustained oscillations in a communications system. The term is used particularly in connection with telephone lines in which *repeaters are incorporated.

singing arc An arc that emits a musical note as a result of impressed oscillations that cause a variation in the heating effect. The oscillations are set up by shunting the arc with an inductance and capacitance in series. *See* speaking arc.

singing flame A jet of hydrogen or carbon monoxide burning in an open tube that, under certain conditions, causes a musical note. This was noticed long ago by many observers, the first one probably Higgens in 1777.

The phenomenon has been widely studied and its real nature was established by Rayleigh who showed that it depends on the intermittent supply of heat by the jet. If the impulses given to the resonator (the air in the tube) by the driving force (the heat supply) occur at the phase of maximum displacement or condensation of the sound, the note is maintained. For the maintenance of the vibrations the length of the gas tube as well as its position in the air tube should be adjusted such that a condensation at the jet will travel down and back and arrive in phase with a condensation in the air tube. In fact, there should be stationary waves in the gas tube as well as in the air tube. This means that if the gas tube is a short narrow one terminating below in a reservoir, its length ought to be such that it has a displacement node at the jet and displacement antinode at the junction with the reservoir, i.e. the best lengths are $\lambda/4$, $3\lambda/4$, $5\lambda/4$, ..., where λ is the wavelength of sound in the gas. It should be noted that this condition of resonance differs from the ordinary cases of forced vibration where the vibration is best encouraged if the driving force is a maximum where the condensation is zero (*see* forced vibrations; resonance). The difference in the present case arises from the fact that the heat supply is intermittent and not sinusoidal.

E. G. Richardson has carried out experiments on singing flames that verify Rayleigh's theory. The flame of a König manometric capsule (*see* manometric flame) attached to a singing tube on a level with the jet was viewed simultaneously with the singing flame itself in a stroboscope. It was found that the two flames vibrate in the same phase as the theory requires, heat being given to the air at each condensation. Stationary waves were formed both in the air tube and in the gas tube.

single crystal A crystal that may be either *perfect or *mosaic but in which the atomic planes of the same kind are sufficiently parallel to diffract a collimated beam of incident radiation cooperatively and thus to give single spots in the diffraction pattern.

single-current system A telegraph system in which the transmission of signals is carried out by means of an electric current that flows in one direction only. *Compare* double-current system.

single-disc winding *Se* disc winding.

single-phase Of an electrical system or apparatus. Having only one alternating voltage. *Compare* polyphase system.

single-pole *See* number of poles.

single scattering *See* scattering.

single-shot multivibrator *See* monostable.

single-sideband transmission (SST) The transmission of one only of the two *sidebands produced by the *amplitude modulation of a *carrier wave. The carrier and the other sideband are usually suppressed at the transmitter. At the receiver, it is necessary to reintroduce the carrier artificially by combining the sideband with a locally generated oscillation. The frequency of the latter should be as nearly as possible equal to that of the original carrier, but this requirement is not so stringent as in the case of *double-sideband transmission. The main advantages of SST over the method in which the carrier and the two sidebands are transmitted are the reduction in transmitter power (since the carrier and one sideband are not transmitted) and the reduction of the *bandwidth required for the transmission of signals within a specified *frequency band.

singularity *See* Schwarzschild radius.

sink *See* source.

sintering Heating fine particles so that they soften and adhere.

sinusoidal Of a periodic quantity. Having a *waveform that is the same as that of a sine function. It is only the shape of the graph that is significant. For example, two e.m.f.s represented by:
$$e_1 = E_1 \sin \omega t \text{ and } e_2 = E_2 \cos \omega t,$$
would both be described as sinusoidal e.m.f.s, and both would be classed under the heading of *sine waves*.

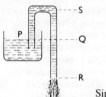

Siphon

siphon An inverted U-tube with one limb longer than the other. The shorter limb dips into a liquid and, when the siphon is completely full of the liquid, it emerges from the lower end. This is a convenient method for removing water from a receptacle that cannot be emptied conveniently in any other way. The simple explanation for its action is that the pressures at P and Q are equal, and that the excess pressure due to the column of liquid QR causes the liquid flow. Since the liquid has to rise a distance PS, this must be less than the barometric height of the liquid. Recent studies have shown that a siphon will work satisfactorily if PS is greater than the barometric height, and will indeed work in a vacuum, provided that the liquid is free from dissolved gas. This suggests that the cohesive forces between molecules also contribute to the action.

siphon barometer *See* barometer.

siphon recorder An obsolete device made by Lord Kelvin, consisting of a very small coil through which the signal current passes. The coil is suspended between the poles of a powerful magnet and is deflected against the twist of the fine suspension wire when the current flows. The small movements of the coil are seen by reflecting a light beam from a mirror fixed to the coil. Continuous records are obtained by attaching to the coil a small glass pen in the form of a siphon, which marks a paper strip drawn beneath the siphon.

siren The use of the siren seems to derive from John Robinson, who described the arrangement in an article on "Temperament of the Musical Scale", in the *Encyclopedia Britannica* (1801). The arrangement was improved by Baron Charles Cagniard de la Tour in 1819. It consisted of a rotating disc perforated with a ring of equally spaced holes mounted in front of a jet of air from a tube. A number of puffs of air are released as the orifice is opened and closed by the revolving disc. If these follow each other with sufficient rapidity, a tone is produced whose frequency can be calculated from the speed of rotation of the disc and the number of holes in it. This arrangement was further modified by Seebeck, who replaced the jet by a series of perforations on the top of a wind chest. These perforations exactly correspond to those of the disc except they are set at opposite direction to each other so that the air blast drives the disc forward. This greatly increases the intensity of the sound produced as the puffs of air emerge simultaneously from all the performations of the disc when the two rows of holes correspond. An automatic revolution counter is attached recording in tens and hundreds. The note given by the siren is simple in character and hence can be used as a standard of frequency provided that the speed of rotation and the pressure of the blast are kept constant. The rotation of the disc can be made constant by stroboscopic means or by attaching the disc to an axle of a phonic wheel (*see* stroboscope; phonic wheel).

Milne and Fowler devised a special form of siren of the Seebeck type giving pure tones. The jet is rectangular in section and the holes are so shaped that the area of the jet exposed varies sinusoidally.

During recent years much progress has been made in the design of large power sirens for use in fog signalling from lighthouses and lightships for the safety of ships at sea.

Sitter, William de (1872–1934) Dutch astronomer. He made observations with the heliometer for the determination of stellar distances; his work on the satellites of Jupiter led to an accurate determination of their masses. He was responsible for bringing the Einstein papers on the theory of *relativity to the notice of British scientists (1914–1918). He considered the astronomical effects of the theory; he also worked on fundamental astronomical constants and the variability of the rate of the earth's rotation.

SI units (Système International d'Unités). The internationally agreed system of *coherent units that is now in use for all scientific and most technological purposes in many countries (including the UK). SI units are based on the *MKS system and replace the units used in the *CGS system and the f.p.s. system (Imperial units). SI units are now of two kinds: *base units* and *derived units*. There are seven base units for the seven dimensionally independent physical quantities shown in the Appendix, Table 2. The base units are arbitrarily defined in terms of reproducible physical phenomena or prototypes. Table 3 gives the derived units with special names and symbols. Two units, the radian and steradian, were originally given the status of *supplementary units*, but are now regarded as dimensionless derived units. All these units are defined in this dictionary in their alphabetical place. The SI unit of any other quantity is derived by multiplication and/or division of the base units without introducing any numerical factors.

SI units are used with a set of prefixes to form decimal multiples and submultiples of the units. The Appendix, Table 4 lists these prefixes and their symbols.

A number of conventions are observed in the printing or writing of SI units. Symbols for a prefix are written next to the symbol for the unit without a space, e.g. cm. A space is left between symbols for units in derived units, e.g. N m for newton metre. The letter *s* is never added to a symbol to indicate a plural: 10 ms indicates ten milliseconds, not ten metres. Compound prefixes are never used, e.g. nm for 10^{-9} metre is correct, mμm is incorrect. A symbol for a unit with a prefix attached is regarded as a single symbol, which can be raised to a power without using brackets, e.g. cm^2 means $(0.01 \text{ m})^2$ and not 0.01 m^2.

The word gram has a special place in SI units; although it is not an SI unit itself, prefixes are attached to the symbol g and not to kg, e.g. 10^3 kg is written Mg and not kkg.

When writing numbers with SI units, the digits are arranged in groups of three but a space rather than a comma is placed between each group; e.g. 10^5 is written 100 000 not 100,000 and 10^{-5} is written 0.000 01. If a number consists of only 4 digits it may be written without spaces, e.g. 1000 or 0.0001.

Certain units outside the SI system have been retained because of their practical importance (e.g. day, degree (of arc), tonne) or their use in specialized fields (e.g. bar, parsec, electronvolt). SI prefixes can be attached to these units, and in some cases compound units can be formed from these non-SI units and SI units.

Six *See* maximum and minimum thermometer.

skew distribution *See* frequency distribution.

skiatron A cathode-ray tube in which the usual coating of fluorescent substances is replaced by a screen of crystals of alkali metal halides, which become darkened under electron bombardment. It can be arranged so that the trace remains on the screen until erased.

skin effect A nonuniform distribution of electric current over the cross section of a conductor when carrying an alternating current. The current density is greater at the surface of the conductor than at its centre. It is due to electromagnetic (inductive) effects and becomes more pronounced as the frequency of the current is increased. It results in a greater *I^2R loss in the conductor than that occurring when the current is uniformly distributed. In consequence, the *effective resistance of a conductor when carrying alternating current is greater than the true resistance, and the high-frequency resistance is greater than its d.c. or low-frequency resistance. With very high frequencies hollow conductors may be used, or stranded conductors may be employed. These stranded conductors may be formed of many fine filaments (litzendraht wire).

skin friction The resistance or drag experienced by a body in relative motion with a fluid, due to the *laminar flow of the fluid and the large rate of shear of the fluid close to the body boundaries.

sky wave *See* ionosphere.

slepton *See* supersymmetry.

sliding friction *See* friction.

slip Of an *induction motor. The ratio of the difference between the *synchronous speed and actual speed of the motor to the synchronous speed. It is usually expressed as a percentage. Typical values for normal types of motors used industrially, when running at full load, are from about 2% to 5%, the smaller values being for the larger motors and vice versa.

slip bands The result of inhomogeneous plastic deformation in a crystalline substance. Gliding occurs discontinuously on particular lattice planes, parallel to *glide planes, these being more or less equally spaced at about 10^4 *unit cell lengths apart.

slip ring *Syn.* collector ring. A ring, usually made of copper, that is connected to and rotates with a winding (as, for example, in certain types of electrical machines), so that the winding may be connected to an external circuit by means of a *brush or brushes resting on the surface of the ring.

slip-ring rotor *See* induction motor.

slope of ray Symbol: d. The acute angle a ray makes with the axis of a system.

slope resistance *Syn.* electrode a.c. resistance; electrode differential resistance. Of a specified electrode in an electronic device. The ratio of a very small change in the voltage applied to the particular electrode to the corresponding change in the current in that electrode, the voltages of all other electrodes being maintained constant at known values. For example, the *collector slope resistance is given by:

$$r_c = \partial V_c / \partial I_c,$$

where V_c is the collector voltage and I_c is the collector current; the base and emitter voltages are held constant.

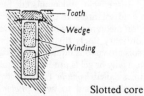

Slotted core

slot One of the long narrow gaps that are formed in the magnetic *core of an electrical machine and house the conductors forming the winding. The part of the core situated between two consecutive slots is called a tooth.

slotted core The core of an electrical machine that has *slots in which are housed the conductors of the winding. *Compare* smooth-core armature.

slot wedge The wedge, commonly made of wood, that holds the conductors in a *slot of the slotted core of an electrical machine. *See diagram under* slot.

slow-break switch A switch in which the speed of breaking depends upon that at which the operating handle, lever, etc., is moved. *Compare* quick-break switch.

slow-down density Symbol: q. A measure of the rate at which *neutrons lose energy by collisions in a *nuclear reactor, expressed by the number of neutrons per unit volume falling below a certain energy per unit time. *See* Fermi age theory.

slow neutron A neutron with a kinetic energy that does not exceed a few electronvolts. The term is sometimes loosely applied to a *thermal neutron.

slow vibration direction The direction of the electric vector of the ray of light that travels with least velocity in a crystal and therefore corresponds to the largest refractive index.

slug 1. *Syn.* geepound. The unit of mass to which a force of one pound-force gives an acceleration of one foot second^{-2}. Thus one slug is equivalent to g pounds.
2. The metre slug has an acceleration of one m s^{-2} when continuously acted on by a force of one kilogram-force.
3. A massive copper cylinder around the core of an electromagnetically operated *relay to retard the rate of change of magnetic flux in the core. If the slug is at the armature end of the core, the operation of the relay is slowed at both make and break; if at the further end (heel), only the release of the relay is slowed down.

small-angle scattering *See* low-angle scattering.

small-signal parameters If the behaviour of an electronic device is represented by instantaneous values of current and voltage appearing at the terminals of a four-terminal *network, the small-signal parameters are the coefficients of the equations representing this behaviour for small values of input. *See also* transistor parameters.

smart fluid *See* electrorheological fluid.

smectic structure A state intermediate between the crystalline and the liquid, in which the lengths of the molecules are parallel and in which their ends lie in a succession of parallel (but not necessarily equally spaced) planes, but in which no further regularities exist. *Compare* nematic structure.

Smith and Menzies' vapour pressure measurements A dynamic method for determining the saturation vapour pressure of various substances. The specimen is placed in the sphere A beside a thermometer T in some liquid in a test tube B, which is joined to a pump and a manometer. Having established a defi-

Smith–Helmholtz law

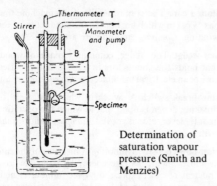

Determination of saturation vapour pressure (Smith and Menzies)

nite pressure in B, the temperature of the bath is slowly raised. When the vapour pressure of the specimen in A becomes equal to the external pressure on the surface of the liquid in B, any further rise in temperature makes the liquid bubble through the liquid in the bath. This temperature is read by the thermometer and the pressure by the manometer.

Smith–Helmholtz law (or **equation**) *See* Lagrange law.

smoke A visible suspension of fine particles, usually smaller than 1 μm, in a gas and generally derived from combustion or from a chemical reaction.

smooth-core armature An armature *core in an electrical machine that is not provided with *slots for housing the conductors of the winding. The conductors are kept in place on the smooth surface of the armature core by means of steel binding wires.

smoothing circuit A circuit used for the purpose of reducing the amplitude of a *ripple. It is commonly a form of low-pass *filter, but may consist of a single inductance.

smoothness A body is said to be smooth when the only force that can be exerted at any point of its surface by another body touching it there is a force along the common normal to the surfaces of the two bodies. A smooth body is a useful concept in theoretical mechanics although no real body is perfectly smooth.

Snell, Willebrord (1591–1626) Dutch astronomer and mathematician. He discovered the law of refraction (1621). *See* Snell's law.

Snellen test type *See* visual acuity.

Snell's law The law of *refraction:
$$\sin i / \sin i' = C,$$
where i is the angle of incidence, i' is the angle of refraction, and C is a constant, now known to be the ratio, n/n', of the *refractive indices of the initial medium and the refracting medium. Also known as Descartes' law.

snow 1. Crystals of ice, forming loose structures usually of a hexagonal pattern, produced in the atmosphere by direct condensation of water vapour

to the solid phase. The flakes form on suitable minute nuclei, probably clay particles. *See also* hail.
2. An unwanted pattern resembling falling snow that appears on a TV or radar screen, usually when the received signal is absent or weak. (With a colour TV the pattern is coloured.) It is caused by electrical *noise in the receiver.

S/N ratio Abbreviation for signal-to-noise ratio.

Soddy, Frederick (1877–1956) British scientist, whose main work was in connection with *radioactivity (at first largely with Rutherford), as a result of which the disintegration theory of the radioactive substances was developed, leading to the recognition of the existence of isotopes (1912). *See* Soddy and Fajans' rule.

Soddy and Fajans' rule *Syn.* displacement rule. A rule concerned with disintegration of radioactive nuclides. It states that the emission of an α-particle causes a decrease of two in atomic number, and the emission of a β-particle increases the atomic number by unity, with consequent changes in the position of the substance in the periodic table.

soft-iron instrument *See* moving-iron instruments.

soft radiation *Ionizing radiation with a low degree of penetration. The adjective *soft* is most commonly applied to X-rays of relatively long wavelength. *Compare* hard radiation.

soft-vacuum tube A vacuum tube in which the degree of the vacuum is such that ionization of the residual gas influences the electrical characteristics of the tube. *Compare* hard-vacuum tube; gas-filled tube.

software The *programs associated with a *computer system. There are two basic forms. *Systems software* is an essential accompaniment to the *hardware of the computer system and includes programs (such as the *operating system*) that are essential to the effective use of the system. *Applications software* relates to the role that the computer system plays within a given organization.

Sohncke's law The stress normal to a cleavage plane that must be applied to produce a fracture in a crystal is a characteristic constant for the crystalline substance.

sol A colloidal suspension. **1.** There are two kinds of liquid sols: suspensoid and emulsoid. Suspensoid sols, e.g. a gold sol, are scarcely more viscous than the dispersion medium and when the particles coagulate, an ordinary finely divided solid is formed. Emulsoid sols are viscous liquids and if coagulation takes place, a jelly or gel is produced. Emulsoid sols are sometimes called *lyophilic* (solvent loving) and suspensoid sols are said to be *lyophobic* (solvent hating). These latter terms are Freundlich's improvements on Perrin's *hydrophilic* and *hydrophobic*. The addition of a small quantity of an emulsoid sol to a suspensoid sol often increases its stability. Emulsoid sols precipitated by the addition of an electrolyte can again be obtained as sols by washing the electrolyte away with more dispersion medium. This reversing cannot be done with suspensoid sols. (*See also* reversible colloids; emulsion.)

2. A suspension of colloidal particles in a gas is called an *aerosol.

solar cell Any device that uses solar radiation to drive an electric current. One type, used to power equipment in spacecraft and artificial *satellites, is essentially a *semiconductor in which a voltage is set up across the p-n junction when photons from the sun fall on the surface (*see* photovoltaic cell). Other types of solar cells, used in desert regions, water-heating systems, etc., consist of complex *thermopiles, one set of junctions being illuminated by solar radiation. A *solar battery* consists of several solar cells. A *solar panel* is a large flat array of solar cells attached to the outside of a spacecraft or satellite and orientated to receive the maximum solar radiation.

solar constant The rate of reception of solar energy by unit area at a specified distance from the sun, the radiation striking the surface normally. A recent measurement of the solar constant at the earth's mean distance from the sun gives a value of

$$1.353 \ (\pm \ 1.5\%) \ \text{kW m}^{-2}.$$

Ground measurements (using *pyrheliometers) must be corrected for atmospheric absorption. Ideally, measurements should be made from a satellite.

solar energy *See* renewable energy sources; solar cell.

solar flare A sudden bright disturbance in the upper atmosphere of the sun, representing an explosive release of particles and radiation. Solar flares are associated with *sunspots and cause magnetic and radio disturbances on earth. *See also* solar wind; cosmic rays.

solarimeter *Syn.* pyranometer. An instrument for measuring the total solar radiation intensity (sun plus sky) received on a horizontal surface. In the *Moll–Gorczynski solarimeter* a *thermopile is used to measure the intensity. A blackened surface, consisting of alternate thin strips of Manganin and constantan, has one set of junctions along the centre line of the surface (the active junctions) while the other junctions are in good thermal contact with the supporting posts. The posts are electrically but not thermally insulated from the base so that this set of junctions remains at the ambient temperature. Incoming radiation raises the temperature of the thin central junctions above the ambient temperature. This temperature difference results in an e.m.f., which is a measure of the radiation intensity and is independent of ambient temperature, wind velocity, etc. In the *Eppley pyranometer*, a central white disc is surrounded by a concentric black ring. Exposure to solar radiation causes the black ring to rise in temperature, the difference between the black and white surfaces being measured by thermocouples. The resulting signal is proportional to radiation intensity. *See also* net radiometer; pyrheliometer.

solarization 1. *Syn.* Sabattier effect. The photographic technique by which a partly reversed final image is obtained by exposing the negative to light (to which the film is sensitive) during developing. In the areas where the negative image had started to form, the emulsion is desensitized so that the additional exposure only affects shadow areas of the original negative, which will consequently have a greater density than before the light flash. Solarization can also be done during printing.

2. *See* reversal (photographic).

solar mass *See* solar units.

solar panel *See* solar cell.

solar time *See* time.

solar units A set of units, including *solar mass* (symbol: $M_\odot$), *solar radius* (symbol: $R_\odot$), and *solar luminosity* (symbol $L_\odot$), in which certain properties of the sun such as the mass, radius, and luminosity are taken as unity; the same properties of other stars can then be compared to those of the sun.

solar wind A stream of ionized particles, mainly protons and electrons, that flow from the sun in all directions. The average particle energy is much lower than that of *cosmic rays, the velocity being between 250–900 km s^{-1} at the distance of the earth's orbit. The number of particles and their velocity increase following solar activity such as sunspots and solar flares. The solar wind causes the shape of the earth's magnetic field to be unsymmetrical, the lines of force being considerably extended in the direction of the wind, beyond the earth. Changes in the intensity of the wind produce fluctuations in the magnetic field, causing magnetic storms and affecting radio communications. Some of the charged particles become trapped in the magnetic field, spiralling along the lines of force from one magnetic pole to another. The *aurora results from excitation of air molecules by these particles. The *Van Allen belts also contain particles from the solar wind.

soldering The process of joining two metal parts together by an alloy of lower melting point than either. The metals to be soldered are first cleaned (e.g. with a wire brush) and then heated with a soldering iron (or a gas flame, if the work is large). A flux of the zinc chloride or resin type is applied to clean the hot surfaces chemically so that the solder, which is next applied, will melt and wet them. Ordinary solder (soft solder) is an alloy mainly of tin and lead and melts in the range of 200–250 °C.

Brazing is a similar process carried out with a gas torch and solders that are grades of brass with melting points of 850–900 °C. A brazed joint is stronger than a soft-soldered one and is sometimes described as hard soldered.

Silver soldering or silver brazing is a similar process using one of a number of alloys of silver with melting points between 630 and 830 °C. Some soldering alloys contain their own flux; these are called resin-cored solders. (*See* alloy.)

Soleil compensator If two equal plates of quartz, one cut parallel to the axis and the other at right angles, are placed in contact, the combination acts as a plate of zero thickness as regards the alteration of phase between ordinary and extraordinary rays. If one of the slabs is cut so as to form two narrow angle wedges, sliding one wedge over the other produces a variable thickness parallel slab, and in combination with the other slab can be arranged to alter the phase relationships between the ordinary and extraordinary rays (e.g. ¼ or ½ wave plates).

solenoid A coil of wire with a length that is large compared with its diameter. At a point inside the solenoid and on its axis, where the ends subtend semiangles θ_1, θ_2, the magnetic flux density B is given by:
$$B = \tfrac{1}{2}\mu_0 nI(\cos \theta_1 + \cos \theta_2),$$
where n is the number of turns per unit length, I the current, and μ_0 the *magnetic constant. (The formula ignores end effects.)

solenoidal A *vector field that has no divergence in a region of space is solenoidal in that region. In such a region, the lines of force (or of flow) either form closed curves (e.g. the magnetic field of a current) or terminate at infinity or on bounding surfaces (e.g. the electric field between capacitor plates).

solid angle Symbol: Ω or ω. An area is said to subtend in three dimensions a solid angle at an outside point in an analogous manner to the subtense by a line at a noncollinear point. The solid angle is measured by the area subtended (by projection) on a sphere of unit radius or by the ratio of the area (A) intercepted on a sphere of radius r to the square of the radius (A/r^2). The unit of solid angle is the *steradian. The solid angle completely surrounding a point is 4π steradians. If a small area (dA) is at a distance R from a point and its normal makes an angle θ with a line drawn to the point, the solid angle formed by the area and point is
$$(\mathrm{d}A \cos \theta)/R^2.$$

solidification curve *See* ice line.

solid solution A homogeneous mixture of two solids whose composition may vary within certain limits. *See* alloy.

solid-state device An electronic device consisting chiefly or exclusively of *semiconducting materials or components.

solid-state memory *See* semiconductor memory.

solid-state physics The branch of physics concerned with the structure and properties of solids and the phenomena associated with solids. These phenomena include electrical conductivity, especially in *semiconductors, *superconductivity, *photoconductivity, *photoelectric effect, and *field emission. The properties of solids and the associated phenomena are often dependent on the structure of the solid.

solid-state relay *See* relay.

soliton A stable particle-like solitary wave state that is a solution of certain propagation equations. Solitons are thought to occur in many areas of physics and applied mathematics, such as plasma physics, fluid mechanics, lasers, optics, solid-state physics, and elementary-particle physics. Strictly, solitons occur only when two solitary waves do not change their form after they collide. They are most commonly produced in one space dimension, as in electron conduction in certain conducting one-dimensional polymers. In particle physics, a magnetic monopole can be regarded as an example of a soliton.

solstice 1. One of the two points at which the ecliptic is furthest north or south of the celestial equator. (*See* celestial sphere.)
2. One of the two days of the year when the sun is at these points. The *summer solstice* occurs about June 21 in the northern hemisphere (December 22 southern hemisphere) and is the day with the greatest number of daylight hours. The *winter solstice* occurs about December 22 in the northern hemisphere (June 21 southern hemisphere) and is the day with the least number of daylight hours. *Compare* equinoxes.

solute A substance dissolved in a pure liquid; the pure liquid is called the *solvent*, and the intimate mixture produced is called the *solution*.

solution *See* solute.

solution pressure 1. The osmotic pressure of a solution in equilibrium with undissolved solid.
2. *See* electrolytic solution pressure.

solvated Denoting a particle in solution associated with a definite portion of the solvent to the exclusion of other particles.

Solvay conferences A series of international conferences on theoretical physics held under the auspices of Ernest Solvay (1838–1922) a Belgian manufacturing chemist. The first conference (1911) established the failure of *classical physics and the need for a *quantum theory.

solvent *See* solute.

Sommerfeld, Arnold (1868–1951) German physicist who made a major contribution to *quantum theory and its application of spectral analysis. He also evolved a theory of the electron in metals.

sonar A contraction of *so*und *na*vigation *r*anging. A method of locating underwater objects by transmitting an ultrasonic pulse and detecting the reflected pulse. The time taken for the pulse to travel to the object and return gives an indication of the depth of the object. *See* ultrasonics; fathometer.

sonde A small *telemetry system in a balloon, rocket, or satellite, used in meteorology, astronomy, etc. *See also* radiosonde system.

sone A unit of loudness. The loudness L in sones is defined by the formula:

$$10 \log_{10} L = (P - 40)\log_{10} 2,$$

where P is the equivalent loudness measured in *phons. The scale has been chosen so that a sound of x sones seems to the listener to be k times as loud as a sound of x/k sones. Experiment has shown this to be justified for loudness between $1/4$ and 250 sones.

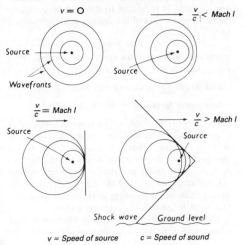

Production of sonic boom

sonic boom The noise originating from the backward projected *shock waves set up by an aircraft travelling at greater than the speed of sound (Mach 1). A stationary source of sound emits a series of concentric *wavefronts, the radius of which will increase with time. For a moving sound source, the wavefronts crowd together in the direction of motion until, at Mach 1, they become tangential to a line perpendicular to the direction of motion. At speeds above Mach 1, two conical lines, tangential to the wavefronts, delineate the shock waves set up by atmospheric-pressure discontinuities. In level flight, the intersection of the shock-wave cone with the ground forms a hyperbola at all points on which the sonic boom is heard simultaneously. A double report is heard as both the nose and tail of the aircraft pass through the sound barrier.

sonic depth finder *See* fathometer; sonar.

sonometer *See* monochord.

sorption *See* absorption.

sorption pump A type of vacuum pump that relies for its action on the *adsorption of gas by a material, usually charcoal or a molecular sieve (*see* zeolites). The adsorbent is contained in a bulb connected to the system and cooled by liquid air. Pumps of this type are used at fairly high pressures in place of rotary pumps and can also be used at low pressures ($<10^{-6}$ Pa) to remove small quantities of gas. *See also* pumps, vacuum.

sound Periodic mechanical vibrations of a medium by means of which sound energy is carried through that medium. (Sound cannot travel through a vacuum.) The word sound is also applied to the sensation felt when such vibrations fall on the ear. As the human ear has a restricted range of perception of *pitch, sound is strictly limited to those vibrations whose frequencies lie between about 20 and 20 000 hertz (audiofrequencies). There is, however, no essential objective peculiarity about those vibrations whose frequency lies below or above this region; these infrasonic and ultrasonic vibrations have the same basic properties as sound waves and can be included in the same study. (*See also* ultrasonics.)

The source of sound executes vibrations. In order to do so, it must possess *elasticity, in the sense that when it is displaced from its position of rest and is released, a force comes into play directed towards that position. It must also possess *inertia, in the sense that in returning, it overshoots its equilibrium position and oscillates to and fro.

The propagation of the sound into the surrounding medium involves portions of the medium in vibration, but because of a progressive lag in the taking up of these vibrations, the disturbance is propagated with a finite speed known as the *speed of sound. Sound travels through a fluid medium as a *longitudinal wave motion. In solids longitudinal, *transverse, and *torsional waves all occur. Finally the waves are picked up by receivers, which may be of a biological or physical nature.

The study of sound finds a number of applications: in meteorology and sound ranging (as atmospheric acoustics), in depth sounding and the detection of submarine objects, and in prospecting by means of echoes; in architecture, as acoustics of buildings; in the design of musical instruments, and as the scientific basis of music and voice production (*phonetics).

sound absorption coefficient *See* acoustic absorption coefficient.

soundboard A board of substantial dimensions with very few, if any, natural resonances, coupled to the strings of a piano, to permit a more suitable transfer of energy from the vibrating strings to the air. The large vibrating area of the soundboard obviously increases the rate of energy emission. The strings are stretched between bridges, one bridge being mounted on the frame of the piano, the other on the soundboard. The energy of the string is transmitted to the soundboard via the bridge, which vibrates in sympathy with the string, but at twice the string frequency. The reverse mechanism is applicable between the bridge and the soundboard. The string is normally struck in such a position to remove the dissonant 7th and 9th *partials. This position also helps to cancel out any natural vibrations of the soundboard. Such excitation was found empirically by the makers to give the best tone production. To maintain the vibrations in phase over the soundboard, a number of bars are fitted to the back of the board. The sound quality produced by a piano is dependent to a large extent on the material and construction of the soundboard. Unlike the body of a violin, the pianoforte loses quality with increasing age. This is generally attributed to the soundboard which, under the string tension, tends to unbend from the original curved form. The design of soundboards is mainly empirical.

sound-energy flux *Syn.* sound power, or, more generally, acoustic power. Symbol P or P_a. The rate of flow of sound energy. For a plane or spherical progressive wave with a speed c in a medium of density ρ it can be shown that at any point:

$$P = p^2 A/\rho c,$$

where p is the root-mean-square *sound pressure at that point and A is the area through which the flux passes. Sound-energy flux is measured in *watts.

sound-energy reflection coefficient *See* acoustic absorption coefficient.

sounding balloons *Syn.* balloon sondes. Small balloons used for carrying recording instruments into the earth's atmosphere.

sound insulation There are three main ways in which unwanted sound can enter a room: (*a*) by passage through the air; (*b*) by transmission through the structure of the building; (*c*) by the diaphragm action of floor, walls, and ceiling. Airborne sounds may be prevented by massive walls, and special attention must be paid to small holes through which the sound can pass. It has been shown that the sound entering a room through a very small crack is likely to be greater than that entering through a large solid wall. Thus, windows and doors should fit tightly and be seated into rubber or felt lining. Ventilation ducts tend to act as speaking tubes and easily carry sound from room to room. Separate ducts should be used for each room and should either be lined with insulating material or a felt-lined baffle inserted in each pipe.

Insulation from structure-borne vibrations necessitates breaking the continuity of the structure. Machinery must be bedded on springy material such as cork or rubber and bolts should be bushed with insulating materials. Similarly, special materials can be placed between walls and floor, under foundations and between steel girders to prevent transmission of structural vibrations. Heating and ventilation pipes should be set in insulating material and pass through cavity walls if possible. A sound-proof floor is made double, the inner floor being floated on insulators and slag-wool placed between the two.

Unless a wall is thick, it tends to act as a diaphragm for sound, but this effect and that of structure-borne vibrations can be reduced by using double walls. Insulating material may be placed between the walls but they should be completely isolated from each other and ties between them must be avoided if possible. Double glazing of windows considerably improves their insulating properties, but the glass should be set in absorbent material and the air space made such as to avoid coupling between the two panes. For acoustical laboratories where sound insulation must be complete, rooms are built on separate foundations with completely unbridged air cavities between them. *See* reverberation chamber; dead room.

sound intensity Symbol: I or L. The rate of flow of sound energy through unit area normal to the direction of flow. It is related to the root-mean-square *sound pressure p and the density ρ of the medium by:

$$I = p^2/\rho c,$$

where c is the speed of the sound. The SI unit is watt per square metre.

sound power *See* sound-energy flux.

sound-power level *See* power-level difference.

sound pressure Symbol: p. The instantaneous value of the periodic portion of the pressure at a particular point in a medium that is transmitting sound. It is that part of the pressure that is due to the propagation of sound in the medium and has an average value of zero over a period of time. The *root-mean-square value of the pressure level is often used. It is measured in pascals or sometimes bars.

The more general term *acoustic pressure* is used when referring to the whole range of acoustic waves.

sound-pressure level Symbol: L_p. A dimensionless quantity given by the natural logarithm of the ratio of the *sound pressure, p, to a reference sound pressure, p_0, i.e. by:

$$\log_e(p/p_0) = \log_e 10 \times \log_{10}(p/p_0).$$

p_0 is either stated or taken to be 2×10^{-5} Pa in air and 0.1 Pa in water. Sound pressure levels are measured in *decibels (dB) or *nepers (Np):

$L_p = 1$ dB when

$$20 \log_{10}(p/p_0) = 1;$$

1 dB $= (\log_e 10)/20$ Np.

sound spectrum *See* quality.

soundtrack A track at one side of a *videotape or cine film on which sound signals can be recorded in order to provide simultaneous reproduction of the sound associated with the vision projection. The soundtrack on film consists of variations in width or density of a silver image, ideally related in linear fashion in frequency and amplitude to the original sound. In reproduction, a beam of light projected through the sound track is amplitude-modulated by these variations. The modulation is converted into an audiofrequency signal by a photocell. The signals are amplified sufficiently to operate a loudspeaker system. In practice, the volume range of the recorded signal is compressed to a range of about 40 to 50 decibels, particularly to raise the lower sound levels above the inherent noise level of the system. The frequency range normally recorded is from 50 to 12 000 hertz, and in laboratory copies of the recording, an output characteristic flat to ±5 decibels for constant amplitude input within this frequency range, is obtainable.

The sound intelligence may be recorded as variations either in width or of density using constant amplitude of the soundtrack. The variations in width may be either single-sided or double, the latter arrangement being push-pull which, when associated with the appropriate transducer, reduces most of the even harmonic distortion. In recording with the variable width method, an oscilloscope gives a sharp line, whose length varies with the sound intelligence. A push-pull or single-sided track is easily obtained with such a system. In the variable density method, a light valve consisting of two very thin metal ribbons carrying the audiofrequency currents, is placed in a magnetic field. The varying attraction between the ribbons alters the gap through which the light passes. The light is then focused onto the undeveloped film. Variable light sources, particularly luminous electric discharges in gases, are often used in the production of news films where a high standard of sound recording is not required. Other methods of modulating a light beam are (i) the Kerr electro-optical effect; (ii) the Faraday magneto-optical effect; (iii) the use of a quartz piezoelectric resonator between crossed Nicol prisms, the exciting oscillator being modulated with the audiofrequency intelligence.

There is an upper limit to the frequencies that may be recorded with this system. The controlling factors are: (*a*) The size of the light slits used in recording and replay. For a given film speed the smaller the slit is, the higher the frequencies that may be reproduced. Diffraction effects place a limit on the minimum width. Suitable, near-monochromatic light and more recently, the use of shorter wavelengths, such as ultraviolet, displaces the diffraction effect to smaller slits. (*b*) The emulsion properties of the film limit the resolution obtainable. (*c*) The speed of movement of the film. The greater this is, the higher the frequencies reproducible, within limits. In cinema pictures a standard number of frames per second has been adopted. (*d*) The time constant of operation of the photoelectric system and recording apparatus. (*e*) The focusing and alignment of the optical systems in both recording and replay.

The amplitude range reproducible is limited by the following main factors: (*a*) In variable density recording the emulsion gradation range is limited, particularly the linear portion. Grain size limits the lowest amplitudes. In the variable width system, the scattering of light at scratches on the clear parts of the track produces a noise background, limiting the lowest sound amplitudes that may be used. Reducing the width of the clear portion during quiet passages limits the noise level. In the variable-density system, the soundtrack is darkened during quiet passages. (*b*) The linear range of operation of the photoelectric system and recording apparatus.

Each system has its own difficulties; the variable density is subject to waveform distortion, particularly at high frequencies; the variable width to the necessity of a greater control in exposure for clean and true outline. Recording at approximately twice the final size on film and reducing to normal, limits some of the irregularities. Some difficulty with sound film results from the noncontinuous motion of the film required for projection of the visual picture, and the necessity of smoothing the variations to regular continuous motion through the sound optical system. In practice, sound-film engineers aim to reduce the flutter to less than $\pm 0.5\%$ at all frequencies. Mechanical and electrical filtering are used. Some systems confine the flutter to one selected frequency band and filter out in the audio-amplifying system.

sound wave *See* sound.

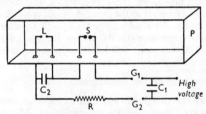

Sound-wave photography

sound-wave photography When a spark is produced in air a sound wave is associated with it that spreads out in a thin spherical shell in which the air has a high density and refractive index. If this sound pulse is illuminated instantaneously, it causes a shadow to be cast on a suitably placed screen.

Although similar to the *schlieren method the apparatus requires no lenses or mirrors. The sound wave is produced by the spark gap S and the illumination comes from another spark gap L. If the shadow caused by a spherical sound pulse from S is to be cast on the photographic plate P, the light pulse must start a fraction of a second after the sound pulse. This is done by putting the two gaps in series in the same circuit with a capacitor C_2 in parallel with L. The delay of L behind S is controlled by the value of C_2 and the resistance R. An induction coil producing about 100 000 volts is used to actuate the spark gaps, which are set off by lowering glass plates into the trigger gaps G_1 and G_2. The spark at L occurs between electrodes of magnesium wire 3 cm apart in an enclosed tube. The sound wave from L, being slower than the light, produces no effect. Light from S is screened from the plate by discs surrounding the electrodes. Reflection and refraction of sound can be studied by placing mirrors or lenses near S and using a suitable delay on L. The method has been used to study the acoustics of buildings.

source 1. The point at which lines of flux originate in a *vector field; e.g. a point at which there is a positive electric charge in an electrostatic field; a point at which a liquid enters a velocity field in hydrodynamics. A *sink* is a point at which lines of flux terminate.

2. In classical hydrodynamics, a point at which fluid is continually emitted and from which the flow is radial and uniform. A negative source is called a *sink*. The strength of a source (m) is the quantity of fluid emitted in unit time. The quantity emitted is sometimes expressed as $4\pi m'$; the symbol m' then defines the strength.

3. The electrode in a *field-effect transistor that supplies charge *carriers (electrons or holes) to the interelectrode space.

4. Any energy-producing device, e.g. current source, power source.

5. An object that emits light, such as the sun, an electric light, a gas-discharge tube, an arc, or a spark.

source impedance The impedance of any energy source presented to the input terminals of a circuit or device. An ideal voltage source will have zero source impedance (i.e. $dV_s/dI_s = 0$), whereas an ideal current source will have infinite source impedance (i.e. $dV_s/dI_s = \infty$).

space centrode *See* instantaneous centre.

space-charge region In any device, a region in which the net *charge density is significantly different from zero. For example, in a *semiconductor or *thermionic valve, space-charge regions can exist in equilibrium under zero applied-bias condition, forming potential barriers; these barriers must be overcome by the applied bias before current flows.

space cone *See* instantaneous axis.

space group A set of operations (rotation about an axis, reflection across a plane, translation, or combinations of these) that when carried out on a periodic arrangement of points in space brings the system of points to self-coincidence.

space inversion *See* parity.

space lattice *See* Bravais lattice.

space quantization *See* spin.

space-reflection symmetry *See* parity.

space-time continuum *See* four-dimensional continuum.

spacing, interplanar The perpendicular distance apart of similar planes of atoms forming a parallel set of planes in a crystal.

spallation A particularly vigorous nuclear reaction caused by a bombardment of a target by high-energy particles, resulting in the emission of a number of nucleons.

spark A visible disruptive discharge of electricity between two places at opposite high potential. It is preceded by ionization of the path. There is a rapid heating effect of the air through which the spark passes, which creates a sharp crackling noise. The distance a spark will travel is determined by the shape of the electrodes and the p.d. between them.

spark chamber A device in which the tracks of charged particles are made visible and their location in space accurately recorded. It was developed from the *spark counter. It consists essentially of a stack of narrowly spaced thin metal plates or grids in a gaseous atmosphere, partially surrounded by one or more auxiliary particle detectors. Any charged particle detected in one of the auxiliary devices triggers the rapid application of a high-voltage pulse to the stack of plates. The passage of the particle through the plates is marked by a series of spark discharges along its path. The track is recorded electronically or photographically. Subsequent events, such as collisions and disintegrations, can also be recorded. Use of the auxiliary devices leads to a select triggering of the spark chamber for a particular type of energy or radiation.

spark channel The path of a spark discharge between two electrodes after the gap has been bridged by a streamer. The channel becomes established by ionization and may persist for some time, although it may not be the shortest possible path.

spark counter A type of particle detector used in the detection and measurement of heavily ionizing particles. The counter consists of a pair of electrodes in the form of a wire or mesh anode in close proximity to a metal-plate cathode. A high potential difference is applied across the electrodes, its value being just less than that required to cause a discharge across the air gap. If a charged particle approaches the anode, the field between the electrodes is increased sufficiently to cause a spark discharge. At the moment of discharge the anode potential drops significantly. Particles may be detected by the noise as sparks occur. The number of particles may be measured by photography or counting circuits designed to respond to the change in voltage across the anode load resistor.

Spark counters can be made to be completely unaffected by beta particles and X- or gamma rays and are therefore very useful in detecting leakage from sealed containers containing radioactive sources (such as radium) that emit α-particles. The very short distance travelled by α-particles means that they are normally completely absorbed in the container and are only observed if leakage occurs.

spark discharge *See* conduction in gases.

spark gap An arrangement of electrodes specially designed so that a disruptive discharge takes place between them when the applied voltage exceeds a predetermined value.

sparkover *See* flashover.

sparkover voltage *See* flashover voltage.

spark photography Any form of photography in which a spark provides the illumination. In spark photography the camera lens is left open and the duration of the spark is controlled to give the correct exposure. *See* schlieren method.

spark spectrum The spectrum of an electric *spark, such as that generated across the points of an induction coil. This consists principally of the spectral lines of air; if a capacitor is placed in parallel with a spark gap, the spectrum is rich in the lines of the metal of the electrodes. As compared with the arc spectrum some lines appear with increased intensity. Spark lines are due to ionized atoms.

Sparrow's criterion *See* resolving power.

spatial filtering A technique whereby an optical image can be improved by filtering (i.e. removing) certain spatial frequencies from it (*see* spatial period). The image may, for example, be a micrograph or one transmitted from a planetary spaceprobe. It is modified so that the information of interest is made more accessible. A mask is used to filter the required components from the diffraction pattern of the image. For example, removal of low spatial frequencies (with a high-pass filter) will enhance the sharp edges in an image at the expense of regions in which the intensity is uniform or changing only slowly. Other filters can enhance contrast or remove extraneous lines from a composite picture or patterns of dots from halftone pictures.

spatial frequency *See* spatial period.

spatial period The distance over which a regular pattern repeats itself. The *spatial frequency* of the pattern is the reciprocal of the spatial period. The pattern may, for example, be a series of lines of equal width and spacing as found on a diffraction grating or it may be a diffraction pattern.

speaker Short for loudspeaker.

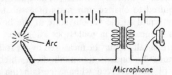

Speaking arc

speaking arc A method for reproducing sounds by the thermal effects of a current. A microphone is used with a transformer to vary the current in an arc circuit. The variations in heat evolved cause pressure variations in the surrounding air, and sounds made into the microphone are reproduced with considerable intensity. The arc acts as an amplifier of the small induced current supplied.

special relativity *See* relativity.

specific The use of the adjective *specific* to qualify the name of an extensive physical property is now restricted to the meaning "per unit mass". For example, the specific heat capacity of a substance is the *heat capacity per unit mass (per kilogram in SI units).

When the physical quantity is denoted by a capital letter (e.g. L for the latent heat), the specific quantity is denoted by the corresponding lower-case letter (l for the specific latent heat).

Formerly the word had other meanings. For example, in the expression *specific gravity* it denoted a ratio of the magnitude of a property of a substance to the magnitude of the same property of a reference substance (in this case water, except that the specific gravities of gases were sometimes referred to hydrogen). In the expression *specific resistance*, the word indicated the value of a general physical property restricted to a particular quantity of a particular substance. These and other uses are now deprecated in favour of the meaning "per unit mass". Specific gravity is now replaced by relative *density, specific resistance by *resistivity.

specific activity Symbol: a. The *activity per unit mass of a radionuclide.

specific charge The ratio of the charge on an *elementary particle to the mass of that particle; the charge per unit mass of a particle. *See also* e/m.

specific conductance The former name for *conductivity (electrical).

specific gravity The former name for relative *density.

specific-gravity bottle *See* relative-density bottle.

specific heat capacity Symbol: c. The *heat capacity per unit mass; the quantity of heat required to raise the temperature of one kilogram of a substance by one kelvin. It is measured in $J\,kg^{-1}\,K^{-1}$.

For a solid, the chief experimental methods for the determination of the specific heat capacity are (*a*) the *method of mixtures; (*b*) using the *Bunsen ice calorimeter; (*c*) the method of *Magnus for high temperatures; (*d*) the method of *Nernst and Lindemann for low temperatures. A theory of the variation of the specific heat capacities of solids with temperature has been given by *Debye using the quantum theory. (*See also* Einstein; Dulong and Petit's law.) For a liquid, the specific heat capacity may be measured by any of the usual calorimetric methods. In the case of water, the specific heat capacity was accurately determined by *Rowland, *Callendar and Barnes, and *Jaeger and Steinwehr. (*See also* cooling method.) All methods for solids and liquids determine the value at constant pressure. For a gas, there are two principal specific heat capacities depending on the way in which the temperature is increased. If the pressure is kept constant the specific heat capacity at constant pressure, c_p, is obtained; if the volume is kept constant, the specific heat capacity at constant volume, c_V, is obtained. c_p may be determined by the methods due to (*a*) *Regnault; (*b*) *Scheel and Heuse at low temperatures; (*c*) *Holborn and Henning at high temperatures. c_V may be determined by (*a*) the *differential steam calorimeter; (*b*) the *explosion method at high temperatures; (*c*) *Eucken's experiments at low temperatures. The specific heat capacity at constant

pressure always exceeds that at constant volume by the work done in expansion. For a solid:
$$c_p - c_V = 9\alpha^2 vT/K,$$
where α is the coefficient of linear expansion, v the specific volume, K the compressibility, and T the thermodynamic temperature. Using *Grüneisen's law that α is proportional to c_p this equation may be written:
$$c_p - c_V = Ac_p^2 T,$$
where A is a constant since K and v vary little with temperature. For an ideal gas in which the internal energy is independent of the volume,
$$c_p - c_V = nR,$$
where n is the number of moles per kilogram and R is the *molar gas constant. These equations are derived from the general thermodynamic equation:
$$c_p - c_V = T\left\{\frac{\partial p}{\partial T}\right\}_V \left\{\frac{\partial v}{\partial T}\right\}_p,$$
while
$$\left(\frac{\partial c_V}{\partial v}\right)_T = T\left(\frac{\partial^2 p}{\partial T^2}\right)_V$$
and
$$\left(\frac{\partial c_p}{\partial p}\right)_T = -T\left(\frac{\partial^2 p}{\partial T^2}\right)_p$$
give the variation of specific heat capacities with volume and pressure. On the kinetic theory the *molar heat capacity at constant volume of a gas is given by $F \times R/2$ assuming the equipartition of energy, where F is the number of *degrees of freedom. Although this result is by no means true for real gases, the specific heat capacity does increase greatly with the atomicity of the molecule. The ratio c_p/c_V of a gas always exceeds unity and is a constant denoted by the symbol γ. For a saturated vapour the specific heat capacity may sometimes be negative. (*See* negative specific heat capacity; calorimetry.)

specific impulse The ratio of the thrust (in kg) produced by a rocket to the rate of fuel consumption ($kg\,s^{-1}$). It can be considered as the length of time 1 kg of propellant (fuel plus oxidant) would last if expended at a rate producing 1 kg of thrust.

specific inductive capacity A former name for relative *permittivity.

specific latent heat of fusion *Syn.* enthalpy of melting. Symbol: l_f or ΔH_s^1. The quantity of heat required to change the state of unit mass of a substance from the solid to the liquid state at the melting point. It is measured in joules per kilogram. Experimentally it may be measured by: (1) *Bunsen ice calorimeter, in which a known amount of the liquid is added to the calorimeter where it solidifies. Part of the heat used to melt the ice in the instrument is the latent heat evolved when the substance solidifies. (2) The *method of mixtures, in which the solid in melting produces a fall in temperature in the liquid in the calorimeter into which the solid is

placed. (3) The method of cooling, in which the time for complete freezing is determined. (*See* Tammann and Jüttner.) (4) The electrical method, which consists of measuring the electrical work required to heat the substance below its melting point to a temperature above the melting point. (*See* Dickinson, Harper, and Osborn.) (5) Indirect methods (*a*) using the *Clausius–Clapeyron equation:
$$l_f = T(dp/dT)(v_1 - v_2),$$
where v_1 and v_2 are the specific volumes of liquid and solid respectively at the melting point T. dp/dT is found from the known variation of the melting point of the substance with the applied pressure. (*b*) by finding the lowering in freezing point ΔT_0 produced by adding a known amount of solute to the substance since:
$$\Delta T_0 = (n_2/n_1)(RT_0^2/l_f),$$
where n_1 and n_2 are the molar concentrations of solvent and solute respectively and T_0 is the melting point of the pure solvent.

According to *statistical mechanics, the change in *entropy on melting one mole of an ideal solid, in the absence of any other changes such as expansion, would be equal to the *molar gas constant R. This gives $l_f/T = nR$ where n is the number of moles per kilogram. By experiment it is found that:
$$l_f/T = (1.35 \pm 0.4)nR.$$

specific latent heat of sublimation *Syn.* enthalpy of sublimation. Symbol: l_s or ΔH_s^g. The quantity of heat required to change unit mass of a substance from the solid to the vapour state without change of temperature. *See* calorimetry.

specific latent heat of vaporization *Syn.* enthalpy of evaporation. Symbol: l_v or ΔH_1^g. The quantity of heat required to change unit mass of a substance from the liquid to the vapour state at the boiling point. It is measured in joules per kilogram.

Experimentally it may be measured by the following methods: **1.** Condensation methods, in which the amount of heat evolved when a certain mass of vapour condenses is measured, as in *Berthelot's apparatus. **2.** Evaporation methods, in which the amount of heat required to vaporize a given mass of liquid is measured. *Henning supplied the heat electrically and determined the specific latent heat of vaporization of water for temperatures between 30 °C and 180 °C. *Simon and Lange used this type of method for determining the specific latent heat of liquid hydrogen. **3.** Indirect method using the thermodynamic relation:
$$\frac{dl_v}{dT} - \frac{l_v}{T} = c_2 - c_1 - \frac{l_v}{v_2 - v_1}\left\{\left(\frac{\partial v_2}{\partial T}\right)_p - \left(\frac{\partial v_1}{\partial T}\right)_p\right\},$$
where v_1 and v_2 are the specific volumes of the liquid and vapour respectively at the boiling point T. (dp/dT) may be found from the curve showing the variation of boiling point with pressure.

The value of l_v decreases as the temperature increases, eventually becoming zero at the *critical temperature. For all substances the variation of specific latent heat with temperature is given by the thermodynamic formula:

$$l_v = T \frac{dp}{dT} (v_2 - v_1),$$

where c_1 and c_2 are the *specific heat capacities of the liquid and vapour respectively. An empirical formula due to Thiesen gives:

$$l_v \propto (T_c - T)^{1/3},$$

where T_c is the critical temperature. *See* Trouton's rule.

specific optical rotary power Symbol: α_D. A measure of the *optical activity of a substance in solution. It is given by the equation:

$$\alpha_D = \alpha V/ml,$$

where α is the angle of rotation of the plane of polarization when light traverses a pathlength l of a solution containing a mass of substance m in a volume V. The units are $m^2\,kg^{-1}$.

specific reluctance The former name for *reluctivity.

specific resistance The former name for *resistivity.

specific rotation *See* rotation of plane of polarization.

specific volume Symbol: v. The volume of unit mass of a substance; the reciprocal of density.

spectacle lenses Thin lenses used to correct *refraction of the eye. The main forms are symmetrical, plano, meniscus, spherocylindrical, and toric: the first three are spherical, correcting spherical errors, and the last two for correcting *astigmatism. The powers in dioptres of lenses are the reciprocals of focal distances expressed in metres (back vertex values preferably). When there are defects of the external muscles, the lenses may require decentring or may be worked on prisms. The tests for power are by neutralization (using trial case lenses, etc.), or by special focimeters, tests that may also include those of *centration. Generally spectacle glass has a standard refractive index 1.523 (n_D) and mean dispersion ($n_F - n_C$) of 0.0093 so that duplication of lens working tools, lens measures, etc., is reduced. Manufacturers make in quantity *uncut* lenses (spheres or spherocylindrical lenses – sph-cyls.), which are sold to prescription houses for cutting the edge to shape and size and also edging to shape with appropriate centration. Dispensing opticians measure faces and fit glazed frames to the spectacle wearers. Bifocal and multiple focus lenses along with protective filters are special types. The *blooming of lenses cuts down unwanted reflections. Spectacle magnification is the ratio of the size of the retinal image when the eye is corrected to that when uncorrected (i.e. [1/(1 − dF)]). (*See* Wollaston.)

spectral 1. Of or relating to a spectrum.
2. Of or relating to a particular frequency or wavelength.

spectral class *See* stellar spectra.

spectral lines *See* spectrum; hyperfine structure of spectral lines.

spectral luminous efficacy *See* luminous efficacy.

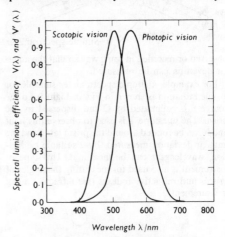

Spectral luminous efficiency curves

spectral luminous efficiency Symbol: $V(\lambda)$ (for photopic vision); $V'(\lambda)$ (for scotopic vision). A ratio expressing the ability of the eye to judge the power of *monochromatic radiation of a particular wavelength, λ. It is equal to the ratio of the *radiant flux at a wavelength λ_m to that producing an equal sensation of *luminosity at the wavelength of interest λ. The comparison is made under standard photometric conditions and λ_m is chosen so that the maximum value of the ratio is unity. The spectral luminous efficiency varies with wavelength and is different for *photopic vision than for *scotopic vision. This is seen in the diagram: the *Purkinje effect, in which the eye becomes relatively more sensitive to blue at fainter illuminations, is revealed by a shift in the maximum of the curve, from 555 nm for bright sources (photopic vision) to 505 nm for faint sources (scotopic vision).

The curves in the diagram apply for most observers and an observer whose spectral sensitivity curve conforms to either of these curves is called a *CIE standard photometric observer. Spectral luminous efficiencies are used in the definition of *luminous flux.

spectral type *See* stellar spectra.

spectrograph *See* spectrometer.

spectrographic analysis *See* spectrometer.

spectrometer 1. An instrument for producing, examining, or recording a *spectrum. When an emission spectrum is investigated, the radiation from the source is passed through a *collimator, which produces a parallel beam of radiation. This is deviated and dispersed by a prism or *diffraction grating and the angular deviation depends on the wavelengths present. The refracted or diffracted radiation is then

spectrophotometer

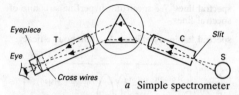

a Simple spectrometer

observed or recorded in some way so that the angular deviation can be measured.

For example, a simple spectrometer suitable for visible radiation is shown in Fig. *a*. Light from the source S is collimated by C and dispersed by the prism. The telescope T is used to observe this light and it can be rotated around the prism table and the angular deviation measured. By a suitable calibration, wavelengths can be measured. This type of instrument is also used for measuring the angles of prisms and refractive indexes (*see* refractive-index measurement).

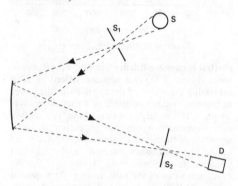

b Spectrometer using concave diffraction grating

Many modern spectrometers use diffraction gratings. Fig. *b* shows a spectrometer employing a concave grating. Radiation enters at the slit S_1, and is reflected by the grating G through S_2 onto a detector D, such as a *photomultiplier. In this instrument both slits are kept constant and the grating is rotated through a range of angles. A collimator is not essential in this type of instrument because the grating is ruled on a concave mirror and focuses the radiation. At each particular angle of the grating, radiation of a particular wavelength is focused onto the exit slit. A graph of the angle of the grating (as abscissa) against the response of the photomultiplier can, with suitable calibration, give a curve of wavelength against intensity of radiation. The wavelength scale could be calculated absolutely but is usually calibrated with standard wavelengths from pure substances. *See* Rowland grating.

When an absorption spectrum is investigated, a source of radiation with a continuous spectrum is usually used. The radiation may be passed through the sample and then into the spectrometer so that the distribution of intensity with wavelength is found. Alternatively, the radiation can be passed into the spectrometer so that one particular wave-

length of radiation is selected and passed through the sample onto the detector. The wavelength can be varied by changing the position of the grating or prism. In this way the spectrometer is used to isolate one wavelength of radiation, and is called a *monochromator*. Spectrometers are also used in studying *electron spin resonance and *nuclear magnetic resonance.

Any spectrometer that records a spectrum is called a *spectrograph*; the record is called a *spectrogram*. The design of spectrographs depends on their application and the wavelength of radiation for which they are used. In the visible region glass prisms and transmission gratings can be used to disperse that radiation. Infrared radiation is absorbed by glass and sometimes rock-salt prisms are used in infrared spectrometers. Gratings are usually used in ultraviolet and X-ray spectrometers.

The instruments used in detection of the radiation consist either of an electronic imaging device or a photographic plate, sensitive to the wavelengths under investigation. Highly sensitive electronic devices, such as *CCD detectors, can detect light, ultraviolet, and often X-rays; the information obtained is fed to a computer for analysis. Other electronic detectors include *bolometers, and devices based on *photoconductivity or the *photovoltaic effect. A variety of sources of radiation are used in producing spectra. Sources used for emission spectroscopy include flames, *gas-discharge tubes, and sparks. X-ray emission is induced by electron bombardment (or photon bombardment) of solids; in absorption spectroscopy, sources of radiation with continuous spectra are used. In the infrared and visible regions these are produced by hot solids. Ultraviolet and visible radiation can be produced by certain gas-discharge tubes. *Synchrotron radiation is a convenient source of both ultraviolet and X-radiation.

Spectrometers are often called *spectroscopes*. A *spectrophotometer* is an instrument for measuring the intensity of each wavelength in a spectrum, and the term is often used synonymously with spectrometer, especially for instruments used for visible and ultraviolet radiation. (*See also* spectroscopy.)

2. A similar instrument for determining the distribution of energies in a beam of particles, such as electrons (*see* electron spectroscopy) or ions (*see* mass spectrum).

spectrophotometer *See* spectrometer.

spectroscope *See* spectrometer.

spectroscopy The technique of producing *spectra, analysing their constituent wavelengths, and using them for chemical analysis or the determination of energy levels and molecular structure. A spectrum is formed by the emission or absorption of electromagnetic radiation accompanying changes between the quantum states of atoms and molecules. The frequency of the radiation depends on the type of states involved and spectroscopic techniques are used over a very wide range of frequencies and yield a wide variety of information.

Gamma-ray spectroscopy involves the measurement of the distribution of energies in *gamma rays emitted from nuclei. The energy levels of nuclei are also found from a spectrum of absorbed gamma rays in the *Mössbauer effect. *X-ray emission occurs when orbital electrons make transitions from outer orbits to vacant inner orbits. It gives information on electronic states in atoms and molecules, and is often used for studying *energy bands in solids. The absorption and emission of ultraviolet radiation are also used for the determination of electronic states. In the far *ultraviolet region it is associated with electronic transitions in ions and *ultraviolet spectroscopy* is applied to the study of discharges (*see* gas-discharge tube). Ultraviolet and visible radiation is also emitted and absorbed by changes in the electron states of atoms and molecules. Spectroscopy of this region is used to determine the energies of valence electrons. The spectra of molecules show band structure due to associated vibrational and rotational states and *ionization potentials can also be determined (*see* Rydberg spectrum). Visible and ultraviolet spectroscopy is a widely used analytical technique. Elements and compounds present in a sample can be detected by the presence of characteristic lines in their emission or absorption spectrum. A quantitative determination can often be made from the intensities of the lines. The techniques are also applied to the detection of species in discharges and flames. (*See also* flash photolysis.)

At longer wavelengths, in the near *infrared region, the photon energies correspond to changes between vibrational states of molecules. *Infrared spectroscopy* gives information on the vibration at frequencies and force constants of chemical bonds and on the potential energy curves of molecules (*see* Morse equation).

Certain groups of atoms in molecules tend to absorb at characteristic frequencies and infrared spectroscopy is widely used by chemists for identifying compounds. Far infrared and *microwave radiation is absorbed during changes in the rotational states of molecules. Spectroscopy in this region can give moments of inertia of molecules and their bond lengths and shapes. *Microwave spectroscopy* is particularly useful in the study of large molecules.

At still lower frequencies (and lower photon energies) the states studied are those due to the *spin of electrons and nuclei in applied magnetic fields. *Electron spin resonance spectroscopy is used to study paramagnetic materials and *nuclear magnetic resonance spectroscopy gives nuclear magnetic moments. *See also* spectrometer; Stark effect; Zeeman effects; Raman effect; electron spectroscopy; hydrogen spectrum.

spectrum 1. Any particular distribution of *electromagnetic radiation, such as the display of colours (violet, blue, green, yellow, orange, red) produced when white light is dispersed by a prism or *diffraction grating. The term is also applied to a plot of the intensity of electromagnetic radiation against wavelength, frequency, or quantum energy, or to a photographic or electronically produced record of dispersed electromagnetic radiation.

Spectra can be obtained with a *spectrometer or spectrophotometer. The spectrum is characteristic of the radiation itself in that it specifies the wavelengths that are present and their intensities. It is also characteristic of the substance that is emitting or absorbing the radiation.

An *emission spectrum* is the spectrum of radiation emitted from a substance as a result of changes in the quantum states of its constituent atoms or molecules. The changes occur from an *excited state to a state of lower energy, often the *ground state. Emission spectra are produced as a result of excitation of atoms by some source of energy, such as heat, electron bombardment, bombardment with X-rays, etc. An *absorption spectrum* is a spectrum of radiation after some of it has been absorbed by matter. In this case radiation with continuous distribution of quantum energy (wavelength or frequency) passes through the substance and particular regions of this radiation are absorbed. The absorbed radiation excites atoms from the ground state to an excited state. Absorption spectra are usually simpler than emission spectra.

Spectra are also classified according to the way in which the intensity of radiation varies with wavelength. A spectrum in which there is a continuous region of radiation emitted or absorbed is a *continuous spectrum*. An example is the spectrum of visible and infrared radiation emitted by a *black body. The spectra of gaseous atoms often contain a number of sharp bright lines of emitted radiation or a number of dark lines on a continuous background due to absorption. Such spectra are called *line spectra* and result because the atoms make transitions between states of definite energy. *Band spectra* contain a series of regularly spaced lines that are very close together and may not be resolved by the apparatus used to disperse the radiation. They are characteristic of emission or absorption by molecules.

Spectra are also classified according to whether the radiation is in the X-ray, ultraviolet, visible, infrared, or microwave region of the spectrum and according to whether the process producing or absorbing the radiation involves a change in electronic, vibrational, or rotational states. Thus in the visible and ultraviolet regions, changes occur between electronic states and an *electronic spectrum* is obtained. These are line spectra in the case of atoms and band spectra in the case of molecules. The bands are formed because in making the transition from one electronic state to another, the molecule may end up in any of a number of possible vibrational states, each differing in energy. In the near-infrared region spectra result from changes from a rotational state in one vibrational state to a rotational state in another vibrational state. A spectrum of this type is called a *vibration-rotation spectrum*. Changes between rotational states in the same vibrational state lead to absorption or emission of radiation in the far-infrared and *microwave regions and the formation of a *rotation spectrum*. At the other end of the scale, ultraviolet spectra are electronic spectra of ions. X-ray spectra involve changes of electronic states of inner tightly bound electrons. (*See* X-rays.)

2. Any distribution of energies, momenta, velocities, etc., in a system of particles, as in a mass spectrum (*see* mass spectrum) or an electron spectrum (*see* electron spectroscopy).

3. *See* quality (in sound).

spectrum analyser An instrument for measuring the energy distribution with frequency for any waveform. It may be used to determine the frequency response and distortion of any transmission system by comparison of input and output waveforms. *Multichannel analysers are commonly used as spectrum analysers.

speculum A copper-tin alloy (Cu 67%, Sn 33%) used for metal mirrors and reflection diffraction gratings. It takes a high polish and does not tarnish readily.

speed 1. The rate of increase of distance travelled, with time. (*Compare* velocity.)

2. A value specifying the sensitivity of a photographic material to light.

3. Of a vacuum pump. The volume of gas extracted from a vessel per second, the volume being measured at the instantaneous pressure at the mouth of the pump.

4. A measure of the light-transmitting power of a lens, commonly indicated by its *f-number: as the f-number decreases, the speed increases. A *fast lens* has a wider than normal maximum aperture for its type, a *slow lens* has a smaller one.

speed of light in vacuum *Syn.* speed of electromagnetic radiation (in vacuum). (Use of the word velocity rather than speed is discouraged.) Symbol: c. A fundamental constant now defined as

$$2.997\ 924\ 58 \times 10^8 \text{ m s}^{-1} \text{ (exactly).}$$

This value has been recommended since 1975 for universal use. It is the speed at which not only light but all *electromagnetic radiation travels in a vacuum. The speed decreases when the radiation enters a material medium.

The Römer method of measurement (1676) was based on the observation of the times of the eclipses of the satellites of Jupiter and the delay of appearance due to the greater distance of travel of light across the diameter of the earth's orbit. Bradley's method (1727) consisted in measuring the angle of aberration due to the orbital motion of the earth. (*See* aberration of light.) These two methods are classed as the astronomical methods. Terrestrial methods include those of Fizeau (1849), which was a cogwheel method used also by Cornu, Young, and Forbes; of Foucault (1850), a rotating mirror method considerably improved by Michelson (1926) and extended from use in the measurement of speed over long distances of air to the measurement of the

speed in an evacuated pipe (1931). Kerr-cell methods developed by Karolus and Mittelstaedt (1925), Anderson (1941), Bergstrand (1950) and others have permitted a much higher frequency of light impulse than the previous methods, so that measurements could be made with much smaller distances of light travel. Radio and radar methods agree with the more recent of the optical determinations.

The precision of the measurements has made possible the redefinition of the *metre, which is now expressed in terms of the distance travelled in a given time (the value given above being exact by definition). *See also* relativity (special theory).

speed of sound Symbol: c. The speed of sound in dry air at STP is 331.4 m s^{-1}. In sea water it is 1540 m s^{-1} and in fresh water 1410 m s^{-1}. The transmission of sound may involve either longitudinal, transverse, or torsional wave motion, according to the medium. The speed with which the waves travel is dependent on the fundamental physical quantities, elasticity and density. The speed of sound of small amplitude elastic waves in any extended medium is given by the equation:

$$c = \sqrt{(K/\rho)},$$

where K is the appropriate *elastic constant and ρ is the normal density of the medium. For fluid media such as gases or liquids K is the bulk *modulus of elasticity while for solids the Young, axial, or shear modulus must be introduced.

In general, there are two elasticities for any fluid medium, the adiabatic and the isothermal, and the ratio between them is equal to the ratio between the two specific heat capacities (γ). For gases, Newton derived the speed of sound by substituting the isothermal elasticity (= p, the pressure of gas) in the equation, so that $c = \sqrt{p/\rho}$. This value was lower than the observed one in the atmosphere. Laplace pointed out that the compressions and rarefactions in sound waves take place so rapidly that the changes are adiabatic and therefore the speed of sound $c = \sqrt{(\gamma p/\rho)}$. This equation fits much more closely to the observed value.

From the above equation, the speed of sound in a gas depends upon the temperature in accordance with the relation:

$$c_\theta = c_0 \sqrt{(1 + \alpha\theta)},$$

where c_0 and c_θ are the speeds of sound at 0 °C and θ °C and α is the coefficient of expansion of the gas. Also, since p/ρ is constant at constant temperature for any gas, the speed is independent of the pressure. At very high pressures where deviations from Boyle's law are noticeable, this is not strictly true. In addition, the speed is independent of the frequency. At high frequencies, however, there is a noticeable dispersion for some gases (*see* dispersion of sound). The speed depends upon the nature of the gas since γ and p are involved in the equation. Hence, moisture and impurities affect the speed.

When sound is transmitted through an infinite medium, it may spread spherically or may theoretically remain in the form of a parallel beam of plane waves. In either case there is an attenuation of intensity as the distance of the source of sound increases and this may affect the speed, especially near a source of large amplitude.

The measurement of speed of sound in air may be done in direct free open air or in air confined in a tube. An early method was that of certain members of the Paris Academy in 1738 by what is known as the method of reciprocal firing (to eliminate the wind correction) and the time was measured by chronometers. Regnault improved the method by using an electrical recording by which the personal equation of the observer is eliminated. The method is open to error on account of the personal equation of the instrument, although it is constant, and the use of intense sounds will give high speeds, especially near the source. All these causes of error are eliminated by making the measurements in pipes or tubes; wind effect is avoided and the conditions of temperature and humidity are controllable. The measurement of the speed of sound may be carried out by the direct method of measuring the distance and the time as was done by Regnault using his electrical recorder, or by a *standing wave method using the principle of *Kundt's tube.

For liquids, either fresh or salt water is the only one with which any large scale method of measuring speed of sound is practicable. The value is of great significance in connection with submarine signalling and detection and under-water survey. Kundt's tube method has also been applied to the determination of speed in liquids. Among the tube methods is Wertheim's organ pipes, which are blown when filled with water thus enabling the velocity to be estimated.

In solids longitudinal, as well as torsional and transverse, waves are set up. The speeds of propagation consequently are different as determined by the elastic constants. The most important cases are: (i) Transverse waves in wires: $c = \sqrt{(T/m)}$; T is the tension and m is the mass per unit length. The elastic properties of the material are disregarded (see stretched string). (ii) Transverse waves in bars: this involves the Young modulus E but the speed depends on the frequency (see bars in transverse vibrations). (iii) Longitudinal waves in wires and bars: $c = \sqrt{(E/\rho)}$. (iv) Torsional vibrations in wires or bars: $c = \sqrt{(\mu/\rho)}$, where μ is the rigidity coefficient.

In all these cases the solid is restricted in certain dimensions, e.g. the diameter relative to the length. For an unlimited solid medium the vibrations are, however, very complex. This is exemplified in the case of earthquake disturbances. During recent years the speed of sound in strata of different compositions (mineral ores) has been determined and applied to *echo prospecting.

spheradian A former name for the *steradian.

sphere gap A spark gap having spherical electrodes. A sphere gap can be used to measure extra high voltages with great reliability.

spherical aberration An *aberration that can occur in optical systems. When rays are traced after reflec-

tion or refraction at large-aperture surfaces (commonly spherical), they do not unite accurately at a focus. When the outer zones focus within the paraxial focus, the spherical aberration is said to be positive (see caustic curve). First-order theory of spherical aberration is based on the employment of the first two terms in the expansion of the sine (Gauss, Seidel). Longitudinal spherical aberration is the axial distance between the paraxial focus and the intersection of the ray from a particular zone with the axis. For incident parallel rays, the lens of least aberration has radii roughly 1:6 in order. The aberration for this order varies as the square of the aperture. For higher order theory, the aberration shows dependence on the fourth power of the aperture.

The term spherical aberration is also used to embrace all the aberrations due to the sphericity (form) of surfaces, e.g. primary spherical aberration, coma, radial astigmatism, curvature of field, distortion (*Seidel aberrations). See aberrations of optical systems.

spherical lenses (and **surfaces**) Surfaces that are portions of spheres are mechanically easier to produce than others, such as ellipsoid, etc. Such surfaces are therefore most commonly used, although the employment of aspherical or deformed spherical surfaces is increasing. Most of the problems of lens design are concerned with correcting the *aberrations produced. For any particular power (focal length), the forms of spherical lenses are equi-bi, plano, periscopic, and deep meniscus. The term spherical lens is also used to describe a complete sphere of a refracting medium acting as a thick lens – a sphere lens.

spherocylindrical lens See cylindrical lens.

spheroid 1. Prolate spheroid; a solid traced by the revolution of an ellipse about its major axis. Volume = $(4/3)\pi ab^2$.

2. Oblate spheroid: a solid traced by the revolution of an ellipse about its minor axis. Volume = $(4/3)\pi a^2 b$. ($2a$ and $2b$ are the lengths of the major and minor axes respectively of the ellipse.)

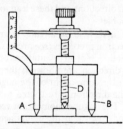

Spherometer

spherometer An instrument for measuring the curvature of the surfaces of lenses and mirrors. The simplest instrument consists of a three-legged table carrying a micrometer screw. The points A, B, C (of which only A and B are shown in the diagram) form an equilateral triangle of side a, and D is always equidistant from A, B, and C. To measure the radius

spider

of a spherical surface, the instrument is stood on the surface and the central screw is adjusted until A, B, C, and D all touch it. The instrument is then transferred to an accurately plane surface and the micrometer screw adjusted until A, B, C, and D are coplanar. If D was displaced through a distance H in this adjustment, the radius R of the spherical surface is given by:

$$R = \frac{a^2}{6H} + \frac{H}{2}$$

spider A rotating part of an electrical machine that has spokes or the equivalent and that supports on the shaft another rotating part such as an armature, commutator, or slip ring (armature spider, commutator spider, slip-ring spider, respectively).

spike A current or voltage transient of extremely short duration.

spin The intrinsic angular momentum of an *elementary particle or group of particles. Bohr's theory of the *atom predicts that lines in the spectra of the alkali metals should be single. In fact, they consist of closely spaced *doublets. To explain this Pauli, in 1925, suggested that each electron could exist in two states with the same orbital motion. Uhlenbeck and Goudsmit interpreted these states as due to the spin of the electron about an axis. The electron is assumed to have an intrinsic angular momentum in addition to any angular momentum due to its orbital motion. This intrinsic angular momentum is called spin. It is quantized in values of:

$$\sqrt{s(s + 1)}h/2\pi,$$

where s is the *spin quantum number* and h the *Planck constant. For an electron the component of spin in a given direction can have values of $+\frac{1}{2}$ and $-\frac{1}{2}$, leading to the two possible states. An electron with spin behaves like a small magnet with an intrinsic *magnetic moment (*see also* magneton). The two states of different energy result from interaction between the magnetic field due to the electron's spin and that caused by its orbital motion. There are two closely spaced states resulting from the two possible spin directions and these lead to the two lines in the doublet.

In an external magnetic field the angular momentum vector of the electron precesses (*see* precession) around the field direction. Not all orientations of the vector to the field direction are allowed; there is quantization so that the component of the angular momentum along the direction is restricted to certain values of $h/2\pi$. The angular momentum vector has allowed directions such that the component is $m_s(h/2\pi)$, where m_s is the *magnetic spin quantum number*. For a given value of s, m_s has the values s, $(s-1)$, ... $-s$. For example, when $s = 1$, m_s is 1, 0, and -1. The electron has a spin of $\frac{1}{2}$ and thus m_s is $+\frac{1}{2}$ and $-\frac{1}{2}$. Thus the components of its spin angular momentum along the field direction are $\pm\frac{1}{2}(h/2\pi)$. This phenomenon is called *space quantization*.

The resultant spin of a number of particles is the vector sum of the spins (s) of the individual particles and is given the symbol S. For example, in an atom two electrons with spins of $\frac{1}{2}$ could combine to give a resultant spin of $S = \frac{1}{2} + \frac{1}{2} = 1$ or a resultant of $S = \frac{1}{2} - \frac{1}{2} = 0$.

Alternative symbols used for spin are J (for elementary particles) and I (for nuclei). Most elementary particles have a nonzero spin, which may either be integral or half integral. (*See* boson; fermion.) The spin of a nucleus is the resultant of the spins of its constituent nucleons. For example, ^{1}H has $I = \frac{1}{2}$, ^{2}H has $I = 1$, ^{10}B has $I = 3$, ^{13}C has $I = \frac{1}{2}$, ^{14}N has $I = 1$, and ^{17}O has $I = 5/2$. *See also* quantum mechanics; gyromagnetic ratio; nuclear magnetic resonance

spin glass A type of alloy containing a small proportion (0.1–10%) of a magnetic metal, such as iron or magnesium, and a nonmagnetic metal, such as gold or copper, in which the magnetic-metal atoms are distributed randomly in the crystal lattice. Examples include AuFe and CuMn. These alloys are 'glasses' in the sense that they have a random distribution of magnetic atoms; the 'spin' refers to the magnetic property of the atoms. Spin glasses have complicated magnetic behaviour, which is difficult to treat by conventional theory because the random distribution of magnetic atoms means that there is a lack of regular order in the lattice.

spinning The process of shaping rapidly rotating metal objects by applying forces to them by means of stationary tools, without cutting away any of the metal.

spin-paired electrons Two electrons with opposing *spins in an *atomic orbital. *See also* Pauli exclusion principle.

spin quantum number Symbol: s, m_s, J, I. *See* spin.

spin–statistics theorem A theorem in relativistic *quantum field theory stating that half-integer *spins can only be quantized consistently if they obey Fermi–Dirac statistics and integer spins can only be quantized consistently if they obey Bose–Einstein statistics (*see* quantum statistics). This theorem, which has been rigorously proved in many ways, provides the basis for the *Pauli exclusion principle.

spinthariscope An apparatus designed by Crookes for observing the scintillations produced by alpha particles from radioactive substances. It consists of a short tube containing a speck of the radioactive substance, a phosphorescent zinc sulphide screen, and a magnifying lens. The alpha particles cause minute flashes of light to appear on the screen. By measurement of the area of screen subjected to radiation, the instrument can be made roughly quantitative. *See* scintillation counter.

spirit level An instrument for determining the direction of the horizontal and, sometimes, for measuring small angles to the horizontal. The indicating device

536

is a bubble trapped under the downward concave upper surface of a glass vessel. If the framework holding the vessel is accurately horizontal, the bubble will be in the centre.

split-phase motor A type of single-phase a.c. *induction motor made self-starting by the provision of a starting winding in addition to the normal running winding. The two windings are wound on the stator and have a separation of 90 *electrical degrees. At starting and during acceleration, the two windings are connected in parallel to the a.c. supply and since the starting winding has a greater ratio of resistance to reactance than the running winding, there is a phase difference between the two currents. Hence the operation during acceleration approximates to that of a two-phase motor. The starting winding is disconnected, either by hand or automatically (e.g. by means of a centrifugally operated switch) when a certain speed has been reached. This type of motor is commonly of the fractional horsepower type and it usually has a squirrel-cage rotor. Fundamentally, a single-phase motor is not self-starting and special means to render it so must be provided as in the above type. *Compare* capacitor motor.

spontaneous compactification *See* Kaluza–Klein theory.

spontaneous fission (S.F.) Nuclear *fission that takes place independently of external circumstances and is not initiated by the impact of a neutron, an energetic particle, or a photon. It is a form of *radioactivity and obeys the exponential decay law. It occurs in nuclides of very high mass and in some cases when the *mass number is 250 or more it is the principal mode of decay.

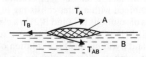

Spreading coefficient

spreading coefficient Imagine a drop of a liquid, A, instantaneously resting on the surface of another liquid, B, with which it is immiscible, as shown in the figure. Let the edge of the drop be straight and consider the three forces acting on unit length. These are numerically equal to the surface tensions T_A and T_B of the liquid and their interfacial surface tension T_{AB}. The condition for equilbrium is that these three forces shall balance and they can therefore be represented in magnitude and direction by the three sides of a (closed) triangle, known, in this case, as *Neumann's triangle*. If the drop does not spread, $T_B < T_A + T_{AB}$.

The quantity $(T_B - T_A - T_{AB})$ is called the spreading coefficient and if this is positive, the drop will spread.

(Spreading coefficient) = (adhesional work) – (cohesional work); adhesional work (W_A) is defined

as the work done in separating unit area of surface where two substances meet; thus,
$$W_A = T_A + T_B - T_{AB}.$$
Cohesional work (W_C) is defined as the work done in separating a column of liquid of unit area into two parts; thus
$$W_C = 2T_A.$$
If the drop is volatile it can spread by evaporating and then condensing on the other liquid's surface, e.g. carbon disulphide spreads on water even though the spreading coefficient is negative.

spreading resistance Of a *semiconductor device. The component of resistance due to the bulk of the semiconductor away from the junctions and contacts.

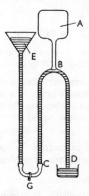

Sprengel air pump

Sprengel air pump This mercurial air pump consists of a tube CBD connected to a mercury reservoir E, through a pinch cock G. The vessel A to be exhausted is joined to the tube at B. Mercury, passing the branch at B, separates into drops; air from the vessel A fills the space between the drops and is dragged down by the descending mercury, the rate of flow of which is regulated by the pinch cock G.

spring balance A device with which a force is measured by the extension produced in a helical spring. It is used in weighing. The extension produced is directly proportional to the force; care must however be taken that the spring is not overstrained so that it takes a permanent set. *See also* balance.

spur 1. The spur, or trace, of an operator is the sum of the diagonal constituents of the *matrix that represents it.
2. *See* ionizing radiation.

spurious MOST An unwanted MOS *field-effect transistor formed in *MOS integrated circuits by the interconnections between parts of the circuit.

spurious response An unwanted output from an electronic circuit, device, or transducer in the absense of an input signal or as a result of the presence of an unwanted input signal.

sputtering The evaporation of particles of the cathode in a *gas-discharge tube during the discharge of electricity, due to bombardment by positive ions. It can be used to coat a nonconductor close to the cathode with a thin adhesive metallic film. Typically the gas pressure is between 150 and 1.5 Pa and the voltage between cathode and anode is between 1 and 20 kV.

square degree A unit of solid angle based on the degree and equal to $(\pi/180)^2$ *steradians. It is the solid angle subtended by a square whose sides each subtend 1°.

square-law detector A *detector that gives an output voltage proportional to the square of the input voltage. Small variations in input voltage are easily detected because of the relatively large output-voltage variations.

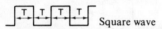 Square wave

square wave A pulse train consisting of rectangular pulses with *mark-space ratio equal to unity.

square-wave response In general, the response of any circuit or device to a square-wave input. In a *camera tube, using a test pattern of alternate black and white bars, it is the peak-to-peak signal amplitude obtained from the camera.

squaring-on The operation of making the optical axes of the lenses of an instrument coincident.

squark *See* supersymmetry.

squegging oscillator An *oscillator in which the oscillations build up in amplitude to a peak value then fall to zero. It is a type of *relaxation oscillator in which the relaxation oscillations modulate the high-frequency main oscillations. A special form is the *blocking oscillator, which may be used as a *pulse generator.

squid (*s*uperconducting *q*uantum *i*nterference *d*evice) Any of a family of superconducting devices that are capable of measuring extremely small magnetic fields, voltages, and currents. Their action is based on the d.c. Josephson current flowing across *Josephson junctions in certain configurations, the current in such a device being highly sensitive to an external magnetic field.

squirrel-cage rotor *See* induction motor.

SST Abbreviation for *single sideband transmission.

SSI Abbreviation for small-scale integration. *See* integrated circuit.

stability *See* equilibrium.

stable circuit A circuit that does not produce any unwanted oscillations under any operating conditions.

stable equilibrium *See* equilibrium.

stainless steel Those steels showing a high resistance to corrosion by the atmosphere or mild chemical

reagents. They are of two main types – martensite with 12–14% chromium, and austenite with 11–25% chromium and 7–36% nickel. Suitable mechanical properties are obtained by the inclusion of up to 0.4% carbon, and sometimes small quantities of other metals such as manganese are included.

stalagmometer A device for measuring the size of liquid drops, e.g. in the determination of surface tension by the *dropweight method.

Stalloy Proprietary alloy of iron with about 4% silicon. It is frequently used for transformer cores and telephone receiver diaphragms, though the hysteresis loss is high.

standard atmosphere An internationally established reference for pressure, defined as 101 325 pascals. Although this was formerly used as a unit of pressure – the *atmosphere*, symbol: atm – the standard atmosphere should not now be regarded as a unit. Atmospheric pressure fluctuates about the standard value.

standard cell An electric cell used as a voltage-reference standard. *See* Clark cell; Weston standard cell.

standard deviation *See* deviation.

standard illuminant An illuminating source set up for standard colorimetry, i.e. to enable measurements of the colours of non-self-luminous samples to be determined. The samples are illuminated at 45° to the normal and viewed normally. Three standards are prescribed:
Illuminant A. 2848 K colour temperature. 500 watt, 100 volt, single coil, uniplanar, class A1 tungsten filament gas-filled projector lamp.
Illuminant B. 4800 K colour temperature. Illuminant A combined with a filter consisting of 10 mm thick layers of solutions B(1) and B(2) in a clear glass double cell. B(1) is a solution of copper sulphate, mannitol, and pyridine; B(2) is a solution of cobalt ammonium sulphate, copper sulphate, and sulphuric acid.
Illuminant C. 6600 K colour temperature (corresponds approximately to light from an overcast sky). As for illuminant B, but with solutions C(1) and C (2), increasing the proportions of solids employed in comparison with liquids.

standardization 1. The process of relating a physical magnitude (e.g. a weight), or the indication of a meter (e.g. one supposedly reading current) to the standard unit of that quantity.
2. Establishment of an international, national, or industrial agreement concerning the specification or production of electronic, electrical, mechanical, and other types of components, or of equipment in general. Among other advantages, this greatly increases the ability to interchange components and devices.

standard model The model used for fundamental interactions in particle physics. It is a combination of *quantum chromodynamics, which describes the *strong interaction, *electroweak theory, which gives a unified description of the *electromagnetic and *weak interactions, and the general theory of *relativity, which describes classical *gravitational interactions. Although the standard model, in principle, gives a complete description of all known phenomena, it is regarded as incomplete because it has many arbitrary features and does not contain a quantum theory of the gravitational interaction. There is felt to be a need for a more fundamental theory such as a *unified field theory or a *string theory to remove the arbitrariness of the standard model and also give a quantum theory of gravity. *See also* broken symmetry; grand unified theory.

standard pile A body of *moderator, usually graphite, containing a neutron source and having a known *neutron flux density at specified positions. It can thus be used as a standard.

standard temperature and pressure (STP) A standard condition for the reduction of gas temperatures and pressures. Standard temperature is now 298.15 K; it was formerly 273.15 K, i.e. 0 °C. The standard pressure for gases was formerly 101 325 pascals. It is now recommended that the standard pressure for reporting thermodynamic data be 10^5 Pa (1 bar); normal boiling points however may still be reported on the basis of a pressure of 101 325 Pa.

standing wave *Syn.* stationary wave. A wave incident normally on a boundary of the transmitting medium is reflected either wholly or partially according to the boundary conditions, the reflected wave being superimposed on the incident wave and thereby creating an interference pattern of *nodes and antinodes. If the incident wave is represented by:

$$\xi = a \sin (2\pi/\lambda)(ct - x),$$

then for total reflection from a rigid boundary the reflected wave would be:

$$\zeta = -a \sin (2\pi/\lambda)(ct + x),$$

and the combination would be:

$$\xi = a \sin (2\pi/\lambda)(ct - x) - a \sin (2\pi/\lambda)(ct + x),$$

i.e.

$$\xi = -2a \sin (2\pi/\lambda)x \cos (2\pi/\lambda)ct.$$

This is a standing wave, i.e. a wave that remains stationary, the displacement being always zero at $x = 0, \lambda/2, \lambda, 3\lambda/2$, etc. (nodes), and vibrating with amplitude $2a$ at $x = \lambda/4, 3\lambda/4, 5\lambda/4$, etc. (antinodes).

In contrast to the above case, at a free boundary the phase of the reflected wave is the same as that of the incident wave, thus creating an antinode at the boundary. The standing-wave pattern has, however, the same spacings between nodes and antinodes as before.

In practice the reflection may be only partial, thus creating minimum deflections at the nodes instead of the zero deflections obtained with perfect reflection. Examples of stationary wave patterns may be obtained by means of a stretched string with one end

coupled to an electrically operated tuning fork. (*See* Melde's experiment.)

Each kind of wave has two types of disturbance. Electromagnetic waves have electric and magnetic fields; sound has pressure variations and particle displacements; surface waves in liquids have longitudinal and transverse displacements; stretched strings have particle displacements and variations of tension. Generally the nodes for one type of disturbance are antinodes for the other and the energy density (averaged over a cycle) is uniform in a system of standing waves, energy being interchanged between the two types of antinode during a cycle. Detectors are often sensitive to only one type of disturbance; for example, a photographic emulsion is exposed in the electric antinodes but not in the magnetic ones (*see* Lippmann process of colour photography). In the case of surface waves the transverse (vertical) nodes are not fixed but oscillate in the direction of the component progressive waves.

Stanhope lens A thick biconvex lens magnifier, with front surface of radius two-thirds the thickness and a concentric back surface with radius one-third the thickness (glass assumed). The object to be viewed is put in contact with the front surface.

Stanton number Symbol: *St*. A dimensionless parameter defined by the equation:

$$St = h/\rho c_p v,$$

where h is the *heat-transfer coefficient, ρ is the density of the fluid, c_p is the specific heat capacity, and v is the speed of flow. It is equal to the *Nusselt number divided by the product of the *Reynolds number and the *Prandtl number. *See* convection (of heat).

star 1. A self-luminous celestial body. Its energy is generated by *nuclear-fusion reactions in its core, and this energy is transported to the surface where it is radiated away into space. The inwardly directed gravitational force in a star balances the outwardly directed gas and radiation pressure, maintaining it in a state of hydrostatic equilibrium. As the star ages its interior structure and chemical composition change. *See also* magnitude; stellar energy; stellar spectra; Hertzsprung–Russell diagram; white dwarf; neutron star; black hole; celestial sphere.

2. For optical testing. A natural or artificial source of light subtending a very small angle at the centre of the entrance pupil of an instrument. (*See* star tests.)

3. In nuclear emulsions, cloud, and bubble chambers. A point from which several tracks radiate, indicating a multiple disintegration of an atomic nucleus.

star connection A method of connection used in *polyphase a.c. working in which the windings of a transformer, a.c. machine, etc., each have one end connected to a common junction the latter being called the *star point*. In three-phase working, the windings may be represented by the symbol Y or T and hence it is also known as the Y or T connection.

Compare mesh connection; delta connection. *See also* zig-zag connection.

star-delta switching starter A special switching starter for use with a three-phase *induction motor. With the switch in the starting position, the stator windings of the motor are connected in star (*see* star connection) and with the switch in the running position they are connected in delta (*see* delta connection). With this method, the supply current at the instant of starting is one-third of that which would be obtained if the motor were switched directly onto the supply with its windings connected in delta. It is necessary, however, for the six separate ends of the stator windings to be available and the motor must be designed for normal running with the delta connection.

Stark, Johannes (1874–1957) German physicist who studied theory of radiation and atomic theory. He discovered the *Doppler effect in positive rays and the effect on the wavelength of light of the application of a strong transverse electric field. (*See* Stark effect.)

Stark effect The wavelength of light emitted by atoms is altered by the application of a strong transverse electric field to the source, the spectrum lines being split up into a number of sharply defined components (Stark, 1913). The displacements are symmetrical about the position of the undisplaced line, and are proportional to the field strength up to about 100 000 volts per cm. The view of the lines parallel to the field is called the *longitudinal Stark effect*, the view (of the polarization) at right angles to the field is the *lateral Stark effect*.

Stark–Einstein equation The formula for the energy per mole E absorbed in a photochemical reaction. If f is the frequency of the absorbed radiation $E = hLf$ where h is the *Planck constant and L the *Avogadro constant.

starter *See* motor starter.

star tests Tests for optical instruments employing a very small source. They include tests for centring, squaring-on, chromatism, spherical aberration, etc.; the *Airy disc diffraction pattern is critically examined around the region of the focus (telescope and microscope objectives).

stat- A prefix that, when attached to the name of a practical electrical unit, denotes the corresponding unit in the CGS-electrostatic system. This system of units is no longer employed.

static 1. Electrical disturbance in a radio or TV system, such as a crackling or hissing sound at a loudspeaker. It is produced by electrostatic induction arising from atmospheric conditions, such as lightning flashes.

2. Electric sparks or crackling produced by friction.

3. Not changing or incapable of being changed over a period of time; undisturbed or causing no disturbance or movement.

static balancer *See* balancer.

static characteristic *See* characteristic.

static friction *See* friction.

statics The branch of *mechanics dealing with bodies at rest relative to some given frame of reference, with the forces between them, and with the equilibrium of the system. *Hydrostatics* is a branch of statics dealing with the equilibrium of fluids and with their stationary interactions (e.g. pressure, flotation) with solid bodies. Statics is sometimes regarded as a branch of *dynamics. *Compare* kinetics.

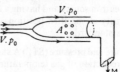

Static tube

static tube An instrument for measuring the static, or undisturbed, pressure in a fluid flow. It consists of a thin tube connected at one end to a manometer, the other end being rounded to minimize the disturbance caused by the tube and pointing upstream. At some distance back from the nose a number of holes (A) are drilled through the tube (*see* diagram). The distance from the nose is such that the flow lines are parallel to the axis of the tube at the holes; the pressure measured at the manometer M is then independent of fluid velocity – the static pressure. *Compare* Pitot tube.

stationary orbit *See* geostationary orbit.

stationary state In the *quantum theory, or *quantum mechanics. The state of an atom or other system that is fixed, or determined, by a given set of quantum numbers. It is one of the various quantum states that can be assumed by an atom.

stationary-time principle *See* Fermat's principle.

stationary value When a quantity y, in a curve having an equation $y = f(x)$, is plotted against x, the values of y, at those points on the graph where the tangent is parallel to the x axis, are called stationary values. Maxima and minima are instances of stationary values.

stationary wave *See* standing wave.

statistical error A fluctuation from an average value that arises from the randomness of the associated event. The event may, for example, be radioactive decay or electron emission. If the average value is n then the statistical error is of the order of $\sqrt{n}$.

statistical mechanics The theory in which the properties of macroscopic systems are predicted by the statistical behaviour of their constituent particles. For example, if a large collection of molecules is considered, its total energy is the sum of all the individual energies of the molecules. These in turn are energies of vibration, rotation, and translation, and electronic energy. According to *quantum theory a molecule can only have certain allowed energies; it can be thought of as occupying any of a number of *energy levels. Consequently the system as a whole can also have any number of possible energy levels, $E_1, E_2, E_3, \ldots$ If a large collection of systems is considered, each containing the same amount of substance, there will be a distribution of systems over energy levels: N_1 will have energy E_1, N_2 energy E_2, etc. According to the Maxwell–Boltzmann distribution law:

$$\frac{N_i}{N} = \frac{g_i e^{-E_i/kT}}{\sum_i g_i e^{-E_i/kT}},$$

where N_i systems have energy E_i, N is the total number of systems, and g_i is the *statistical weight of this energy level. The expression:

$$\sum_i g_i e^{-E_i/kT}$$

is called the *partition function* and has the symbol Z. A collection of systems of this type is called a *canonical assembly* or *ensemble*. The average energy of a system, E, is $\Sigma N_i E_i / \Sigma N_i$ and consequently:

$$E = kT^2 \, (\partial \log_e Z / \partial T).$$

In statistical mechanics it is assumed that this average instantaneous value of a property over a large number of systems is the same as the average value of this property for one system over a period of time. Thus the expression gives the *internal energy of the system. The partition function of the canonical assembly is related to the energy levels of the individual molecules by the equations:

$$Z = z^L \text{ and } z = \Sigma_j g_j \exp(-\varepsilon_j / kT).$$

Here z is the partition function of the assembly of molecules with energy $\varepsilon_1, \varepsilon_2, \varepsilon_3$, etc. Usually the partition function of 1 mole is considered and L is the *Avogadro constant. In principle, statistical mechanics can be used to obtain thermodynamic properties of a system from a knowledge of the energy levels of its components. However, in many cases it is difficult to evaluate the partition functions because of interactions between the particles. *See also* quantum statistics.

statistical weight *Syn.* degeneracy. Symbol: g. If a system has a number of possible quantized states and more than one distinct state has the same energy level, this energy level is said to be *degenerate* and its statistical weight is the number of states having that energy level. *See also* statistical mechanics.

stator The portion of an electrical machine that includes the nonrotating magnetic parts and the windings associated with them. The term is normally used only in connection with a.c. machines. *Compare* rotor.

steady state The state reached in a system under steady conditions. The steady state is obtained after all *transients, produced by one or more recent changes in existing conditions, have died away.

steady-state theory A theory in *cosmology postulating that the universe has always existed and will continue to exist in a steady state such that the average density of matter does not vary with distance or time. To offset the change in density of the *expanding universe, matter must be created in the space left by the receding stars and galaxies at a rate of about 10^{-43} kg m^{-3} s^{-1}. Evidence to support this theory (obtained from radiation received from outer space) must show that the density remains constant to the farthest depths of space thus indicating that density has not increased or decreased at any time. (The radiation received from stars, say 5000×10^6 light years away, has taken 5000×10^6 years to reach us, and therefore reflects conditions at that time.) So far no such conclusive evidence has been obtained. *Compare* big-bang theory.

steam calorimeter A calorimeter in which the amount of heat supplied is calculated from the mass of steam condensed on the body under test. *See* Joly's steam calorimeter.

steam engine A machine that takes heat from a steam boiler, performs external work, and rejects a smaller amount of heat to a condenser. *See* Rankine cycle.

steam line The curve showing the variation of the boiling point of water with pressure, i.e. the variation of the saturation pressure of water vapour with pressure.

steam point The temperature of equilibrium between the liquid and vapour phases of water at standard pressure. Its former importance was as the upper fixed point on the Celsius scale of temperature. Thermodynamic temperature is now based on the triple point of water, and its value (273.16 K) has been chosen to make the steam point equal to 100 °C within the limits of experimental measurement. At a pressure p (in millimetres of mercury) the steam point has a value of:

$$100 + 0.0367(p - 760) - 2.3 \times 10^{-5} \, (p - 760)^2.$$

This temperature is the boiling point of water at a given pressure. *Compare* ice point.

steam turbine *See* turbine.

steel *See* alloy; stainless steel.

steelyard *See* balance.

steerable aerial A *directive aerial whose direction of maximum radiation or sensitivity can be altered. This can be achieved mechanically, as when a dish aerial is tilted and/or rotated, or electronically, as in phased-array *radar.

Stefan, Joseph (1835–1893) Austrian physicist, whose work was concerned primarily with radiation (*see* Stefan–Boltzmann law), the kinetic theory, and hydrodynamics.

Stefan–Boltzmann constant

Stefan–Boltzmann constant Formerly Stefan's constant. *See* Stefan–Boltzmann law.

Stefan–Boltzmann law *Syn.* Stefan's law. A formula relating the radiant flux per unit area emitted by a *black body (*radiant exitance) to the temperature. It has the form $M_e = \sigma T^4$, where M_e is the radiant exitance and σ is the *Stefan–Boltzmann constant*:

$$\sigma = 2\pi^5 k^4 / 15 h^3 c^2,$$

where k is the Boltzmann constant, c the speed of light in vacuum, and h is the *Planck constant. σ has the value

$$5.670\ 51 \times 10^{-8}\ \text{W m}^{-2}\ \text{K}^{-4}.$$

The law was originally deduced by Stefan from the results of experiments by *Dulong and Petit. Boltzmann later gave a proof based on *thermodynamics.

Stefan's constant Former name for Stefan–Boltzmann constant. *See* Stefan–Boltzmann law.

Stefan's law Former name for Stefan–Boltzmann law.

Steinberger, Jack (*b.* 1921) German-born Amer. physicist. A professor at Columbia University (1950–1971), he moved to CERN in 1968. With Leon Lederman and Melvin Schwartz he devised an experiment at the Brookhaven National Laboratory that established the existence of the muon neutrino. For this work they shared a Nobel prize (1988).

Steinmetz hysteresis law An empirical equation giving hysteresis loss for iron and some steel, formulated in 1891 by C. P. Steinmetz (1865–1923). For iron the loss is given as $kB_{max}^{1.6}$ where B_{max} is the maximum *magnetic flux density and k is a constant depending on the material; k is known as the *Steinmetz coefficient*.

stellar energy Stars emit radiation of great intensity for enormous periods of time, for example the sun radiates with a power about 3.6×10^{26} W, equivalent to $4 \times 10^9\ \text{kg s}^{-1}$, and is believed to have emitted for at least 5×10^9 years. *Helmholtz calculated that the sun could have radiated for about thirty million years at the present rate as a result of gravitational contraction. The study of *nuclear fusion, starting about 1937, showed that the major part of stellar energy must be generated by such nuclear processes, but that gravitational contraction is needed to generate the high temperatures (of order 10^7 K) at which fusion can occur, and for certain later stages of evolution.

The primary process of thermonuclear fusion in stars such as the sun is the *weak interaction between protons:

$$^1\text{H} + {}^1\text{H} \rightarrow {}^2\text{H} + e^+ + \nu.$$

Then:

$$^2\text{H} + {}^1\text{H} \rightarrow {}^3\text{He} + \gamma;$$
$$^3\text{He} + {}^3\text{He} \rightarrow {}^4\text{He} + 2{}^1\text{H}.$$

Hence four protons have interacted to form a helium nucleus, two positrons, two neutrinos, and electromagnetic radiation. The positrons interact with electrons to give *annihilation radiation. The overall effect is equivalent to the conversion of hydrogen into helium, with energy release (excluding neutrinos) of about 25 MeV per helium nucleus produced. (*See also* carbon cycle.)

At higher temperatures other fusion reactions occur. One process leads to the production of boron-8, which decays to helium, positrons, and neutrinos. The latter have energy high enough to be detected on earth, although experiments to detect the process

$$^{37}\text{Cl}(\nu,e^-)^{37}\text{Ar}$$

have so far had negative results.

At temperatures of the order 10^8 K helium nuclei will fuse to produce ^{12}C and ^{16}O, while at about 10^9 K heavier nuclei, especially iron, may be formed.

stellar evolution *See* Hertzsprung–Russell diagram.

stellar interferometer *See* Michelson.

stellar magnitude *See* magnitude.

stellar spectra Stars emit radiation over a wide range of wavelengths, the maximum amount of energy being emitted at a particular wavelength. This wavelength will occur at the red end of the visible spectrum if the energy emitted is not very high (a cool star). An energetic star (a hot star) emits at the blue end of the spectrum.

Various groups of emission and absorption lines appear in a star's spectrum depending on temperature. Stars are usually classified according to their spectra and can be grouped into *spectral types*. (*See* Table. Note: All data given here are for typical *main-sequence stars. The colour description is conventional as the sensation is greatly affected by contrast. The sun is classified as yellow, although it looks blue-green in contrast to the hottest filament lamps, which are normally seen as white although their temperature is about that of a "red" star.) At very high temperatures over 50 000 K, only ionized gases exist. At lower temperatures, metals appear and in even cooler stars elementary molecules can survive. Stellar spectra therefore indicate not only temperature but chemical composition. The *redshift of the *Fraunhofer lines in absorption spectra is used in calculating the velocity of recession of stars.

St. Elmo's fire The brush discharge from the pointed parts of ships or aircraft when in a strong atmospheric electrical field.

STEM Abbreviation for scanning-transmission *electron microscope.

step-down, step-up transformer *See* transformer.

stepped-index device *See* fibre-optics system.

stepped leader stroke Of lightning. *See* leader stroke.

stellar spectra

Spectral type (Harvard classification)	Colour	Typical temperature/ kK	Typical absolute magnitude	Typical mass (sun = 1)	Typical radius (sun = 1)	Characteristics
O	blue	40	−7	40	20	ionized helium in absorption.
B	blue–white	20	−2	10	5	strong neutral helium; hydrogen developing.
A	white	9	+1	3	2	very strong hydrogen, then decreasing; ionized calcium increasing.
F	white–yellow	7	+3	1·5	1·4	ionized calcium well developed; metals developing.
G	yellow	6	+5	1	1	ionized calcium strong; iron and other metals strong; hydrogen weak
K	orange	4·5	+7	0·8	0·8	metals strong; some molecules (CH, CN) developing.
M	red	3	+10	0·2	0·3	many molecules; titanium oxide very strong.

Additional classes:
W   Wolf–Rayet stars; subdivided into WC, WN according to dominance of carbon, nitrogen in spectrum. Very hot stars (80 kK) of twice solar radius; typical absolute magnitude −4·1.
R   Strong bands of CN; C_2 increasing.
N   CN decreasing; C_2 very strong.
S   zirconium oxide bands. These last three classes are cool stars (2·5 to 3 kK).

step wedge A block or sheet of material having a series of layers successively more opaque to a given radiation in steps of definite value. Step wedges are employed in *photometry (to measure *luminous flux), in X-ray studies (to compare radiographic effects under various conditions), etc.

steradian Symbol: sr. A unit of *solid angle. One steradian is subtended at the centre of a sphere of radius r by a portion of its surface of area r^2:

$$1 \text{ sphere} = 4\pi \text{ sr}.$$

The steradian is a dimensionless derived *SI unit.

stereocomparator A *stereoscope provided with stereoscopic scales for measuring distances in depth from stereoscopic pictures obtained from a wide base (Pulfrich, 1901).

stereographic projection A projection obtained by assuming a crystal to be placed within a sphere from the centre C of which normals are drawn to the crystal faces; these normals intersect the sphere in points that are then joined to a pole P on the surface of the sphere. The straight lines thus obtained intersect a plane through C normal to CP to give the stereographic projection. The term has a wide mathematical significance that covers the projection of any points or figures on the surface of a sphere, from a pole P on the surface onto the plane through the centre C normal to PC.

stereophonic reproduction Reproduction of sound so as to give an illusion of location and direction from which a sound has originated. A normal person can determine the direction from which a sound is coming by distinguishing differences in arrival times of sound waves at the two ears (*see* binaural location). The single-channel sound system gives no illusion of sound location, all sounds appearing to come from a single source – the loudspeaker system. Multiple loudspeakers or a diffusion of the sound by broken reflectors give only a small improvement on the single sound source. Two or more channels of communication are essential for an illusion of sounds located in space. (*See* gramophone.)

stereopower *See* total relief.

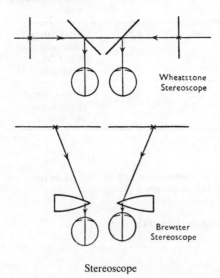

Wheatstone
Stereoscope

Brewster
Stereoscope

Stereoscope

stereoscope The impression of depth (sensation of relief, plasticity, stereoscopy), arises during binocular vision because the two eyes receive slightly different images of the same object (or distribution of objects at different distances). The effects are exag-

544

gerated if the baseline between the two visual viewpoints is increased. If photographs of solid objects, taken from appropriate viewpoints, are presented to the two eyes, stereoscopic effects are produced. The instrument for effecting this is the stereoscope. The Wheatstone type uses two plane mirrors; the Brewster instrument (common form) uses spheroprisms so that two separated pictures (stereograms) can be fused while their separation is greater than the interocular distance. By placing the pictures in the focal plane of the lenses the pictures can be viewed without accommodation. Stereoscopes have a wide application (educational, training binocular vision, etc.). The *stereoscopic rangefinder* uses a wide base to increase the range of stereoscopic vision and by means of a suitable scale and mark, distances on a scale apparently fading into the distance can be read off directly. *Stereotelescopes* permit of enhanced stereoscopic effects. (*See* Stratton pseudoscope; microscope.)

stereoscopic acuity The angle subtended by the baseline at a distant object is the parallax of the object. The difference of the binocular parallaxes is sometimes referred to as the *stereoscopic parallax*. If b is the length of the base line, R is the distance of an object, $(R + \delta R)$ the distance of an object a little farther away or nearer, then the binocular parallax is b/R and the stereoparallax $b \cdot \delta R/R^2$. The smallest detectable depth by the two unaided eyes expressed as steroparallax is called the stereoscopic acuity (of the order of 5 to 10 seconds of arc).

stereoscopic microscope *See* microscope.

stereoscopic parallax *See* stereoscopic acuity.

stereoscopic rangefinder *See* stereoscope.

stereotelescope *See* stereoscope.

Stern, Otto (1888–1969) German physicist, who worked with Gerlach on the determination of the magnetic moment of atoms (*see* Stern–Gerlach experiment). He also (with Estermann) furnished evidence of the wave nature of matter by demonstrating that whole molecules (e.g. hydrogen and helium) are capable of refraction on reflection from crystal surfaces.

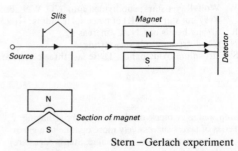

Stern – Gerlach experiment

Stern–Gerlach experiment An experiment, first performed by Stern and Gerlach (1921), that demonstrates the existence of the magnetic moment of an electron, particularly that due to its spin. When a sharply bounded stream of atoms is shot through a nonuniform magnetic field, the stream is split up into distinct components, dependent on the magnetic properties of the atoms. The atoms take up definite orientations relative to the field, and, in consequence of its nonuniformity, are deflected by different amounts. The experiment provides proof that there exist only certain permitted orientations; otherwise instead of splitting up, there would merely be a broadening of the beam since the atoms would be randomly orientated. This in turn provides proof of the quantization of the angular momentum of the electron. *See* spin.

stiffness 1. (acoustic) *See* reactance. **2.** (mechnical) *See* reactance.

stilb Symbol: sb. A unit of *luminance equal to 1 *candela per square centimetre. 1 stilb = 10^4 *nits. *See also* photometry.

stimulated emission *See* laser.

STM Abbreviation for *scanning tunnelling microscope.

stochastic process A process resulting from the random behaviour of its generators.

Stockbridge vibration-damper *See* vibration-damper.

stokes Symbol: St. The *CGS unit of *kinematic viscosity. 1 St = 10^{-4} metre squared per second.

Stokes, Sir George Gabriel (1819–1903) Brit. mathematician and physicist. He made substantial contributions to many branches of physics, and especially to the science of hydrodynamics. In particular, he studied the passage of waves through various media and the transformations that take place, the motion of elastic solids, and fluid motion (*see* Stokes's law of fluid resistance). In the field of sound, he investigated the effect of wind and of the nature of the gas on intensity; and in light he studied aberration, polarization, diffraction, and the optical properties of glasses. (*See* Stokes's law of fluorescent light; Stokes's lens; stream function.)

Stokes–Kirchhoff equation *See* absorption of sound.

Stokes's law 1. Of fluid resistance. The drag D of a sphere of radius r moving with a velocity V through a fluid of infinite extent is $D = 6\pi\eta rV$, where η is the *viscosity. The law holds only for a restricted range of conditions but has been the basis for an experimental determination of the viscosity of fluids. (*See* Reynolds number; sedimentation.)

2. Of fluorescent light. The wavelength of light emitted during fluorescence is longer than that of the absorbed light (Stokes, 1852). (*See* fluorescence.) When Einstein (1905) examined Planck's hypothesis

of quanta to explain *black-body radiation he cited Stokes's law as his first example to illustrate the argument that electromagnetic radiation generally interacts with matter in quanta.

Stokes's lens A variable-power cylindrical lens consisting of combinations of cylindrical lenses whose axes can be crossed to varying degrees. *See* crossed cylinder.

Stokes's stream function *See* stream function.

Stoney, George Johnstone (1826–1911) Irish physicist who was the first to calculate the charge on the *electron (a word that he introduced).

stop A perforated screen or diaphragm that limits the width of pencils of light traversing a system (*aperture stop*) or limits the field of view (*field stop*). Sometimes stops are used to reduce the blur of *spherical aberration, at other times to reduce the illumination of the image, to cut off the indistinct portions of a field, to prevent reflections from the inside of the tube, etc. *See* apertures and stops in optical systems; entrance pupil; exit pupil.

stop number *See* f-number.

stopper *See* parasitic oscillations; channel stopper.

stopping power A measure of the effect of a substance upon the kinetic energy E, of a charged particle passing through it.

1. The *linear stopping power*, S_i, is the energy loss per unit distance: $S_i = -dE/dx$, expressed in MeV/cm or keV/m.

2. The *mass stopping power*, S_m, for a substance of density, ρ, is the energy loss per unit surface density:
$$S_m = S_i/\rho = S_i L/nA,$$
where L is the Avogadro constant, A the relative atomic mass (atomic weight), and n is the number of atoms per unit volume.

3. The *atomic stopping power*, S_a, is the energy loss per atom per unit area of the substance normal to the motion of the charged particle:
$$S_a = S_i/n = S_m A/L.$$
Stopping power is often expressed relative to that of a standard substance, such as air or aluminium.

storage capacity *See* capacity.

storage device A device in a *computer system that can receive data and retain it for subsequent use. Storage devices are either *backing store or semiconductor devices used as *main memory, and vary widely in the amount of data that can be stored and the speed of access to a particular item.

storage oscilloscope An oscilloscope that can capture a signal, especially a fast nonrepetitive signal, and continue to display it until reset. A digital storage oscilloscope samples the incoming signal, stores these samples, and displays them. Other storage oscilloscopes use a special cathode-ray tube that retains the image by mapping it as a charge pattern on a target electrode behind the screen. The pattern then modulates the electron beam to give a picture of the captured signal. *See also* storage tube.

storage ring *See* accelerator.

storage time 1. In general, the time for which information may be stored in any device, without significant loss of information.

2. Of a *p-n junction. Under conditions of *forward bias, charge is stored in a p-n junction in the form of a net concentration of excess minority *carriers, which are injected across the junction; this is known as *carrier storage*. When reverse bias is applied to the junction the current does not change instantaneously to the small reverse leakage value, but is maintained until the stored charge is removed. The storage time is the time required for the excess minority charges to be removed and hence is the time interval observed between application of the reverse bias and cessation of a reverse current surge.

storage tube An *electron tube in which information may be stored for a determined and controllable time, and extracted as required. Various operating principles are used in storage tubes and many different types exist.

The most common types of tube in general use are *charge storage tubes* in which the information is stored as a pattern of electrostatic charges. The information may appear as a visual display as in the storage *cathode-ray tube or may be extracted as an electronic signal as in the electrostatic charge storage tube.

Both types of tube depend on *secondary emission of electrons from a target plate. The information to be stored is used to modulate the intensity of an electron beam that scans the target. Secondary emission of electrons occurs and an electrostatic image is left on the target. The numbers of secondary electrons emitted from each of the tiny storage elements making up the target plate is a function of the energy of the beam, the beam intensity, and the tube design. If more secondary electrons are produced than fall on the target, a positive charge image is produced; if fewer, then a negative charge results. The information-modulated electron beam is called the writing beam.

Storage cathode-ray tubes are usually designed to have a positive charge image, and the information is extracted by *flooding* the target with electrons, which pass through the target to the screen. The deposited charge modulates the numbers of electrons reaching the screen and the light image produced is therefore proportional to the original information.

Electrostatic-charge storage tubes utilize an unmodulated electron beam (the reading beam), which scans the target and is modulated by the pattern of stored charges. The beam is collected on a collector electrode, the output signal being proportional to the original input material. The reading and writing beams may be produced by the same electron gun (single-gun tubes) or by separate electron guns (two-gun tubes).

Other effects utilized in storage tubes include *photoconductivity and also the related bombardment conductivity, in which the conductivity of a material is temporarily increased by bombardment of a target with light quanta or an electron beam. The target consists of a photoconductive material in contact with a back-plate electrode. The target is exposed to a light image or a modulated electron beam and the conductivity of small areas of the target is proportional to the intensity of the incident radiation. An unmodulated electron beam is used to scan the target and discharge each small area through the back-plate electrode. Variations in the current through a load resistor in series with the electrode are observed and are proportional to the input information. Photoconductivity is extensively used in television camera tubes.

Storage tubes have also been developed that depend on *photoemission, *luminescence, and photochromism (*see* photochromic substance) for their operation, but these are less common.

STP Abbreviation for *standard temperature and pressure.

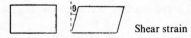

Shear strain

strain The change of volume and/or shape of a body, or part of a body, due to applied stresses. The three simplest strains are as follows. (1) *Linear* (or *longitudinal*) *strain*: the change in length per unit length (e.g. when stretching a wire). (2) *Volume* (or *bulk*) *strain*: the change in volume per unit volume (e.g. when a hydrostatic pressure is applied to a body). (3) *Shear strain* (or *shear*): angular deformation without change in volume (e.g. a rectangular block strained so that two opposite faces become parallelograms, the others not changing shape). The radian measure of the change in angle, θ, at one corner is a measure of the strain (*see* diagram). θ is small in practice and is equal to the tangential displacement of two planes unit distance apart. *See* homogeneous strain; stress.

strain components *See* homogeneous strain; stress components.

strain ellipsoid *See* homogeneous strain.

strain gauge An instrument for measuring strain at the surface of a solid body. The types used include: (1) resistance strain gauge in which a few centimetres of fine wire or a semiconductor are stuck to the surface and changes in length cause the electric resistance of the gauge to change; (2) variable inductance gauge. An iron armature associated with the magnetic circuit of an iron-cored coil is moved by the strain and alters the coil's inductance; (3) variable capacitance gauge. The distance between the plates of a capacitor, and hence the capacitance, are altered by the strain; (4) acoustic strain gauge. A taut wire is held between the gauge points, which are firmly clamped against the test surface. Any strain in the surface alters the distance between the gauge points and hence the frequency of vibration of the wire is changed. (5) Miscellaneous methods include utilization of the piezoelectric and magnetostriction effects. (*See also* extensometer.)

strain insulator *See* tension insulator.

strange matter Matter composed of a mixture of up, down, and strange *quarks (rather than the up and down quarks of protons and neutrons). It has been postulated that stable nuggets of strange matter might have been formed in the extreme conditions of the big bang. Some astrophysicists have also suggested that it could be formed in supernova explosions. There is also a possibility that drops of strange matter (known as *S-drops*) could be produced in particle accelerators by bombarding a target with a beam of heavy nuclei. So far, no evidence for its existence has been found.

strangeness Symbol: *S*. A *quantum number associated with *elementary particles (specifically *hadrons) that is conserved in *strong and *electromagnetic interactions but not in *weak interactions. *S* changes by ±1 in weak interactions. Its existence was postulated in order to explain the fact that some elementary particles (e.g. kaons, Σ, and Λ) that were expected to decay very rapidly by strong interaction (since they could do so without violating any of the known conservation laws), had much longer lifetimes than expected. Strangeness is associated with the presence of one or more strange *quarks in the particle. The strange quark (s) has strangeness –1 and the antiquark (s̄) has strangeness +1. All other quark flavours have strangeness 0. The strangeness of a particle is the sum of the number of s̄ quarks minus the number of s quarks.

stratosphere One of the earth's *atmospheric layers, lying between the *troposphere and the *mesosphere. The air pressure is very low, being below 1 millibar at a height of 48–50 km, compared with about 1 bar at sea level. The temperature in the middle and higher latitudes between 12–20 km is almost constant; it then increases with height at about 1–3 °C/km. The maximum temperature, (0–10 °C), corresponds to the upper boundary of the stratosphere, called the *stratopause*, lying at a height of 50–60 km above the earth. The warmth of this layer is due to absorption of ultraviolet radiation by ozone in the atmosphere. The greatest ozone concentration occurs at about 55 km and this altitude marks the upper limit of the *ozone layer* (or ozonosphere).

Stratton pseudoscope A form of Wheatstone *stereoscope in which the mirrors reverse the right- and left-eye views of a distant object, and so produce a reversed stereoscopic effect.

stray capacitance Any capacitance in a circuit or device due to interconnections, electrodes, or the proximity of elements in the circuit in addition to the intentional capacitance provided.

stray load loss The total of the extra losses that occur in an electrical machine or transformer when on load, due to changes in the magnetic flux distribution and to *eddy currents caused by the load current. The term includes all the losses of this type wherever they may occur. Thus, for example, in a transformer it includes the losses due to eddy currents in any surrounding metal (e.g. the tank) caused by stray magnetic fields when on load. For a large synchronous alternating-current generator, the stray load loss may exceed 15% of the total losses at full load.

stream function *Syn.* current-function. Symbol: ψ. In two-dimensional motion of a fluid the stream function at any point P is defined as the flow across a curve AP where A is a fixed point in the two-dimensional plane. If the plane is the Cartesian plane xy then the component velocities at any point (x, y) are:
$$V_x = -\partial\psi/\partial y \text{ and } V_y = \partial\psi/\partial x.$$
In axisymmetrical three-dimensional motion of a fluid – motion that is the same in every plane through a certain axis of symmetry – there is a similar function (ψ) whose value at a point P is defined as $1/2\pi$ of the flow out of a surface of revolution formed by rotating a curve AP about the axis, where A is any fixed point in the meridian plane of P. This function is called *Stokes's stream function*. In both cases the curves: ψ = constant, give the streamlines of the fluid motion.

streaming *See* molecular flow.

streaming potential (Quincke, 1859) A difference of potential set up between the two sides of a porous material such as clay when water is forced through. It is the potential set up between the ends of a capillary tube when an electrolyte is forced through it. This may be regarded as the reverse of *electrosmosis.

streamline A line drawn in a fluid so that the tangent at any point is in the direction of the fluid velocity at that point. The aggregate of streamlines at any instant of time forms the flow pattern. The tube formed by drawing the streamlines through every point of a closed curve in the fluid is called a *stream tube*.

strength of sound *See* impedance (acoustic).

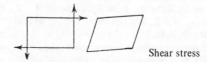

Shear stress

stress A system of forces in equilibrium producing *strain in a body or part of a body. The stresses may be regarded as the forces applied to deform the body or as the equal and opposite forces with which the body resists. In all cases a stress is measured as a force per unit area. The simplest stresses are: (1) *tensional* or *compressive stress* (i.e. *normal stress*), e.g.

stress components

the force per unit area of cross section applied to each end of a rod to extend or compress it; (2) *hydrostatic pressure*, e.g. the force per unit area applied to a body by immersion in a fluid; (3) *shear stress*, e.g. the system of four tangential stresses applied to the surfaces of a rectangular block (each force being parallel to one edge) tending to strain it so that two of the sides become identical parallelograms, the others being unaltered in shape. *See* strain, for the effects of these stresses; modulus of elasticity, for the relations between stresses and strains; stress components, for further definitions.

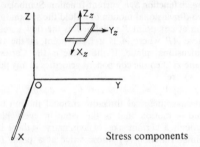

Stress components

stress components The internal forces per unit area arising between contiguous parts of a body due to applied surface and body forces. Consider an infinitesimal plane area within the body; the force exerted by the matter on one side to that on the other can be resolved into components normal and tangential to the area that are called the *normal stress component* and the *shear stress component* respectively, at the area. Except in the special case of hydrostatic pressure, the stress at a point depends on the orientation of the area used in defining it.

If the component stresses at three infinitesimal areas at the point, each being parallel to one of the planes defined by Cartesian axes outside the body, are known, it is possible to calculate the stress across an infinitesimal area at the point however orientated. The tangential component in the plane of the infinitesimal area perpendicular to OZ (see diagram) may be resolved into components X_z and Y_z (the subscript indicates the axis perpendicular to the plane across which the force acts). The system of 9 stress components (3 normal and 6 shear) finally reduces to 6, for by considering the equilibrium of a small rectangular solid at the point, it is found that $X_y = Y_x$, $Z_x = X_z$, $Y_z = Z_y$. These six, X_x, Y_y, Z_z and X_y, Y_z, Z_x are the stress components at the point (x, y, z).

If the coordinate axes are rotated to new positions OX'Y'Z', a position can be found at which $X'_{y'} = Y'_{z'} = Z'_{x'} = 0$ and the values of $X'_{x'}$, $Y'_{y'}$, and $Z'_{z'}$ are known as the *principal stresses* and their directions as the *axes of stress*. In an isotropic solid, the axes of stress will also be the axes of strain (*see* homogeneous strain) so that if *Hooke's law and the principle of *superposition of vibrations are obeyed by the solid, and if the coordinate axes are parallel to the axes of stress and strain, the strain components will be:

$$s_x = \frac{1}{E}\left\{X_x - \sigma(Y_y - Z_z)\right\}$$

$$s_y = \frac{1}{E}\left\{Y_y - \sigma(Z_z - X_x)\right\}$$

$$s_z = \frac{1}{E}\left\{Z_z - \sigma(X_x - Y_y)\right\}$$

where E is Young's modulus (*see* modulus of elasticity) and σ is the *Poisson ratio. *See* stress; force.

stretched string The theory of *transverse vibrations in a stretched string assumes it to be uniform, perfectly flexible, and of practically constant length while in vibration. These conditions are nearly fulfilled when a long thin wire is stretched between two rigid supports. *Standing waves set up in a stretched string can be considered to be due to the superposition of two transverse progressive waves travelling in opposite directions with velocity $v = \sqrt{(T/m)}$, where T is the tension and m the mass per unit length. Ideally, in the fundamental mode there is a single loop with an antinode at the centre and nodes at the ends. In this case the wavelength is twice the length of the string l, and the fundamental frequency is given by:

$$f = \sqrt{(T/m)}(2l)^{-1}.$$

The various *partials are produced when the string vibrates in several loops. All the partials are *harmonic, their frequencies being obtained by multiplying that of the fundamental by the number of loops on the string. A string may vibrate with several partials at the same time, their number and magnitude depending on the method of excitation. In practice the ends are not perfect displacement nodes, thus the quality of a musical note is affected. The vibrations of a string may be observed photographically or by stroboscopic methods.

A transversely vibrating string is used as a source of sound in many musical instruments when it is attached to a *soundboard that increases its ability to radiate sound energy. In the harp the method of excitation is by plucking the strings with the fingers. In the piano the string is struck by a hammer and in the violin it is excited by bowing. The different sound of the same note from these instruments is due to different partials being excited by each method of excitation. A string may be kept vibrating continuously if it is excited electromagnetically by placing a magnet near a wire carrying an alternating current of suitable frequency (*see* electromagnetically maintained vibration). For laboratory work the string is used in the form of a *monchord to measure the frequency of a source of sound.

striations Grooves or furrows on a surface, such as arise on crystal faces from repeated lamellar twinning, and the pattern in the powder of a Kundt's tube; hence the periodic structure of the positive column of a *gas-discharge tube.

strike note The note of a bell whose pitch is assigned to the bell. This note is generally an octave below the 5th partial. When a bell is struck this note is the loudest heard, but it decays more rapidly than the other partials. In a good bell the strike note is made coincident with, or harmonic to, the 2nd partial, even if the other partials are inharmonic to each other (see bell sounds). There is no satisfactory explanation of the occurrence of the strike note. It cannot be elicited from the bell by resonance, nor will it beat with a tone of nearly the same frequency. The abnormal intensity would suggest that the origin of the tone is other than subjective, although one suggestion is that it may be formed in the ear by the large forcing of the original stroke. It may also be due to an intermittent contact between the clapper and the bell.

string efficiency Of an insulator chain (see suspension insulator) when used in an a.c. system. Owing to the earth capacitances of the individual units in the chain, the voltage applied across the complete chain (or string) is not divided equally between the units. As a consequence of this, the *flashover voltage of the complete string is less than that of one unit multiplied by the number of units. The string efficiency is defined by

$$\frac{\text{flashover voltage of the string}}{\text{flashover voltage of a unit} \times \text{no. of units}} \times 100\%.$$

Values of the order of 80% are common for suspension insulators of normal type.

string galvanometer See Einthoven galvanometer.

string theory A theory of *elementary particles based on the idea that the fundamental entities are not point-like particles, but finite lines (strings) or closed loops formed by strings. The original idea was that an elementary particle was the result of a standing wave in a string.

A considerable amount of theoretical effort has been devoted to developing string theories. In particular, to combining the idea of strings with that of *supersymmetry, which leads to the idea of superstrings. This theory may be a more useful route to a unified theory of fundamental interactions than *quantum field theory because it probably avoids the infinities that arise when gravitational interactions are introduced into field theories. Thus, superstring theory inevitably leads to particles of spin 2, identified as *gravitons. String theory also shows why particles violate *parity conservation in weak interactions.

Superstring theories involve the idea of higher dimensional spaces: 10 dimensions for fermions and 26 dimensions for bosons. It has been suggested that there are the normal 4 space–time dimensions, with the extra dimensions being tightly 'curled' up.

There is no direct experimental evidence for superstrings. They are thought to have a length of about 10^{-35} m and energies of 10^{19} GeV, which is well above the energy of any accelerator. An extension of the theory postulates that the fundamental entities are not simply one-dimensional but two-dimensional, i.e. they are supermembranes.

strip core Some magnetic materials, such as the silicon steels, have their magnetic properties improved when they have been subjected to a rolling treatment. This process, however, gives grain orientation, and different properties with, and across, the grain. When making a core of these materials, a continuous strip has to be used, and is wound around a former and then cut across to break the electrical continuity of the core and to facilitate insertion of the windings.

stripping A *nuclear reaction in which a *nucleon of the bombarding nucleus is captured by the struck nucleus without the nuclei merging to form a compound nucleus. The *Oppenheimer–Phillips (O–P) process* is a reaction in which a low-energy deuteron gives its neutron to a nucleus without entering it.

Stroboscope disc

stroboscope If a body is in motion and is illuminated for a sufficiently short period of time, it will appear to be motionless. If the motion is periodic, then by suitable devices the body may be exposed or illuminated at successive instants while it is passing through the same space intervals in the same direction. These devices, which give appearance of rest or slow motion to bodies in rapid motion, are known as stroboscopes. This can be simply illustrated in the case of a tuning fork. Let the tuning-fork prongs carry two diaphragms that overlap and both are slotted so that when the prongs are at rest the two slots are in line with the beam of light passing through. This light is allowed to fall on a rotating disc that contains a series of concentric rings, filled with equidistant black triangles (see diagrams). If the fork is vibrating and the disc is rotating in such a way that any ring of triangles appears stationary, then for that ring each triangle succeeds its neighbour in the period between two flashes. It may happen also to be stationary if the ring has traversed the spaces of 2, or 3, or 4, etc. . . . of triangles in the interval between two successive flashes of light.

If f is the frequency of illumination (or otherwise the frequency of the fork), r is the number of revolutions of the disc per second and m is the number of triangles in one stationary ring then $f = mr$, or $mr/2$, or $mr/3$, etc. . . . The patterns observed for the first and second cases are shown in the diagrams. Any ambiguity involved can be eliminated by gradually increasing the speed of rotation of the disc. For a small difference between f and mr, f being less than mr, the pattern revolves with the wheel, while if f is greater than mr the pattern revolves against it. This method can be easily applied to determine the frequency of an electrically maintained fork.

stroke

In a later form of the stroboscope the source of light itself flashes periodically, a discharge lamp being used. The flashing of the lamp can be controlled up to a frequency of the order of 10 000 hertz by varying an oscillatory circuit consisting of a capacitor in parallel with the lamp or a resistance in series with it.

stroke *See* leader stroke; lightning stroke.

stroke pulse A *pulse used for timing or indicating purposes in computing systems and *radar. It may be used as a form of *gate only allowing a *signal to pass during the *pulse duration, or be superimposed on a signal as a marker (as in radar systems).

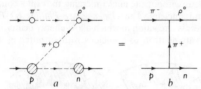

Strong interaction between a proton and negative pion

strong interactions Interactions between *elementary particles involving the strong interaction force. This force is about one hundred times greater than the *electromagnetic force between charged elementary particles. However, it is a short-range force – it is only important for particles separated by a distance of less than about 10^{-15} m – and is the force that holds hadrons together as well as nucleons in atomic nuclei. Just as the electromagnetic interaction between charged particles may be described in terms of the exchange of virtual photons (*see* virtual particle), so strong interactions between *hadrons may be described in terms of the exchange of virtual hadrons. In strong interactions any hadron can act as the exchanged particle providing certain *quantum numbers are conserved. These quantum numbers are the total angular momentum, charge, *baryon number, *isospin (both I and I_3), *strangeness, *parity, *charge conjugation parity, and *G-parity. Strong interactions are investigated experimentally by observing how beams of high-energy hadrons are scattered when they collide with other hadrons. Two hadrons colliding at high energy will only remain near to each other for a very short time. However, during the collision they may come sufficiently close to each other for a strong interaction to occur by the exchange of a virtual particle. As a result of this interaction, the two colliding particles will be deflected (scattered) from their original paths. If the virtual hadron exchanged during the interaction carries some quantum numbers from one particle to the other, the particles found after the collision may differ from those before it. Sometimes the number of particles is increased in a collision. An example of a strong interaction process is

$$\pi^- p \rightarrow \rho^\circ n.$$

Here the colliding π^- and proton become a ρ° and neutron after the collision. This process is thought to take place by the proton emitting a virtual π^+, which combines with the π^- to give a ρ° as illustrated in Fig. *a* below. This diagram is often simplified to the form shown in Fig. *b*. Where all the necessary quantum numbers can be conserved, elementary particles can decay by a strong interaction. An example of this is the decay of the ρ-meson into two pions. Here the ρ-meson can be thought of as breaking up into two virtual pions that are bound by the strong interaction force. However, since this can occur without violating the law of *conservation of mass and energy (the mass of two pions is less than the mass of a ρ) these pions can become physical particles and separate from each other.

In high-energy hadron–hadron interactions, the number of hadrons produced increases approximately logarithmically with the total centre of mass energy, reaching about 50 particles for proton–antiproton collisions at 900 GeV, for example. In some of these collisions, two oppositely directed collimated *jets* of hadrons are produced, which are interpreted as being due to an underlying interaction involving the exchange of an energetic gluon between, for example, a quark from the proton and antiquark from the antiproton. The scattered quark and antiquark cannot exist as free particles but instead *fragment* into a large number of hadrons (mostly pions and kaons) travelling approximately along the original quark or antiquark direction. This results in collimated jets of hadrons which can be detected experimentally. Studies of this and other similar processes are in good agreement with QCD predictions. *See also* bootstrap theory; exchange force; Regge pole model; resonances.

structure-sensitive properties Properties that depend on the *crystal texture (not on the atomic arrangement as such).

strut A long, thin, naturally straight rod, subject to a thrust. The theory of their buckling was first given by Euler.

Strutt, John William *See* Rayleigh, Lord.

Strutt, Robert John *See* Rayleigh, Lord.

Sturgeon, William (1783–1850) Brit. physicist who made significant contributions to the early science of electricity, including his inventions of the electromagnet (1821), the moving-coil galvanometer (1836), and the commutator for an electric motor (1836).

Sturm, conoid of *See* astigmatism.

St. Venant's principle (1885) The strains produced in a body by the application, to a small part of its surface, of a system of forces statically equivalent to zero force and zero couple are of negligible magnitude at distances large compared with the linear dimensions of the part. E.g. if a wire is stretched by in-line forces applied to its ends, the strains are simple longitudinal extension and lateral contraction at distances greater than a few wire-radii from the chucks, but are exceedingly complicated at and near them.

SU₃ *See* unitary symmetry.

subatomic Denoting a constituent particle of the atom (a *nucleon or electron) or a process occurring within atoms. Subatomic processes include *radioactivity, *X-ray production, and *absorption of radiation.

subcarrier A *carrier wave used to modulate another carrier wave.

subcritical *See* chain reaction.

subgroup A set of operations that forms part of a larger group, but that is complete in itself. The term is used in spacegroup theory.

subharmonic vibration A vibration of a frequency that is a whole-number submultiple of the fundamental frequency. Subharmonic vibrations (at half the frequency of the fundamental) occur, for example, when a tuning fork is sounded on a board; this has been found to be due to intermittent contact of the fork with the board.

sublimation The direct transition from solid to vapour, or conversely, without any liquid phase being involved. *See* hoar-frost line.

sublimation curve *See* hoar-frost line.

submillimetre waves Radio waves with a wavelength ranging from 1 mm down to about 0.1 mm, i.e. with a frequency between 300 and about 3000 GHz. They are of particular interest in radio astronomy because of the large number of molecular emission lines to be found in this wavelength range.

subshell *See* atomic orbital; electron shell.

subsonic Denoting an object, airflow, etc., moving at less than the speed of sound, i.e. at less than Mach 1.

substandard A standard measuring device not quite as accurate as the primary standard, used as the intermediate link between the primary standard and the device being calibrated or checked.

substation A complete assemblage of plant, equipment, and the necessary buildings at a place where electrical power is received (from one or more *power stations) for conversion (e.g. from alternating current to direct current by means of *rectifiers, rotary *converters, etc.), for stepping-up or down by means of transformers, or for control (e.g. by means of switchgear, etc.).

substrate 1. A single body of material on or in which one or more electronic circuit elements or integrated circuits are fabricated. It may be a semiconductor crystal, which plays an active role in the device, or it may act as a support, as with the insulating layer in a printed-circuit board.

2. Any surface or layer used as a basis for some process.

subtractive process A process by which colours can be produced or reproduced by mixing absorbing media (or *filters) of three different dyes or pigments, called *subtractive primaries*. The colour of light reflected by (or passing through) the mixture is determined by the *absorption, or subtraction, of specific colours by each medium. The three dyes or pigments are usually yellow, magenta, and cyan (greenish-blue), an approximately equal mixture of which will appear black. *Compare* additive process. *See* colour photography.

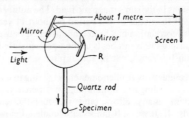

Sucksmith's method of measuring susceptibility

Sucksmith's method For measuring the *susceptibility of paramagnetic and ferromagnetic materials. A ring of phosphor-bronze R is fixed at the top and is deformed by the magnetic force acting on a specimen of a few cubic millimetres supported at the end of a quartz rod fastened to the bottom of the ring. The specimen is held in a nonhomogeneous magnetic field. Deformation of the ring is recorded by light reflected from a system of two mirrors fastened to the ring, deformations being kept small.

sulphation The undesirable formation of insoluble lead sulphate in a lead-acid storage cell when it is left for a time in an undischarged condition. It reduces the capacity and efficiency.

summation instrument A single instrument for measuring the aggregate of the current, power, or energy in a number of separate circuits.

summation tone If two intense tones are sounded simultaneously, other tones may be heard. One consists of a tone whose frequency is the sum of the frequencies of the primaries and is known as the summation tone. *See* combination tones.

summer solstice *See* solstice.

sundial An instrument for measuring time by observing the position of a shadow due to the sun cast by a fixed object on a graduated surface. Since the speed of the sun across the heavens is not uniform every day, the apparent time given by the sun differs from the mean time as given by a clock; the equation of *time given by the calendar gives the difference between them. If the graduated surface is horizontal, the rod that casts the shadow must point in the direction parallel to the earth's axis: the angle it makes with the dial is that given by the latitude of the place and it must also point true North. In the case of altitude dials, the time is determined by the altitude of the sun above the horizon: the object casting the shadow usually consists of a piece of metal with a hole in it and is set in different positions according to the time of the year.

sunspots Dark patches seen on the sun's *photosphere usually in groups and with a lifetime of several weeks. All but the smallest have a dark inner region (the umbra) surrounded by a less dark edge (the penumbra). Sunspots are regions of relatively cool gas and their presence is connected with local variations in the sun's magnetic field. The number of sunspots fluctuates over a period of about 11 years. A close correlation has been demonstrated between the sunspot cycle and beats in the tides generated in the sun by Venus, earth, and Jupiter.

superconductivity A phenomenon occurring in many metals including tin, aluminium, zinc, mercury, and cadmium, many alloys, and intermetallic compounds. If these substances are cooled below a *transition temperature, T_c, the electrical resistance vanishes. For pure metals the transition temperature is usually a few kelvin but for some compounds much higher values are obtained. Current research is aimed at developing superconductors with T_c near to, or even above, room temperatures.

If a superconducting metal consists of several isotopes, the different isotopes will have different transition temperatures. The isotopic mass number, A, can be related to the transition temperature: $T_c \sqrt{A}$ = constant. T_c also depends on pressure P such that the derivative dT_c/dP is constant. This behaviour does not apply to all superconductors. At 1 atmosphere, bismuth is not a superconductor. Between 20 000 and 40 000 atmospheres, bismuth becomes superconducting at T_c = 7 K. The change in specific heat capacity with temperature of a superconductor below T_c differs markedly from its behaviour above T_c, varying as $a \exp(-bT_c/T)$, where a and b are constants.

Since a compound of two nonsuperconducting metals can be superconducting, the phenomenon is not a property of the atom but of the free electrons in the metal. In the superconducting state the electrons do not move independently. There is a dynamic pairing of electrons (a *Cooper pair*) such that if the quantum state with *wavenumber σ and *spin ½ is occupied by an electron, then so is the state with wavenumber –σ and spin –½. These pairs are superimposed in phase. The two electrons interact through lattice vibrations, and the formation of these bound pairs is not prevented by the presence of other electrons. Cooper pairs are the basis of the *BCS theory* (1957, named after Bardeen, Cooper, and Schrieffer). This accounts for many of the properties of conventional superconductors but is less successful for the recently discovered high-temperature superconductors; these rely on *heavy-fermion systems in which the transition temperature can be as high as 100 K. The practical advantage of these *high-temperature superconductors* is that they can operate at liquid-nitrogen temperatures rather than the liquid-helium temperatures required by BCS superconductors. An example of such a superconductor is $YBa_2Cu_3O_{1-7}$. The theory of these superconductors has not been established yet but various models have been proposed and tested.

The magnetic behaviour of superconductors is extremely complicated. When a superconductor, in a weak magnetic field, is cooled below its transition temperature, the magnetic flux inside the substance is expelled except for a thin surface layer. This is the *Meissner effect*. A bar magnet dropped onto such a conductor will be repelled and will hover above it, exhibiting *levitation*. The Meissner effect implies that a superconductor exhibits perfect *diamagnetism. This in turn implies the existence of a large energy gap between ground state and first excited state so that all the superconducting electrons are in a particular ground state; this is possible for Cooper pairs. Superconductivity can be destroyed by a magnetic field – either an external field or one produced by a current flowing in the metal. This is used in the *cryotron. A current induced in a closed ring of superconducting material by a magnetic field will continue to flow after the removal of the field for a considerable time, without diminution in strength, if the temperature is kept below the transition temperature. This effect has been used in *superconducting magnets* in which very large magnetic field strengths can be produced without the expenditure of large amounts of electrical power or the production of heat. The superconducting electrons form an *energy band below that of the normal conduction band and do not take part in heat conduction. Hence the thermal conductivity of metals is usually less in the superconducting state. At very low temperatures, however, it may rise because of increased *phonon conductivity.

See also Josephson effect.

supercooling The process by which liquids, by slow and continuous cooling, are reduced to a temperature below the normal freezing point. A supercooled liquid is a *metastable state and the introduction of the smallest quantity of the solid at once starts solidification. Small mechanical disturbance may also serve to initiate solidification and once this has started it will continue with the evolution of heat until the normal freezing point is reached, when subsequently further solidification will take place only as heat is lost from the liquid. *Compare* superheating.

Clouds very commonly consist of supercooled water droplets. These may freeze on contact with solids, causing *icing* on aircraft or *glazed frost* at ground level. (*See also* hail.) If ice particles form on suitable nuclei in a cloud of supercooled droplets, they grow at the expense of the latter to form *snow flakes because of the difference in saturation vapour pressure over solid and supercooled liquid surfaces (*see* diagram at triple point).

supercritical *See* chain reaction.

superfluid A fluid that flows without any resistance. The electrons in *superconductivity constitute a superfluid. Another important case is that of liquid helium II, that is the form of liquid helium (of the common isotope ^{4}He) below a certain transition temperature dependent upon pressure. This has its highest value, 2.186 K, at the equilibrium vapour pressure, and is called the *lambda point because of the shape of the peak in the graph of specific heat capacity against temperature. The phenomenon is associated with exceptionally high thermal conductivity, which rises to 10^6 times the value above the lambda point.

Superfluidity is attributed to *Bose–Einstein condensation*, when a large number of helium atoms (which are *bosons) are in the translational state of lowest energy. These atoms constitute a superfluid that is homogeneously mixed with normal fluid consisting of atoms with higher translational kinetic energies. *See also* second sound.

Atoms of the rare isotope ^{3}He are *fermions since each nucleus contains an odd number (one) of neutrons, thus this isotope behaves very differently. At low temperatures the atoms form weakly associated pairs joined by the interaction of their nuclear magnetic moments. These pairs act as bosons and at temperatures of the order of millikelvin begin to behave in a manner analogous to liquid helium II.

supergiant *See* Hertzsprung–Russell diagram.

supergravity An unproved *unified field theory involving *supersymmetry that encompasses all four fundamental *interactions. By means of introducing supersymmetry the number of infinities in the calculations is fewer than in other quantum-theory based unified theories that include the gravitational interaction, but the theory still contains infinities that cannot be removed. Some theorists believe that a unified theory to include gravity is unlikely to be a *quantum field theory and seek instead a theory based on superstrings (*see* string theory).

superheater A component of a steam generator in which saturated steam from the boiler is raised in temperature at constant pressure before being admitted to the cylinder or turbine. The unsaturated steam is said to be *dry* or *superheated*. *See* Rankine cycle.

superheating The heating of a liquid above its normal boiling point without boiling occurring. *See also* supercooling.

superheterodyne receiver The most widely used type of *radio receiver, in which the incoming signal is fed into a *mixer and mixed with a locally generated signal from a *local oscillator*. The output consists of a signal of *carrier frequency equal to the difference between the locally generated signal and the carrier frequencies, but containing all the original modulation. This signal (the *intermediate frequency* (or i.f.) signal) is amplified and detected in an i.f. amplifier, and passed to the audiofrequency amplifier. The advantages of superheterodyne reception are direct-

ly attributable to the use of the intermediate frequency. High-gain amplification is very much easier at intermediate frequencies than at radio frequencies. The use of the intermediate frequency allows much greater selectivity and hence easier elimination of any unwanted signals, since it is a difference signal and the percentage difference between the signal and other signals is much greater than at the original radio frequency.

superhigh frequency (SHF) *See* frequency bands.

superlattice 1. *See* solid solution.
    2. In electronics, a semiconducting crystal in which two semiconductors with different electronic properties are interleaved in ultrathin layers. Such crystals are produced by depositing the two materials in alternating layers or by introducing impurities into layers of a single semiconductor; viable devices have only recently been produced. The two semiconductor materials are selected so that the energy difference (or band gap) between their valence and conduction bands is different (*see* energy bands). An example is a superlattice consisting of gallium arsenide and gallium aluminium arsenide, the latter having the larger band gap. The great advantage of a superlattice device is that its electronic and optical properties can be tailored to a particular function by an appropriate choice of the semiconductors and the width and number of the layers.

Supermalloy A magnetic alloy of Fe, Ni, and Mo. It is similar to *permalloy but has a higher permeability.

supermembrane *See* string theory.

supernova A star that explodes as a result of instabilities following the exhaustion of its nuclear fuel. The explosion involves an enormous energy release: a supernova can become 10^9 times as bright as the sun (reaching an absolute *magnitude of –17). All or most of the star's matter is ejected at relativistic speeds, forming an expanding shell of debris called a *supernova remnant*. An example of a supernova remnant is the Crab nebula; such a nebula can radiate light, radio waves, and X-rays, for thousands of years. A *pulsar can be formed at the core of the supernova. *Compare* nova.

superposition of vibrations Two or more vibrations may be superimposed upon each other to give a single complex vibration whose displacement at any instant is given by the sum of the instantaneous displacements of the individual vibrations. This is the *superposition principle. For example, two vibrations of the same frequency, but differing in amplitude and phase, may be combined to form a resultant vibration of the same frequency as the component vibrations, and of amplitude and phase that are both functions of the component amplitudes and phases.

Expressed mathematically, if the component vibrations are:

$$y_1 = a_1 \sin(2\pi nt + \delta_1),$$

superposition principle

and
$$y_2 = a_2 \sin(2\pi nt + \delta_2),$$
then the resultant vibration, y, is given by:
$$y = y_1 + y_2 = A \sin(2\pi nt + \Delta),$$
where
$$A = \sqrt{(a_1{}^2 + a_2{}^2 + 2a_1 a_2 \cos(\delta_1 - \delta_2))},$$
and
$$\tan\Delta = \frac{a_1 \sin\delta_1 + a_2 \sin\delta_2}{a_1 \cos\delta_1 + a_2 \cos\delta_2}.$$

superposition principle A principle that holds generally in physics wherever linear phenomena occur. In elasticity, the principle states that each stress is accompanied by the same strains whether it acts alone or in conjunction with others; it is true so long as the total stress does not exceed the limit of proportionality. In vibrations and wave motion the principle asserts that one set of vibrations or waves is unaffected by the presence of another set, e.g. two sets of ripples on water will pass through one another without mutual interaction so that the resultant disturbance at any point traversed by both sets of waves is the sum of the two component disturbances. (*See* superposition of vibrations.)

super-regenerative reception A method of reception of ultrahigh frequencies by means of a radio receiver employing an oscillating *detector, the oscillations of which are periodically stopped (or quenched) at a frequency dependent on the input frequency. A particular characteristic of this method of reception is the very great amplification that can be obtained, but the selectivity is rather poor compared to a *superheterodyne receiver.

supersaturated vapour A vapour, the pressure of which exceeds the saturation vapour pressure at that temperature. It is unstable and condensation occurs in the presence of suitable nuclei or surfaces.

supersonic flow The movement of a fluid at a speed exceeding the speed of sound in a fluid. In such a case, changes of density in the flow can no longer be neglected. (*See* compressible flow.) As the speed of an object moving through a fluid is increased through the speed of sound and beyond, the resistance (drag) rises due to the formation of *shock waves (*see also* sonic boom). This drag is very sensitive to roughness of the surface of the projectile so that lifting surfaces employed in aircraft designed for supersonic speeds must be made both fine and very smooth. Supersonic flow may also supervene when a reservoir of air at high pressure is discharged through a nozzle so that the air is let into the atmosphere without energy loss. The throat of such a nozzle may be used as a miniature wind tunnel for measuring the drag of models at supersonic speeds.

supersonic frequency A former term for a frequency that is too high to be audible, i.e. that lies above the *audiofrequency range; replaced by *ultrasonic*.

superstring *See* string theory.

supersymmetry A *symmetry that includes both *bosons and *fermions. In the simplest supersymmetry theories, every boson has a corresponding fermion partner and every fermion has a corresponding boson partner. The boson partners of existing fermions have names formed by adding "s" to the beginning of the fermions' names, e.g. *selectron*, *squark*, and *slepton*. The fermion partners of existing bosons have names formed by replacing "on" at the end of the word with "ino", e.g. *gluino*, *photino*, *wino*, and *zino*.

The infinities which afflict calculations in *relativistic quantum field theories are less severe in supersymmetric theories than non-supersymmetric theories due to cancellations between infinities of bosons and fermions. Although there is no experimental evidence for supersymmetry, it may be important in the search for a unified theory of the fundamental interactions. *See also* string theory; supergravity.

supported end The end of a structure, e.g. a beam, which is held by a knife-edge but is not restrained in any other way. In the case of a beam, at the end under consideration, the displacement, y, is always zero, but the slope $\mathrm{d}y/\mathrm{d}x$ (where x is measured along the beam) may not be zero.

suppressed-zero instrument *Syn.* set-up scale instrument. An electrical instrument in which the moving part does not begin to deflect until the deflecting force (or torque) reaches a predetermined value.

suppressor grid *See* thermionic valve.

surface acoustic wave devices (SAW devices) Miniature devices that are used for signal processing and employ an analogue rather than a digital representation of information. In such a device, an ultrasonic wave propagates along the plane surface of a solid. The frequency ranges from a few megahertz to a few gigahertz so high transmission rates are possible. Electric signals can be converted to surface acoustic waves (and vice versa) by means of transducers based on the *piezoelectric effect. Piezoelectric crystals are therefore used as the substrate along which the waves travel. SAW devices can be fabricated to perform a variety of functions; they are particularly important as filters (*SAW filters*).

surface-barrier transistor A *transistor in which the usual p-n junctions are replaced by metal-semiconductor contacts called Schottky barriers (*see* Schottky diode). Carrier storage under saturation conditions is zero with the Schottky barriers (*see* storage time) and the transistors are useful for high-frequency switching applications.

surface-charge transistor *See* charge-transfer device.

surface colour Coloured light reflected by a surface as distinct from the more common body colour that arises from reflection after some penetration into the medium. Transmitted light by bodies showing surface colour is complementary to the reflected colour.

554

surface density *Syn.* superficial density. The quantity per unit area of anything distributed on a surface.

surface energy The energy per unit area of exposed surface. The (total) surface energy in general exceeds the *surface tension, which is the *free* surface energy, concerned in isothermal changes.

surface film 1. *Syn.* monolayer; monofilm. A film formed by insoluble oils spreading on water. The molecules in such a film form a monomolecular layer with any excess in the form of globules. Since the molecules are usually orientated in some simple manner in these films, their properties give information as to the shape and structure of the molecules. The films behave in a manner analogous to two-dimensional gases, liquids, and solids. For example, if there are not enough molecules to fill the available surface, they move about like the molecules of a two-dimensional gas exerting a force per unit length on the boundary called *surface pressure* by analogy with an ordinary gas pressure. If the area of the water surface available to the molecules is reduced until all the molecules touch, the film is said to have condensed and to be a liquid film; at this stage the film vigorously resists further compression. The critical area, at which the film has just liquefied, can be measured and, for example, the area occupied by each molecule can be deduced. For the fatty acids this area is found to be the same no matter how many $-CH_2-$ groups there are in the chain; thus we deduce that the chains are more or less vertical and their cross section is therefore known. Solid films, which can be blown about on the surface and resist changes in shape, are also known.

2. Surface film of a solution. The concentration of a solute is different in the surface layer from that in the bulk of the solution. (*See* Gibbs's adsorption formula.)

3. Surface films formed on solids include: (i) The *Beilby layer formed by polishing. (ii) The films formed by the spreading of liquids on solids. (iii) The gaseous films adsorbed on solids that are important in connection with catalysis, with the protection of a metal surface by the formation of an oxide layer (e.g. aluminium) and with the necessity for thorough cleaning of metals used *in vacuo* for the photoelectric effect, etc.

surface leakage A leakage current due to charges flowing at the surfaces of a material, rather than in the bulk of the material.

surface of buoyancy *See* metacentre.

surface of discontinuity The surface lying in a fluid across which there is a discontinuity of the fluid velocity, often set up in the *wake of an obstacle. The surface is unstable and exists for only a short period after which it breaks up and forms vortices.

surface passivation The application or growth of protective layers on the surface of a *semiconductor, to reduce *surface-leakage effects.

surface pressure *See* surface film.

surface resistivity The resistance between two opposite sides of a unit square of the surface of a material. Its reciprocal is the *surface conductivity*.

surface tension Symbol: σ, γ. Intermolecular forces are repulsive at small separations, decrease to zero at about 10^{-10} m, then become attractive at slightly greater distances, rise to a maximum and then fall off rapidly to zero. Liquids are usually under pressure so any molecule in the interior must be, on average, repelled by the molecules on each side. The separation between near neighbours is therefore slightly less than that for zero interaction, in the region of repulsion. In the plane of the surface however the separation of near neighbours is greater, i.e. in the region of attractive forces increasing with distance. Consequently, the liquid surface is in tension and contracts as far as is possible; thus a small free drop is spherical. The increased separation in the surface results from the fact that surface molecules have fewer near neighbours, and so have less negative potential energy, than those in the interior. The work done in creating unit area of surface against this tension at constant temperature is called the *free surface energy*. If the liquid surface is assumed to be in tension in all directions so that the force required to hold the straight edge of a plane liquid surface is γ (usually expressed in newtons per metre), then γ is called the surface tension. If the surface is stretched isothermally so that the edge moves unit distance so creating unit area of new surface, the work done is γ (in joules). Thus the surface tension expressed in $N \, m^{-1}$ is numerically and dimensionally equal to the free surface energy in $J \, m^{-2}$ but is not the total surface energy.

Small additions of foreign substances often profoundly affect the surface tension. (*See* Gibbs's adsorption formula.)

Due to the surface tension there is a pressure difference between the two sides of a liquid surface equal to

$$\gamma(1/R_1 + 1/R_2),$$

where R_1 and R_2 are the radii of curvature of two perpendicular normal sections. (*See* principal radii.) Thus the pressure inside a soap bubble (which has *two* surfaces of radius R) is $4\gamma/R$.

surface wave 1. *See* ripples; water waves.
2. *See* ground wave.

surfactant A substance, such as soap, that changes the surface tension of a liquid, especially of water.

surge An abnormal transient electrical disturbance in a conductor. Surges result, for example, from lightning, sudden faults in electrical equipment or transmission lines, or switching operations.

surge-current indicator *Syn.* magnetic link. An apparatus used for determining the approximate value of a *surge current that has flowed in a conductor due, for example, to lightning. Fundamentally, it consists of a specimen of steel, having a high value of *remanence, which is mounted close to the conductor concerned. Changes in the magnetic condition of the specimen are related to the magnitudes of the surge currents.

surge diverter A device connected between one electric conductor of a system and earth to divert to earth the major part of any *surge of excessively high voltage. It usually consists of one or more *spark gaps connected in series with a material, the electrical resistance of which decreases as the voltage increases. This special material (of which silicon carbide commonly forms a basis) assists the spark gaps in clearing themselves and in interrupting the *follow current after the surge has been discharged.

surge generator See impulse generator.

surge impedence Of an electrical *transmission line. The voltage divided by the current in a *surge propagated along the line as a *travelling wave. It is given approximately by $\sqrt{(L/C)}$, where L and C are respectively the inductance and capacitance for the same length of line (e.g. per unit length). The expression is exact for the hypothetical case of a line without losses. *Compare* characteristic impedence.

surge modifier An apparatus connected to a conductor of a power or commuication system to modify the waveshape of *surges that may occur on the conductor, with the object of affording protection to other electrical apparatus connected to the system. For example, a *travelling wave having a steep wavefront produces very high electric stresses in the end turns of a transformer winding connected to the line. Fitting a surge modifier reduces the steepness of the wavefront and hence also the stresses in the winding. A surge modifier may be equipped with some form of resistance in order to dissipate part of the energy of the travelling wave, in which case it is described as a *surge absorber*.

susceptance Symbol: B. The imaginary part of the *admittance, Y, i.e.
$$Y = G + iB,$$
where G is the *conductance. It is the reciprocal of the reactance and is measured in *siemens.

susceptibility 1. (magnetic) Symbol: χ_m. The quantity $\mu_r - 1$, where μ_r is the relative *permeability.
2. (electric) Symbol: χ_e. The quantity $\varepsilon_r - 1$, where ε_r is the relative *permittivity.

suspension insulator A type of insulator used to support a line conductor in an overhead transmision-line system especially at voltages above about 33 kV. It consists of a number of separate insulating units arranged in series so that each one hangs vertically below the next above it. The complete string of units has a certain degree of flexibility. See tension insulator; string efficiency.

suspensoid sol See sol.

Sutherland's formula One of several formulae proposed to show the variation of viscosity η of a gas with thermodynamic temperature T:
$$\eta = \eta_0(T/273)^{3/2}(273 + k)/(T + k),$$

where k is constant for a given gas, and η_0 is the viscosity at 0 °C.

SVP Abbreviation for *saturated vapour pressure.

Swan, Sir Joseph Wilson (1828–1914) Brit. inventor and industrialist who, independently of Thomas Edison, developed the carbon-filament incandescent electric light bulb (1878).

sweep See time base.

sweep generator Syn. time-base generator. See time base.

swing The limits of the values of a varying electrical parameter.

switch A device for opening or closing a circuit, or for changing its operating conditions between specified levels. It is also used to select from two or more components, circuits, etc., the desired element for a particular mode of operation. Switches may consist of a mechanical device, such as a *circuit breaker, or a solid-state device such as a *transistor, *Schottky diode, or *field-effect transistor.

sylphon Flexible metal bellows usually made by spinning from a single piece of metal. They were used by Bridgman for the determination of the compressibility of liquids that were sealed up in the sylphon and the whole exposed to external hydrostatic pressure. Compression of the liquid is shown by shortening of the sylphon.

symbols See Appendix, Table 12 for a list of the principal symbols used in physics and Table 6 for symbols used in electronics. See also SI units.

symmetrical components An unsymmetrical three-phase system of voltages can, in general, be considered as the superposition of the following three component systems of voltages: (i) Positive-sequence component, consisting of three equal voltages with relative phase differences of 120 degrees and having the normal (i.e. positive) *phase sequence. (ii) Negative-sequence component, consisting of three equal voltages with relative phase differences of 120 degrees and having the reverse (i.e. negative) phase-sequence. (iii) Zero-sequence component, consisting of three equal voltages that are all in phase.

An unbalanced three-phase system of currents can be dealt with in the same manner. The use of symmetrical components usually greatly simplifies the calculations of three-phase systems under unbalanced conditions and especially under fault conditions. The ratio of the negative-sequence component to the positive-sequence component is called the *unsymmetry factor* (for voltages) or *unbalance factor* (for currents).

symmetrical (and asymmetrical) deflection Of the beam in a *cathode-ray tube employing electrostatic deflection. In this method of deflecting the beam of electrons, deflecting voltages, which are always equal in magnitude and opposite in sign, are applied to the two deflector plates so that the potential, with respect to the final accelerator (final anode), of a point midway between the two deflector plates is constant. In asymmetrical deflection, one deflector plate is maintained at a potential fixed relative to that of the final accelerator and the deflecting voltage is applied to the other. With this method of deflection, *deflection defocusing and *trapezium distortion may be serious, whereas symmetrical deflection is better in these respects.

symmetry The set of invariances of a system. A symmetry operation on a system is an operation that does not change the system. It is studied mathematically using *group theory. Some symmetries are directly physical, for example reflections and rotations for molecules and translations in crystal lattices. More abstract symmetries involve changing properties, as in the *CPT theorem and the symmetries associated with *gauge theories.

symmetry operation of the first kind An operation that, when performed upon a system of points or a rigid body, brings it to coincidence with a congruent system.

symmetry operation of the second kind An operation that, when performed upon a system of points or a rigid body, brings it to coincidence with an enantiomorphous system. *See* enantiomorphy.

synchrocyclotron *Syn.* frequency-modulated cyclotron. A modification of the *cyclotron in which the magnetic field remains constant but the frequency of the accelerating electric field is slowly decreased. In the cyclotron the time to complete a semicircle is proportional to the mass of the particle. As the velocity becomes relativistic an increase in mass occurs and as a result the particle gets out of phase with the alternating electric field. To counteract this, the alternating frequency in the synchrocyclotron is slowly decreased so that the particles remain in phase with the field. Energies up to 700 MeV for protons can be obtained. *Compare* synchrotron.

synchronism The condition that exists when two or more a.c. machines or sources of a.c. supply have exactly the same frequency and are in phase.

synchronous alternating-current generator *Syn.* alternator; synchronous generator. An electrical machine for generating alternating current. It has a number of *field magnets, which are usually excited by means of *field coils carrying direct current obtained from an independent source. The frequency of the generated e.m.f.s and currents is determined by the number of magnetic poles in the machine and the speed at which it is driven (*see* synchronous speed). This type of generator can operate and deliver its output independently of any other source of alternating current. (*Compare* induction generator.) Power stations usually employ generators of this type.

synchronous capacitor A *synchronous motor designed to run unloaded and overexcited so as to take a *leading current at low *power factor from the supply system and thus to be electrically equivalent to a capacitor. In this form it is used extensively for *power-factor improvement. If means are provided for varying the excitation from high to low values, the power factor may be varied from leading, through unity to lagging respectively, and the motor may be made to simulate a capacitive or inductive reactor as required. In this form, it is commonly employed in a transmission system for the regulation of voltage and may be described as a *synchronous phase modifier*.

synchronous clock A *clock in which a *synchronous motor drives the mechanism that advances the hands. The time-keeping is determined entirely by the frequency of the a.c. electricity supply to which the motor is connected.

synchronous converter *See* converter.

synchronous generator *See* synchronous alternating-current generator.

synchronous impedence Of a *synchronous alternating-current generator. The fictitious *impedance sometimes used in calculations of voltage regulation and parallel working of machines. Its value in ohms is obtained by dividing the open-circuit e.m.f. in volts by the sustained short-circuit current in amperes, both corresponding to the same value of field excitation. It is a fictitious impedance since *armature reaction is considered as having an effect similar to that of an inductive *reactance. Its usefulness in calculations is seriously impaired by the fact that it is not constant in magnitude since it varies, owing to magnetic saturation with the excitation. The reactive component of the synchronous impedance is called the *synchronous reactance*.

synchronous induction motor Basically, an *induction motor having a slip-ring rotor and a direct-coupled d.c. exciter. It is started as a normal induction motor with external resistances connected in the rotor circuit, and when these resistances have been reduced to zero and the motor is running with a small *slip, direct current from the exciter is injected into the rotor circuit. The motor then runs at *synchronous speed as a normal *synchronous motor. Sometimes, the primary, or a.c. winding is mounted on the rotor and the d.c. excitation is supplied to the secondary winding on the stator. This motor combines the advantages of the constant speed and high or leading *power factor of a normal synchronous motor with the advantage of the high starting torque of a slip-ring induction motor. The design is, however, a compromise since the requirements for good design in induction and synchronous motors conflict with one another in many particulars.

synchronous logic If the timing of all the operations in an electronic *logic circuit is controlled by and hence is in synchronism with externally generated *clock pulses, the system is said to be synchronous. If the operations are allowed to proceed without

synchronous motor

controlling clock pulses the system is said to be *asynchronous*.

synchronous motor An a.c. electric motor, the mean running speed of which is independent of the load and is determined by the number of its magnetic poles and the frequency of the electric supply. (*See* synchronous speed.) A typical industrial motor of this type consists of a stator carrying a winding that, when connected to the a.c. supply, produces a magnetic field that rotates in space, and a rotor excited by direct current. The rotor is, fundamentally, an electromagnet that locks with the field produced by the stator and rotates at the same speed as the field. By overexciting the rotor, the motor can be made to operate at a leading *power factor and this is a valuable property since it assists in the *power-factor improvement of an industrial load. Special arrangements have to be made for starting since a plain synchronous motor is, fundamentally, not self-starting. *See also* synchronous capacitor; synchronous induction motor; self-starting synchronous motor. *Compare* asynchronous motor.

synchronous orbit *See* geostationary orbit.

synchronous phase modifier *See* synchronous capacitor.

synchronous reactance *See* synchronous impedance.

synchronous speed The speed of rotation of the magnetic flux in an a.c. machine. It is given by the ratio f/p r.p.s., where f = frequency in hertz of the a.c. supply, and p = number of pairs of magnetic poles for which the a.c. winding has been designed.

synchronous vibrator *See* vibrator.

synchroscope A special form of power-factor meter for indicating the phase relation of the voltages of two alternators, or two a.c. supply systems, that are to be operated in parallel.

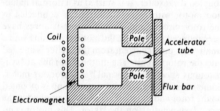

a   Cross section of a synchrotron

synchrotron *Syn.* electron synchrotron. A cyclic *accelerator that is based on the *betatron but uses a constant-frequency electric field in addition to a changing magnetic field. As in the betatron, the increasing magnetic flux density B counteracts the relativistic increase in mass at high velocities. Fig. *a* shows a section of a synchrotron. The steel flux bars become magnetically saturated very early in the magnetic cycle and further time variation in B takes place in the region of the electron orbit. Fig. *b* shows the circular vacuum tube. A high-frequency electric

field from a radio-frequency oscillator such as a *klystron is applied across a gap in a metallic cavity inside the chamber. The frequency is in synchronism with the constant angular frequency of the particles so that the electrons are accelerated inside the cavity. There are usually several RF cavities interspersed between the magnets that bend the particle beam and that focus it.

Initially the machine acts as a betatron until the electrons have an energy of several MeV. The high-frequency field is then applied while the magnetic field is increasing. At the required energy the magnetic flux condition for a stable orbit is destroyed and the electrons are deflected from the path. Very high energies in the GeV range have been achieved. *See also* proton synchrotron; synchrotron radiation.

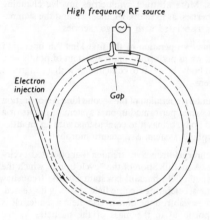

b   The vacuum tube of a synchrotron

synchrotron radiation *Syn.* magnetobremsstrahlung. Electromagnetic radiation emitted by high-energy charged particles moving at relativistic speeds in a strong magnetic field. The radiation is emitted as the particles follow a circular path through the field. Such emission occurs, for example, in *synchrotrons and *accelerator storage rings. There is a smooth distribution of wavelengths, ranging from microwaves to hard X-rays. The exact profile of the spectrum depends on the radius of the particle orbit and the particle energy. It is not *thermal radiation but is strongly polarized.

Synchrotron radiation can be a significant problem, in terms of energy loss from the charged particles. The radiation can, however, be used as a tool. Synchrotron sources are being built to produce intense beams of radiation, primarily of X-ray and UV frequencies.

Because many regions of the universe are associated with very high magnetic fields, the radiation emitted from electrons moving in these regions is called synchrotron radiation. It is the mechanism thought most likely to explain radio emission from extragalactic *radio sources and the emission from *supernova remnants.

synclastic *See* principal radii.

syneresis The reverse of *imbibition.

synoptic chart A map showing wind, barometric pressure, etc., at a particular time. It is used in weather forecasting.

systematic error *See* errors of measurement.

Système International d'Unités *See* SI units.

systems software *See* software.

syzygy The moon (or a planet) is in syzygy when it is in conjunction (i.e. when it has the same celestial longitude as the sun) or when it is in *opposition.

Szilard, Leo (1898–1964) Hungarian-born Amer. physicist who worked with *Fermi on the first atomic pile. He collaborated with Teller in persuading Einstein to write a letter to President Roosevelt in 1939 advising him of the feasibility of constructing an atom bomb.

Szilard–Chalmers effect In certain circumstances an active isotope formed in a nuclear reaction without change of atomic number can be separated chemically from the original irradiated material. The product nucleus suffers *Compton recoil with sufficient energy to break the chemical bond between the affected atom and the rest of the molecule.

T

tachometer An instrument for measuring angular speeds. The many types available include: (i) Those depending on centripetal force. E.g. a circular metal disc is carried on the shaft so as to be coaxial with it; it is pivoted so that its plane is not necessarily perpendicular to the shaft. As the shaft rotates, the disc tends to set itself with its plane perpendicular to the shaft but this movement is constrained by a spring. The actual movement depends on the angular velocity of the shaft and is magnified and communicated to a pointer. (ii) Instruments in which the number of revolutions of the shaft is automatically and repeatedly counted for small equal intervals of time. (iii) Instruments based on the *eddy current drag exerted on neighbouring pieces of metal by a magnet rotating with the shaft. (iv) Electrical instruments similar to a dynamo. (v) Instruments based on the viscous drag exerted by a whirling fluid. (vi) Resonance tachometers in which a vibrating reed indicates that a particular frequency of revolution has been reached.

tachyon A particle postulated to move with velocity greater than that of the speed of electromagnetic radiation. Of the two properties rest mass and energy, one must be real and the other imaginary. If it exists, it may be detected through the emission of *Cerenkov radiation.

Talbot's law The apparent intensity, I, of a light source, flashing at a frequency greater than 10 hertz, is given by:
$$I = I_0(t/t_0),$$
where I_0 is the actual intensity, t is the duration of the flash, and t_0 is the total time. The light appears steady due to the persistence of vision. *Compare* Blondel–Rey law.

Tamm, Igor Yevgenyevich (1895–1971) Soviet physicist who worked in Moscow. He shared the 1958 Nobel prize for physics with Ilya Frank and Pavel Cerenkov for his work in explaining *Cerenkov radiation.

Tammann and Jüttner's measurement of specific latent heat The specific latent heat of fusion was determined by observing the time (t) required for complete freezing, by heating the substance above its melting point and plotting the cooling curve. If $d\theta/dt$ is the rate of cooling given by the slope of the curve when freezing just commences, then
$$L = At \; d\theta/dt,$$
where A is a constant.

Tammann point The temperature of a solid at which the molecules or atoms become noticeably mobile; it is about half-way between absolute zero and the melting point. An exact value can rarely be given.

tandem generator A modification of the *Van de Graaff generator in which a doubling of the energy of the particles is achieved for the same accelerating potential. Negative ions are accelerated from earth potential, the electrons are stripped off, and the resulting positive ions are accelerated back to earth potential. This is produced by connecting two generators in series.

tangent galvanometer A *galvanometer in which a short magnetic needle similar to that of a *magnetometer is suspended at the centre of a narrow, circular coil of wire of radius large compared with the length of the needle. The plane of the coil is placed along the magnetic meridian so that a current in the coil deflects the needle against the controlling couple of the earth's magnetic field. The tangent law of the magnetometer applies, i.e. the current through the coil is proportional to tan θ, where θ is the angle of deflection of the needle. Thus, $i = k \tan \theta$, where k is a constant called the *reduction factor* of the galvanometer. The galvanometer can be used to compare currents or to determine the earth's field if the current is known.

tangential coma *See* coma.

tangential foci A *meridian plane is also called a tangential plane. Rays in a meridian plane focus at a point in the plane, the tangential focus. Through this focus, a focal line perpendicular to the plane is formed by all the other rays striking the surface or lens obliquely.

tangential-polar equation An equation of a curve given in terms of the distance of a point P on the

tangential surface

curve from some reference point P_0 and the perpendicular distance of P_0 from the tangent to the curve at P.

tangential surface *See* astigmatic (focal) surfaces; meridian (plane, rays, section).

tangent law *Syn.* Helmholtz tangent law. The law:
$$ny \tan \alpha = n'y' \tan \alpha',$$
where n is the refractive index of object space, y the size of object, α the angle of inclination of a paraxial ray from an axial point on the object. The symbols with a dash refer to the image side.

tap changer *Syn.* ratio adjuster. A device for changing the ratio of a *transformer by selecting *tappings. If it is designed for operation only when the transformer is disconnected from the electric supply, it is called an off-circuit tap changer, whereas if it is designed for operation when the transformer is connected to the supply (whether or not the transformer is on load) it is called an on-load tap changer.

tape *See* magnetic tape; paper tape.

tape punch *See* paper tape.

tape reader *See* paper tape; magnetic tape unit.

tape recorder *See* magnetic recording.

tapping A conductor, usually a wire, that makes an electrical connection with a point between the ends of a winding or coil. In a *transformer, it may be described as plus or minus according to whether it is designed so that by its use more or fewer turns respectively are included in the active portion of that winding than the number of turns that corresponds to the *service voltage ratio. *See* tap changer.

target 1. A substance deliberately subjected to ionizing radiation.
   2. A surface (generally an electrode) on which high-energy electrons impinge, and which becomes a source of secondary radiation (e.g. fluorescence, X-rays).
   3. An object detected by the reflection of infrared radiation, radar, etc.

tauon *Syn.* tau particle. Symbol: τ. A negatively charged *elementary particle, a *lepton, of considerable mass (1784 MeV, i.e. 3560 times the electron mass). It has a mean life of 3×10^{-13} second and three principal decay modes. Its antiparticle is called the *positive tauon*. The tauon is assumed to have an associated neutrino, ν_τ.

Taylor series A mathematical expansion of a continuous function f(x) in terms of an approximating polynomial function. If a is a fixed value of x and $f'(a), f''(a) \ldots$ are successive derivatives of f(x) evaluated at a then:
$$f(x) = f(a) + f'(a)(x - a) + f''(a) \frac{(x - a)^2}{2!}$$
$$+ \cdots + f^n(a) \frac{(x - a)^n}{n!} + R_n,$$

where R_n is the remainder after n terms. If R_n is ignored, then the resulting polynomial is an approximation of the original function f(x). Hence
$$\log(1 + x) = x - \frac{x^2}{2} + \frac{x^3}{3} - \frac{x^4}{4} + \cdots$$
$$+ (-1)^{n+1} \frac{x^n}{n} \cdots.$$

This is the *logarithmic series*.

The *Maclaurin series* is obtained when $a = 0$, i.e.
$$f(x) = f(0) + f'(0)x + f''(0) \frac{x^2}{2!} + \cdots.$$

Expansions for sin x and cos x can then be found:
$$\sin x = x - \frac{x^3}{3!} + \frac{x^5}{5!} - \frac{x^7}{7!} + \cdots$$
$$\cos x = 1 - \frac{x^2}{2!} + \frac{x^4}{4!} - \frac{x^6}{6!} + \cdots.$$

teaser transformer *See* Scott connection.

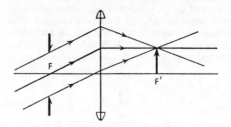

Telecentric

telecentric If the centre of the *entrance pupil lies at the anterior focus of a lens, the chief ray of a pencil traversing the lens will emerge parallel to the axis. Any error of focus will show the centres of the blur circles at the same distance from the axis. Thus, a measuring scale placed near the image will show less error if it is slightly out of focus compared with the image. By placing a rear stop (or *exit pupil) at the rear focus, then the chief ray of a pencil from an extra axial point will be incident parallel to the axis. A scale that may not be in its correct object plane will therefore show less error where the measurement of an object size is under consideration.

telecommunications The study and practice of the transfer of information by any kind of electromagnetic system, e.g. wire, radio waves, etc. There are many types of telecommunications system including telephony, *television, *radio, and communication *satellites.

telegraphy A *telecommunications system used for the transmission of graphical symbols such as letters or numbers, using a signal code. International Morse code is a very well known example of such a code.

telemeter An instrument for providing the value of a quantity by remote control process. *See* telemetry.

telemetry A means of making measurements, in which the measured quantity is distant from the recording apparatus and the data are transmitted over a particular *telecommunications system from the measuring position to the recording position. Particular examples of telemetry systems are space exploration and physiological monitoring in hospitals.

telephony A communications system designed to transmit speech and other information such as fax and electronic mail. A complete system consists of all the circuits, switching apparatus, and other equipment necessary to establish a *communications channel between any two users. The communication may take place along suitable electric cables or optical fibres, and/or by means of radio links (as in cellular radio) or satellite links.

The use of *digital circuits and signals rather than *analogue circuits and signals to transmit the information provides a faster more reliable telephone connection; *analogue/digital converters are used to convert speech into digital form. Conversion to digital systems is under way worldwide. In the case of an analogue system, digital input and output (e.g. from a computer) is achieved using *modems.

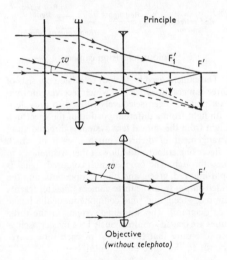

Principle

Objective
(without telephoto)

Telephoto lens

telephoto lens A photographic lens that can produce a large image of a distant object with normal camera extensions. It has a large effective focal length, and hence a narrow angle of view and small depth of field. It consists of a converging lens followed by a diverging lens after the manner of the Galilean telescope (*see* refracting telescope), but with a separation such that convergence to a real image occurs. Appreciable magnifications can thereby be obtained with constant back focal length, since the second

principal plane of the combination can be placed at relatively great distances in front of the back lens. The *telephoto effect* is the ratio of the principal focal length to the back focal length. Modern telephoto lenses are fixed focus; the earlier patterns were telephoto attachments to ordinary lenses with a variable separation, and a scale attachment to give the magnifying effect, i.e. ratio of image size with the attachment to that without. *See also* zoom lens.

teleprinter A form of typewriter used for converting keyboard information into electrical signals and vice versa. Teleprinters are used in *telex systems.

telescope 1. An optical device for collecting light in order to form images of distant objects. Its light-gathering power allows fainter objects to be discerned than with the unaided eye. The images of distant extended objects, such as the moon, are magnified (*see* magnifying power). Point sources, such as stars, are easier to distinguish (*see* resolving power).

An optical telescope can either be a refracting or a reflecting system. A *refracting telescope consists essentially of two lens systems, the *objective and the *eyepiece. The objective is always a convex lens of long focal length; the eyepiece has a short focus and is either convex or concave depending on the type of telescope. Chromatic and spherical *aberration, coma, etc., are severe in these telescopes, especially if a large-diameter objective is used, and must therefore be corrected. These aberrations are nonexistent or of little importance in a *reflecting telescope in which a large-aperture concave mirror, usually paraboloid, is used to collect and focus light. These telescopes are employed mainly in astronomy.

The *Schmidt telescope is a combination of a lens and a mirror by which spherical aberration and coma are reduced to a minimum and a large field of view is possible. The *Maksutov telescope uses a similar system.

2. Any astronomical device by which radiation from a particular region of the spectrum is collected and brought to a suitable recording system, usually electronic, for analysis. Ground-based *radio telescopes and infrared telescopes collect radiation penetrating the radio and infrared *atmospheric windows. Satellite-borne instruments must be used to detect ultraviolet radiation, X-rays, and gamma rays from space.

television A *telecommunications system in which visual and aural information is transmitted for reproduction at a receiver. The basic elements of the system are as follows: *television cameras plus microphones to convert the information into electrical signals, i.e. into *video* and *audio signals*; amplifying, control, and transmission circuits to transmit the information; broadcast information using a modulated radio-frequency *carrier wave; a *television receiver* that detects the signals and produces an image on the screen of a specially designed cathode-ray tube.

television camera

The information on the target in the television *camera tube is extracted by *scanning, and the spot on the screen of the receiver tube is scanned in synchronism with it to produce the final image. A process of rectilinear scanning is used in which the electron beam traverses the target area in both the horizontal and vertical directions; this is known as *raster scanning*. *Sawtooth waveforms are used to produce the deflections of the beam and in both the camera and receiver the portion of the waveform during which the beam returns to the start of the scan is blanked out. This is the beam retrace or *flyback* period. For the maximum information to be obtained from the target area the number of horizontal scans is made larger than the number of vertical scans so that as much of the target area of the receiver is covered as possible. Each horizontal traverse is a *line* and the repetition rate is the *line frequency*.

The vertical direction is the *field*, the number of vertical scans per second being the *field frequency*. Each vertical scan is a *raster*. If the entire picture is produced in a single raster, the scanning is called *sequential scanning*. Most broadcast television systems use a system of *interlaced scanning*. In this system the lines of successive rasters are not superimposed on each other, but are interlaced, and two rasters constitute a complete picture or *frame*. The number of complete pictures per second is the *frame frequency* or *picture frequency*, and is commonly half the number of rasters per second or *field frequency*. The field frequency needs to be relatively slow to allow as many horizontal lines as possible, but sufficiently fast to eliminate flicker. Various compromises are used. European television systems use a 50 hertz field frequency (25 Hz frame frequency) system with 625 lines per frame. American television uses a 60 Hz field frequency and 525 lines per frame. *Definition* in television is a measure of the resolution of the system, which in turn depends on the number of lines per frame. High-definition systems have more lines. Some *closed-circuit television systems use as many as 2000 lines per frame.

Positive or *negative transmission* may be employed for transmitting the video signals. A particular value of the modulated carrier wave represents the black level of the picture. In positive transmission an increase in amplitude or frequency above this value is proportional to the light intensity. In negative transmission the carrier wave value decreases below the black level in proportion to the light intensity.

The basic television system transmits images in black and white only (monochrome television). *Colour television is now widely used, and the broadcast signal is received on special colour receivers. Modern monochrome receivers also use the broadcast colour signal but the image produced is black and white.

television camera The device used in a *television system to convert the optical images from a lens into electrical *video signals*. The optical image formed by the lens system of the camera falls onto photosensi-

tive material. This is scanned, usually by a low-velocity electron beam, and the resulting output is modulated with video information obtained from the target area. The camera consists of three major parts housed in one container: optical lens system, *camera tube, and preamplifier. The resulting output is further amplified and transmitted in the broadcasting network. Some cameras are self-contained, with the amplifier and transmitter in the same container. These cameras are usually employed in a closed-circuit system, or for special applications. Several types of camera tube have been developed; the major differences between various types of tubes are in the composition of the photosensitive material used, and the means of extracting the electrical information produced.

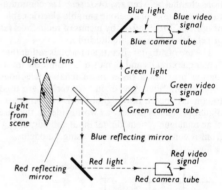

Colour-television camera

The camera used in *colour television consists of three camera tubes, each of which receives information that has been selectively filtered to provide it with light from a different portion of the spectrum. Light from the optical lens system is directed at an arrangement of dichroic mirrors, each of which reflects one colour and allows other frequencies to pass through. The original multicoloured signal is split into red, green, and blue components, and the video output from the three camera tubes represents the red, green, and blue components of the image (*see* diagram). The scanning systems in the three tubes are driven simultaneously by a master oscillator to ensure that the output of each tube corresponds to the same image point.

television receiver *See* television.

telex A telegraphy system in which the transmitter and receiver are combined in a typewriting device (a *teleprinter), which converts keyboard information into electrical signals and vice versa. In the USA the system is called TWX.

Teller, Edward (*b.* 1908) Hungarian-born Amer. nuclear physicist. He contributed to the understanding of stellar energy and was responsible for the programme to build the first H-bomb, having worked on the fission bomb. With *Szilard he persuaded Einstein to write to Roosevelt in 1939 advising him that an atom bomb was feasible.

telluric spectrum line One of several absorption lines seen in the spectrum of the sun and stars that originate in the atmosphere of the earth itself. The lines are due to absorption, especially by oxygen and water vapour.

TE modes *See* waveguide.

temper Of steel. *See* alloy; tempering.

temperament The adjustment of tuning of the notes of a keyboard instrument to give a near diatonic scale (*see* musical scale) for all keys. The keyboard instrument traditionally has thirteen finger keys to the octave. If diatonic scales were built up on all keys, the number of finger keys to the octave would be very large. The first table of frequencies illustrates this.

Diatonic scale of C		Diatonic scale of D
264	C	
297	D	297
330	E	334
352	F	
	F♯	371
396	G	396
440	A	445
495	B	495
528	C	
	C♯	557
	D	594

To improve the accuracy of tuning, keyboards with an increased number of finger keys have been introduced, but with very limited success, e.g. the Bosanquet keyboard has fifty-three keys to the octave. Two solutions have been tried: (*a*) the *mean tone scale and (*b*) the *equitempered scale, of which the latter is now used almost to the exclusion of the former. In the mean tone scale the intervals of major thirds are made accurate, the other intervals being adapted. Six major and three minor scales are tolerable but beyond these the out-of-tuneness is bad. To overcome this disadvantage, a movable keyboard has been proposed that operates on different sets of notes depending upon the key in which the music is played. In the equitempered scale the mistuning is evened out, the octave being divided into twelve intervals with a semitone ratio of $2^{1/12}$. Only the octaves are true but the errors are not such that an ear accustomed to the system feels distress. Such a tuning permits the use of all keys, and modulation during a performance. The origin of the equitempered scale is obscure but it is said that it was proposed by Aristoxenus about 350 B.C. The second table gives a comparison between the scales of just intonation (diatonic) and equitemperament. A (440 hertz) has been taken as the standard in both scales quoted.

Just intonation		Equal temperament
264	C	261.6
297	D	293.7
330	E	329.6
352	F	349.2
396	G	392.0
440	A	440
495	B	493.9
528	C	523.3

In actual practice, because of the tolerance of the ear, tuning will only approximate to these frequencies. The equitempered scale tends to be used in place of the diatonic scale in present-day music. Microtones, whole tone scale, and atonality are dependent upon equal temperament. It has been often stated that performers on instruments without a keyboard and fixed tuning, e.g. violin, trombone, and human voice, use the scale of just intonation, adjusting the intervals to the appropriate value for any key in use. Tests have shown that generally the intonation used varies more from just intonation than does equal temperament.

temperature Symbol: *T*. The property of an object that determines the direction of heat flow when the object is brought into thermal contact with other objects: heat flows from regions of higher to those of lower temperatures. (*See* thermodynamics (second law).)

This definition merely allows one to place in order of temperature a number of objects. To assign numerical values, it is necessary to establish a scale of temperature. The only temperature scale now in use for scientific purposes is the *International Practical Temperature Scale, temperature being expressed in *degrees Celsius or *kelvins. Temperature is a measure of the kinetic energy of the molecules, atoms, or ions of which a body or substance is composed. The *thermodynamic temperature of a body is now treated as a physical quantity and is measured in kelvins.

The measurement of low and moderate temperatures (roughly up to 500 °C) is usually classed as *thermometry* while *pyrometry* covers the high-temperature ranges.

temperature coefficient of resistance For any material, the small change in the resistance of the material for changes in the thermodynamic temperature of the material.

At a given Celsius temperature *t* the resistance of a body, R_t, can be expressed as a power series in *t*. For a moderate range of temperature,

$$R_t = R_0 (1 + \alpha t + \beta t^2),$$

where R_0 is the resistance at 0 °C and α and β are constants characteristic of the material. For limited range of *t* or moderate precision, β is usually negligible and α is called the temperature coefficient of resistance. It is measured in reciprocal degrees. For pure metals α is typically about 3.5×10^{-3} °C^{-1}; for alloys it is generally much less, and for some specimens of *constantan it may have a small negative value.

In general, conductors have a positive coefficient of resistance, semiconductors and insulators have a negative coefficient of resistance. This may be most easily explained by considering the energy distribution of electrons in the material (*see* energy bands). Conductors always have quantum states available for conduction, and increasing the temperature above absolute zero causes ions to vibrate less uniformly about their lattice positions and scatter the conduction electrons more as they drift through the material. This has the effect of increasing the resistance.

In semiconductors and insulators, where a forbidden band exists between the valence bands and the conduction bands, increasing the temperature increases the number of electrons that can cross the forbidden band and decreases the resistance. Scattering by ions will normally make only a small contribution to the resistance.

temperature detector *See* embedded temperature-detector.

temperature gradient The rate of change of temperature in a substance with distance. In the calculus notation it is denoted by dT/dx in the direction of the x axis.

temperature inversion 1. A phenomenon that occurs in the troposphere when the rate of decrease in temperature with height is less than the *adiabatic lapse rate.
2. A level in the atmosphere in which the temperature gradient changes sign.

tempering A process intended to alter the hardness of a metal, already heat-treated, by heating it to a lower temperature than that of the first treatment. *See* alloy.

tensile strength The *ultimate strength of a material as measured under tension.

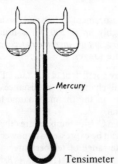

Tensimeter

tensimeter A device for measuring differences in vapour pressure. In the diagram water may be put in one bulb and an organic liquid in the other. Air is removed and the bulbs sealed off. The apparatus is then placed in a thermostat and the differential manometer enables the vapour pressure difference to be determined at the temperature of the thermostat. Since the vapour pressure of water is accurately known, the vapour pressure of the other liquid may be determined.

tensiometer A *torsion balance with a horizontal wire used for measuring the force needed to pull a flat horizontal ring of platinum-iridium from a liquid surface. It is used for determining the surface tension of liquids.

tension insulator *Syn.* strain insulator. A type of insulator used to support a line conductor in an overhead transmission-line system and designed to transmit the whole of the tension in the line conductor to the tower, pole, or other support. *Compare* suspension insulator.

tensor A point in a space of n dimensions can be specified by a set of n coordinates, $x_1, x_2, \ldots x_n$. The set can be written in the form x_i where i takes the values 1, 2, 3, $\ldots n$. If a transformation of coordinates is made, the x_i becomes x'_i in the new coordinate system. A set of n magnitudes can be formed:

$$A_1, A_2, \ldots A_n,$$

each one being a function of the original coordinates x_i. The set is denoted A_i. When the transformation occurs these change to A'_i. Such a set is a tensor if certain transformation laws hold.

Thus, if the transformation law:

$$A'_i = \partial x_j / \partial x'_i A_j,$$

holds, where each value of A_i is obtained by summing over $(j = 0, 1, \ldots n)$, it is called a *covariant tensor* of rank 1. A *contravariant tensor* of rank 1 is defined by the rule:

$$A'_i = \partial x'_i / \partial x_j A^j.$$

Here in A^j the j is not an exponent but a conventional way of denoting the components of the tensor. A tensor of rank r has n to the power r components. A tensor of rank zero is a *scalar and a tensor of rank 1 is a *vector. Tensor analysis is a generalization of vector analysis, although a tensor cannot be visualized and is an abstract mathematical entity.

If there is a tensor equation for a physical system in one coordinate system, it also holds on transformation to another coordinate system. This accounts for the importance of tensor analysis in physics, where it is used to express physical laws that remain valid on passing from one coordinate system to another.

The *metric tensor is g_{ik} in the relationship:

$$(ds)^2 = g_{ik} dx^i dx^j,$$

where $(ds)^2$ is the square of the distance between points x and $x + dx$ and the summation convention is used. The metric tensor of Riemannian space is used to represent gravitational potentials in the general theory of *relativity.

tenth-metre A unit of length formerly used in specification of wavelengths: equal to 10^{-10} metre.

tephigram A temperature-entropy diagram. It is used, for example, in recording the dynamical state of the atmosphere.

tera- Symbol: T. A prefix denoting 10^{12}; for example, one terametre (1 Tm) is equal to 10^{12} metres.

terminal 1. A device remote from and linked to a *computer, providing *input/output facilities. The most common terminal is a *visual-display unit paired with a keyboard. This has an additional built-in capacity to store and manipulate data, and is thus described as an *intelligent terminal. See also* time sharing.
2. Any of the points on an electronic circuit or device at which interconnecting leads may be attached and signals fed in and out.

terminal velocity The velocity with which a body moves relative to a fluid if the resultant force on the body is zero. From *Stokes's law the terminal velocity of a sphere falling in a fluid under gravity is:
$$2(\sigma - \rho)r^2g/9\eta,$$
where σ is the body density, ρ is the fluid density, r is the radius of the sphere, and η is the coefficient of the viscosity. *See also* Reynolds number.

termination 1. Of a *transmission line. A load impedance placed at the end of the transmission line to ensure *impedance matching and prevent unwanted reflection. **2.** *See* air termination; earth terminations.

terrestrial magnetism *See* geomagnetism.

tertiary winding An additional secondary winding on a *transformer. Some of its uses are: (i) Suppression of a third, ninth, fifteenth, etc. harmonics (i.e. *triplen harmonics) in the e.m.f.s of three-phase transformers in which the primary and secondary windings are both star-connected (*see* star connection). For this purpose the tertiary winding is delta-connected (*see* delta connection). This arrangement also improves the *regulation when the load is unbalanced. (ii) Supplying a load at a voltage different from that of the normal secondary or supplying a load that must be kept electrically insulated from that of the normal secondary. (iii) Supplying apparatus used for *power-factor improvement or control. (iv) Interconnecting supply systems that operate at different voltages.

tesla Symbol: T. The *SI unit of *magnetic flux density, defined as one weber of magnetic flux per metre squared.

Tesla, Nikola (1870–1943) Croatian-born Amer. physicist and inventor who developed a successful a.c. system and is remembered for the *Tesla coil. The SI unit of *magnetic flux density is named after him.

Tesla coil An apparatus for generating very high frequency currents at high potential. An induction coil or spark I, discharging across a spark gap, feeds the primary P of a transformer through two large capacitors C. The transformer is wound on a large open frame, the primary having only a few turns. The very high potentials obtained are at such high frequency that they have little physiological effect, and sparks of several feet in length can safely be drawn through the body. The frequency is also so high that nodes and antinodes of potential (*see*

lecher wires) are set up along the secondary coil. Tesla coils are used to test vacuum systems for leakage. If gas is entering the system, a Tesla coil will induce a *glow discharge in the gas as it enters the system.

tetragonal system *See* crystal systems.

tetrode Any electronic device with four electrodes, in particular a *thermionic valve in which an auxiliary grid is used. This grid is usually a *screen grid* designed to decrease the anode-grid capacitance and hence increase the resistance to high-frequency currents. It may also be used to decrease the anode-cathode resistance or to inject an a.c. potential into the main electron stream to modulate it.

TFT Abbreviation for *thin-film transistor.

theodolite A telescope fitted with spirit levels and angular scales for measuring altitude and azimuth; it is used in surveying, etc.

theorem 1. A universal or general proposition or statement, not self-evident (*compare* axiom) but demonstrable by argument.
2. A proposition embodying merely something to be proved as distinct from a problem that embodies something to be done.

theorem of parallel axes If the moment of inertia of a body of mass M about an axis through the centre of mass is I, the moment of inertia about a parallel axis distance h from the first axis is $I + Mh^2$. If the radius of gyration is k about the first axis, it is $\sqrt{(k^2 + h^2)}$ about the second.

theory of exchanges *See* Prévost's theory of exchanges.

therm A unit of heat used in the gas industry, equal to 100 000 *British thermal units.

thermal agitation The random movement of the molecules of a substance, the total energy of which (kinetic and potential) is the *internal energy.

thermal capacity A former name for *heat capacity.

thermal conductance *See* heat-transfer coefficient.

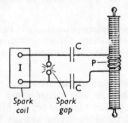

Tesla coil

thermal conductivity Symbol: λ, K, or k. The rate at which heat passes through a small area A inside the body is given by:
$$dQ/dT = -\lambda A \, dT/dx,$$
where dT/dx is the temperature gradient normal to the area A in the direction of heat flow, and λ is the thermal conductivity of the body at temperature T.

thermal diffusivity

The units are $J\,s^{-1}\,m^{-1}\,K^{-1}$. For good conducting solids, the thermal conductivity may be determined by the methods due to (i) Searle; (ii) Forbes; (iii) Lees for low temperatures; (iv) Jaeger and Diesselhorst for high temperatures. For crystals, the method due to Eucken may be used, while for a comparison of thermal conductivities the method due to Ingen-Hausz may be employed. For bad conducting solids and liquids, the Lees disc method is used. In the case of a liquid the heating must be done from above to prevent the transfer of heat by convection. For a gas, the conductivity may be determined using coaxial cylinders as used by Schleiermacher and by Gregory and Archer (*see* separate entries for these methods).

On the *kinetic theory, the thermal conductivity of a gas is given by:
$$\lambda = \tfrac{1}{3}\rho \bar{c} L C_v,$$
so that the conductivity is independent of the pressure. This is true for moderate pressures but at very low pressures the conductivity becomes proportional to the pressure, a fact made use of in the measurement of low pressures by the *Pirani gauge. The thermal conductivity of a solid metal is related to the electrical conductivity by the *Wiedemann–Franz–Lorenz law.

thermal diffusivity Symbol: a. The quantity defined by the expression $\lambda/\rho c_p$, where λ is the *thermal conductivity, ρ is the density, and c_p is the specific heat capacity at constant pressure. It has the units metre squared per second.

thermal effusion *See* thermal transpiration.

thermal equilibrium The condition of a system in which the net rate of exchange of heat between the components is zero.

thermal imaging The production of images using the infrared radiation emitted by objects. The radiation is usually collected by a two-dimensional array of infrared detectors, whose electrical signals are extracted by a scanning method and converted into a visible image.

thermal instrument An instrument that depends for its action upon the heating effect of an electric current.

thermalize To bring neutrons into thermal equilibrium with their surroundings. *Thermal neutrons can be produced by passing *fast neutrons through a *moderator.

thermal neutrons Neutrons that are approximately in thermal equilibrium with matter within which they are diffusing. They roughly obey the Maxwell *distribution of velocities, giving average kinetic energy $(3/2)kT$, most probable kinetic energy $(1/2)kT$, and most probable speed $v = \sqrt{(2kT/m)}$, where T is the thermodynamic temperature of the matter, k is the Boltzmann constant, and m is the mass of the neutron. At 20 °C v is 2200 m s⁻¹, and the value of kT is 4.05×10^{-21} J (0.0253 eV). Values of the *cross sections for nuclear reactions are commonly tabulated for the standard speed 2200 m s⁻¹. Thermal neutrons are sometimes loosely called *slow neutrons. *See also* thermalize.

thermal noise *See* noise.

thermal power station 1. A *power station in which electrical power is produced by combustion of a fossil fuel such as coal, coke, oil, etc.

2. A nuclear power station in which electrical power is produced by a thermal reactor (*see* nuclear reactors).

thermal radiation At all temperatures bodies emit radiant energy whose quantity and quality depend on the thermodynamic temperature T of the body. For any given body the radiation that depends only on temperature is known as thermal radiation. It is excited by the *thermal agitation of molecules or atoms, and its spectrum is continuous from the far infrared to the extreme ultraviolet. *See* black-body radiation.

thermal reactor *See* nuclear reactors.

thermal resistance *See* resistance (thermal).

thermal shield *See* nuclear reactors.

thermal transpiration *Syn.* thermal effusion. A phenomenon occurring when a temperature gradient exists in a tube containing gas at such a pressure that the *mean free path of the molecules is not negligible compared to the tube diameter. The pressure then is no longer uniform but is greatest at the high-temperature end. The term also applies to the case of a gas contained in two vessels at different temperatures connected by a porous medium whose holes are small compared with the mean free path of the molecules. In the steady state, the ratio of the pressures in the vessels equals the ratio of the square roots of the thermodynamic temperatures of the vessels.

thermionic cathode A *cathode that provides a source of electrons due to *thermionic emission. *Compare* photocathode.

thermionic emission The spontaneous emission of electrons from solids and liquids observed at high temperatures. According to the *Fermi–Dirac distribution function, a small number of the valence electrons have kinetic energies larger in magnitude than the negative potential energy within the substance, thus their total energies are positive. Such high-energy electrons can leave the surface spontaneously. The current emitted rises sharply with temperature (*see* Richardson–Dushman equation). The effect was first noticed by Edison and was formerly called the *Edison effect*.

In addition to electrons both positive and negative atomic or molecular ions may be emitted as a result of the presence of impurities within the substance or on its surface. Such particles and electrons are collectively called *thermions*.

See also Schottky effect.

thermionics The study of the emission of electrons from heated bodies and their subsequent behaviour and control, especially in vacuo as in the *thermionic valve.

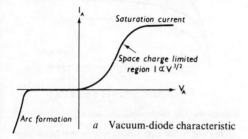

a Vacuum-diode characteristic

thermionic valve Any of a group of electronic devices once extensively used for a variety of purposes. Each one is a multielectrode evacuated *electron tube, containing a *thermionic cathode as the source of electrons. Thermionic valves containing three or more electrodes are capable of voltage amplification, the current flowing through the valve between two electrodes (usually the anode and the cathode) being modulated by a voltage applied to one or more of the other electrodes. Thermionic valves have rectifying characteristics, i.e. current will flow in one direction only (the forward direction) when positive potential is applied to the anode.

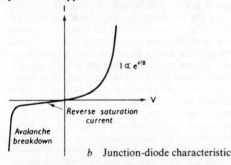

b Junction-diode characteristic

The simplest type of thermionic valve is the *diode*, which has been most often used in rectifying circuits. Electrons are released from the cathode by *thermionic emission. The cathode may be heated by a current passing through it (*direct heating*) or by surrounding it with a *heater coil (*indirect heating*). Under zero-bias conditions electrons released by the cathode form a *space-charge region in the vacuum surrounding the cathode, and exist in dynamic equilibrium with the electrons being emitted. If a positive potential is applied to the anode, electrons are attracted across the valve to the anode and current flows. The maximum available current (the *saturation current) varies as:

$$I_{sat} = AT^2 e^{-B/T},$$

where A and B are constant and T is the thermodynamic temperature of the cathode. The current does not rise rapidly to the saturation value as the anode voltage is increased, but is limited by the mutual

repulsion of electrons in the interelectrode region. This is the *space-charge limited* portion of the characteristic and the current obeys Child's law approximately where:

$$I \propto V_A{}^{3/2}.$$

V_A is the anode voltage. The motion of the electrons may also be affected by the magnetic field associated with the current flowing in the heater, and they will be deflected from a linear path. This effect is the *magnetron effect*, and contributes to the delay in reaching the saturation current.

Under conditions of reverse bias (*see* reverse direction), no current flows in the valve until the field across the valve is sufficient to cause *field emission from the anode or *arc formation, when *breakdown of the device occurs. The characteristics of a simple diode are shown (Fig. *a*). This can be compared with the characteristics of a simple *p-n junction diode, which is the solid-state analogue of the device (Fig. *b*).

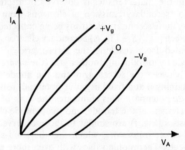

c Anode characteristics of a triode

The diode characteristic can be modified by interposing extra electrodes, called *grids* as they are usually in the form of a wire mesh, between the anode and the cathode. The simplest of these is the *triode with only one extra electrode, a *control grid*. Application of a voltage to the grid affects the electric field at the cathode, and hence the current flowing in the valve. A family of characteristics is generated for different values of grid voltage, similar in shape to the diode characteristic. The anode current, at a given value of anode voltage, is a function of grid voltage. Amplification may thus be achieved by feeding a varying voltage to the grid; comparatively small changes of grid voltage cause large changes in the anode current. In normal operation the grid is held at a negative potential, and therefore no current flows in the grid, as no electrons are collected by it. Anode and transfer characteristics of a triode are shown in Figs. *c* and *d*. Triodes have been extensively used in amplifying and oscillatory circuits.

A disadvantage of the triode is the large grid-anode capacitance, which allows a.c. transmission, and extra electrodes have been added to reduce this effect. Such valves are called *screen-grid valves*, the simplest of which are the *tetrode and *pentode. The tetrode has one extra grid electrode, the screen grid, placed between the control grid and the anode

thermions

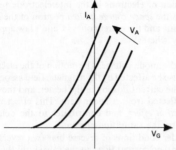

d Transfer characteristics of a triode

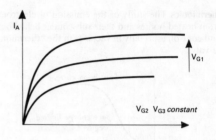

f Characteristics of a pentode

and held at a fixed positive potential. Some electrons will be collected by the screen grid, the number of electrons being a function of anode voltage. At high anode voltages the majority of electrons pass through the screen grid to the anode. An undesirable kink in the characteristics is observed in a tetrode due to *secondary emission of electrons from the anode, these secondary electrons being collected by the screen grid. (*See* Fig. *e*.) Secondary electrons are prevented from reaching the screen grid in the pentode by introducing another grid (the *suppressor grid*) between the screen grid and the anode and maintaining it at a fixed negative potential (usually cathode potential). This eliminates the kink of the characteristic of the tetrode, and the pentode characteristics (Fig. *f*) are similar to those observed in *field-effect transistors, which are the solid-state analogues. Thermionic valves with even more electrodes (*hexode, *octode, etc.) have been designed to produce particular characteristics and also multipurpose valves with an arrangement of electrodes such that the functions of several simpler valves are combined in a single envelope (diode-triode, etc.).

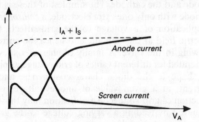

e Characteristics of a typical tetrode

Thermionic valves have been almost completely replaced by their solid-state equivalents. In applications requiring high voltages and currents, valves are still used, but these are special-purpose valves such as *cathode-ray tubes, *magnetrons, and *klystrons. For most applications, solid-state devices such as the p-n junction *diode, bipolar junction *transistor, and field-effect transistor (frequently in the form of *integrated circuits) have the advantages of small physical size, cheapness, robustness, and safety as the power required is very much less than for valves.

thermions *See* thermionic emission.

thermistor A *semiconductor device that has a large negative *temperature coefficient of resistance, and can be used for temperature measurement, or as a controlling element in electronic control circuits. It can also be used to compensate for temperature variations in other components.

thermoammeter An *ammeter that measures a current (a.c. or d.c.) in terms of its heating effect. For instance, the heating effect can be used to raise the temperature of a thermocouple connected to a sensitive galvanometer.

thermocouple An electrical circuit consisting of two dissimilar metals (or a metal and a semiconductor) joined at each end, in which an e.m.f. is produced when the two junctions are at different temperatures. This is due to the Seebeck effect (*see* thermoelectric effects). The thermocouple is often used as a temperature-measuring device: it can be used over a very wide temperature range, and is a conveniently usable device that can give temperatures at a very small point, and present readings at a considerable distance away if necessary. For the measurement of temperatures up to about 500 °C copper/constantan or iron/constantan thermocouples are used; at temperatures up to 1500 °C chromel/alumel or platinum/platinum plus 10% rhodium alloy is used, and at still higher temperatures, iridium/iridium-rhodium alloy.

The sensitivity of a thermocouple instrument is increased by connecting a number of junctions, in series, forming a *thermopile.

thermocouple pyrometer or thermometer A *pyrometer that depends for its action upon changes of e.m.f. that occur in a *thermocouple with changes of temperature.

thermodynamic potential A measure of the energy level of a system that represents the amount of work obtainable when the system undergoes a change. The main types of potential are the *internal energy (U), the *Helmholtz function defined as $(U - TS)$, the *enthalpy given by $(U + pV)$, and the *Gibbs function defined as $(U + pV - TS)$; the latter is sometimes referred to as the thermodynamic potential. This latter quantity is the maximum amount of external work that may be obtained during a reversible change at constant temperature and pressure. For a change at constant pressure and temperature this thermodynamic potential either decreases or, in the limit for a reversible change, remains constant. Hence a system in which both temperature and pressure are constant is in equilibrium when its thermodynamic potential $(U + pV - TS)$ is a minimum.

thermodynamics The study of the interrelation between heat, work, and internal energy. The thermodynamic state of a body is defined in terms of certain thermodynamic variables, for example, in the case of a simple homogeneous body such as a gas or a solid, in terms of pressure, volume, and temperature. There is generally an *equation of state:

$$f(p, V, T) = 0,$$

according to which the condition of a simple system is determined by fixing the values of two of the variables.

It must be borne in mind that p, V, and T are statistical quantities, for thermodynamics deals with systems consisting of very large numbers of particles and not with the behaviour of individual molecules. The *kinetic theory of heat is an attempt to relate the thermodynamic variables with the dynamic variables of the individual molecules. Thermodynamics is only concerned with changes of energy and not with the mechanism by which that change is brought about and its methods and results are therefore very general and independent of a particular mechanism. Thermodynamics is based on two fundamental laws, the 1st and 2nd laws, to which is sometimes added a 3rd law, more generally known as the *Nernst heat theorem (see also zeroth law of thermodynamics).

1st law of thermodynamics. This states simply that heat is a process of energy transfer and that in a closed system the total amount of energy of all kinds is constant. It is therefore the application of the principle of conservation of energy to include heat transfer. An alternative statement of the law is that it is impossible to construct a continuously operating machine that does work without obtaining energy from an external source. To obtain a mathematical interpretation of the law it is necessary to introduce the internal or thermodynamic energy U of the system as a function of the thermodynamic variables:

$$\delta Q = dU + \delta W,$$

where δQ is the heat absorbed by the system, dU is the increase in internal energy, δW is the work done by the system.

2nd law of thermodynamics. This deals with the question of the direction in which any chemical or physical process involving energy takes place. The formulation due to Lord Kelvin is that "it is impossible to construct a continuously operating machine which does mechanical work and which cools a source of heat without producing any other effects". The truth of this law is established by the correctness of the many deductions made from it that are capable of direct test. In nature, heat is never found to proceed up a temperature gradient of its own accord, this being one special case of the general truth expressed by the law, which may in fact be stated in the form: no self-acting machine can transfer heat continuously from a colder to a hotter body and produce no other external effect.

*Carnot's theorem, based on the 2nd law, states that the efficiency of a perfectly reversible engine, operating between given temperatures, cannot be exceeded by any other engine working under the same conditions, and consideration of the *Carnot cycle leads to the concept of *thermodynamic temperature. When a working substance is taken through a complete Carnot cycle the total change in *entropy of the universe is zero, and the entropy, S, of a substance is a definite function of its condition, just as its pressure, volume, temperature, or its internal energy. This result is a direct deduction from the second law, and may be regarded as a statement of that law. Thus for a perfectly reversible process:

$$\delta Q = TdS = dU + \delta W.$$

If, as is usually the case, the only external work results from a uniform pressure p, then:

$$TdS = dU + pdV.$$

This is a convenient mathematical statement of the first and second laws taken together.

For the third law of thermodynamics, see Nernst heat theorem. See also enthalpy; Gibbs function; Helmholtz function.

thermodynamic temperature Temperature measurement has been based upon many properties of substances, for example the expansion of a gas, the change in resistance, or the brightness of hot bodies. Kelvin was the first to propose a thermodynamic scale of temperature (1848) in which changes of temperature are independent of the working substance used.

Temperatures on this scale are defined so that if a reversible engine working on a *Carnot cycle takes up a quantity of heat q_1 at a temperature T_1 and rejects a quantity q_2 at T_2, then:

$$T_1/T_2 = q_1/q_2.$$

Kelvin showed that the scale so defined was identical to one based upon an *ideal gas. Using this concept, thermodynamic temperature is regarded as a physical quantity that can be expressed in the unit called the *kelvin, the triple point of water being defined as 273.16 kelvin. In practice thermodynamic temperature is measured on the *International Practical Temperature Scale.

thermoelectric effects A series of phenomena occurring when temperature differences exist in an electrical circuit.

(1) *Seebeck effect*: If two different metals (or a metal and a semiconductor) are joined and the two junctions are kept at different temperatures, an electromotive force is developed in the circuit. The circuit constitutes a *thermocouple. The e.m.f. is not affected by the presence of other junctions in the circuit if they are all maintained at the same temperature. The e.m.f. is given by the equation:

thermoelectric generator

$$E = \alpha + \beta\theta + \gamma\theta^2,$$

where θ is the temperature difference between the hot and cold junctions, and α, β, and γ are constants that depend on the substances comprising the circuit. γ is normally quite small so that for a small temperature difference the e.m.f. change is directly proportional to this difference.

(2) *Peltier effect: If a current is passed through a metallic junction or a metal-semiconductor junction, the junction is either warmed or cooled according to the direction of flow.

(3) Kelvin (or Thomson) effects: A potential difference is developed between different parts of a single conductor if there is a temperature difference between them. Also if a current passes through a wire in which a temperature gradient exists, this current causes a flow of heat from one part to the other – the direction of flow depending on the substance concerned. These Kelvin (or Thomson) effects are infinitesimal in the case of lead, so that it is often used as a reference metal in measuring the effects in other metals. See Thomson effects.

thermoelectric generator A device that converts heat directly into electric power by means of a *thermoelectric effect. An e.m.f. is developed between the junctions of a thermocouple, or between two regions of a metal, when a temperature difference exists between them. The e.m.f. is used to power an external circuit. The hot junction or region can be heated, for example, by the decay of a radionuclide.

thermoelectric series A series of metals arranged so that if a thermocouple is made from two of them, current flows at the hot junction from the metal occurring earlier in the series to the other metal.

thermogalvanometer See galvanometer.

thermograph A recording thermometer. See Bourdon tube.

thermoluminescence See luminescence.

thermomagnetic effect See magnetocaloric effect.

thermometer A device for measuring the temperature of a body, from a measurement of some property of the thermometric substance that depends on temperature. Various types are used for special purposes including (1) mercury in glass; (2) mercury in steel; (3) Beckmann; (4) clinical; (5) maximum and minimum; (6) bimetallic strip; (7) constant volume gas; (8) differential air; (9) vapour pressure; (10) platinum resistance (see separate entries). See also pyrometer; thermometric fluid; thermoelectric effects (Seebeck effect).

thermometric fluid A fluid suitable for use in a thermometer. It should possess as many as possible of the following properties:

(a) Marked degree of expansion for small temperature rise.

(b) Uniform expansion rate referred to the *International Practical Temperature Scale.

(c) Good thermal conductivity.

(d) Chemical stability.

(e) High boiling point (if liquid).

(f) Low freezing point (if liquid) or liquefaction point (if gas).

thermometry See temperature.

thermonuclear reaction See nuclear fusion.

thermophone A source of sound whose action depends on the heating of the air surrounding a conductor through which an alternating current is passed. The conductor is a strip of platinum or gold about 0.7 μm thick mounted between two terminal blocks. Alternating current passing through this strip causes periodic variations in its temperature and expansions and contractions of the air surrounding it. The corresponding pressure variations are radiated in the form of sound waves. The heating strip must have a very small heat capacity so that its temperature will accurately follow rapid current variations. The sound output of the thermophone is low but may be amplified with a resonator. When alternating current is superimposed on direct current, the sound has essentially the frequency of the a.c., but when a.c. alone is passed through the strip, the frequency of the sound is double that of the current. A thermophone is often used in a cavity filled with hydrogen, since the wavelength of a sound is much longer in hydrogen than in air so that there is less chance of *standing waves being produced for a given cavity size. The thermophone may be used as a source of sound in the pressure calibration of a microphone. A block of metal having a cylindrical cavity is fixed to the face of the microphone and the heating strip is placed in the enclosure so formed. The sound pressure produced may be calculated and the microphone output voltage measured.

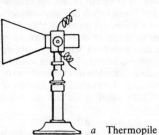

a Thermopile

thermopile A device consisting essentially of a large number of *thermocouples connected in series to give an easily measurable e.m.f. when heat radiation is allowed to fall on one set of junctions, the other set being shielded from the radiation. The metals are in the form of short fat rods for compactness, a trumpet being mounted to direct the radiation onto the hot junctions (Fig. a). The cold junctions either lie at the back of the apparatus or are protected from the incoming radiation in some other way (see solarimeter); they are at air temperature throughout. Such an instrument gives a current that is easily detectable.

One such arrangement of the bars is evident from the diagram (Fig. *b*), where thick lines represent mica insulation. If the bars marked A are of antimony and those marked B are bismuth, a current will flow from C to D through the bars in the directions shown by the arrows.

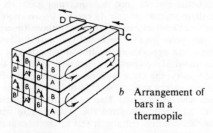

b Arrangement of bars in a thermopile

Thermos flask *Trademark* A commercial form of *Dewar vessel usually used for keeping hot things hot. A metal outer casing is used for protection against breakage.

thermosphere *Syn.* chemosphere. An *atmospheric layer of the earth, lying between the *mesosphere and the *exosphere, extending to an altitude of about 400 km. The temperature increases from approximately –50 °C at the mesopause to over 1000 °C in the highest layers. The pressure is extremely low decreasing from 10^{-6} bars to approximately 10^{-11} bars. The *ionosphere lies in and beyond this region, the ionization and absorption processes causing the high temperatures. *Auroras occur in the thermosphere at altitudes between 100–300 km.

thermostat A device that responds to changes of temperature and automatically actuates a mechanical valve, electric switch, etc. Common types depend for their action upon the variation with temperature of the expansion of a metal rod, the shape of a spring, or the pressure (and/or volume) of a gas. Thermostats are usually employed to regulate the supply of heat in situations where a substantially constant temperature is to be maintained.

theta pinch *See* fusion reactor.

thick-film circuit A circuit usually consisting only of interconnections and *passive components (e.g. resistors and inductors) that is fabricated on a film, up to about 20 μm thick, deposited on a glass or ceramic substrate. The film itself is a suitable glaze or cement, such as a ceramic/metal alloy. *Active components or devices on silicon chips may be wire-bonded to the thick-film circuit to produce a form of hybrid *integrated circuit.

thick lens (and **mirror**) A real lens, as distinct from a hypothetical infinitely thin lens. In a thick lens the separation between the two surfaces cannot be ignored; this may require the calculation of focal lengths, positions of principal planes, etc. A system involving lenticular refraction and reflection with thickness and separation to be taken into account, is technically described as a thick mirror.

thin-film circuit A circuit consisting of interconnections and components deposited on a glass or ceramic substrate usually by *vacuum evaporation or *sputtering. The deposited layers are up to a few micrometers thick. Components are usually passive, but active components such as the *thin-film transistor have been made.

thin-film transistor (TFT) An insulated-gate *field-effect transistor constructed by thin-film circuit techniques (i.e. *vacuum evaporation or *sputtering) on an insulating substrate rather than a semiconductor chip. The insulating substrate leads to fast switching speeds. The technique was originally used to produce discrete cadmium sulphide transistors but is now used mainly to construct silicon-on-sapphire MOS logic circuits.

thixotropy As defined by Freundlich, thixotropy is the reversible transformation of a gel to a sol, induced by mechanical vibration.

A gel possesses a structure that confers elasticity upon it. Agitation breaks down this structure but in thixotropic bodies, of which paints and printing inks are examples, the gel resets, on standing. (This *false body* is obviously a desirable property in a brushing paint.) If measurements of apparent viscosity are made on such substances and a graph of mean rate of shear against flow plotted, it is found that the curve obtained for increasing shear lies above that for decreasing shear, exhibiting mechanical hysteresis. The area between the curves represents the work done in disrupting the structure.

The phenomenon must be distinguished from the transformation that may be produced in a gel by change of temperature, for example, in warming and recooling a jelly.

Inverse thixotropy, which in special cases is called *dilatancy, is an isothermal reversible increase in viscosity with increasing rate of shear, e.g. wet sand becoming dry if walked on (but note that quicksand is thixotropic).

Thomson, Sir George Paget (1892–1975) Brit. physicist (son of J. J. Thomson). He was a professor at Imperial College and later at Cambridge. He discovered electron diffraction by crystals, for which he shared a Nobel prize (1937) with Clinton Davisson.

Thomson, Sir Joseph John (1856–1940) Brit. physicist. The creator of the nuclear physics school at the Cavendish Laboratory in Cambridge, "J.J.T." had earlier covered a wide field including very important researches on discharge phenomena in gases, culminating in his demonstration of the corpuscular nature of cathode rays and the existence of the electron with an unvarying negative electrical charge (1897). He later applied deflection methods to positive rays and sorted the constituents of a mixture into positive-ray parabolas, each corresponding to one definite ratio of charge to mass. This led to Aston's work on isotopes and the development of the mass spec-

Thomson, Sir William

trograph. He was succeeded at the Cavendish Laboratory in 1919 by Ernest (later, Lord) Rutherford, who had previously worked under him.

Thomson's ideas on the structure of the atom early in the century were superseded by Rutherford's nuclear model. (*See* atom.)

Thomson, Sir William *See* Kelvin, Lord.

Thomson coefficient *See* Thomson effects.

Thomson effects *Syn.* Kelvin effects. A potential difference is developed between different parts of a metal if there is a difference of temperature between them. If two points in a metal differ in temperature by an amount dT, the electromotive force in this element of metal is σdT, where σ is the *Thomson coefficient*. The quantity σ is positive when directed from points of lower to points of higher temperature. This is analogous with the flow of liquid in a tube heated at a point along its length; σ may be positive or negative and corresponds to positive or negative specific heat capacity. *See also* thermo-electric effects.

Thomson scattering *See* scattering.

thorium-lead dating *See* dating.

thorium series *See* radioactive series.

three-body problem The most important example of the *n*-body problem. The determination of the positions and motions of three bodies in a mutual gravitational field. No analytical solution can be obtained (except in special cases, e.g. where the bodies occupy the vertices of an equilateral triangle), but positions are determined from their previous values.

three-colour process *See* additive process; subtractive process.

three-phase Of an electrical system or device. Having three equal alternating voltages between which there are relative *phase differences of 120°. *See* polyphase system.

three-wire system A system of distribution of electric power used with direct current or with single-phase alternating current. It comprises three conductors: two *outers* and a middle (or neutral) wire and the consuming apparatus is connected between the middle wire and either outer conductor, or between the two outer conductors, different pieces of apparatus being connected by any one of these methods independently of the other apparatus. The voltage between the outer conductors is nominally twice the voltage between the middle wire and either outer conductor. The middle wire carries only the difference (vector difference with alternating current) between the currents in the outer conductors and is usually connected to earth. Its advantages over the *two-wire system are increased overall efficiency and the availability of two different voltages.

threshold energy The least energy required to bring about a certain process, in particular a reaction in nuclear or particle physics. It is often important to distinguish between the energies required in the laboratory and in centre-of-mass coordinates. For example, consider the collision of two identical particles, each of *rest mass m_0, that generates a particle-antiparticle pair. In centre-of-mass coordinates the least total energy required is the rest energy of the pair, $2m_0c^2$. The initial two particles have equal and opposite momenta and come to rest at threshold. If, however, one particle is at rest in the laboratory, the other must be projected at it with kinetic energy at least $6m_0c^2$ in order that momentum shall be conserved. Thus an electron that generates an electron–positron pair by colliding with another electron initially at rest in the laboratory must have a kinetic energy at least 3.06 MeV. A proton must have a kinetic energy at least 5.63 GeV if it is to generate a proton-antiproton pair by colliding with a stationary proton.

threshold frequency The minimum frequency giving rise to a particular phenomenon, such as *photoconductivity or the *photoelectric effect.

threshold of feeling That minimum intensity level of a sound wave that gives rise to a feeling of discomfort. It is reached at a loudness of about 120 phons (equal to 120 decibels at 1 kHz).

threshold of hearing That minimum intensity level of a sound wave that is audible. It occurs at a loudness of about 4 phons (equal to 4 decibels at 1 kHz).

threshold voltage The voltage at which a particular characteristic of an electronic device first occurs. For an insulated-gate *field-effect transistor, it is the voltage at which *channel formation occurs.

throwing power That quality of an electroplating bath that allows a reasonably even electrodeposit being made on a cathode of irregular shape without the use of a specially shaped anode. It is favoured by low electrolyte resistance and high cathodic polarization.

Thury system A system of transmission of electrical power using direct current at high voltage. The current is kept constant and variations in power are allowed for by varying the voltage of transmission. The lines are supplied at the generating station by *series-wound machines, which are all connected in series. The electrical power is utilized by means of series-wound motors connected in series to the lines. These motors usually drive generators that can supply direct or alternating current at any desired voltage. All the series-wound generators and motors have to be insulated from earth to withstand the maximum voltage of transmission. A particular disadvantage is that the line losses are constant at all loads and hence the efficiency is low at low loads.

thyratron *Syn.* gas-filled relay. A gas-filled *electron tube with three electrodes in which the voltage on one electrode (the grid) controls the starting of the discharge in the tube. A positive potential is applied to the anode, the potential being greater than the *ionization potential of the gas. A negative potential is applied to the grid, and, if sufficiently large, this neutralizes the effect of the anode potential at the thermionic cathode and prevents any current flowing. If the grid voltage is made less negative, the field at the cathode increases until the discharge starts. This is called *striking* the tube. Once the discharge has started, the grid has no further effect on the anode current even if the voltage is made very negative. At the instant of striking, the anode potential falls to approximately the ionization potential of the gas, and the discharge may only be stopped by reducing the anode potential below this value.

Thyratrons were formerly often used as relays and for counting radioactive particles. They have now been largely superceded by their solid-state analogue, the *silicon controlled rectifier.

thyristor *See* silicon controlled rectifier.

tidal energy *See* renewable energy sources.

tight-coupled inductor A very efficient inductor, in which the square of the mutual inductance is only slightly less than the product of the self inductances.

timbre The distinguishing quality, other than pitch or intensity, of a note produced by a musical instrument, voice, etc. The *quality is generally stated to be dependent upon the relative amplitude and number of the *partials and possibly of the *formant.

time Symbol: *t*. A fundamental quantity usually indicating duration – a period or interval of time – or a precise moment. Time is measured in *seconds in *SI units, but the minute, hour, day, and year may still be used. One day is defined as 24 hours, i.e. as 86 400 seconds. The length of the year depends on how it is measured. The practical measurement of time formerly depended on determining the period of rotation of the earth relative to the astronomical bodies. An atomic *clock and other types of clocks are now used for more precise measurements of time

Apparent solar time is measured by successive intervals between transits of the sun across the meridian, and is shown on a sundial. *Mean solar time* averages out this interval over the course of one year, and is measured with reference to the motion of the *mean sun*. This is a point that moves uniformly around the celestial equator (*see* celestial sphere) in the same total time as the real sun takes in its apparent motion (which is not uniform) round the ecliptic. The difference on any day between apparent and mean solar time, up to 16 minutes, is known as the *equation of time*. The provision of a uniform timescale using mean solar time was based on the assumption that the earth's rotation rate is constant. The rotation rate is now known to vary very slightly and irregularly.

Sidereal time, used in astronomy, is measured in terms of the *sidereal day*; this is the interval between successive transits of a point (related to the vernal

*equinox) across a given meridian. It is thus also based on the earth's rotation rate. The sidereal day is about 3 minutes 56 seconds shorter than the 24-hour day. A star will be on the meridian of an observatory when the local sidereal time becomes equal to the star's right ascension (*see* celestial sphere).

Greenwich Mean Time (GMT) is determined from transits of certain stars across the Prime Meridian (of zero longitude) at Greenwich. The coordinates of the stars are known, and this allows a correction to be made to the time given by a sidereal clock and hence obtain GMT.

Universal coordinated time is the mean of that provided by sets of atomic clocks distributed around the world. It thus provides a timescale that changes uniformly. It is kept within one second of GMT by the insertion or deletion of *leap seconds* when the earth's irregular rotation necessitates this (usually made at the end of December).

The *year* is the time taken by the earth to make one complete orbit of the sun, measured with respect to a given point. The *sidereal year* is measured with respect to a particular star regarded as fixed in position; it is equal to about 365.256 36 days. The *tropical year* is the interval between successive passages of the sun through the vernal equinox; it is equal to about 365.242 19 days. Since the equinox is not a fixed point, the tropical and sidereal years are different.

The *calendar year* (or *civil year*) is adjusted so that its average length is very close to that of the tropical year; for practical reasons the calendar year must contain a whole number of days. In the *Gregorian calendar* used almost worldwide, the calendar year has 365 days, plus an extra day in leap years, but the century years are not leap years unless the number of the century is exactly divisible by 400. On average, the calendar year is equal to 365.2425 days. The *Julian calendar* made all century years into leap years, so that the calendar year was exactly 365.25 days, and by 1752, when the Gregorian calendar was adopted in England, September 2 had to be followed by September 14 because of the accumulated error.

See also arrow of time; time reversal.

time base A voltage that is a predetermined function of time and is used to deflect the electron beam of a *cathode-ray tube so that the luminous spot traverses the screen in the desired manner. One complete traverse of the screen is called a *sweep* (or sometimes a time base). The most common type of time base is one that produces a linear sweep; a *sawtooth waveform is usually employed for this so that at the end of each useful trace the luminous spot is returned rapidly to its starting point. The return of the spot to its starting point is called the *flyback*.

time constant Physical quantities such as voltage, current, and temperature sometimes decrease with time in such a manner that, at any instant, the rate of change of the quantity is equal to minus the value of the quantity divided by a constant (called the time constant). Thus, in mathematical terms,

$$-(\mathrm{d}v/\mathrm{d}t) = v/T, \qquad (i)$$

where $\mathrm{d}v/\mathrm{d}t$ = the instantaneous rate of change of the quantity, v = the instantaneous value of the quantity, T = the time constant. In this case, the time constant is the time taken for the quantity to decrease to $1/e$ (approximately 0.368) of its initial value. Alternatively, a quantity may increase with time in such a manner that, at any instant, the rate of change of the quantity is equal to the difference between the ultimate and instantaneous values of the quantity divided by a constant (also called the time constant). This may be represented by:

$$\mathrm{d}v/\mathrm{d}t = (V - v)/T, \qquad (ii)$$

where $\mathrm{d}v/\mathrm{d}t$ = the instantaneous rate of change of the quantity, v = the instantaneous value of the quantity, V = the ultimate value of the quantity (a constant), T = the time constant. In this case, the time constant is the time taken for the quantity to increase from zero to $1 - 1/e$ (approximately 0.632) of its ultimate value.

The time constant is particularly important in connection with electrical circuits. For example, in a circuit containing either (*a*) a resistance in series with a capacitance, or (*b*) a resistance in series with an inductance, which is connected suddenly to a constant-voltage d.c. supply, the component voltages, current, and charge (as appropriate) vary as in (i) or (ii) above. The time constant in seconds for circuit (*a*) may be calculated by multiplying the resistance by the capacitance, and for circuit (*b*) by dividing the inductance by the resistance. (Note. e = 2.718..., the base of natural logarithms.) *See also* transducer.

time delay *See* time lag.

time dilation *See* relativity.

time-division multiplexing A method of *multiplex operation in which each of the input signals is sampled and transmitted sequentially; the transmission channel is thus shared by allocating specific time intervals to each signal. The time of switching from one to the next must be such that each signal is sampled many times in the course of a cycle. *Pulse modulation is frequently used for time-division multiplexing.

time lag *Syn.* time delay. The time that elapses between the closing of one circuit in a circuit breaker, relay, or similar apparatus, and the response of the current in the main circuit. If the delay time is independent of the magnitude of the quantity (e.g. overcurrent) that causes operation, it is a definite time lag (the delay time may be adjustable). If the delay time is inversely dependent upon the magnitude of the quantity that causes operation, it is an inverse time lag. (*See* overcurrent release.)

time-lapse photography A means of obtaining a speeded-up version of a slow process, such as a flower opening, by recording single exposures, taken at regular intervals, on ciné film without moving the camera and then projecting the film at normal speed.

time reversal Symbol: T. The substitution of time *t* by time *–t*, the symmetry of which is known as *T invariance*. As with CP violation (*see* CP invariance), T violation occurs in *weak interactions involving kaon decay. *See also* CPT theorem.

time sharing A technique whereby the processing time of a *computer is shared among several jobs by means of rapid switching between them. For example, in a *multiaccess system* a number of computer users communicate with the machine seemingly simultaneously via individual *terminals. The speed of the machine gives each user the impression that he has sole use of it. *See also* interactive.

time switch A switch that incorporates a type of clock mechanism for making and/or breaking an electric circuit at times that are predetermined by the setting of the mechanism.

time to flashover *See* impulse voltage (or current).

time to puncture *See* impulse voltage (or current).

Ting, Samuel Chao Chung (*b.* 1936) Amer. physicist. Ting worked on the synchrotron at the Brookhaven National Laboratory, New York, where he discovered (1974) the *J/psi particle in experiments involving proton collisions with beryllium. He shared the 1976 Nobel prize with Burton *Richter, who had independently made the same discovery.

tint *See* colour.

T invariance *See* time reversal.

Tizard, Sir Henry (1885–1959) Brit. physical chemist who did valuable research on aircraft fuels. He is best known for his work in organizing scientists in World War II, especially in the development of *radar.

TM modes *See* waveguide.

T-number A modified (increased) *f-number that a real lens would have to have in order to transmit an amount of light corresponding to its calculated f-number. The T-number takes into account the reflection and absorption losses that occur in practice with a lens.

toe and heel *See* brush.

Toepler points *See* negative principal points.

Toepler pump A mercury-in-glass pump, worked by hand. It was originally used for producing high vacua but is now obsolescent for that purpose due to its slow speed.

tokamak An arrangement for confining plasma in a *fusion reactor, which shows promise for achieving controlled energy release from nuclear fusion. Devised in the USSR in the 1960s, the device combines two principles: the torus, which has no ends through which the plasma can escape, and a magnetic field, which spirals around this toroidal plasma.

A spiral (or helical) magnetic field is achieved by adding together magnetic fields produced by toroi-

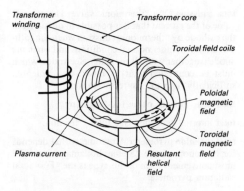

Transformer winding

Transformer core

Toroidal field coils

Poloidal magnetic field

Toroidal magnetic field

Plasma current

Resultant helical field

Principal magnetic fields of a tokamak

dal coils (along the plasma) with the magnetic field (around the plasma) resulting from the strong plasma current driven by a transformer core. The principle of the tokamak is illustrated in the diagram.

Whether a demonstration power-producing tokamak will be built will depend upon the results of major experiments being carried out in the UK, USA, Japan, and the Soviet Union.

tolerance The maximum permissible error or variation permitted in a component, etc.; e.g. ±0.05 mm in the divisions of the millimetre scale is permissible if subdivisions of the millimetre are made by eye estimations.

Tolman and Stewart effect When a metal rod in rapid longitudinal motion is suddenly stopped, it becomes negatively charged at the forward end, suggesting that electrons pile up at the front.

tomography A technique for producing a three-dimensional image of the internal structure of a specimen. The most widely used method involves X-rays and is used in medicine. A narrow beam of X-rays is swept across an area of the body and the signals from a range of surrounding detectors are recorded. It is possible to construct, by computer, an image of the cross section of the body and, by scanning the beam, to produce a three-dimensional image of the internal structure. This technique was developed in the early 1970s by Godfrey Hounsfield and is called *computerized tomography* (*CT*) or *computerized axial tomography* (*CAT*).

Since then, various other tomographic techniques have been developed. *NMR tomography* (or *nuclear magnetic resonance tomography*) uses *nuclear magnetic resonance signals to produce the image. It has the advantages that the dangers of X-rays are avoided and that structures that are transparent to X-rays can be revealed. Another technique, *positron-emission tomography* (or *PET*), involves emission of positrons from a radionuclide injected into the body. Annihilation of positrons and electrons causes emission of gamma rays, which are recorded by the external detectors. PET is used particularly for investigating brain and heart conditions.

Tomonaga, Sin-Itiro (1906–1979) Japanese theoretical physicist who worked at Tokyo. Tomonaga was one of the first to introduce the idea of *virtual particles and to develop a theory of *quantum electrodynamics. He shared the 1965 Nobel prize for physics with Richard Feynman and Julian Schwinger.

tone 1. An audible note containing no *partials.

2. The quality of a musical sound, e.g. loud tone, soft tone, thin tone, poor tone, good tone; also tone colour.

3. The interval of a major second, e.g. C to D, as opposed to a *semitone. Any of the larger intervals in the diatonic scale (*see* musical scales).

tone control A means of adjusting the relative frequency response of an audiofrequency amplifier used in the reception or production of sound in order to achieve a more pleasing result.

tonne *Syn.* metric ton. Symbol: t. A commonly used metric unit of mass for engineering problems. 1 t = 1000 kg. It differs in size from an imperial ton (1016 kg) by about $1\frac{1}{2}\%$.

tooth *See* slot.

topology The branch of geometry concerned with continuity in geometrical figures, i.e. with those properties of a figure that remain unchanged (*topologically equivalent*) when a figure is bent, stretched, or shrunk, but not when it is torn or deformed by the fusion of points on it. For example, a torus and a cup are topologically equivalent (because the cup's handle and the torus both have a single hole through them), whereas a torus and a sphere are not. Topology has many applications, including the theory of liquid crystals and the classification of *magnetic monopoles in *gauge theory.

toric (toroidal) lenses (and **surfaces**) The most general definition of a toroidal surface is that it is a surface generated by the rotation of an arc of a circle about a line in its own plane. From a technical point of view the special toric surfaces so defined, viz. spheres and cylinders, are considered separately, and the word *toric* is used with a restricted meaning for the remainder. Again, from a practical point of view, only those toric surfaces that have the same aspect of curvature (but different in the various meridians) are considered, i.e. the axis of revolution lies on the same side as the centre of the generating circle. Spectacle toric lenses are made with so-called base curves to reduce the number of working tools. The base curve is the lowest power of the surface. The astigmatic difference, which it is the main function of the lens to produce, is the difference in power between the meridian of the lowest power (base) and the meridian at right angles. Toric lenses are produced to provide a choice of form for astigmatic lenses in an analogous way to which deep meniscus lenses provide better peripheral vision in the case of the spherical lenses. Ideally, they should provide the same astigmatic difference at the periphery in oblique vision to that which obtains at the centre.

toroid *See* anchor ring.

toroidal winding *See* ring winding.

torque *Syn.* moment of a force about an axis or of a couple. *See* moment.

torquemeter *Syn.* dynamometer. An apparatus for measuring the torque exerted by the rotating part of a prime mover, electric motor, etc. It may incorporate a brake to absorb the mechanical output of the machine, etc., in which case it is sometimes called a brake dynamometer.

torr A unit of pressure formerly used in vacuum technology. It is equal to 133.322 pascals.

Torricelli, Evangelista (1608–1647) Italian physicist remembered for his work on *barometers and his law of fluid flow. He also built a primitive microscope.

Torricellian vacuum *See* barometer.

Torricelli's law The velocity of efflux of a fluid from an orifice in a reservoir is $\sqrt{(2gH)}$, where H is the depth of the orifice – more precisely of the *vena contracta below the free surface of the reservoir.

torsional hysteresis *See* hysteresis.

torsional vibrations Vibrations in a body, usually a cylindrical bar or tube, in which the displacement is in the form of a twist due to the application of an alternating torque to one end of the body, the other end being clamped.

torsional waves Waves formed in a medium, usually a cylindrical bar or tube, as the result of the application of torsional vibrations to one or more parts of the medium. The velocity, c, of torsional waves is given by $c = \sqrt{(G/\rho)}$, where G is the modulus of rigidity (*see* modulus of elasticity), and ρ is the density.

torsion balance A very sensitive *balance consisting of an arm attached to a fibre; when a force is applied to the arm, the fibre twists until the torque on the arm is balanced by that in the fibre. Since the latter is proportional to the angle of twist, this quantity is determined by the use of a light pointer. Torsion balances are employed in the measurement of small forces such as those associated with surface tension and static charges.

torsion meter *See* dynamometer.

total angular momentum quantum number A number, integral or half-integral, that characterizes the total angular momentum (*orbital angular momentum plus *spin) of an atom, nucleus, particle, etc. The symbol j is used for a single entity, J or j_i for a whole system.

total curvature *Syn.* Gaussian curvature. *See* principal radii.

total emission Of a *thermionic valve. The greatest value of the current that it is possible to obtain from the cathode by *thermionic emission when the cathode is heated under normal conditions and when the anode (together with all the other electrodes, which must be connected to it) is raised to a potential sufficiently high to ensure saturation.

total heat Former name for *enthalpy.

total internal reflection *See* total reflection.

total radiation pyrometer A pyrometer that depends for its action on the *Stefan–Boltzmann law. The most convenient form of this type is the *Féry total radiation pyrometer.

total reflection *Syn.* total internal reflection. A phenomenon occurring when light strikes the surface of an optically less dense medium at an angle of incidence greater than the *critical angle: instead of emerging into the less dense medium, it is reflected back into the optically denser (incident) medium. Total-reflection prisms have a wide application in instruments for changing direction of rays, for producing lateral inversion, or for completing inversion of the image. (*See* prism; Porro prism; Dove prism; Amici prism.) Although the reflection is total, the light actually penetrates a small distance into the rarer medium.

Total reflection also occurs with other waves: e.g. with sound, at moderately oblique incidence from air to water, and with X-rays, at almost grazing incidence from air to a solid or liquid.

total relief Sometimes called stereo-power of a binocular instrument. The product of base magnification and telescope magnification. (The base magnification is the ratio of the interobjective distance to the interocular distance.)

tourmaline A mineral crystal that exhibits dichroic properties, i.e. doubly refracting with selective absorption of the ordinary ray while transmitting the extraordinary ray. A pair of tourmalines can be adapted to serve as analyser and polarizer (tourmaline clips) for rough work with small fields.

Townes, Charles Hard (*b.* 1915) Amer. physicist. A professor at Columbia and later at the University of California (Berkeley), he developed the maser (1953), for which he shared a Nobel prize (1964) with Nicolai Basov and Alexandr Prokhorov, who had developed it independently in the Soviet Union in 1955. Townes predicted the development of the laser in 1958, but failed to construct a working laser before *Maiman (in 1960).

Townsend, Sir John Sealy Edward (1868–1957) Irish physicist who is remembered for his work on ionization and conduction in gases.

Townsend discharge *See* gas-discharge tube.

trace 1. A figure traced out by the luminous spot on the screen of a *cathode-ray tube.
   2. *See* spur.

tracer A substance introduced into a moving system so as to be able to follow part of the action of the system. The substance may be radioactive, or have another characteristic property, and is followed with the aid of an appropriate detector. *See* radioactive tracer.

track An illuminated track of an ionizing particle created in a *cloud chamber, on a photographic film, or in a *bubble chamber.

tracking 1. Of any two electronic devices or circuits. An arrangement by which an electrical parameter of one device or circuit varies in sympathy with the same or another parameter of the second one, when both are subjected to a common stimulus. In particular, it is the maintenance of a constant difference in the *resonant frequencies of two ganged tuned circuits.
2. The formation of unwanted electrically conducting paths (often by carbonization) on the surface of solid dielectrics and insulators, when subjected to high electrical fields.

track-return system A system for the distribution of electric power to trains and vehicles, in which the track rails are utilized as an uninsulated return conductor. *Compare* insulated-return system.

traction meter *See* dynamometer.

trammel A device having a rigid bar that moves so that each of two of its points is constrained, by suitable guides, to move along one of two perpendicular lines. It is used (*a*) in drawing ellipses and (*b*) in the *Rowland mounting of a concave *diffraction grating.

transconductance *See* mutual conductance.

transducer *Syn.* sensor. Any device for converting a nonelectrical signal into electrical signals (or vice versa), the variations in the electrical signal being a function of the input. Transducers are used as measuring instruments and in the electroacoustic field, the term being applied to gramophone pick-ups, microphones, and loudspeakers.
The physical quantity measured by the transducer is the *measurand*, the portion of the transducer in which the output originates is the *transduction element*, and the nature of the operation is the *transduction principle*. The device in the transducer that responds directly to the measurand is the *sensing element* and the upper and lower limits of the measurand value for which the transducer provides a useful output is the *dynamic range*.
Several basic transduction elements can be used in transducers for different measurands. They include capacitive, electromagnetic, inductive, photoconductive, photovoltaic, and piezoelectric elements. Most transducers require external electrical excitation for their operation; exceptions are self-excited transducers such as piezoelectric crystals, photovoltaic, and electromagnetic types.
Most transducers provide linear (analogue) output, i.e. the output is a continuous function of the measurand, but some provide digital output in the form of discrete values. Most transducers are designed to provide output that is a linear function of the measurand as this allows easier data handling. If the measurand varies over a stated frequency range, the output of the transducer varies with frequency. The frequency response of the transducer is the change with frequency of the output/measurand amplitude ratio. The portion of the response curve over which attenuation of the measurand is significant is the *rejection band* or *attenuation band* of the transducer. The time taken for a change in output to reach 63% of the final steady value in response to a change in the measurand is the *time constant* of the transducer.

transductor *Syn.* saturable reactor. A device that is used in control circuits and consists of a number of windings on a magnetic core. Usually a standing current in one winding is adjusted to bring the core to a magnetic state in which small changes in the current of one of the other windings (the signal winding) can control large powers in coupled circuits. Variations in the control circuit supplying the signal winding must be slow relative to the frequency of supply current, but it is possible to use up to 2000 hertz and thus to control signals in the lower audiofrequency range. The instrument is of robust construction, without moving parts, and in consequence has been widely used in aircraft.

transfer characteristic The relation between the current (or the voltage) at one electrode of an amplifier, transducer, or other electronic device or network and the voltage (or current) at a different electrode; it is usually shown in graphical form. The tangent at any point on a given transfer-characteristic curve gives the associated *transfer parameter* at that point. The transfer parameter most commonly used is the *mutual conductance, determined from the characteristic relating output current and input voltage.

transference number *Syn.* transport number. *See* migration of ions.

transformation *See* transition.

transformer An apparatus without moving parts for transforming electrical power at one alternating voltage into electrical power at another (usually different) alternating voltage, without change of frequency. It depends for its action upon mutual induction (*see* electromagnetic induction) and consists essentially of two electric circuits coupled together magnetically; the usual construction is of two coils (or windings) with a laminated core suitably arranged between them. One of these circuits, called the *primary*, receives power from an a.c. supply at one voltage, and the other circuit, called the *secondary*, delivers power to the load at (usually) a different voltage.
If core losses are ignored, the ratio of primary to secondary voltage is equal to the ratio, n, of the number of turns in the primary winding to the number in the secondary winding. The transformer

transformer coupling

is described as *step-up* or *step-down* according to whether the secondary voltage is respectively greater or less than the primary voltage. Apart from the property of voltage transformation, it has the property of current transformation: if core losses are ignored, the ratio of primary to secondary current is equal to $1/n$. *See* autotransformer; voltage transformer; current transformer; core-type transformer; shell-type transformer.

transformer coupling *See* interstage coupling.

transients Temporary disturbances in a system resulting from the sudden incidence of an impulse voltage (or current) or the application or removal of a driving force. The form of any such transient is characteristic of the system, but its magnitude is a function of the magnitude of the impulse or driving force. The persistence of the transient is controlled by the dissipative components of the system. An example of the production of transients is that of *forced vibrations.

For a mechanical system, a transient may be discerned in the particle displacement, the particle velocity, and the pressure at any given point; the corresponding quantities in an electrical system are the charge, the current, and the voltage at a particular point. *Consonants are examples of transient sounds.

a Point contact transistor

transistor A multielectrode *semiconductor device in which the current flowing between two specified electrodes is modulated by the voltage or current applied to one or more specified electrodes. The semiconductor material is usually silicon. Transistors are small robust cheap devices requiring small supply voltages. They have replaced *thermionic valves as the general-purpose *active components in electronic circuits except for some specialized uses.

The first transistors were invented in 1948. These were *point-contact transistors*, which are now obsolete. The point contact transistor consisted of a small crystal of semiconductor (usually germanium) with two rectifying point contacts attached in close proximity to each other and a single large-area ohmic contact at some distance from the point contacts (Fig. *a*). On the application of small voltages, current flowed in the device and could be modulated by signals fed onto the ohmic contact. The *junction transistor* was developed in 1949 and its action was fully described by Shockley in 1950. *Field-effect transistors were developed more recently and have a different principle of operation to the point contact or junction transistors.

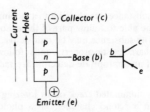

b A *p–n–p* bipolar junction transistor

Modern transistors fall into two main classes: bipolar devices, which depend on the flow of both *minority and *majority carriers through the device, and unipolar transistors (*see* field-effect transistors) in which current is carried by majority carriers only.

Bipolar junction transistors (usually simply called *transistors*). The basic device consists of two *p-n junctions in close proximity, with either the n or p regions common to both junctions (Figs. *b* and *c*) and so forming either *p-n-p* or *n-p-n* transistors respectively; the latter is most commonly used. The central region is called the *base* and the electrode attached to it is the *base electrode*.

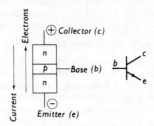

c An *n–p–n* bipolar junction transistor

The *energy bands of an n-p-n transistor under zero-bias conditions are shown in Fig. *d*. If a voltage is applied across the transistor, one junction becomes *forward biased and the other junction *reverse biased. Current flows across the forward-biased junction; electrons (the majority carriers) from the n-type region cross into the base, and *holes (the minority carriers) from the base cross into the n-region. This junction is called the emitter-base junction, the n-region being the *emitter*. The electrons entering the base diffuse across it towards the reverse-biased junction. Once they enter the *depletion layers associated with this junction, they are swept across into the other n-region, which is called the *collector*, entering the lower energy bands associated with this electrode.

The hole concentration in the base falls due to holes entering the emitter across the forward-biased junction and to holes recombining with injected electrons from the emitter as they diffuse across the base. This will reduce the forward voltage across the junction until the current ceases. If the base region is connected to a suitable point in the circuit so that the emitter-base junction remains forward biased, electrons flow out of the base region to maintain the hole concentration and current continues to flow

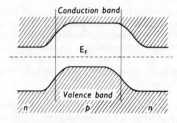

d Energy bands in an *n–p–n* transistor under zero bias

across the device, from collector to emitter (i.e. in the opposite direction to the flow of electrons). The total current flowing will be related by:
$$I_e = I_b + I_c,$$
where I_e is emitter current, I_b base current, and I_c collector current.

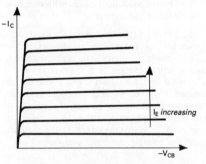

e Common-base output characteristics of a junction transistor

Voltage amplification is possible using the emitter as the input terminal. This operation is called *common base connection; the characteristics are shown in Fig. *e*. For efficient amplification, the collector current must be as nearly equal to the emitter current as possible; this may be achieved by reducing the base current to as low a value as is practicable. Several factors contribute to the base current, the most important being holes crossing into the emitter and holes recombining with electrons in the base. Increasing the *injection efficiency of the emitter reduces the hole current as most of the current across the junction is due to electrons from the emitter. This is achieved by making the concentration of charge carriers in the emitter much larger than in the base, i.e. a high *doping level* is used in the emitter. Compared to the base the emitter is therefore an n⁺ semiconductor. The recombination current in the base is reduced by making the base region narrower than the diffusion length of the minority electrons so that most of them reach the collector before recombination occurs. The concentration of carriers through the base may also be altered by varying the doping level through the base region (Fig. *f*). This alters the energy levels throughout the base as in Fig. *g*. Since the Fermi level must be constant throughout the material, the conduction

bands bend and assist the passage of electrons through the material. Such a transistor is called a *graded-base* transistor.

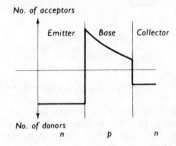

f Variation of doping level in a graded-base transistor

The *common-base current gain* or *collector efficiency*, α, is given by I_c/I_e, and is a function of a particular device. The base current I_b is equal to $(1-\alpha)I_e$. For any given transistor the ratio:
$$I_e:I_b:I_c = 1:(1-\alpha):\alpha,$$
is constant.

Current amplification is possible by applying the input signal to the base rather than to the emitter. Provided the emitter-base junction is always forward biased, carriers travel through the device as described above, but a variation in the base current causes a corresponding change in the collector current to maintain the ratio $\alpha:(1-\alpha)$. This is known as *common emitter connection, and is the most usual method of operating the device. The characteristics are shown in Fig. *h*. The *beta current gain factor is defined as:
$$\beta = I_c/I_b = \alpha/(1-\alpha).$$
Since α can approach unity, β can be very large. Large collector currents are therefore generated by small base currents. This leads to saturation of the transistor as the current that can flow out of the collector is limited by the components in the external circuit. When the collector current is saturated, the collector-emitter voltage drops to a small value; this is the *saturation voltage of the device. It is dependent on the value of base current and the external circuit components. If the base current falls, the transistor ceases to be saturated when $\beta I_b < V/R_L$ (where R_L is a load resistor connected between the collector and the power supply V) and the collector voltage rises to a value determined by R_L. The transistor may therefore be used as a switch by driving it into saturation using the base current as the driver. It has many switching applications in *digital circuits (*see* transistor-transistor logic; diode transistor logic).

One of the most important differences between holes and electrons is the difference between their mobilities, electrons being about three times as mobile as holes. This makes devices depending mainly on electron flow much faster in operation than those depending on the flow of holes, and capable of being used at higher frequencies. This accounts for some

579

transistor parameters

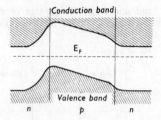

g Energy levels in a graded-base transistor

of the small differences between n-p-n and p-n-p transistors. However, in principle they are the same, except that holes replace electrons in the above description.

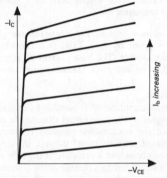

h Common-emitter output characteristics of a junction transistor

Transistor types. Bipolar junction transistors rapidly replaced the original point contact transistors. The different types of junction transistor are classified by the manufacturing method used, many slightly different types of transistor having been produced. Three basic junction-forming methods have been used; *grown junctions (an obsolete method), *alloyed junctions, and *diffused junctions. The most commonly used method is the diffused junction technique.

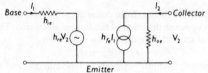

a Two port network

transistor parameters A transistor is a nonlinear device whose behaviour is difficult to represent exactly by a set of mathematical equations. When designing transistor circuits the behaviour of the transistor is represented approximately by *equivalent circuits that act as models of the device. The particular equivalent circuit used will be the one that is most appropriate for the type of circuit being designed (i.e. for use with large signals, small signals, as switches, etc.).

(i) *Matrix parameters.* The transistor is represented by an equivalent circuit with two input terminals and two output terminals (Fig. *a*). This is a quadri-

580

pole network. Over a small portion of its operating characteristic the device is assumed to behave linearly. This is particularly true in small-signal operation, but is only an approximation for large-signal operation. The input and output voltages and currents are related by two simultaneous equations of the general matrix form:

$$[A] = [p][B],$$

where A and B represent current or voltage and p the particular transistor parameter used. The most common matrix parameters are the $h, y, z,$ and g parameters.

b Hybrid equivalent circuit with common-emitter connection

The h or *hybrid* parameters are the most commonly used for bipolar junction *transistors; they are so called because the dimensions are mixed and a true matrix representation is not possible. A typical equivalent circuit is shown in Fig. *b* with the transistor used in the *common emitter configuration. The transistor is assumed to consist of a voltage source with series impedance at the input and a current source with shunt impedance at the output. The equations relating the input and output are:

$$V_1 = h_{11}I_1 + h_{12}V_2;$$
$$I_2 = h_{21}I_1 + h_{22}V_2.$$

The values of the constants h_{11} etc. are characteristic of the transistor used and may be measured. If the output terminals are short-circuited ($V_2 = 0$) then:

$$h_{11} = V_1/I_1 \quad \text{(an impedance)};$$
$$h_{21} = I_2/I_1 \quad \text{(dimensionless)}.$$

If the input terminals are open circuited ($I_1 = 0$) then

$$h_{12} = V_1/V_2 \quad \text{(dimensionless)};$$
$$h_{22} = I_2/V_2 \quad \text{(an admittance)}.$$

A standard nomenclature has been adopted:

$h_i = h_{11}$ = input impedance with output short-circuited;

$h_r = h_{12}$ = reverse voltage feedback ratio;

$h_f = h_{21}$ = forward current transfer ratio;

$h_o = h_{22}$ = output admittance with input open-circuited.

The method of connecting the transistor in the circuit is indicated by a second subscript. These are b for *common base connection, e for common emitter connection, and c for *common collector connections; e.g. h_{fe} = common emitter forward-current transfer ratio.

The y parameters are frequently used for circuits containing *field-effect transistors. The equivalent circuit is a current source with shunt impedance at both input and output (Fig. *c*). The simultaneous equations corresponding to the circuit are:

$$[I] = [y][V],$$

where y has dimensions of admittance.

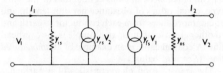

c   *y*-Equivalent circuit with common-source connection

Other parameters used are z parameters, which are impedance parameters and are the inverse of y parameters (voltage sources with series impedances); and g parameters, which are the inverse of h parameters.

If the circuit is for use with small signals, the *small-signal parameters* are written h_{fe}, etc.; if for use with large signals, capitals are used for the subscripts: h_{FE}, etc.

(ii) *Other equivalent circuits.* These are made up of components relating to the actual physical nature of the device rather than the more abstract linear networks described above. The most common circuit that is useful for high-frequency circuits is the hybrid-π model (Fig. *d*). The components are labelled with subscripts that are unprimed and primed. The letters *b*, *c*, and *e* represent the transistor regions (base, etc.). The unprimed subscripts represent the external portions of the region (connections and some semiconductor material) while the primed subscripts represent the intrinsic or internal portion of the electrode that is close to the junction and actually involved in the transistor action. Thus $r_{bb'}$ represents the base *spreading resistance and $C_{b'e}$ represents the intrinsic base to emitter terminal capacitance.

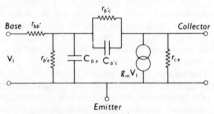

d   Hybrid-π equivalent circuit with common-emitter connection

transistor–transistor logic (TTL) A family of high-speed *logic circuits similar in operation to a *diode transistor logic circuit but the the cluster of *diodes at the input replaced by a multiemitter *transistor; usually the output stage is *push-pull. A typical circuit (a three-input NAND circuit) is shown in Fig. *a*. If the input levels are all high, the emitter-base junctions of the input transistor T_1 will all be reverse biased, and current through the base will flow through the forward-biased collector junction to the phase-splitting transistor T_2, which will switch on. Current flows through T_2 to T_4 and turns on T_4. T_3 will remain off as current is shunted away

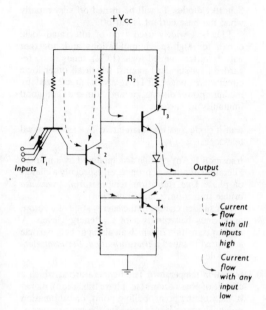

a   Three-input NAND circuit

from the base, and the output voltage will be low. If any one or more of the inputs is low, the emitter-base junction of T_1 will be forward biased, and current flows out of the base through the emitter of T_1. The current is therefore diverted away from T_2, and T_2 and T_4 will be turned off. The current through R_2 flows into the base of T_3 and T_3 is switched on and the output voltage is high. The output voltage will change rapidly when the input conditions change, since the transistors drive the level in both directions.

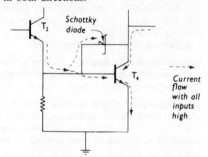

b   Schottky TTL circuit

The biggest limitation in speed is caused by the *delay time due to hole storage in the saturated output transistor T_4. The speed may be improved by adding a *Schottky diode with low *forward bias, across the base-collector junction of T_4. This is called *Schottky TTL* and part of the circuit is shown in Fig. *b*. This diode prevents T_4 saturating. Hole storage therefore does not occur in the collector-base junction, and since it is not a feature of

transit circle

Schottky diodes, T_4 will be turned off very rapidly when the base current is cut off.

TTL is a widely used type of integrated logic circuit for high-speed applications and, together with *emitter-coupled logic (ECL), tends to be regarded as a standard against which all other logic circuits are judged. TTL is also characterized by medium power dissipation and *fan-out, and good immunity to *noise.

transit circle *Syn.* meridian circle. *See* astronomical telescope.

transition 1. Any change accompanied by a marked alteration of physical properties, especially a change of phase (*see* transition temperature). (*See also* bridge transition; shunt transition).

2. A sudden change in the energy state of an atom or nucleus between two of its *energy levels. A nuclear transition in which an alpha or beta particle is emitted is called a *transformation*. *See also* selection rules.

transition temperature The temperature at which a change of phase occurs (*see* phase transition), namely the freezing point, boiling point, or sublimation point. The term is also used for the temperature at which a substance becomes superconducting.

transit time In an electronic device, the time taken for a charge *carrier to pass directly from one specified point to another under given operating conditions.

translation The movement of a body or system in such a way that all points are moved in parallel directions through equal distances.

transmission coefficient 1. Symbol: τ. The ratio P_{tr}/P_0, where P_0 and P_{tr} are the *sound-energy flux (or more generally the acoustic power) incident on and transmitted by a body respectively. The *dissipation factor* is $(\alpha_a - \tau)$ where α_a is the *acoustic absorption coefficient.

2. Former name for *transmittance.

transmission density *Syn.* optical density. Symbol: D. The logarithm to base ten of the reciprocal of the *transmittance, i.e. $D = \log_{10} 1/\tau$.

transmission line 1. *Syn.* power line. An electric line, often an overhead wire not surrounded by insulation, that conveys electric power from a power station or substation to other stations or substations.

2. An electric cable or *waveguide that conveys electric signals from one point in a communications system to another, and forms a continuous path between the points.

3. *Syn.* feeder. The one or more conductors – wires or waveguides – that connect an aerial to a transmitter or receiver, the conductor(s) being substantially nonradiative.

Any of the above transmission lines is described as *uniform* if its electrical parameters, e.g. series resistance, are distributed uniformly along its

length. A *balanced* line has conductors of the same type, equal values of resistance per unit length, and equal impedances from each conductor to earth and to other electric circuits; an example is the overhead line. A coaxial cable is an example of an *unbalanced* line. (*See also* balun.)

The transmission characteristics of a particular transmission line are usually frequency-dependent. The characteristics throughout a given frequency band can be improved by the addition of inductance to a line; this is called *loading*.

The most efficient transfer of signal power from a transmission line to the *load occurs when the line is terminated by a load impedance equal to the *characteristic impedance, Z_0, of the transmission line. All the power travelling down the line is then absorbed by the load and none is reflected; the line is said to be *matched*. In a uniform line carrying high (e.g. radio) frequencies, Z_0 tends to the value $\sqrt{(L/C)}$, i.e. a pure resistance, where L and C are the inductance and capacitance per unit length of the line.

See also travelling wave.

transmission loss In any communications or acoustics system, the power at a point remote from the source is, in general, different from the power at a point nearer the source. The ratio of the two powers, expressed in *decibels, is the transmission loss between the two points.

transmission modes *See* waveguide.

transmissivity A measure of the ability of a material to transmit radiation as measured by the *internal transmittance of a layer of substance when the path of radiation is of unit length and the boundaries of the material have no influence.

transmittance Symbol: τ. A measure of the ability of a body or substance to transmit electromagnetic radiation, as expressed by the ratio Φ_{tr}/Φ_0 of the *radiant flux or *luminous flux transmitted to the incident flux.

Translucent bodies transmit light by diffuse transmission – the path of the light is independent, on the macroscopic scale, of the laws of *refraction. Transparent bodies on the other hand have regular transmission with no diffusion of light. In general, a body exhibits mixed transmission and the total transmittance can then be divided into a regular transmittance (τ_r) and a diffuse transmittance (τ_d), where $\tau = \tau_r + \tau_d$. *See also* internal transmittance; transmission density.

transmittancy A measure of the ability of a solution to transmit radiation, as determined by the ratio of the *transmittance of the solution to the transmittance of the solvent. The solution can be a solid or liquid solution and the definition applies when the thickness of the solution is the same as that of the solvent.

transmitter In any communications system, the device, apparatus, or circuits by means of which the signal is transmitted to the receiving parts of the system. *See also* aerial.

transmutation The formation of one element from another, generally naturally as a result of radioactive decay or artificially as a result of radioactive bombardment with particles or electromagnetic radiation.

transponder A combined transmitter and receiver system that automatically transmits a signal on reception of a predetermined *trigger. The trigger is usually in the form of a pulse, which must have a minimum amplitude.

transport number *Syn.* transference number. *See* migration of ions.

transport phenomena The class of phenomena due to the transfer of mass, momentum, or energy in a system as a result of molecular agitation, including such properties as thermal conduction, viscosity, diffusion, etc.

transposition Of a *transmission line. The manner in which the line conductors have their positions interchanged at points along the line separated by equal distances, for the purpose of reducing the effects of asymmetry. It is particularly useful for reducing the radiation from a line operating at radio frequencies and also for reducing the interaction between lines that run parallel to one another and in close proximity such as that between power and communication lines.

transuranic elements Elements having an *atomic number greater than 92, lying beyond uranium in the *periodic table. *See* radioactivity.

transverse mass The relativistic mass in a direction perpendicular to the motion of the particle relative to the observer (*see* relativity). It is given by:
$$m_0 / \sqrt{(1 - \beta^2)},$$
for a particle of rest mass m_0 moving with relative speed $\beta = v/c$, expressed as a fraction of the speed of light. *Compare* longitudinal mass.

transverse vibrations Vibrations in which the displacement is perpendicular to the main axis or direction of the vibrating body or system and so to the direction in which waves are travelling. Typical examples are the vibrations of a string as in sounding a note on the piano, and the vibrations of a tuning fork.

transverse waves Waves in which the displacement of the transmitting medium is perpendicular to the direction of propagation. The velocity, c, of transverse waves along a string is given by $c = \sqrt{T/m}$, where T is the tension in the string and m is the mass per unit length. The velocity of transverse waves along a bar or plate does not lend itself to a simple formula. Examples of transverse wave motion are light and electromagnetic radiation in general. Waves on the surface of liquids are half transverse and half longitudinal.

trapezium distortion In a *cathode-ray tube, distortion arises when the *trace on the screen does not faithfully reproduce the waveform of the voltage applied to a pair of deflector plates. If the voltages applied to the two pairs of deflector plates are periodic and of constant amplitude, the resulting trace should be confined within a rectangle on the screen when the relative phase displacement or the frequencies of the deflecting voltages are varied. In practice the trace, in these circumstances, is found to be contained with a trapezium instead of within a rectangle. This effect is known as trapezium distortion. Usually, the major cause is interaction between the deflector plates, which makes the effective *deflection sensitivity of one pair dependent upon the voltage applied to the other pair.

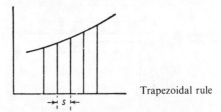
Trapezoidal rule

trapezoidal rule For finding the area underneath a curve. Any number of ordinates are erected at equal distances s apart (*see* diagram), then the area intercepted between the curve, the base, and the first and last ordinates is given by: area = $s \times$ (half the first ordinate plus half the last ordinate plus the sum of the intermediate ordinates). This rule assumes each section of the curve of length s to be a straight line. The accuracy of the result is greatly enhanced by increasing the number of ordinates in the given range.

travelling microscope A *microscope with a low magnifying power (normally about $\times 10$) and a *graticule placed in the plane of the eyepiece. It is mounted on rails to enable it to travel in the horizontal or vertical, or both, and is used to make very accurate determinations of length, e.g. on a photographic plate. Measurements are correct to 0.01 mm or better, over a distance of perhaps 0.2 m or more.

travelling wave Consider (hypothetically) a loss-free uniform *transmission line of infinite length situated in a medium of relative permittivity ε_r and relative permeability μ_r. If this line is connected to a sinusoidal a.c. supply at its sending end, electric power is transmitted along the line and the instantaneous values of the current and voltage at all points along the line are distributed in space as sine waves. As time increases, these sine-wave space distributions travel in the direction from the sending end to the receiving end with a velocity given by $c/\sqrt{(\varepsilon_r \mu_r)}$, where c is the speed of light. Either of these sine waves may be regarded as a travelling wave (of current or voltage respectively). More generally, it is an electromagnetic wave that travels along, and is guided by, a transmission line: sinusoidal conditions are not implicit (e.g. a surge propagated along a conductor or transmission line is also a travelling wave). Losses in the line cause a reduction in the velocity given above and also give rise to *attenuation. The above also applies to a line of finite length if it is terminated at its receiving end with an impedance equal to its *characteristic impedance. If an impedance discontinuity occurs at any point in the line (i.e. if there is an abrupt change in the characteristic impedance), reflection takes place at that point and the initial travelling wave (incident wave) is divided into the transmitted wave that travels towards the receiving end and the reflected wave that travels from the point towards the sending end.

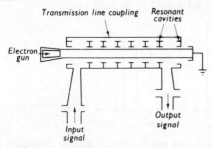

Travelling wave tube

travelling-wave tube An *electron tube that depends on the interaction of a velocity-modulated electron beam (see velocity modulation) with a radio-frequency electromagnetic field to produce amplification at microwaves frequencies. One form of tube is a modification of the multicavity *klystron tube. Several resonant cavities are placed along the length of the tube and coupled together with a *transmission line. The electron beam is velocity modulated by the radio-frequency input signal at the first resonant cavity, and induces radio-frequency voltages in each subsequent cavity. If the spacing of the cavities is correctly adjusted, the voltages at each cavity induced by the modulated electron beam are in phase and travel along the transmission line to the output, with an additive effect, so that the power output is much greater than the power input. Some power travels in the other direction along the transmission line (the backward waves) but these contributions are designed to be out of phase with each other and cancel out. The effect of the transmission-line coupling is to lower the *Q-factor of each cavity and results in a greater bandwidth for the amplifier, compared to a klystron.

The *backward-wave oscillator* is a modification of the travelling wave tube in which the backward waves are in phase with each other and provide positive *feedback to the input cavity, allowing the tube to oscillate.

Trevelyan rocker Vibrations can be set up in a solid or a volume of gas by the intermittent periodic communication of heat to some part of the system (see singing flame). The Trevelyan rocker is one of the earliest examples of heat-maintained vibrations. It consists of a prismatic block of copper having one edge grooved to form two adjacent parallel ridges. The prism rests with its grooved edge on a block of lead, the other point of support being a knob at the end of a thin round handle. When the prism is heated and placed on the lead block, it begins to vibrate, the weight being carried alternately on one or other of the two ridges. The explanation of this vibration was given by Leslie. It is due to the expansion of the cold block at the point of contact with the hot prism, the rocking being caused by the inequality of the inertia of the portions of the rocker on opposite sides of the ridge. Trevelyan found it necessary to have the ridges very smooth and clean but the lead to have a rough surface. Leslie's explanation was put under investigation by E. G. Richardson who used a stroboscopic method (see stroboscope) to measure both the frequency and the amplitude of the vibration. The results of his experiments can be summarized as follows: (1) f^2 is fairly constant where f is the frequency and a is the amplitude of the vibration. This is in agreement with the gravity theory. (2) The alteration of the length of the rocker handle as well as its nature (material) does not affect the average value of f^2a nor the range of the frequency. (3) The amplitudes are greater at higher temperatures although they may be dependent more on the excess of temperature of the copper over that of the lead rather than on the actual temperature of the former.

triac See silicon controlled rectifier.

triangle of forces A geometrical construction expressing the combined effect of two forces acting on the same particle. See vector.

triangle of vectors A geometrical construction expressing the law of addition of two *vectors.

triboelectricity See frictional electricity.

tribology The study of *friction, lubrication, and wear of surfaces in relative motion.

triboluminescence *Luminescence resulting from *friction.

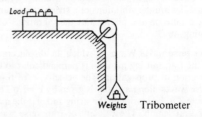

Tribometer

tribometer A loaded sled subject to a measurable horizontal force used in an elementary determination of coefficients of friction.

trichromatic theory of vision See colour vision.

trickle charge A continuous charge given to an accumulator or battery to maintain the fully charged condition. The charging current, which is relatively small, is adjusted to a value that compensates for the losses caused by local action.

triclinic system *See* crystal systems.

trigger Any stimulus that initiates operation of an electronic circuit or device. In general, the response to a trigger continues after the cessation of the stimulus.

trimmer *Syn.* trimming capacitor. A variable capacitor of relatively small maximum capacitance used in parallel with a fixed capacitor to enable the capacitance of the combination to be finely adjusted.

Trinitron *See* colour picture tube.

triode Any electronic device with three electrodes, such as bipolar junction *transistors, *thyratrons, etc. The term is usually applied to a three-electrode evacuated *thermionic valve.

triode-hexode A multiple (or multielectrode) *thermionic valve containing a triode and a hexode. The triode portion is normally used as an oscillator and the hexode as a *mixer or frequency-changer.

trip coil *See* tripping device.

triple mirror A total reflecting prism corresponding to one formed by cutting off the corner of a cube by a plane making equal angles with the cube faces. (Also three plane mirrors forming the angles at a corner of a cube.) Any ray entering emerges in a parallel direction after reflection at all three faces. It is used in signalling.

triplen harmonic A *harmonic, of which the frequency is an integral multiple of three times that of the fundamental.

triple point The point on a pressure–temperature diagram representing the condition that the solid, liquid, and vapour phases can be together in equilibrium. The diagram for water is unusual in that the line showing equilibrium between solid and liquid has a negative slope. This is because water has the rare property of contracting on melting, so $(v_2 - v_1)$ is negative in the *Clausius-Clapeyron equation. The lower equilibrium pressure for hoar frost compared with supercooled water is significant in the formation of *snow.

The temperature of the triple point for pure water is defined to be 273.16 kelvin. This is the fixed point for the *thermodynamic temperature scale and it, together with the triple points of hydrogen and oxygen, are fixed points for the *International Practical Temperature Scale.

For some substances, notably carbon dioxide, the pressure at the triple point is above normal atmospheric pressure. These substances only have the liquid phase when in closed vessels at high pressures.

tripping device A device that when actuated, sets free the restraining mechanism of a *circuit breaker,

causing the latter to break the circuit. Some common types are hand operated; others are operated electromagnetically by means of, for example, a *trip coil*, which forms part of the tripping device, and consists of a coil and a movable plunger or armature.

tristimulus values *See* chromaticity.

tritanopia *See* colour vision.

tritium The radioactive isotope of hydrogen with mass number 3. Symbol: ^{3}H or T.

triton The nucleus of a *tritium atom.

tropical year *See* time.

tropopause *See* troposphere.

troposphere The lowest of the earth's *atmospheric layers, lying below the *stratosphere. The thickness varies over the earth's surface and also shows diurnal and seasonal changes. It is between 10–12 km in middle latitudes, about 15–16 km at the equator, and 7 km at the poles. The air temperature normally decreases with height at approximately 6.5 °C/km. The troposphere is the most turbulent atmospheric layer, cloud precipitation and other meteorological phenomena occurring here. The *tropopause* is the boundary between troposphere and stratosphere, marking an abrupt change in the temperature variation with height. It consists of one or more layers. The temperature of the tropopause is between –45 and –65 °C.

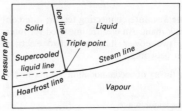

Water

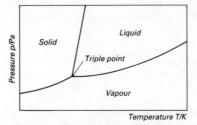

Typical pure substance

Triple point

Trouton–Noble experiment A plate capacitor is suspended so that it is free to turn. Suppose that the capacitor and its mount move through space with velocity v. The total electric and magnetic energy of the capacitor can be shown to involve a term including v^2/c^2 times the cosine of the angle θ the plates

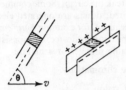

Trouton-Noble
experiment
(ether drift)

make with the direction of the velocity of motion. Thus it would be expected to set itself at right angles to the direction of motion to make the total energy as small as possible. The experiment was suggested as a means for detecting velocity drift through the *ether, as a parallel with the *Michelson–Morley experiment. That no effect is observed is explained by the theory of *relativity.

Trouton's rule The molar *latent heat of vaporization L at the standard boiling point T_B on the thermodynamic temperature scale is given for most pure substances by:

$$L/T_B = (10.5 \pm 1.5)R = (88 \pm 12) \text{ J mol}^{-1}\text{K}^{-1},$$

where R is the molar gas constant. Those substances that boil at very low temperatures (isotopes of hydrogen and helium) give much lower values because the *zero-point energy is relatively very high. For He-4 the value is about a quarter, and for He-3 about one-eighth of Trouton's value. The quantity represents the change of *entropy on evaporation per mole.

trunk feeder *Syn.* interconnecting feeder; interconnector; trunk main. An electric line used for interconnecting two power stations or two electrical distribution networks.

Tscherning theory of accommodation *See* accommodation.

tsunami *See* water wave.

TTL Abbreviation for *transistor–transistor logic.

tube *Syn.* valve. Abbreviation of electron tube. The term is more common in the US than Britain.

Tube of flux

tube of flux *Syn.* tube of force; field tube. The vector field of an electric field can be divided into tubes that are bundles of lines of electric field strength. The lateral surface of the tube is made up of lines of force so that the field strength at any point of that surface is tangential to it and the *flux is the same through all cross sections. The field strength at any part of the tube is inversely proportional to the section at that place, taken normal to the lines, the tube narrowing as the field strength increases. A unit tube is one through which unit flux flows. An analo-

gous concept can be applied to magnetic flux density.

tube of force *See* tube of flux.

tuned circuit A *resonant circuit whose resonant frequency can be varied. The process of adjusting the circuit to a condition of *resonance is known as *tuning*, and may be carried out by adjusting the capacitance of the circuit (capacitive tuning) or by adjusting the inductance of the circuit (inductive tuning). Sometimes the inductance is varied by altering the *reluctance of the magnetic circuit of the coil (by moving a ferromagnetic core relative to the coil). This latter form of inductive tuning is known as permeability tuning.

tungsten lamp A lamp with a tungsten filament. Incandescent tungsten is a slightly selective radiator serving as a grey body and as a standard radiation source (tungsten substandard filament lamps). *See also* quartz-iodine lamp.

tuning *See* tuned circuit.

tuning fork A suitably proportioned metal bar bent into the shape of a U and mounted upon a stem at the base of the U. If excited by bowing, striking, or pressing the prongs together, a note consisting almost entirely of a fundamental is faintly heard. Overexcitation elicits a few very weak, mostly inharmonic, upper *partials. The tuning fork maintains a constant frequency for long periods, the frequency varying only very slightly with amplitude of vibrations and temperature. Forks, which have been made for frequencies as high as 90 kHz, are generally made of steel, invar, or elinvar. The vibrating system is a dynamically balanced one – the prongs moving in together and out together. Because of the small size of the prongs there is little output of sound direct to the air, and because of the method of vibration there is only a very small longitudinal component of vibration down the stem. The damping of the vibrating system is therefore very small. The use of resonators coupled to the fork through the stem increases the sound output and also only reproduces the fundamental since the upper partials of the fork and resonator are not similar.

The tuning fork has many applications. It is used as an easily carried standard of pitch in the tuning of musical instruments. It is also used in a musical instrument in place of strings as a practice piano with small sound output. If the fork is maintained at constant amplitude in a constant temperature enclosure, it can be used as a standard of time with an accuracy of between 1 in 10^4 and 1 in 10^6. The fork is generally maintained in vibration by means of an electronic circuit; electromagnetic coils placed around or near the two prongs, but not touching them, are suitably connected. Suitably placed and designed contacts with a d.c. supply are sometimes used to maintain the fork in vibration. The accurate placing of the contacts, and the use of a suitable inductance in the circuit, are essential for constant frequency output. It is found that the amplitude/frequency curve has a minimum, and electrically maintained forks are usually operated at the centre of the minimum.

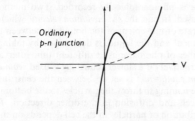

Characteristic of a tunnel diode

tunnel diode *Syn.* Esaki diode. A *p-n junction *diode that has extremely high *doping levels on each side of the junction. As a result electrons can tunnel across the junction (*see* tunnel effect) in the forward direction, i.e. with positive voltage applied to the p-region. With increasing forward bias on the junction, the tunnel effect contributes less and less, producing on the diode characteristic a negative-resistance portion; finally the forward-voltage characteristic resembles that of an ordinary p-n junction (*see* diagram). Tunnelling also occurs in the *reverse direction, producing a large reverse current. This is similar to what occurs in a *Zener diode, but the effective Zener breakdown voltage can be considered to occur at a small positive voltage.

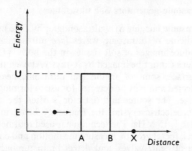

Particle approaching a potential barrier

tunnel effect The movement of particles through barriers that, on classical theory, they would have insufficient energy to surmount. Classically, if a particle moves in one direction with kinetic energy E and approaches a potential-energy barrier of height U, it can get over it by converting some of its kinetic energy into potential energy. If $E < U$ the probability of finding the particle at a point X (*see* diagram) would be zero. However, even when $E < U$ the particle can tunnel through the barrier. This is explained by wave mechanics. If the *wave function, ψ, of the particle is considered inside the region of the barrier, i.e. taking point A as $x = 0$, the *Schrödinger wave equation has the form:

$$\frac{d^2\psi}{dx^2} + \frac{8\pi^2 m}{h^2}(E - U)\psi = 0,$$

and a general solution:

$$\psi = A \exp[(2\pi i x/h)\sqrt{2m(E - U)}],$$

where A is a constant, m the particle mass, h the Planck constant, x the distance into the barrier, and $i = \sqrt{-1}$. If $E < U$ the solution is:

$$\psi = A \exp[-(2\pi x/h)\sqrt{2m(U - E)}].$$

Provided the barrier is not infinitely thick or wide there is thus a probability of the particle crossing this region and reaching X. The effect is too unlikely to occur in macroscopic systems but is the basis of *alpha decay and *field emission.

turbine An engine in which a shaft is rotated by fluid impinging upon a system of blades or buckets mounted upon it. *Water turbines* are used in *hydroelectric power stations; *steam turbines* are used for ship propulsion and for electric power stations using coal, oil, nuclear, or other fuel; *gas turbines* have been used particularly for aircraft propulsion but have other applications. According to the system of nozzles or blades fixed upon the framework of a steam turbine, it is conventionally classified as an *impulse turbine, reaction turbine* or *impulse-reaction turbine.*

turbo-alternator An alternator (e.g. a *synchronous alternating-current generator) intended for steam-turbine drive. Turbo-alternators are employed in thermal power stations for generating alternating current. For a 50 hertz supply, machines having two poles and running at 3000 r.p.m. are built with ratings up to about 50 000 kVA. For higher ratings, four poles are used and the speed is 1500 r.p.m.

turbulence The state of a fluid possessing a nonregular motion such that the velocity at any point may vary in both direction and magnitude with time. Turbulent motion is sometimes called sinuous motion and is accompanied by the formation of eddies and the rapid interchange of momentum in the fluid. The change from laminar to turbulent motion occurs at a critical value of the *Reynolds number. The drag resistance to a body for turbulent flow is proportional to the square of the velocity; resistance for laminar flow is proportional to the velocity.

turns ratio *See* ratio.

tweeter A loudspeaker of small dimensions designed to reproduce sounds of relatively high frequency in a hi-fi system. *Compare* woofer.

twin cable *See* paired cable.

twinning The growing together of two crystals of the same kind in such a way that one individual may be brought to coincidence with the other by reflection across a plane (the *twin plane*) or by rotation about an axis (the *twin axis*). Twinning can also be brought about by mechanical deformation involving *glide.

twin paradox A concept that has been discussed in terms of both the special and general theories of *relativity. It is supposed that a pair of identical twins are initially together in an inertial system. One twin is then accelerated to a high speed and departs for a long space journey, finally returning to rest beside the other. It is predicted that the twin who has made the journey will have aged less than the one who has remained at home.

twisted pair A pair of insulated wires, twisted together to form a *transmission line, whose characteristic *impedance depends on the physical dimensions of the component wires. Twisted pairs are often used in high-frequency circuits as an alternative to *coaxial cable. *See* paired cable.

two-phase *Syn.* quarter-phase. Of an electrical system or device. Having two equal alternating voltages between which there is a relative *phase difference of 90°. *See* polyphase system.

two-wire system A system of distributon of electrical power used with direct current or with single-phase alternating current. It comprises two conductors only, the electrical apparatus being connected between them. *Compare* earth-return system; three-wire system.

Twyman–Green apparatus *See* interference of light.

Tyndall, John (1820–1893) Irish physicist who investigated thermal conductivity in metals and the *scattering of light by gases. He suggested that the blue of the sky is caused by scattering of sunlight by small particles of water in the upper atmosphere.

Tyndall effect *See* Rayleigh scattering.

type metal An alloy of lead, antimony, and tin, sometimes also with copper. A typical composition is Pb 66, Sb 16, Sn 18. Pure antimony expands on freezing; type metal shrinks only a little and it makes sharp casts. Readily melting, it sets to give a hard surface, resistant to air, water, oils, and inks, and to alkaline solutions such as are used in the cleansing of type.

U

UHF Abbreviation for ultrahigh frequency. *See* frequency bands.

Uhlenbeck, George Eugene (*b.* 1900) Dutch-born Amer. physicist noted for his collaboration with Samuel Goudsmit in their discovery of electron spin (1925).

ultimate strength The limiting stress (in terms of force per original unit area of cross section) at which a material completely breaks down (i.e. fractures or crushes).

ultracentrifuge A centrifuge operating at a very high angular speed, suitable for use with colloidal solutions. It is used to separate colloidal particles and also can be used to estimate their size and measure the relative molecular masses of very large molecules, such as proteins. Quantitative instruments have a transparent cell and the formation of sediment is photographically recorded. Two methods are used. In one, the *sedimentation velocity*, which is the rate of movement of the boundary between the sediment and the liquid, is determined by taking a number of photographs at different time intervals. Alternatively if a slower centrifuge is used *sedimentation equilibrium* is set up between the centrifugal force tending to throw the particles to the bottom of the cell and diffusion in the other direction. The distribution of particles in the cell depends on their size.

ultrahigh frequency (UHF) *See* frequency bands.

ultramicrobalance A very sensitive balance for accurate weighing to 10^{-8} grams.

ultramicroscopy A technique in optical microscopy for viewing particles that are smaller than those resolvable by ordinary methods. An intense narrow-slit beam is passed through a medium under investigation (e.g. colloidal particles of order 10^{-5} to 10^{-6} mm), transversely to the axis of the observing microscope; the light is then diffracted into the microscope objective. Dark-field illumination (*see* microscope) can also be used as a form of ultramicroscopy.

ultrasonic detectors *See* ultrasonics.

ultrasonic generators *See* ultrasonics.

ultrasonic imaging *Syn.* ultrasonography. The use in medicine of ultrasonic waves (*see* ultrasonics) to produce images of soft tissues in the body. (These tissues cannot be imaged by X-rays.) At each tissue interface, some of an incident ultrasonic beam is reflected and may be detected for use in forming the image. The source and detector is often the same *piezoelectric crystal, the frequency being in the range 1–15 MHz. The technique is used routinely to check foetal growth, measure the motion and condition of heart valves, and detect brain tumours.

ultrasonics The study and application of mechanical vibrations with frequencies beyond the limits of hearing of the human ear, i.e. with frequencies about 20 kilohertz and upwards. There is no theoretical upper limit to the ultrasonic frequency, although ultrasonic applications are usually restricted to about 20 MHz. Ultrasonic waves have the same basic properties as sound waves, both being examples of acoustic waves (*see* sound).

The main methods of generation of ultrasonic frequencies are: (1) *Mechanical generators*: e.g. *Galton whistle and *Hartmann generator. (2) *Magnetostriction generators*: These are based on the principle of the *magnetostriction effect. (3) *Piezoelectric generators*. As in the case of magnetostriction oscillators, the high-frequency electric oscillations are transformed into powerful mechanical oscillations by piezoelectric crystals (*see* piezoelectric oscillators). (4) *Thermal generators*: These are now hardly used, but ultrasonic waves can be produced by means of a spark gap fed by a damped oscillatory circuit or by a direct current arc with an alternating current superimposed on it. The defect of such methods is the instability of the frequency as well as the amplitude.

Since ultrasonic waves are inaudible, it is necessary to use physical apparatus to detect their existence and measure their amplitude. Some of the methods used in the audible range for measuring wavelength and absorption of sound can be used in the lower ultrasonic region, while they may fail at higher ultrasonic frequencies where the indicator may be too large compared with the ultrasonic vibration itself and thus causes disturbance in the propagation. On the other hand, the optical methods used now are specially suitable for high ultrasonic frequencies to make the waves visible and to measure their length and amplitude with great accuracy. The methods can be summarized as follows:

Mechanical methods: (i) Fine particles such as dust, powdered coke, or bubbles in a liquid or tobacco smoke or mist, produced by a substance like salammoniac, may be used in tubes to render the standing ultrasonic waves visible. (ii) Radiation pressure measurements are taken by radiometers in the form of a torsion balance (*see* radiation pressure). (The *Rayleigh disc and the manometric capsule (*see* manometric flame) are good mechanical receivers and detectors for audible sound, but for ultrasonics, their dimensions are too great compared with these waves.)

Thermal methods: These are based on the principle of the *hot-wire microphone. A thin platinum wire placed between the ultrasonic source and the reflector will be cooled. As the wire is traversed along a standing wave pattern it is cooled more in the displacement antinode where there is much movement than in the nodes where the particles of the medium are still. By such a method velocity and absorption of ultrasonics can be measured.

Electrical methods: The only electrical receiver suitable for ultrasonic vibrations is the piezoelectric quartz crystal, since all other forms of microphones are insensitive for high frequencies. If a wave meets the piezoelectric crystal in one of its vibrational directions and has the same frequency as the natural frequency of the crystal the latter is set into mechanical vibration, causing electric charges on its surfaces or alternatively alternating potential differences. Langevin was the first to suggest the piezoelectric receivers and used them in his echo-sounding system (*see* Langevin–Florissen sounder). Since it is in general very difficult to get two crystals, one to be a source and the other a receiver, of exactly the same frequency, it is necessary to make use of the method of electrically tuning the receiving quartz to the frequency of the incoming wave.

Instead of using a separate crystal as a detector, the same transmitting crystal could be used as a receiving one by allowing the waves to fall on a plane reflector. As the position of the reflector is changed, there is a periodical varying reaction of the reflected sound waves on the transmitting crystal. These variations could be measured by change of plate current actuating the crystal or the variation of potential difference across the crystal. This method due to Pierce is a very sensitive one for measurements of wavelengths. According to Hubbard the gas or liquid column between the quartz surface and the reflecting plate may be regarded as a mechanical system, the apparent impedance of which changes periodically with the change of the distance of the reflector.

Optical methods: (i) *Striation methods or dark-field method*: These depend upon the change of the refractive index of a light beam passing transversely across a standing sound wave. The light is brought to a focused image at a screen. The points in the sound wave at which a change of refraction takes place divert the light, so that a picture is formed of the standing waves in which pressure nodes are dark and antinodes are bright.

(ii) *Diffraction methods*: Lucas and Biquard showed that diffraction of light occurs in a liquid traversed by ultrasonic waves. The series of compressions and rarefactions that constitute ultrasonic waves in a liquid act as a grating of light and cause optical diffraction. A narrow parallel beam of laser light, wavelength λ, emerging from a slit passes through a trough of liquid at right angles to the ultrasonic beam. Looking along the laser beam a central image of the slit is seen together with its diffraction images. If λ_s is the wavelength of the sound in the liquid and α_k is the angle of diffraction for the kth order, then:

$$\sin \alpha_k = k\lambda/\lambda_s.$$

It is a very accurate method for measuring the wavelength of sound and hence the speed of sound at ultrasonic frequency in liquids.

The applications of ultrasonics are numerous in many branches of applied and pure science. They include *sonar, *ultrasonic imaging, the formation of emulsions, the coagulation of smokes, the degassing of melts, *flaw detection, cleaning small objects by vibrating them in a solvent, welding dissimilar materials, and soldering aluminium.

Formerly the term *supersonic* was sometimes used to mean ultrasonic. According to modern practice supersonic is only used to refer to speeds greater than that of sound.

ultrasonography See ultrasonic imaging.

ultrasound Acoustic waves of ultrasonic frequency. *See* ultrasonics.

ultraviolet astronomy See ultraviolet radiation.

ultraviolet catastrophe See quantum theory.

ultraviolet microscopy See resolving power; ultraviolet radiation.

ultraviolet radiation (UV radiation) Ritter (1801) demonstrated with silver chloride that solar radiation included chemically active (actinic) rays just beyond the violet end of the spectrum. Since then a range of invisible radiations produced by other sources, overlapping the long wave X-rays, have been discovered. Ultraviolet radiation is grouped according to wavelength into long or near ultraviolet (wavelengths 397–300 nm), short (300–185 nm), Schumann region (185–120 nm), and extreme ultraviolet (120–13 nm).

Substances that may be transparent to light absorb strongly as the wavelength is decreased in the ultraviolet, e.g. Crookes "A" at 360 nm, ordinary crown 300 nm, vita glass 252 nm, quartz 180 nm, fluorspar 120 nm. In the extreme ultraviolet, most substances become opaque or show selective absorption, and even small paths in air or gas at low pressure may show considerable absorption. Atmospheric absorption by ozone in the *stratosphere restricts the solar ultraviolet at about 300 nm. Incandescent bodies even at high temperature yield a relatively small proportion of ultraviolet to total radiation. Arc and spark discharges and vacuum-tube discharge with enclosures of transparent media (quartz) are the main sources.

UV radiation can be detected and investigated by its imaging action on photographic film and plates, and also by the *photoelectric effects and *fluorescence that it produces. In *ultraviolet spectroscopy*, according to the region investigated, spectrographs with quartz or fluorite lenses and prisms, reflecting gratings enclosed in vacuum, plastic windows, and special photographic plates are used. The UV band includes many of the resonant absorption and emission lines (resulting from transitions of electrons to and from atomic ground states) of most of the more abundant chemical elements. *Ultraviolet microscopy* uses special quartz lenses and commonly works with a monochromatic source (wavelength 275 nm). This enables higher resolution and magnifications (3600 times) than can be obtained with visual observation. *Ultraviolet astronomy* can only be carried out using satellites, or balloons or rockets, due to strong absorption in the earth's atmosphere. Celestial sources include high-temperature stars, the outer atmospheres of cooler stars, and interstellar gas.

ultraviolet spectroscopy *See* ultraviolet radiation.

umbra A region of complete shade within a shadow. Geometrically no rays can enter from any source present. By drawing straight lines from any point within an umbra, and passing outside the obstacle, none of them should strike a source. This serves as a test for umbra. *See* shadows.

Umklapp process A collision process between *phonons, or between phonons and electrons. The crystal momentum is not conserved (no conservation principle applies). The process is responsible for thermal resistance in nonconducting materials.

unbalance factor In a three-phase system. *See* symmetrical components.

uncertainty principle *Syn.* Heisenberg uncertainty relation; indeterminacy principle. The principle, enunciated by Heisenberg, that the product of the uncertainty in the measured value of a component of momentum (p_x) and the uncertainty in the corresponding coordinate of position (x) is of the same order of magnitude as the *Planck constant. In its most precise form:

$$\Delta p_x \times \Delta x \gg h/4\pi,$$

where Δx represents the root-mean-square value of the uncertainty. For most purposes one can assume:

$$\Delta p_x \times \Delta x \simeq h/2\pi.$$

The principle can be derived exactly from *quantum mechanics but is most easily understood as a consequence of the fact that any measurement of a system must disturb the system under investigation, with a resulting lack of precision in measurement. For example, if it were possible to see an electron and thus measure its position, photons would have to be reflected from the electron. If a single photon could be used and detected with a microscope, the collision between the electron and photon would change the electron's momentum (*see* Compton effect). The accuracy of the position determination is increased at shorter photon wavelengths because the *resolving power of the microscope increases with shorter wavelengths. However, the momentum of a photon also increases with shorter wavelengths and increased precision in measurement of position is gained at the expense of loss of precision in the measurement of the electron's momentum.

An obvious inference from the principle is that when the position is known with absolute accuracy (i.e. $\Delta x \rightarrow 0$) the uncertainty Δp_x is infinitely large and p cannot be known.

A similar relationship applies to the determination of energy and time, thus:

$$\Delta E \times \Delta t \gg h/4\pi.$$

The effects of the uncertainty principle are not apparent with large systems because of the small size of h. However the principle is of fundamental importance in the behaviour of systems on the atomic scale. For example, the principle explains the inherent width of spectral lines; if the lifetime of an atom in an *excited state is very short there is a large uncertainty in its energy and a line resulting from a transition is broad.

One consequence of the uncertainty principle is that it is impossible fully to predict the behaviour of a system and the macroscopic principle of *causality cannot apply at the atomic level. *Quantum mechanics gives a statistical description of the behaviour of physical systems.

undercurrent release A *tripping device that operates when the current falls below a predetermined value (usually adjustable). A current that causes the release to operate is called an undercurrent. *Compare* overcurrent release.

underdamped *See* damped.

undervoltage release A *tripping device that operates when the voltage falls below a predetermined value. It is sometimes used to prevent the closing of a *circuit breaker when the supply voltage is less than a predetermined value, and in this application it is described as an undervoltage, no-close release.

uniaxial crystals Crystals belonging to the tetragonal, rhombohedral, and hexagonal systems. They are double refracting except for light travelling through them in the direction of the principal crystallographic axis. In this direction only they are singly refracting. Iceland spar (calcite) and quartz are uniaxial; the former is a negative crystal (the ordinary refractive index is greater than the extraordinary), the latter is a positive crystal. *See* calcite; double refraction.

unidirectional current An electric current that, although not necessarily constant in magnitude, flows in one direction only. *Compare* pulsating current.

unified atomic mass unit *See* atomic mass unit.

unified field theory A field theory that seeks to unite the properties of gravitational, electromagnetic, weak, and strong interactions, so that a single set of field equations can be used to predict all their characteristics. At present there are unsolved problems in using field theory to encompass the fundamental interactions, in particular the gravitational interaction. *See also* grand unified theory; string theory.

unifilar suspension A type of suspension used in electrical instruments in which the moving part is suspended on a single thread, wire, or strip. The controlling (or restoring) torque is produced by the twisting of the thread, etc., which takes place when the moving part is moved from its initial position of rest. The *quadrant electrometer is an instrument in which this type of suspension is commonly employed. *Compare* bifilar suspension.

uniform temperature enclosure An enclosure whose walls are maintained at the same constant temperature. The amount and kind of radiation within such an enclosure is independent of the nature of the walls and the contents of the enclosure, depending only on the temperature of the walls. Such radiation is called full, complete, cavity, or black-body radiation since the radiation inside the enclosure is identical with that emitted by a *black body at the same temperature.

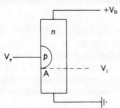

a Unijunction transistor

unijunction transistor *Syn.* double-base diode. A *transistor consisting of a bar of lightly doped (high-

resistivity) *semiconductor, usually n-type, with a region of highly doped semiconductor of opposite polarity located near the centre (Fig. *a*). Ohmic contacts are formed to each end of the bar (base 1 and base 2) and the central region (the emitter).

If a voltage V_b is applied across the bar a potential drop will be present through the bar. Let the potential on the less positive side of the junction at point A be V_1. If a voltage V_e is applied to the emitter, for $V_e < V_1$ the junction will be *reverse biased and very little current will flow. If V_e is increased to V_1, the junction will become *forward biased at point A. *Holes will be injected into the bar and attracted to the less positive terminal (base 1), lowering the resistivity between A and base 1. Point A becomes less positive and more of the junction becomes forward biased. This causes the emitter current, I_e, to increase rapidly. As I_e increases, the decreased resistivity causes V_e to drop and the device exhibits negative resistance. The typical characteristics are shown in Fig. *b*. The emitter voltage V_2 is the point at which the device ceases to show negative resistance. The switching time between V_1 (at which the device starts to conduct) and V_2 depends on the device geometry and the biasing voltage V_b. As V_b is increased, V_1 increases and the emitter current at V_2 also increases.

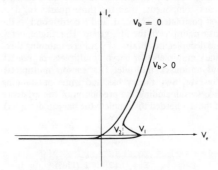

b Unijunction transistor characteristics

The most common use of the unijunction transistor is in *relaxation oscillator circuits.

unipolar transistor A *transistor in which current flow is due to the movement of *majority carriers only. *See* field-effect transistor.

unitary symmetry A generalization of *isospin theory. In *group theory, isospin is concerned with a group called SU_2 (the special unitary group of 2×2 matrices). Unitary symmetry is concerned with a group called SU_3. It predicts that as far as *strong interactions are concerned *elementary particles can be grouped into multiplets containing 1, 8, 10, or 27 particles and that the particles in each multiplet may be regarded as different states of the same particle. Unitary symmetry multiplets contain one or more isospin multiplets. All particles in a multi-

591

plet have the same spin (J), parity (P), and baryon number (B).

The multiplets are most clearly illustrated by plotting the constituent particles on a graph of *hypercharge (Y) against isospin quantum number (I_3). Examples of a meson and baryon octet (multiplet of 8 particles) and a baryon decuplet (10 particle multiplet) are illustrated in Figs. *a* and *b*.

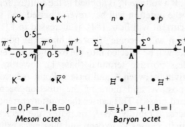

$$J=0, P=-1, B=0 \qquad J=\tfrac{1}{2}, P=+1, B=1$$
Meson octet   Baryon octet

a   Meson and baryon octets

The predictions of SU$_3$ theory do not agree with experiment as well as those of isospin. If it were an exact symmetry of elementary particles, all particles in a multiplet should have the same mass; this is far from true. In addition, in the case of *quarks, the resulting symmetry is badly broken now that hadrons have been discovered with more massive quark components than the three quarks (u, d, s) first postulated. The u, d, and s correspond to the basic multiplet of the SU$_3$ group. The singlet, octet, and decuplet are obtained by either combining three quark multiplets (for baryon multiplets) or a quark and antiquark multiplet (for meson multiplets). However, SU$_3$ theory has had some outstanding successes, including the prediction of the existence of the Ω^- (needed to complete the baryon decuplet).

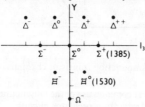

$$J=3/2, P=+1, B=1$$

b   Baryon decuplet

unit cell The smallest crystal unit possessing the entire symmetry of the whole periodic structure. It is defined in terms of six elements or parameters: a, b, c, the lengths of the cell edges (taken as axes), and α, β, γ, the angles between the axial directions.

unit planes (and **points**) Conjugate planes of an optical system for which the lateral magnification is $+1$ (erect). *See* principal, cardinal, equivalent planes.

unit pole A magnetic pole that when placed 1 cm from an equal pole, in a vacuum, is repelled with the

force of one dyne. This CGS unit is no longer in use. *See also* magnetic monopole.

units Since its earliest beginnings physical science has been beset by entirely unnecessary complications arising from the variety and diversity of the units that have been used to express the magnitudes of physical quantities. These complications have finally been resolved by the adoption of the internationally agreed system of units known as *SI units (Système Internationale d'Unités). This system, which has been adopted by the scientific communities in most countries of the world (including the UK), is based on the *metre-kilogram-second (MKS) system and replaces the *CGS system and the *foot-pound-second (f.p.s.) system. SI units make quite clear the relationship between a physical quantity (say length l) and the units in which it is expressed, i.e.

Physical quantity = numerical value × unit.

Or, in the case of length, $l = n$ m, where n is a numerical value and m is the agreed symbol for the SI unit of length, the metre. This equation should be treated algebraically, e.g. the axis of a graph or a column of a table giving numerical values in metres should be labelled $l/$m. If the axis or column gives the values of $1/l$, it should be labelled m$/l$. *See also* rationalization of electric and magnetic quantities.

unit switch *See* contactor.

unit system A system that transfers the position of an image with unit magnification (± 1), e.g. erecting systems of terrestrial eyepieces, the Toepler points of a single lens. *See* negative principal points.

unit vector *Vectors, usually written i, j, and k, that have unit length and lie along the x-, y-, and z-axes, respectively. A vector function, F, can therefore be written

$$F = xi + yj + zk.$$

As the angles between these vectors are 90°, the scalar products can be given as:

$$i \cdot i = i^2 \cos 0 = 1, \text{ since } i = 1$$
$$i \cdot j = ij \cos 90° = 0, \text{ etc.}$$

The vector products are:

$$i \times i = i^2 \sin 0 = 0$$
$$i \times j = ij \sin 90° = 1, \text{ etc.}$$

universal coordinated time *See* time.

universal gas constant Former name for *molar gas constant.

universal motor An electric motor that is suitable for use with direct current or alternating current. It incorporates a *commutator and is usually *series-wound machine. Small motors of this type are (fractional horse-power) are commonly used in electrical appliances such as vacuum cleaners, portable drills, etc.

universal shunt A galvanometer shunt due to Ayrton and Mather. It is tapped so that it can pass 1/10, 1/100, 1/1000, etc., of the main current through the galvanometer, whatever resistance this has.

unpitched sound A sound with no definite pitch, composed of a wide spread of frequencies.

unsaturated vapour A vapour at a certain temperature that contains less than the equilibrium amount of the substance in the gaseous phase. Such a vapour may undergo slight isothermal compression without condensation occurring, and obeys the *ideal gas laws approximately.

unstable equilibrium See equilibrium.

unstable oscillation Any oscillation – mechanical, electrical, etc. – that tends to increase in amplitude with time.

unsymmetry factor In a three-phase system. See symmetrical components.

Unwin coefficients See Reynolds' law.

upper atmosphere The outer layers of the gaseous envelope surrounding the earth above about 30 km. It includes part of the *stratosphere, the *mesosphere, *thermosphere, and *exosphere, and therefore contains the *ionosphere which extends from the mesosphere into part of the exosphere. It is the region of the earth's atmosphere that is inaccessible to direct observation by balloons and information must be obtained from artificial *satellites and space probes.

uranium–lead dating See dating.

uranium series See radioactive series.

Urey, Harold Clayton (1893–1981) Amer. chemist who worked on atomic and molecular structure, especially on isotopes. He was the first to isolate heavy water.

UV Abbreviation for ultraviolet.

V

vacancy A position in a crystal lattice that is not occupied by an atomic nucleus. A vacancy should not be confused with a *hole. See defect.

vacuum 1. Strictly, a physical state totally devoid of particles – either particles of matter or photons of radiation (*compare* free space). Such a state does not exist in practice.
2. In vacuum technology, any space in which the pressure is below normal atmospheric pressure. The degree or quality of the vacuum attained – coarse, medium, high, very high, ultrahigh – is indicated by the total pressure of the residual gases. The highest vacuums achieved on earth have a pressure of less than 10^{-8} N m^{-2}.

vacuum evaporation A technique used for producing a coating of one solid on another, usually a coating of metal or semiconductor. Evaporation occurs from a solid or liquid at high temperature in a vacuum. At low pressures atoms leaving the hot surface suffer few collisions in the gas and can travel directly to a nearby cool surface where they condense, thus forming a thin film. See also thin-film circuit.

vacuum flask See Dewar vessel.

vacuum gauge See pressure gauges.

vacuum pump See pumps, vacuum.

vacuum state The ground state in a relativistic *quantum field theory. A vacuum state does not mean a state of nothing. As a consequence of *quantum mechanics, the vacuum state has a zero-point energy, which gives rise to *vacuum fluctuations*. The existence of vacuum fluctuations has observable consequences in *quantum electrodynamics.

vacuum tube An *electron tube evacuated to a sufficiently low pressure that its electrical characteristics are independent of any residual gas. *Compare* gas-filled tube.

vacuum ultraviolet Part of the ultraviolet region of the spectrum, in which the radiation is absorbed by air and experiments have thus to be performed in a vacuum. It is the part of the ultraviolet region having a wavelength less than about 200 nanometres and the highest quantum energies.

valence band See energy band.

valence electrons Electrons in the outermost shell of an atom that are involved in chemical changes. See also energy bands.

valency The number of hydrogen (or equivalent) atoms that an atom will combine with or displace.

valve *Syn.* electron tube. A device in which two or more electrodes are enclosed in an envelope usually of glass, one of the electrodes being a primary source of electrons. The electrons are most often provided by thermionic emission (*thermionic valve), and the device may be either evacuated (*vacuum tube) or gas-filled (*gas-filled tube). The name derives from the rectifying properties (*see* rectifier) of the devices, i.e. current flows in one direction only. The word valve is becoming obsolete and is being replaced by the term electron tube.

valve voltmeter An *amplifier using *thermionic valves, with a measuring instrument in the output circuit. The voltage to be measured is applied to the input circuit of the amplifier which has very high input impedance (since current does not flow in the grid circuit). Such instruments are used to measure both d.c. and a.c. voltages. They have now been generally superseded by *digital voltmeters.

Van Allen, James Alfred (b. 1914) Amer. astrophysicist who first detected the radiation belts that bear his name from information collected by the satellites Explorer I and III.

Van Allen belts *Syn.* radiation belts. Two belts of energetic charged particles, mainly electrons and protons, lying around the earth, across the equatorial plane. The particles are confined there by the

Van Arkel and De Boer's process

earth's magnetic field; they oscillate back and forth between the magnetic poles, spiralling around the field lines and emitting radiation as they do so. The existence of the radiation belts was suggested in 1958 by J. A. Van Allen from data obtained from an artificial *satellite. The inner belt is approximately 1000 to 5000 km above the equator and contains more energetic particles. These are electrons and protons captured from the *solar wind or produced as secondary products of *cosmic-ray bombardment in the upper atmosphere. The outer belt is approximately 15 000 to 25 000 km above the equator, curving down towards the magnetic poles. It contains mainly electrons from the solar wind, their energy and number fluctuating with solar activity.

Van Arkel and De Boer's process A method of producing certain elements, such as boron, silicon, and titanium, in a very pure state. The volatile iodide vapour is decomposed to the element and iodine vapour, on an incandescent filament in a vacuum. A similar process occurs in tungsten-halide lamps (used in car headlamps). Tungsten, evaporated from the filament, combines with iodine and forms tungsten iodide which then decomposes on the filament, thus keeping the bulb clear.

Van de Graaff, Robert Jemison (1901–1967) Amer. physicist remembered for the high-voltage generator that bears his name. It was developed during studies on the mobility of gaseous ions and was an improvement on the *Wimshurst machine then in use.

Van de Graaff accelerator A type of *accelerator in which the high-voltage terminal of a *Van de Graaff generator is used as a source of charged particles.

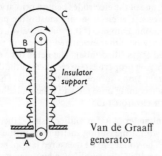

Van de Graaff
generator

Van de Graaff generator A high-voltage electrostatic generator that can produce potentials of millions of volts. It consists essentially of an endless insulated fabric belt moving vertically (*see* diagram). A charge is applied to the belt by the needle points A from an external source of up to 100 kV. This charge is carried continuously up into a large hollow sphere C where collector points B remove it. The potential of the sphere continues to increase (no charge residing on the interior) and is limited only by the leakage rate of the supporting insulators and the surrounding gas. The fabric belt may be replaced by a series of metal beads connected by insulating string. This

system, called the *pelletron*, has been developed to enable a voltage of 30 MV to be produced by two generators in series. This device is called a *tandem generator. *See also* Van de Graaff accelerator.

van der Meer, Simon (*b.* 1925) Dutch physicist and engineer. Senior engineer at the European Organization for Nuclear Research (formerly CERN), he shared a Nobel prize (1984) with Carlo Rubbia for their experiments at CERN (with colliding beams) that established the existence of W and Z particles.

van der Waals, Johannes Diderik (1837–1923) Dutch physicist, who worked on the liquefaction of gases and whose paper "On the continuity of the liquid and gaseous states" had a profound influence on liquefaction technique.

van der Waals equation of state The first equation of state proposed to describe both the gaseous and liquid phases of a substance (1873), which remains the most useful equation for the approximate analysis of fluid properties. It can be written for one mole of fluid:
$$(p + a/V^2)(V - b) = RT,$$
where p is pressure, V volume, T the thermodynamic temperature, R the molar gas constant, and a and b are constants characteristic of a given substance. The quantity b allows for the volume effectively occupied by molecules because of the very short-range repulsive forces and a/V^2 allows for the short-range attractive forces.

The equation gives a more accurate account of the properties of gases than does the *ideal gas equation. It predicts a critical point at which the value of RT/pV is 2.67, compared with values in the range 3 to 4 found by experiment for real fluids. It is less accurate for the liquid phase but is interesting in that it predicts the capacity of liquids to resist negative pressures, and the existence of supercooled vapours and superheated liquids. Since there are only three arbitrary constants a, b, and R, the law of corresponding states applies and the equation may be written in the reduced form (*see* reduced equation of state),
$$(\propto + 3/\beta^2)(3\beta - 1) = 8\gamma,$$
where α, β, and γ are the reduced values of the pressure, volume, and temperature. *See also* equations of state.

van der Waals forces Intermolecular and interatomic forces other than *covalent and *electrovalent bonds. These forces are all electrostatic in origin. If two identical molecules have permanent *dipole moments and are in random thermal motion then some of their relative orientations cause repulsion and some attraction. On average there will be a net attraction (*orientation interaction*) and the potential energy of the molecules is given by:
$$E_p = -2\mu^4/3r^6kT,$$
where μ is the dipole moment, r the molecular separation, k the *Boltzmann constant, T the thermodynamic temperature, and ε_0 is the electric constant.

A molecule with a permanent dipole can also induce a dipole in a similar neighbouring molecule and cause mutual attraction (*induction interaction*). The potential energy due to this type of interaction is given by:

$$E_p = -2\alpha\mu^6/r^6,$$

where α is the *molecular polarization. These dipole-dipole and dipole-induced dipole interactions cannot occur between atoms.

The van der Waals forces between single atoms are called *dispersion forces* and they arise because of small instantaneous dipole moments in the atoms themselves. For example hydrogen, with one proton and one electron, has a symmetrical electron cloud about the nucleus. However, at any instant there is an asymmetric charge distribution that depends on the position of the electron. Thus the atom can be considered to have a fluctuating rotating dipole moment. There is no net attraction between dipoles of neighbouring atoms because there is insufficient time for orientation to occur. However, an instantaneous dipole in one atom can polarize a similar neighbouring atom and the potential energy of the two is given by:

$$E_p = -h\nu_0\alpha^6/r^6,$$

where h is the *Planck constant and ν_0 the frequency of oscillation of the charge distribution.

The three types of interaction all have the form: $E_p = -A/r^6$ where A is a constant for a particular atom or molecule. The full expression for potential energy includes a term representing repulsion between the atoms:

$$E_p = B/r^6 - A/r^n,$$

where B is a constant and n lies between 9 and 12. E_p is called the *Lennard-Jones potential*. Van der Waals forces are responsible for departures from *ideal gas behaviour in real gases. They are also the forces between molecules or atoms in liquids and in non-ionic solids.

vane anemometer See anemometer.

vane radiometer See Nichols's vane radiometer.

van't Hoff factor Symbol: i. A factor introduced into the equation for osmotic pressure (*see* osmosis) to account for the fact that electrolyte solutions deviate from this equation and have a higher osmotic pressure than that expected. This arises because the osmotic pressure of a solution depends on the number of entities (molecules, atoms, ions) present and in electrolyte solutions *dissociation into ions occurs. The factor is the ratio of the actual number of entities present in the solution to the number that would be present if no dissociation occurred. The equation for the osmotic pressure (Π) of electrolytes is $\Pi V = iRT$.

van't Hoff's equation See isochore of reaction.

Van Vleck, John Hasbrouck (1899–1980) Amer. physicist. A professor at Harvard, he is known for his work on the quantum mechanical theory of magnetism and the phenomenon known as Van Vleck *paramagnetism. In 1977 he shared a Nobel prize with Philip Anderson and Nevill Molt for their work on the electronic structure of magnetic and disordered systems.

Van Vleck paramagnetism See paramagnetism.

vaporization A change from solid or liquid to vapour. The rate of vaporization per unit surface area of solid or liquid at temperature T is given by

$$dm/dt = \alpha p\sqrt{(M/2\pi RT)},$$

where p is the vapour pressure of the vapour of relative molecular mass M, R is the molar gas constant, and α is the *vaporization coefficient*, which is necessarily less than unity. It may be measured by determinations of the rate of vaporization at temperatures at which the vapour pressure is known.

vapour A substance, in gaseous form but below its *critical temperature so that it could be liquefied by pressure alone, without cooling to a lower temperature.

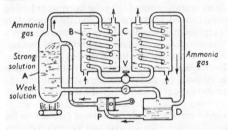

Ammonia-absorption refrigerator

vapour-absorption refrigerator An early type of refrigerator consisting of a boiler A containing a strong aqueous solution of ammonia. On heating this, ammonia gas passes into the water-cooled coils in B where it condenses, the liquid then passing through a valve V to a spiral in the refrigerating chamber C. The liquid ammonia evaporates and cools the brine in C. The gas is then absorbed by dilute ammonia solution in the absorber D, the concentrated solution being pumped into the boiler via P. The valve V is adjusted to maintain the desired difference of pressure between the two sides.

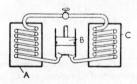

Vapour-compression refrigerator

vapour-compression refrigerator A *refrigerator having as its working substance one with a large vapour pressure at moderate temperatures. The refrigerant evaporates in a coil immersed in a brine bath A, taking its latent heat from the brine. The vapour is removed by the pump B and compressed to the liquid in the coil immersed in the vessel C which contains cold water to remove the latent heat

vapour concentration

liberated by the liquefaction. The cycle is completed by the liquid passing through a valve to the spiral in A.

vapour concentration *See* humidity.

vapour density The mass per unit volume of a vapour under specified conditions of temperature and pressure. It is sometimes defined relative to the vapour density of hydrogen under the same conditions of temperature and pressure. It may be determined experimentally by Victor Meyer's method (*see* Meyer).

vapour pressure The pressure exerted by a *vapour. For a vapour in equilibrium with its liquid, the vapour pressure depends only on the temperature of the liquid for a plane surface and is known as the saturated vapour pressure (SVP) at that temperature.

The SVP of a liquid may be measured by the direct or static method or by the dynamic or boiling point method. In the static method, the temperature is kept constant and the SVP determined either by a manometer as in Regnault's method or by measuring the density of the saturated vapour (*see* Fairbairn and Tate's measurements). In the dynamic method, the pressure is kept constant and the temperature at which the liquid boils is determined, use being made of the fact that a liquid boils when its vapour pressure equals the external pressure on the liquid surface. (*See* Ramsay and Young's measurement; Smith and Menzies' vapour pressure measurements.) For the measurement of the vapour pressure of a metal at its freezing point *see* Egerton's effusion method.

The variation of vapour pressure with temperature over a small range is given by the Kirchhoff formula,

$$\log p = A - B/T - C \log T,$$

where A, B, and C are constants and T the thermodynamic temperature. The general equation for vapour pressure is

$$\log p = -\frac{L_0}{RT} + \frac{5}{2} \log T$$

$$+ \int_0^T \frac{dT}{RT^2} \int_0^T (C_1 - C_p)\, dT + i,$$

where L_0 is the molar latent heat of vaporization at absolute zero, C_1 is the internal molar heat capacity due to rotations and vibrations, and C_p is the molar heat capacity of the solid; i is an integration constant known as the *chemical constant.

vapour-pressure thermometer A thermometer that uses the fact that the (saturation) vapour pressure of a liquid is a function only of temperature, in order to measure temperatures. Thermometers of this type are the most reliable for the measurement of temperatures below the boiling point of helium (−268 °C). For this application, a small bulb, a (*see* diagram), containing the liquid, is connected by glass capillary

596

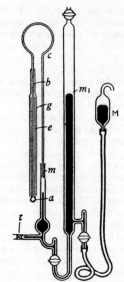

Vapour-pressure thermometer

tubes b, c to the mercury manometer m, m_1 whose levels are altered by means of the reservoir M. Uniformity of temperature is achieved by surrounding the bulb a and part of the tube b by a copper tube e, and radiation of heat from the bulb and from the tube b is prevented by surrounding it with a glass tube g closed at the top. To set up the apparatus the level m is lowered, the system evacuated and the gas admitted at t. By raising M the gas is compressed and part of it liquefies in the bulb a which is then immersed in the bath whose temperature is required, the vapour pressure being measured on the manometer. From tables giving the vapour pressure of the liquid at various temperatures the corresponding temperature may be found. He, NH_3, SO_2, CO_2, CH_4, C_2H_4, O_2, and H_2 may be used as the working liquid in the thermometer; their great disadvantage is that their range is very limited since very high or very low pressures cannot be conveniently measured. Low pressures are measured with a *Pirani gauge, a correction being made for the fact that, due to *thermal transpiration, the pressure in the bulb differs from that in the manometer.

Simple vapour-pressure thermometers with a Bourdon-type indicator (*see* Bourdon tube) are commonly used where an open scale is required at the upper part of the range, e.g. in dashboard operation to indicate radiator temperature of vehicles.

var A unit of *power identical to the *watt but used for the reactive power of an alternating current. One var is one volt times one ampere.

varactor A semiconductor *diode, operated with reverse bias so that it acts as a voltage-dependent capacitor. The *depletion layer at the junction acts as the dielectric, the n- and p-regions act as the plates. A diode used in this way is usually designed to have an unusually large capacitance. The depletion-layer width, and therefore the capacitance, depend on the voltage across the junction. If the semiconductor type changes abruptly from n-type to p-type then $C \propto V^{-1/2}$.

If it changes gradually (linearly graded junctions) then $C \propto V^{-1/3}$. A *Schottky diode can be used as a varactor in a similar way.

varactor tuning A means of tuning employed in receivers (e.g. television receivers) in which *varactors are used as the variable-capacitance elements.

variable-focus condenser See Abbe condenser.

variable mu valve A *thermionic valve whose *amplification factor varies with the grid voltage. This is usually achieved by constructing the grid from nonuniformly spaced wires.

variable star A star whose physical properties, most noticeably brightness, vary either regularly or irregularly with time. The variation in an *intrinsic variable* is caused by changes in internal conditions. A *pulsating star* is of this kind, the light variation being due to expansion and contraction of the surface layers of the star. The variation can also be due to external causes. Two stars rotating around a common centre of gravity (*binary stars) can sometimes eclipse one another if the plane of the orbit lies in the line of sight. These are known as *eclipsing binaries*. *Novas are an example of *cataclysmic variables*, the brightness increasing by an enormous amount in a very short time.

variance The square of the standard *deviation.

variation See geomagnetism.

Varignon's theorem See moment.

variometer A variable inductor that usually consists of a fixed coil connected in series with a movable coil so that by moving (rotating) the latter the coupling between the two coils, and hence also the self inductance of the series combination, may be varied.

varistor A *resistor with characteristics that do not follow Ohm's law. It may be formed from a semiconductor *diode. A symmetrical varistor consists of two diodes connected in parallel with opposite polarity. This arrangement exhibits the forward current-voltage characteristic of a diode in either direction of applied voltage, and may be used as a *voltage limiter*.

V-beam radar See radar.

VDU Abbreviation for *visual-display unit.

vector A quantity with magnitude and direction that can be represented by a line whose length is proportional to the magnitude and whose direction is that of the vector. It can be represented by three components in a rectangular coordinate system, which may be expressed as a matrix. (*See also* unit vector.)

A true vector, or *polar vector*, involves a displacement or virtual displacement. Polar vectors include velocity, acceleration, force, electric and magnetic field strength. They are odd functions, that is the signs of their components are reversed on reversing the coordinate axes. Their dimensions include length to an odd power.

A *pseudovector*, or *axial vector*, involves the orientation of an axis in space. The direction is conventionally obtained in a right-handed system by sighting along the axis so that the rotation appears clockwise. Pseudovectors include angular velocity, vector area, and magnetic flux density. They are even functions, that is the signs of their components are unchanged on reversing the coordinate axes. Their dimensions include length to an even power. A *free vector* is an axial vector, such as a couple, where the direction of the axis is given but it lacks location.

Polar vectors and axial vectors obey the same laws of vector analysis.

Vector analysis is a branch of mathematics by means of which vectors may be handled in a way consistent with the physical problems in which they occur.

(*a*) *Vector addition.* The law of vector addition is that if the two vectors *A* and *B* are represented in magnitude and direction by the adjacent sides of a parallelogram, the diagonal represents the vector sum (*A* + *B*) in magnitude and direction. Obviously the *vector triangle* also shown is equivalent to the parallelogram construction. Forces, velocities, etc., combine in this way. (*See* polygon of vectors.)

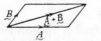

(*b*) *Vector multiplication.* There are two ways of multiplying vectors. (i) The *scalar product* of two vectors equals the product of their magnitudes and the cosine of the angle between them, and is a scalar quantity. For example, the amount of work done by a force when its point of application is moved through a given displacement is the scalar product of the vectors representing the force and the displacement. (ii) The *vector product* of two vectors *A* and *B* is defined as a pseudovector of magnitude $AB \sin \theta$, having a direction perpendicular to the plane containing them. The sense of the product along this perpendicular is defined by the rule: if *A* is turned towards *B* through the smaller angle, this rotation appears clockwise when sighting along the direction of the vector product.

Notations of vector analysis:

Scalar product: *AB*; (*AB*); *A* · *B*; (the last is read as "*A* dot *B*").

Vector product: [*AB*]; *A* × *B* (read as "*A* cross *B*"). (The dot and cross notation is specially recommended.) Vectors should be distinguished from scalars by printing the symbols in bold italic letters.

vector analysis See vector.

vector field A *field, such as a gravitational field or

vector potential

magnetic field, in which the magnitude and direction of the vector quantity are one-valued functions of position. They can be mapped by curved lines whose direction at any point is that of the vector and whose density (i.e. number per unit area crossing an infinitesimally small area perpendicular to the lines) is proportional to the magnitude of the vector at the point. These lines are called lines of flux (or force).

vector potential *See* magnetic vector potential.

vector product *See* vector.

Vegard's law To a first approximation the lattice parameters of a primary solid solution vary linearly with the atomic percentage of the solute element.

velocity 1. Linear velocity, symbol: v. The average velocity is the displacement divided by the time taken. Instantaneous velocity is the rate of change of displacement. Linear velocity is a polar *vector. The magnitude of the instantaneous velocity is called speed, which is a scalar. In precise technical usage velocity and speed are clearly distinguished, the former only being used when the direction of motion is specified. The distinction is particularly significant in the case of a body moving with constant speed in a curved path. As the direction is changing the velocity is not constant although its magnitude is, so the body has an *acceleration and must be subject to a resultant force (*see* Newton's laws of motion). The term velocity is however often used loosely to mean speed.

2. *See* angular velocity.

velocity modulation If a beam of electrons passes through a sharply defined region, such as a *cavity resonator, and is subjected to a radio-frequency field, the individual electrons will be retarded or accelerated according to the half-cycle prevailing when they enter the region. If retarded, the electrons following will catch up (and the converse) so that the total effect will be a *bunching* of the electron beam into a series of pulses similar to the rarefactions and compressions of a sound wave. Such a beam is said to be velocity modulated. Velocity modulation is employed for the amplification and generation of microwave frequencies in *electron tubes such as the *klystron and *travelling-wave tube.

velocity of light *See* speed of light (the more correct term).

velocity of sound *See* speed of sound (the more correct term).

velocity potential If the velocity of a point (x, y, z) of a fluid has components (Cartesian) u, v, w, and there exists a scalar function ϕ such that:

$$u = -\partial\phi/\partial x$$
$$v = -\partial\phi/\partial y, \quad w = -\partial\phi/\partial z,$$

then the motion is irrotational and ϕ is called the velocity potential. The negative sign is conventional and is sometimes omitted.

velocity ratio *See* machine.

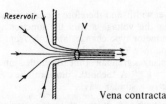

Vena contracta

vena contracta When a jet issues from an orifice in a reservoir tank the change in direction of the stream lines is not completed at the orifice but continues past the orifice causing the subsequent jet section to be smaller than that of the orifice. The point of the jet where the contraction is complete is called the vena contracta, V, and at this point the stream lines are all parallel and the pressure of the fluid is that of the surrounding medium except for a small effect of surface tension.

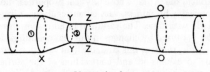

Venturi tube

venturi tube A device used for measuring the quantity of fluid flowing through a pipe. The principal features are: the inlet section, XY (*see* diagram), which converges to the throat, YZ, which consists of a short straight portion of the pipe. From the throat the pipe diverges again, ZO, usually to the original size. The quantity of fluid flowing per second through the pipe is given by:

$$Q = \frac{A_1 A_2}{\sqrt{A_1^2 - A_2^2}} \sqrt{\frac{2(p_1 - p_2)}{\rho}}$$

where A is the area of cross section, p the pressure (static) of the fluid, ρ the fluid density; the suffixes 1 and 2 refer to the inlet and throat respectively.

Verdet constant *See* Faraday effect.

vergence The convergence and divergence of rays. Reciprocal distances measure vergence. *Reduced vergence* is the reciprocal of a *reduced distance. The change of vergence of rays is equal to the focal power of an optical element. If L = reduced object vergence, F = reduced power, L' = reduced image vergence, then $L + F = L'$, where $F = (n' - n)/r$, for a single surface. This formula holds for reflection or refraction through any media whose refractive indexes are known. Further, it holds for single surface refraction or for vergences referred to principal planes.

vernal equinox *See* equinoxes.

Vernier, Pierre (*c.* 1580–1637) French mathematician and soldier who invented the *vernier scale.

vernier callipers Sliding *callipers whose scale is fitted with a *vernier scale.

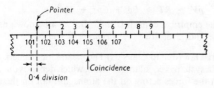

Vernier scale

vernier scale A short scale sliding on the main scale of a length- or angle-measuring instrument. It is used for determining the fraction of the smallest interval into which the main scale is divided by the instrument pointer, which is the zero division of the vernier scale. For example, in the diagram the vernier scale has 10 intervals equal to 9 intervals on the main scale. The zero division of the vernier (the pointer) is seen (by estimating 1/10ths) to be reading 101.4 divisions on the main scale. The fractional part of the reading is given with greater certainty by the 4th division of the vernier being exactly in line with a division on the main scale.

Angular verniers which divide a degree into minutes work on the same *principle of coincidence.*

versor A factor of a *vector determining its geometrical direction, which when multiplied by the *tensor gives the complete expression for the vector.

vertex focal length The distance measured from the last surface of a thick lens or combination of lenses to the principal focus. Its reciprocal is the *vertex power.* The back vertex focal length (shortened to back focal length) is of particular importance in *spectacle lenses and photographic objectives.

vertex polarization *See* horizontal polarization.

very high frequency (VHF) *See* frequency bands.

very low frequency (VLF) *See* frequency bands.

VHF Abbreviation for very high frequency. *See* frequency bands.

vibration The rapid to and fro motion characteristic of an elastic solid (e.g. a tuning fork) or a fluid medium influenced by such a solid. The time occupied in each to and fro motion is constant and is called the *period. The frequency is the number of vibrations per unit time and is, therefore, the inverse of the period. *See* column of air; stretched string.

vibrational quantum number *See* energy level.

vibration-damper In an overhead electric line. A device attached to a line conductor to prevent vibrations (produced in the conductor by wind) from reaching the part of the conductor that is at the clamp or other support. It thus prevents the damage to the conductor (breaking by fatigue) which such vibrations can cause. A common type is the Stockbridge vibration-damper which consists of two weights attached to the ends of a straight piece of stranded cable 1 or 2 ft. in length. The stranded cable is clamped at its mid-point to the line conductor. The energy of vibration is largely absorbed by

this piece of stranded cable and the vibrations are rapidly damped out.

vibration galvanometer A tuned current-detecting instrument for use in alternating-current bridge

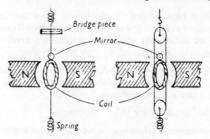

Vibration galvanometer

measurements in place of a telephone receiver. Vibration galvanometers are usually of the moving-coil type (though moving-magnet types are possible) and can be tuned for frequencies between 5 and 1000 hertz. The coil is suspended on a single or double strand of wire, under tension as in the diagram. Tuning is carried out by varying the spring tension or the position of the bridge piece.

vibration of a magnet The time of oscillation of a freely suspended bar magnet in a magnetic field is given by:
$$T = 2\pi\sqrt{(I/MH)},$$
for oscillations of small amplitude, where T is the time of one complete oscillation, I is the moment of inertia of the magnet, M is the moment of the magnet, and H is the field strength. For a rectangular magnet of length l and breadth b:
$$I = m(l^2 + b^2)/12$$
and for a cylindrical bar magnet of length l and radius r:
$$I = m(l^2/12 + r^2/4).$$
The vibration of a magnet can easily be used to compare field strengths as it will be seen that $H_1/H_2 = T_2^2/T_1^2$, or, if n_1 and n_2 are the number of vibrations noted in the same period of time,
$$H_1/H_2 = n_1^2/n_2^2.$$

vibration–rotation spectrum *See* spectrum.

vibrator A device for producing an alternating current by periodically interrupting or reversing the current obtained from a direct-current source. It is operated electromagnetically and has a vibrating armature that alternately makes and breaks one or more pairs of contacts. It is most commonly used in a power-supply unit that is required to produce direct current at high voltage from a low-voltage d.c. source, such as a battery. The vibrator produces a low-voltage a.c. supply that a transformer converts into a high-voltage a.c. supply, and the latter is then rectified to produce a high-voltage d.c. supply. The rectification may be carried out by means of a *rectifier or the vibrator may be fitted with additional contacts which are used to reverse the connections to the secondary winding of the transformer synchronously with the reversals of current in the primary winding, so that a separate rectifier is not required. This latter type of vibrator is called a *synchronous vibrator*.

vicinal faces Smooth bright crystal faces that are inclined at only a few minutes of arc to crystal faces of low indices; the indices of a vicinal face therefore can only be expressed by very high numbers.

Victor Meyer's apparatus An apparatus, not capable of very high accuracy, enabling the vapour density of a liquid to be deduced. *See* Meyer.

video amplifier *See* video frequency.

videocassette, videocassette recorder *See* videotape.

video frequency The frequency of any component of the signal produced by a *television camera. Video frequencies lie between 10 Hz and 2 MHz. An amplifier designed to amplify video-frequency signals is called a *video amplifier*.

video signal *See* television camera.

videotape A form of magnetic tape that is suitable for use with a *television camera. Simultaneous recording of the video signal from the TV camera and the audio signal from the microphone system is carried out on separate tracks on the videotape. Many TV programmes are recorded on videotape before they are transmitted.

A form of videotape recorder is available for use with domestic TV receivers. The tape is enclosed in a container for protection and easy handling; the package is called a *videocassette*. The *videocassette recorder* (VCR) can be used to record a TV programme on tape, which can then be subsequently replayed directly into the TV receiver, or it can replay a prerecorded videocassette.

vidicon *See* camera tube.

vignetting The progressive reduction in the cross-sectional area of a beam of light passing through an optical system as the obliquity of the beam is increased. It is due to obstruction of the beam by mechanical apertures, lens mounts, etc., of the system.

Villari reversal In iron and steel, longitudinal tension increases magnetization when the magnetizing field is weak, but the reverse effect occurs in strong fields (E. Villari, 1868).

virgin neutrons Neutrons from any source that have not been involved in any collisions.

virial In a system in which an atom having coordinates (x,y,z) is acted upon by a force having components X, Y, Z parallel to these axes, the virial is defined as the average value with respect to time of the sum of all expressions of the form:
$$-\tfrac{1}{2}(xX + yY + zZ).$$

virial expansion A relation expressing the behaviour of a real gas:

$$pV = RT + Bp + Cp^2 + Dp^3 + \ldots.$$

The empirical constants $B, C, D \ldots$ are known as the 2nd, 3rd, 4th, ... *virial coefficients*.

virial law (due to Clausius) The mean kinetic energy of a system is equal to its *virial, which depends only on the forces acting on the atoms and not on their motions.

virtual cathode In a *thermionic valve. The surface, situated in the *space-charge region, at which the potential is a mathematical minimum and the potential gradient (or electric force) is zero. It acts as if it were a source of electrons.

virtual image (and **object**) When rays only appear to be diverging from an image behind a mirror, or from an image on the same side as the object in the case of the lens, the image is said to be virtual. For concave mirrors and convex lenses, virtual images are produced when the real object lies between the focal point and the optical element or first principal plane. Virtual images cannot be received on screens. A virtual object requires that the incident rays are converging to a real image but before they focus, the mirror or lens intercepts the converging pencils.

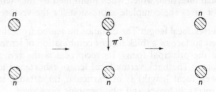

Interaction of neutrons by exchange of a virtual pion

virtual particle Because of the *uncertainty principle it is possible for the law of *conservation of mass and energy to be broken by an amount ΔE providing this only occurs for a time Δt such that:
$$\Delta E \Delta t \le h/4\pi.$$

This makes it possible for particles to be created for short periods of time when their creation would normally violate conservation of energy. These particles are called virtual particles. The electrostatic force between charged particles may be described in terms of the emission and absorption of virtual photons by the particles. Similarly the *strong interaction between nucleons may be thought of as being due to the emission and absorption of virtual hadrons. The diagram below illustrates how two neutrons can interact by the exchange of virtual π°. The law of conservation of energy is broken for a time Δt, where:

$$\Delta t \le \frac{h}{4\pi c^2 m_\pi},$$

m_ρ being the mass of the pion. Other conservation laws such as those applying to angular momentum, *isospin, etc., cannot be violated even for short periods of time.

virtual-work principle A principle much used in solving problems involving forces in equilibrium: it states that a system with workless constraints is in equilibrium under applied forces if, and only if, zero (virtual) work is done by the applied forces in an arbitrary infinitesimal displacement satisfying the constraints. *See* constrain.

viscometer A device for measuring the viscosity of a fluid (liquid or gas). The main types include those based on (i) the flow of fluids through capillary tubes (Poiseuille, Ostwald, Rankine); (ii) the time of fall of a sphere through a liquid (Stokes); (iii) the torque needed to keep two coaxial cylinders in rotation when the space between is filled with the liquid under test (Couette, Hatschek, Searle); (iv) the rate of damping of a vibrating body by the fluid (Maxwell, Stokes). *See also* Ostwald viscometer; Couette viscometer.

viscosity The property of fluids by virtue of which they offer a resistance to flow for low values of the *Reynolds number. Newton's law of viscous flow for streamline, as opposed to turbulent, liquid motion is:
$$F = \eta A \, dv/dx,$$
where F is the tangential force between two parallel layers of liquid of area A, dx apart, moving with a relative velocity dv. The quantity η is called the *coefficient of viscosity* (or just the viscosity) of the liquid and is measured in $kg\, m^{-1}\, s^{-1}$ or $N\, s\, m^{-2}$. Viscosity of a liquid usually decreases with temperature but that of a gas increases.

A very large number of liquids obey Newton's law in that the viscosity is independent of the velocity gradient (dv/dx); these are called *Newtonian fluids. (*See also* anomalous viscosity; kinematic viscosity.)

viscosity gauge *See* molecular gauge.

viscosity manometer *See* molecular gauge.

viscosity of a gas On the *kinetic theory, the viscosity of a gas is given by the equation:
$$\eta = \tfrac{1}{3}\rho \bar{C} L,$$
where ρ is the density, $\bar{C}$ the mean speed, and L the mean free path. Since the product (ρL) is independent of the pressure, on this theory the viscosity of a gas should be independent of the pressure (Maxwell's law), a fact that is found to be true over a large range of pressures. At very low pressures, however, the law breaks down, the effective viscosity becoming proportional to the pressure, a fact which is made use of in the design of low-pressure manometers. The mass of gas flowing per second, Q, through a tube of length l and radius a under a pressure difference $(p_1 - p_2)$ is given by
$$Q = \frac{\pi}{2} \frac{a^3}{l} \frac{p_1 - p_2}{\alpha} \sqrt{\frac{2\pi M}{RT}},$$
when the pressure is extremely low, α being the *accommodation coefficient and M the molecular weight of the gas. The flow at low pressures is thus quite different from that at ordinary pressures where the flow depends on a^4 and $(p_1{}^2 - p_2{}^2)$.

viscous damping Damping in which the opposing force is proportional to velocity, as with the damping resulting from *viscosity of a fluid or from *eddy currents.

viscous flow *See* Newton's laws of fluid friction; viscosity.

visible spectrum The continuous *spectrum of visible radiation, i.e. radiation lying in the wavelength range between approximately 380 and 780 nm. It is seen in the *rainbow and in the display of colours produced when a beam of white light is dispersed by a prism or *diffraction grating. There is a continuous variation of wavelength but six colours are usually distinguished: violet, blue, green, yellow, orange, and red; red is the component of longest wavelength. The value of the longest wavelength to which the eye is sensitive depends on the brightness.

visual acuity Keenness of vision generally measured by letters of standard dimensions (Snellen test type). The limbs of the letters are one-fifth the total height of the letter, and the width four-fifths or five-fifths of the height. Normal vision (Snellen) is considered to hold when the vertical height is five minutes of angle. Letters of different sizes are marked according to the distance that a normal eye could detect them. When testing, the letters are placed 6 metres away; the smallest line of letters that can be read is then recorded. Suppose the letter marked 18 metres can only be read at 6 metres, then the visual acuity is expressed as 6/18.

visual angle The angle subtended by an object at the nodal point of the eye.

visual-display unit (VDU) A device that displays the output from a *computer temporarily on a screen, generally that of a *cathode-ray tube. The information may be in the form of letters, numbers, and other characters or may be graphical, e.g. diagrams, graphs, etc. A VDU is usually paired with a keyboard, by which information can be fed into the computer, and there are often other input devices such as a mouse or light pen.

VLF Abbreviation for very-low frequency. *See* frequency bands.

VLSI *See* integrated circuit.

V-number *See* Abbe number.

voice frequency *See* audiofrequency.

Voigt effect The *double refraction produced when light traverses a vapour acted on by a transverse magnetic field (Voigt, 1902), the vapour acting as a *uniaxial crystal with axis parallel to the field direction.

volt Symbol: V. The *SI unit of electric *potential, *potential difference, and *electromotive force, defined as the potential difference between two points

Volta, Count Alessandro

on a conductor carrying a current of one ampere when the power dissipated is one watt. In practice voltages are measured by comparison with the electromotive force of a *Weston standard cell using a *potentiometer.

Volta, Count Alessandro (1745–1827) Italian physicist. The inventor of the voltaic pile and of the electroscope; he played an important part in the early development of current electricity; the *volt* was named after him.

voltage Symbol: V. A term loosely used for the potential difference between two specified points in a circuit or device or for electromotive force. It is expressed in volts. *See also* active voltage; reactive voltage.

voltage amplifier *See* amplifier.

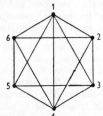

Hexagon voltage

voltage between lines *Syn.* line voltage; voltage between phases. Of an electrical power system. The voltage between the two lines of a *single-phase system, or between any two lines of a symmetrical *three-phase system, consider the lines to be arranged at the corners of a regular hexagon in correct order of phase sequence round the periphery, as shown. Then, the voltage between lines (or *hexagon voltage* or *mesh voltage*) is the voltage between any two consecutive lines (e.g. between 1 and 2). The voltage between alternate lines (e.g. between 1 and 3) is called the *delta voltage*, and the voltage between opposite lines (e.g. between 1 and 4) is called the *diametral voltage*.

voltage between phases *See* voltage between lines.

voltage divider *See* potential divider.

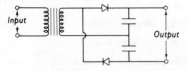

Voltage doubler

voltage doubler An arrangement of two rectifiers to give double the voltage output of a single rectifier. The diagram shows a typical circuit for *diode rectifiers.

voltage drop 1. (general) The voltage between any two specified points of an electrical conductor (such

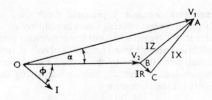

Vector diagram for a transformer supplying
a lagging load

as the terminals of a circuit element or component) due to the flow of current between them. The voltage drop is equal to the product of the current and the resistance between the two points (for direct current) or the product of the current and the impedance between the two points (for alternating current). In the case of a.c. the product of the current and the resistance gives the *resistance drop*, which is in phase with the current, whereas the product of the current and the reactance gives the *reactance drop*, which is in *quadrature with the current.

2. In transformers. A *transformer may be considered as having a series internal impedance (Z) composed of a resistance (R) in series with an inductive reactance (X). When the transformer is supplying a load, this impedance causes the voltage (V_2) at the load terminals to be, in general, different from the internal voltage (V_1). The voltage drop (or impedance drop) is defined as the arithmetic difference between V_1 and V_2. For some conditions of loading, V_2 may be greater than V_1, in which case the voltage drop is negative and is described as a *voltage rise* (or impedance rise). In the vector diagram, the *lagging load takes a current (I) lagging the voltage V_2 by an angle φ. BC (equal to IR volts) is the resistance voltage that is in phase with the current I (BC is parallel to OI). CA (equal to IX volts) is the reactance voltage that leads the current I by 90° (CA is at right angles to OI). BA (equal to IZ volts) is the impedance voltage that is the vector sum of BC and CA. Since the angle α is usually less than about 5° it can be shown that, to a very close approximation, the voltage drop is equal to:
$$IR \cos φ \pm IX \sin φ,$$
the negative sign being used when I is a *leading current. $IR \cos φ$ is the voltage drop due to resistance (the resistance drop). $IX \sin φ$ is the voltage drop (or rise if negative) due to reactance (the reactance drop or rise). In machines, the impedance drop, resistance drop, and reactance drop are less than the impedance voltage, resistance voltage, and reactance voltage respectively. The above may also be applied to a.c. generators. (*See* inherent regulation; regulation.)

3. *See* anode drop; gas-discharge tube.

voltage feedback *See* feedback.

voltage limiter *See* varistor.

voltage ratio *See* ratio.

voltage regulation *See* regulation.

voltage regulator diode *See* Zener diode.

voltage stabilizer A device or circuit designed to maintain a voltage at its output terminals that is

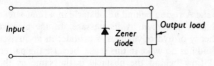

a　Zener-diode circuit

substantially constant and independent of either variations in the input voltage or in the load current. Fig. *a* shows a typical circuit using a *Zener diode and Fig. *b* shows a stabilizer in which the load impedance is in series with the circuit.

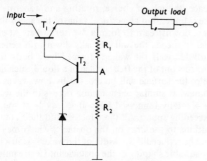

b　Series stabilizer circuit

voltage to neutral Of an electrical power system. In three-phase and six-phase systems, the voltage between any line conductor and the *neutral point of the system. It has more than one value unless the system is symmetrical. *Compare* voltage between lines; phase voltage. *See* polyphase system.

voltage transformer *Syn.* potential transformer. An instrument transformer utilizing the voltage-transformation property of a transformer. The primary winding is connected to the main circuit and the secondary winding is connected to a measuring instrument (e.g. voltmeter). Voltage transformers are extensively used to extend the range of a.c. instruments and to isolate instruments from high-voltage circuits.

voltaic cell *See* cell.

voltameter Former name for coulombmeter.

volt-amperes Symbol: VA. The unit of apparent electric *power, defined as the product of *root-mean-square values of voltage and current in an alternating-current circuit. *See* active volt-amperes; reactive volt-amperes.

Volta's pile An early (1800) battery consisting of a pile of plates of zinc and copper separated by pieces of felt moistened with dilute sulphuric acid.

voltmeter A device for measuring *potential differences. Voltmeters in common use include *digital voltmeters, *cathode-ray oscilloscopes, and d.c. instruments such as permanent-magnet moving-coil devices. Voltmeters should cause no appreciable disturbance in the circuit to which they are connected. They therefore require very high input impedances so that very little current is taken from the circuit. Digital voltmeters and oscilloscopes comply with this requirement, but a large series impedance is used with moving-coil instruments to increase their input impedances.

volume 1. Symbol: *V*. The amount of space occupied by a body. It is measured in cubic metres or litres.
　2. The general loudness of sounds, or the magnitude of transmitted audiofrequency signals giving rise to sounds. *See* automatic gain control; volume compressors (and expanders).

volume charge density *Syn.* volume density of charge. *See* charge density.

volume compressors (and **expanders**) A compressor is an electrical device that automatically reduces the range of amplitude variations of an audiofrequency (a.f.) signal in a transmission system: it decreases the amplification when the signal amplitude exceeds a predetermined value and increases the amplification when the signal amplitude is less than a second predetermined value. A volume expander is a device that produces the opposite effect to a compressor, i.e. it automatically extends the range of the amplitude variations of the transmitted a.f. signal. With suitable design, an expander included in one part of the system can be made to compensate for the effect of a compressor in another part of the system.
　In recording sound on a film track or on a gramophone record, a compressor may be used effectively to reduce the volume range of the recorded sound. An expander may then be used in the sound-reproducing apparatus to restore the original volume range. The *signal-to-noise ratio of a transmission system (e.g. radio-telephone transmission system) can be improved by using a compressor at the transmitter and an expander at the receiver. A compressor and an expander used in this manner are together described as a *compandor*.

volume density of charge *Syn.* volume charge density. *See* charge.

volume elasticity *See* modulus of elasticity.

volume expander *See* volume compressors (and expanders).

volume strain *Syn.* bulk strain. *See* strain.

von Laue, Max *See* Laue, Max von.

vortex *Syn.* vortex filament. A fluid contained within a tube formed by drawing, through every point of a small closed curve in the fluid, the lines for which the tangent at any point of the line is in the direction of the instantaneous axis of rotation of the fluid at that point. These lines are called the vortex lines. The strength of a vortex is defined as the circulation round any circuit embracing the vortex tube and is constant for all time (*see* circulation). The vortex tube cannot begin or end at an interior point of the fluid but only at the fluid boundaries or at a free surface. *See* vortex street.

Vortex street

vortex street When an obstacle such as a cylinder is set with its axis perpendicular to the direction of motion of a fluid and the relative velocity exceeds that critical value determined by the *Reynolds number exceeding 30, the vortices formed by the rolling up of the two *surfaces of discontinuity in the wake arrange themselves into two equispaced parallel rows, forming an *avenue* or *street* in which the one set lies intermediate in station to the other. Von Kármán showed that the system will continue in stable formation as it marches downstream, i.e. preserving the relative spacing, provided $h/l = 0.28$ and $a = l/2$, where h is the lateral and l the longitudinal spacing in the street. This result has been confirmed experimentally.

The vortices are initiated periodically on either side of the obstacle as the system proceeds downstream, thereby setting up cross forces upon it which modify its resistance and may give rise to forced oscillation of the obstacle across the direction of flow. (*See* aeolian tone.)

vortical field *See* curl.

vorticity In the three-dimensional motion of a fluid the velocity of an element of fluid at the point (x, y, z) has Cartesian components u, v, w. The general motion of the element is three-part: (*a*) a general translation; (*b*) a pure strain motion; (*c*) a rotational motion of the whole element about an instantaneous axis. The component angular velocities in (*c*) are given by $\frac{1}{2}\xi$, $\frac{1}{2}\eta$, $\frac{1}{2}\zeta$, where:

$$\xi = \frac{\partial w}{\partial y} - \frac{\partial v}{\partial z}, \eta = \frac{\partial u}{\partial z} - \frac{\partial w}{\partial x}, \zeta = \frac{\partial v}{\partial x} - \frac{\partial u}{\partial y}.$$

The vector having the components ξ, η, ζ is called the vorticity at the point (x,y,z).

V-ring Of a *commutator fitted to electrical machines. (*a*) Metal V-ring. *Syn.* metal V-collar. A metal ring having a V-section that is used to clamp the commutator bars. (*b*) Mica V-ring. *Syn.* mica cone. A ring made out of a mica compound, such as micanite, having a V-section that is used to insulate the commutator bars from the metal V-ring.

W

Wadsworth prism An equilateral glass prism with a plane mirror at 45° with the base. A ray passing

through the prism at minimum deviation is reflected by the mirror to 90° deviation from the incident ray.

wafer *See* chip.

wake A region behind an obstacle in a fluid stream that is disturbed by its passage. In the classical theory of hydrodynamics, a cylinder in an inviscid fluid has no wake in this sense, because the pattern of flow behind the obstacle is the same as that upstream. It is then a consequence of *Bernouilli's theorem that the forces on the obstacle at opposite ends of a diameter balance, so that the total resistance (drag) is nil. In a real fluid, even at low speeds, viscous forces in the *boundary layer produce skin friction and as the speed increases, vortices are formed behind it. At first these remain stationary in space, though, of course, rotating and adhering to the cylinder, but at a critical value of the speed they pass downstream to form a *vortex street. At still higher speeds the wake breaks down into general turbulence. If the wake is formed by a body that projects out of the free surface of a liquid, such as a ship, the surface will be crossed by bow waves. A somewhat similar phenomenon occurs in the wake of a totally immersed body if it travels at speeds exceeding sound. All these factors add their contributions to the drag on the obstacle, which may in fact be reduced if the width of the wake is restricted by suitable shaping of the obstacle into a streamline form.

Walden's rule For a given electrolyte the product of its equivalent conductance at infinite dilution in a solvent, and the viscosity of that solvent, is approximately constant.

wall effect In general, any significant effect of the inside wall of a container or reaction vessel on the behaviour of the enclosed system. Some examples are: (i) The contribution to the current in an *ionization chamber made by electrons that are liberated from the inside walls rather than from the enclosed gas. (ii) The loss of ionization in a *counter when some of the energy of the primary ionizing radiation is absorbed in the chamber wall rather than in the enclosed gas. (iii) The effect of the wall in a *gas-discharge tube in promoting the recombination of positive ions and electrons to neutral molecules or atoms. (iv) The effect of the wall in a gas-discharge tube or in *photolysis reactions in promoting the recombination of *free radicals.

wall energy The energy per unit area of the boundary between the *domains in a ferromagnetic material (*see* ferromagnetism).

Walton, Ernest Thomas Sinton (*b.* 1903) Irish physicist who collaborated with *Cockcroft in being the first to disintegrate lithium nuclei by proton bombardment, for which they shared the 1951 Nobel prize.

Wanner optical pyrometer *See* polarizing pyrometer.

warble tone A tone in which the frequency varies cyclically between two limits. The frequency variation is usually small compared with the actual frequency of the note and the warble occurs several times per second. A warble tone can be produced from an oscillator by using a small variable capacitor in the tuned circuit. This type of note is extensively used in *reverberation chambers since it enables a measurably uniform sound field to be produced by eliminating *standing wave patterns.

Ward–Leonard system A method of controlling the speed and, if required, the direction of rotation of a d.c. motor. The motor armature is supplied directly from the armature of a separate d.c. generator and both machines are separately excited (see excitation). The whole of the control is effected by varying the field current and hence the voltage of the generator. If this field current is increased, the voltage increases and the speed of the motor increases. If the field current is reversed, the voltage reverses and the direction of rotation of the motor reverses. The d.c. generator is driven by a separate d.c. or a.c. motor. The method provides a wide speed range and fine control. It is much used for large reversing rolling mill motors, for colliery winding, and for paper making. A modification is made in the *Ilgner system.

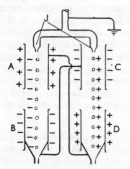

Water dropper

water dropper A simple generator of electrostatic charges. Water drops from a double jet, J (see diagram), through two cylindrical metal cans A, C, into two lower cans B, D, provided with metal funnels as shown. The cans are cross connected. A small charge (say) is given to the can A. The jets are now in a positive field and a current flows to earth leaving them negatively charged. Thus a series of negatively charged drops falls into can B and there discharges against the funnel, imparting a negative charge to C. The operation is now progressive on both sides of the apparatus and a large charge accumulates. The source of energy is that derived from the gravitational force on the drops.

water equivalent The mass of water that would have the same *heat capacity as a given body. It is numerically equal to the product of the body's mass and its specific heat capacity.

Waterston, John James (1811–1883) Scottish physicist who developed a kinetic theory of gases which he sent to the Royal Society in 1845 but which was overlooked until it was discovered by Lord *Rayleigh in 1892. This theory largely anticipated the work of *Clausius (1857).

water waves Waves may be set up in the free surface of a liquid or in the interface between two liquids by a disturbance of the plane surface, as when the wind blows over a sea or lake. The restoring force is gravity, hence the name *gravity wave*.

In the simplest form of surface wave that can be set up in deep water, the individual particles in the surface trace out circles of radius r with a frequency f. From the aspect of an observer travelling with the waves at their velocity c, the flow is steady and *Bernouilli's theorem can be applied. Relative to such an observer, the particle velocities in crests and troughs are, respectively:
$$U_1 = c - 2\pi rf \text{ and } U_2 = c + 2\pi rf.$$
Bernoulli's theorem then gives:
$$U_2{}^2 - U_1{}^2 = 4gr,$$
where $2r$ is the vertical difference in level between a crest and a trough, whence:
$$c = g/2\pi f = \sqrt{(g\lambda/2\pi)},$$
where the wavelength $\lambda = c/f$.

The wave speed thus varies with the wavelength and in mixed waves, dispersion ensues, the group speed being half the phase speed.

The above treatment ignores surface tension. This factor is important for short waves and modifies their velocity. (See ripples.)

The amplitude A_x of the wave at depth x is given by:
$$A_x = A_0 \exp(-2\pi x/\lambda),$$
where A_0 is the amplitude at the surface.

For a liquid that is not deep compared with the wavelength, the speed of the wave is affected by the depth. For very shallow liquid of depth h the speed is given by:
$$c = \sqrt{hg}.$$
An earthquake under the ocean can generate a large wave of extremely long wavelength. As this reaches shallow water the speed decreases and the amplitude increases correspondingly. The wave, which can be very destructive, is known by the Japanese name *tsunami*.

Watson-Watt, Sir Robert Alexander (1892–1973) Scottish physicist who led the team of scientists responsible for the development of *radar.

watt Symbol: W. The *SI unit of *power (mechanical, thermal, and electrical), defined as the power when work of one joule is done in one second, or an equal heat transfer occurs in one second. In electrical circuits one watt is the product of one ampere and one volt.

Watt, James (1736–1819) Scottish engineer, who made fundamental improvements in the steam engine and by inventing a separate condenser reduced the operating cost by 75%. He also invented in 1781 a sun and planet gear to produce a rotation as against a reciprocating motion engine (because the crank had been patented) and he devised parallel motion linkage to provide the connection between piston and beam. He was the inventor of the copying press.

watt-hour A unit of work or energy, equal to one watt operating for one hour (equal to 3.6×10^3 joules).

watt-hour meter *Syn.* integrating wattmeter; recording wattmeter. An integrating meter for measuring electrical work, expressed in watt-hours or, more usually, in kilowatt-hours.

wattless component 1. Of current. *See* reactive current.

2. Of volt-amperes. *See* reactive volt-amperes.

3. Of voltage. *See* reactive voltage.

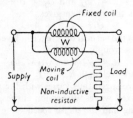

Connections for single phase electrodynamic wattmeter

wattmeter An instrument used for measuring electric *power – active power – and having a scale graduated in watts, multiples of a watt or submultiples of a watt. In the electrodynamic type (*see* electrodynamic instrument) in common use, a fixed coil (*see* diagram) forms the current circuit, which is connected directly in series with the main circuit (or in the secondary circuit of a current transformer, the primary of which is connected in series with the main circuit). The voltage circuit, consisting of the moving coil in series with a noninductive resistor, is connected across the main circuit so that the current in the moving coil is proportional to the voltage of the circuit. The moving coil with its attached pointer rotates from its initial position of rest against a torque provided by control springs, and the final steady deflection depends upon the power in the main circuit. A *damper of the air friction type is provided. This type of wattmeter may be used with d.c. or a.c. circuits. Less common is the induction type (*see* induction instrument) that is suitable only for circuits in which the frequency and voltage are substantially constant. The electrostatic wattmeter consists of a *quadrant electrometer arranged in a special circuit with one or more noninductive resistors to enable it to measure power directly. It is mainly used in research and standardizing laboratories, and is the standard method of measuring power in the testing of watt-hour meters and the calibra-

tion of wattmeters at the National Physical Laboratories.

wave A time-varying quantity that is also a function of position. It is a disturbance, either continuous (e.g. sinusoidal) or transient, travelling through a medium by virtue of the elastic and inertia factors of the medium, or of magnetic and electric properties of space, the resulting displacements (mechanical, electric, etc.) of the medium being relatively small and returning to zero when the disturbance has passed. The general impression therefore in the transmission of a wave is that the particles of the medium vibrate relatively to each other in such a way that the wave appears to travel bodily forward with a velocity given by the velocity of the wave motion. *See* diffraction of light; diffraction of sound; travelling wave; standing wave.

wave analyser An instrument for the resolution of a given waveform into its fundamental and harmonic components. The analysis may be made manually or it may be made automatically according to the design of the instrument. The result is expressed in the form of frequencies and amplitudes of the various components. Instruments most frequently used for wave analysis are *spectrum analysers, frequency analysers, and heterodyne instruments.

With the last type the wave under investigation is made to modulate a carrier wave generated internally. The frequency of the carrier is adjusted until the sideband created by a component of the wave reaches a frequency to which a highly selective amplifier is tuned. The signal is fed into this amplifier and the output measured by a voltmeter. A search is made over a suitable range of carrier frequencies in order to obtain the frequencies and relative amplitudes of all components of the wave under examination.

waveband A range of wavelengths in the electromagnetic spectrum, defined according to some property of the radiation, or some requirement or functional aspect of a detecting or transmitting system.

wave energy *See* renewable energy sources.

wave equation The partial differential equation:

$$\frac{\partial^2 U}{\partial x^2} + \frac{\partial^2 U}{\partial y^2} + \frac{\partial^2 U}{\partial z^2} = \frac{1}{c^2} \frac{\partial^2 U}{\partial t^2}$$

(or its counterpart in one or two dimensions or in other coordinates), the solution of which represents the propagation of displacements U as waves with speed c. *See* wave mechanics.

waveform *Syn.* waveshape. Of a periodic quantity (*see* period). The shape of the graph obtained by plotting the instantaneous values of the quantity against time. The waveform is usually described as being distorted if it is not *sinusoidal. In acoustics the waveform determines the *quality of the sound.

wavefront 1. A surface over which the oscillations in a wave have the same phase. The surface is normal to rays in isotropic media; in doubly refracting media a pair of wavefronts progress (forming the *wave surface*) and it is only for the ordinary wave that the wavefront is normal to the ordinary ray. The optical path between two successive positions of a wavefront, measured along rays, is constant.

2. *See* impulse voltage (or current).

wave function Symbol: ψ. A mathematical quantity analogous to the amplitude of a wave which appears in the equations of *wave mechanics, particularly the *Schrödinger wave equation. The most generally accepted interpretation is that of *Born according to whom $|\psi|^2 dV$ represents the probability that a particle is located within the volume element dV (*see* de Broglie waves). ψ is often a complex quantity.

The analogy between ψ and the amplitude of a wave is purely formal. There is no macroscopic physical quantity with which ψ can be identified (in contrast with, for example, the amplitude of an electromagnetic wave, which is expressed in terms of electric and magnetic field intensities).

In general there is an infinite number of wave functions satisfying a wave equation but only some of these will satisfy the boundary conditions. ψ must be finite and single-valued at every point, and the spatial derivatives must be continuous at an interface. For a particle subject to a law of conservation of numbers (*see* fermion), the integral of $|\psi|^2 dV$ over all space must remain equal to 1, since this is the probability that it exists somewhere. To satisfy this condition the wave equation must be of the first order in $(d\psi/dt)$. (*See* Dirac wave equation. *Compare* Klein–Gordon equation.) Wave functions obtained when these conditions are applied are called *proper wave functions* and form a set of *characteristic functions of the Schrödinger wave equation. These are often called *eigenfunctions* and correspond to a set of fixed energy values in which the system may exist, called *eigenvalues* (proper values). Energy eigenfunctions describe *stationary states of the system.

For certain bound states of a system the eigenfunctions do not change sign on reversing the coordinate axes. These states are said to have *even *parity*. For other states the sign changes on space reversal and the parity is said to be *odd*.

waveguide A hollow metal conductor containing a dielectric (usually air) down which *travelling waves are propagated. Waveguides are thus used as *transmission lines, especially for UHF radio waves. They have much lower attenuation than coaxial cables at these high frequencies, and also have a higher power-carrying capacity and a simpler construction. The conductor is usually of rectangular or circular cross section, but irregular shapes are also used for special applications.

Electromagnetic waves can be excited in a waveguide by the electric and magnetic fields associated with waves present in another device, such as a *cavity resonator or *microwave tube. Source and waveguide are connected so as to achieve the optimum transfer of energy. Energy can be extracted in a similar manner. Energy may also be transferred by using a probe to which a voltage is applied or a coil that carries a current. Again, energy can be similarly extracted.

The electromagnetic wave in a waveguide can have an infinite number of *transmission modes*, characterized by the electric and magnetic field patterns of the wave. In general these modes are of two kinds: in *transverse electric (TE) modes* the electric vector is always perpendicular to the direction of propagation; in *transverse magnetic (TM) modes* the magnetic vector is perpendicular to the propagation direction. Physical constraints and the frequency of the wave usually limit the number of modes. For each mode there is a cut-off frequency, determined by the size and shape of the guide; waves below this frequency cannot be propagated by that mode. For any transmission frequency it is usually possible to choose the dimensions of the waveguide so that only one mode is above the cut-off frequency and all other modes are rapidly attenuated. In a rectangular guide this *dominant mode* is a TE mode for which the cut-off wavelength is twice the wide dimension.

A waveguide is a completely shielded transmission line, and may be bent and twisted with no radiation loss as long as the cross section remains uniform. A change in dimensions amounts to a change in characteristic impedance.

wavelength Symbol: λ. The least distance in a progressive wave between two surfaces with the same phase. If v is the *phase speed and ν the frequency, the wavelength is given by $v = \nu\lambda$. For electromagnetic waves the phase speed and wavelength in a material medium are equal to their values in free space divided by the *refractive index. Particular regions in the spectrum and the wavelengths of spectral lines are normally specified for free space.

Optical wavelengths are measured absolutely using interferometers or diffraction gratings, or comparatively using a prism *spectrometer.

The wavelength can only have an exact value for an infinite wave train. If an atomic body emits a quantum in the form of a train of waves of duration τ the fractional uncertainty of the wavelength, $\Delta\lambda/\lambda$, is approximately $\lambda/2\pi c\tau$, where c is the speed in free space. This is associated with the indeterminacy of the energy given by the *uncertainty principle.

See also Doppler effect; interference of light; interference of sound; redshift.

wavelength constant For a plane progressive wave transmitting a sinusoidal vibration. The wavelength constant (β) is the phase difference between two points along the direction of propagation, unit distance apart. It is therefore equal to $2\pi/\lambda$, where λ is the wavelength. *See* propagation coefficient.

wave-making resistance The portion of drag resistance of a floating body, moving on a fluid, due to the dissipation of energy in the formation of waves.

The presence of this resistance determines the ideal shape of the hull of a ship.

wave mechanics One of the forms of *quantum mechanics, due to Louis de Broglie and extended by E. Schrödinger, P. A. M. Dirac, and many others. It originated in the suggestion that light consists of corpuscles as well as of waves and the consequent suggestion that all elementary particles are associated with waves. In its essentials it is virtually a broadening of the scope of the old analogy between geometrical optics and mechanics (de *Maupertuis and *Hamilton). Characteristic of it is the *wave function which is expressed as an expansion of the Fourier kind in terms of so-called eigenfunctions.

wavemeter An apparatus for measuring the frequency or wavelength of a radio wave. It consists essentially of a capacitively *tuned circuit and a current-detecting instrument. The variable capacitor is calibrated in terms of frequency or wavelength. A current maximum is obtained when the resonant frequency of the circuit corresponds to the frequency of the radio wave.

wave motion The process of transmitting *waves. Wave motion appears naturally in many different forms, the principal ones being surface waves (e.g. water waves), longitudinal waves (e.g. sound), transverse waves (e.g. electromagnetic radiation and waves in a vibrating string), and torsional waves (e.g. waves in a bar due to torsional vibrations at one end). All types are governed by a single equation, the equation of wave motion (*see* progressive wave), and have the property of transmitting energy over considerable distances.

wavenumber Symbol: σ. The reciprocal of the *wavelength, i.e. the number of waves per unit path length. It is expressed in m^{-1}. $\sigma = 1/\lambda = f/c$, where λ is the wavelength, f the frequency, and c the speed. The *angular wavenumber*, symbol: k, is given by $2\pi\sigma$, i.e. $2\pi/\lambda$. This is the magnitude of the *angular wave vector* or *propagation vector*, symbol: $\mathbf{k}$.

wave packet *See* wavetrain.

wave-particle duality The phenomenon whereby electromagnetic radiation and particles can exhibit either wave-like or particle-like behaviour, but not both. *See* corpuscular theory; complementarity.

wave plate *See* half-wave plate; quarter-wave plate.

waveshape *See* waveform.

wave surface *See* wavefront.

wavetail *See* impulse voltage (or current).

wave theory of light *See* corpuscular theory.

wavetrain A succession of waves, especially a group of waves of limited duration (also called a *wave packet*).

wavetrap A *tuned circuit, usually a *rejector, incorporated in a radio receiver to reduce interference at a particular radio frequency.

wax-block photometer *See* Joly photometer.

weak interaction A kind of interaction between *elementary particles that is weaker than the *strong interaction force by a factor of about 10^{12}. When strong interactions can occur in reactions involving elementary particles the weak interactions are usually unobservable. However, sometimes strong and *electromagnetic interactions are prevented because they would violate the conservation of some *quantum number (e.g. *strangeness) that has to be conserved in such reactions. When this happens weak interactions may still occur. The weak interaction operates over an extremely short range (about 2×10^{-18} m). It is mediated by the exchange of a very heavy particle (a gauge boson; *see* gauge theory) that may be the charged W^+ or W^- particle (mass about 80 GeV/c^2) or the neutral Z^0 particle (mass about 91 GeV/c^2). The gauge bosons that mediate the weak interactions are analogous to the photon that mediates the *electromagnetic interaction. Weak interactions mediated by W particles involve a change in the charge and hence the identity of the reacting particle. The neutral Z^0 does not lead to such a change in identity. Both sorts of weak interaction can violate *parity.

Most of the long-lived elementary particles decay as a result of weak interactions. For example, the kaon decay

$$K^+ \rightarrow \mu^+ + \nu_\mu$$

may be thought of as being due to the annihilation of the u quark and the $\bar{s}$ antiquark in the K^+ to produce a virtual W^+ boson, which then converts into a positive muon and a neutrino. This decay process cannot take place by a strong or electromagnetic interaction because strangeness is not conserved. *Beta decay is the most common example of a weak interaction decay. Because it is so weak, particles that can only decay by weak interactions do so relatively slowly, i.e. they have relatively long lifetimes.

Important examples of weak interactions other than decay processes are given under *Reines' and Cowan's experiments and *stellar energy. As the probability of weak interaction increases with energy other processes are observed in *bubble chambers using particles of very high energy from *accelerators.

Understanding of weak interactions is based on the *electroweak theory, in which it is proposed that the weak and electromagnetic interactions are different manifestations of a single underlying force known as the electroweak force. Many of the predictions of the theory have been confirmed experimentally.

weakly interacting massive particle (WIMP) *See* missing mass.

weber Symbol: Wb. The *SI unit of *magnetic flux, defined as the flux that, linking a circuit of one turn, produces an electromotive force of one volt when the flux is reduced to zero at a uniform rate in one second. 1 Wb = 10^8 maxwell.

Weber, Wilhelm Edward (1804–1891) German physicist who studied magnetism and played an important part in the development of a system of absolute electrical units. The *SI unit of magnetic flux was named after him.

Weber–Fechner law To make a sensation (such as *loudness or brightness) increase in arithmetical progression, the stimulus must increase in geometrical progression.

Weber number A dimensionless parameter used in the study of the formation of bubbles, and given by the function:

$$v(l\rho/\sigma)^{1/2},$$

where v is velocity, ρ is density, σ is surface tension, and l represents some characteristic length (e.g. radius or diameter). Occasionally the Weber number is defined as the square of this function.

Weber's law A just perceptible difference of sensation is proportional to the intensity of the stimulus.

wedge A strip of material, such as a photographic plate or piece of gelatin, that shows a gradation of transmission from clear to opaque along its length. The gradation may be continuous or in steps, and may be in neutral tones or a single colour.

Wehnelt cathode In 1904, Wehnelt discovered that oxides of barium, strontium, and calcium when coated on platinum are very good emitters of electrons. Almost all thermionic valves use coated cathodes, often the coating being on a cylinder raised to operating temperature by radiation from an enclosed independent heating filament. Such cathodes can operate at a much lower temperature than clean metals and the electron tubes in which they are used are called dull emitters.

Wehnelt interrupter An electrolytic interrupter consisting of a lead plate, and a platinum wire sheathed by glass or porcelain so that only its tip is exposed, these being immersed in 30% sulphuric acid. This apparatus is put in series with the primary of an induction coil, the platinum point being the anode. By a combination of gas evolution and vaporization under the high current density the circuit is rapidly broken and remade at the point, very high frequencies of interruption being possible.

weight 1. In mechanics, a term used loosely with different meanings that are often confused. (*a*) The actual force of gravitation acting on a body near to the surface of the earth or another astronomical body. (*b*) The apparent force of gravitation equal to the mass of the body times the acceleration of *free fall *g*; this differs slightly from the former since *g* is measured with respect to a nearby point on the surface of the rotating planet, not with respect to the centre. (*c*) For a body remaining supported on the surface, the force exerted by the body on the support; this is equal to (*b*) but acts on a different body at a different point by a different mechanism. (*d*) The force exerted by the support on the body in the last case; this force is sometimes called the "reaction" to

the weight and is often wrongly supposed to be related to the force of gravity by Newton's third law of motion. (*e*) The mass of a body, in particular a standard mass used in weighing. In all cases except (*e*) weight means a force so it is measured in newtons. *See also* weightlessness.

2. *See* weighted mean.

weighted mean Of a number of values $x_1, x_2, x_3 \ldots x_n$. The quantity given by:

$$\frac{w_1 x_1 + w_2 x_2 + \cdots w_n x_n}{w_1 + w_2 + \cdots w_n},$$

where w is the *weight* of an observation; w is a measure of the reliability of the corresponding x, and can either be assigned intuitively or calculated from $w = (\text{probable error})^{-2}$.

weightlessness The condition of a body in *free fall. Since there is no support the body is not acted upon by a force, as described in *weight 1 (*d*). The condition of weightlessness does not imply that the body is not subject to gravitation, but that no other force acts on it. Since gravity acts uniformly throughout a body it does not by itself cause any stress. Any force exerted by a support on the surface of a body necessarily causes deformation, so the normal condition of living organisms on earth is one of stress, which is removed when in free fall in an orbit.

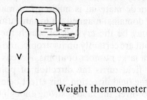

Weight thermometer

weight thermometer A vessel V made of glass or fused silica, used to determine the coefficient of expansion of a liquid by finding the masses of liquid required to fill the thermometer at different temperatures θ_1 and θ_2 respectively. Then

$$\frac{m_1}{m_2} = \frac{1 + \alpha(\theta_2 - \theta_1)}{1 + \gamma(\theta_2 - \theta_1)},$$

where α is the coefficient of absolute expansion of the liquid and γ is the coefficient of cubical expansion of the material of the thermometer. The coefficient of apparent expansion is thus given by:

$$\alpha_a = \frac{m_1 - m_2}{m_2(\theta_2 - \theta_1)}$$

$$= \frac{\text{Mass expelled}}{\text{Mass remaining} \times \text{Rise of temperature}}.$$

Weinberg, Steven (*b*. 1933) Amer. physicist; professor at the Massachusetts Institute of Technology and at Harvard University. He shared the 1979 Nobel prize with Sheldon Glashow and Abdus

Weinberg–Salam model

Salam for work done on the *electroweak theory of particle interactions.

Weinberg–Salam model *See* electroweak theory.

Weiss constant *See* Curie–Weiss law.

Weissenberg photography A crystal-diffraction method in which a single crystal is allowed to rotate about an axis normal to the incident beam of monochromatic radiation, while the cylindrical photographic film moves in a synchronized way, to and fro parallel to the rotation axis, screens being arranged so that only one layer line is recorded at a time. It is used for the measurement of intensities of diffracted spectra.

Weiss magneton Weiss discovered that the susceptibilities of both iron and nickel, extrapolated to absolute zero, were multiples of 1123.5, and suggested that the molar susceptibility was always a multiple of this number. On this basis the magnetic moment of a single atom would be 1.87×10^{-24} J T^{-1}; this value is called the Weiss magneton. No supporting evidence for this value has ever been found and it was soon to be replaced by that proposed by *Bohr (1913) on the basis of the quantum theory. The Bohr *magneton is approximately five times that of Weiss.

Weiss' theory of magnetism Weiss supposed that a ferromagnetic material is made up of many small regions (*domains) magnetized to saturation. These regions may be the crystals of which the body is formed, but are certainly nonisotropic, and are composed of a large number of atoms. A weak external magnetic field turns the direction of polarization towards the field direction, this effect being reversible. If, however, the field is strong enough there is a sudden irreversible change in the direction of polarization throughout the whole region (*see* Barkhausen effect). This type of change may perhaps cause a subsequent change in adjacent domains. *See also* ferromagnetism; Curie–Weiss law.

welding The process of joining two pieces of the same metal together using (*a*) no extra metal at all, or (*b*) extra metal of the same kind as the parts being joined.

The methods include: (i) Hammering together two pieces of metal, e.g. lead, platinum need no heating, but steel needs to be hammered while hot and also sprinkled with a flux. (ii) Resistance butt-welding in which a heavy electric current passes across the junction and develops a high temperature at the points of contact where the electric current is greatest. (iii) Spot welding for sheet metal. This is similar to (ii) and can be used to produce a row of small welded spots – like a line of rivets. Seam welding is a further development of spot welding. (iv) Fusion welding using an oxy-acetylene flame, a filler-rod of the same material as the metals being joined and a flux. (Iron and steel need no flux.) (v) Arc welding which is a form of fusion welding in which an electric arc is developed between the filler-rod which is coated with flux, and the job.

well *See* potential well.

well-behaved *See* characteristic function.

well counter A radiation counter used in connection with radioactive fluids. The fluid is held in a cylindrical container within the detecting device.

Wertheim effects *See* Wiedemann effects.

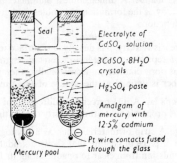

Weston standard cell

Weston standard cell *Syn.* cadmium cell. A cell that is a portable standard of electromotive force and is used for calibrating potentiometers and hence all other voltage-measuring instruments. It is constructed in an H-shaped glass vessel, the constituents being shown in the diagram. The cell has a very low temperature coefficient of e.m.f. Its e.m.f. at $t\ °C$ is given as:
$$E_t = 1.018\ 58 - 4.06 \times 10^{-5}(t - 20)$$
$$- 9.5 \times 10^{-7}(t - 20)^2 + 1 \times 10^{-8}(t - 20)^3.$$
See Clark cell; Josephson effect.

Westphal's measurement of the Stefan–Boltzmann constant (1913) Westphal determined the *Stefan–Boltzmann constant at ordinary temperatures from observations on the work done electrically on a sphere, first polished, then blackened, in order to maintain the temperature T of the sphere constant when the sphere was suspended in a large evacuated enclosure maintained at a lower temperature (T_0). The Stefan–Boltzmann constant σ is then given by:
$$\sigma = \frac{w_2 - w_1}{A(e_2 - e_1)(T^4 - T_0^4)},$$
where A is the area of the sphere and e_1 and e_2 are the emissivities of the polished and blackened sphere; w_1 and w_2 are the electrical works done on the polished and blackened spheres respectively. The emissivities are determined by comparing the total emission from the sphere with that of a black body by bringing them in turn before a radiometer.

wet and dry bulb hygrometer A simple hygrometer consisting of an ordinary thermometer side by side with another thermometer whose bulb is surrounded by fibres dipping into water. Tables are used to calculate the relative humidity of the atmosphere from the difference in the thermometer readings. These tables are based on Regnault's formula,

$$p_w - p = a(t - t_w)B,$$

where p is the actual pressure at temperature t, p_w is the SVP at the temperature t_w of the wet bulb, and B is atmospheric pressure. Pressures are measured in mmHg. The value of a depends on the wind speed, but is constant for all speeds above 2.5 m/s. The tables give the value for the relative humidity for a given value of t and of the difference $(t - t_w)$ since the relative humidity is the ratio of p (determined by the above equation) to the SVP at t °C. Two instruments of this type are *Mason's hygrometer and the Assmann psychrometer. *See also* Apjohn's formula.

wetting A surface is said to be completely wet by a liquid if the contact angle is zero and incompletely wet if it is a nonzero angle. However, many writers say that if the angle is greater than 90° the surface is unwettable, e.g. paraffin wax and water.

Wheatstone, Sir Charles (1802–1875) Brit. physicist, founder of modern telegraphy. He determined the speed of electric discharge in conductors by means of a revolving mirror and in conjunction with W. F. Cooke took out a patent for the electric telegraph in 1837. He investigated the transmission of sound in solids, furnished an explanation of *Chladni's figures and invented the concertina. He also studied optics, in particular the eye, vision and colour, and was the first to use (1847) the *Wheatstone bridge, which was devised in 1833 by S. H. Christie.

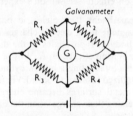

Galvanometer

a Wheatstone bridge circuit

Wheatstone bridge A *bridge used to measure resistance. A network of resistors is arranged as in Fig. *a*, R_1 and R_2 being the unknown and the reference resistance. When the galvanometer shows no deflection then the currents in the four arms are balanced so that:

$$R_1/R_2 = R_3/R_4.$$

The two arms R_3 and R_4 may be sections l_1, l_2 of a uniform resistance wire tapped off by a slider (Fig. *b*). Then:

$$R_1/R_2 = l_1/l_2.$$

Several forms of Wheatstone bridge exist: *see*, for example, post office box; Carey–Foster bridge; Kelvin double bridge.

white dwarf Any of a large class of very faint stars that are thought to be low-mass stars in the last stage of stellar evolution. Their mass lies below the *Chandrasekhar limit* (about 1.4 solar masses). Their nucle-

ar fuel (hydrogen) has been completely exhausted and they have undergone *gravitational collapse to form small but very dense bodies consisting of helium nuclei and a degenerate gas of electrons. A man on the surface of a white dwarf would weigh about 10^{10} N. *See also* pulsar; black hole; Hertzsprung–Russell diagram.

white light Light, such as daylight, containing all wavelengths of the visible spectrum at normal intensities so that no coloration is apparent.

white noise *See* noise.

wide-angle lens A camera lens with a relatively short focal length compared with a standard lens and a large field of view – typically 80° to 100° for still cameras.

Wiedemann effects *Syn.* Wertheim effects. Circular magnetic effects: (1) the twist produced in a rod due to interaction of longitudinal and circular magnetic fields; (2) the longitudinal magnetization produced by twisting a circular magnetized rod; and (3) the circular magnetic field produced by twisting a longitudinally magnetized rod.

Wiedemann–Franz–Lorenz law An approximate relationship between thermal and electrical conductivities of metals and alloys that holds roughly over a wide range of temperatures, but becomes less accurate at very low temperatures:

$$\lambda/\sigma T = (2.0 \pm 0.5) \times 10^{-8} \text{ V}^2\text{K}^{-2},$$

where λ is the thermal conductivity, σ is the electrical conductivity, and T is the thermodynamic temperature.

The relationship $\lambda/\sigma \simeq$ constant, at constant T, was published in 1853 by Wiedemann and Franz. Later work has established the more general form of the law over a wide range of temperatures. The first theory of conduction by electrons given by *Drude (1900) was by chance consistent with the relationship, two major errors cancelling. *Sommerfeld gave a consistent theory in 1928.

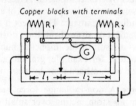

Copper blocks with terminals

b Wheatstone bridge

Wien, Wilhelm (1864–1928) German physicist who became a professor at Munich. He studied *blackbody radiation (*see* Wien displacement law; Wien radiation law) for which he was awarded a Nobel prize (1911). He also studied ionic conductivity (*see* Wien effect).

Wien bridge A four-arm a.c. *bridge that can be used to measure capacitance, inductance, or *power factor.

Wien displacement law

Wien displacement law Although the law for the spectral distribution of radiation from a *black body could not be deduced until the work of *Planck (1900), Wien showed by a thermodynamic argument (1893) that the thermodynamic temperature T must appear in the law only in a function of λT, where λ is the wavelength. Hence if the peak of the distribution as a function of wavelength at T_1 occurs at λ_1, then when the temperature is changed to T_2 the peak is displaced to λ_2, where
$$\lambda_1 T_1 = \lambda_2 T_2.$$
This relationship was later confirmed by *Planck's formula. *See also* Wien radiation law.

Wien effect The increase in conductivity of an electrolyte under a high voltage gradient (of the order of 2 MV m^{-1}). Under this condition, the rate of movement of an ion in solution is such that it passes completely out of its ionic atmosphere during the *relaxation time, and is thus free from the retarding effect normally encountered.

Wien radiation law By combining the *Wien displacement law with empirical results Wien was able (1896) to deduce an approximate form of the law later formulated precisely by *Planck for the energy distribution of *black body radiation:
$$M_{e\lambda} = \frac{c_1}{\lambda^5} e^{-c_2/\lambda T}.$$

c_1 and c_2 have the same values as they have in Planck's formula. The formula applies when λT is small (< 0.002 m K). *Compare* Rayleigh–Jeans formula.

Wigner, Eugene Paul (*b.* 1902) Hungarian-born Amer. physicist who has contributed to the theory of nuclear resonance and the conservation of angular momentum of electron spin. He worked with *Fermi on the construction of the first atomic pile. He was awarded a share in the 1963 Nobel prize. *See also* Wigner effect; Wigner force.

Wigner effect *Syn.* discomposition effect. A change in the physical or chemical properties of a solid as a result of radiation damage. The effect is caused by the displacement of atoms from their normal lattice positions as a result of the impact of nuclear particles. In reactors, for example, graphite changes its size because of bombardment by neutrons.

Wigner force A non-exchange force between the nucleons in an atom, acting over small distances.

Wigner nuclides *See* mirror nuclides.

Wilkins, Maurice Hugh Frederick (*b.* 1910) New Zealand physicist who applied X-ray crystallography to the structure of biological molecules. With Watson and Crick he unravelled the structure of DNA, sharing with them a Nobel prize (1962).

Wilson, Charles Thomas Rees (1869–1961) Scots physicist. He devised the Wilson cloud chamber by which tracks of charged particles are made visible and can be photographed (*see* cloud chamber). For

this work he was awarded a Nobel prize (1927). He also studied atmospheric electricity and designed an improved form of gold-leaf electrometer for this purpose.

Wilson, Kenneth (*b.* 1937) Amer. theoretical physicist who received the 1982 Nobel prize for physics for his work on the theory of *phase transitions. Wilson developed a theory that could be applied near the critical point. His original work was on the ferromagnetic–paramagnetic transition using a technique of taking a block of atoms and calculating their magnetic properties, then calculating the effect of several blocks forming a larger block, and so on. The method involves using *renormalization and, although originally used for magnetic properties, has been extended to other phase transitions, including interactions of elementary particles.

Wilson, Robert Woodrow (*b.* 1936) Amer. physicist who worked at Bell Telephone Laboratories. He shared the 1978 Nobel prize with Arno Penzias for their discovery (1964) of *microwave background radiation.

Wilson cloud chamber *See* cloud chamber.

Wilson effect When an insulating material is moved through a region of magnetic flux an induced potential difference is set up across the material. Because the creation of an electric current is inhibited by the nonconducting properties of the material, it becomes electrically polarized, a phenomenon known as the Wilson effect. *See* Wilson's experiment.

Wilson's experiment If a dielectric body moves across a magnetic field, electric charges are set up across the body at right angles to the direction of motion (*see* Wilson effect). H. A. Wilson first demonstrated this by rotating a hollow cylinder of dielectric (metallized inside and outside) in a magnetic field. The metal surfaces became charged as expected, reversing their charge when the direction of rotation changed.

WIMP Weakly interacting massive particle. *See* missing mass.

Wimshurst machine An early electrostatic generator consisting of two parallel plates with tinfoil sectors arranged radially. There are two cross-connectors with tinsel wipers, and a system of pointed combs to act as collectors of the induced electrostatic charges (Fig. *a*). The operation is identical with the *Kelvin replenisher if the number of collectors and inductors is considered to be increased and if both systems rotate in opposite directions to double the number of operations in a given time. The diagrammatic representation (Fig. *b*) will make the operation clear. Capacitors are usually fitted to increase the capacity and to give a stronger spark, and in the larger machines more than one pair of plates may be used.

windage loss The power loss, usually expressed in watts, that occurs in an electrical machine as a result of the motion imparted to the gas or vapour (commonly air) surrounding the moving parts, by the latter. This loss is inherent in all electrical machines since ventilation is required for cooling purposes. Some modern machines use hydrogen instead of air as the cooling medium since, among its other advantages, it enables the windage loss to be reduced considerably.

wind energy *See* renewable energy sources.

winding Of an electrical machine, transformer, or other piece of apparatus. A complete group of insulated conductors designed to produce a magnetic field or to be acted upon by a magnetic field. A winding may consist of a number of separate conductors connected together electrically at their ends or may consist of a single conductor (wire or strip) that has been shaped or bent to form a number of loops or turns.

wind instruments *See* column of air.

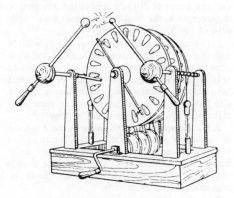

a Wimshurst machine

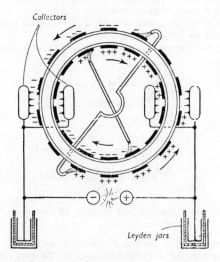

b Diagrammatic representation of Wimshurst machine

window 1. When a medium, generally opaque to the passage of radiation, selectively transmits a particular small range of radiation it is said to operate as a window in that range. (*See* atmospheric windows.)

2. The thin sheet of material (often mica) covering the end of a radiation detector or counter, through which the radiation is received.

3. The winding space of a transformer.

wind tunnel Essentially a hollow tube through which a uniform flow of air is passed. A model, constructed to scale of all or part of an aircraft or missile is suspended in the air stream. From the observations made upon the model the behaviour of the aircraft in normal flight can be estimated. There is *dynamic similarity between the model and aircraft when the *Reynolds number is the same in both cases. This condition cannot, in general, be satisfied, due to the large velocity of air flow required in the wind tunnel; and a method is adopted whereby the results of observations made upon the model at small values of the Reynolds number are extrapolated to the higher values required for the full scale. The equality of the Reynolds number may be achieved by using a compressed air tunnel, the air being at a pressure of up to 20 atmospheres. The effect of the pressure is to increase the density and so to reduce the value of the *kinematic viscosity in nearly the same proportion. The air stream is produced by a fan and it is necessary to correct for the effect of *turbulence in the stream caused by the rotation of the fan. Irregularities in the flow of the air are also caused by surface friction at the boundaries, variations in sectional area, and changes in direction, of the tunnel. Allowance must be made for these in subsequent calculations. Wind tunnels for supersonic flow usually take the form of a *Laval nozzle into which compressed air is discharged, the model being placed in the section of maximum contraction.

winter solstice *See* solstice.

wires *See* guard wires; Price's guard wire.

wire-wound resistor *See* resistor.

wobbulator A *signal generator whose output frequency can be automatically varied over a definite range of values. It is used in testing the frequency response of electronic circuits and devices.

wolf *See* mean tone scale.

wolf note In bowed string instruments certain notes are difficult if not impossible to produce by bowing, for the bow does not then adequately bite the string. At such notes the whole body of the instrument vibrates to an unusual degree. The notes produced are similar to a howl and are known as wolf notes. Makers of string instruments aim to have the wolf note below the frequency of the open lowest string. Experiments have shown that at a wolf note the belly of the instrument vibrates almost entirely in *simple harmonic motion. At frequencies other than the wolf note the belly has a complex vibration. However, it is observed that the wolf note is usually

Wollaston, William Hyde

found at the upper harmonics of the wooden system and not at the fundamental. Raman has based his theory of the production of the wolf note on the fact that for a string bowed at one end more bowing pressure is required for the fundamental than for the octave. At the wolf note it is suggested that the string sounds its fundamental initially, but that the drain of energy to the belly is such that the pressure between string and bow becomes inadequate to maintain the fundamental and the vibration changes to one where the octave is predominant. The wood vibrations then cease and the pressure becomes sufficient to re-excite the fundamental, and the process repeats. The cyclic changes of octave to fundamental in the string vibrations have been observed.

Wollaston, William Hyde (1766–1828) Brit. chemist and philosopher. He observed solar absorption lines (later called Fraunhofer lines) in 1802. The *Wollaston magnifier* or *doublet* consists of two hemispheres of glass separated by a small central stop. The Wollaston form of spectacle lens is highly meniscus in order to reduce peripheral oblique astigmatism (contrasted with the Ostwalt form, also meniscus, but less steeply curved, correcting for the same aberration). He also described the *reflecting goniometer* (1809) and the *camera lucida* (1812). He made a special study of the platinum metals and was the first to detect palladium (1804) and rhodium (1805) and to show the elementary nature of columbium and titanium.

Wollaston prism

Wollaston prism A polarizing beam splitter made from two prisms of calcite or quartz, cemented or in optical contact along their diagonals. Their optic axes are arranged so as to separate the ordinary and extraordinary components of a ray of unpolarized light at the diagonal interface (*see* double refraction). The diagram shows the refractive action of a calcite device. For prism P the optic axis is parallel to AB; for prism Q it is normal to the plane of the paper. The angle of deviation of the emerging beams is determined by the wedge angle between AB and the diagonal. The two beams are orthogonally polarized.

Wollaston's method *See* refractive-index measurement.

Wollaston wire Exceedingly fine wire produced by encasing platinum wire in a silver sheath, drawing them together, and then dissolving away the silver by acid. The platinum can be reduced to a diameter of 1 μm. It is used for electroscope wires, microfuses, and hot-wire instruments.

Womersley *See* explosion method.

Wood's glass A glass with the unusual property of having a high transmission factor in the ultraviolet range of the spectrum, but being relatively opaque to visible radiation.

Wood's metal A bismuth, lead, tin, cadmium alloy with a melting point about 70 °C.

woofer A loudspeaker of large dimensions, used to reproduce sounds of relatively low frequency in a hi-fi system. *Compare* tweeter.

word A string of *bits used to store an item of information in a *computer. Word length depends on the machine but typically consists of 32 or 16 bits.

work Symbol: W; unit: joule. **1.** If a constant force F acts at a point on a body while it undergoes a displacement D, the work done by the body that exerts the force is the scalar product $F \cdot D$, i.e. $W = FD \cos \theta$, where θ is the angle between F and D. This can be expressed as the product of the force times the component of displacement in the direction of the force, or as the displacement times the component of the force in that direction.

2. If a constant torque G acts on a body while it undergoes a rotation through an angle H, the work done is the scalar product $G \cdot H$ (H measured in radians). If the torque and angular displacement have the same direction this gives the simple product GH. If the axes of torque and rotation are perpendicular (as in *precession) no work is done.

3. If a surface is displaced sweeping out a volume ΔV against a pressure p, the work done is $p\Delta V$.

4. If a charge Q is displaced between two points with potential difference U, the work done electrically is QU. If work is in joules and charge in coulombs, then U is in volts.

5. Work may also be done in changes of the state of magnetization or electrification of a body. For example, a magnetized paramagnetic substance does work at the expense of its internal energy on demagnetizing itself on removal of an external field (*see* adiabatic demagnetization).

See also energy; power; virtual-work principle.

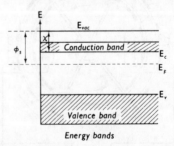

Work function and electron affinity of a semiconductor

work function Symbol: Φ. The difference in energy between the *Fermi level of a solid and the energy of the free space outside the solid (the vacuum level). At the absolute zero of temperature the work function is the minimum energy required to remove an electron from a solid. In a metal there is a contribution to the work function from the *image potential that an electron would experience outside the metal.

In a *semiconductor (*see* diagram) the *electron affinity* (symbol: χ) is defined by the energy difference between the vacuum level and the bottom of the conduction band.

Work function and electron affinity are usually defined as energies and measured in electronvolts, although volts are sometimes used.

work hardening The hardening of a metal when it is strained considerably by a stress above the *elastic limit. It is attributed principally to the locking together of dislocations (*see* defect). The *ductility and *malleability of a metal can be restored after work hardening by *annealing.

wound *See* compound-wound, series-wound, and shunt-wound machines.

wow An undesirable form of *frequency modulation heard in the reproduction of high-fidelity sound and characterized by variations in pitch up to about 10 Hz. In the case of a gramophone record it is often due to nonuniform rotation of the turntable. *See also* flutter.

W particle (or **W boson**) The extremely massive charged particle, symbol: W^+ or W^- (antiparticles), that mediates certain types of *weak interaction. The neutral *Z particle* (or *Z boson*), symbol: Z^0, mediates the other types. Both are gauge bosons (*see* gauge theory). The W and Z particles were first detected at CERN (1983) by studying collisions between protons and antiprotons with total energy 540 GeV in centre-of-mass coordinates. The rest masses were determined as about 80 GeV/c^2 and 90 GeV/c^2 for the W and Z particles respectively.

wrench A force together with a couple whose axis is the line of action of the force. The quotient of the couple by the force is the pitch of the wrench and has the dimensions of a length. The magnitude of the force is the intensity and the line of action of the force is the axis of the wrench. In general, any system of forces can be reduced to a wrench.

write The operation of entering information into a storage location.

write-time Of a storage component or memory. The time taken to enter information into a storage location.

Wu, Chien Shiung (*b*. 1912) Chinese-born Amer. physicist who became professor of physics at Columbia University. One of the world's leading experimental physicists, she is noted for her demonstration (1956) that *parity is not conserved in beta decay – a theory advanced by Tsung Dao Lee and Chen Ning Yang.

X

xerography A photographic process in which the image is formed by electrical effects rather than chemical effects. Ultraviolet radiation is passed through, for example, a document to be copied and falls on an electrostatically charged plate, usually coated with selenium. This is discharged to an extent that depends on the intensity of the incident radiation. A powder with an opposite electric charge is then sprayed on the plate and sticks to the "dark" areas where the plate has not been discharged by the radiation. The powder, which is a mixture of graphite and a thermoplastic resin, is then transferred from the plate to a charged paper where it is fixed by heat treatment.

xeroradiography Radiography in which the X-ray image is produced by *xerography rather than by a normal photographic process. This is useful when low-energy X-rays are used since lower exposure times are possible.

xi particle An *elementary particle classified as a *hyperon.

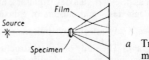

a Transmission method

X-ray analysis Analysis of the structure of crystalline substances, or of those that have crystalline phases, based on the diffraction of *X-rays. X-rays are diffracted by crystals in a manner dependent on the wavelength of the rays and the *Bravais lattice of the crystal. The analytical method adopted depends on the form in which the substance is available. With large crystals the *Laue method can provide useful characterization, but more frequently the crystal is rotated when mounted at the centre of a cylindrical film, thus bringing successive sets of crystalline planes into position. The *Debye–Scherrer ring or powder method is used when the specimen consists of a number of small crystals. Because of the number of crystals, randomly distributed, some are usually available in each plane to diffract the X-ray beam. The plate may be set up as in Fig. *a* or *b*, or may be rotated in a cylindrical camera, which is the most usual method.

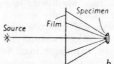

b Back reflection method

X-ray analysis can be used to investigate alloy systems and alloy transformations, to study atomic distribution in crystalline materials, for chemical analysis under certain limited conditions, and to study anisotropic behaviour of crystalline material

X-ray astronomy

under deformation. *See also* Bragg's law; X-ray spectrum.

X-ray astronomy The study of X-ray emission from astronomical sources both in and beyond our Galaxy. Since X-rays are absorbed by the earth's atmosphere, observations must be made at altitudes above about 150 km using instruments mounted in satellites, rockets, and balloons. The X-rays may be *thermal radiation produced from very high temperature gas (about 10^6 to 10^8 K), or nonthermal X-rays arising from interaction of high-energy electrons with a magnetic field (*synchrotron radiation) or with low-energy photons (inverse *Compton effect). The X-rays may be detected, recorded, and analysed by various instruments, including *proportional counters, *CCDs (charge-coupled devices), and grazing-incidence X-ray telescopes.

The most common and luminous sources of X-rays in our Galaxy are *X-ray binaries*, in which gas is flowing from a normal star to a close companion – either a *white dwarf, *neutron star, or even a *black hole. *Supernova remnants, such as the Crab nebula, are another source. Fainter but intrinsically much more powerful X-ray emission is detected from many extragalactic objects, especially active galaxies – such as *Seyfert galaxies, *quasars, and powerful radio galaxies – and clusters of galaxies.

X-ray binary *See* X-ray astronomy; pulsar.

X-ray crystallography The study of crystal structure, texture, and behaviour, and the identification of crystals, by methods involving the use of X-rays. *See* X-ray analysis.

X-ray diffraction *See* X-rays; X-ray analysis.

X-ray microscopy *See* microscope.

X-rays *Syn.* Röntgen rays. There is no universally agreed classification for electromagnetic radiations of high quantum energy. The name X-radiation was given by Röntgen in 1895 to the unidentified radiation that he observed when electrons of high energy (10^3 eV upwards) strike matter. For many years research was limited to quantum energies less than about 10^5 eV, so the term X-ray came to be used by some workers for radiations with quantum energies up to about this value (corresponding to a wavelength about 10^{-11} m). Thus radiations in this range produced by other mechanisms are sometimes called X-rays. (*See* synchrotron radiation.) In 1896 Becquerel discovered those rays from radioactive substances called *gamma rays, which were later shown to originate in the nucleus. Early research detected only those radiations with quantum energies of the order 10^5 to 10^6 eV, hence it was said that gamma rays were of shorter wavelength than X-rays; the term gamma ray is thus sometimes applied to a radiation of this very high quantum energy irrespective of origin.

Such a distinction between X- and γ-rays is now invalid: nuclei often emit radiations of much lower quantum energy than 10^5 eV while machines have been developed giving very much shorter wavelength radiation than that emitted from any nucleus. In this article the term X-ray is used to mean radiation produced in a machine in which electrons are accelerated to any high energy and strike a target.

An *X-ray tube* is an evacuated vessel containing an *electron gun in which electrons from a hot filament are focused onto a *target*, or *anticathode*. This is usually a metal of high melting point, tungsten being particularly suitable. The target is cooled either by air or oil. X-rays are emitted from the area, typically a few square millimetres, struck by the electron beam. In the side of the tube there is a window of material with a low atomic number to transmit the X-rays.

The spectrum of X-rays emitted from an X-ray tube consists of lines characteristic of the target material and a continuum that has a short-wave limit λ_m, given by hc/eV, where h is the Planck constant, c the speed of light, e the electron charge, and V the potential difference across the tube. The *characteristic X-rays* are caused by the transitions of electrons between the various shells of the atom; the continuous band of wavelengths is caused by the acceleration of electrons in the vicinity of nuclei.

X-rays can be reflected, refracted, and polarized by suitable materials. They also show interference and diffraction effects. The wavelengths of X-rays were first determined (Bragg 1911) using a crystal as a three-dimensional diffraction grating. The atoms in regular array in a crystal act as point sources when placed in an X-ray beam and constructive interference takes place in preferential directions dependent on wavelength, the crystal, and its setting. The wavelength is typically 10^{-10} to 10^{-11} metre. In *quantum theory the energy, E, of X-ray photons is related to the frequency ν, by the expression: $E = h\nu$.

X-rays ionize gases but the process is a secondary one caused by the electrons set free when X-rays interact with matter. The ionizing effects produced in body tissue can be harmful to health; under strict control, however, they are used in the treatment of cancer. X-rays penetrate matter to a degree dependent on the wavelength of the rays. In general, *hard X-rays* (small wavelengths) are able to penetrate a given substance more easily than *soft X-rays* (longer wavelengths). The intensity of homogeneous radiation transmitted through a given material of thickness t, is related to the initial intensity I_0, by the relation:

$$I = I_0 e^{-\mu t},$$

where μ is the *absorption coefficient* of the material. The incident beam is weakened by (a) scattering caused by atoms, (b) the *photoelectric effect, and (c) the *Compton effect. In the latter two processes, free electrons are ejected from the atoms of the material. Photographic plates are blackened by X-rays to a degree dependent on the intensity of the radiation and the wavelength.

The intensities of X-ray beams may be compared by effecting a measure of the ionization currents produced in an *ionization chamber, or by determining the blackening produced in a photographic plate. The unit of X-ray dose used in health physics is related to the ionizing effects of the radiation. (*See* roentgen; gray; sievert.)

X-rays are widely used for (*a*) investigating flaws in structures (radiography), (*b*) diagnostic purposes (radiography), (*c*) therapeutic purposes (destruction of diseased tissue), and (*d*) investigating crystal structure (X-ray crystallography).

X-ray spectrum When X-rays are scattered by atomic centres arranged at regular intervals, interference phenomena occur, crystals providing gratings of suitable small interval. The interference effects may be used to provide a spectrum of the beam of X-rays, since, according to *Bragg's law, the angle of reflection of X-rays from a crystal depends on the wavelength of the rays. For lower-energy X-rays mechanically ruled gratings can be used. Each chemical element emits *characteristic X-rays* in sharply defined groups in more widely separated regions. They are known as the K, L, M, N, etc., series. There is a regular displacement of the lines of any series towards shorter wavelengths as the atomic number of the element concerned increases. *See* Moseley's law; spectroscopy; X-rays.

X-ray tube *See* X-rays.

X-unit *Syn.* X-ray unit. A small unit of length equal to 10^{-13} metre, formerly used to describe short ultraviolet and X-ray wavelengths.

x-y recorder A recording instrument in which a graph is produced showing the relationship between two varying electric currents or potentials. One signal moves the pen in the direction of the x-axis and the other independently moves it along the y-axis.

Y

Yagi aerial A sharply directional *aerial array used especially for television that consists of one or two dipoles, connected to the transmitting or receiving circuits, a parallel reflector, and a series of directors. The directors are parallel and spaced from 0.15 to 0.25 of a wavelength apart such that, in transmission, energy is absorbed from the field of the dipole and re-radiated so as to reinforce the field in a forward direction and oppose it in the reverse direction; in receiving, the signals are focused onto the dipole.

Yang, Chen Ning (*b.* 1922) Chinese-born Amer. particle physicist who, together with *Lee, first showed that *parity is not conserved in *weak interactions, for which they shared the 1957 Nobel prize.

Yang–Mills theory *See* gauge theory.

yard The Imperial standard yard was formerly defined as the distance at 62 °F between the central traverse lines on two gold plugs in a bronze bar kept by the Board of Trade. It is now defined as 0.9144 metre.

Y connection *See* star connection.

year *See* time.

yield point A point on a graph of *stress versus *strain for a material at which the strain becomes dependent on time and the material begins to flow. *See* yield value.

yield stress The minimum stress for *creep to take place. Below this value any deformation produced by an external force will be purely elastic.

yield value The minimum value of stress that must be applied to a material in order that it shall flow. *See* plasticity; Bingham fluid.

YIG Abbreviation for yttrium iron garnet. A synthetic *ferrite widely used for microwave applications. The magnetic properties are altered by the amount of trace elements present.

yoke A piece of ferromagnetic material that is used to connect permanently two or more magnetic *cores and thus complete a magnetic circuit without surrounding it by a winding of any type.

Young, Thomas (1773–1829) Brit. physician and physicist; he discovered the *interference of light and developed a wave theory of light which was later extended by Fresnel; he proved that accommodation of the eye was effected by change of curvature of the crystalline lens. He first observed ocular astigmatism and introduced a trichromatic theory of colour vision developed later by Helmholtz (Young–Helmholtz theory; *see* colour vision).

Young–Helmholtz laws Two mainly empirical laws serving generally to determine the motion of bowed strings. They may be stated as follows: (i) No overtone is present that would have a node at the point of excitation. (ii) When a string is bowed at an aliquot point $(1/k)$, the part of the string immediately under the bow moves to and fro with constant velocities whose ratio is equal to the ratio $1/(k-1)$ of the segments into which the string is divided by the point in question. The smaller of these two velocities has the same direction as that of the bow, and is equal to it.

The first law applies to strings stretched between sharp bridges in whatever way excited. The second law refers to the motion of a bowed point, but only considers the case of a string bowed $\frac{1}{2}, \frac{1}{3}, \frac{1}{4}$, etc., of the length from the end. A third law, more general than the second, is the *Krigar–Menzel law*, which states that when a string is bowed at any rational point p/q, where p and q are primaries to each other, the part of the string immediately under the bow moves to and fro with two constant velocities whose ratio depends only on q, and is $1/(q-1)$.

Young–Helmholtz theory

Young–Helmholtz theory *See* colour vision.

Young modulus *See* modulus of elasticity.

y-parameters *See* transistor parameters.

Yukawa, Hideki (*b.* 1907) Japanese physicist who became a professor at Kyoto University. He predicted the existence of the *meson in 1935. At first this was thought to have been confirmed by Anderson and Neddermeyer's discovery in 1938 of the particle now called the *muon, but it was not until 1947 that *Powell discovered the *pion, which exactly fitted Yukawa's hypothesis. Yukawa was the first Japanese physicist to win the Nobel prize (1949).

Yukawa potential A potential used by Yukawa to explain the forces between nucleons. Two particles of equal and opposite charge are attracted by the *electromagnetic interaction and their mutual potential energy is $-e^2/r$, where e is their charge and r their distance apart. In the nucleus, stronger short-range forces are acting (*see* strong interactions) and Yukawa assumed that their potential energy varied according to $e^{-\mu r}/r$ rather than $1/r$, where μ is a constant. The interaction is assumed to be caused by the virtual production of a boson that is exchanged between the nucleons. The quantity $\mu = 2\pi m_0 c/h$, where m_0 is the mass of the particle, c the speed of light, and h is the Planck constant. Equating the reciprocal of μ with the effective range of the strong interaction ($\simeq 1.2 \times 10^{-15}$ m) gives $m_0 \simeq m_p/6$, where m_p is the proton mass. This is consistent with the mass of the *pion.

Z

Zamboni's pile A variant of *Volta's pile using gold and silver foil as alternate metals.

Zeeman, Pieter (1865–1943) Dutch physicist, who discovered (1896) the splitting of spectral lines when light from a sodium flame is subjected to a strong transverse magnetic field (*see* Zeeman effects). He studied (1915) the propagation of light in media moving in the laboratory and obtained results more consistent with the theory given by *Lorentz based upon electromagnetism than with the earlier theory of *Fresnel based upon the concept of the *ether. He shared a Nobel prize (1902) with Lorentz.

Zeeman effects The original Zeeman effect (1896) was the broadening of sodium lines when a sodium flame is placed in a moderately strong magnetic field, transverse to the source of light. Later, it was established that each line is split into components: (*a*) two components circularly polarized when viewed parallel to the field; (*b*) three components plane polarized when viewed at right angles to the field (normal triplets); (*c*) multicomponent systems arising from multiplet components, e.g. D lines of sodium (*anomalous Zeeman effect*). If the field is applied along the beam of light, the undisplaced line

is absent. The *inverse Zeeman effect* refers to the absorption of unpolarized light by vapours placed in a strong magnetic field.

The Zeeman effect occurs because the energies of individual electron states depend on their inclination to the direction of the magnetic field, and because quantum energy requirements impose conditions such that the plane of an electron orbit can only set itself at certain definite angles to the applied field. These angles are such that the projection of the total angular momentum on the field direction is an integral multiple of $h/2\pi$ (h is the *Planck constant). The Zeeman effect is observed with moderately strong fields where the precession of the orbital angular momentum and the spin angular momentum (*see* spin) of the electrons about each other is much faster than the total precession around the field direction. For stronger fields the *Paschen–Back effect predominates. The normal Zeeman effect is observed when the conditions are such that the *Landé factor is unity, otherwise the anomalous effect is found. This anomaly was one of the factors contributing to the discovery of electron spin. *See also* interference of light.

Zener, Clarence Melvin (*b.* 1905) Amer. physicist who has contributed considerably to the theory and practice of solid-state physics. *See* Zener breakdown; Zener diode.

Zener breakdown A type of *breakdown, observed in reverse-biased p-n junctions in which very high *doping levels exist: the built-in potential across the junction is therefore high and the *depletion layer narrow. The application of a small reverse voltage is sufficient to cause the electrons to tunnel directly from the valence band to the conduction band (*see* energy bands; tunnel effect). No multiplication of charge *carriers occurs (*compare* avalanche breakdown) and the breakdown is reversible since the dielectric properties of the material remain unchanged. A very sharp increase in the reverse current is observed at the breakdown potential. Avalanche breakdown is the dominant breakdown mechanism in most semiconductor devices, except those with high doping levels. *See also* Zener diode.

Zener diode A p-n junction *diode that has sufficiently high *doping levels on each side of the junction for *Zener breakdown to occur. The diode thus has a well-defined reverse *breakdown voltage (about a few volts) and can be used as a voltage regulator. It behaves like a normal diode in the forward direction.

The term is also applied to less highly doped diodes that have higher values of breakdown voltage (up to 200 V) and whose characteristics depend on *avalanche breakdown.

zenith The point, infinitely distant, above an observer on the earth's surface, in the direction in which gravity acts. *See* celestial sphere.

zenith distance *See* celestial sphere.

zeolites *Syn.* molecular sieves. A class of natural or synthetic compounds that are alumino-silicates of sodium, potassium, and calcium. They have an open cage-like crystal structure and contain water molecules that can be removed by heating. Anhydrous zeolites have cavities in their crystal structure that can trap molecules of suitable size. For this reason they are used as selective absorbers for separating mixtures of liquids and gases. Their other main application is in *ion exchange.

Zernicke, Fritz (1888–1966) Dutch physicist who developed the phase-contrast *microscope.

zero error *See* index error.

zero-point energy In classical physics it was assumed that at absolute zero all particles would be at rest, hence the translational and rotational kinetic energies and the vibrational energy of molecules would be zero. *Quantum theory shows that the lowest energy state of a system is often nonzero, hence although at absolute zero all particles would be in the lowest energy states available their energies may be significant.

From the *uncertainty principle a particle cannot have zero momentum unless the uncertainty in its position is infinite. Thus the translational kinetic energy of molecules at absolute zero can only be zero for an *ideal gas, which is in principle realized by extrapolation to infinite volume.

In a condensed substance each atom can be considered as being equivalent to three linear oscillators that are very nearly simple harmonic except at high temperatures. By *wave mechanics the lowest energy of a linear simple-harmonic oscillator of frequency ν is $\frac{1}{2}h\nu$ where h is the Planck constant. The *Debye theory of specific heat capacities shows from this that the zero-point energy of a mole of solid resulting from vibrations is $9R\Theta_D/8$, where R is the *molar gas constant and Θ_D is the Debye characteristic temperature. The zero-point energy of the valence electrons in a solid or liquid is relatively very large (*see* energy bands). The zero-point energy affects the *saturated vapour pressure and the *latent heats of vaporization and sublimation (*see* Trouton's rule). The existence of zero-point energy in crystals may be verified by investigation of the temperature variation of the intensity of X-rays reflected from the crystal.
See also vacuum state.

zeroth law of thermodynamics The law of *thermodynamics that is fundamental to the three basic laws of thermodynamics (hence its name). It states that if two bodies are each in thermal equilibrium with a third body, then all three bodies are in thermal equilibrium with each other.

zeta pinch *See* fusion reactor.

zeta potential *Syn.* electrokinetic potential. According to Stern, the electrical layer at the boundary between solid and liquid is composed of two parts. In one, approximately one ion thick, there is a sharp

potential fall, the other part showing a gradual rise or fall of potential and extending some way into the liquid. This part of the potential change is called the zeta potential, and is involved in electrosmosis and similar effects. The ζ potential in millivolts is given by $4\pi\eta\nu/\varepsilon_r E$, where E is the electric field, ν the liquid velocity, η is the viscosity of the liquid medium, and ε_r its relative permittivity.

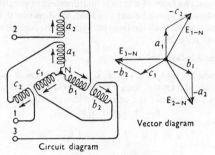

Circuit diagram   Vector diagram

Zig-zag connection

zig-zag connection *Syn.* interconnected star connection. Of the windings of a three-phase alternator, transformer, or inductive reactor. Each phase is wound in two equal sections which are connected as shown in the diagram. The sections a_1 and a_2 are associated with the same magnetic field and the e.m.f.s in them are equal and in phase; similarly for sections b_1 and b_2 and sections c_1 and c_2. The e.m.f.s in the pairs of sections a, b, and c have relative phase differences of 120° (in a transformer or reactor the pairs of sections are situated on the separate limbs of a three-limbed core or upon three separate cores). The total e.m.f. between, for example, line 1 and the *neutral point N is the vector sum of the e.m.f. in a_1 and the reversed e.m.f. in c_2, i.e. the vector sum of two equal e.m.f.s differing in phase by 60°. This connection avoids the undesirable effects of *triplen harmonics that may be present in the e.m.f.s of the individual sections. In transformers and alternators, it improves conditions when working with unbalanced loads. For these reasons it is sometimes used when supplying three-phase rectifiers. Reactors employing this connection are used for obtaining an artificial neutral point in three-phase systems. Compared with the normal *star connection this method of connection results in a reduced output (or rating) for a given amount of active material in the apparatus, and this is its main disadvantage.

Zinn, Walter Henry (*b.* 1906) Canadian-born Amer. physicist. Zinn did most of his work at Columbia University, where he collaborated with Leo Szilard. Together they demonstrated that uranium could undergo fission when bombarded with neutrons. After World War II, Zinn worked on the design of nuclear reactors and was responsible for building the first breeder reactor (1951).

Zisman apparatus An apparatus for measuring contact potential differences between solid/solid or solid/liquid interfaces. The two media whose contact p.d. is to be determined are arranged to form a parallel-plate capacitor the capacitance of which is changed periodically by vibrating one of the plates. This produces an alternating current that can be detected after audiofrequency amplification.

z-modulation *See* intensity modulation.

zone The faces of a crystal that intersect in parallel edges are said to belong to a zone, the direction that is common to all the faces being called the *zone-axis*. *See* zone law.

zone indices *See* zone law.

zone law If $(h_1 \ k_1 \ l_1) (h_2 \ k_2 \ l_2)$ are the indices of two faces of a crystal, then a third face $(h_3 \ k_3 \ l_3)$ will lie in the *zone of the other two if, and only if,
$$h_3 u + k_3 v + l_3 w = 0,$$
where
$$u = k_1 l_2 - k_2 l_1, \ v = l_1 h_2 - l_2 h_1,$$
and
$$w = h_1 k_2 - h_2 k_1,$$
(uvw) are then the *zone indices*.

zone of silence Associated with the sound caused by large explosions is a zone of silence where no sound is heard, while at greater distances from the source the sound is clearly heard again over a considerable area. The direct sound travelling with normal speed is heard within an area up to about sixty miles from the source. Beyond this is a zone of silence and then a large area sometimes hundreds of miles away where the sound is heard a considerable time after its propagation. This suggests that the sound waves travel into the upper atmosphere and are reflected or refracted back into the outer audible zone. There were doubts about the reasons for this phenomenon for some time but it is known that the temperature of the atmosphere decreases with height for some distance, it then remains steady and beyond this it increases with height (*see* atmospheric layers). This causes a bending of the sound waves and they subsequently return to the earth at a considerable distance from the source. The zones round the source are often irregular in shape and it is suggested that winds in the *stratosphere also affect the areas of abnormal audibility. A similar effect on a much smaller scale has also been observed with fog signals, but this is due to quite different causes. Near the earth's surface the wind speed increases with height causing the sound waves travelling against the wind to be concave upwards. Higher up, however, there is a reversal of wind velocity gradient causing the sound to become concave downwards and return to

the earth on the windward side of the source beyond the region of normal audibility. *See* refraction of sound.

zone plate *See* diffraction of light.

zone refining A technique for producing very pure materials, used in the manufacture of *semiconductors. The sample is in the form of a bar and a small portion of it is melted by induction heating, electron bombardment, or by a resistance coil. The sample is slowly moved past the heater so that the molten zone passes along the length of the bar. Impurities in the material tend to concentrate in the melt and are thus segregated at one end of the sample. By using a large number of heaters it is possible to reduce the impurity concentration to 1 part in 10^{10}.

zoom lens A lens system consisting of converging and diverging elements, one or more of which can be moved so that the focal length, which depends on the separation of the elements, can be continuously adjusted. Variation of the focal length without changing the sharpness of the image can be achieved by connecting two or more of the elements together so that they both move through the same distance, yet keep the image sharp at the different focal lengths. This optical compensation is shown in the diagram. It is also desirable that the *f-number should not need resetting as the focal length changes. To avoid this the lens system is usually in two parts: the basic imaging system, for which the f-number remains constant, and a variable-focus attachment.

z-parameter *See* transistor parameters.

Z particle *See* W particle.

zwitterions *Syn.* ampholyte ions. Ionized molecules that have both positive and negative charge. Zwitterions are formed by substances such as amino acids that have a basic and acidic group in the same molecule.

Zworykin, Vladimir Kosma (1889–1982) Russian-born Amer. physicist who worked mostly in America. He was a pioneer of many advances in electron optics and the inventor (1923) of the first electronic scanning *television camera (the iconoscope). Zworkin also worked on the transmission of pictures, an early version being produced in 1923. He also produced an electron-image tube and electron multiplier. James *Hillier joined his research group at the Radio Corporation of America in 1940, and it was here that Hillier constructed his electron microscope. *See also* Baird, John Logie.

APPENDIX

Table 1 **Conversion Factors**
SI, c.g.s., and f.p.s. units

Length	m	cm	in	ft	yd
1 metre	1	100	39·3701	3·280 84	1·093 61
1 centimetre	0·01	1	0·393 701	0·032 808 4	0·010 936 1
1 inch	0·0254	2·54	1	0·083 333 3	0·027 777 8
1 foot	0·3048	30·48	12	1	0·333 333
1 yard	0·9144	91·44	36	3	1

	km	mile	n. mile
1 kilometre	1	0·621 371	0·539 957
1 mile	1·609 34	1	0·868 976
1 nautical mile	1·852 00	1·150 78	1

1 light-year = $9·460\ 528 \times 10^{15}$ metres = $5·878\ 514 \times 10^{12}$ miles.
1 astronomical unit = $1·495\ 979 \times 10^{11}$ metres.
1 parsec = $3·085\ 677 \times 10^{16}$ metres = 3·261 633 light-years.

Area	m^2	cm^2	in^2	ft^2
1 square metre	1	10^4	1550	10·7639
1 square centimetre	10^{-4}	1	0·155	$1·076\ 39 \times 10^{-3}$
1 square inch	$6·4516 \times 10^{-4}$	6·4516	1	$6·944\ 44 \times 10^{-3}$
1 square foot	$9·2903 \times 10^{-2}$	929·03	144	1

	m^2	km^2	yd^2	mi^2	acre
1 square metre	1	10^{-6}	1·195 99	$3·860\ 19 \times 10^{-7}$	$2·471\ 05 \times 10^{-4}$
1 square kilometre	10^6	1	$1·195\ 99 \times 10^6$	0·386 019	247·105
1 square yard	0·836 127	$8·361\ 27 \times 10^{-7}$	1	$3·228\ 31 \times 10^{-7}$	$2·066\ 12 \times 10^{-4}$
1 square mile	$2·589\ 99 \times 10^6$	2·589 99	$3·0976 \times 10^6$	1	640
1 acre	$4·046\ 86 \times 10^3$	$4·046\ 86 \times 10^{-3}$	4840	$1·5625 \times 10^{-3}$	1

1 are = 100 square metres.
1 hectare = 10 000 square metres = 2·471 05 acres.

Volume	m^3	cm^3	in^3	ft^3	gal
1 cubic metre	1	10^6	$6·102\ 36 \times 10^4$	35·3146	219·969
1 cubic centimetre	10^{-6}	1	0·061 023 6	$3·531\ 46 \times 10^{-5}$	$2·199\ 69 \times 10^{-4}$
1 cubic inch	$1·638\ 71 \times 10^{-5}$	16·3871	1	$5·787\ 04 \times 10^{-4}$	$3·604\ 64 \times 10^{-3}$
1 cubic foot	0·028 316 8	28 316·8	1728	1	6·228 82
1 gallon (UK)	$4·546\ 09 \times 10^{-3}$	4546·09	277·42	0·160 544	1

1 gallon (US) = 0·832 68 gallon (UK).
1 cubic yard = 0·764 555 cubic metre.
The *litre* is now recognized as a special name for a cubic decimetre, but is not used to express high precision measurements.

623

tables

Table 1 **Conversion Factors** (*continued*)

Velocity	m s^{-1}	km h^{-1}	mile h^{-1}	ft s^{-1}
1 metre per second	1	3·6	2·236 94	3·280 84
1 kilometre per hour	0·277 778	1	0·621 371	0·911 346
1 mile per hour	0·447 04	1·609 344	1	1·466 67
1 foot per second	0·3048	1·097 28	0·681 817	1

1 knot = 1 nautical mile per hour = 0·514 444 metre per second.

Mass	kg	g	lb	long ton
1 kilogram	1	1000	2·204 62	9·842 07 × 10^{-4}
1 gram	10^{-3}	1	2·204 62 × 10^{-3}	9·842 07 × 10^{-7}
1 pound	0·453 592	453·592	1	4·464 29 × 10^{-4}
1 long ton	1016·047	1·016 047 × 10^6	2240	1

1 slug = 14·5939 kg = 32·174 lb.

Density	kg m^{-3}	g cm^{-3}	lb ft^{-3}	lb in^{-3}
1 kilogram per cubic metre	1	10^{-3}	0·062 428	3·612 73 × 10^{-5}
1 gram per cubic centimetre	1000	1	62·428	3·612 73 × 10^{-2}
1 pound per cubic foot	16·0185	0·016 018 5	1	5·787 04 × 10^{-4}
1 pound per cubic inch	2·767 99 × 10^4	27·6799	1728	1

1 lb/gal (UK) = 0·099 776 3 kg dm^{-3}.

Force	N	kg	dyne	poundal	lb
1 newton	1	0·101 972	10^5	7·233 00	0·224 809
1 kilogram force	9·806 65	1	9·806 65 × 10^5	70·9316	2·204 62
1 dyne	10^{-5}	1·019 72 × 10^{-6}	1	7·233 00 × 10^{-5}	2·248 09 × 10^{-6}
1 poundal	0·138 255	1·409 81 × 10^{-2}	1·382 55 × 10^4	1	0·031 081
1 pound force	4·448 22	0·453 592	4·448 23 × 10^5	32·174	1

Pressure	Pa	kg/cm^2	lb/in^2	atm
1 pascal	1	1·019 72 × 10^{-5}	1·450 38 × 10^{-4}	9·869 23 × 10^{-6}
1 kilogram per square centimetre	980·665 × 10^2	1	14·2234	0·967 841
1 pound per square inch	6·894 76 × 10^3	0·070 306 8	1	0·068 046
1 atmosphere	1·013 25 × 10^5	1·033 23	14·6959	1

1 **pascal** = 10 dynes per square centimetre.
1 bar = 10^5 **pascals** = 0·986 923 atmosphere.
1 torr = 133·322 **pascals** = 1/760 atmosphere.
1 atmosphere = 760 mmHg = 29·92 in Hg = 33·90 ft water (all at 0 °C).

Table 1 **Conversion Factors** (*concluded*)

Work and Energy	J	cal$_{IT}$	kW hr	btu$_{IT}$
1 joule	1	0·238 846	$2·777\,78 \times 10^{-7}$	$9·478\,13 \times 10^{-4}$
1 calorie (IT)	4·1868	1	$1·163\,00 \times 10^{-6}$	$3·968\,31 \times 10^{-3}$
1 kilowatt hour	$3·6 \times 10^{6}$	$8·598\,45 \times 10^{5}$	1	3412·14
1 British Thermal Unit (IT)	1055·06	251·997	$2·930\,71 \times 10^{-4}$	1

1 joule = 1 newton metre = 1 watt second = 10^{7} erg = 0·737 561 ft lb.
1 electronvolt = $1·602\,10 \times 10^{-19}$ joule.

Table 2 **Base SI Units**

Physical quantity	Name	Symbol
length	metre	m
mass	kilogram	kg
time	second	s
electric current	ampere	A
thermodynamic temperature	kelvin	K
amount of substance	mole	mol
luminous intensity	candela	cd

Table 4 **Prefixes Used with SI Units**

Factor	Name of Prefix	Symbol	Factor	Name of Prefix	Symbol
10	deca-	da	10^{-1}	deci-	d
10^{2}	hecto-	h	10^{-2}	centi-	c
10^{3}	kilo-	k	10^{-3}	milli-	m
10^{6}	mega-	M	10^{-6}	micro-	μ
10^{9}	giga-	G	10^{-9}	nano-	n
10^{12}	tera-	T	10^{-12}	pico-	p
10^{15}	peta-	P	10^{-15}	femto-	f
10^{18}	exa-	E	10^{-18}	atto-	a

Table 3 **Derived SI Units with Special Names**

Physical quantity	Name	Symbol
frequency	hertz	Hz
force	newton	N
pressure, stress	pascal	Pa
energy, work, quantity of heat	joule	J
power	watt	W
electric charge, quantity of electricity	coulomb	C
electric potential, potential difference, tension, electro-motive force	volt	V
electric capacitance	farad	F
electric resistance	ohm	Ω
electric conductance	siemens	S
flux of magnetic induction, magnetic flux	weber	Wb
magnetic flux density, magnetic induction	tesla	T
inductance	henry	H
Celsius temperature	degree Celsius	°C
luminous flux	lumen	lm
illuminance	lux	lx
activity (of a radionuclide)	becquerel	Bq
absorbed dose, specific energy imparted, kerma, absorbed dose index	gray	Gy
dose equivalent	sievert	Sv
plane angle	radian	rad
solid angle	steradian	sr

625

tables

Table 5 Fundamental Constants

Constant	Symbol	Value (with estimated error)
speed of light	c	2.99792458×10^8 m s^{-1} exact by definition
magnetic constant (permeability of free space)	μ_0	$4\pi \times 10^{-7} = 1.25663706144 \times 10^{-6}$ H m^{-1}
electric constant (permittivity of free space)	$\varepsilon_0 = \mu_0^{-1}c^{-2}$	$8.854187817 \times 10^{-12}$ F m^{-1}
charge of electron or proton	e	$\pm 1.60217733 \times 10^{-19}$ C
rest mass of electron	m_e	$9.1093897 \times 10^{-31}$ kg
rest mass of proton	m_p	$1.6726231 \times 10^{-27}$ kg
rest mass of neutron	m_n	1.674929×10^{-27} kg
electronic radius	$r_e = \dfrac{e^2}{4\pi\varepsilon_0 m_e c^2}$	$2.81794092 \times 10^{-15}$ m
Planck constant	h	6.626076×10^{-34} J s
Boltzmann constant	$k = \dfrac{R}{L}$	1.380658×10^{-23} J K^{-1}
Avogadro constant	L, N_A	6.0221367×10^{23} mol^{-1}
Loschmidt constant	N_L, n_o	2.686763×10^{25} m^{-3}
molar gas constant	$R = Lk$	8.314510 J K^{-1} mol^{-1}
Faraday constant	$F = Le$	9.6484531×10^4 C mol^{-1}
Stefan-Boltzmann constant	$\sigma = \dfrac{2\pi^5 k^4}{15h^3 c^2}$	5.67051×10^{-8} W m^{-2} K^{-4}
fine structure constant	$\alpha = \dfrac{e^2}{2\varepsilon_0 hc}$	7.2973531×10^{-3}
Rydberg constant	$R = \dfrac{m_e e^4}{8\varepsilon_0^2 h^3 c}$	1.0973731534×10^7 m^{-1}
gravitational constant	G	6.67259×10^{-11} N m^2 kg^{-2}
acceleration of free fall (standard value)	g_n	9.80665 m s^{-2}

Table 6 Symbols used in Electronics

Device or concept	Symbol	Device or concept	Symbol
Qualifying graphical symbols			
alternating current		ionizing radiation	
variability (noninherent)		positive-going pulse	
variability in steps		negative-going pulse	
thermal effect		pulse of a.c.	
electromagnetic effect		positive-going step function	
radiation, electromagnetic nonionizing		negative-going step function	
coherent radiation		fault	

Table 6   **Symbols used in Electronics** (*concluded*)

Device or concept	Symbol	Device or concept	Symbol
Graphical symbols			
connection of conductors	●	photodiode	
terminal (circle may be filled in)	○	pnp transistor	
junction of conductors		npn transistor	
plug & socket (male & female)		JFET, n-type channel	
earth		JFET, p-type channel	
primary cell or accumulator (longer line represents +ve pole)		IGFET, enhancement type, single gate, p-type channel without substrate connection	
battery of accumulators or primary cells		amplifier, general symbol	
switch, general symbol; make contact		AND gate, general symbol	
resistor, general symbol (first form preferred)		OR gate, general symbol	
variable resistor		inverter (NOT gate)	
resistor with sliding contact			
capacitor, general symbol (first form preferred)		NAND gate (negated AND)	
inductor, coil, winding, choke, general symbol		NOR gate (negated OR)	
inductor with magnetic core		exclusive-OR gate	
transformer, 2 windings			
piezoelectric crystal, 2 electrodes		indicating instrument (first form) & recording instrument; asterisk is replaced by symbol of unit of quantity being measured (e.g. V for voltmeter, A for ammeter, or by some other appropriate symbol)	
semiconductor diode, general symbol			
light-emitting diode, general symbol		antenna, general symbol	

627

Table 7 **Spectrum of Electromagnetic Radiation**

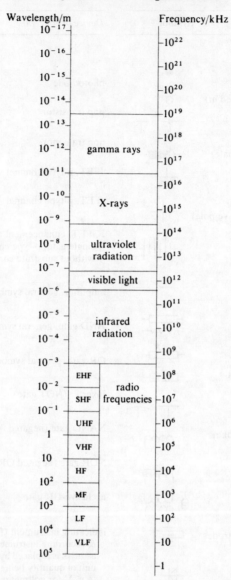

Table 8　**Elementary Particles**

	Particle	Quark content	Mass MeV/c^2	Isospin I	J^{PC}	Lifetime/s
gauge bosons	γ		0		1^-	stable
	$W^{\pm}$		80000		1	
	Z^0		91000		1	
leptons	ν		0	$\frac{1}{2}$		stable
	e		0.511	$\frac{1}{2}$		stable
	μ		105.7	$\frac{1}{2}$		2.2×10^{-6}
	τ		1784.1	$\frac{1}{2}$		3.0×10^{-13}
mesons	$\pi^{\pm}$	$u\bar{d},\bar{u}d$	139.6	1	0^-	2.6×10^{-8}
	π°	$u\bar{u},d\bar{d}$	105.7	1	0^{-+}	8.4×10^{-17}
	$K^{\pm}$	$u\bar{s},s\bar{u}$	493.6	$\frac{1}{2}$	0^-	1.2×10^{-8}
	K°	$d\bar{s}$	497.7	$\frac{1}{2}$	0^-	
	K°_S		497.7	$\frac{1}{2}$	0^-	8.9×10^{-11}
	K°_L		497.7	$\frac{1}{2}$	0^-	5.2×10^{-8}
	η°	$u\bar{u},d\bar{d},s\bar{s}$	548.8	0	0^{-+}	
	$D^{\pm}$	$c\bar{d},d\bar{c}$	1869	$\frac{1}{2}$	0^-	1.1×10^{-12}
	D°	$c\bar{u}$	1865	$\frac{1}{2}$	0^-	4×10^{-13}
	$D^{\pm}_s$	$c\bar{s},s\bar{c}$	1969	0	0^-	4×10^{-13}
	$B^{\pm}$	$u\bar{b}, b\bar{u}$	5278	$\frac{1}{2}$	0^-	1×10^{-12}
	B°	$d\bar{b}$	5279	$\frac{1}{2}$	0^-	1×10^{-12}
baryons	p	uud	938.3	$\frac{1}{2}$	$\frac{1}{2}^+$	stable
	n	udd	939.6	$\frac{1}{2}$	$\frac{1}{2}^+$	896
	Λ°	uds	1115.6	0	$\frac{1}{2}^+$	2.6×10^{-10}
	Σ^+	uus	1189.4	1	$\frac{1}{2}^+$	8.0×10^{-10}
	Σ°	uds	1192.5	1	$\frac{1}{2}^+$	7.4×10^{-20}
	Σ^-	dds	1197.4	1	$\frac{1}{2}^+$	1.5×10^{-10}
	Ξ°	uss	1314.9	$\frac{1}{2}$	$\frac{1}{2}^+$	2.9×10^{-10}
	Ξ^-	dss	1321.3	$\frac{1}{2}$	$\frac{1}{2}^+$	1.7×10^{-10}
	Ω^-	sss	1672.5	0	$\frac{3}{2}^+$	1.3×10^{-10}
	Λ^+_c	udc	2285	0	$\frac{1}{2}^+$	2×10^{-13}

Table 9　**The Chemical Elements**

Melting points and boiling points are for a pressure of one standard atmosphere.

Element	Symbol	Atomic number	Relative atomic mass	Melting point/°C	Boiling point/°C	Relative density
actinium	Ac	89	227	1050	3200 (est.)	10·07
aluminium	Al	13	26·9815	660·2	2467	2·699
americium	Am	95	243	995	2607 (est.)	13·67
antimony	Sb	51	121·75	630·5	1640	6·684
argon	Ar	18	39·948	−189·2	−185·7	($\rho = 1.78$ kg m^{-3})
arsenic	As	33	74·9216	817 (grey) (at 3 MPa.)	sublimes at 613 (grey)	5·73 (grey)
astatine	At	85	210	250	350	
barium	Ba	56	137·34	725	1140	3·5
berkelium	Bk	97	247			14 (est.)
beryllium	Be	4	9·012 18	1278	2970	1·85
bismuth	Bi	83	208·9806	271·3	1560	9·75
boron	B	5	10·81	2300	2550	2·34 (crystalline) 2·37 (amorphous)
bromine	Br	35	79·904	−7·2	58·78	3·12 ($\rho_{gas} = 7.59$ kg m^{-3})

Table 9   **The Chemical Elements** (*continued*)

Element	Symbol	Atomic number	Relative atomic mass	Melting point/°C	Boiling point/°C	Relative density
cadmium	Cd	48	112·4	320·9	765	8·65
caesium (or cesium)	Cs	55	132·9055	28·5	690	1·87
calcium	Ca	20	40·08	842–848	1487	1·55
californium	Cf	98	251			
carbon	C	6	12·011 15	sublimes above 3500	4827	1·8–2·1 (amorphous) 1·9–2·3 (graphite) 3·1–3·5 (diamond)
cerium	Ce	58	140·120	795	3468	6·77
chlorine	Cl	17	35·453	−100·98	−34·6	($\rho = 3\cdot124$ kg m^{-3})
chromium	Cr	24	51·996	1890	2482	7·19
cobalt	Co	27	58·9332	1495	2870	8·9
columbium	Cb	41	(see niobium)			
copper	Cu	29	63·546	1083	2595	8·96
curium	Cm	96	247	1340 (approx.)		13·51 (calc.)
dysprosium	Dy	66	162·50	1407	2335	8·56
einsteinium	Es	99	254			
element 104		104				
element 105		105				
erbium	Er	68	167·26	1522	2510	9·045
europium	Eu	63	151·96	826	1439	5·25
fermium	Fm	100	257			
fluorine	F	9	18·9984	−219·62	−188·14	($\rho = 1\cdot696$ kg m^{-3})
francium	Fr	87	223	30	650	
gadolinium	Gd	64	157·25	1312	3000 (approx.)	7·898
gallium	Ga	31	69·72	29·78	2403	5·91
germanium	Ge	32	72·59	937·4	2830	5·32
gold	Au	79	196·9665	1063	2660	19·30
hafnium	Hf	72	178·49	2150	5400	13·31
helium	He	2	4·0026	−272·2	−268·6	($\rho = 0\cdot178$ kg m^{-3})
holmium	Ho	67	164·9303	1461	2600	8·803
hydrogen	H	1	1·007 97	−259·14	−252·5	($\rho = 0\cdot0899$ kg m^{-3})
indium	In	49	114·82	156·61	2000 (approx.)	7·31
iodine	I	53	126·9045	113·5	184·35	4·93
iridium	Ir	77	192·22	2410	4130	22·42
iron	Fe	26	55·847	1539	2800	7·90
krypton	Kr	36	83·80	−111·9	−107·1	($\rho = 3\cdot733$ kg m^{-3})
lanthanum	La	57	138·9055	920	3454	6·17
lawrencium	Lr	103	257			
lead	Pb	82	207·19	327·3	1750	11·3
lithium	Li	3	6·941	179	1317	0·534
lutetium	Lu	71	174·97	1656	3315	9·835
magnesium	Mg	12	24·305	651	1107	1·738
manganese	Mn	25	54·938	1244	2097	7·21–7·44 (depending on allotrope)
mendelevium	Md	101	256			
mercury	Hg	80	200·59	−38·87	356·58	13·55
molybdenum	Mo	42	95·94	2610	5560	10·22
neodymium	Nd	60	144·24	1024	3127	6·80 and 7·00 (depending on allotrope)
neon	Ne	10	20·179	−248·67	−246·05	($\rho = 0\cdot8999$ kg m^{-3})
neptunium	Np	93	237·0482	640	3902 (est.)	20·25
nickel	Ni	28	58·71	1453	2732	8·90

Table 9 **The Chemical Elements** (*continued*)

Element	Symbol	Atomic number	Relative atomic mass	Melting point/°C	Boiling point/°C	Relative density
niobium	Nb	41	92·9064	2468	4927	8·57
nitrogen	N	7	14·0067	− 209·86	− 195·8	($\rho = 1·251$ kg m^{-3})
nobelium	No	102	256			
osmium	Os	76	190·2	3045 (approx.)	5027 (approx.)	22·57
oxygen	O	8	15·9994	− 218·4	− 182·96	($\rho = 1·429$ kg m^{-3})
palladium	Pd	46	106·4	641	3327	12·02
phosphorous	P	15	30·9738	44·1 (white)	280 (white)	1·82 (white) 2·20 (red)
platinum	Pt	78	195·09	1769	3800	21·45
plutonium	Pu	94	244	641	3327	19·84 (α form)
polonium	Po	84	209	254	962	9·40
potassium	K	19	39·102	63·65	774	0·86
praseodymium	Pr	59	140·9077	931	3212	6·77
promethium	Pm	61	147	1080 (approx.)	2460 (approx.)	
protactinium	Pa	91	231·0359	1200	4000	15·4 (calc.)
radium	Ra	88	226·0254	700	1140	5 (approx.)
radon	Rn	86	222	− 71	− 61·8	($\rho = 9·73$ kg m^{-3})
rhenium	Re	75	186·2	3180	5627 (est.)	21·0
rhodium	Rh	45	102·9055	1966	3727 (approx.)	12·4
rubidium	Rb	37	85·4678	38·89	688	1·53 (solid) 1·47 (liquid)
ruthenium	Ru	44	101·07	2310	3900	12·41
samarium	Sm	62	150·35	1072	1778	7·5
scandium	Sc	21	44·9559	1539	2832	2·99
selenium	Se	34	78·96	217 (grey)	684·9 (grey)	4·79 (grey)
silicon	Si	14	28·086	1410	2355	2·33
silver	Ag	47	107·868	961·93	2212	10·5
sodium	Na	11	22·9898	97·81	892	0·97
strontium	Sr	38	87·62	769	1384	2·54
sulphur	S	16	32·064	112·8 (rhombic) 119·0 (monoclinic)	444·6	2·07 (rhombic) 1·96 (monoclinic)
tantalum	Ta	73	180·9479	2996	5425	16·65
technetium	Tc	43	98·9062	2200 (approx.)	5030	11·5 (approx.)
tellurium	Te	52	127·6	449·5	989·8	6·24
terbium	Tb	65	158·9254	1360	3041	8·234
thallium	Tl	81	204·37	303·5	1457	11·85
thorium	Th	90	232·0381	1750	3800 (approx.)	11·72
thulium	Tm	69	168·9342	1545	1727	9·31
tin	Sn	50	118·69	231·89	2270	5·75 (grey) 7·31 (white)
titanium	Ti	22	47·90	1675	3620	4·54
tungsten	W	74	183·85	3410	5927	19·3
uranium	U	92	238·029	1132	3818	18·95 (approx.)
vanadium	V	23	50·9414	1890	3380	6·1
wolfram	W	74	(see tungsten)			
xenon	Xe	54	131·30	− 111·9	− 107·1	($\rho = 5·887$ kg m^{-3})

tables

Table 9 **The Chemical Elements** (*concluded*)

Element	Symbol	Atomic number	Relative atomic mass	Melting point/°C	Boiling point/°C	Relative density
ytterbium	Yb	70	173·04	824	1193	6·97 or 6·54
yttrium	Y	39	88·9059	1523	3337	4·46
zinc	Zn	30	65·37	419·58	907	7·133
zirconium	Zr	40	91·22	1852	4377	6·51

Table 10 **Periodic Table of the Elements**

1A	2A	3B	4B	5B	6B	7B		8		1B	2B	3A	4A	5A	6A	7A	0
1 H																	2 He
3 Li	4 Be											5 B	6 C	7 N	8 O	9 F	10 Ne
11 Na	12 Mg				← transition elements →							13 Al	14 Si	15 P	16 S	17 Cl	18 Ar
19 K	20 Ca	21 Sc	22 Ti	23 V	24 Cr	25 Mn	26 Fe	27 Co	28 Ni	29 Cu	30 Zn	31 Ga	32 Ge	33 As	34 Se	35 Br	36 Kr
37 Rb	38 Sr	39 Y	40 Zr	41 Nb	42 Mo	43 Tc	44 Ru	45 Rh	46 Pd	47 Ag	48 Cd	49 In	50 Sn	51 Sb	52 Te	53 I	54 Xe
55 Cs	56 Ba	57* La	72 Hf	73 Ta	74 W	75 Re	76 Os	77 Ir	78 Pt	79 Au	80 Hg	81 Tl	82 Pb	83 Bi	84 Po	85 At	86 Rn
87 Fr	88 Ra	89† Ac															

*lanthanides		57 La	58 Ce	59 Pr	60 Nd	61 Pm	62 Sm	63 Eu	64 Gd	65 Tb	66 Dy	67 Ho	68 Er	69 Tm	70 Yb	71 Lu
†actinides		89 Ac	90 Th	91 Pa	92 U	93 Np	94 Pu	95 Am	96 Cm	97 Bk	98 Cf	99 Es	100 Fm	101 Md	102 No	103 Lr

Table 11   **The Radioactive Decay Series**

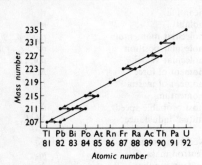

The actinium series

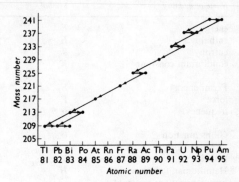

The neptunium series

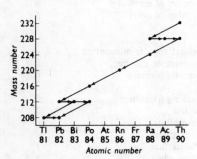

The thorium series

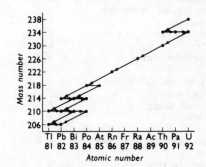

The uranium series

Table 12   **Symbols for Physical Quantities**

Name of quantity	Symbol	Name of quantity	Symbol
absorptance	α	conductivity	γ, σ
acceleration	$\boldsymbol{a}$	cross section	σ
activity, radioactivity	A	cubic expansion coefficient	α_v
admittance	Y	Curie temperature	T_C
amount of substance	n		
angular acceleration	α	decay constant	λ
angular frequency	ω	density	ρ
angular momentum	L		
angular velocity	ω	efficiency	η
area	A, S	electric charge	Q
atomic mass constant	m_u	electric current	I
atomic number, proton		electric current density	$j, \boldsymbol{J}$
number	Z	electric dipole moment	p
		electric displacement	D
Bragg angle	θ	electric field strength	E
bulk modulus	K	electric flux	Ψ
		electric polarization	P
capacitance	C	electric potential	V
characteristic temperature	Θ	electric susceptibility	χ_e
charge density	ρ	electromotive force	E
coefficient of friction	μ	electron mass	m, m_e
compressibility	κ	elementary charge, charge of	
concentration	c	proton	e
conductance	G	emissivity	ε

633

tables

Table 12 **Symbols for Physical Quantities** (*continued*)

Name of quantity	Symbol	Name of quantity	Symbol
energy	E	mean life	τ
enthalpy	H	molality	m_A
entropy	S	molecular momentum	$p(p_x, p_y, p_z)$
equilibrium constant	K	molecular position	$r(r_x, r_y, r_z)$
		molecular velocity	$u(u_x, u_y, u_z)$
Fermi energy	E_F, ε_F	moment of force	M
force	F	moment of inertia	I
frequency	ν, f	momentum	p
		most probable speed	$\hat{u}$
Gibbs function	G	mutual inductance	M, L_{12}
half-life	$T_{1/2}, t_{1/2}$	Néel temperature	T_N
Hamiltonian function	H	neutron mass	m_n
heat capacity: at constant		neutron number	N
pressure	C_p	nuclear magneton	μ_N
heat capacity: at constant		number of molecules	N
volume	C_v	number of turns	N
heat flow rate	$\ominus$		
height	h		
Helmholtz function	A, F	orbital angular momentum	
		quantum number	L, l_1
illuminance, illumination	E_V, E	osmotic pressure	Π
impedance	Z		
internal energy	U	packing fraction	f
irradiance	E_e, E	period	T
		permeability	μ
Joule–Thomson coefficient	μ, μ_{JT}	permittivity	ε
		Planck function	Y
kinematic viscosity	ν	plane angle	$\alpha, \beta, \gamma, \theta, \varphi$
kinetic energy	T, E_k, K	polarizability	α, γ
		position vector, radius vector	r
Lagrangian function	L	potential difference	U, V
linear absorption coefficient	a	potential energy	E_p, V, Φ
linear attenuation (extinction)		power	P
coefficient	μ	pressure	p
linear expansion coefficient	α	principal quantum number	n
linear strain (relative		propagation coefficient	P, γ
elongation)	ε	proton mass	m_p
loss angle	δ	proton number, atomic	
luminance	L_V, L	number	Z
luminous exitance	M_V, M	quantity of heat	Q
luminous flux	Φ_V, Φ	quantum number of	
luminous intensity	I_V, I	electron spin	S
		quantum number of nuclear	
magnetic field strength	H	spin	I
magnetic flux	$\ominus$	quantum number of total	
magnetic flux density, magnetic		angular momentum	N
induction	B	quantum number of	
magnetic moment	m	vibrational mode	v
magnetic moment of particle	μ		
magnetic quantum number	M, m_i	radiance	L_e, L
magnetic susceptibility	χ, χ_m	radiant exitance	M_e, M
magnetization	M	radiant flux, radiant power	Φ_e, Φ
magnetomotive force	F_m	radiant intensity	I_e, I
mass	m	radius	r
mass excess	Δ	ratio of heat capacities, C_p/C_V	γ, κ
mass number, nucleon		reactance	X
number	A	reduced mass	μ
mean free path	λ, l	reflectance	ρ

Table 12   **Symbols for Physical Quantities** (*concluded*)

Name of quantity	Symbol	Name of quantity	Symbol
refractive index	n	surface tension	γ, σ
relative atomic mass	A_r	susceptance	B
relative density	d		
relative permeability	μ_r	temperature	T, t
relative permittivity (dielectric constant)	ε_r	thermal conductivity	λ
		thermal diffusion factor	α_T
relaxation time	τ	thermal diffusion ratio	k_T
resistance	R	thermal diffusivity	a
resistivity	ρ	thermodynamic temperature	T
Reynolds number	Re	time	t
rotational quantum number	J, K	torque	T
		transmission coefficient (acoustics)	τ
self-inductance	L		
shear modulus	G	velocity	v
solid angle	Ω, ω	vibrational quantum number	v
specific heat capacity: at constant pressure	c_p	viscosity	η
		volume	V, v
specific heat capacity: at constant volume	c_V	volume (bulk) strain	θ
specific volume	v	wavelength	λ
speed	u, v	wavenumber	σ
spin quantum number	S, s	weight	G, W
strain, linear	ε, e	work	W
strain, shear	γ	work function	Φ
stress, shear	τ		
surface charge density	σ	Young modulus (modulus of elasticity)	E

Table 13   **The Greek Alphabet**

Letters		Name
A	α	alpha
B	β	beta
Γ	γ	gamma
Δ	δ	delta
E	ε	epsilon
Z	ζ	zeta
H	η	eta
Θ	θ	theta
I	ι	iota
K	κ	kappa
Λ	λ	lambda
M	μ	mu
N	ν	nu
Ξ	ξ	xi
O	o	omikron
Π	π	pi
P	ρ	rho
Σ	σ	sigma
T	τ	tau
Y	υ	upsilon
Φ	ϕ	phi
X	χ	khi
Ψ	ψ	psi
Ω	ω	omega

tables

Table 14 Nobel Prizewinners in Physics

1901	W Röntgen	1938	E Fermi		L Néel
1902	H Antoon Lorentz	1939	E Lawrence	1971	D Gabor
	P Zeeman	1943	O Stern	1972	J Bardeen
1903	A Becquerel	1944	I Rabi		L N Cooper
	P Curie	1945	W Pauli		J R Schrieffer
	M Curie	1946	P Bridgman	1973	L Esaki
1904	Lord Rayleigh	1947	Sir E Appleton		I Giaever
1905	P Lenard	1948	P Blackett		B Josephson
1906	Sir J J Thomson	1949	H Yukawa	1974	Sir M Ryle
1907	A A Michelson	1950	C Powell		A Hewish
1908	G Lippmann	1951	Sir J Cockcroft	1975	J Rainwater
1909	G Marconi		E Walton		A Bohr
	K Braun	1952	F Bloch		B Mottelson
1910	J van der Waals		E Purcell	1976	B Richter
1911	W Wien	1953	F Zernike		S Ting
1912	N G Dalén	1954	M Born	1977	P W Anderson
1913	H Kamerlingh Onnes		W Bothe		Sir N F Mott
1914	M von Laue	1955	W Lamb, Jr		J H van Vleck
1915	Sir W Bragg		P Kusch	1978	P L Kapitsa
	Sir L Bragg	1956	W Shockley		A A Penzias
1916	(No Award)		J Bardeen		R W Wilson
1917	C Barkla		W Brattain	1979	S L Glashow
1918	M Planck	1957	Tsung-Dao Lee		A Salam
1919	J Stark		C N Yang		S Weinberg
1920	C Guillaume	1958	P A Cherenkov	1980	J Cronin
1921	A Einstein		I M Frank		V Fitch
1922	N Bohr		I Y Tamm	1981	K Siegbahn
1923	R Millikan	1959	E Segrè		N Bloembergen
1924	K Siegbahn		O Chamberlain		A Schawlow
1925	J Franck	1960	D Glaser	1982	K G Wilson
	G Hertz	1961	R Hofstadter	1983	S Chandrasekhar
1926	J Perrin		R Mössbauer		W Fowler
1927	A H Compton	1962	L D Landau	1984	C Rubbia
	C Wilson	1963	J H D Jensen		S van der Meer
1928	Sir O Richardson		M G Mayer	1985	K von Klitzing
1929	Prince L de Broglie		E P Wigner	1986	E Ruska
1930	Sir C Raman	1964	C H Townes		G Binnig
1931	(No Award)		N G Basov		H Rohrer
1932	W Heisenberg		A M Prokhorov	1987	A Müller
1933	P A M Dirac	1965	J S Schwinger		G Bednorz
	E Schrödinger		R P Feynman	1988	L Lederman
1934	(No Award)		S Tomonaga		M Schwartz
1935	Sir J Chadwick	1966	A Kastler		J Steinberger
1936	V Hess	1967	H A Bethe	1989	H Dehmelt
	C Anderson	1968	L W Alvarez		W Paulm
1937	C Davisson	1969	M Gell-Mann		N Ramsey
	Sir G P Thomson	1970	H Alvén		